Anesthesia, Analgesia, and Pain Management

FOR VETERINARY TECHNICIANS

FIRST EDITION

Janet Amundson Romich

Christopher L. Norkus

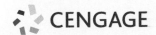

CENGAGE

Australia • Brazil • Canada • Mexico • Singapore • United Kingdom • United States

Anesthesia, Analgesia, and Pain Management for Veterinary Technicians, **First Edition**
Janet Amundson Romich and Christopher L. Norkus

SVP, Higher Education & Skills Product: Erin Joyner

VP, Higher Education & Skills Product: Mike Schenk

Product Director: Matthew Seeley

Senior Product Manager: Stephen Smith

Product Assistant: Dallas Wilkes

Learning Designer: Debbie Bordeaux

Content Manager: Emily Olsen

Digital Delivery Lead: Lisa Christopher

Marketing Manager: Courtney Cozzy

IP Analyst: Ashely Maynard

IP Project Manager: Nick Barrows

Production Service: MPS Limited

Cover Design: Angela Sheehan

Cover Image Source: urbancow/iStock.com

For product information and technology assistance, contact us at
**Cengage Customer & Sales Support, 1-800-354-9706
or support.cengage.com.**

For permission to use material from this text or product, submit all requests online at **www.cengage.com/permissions.**

Library of Congress Control Number: 2021935987

ISBN: 978-1-2857-3740-9

Cengage
200 Pier 4 Boulevard
Boston, MA 02210
USA

Cengage is a leading provider of customized learning solutions with employees residing in nearly 40 different countries and sales in more than 125 countries around the world. Find your local representative at **www.cengage.com.**

To learn more about Cengage platforms and services, register or access your online learning solution, or purchase materials for your course, visit **www.cengage.com.**

Notice to the Reader

Publisher does not warrant or guarantee any of the products described herein or perform any independent analysis in connection with any of the product information contained herein. Publisher does not assume, and expressly disclaims, any obligation to obtain and include information other than that provided to it by the manufacturer. The reader is expressly warned to consider and adopt all safety precautions that might be indicated by the activities described herein and to avoid all potential hazards. By following the instructions contained herein, the reader willingly assumes all risks in connection with such instructions. The publisher makes no representations or warranties of any kind, including but not limited to, the warranties of fitness for particular purpose or merchantability, nor are any such representations implied with respect to the material set forth herein, and the publisher takes no responsibility with respect to such material. The publisher shall not be liable for any special, consequential, or exemplary damages resulting, in whole or part, from the readers' use of, or reliance upon, this material.

Printed at CLDPC, USA, 07-21

BRIEF CONTENTS

CONTENTS

PREFACE

The practice of anesthesia is an art. Art motivates us as veterinary professionals to have compassion for patients that suffer from conditions that need our intervention. The art of anesthesia can take on many forms. General anesthesia, a controlled unconsciousness, ensures patient safety by alleviating their overall perception of pain and movement so that a surgical incision can be made or a bone fracture repaired. In contrast, local anesthesia reduces pain perception in a limited area such as an injection of a local anesthetic around a laceration so that it can be cleaned and sutured in a sedated patient. For some procedures, local anesthesia may be faster, safer, and less stressful to patients and veterinary staff versus placing the animal under full general anesthesia. For other procedures, general anesthesia is required to ensure adequate reduction of pain sensation to patients and safety to all parties. The beauty of anesthesia is that it takes on many forms that require a series of decisions to be made by well-trained, competent anesthetists.

The practice of analgesia and pain management is also an art. Veterinary professionals do not want their patients to suffer from conditions that cause them pain. The first step in providing effective analgesia and pain management to animals involves the challenging task of identifying unpleasant sensory and emotional experience in stoic, nonverbal patients. It is essential to recognize pain in our patients because if it is not recognized, then it is unlikely to be treated. It is also important to assess the intensity of pain because failure to identify the intensity of pain will result in the selection of inappropriately potent analgesic drugs, cause doubts about the effectiveness of the administered drug dose, and result in less than optimal pain management. There have been great strides in improving the recognition of pain in animals, which have led to the development of ethical standards of care based on effective therapeutic options for veterinary patients. The goal of providing analgesia and pain management to animals is to identify pain as early as possible, create a treatment plan, and explain the treatment plan and expected outcomes to owners in terms they understand. The success of analgesia and pain management requires a series of decisions to be made by well-trained, competent veterinary technicians.

Although the practice of anesthesia, analgesia, and pain management is an art, it is soundly based on science. Anesthesia is a scientific gift to our patients that has progressed over time from immobilizing a patient so a surgery can be performed to providing optimal perioperative outcomes and survival of our patients. No one anesthesia protocol is appropriate for all species or procedures; therefore, to effectively practice the art of anesthesia we must continue to learn new scientific-based advances in veterinary medicine and apply those new advances to our daily clinical practice. The same holds true for analgesia and pain management. The role of veterinary professionals is to minimize pain in their patients in order to maximize the human-animal bond and to promote the health and well-being of both people and animals. Analgesic and pain management protocols are designed to meet the needs of individual patients; therefore, to effectively provide pain relief for our patients we must avoid a cookie-cutter approach to managing pain and instead rely on continuing education to provide the most current knowledge regarding therapeutics and pain management techniques. The goal for veterinary technicians is to understand the fundamentals of anesthesia, analgesia, and pain management techniques so they can apply them to provide the best patient care plans possible to ensure each patient receives the most appropriate and best possible care. It is the aim of the authors of this textbook to help veterinary technicians reach their goal of becoming competent in the practice of anesthesia, analgesia, and pain management as they continue to learn and utilize new information in this challenging yet rewarding field of veterinary medicine.

SUPPLEMENTAL MATERIALS

Additional instructor and student resources for this product are available online. Instructor assets include an Instructor's Manual, Educator's Guide, PowerPoint® slides, and a test bank powered by Cognero®. Student assets include a student solution manual and data sets. Sign up or sign in at www.cengage.com to search for and access this product and its online resources.

Instructor's Manual

The Instructor's Manual provides the following for each chapter: Chapter Objectives, Key Terms, and a Chapter Outline. Also included are additional activities and discussions to help guide classroom instruction.

The Solution and Answer Guide provides answers to the Review Questions, Case Studies, and Critical Thinking Questions.

PowerPoint® Lecture Slides

Customizable Microsoft PowerPoint® Presentations focus on key points for each chapter. The slides allow instructors to tailor the course to meet the needs of the individual class.

Cognero Test Banks

Cognero is a flexible, online system that allows instructors to author, edit, and mange test bank content from Cengage; create multiple test versions in an instant; and deliver tests from the instructor's LMS, classroom, or wherever the instructor desires. The test bank contains more than 500 questions in multiple-choice format.

ABOUT THE EDITORS

Janet Amundson Romich, DVM, MS

Dr. Janet Romich started her veterinary education by receiving her Bachelor of Science degree in Animal Science from the University of Wisconsin–River Falls and then her Doctor of Veterinary Medicine from the University of Wisconsin–Madison. She continued her studies at the University of Wisconsin–Madison earning a Master of Science degree which focused on Food and Drug Administration (FDA) drug research. Dr. Romich has taught a variety of veterinary technician and science courses at Madison Area Technical College in Madison, Wisconsin, and was honored with the Distinguished Teacher Award for her use of technology in the classroom, advisory and professional activities, publication list, and fundraising efforts. She also received the Wisconsin Veterinary Technician Association's Veterinarian of the Year Award for her work in teaching and mentoring veterinary technician students. Dr. Romich has authored the textbooks *An Illustrated Guide to Veterinary Medical Terminology*, with MindTap®, *Fundamentals of Pharmacology for Veterinary Technicians*, with MindTap®, and *Understanding Zoonotic Diseases*, as well as coauthored *Delmar's Veterinary Technician Dictionary*. Dr. Romich remains active in veterinary medicine through her clinical relief practice and work on an Institutional Animal Care and Use Committee for a hospital research facility.

Christopher L. Norkus, DVM, DACVAA, CVPP, DACVECC

Dr. Chris Norkus began his veterinary career as a veterinary technician, obtaining duel Veterinary Technician Specialist (VTS) status in both emergency-critical care and anesthesia. He completed his undergraduate education at the University of Massachusetts-Amherst and proceeded to veterinary school at Ross University School of Veterinary Medicine and Tufts Cummings School of Veterinary Medicine. He then completed a prestigious internship at The Animal Medical Center in New York City followed by a residency in anesthesiology at Kansas State University. Following this, Dr. Norkus became board-certified as a Diplomate of the American College of Veterinary Anesthesia and Analgesia and was credentialed as a Certified Veterinary Pain Practitioner in 2015. Dr. Norkus then completed a second 3-year residency in emergency medicine and critical care at Allegheny Veterinary Emergency Trauma & Specialty in Pittsburgh, Pennsylvania, and became board certified as a Diplomate of the American College of Veterinary Emergency and Critical Care in 2018. He is currently the Medical Director of Anesthesiology, Emergency and Critical Care at VCA Veterinary Specialists of CT in West Hartford, Connecticut. To date, Dr. Norkus has published more than two dozen peer-reviewed publications and book chapters and is the editor for Veterinary Technician's Manual for Small Animal Emergency and Critical Care. He lectures regularly for both regional and national audiences.

ABOUT THE CONTRIBUTING AUTHORS

Emma Archer, RVN, Dip AVN (Surgical), VTS (Anesthesia/Analgesia)

Emma qualified as a veterinary nurse in the United Kingdom in 2001 and spent a year in general practice before joining the Animal Health Trust in 2002 as a theatre nurse. She gained the Diploma in Advanced Nursing (Surg) in 2006 and then in 2008 became their Anesthesia Technician. In 2010, she became a Veterinary Technician Specialist in Anesthesia and a member of the Academy of Veterinary Technician Anesthetists. Since 2013 she has been working as a locum Anesthesia Technician at the Animal Health Trust. Emma continues her passion for providing continuing education for nurses by organizing and speaking at Continuing Professional Day courses at the Animal Health Trust and lecturing at BSAVA congress and at local colleges, as well as writing for various nursing journals.

Kristen Cooley BA, CVT, VTS (Anesthesia/Analgesia)

Kristen is a member of the Academy of Veterinary Technician Anesthetists through which she is recognized as a veterinary technician specialist in anesthesia. For many years, Kristen worked as an Instructional Specialist at the University of Wisconsin Veterinary Medical Teaching Hospital and is currently a veterinary cannabis consultant for Earthwise Veterinary Cannabis Consulting as well as an independent veterinary anesthesia and pain management consultant. Kristen is a member of both the Wisconsin Veterinary Technician Association (past president) and the Academy of Veterinary Technician Anesthetists (conference committee chair). She has presented at several conferences including Central Veterinary Conference, Western Veterinary Conference, and International Veterinary Emergency and Critical Care Symposium. Kristen is also an author of several anesthesia related textbook chapters.

Lorelei D'Avolio, LVT, VTS (Clinical Practice-Exotic Companion Animal), CVPM Practice Manager

Lorelei received her BA in journalism from Boston University; however, her lifelong love of animals, especially birds, eventually pulled her away from writing into the field of veterinary medicine. After graduating from the Veterinary Technology program at Mercy College in 2001, Lorelei has been working exclusively with birds and exotic pets and is currently with The Center for Avian & Exotic Medicine in New York as the practice manager and nursing team member since it opened in 2004. In addition to being an outstanding nurse with many years of experience, she has also been an adjunct instructor of avian and exotic medicine at a local veterinary technology program, plus lectured at several national and international veterinary symposiums. She also has authored text book chapters, journal articles, and edited several professional publications. Lorelei was a founding member and is a past president of the Academy of Veterinary Technicians in Clinical Practice (AVTCP). The AVTCP is the ninth officially recognized veterinary technician specialty (VTS) and is modeled after the American Board of Veterinary Practitioners (ABVP). In 2018, Lorelei earned the status of Certified Veterinary Practice Manager (CVPM) through the VHMA (Veterinary Hospital Managers Association).

Trish Farry, CVN, VTS (ECC, Anesthesia), TAA, GCHEd

Trish is an Australian certified veterinary nurse with specialist qualifications in Emergency/Critical Care and Anesthesia/Analgesia. She is an associate lecturer and clinical instructor at The University of Queensland, and co-coordinates final year and postgraduate subjects in the Bachelor of Veterinary Technology program. Her areas of teaching include emergency medicine, anesthesia, analgesia, and clinical practices for undergraduate veterinary and veterinary technology students. Trish has published numerous textbook chapters and journal articles and is regularly asked to lecture at national and international conferences. Her professional positions include President of the Academy of Emergency and Critical Care Technicians (AVECCT), and past board member of the Academy of Veterinary Technician

Anesthetists (AVTAA) and the Board of Regents for the International Veterinary Academy of Pain Management (IVAPM). In 2018, she was honored to receive the VNCA Veterinary Nurse of the Year and the AVECCT Specialty Technician of the Year awards.

Mary Ellen Goldberg BS, CVT, SRA, CCRVN, CVPP, VTS-Lab Animal Medicine (Research Anesthesia), VTS-Physical Rehabilitation, VTS-H (Anesthesia/Analgesia)

Mary Ellen is a graduate of Harcum College and the University of Pennsylvania and has worked in various aspects of veterinary medicine from small animal and equine to mixed practice to zoo animal medicine and laboratory animal medicine as well as pharmaceutical research. She worked at Virginia Commonwealth University in the Division of Animal Resources and for research scientists and was a member of their IACUC. She has been an instructor of Anesthesia and Pain Management at VetMedTeam since 2003. She is a Certified Veterinary Pain Practitioner and the Exam Chair for the International Veterinary Academy of Pain Management (IVAPM). Mary Ellen is also a Surgical Research Anesthetist certified through the Academy of Surgical Research. She is the Exam Chair for APRVT (Academy of Physical Rehabilitation Veterinary Technicians) and for the Academy of Laboratory Animal Veterinary Technicians and Nurses (APRVT). Currently, she is a staff member at the Canine Rehabilitation Institute, as a Certified Canine Rehabilitation Veterinary Nurse (CCRVN). She is an IACUC member at the Manheimmer Foundation, Inc. and teaches their residents and interns about non-human primate anesthesia/analgesia. Mary Ellen received NAVTA's Veterinary Technician of the Year 2017. In 2019, she received the honorary VTS-Anesthesia/Analgesia from the Academy of Veterinary Technician Anesthesia and Analgesia. Mary Ellen has authored several books on anesthesia, pain management, and rehabilitation as well as speaking at national meetings and conducting continuing education to organizational groups on these topics.

Kristen Hagler, BS (An. Phys), RVT, VTS (Physical Rehabilitation), CCRP, CVPP, VCC, OACM, CBW

Kristen received her undergraduate degree at Sonoma State University (BS in Biology with a concentration in Animal Physiology) and passed the California Registered Veterinary Technician licensing examination in 2002. She then completed the University of Tennessee's Certificate Program in Canine Physical Rehabilitation in 2005. She has completed coursework in canine massage therapy through the Caninology® massage therapy program in 2009 and in 2012 completed coursework to become a Certified Osteoarthritis Case Manager (COCM) as well as being awarded a Certified Veterinary Pain Practitioner (CVPP) through the International Veterinary Association of Pain Management (IVAPM). Her business Golden Gait Canine was started in 2013 to help provide physical rehabilitation services to veterinarians in the community in Marin, Sonoma, and San Francisco counties. She is an active participant with the American Association of Rehabilitation

Veterinarians (rehabvets.org) promoting veterinary physical rehabilitation. Kristen is a charter member of the Academy of Physical Rehabilitation Veterinary Technicians recognized by NAVTA in 2017 and is a VTS (Physical Rehabilitation). She is the recipient of the 2018 California Registered Veterinary Technician of the Year Award.

Janel Holden LVT, VTS (Anesthesia/Analgesia)

Janel is a licensed veterinary technician at Washington State University Veterinary Teaching Hospital where she spent her first 5 years working in equine surgery and radiology. For the past 15 years she has worked specifically in the anesthesia department teaching fourth year veterinary students how to safely anesthetize and provide pain management to a variety of small and large animal species. In addition to her clinical work, Janel also serves as the Director of the Washington State University Veterinary Technician Internship Program. In 2011, Janel received her Veterinary Technician Specialist in Anesthesia and Analgesia. Over the years she has been able to present several continuing education lectures and labs to veterinary professionals on anesthesia and analgesia topics and techniques. She is a contributing author for pain management and anesthesia textbooks for veterinary technicians.

Rebecca Johnson, DVM, PhD, DACVAA

Dr. Rebecca Johnson is a Clinical Associate Professor of Anesthesia and Pain Management at the University of Wisconsin–Madison. She earned her Doctor of Veterinary Medicine (DVM), cum laude, from The Ohio State University and completed a Small Animal Rotating Internship at Metropolitan Veterinary Hospital, Akron, Ohio. Dr. Johnson completed her Residency Program in Veterinary Anesthesiology and earned her PhD in Veterinary Science (emphasis: Respiratory Neurobiology) and Master of Science in Veterinary Science at the University of Wisconsin–Madison. Her main research goal is the study of new and more efficacious means of providing anesthesia and analgesia to exotic and laboratory animal species. Her lab focuses on improving pharmacologic options and techniques to improve animal care in client owned pets as well as research animals. She also is involved in continuing clinical education of veterinary technicians and practicing veterinarians.

Teri (Raffel) Kleist, CVT, VTS (Surg)

Teri Kleist earned her AAS from Madison College in 1981. Her career has included private practice employment, a small animal surgery department position at Purdue University Veterinary School, as well as spending eight years at University of Wisconsin Veterinary Care Teaching Hospital in the small animal and large animal surgery departments. Her 30-year career at Madison College Veterinary Technician program has included positions as a laboratory coordinator as well as a supervisory position for student workers, animal care, and program management. Her professional involvement includes holding the presidential trilogy for the Wisconsin Veterinary Technician Association, Legislative Committee Chair for WVTA, multiple board positions for NAVTA including member at large, president elect, and president. Her involvement in the creation, development, and administration of the VTNE includes both item writing and acting as the VTNE Committee Chair for AAVSB. She was a charter member of the Academy of Veterinary Surgical Technicians and has held multiple positions within the organization including secretary, exam chair, and the president trilogy. Additionally her professional interactions include being the Technician Program Chair for the Surgical Summit of the American College of Veterinary Surgeons. Teri has authored articles in peer-reviewed journals, chapters in various textbooks, co-author of a surgical patient care textbook for veterinary technicians, and presentations at national and international conferences.

Katrina Lafferty, CVT, VTS (Anesthesia/ Analgesia)

Katrina is a 2001 graduate of DePaul University with a Bachelor of Fine Arts in Theatre. She received her degree in Veterinary Technology in 2005 and earned her veterinary technician specialty (VTS) in Anesthesia in 2009. Katrina spent 2005 through 2016 as a Senior Technician in the Anesthesia and Pain Management Department at the University of Wisconsin–Madison School of Veterinary Medicine. In 2016, she moved to the Wisconsin National Primate Research Center as a lead technician in the Surgery and Anesthesia Department. In her current position Katrina is responsible for improving and innovating the anesthetic and analgesic care of a number of nonhuman primate species. Katrina is involved in the education of all members of the veterinary community, and she has written numerous articles and textbook chapters. She has presented at over 40 continuing education seminars. Her passion is anesthesia and pain management, most particularly in exotic species.

Anita Parkin, RVN, AVN Dip (Surgery & ECC), VTS (Anesthesia/Analgesia), CVPP, TAE

Anita has been a veterinary nurse for over 25 years. She started working at Veterinary Specialist Services in Australia in 2001 as a medical and surgical nurse. In 2009, Anita became an Accredited Veterinary Nurse through the VNCA and completed her Veterinary Technician Specialty in Anesthesia. In 2011, Anita was the first Australian Veterinary professional to pass the International Veterinary Academy of Pain Management (IVAPM) exam. In 2013, she completed her diploma in Emergency and Critical Care, recertified in her diploma in Surgery, and was re-accredited through the VNCA (Veterinary Nurses Council of Australia). In 2018, Anita joined the Board of Directors and the Continuing Professional Day committee with the VNCA. The skills and experience she has gained over the last 25 plus years are now being passed on to veterinary nurses throughout Australia and internationally through lectures, workshops, and in-clinic hands-on training. Anita's special interests include critical care, challenging anesthetics, pain management, and surgical nursing.

Trish Roehling, CVT, VTS (Anesthesia/ Analgesia)

Trisha Roehling graduated from Madison Area Technical College with her AAS degree in Veterinary Technology. She was accepted to the Academy of Veterinary Technicians in Anesthesia and Analgesia (AVTAA) in 2012. Trisha has worked at the University of Wisconsin–Madison School of Veterinary Medicine in the Large Animal Surgery Department and as a senior technician in the Anesthesia and Pain Management Department. In 2017, she moved to the Wisconsin National Primate Research Center and is currently working as an anesthesia/surgery technician with their veterinary service. Trish also works part time as a veterinary technician in private practice at Columbus Countryside Veterinary Clinic in Columbus, Wisconsin. Trisha has served as the president, treasurer, and NAVTA liaison for the Wisconsin Veterinary Technician Association. She also assists the AVTAA as part of the application credentials committee and as an application mentor. Trish's first love is large animal anesthesia, and she has written on both equine and bovine anesthesia. She continues to learn new ways to provide safe and adequate anesthesia and pain management to her large animal patients.

Kim Spelts, BS, CVT, VTS (Anesthesia/ Analgesia)

Kim originally received a BS in Aerospace Engineering Sciences from the University of Colorado, Boulder. After spending time working for NASA and then in the corporate world, she went back to school and earned her AAS in Veterinary Technology and achieved CVT status. She initially worked as an emergency technician, then joined the Anesthesia Department at Colorado State University's Veterinary Teaching Hospital where she trained veterinary students, interns, residents, and other technicians in anesthesia. She became a Veterinary Technician Specialist in Anesthesia/Analgesia in 2004. She eventually left CSU and was instrumental in launching the rehabilitation and anesthesia components of a specialty practice in Colorado Springs. Kim started her own business, PEAK Veterinary Anesthesia Services, in 2013. She strives to advance anesthesia knowledge and standards of care through advanced training and case management. Kim has lectured at many conferences to veterinary technicians and veterinarians alike to further their knowledge of anesthesia and pain management. She was honored in 2008 with a cover photo and interview for Veterinary Technician magazine, and she was named the Western Veterinary Conference Veterinary Technician Continuing Educator of the Year in 2011. Kim was selected President-Elect of the Academy of Veterinary Technician Anesthetists in 2014, has coauthored the AAHA Anesthesia Guidelines for Dogs and Cats, and is a contributing author to several anesthesia and pain management textbooks.

Patricia R. Zehna, RVT, CPT1

Patricia received her RVT from Cedar Valley College and is currently an instructor of veterinary technology at Central Coast College in Salinas, California. After a varied career as a RVT for Guide Dogs for the Blind and in private veterinary practice as a surgical and ICU technician and RVT trainer, she is especially passionate about emergency and critical care, pain management, and anesthesia. From 2017 to 2018 Patricia was a NAVTA National District Representative Zone X. She currently serves on the Board of Directors of the California Registered Veterinary Technicians Association (CaRVTA) and the Human Animal Bond Association (HABA) and is the Executive Secretary and Treasurer of the National Veterinary Professionals Union.

ACKNOWLEDGMENTS

Benita Altier, LVT, VTS (Dentistry)
Pawsitive Dental Education LLC
Easton, WA

Jan Barnett, DVM
Program Director, Veterinary Technology Program
Tulsa Community College
Tulsa, OK

Jody Bearman, DVM, CVA, CVCH, CVSMT
AnShen Holistic Veterinary Care
Madison, WI

Stephen Cital, RVT, RLAT, SRA, VTS-LAM (Research Anes.)
Chief Operating Officer Veterinary Anesthesia Nerds and Veterinary Cannabis Academy
Executive Director for the Academy of Laboratory Animal Veterinary Technicians and Nurses Chair Elect for the Committee on Veterinary Technician Specialties

Lisa Ebner, DVM, MS, DACVAA, CVA
Associate Professor of Veterinary Anesthesiology
Lincoln Memorial University
Harrogate, TN

Caroline Mead, CVT, RLATg
Translational Research Center
Arizona Heart Foundation
Phoenix, AZ

Lianne Nelson, DVM, BSc
Education Advisor and Lead Instructor
Veterinary Technician, Veterinary Assistant
Hopewell Cape, NB

Disclaimer

This textbook is designed to teach students the theories and methods of anesthesia and pain management in a variety of animal species. It is not meant to serve as a drug formulary. Drug dosages listed are those used by select authors or published in referenced publications.

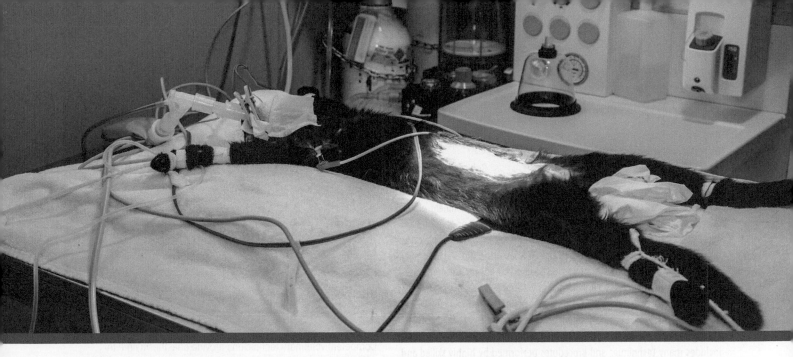

CHAPTER 1

Introduction to Anesthesia

Trish Farry, *CVN, VTS (ECC), VTS (Anesthesia/Analgesia)*

LEARNING OBJECTIVES

Upon completion of this chapter, it is expected that the reader should be able to:

1.1 Recognize the importance of and need for well-trained professionals in the field of veterinary anesthesia

1.2 Describe the veterinary technician's role in anesthesia

1.3 Identify the role veterinary anesthesia organizations have in enhancing the practice of veterinary anesthesia

1.4 Define common terminology used in the field of veterinary anesthesia

KEY TERMS

American Society of Anesthesiologists (ASA) physical status

Analgesia

Anesthesia

Anesthetic depth

Balanced anesthesia

Dissociative anesthesia

General anesthesia

Local anesthesia

Regional anesthesia

Sedation

Surgical anesthesia

Tranquilization

INTRODUCTION

"There are no safe anesthetic agents; there are no safe anesthetic procedures; there are only safe anesthetists."

Robert M. Smith, MD

The word **anesthesia**, derived from the Greek word *anaisthesia* meaning "without feeling or sensibility," was introduced by the prominent physician Oliver Wendell Holmes in the 1840s to describe part or entire loss of sensation to the body. Anesthesia has evolved since then, and the accompanying changes have made the practice of anesthesia more than simply giving drugs for immobilization so that surgeries and painful procedures can be performed. Increased knowledge of veterinary anatomy, physiology, and pharmacology, along with development of specialized techniques, has improved delivery of anesthetic agents to veterinary patients as well as the management of patients under anesthesia during surgery. Today, anesthesia includes many techniques and procedures performed by highly skilled and trained veterinarians and veterinary technicians.

This text provides the fundamentals of anesthesia for students of veterinary technology. The techniques and information presented will equip students to provide the best possible care for their patients. Veterinary technicians competent in the practice of anesthesia must know the anatomy and physiology of veterinary patients and understand how medications administered and equipment used in practice will affect patients. Veterinary technicians must also perform physical assessments of their patients, take accurate and detailed medical histories, and care for and advocate for their patients by being keen observers of patients' health status and reactions to drugs given to them. Veterinary technicians must work together with the veterinarian and client to formulate an individualized anesthesia and analgesia plan and thereby ensure that each patient receives the most appropriate and best possible care.

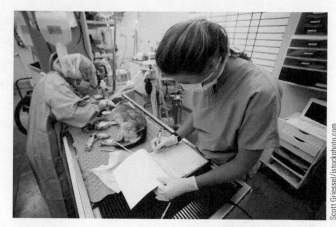

FIGURE 1-1 The veterinary technician's role in anesthesia. Competent veterinary technicians provide anesthesia care to patients in clinical practice by preparing, operating, and maintaining anesthetic equipment; administering anesthetic drugs under the veterinarian's direction; and monitoring and recovering patients of a variety of animal species.

Veterinary technician duties are delegated under veterinarian supervision and are under the authority of the state and/or local veterinary exam boards.[1] For veterinary technicians performing anesthesia, proper delegation typically assumes that personnel are sufficiently trained and evaluated by the delegating veterinarian to possess the essential knowledge and psychomotor and critical thinking skills to perform anesthetic procedures that provide positive patient outcomes. Most states require that veterinarians provide direct or immediate supervision for veterinary technicians performing anesthetic procedures.[1]

Tech Tip

To safely perform anesthesia, veterinary professionals need a thorough knowledge of physiology, pharmacology, and anesthetic equipment function. This knowledge will help the entire anesthesia team understand how the patient's physical condition may affect its physiological responses to anesthesia and surgery.

Tech Tip

Depending on state or national legislature, veterinary technicians may have varying degrees of responsibility and may be able to perform various anesthesia-related procedures. For example, veterinary technicians can perform an epidural on a patient in some countries but not in others. Veterinary technicians should always check with local agencies to ensure they are conforming to the scope of practice for their locale.

THE VETERINARY TECHNICIAN'S ROLE IN ANESTHESIA

Veterinary technicians are the individuals who most commonly provide anesthesia care to patients in clinical practice. Veterinary technicians play a pivotal role in the practice of anesthesia by preparing, operating, and maintaining anesthetic equipment; administering anesthetic drugs under the veterinarian's direction; and monitoring and recovering patients of a variety of species (specific topics are described in their respective chapters) (Figure 1-1). These tasks require an advanced knowledge and skill level not only to monitor and assess the patient, but also to initiate or suggest an appropriate response to the changing physiological status of an anesthesia patient and complications that may arise.

Although the veterinarian has final authority regarding the diagnosis and treatment of patients, the veterinary technician is often the person who administers anesthesia to the patient and monitors their response to the anesthetic agent. Being the patient's advocate is therefore another critical part of the veterinary technician's job and one that should not be taken lightly. Advocating for proper analgesia and nursing care following a procedure improves both the patient's experience and the success of every veterinary anesthesia case.

As part of their educational training, veterinary technician students must obtain the knowledge and skills necessary to practice anesthesia safely and effectively. The following actions will guide students toward this goal:

- *Commit normal values to memory.* Knowing normal vital signs and indicators of **anesthetic depth** (the degree to which the central nervous system (CNS) is depressed by a general anesthetic agent) is the only way to determine if a particular value is abnormal. Without this knowledge, veterinary technicians cannot adjust the level of anesthetic agent given.
- *Concentrate on changes in anesthetized patients.* Being a good observer and a detail-oriented person is important so that subtle changes in the anesthetized patient are noticed. Identifying subtle changes and assessing their importance may prevent complications.
- *Communicate concerns to the veterinarian.* All professionals involved with anesthesia need to know the status of the patient at all times.
- *Contemplate how the mechanisms of action of anesthetic and analgesic drugs will impact the patient.* Knowing how a drug works in the species being anesthetized, at which point in the anesthetic protocol the drug is administered, and its benefits and risks to the patient are important parts of the veterinary technician's job. Being able to anticipate a patient's response to a drug, both desired and potentially adverse, is key to safe and effective drug administration to anesthesia patients.
- *Calculate drug doses and fluid rates carefully.* All drug dosages used and calculations performed are approved by the veterinarian; however, accurate calculations serve as a check-and-balance system among professionals.
- *Consider the limitations of the equipment.* Never rely solely on instrument values when monitoring anesthetized patients, but rather rely on interpretation of these values in conjunction with the patient's parameters obtained during the procedure. For example, if the electrocardiogram (ECG) determines that the patient's heart rate is 30 beats per minute (bpm) yet you auscult a heart rate of 100 bpm and palpate a strong, regular pulse rate, trust your assessment of the patient rather than the instrument.
- *Care for and fully understand how anesthesia equipment works.* Without a thorough understanding of the anesthesia machine and the equipment used to deliver anesthesia, errors in the machine setup or the inappropriate use of equipment can occur. Anesthesia machines also need regular maintenance. Many monitors need regular calibration to ensure accurate display of information used to assess cardiopulmonary status of anesthetized patients. Not properly maintaining and calibrating equipment may put the anesthesia patient at risk. Data or values displayed by equipment that has not been maintained correctly may be erroneous.
- *Continue to expand your knowledge base.* Staying inquisitive, pursuing continuing education (CE), and learning from your (and others') successes (and sometime failures) will ensure that you continually improve your skills and knowledge.

Tech Tip

Anesthesia equipment should calibrated and inspected regularly according to manufacturers' recommendations. Calibration of equipment should be documented and part of the clinic's records.

ANESTHESIA ASSOCIATIONS

Anesthesiology is a service-oriented specialty that helps other fields such as surgery and internal medicine enhance the well-being of their patients. To advance the practice of anesthesiology, both veterinarians and veterinary technicians have formed specialty organizations for professionals interested in advanced training in anesthesia.

Veterinarian Association

Veterinary anesthesiologists are veterinarians who are recognized and respected for their role in providing the highest quality of anesthesia service to veterinary patients. The American College of Veterinary Anesthesia and Analgesia (ACVAA) is the American Veterinary Medical Association's (AVMA)—recognized, not-for-profit specialty board founded in 1975 that sets the standards for advanced professionalism and anesthesia excellence in veterinary anesthesiology.[2] The ACVAA is the governing college that grants credentials to veterinarians as specialists in anesthesia in the United States; its mission statement consists of goals "to promote the highest standards of clinical practice of veterinary anesthesia and analgesia, define criteria for designating veterinarians with advanced training as specialists in the clinical practice of veterinary anesthesiology, issue certificates to those meeting these criteria, maintain a list of such veterinarians, and advance scientific research and education in veterinary anesthesiology and analgesia."[2] Only licensed veterinarians who have successfully completed an internship and 3-year residency, completed and published research, received letters of recommendation, and then passed a written board examination and written Clinical Competency Exam by the ACVAA are Diplomates of the American College of Veterinary Anesthesia and Analgesia and have earned the right to be called Board Certified Specialists in Veterinary Anesthesia and Analgesia®.[2]

To provide optimal patient care, the ACVAA developed Small Animal Monitoring Guidelines and Guidelines for Anesthesia in Horses (see online resources). The goal of these guidelines, in addition to ensuring a knowledgeable anesthesia team, is to improve the level of anesthesia care for veterinary patients.

Veterinary Technician Association

The importance of veterinary technicians' role in the accurate administration and monitoring of anesthesia has been demonstrated by the formation of a specialty group that certifies each technician as a Veterinary Technician Specialist (VTS) in Anesthesia/Analgesia. To become a VTS (Anesthesia/Analgesia) and member of the Academy of Veterinary Technicians in Anesthesia and Analgesia (AVTAA), veterinary technicians must complete formal education that leads to becoming a credentialed veterinary technician, complete and document extensive postgraduate experience in high-quality anesthesia and analgesia, receive letters of recommendation, obtain and document continuing education in anesthesia, submit case reports and case logs, and demonstrate advanced psychomotor skills and submit all of these as part of the application process. Once the application is accepted, veterinary technicians must subsequently pass a multiple-choice and practical examination. Veterinary technicians who become certified as a VTS (Anesthesia/Analgesia) demonstrate advanced knowledge in the care and management of anesthesia cases, promote patient safety, and provide excellence in anesthesia care and competence in this continually evolving specialty. The VTS (Anesthesia/Analgesia) title is similar to the title of nurse anesthetist in human medicine.

The vision of VTS (Anesthesia) began in 1996, when the Veterinary Technician Anesthetist Society (VTAS) petitioned the National Association of Veterinary Technicians in America—Committee on Veterinary Technician Specialties (NAVTA-CVTS) for recognition as a certifying body of veterinary technician specialists in anesthesia.[3] In 1998, a formal petition was submitted to NAVTA-CVTS requesting recognition of the Academy of Veterinary Technicians in Anesthesia (AVTA) by NAVTA as a certifying body for veterinary technician specialists in anesthesia (VTS [Anesthesia]).[3] The AVTA was officially recognized by the NAVTA-CVTS in January 1999 and was the second organization to be recognized by NAVTA-CVTS.[3] In 2014, the American College of Veterinary Anesthesia (ACVA) officially changed its name to the American College of Veterinary Anesthesia and Analgesia (ACVAA). In concert with this, the AVTA changed its name to the Academy of Veterinary Technicians in Anesthesia and Analgesia (AVTAA) in 2015 with the continued goal of promoting patient safety, consumer protection, professionalism, and excellence in the provision of anesthesia and analgesia to veterinary patients.

Other Organizations

In addition to the anesthesia specialty associations, other veterinary organizations support the practice of veterinary anesthesia and pain management through the development of standard of care guidelines and publication of research and clinical recommendations for veterinary professionals. For example, the American Animal Hospital Association's (AAHA) Anesthesia Guidelines for Dogs and Cats provides recommendations for preanesthetic patient evaluation; selection of premedication, induction, and maintenance agents as well as equipment; and standards of care for monitoring and recovering anesthetized patients.[4] The AVMA provides its members with notification of anesthesia guidelines, publishes recommendations on procedures that require anesthesia and consent forms for anesthesia procedures, and develops client brochures to educate owners about anesthesia and its risks.[5] The International Veterinary Academy of Pain Management (IVAPM) offers an interdisciplinary certification program (Certified Veterinary Pain Practitioner [CVPP])[6] for veterinarians and veterinary technicians. To become a CVPP, a veterinary health professional must have a minimum of 5 years of full-time practice and at least 2 years of clinical experience working with animals in pain, obtain continuing education in pain management, submit letters of recommendation, document ability to perform skills related to pain management, submit case reports, and pass a written exam.[6] The North American Veterinary Anesthesia Society (NAVAS) is an organization that works in collaboration with the existing ACVAA/European College Veterinary Anaesthesia and Analgesia (ECVAA) Specialty Colleges and AVTAA "to strengthen and advance international efforts to improve and achieve quality veterinary anesthesia and analgesia care on a global scale."[7] NAVAS develops guidelines based on scientific publications and textbooks to help veterinary professionals and caregivers advance and improve the safe administration of anesthesia and analgesia to all patients.[7] NAVAS's mission parallels that of the ACVAA/ECVAA and their board-certified diplomates, who are recognized by the AVMA as specialists in these fields of clinical medicine in North America.[7]

ANESTHESIA TERMINOLOGY

To communicate and practice effectively with professionals, veterinary technicians must be familiar with common terms used in anesthesia, including those that describe the types of anesthesia and categories of drugs used to anesthetize patients. Knowing which category a drug belongs to will help veterinary technicians know when in the anesthetic protocol the drugs is used (premedication, inductions, maintenance, or recovery), the drug's desired effect (sedation, analgesia, local versus general anesthesia, etc.), and the drug's potential risks (cardiovascular and/or respiratory depression, alteration of nervous system function, vomiting and potential aspiration of vomitus, etc.). In addition, the use of correct terminology reflects technicians' professionalism and an understanding of the practice of veterinary anesthesia.

Some of the many terms used in describing the effects of anesthetic drugs and related procedures are as follows:

- **General anesthesia**: A reversible state of unconsciousness produced by anesthetic agents, with absence of pain perception over the entire body and a greater or lesser degree of muscular relaxation; the drugs producing this state are most commonly administered by inhalation, intravenously, or intramuscularly.
- **Surgical anesthesia**: The degree of anesthesia at which a response to surgical stimulus does not occur.
- **Balanced anesthesia**: A theory in which general anesthesia is produced by administering several drugs with the goal of exploiting each drug's positive actions while avoiding potential adverse effects associated with large doses of a single drug. The philosophy encourages the use of several agents, each designed to affect a different function (e.g., providing muscle relaxation and amnesia or blocking pain and motor function).
- **Local anesthesia**: Anesthesia produced in a limited area, as by injection of a local anesthetic, topical application, or freezing. Examples include infiltration, topical, regional block, spinal, and epidural techniques.
- **Regional anesthesia**: Anesthesia caused by interrupting the sensory nerve conductivity of any region of the body. Regional anesthesia may be produced by a **field block** (encircling the operative field by means of injections of a local anesthetic) or by a **nerve block** (making multiple injections in close proximity to the nerves supplying the area).
- **Dissociative anesthesia**: Anesthesia that produces a catalepsy-like state, in which the patient feels dissociated from its environment, and good somatic analgesia but poor visceral analgesia.
- **Analgesia**: Relief of pain without loss of consciousness; absence of pain or noxious stimulation.
- **Sedation**: The state characterized by CNS depression accompanied by sleepiness/drowsiness and some degree of relaxation.
- **Tranquilization**: Administration of any of a group of compounds that calm and relax an anxious patient but do not induce sleep.
- **American Society of Anesthesiologists (ASA) physical status**: The ASA physical classification system rating patient risk during anesthesia based on the patient's health (see Chapter 3, Cardiovascular and Respiratory Physiology Review).

SUMMARY

No single anesthesia protocol is appropriate for all species or procedures; therefore, properly trained veterinary anesthetists are essential to successful anesthesia in patients. Drugs and techniques should be selected on an individual patient basis that focuses on identifying a list of potential patient problems and management concerns. For example, a patient undergoing surgery for a cruciate ligament repair will have different anesthetic and analgesic needs from a patient that has heart failure and is having a tumor surgically removed. A geriatric dog with kidney disease will have a different patient problem list and different management concerns from a healthy kitten being neutered or a racehorse with a fractured leg or a bird that is egg bound. The veterinary technician's job is to understand veterinary anatomy, physiology, and pharmacology as it relates to the practice of anesthesia as well as to identify concerns associated with anesthesia and thereby customize an anesthesia plan and anticipate potential complications. Due to the virtually endless supply of information provided by the Internet, clients are well informed and demand up-to-date procedures for their animals as well as knowledgeable and skilled veterinary professionals to care for them. It is therefore important for veterinary technicians to continue learning about new drugs and techniques used for anesthesia. In modern medicine, including anesthesia, veterinary technicians have access to immense amounts of information as well as ongoing updates in veterinary anesthesia. With these advantages comes the responsibility of using this information to provide optimal care for veterinary patients.

CRITICAL THINKING POINTS

- Ever-expanding knowledge in the field of veterinary anesthesiology makes it crucial for professionals to stay informed on the development of new drugs, equipment, and techniques that will provide the best patient care in the perioperative period.

- The role of veterinary technicians in preparing, operating, and maintaining anesthetic equipment; administering anesthetic drugs; and monitoring and recovering patients of a variety of species is pivotal in providing effective anesthesia in veterinary patients.

- Veterinary technician duties are delegated under veterinarian supervision and are under the authority of the state and/or local veterinary exam board. Most states require that veterinarians provide direct or immediate supervision for veterinary technicians performing anesthetic procedures.

- Being familiar with professional veterinary associations and their anesthesia recommendations is key to providing optimal care for patients.

- Using correct anesthetic terminology displays professionalism and an understanding of the practice of veterinary anesthesia.

REVIEW QUESTIONS

1. What is the correct definition of the term *sedation*?
   a. A drug-induced reversible state of unconsciousness and loss of sensation
   b. The state characterized by CNS depression accompanied by sleepiness/drowsiness and some degree of relaxation
   c. A drug-induced state of calm in which anxiety is relieved the patient is relaxed but remains aware of surroundings
   d. The loss of sensation in a discrete area of the body caused by the administration of a local anesthetic agent

2. What is the correct definition for the term *tranquilization*?
   a. A drug-induced reversible sate of unconsciousness and loss of sensation
   b. A drug-induced CNS depression and drowsiness in which the patient is generally unaware of surroundings but will respond to painful stimuli
   c. The administration of any of a group of compounds that calm and relax an anxious patient but do not induce sleep
   d. The loss of sensation in a discrete area of the body caused by the administration of a local anesthetic agent

3. What do the letters ACVAA stand for?
   a. American Congress for Veterinary Anesthesia and Analgesia
   b. Association Certification in Veterinary Anesthesia and Analgesia
   c. American College of Veterinary Anesthesia and Analgesia
   d. American Congress for Veterinary Analgesia and Anesthesia

4. The Academy of Veterinary Technicians in Anesthesia petitioned which veterinary technician organization in 1998 for recognition as the certifying body for veterinary technician specialists in anesthesia?
   a. Veterinary Support Personnel Network
   b. National Association of Veterinary Technicians in America— Committee on Veterinary Technician Specialties
   c. VetMedTeam—Anesthesia Technicians
   d. Veterinary Anaesthesia School for Technicians

5. Which two organizations write anesthesia guidelines?
   a. AVMA and AAHA
   b. AVMA and NAVTA
   c. IVAPM and ACVAA
   d. IVECCS and IVAPM

6. The job of working in veterinary anesthesia entails
   a. monitoring and recovering the anesthesia patient, caring for all anesthetic equipment, and administering certain medications.
   b. preparing and monitoring surgical patients, autoclaving surgical equipment, and discharging patients after they recover from anesthesia.
   c. monitoring and recovering the anesthesia patient, choosing which medications to administer, and preparing surgical summary reports.
   d. choosing which medications to administer, recording vital signs of the patient while under anesthesia, and disinfecting surgical equipment.

7. Why is it important to have anesthetic equipment calibrated regularly?

   a. Without calibration there is no guarantee that everyone has been properly trained in the use of anesthesia equipment.
   b. Without calibration the equipment will require extensive repairs the next time it is serviced.
   c. Without calibration unqualified veterinary staff may manually alter the anesthetic level.
   d. Without calibration the equipment may give incorrect readings and/or operate incorrectly.

8. What does ASA stand for?

   a. Anesthetic Society of Anesthetists
   b. Association for Safe Anesthesia
   c. American Society of Anesthesiologists
   d. Anesthesia Safety for Animals

9. What does *balanced anesthesia* mean?

   a. Anesthetic drugs are weighed on a scale before being administered to the patient because many drugs are administered during the anesthetic event
   b. General anesthesia produced by administering several drugs with the goal of utilizing each drug's positive actions while avoiding potential adverse effects associated with large doses of a single drug
   c. Anesthesia produced by introducing the anesthetic agent into the rectum in an attempt to use varying administration routes
   d. General inhalation anesthesia in which there is no rebreathing of the expired gases

10. What does the term *analgesia* mean?

    a. Relief of pain without loss of consciousness; absence of pain or noxious stimulation
    b. Injection of an agent into the spinal canal, generally either into the subarachnoid or epidural space
    c. A substance used for the control of excessive reaction to or overdosing with a narcotic
    d. The combining of pain-relieving agents that act on different mechanisms of nociceptive modulation to enhance additive and synergistic effects.

11. To perform ocular procedures, such as corneal foreign body removal, proparacaine drops are applied to desensitize the cornea. Which type of anesthesia does this describe?

    a. Regional
    b. Surgical
    c. Local
    d. Dissociative

12. By which routes is general anesthesia is most commonly administered?

    a. Intravenously, intramuscularly, and subcutaneously
    b. Intravenously, subcutaneously, and intraperitoneally
    c. Inhalation, intramuscularly, and intraperitoneally
    d. Inhalation, intravenously, and intramuscularly

13. A dog is given anesthetic agents and becomes unconscious and does not have any pain perception. Which type of anesthesia does this describe?

    a. Dissociative
    b. General
    c. Regional
    d. Local

14. A sedated horse is given a maxillary nerve block in order to perform dental work on the upper incisors. Which type of anesthesia is a nerve block?

    a. General
    b. Regional
    c. Dissociative
    d. Tranquilization

15. Which of the following organizations offers the Certified Veterinary Pain Practitioner for veterinarians and veterinary technicians?

    a. AVMA
    b. IVAPM
    c. AAHA
    d. NAVAS

Case Study

1. You have accepted a position at a small animal veterinary hospital. One of your duties according to your job description will be to work on surgical cases assisting with anesthesia. List your top four responsibilities in assisting with anesthesia.

Critical Thinking Questions

1. Jada is a veterinary technician student who has done well in her previous courses yet is anxious about taking the anesthesia course because is it one of the more advanced courses in her program. To be successful, she knows she will need to utilize information and skills she learned in previous courses and apply it, along with content she is currently learning, to perform advanced skills such as administering anesthesia to patients. As she reads the introductory chapter to prepare for her anesthesia class, she begins to feel overwhelmed with the amount of information she will need to learn to become a competent anesthetist. Jada texts one of her classmates to ask if she is the only one stressed about taking the anesthesia course. Jada learns that many other students are also concerned about whether they will be able to learn all of the information needed to administer anesthesia safely to their patients. Jada and her classmates decide to have a group chat about how they can be better prepared for their anesthesia class and less

anxious when they have to perform anesthesia on patients. What are some things Jada and her classmates can do now to help them prepare for their anesthesia class?

2. A client calls the clinic to make an appointment to have her cat spayed. As the veterinary technician responsible for admitting surgical/anesthesia cases, you begin to explain the dropoff procedure for the surgery day (i.e., signing the informed consent form and describing the risks involved with anesthesia to the client). The client tells you that she is usually in a hurry in the morning and needs to "simply drop off the cat to be put to sleep so she can get spayed." How can you quickly educate the client about the process of anesthetizing this cat using professional language that is also clear to the client?

ENDNOTES

1. Staff research, Duties of veterinary technicians and assistants, American Veterinary Medical Association, https://www.avma.org/advocacy/state-local-issues/duties-veterinary-technicians-and-assistants, May 2019.
2. American College of Veterinary Anesthesia and Analgesia, www.acvaa.org, 2020, accessed May 19, 2020.
3. Academy of Veterinary Technicians in Anesthesia and Analgesia, https://www.avtaa-vts.org/, 2019.
4. Grubb, T., Sager, J., Gaynor, J. S., Montgomery, E., Parker, J. A., Shafford, H., & Tearney, C. (2020). 2020 AAHA anesthesia and monitoring guidelines for dogs and cats. *Journal of the American Animal Hospital Association*, *56*(2), 59–82.
5. American Veterinary Medical Association, www.avma.org, 2020, accessed May 19, 2020.
6. International Veterinary Academy of Pain Management, www.ivapm.org, 2020.
7. North American Veterinary Anesthesia Society, https://www.mynavas.org/, 2020.

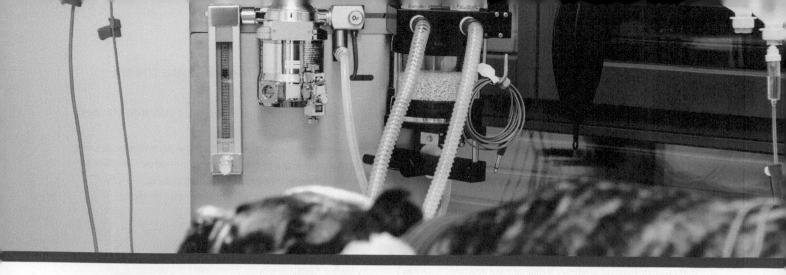

CHAPTER 2
The Anesthesia Machine

Kristen Cooley, BA, CVT, VTS (Anesthesia/Analgesia)

Teri Raffel Kleist, CVT, VTS (Surgery)

LEARNING OBJECTIVES

Upon completion of this chapter, it is expected that the reader should be able to:

2.1 Describe different types of oxygen sources

2.2 Outline compressed gas cylinder safety

2.3 Identify the fundamental components of the anesthesia machine

2.4 Explain the function of each fundamental component of the anesthesia machine

2.5 Describe the safe use of each fundamental component of the anesthesia machine

2.6 Identify the components of the machine's breathing system

2.7 Explain the function of each component of the machine's breathing system

2.8 Describe the safe use of each component of the machine's breathing system

2.9 Recognize the different types of breathing circuits and which patients use each type

2.10 Select appropriate oxygen flow rates based on patient needs

2.11 Outline proper anesthesia machine maintenance

2.12 Explain how to pressure check the anesthesia machine

KEY TERMS

Active scavenging systems

Adjustable pressure limiting (APL) valve

Barotrauma

Carbon dioxide (CO_2) absorber

Closed circle system

Coaxial

Common gas outlet

Expiratory pause

Flowmeter

Fresh gas inlet

Hanger yokes

High-pressure system

Hypoxic mixture

Intermediate-pressure system

Low-flow circle system

Low-pressure system

Mechanical dead space

Metabolic oxygen consumption

Non-rebreathing systems

Oxygen concentrator

Oxygen flush valve (oxygen fast flush valve)

Passive scavenging systems

Pin Index Safety System (PISS)

Pop-off valve

Pressure manometer

Pressure regulators (pressure reducing valves)

Reservoir bag

Semiclosed circle system

Splitting ration

Tidal volume

T-piece

Unidirectional valves

Vaporizers

Volatile

Y-piece (wye-piece)

Yoke block

INTRODUCTION

Inhalant (gas) anesthesia is the basis of many anesthetic protocols in veterinary medicine. An anesthesia machine delivers precise amounts of inhalant anesthetic agent to the patient. Although the basic components and functions of all anesthetic machines are similar, they can vary in design, from simple mobile devices to complex workstations with monitoring systems. All anesthesia machines have a gas source (for agents such as oxygen, medical grade air, and nitrous oxide), pressure regulator, flowmeter, and vaporizer that are connected to a breathing system for delivery of inhalant anesthetic agent to the patient. The anesthesia machine's function is to direct a controlled flow rate of oxygen to a vaporizer, where a liquid anesthetic agent is introduced into a carrier gas (e.g., oxygen, oxygen and medical grade air, oxygen and nitrous oxide), which is then delivered at a controlled flow rate to the patient via a breathing circuit. Carbon dioxide (CO_2) is then removed from exhaled gases through a CO_2 absorbent and moved away from the patient via a scavenging system or is moved away from the patient if the gases are recirculated. Understanding the parts of the anesthetic machine is necessary for its safe and effective use and is the goal of this chapter.

THE ANESTHESIA MACHINE

The anesthesia machine supplies both oxygen and inhalant anesthetic agent to the patient, removes carbon dioxide from the breathing system, and facilitates controlled ventilation. The anesthesia machine consists of a series of components that work together to deliver controlled amounts of oxygen and inhalant anesthetic agent into a breathing system. The fundamental components of the machine are involved in delivering controlled amounts of gas and include

- compressed medical gas source (oxygen and other medical gases such as N_2O);
- pressure regulator (pressure reducing valve);
- flowmeter;
- vaporizer; and
- oxygen flush valve.

Other components of the machine make up the breathing system and include

- common gas outlet (also called the fresh gas inlet);
- unidirectional valves;
- breathing hoses;
- reservoir bag;
- pop-off valve;
- pressure manometer;
- carbon dioxide absorber; and
- scavenging system.

The pressure of gases in the anesthesia machine varies depending on their location. There are high, intermediate, and low-pressure systems in different areas of the anesthesia machine. The components associated with each system are grouped together based on the amount of pressure present. The **high-pressure system** takes gases at O_2 cylinder pressure and reduces and regulates them so they can be used safely within the machine.[1] The high-pressure system includes the compressed gas cylinder or piped oxygen, **hanger yokes**, high-pressure hoses, pressure gauges, and pressure regulators (also known as the pressure reducing valve). Pressures in the high-pressure system may be as high as 2200 psi (Figure 2-1a).

The **intermediate-pressure system** receives gas from the pressure regulator and moves it into the flush valve and flowmeter. Oxygen must flow through a pressure regulator to reduce pressures to a level that can be used safely within the anesthesia machine and used at a constant pressure level regardless of what pressure is upstream (e.g., in an oxygen cylinder). Once in the machine, oxygen is piped to the flowmeter and the oxygen flush valve. The pressure in the intermediate system range is adjustable but is typically set between 50 and 55 psi (pounds per square inch) in most anesthesia machines with one regulator (the user can change this pressure level within the anesthesia machine, which accounts for the wide range (35–75 psi) in pressure, although it is often left at the manufacturer setting). Newer anesthesia machines may have a secondary regulator that reduces the pressure even lower for added safety (Figures 2-1b and 2-1c)[2].

The **low-pressure system** is comprised of the components downstream (closer to the patient) of the flow control valve. These components consist of the flowmeter, piping from the flowmeter to the vaporizer, the vaporizer, the channel from the vaporizer to the common gas outlet, and the channel from the common gas outlet to the breathing system. Pressures within this system are near atmospheric pressure and will vary once they reach the breathing system; for example, between 0 and 30 cm H_2O (Figures 2-1d and 2-1e).[1]

Tech Tip

Working pressures of compressed gas cylinders and anesthetic machines are typically expressed in pounds per square inch (psi) while pressures within the breathing system are expressed in centimeters of water (cm H_2O).

Fundamental Components of the Anesthesia Machine

The fundamental components of the anesthesia machine are the compressed medical gas source, the pressure regulator (pressure reducing valve), flowmeter, vaporizer, and oxygen flush valve. They are considered the fundamental components because they are of central importance to delivering controlled amounts of oxygen and inhalant anesthetic agent into a breathing system.

Safety Alert

Anesthesia machines used in human medicine are required to meet design and safety standards established by the American Society for Testing and Materials (ASTM) International and the Canadian Standards Association (CSA); however, those designed for veterinary use do not have to meet any safety standards other than those associated with hazards to the anesthetist. It is critical for the veterinary anesthetist to understand the function of each component of the anesthesia machine to ensure the machine has been designed to operate safely.

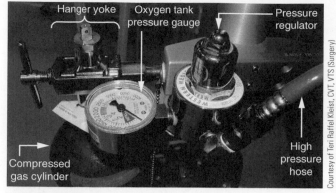

(a)

Hanger yoke Oxygen tank pressure gauge Pressure regulator

Compressed gas cylinder

High pressure hose

Courtesy of Teri Raffel Kleist, CVT, VTS (Surgery)

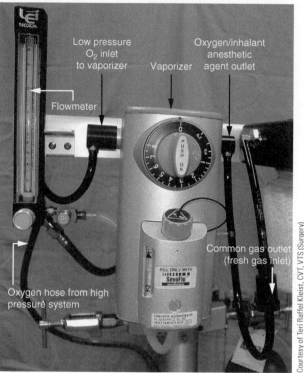

(b)

Oxygen pipeline to vaporizer

Oxygen pipeline from local tank

Oxygen pipeline from large/central tank

Oxygen flush valve

Flowmeter knob

Courtesy of Teri Raffel Kleist, CVT, VTS (Surgery)

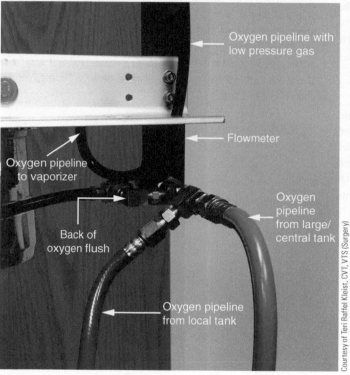

(c)

Oxygen pipeline with low pressure gas

Flowmeter

Oxygen pipeline to vaporizer

Back of oxygen flush

Oxygen pipeline from large/central tank

Oxygen pipeline from local tank

Courtesy of Teri Raffel Kleist, CVT, VTS (Surgery)

(d)

Low pressure O₂ inlet to vaporizer Vaporizer Oxygen/inhalant anesthetic agent outlet

Flowmeter

Common gas outlet (fresh gas inlet)

Oxygen hose from high pressure system

Courtesy of Teri Raffel Kleist, CVT, VTS (Surgery)

FIGURE 2-1 Anesthesia Machine Pressure Systems (a) The high-pressure system includes the compressed gas cylinder or piped oxygen, hanger yoke/yoke block, high-pressure hoses, pressure gauges, and pressure regulators. Pressures in the high-pressure system may be as high as 2200 psi. (b) The intermediate-pressure system receives gas from the pressure regulator and moves it into the flush valve and flowmeter. It includes the pipeline inlet connections, pipeline pressure gauge (may not be present on portable anesthetic machines), conduits from pipeline to flowmeter, conduits from regulator to flowmeter, flowmeter assembly, oxygen fail-safe valve, oxygen supply failure valve, and oxygen flush valve. The pressures in the intermediate system typically range from 50 to 55 psi in most anesthesia machines with one regulator. (c) Rear view of the intermediate-pressure system connections. (d) The low-pressure system is where the oxygen and volatile anesthetic agent join together and includes the flowmeter, piping from the flowmeter to the vaporizer, the vaporizer, the channel from the vaporizer to the common gas outlet, and the channel from the common gas outlet to the breathing system. Pressures within this system are near ambient pressure and will vary once they reach the breathing system; between 0 and 30 cm H₂O.

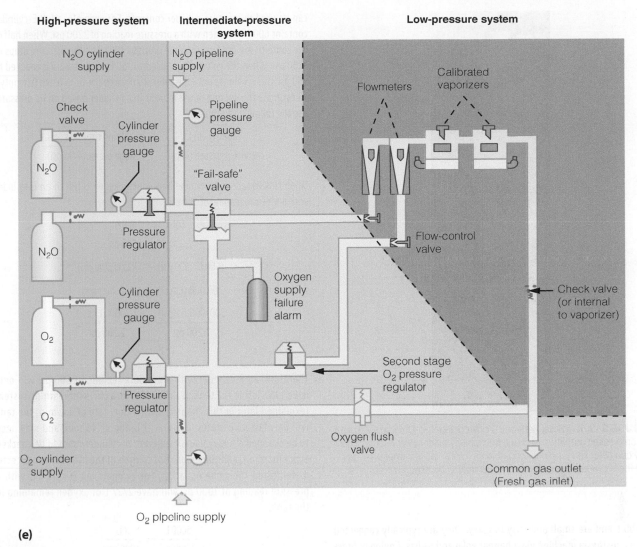

High-pressure system **Intermediate-pressure system** **Low-pressure system**

(e)

FIGURE 2-1 (*Continued*) (e) Pressure system labeled diagram of an anesthesia machine outfitted with a circle or rebreathing system.

Compressed Medical Gas Sources

Compressed medical gases create the pressures within the anesthetic machine. Anesthesia machines are designed to deliver **volatile** (easily evaporate at room temperature) anesthetics in the presence of compressed gas (mainly oxygen). Because many inhalant anesthetic agents cause some degree of respiratory depression, high oxygen concentrations (>30%) are used to help the patient deliver adequate amounts of oxygen to the tissues. While other medical gases (such as nitrous oxide or medical grade air) can be delivered with oxygen, oxygen must always be part of the anesthetic gas to avoid hypoxic mixtures. Inadequate amounts of oxygen can lead to hypoxemia and asphyxiation. Many years ago nitrous oxide (N_2O) was commonly used in veterinary anesthesia. Despite its benefits of providing analgesia and cardiovascular stability, the use of N_2O has dramatically decreased due to human abuse potential, additional cost, and its inability to produce general anesthesia in dogs and cats.

Oxygen gas can be delivered to the anesthetic machine in several ways: compressed gas in a metal cylinder, an oxygen concentrator, or liquid oxygen through a central gas supply system.

Compressed Gas Cylinders

Most medical gases are stored in metal cylinders under pressure. Compression, or pressure, of the gas is necessary to fit a large amount of it into a small container. Compressed gas cylinders are beneficial because they provide the pressure necessary to run a variety of types of anesthesia equipment. Small cylinders are portable and can be easily connected to an anesthesia machine for transport. Compressed gas cylinders are the most common way oxygen and other medical gases are supplied in veterinary medicine. The disadvantages of this system are that compressed gas cylinders can be dangerous if not used correctly and they only provide a finite amount of gas.

Many cylinder sizes are available, but the most commonly used sizes are E and H (Figure 2-2). E-cylinders have an approximate volume

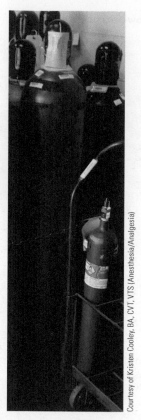

FIGURE 2–2 H-cylinder (left) and E-cylinder (right) both contain 2200 psi of compressed gas when new and full but they differ in the volume of gas they can hold. The E-cylinder can hold up to 600 mL of compressed gas and the H-cylinder can hold 10 times as much (6000 mL).

of 600 L and are small and easy to carry. They are typically connected to an anesthesia machine via a hanger yoke and utilize a nylon or brass washer to form a seal. These tanks are recommended for use during patient transport, as part of a portable anesthetic machine configuration, or in an emergency situation. One full back-up E-cylinder should always be on hand.[1] E-cylinders should be checked before every surgery and replaced when the pressure gauge is below 500 psi (if heavy use is expected). If light use or a short duration of use is anticipated, a tank that has a pressure reading below 200 psi may be used but must be closely monitored to avoid running out of oxygen. The H-cylinder can hold up to 6000 L; therefore, it is large and heavy. These cylinders are typically stationary and often stored away from the machine and piped in for use. Banks of H-cylinders can be connected in tandem and piped throughout the hospital. The anesthesia machine can access the H-cylinder system via quick-release connections. Regardless of size, all oxygen tanks should be secured with two chains or other means to prevent them from tipping over. Free-standing tanks are absolutely prohibited.

Full oxygen cylinders contain 2200 psi of gas regardless of their size. Pressure within the cylinder decreases as gas volume is depleted. As gas is released more space is available for the gas molecules to move within the cylinder; therefore the pressure reading decreases. The pressure reading

can be used to measure cylinder contents. For example, a full E-cylinder contains 600 L of oxygen with a pressure reading of 2200 psi. When half of the contents are used, the volume drops to 300 L and the pressure drops to 1100 psi. When the volume is reduced to 150 L, the pressure is reduced to 550 psi; this continues until the volume and pressure both reach 0 (empty). To estimate the volume of gas in *liters* in a cylinder based on its pressure, use the following formula:

$$\frac{\text{capacity (in L)}}{\text{service pressure (in psi)}} = \frac{\text{remaining contents (in L)}}{\text{gauge pressure (in psi)}}$$

Using this equation, determine how much oxygen is left in an E-cylinder with a pressure reading of 500 psi.

$$\frac{600\,L}{2{,}200\,\text{psi}} = \frac{X\,L}{500\,\text{psi}}$$

$$(600\,L)\,(500\,\text{psi}) = (X\,L)\,(2{,}200\,\text{psi})$$

$$300{,}000\,(L)(\text{psi}) = 2{,}200\,X\,(L)\,(\text{psi})$$

$$\frac{300{,}000\,(L)(\text{psi})}{2{,}200\,\text{psi}} = \frac{2{,}200\,X\,(L)(\text{psi})}{2{,}200\,\text{psi}}$$

$$136\,L = X$$

Therefore, an E-cylinder with a pressure reading of 500 psi will only have 136 L left in the tank. If a patient is on a circle rebreathing system (explained later in this chapter) with a flow rate of 3 L/min, the tank will only last 45 minutes (136 L ÷ 3 L/min = 45 min) and may need to be changed if heavy use is expected. H-cylinders should be checked every morning, and the amount of oxygen in the cylinder can be estimated using the same equation. For example, an H-cylinder with a pressure reading of 1000 psi will have 2727 L of oxygen remaining in the tank.

$$\frac{6000\,L}{2{,}200\,\text{psi}} = \frac{X\,L}{1000\,\text{psi}}$$

$$(6000\,L)\,(1000\,\text{psi}) = (X\,L)\,(2{,}200\,\text{psi})$$

$$6{,}000{,}000\,(L)\,(\text{psi}) = 2{,}200\,X\,(L)\,(\text{psi})$$

$$\frac{6{,}000{,}000\,(L)(\text{psi})}{2{,}200\,\text{psi}} = \frac{2{,}200\,X\,(L)(\text{psi})}{2{,}200\,\text{psi}}$$

$$2{,}727\,L = X$$

Tech Tip 🐾

Oxygen levels should be checked regularly to ensure that enough oxygen is available for use during the entire procedure. Low oxygen alarms are available to alert the veterinary technician when oxygen reaches a critical level and the cylinder should be replaced (but often are not used unless it is included in an anesthesia machine or built into the central supply; therefore, the average anesthesia machine will not sound an alarm when oxygen levels are low).

TABLE 2-1 Color Coding of Medical Gases

Gas	Formula	United States	International
Oxygen	O_2	Green	White
Carbon dioxide	CO_2	Gray	Gray
Nitrous oxide	N_2O	Blue	Blue
Helium	He	Brown	Brown
Nitrogen	N_2	Black	Black
Air	---	Yellow	Black & White

Data from American Society for Testing and Materials (ASTM) International

Compressed gas cylinders, hoses, and connections are color coded for safety to prevent inadvertent connection of the wrong gas (Table 2-1). A **pin index safety system (PISS)** is also present on E-cylinders and consists of a series of holes located just beneath the outlet port on the neck of the compressed gas cylinder; corresponding pins are located on the hanger yoke or pressure regulator (Figures 2-3a and 2-3b). Unless the pins and holes are aligned, the cylinder will not seat properly. A hanger yoke or pressure regulator may be without pins and will seat on any cylinder; care must always be taken to assure that these connections are correct.[3] Sometimes a **yoke block** is used when a compressed gas cylinder is removed from the hanger yoke (Figure 2-3c). The yoke block prevents dust and dirt from entering the yoke area. Manufacturers often chain yoke blocks to the machine.

Compressed gas cylinders can be opened with a short counterclockwise turn. This is accomplished by use of a handle, wrench, or handwheel located near the neck of the cylinder. The same component can be turned clockwise to close the cylinder. Large cylinders (H) have handwheels whereas smaller cylinders (E) use a wrench to open and close the valve. Wrenches should be attached to the anesthesia machine near the cylinder for ease of use.

Oxygen Concentrators

An **oxygen concentrator** can also supply oxygen to an anesthesia machine when obtaining and storing compressed gas cylinders is inconvenient, expensive, or impossible (Figure 2-4). The oxygen concentrator collects room air and separates the oxygen from the other gases; namely, nitrogen, argon, and carbon dioxide. Oxygen concentrators are a cost-effective way to supply oxygen without having the risk of housing compressed gas cylinders. Oxygen concentrators require regular maintenance to ensure that appropriate levels of oxygen are being delivered. Pressures exiting the oxygen concentrator tend to be lower than those exiting a compressed gas cylinder; therefore, a single unit oxygen concentrator will generate enough pressure to effectively run an anesthesia machine, but not a mechanical ventilator (typically 50–55 psi is required). Based on the concern that oxygen concentrators may not produce a reliable concentration of oxygen and may not generate adequate pressure, these units are not used frequently in veterinary medicine.

Central Gas Supply Systems

Some larger hospitals may have a central supply of oxygen that is piped to various areas of the hospital. This may be a liquid supply of oxygen or a bank of cylinders connected to a common manifold that converts multiple cylinders to a continuous supply chain (Figures 2-5a through 2-5c). An advantage of this system is the continuous supply of oxygen. Like any complicated system, proper maintenance is essential and for central gas systems; maintenance teams often include medical engineers and facility managers.

Veterinary clinics and hospitals should choose their oxygen source based on need. Small practices may use an oxygen concentrator or

Safety Alert

Cylinder Safety:

- Cylinder valves, pressure regulators, gauges, or fittings should never come into contact with combustible substances, greases, oils, or lubricants. Oxygen supports combustion, and these substances become highly combustible in the presence of 100% oxygen.
- Cylinders should be kept away from extreme changes in temperature (below −7°F or above 125°F) to prevent fluctuations in the tank pressure.
- Only those regulators, hoses, and gauges designed for the type of gas in the tank should be used. They are designed for use with specific gases at prescribed pressure ranges. The use of an incorrect regulator or gauge may cause the gas to ignite.
- Valves should be kept closed when the cylinder is not in use to avoid pollution, combustion, and wasting of the gas. Cylinders should be turned off with minimal force to avoid jamming the valve in the closed position and damaging the threads.

- Cylinders should only be transported using a cart or carrier made specifically for the transport of gas cylinders. To avoid turning the cylinder into a high-speed rocket, do not drop, drag, or roll cylinders (which can damage the cylinder) and do not lift cylinders by their necks or valve caps (the neck is the weakest point of the unit, and lifting the cylinder by the neck can damage it and cause release of unregulated, very high-pressure gas).
- Cylinders should always be kept properly secured to prevent them from falling or being knocked over. Do not allow them to sit upright unsecured. If there is no way to secure a cylinder in an upright position, it is safer to lay it on its side. Damaged cylinders can be very hazardous because the unregulated release of high-pressure gas can turn the cylinder into a dangerous projectile.
- An empty cylinder is never out of gas and should be treated the same way as a full cylinder to avoid injury.[1, 3]

(a) **(b)** **(c)**

FIGURE 2–3 The pin index safety system (PISS) uses precisely placed holes on the neck of the gas cylinder that correspond to pins located in the hanger yoke specific to that type of gas. The PISS is in place to prevent the wrong gas from being connected to the anesthesia machine. (a) shows the nitrous oxide PISS and (b) shows the oxygen PISS. Notice the difference in the spacing of the pins. (c) an example of a yoke block.

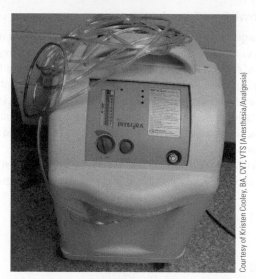

FIGURE 2–4 An oxygen concentrator is a device that concentrates oxygen from the ambient air to deliver an enriched oxygen mixture to the anesthesia machine. It is a cost-effective way to provide oxygen in situations where the presence of compressed gas cylinders is undesirable.

E-cylinders because the amount of oxygen they consume for anesthesia and nursing care is minimal. Larger clinics may use one or two H-cylinders piped to the operating room, whereas even larger clinics may use a bank of cylinders with a common manifold. Teaching hospitals and universities may use a large storage tank of liquid oxygen because their oxygen needs are high and constant.

Pressure Regulators

Pressure regulators (also known as **pressure reducing valves**) control the pressure at a steady level and reduce the compressed gas cylinder's pressure (2200 psi when new) to a level that is appropriate and safe for use inside the anesthesia machine (e.g., 55 psi). The pressure regulator should have two gauges, one that reads the tank pressure and one that reads the line pressure (which may not always be available or visible to the user) (Figure 2-6). Each type of medical gas supplied to the anesthesia machine should have a regulator specially designed for that particular gas.[3] Pressure regulators are set to drop pressure from as high as 2200 psi down to 35–50 psi. Secondary

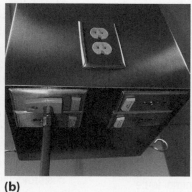

(a) **(b)** **(c)**

FIGURE 2–5 A central gas supply utilizes ceiling drops or wall connections to access the oxygen hospital-wide. (a) a ceiling drop column, (b) an oxygen hose plugged into the column, (c) an oxygen hose plugged into a wall outlet.

FIGURE 2-6 The pressure regulator reduces the pressure of the gas entering the anesthesia machine.

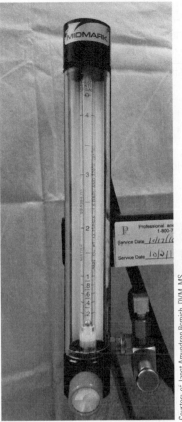

FIGURE 2-7 The oxygen flowmeter controls the flow of oxygen through the anesthesia machine.

regulators may be present to drop those pressures even lower, to 12–16 psi. The line pressure gauge should read 55 psi for use with most anesthesia machines and ventilators.[2]

Flowmeter

The **flowmeter** is located downstream from the pressure regulator. It controls the flow of oxygen through the vaporizer to the breathing system at a specific flow rate shown in liters per minute (L/min) or milliliters per minute (mL/min) (Figure 2-7). Flowmeters are tapered gas tubes with a float or ball. Gas flows in the bottom of the tube and out the top. While most flowmeters deliver oxygen in liters per minute, some flowmeters may express rates below 1 L/min in milliliters per minute (mL/min) for improved accuracy during low-flow anesthesia. Flowmeter knobs are connected to a needle valve that controls the rate of oxygen flow. When the valve is opened, oxygen flows into a vertical tube and raises the indicator ball or float to a height proportional to the flow of gases (Figure 2-8a). The indicator should be read at the point where there is the greatest resistance to flow; a ball indicator is read at the center (Figure 2-8b) whereas a float is read at the top (Figure 2-8c). The flowmeter should be gently turned off at the end of the day or when not in use to prevent wasting oxygen and

damage to the flow tube (if the flowmeter is on when the compressed gas cylinder is opened, the indicator may lodge at the top of the tube or even break the tube due to the sudden surge of oxygen). Closing the flowmeter too tightly may damage the needle valve. Compressed gas cylinders should be closed and machines disconnected from the central oxygen supply to prevent any needless oxygen consumption from small leaks or an active flowmeter.

Tech Tip 🐾

The distance between markings on the flowmeter is not equally spaced throughout the tube. For example, the markings may be closer together between 0 and 1000 mL than those between 1000 and 2000 mL.

Vaporizers

Vaporizers convert anesthetic agents, like isoflurane or sevoflurane, from a volatile liquid into a vapor that is delivered in the carrier gas (e.g., oxygen).

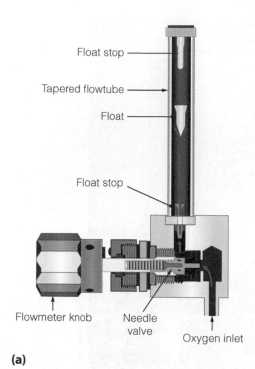

Float stop

Tapered flowtube

Float

Float stop

Flowmeter knob

Needle valve

Oxygen inlet

(a)

Courtesy of Teri Raffel Kleist, CVT, VTS (Surgery)

(b)

Courtesy of Teri Raffel Kleist, CVT, VTS (Surgery)

(c)

FIGURE 2–8 Flowmeters. (a) Internal view of a flowmeter. A flowmeter is a uniformly tapered gas tube with a measurement scale and float. A flowmeter knob is connected to a needle valve that controls the rate of oxygen flow. The flowmeter is positioned vertically with the smallest diameter end of the tapered gas tube at the bottom (this is the gas inlet). The float is located inside the gas tube and is made so its diameter is nearly identical to the gas tube's inlet diameter. When gas is introduced into the tube, the float is lifted from its initial position and rises to a height proportional to gas flow. (b) A flowmeter with a ball indicator is read at the center. (c) A flowmeter with a float is read at the top.

The vaporizer can be adjusted to change the amount of anesthetic agent added to the breathing system (Figure 2-9). Most vaporizers in use today are agent-specific, precision vaporizers that are located outside of the breathing system (Figure 2-10). Delivery of inhalant anesthetic agent with a nonprecision, uncompensated vaporizer is associated with a high degree of risk[1] and is not recommended for general use in veterinary medicine. Accurate delivery of inhalant anesthetic agents is essential to patient safety.

Modern vaporizers are classified as variable bypass, which means the user can adjust how much oxygen passes through the variable bypass chamber and how much passes through the vaporizing chamber; therefore, some oxygen flows into the vaporizing chamber and picks up anesthetic agent while some does not (Figure 2-11). A precise amount of oxygen is directed through the vaporization chamber, where it picks up anesthetic vapor, and the rest of the oxygen bypasses the chamber and does not pick up anesthetic vapor. These gases join together just prior to exiting the vaporizer (see Figure 2-9). The ratio of the gas that picks up inhalant anesthetic agent to the gas that does not pick up inhalant anesthetic agent determines the concentration of gas leaving the vaporizer and is called the **splitting ratio**. The amount of oxygen that is diverted into the vaporizing chamber alters how much inhalant anesthetic agent is picked up. The concentration of inhalant anesthetic agent can be established by setting the dial at the desired level.[3]

Concentration dial

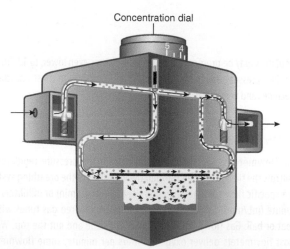

FIGURE 2–9 Inside the anesthetic gas vaporizer: The green circles represent fresh gas (oxygen) and the purple circles represent anesthetic gas (e.g., isoflurane, sevoflurane). When the vaporizer is in the off position, the oxygen flows through the "oxygen only" chamber and exits the vaporizer as just fresh oxygen. If the vaporizer is in the on position, a controlled amount of oxygen is sent to pick up molecules of anesthetic and carry them out of the vaporizer, where they are mixed with 100% oxygen to become the fresh gas mixture. This fresh gas mixture continues on to the breathing circuit and patient.

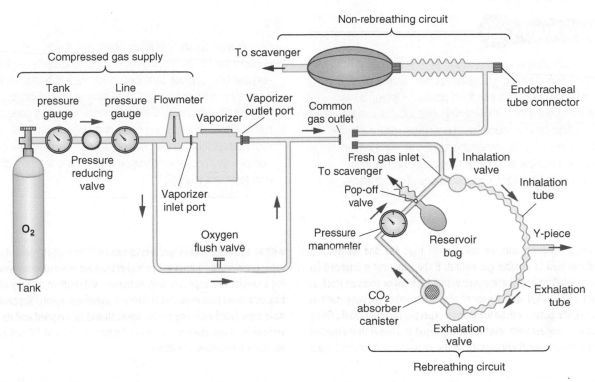

FIGURE 2-10 Vaporizers can be located inside of or outside of the breathing circuit. Vaporizers-out-of-the-circuit (VOC) are located outside of the breathing circuit. Contemporary precision vaporizers are all located out of the circuit.

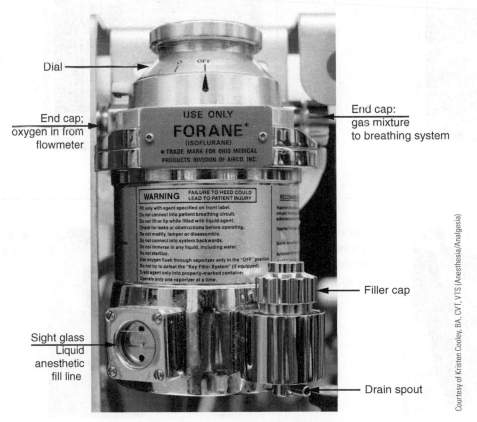

FIGURE 2-11 Parts of a vaporizer. This vaporizer delivers isoflurane and is set in the off position.

Safety Alert

The concentration set on the vaporizer may not be the actual concentration of inhalant anesthetic agent delivered to the patient if the vaporizer is in a tilted position. Tipping some vaporizers greater than 35 degrees while the dial is in the "on" position can introduce liquid anesthetic agent into the bypass chamber (with some vaporizers, this may happen even if the vaporizer is off). This may cause higher than expected, and even fatal, concentrations of anesthetic to

be delivered. Sloshing of liquid anesthetic during transport is also a risk for introducing liquid anesthetic agent into the "oxygen only" bypass chamber; therefore, it is best to drain a vaporizer prior to transport (e.g., moving a machine to another clinic location). If either of these scenarios occurs, run oxygen through the vaporizer at a high rate with the vaporizer turned to a low percent; after 15 minutes check the output with a gas/agent analyzer or contact a service technician to ensure it is safe.[1]

Filling a vaporizer with an agent other than the one for which it is labeled can lead to variable gas output. If the vaporizer is intended for use with an inhalant anesthetic agent with a high vapor pressure (such as isoflurane) and is filled with an agent of a lower vapor pressure (such as sevoflurane), the output will be lower than expected (in other words, filling an isoflurane vaporizer with sevoflurane will result in a lower than expected output of inhalant anesthetic agent).[1] If the vaporizer is intended for use

with an agent of a lower vapor pressure and is filled with an agent of a higher vapor pressure, the output will be higher than expected (in other words, filling a sevoflurane vaporizer with isoflurane will result in delivery of higher concentrations than expected of inhalant anesthetic agent).[1] Vaporizers that have been filled with the wrong agent should be emptied and its output verified by a gas analyzer or service technician. How to fill and empty a vaporizer is described as follows:

◀ **1.** Turn vaporizer off and shut off incoming oxygen flow.

Courtesy of Teri Raffel Kleist, CVT, VTS (Surgery)

(a)

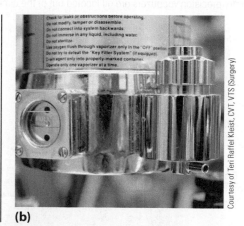

◀ **2.** Add more liquid inhalant anesthetic agent to the vaporizer if the amount of liquid inhalant anesthetic agent in the vaporizer window is less than $\frac{1}{4}$ full.

Courtesy of Teri Raffel Kleist, CVT, VTS (Surgery)

(b)

◀ Some vaporizers have a well that has a key built into the cap.

Courtesy of Teri Raffel Kleist, CVT, VTS (Surgery)

(c)

Courtesy of Teri Raffel Kleist, CVT, VTS (Surgery)

(d)

◀ Some vaporizers have a simple cap.

Courtesy of Teri Raffel Kleist, CVT, VTS (Surgery)

(e)

(f)

◀ **3.** Properly align the adapter for the inhalant anesthetic agent bottle. Ensure that the larger slot fits over the large tab on the bottle collar.

Courtesy of Teri Raffel Kleist, CVT, VTS (Surgery)

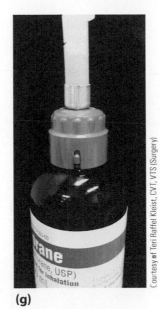

(g)

◀ **4.** Ensure the smaller slot fits over the smaller tab on the bottle collar.

Courtesy of Teri Raffel Kleist, CVT, VTS (Surgery)

(h)

◀ **5.** Pour liquid inhalant anesthetic agent into the well until the desired level appears on the vaporizer window. Be sure not to overfill the well.

Courtesy of Teri Raffel Kleist, CVT, VTS (Surgery)

(i)

◀ **6.** An empty bottle should always be kept in the event a vaporizer needs to be emptied.

Courtesy of Teri Raffel Kleist, CVT, VTS (Surgery)

(j)

◀ **7.** Vaporizers with a well require a key to empty the vaporizer; the key can be found in the cap of the vaporizer.

Courtesy of Teri Raffel Kleist, CVT, VTS (Surgery)

(k)

◀ Placing the cap upside down on the well will allow the drain to be opened.

Courtesy of Janet Amundson Romich, DVM, MS

(l)

Courtesy of Teri Raffel Kleist, CVT, VTS (Surgery)

◀ **8.** Vaporizers with a simple cap require the use of a screwdriver to loosen the screw to allow the drain to open.

(m)

Courtesy of Teri Raffel Kleist, CVT, VTS (Surgery)

◀ **9.** Some vaporizers have a screw near the bottom of the vaporizer that needs to be loosened to open the drain.

(n)

Courtesy of Teri Raffel Kleist, CVT, VTS (Surgery)

◀ **10.** Place the empty bottle below the spout, before the drain is opened, to capture any liquid inhalant anesthetic agent that drains out.

FIGURE 2–12 Maintaining a vaporizer includes properly filling and emptying it. The details of filling and emptying a vaporizer are described in the text.

It is essential to patient health to maintain the vaporizer according to manufacturer's standards. Yearly cleaning and calibration is strongly recommended.[4] Vaporizers cannot be calibrated or cleaned in the field because doing so requires disassembly. In-field "calibration" simply means checking the output of the vaporizer to make sure it is within tolerance levels and specifications ($+/-$ 10–20% of the setting). Any vaporizer outside of tolerance requires service that includes cleaning, replacement of worn parts, and recalibration.

Tech Tip

To track and properly schedule regular maintenance of the anesthetic machine, clinic staff should keep records of vaporizer output checks performed by service technicians.

Oxygen Flush Valve

The **oxygen flush valve** (also known as the **oxygen fast flush valve**) is a button that can be utilized to supply large volumes of oxygen quickly to the patient (Figure 2-13). When depressed, this valve rapidly delivers large volumes of 100% oxygen to the breathing circuit and bypasses the flowmeter and the vaporizer. Delivering oxygen with the oxygen flush valve provides oxygen *only* and must be used judiciously because the oxygen is delivered at relatively high flow rates (35–75 L/min) that could cause lung injury. The oxygen flush valve may be used to fill an empty reservoir bag but dilutes the concentration of inhalant anesthetic agent within the system and can lead to a light level of anesthesia. In some emergency situations, this is desired, but the oxygen flush valve should be used with caution to avoid over pressurization and **barotrauma** (damage to the lung from over pressurization and overstretching of the delicate tissue).

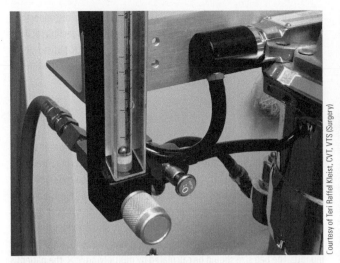

FIGURE 2-13 Oxygen flush valve supplies large volumes of oxygen quickly to the patient.

Do not use the oxygen flush valve when the pop-off valve is closed, when using a non-rebreathing system, or when connected to a small patient.[1]

Safety Alert

The oxygen flush valve delivers 100% oxygen very rapidly (35–75 L/min) and should not be used when the machine is connected to a patient because it can damage delicate lung tissue. Since the oxygen flush valve bypasses the vaporizer and delivers pure oxygen, it dilutes the concentration of inhalant anesthetic agent in the breathing system and can lead to a light level of anesthesia.

Other Components of the Anesthesia Machine

The other parts of the anesthesia machine help facilitate the delivery of the gas mixture to the patient and enable the removal of carbon dioxide and waste gases from the circuit and patient. These other components of the anesthesia machine include the common gas outlet, unidirectional valves, breathing hoses, reservoir bag, pop-off valve, pressure manometer, carbon dioxide absorber, and scavenging system.

Common Gas Outlet

The point at which oxygen and anesthetic gas enter the breathing circuit is called the **common gas outlet** (also called the **fresh gas inlet**). Gas in the common gas outlet has gone through the oxygen source, pressure regulator, flowmeter, and vaporizer. On some modern anesthesia machines, the common gas outlet can be easily manipulated to accept a non-rebreathing circuit (see explanation later in this chapter) (Figure 2-14).

Unidirectional Valves

The **unidirectional** or **one-way valves** direct the flow of gases in a circle to prevent the direct return of carbon dioxide to the patient. These valves help move

FIGURE 2-14 The common gas outlet (also called the fresh gas inlet) is the point at which oxygen and anesthetic gas enter the breathing circuit.

fresh gases toward the patient and exhaled gases away from the patient. In a circle rebreathing system using unidirectional valves, exhaled gases must first pass through the **CO_2 absorber**, a device containing absorbent chemicals that removes CO_2 from the exhaled gases before they are reintroduced to the patient. In a typical rebreathing system, after exiting the vaporizer, the gases enter the breathing circuit at the common gas outlet. From the common gas outlet, the gases flow from underneath into the inspiratory unidirectional valve, raising the flutter disc as they pass around it. The gases then travel through the inspiratory breathing hose of the circuit, down the endotracheal tube, and into the patient's lungs. Once in the lungs, oxygen and inhalant anesthetic agent diffuse from the alveoli into the blood while carbon dioxide and waste anesthetic diffuse from the blood into the alveoli. These "used" gases exit the patient and travel through the expiratory breathing hose and enter the expiratory unidirectional valve from below, raising the flutter disc as they pass around it. The gases then continue on to the CO_2 absorber and reservoir bag or the pop-off valve and scavenging system and then back to the patient. Unidirectional valves and their associated flutter disc contribute resistance to breathing and should be inspected regularly. Damaged discs or domes should be replaced to ensure proper function and to avoid rebreathing of CO_2 (Figure 2-15).[1]

Tech Tip

During *inspiration*, the inspiratory unidirectional valve opens, and gas moves from the common gas outlet through the valve and into the inspiratory breathing hose. During *expiration*, the inspiratory unidirectional valve closes, preventing exhaled gas from entering the inspiratory breathing hose.

During *expiration*, the expiratory unidirectional valve opens, gas moves from the patient to the expiratory breathing hose, through the expiratory unidirectional valve, and into the reservoir bag. During *inspiration*, the expiratory unidirectional valve closes, preventing gas from entering the expiratory breathing hose and guiding it from the reservoir bag to the CO_2 absorber. The unidirectional valves help to create a direction for gases to flow.

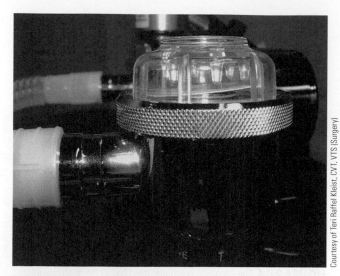

FIGURE 2–15 Inspiratory one-way valve. Notice the flutter disc in the up position as would occur with an inspiration.

Breathing Hoses

Breathing hoses are one of the dynamic components of the anesthesia machine. These corrugated plastic or rubber hoses carry gases to (via the inspiratory breathing hose) and from (via the expiratory breathing hose) the patient. The inspiratory and expiratory breathing hoses are connected by a **Y-piece** (also called a **wye piece**) that also connects to the endotracheal tube (or facemask) (Figure 2-16a). Breathing hoses vary in length, volume, and internal diameter. Breathing hoses are corrugated to prevent kinking, bending, or crushing and to allow for expansion of the breathing circuit (Figure 2-16b). Some breathing hoses are **coaxial** (tube within a tube) to reduce bulk, make them less cumbersome, and help them warm the cold, fresh inspired gases with warm expired gases as they bathe the inner tube. Coaxial rebreathing hoses are often referred to as Universal F circuits and should not be confused with other coaxial non-rebreathing systems (Bain). The Universal F circuit connects to the machine's inspiratory and expiratory ports and utilizes the unidirectional valves and CO_2 absorber, whereas the non-rebreathing system (Bain) does not (Figure 2-16c). Disadvantages of coaxial breathing tubes include breaking or kinking of the inner tube or the inner tube disconnecting from either end. The inspiratory inner tube of

coaxial circuits should be regularly inspected for cracks or disconnections (described later in this chapter in the Anesthesia Machine Inspection section).

Different hose configurations are available for specific patient populations and personal preference. These configurations are outlined in the breathing system section later in this chapter.

Reservoir Bag

The **reservoir bag** attaches to a port located on the CO_2 absorber side of the circle near the expiratory unidirectional valve (the term *rebreathing bag* is sometimes used for rebreathing systems instead of *reservoir bag*). The reservoir bag provides extra gas holding capacity within the system during exhalation. This allows the patient to take large breaths and pull gas from the reservoir volume within the bag. It also provides a way for the anesthetist to assist the patient's respirations and to monitor breathing rate. The bag size should be calculated by multiplying 5 × the patient's **tidal volume** (10–15 mL/kg) and should not allow pressures in excess of 30 cm H_2O to accumulate within the patient.[1, 5] The bag's volume should exceed the patient's inspiratory capacity so that a spontaneous deep breath does not cause it to collapse completely.[3] Appropriately sized bags allow for monitoring of breathing rate and manual ventilation support (Table 2-2, Figure 2-17). A bag that is too large can impede monitoring and adds too much volume to the circuit. This slows any desired changes in inspired gas concentrations when vaporizer settings are changed. A bag that is too small does not provide adequate gas supply to meet the maximum inspiratory volume and will collapse during inspiration and overinflate during expiration. When the calculation for bag size results in a bag size that is not available (e.g., 737 mL), choose the next largest bag size available (in this case 1 L).

Pop-off or Adjustable Pressure Limiting Valve

The **pop-off valve** or **adjustable pressure limiting (APL) valve** is a safety valve that allows excess gases to vent from the breathing circuit to avoid a build-up of pressure within the system. It also transfers waste gases to the scavenging system for removal. This valve is often located near the expiratory unidirectional valves or the reservoir bag. The pop-off valve

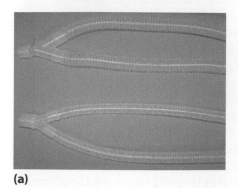
(a)

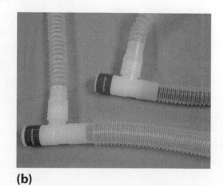
(b)

(c)

FIGURE 2–16 Breathing hoses. (a) Standard adult diameter Y hose (top) and pediatric diameter Y hose (bottom). (b) Adult and pediatric diameter Universal F/coaxial rebreathing hose. (c) Inside view of coaxial breathing hose.

TABLE 2-2 Reservoir (Rebreathing) Bag Size

Weight of Patient	Recommended Bag Size (5* × TV**)
1-4 kg	¼ L bag (½ L can be used when ¼ L not available)
5 kg	½ L bag
5.5–11 kg	1 L bag
11.5–22 kg	2 L bag
22.5–33 kg	3 L bag
33.5–44 kg	4 L bag (5 L bag can be used if 4 L is not available)
44.5–55 kg	5 L bag
100–150 kg	15 L bag
150 + kg	30 L bag

*There is a range in references used to determine reservoir bag size. Some veterinary professionals use the equation 6 × TV. It is wise to check with the veterinarian prior to reservoir bag determination.

**TV = Amount of air exchanged during normal respiration (air inhaled and exhaled in one breath); 10–15 mL/kg

FIGURE 2-17 A variety of reservoir bags. From left to right: 0.25 L, 0.5 L, 1.0 L, 2.0 L, 3.0 L, 4.0 L, 5.0 L.

will vent pressure at 1–2 cm H_2O and will leave a small amount of pressure in the system to allow for a passively full reservoir bag. It is important to keep the pop-off valve in the open position unless manual or mechanical ventilation is taking place to avoid dangerously high increases in pressure or volume of gases in the circuit (Figure 2-18a). If the valve is left closed, unsafe increases in pressure will occur within the circuit that could damage a patient's lungs due to overdistention and cause rupture of alveoli with subsequent cardiovascular collapse (Figure 2-18b). Increased intrathoracic pressure may also reduce the patient's ability to expire gas, decrease cardiac output, and, if not corrected, lead to death. A pop-off occlusion or safety pop-off button is available (SurgiVet, Matrx) (Figure 2-18c). This button allows the anesthetist to easily close the pop-off valve and to deliver a positive pressure manual breath by pressing down on a button. Once that button is released, the valve opens and pressure is allowed to vent. This type of valve drastically reduces the incidence of complications from increased circuit pressures that may develop from the pop-off valve being left closed.

Pressure Manometer

The **pressure manometer** is a pressure gauge located after the expiratory unidirectional valve. It is calibrated in cm H_2O (centimeters of water) or mm Hg (millimeters of mercury) and reflects the pressure in the breathing system and the patient's lungs. The pressure is influenced by the oxygen flow and the pop-off valve position. An open pop-off valve will readily release excess pressure and gas into the scavenging system whereas a closed pop-off valve will allow pressure to build up. The pressure manometer is also used to assess patient pressures during manual or mechanical ventilation. Ten to twenty centimeters of H_2O (7.4–14.8 mm Hg) is sufficient pressure to inflate normal lungs of small animals. As mentioned previously, excessive pressure within the circuit may lead to barotrauma and/or a decrease in cardiac function and possibly death.[1] Frequent monitoring of the pressure in the system is the responsibility of the anesthetist and is critical during every anesthetic procedure (Figure 2-19).

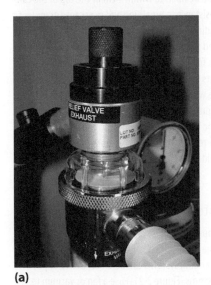
(a)

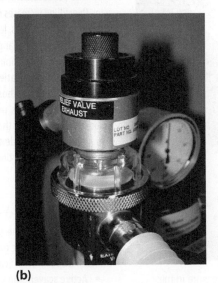
(b)

(c)

FIGURE 2-18 Adjustable pressure limiting or pop-off valve. (a) Pop-off valve is in the open position; (b) pop-off valve is in the closed position; (c) pop-off safety button.

Carbon Dioxide Absorber

One of the functions of the anesthesia machine is to remove carbon dioxide from the breathing circuit. To do this, exhaled gases enter a canister filled with absorbent granules that remove carbon dioxide. The CO_2-free gases are then recirculated to the inspiratory unidirectional valve to be rebreathed. The canister contains absorbent granules such as soda lime (4% sodium hydroxide, 1% potassium hydroxide, 14–19% water and sufficient calcium hydroxide to equal 100%) with a small amount of pH

FIGURE 2–20 Fresh carbon dioxide absorbent (white) on the bottom and spent carbon dioxide absorbent (purple) on the top of the canister.

sensitive indicator dye.[3] The granules chemically absorb acidic carbon dioxide and neutralize it with a base. This exothermic reaction produces heat, water, and carbonate (limestone). Activity of the absorbent granules can be observed by moisture and heat production in the canister. The pH sensitive indicator dye changes the white granules to purple once they become exhausted (Figure 2-20). The color change will dissipate after a few hours with standard absorbent; however, a CO_2 absorbent is available that contains a permanent dye that does not change color. The granules must be fresh for proper operation and to prevent the patient from rebreathing CO_2. When granules are saturated (exhausted), they become very hard and brittle. Fresh granules are chalky and crush easily with finger pressure. All CO_2 absorber canisters contain a screen to prevent absorbent granules from entering the breathing circuit.

The CO_2 absorber should be filled to within 1 cm of the top of the canister. Overfilling increases resistance to breathing and may cause granules to lodge under the gaskets, leading to leaks in the anesthetic system. Standard absorbent should be changed after 8–12 hours of use for small animals and approximately after 3 hours for horses or when two-thirds of the granules have changed color. Patients with higher tidal volumes will exhaust the absorbent faster than those with smaller tidal volumes. Scheduled changing of the absorbent is recommended, but inspection of the absorbent's color before every anesthetic event is very important. Exhausted granules will allow carbon dioxide to reenter the breathing system and may lead to rebreathing of carbon dioxide, resulting in hypercapnia and subsequent respiratory acidemia.

Scavenging System

Hoses connected to the pop-off valve or other breathing circuit outlet move waste anesthetic gas away from the anesthesia machine to a container or disposal area. Scavenging systems can be either active or passive.

- **Active scavenging systems** (Figure 2-21a) use a fan or vacuum to remove waste gas after it has passively moved from the pop-off valve

FIGURE 2–19 Pressure manometer reflects the pressure in the breathing system and the patient's lungs.

(a)

Courtesy of Kristen Cooley, BA, CVT, VTS (Anesthesia/Analgesia)

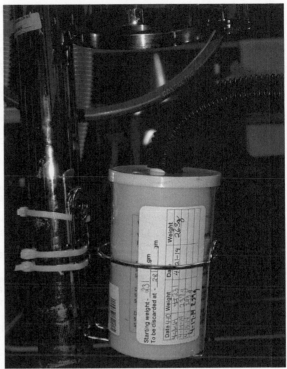

(b)

Courtesy of Teri Raffel Kleist, CVT, VTS (Surgery)

FIGURE 2-21 Scavenging systems. (a) Active scavenge system, (b) passive scavenge system.

to the scavenging duct system. An atmospheric interface is used in an active system to draw room air. This action keeps the system from actively pulling gas from the pop-off valve and causing the reservoir bag to collapse.

- **Passive scavenging systems** (Figure 2-21b) rely on positive pressure generated during patient exhalation and gravity to move gases into a disposal assembly.[1] An example of a passive system is an activated charcoal canister that absorbs anesthetic gases. An activated charcoal canister has a simple construction and is lightweight, making it easy to move around with the anesthetic machine. It is weighed regularly and disposed of once it gains a specific amount of weight (usually 50 g for canisters such as F-air® and Anesorb®). The holes at the bottom of the canister should not be covered or occluded so that filtered air is allowed to escape; otherwise pressure can build up within the canister and breathing circuit. These systems are easy to use and convenient but must be disposed of in a manner recommended by the manufacturer and any local disposal laws. Another example of a passive scavenging system is a tube that is directed toward a non-recirculating ventilation exhaust duct.[6]

Tech Tip 🐾

It is important that scavenging canisters not sit upright on a solid surface (e.g., table top, floor). The venting system on the bottom of the canister must have uninhibited flow to allow escape of the oxygen leaving the canister.

Exposure to waste anesthetic gas has become increasingly important over the past 20 years. There is no conclusive evidence that supports the idea that small amounts of waste anesthetic gas cause a specific health problem; however, there is evidence suggesting that proper removal of waste anesthetic gas will improve the occupational health of the veterinary team.[1] The recommended exposure limits put forth by the National Institute for Occupational Safety and Health (NIOSH) concerning halogenated anesthetics (e.g., sevoflurane and isoflurane) is two parts per million (ppm). All personnel working with or around anesthetic gas should be aware of the potential risks associated with exposure to waste gas and minimize such exposure by following the guidelines listed.[1, 3, 5] These are listed in Appendix A and are summarized in Figure 2-22.

Breathing Systems

Anesthetic breathing systems facilitate the delivery of oxygen and inhalant anesthetic agents to the patient, remove carbon dioxide from exhaled gases, and provide a means to assist ventilation. The breathing system consists of the machine components after the common gas outlet (fresh gas inlet) and involves those pieces and parts in which a breath passes through: unidirectional valves, breathing hoses, reservoir bag, pop-off valve, CO_2 absorber, and scavenging system. The breathing circuit consists of the removable breathing hoses and reservoir bag, which are customized and changed for each patient based on their lean body weight. Breathing circuits should be as short as practical for the procedure with few bends or restrictions in the path of gas flow.[1] The length of the breathing hoses adds resistance to breathing but does not

SCAVENGE SAFETY CHECKLIST

✓ Always use a scavenging system with all anesthesia machines and breathing circuits. This includes using an activated charcoal canister during transport.

✓ Rooms where inhalants are used should be well ventilated with at least 15 air exchanges per hour.

✓ Anesthesia machines should be leak tested daily and the breathing circuits should be leak tested prior to each case. This equipment should be as leak-free as possible and comply with the established criteria of less than 300 mL/min at 30 cm H_2O for circle systems with the pop-off valve closed.

✓ A log should be kept of machine maintenance and vaporizer calibration checks.

✓ Always use an anti-spill adapter or key filler when filling vaporizers to minimize spillage.

✓ Periodically monitor waste anesthetic gas in the operating room and induction and recovery areas using an anesthetic detection badge to ensure that proper scavenging in taking place.

✓ Scavenging system should be checked regularly to make sure that tubing is not blocked, canisters are still viable, and fans and suction are at the appropriate level.

✓ Start the flow of anesthetic gas only after intubating the patient and checking the endotracheal tube for a proper seal.

✓ Run 100% oxygen through the system at the end of a case while the patient is still connected, or disconnect from patient, place your hand over the Y-piece and flush the anesthetic gas out to scavenge.

FIGURE 2–22 An example of a scavenge safety checklist.

contribute to dead space as long as the unidirectional valves are working properly. Mechanical dead space is the area where bidirectional flow takes place yet no gas exchange occurs. It consists of the endotracheal tube (ETT) extending beyond the patient's incisors, the Y-piece and anything between the Y-piece and the ETT, and patient monitors (CO_2 monitor adapter, swivel adapter, elbow) (see Chapter 3, Cardiovascular and Respiratory Physiology Review). Mechanical dead space within the breathing circuit and ETT should be minimized to avoid the rebreathing of expired carbon dioxide.

Tech Tip

Ways to minimize mechanical dead space include minimizing the connectors attached to the ETT, making sure that the ETT is not excessively long, inspecting all anesthetic machines and systems regularly (paying particular attention to valve function and inner hose integrity), using no more than one monitor adaptor, and choosing the proper circuit to use for the size of the patient.

The type of breathing system used is determined based upon the weight of the patient in kilograms. Although these ranges are subjective and are often quoted differently by various professionals, most patients weighing more than 7 kg use the rebreathing system, while those weighing less than 7 kg use the non-rebreathing system. Some veterinary professionals will refer to a "gray zone": for patients weighing between 3 and 7 kg, use a pediatric rebreathing system and for patients weighing less than 3 kg, use a non-rebreathing system. The types of breathing systems are summarized in Table 2-3.

Tech Tip

In rebreathing systems, CO_2 is removed from exhaled gases and they are recirculated to the patient. The patient rebreathes the CO_2-free exhaled gases along with the newly added fresh oxygen and inhalant anesthetic agent from the anesthesia machine.

Breathing systems are divided into rebreathing and non-rebreathing systems based on whether or not exhaled gases are reintroduced to the patient. Rebreathing systems remove carbon dioxide from exhaled gases as they pass through the CO_2 absorber. Once carbon dioxide is removed, all or some of the exhaled gases go back to the patient, which allows rebreathing of gases. Non-rebreathing systems rely on higher fresh gas flow rates to remove carbon dioxide from the system, which eliminates the need for a CO_2 absorber.

Tech Tip

Rebreathing systems remove CO_2 via a CO_2 absorber; non-rebreathing systems remove CO_2 via high fresh gas flow rates.

Rebreathing Systems

The most common rebreathing system is the circle system, which utilizes inspiratory and expiratory unidirectional valves to keep gases moving one direction in a circle. This system also has a method to remove carbon dioxide. The circle system prevents rebreathing of CO_2 but allows

TABLE 2-3 Summary of Breathing Systems

Breathing System	Principles	Example*	Advantages	Disadvantages

Rebreathing
• Parts include
  ○ Common gas outlet (fresh gas inlet)
  ○ Unidirectional valves
  ○ Breathing hose
  ○ Y-piece connector
  ○ Pop-off (APL) valve
  ○ Reservoir bag
  ○ CO_2 absorber
  ○ Pressure manometer
	• Allows rebreathing of exhaled gases
	• CO_2 removed chemically (via absorbent granules)
	• Fresh gas mixture added continually
	• Used for patients weighing more than 7 kg
	• Resistance to movement of the gas mixture is increased
	• More economical because gas flow rates are lower
	• Changes in anesthetic depth relatively slow due to lower flow rate of gas mixture
		Closed circle system (oxygen flow rates 5–10 mL/kg/min)	• Economical
			• Helps regulate body temperature through conservation of heat and moisture
			• Conserves inhalant anesthetic agent
			• Conserves oxygen
			• Environmentally safer
				• Has more resistance to ventilation
				• Unable to rapidly change anesthetic concentration (need to increase gas flow rate and vaporizer setting simultaneously
				• CO_2 removal dependent on chemical absorption (CO_2 absorbent must be checked before all procedures)

		Low-flow circle system (oxygen flow rates 10–25 mL/kg/min)	• Economical • Helps regulate body temperature through conservation of heat and moisture • Reduces waste gas	• Slow induction (low-flow technique is best for maintaining anesthesia) • Slower changes in patient anesthetic depth
		Semiclosed circle system (oxygen flow rates 25–50 mL/kg/min)	• Rapid changes in inhalant anesthetic agent concentration • Less reliance on CO_2 absorbent to remove CO_2	• Higher operating cost • Excess waste gas produced due to higher fresh gas flow • Increased loss of body heat

Non-Rebreathing
• Parts include
  ○ Fresh gas breathing hoses
  ○ Straight connector
  ○ Exhalation hoses
  ○ Excess gas venting system
  ○ Reservoir bag
	• Little to no exhaled gases are recirculated
	• CO_2 removed physically (via fresh gas flow)
	• Gases removed by scavenging system
	• Offers little resistance to the patient when breathing because it has less valves and parts (advantageous to smaller patients)
	• Breath by breath response to vaporizer changes
	• Used in patients weighing less than 7 kg
		Bain or Modified Jackson-Rees (oxygen flow rates 200 mL/kg/min)	Bain:
			• Reduces resistance to breathing for small, spontaneously breathing patients
			• Produces rapid changes in inspired inhalant anesthetic agent concentration due to relatively small volume and increased flow rates (therefore, depth of anesthesia can be changed rapidly)
			• More precise inhalant anesthetic agent delivery
				• Less economical because gas flow rates are higher
				Bain:
				• Possible disconnection or kinking of the inner fresh gas tube
				• Risk of rebreathing expired gases due to inadequate fresh gas flows.
				• Inability to monitor respiration or assist breathing with a Bain circuit without a pop-off valve and reservoir bag

(Continued)

TABLE 2-3 Summary of Breathing Systems (*Continued*)

Breathing System	Principles	Example*	Advantages	Disadvantages
	• Produces less strain on the endotracheal tube		• Fresh gases are warmed by the surrounding exhaled gases • Partial rebreathing of gases conserves moisture • Unit is lightweight, reusable • Less cumbersome in patients having dental procedures or head and neck surgery Modified Jackson-Rees: • Lack of movable parts (other than the pop-off valve) • Presence of a reservoir bag that allows the anesthetist to monitor breathing and offer assisted breaths when needed. • Unit is lightweight, has minimal dead space, offers minimal resistance to breathing, and is ideal for small, spontaneously breathing patients.	Modified Jackson-Rees: • High flow rate of fresh gases dries and slightly cools patient's respiratory tract

*These ranges are suggested starting points; therefore, different resources may state different values. Keep in mind that there is some room for error for these ranges.

partial rebreathing of other exhaled gases (Figure 2-23). The circle system consists of the

- common gas outlet (also called the fresh gas inlet);
- inspiratory unidirectional valve;
- breathing hoses;
- Y-piece;
- expiratory unidirectional valve;
- pop-off valve;
- pressure manometer;
- reservoir bag; and
- CO_2 absorber.

The circle system provides an economical way to deliver inhalant anesthetic agents because after expired gases pass through the CO_2 absorber for carbon dioxide removal, they are then recirculated back to the patient. The rebreathed gases are warmer and contain more humidity than fresh gas; this increased temperature may help maintain normal body temperature.

Circle system breathing hoses are available in adult, pediatric, and large animal sizes that differ only in their internal diameters. Pediatric hoses have a 15 mm internal diameter and are intended for use on veterinary patients less than 7 kg. The pediatric diameter hoses decrease the total volume of the breathing system and allow for more rapid changes in anesthetic concentrations. Adult rebreathing hoses have an internal diameter of 22 mm and are recommended for use in patients weighing between 7 and 135 kg. Large animal breathing hoses have an internal diameter of 50 mm and are recommended for patients weighing greater than 135 kg (see Chapter 19, Anesthesia and Analgesia of Large Animal and Food Producing Species). Special adapters are available to connect endotracheal tubes to the large diameter Y-pieces used in circle systems. It is recommended that the internal diameter of the breathing hoses be larger than the internal diameter of the endotracheal tube to avoid unnecessary resistance to breathing.[1]

Excess gas is vented out through the pop-off (APL) scavenging system

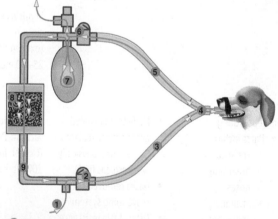

1. Fresh gas enters the circle from the common gas outlet of the anesthetic machine...

2. ...flows through the inspiratory unidirectional valve...

3. ...then flows through the inspiratory breathing tube...

4. ...through the Y-piece to the patient...

5. ...from the patient through the expiratory brething tube...

6. ...through the expiratory unidirectional valve...

7. ...in & out of the reservoir bag...

8. ...through the absorbent canister where CO_2 is removed...

9. ...and then back toward the patient.

FIGURE 2–23 Circle rebreathing system utilizes inspiratory and expiratory unidirectional valves to keep gases moving one way in a circle, has a method to remove CO_2, and prevents rebreathing of CO_2 but allows partial rebreathing of other exhaled gases.

Recommendations for fresh gas flow rates in circle systems vary depending on the source. Personal preference along with experience also determines how flow rates are set. Flow rates in circle systems can be categorized into closed, low-flow, and semiclosed circle systems. These systems are based on the flow of oxygen in relation to the metabolic needs of the patient and *do not* reflect the position of the pop-off valve. Anecdotally, the position of the pop-off valve can be adjusted during alternative flow states to keep the reservoir bag passively full. Extreme vigilance is essential whenever manipulating the pop-off valve, and this method requires an attentive, dedicated anesthetist.

Closed Circle System

In the **closed circle system**, the oxygen flow rate approximates the patient's **metabolic oxygen consumption**, which varies based on the metabolic state of the patient as well as body weight, surface area, temperature, underlying disease condition, and anesthetic drugs used. Published recommendations for fresh gas flow rates in a closed circle system vary from 5 to 10 mL/kg/min. Closed circle systems reduce waste anesthetic gas and conserve body temperature. However, they require early intervention to allow for depth changes to match anesthetic needs (see Chapter 10, The Maintenance Period).

Low-Flow Circle System

In the **low-flow circle system**, the oxygen flow rate is greater than the patient's metabolic oxygen needs and ranges from 10 to 25 mL/kg/min.[7] A typical low-flow circle system uses oxygen flow rates of 10–15 mL/kg/min. The advantages of the low-flow circle system are similar to those with the use of a closed circle system, including economy and retention of body heat and moisture. The disadvantages of lower flow rates are slower changes in the concentration of inhalant anesthetic agent and therefore slower changes in the patient's anesthetic depth. Low-flow techniques can be instituted as maintenance anesthetic flow rates and are helpful in counteracting hypothermia (along with supplemental heating).

Semiclosed Circle System

The **semiclosed circle system** utilizes fresh gas flow rates that exceed the metabolic oxygen consumption rate of the patient and range from 25 to 50 mL/kg/min (or greater). These higher flow rates mean that an excess of waste anesthetic gas is produced and must be vented through the pop-off valve and ushered to the scavenging system. The semiclosed circle system is the most commonly used system in veterinary medicine. Advantages include faster changes in inhalant anesthetic agent concentrations and less reliance on absorbents to remove carbon dioxide. Disadvantages include higher operating cost, more waste anesthetic gas pushed out to the scavenging system, and loss of body heat from the dry incoming fresh gas. Many practices use the arbitrary setting of 1 L/min oxygen flow rate for maintenance of anesthesia. Vaporizers are calibrated to be accurate between certain flowmeter settings, such as 0.5 L/min to 4 L/min, and because most practices probably don't know the manufacturer's recommended flow rate settings for vaporizer accuracy it seems "safest" to use the default setting of 1 L/min. This technique is often not as economical as calculating the oxygen flow rate and can compound the occurrence of excess waste anesthetic gas and loss of body heat.

Non-Rebreathing Systems

Non-rebreathing systems utilize high fresh gas flow rates to avoid rebreathing of expired gases. This system bypasses the unidirectional valves and CO_2 absorber to reduce the resistance to breathing for small (<7 kg), spontaneously breathing patients. Non-rebreathing systems minimize dead space, and the relatively small volume and high flow rates allow for rapid changes in inspired anesthetic concentrations (Figure 2-24). The high fresh gas flow rates are very drying and potentially cooling to the patient's respiratory tract but are needed to minimize rebreathing of expired gases. The parts of the non-rebreathing system include

- fresh gas breathing hoses;
- straight connector;
- exhalation hoses;
- excess gas venting system; and
- reservoir bag.

Most systems have fresh gases entering the breathing hose at a point near the patient's mouth via fresh gas hoses. This allows the fresh gas inflow to push the CO_2-rich expired breath to the exhalation hoses toward the reservoir bag and then to the scavenging system. When the patient inspires, the inspired gas comes from both the fresh gas hose and exhalation hose; therefore, a patient may rebreathe some exhaled gases despite high fresh gas flow rates. Continuous monitoring of end tidal carbon dioxide levels using a capnograph (see Chapter 7, Anesthesia Monitoring) is important when using non-rebreathing systems; however, their accuracy may be altered due to high fresh gas flow.

Fresh gas flow rates for the non-rebreathing system begin at 200 mL/kg/min for patients weighing less than 7 kg. Fresh gas flow rates below these parameters allow rebreathing of gases and may lead to the improper removal of expired carbon dioxide. When using a non-rebreathing system, the oxygen flush valve should not be used as it could cause rapid overpressurization of the lungs due to high oxygen flow moving through this valve.

Non-rebreathing systems are sometimes referred to as Mapleson systems and are classified based on the location of the different components. There are Mapleson systems A–F, but the most commonly used ones in veterinary medicine include the Mapleson F system (also known as Modified Jackson-Rees) and the Modified Mapleson D system (also known as Bain).

Bain System

The Bain system is a coaxial version of the Mapleson D system. The fresh gas tube enters the circuit near the reservoir bag and fresh gas flows through the inner tube for delivery at the patient end of the circuit. When the patient exhales, fresh gas and expired gas enter the tube, thereby venting any excess through the pop-off valve. During the next inhalation, the patient receives a mixture of gas: some fresh and some expired. High fresh gas flow rates are essential to minimize rebreathing of inspired gases. The exhaled gases are vented through a pop-off valve located near the reservoir bag. Some Bain systems do not have a reservoir bag and pop-off valve. These systems are simply a fresh gas tube and an expiratory tube, where the expiratory limb acts as a reservoir for gases.

Advantages of the Bain circuit are as follows: less resistance of breathing, faster anesthetic depth changes, more precise inhalant anesthetic agent delivery, fresh gases warmed by the surrounding exhaled gases, conservation of moisture due to partial rebreathing of gases; and the unit's light weight, reusability, and less cumbersome form in patients having dental procedures or head and neck surgery. Disadvantages include possible disconnection or kinking of the inner fresh gas tube and the risk of rebreathing expired gases due to inadequate fresh gas flow rates. In the case of the Bain circuit without a pop-off valve and reservoir bag, the main disadvantage is the inability to monitor respiration or assist breathing (Figure 2-25).

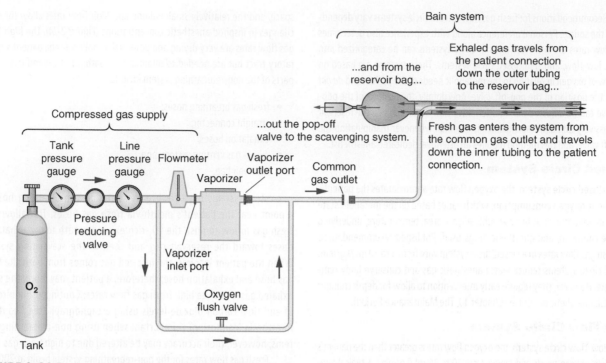

FIGURE 2-24 For the non-rebreathing system, the breathing circuit completely bypasses the unidirectional/one-way valves and carbon dioxide absorber. The non-rebreathing circuit only utilizes the oxygen source, flowmeter, and vaporizer by connecting to the machine at the fresh gas outlet (some connect directly to the outlet port on the vaporizer).

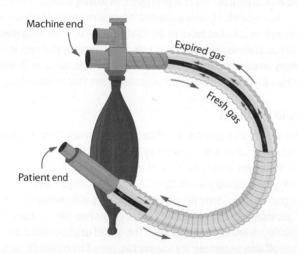

FIGURE 2-25 The Bain circuit is a coaxial non-rebreathing circuit in which fresh gas tube enters the circuit near the reservoir bag and fresh gas flows through the inner tube to be delivered at the patient end of the circuit. When the patient exhales, fresh gas and expired gas enter the tubing, venting any excess through the pop-off valve. During the next inhalation, the patient receives a mixture of gas, some fresh and some expired.

Modified Jackson-Rees System

The Mapleson F or Modified Jackson-Rees system is a non-rebreathing system with a **T-piece** (fresh gas port that sits at a 45–90° angle to the patient end of the circuit) to facilitate delivery of fresh gas near the patient and a reservoir bag

and a pop-off valve at the distal end of the circuit. During spontaneous ventilation, exhaled gases pass through the expiratory tube and mix with incoming fresh gases. The **expiratory pause** (break in breathing following expiration) allows the fresh gas to push the exhaled gases through the expiratory limb of the circuit. When the patient inhales, the gas mixture comes from the fresh gas flow and expiratory limb, including the reservoir bag.

Advantages of this system include the lack of movable parts (other than the pop-off valve) and the presence of a reservoir bag, which allows the anesthetist to monitor breathing and offer assisted breaths when needed. In addition, the unit is lightweight, has minimal dead space, offers minimal resistance to breathing, and is ideal for small, spontaneously breathing patients. A disadvantage of the Modified Jackson-Rees system is that high fresh gas flow rates can contribute to hypothermia; however, the length of the procedure rather than the type of anesthetic circuit used is more influential in heat loss[8] (Figures 2-26a and 2-26b).

MACHINE MAINTENANCE

The anesthesia machine requires basic routine maintenance to keep it functioning properly (Appendix B). A veterinary technician can easily perform much of the routine maintenance of the anesthesia machine. The machine should be inspected by a trusted and qualified professional at least once a year.

Machine cleaning involves removing the canister of the CO_2 absorber, dome caps, gaskets, and flutter discs and cleaning them with mild soapy disinfectant water and drying them thoroughly. The inside of the domes

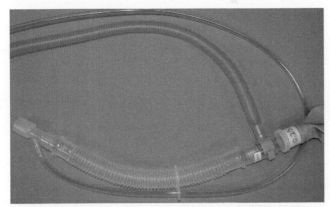

FIGURE 2-26 The Modified Jackson-Rees circuit is a non-rebreathing circuit with a T-piece to facilitate delivery of fresh gas near the patient (a) and a reservoir bag and a pop-off valve at the distal end of the circuit (b). During spontaneous ventilation, exhaled gases pass through the expiratory tube and mix with incoming fresh gases.

and the CO_2 absorber gasket should be wiped down with a damp cloth and allowed to dry. Wipe down the exterior of the machine with a mild soap and water if it gets soiled. The outside of the vaporizer should be wiped down, as should all other machine surfaces.

Safety Alert

The anesthesia machine should be pressure checked or leak tested prior to every anesthetic case (it is generally accepted that pressure checked and leak tested can be used interchangeably). Testing is done when a machine is being pressurized to detect a leak prior to use. A machine with a leak exposes personnel to anesthetic gas and should be avoided.

Anesthesia Machine Daily Check List

The anesthesia machine should be inspected every day and pressure checked before being used on a patient There are no recognized standards for maintaining and inspecting veterinary anesthetic machines; however,

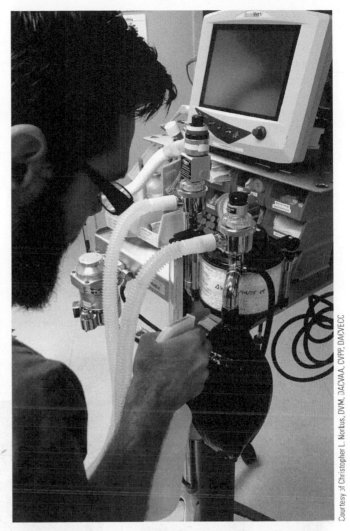

FIGURE 2-27 Spraying machine with soapy water to detect leaks. A leak in an anesthesia machine can be detected using a spray bottle containing soapy water. When spraying the area with the leak with soapy water, bubbles will form so that the leak can be identified and then repaired.

the following serves as a guideline for maintaining a properly functioning anesthetic machine. This anesthesia machine inspection should be done before every case and includes the following:

- Check the oxygen source to make sure there is enough oxygen available.
  - The line pressure is between 40–60 psi (50–55 psi if running a mechanical ventilator).
  - If there are no detectable leaks in connections (many times these leaks are audible due to the high pressures of the oxygen source; however, sometimes they are not). If a leak is suspected but not heard, spray all connections down with a mild soapy liquid and look for the presence of bubbles (Figure 2-27).
  - If using an oxygen concentrator, make sure the output is at least 93%.
- Turn on the flowmeter to make sure indicator moves freely within the tube. Turn it off gently.

- Depress oxygen flush valve to make sure it is working properly.
  - Oxygen should stop flowing as soon as the button is released, if it does not it must be serviced
- Ensure that the vaporizer is filled and contains the correct inhalant anesthetic agent. The funnel fill cap should be tightened. If there is a drain key, it should also be tightened. Check vaporizer connections.
- Replace CO_2 absorbent as needed. Be sure not overfill the canister. The inside of the CO_2 gasket can be wiped down with a damp cloth and allowed to dry. The CO_2 canister should be filled to within 1 centimeter of the top of the canister and the cover closed finger tight. Check the CO_2 canister for a proper seal.
- The scavenging system hoses should be connected to the pop-off valve and the scavenge assembly; if using a charcoal canister such as F-air®, the canister should be weighed and discarded if overweight. It is recommended to change canisters after 12 hours of use or when there is a 50-gram weight increase (making sure you weigh canisters before use will allow you to determine when a 50-gram increase has occurred).
- Wash breathing tubes after every use with 0.2% chlorhexidine solution. Rinse and hang to dry.

- Clean reservoir bag after every use with 0.2% chlorhexidine solution. Hang it to dry.
- If one-way valves have excessive moisture, remove dome and wipe dry. The unidirectional valve domes should be checked for chips or cracks. Dome caps should thread easily; the gasket or O-ring should be in good condition. Flutter discs should be flat and free from defects.
  - If any of these components are damaged they should be replaced immediately. Poorly functioning unidirectional valves can make breathing difficult for the patient and may lead to a dangerous increase in carbon dioxide levels.
- The end caps should be secure and the silicone hoses should be in good condition.
  - Rubber hoses (black) are prone to cracking. Cut hose below the crack and reconnect (do not tape!) or replace with silicone tubing.
- The fresh gas inlet should be firmly seated in the machine.
- The pop-off valve should open and close easily; the pop-off occlusion valve button should depress and release with ease.
- Perform a pressure check or leak test. To test the pressure on an anesthesia machine, follow the steps described in Figure 2-28 a through 2-28h:

(a)

Attach breathing hoses and reservoir bag to the anesthesia machine.
▲ Plug in oxygen hose/turn on oxygen source and ensure flowmeter is turned off.

(b)

▲ Close the pop-off valve.

(c)

▲ Occlude the Y-(wye) piece of the breathing circuit with your thumb or stopper. The use of a stopper provides standard closure compared to the palm of the hand.

(d)

▲ Gently depress the oxygen flush valve.

(e)

▲ To pressurize the breathing system to 30 cm H_2O.

(f)

▲ Set the O_2 flowmeter at 200 mL/min. Observe the manometer for decreasing pressure.

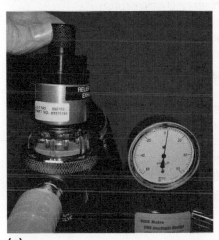

(g)

▲ If the pressure holds or rises very slowly, there is no leak.

(h)

▲ Before removing your finger or stopper from the breathing circuit, open the pop-off valve to ensure that it is working correctly. Keep the pop-off valve open after the test is finished. If pressure drops, there is a leak greater than 200 mL/min and it should be identified and resolved.

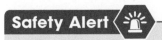

FIGURE 2-28 The steps to performing a machine pressure check or leak test are shown here and described in the text.

Tech Tip 🐾

There are slight variations to how pressure checks/leak tests are performed. In addition to the description in Figures 2-28a through 2-28h, some anesthetists take the pressure up to 30 cm H_2O for approximately 30 seconds. If the pressure drops more than 1 cm H_2O per second, then the flowmeter is slowly titrated up to 300 mL/min (in an effort to go to lowest setting that will stop the leak). As long as a leak is stopped at 300 mL/min or below, most anesthetists are comfortable using the anesthesia machine if all the possible sources of a leak have been inspected.

Safety Alert 📢

The last step of pressure checking the anesthesia machine is opening the pop-off valve. This serves three functions: to release pressure from the system, to ensure that the pop-off valve is working properly, and to leave the valve open when the machine is connected to a patient.

SUMMARY

The anesthesia machine is an important piece of equipment whose operation can easily be mastered once all of its parts are understood (Table 2-4). From the fundamental components that provide the oxygen and inhalant anesthetic agent to the parts of the breathing system that carry the gases to the patient and facilitate the removal of waste gases, all anesthetic machines have the same basic design, which makes the principles of their use universal. Anesthesia machines direct a controlled flow rate of oxygen to a vaporizer, where a liquid anesthetic agent is introduced into a carrier gas, which is then delivered at a controlled flow rate to the patient via a breathing circuit. Carbon dioxide is removed from exhaled gases, which are moved away from the patient and either recirculated to the patient or removed via a scavenging system. Like any piece of equipment, the anesthesia machine requires regular maintenance and quality checks to ensure it is safe and free from leaks. Checklists and maintenance logs are helpful to keep track of these events. Understanding how the anesthesia machine works and maintaining it so that it functions properly every time it is used will help the veterinary team avoid anesthetic accidents.

TABLE 2-4 Summary of Parts of the Anesthesia Machine

Anesthesia Machine Part	Function	Comments
Oxygen Source 1. Compressed gas cylinder	Supplies O_2 and other medical gases (N_2O) to the anesthesia machine	• Pressure and volume of cylinder varies with each compressed gas being used • "E" and "H" most common types • Attached to the anesthesia machine via hanger yoke • Compressed gas cylinders are color coded to identify the gas inside (see Table 2-1)
2. Oxygen concentrator	Supplies O_2 to the anesthesia machine	• Cost effective • No risks to personnel related to storage of compressed gas • Requires regular maintenance (to ensure O_2 levels) • May not produce a reliable concentration of oxygen • May not generate enough pressure to run a mechanical ventilator
3. Central gas supply systems	Supplies O_2 and other medical gases (medical grade air) to the anesthesia machine	• May be a liquid supply of O_2 or band of cylinders connected to a common manifold • Continuous supply of O_2 • Proper maintenance may include medical engineers and facility managers
Pressure regulator (pressure reducing valve)	Decreases and maintains gases at safe operating level	• Should have two gauges, one that reads the tank pressure and one that reads the line pressure • Each medical gas supplied to the anesthesia machine should have a regulator specially designed for that particular gas
Pressure gauge (not the pressure gauge on the anesthetic machine that notes pressure within the breathing circuit, but rather the pressure gauge for the intermediate pressure system)	Indicates pressure of the compressed gas cylinder side of the pressure regulator	• Should read 55 psi for use with most anesthesia machines and ventilators
Flowmeter	Delivers specific flow rate of medical gases	• Turned off when not in use to prevent sudden build-up of pressure in the glass tube when gas flow is turned on • Ball indicator is read at the center whereas a float indicator is read at the top • Should be gently turned off to prevent wasting oxygen and damage to the flow tube

Vaporizer	Stores liquid anesthetic agent and mixes anesthetic agent and medical gases (O_2, N_2O) as it flows to the patient	• Precision
		  ○ Delivers an exact concentration of anesthetic agent
		  ○ Flow, temperature, and backpressure compensated
		  ○ Expensive
		  ○ Calibrated for each anesthetic agent
		  ○ Used as VOC (vaporizer out of circle) because it is not located in the breathing circuit
		    ■ Oxygen flows from the flowmeter into the vaporizer before entering the breathing circuit. Precision vaporizers have a VOC configuration because when resistance to gas flow is high, respiratory drive is not enough to push gases through the vaporizer
		  ○ Examples include Tec, Vapor, and Ohio

(Continued)

TABLE 2-4 Summary of Parts of the Anesthesia Machine (*Continued*)

Anesthesia Machine Part	Function	Comments
Oxygen flush valve	Rapidly supplies large volume of O_2	• Should be used with caution to avoid overpressurization and barotrauma • Fills empty reservoir bag with 100% oxygen, thereby diluting inhalant anesthetic agent concentration and can lead to a light level of anesthesia • Do not use when pop-off valve is closed, when using a non-rebreathing system, or when connected to a small patient
Common gas outlet (also called the fresh gas inlet)	Point where O_2/medical gas (N_2O)/inhalant anesthetic agent mixture exits the anesthetic machine or enters the breathing circuit	• Used with either rebreathing or non-rebreathing system • Usually located on inspiratory side of rebreathing system • Gas in the common gas outlet has gone through the oxygen source, pressure regulator, flowmeter, and vaporizer • Minimizes dilution of fresh gas with expired gases or loss of fresh gas through the pop-off valve
Unidirectional valves (also called one-way valves)	Prevent expired gases from being recirculated directly to the patient by facilitating the movement of fresh gases toward the patient and exhaled gases away from the patient	• Upon inspiration, the inspiratory unidirectional valve opens, allowing the gas mixture to flow toward the patient • Upon expiration, the expiratory unidirectional valve opens (and inspiratory unidirectional valve closes), allowing expired gases to pass into the CO_2 absorber • Unidirectional valves and their associated flutter discs contribute resistance to breathing and should be inspected regularly
Breathing hoses	Channels medical and anesthetic gases	• Inspiratory and expiratory breathing hoses are connected by a Y- (wye) piece that also connects to the endotracheal tube (or facemask) • Corrugated to prevent kinking, bending, or crushing and to allow for expansion of the breathing circuit • Corrugation increases resistance to breathing • Some are coaxial to reduce bulk, make them less cumbersome, and help them warm cold inspired gases with warm expired gases as they bathe the inner tube
Reservoir (rebreathing) bag	Provides tidal volume for the patient and compliance for the breathing system	• Also used to store gases, monitor respirations, and manually ventilate a patient • Bag size: 5 × tidal volume (10–15 mL/kg) • Bag's pressure should not be more than 30 cm of H_2O when attached to the patient. (Higher volumes are used for machine leak checks.) • Bag that is too large can impede monitoring and adds too much volume to the circuit • Bag that is too small does not provide adequate gas supply to meet the maximum inspiratory volume and it will collapse during inspiration and overinflate during expiration
Pop-off (APL) valve	Allows excess gases to vent from the breathing circuit to avoid a build-up of pressure within the system; also allows anesthetist to increase pressure in the breathing system when needed Transfers waste gases to the scavenging system for removal	• Usually left open when patient is spontaneously breathing • Closed for manual or mechanical ventilation • Valve can be adjusted to supply appropriate pressure • Pop-off occlusion or safety pop-off button is available to deliver a positive pressure manual breath to reduce the incidence of complications from increased circuit pressures if pop-off valve is left closed
Pressure manometer (Patient pressure manometer in the breathing circuit)	Measures the amount of pressure in the breathing circuit and patient's lungs Used to assess patient pressures during manual or mechanical ventilation	• Pressure is influenced by the oxygen flow rate and the pop-off valve position. • 10–20 cm H_2O is sufficient pressure to inflate normal lungs of small animals

(Continued)

TABLE 2-4 Summary of Parts of the Anesthesia Machine (*Continued*)

Anesthesia Machine Part	Function	Comments
CO_2 absorber	Removes CO_2 from expired gases	• Granules in CO_2 absorber will turn purple once exhausted and will return back to normal color when not in use (after some time has passed) • When granules are saturated they become very hard and brittle. Fresh granules are chalky and crush easily with finger pressure • CO_2 absorber canister contains a screen to prevent absorbent granules from entering the breathing circuit • Standard absorbent should be changed after 8 hours of use or when 2/3 of the granules have changed color. • Should be filled to within 1 cm of the top of the canister

Scavenging system	Removes waste anesthetic gases from the anesthetic breathing system and reduces contamination of the environment	• Two types
		 ○ Active
		  ▪ Uses suction created by a vacuum pump or fan to draw gas into the system
		 ○ Passive
		  ▪ Uses positive pressure of the anesthetic machine to push gas into the system
		  ▪ An example is an activated charcoal canister

CRITICAL THINKING POINTS

• Having a thorough understanding of each component of the anesthesia machine is important when providing anesthesia to patients.

• To prevent injury to personnel and patients, veterinary technicians must be aware of the safety concerns surrounding the use of compressed gas and inhalant anesthetic agents.

• Proper maintenance and pressure checking of the anesthesia machine is a critical part of the veterinary technician's job.

• Recognizing the different types of anesthetic breathing circuits and how and when to use them will help individualize anesthetic plans for each patient.

• Understanding why different oxygen flow rates are used will help anticipate advantages and disadvantages of each rate.

REVIEW QUESTIONS

Multiple Choice

1. Which machine component is part of the high-pressure system?
   a. Flowmeter
   b. Oxygen flush valve
   c. Pressure regulator
   d. Scavenging system

2. How is the reservoir bag size determined?
   a. Based on the patient's lean body weight because excess subcutaneous fat can impede inspiratory volume
   b. Multiplying 5 × tidal volume (10–15 mL/kg) and choosing that size or the next largest size bag available
   c. By estimating lung capacity based on chest circumference and body weight
   d. By animal species and multiplying the weight by two-thirds

3. Which component is part of the low-pressure system?
   a. Piping from oxygen tank to machine
   b. Channel from the vaporizer to the common gas outlet
   c. Oxygen drops from the ceiling
   d. Piping from pressure regulator to oxygen flush valve

4. What is the term for a mixture of gases that does not provide enough oxygen for a patient's individual needs?
   a. Volatile mixture
   b. Hypoxic mixture
   c. Compressed mixture
   d. Fresh gas mixture

5. Oxygen cylinders, in the United States, along with all hoses and knobs, are color-coded _____ for safety.
   a. green
   b. white
   c. yellow
   d. black

6. Which type of cylinder is the small, portable oxygen tank that holds around 600 L?
   a. A
   b. H
   c. D
   d. E

7. What is the pressure of gas in a full oxygen cylinder?

   a. 2200 psi
   b. 1700 psi
   c. 200 psi
   d. 700 psi

8. What is the volume of oxygen remaining in an E tank if the pressure reading is 735 psi?

   a. 150 L
   b. 175 L
   c. 200 L
   d. 225 L

9. Cylinder valves, pressure regulators, gauges, or fittings should never come into contact with combustible substances, greases, oils, or lubricants because of which risk?

   a. Fire
   b. Turning the cylinder into a rocket
   c. Wasting expensive gases
   d. Damaging the cylinder neck

10. Where should a float indicator in a flowmeter be read?

    a. The top
    b. At the dot
    c. In the center
    d. At the point

11. What can tipping a vaporizer when it is in the "on" position lead to?

    a. Liquid anesthetic leakage that may lead to machine contamination
    b. Delivery of the maximum percent of anesthetic inhalant agent and patient death
    c. Exposure to waste anesthetic gas
    d. Improper scavenging of liquid material

12. The oxygen flush valve delivers _____ oxygen at _____ pressures.

    a. 95%; high
    b. 100%; low
    c. 100%; very high
    d. 95%; normal

13. When should you absolutely NOT use the oxygen flush valve?

    a. When the pop-off valve is closed, when using a non-rebreathing system, and when connected to a small patient
    b. When the pop-off valve is closed, when using a rebreathing system, and when connected to a larger patient
    c. When the pop-off valve is open, when using a non-rebreathing system, and when connected to a larger patient
    d. When the pop-off valve is open, when using a rebreathing system, and when connected to a smaller patient

14. What part of the anesthesia machine contains gas that has gone through the oxygen source, pressure regulator, flowmeter, and vaporizer?

    a. Oxygen flush valve
    b. Concentrator
    c. Common gas outlet
    d. Cylinder

15. The pressure manometer is located after the expiratory valve and is used to assess patient airway pressure during manual or mechanical breaths. How much pressure is sufficient to properly inflate the normal lungs of a small animal to facilitate gas exchange during a mechanical breath?

    a. 10–15 cm H_2O
    b. 20–30 cm H_2O
    c. 35–40 cm H_2O
    d. 45–50 cm H_2O

16. The exothermic reaction that takes place within the carbon dioxide absorber produces _____ and _____ as well as carbonate.

    a. carbon dioxide, water
    b. limestone, condensation
    c. heat, water
    d. acid, sodium hydroxide

17. What kind of scavenging system is a charcoal canister?

    a. Active
    b. Passive
    c. Vacuum
    d. Dynamic

18. A non-rebreathing system is typically used in which weight patients?

    a. < 7 kg
    b. > 15 kg
    c. 8–12 kg
    d. < 10 kg

19. What is a coaxial circle system also called?

    a. Mapleson
    b. Wye
    c. Jackson-Rees
    d. Universal F

20. What is the breathing system that bypasses the unidirectional valves, pop-off valve, and carbon dioxide absorber?

    a. Non-rebreathing
    b. Circle
    c. Rebreathing
    d. Pediatric

Case Studies

Case Study 1: With the anesthetic machine plugged into an oxygen source, Daisy, a F/S canine patient weighing 34 kg, is attached with an appropriately sized rebreathing circuit yet does not stay anesthetized. Despite turning up the vaporizer setting to deepen the patient's stage of anesthesia, the patient continues to "lighten" up. The only effective way to keep Daisy at the proper plane of anesthesia is through administration of an intravenous induction agent.

1. What could the problem(s) be?
2. How can the problem be fixed?
3. What can be done to prevent this from happening again?

Case Study 2: Tumblebrutus is a 3-year-old M/N Labrador Retriever dog weighing 25 kg. The dog presented on emergency with a broken and bleeding toenail and refuses to let anyone examine it. Tumblebrutus needs to be anesthetized to clean the wound and stop the bleeding. The dog is given a sedative as pre-anesthetic medication. Sedation goes well and induction of anesthesia is smooth. Together the veterinary technician and the veterinarian perform a ring block around the toe using lidocaine to help keep the patient comfortable. Despite the block, the dog is very difficult to keep anesthetized. Vital signs are all high-normal and the patient is breathing well unassisted, so the vaporizer setting is increased along with the oxygen flow to try to deepen the plane of anesthesia. Based on the premedications and local block, it is felt that pain has been adequately addressed, but the dog will not stay anesthetized. Tumblebrutus begins to move and wake up about 10 minutes after induction.

1. What might the problem be?
2. How can it be corrected?
3. How can the anesthetist prevent this from happening again?

Case Study 3: At the end of each day it is the anesthetist's job to check the level of oxygen in the portable anesthesia machine E-cylinders. If they contain less than 500 psi of oxygen, the hospital's protocol is to remove the tank and replace it with a full one. The tank being used has 400 psi of oxygen left, so it is removed and laid down flat on the floor while a connection is made to the full tank. After tightening down the hanger yoke, a wrench is used to turn on the tank and to make sure it is connected correctly. As this is done oxygen begins to whistle out of the tank at very high pressure.

1. What might be the problem?
2. Why might this problem be occurring?

Case Study 4: Lilly, a 7 year-old, F/S Standard Schnauzer who weighs 37.6 lb, is anesthetized for a dental cleaning. While under anesthesia, the patient is stable; however, the reservoir bag on the machine continues to remain more than ¾ full. Lilly's oxygen flow rate was set at the clinic's standard setting of 1 L/min. The patient is being run on a semiclosed system.

1. What could be the problem(s)?
2. How can the problem be fixed?
3. What can be done to prevent this from happening again?

Case Study 5: Before every anesthetic case, it is routine to pressure check the anesthesia machine using the hoses and reservoir bag that will be used on the next case.

1. If the next case is a 25 kg dog, what size reservoir bag should you use for the pressure check (leak test)?

During the pressure check (leak test), it is found that there is a substantial leak. It is so big that the reservoir bag will not fill.

2. What might the problem be?
3. How can this problem be fixed?
4. How can the anesthetist prevent this from happening again?

After fixing the problem, the pressure check (leak test) is repeated. With the pop-off closed and the patient end occluded, the system is inflated to 30 cm H_2O. While watching the pressure manometer, the veterinary technician notices that the pressure is dropping. By turning on the flowmeter it is determined that the machine is leaking 300 mL/min. After tightening the dome caps and checking the CO_2 absorber, the leak is still present. A coworker comes over to help determine the leak source. The coworker sprays the machine with a dilute soapy disinfectant in an attempt to find bubbles produced by the machine's leak, and bubbles are found around the neck of the reservoir bag.

5. What is the problem?
6. How can this situation be fixed?
7. How can the anesthetist prevent this from happening again?

Critical Thinking Questions

1. What should be done if a leak in the reservoir bag was not discovered until the patient was already anesthetized and connected to the anesthesia machine? What are some pros and cons for addressing this issue intraoperatively?

2. You are working with a 75 kg cheetah at the local zoo. You have an E-cylinder oxygen tank with a pressure reading of 1900 psi that contains approximately 500 L of oxygen, and you are running the patient at an oxygen flow rate of 5 L/min. How long will it take before the tank runs out of oxygen?

ENDNOTES

1. Hartsfield, S. M. (2007). Anesthetic machines and breathing systems. In W. J. Tranqulli, J. C. Thrumon, & K. A. Grimm (Eds.), *Lumb & Jones' veterinary anesthesia and analgesia* (4th ed., pp. 463–491). Ames, IA: Blackwell Publishing.

2. Andrews, J. J. (1990). Inhaled anesthetic delivery systems. In R. D. Miller (Ed.), *Anesthesia* (3rd ed., p. 171). New York: Churchill Livingston 1990.

3. Dorsch, J. A., & Dorsch, S. E. (2008). Medical gas cylinders and containers. *Understanding anesthesia equipment* (pp. 9–28). Philadelphia, PA: Lippincott, Williams & Wilkins.

4. Muir W.W. III, & Hubbell, J .A. E. (1995). *Handbook of veterinary anesthesia* (2nd ed.). St. Louis, MO: Mosby.

5. Doherty, T. & Greene, S. A. (2002). In S. A. Greene (Ed.), Anesthesia Circuits. *Veterinary anesthesia and pain management secrets* (pp. 66–69). Philadelphia, PA: Hanley & Belfus.

6. Hall, L. W., Clarke, K. W., & Trim, C. M. (2001). Apparatus for the administration of anaesthetic. In *Veterinary anaesthesia* (10th ed., pp. 208–218). London: W.B. Saunders.

7. Wagner, A .E., & Bednarski, R. M. (1992). Use of low-flow and closed-system anesthesia. *Journal of the American Veterinary Medical Association*, *200*, 1005.

8. Kelly, C. K., Hodgson, D. S., & McMurphy, R. M. (2012). Effect of anesthetic breathing circuit type on thermal loss in cats during inhalation anesthesia for ovariohysterectomy. *Journal of the American Veterinary Medical Association*, *240*(11), 1296–1299. DOI: 10.2460/javma.240.11.1296

CHAPTER 3
Cardiovascular and Respiratory Physiology Review

Trish Farry, CVN, VTS (ECC), VTS (Anesthesia/Analgesia)

LEARNING OBJECTIVES

Upon completion of this chapter, it is expected that the reader should be able to:

3.1 Describe the American Society of Anesthesiologists (ASA) Physical Status Scale and its relationship to determining anesthetic risk

3.2 Identify the structures of the cardiovascular system

3.3 Describe the function of the cardiovascular system

3.4 Identify the structures of the respiratory system

3.5 Describe the function of the respiratory system

3.6 Explain cardiovascular physiology as related to the anesthesia patient

3.7 Explain respiratory physiology as related to the anesthesia patient

KEY TERMS

Afterload
Alveolar dead space
Anatomical dead space
Apnea
Apneustic ventilation
Atelectasis
Atrioventricular (AV) node
Autoregulation of cerebral blood flow
Bradypnea
Bundle of His
Cardiac output (CO)
Cushing reflex
Dead space
Dyspnea
Eucapnia
Eupnea
Expiratory reserve volume (ERV)
Functional residual capacity (FRC)
Hypercapnia
Hyperpnea
Hypocapnia
Hypopnea
Hypoxemia

Hypoxia
Inotropy
Insufflation
Interventricular septum
Inspiratory reserve volume (IRV)
Mean arterial pressure (MAP)
Mechanical dead space
Physiological dead space
Polypnea
Preload
Purkinje fibers
Residual volume (RV)
Sinoatrial (SA) node
Stroke volume (SV)
Systemic vascular resistance (SVR)
Tachypnea
Tidal volume (V_T)
Total lung capacity (TLC)
Ventilation-perfusion mismatch (V/Q mismatch)
Vital capacity (V_C)

INTRODUCTION

Anesthetizing animals involves administering drugs that can significantly reduce cardiovascular and pulmonary function. Since the cardiovascular and respiratory systems work together to provide oxygen and adequate blood flow for delivery of anesthetic agents and oxygen to the patient as well as to eliminate drugs from the animal's body, any alteration to their function can result in complications affecting the outcome of an anesthetic event. To minimize anesthetic risk, the anesthesia team determines if the animal's cardiovascular and respiratory systems are functioning adequately. Therefore, an understanding of how each of these systems works as well as their interaction with each other helps in the evaluation of patients undergoing anesthesia and surgery. Once the patient's health status is determined, the patient is assigned an American Society of Anesthesiologists (ASA) status, which enables the anesthesia team to estimate anesthetic risk, including morbidity and mortality.

The goal of this chapter is to provide a thorough understanding of the cardiovascular and respiratory systems so veterinary technicians can safely deliver anesthesia to their patients. Other anesthetic factors, such as choice of anesthetic agent, anesthetic technique, and duration of anesthesia, also affect patient outcome and will be described in Chapter 5 (Anesthetic and Analgesic Pharmacology), Chapter 8 (Premedication), Chapter 9 (The Induction Period), Chapter 10 (The Maintenance Period), and Chapter 11 (The Recovery Period and Post-Anesthetic Care). Some procedures and surgeries are more risky than others based on how they affect a patient's homeostasis or if they are performed on already compromised patients; those procedures will be described in Chapter 14 (Anesthesia Techniques for Special Cases).

THE RELATIONSHIP OF PHYSIOLOGY, PATIENT HEALTH STATUS, AND MORTALITY IN ANESTHESIA

The patient's health status has been identified as a major risk determinant in all patients undergoing sedation or anesthesia.[1,2] Factors that may affect anesthetic risk include the following:

- *Cardiovascular function.* Drugs used during anesthesia have significant cardiovascular adverse effects. Pre-existent cardiovascular disease or a patient in cardiovascular collapse from hypovolemic shock, for example, increase patient anesthetic risk.
- *Pulmonary function.* Many anesthetic agents (including inhalant anesthetic agents) depress respiratory function, which may lead to the patient's decreased ability to oxygenate and ventilate. These effects may also linger into the anesthetic recovery period. These factors, in the face of concurrent pulmonary disease, will also lead to increased patient anesthetic risk.
- *Kidney or liver function.* Patients with renal or hepatic dysfunction may experience prolonged anesthetic effects since many anesthetic agents are removed from the body via metabolism by the liver and/or excretion by the kidney.
- *Neurologic function.* Anesthetized patients need optimal cerebral blood flow and perfusion because neurologic tissue is intolerant of hypoxia. Intracranial masses, hypertension, and traumatic brain injury as well as inhalant anesthetic agents can impair the autoregulation of cerebral blood flow. **Autoregulation of cerebral blood flow** is a protective mechanism that maintains constant cerebral blood flow via contraction and relaxation of cerebral arterioles in response to changes in cerebral perfusion pressures.

- *Age.* Geriatric animals may have alterations to organ function and concurrent co-morbidities that impact their physiology and thus potential anesthetic outcome. Pediatric animals and neonates may have immature physiologic responses and organ development which may also lead to higher anesthetic risk or different anesthetic considerations[1,2] (Figures 3-1a and 3-1b).
- *Extremes of weight.* There is very little uptake of most drugs in adipose tissue (Figure 3-1c); therefore, it is important to calculate drug doses for obese animals based on an estimate of their lean body weight to avoid giving too much drug. Other body systems can be impacted in obese patients; for example, excess weight can impede oxygenation and ventilation. Patients with less body fat may also become more susceptible to hypothermia (Figure 3-1d), which may decrease drug metabolism and result in a prolonged drug effect in the patient.
- *Breed.* Some breeds have anatomical features that affect anesthesia. For example, brachycephalic breeds have an increased risk of respiratory obstruction (Figure 3-1e), while small breeds may be more prone to hypothermia (Figure 3-1f). In addition, specific breeds may have unique health concerns. For example, a quarter horse may be at risk for the genetic disease hyperkalemic periodic paralysis (HYPP), which may cause the anesthetized horse to develop life-threatening hyperkalemia, cardiac arrhythmias, bradycardia, muscle weakness, collapse, and difficulty breathing.

A preanesthetic physical examination helps veterinary professionals understand the patient history and evaluate current health status (see Appendix C). Because physiology and patient health status can significantly affect the outcome of anesthesia, the ASA developed a physical status classification system that is used in the preanesthetic evaluation of human patients. This ASA Physical Status Scale was developed in 1963 to help the anesthesia team assess their patient's anesthetic risk for morbidity or mortality.[3] Patients with a higher ASA physical status are at greater risk for death and/or anesthetic complications.[4] The ASA Physical Status Scale has been adapted for use in animals to assess anesthetic morbidity and mortality in veterinary patients.

The ASA Physical Status Scale consists of five classes (and a special category) based on patient health and anesthetic risk.[5] The classes are as follows:

- Class I
  - Minimal risk
    Normal healthy patient, no underlying disease. For example, a healthy patient undergoing an ovariohysterectomy or neuter
- Class II
  - Slight risk, mild disease present
  - Patient with slight to mild single systemic disturbance for which the patient is able to compensate. For example, patients with obesity, dental disease, and cranial cruciate rupture
- Class III
  - Moderate risk, obvious disease present affecting one or more body systems
  - Systemic disease or disturbances that are compensated for and are not causing severe functional limitations. For example, patients with chronic kidney disease, controlled diabetes mellitus, and compensated cardiac disease
- Class IV
  - High risk, significantly compromised by disease and functional limitations

○ Current condition is a threat to life. For example, a horse with colic, a dog with gastric dilatation volvulus (GDV), and a cat with congestive heart failure

- Class V
  ○ Extreme risk, patient not expected to survive beyond 24 hours with or without anesthesia. For example, patients from category IV that are in terminal stage of disease (a horse with colic in profound shock, an unresponsive GDV patient, a cat with end-stage heart failure)
- "E" denotes emergency and is assigned to any of the five categories when the anesthesia is an emergency (e.g., IIIE).

Assigning an ASA physical status to patients is essential and requires that the anesthesia team evaluate a patient's physical health to ultimately help determine the most suitable anesthetic agents to use for each patient. The purpose of the grading system is simply to evaluate the degree of a patient's "sickness" or "physical state" before selecting the anesthetic drugs or before performing surgery. Describing patients' preoperative physical status is used for recordkeeping, for communicating between colleagues, and for creating a uniform system for statistical analysis. The grading system is not intended for use as a measure to predict operative risk. Although assigning ASA physical status to a patient is subjective and the number assigned to an individual patient may vary among professionals, it is still useful to perform a physical examination and assess the health status of patients to determine the risk involved with anesthetizing them. The level of risk can then be used to determine which anesthetic agents to use and which types of medical intervention may be needed for each patient.

Tech Tip

Performing the preanesthetic physical examination in the owner's presence allows the veterinary staff to ask the client questions and communicate to them the risks of anesthesia. Assigning ASA physical status to a patient also helps veterinary technicians communicate with the client the level of risk that their pet will take by undergoing anesthesia.

FIGURE 3–1 Factors increasing the risk of anesthesia: (a) A pediatric patient may have immature physiologic responses and organ development, which may also lead to higher anesthetic risk. (b) As animals age, their metabolic rate may decrease, which may increase the depressant effects of anesthetic agents. (c) Obese patients may receive more anesthetic drug than needed if the dose is not calculated based on an estimate of the lean body weight. (d) Underweight patients may be more prone to hypothermia, which may decrease metabolic rate and prolong anesthetic drug effect. (e) Brachycephalic animals are a greater anesthetic risk due to the increased risk of respiratory obstruction. (f) Small breeds are more prone to hypothermia than larger breeds.

THE CARDIOVASCULAR SYSTEM

The heart is an organ that pumps blood throughout the body via blood vessels, which supply oxygen and nutrients to the tissues and remove carbon dioxide and other wastes from tissues. The mammalian heart has four chambers: the left atrium and ventricle and the right atrium and ventricle. The mitral valve separates the left atrium and ventricle, and the tricuspid valve separates the right atrium and ventricle. These valves are known as **atrioventricular (AV) valves**. During circulation, deoxygenated blood returns from the body via the vena cava to the right side of the heart. From the right side of the heart, blood is pumped through the pulmonary semilunar valve into the pulmonary artery to the pulmonary capillaries in the lungs, where it picks up oxygen and carbon dioxide is removed. Blood returns to the heart via the pulmonary veins into the left atrium then through the mitral valve into the left ventricle. Blood is then ejected through the aortic semilunar valve into the aorta and travels to the organs of the body via the systemic vessels (Figure 3-2).

The cardiac cycle is the sequence of events that occurs from the beginning of one heartbeat to the next. There are two phases of the cardiac cycle: diastole and systole. In the diastole phase, the ventricles are relaxed and the heart fills with blood. In the systole phase, the ventricles contract and pump blood to the arteries. One cardiac cycle is completed when the heart fills with blood and the blood is pumped out of the heart (Figure 3-3).

Throughout the cardiac cycle, blood pressure increases and decreases. When atrial pressure exceeds ventricular pressure, the AV valves open and a volume of blood exits the atria and enters the ventricles. When the sinoatrial (SA) node depolarizes, there is atrial contraction, which allows for additional blood volume to leave the atria and enter the ventricles. The period of ventricular relaxation is termed diastole. When the ventricles begin to contract (systole), the AV valves close to prevent backflow into the atria. There is a rapid increase in pressure as the ventricles contract, and when the pressure in the ventricles exceeds the pressure in the aorta and the pulmonary artery, the pulmonary and aortic valves open and ejection of blood to the systemic circulation and lungs occurs.

The electrical activity in the conduction system of the heart is visualized by wave movement on an electrocardiogram (ECG). An ECG is a tracing that shows the changes in voltage and polarity (positive and negative) over time (Figure 3-4). The ECG is the sum of all action potentials produced by each cardiac cell and represents variations in electrical potential caused by excitation of the myocardium.

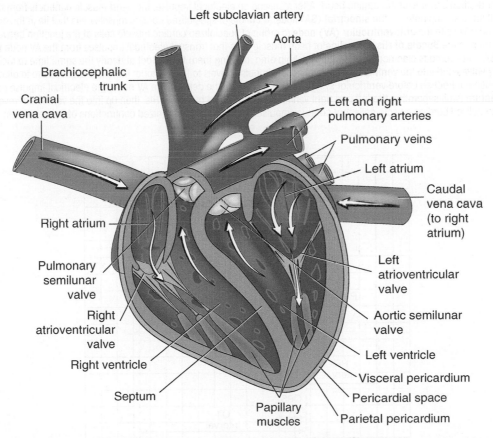

FIGURE 3–2 Internal structures and blood flow through the canine heart. The right atrium receives deoxygenated blood from all tissues except the lungs through the cranial and caudal vena cavae. From the right atrium, blood flows through the tricuspid valve (right atrioventricular valve) into the right ventricle (systemic circulation). The right ventricle contracts and pumps blood through the pulmonary semilunar valves into the pulmonary artery. The pulmonary artery carries blood to the lungs, where carbon dioxide (CO_2) diffuses from blood and oxygen (O_2) diffuses into blood (pulmonary circulation). The left atrium receives oxygenated blood from the lungs through the pulmonary veins. Blood flows through the mitral valve (left atrioventricular valve) into the left ventricle. From the left ventricle, blood leaves the heart through the aortic semilunar valve into the aorta, where it is pumped to all parts of the body except the lungs (systemic circulation).

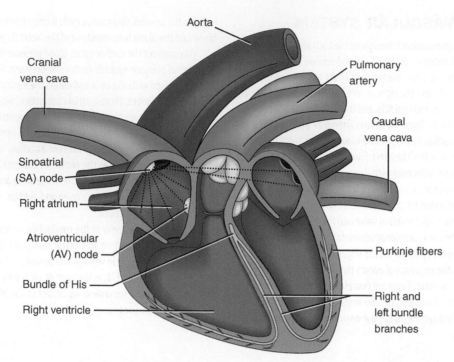

FIGURE 3-3 Conduction systems of the canine heart. After receiving an electrical impulse, the heart muscle contracts from its base to its apex. The electrical impulse originates in the **sinoatrial (SA) node** (a bundle of specialized cardiac muscle cells that lie in the cranial portion of the right atrium) and travels to the **atrioventricular (AV) node** (a group of specialized cardiac muscle cells at the junction between the atrium and ventricle) and then to the **Bundle of His** (collection of heart muscle cells that transmit electrical impulses from the AV node to the apex of the heart). The SA node spreads electrical current through both atria, causing them to contract at nearly the same time to facilitate blood flow through the AV valves and into the ventricles. As the SA node's impulse travels to the AV node, there is a slight pause to allow the atria to complete their systolic contraction before ventricular systole begins. Following the delay at the AV node, the electrical impulse continues down the **interventricular septum** (wall separating the left and right ventricles), through the bundle of His, then up into the ventricular myocardium via the **Purkinje fibers** (specialized cardiac muscle cells that conduct impulses to facilitate synchronized contractions of the ventricles).

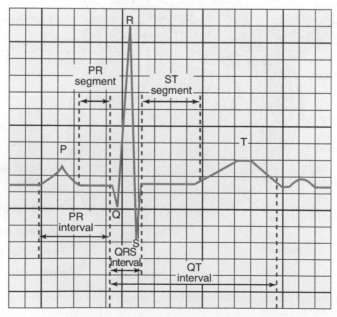

FIGURE 3-4 The anatomy of an electrocardiogram. The first deflection, the P wave, represents atrial depolarization. The PR interval represents conduction through the atria and AV node (affected by parasympathetic tone). The QRS complex represents ventricular depolarization. The QT interval represents ventricular depolarization and repolarization. The ST segment represents end of ventricular depolarization to the onset of ventricular repolarization (the pause between ventricular muscular firing and ventricular muscular repolarization). The T wave represents ventricular repolarization.

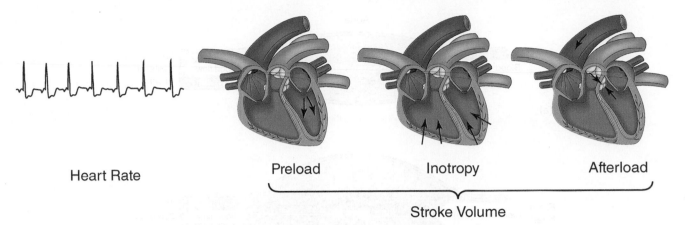

Heart Rate   **Preload**   **Inotropy**   **Afterload**

Stroke Volume

FIGURE 3-5 The components of cardiac output. Cardiac output is determined by heart rate and stroke volume. Stroke volume is determined by preload, afterload, and **inotropy** (cardiac contraction force).

The amount of blood ejected in one cardiac cycle is referred to as **stroke volume** (SV). Determinants of SV include **preload** (the volume of blood entering the right side of the heart), **afterload** (the tension in the left ventricle immediately before the aortic valve opens), and contractility (the strength of heart muscle to contract or shorten) (Figure 3-5). The end of systole occurs when the pressure in the ventricles drops below that of the pressure of the aorta and pulmonary artery, and the valves close.

Blood Pressure

Blood vessels facilitate the delivery of blood to the lungs and tissues and then return it to the heart. Blood pressure is the pressure exerted by circulating blood on the walls of blood vessels. Upon leaving the heart blood flows to arteries, then to smaller distal arteries and arterioles. Arteries have a higher proportion of elastic tissue than other blood vessels, which accounts for their higher pressures. Distal arteries and arterioles have more smooth muscle than arteries, which helps move blood under high pressure to the tissues and provides resistance to help regulate blood flow. Next, blood flows to capillaries, which have single-cell-layer-thick walls and are the exchange sites for oxygen, electrolytes, nutrients, and waste products (like CO_2). Upon leaving the capillaries, blood flows to venules, which are made of fibrous tissue, have thin walls, and collect blood from the capillaries and transport it to the veins. Veins have thin walls with increasing amounts of fibrous tissue as well as smooth muscle and elastic tissue that, together with compression from muscle contraction, help return blood to the heart. Most veins have valves that help prevent backflow of blood.

Determining tissue perfusion is most accurately assessed by **cardiac output** (CO) (the volume of blood ejected by each ventricle per minute). Measuring CO is complex; therefore, blood pressure is the common cardiovascular parameter measured clinically. Since anesthetic agents can cause hypotension and/or decreased CO, understanding the components of maintaining adequate blood pressure is important. Understanding the physiology of hypotension ensures the most appropriate approach to its treatment.

Tech Tip

When administering anesthesia to patients, it is important to understand the relationship of cardiac output to heart rate and stroke volume because inhalational and injectable anesthetic agents can alter cardiac function. Committing the following equation to memory is important: $CO = HR \times SV$.

Mean arterial pressure (MAP) is the average pressure in the arteries during one cardiac cycle and is the product of **systemic vascular resistance** (SVR) (resistance to blood flow by the peripheral circulation.) and CO (see Figure 3-5).

- SVR is the resistance to blood flow by the peripheral circulation.
- CO is the volume of blood ejected by each ventricle per minute and is the product of stroke volume (SV) and heart rate (HR). $CO = SV \times HR$.
- $MAP = SVR \times CO$, and since $CO = SV \times HR$; $MAP = SVR \times SV \times HR$.

In the hypotensive patient, therapy to correct hypotension should be directed at restoring the component most likely affected, whether it be SVR or CO (Figure 3-6).[6]

- For example, if a patient presented to the hospital with acute, significant blood loss, this would require restoration of circulating *volume* to improve CO.
- Conversely, if an anesthetized patient is hypotensive due to vasodilation from isoflurane (a common inhalant anesthetic agent), a drug such as phenylephrine may be required to *vasoconstrict* and improve SVR.

Tech Tip

Hypotension in the anesthesia patient can vary from mild and clinically insignificant to a severe, life-threatening emergency where organ perfusion is in jeopardy. Based on the distribution percent of blood they receive from systemic cardiac output, the organs most affected by reduced tissue perfusion are the kidney, liver, and abdominal organs followed by heart, lungs, and brain.[7]

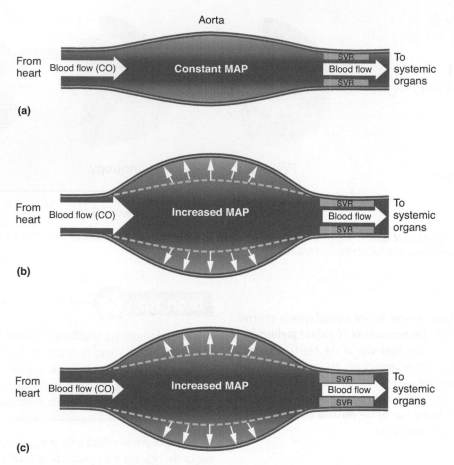

FIGURE 3–6 Mean arterial pressure. (a) Normal relationship among cardiac output, systemic vascular resistance, and mean arterial pressure. (b) When systemic vascular resistance stays the same, an increase in cardiac output leads to an increase in the volume of blood in the aorta and an increase in mean arterial pressure. (c) When systemic vascular resistance increases, a constant cardiac output leads to an increase in the volume of blood in the aorta and an increase in mean arterial pressure.

Heart Rate

Heart rate is the number of times the heart beats per minute. Changes in HR can affect blood flow through the patient and may be caused by many factors. Cardiac arrhythmias may also cause an increase or decrease in heart rate.[6]

- Bradycardia is a slower than normal HR that may or may not affect cardiac output. Causes of bradycardia include:
  - Drug administration. Alpha-2 adrenergic agonists and opioids are drugs well known for their ability to cause a decrease in heart rate. Even though opioids can cause a decrease in heart rate, at clinical dosages they usually have very little effect on blood pressure.
    - If bradycardia is reducing cardiac output, resulting in an overall decrease to blood pressure, then the use of an anticholinergic drug may be warranted.
    - If bradycardia is a result of the use of an alpha-2 adrenergic agonist, anticholinergic drugs should be used judiciously as they may produce hypertension and other adverse effects. Alpha-2 adrenergic agonists are discussed in more detail in Chapter 5 (Anesthetic and Analgesic Pharmacology) and Chapter 8 (Premedication).

  - Medical procedures. Some medical or surgical procedures may produce bradycardia because they stimulate the vagus nerve (the vagus nerve supplies parasympathetic innervation to various organs, including the heart; parasympathetic innervation to the heart lowers heart rate). Examples include **insufflation** (inflating the stomach with air for internal visualization) during endoscopy and ophthalmic procedures that stimulate the oculocardiac reflex.
  - Medical or physiological conditions. Examples of medical conditions that can lower HR include the following:
    - Elevated intracranial pressure may result in a **Cushing reflex**, which is hypertension that results in reflex bradycardia. The hypertension is a protective mechanism that helps preserve brain tissue during periods of poor perfusion.
    - Cardiac arrhythmias that affect the SA node (e.g., sick sinus syndrome) or passage of signals to the AV node or **bundle of His** (e.g., AV block) resulting in a lower HR.
    - Hypothermia may result in bradycardia because it may reduce spontaneous depolarization of the sinoatrial and atrioventricular nodes.
    - Electrolyte abnormalities such as hyperkalemia due to urethral obstruction may result in bradycardia.

- Tachycardia is a rapid HR that may affect CO because the heart does not have time to fill with blood during diastole. Causes of tachycardia include:
  - Disease conditions. Certain disease states, such as hyperthyroidism and pheochromocytoma, produce a rapid HR.
  - Drug administration. Drugs that increase HR are called positive chronotropes. Examples of drugs in this class are epinephrine, dopamine, and anticholinergic drugs such as atropine. Anesthetic drugs such as ketamine can also increase HR.
  - Medical or physiological conditions. Examples of medical/physiological conditions can raise HR include the following:
    - Hypercapnia (increased CO_2 in the blood) will cause stimulation of the CNS, resulting in peripheral vasodilation and tachycardia.
    - Hypoxemia (low partial pressure of dissolved oxygen in the arterial blood) initially causes the body to compensate for insufficient oxygen by producing tachycardia and tachypnea. As hypoxemia worsens, these compensatory mechanisms fail and hypoxia (reduced level of tissue oxygenation) develops, leading to respiratory and cardiac arrest.
    - Physiological response to surgical stimulation/pain/anxiety/urge to urinate, which increases release of cortisol, catecholamines, and inflammatory mediators, which increase HR.[8]
    - Blood loss and hypovolemia may increase HR in an attempt to preserve CO. This is a compensatory mechanism because as blood pressure drops the heart attempts to maintain adequate circulation by increasing HR.
    - Anemia may increase HR due to a decreased hemoglobin concentration that decreases the oxygen content of the blood. To compensate for the lowered oxygen content, HR and SV increase.
    - Arrhythmias such as atrial or ventricular tachycardia.[9]

If the HR is within normal limits but hypotension is present, there may be an issue with SV or SVR.[2] It is important to identify the probable cause of the tachycardia and the source of fluctuations of blood pressure. This will enable the anesthetist, working in conjunction with the attending veterinarian, to formulate a plan for providing the most appropriate treatment.

THE RESPIRATORY SYSTEM

The function of respiration is to oxygenate hemoglobin (Hb) so it can deliver oxygen to the tissues and remove carbon dioxide (CO_2) from the blood. Anesthesia may cause a dosage-dependant depression of ventilation and impact lung function; therefore problems with ventilation and hypoxemia are frequent issues that may need to be addressed in anesthetized patients. An understanding of the functional anatomy and physiology of each patient's respiratory system, as well as terms related to the respiratory system (Table 3-1), is essential for the veterinary technician providing anesthesia support.

The respiratory system is made up of the upper respiratory tract (nasal cavity, pharynx, and larynx), and the lower respiratory tract (trachea, bronchi, and lungs) (Figure 3-7). The diaphragm is the major muscle of respiration, supported by the intercostal and abdominal muscles. Inhaled air

TABLE 3-1 Respiratory Terms

Respiratory Term	Definition
Apnea	Cessation of breathing
Apneustic ventilation	Prolonged inspirations with subsequent short exhalations; patient appears to be holding its breath
Bradypnea	Slow, regular breathing
Dyspnea	Labored breathing
Eupnea	Ordinary and quiet breathing
Hyperpnea	Rapid breathing that may be increased in depth ; overrespiration
Hypopnea	Slow breathing that may be shallow in depth; underrespiration
Polypnea	Rapid, shallow panting
Tachypnea	Rapid breathing that is not labored
Hypoxia	Reduced level of tissue oxygenation; a condition in which the body as a whole or a region of the body is deprived of adequate oxygen supply.
Hypoxemia	Low partial pressure of dissolved oxygen in arterial blood; PaO_2* < 80 mm Hg at sea level
Hypercapnia	Elevated CO_2 tension in blood, $PaCO_2$** > 45 mm Hg
Hypocapnia	Lowered CO_2 tension in blood, $PaCO_2 < 35$ mm Hg
Eucapnia	Normal CO_2 tension in blood, $PaCO_2$ between 35 and 45 mm Hg

* PaO_2 is the partial pressure of oxygen molecules dissolved in the plasma phase of an arterial sample
** $PaCO_2$ is the partial pressure of carbon dioxide molecules dissolved in the plasma phase of an arterial sample

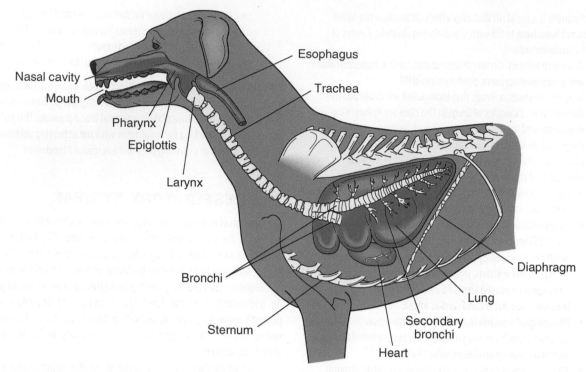

FIGURE 3-7 Structures of the respiratory system.

moves through the larynx and trachea to the lungs via the two bronchi, one supplying each lung. Each main stem bronchus divides into secondary and tertiary bronchi, which terminate into the bronchioles. The bronchioles end in air sacs called alveoli, where diffusion of gases occurs. Capillaries carrying blood are found on the surface of each alveolus. The blood in these capillaries has returned from the body and has low levels of oxygen and high levels of CO_2. Gas exchange occurs when oxygen from air is inhaled into the alveoli and then diffuses into the blood capillary; CO_2 is removed from blood when it diffuses into the alveoli and is exhaled.

The inner surfaces of the alveolar walls are coated with a thin film of water molecules that move freely in all directions within the liquid layer. However, when a liquid gas interface occurs (such as between the water molecules in the fluid coating the alveolar walls and the air in the alveoli), the liquid molecules are attracted to each other more than they are attracted to air molecules.[10] This attraction of water molecules to each other creates a force called surface tension. Surface tension increases as water molecules come closer together, which happens when an animal exhales and the alveoli become smaller and their inner surfaces become close to each other. Surface tension could cause alveoli to collapse and make it difficult to re-expand the alveoli on inspiration.

Alveoli do not collapse under normal circumstances because the alveolar type II cells of the lungs produce surfactant, a substance that reduces surface tension. Surfactant contains phospholipids and different types of surfactant proteins that are either hydrophilic (water-loving) or hydrophobic (water-hating).[3] The main role of surfactant is to prevent collapse of the alveoli by reducing the effort needed to expand the lungs during inspiration and allow gas exchange to take place. Surfactant prevents alveolar collapse during expiration by reducing surface tension (Figure 3-8).

Not all alveoli are functional at any one time. Some alveoli may be collapsed due to surface tension. Some alveoli, particularly those in the upper segments of the lungs, will be underperfused relative to ventilation and some in the dependent parts of the lungs will be overperfused relative to ventilation. This unequal perfusion of the alveoli is due to gravity and is called **ventilation-perfusion mismatch (V/Q mismatch)**.

Anesthetized patients are at increased risk of developing **atelectasis** (alveolar collapse). Development of atelectasis is associated with decreased lung compliance (ability for the lung to distend or stretch), impairment of oxygenation, and lung injury.[11] The adverse effects of atelectasis persist into the postoperative period and can impact patient recovery.[11] Patients may develop atelectasis by three different mechanisms: absorption (gas) atelectasis, compression atelectasis, and surfactant impairment.

- *Absorption atelectasis* occurs due to small airway closure and the inability to keep alveoli distended due to inadequate nitrogen reserves that normally help splint alveoli open. In anesthetized patients, most atelectasis is absorption atelectasis because in veterinary medicine high inspired oxygen concentrations are used and in doing so no nitrogen source is provided.

- Along with absorption atelectasis and positioning, *compression atelectasis* occurs when the pressure distending the alveolus is reduced to a level that allows an alveolus to collapse. The diaphragm separates the thoracic and abdominal cavities and under normal

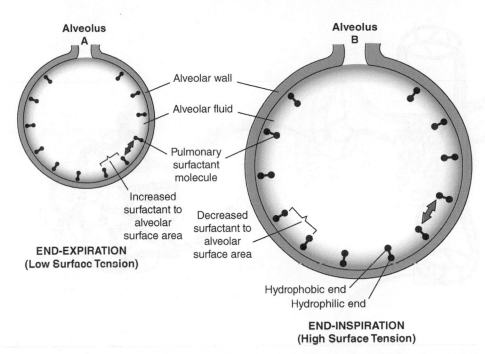

FIGURE 3-8 Relationship of surfactant and surface tension. (a) In the normal lung, the surface tension is low in a small alveolus because the ratio of surfactant to alveolar surface is high. (b) As the alveolus enlarges, the surface tension increases because the ratio of surfactant to alveolar surface decreases.

circumstances produces pressure differences between the chest and abdomen. Under anesthesia, the diaphragm is relaxed and does not maintain pressure differences between the thoracic and abdominal cavities, which can result in alveolar collapse. Similarly, upon anesthetic induction there is a decrease in functional residual capacity (volume of gas remaining in the lungs at the end of a normal expiration) due to patient positioning, which can result in the formation of atelectasis. In addition to patient positioning, other conditions that can cause compression atelectasis include lung tumors and pleural space disease.

- Atelectasis caused by *surfactant impairment* may occur because anesthetic agents may depress the stabilizing function of surfactant; however, surfactant reserve makes this mechanism of developing atelectasis less important. Surfactant impairment is uncommon in veterinary medicine.

Oxygen Transport

Respiration is the diffusion of gases (O_2 and CO_2) between the atmosphere and the cells of the body. Gases move from a region of high partial pressure to a region of low partial pressure down a partial pressure gradient. Oxygen is carried through blood by hemoglobin. How much oxygen is bound to hemoglobin is mainly determined by the partial pressure of O_2 in the hemoglobin solution (a small amount of O_2 is dissolved in the plasma, but this amount is physiologically insignificant). When every oxygen-binding site on all the hemoglobin molecules is bound to oxygen, the blood is 100% saturated

and cannot carry any more oxygen. When half of the sites are filled with oxygen, the blood is 50% saturated. At high partial pressures of O_2, hemoglobin saturation remains high. As oxygen levels decrease, hemoglobin saturation decreases and less oxygen is available to be delivered to cells.

External respiration is the exchange of O_2 and CO_2 between the external environment (atmospheric air) and the blood in the alveolar capillaries (Figure 3-9). Gas exchange in the lungs occurs rapidly across the very thin alveolar and capillary membranes, both of which have a large surface area. In healthy animals, O_2 enters the alveoli and CO_2 exits the alveoli based on partial pressure differences. The partial pressure of O_2 is lower in the alveoli compared to the external environment because oxygen continuously diffuses across the alveolar wall. In contrast, the partial pressure of O_2 in the alveoli is higher than that in the capillaries; therefore, oxygen diffuses into the blood. Once through the alveolar and capillary walls, the oxygen combines with hemoglobin to form oxyhemoglobin and is transported within the bloodstream. CO_2 diffuses from the blood into the alveoli because the concentration (pressure) of CO_2 in the alveoli is at a lower level than in the blood. As an animal breathes, it continuously brings fresh air containing a lot of O_2 and a small amount of CO_2 into the alveoli to continue movement along this concentration gradient.

Internal (or cellular) respiration is gas exchange between the blood and the cells (see Figure 3-9). Diffusion occurs between the blood and cells because the concentration (partial pressure) of O_2 in the blood is at a higher level than that in the cells and thus O_2 moves into the cell. The concentration (pressure) of CO_2 in the blood is at a lower level than in the cells, which creates a concentration gradient. CO_2 is a cellular waste product and diffuses from the tissue to blood. The partial pressure of CO_2 in the capillaries is higher than in the alveoli; therefore, CO_2 diffuses into the alveoli and is exhaled.

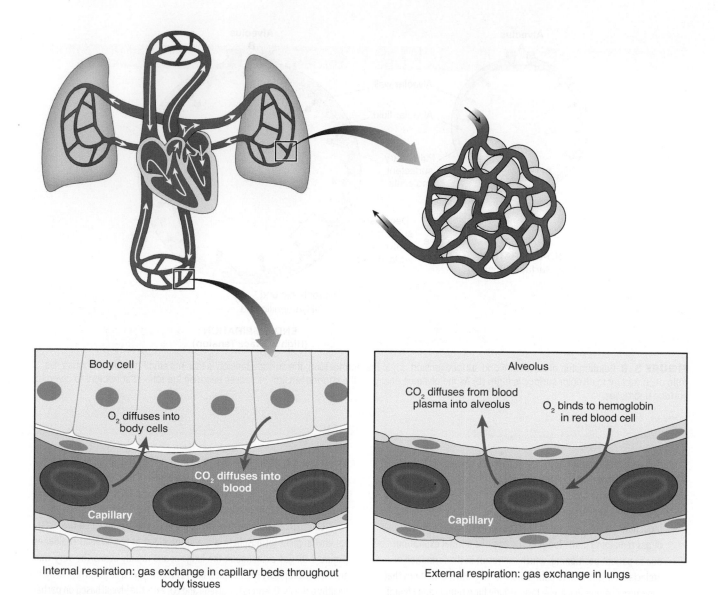

Internal respiration: gas exchange in capillary beds throughout body tissues

Body cell

O_2 diffuses into body cells

CO_2 diffuses into blood

Capillary

External respiration: gas exchange in lungs

Alveolus

CO_2 diffuses from blood plasma into alveolus

O_2 binds to hemoglobin in red blood cell

Capillary

FIGURE 3–9 External and internal respiration. In external respiration, the partial pressure of oxygen is higher in the alveoli than in blood; therefore, oxygen diffuses to blood. The partial pressure of carbon dioxide is higher in blood than in alveoli; therefore, carbon dioxide diffuses into the alveoli. In internal respiration, the partial pressure of oxygen is higher in the blood than in tissue; therefore, oxygen diffuses to the tissue. The partial pressure of carbon dioxide is higher in tissue than in blood; therefore, carbon dioxide diffuses into the blood.

Respiratory Volume

When describing the events that occur during ventilation, a variety of terms are used to subdivide the volumes and capacities within the lungs. These terms are defined in Table 3-2.

Dead Space

Dead space is the volume of inhaled air that does not take part in gas exchange. At inspiration the volume of air occupying the space in the nose, pharynx, larynx, trachea, bronchi, and bronchioles is called **anatomical dead space**. **Alveolar dead space** is the volume of gas in alveoli that are ventilated but not perfused. Together, the anatomical dead space and the alveolar dead space make up the **physiological dead space** (Figure 3-10). Gas within this space does not take part in gas exchange. When the patient

undergoing anesthesia is connected to a breathing system, care must be taken so the apparatus itself does not contribute significantly to the mechanical dead space. **Mechanical dead space** is the area in the breathing circuit or endotracheal tube where bidirectional flow takes place yet no gas exchange occurs. To understand the effects of dead space, consider a swimmer using a snorkel. The volume of the snorkel must be added to the amount of air inhaled and exhaled in one breath (tidal volume) if ventilation is to be maintained adequately. The snorkel contributes dead space, called the mechanical dead space (also known as apparatus dead space). Similarly, the breathing system of an anesthetic machine can contribute to mechanical dead space. This means that the patient must breathe more deeply and/or more rapidly to achieve the same alveolar ventilation as it had before being connected to the anesthesia machine. Care must be taken to ensure that intubation of the trachea and connection of a patient to a breathing system

TABLE 3-2 Lung Volume Terminology

Term and Abbreviation	Definition
Tidal volume (V_T)	Amount of air exchanged during normal respiration (air inhaled and exhaled in one breath)
Inspiratory reserve volume (IRV)	Amount of air inspired over the tidal volume (extra amount that could be inhaled after normal inspiration)
Expiratory reserve volume (ERV)	Amount of air expired over the tidal volume (extra amount that could be exhaled after normal expiration)
Residual volume (RV)	Air remaining in the lungs after a forced expiration (amount of air trapped in alveoli)
Functional residual capacity (FRC)	Amount of air remaining in the lungs after a normal respiration (ERV + RV)
Vital capacity (V_c)	Largest amount of air that can be moved in the lungs (V_T + IRV + ERV)
Total lung capacity (TLC)	Maximum volume to which the lungs can be expanded (V_T + ERV + RV)

Anatomical + Alveolar + Mechanical = Total Dead Space

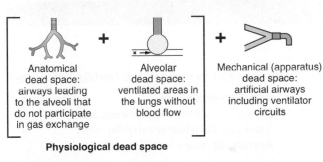

Anatomical dead space: airways leading to the alveoli that do not participate in gas exchange

Alveolar dead space: ventilated areas in the lungs without blood flow

Mechanical (apparatus) dead space: artificial airways including ventilator circuits

Physiological dead space

FIGURE 3-10 Components of dead space. Physiological dead space is made up of the anatomical and alveolar dead space. Patients receiving inhalant anesthetic agents may also have mechanical (apparatus) dead space due to the breathing system of the anesthesia machine. Since gas within this space does not take part in gas exchange, it is important to limit dead space in anesthetized patients. The more dead space that is present, the deeper and/or more rapidly the patient must breathe to achieve the proper level of alveolar ventilation.

do not significantly increase the dead space (Figure 3-11). If an anesthetized patient (likely to have depressed ventilation) is connected to a breathing system that has significant mechanical dead space, hypoventilation and increased CO_2 levels as a result of rebreathing of expired gases are likely to occur. It is important to remember that it is only the functioning alveoli that are involved in gas exchange.

Many anesthetic agents depress pulmonary ventilation by causing a reduced respiratory rate or tidal volume or both. Depression of ventilation will also cause an increase in alveolar CO_2 that results in an increase in blood CO_2 and lowering of pH (carbon dioxide is acidic). The anesthetist may need to support ventilation during anesthesia to prevent hypercapnia. High blood CO_2 may cause

- vasodilation and consequentially hypertension, tachycardia, and increased cardiac output;
- lower blood pH, which can cause subsequent electrolyte changes;
- moderate hypercapnia, which is a respiratory stimulant in the conscious patient, but at high CO_2 levels depresses respiration and can act as an anesthetic in its own right; and
- decreased renal blood flow.

Hypoxemia is decreased partial pressure of oxygen in arterial blood. There are many possible causes of hypoxemia for the anesthetized patient. Hypoventilation is one of the most common causes of hypoxemia. If hypoventilation is ruled out as the cause of hypoxemia, other causes include inadequate fraction of inspired oxygen (FiO_2), ventilation-perfusion mismatch (V/Q mismatch), diffusion impairment (a diffusion barrier diminishing the transfer of gas), or alveolar (right to left) shunting[6] (see Chapter 12, Anesthetic Complications).

Normal room air contains 21% oxygen. When receiving supplemental oxygen (facemask or nasal cannula), the patient is breathing a combination of room air and oxygen. Delivery of oxygen by these means results in an FiO_2 that is always going to be significantly less than 100%.

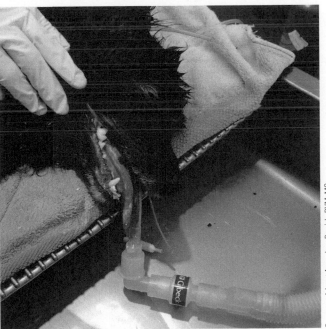

Courtesy of Janet Amundson Romich, DVM, MS

FIGURE 3-11 Mechanical (apparatus) dead space is increased in this patient due to both the endotracheal tube extending beyond the patient's incisors and a monitoring adaptor added between the end of the endotracheal tube and breathing hose. Increased dead space can result in hypoventilation and increased CO_2 levels as a result of rebreathing of expired gases.

SUMMARY

An patient's health status affects anesthetic outcomes; therefore, knowledge of normal cardiovascular and respiratory physiology along with ways to assess a patient's health help determine how anesthetic agents will affect a patient. The ASA Physical Status Scale (see Appendix D) is a tool that helps the anesthesia team predict their patient's anesthetic risk for morbidity or mortality. By using the patient's ASA status and knowing the effects anesthetic agents can have on body systems such as the cardiovascular and respiratory systems, the anesthesia team can develop an individualized plan for each patient in their care.

Each veterinary species may have unique anatomic variations that should be taken into account prior to anesthesia. It is beyond the scope of this text to provide complete comparative anatomy and physiology for all species; however, chapters of this text that discuss specific species will provide basic species-specific anatomy and physiology.

CRITICAL THINKING POINTS

- Understanding the normal anatomy and physiology of the cardiovascular and respiratory systems helps the anesthesia team provide quality care for anesthetized patients and allows for appropriate intervention during emergency situations.

- There are a variety of factors that affect the incidence of anesthetic deaths, including a patient's organ function, age, weight, and breed.

- The ASA Physical Status Scale helps the anesthesia team assess their patient's anesthetic risk for morbidity or mortality. Patients with a higher ASA physical status are at greater risk for death and/or anesthetic complications.

REVIEW QUESTIONS

Multiple Choice

1. During circulation, what structure returns deoxygenated blood from the body to the heart?

    a. Vena cava
    b. Pulmonary artery
    c. Aorta
    d. Lungs

2. What is an electrocardiogram?

    a. A diagnostic method to detect only atrial conditions such as atrial depolarization
    b. A diagnostic method to detect only ventricular conditions such as ventricular depolarization
    c. A tracing that represents variations in electrical potential caused by excitation of the myocardium.
    d. A technique that is performed by placing two fingers at the base of the wrist to feel for a beat

3. What does apneustic ventilation mean?

    a. Rapid, shallow panting
    b. Labored breathing
    c. Elevated CO_2 tension in blood
    d. Prolonged inspirations with subsequent short exhalations

4. How much oxygen does normal room air contain?

    a. 13%
    b. 21%
    c. 44%
    d. 67%

5. Mean arterial pressure (MAP) is the product of which two factors?

    a. Stroke volume and cardiac output
    b. Stroke volume and systemic vascular resistance
    c. Systemic vascular resistance and cardiac output
    d. Heart rate and stroke volume

6. Cardiac output (CO) is the product of which of the following two factors?

    a. Contractility and heart rate
    b. Stroke volume and systemic vascular resistance
    c. Systemic vascular resistance and cardiac output
    d. Heart rate and stroke volume

7. What is the major muscle of respiration?

    a. The diaphragm
    b. The lungs
    c. The deltoid muscles
    d. The abdominal muscles

8. The cardiac cycle begins with depolarization at which structure?

    a. Bundle of His
    b. Interventricular septum
    c. AV node
    d. SA node

9. Which valve separates the left atrium and ventricle?

    a. The tricuspid valve
    b. The mitral valve
    c. The aortic valve
    d. The pulmonary valve

10. Which valve separates the right atrium and ventricle?

    a. The tricuspid valve
    b. The mitral valve
    c. The aortic valve
    d. The pulmonary valve

11. What is the time period of ventricular relaxation?

    a. Systole
    b. Asystole
    c. Diastole
    d. Depolarization

12. What effects on blood CO_2 level and pH are due to depression of ventilation?

   a. An increase in blood CO_2 and raising of pH
   b. No change in blood CO_2 and no change of pH
   c. A decrease in blood CO_2 and raising of pH
   d. An increase in blood CO_2 and lowering of pH

13. If an anesthetized patient is connected to a breathing system that has significant mechanical (apparatus) dead space, which of the following may occur?

   a. Hypoventilation and rebreathing of expired gases
   b. Eupnea and rebreathing of inspired gases
   c. Hyperventilation and rebreathing of expired gases
   d. Hypoventilation and hypocapnia due to rebreathing of inspired gases

14. What is one of the most common causes of hypoxemia?

   a. Hypoventilation
   b. Hyperventilation
   c. Blood loss
   d. Oxygen toxicity

15. When calculating a drug dose for an obese patient, what value should be used?

   a. Their actual weight
   b. Their estimated lean body weight
   c. Their actual weight minus 12%
   d. An empirical formula

16. What term is defined as the resistance to blood flow by the peripheral circulation?

   a. Systemic vascular resistance (SVR)
   b. Mean arterial pressure (MAP)
   c. Cardiac output (CO)
   d. Preload (PL)

17. For what complication are brachycephalic breeds at an increased?

   a. Respiratory obstruction
   b. Cardiac arrhythmia
   c. Cardiac arrest
   d. Respiratory discharge

18. What factor reduces surface tension of the alveoli?

   a. Water molecules in the fluid of the alveoli
   b. Inspiratory volume
   c. Expiratory volume
   d. Surfactant

19. To what type of dead space can the breathing system of the anesthetic machine contribute?

   a. Anatomical dead space
   b. Alveolar dead space
   c. Mechanical (apparatus) dead space
   d. Physiologic dead space

20. Which organs are most affected by reduced tissue perfusion?

   a. Brain and heart
   b. Kidneys and liver
   c. Liver and lungs
   d. Brain and lungs

Case Studies

Case Study 1: A 6-year-old M/N Cavalier King Charles Spaniel dog is brought in for a left front limb mass removal. This breed of dog can often suffer from mitral valve disease, which leads to heart failure. The owner tells you that the veterinarian has diagnosed a low-grade heart murmur in her pet. The dog is not on any medication and has no evidence of exercise intolerance, syncope, or cough.

1. As the veterinarian is preforming a physical examination, you record the patient's history in the medical record. You are the anesthesia technician for this case and would like to have determined the ASA physical status when you approach the veterinarian to discuss the anesthetic protocol for this patient. What do you think is the ASA physical status of this patient (see Appendix D)?
2. Will the ASA physical status impact how the animal is anesthetized?
3. Does the anesthesia team need to be made aware of this patient's pre-existing medical condition and ASA physical status prior to the patient's premedication?

Case Study 2: A 9-year-old F/S domestic longhair (DLH) cat is presented to the clinic for an annual dental check-up. She currently weighs 6 kg and has diabetes mellitus, which is controlled with insulin. The veterinarian has decided that the feline patient will have a dental prophylactic cleaning and possible extractions performed the next day. The results of preanesthetic blood tests are normal.

1. As the anesthesia technician for this case, you would like to have determined the ASA physical status when you approach the veterinarian to discuss the anesthetic protocol for this patient. What ASA physical status do you think this patient is and why?

Case Study 3: A 5-year-old M/N Rottweiler dog has been hit by a car and is presented to the veterinary clinic in lateral recumbency and in shock. The owners inform you that on his previous visit to the veterinarian he weighed 35 kg and has been in good health. The dog is tachycardic, tachypneic, and has very pale mucous membranes and suddenly becomes unresponsive. The veterinarian has diagnosed massive internal hemorrhage in the abdomen and thorax by diagnostic imaging, abdominocentesis, and thoracocentesis. The owners want to do everything they can for their pet and have given authorization for stabilization of shock and surgical intervention.

1. As the veterinary technician for this case, you would like to have determined the ASA physical status when you approach the veterinarian to discuss the anesthetic protocol for this patient. What ASA physical status do you think this patient is and why?

Case Study 4: You are monitoring a healthy 3-year-old F/S Doberman dog weighing 25 kg who is undergoing a gastroscopy procedure for a foreign body removal. In the middle of the procedure the patient has suddenly become bradycardic. The heart rate has decreased from 88 bpm to 48 bpm. Why might this have happened?

Case Study 5: You are monitoring a patient under anesthesia who becomes hypotensive. Being able to recommend the best intervention for the patient requires considering the components of mean arterial pressure.

1. What two components make up MAP?
2. What intervention should be considered if the patient is hypotensive due to significant blood loss?
3. What intervention should be considered if the patient is hypotensive due to vasodilation from an anesthetic agent?

Critical Thinking Questions

1. You have a client that is a human anesthesiologist, and she has a 15-year-old M/N Abyssinian cat that requires general anesthesia for a dental extraction. She has asked you to explain the risks involved with feline anesthesia and to explain the anesthetic protocols in your hospital. What do you tell her?

2. You are working with a new veterinary technician and have been going through some respiratory physiology with him. He is having a difficult time understanding how dead space can be an important factor in respiration and ventilation. How do you explain the concept of dead space to the veterinary technician?

ENDNOTES

1. Brodbelt, D. C., Blissitt, K. J., Hammond, R. A., Neath, P. J., Young, L. E., Pfeiffer, D. U., & Wood, J. L. N. (2008). The risk of death: The confidential enquiry into perioperative small animal fatalities. *Veterinary Anaesthesia and Analgesia, 35*, 365–373.
2. Brodbelt, D. (2009). Perioperative mortality in small animal anaesthesia. *The Veterinary Journal, 182*(2), 152–161.
3. American Society of Anesthesiologists. ASA Physical Status Classification System, www.asahq.org
4. Brodbelt, D., Flaherty, D., & Pettifer, G. (2015). Anesthetic risk and informed consent. In K. G. Grimm, L. Lamont, W. J. Tranquilli, S. Greene, & S. Robertson (Eds.), *Veterinary anesthesia and analgesia: The fifth edition of Lumb and Jones* (pp. 11–21). Ames, IA: Wiley Blackwell.
5. Academy of Veterinary Technician Anesthetists (AVTA), http://www.avta-vts.org/asa-ratings.pml
6. Guedes, A. G. P. (2008). Support. In G. L. Carroll (Ed.), *Small animal anesthesia and analgesia* (pp. 159–163). Ames, IA: Blackwell.
7. Stoelting, R. K., & Miller, R. D. (2007). *Basics of anesthesia* (pp. 49–63). Philadelphia, PA: Churchill Livingstone.
8. Okafor, R. O. S., Remi Adewunmi, B. D., Fadason, S.T., Ayo, J. O., & Muhammed, S. M. (2014). Pathophysiologic mechanisms of pain in animals—A review. *Journal of Veterinary Medicine and Animal Health, 6*(5), 123–130.
9. Muir, W., Hubbell, J., Bednarski, R., & Lerche, P. (2013). Cardiovascular emergencies. In *Handbook of veterinary anesthesia* (5th ed., pp. 521–522). St. Louis: Elsevier.
10. Des Jardins, T. (2013). *Cardiopulmonary anatomy and physiology: Essentials of respiratory care* (6th ed., pp. 100–107). Boston, MA: Cengage Learning.
11. Duggan, M., & Kavanagh, B. (2005). Pulmonary atelectasis: A pathogenic perioperative entity. *Anesthesiology, 102*, 838–854.

CHAPTER 4

Fluid Therapy and Intravenous Catheterization

Anita Parkin, *RVN, AVN Dip (Surgery and ECC), VTS (Anesthesia/Analgesia), CVPP, TAE*

LEARNING OBJECTIVES

Upon completion of this chapter, it is expected that the reader should be able to:

4.1 Summarize fluid distribution in animals

4.2 Discuss the rationale of using different types of intravenous (IV) fluids, including crystalloids, colloids, and blood products

4.3 Describe the various types of IV catheters used in veterinary medicine

4.4 Explain the importance of placing an IV catheter in anesthetized patients

4.5 Describe how to place an IV catheter

4.6 Describe IV catheter management

4.7 Identify different ways to administer fluids during anesthesia

KEY TERMS

Anion

Balanced solution

Blood products

Bolus

Buffer

Cation

Central venous catheter (central line)

Colloid

Cross-matching

Crystalloid

Dehydration

Electrolyte

Fluid overload

French (Fr)

Gauge

Hyperosmolar

Hypertonic

Hypotonic

Hypovolemia

Insensible body water losses

Ion

Isotonic

Maintenance fluid

Oncotic pressure

Ongoing fluid loss phase

Ongoing maintenance phase

Osmolality

Osmolarity

Osmole

Osmotic pressure

Over-the-needle catheter

Over-the-wire catheter

Peel-away catheter

Replacement fluid

Replacement phase

Resuscitation phase

Sensible body water losses

Solute

Shock

Solvent

Thrombogenic

Through-the-needle catheter

Tonicity

INTRODUCTION

The body and its supporting tissues need constant levels of oxygen delivery to perform their metabolic functions and sustain life. As discussed in Chapter 3 (Cardiovascular and Respiratory Physiology Review), a properly functioning cardiovascular and respiratory system is needed to transport oxygen down a concentration gradient from the lungs, to the macrocirculation (large blood vessels), then to the microcirculation (the smallest blood vessels embedded within organ tissues) and then finally into individual tissue cells within an organ. Systemic disease or the use of anesthetic agents that affect cardiopulmonary performance can alter these physiologic systems, which in turn can adversely impact the oxygen delivery system and result in increased patient morbidity and mortality. These patients will often benefit from IV fluid therapy and/or blood products to support cardiac output, the oxygen-carrying capacity of blood, and ultimately oxygen delivery to systemic tissues. Since IV fluids and blood products are essentially drugs that have their own individual indications, benefits, and risks, it is important to understand the types of products available and how they affect individual patients who also are unique in their levels of hydration, circulating intravascular volume, electrolytes, plasma proteins, and number of red blood cells. It is therefore important that the anesthetist, working in conjunction with the attending veterinarian, identify these

factors and construct a plan that is specific to an individual patient. This chapter aims to help the reader gain a deeper understanding of the physiology of fluid therapy, the individual types of IV fluids and blood products available, and the methods surrounding the delivery of IV fluid and blood products.

BODY FLUID BASICS

Water is the primary fluid of the human or animal body and is vital for normal oxygen delivery and cell function. Body water makes up a large part of an animal's total body weight (approximately 60%) and is distributed among three separate "compartments": cells, blood vessels, and tissue spaces, which are described as *intracellular* (within the cell), *intravascular* (within the blood vessels), or *interstitial* (in the tissue spaces between blood vessels and cells), respectively, based on their location. Fluid within the cell is known as *intracellular fluid (ICF)*, while intravascular fluid and interstitial fluid together are classified as *extracellular fluid (ECF)*. About two-thirds of the body's entire water is intracellular and is found mainly in skeletal muscle, blood cells, bone cells, and adipose cells. The remaining one-third of body water is extracellular and is found in circulating intravascular volume (about 25%) and in the interstitial fluid between cells (about 75%) (Figure 4-1).

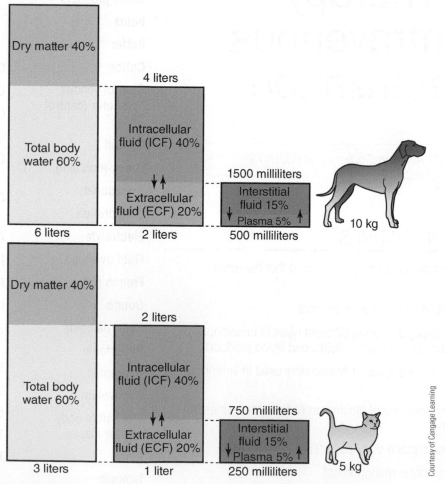

FIGURE 4-1 Fluid Compartments in an adult animal. In adult animals one way to think about total body water is by the 60:40:20 rule. Sixty percent of total body weight is water which is made up of 40% intracellular fluid volume and 20% extracellular fluid volume. The extracellular portion is further divided into intravascular, interstitial, and transcellular water (such as synovial fluid, cerebral spinal fluid, etc.).

Body fluid contains not only water but also important **solutes**, dissolved substances of a solution such as electrolytes (e.g., sodium, potassium). **Electrolytes** are particles or **ions** that can conduct an electrical charge and consist of **cations** (positively charged ions) and **anions** (negatively charged ions). The primary electrolytes and ions in the body are sodium (Na^+), potassium (K^+), chloride (Cl^-), phosphate (PO_4^{-3}), and bicarbonate (HCO_3^-). Sodium is the main extracellular cation and potassium is the primary intracellular cation. Chloride is the primary extracellular anion and phosphate is the primary intracellular anion. Bicarbonate ions are also extracellular (Figure 4-2).

Within the extracellular space, fluid will shift between the intravascular space (circulating blood vessels) and the interstitial space (tissue space between cells) to maintain a fluid equilibrium or balance. This fluid shift or exchange occurs only across the walls of capillaries, not across other blood vessels. The capillary membrane acts as a selectively permeable membrane by permitting free passage of water and small solutes but is relatively impermeable to plasma proteins (e.g., albumin). Under normal conditions fluid constantly leaves the capillary into the interstitial fluid. Interstitial fluid is returned to the vascular system by the lymphatic capillaries. The amount of **osmotic**

pressure (pressure or force that develops when two solutions of different concentrations are separated by a selectively permeable membrane) that develops at the selectively permeable membrane is due to the number of particles rather than the size of those particles. Large numbers of particles in the vascular space draw in water. Fluid shifts only when there is a difference in pressure between the intravascular fluid and the interstitial fluid.

Tech Tip

Remember that sodium and potassium are the predominant cations in animals and therefore are important in regulating overall electrolyte and water balance. For example, if an animal patient has an elevated sodium concentration in the extracellular fluid, osmotic pressure will increase, causing water to move from the intracellular compartment to the extracellular compartment.

	Plasma mmol/L	Interstitial fluid mmol/L	Intracellular fluid mmol/L
Cations			
Sodium (Na^+)	142	145	10
Potassium (K^+)	4	4	150
Calcium (Ca^{+2})	2.5	2.4	4
Magnesium (Mg^{+2})	1	1	34
Anions			
Chloride (Cl^-)	104	117	4
Bicarbonate (HCO_3^-)	24	27	12

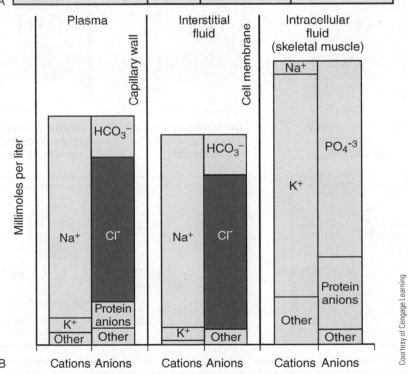

FIGURE 4-2 Ions in the animal's body. (a) Values of select ions in plasma, interstitial fluid, and intracellular fluid. (b) Ion balance in plasma, interstitial fluid, and intracellular fluid.

It is also important to consider that a continuous state of balance exists between the amount of water that is taken in by an animal and the amount of water that is lost during normal physiologic processes. The kidneys are the primary regulators of the body's volume of water within the body. In other words, what comes in (e.g., the patient drinks water) is used to replace what leaves (e.g., the patient urinated, defecated). Fluid is normally lost from the body in many ways, including from urinary loss (urination), gastrointestinal loss (defecation), respiratory (moisture in each breath), and skin sources (evaporation) while fluid intake comes from ingestion of liquids and food. Inappropriate and unexpected fluid loss that can dramatically alter the body's fluid balance can also occur with things like vomiting, diarrhea, draining wounds, shifting of fluid into body cavities where it doesn't belong (e.g., pleural effusion or ascites), and acute or chronic blood loss. When the patient is unable to eat or drink, such as during illness or during general anesthesia, intravenous fluid therapy is necessary to support this fluid balance.

INTRAVENOUS FLUIDS AND BLOOD PRODUCTS

There are three main groups of fluids that can be given IV: crystalloids, synthetic colloids, and blood products. **Crystalloids** are solutions that consist of a sodium or dextrose base dissolved in water and may contain other electrolytes and/or buffers as well. An example of this type of fluid is 0.9% sodium chloride (also referred to as "normal" saline despite being much higher in chloride than the blood) or lactated Ringer's solution (LRS). Crystalloids can be further described as isotonic, hypotonic, or hypertonic solutions (Figure 4-3). Crystalloids are most commonly administered intravenously but are sometimes administered subcutaneously, intraperitoneally, or orally. **Colloids** are different in that these solutions contain microscopically dispersed suspended particles (e.g., starch molecules) that are insoluble. Water and electrolytes are often present in colloids as well. Colloids are only administered intravenously. In the United States, the most common synthetic colloids used in veterinary medicine are heta- or tetrastarches. **Blood products** (sometimes referred to as natural colloids) are unique in that these products contain red blood cells, plasma proteins, or both. Examples of blood products include whole blood, packed red blood cells, plasma, and albumin. Blood products are only administered intravenously.

Crystalloids

Crystalloids, as previously stated, are solutions that consist of a sodium or dextrose base dissolved in water. Some crystalloids may also contain other electrolytes such as chloride, potassium, calcium, and magnesium, and/or buffers such as gluconate, lactate, or acetate. **Buffers** are simply additives that help the fluid resist changes in pH. After a crystalloid is administered intravenously, it immediately enters the circulating intravascular space but approximately only one-quarter of the amount administered remains

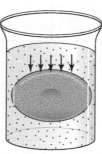
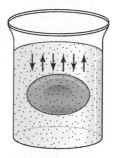
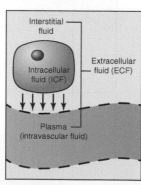
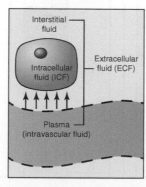
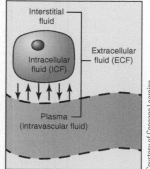

Hypertonic solution (more particles outside cell)

Hypotonic solution (fewer particles outside cell)

Isotonic solution (equal particles inside and outside cell)

FIGURE 4-3 Solution tonicity. Blood cells placed in hypertonic solution lose fluid in an attempt to equalize the osmolality in the cell to the solution. Blood cells placed in hypotonic solution gain fluid in an attempt to equalize the osmolality in the cell to the solution. Blood cells placed in balanced (isotonic) solution have neither a net gain nor loss of fluid.

in the intravascular space (within the blood vessels) after 1 hour because the remainder migrates to the rest of the extracellular space (e.g., the interstitial space). Because of this property, crystalloid fluids replace volume within the intravascular space (e.g., to correct hypovolemia) and are beneficial in correcting deficits within the interstitial space (e.g., dehydration). They are typically inexpensive, have a long shelf life, and are easy to obtain because they are the most widely used fluid type. The main disadvantages of crystalloid fluids are that they provide only a short duration of action in the intravascular space and in large volumes they may cause dilution of plasma proteins, red blood cells, and clotting factors. As with any fluid type, administration of too much crystalloid fluid can result in volume overload.

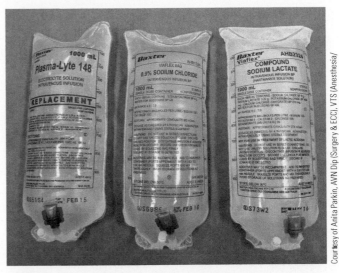

Courtesy of Anita Parkin, AVN Dip (Surgery & ECC), VTS (Anesthesia/ Analgesia), CVPP, TAE

FIGURE 4–4 Examples of Isotonic Crystalloids. Plasma-Lyte® 148 (left), 0.9% NaCl (sodium chloride) (middle), and Hartmann's solution (right).

Tech Tip 🐾

Dehydration and hypovolemia are not the same. **Dehydration** is loss of body water and refers to lack of fluid in the interstitial compartment. **Hypovolemia** is reduction of extracellular fluid volume resulting in decreased tissue perfusion and describes lack of fluid in the intravascular compartment.

Tonicity, the concentration of a solution as compared to another solution, is based on a measurement called **osmolality**, which is the osmotic pressure of a solution based on the number of particles per kilogram of solution. Osmotic pressure is the ability of solutes to attract water and cause osmosis. Not all particles contribute to osmolality. Sodium and glucose provide most of the particles that contribute to osmotic pressure. Normal osmolality of blood and ECF is 290 to 310 mOsm/kg.

The ratio of solutes to **solvent** (the dissolving substance of a solution; water is the biological solvent) in the body is *plasma osmolality*. Cells are affected by the osmolality of the fluid that surrounds them. In the body, *osmolality* measures the number of dissolved particles, regardless of their size, per kilogram of water. Of the particles in body fluids, sodium is the largest contributor to osmolality. Plasma osmolality is expressed in units called **osmoles** or *milliosmoles per kilogram of water*. A solution that has a higher concentration of solutes will have a greater plasma osmolality.

Tech Tip 🐾

Osmolality and osmolarity are two similar, but not identical, terms. Osmolality, usually used to describe fluids inside the body, is the measure of the solute concentration in fluid by weight (kg of solution). **Osmolarity** is the measure of the solute concentration in fluid by volume (L of solution). Because 1 L of water weighs 1 kg, the normal ranges are the same and the terms are often incorrectly used interchangeably.

Crystalloids that have an osmotic pressure similar to plasma are called **isotonic**. Isotonic solutions are considered balanced if their electrolyte concentration and pH are close to that of plasma. **Balanced solutions** are buffered with precursors of bicarbonate and more closely mimic plasma electrolyte composition, particularly in regard to chloride content. Balanced electrolyte solutions are commonly used for intraoperative fluid therapy in healthy patients (Figure 4-4) and as **replacement fluids** (to correct pre-existing dehydration) and **maintenance fluids** (to replace what the body normally loses on its own during the day). Common crystalloid fluid types include the following:

- Isotonic saline (0.9% sodium chloride) is an unbalanced solution that contains only sodium and chloride ions without a buffer. Isotonic saline is used to expand plasma volume and to correct hyponatremia (decreased sodium levels) or metabolic alkalosis due to its chloride content. Isotonic saline is also a better choice than lactated Ringer's solution for patients with cerebral edema. Because of its high sodium content, saline should not be used in patients with heart failure or those that have sodium retention due to liver disease. Isotonic saline, like other crystalloids, causes dilutional effects (reduced potassium, total protein, and packed cell volume) and in large volumes can cause acidifying effects. Large amounts of isotonic saline can potentiate or precipitate hypokalemic and hyperchloremic metabolic (nonrespiratory) acidosis. Solutions that are high in chloride should also be avoided in patients receiving potassium bromide for treatment of seizures.

- Lactated Ringer's solution is a balanced solution that contains sodium and chloride with lactate and other electrolytes such as calcium and potassium. It has a lower sodium content as compared to 0.9% sodium chloride and does not contain magnesium. The lactate molecule found in LRS is a buffer, is metabolized in the liver to bicarbonate, and is helpful in the treatment of acidosis. LRS may not be recommended for use in animals with liver failure or diabetic ketoacidosis because the lactate conversion may not occur, although this concept is controversial and not widely accepted. Neonates can use lactate as an energy source, which makes LRS the preferred fluid choice for young patients. LRS should not be given with blood products because the calcium in LRS will be chelated by the anticoagulant in the blood product, which could lead to clot formation and potentially thrombosis.

- Normosol®-R (Abbott Laboratories) is another balanced solution with less sodium, more potassium, more magnesium, less chloride, and no calcium as compared to LRS. It is an all-purpose replacement fluid and has acetate as a buffer.
- Plasmalyte® (Baxter Laboratories) is another balanced solution that has less chloride, more magnesium, and no calcium as compared to LRS. Plasmalyte® is available in a variety of formulations and is largely identical to Normosol-R®; however, Plasmalyte® A and Plasmalyte® R have the buffers acetate and lactate while Normosol-R® contains acetate and gluconate. Gluconate is converted in skeletal muscle to bicarbonate.

When administering fluid therapy to a neonate, monitor blood glucose levels to determine if glucose should be added to the fluids. Neonates have difficulty maintaining blood glucose levels when fasted due to their limited glycogen stores and inefficient glucose production by the liver.

Crystalloids that have a lower salt concentration than normal body cells are **hypotonic**. Hypotonic crystalloid solutions are fluids with a much lower sodium concentration than extracellular fluid; therefore, they cause water to move into the cells by osmosis, which could lead to cellular swelling. Hypotonic solutions are used to replace free water deficits in the hypernatremic patient. When administering hypotonic solutions, care should be taken to not decrease sodium concentrations too quickly as neurologic adverse effects, including cerebral edema, may develop. Commonly used hypotonic fluid solutions include the following:

- 5% dextrose in water (D₅W), which is given intravenously only (Figure 4-5). The administered dextrose is quickly metabolized and only water remains, which immediately diffuses out of the intravascular space and divides itself between the intracellular and extracellular fluid compartments. Two-thirds of the fluid goes to intracellular fluid and one-third goes to the extracellular fluid (other crystalloids retain about one-third to one-half to in the intravascular space). D₅W is used to treat hypernatremia or as a fluid supplement in patients that cannot tolerate sodium (such as patients with heart failure). D₅W does not contain enough calories or a source of fat or protein to meet an animal's energy requirements and therefore it is not an appropriate substitution for adequate nutrition.
- Other common examples of hypotonic solutions include 0.45% NaCl and 2.5% dextrose in 0.45% NaCl. Custom hypotonic solution can also be made by adding sterile water to balanced crystalloids to address individual needs for specific cases. One potential example of this might be for a patient receiving potassium bromide (KBr) for seizure therapy, in which very low chloride content is required so as to not counteract the KBr.

Crystalloids that have an electrolyte composition that exceeds that of plasma are **hypertonic**. Hypertonic crystalloid fluids have high sodium concentrations that will increase the sodium concentration of the extracellular fluid and thereby will draw fluid from the interstitial and intracellular spaces into the intravascular space. Hypertonic solutions can help to rapidly restore intravascular volume and for this reason can be highly beneficial in the emergency treatment of hypovolemic

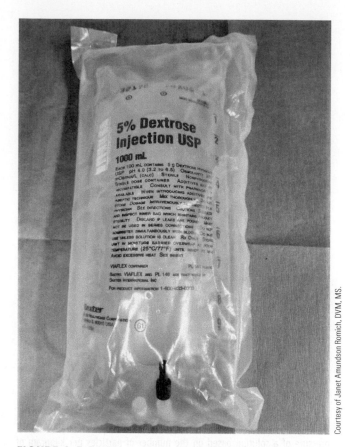

Courtesy of Janet Amundson Romich, DVM, MS.

FIGURE 4–5 D₅W is an example of hypotonic fluid.

shock. However, this principle only works if the interstitial space has fluid to spare, and therefore the patient must be well hydrated or else fluid will be drawn from an already dehydrated space. Hence, hypertonic solutions are useful in restoring intravascular volume in cases that only lack intravascular volume (such as acute blood loss) and not dehydration with concurrent hypovolemia. One additional problem associated with all hypertonic solutions is that they only result in fluid staying in the intravascular space for a short time. Once the solute (dextrose or sodium) diffuses out of the intravascular space, all the fluid that was drawn into the intravascular space diffuses out again. Therefore, the effects of hypertonic solutions are only transient effects; however, they can be very useful "bandages" to be administered in a small volume, create rapid expansion of the intravascular space, and then allow time for the veterinary team to continue treatment. Some volume of a balanced crystalloid should be given following hypertonic fluids to insure that the interstitial space remains hydrated. Hypertonic solutions are often given with colloids for rapid and more sustained resuscitation compared to administering hypertonic solutions alone. Some hypertonic solutions (such as 23% NaCl or 50% dextrose) should be diluted before use to prevent pain on administration and phlebitis (inflammation of the vein). The most common hypertonic solutions include

- 7.2% NaCl (Figure 4-6);
- 23% NaCl (must be diluted before use);
- 20% mannitol; and
- 50% dextrose.

Table 4-1 is a quick reference guide to the solute levels in commercially available electrolyte solutions.

TABLE 4-1 Electrolyte Composition of Commercially Available Crystalloid Fluids

Fluid	Glucose (g/L)	Na⁺ (mEq/L)	Cl⁻ (mEq/L)	K⁺ (mEq/L)	Ca²⁺ (mEq/L)	Mg²⁺ (mEq/L)	Buffer (mEq/L)	Osmolarity (mOsm/L)	pH	Tonicity
5% dextrose	50	0	0	0	0	0	0	253	4.0	Isotonic*
2.5% dextrose in 0.45% NaCl	25	77	77	0	0	0	0	280	4.5	Isotonic*
5% dextrose in 0.45% NaCl	50	77	77	0	0	0	0	405	4.0	Isotonic*
5% dextrose in 0.9% NaCl	100	154	154	0	0	0	0	560	4.0	Isotonic
0.45% NaCl	0	77	77	0	0	0	0	155	5.0	Hypotonic
0.9% NaCl	0	155	155	0	0	0	0	310	5.0	Isotonic
Ringer's Solution	0	147	156	4	4	0	0	310	5.5	Isotonic
Lactated Ringer's Solution (Hartmann's Solution)	0	130	109	4	4	0	28(L**)	273	6.5	Isotonic
Normosol®-R	0	140	98	5	0	3	27 (A) 23 (G)	294	6.4	Isotonic
2.5% dextrose in Lactated Ringer's Solution	25	130	109	4	3	0	28 (L)	398	5.0	Isotonic
5% dextrose in Lactated Ringer's Solution	50	130	109	4	3	0	28 (L)	524	5.0	Isotonic
Plasma-Lyte ®A	0	140	103	10	5	3	47(A) 8 (L)	294	5.5	Isotonic
Plasma-Lyte ®R	0	140	103	10	5	3	47(A) 8 (L)	312	5.5	Isotonic
D₅W	50	0	0	0	0	0	0	253		Hypotonic
0.45% NaCl	0	77	77	0	0	0	0	155		Hypotonic
2.5% dextrose in 0.45% NaCl	25	77	77	0	0	0	0	280		Hypotonic
20% mannitol	200 (M***)	0	0	0	0	0	0	1099	5.3	Hypertonic
50% dextrose	500	0	0	0	0	0	0	2525	4.2	Hypertonic
7.2% NaCl	0	1199	1199	0	0	0	0	2400	5.0	Hypertonic
5% NaCl	0	855	855	0	0	0	0	1710	5.0	Hypertonic

*These solutions are isotonic in the fluid bag; however, dextrose gets metabolized almost immediately in the lining of the blood vessels leaving free water (or 0.45% NaCl depending on the fluid type) which makes it hypotonic in the body.

**Buffers include L = lactate; A = acetate; G = gluconate

***M = mannitol not glucose

Table modified from Plumb, D. (2019). *Plumb's veterinary drug handbook*, 9th ed, p. 1390.

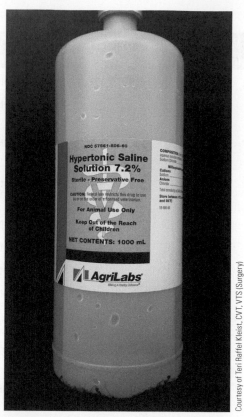

FIGURE 4-6 Hypertonic saline is an example of hypertonic fluid.

Courtesy of Teri Raffel Kleist, CVT, VTS (Surgery)

Tech Tip

The osmolality of *isotonic* solutions is the same as the fluid component of blood and extracellular water; thus, isotonic solutions produce no significant changes in the osmolality of blood. *Hypotonic* solutions have osmolality lower than that of the fluid component of blood. If given in extremely large quantities, they may cause the red blood cells (RBCs) to swell because the osmolality inside the RBCs is greater than that in the blood (causing fluid to diffuse into the RBCs and distending them). The osmolality of *hypertonic* solutions is greater than that of the fluid component of blood and may cause the RBCs to shrink; however, with proper administration volumes, hypertonic solutions can cause fluid to shift into the intravascular space from the extravascular space without causing changes in the RBCs.

Colloids

Colloids are solutions that contain higher molecular weight molecules that contribute to oncotic pressure. **Oncotic pressure** is the osmotic pressure exerted by higher molecular weight molecules in a solution; in other words, it is the pressure exerted mostly by the plasma proteins that helps keep fluid within the intravascular space and opposes interstitial oncotic pressure. Colloids dehydrate the glycocalyx (a gel-like layer that covers the vascular

lumen) and slow filtration but do not completely prevent it. Colloids tend to stay within the vascular space longer than crystalloids and can also weakly expand plasma volume but do so far less effectively than hypertonic solutions. Recall that following administration of a crystalloid, approximately 75% of the administered volume shifts out of the vascular space into the interstitial space. While this is very useful to treat dehydration, it may not be useful is sustaining intravascular volume. Colloids, on the other hand, have a sustained effect within the vascular space for several hours or longer (depending of the type of colloid) and do a poor job of rehydrating the interstitial space. Hence colloids are used for intravascular volume expansion in treating hypovolemia and for treating disease in which hypoproteinemia is present and a low colloid osmotic pressure is suspected. They are a poor choice for addressing patient dehydration.

Synthetic colloids include dextrans and hydroxyethyl starch. The hydroxyethyl starch (HES) family (VetStarch®, Hespan®, Hextend®, Voluven®) (Figure 4-7) is the most widely used synthetic group of colloids in the Unites States at this time. Synthetic colloids such as the HES family have widely fallen out of use in human medicine in Europe and across other parts of the world. Synthetic colloids may increase the risk for acute kidney injury (especially in patients that are septic), can easily result in volume overload, and can cause coagulation abnormalities and rarely anaphylaxis. Pruritus is also reported in humans; however, the incidence of this in dogs and cats is unclear. Once a synthetic colloid has been administered, microscopically it can be found in tissues for weeks and potentially months after administration.

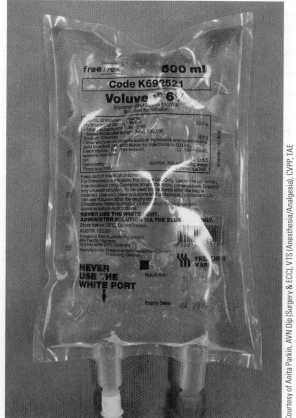

FIGURE 4-7 Voluven® is an example of a colloid.

Courtesy of Anita Parkin, AVN Dip (Surgery & ECC), VTS (Anesthesia/Analgesia), CVPP, TAE

The first HES developed was hetastarch, a combination of hydroxyethyl starch in normal saline that is used to treat hypovolemic shock and hypoproteinemia. Second generations (pentastarches) and third generations (tetrastarches) of HES have since been developed. Tetrastarches are safer than hetastarches and have fewer coagulation-altering effects. The veterinary specific tetrastarch product VetStarch® contains 6% hydroxyethyl starch 130/0.4 in 0.9% NaCl and is labeled as a plasma volume substitute for the treatment and prophylaxis of hypovolemia. The dosage of hydroxyethyl starch depends on the specific type of product that is administered and typically ranges between 2.5 and 50 mL/kg/day, with most products being given between 2.5 and 20 mL/kg/day. As a general recommendation VetStarch® is administered up to 20 mL/kg/day in dogs and cats with the initial 10 to 20 mL infused slowly while the patient is closely monitored for potential anaphylaxis.

Safety Alert

Since 2013, HES has come with a black box warning label issued by the United States Food and Drug Administration (FDA) due to its risk of increased mortality and severe renal injury in critically ill adult human patients, including patients with sepsis. A black box warning label is the strongest warning that the FDA requires and indicates that medical studies have found that the drug carries a significant risk of serious or life-threatening adverse effects. In animals, HES should be used with caution in patients with renal dysfunction, congestive heart failure, or pulmonary edema.

Blood Products

Whole blood contains the entire cellular (erythrocytes, leukocytes, and platelets) and plasma components (proteins, clotting factors) of blood that circulates in the body. Whole blood provides red blood cells for carrying oxygen to tissues as well as plasma volume for expanding the intravascular space. Whole blood can be fresh or stored. If it is used for platelet replacement, whole blood must be used within 4 to 6 hours of collection (e.g., fresh whole blood) to ensure that the platelets are viable. As blood is stored and becomes older, the platelets and potentially the clotting factors become less viable and the blood develops "storage lesions," which increase the risk of adverse effects and decrease its oxygen-carrying capacity.

When whole blood and other blood products are collected, they are mixed with an anticoagulant and often a nutrition source that provides fuel to the cells. Examples of these solutions include citrate phosphate dextrose adenine (CPDA), acid citrate dextrose (ACD), saline adenine glucose mannitol (SAGM), and heparin (heparin does not provide cellular nutrition). Each additive mixture gives whole blood its own unique "shelf life," which influences how long stored whole blood can be kept. Ideally, stored whole blood is used within 21 days from being drawn; however, depending on the additives used, stored blood can be good for as long as 42 days (e.g., blood mixed with SAGM). More commonly, blood is stored no more than 28 to 30 days. Whole blood can be collected in glass vacuum bottles, plastic bags, and plastic syringes. Glass is inert to stored blood, yet is breakable, is more expensive, and requires more storage space than plastic bags. A plastic bag of fresh or stored blood is considered full when it contains 463 grams of blood.

Blood group typing and testing of donors for infectious diseases transmissible via blood is recommended. **Cross-matching** (test performed prior to a blood transfusion in order to determine if the donor's blood is compatible with the blood of an intended recipient) is performed prior to administering blood or blood products to determine if the donor's blood/blood product is compatible with the recipient's blood. Major cross-matching checks the compatibility of the donor's RBCs with the recipient's plasma by detecting antibodies in the recipient's serum that may agglutinate or lyse the donor's erythrocytes. Major cross-matching is most important in all species. Minor cross-matching detects antibodies in the donor plasma directed against recipient erythrocytes and is performed by testing the red cells of the recipient with plasma from the donor. Minor cross-matching is important in species such as cats, in cases where the donor has been previously transfused, and in horses with a history of pregnancy. Compatibility is determined through detection of antibodies present in the recipient against the RBCs of the donor. If antibodies are present, they attach to the donor RBCs after transfusion and can cause hemolytic anemia or death. Dogs and horses do not have naturally occurring antibodies against the important RBC antigens and therefore do not require cross-matching for the first transfusion. Transfusion of blood products sensitizes the RBC antigen; therefore, subsequent transfusions in dogs and horses need to be cross-matched. Cats have naturally occurring alloantibodies to RBC antigens and always need to be typed and cross-matched on the first as well as subsequent transfusions.

Blood is usually collected via the jugular vein and is typically given using a 20 or 23 gauge indwelling catheter. Blood products need to be administered using a filtered blood administration set. Fresh whole blood can be administered initially at a low rate such as 0.25 mL/kg/hr for the first 15 to 30 minutes and then ultimately increased as high as 10 mL/kg/hr IV as long as the patient is tolerating it without an adverse response. Patients receiving a blood transfusion are monitored carefully and vital signs are taken every 5 minutes for at least the first hour and then less frequently after that (e.g., once every 15 minutes).

The most common type of adverse transfusion reaction is known as a *febrile non-hemolytic* reaction. These reactions are due mainly to an immune response signaled by cytokines and other inflammatory mediators in the donor blood. Signs of this type of reaction are most commonly an increase in patient temperature and vomiting. Any increase of 1.8°F or 1°C or more within 1 to 2 hours must be considered a febrile reaction. The treatment for this type of reaction is to stop the transfusion until the cause of the temperature increase is determined. These reactions can be reduced by using blood that has undergone *leukoreduction*, meaning the white blood cells have largely been commercially filtered away. However, in patients under general anesthesia, increased temperature or vomiting is often not observed and thus this reaction type is often overlooked in anesthetized patients.

Tech Tip

As a quick general guide, 1 mL of whole blood per 1 kg of body weight raises the packed cell volume by approximately 1%. This is known as the "rule of 1's".

Allergic reactions are another type of adverse acute reaction to a transfusion. These reactions can be life threatening and show general signs of an allergic reaction, which may include facial swelling, hives, itchiness, restlessness, vomiting, low blood pressure, tachycardia, and difficulty breathing. This type of transfusion reaction requires stopping the transfusion and treating the patient immediately for an allergic reaction (e.g., antihistamines, glucocorticoids, IV fluid therapy, and potentially epinephrine). The "pretreatment" or use of the antihistamine diphenhydramine before a patient receives a transfusion may lessen the incidence of this specific type of reaction.[1] In patients under general anesthesia, however, it can be difficult for the anesthetist to detect this type of transfusion reaction, and unexplained, refractory hypotension is often the earliest clue.

Two other less common but important types of acute transfusion reactions include transfusion-associated circulatory overload (TACO) and transfusion-related acute lung injury (TRALI), which predominantly result in patient respiratory distress from either giving too much volume to the intravascular space (TACO) or an acute lung injury (TRALI). A typical dosage for whole blood for a dog or cat is 5 to 20 mL/kg, but the actual dosage will depend upon the individual patient's needs and responses to administered blood.

Tech Tip

Blood products should not be administered with an IV fluid that contains calcium (e.g., lactated Ringer's solution) because the calcium will inactivate the anticoagulant and may cause the blood to coagulate. As a result, 0.9% NaCl is the only type of IV fluid that used concurrently when giving a transfusion.

Plasma is another type of widely used blood product that contains none of the cellular components (RBCs, WBCs, and platelets) and only contains the plasma components (proteins, clotting factors) of blood that circulates in the body. Because plasma does not contain red blood cells for carrying oxygen to tissues, it is only used to expand the intravascular space and provide a source of albumin, clotting factors, and colloidal osmotic pressure.

Plasma is made by drawing whole blood from a patient and then separating it into individual components such that the plasma component itself can be removed. This process should be done 4 to 6 hours after collection. It can then be stored frozen for long periods of time at −70 to 40°C. Plasma that is frozen for less than one year is known as fresh frozen plasma (FFP). Plasma that has been frozen for longer than one year is known solely as frozen plasma and contains less viable clotting factors (e.g., labile coagulation factors such as fibronectin, fibrinogen, and factors V and VIII). Fresh plasma and fresh frozen plasma can also cause transfusion reactions and should initially be given slowly as with whole blood but can ultimately be given at up to a rate of 10 mL/kg/hr. A typical dosage for plasma in dogs and cats is 2.5–10 mL/kg administered based upon the patient's individual needs and response. Frozen plasma should be used within 6 hours of thawing to maintain the clotting proteins. Adverse reactions to plasma are similar to those reported for whole blood.

Packed red blood cells (pRBCs) are a widely used blood product that contains red blood cells and almost no plasma components (proteins, clotting factors) of blood that circulates in the body. Packed red blood cells are not a source of platelets or white blood cells. Packed red blood cells are never frozen and are refrigerated in the same manner and duration as for whole blood. They can produce the same types of transfusion reactions as

other blood products and should be handled in the same fashion. Typical dosages for pRBCs in dogs and cats range between 5 and 10 mL/kg.

Albumin is the main protein in the blood and constitutes about 50% of the total protein. It serves many functions, including, but not limited to, maintaining the colloid osmotic pressure of blood and acting as a source for drug binding. Without albumin (and its role in maintaining colloid osmotic pressure), intravascular plasma would readily leave this space and accumulate in the interstitial tissues, leading to tissue edema. Transfusions of albumin, which are widely done in human medicine and less commonly done in veterinary medicine, can be helpful in restoring body albumin levels. Adverse reactions to albumin include fever, anorexia, sudden death, and a 1- to 2-week delayed hemolytic reaction that can be fatal. Currently, canine serum albumin and human albumin products are available but are infrequently used.

INTRAVENOUS CATHETERIZATION

To administer fluids, blood products, and anesthetic drugs intravenously, a catheter is placed into a vein because using a catheter produces fewer complications than using a needle. Placement of an IV catheter is a standard of care in veterinary medicine for patients undergoing general anesthesia and potentially sedation. IV catheters are beneficial because they

- facilitate the delivery of IV anesthetic agents;
- prevent potentially harmful perivascular drug administration;
- provide access for rapid drug/fluid administration in emergency situations; and
- allow for continuous fluid or blood product administration with less injury to the patient's blood vessel.

Safety Alert

Hypodermic needles, including catheters, should never be recapped once the protective sleeve is removed. This prevents accidental needle stick injury and potential risk for zoonotic infection between patient and caregiver.

Intravenous catheters are available in a variety of shapes, sizes, and materials to fit the diverse needs of veterinary patients. Catheter selection is based on patient size, expected length of time the catheter will be in place, whether blood samples are expected to be removed through the catheter, and the rate, volume, and type of fluid administered. Other factors such as cost, experience, and preference of the veterinary professional placing the IV catheter are also considered.

Catheter Size, Length, and Lumen

Catheter size is expressed in terms of outer diameter. Two units of measure are used for catheter size: gauge and French units. Catheters with larger **gauge** numbers have narrower diameters than those with smaller gauge numbers. For example, a 24 gauge catheter has a narrower diameter than an 18 gauge catheter. **French (Fr)** size is approximately equivalent to a catheter's circumference and is a series of whole numbers that represents three times the catheter's outer diameter in millimeters. For example, a size 3 French catheter has a circumference of 3.14 mm and an outer diameter of 1 mm (3 $\times$ 1 mm = 3). French size is mainly used for urinary catheters and feeding tubes, whereas gauge is usually used for IV catheters. IV catheters as large as 10 gauge and as small as 26 gauge are available for use in veterinary patients (Figure 4-8a).

Catheter size (both internal diameter and length) profoundly affects IV anesthesia administration rate and can impact the incidence of unwanted adverse effects. For example, a narrower IV catheter with a smaller diameter (in relation to the vein) produces less vein inflammation; however, it will dramatically increase resistance to flow and slow the rate of fluid passage though the catheter. A wider IV catheter (relative to the size of the vessel) will have less resistance and therefore deliver flow of fluid more rapidly to the patient, yet it increases the incidence of phlebitis due to contact with the vessel wall and increases patient discomfort. Increased catheter length increases resistance to flow, slows administration rates, and increases vessel inflammation; however, increased length makes the catheter more stable and less likely to become dislodged. For example, a 16 gauge catheter can deliver fluids more rapidly than a 22 gauge catheter because it has a wider diameter. A shorter length catheter can deliver fluids more rapidly than a longer length catheter but will become dislodged more easily. Therefore, clinically, more rapid and larger volumes should be administered with a large diameter (small gauge), shorter length catheter. This is commonly done in emergency settings. However, in stable patients for elective procedures or in patients where the catheter is expected to be present for a prolonged period, a longer length, smaller gauge catheter can be used to increase patient comfort.

Tech Tip

Gauge is a descending scale (large number = small diameter); in contrast, the French scale is ascending (large number = large diameter).

Catheters can be single- or multi-lumen. Single-lumen catheters, primarily used for peripheral vein catheterization, have one lumen (internal channel) that resides in the vessel where the fluids combine and may have several ports that connect at the entrance of the catheter (hub). Multi-lumen catheters have a central, cylindrical tube that is internally divided into two (double lumen) or three (triple lumen) separate lumens that do not communicate with each other and run within the outer shell of the catheter (Figure 4-8b). The exit sites of the various lumens in multi-lumen catheters are at different locations at the distal end of the catheter shaft so the infusions do not mix prior to delivery into the vein. This configuration allows a different intravenous infusion to be connected to each lumen and the patient to receive several different infusions with only one access site. Multi-lumen catheters are available in both over-the-needle and through-the-needle style (see description below) and are generally preferred for **central venous catheters** (an indwelling catheter placed into a central vein, typically the jugular or caudal vena cava, for long term fluid administration; also called a "central line"). The type of multi-lumen catheter to use depends on the functional requirements of the catheter. For example, if frequent blood sampling and IV fluid administration are required, double-lumen catheters may be warranted. A triple lumen may be a better choice if the patient also needs total parenteral nutrition (TPN) in addition to blood sampling and IV fluid administration. Keep in mind that the more lines a catheter has, the narrower the lines are; this will affect the infusion rate.

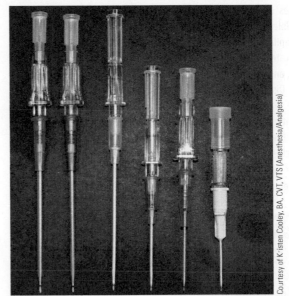

(a)

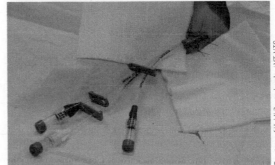

(b)

(c)

FIGURE 4-8 Types of catheters. (a) Over-the-needle single lumen IV catheters are available in a variety of diameters (10–26 gauge are available for use in veterinary patients) and lengths (¾ to 5 ½ inches; 19 to 139.7 mm). The choice of catheter to use depends on the vein size; amount, rate, and viscosity of the fluid/drugs to be administered; and the length of time the catheter will remain in the patient. This photo shows a variety of different gauges and lengths of peripheral IV catheters. From left to right: 14 G, 45 mm (orange); 16 G, 45 mm (gray); 18 G, 45 mm (green); 20 G, 25 mm (pink); 22 G, 25 mm (blue); 24 G, 19 mm (yellow). (b) Multi-lumen (triple-lumen) catheter used to sample blood and administer fluids/medications to a patient. Each catheter tubing enters into the blue hub individually so that solutions do not mix prior to administration to the patient. (c) An example of a multi-lumen catheter kit.

Advantages of multi-lumen catheters include the ability to administer incompatible solutions simultaneously, to perform repeated atraumatic blood sampling, to administer irritant solutions (TPN, hypertonic solutions, and medications known to cause phlebitis) while administering fluids continuously, and to perform hemodynamic monitoring (e.g., central venous pressure). Disadvantages of multi-lumen catheters are that they are more expensive than single-lumen catheters, are difficult to place, and usually require patients to be sedated or under general anesthesia if they are not gravely ill (because it usually takes approximately 10 to 20 minutes to fully place a multi-lumen catheter). Multi-lumen catheters can be purchased individually or as a kit that contains all the components necessary for insertion (e.g., catheter, dilators, local anesthetic, scalpel, and syringe) (Figure 4-8c).

> **Tech Tip** 🐾
>
> Multi-lumen catheters are useful in the critical care setting because only one catheter is placed but that single catheter provides the same functions as two to three separately introduced single-lumen catheters.

Types of Intravenous Catheters

There are three general intravenous catheter types used in veterinary anesthesia: over-the-needle, through-the-needle, and wire-guided. Butterfly catheters are for short-term use and are described in Figure 4-9a.

Over-the-needle catheters have a stylet with a sharp needle tip that is surrounded and covered by a catheter. The needle is used to penetrate the skin, and then the surrounding catheter is advanced by the anesthetist into a blood vessel (Figure 4-9b). Once the catheter is in the blood vessel, the sharp stylet is removed and discarded. Over-the-needle catheters are typically placed in peripheral veins (e.g., cephalic vein, saphenous vein) and are generally for shorter term use (such as hours to days). These catheters are also commonly used for arterial catheterization for invasive blood pressure monitoring. Disadvantages of the over-the-needle catheters are that the tip of the catheter can become blunted or damaged during insertion and the catheter may kink easily because it is made from rigid materials with poor flexibility and memory. These catheters are typically short in length, can become dislodged easily, and do not facilitate repeated blood sampling. They are often made of material that does not promote long-term usage and therefore are typically maintained for less than five days and then removed. Over-the-needle catheters range in size (width) from 26 to 16 gauge and in length from ¾ to 5½ inches.

Through-the-needle catheters have a larger gauge needle through which a smaller gauge catheter is fed into the blood vessel through the middle of the needle. Through-the-needle catheters range in size (width) from 18 to 16 gauge and in length from 5¼ to 12 inches. They require a large gauge needle (14 gauge or larger) to be placed percutaneously as an access port for the catheter itself. The smaller gauge catheter is then fed through the needle and into the vessel (Figure 4-9c). The needle is then pulled out of the vessel and a needle guard is snapped closed around it to prevent the needle from lacerating the patient or the catheter. These catheters are commonly used for placement within central veins such as the jugular vein or caudal vena cava. They are longer in length and are less likely to become dislodged. The advantages of these catheters are that they allow for repeated blood sampling and, when placed into a central vein, allow for the use of **hyperosmolar** solutions (> 600 mOsm/L) (e.g., high-concentration dextrose, TPN solutions), parenteral nutrition, potentially irritating drugs known to cause phlebitis and tissue sloughing, and central venous pressure (CVP) measurement. The disadvantages of this type of catheter include the technical difficulty of their placement, the ease at which they kink, and increased tissue trauma (potentially resulting in cellulitis or hematoma development) because they require the use of a large gauge needle. These catheters can be left in place for prolonged periods of time (up to a week) if well maintained.

A **peel-away catheter** is a special type of through-the-needle catheter that uses a large gauge needle and a peel-away sheath to facilitate catheter placement. Once the needle and plastic sheath are placed percutaneously into the vessel, the needle is removed, leaving the sheath in the vessel. The IV catheter is fed through the sheath and is then peeled away, leaving only the catheter in the vessel. This method is less cumbersome because the needle is removed; however, the needle hole used to introduce the IV catheter is large (leading to potential tissue trauma), and once the large bore needle is removed, the plastic sheath cannot move (movement will prevent the catheter from threading into the vein).

Over-the-wire catheters have a kit and guide wire that is used to establish a path for the intravenous catheter (Figure 4-9e). To place this type of catheter, the Seldinger insertion technique is used (Figure 4-9e). After performing a sterile catheter prep, the small gauge needle (provided in the catheter kit) is introduced into the vein and a flexible sterile guide wire is threaded through the needle. This guide wire has a J-shaped end that must be straightened when inserted through the needle and then bends back to the J shape after it is inside the vein to lessen the chance of vessel trauma. The needle is then withdrawn, with the guide wire left in place, and a rigid dilator is passed over the guide wire to dilate the entry site through the skin and vein to further ease catheter placement. The dilator is then withdrawn, with the guide wire again left in the blood vessel, and a new catheter is advanced over the guide wire into the blood vessel. At this time the guide wire is removed and discarded and the catheter is secured in place.

The disadvantage of using over-the-wire catheters and the Seldinger technique is that proper placement requires several steps and a high skill level. Bleeding or initial hematoma formation can also occur while establishing the initial needle stick into the blood vessel. However, over-the-wire catheters and the Seldinger technique have many advantages: they allow for placement of a long catheter and, when performed correctly, they ensure easy passage of a catheter with minimal trauma to tissue. Additionally, these catheters lend themselves to repeated blood sampling and, when placed into a central vein, allow for the use of hyperosmolar solutions (> 600 mOsm/L), parenteral nutrition, potentially irritating drugs known to cause phlebitis and tissue sloughing, and central venous pressure (CVP) measurement. These catheters can be left in place for prolonged periods of time (e.g., days to weeks) if well maintained.

Catheter Material

Catheters can be made from polypropylene, polyvinyl chloride (PVC), polyethylene, polyurethane, silicone rubber (silastic), polytetrafluoroethylene (Teflon®), or a combination of these compounds. All catheters are made from

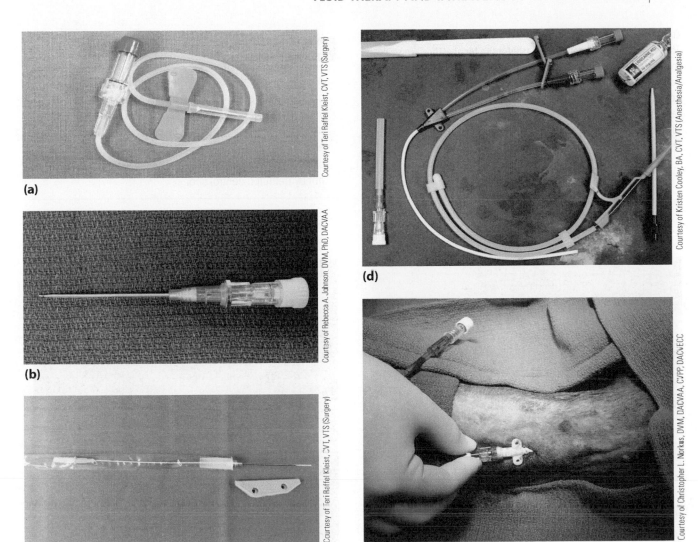

FIGURE 4-9 Types of catheters. (a) A butterfly catheter is a small gauge needle (ranging in size from 25 to 18 G), with a plastic set of "wings" just below the needle hub that make holding the needle easier and more stable, and plastic tubing of various lengths extends behind the needle. When securing a butterfly catheter in place, it is taped directly behind the catheter wings. It is for short-term use such as during short anesthetic procedures, blood collection, or medication administration. Butterfly catheters are easy to place but are the least stable, can easily puncture the blood vessel wall (causing subcutaneous infiltration of fluids), and are difficult to maintain especially in animals that are able to move. (b) Over-the-needle peripheral IV catheters have the catheter mounted on the needle (stylet) that is removed once the catheter is in the vessel. Over-the-needle IV catheters can become burred during insertion and may bend or break between the hub and shaft; however, they are commonly used in veterinary practice. (c) A through-the-needle catheter. The needle is placed percutaneously into the vessel, and at that point the catheter is then fed into the vessel through the needle. Once the needle is removed, the protective sheath is placed over the needle. (d) An over-the-wire catheter consists of a needle, wire, dilator, and catheter. The J-shaped end prevents the wire from having a sharp tip which can damage the vessel. (e) Over-the-wire IV catheters use a wire that is inserted into the vessel and then the IV catheter is fed over the wire into the blood vessel. This IV catheter is in place and ready to be secured to the patient. The IV catheter is sutured to the patient and an occlusive dressing is placed over the neck.

materials that are **thrombogenic** (meaning they tend to initiate blood clot formation), but some catheter materials are less thrombogenic than others. Silicone and polyurethane catheters are minimally thrombogenic; polytetrafluoroethylene catheters are mildly thrombogenic; and polyvinyl chloride, polypropylene, and polyethylene tend to be the most reactive and thrombogenic.[2] The type of intravenous catheter material used is based on the length of time the catheter must stay in place. For example, if an intravenous catheter is required to be in place longer than a few days, a polyurethane (less thrombogenic) catheter would be a good choice.

Placement Sites for Intravenous Catheters

The ideal site for intravenous catheter placement should minimize patient stress while considering the type and length of treatment, disease condition present, and accessibility of the vessel. For example, a lateral saphenous catheter in a dog with diarrhea may become contaminated with fecal material. A cephalic intravenous catheter can be contaminated with drool or vomitus. IV catheters should be placed into the most convenient and easily accessible vessel that allows stress-free handling. The cephalic vein is a popular IV catheter site for anesthesia in small animals because this vessel is readily accessible, relatively large, and the catheter is easy to secure in place. However, in large animals, the jugular vein is more commonly used. Various venous access sites and their advantages/disadvantages are illustrated in Figures 4-10 through 4-15. Further discussion on venipuncture and catheter sites can be found in individual species chapters.

Tech Tip

A catheter is considered central based on the position of the tip of the catheter. The most distal end of central venous catheters terminates in a central vessel within the chest cavity (cranial or caudal vena cava). A jugular catheter can be a central venous catheter as long as it is the correct length to reach the vena cava. A jugular catheter is not by default a central line.

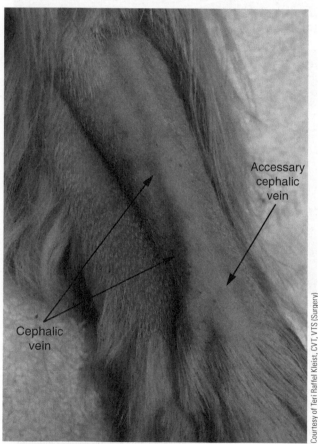

Accessary cephalic vein

Cephalic vein

Courtesy of Teri Raffel Kleist, CVT, VTS (Surgery)

FIGURE 4-10 The cephalic vein is used in dogs, cats, rabbits, and ferrets. It is located in the forelimb from the carpus to elbow, typically running medial to lateral. The cephalic vein is surrounded by muscles and tendons of the forelimb. Advantages of its use are that it is easy to access and easy to secure. It is contraindicated in aggressive animals and in vomiting/drooling/chewing/licking animals. The accessory cephalic vein is used in dogs. It is located in the most medial and distal portion of the cephalic vein; it runs very medially and tends to be short. The accessory cephalic vein is surrounded by the carpus. An advantage of its use is that it is easy to visualize in dogs. The disadvantages of its use are that it has a difficult angle and the catheter may not feed past the bifurcation of vein.

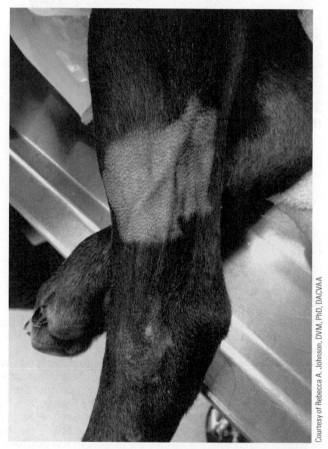

Courtesy of Rebecca A. Johnson, DVM, PhD, DACVAA

FIGURE 4-11 The lateral saphenous or tarsal vein is used in dogs; it is possible to use in cats but difficult to access. It is located on the lateral hind limb proximal to the hock and runs diagonally dorsal to ventral (the stifle is considered dorsal whereas the semimembranosis muscle is ventral). The lateral saphenous vein is surrounded by the common calcaneal tendon and other smaller tendons. Advantages of its use are safer placement (for veterinary technician) in mouthy or aggressive animals and it is easy to secure. Disadvantages of its use are that down animals and those with diarrhea or those that are urinating frequently can contaminate the catheter and the catheter can be uncomfortable if secured incorrectly; it is not generally used in the cat.

Courtesy of Rebecca A. Johnson, DVM, PhD, DACVAA

FIGURE 4-12 The medial tarsal vein is used in dogs. It is located in the distal portion of the lateral saphenous over the dorsal aspect of the hind limb and over the metatarsals. The medial tarsal vein is surround by the metatarsals, tendons, and dorsal pedal artery. An advantage of its use is that it is a good option when all other vessels have failed. Disadvantages are that it may be difficult to palpate and is difficult to secure.

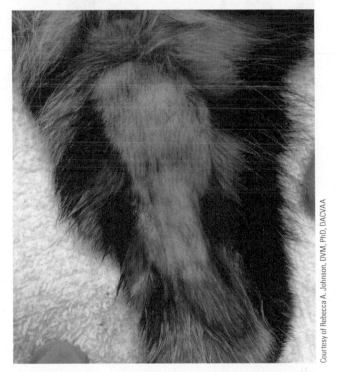

Courtesy of Rebecca A. Johnson, DVM, PhD, DACVAA

FIGURE 4-13 The medial saphenous vein is used in dogs, cats, and ferrets. It is located on the medial aspect of the hind limb from the hock to the proximal aspect of the thigh (distal portion is more superficial). The medial saphenous vein is a superficial vein; its underlying structures include muscles. Advantages of using this vein are that it is easy to locate and easy to see. Disadvantages are hematoma formation; vessel is not very stable, it is small and can be a challenge to secure; and it is not recommended in animals that are down, have diarrhea, or are urinating frequently because the catheter can become easily soiled.

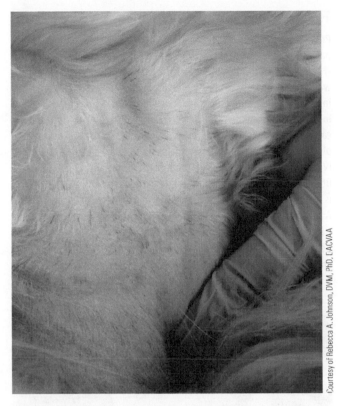

Courtesy of Rebecca A. Johnson, DVM, PhD, DACVAA

FIGURE 4-14 The jugular vein is used in dogs and cats (its use in large and food producing animals is described in Chapter 19 [Anesthesia and Analgesia of Large Animals and Food Producing Species]). It is located in the ventral neck, lateral to the trachea, and medial to the brachiocephalicus muscle; it runs in a line from the ramus of the mandible to the thoracic inlet. The jugular vein is surrounded by the trachea, sternomastoidius muscle, and skin. Advantages of its use are that it is a large vessel and good for central venous access. It is contraindicated for use in patients with thrombocytopenia due to the potential for bleeding and hematoma formation.

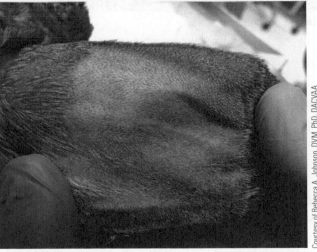

Courtesy of Rebecca A. Johnson, DVM, PhD, DACVAA

FIGURE 4-15 The marginal ear vein is used in dogs (floppy-eared) and rabbits. It is located on the periphery of the pinna; note the marginal ear vein is peripherally located while the auricular artery is centrally located. An advantage to its use is that it is easy to visualize once hair is removed. Disadvantages are that only small gauge catheters tend to fit in it, hematoma formation is possible, and it can be difficult to secure.

Preparation and Placement of IV Catheters

Prior to intravenous catheter placement, the materials needed to prepare the site and secure the IV catheter in place should be gathered (Figures 4-16a through 4-16c). The veterinary technician begins by washing their hands and putting on protective gloves. Hair or fur at the intended vessel location may be clipped and is then shaved. The use of a local anesthesia cream (e.g., EMLA® cream, which contains a mix of lidocaine and prilocaine that, in humans, penetrates skin to a maximum depth of 5 mm) may be useful to lessen the pain caused by inserting a needle or IV catheter (Figure 4-16d). Such products should be applied 30 to 60 minutes before attempted IV catheter placement so their full anesthetic effect is realized. Next, the site is gently cleaned and prepped for up to 5 minutes. Aseptic preparation of the area involves using either a dilute chlorhexidine (Nolvasan®, Virosan®) or povidone iodine (Betadine®, Povidone®) solution in a target pattern, taking care to not touch the catheter site with any part of a gauze square that has touched the nonshaved hair or fur (Figure 4-17). Seventy percent isopropyl alcohol or sterile saline is then used to remove the aseptic solution. It is important to avoid abrading the skin with aggressive scrubbing of the site as skin abrasions are irritating, painful, and may cause the patient to further traumatize the area or lead to contamination of the site. Once the area is clean, it should be dried with sterile gauze so that tape will stick to the patient's skin (i.e., tape will not stick to wet skin). The veterinary technician then puts on sterile gloves for the IV catheter placement. Not everyone places IV catheters using sterile gloves, but attention to aseptic technique is recommended. IV catheter technique is summarized in Appendix E.

Sometimes a "pilot" hole is made with a needle or a scalpel blade prior to insertion of the IV catheter to reduce drag as the catheter is moved through the skin and improve visualization of the blood vessel. A pilot hole is made by lifting the skin over the vessel and making a very small (1–2 mm) incision, paying careful attention not to cut into the blood vessel itself. It is important to understand, however, that some state laws prevent veterinary technicians from performing this technique; a check of state laws will ensure adherence to legal regulations.

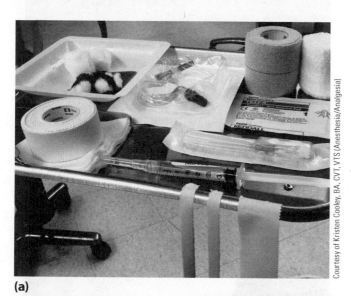

(a)

Courtesy of Kristen Cooley, BA, CVT, VTS (Anesthesia/Analgesia)

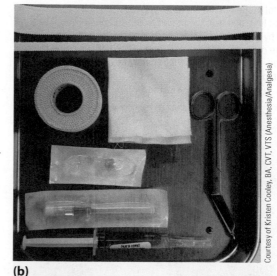

(b)

Courtesy of Kristen Cooley, BA, CVT, VTS (Anesthesia/Analgesia)

(c)

Courtesy of Janet Amundson Romich, DVM, MS

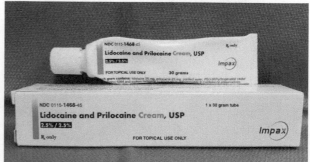

(d)

Courtesy of Janet Amundson Romich, DVM, MS

FIGURE 4-16 Preparing to place an IV catheter. (a) Supplies used for a sterile/long-term catheter placement (in order clockwise from upper left corner to lower left corner: Cotton balls with antiseptic scrub and alcohol, a T-port, Elasticon® or Vetrap®, Kling®/stretchy gauze, a Telfa® pad or other nonstick wound dressing, IV catheter, saline flush, pre-torn tape for securing in the catheter, tape roll, extra gauze, and bandage scissors. (b) Supplies used for a non-sterile short-term anesthesia catheter: Pre-torn tape across the top, and then clockwise from upper left corner to lower left corner: roll of tape (in case you need new pieces), alcohol gauze to wipe the site and remove skin flakes and tiny hairs, a dry gauze to dry the area so the tape can stick, bandage scissors, saline flush, an IV catheter, and an injection cap/PRN (in the middle of the tray). (c) Courtesy tabs on the end of tape make it easier to find the tape-end. (d) A topical anesthetic cream which contains a mix of lidocaine and prilocaine may be useful to lessen the pain caused by inserting a needle or IV catheter.

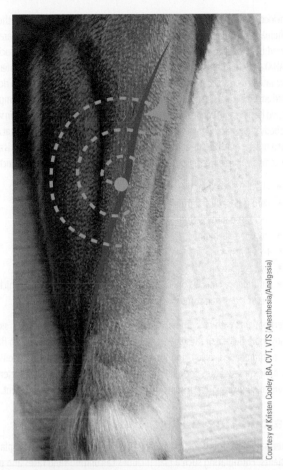

Courtesy of Kristen Cooley BA, CVT, VTS (Anesthesia/Analgesia)

FIGURE 4-17 A generous amount of fur is clipped and a target pattern is used when prepping the IV catheter site. Once the gauze has touched the fur at the periphery of the clip, it is discarded.

Equipment for Sterile Peripheral Catheter Placement (see Figures 4-16a and 4-16b)

- Scrub:
  - Chlorhexidine or povidone iodine solution
  - 70% alcohol or sterile saline
- Pre-torn tape:
  - Two narrow pieces and one thick piece long enough to encircle the limb 1.5 times
  - Fold courtesy tabs on one end (see Figure 4-16c)
- Telfa® pad or other nonstick dressing
- Saline flush
- Injection cap or T-port
- Kling® or stretchy gauze
- Elasticon® or Vetrap® unrolled and then rerolled
- Tape and permanent marker to mark date of placement
- IV catheter
- Extra gauze
- Scissors
- Sterile gloves

IV Catheter Care and Management

Placing an IV catheter through skin damages the body's protective outer barrier, which may allow microbes to enter the underlying tissue or blood vessel. Good catheter care avoids local and systemic infection, prolongs the life of the catheter, and prevents IV catheter dislodgement. Maintaining patent IV catheters is also important because replacing IV catheters is costly, is stressful to both the patient and staff, and increases the chances of thrombophlebitis when multiple veins must be accessed. Effective IV catheter care starts with following good hygiene and proper aseptic technique every time an IV catheter is placed or assessed.

Some patients may develop phlebitis, inflammation, or discomfort due to complications from placing and maintaining the IV catheter. Good nursing care and a watchful eye help prevent any IV catheter complications. Guidelines for the care and maintenance of IV catheters include the following:

- Whether using gravity flow or an infusion pump (described later in this chapter), check IV catheters for patency every 4 to 6 hours when a patient is receiving IV fluids. If fluids are not moving, the IV catheter may contain a clot or may be kinked and need to be replaced. The IV catheter itself may have a tear/hole in it or may have pulled out of the vein, causing fluids to be administered perivascularly. If the IV catheter site is hot, swollen, red, or painful, then phlebitis may be developing. Frequent IV catheter checks allow personnel to address concerns earlier, which may reduce serious consequences.
- Check the patient's digits for swelling or coolness, which may indicate that the catheter bandage is too tight (Figure 4-18). Take care when wrapping a bandage to protect an IV catheter as folded skin and pressure points can lead to discomfort and cause the patient to lick or chew at the area and thereby dislodge the catheter.
- Check the IV catheter bandage for moisture and change if needed. IV catheter bandages should be changed every 24 hours or if they become wet or soiled. A wet bandage wicks bacteria, which can lead to infection. Evaluate the site for redness or swelling, which may indicate irritation or the start of an infection.
- Check the IV catheter site both proximally and distally to the insertion site. If the patient is uncomfortable, unwrap the catheter and inspect it closely; replace the IV catheter if the insertion site shows signs of infection or inflammation.
- If the animal is not on fluid therapy, flush the IV catheter with saline/crystalloid fluids every 4 to 6 hours to maintain a patent line. Historically, heparin was often added to saline in an effort to prevent catheter clotting. Today this practice is no longer recommended in veterinary or human medicine because we now know that is the act of flushing itself rather than the presence of an anticoagulant that prevents catheter clotting.
- Check the IV catheter before and after any IV drug is administered. If multiple medications are given, use a sufficient amount of fluid between injections to prevent any precipitate from plugging the catheter. Flush with saline/crystalloid fluid after IV drug administration to ensure the entire amount of drug was administered.
- Pain on injection may be due to the pH of the drug or it may indicate phlebitis or questionable patency of the IV catheter. Note any discomfort the patient displays, check the IV catheter site, and alert the veterinarian about any abnormalities.

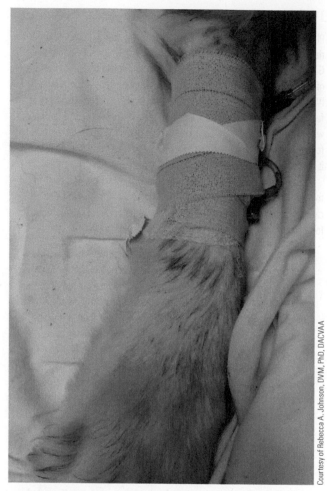

Courtesy of Rebecca A. Johnson, DVM, PhD, DACVAA

FIGURE 4-18 When a catheter is taped too tightly, the paw can swell from a lack of venous return.

- Place a leash over the shoulder or use a harness in patients with jugular catheters to avoid irritating the area or inadvertently dislodging the catheter.
- When changing an IV catheter, keep the old one in place until the new one is secured; doing so maintains venous access at all times.

DEVELOPING A PERIOPERATIVE FLUID ADMINISTRATION PLAN

The volume and rate of IV fluid (crystalloids, colloids, or blood products) given to a patient before, during, or after general anesthesia depends on many factors, including the patient's current level of hydration; need for intravascular volume replacement; potential for continued or future fluid loss; need for red blood cells and improved oxygen-carrying capacity, clotting factor replacement, and colloidal oncotic pressure (COP) support; and numerous other patient-specific factors such as concurrent disease states, acid-base balance, and electrolyte levels.

A complete discussion of the theory behind fluid therapy is beyond the scope of this chapter; however, it is widely accepted that perioperative fluid administration in healthy animals is sufficient to ensure adequate hydration and circulating plasma volume and supports cardiovascular stroke volume, cardiac output, and blood pressure. Perioperative fluid administration along with an IV catheter has become the standard of care in veterinary medicine and should be provided to almost *all* patients undergoing general anaesthesia. Modern theories in human medicine that have been adapted to veterinary medicine and are supported by recent recommendations by the American Animal Hospital Association (AAHA) and American Association of Feline Practitioners (AAFP) cite a maintenance fluid rate of a crystalloid solution at 3 mL/kg/hr in cats and 5 mL/kg/hr in dogs to avoid adverse effects associated with overhydration such as tissue edema, impaired wound healing, and congestive heart failure.[3] **Fluid overload**, a condition in which the body's fluid requirements are met and the administration of fluid occurs at a rate that is greater than the rate at which the body can use or eliminate the fluid, is a serious complication caused by excessive fluid delivery. Signs of fluid overload include

- chemosis (conjunctival edema);
- serous nasal discharge;
- increased body weight;
- dyspnea/tachypnea;
- tachycardia;
- hypoxemia; and
- pulmonary crackles on auscultation and pulmonary edema, which occur later than some of the other signs.

Select patients with concurrent co-morbidities may require less or often significantly more fluid than these general guidelines. Just as giving too much fluid is not recommended, giving too little perioperative fluid can have serious and profound consequences, including death. Therefore, it is critical for the attending veterinarian to prescribe an appropriate fluid plan for the individual patient.

Calculating a Fluid Plan

Veterinary technicians need to understand how to formulate a fluid plan for patients. Formulating such a plan includes calculating fluid rates and properly assessing, administering, and monitoring fluid therapy. The attending veterinarian should calculate and approve an individual fluid plan for every case undergoing anesthesia (as well as every case receiving fluid therapy in the hospital). To get to a precise fluid rate for a given patient, several components must be considered individually and then totaled:

- **The resuscitation phase:** Every patient should initially be evaluated for signs of cardiovascular collapse and **shock** (any state where the body's consumption of oxygen is not met by the body's delivery of oxygen). Typically these signs include weakness and lethargy, tachycardia in dogs and bradycardia in cats, pale to white mucous membranes, weak pulses, cold extremities, low blood pressure, hypothermia (especially in cats), and potentially tachypnea. Those patients in hypovolemic, septic, or distributive shock should receive fluid therapy directed initially at patient resuscitation to get them out of a state of cardiovascular collapse. However, fluid therapy is contraindicated in patients in cardiogenic shock (e.g., heart failure). This phase of therapy is known as the resuscitation phase, and its goal is to expand intravascular volume. In this phase, patients receive a large volume of fluid (e.g., crystalloid, colloid, blood products, or some combination thereof) rapidly over a short period of time (e.g., from 5 to 30 minutes). Most commonly, patients may receive up to their full blood volume in crystalloids (90 mL/kg for dogs and horses and 55 mL/kg for cats) in one hour. Typically ¼ to ⅓ of this volume is administered rapidly as a **bolus** (volume of fluid given rapidly) over 10 to 20 minutes and then the patient is reevaluated. This crystalloid bolus usually equates to 20 mL/kg for dogs and 15 mL/kg for cats.

Once the patient is out of cardiovascular collapse and shock (or if the patient was never in it to begin with), the next phase is addressed.

- **The replacement phase:** The next phase of fluid therapy is the replacement phase, which serves to correct dehydration and replace water previously lost. The amount of fluid needed to rehydrate a patient is based on estimating the percent of dehydration (Table 4-2). Dehydration is estimated between 5 and 12% of body weight. Patients having < 3–5% dehydration are described as having mild dehydration and do not show obvious signs on physical exam. Patients with a history of fluid loss, such as vomiting, diarrhea, excessive drooling, lack of eating, and weight loss, automatically fit into this category. Patients that are between 6 and 8% dehydrated are described as moderately dehydrated and upon physical exam may show signs such as moderate loss of skin turgor, dry or tacky mucous membranes, lethargy, and potentially a weak and rapid pulse. Patients that are >10% dehydrated have marked loss of skin turgor, extremely dry mucous membranes, dry sunken eyes, delayed capillary refill time, tachycardia, weak/thready pulses, hypotension, and altered state of consciousness. These patients are in a state of cardiovascular collapse because of their dehydration and should initially be treated with resuscitation phase interventions. Death results without immediate intervention in patients with >12% dehydration.

In theory, to calculate the amount of fluid a patient requires to correct dehydration, the rule is to take the estimated percent dehydration (e.g., 5% = 0.05, 7% = 0.07, 10% = 0.10) and multiply it times the weight of the animal in kilograms. This calculation gives the fluid deficit in *liters*. To convert this value to milliliters, the deficit in liters should be multiplied by 1000. For example, a 30 kg dog is estimated to be 7% dehydrated. This dog would need to receive 2100 mL of crystalloid fluid to correct its dehydration because 30 kg × 0.07 × 1000 = 2100 mL. It is important to understand that this calculation is entirely an estimate and is not precise. Often veterinarians underestimate the true degree of how dehydrated a patient actually is and do not give enough replacement fluid. If the patient became dehydrated rapidly or is in cardiovascular collapse/shock, the patient can typically have its replacement phase and correction of dehydration done rapidly over 6 to 12 hours with proper attention to electrolyte values. For example, if the patient is dehydrated and also hypernatremic, care should be taken and adjustments made about how quickly fluids are given. If the patient became dehydrated over a prolonged period (e.g., weeks to months), then the replacement phase and correction of dehydration should be done more slowly over a 24-hour period. This amount of fluid is then combined with the next component of fluid therapy to help determine an hourly fluid rate.

TABLE 4-2 Estimating Level of Dehydration

Dehydration Percentage	Physical Examination Findings

< 3–5%	• No obvious clinical signs or physical exam findings
	• Patients with a history of fluid loss
	  ○ vomiting
	  ○ diarrhea
	  ○ excessive drooling
	  ○ decreased eating or drinking
	  ○ weight loss

5%	• Physical exam may show the following signs:
	  ○ tacky mucous membranes
	  ○ abdomen feels "doughy" (free peritoneal fluid has be reabsorbed)

6–8%	• Physical exam may show the following signs:
	  ○ moderate loss of skin turgor
	  ○ dry or tacky mucous membranes
	  ○ lethargy
	  ○ potentially a weak and rapid pulse

10–12%	• Physical exam may show the following signs:
	  ○ marked loss of skin turgor
	  ○ extremely dry mucous membranes
	  ○ dry sunken eyes
	  ○ delayed capillary refill time
	  ○ tachycardia
	  ○ weak/thready pulses
	  ○ hypotension
	  ○ altered state of consciousness

> 12%	• Without intervention, death results

- **The ongoing maintenance phase:** The ongoing maintenance phase is used to replace body water lost continuously during daily normal body functions. Maintenance fluid volumes can be determined based on the amount of fluid expected to be lost from sensible and insensible losses. **Sensible body water losses** occur in urine and feces and can actually be measured. **Insensible body water losses** are the result of normal metabolic processes but are not easily measured; such losses occur through sweating, breathing, and mucous membrane evaporation. A typical daily rate for maintenance fluid is 60 mL/kg/day in adult dogs and cats and 110 mL/kg/day in pediatric patients. There are numerous other formulas that are available; for example, in patients weighing more than 2 kg the following equation can also be used:

$$[(30 \times \text{body weight in kilograms}) + 70]$$

These values will replace both sensible and insensible body water losses.

> ### Tech Tip 🐾
>
> Regardless of which formula for maintenance fluid rate is used, the fundamental component of fluid administration is diligent patient monitoring and assessment to ensure the patient is benefiting from the fluids and, if necessary, making adjustments to the fluid plan.

- **The Ongoing Fluid Loss Phase:** The ongoing fluid loss phase replaces body water lost through things like continued vomiting, diarrhea, oozing skin, and excessive drooling. Patients that are losing additional fluid amounts due to vomiting or diarrhea need to have this fluid loss replaced or they will quickly become dehydrated again. The amount of fluid lost (e.g., 100 mL lost from vomiting) is typically estimated but can be calculated by measuring or weighing the loss. This volume is then added to the other volumes described earlier.

The following example applies all of the fluid therapy phases together. Molly, an 8-month-old spayed female pit bull terrier puppy presents to the hospital for parvovirus enteritis. She has had significant vomiting and diarrhea and is in a state of cardiovascular collapse upon presentation. She is estimated to be 12% dehydrated and weighs 20 kg.

1. *Resuscitation phase:* Molly is given 20 mL/kg (20 kg $\times$ 20 mL/kg = 400 mL) IV LRS as a bolus over 20 minutes. When this does not resolve her cardiovascular collapse/shock, she is given another 20 mL/kg (400mL) bolus over 20 minutes. Once stable, other phases are considered.
2. *Replacement phase:* Molly was assessed to be 12% dehydrated. This equated to 0.12 $\times$ 20 kg $\times$ 1000 mL = 2400 mL. Molly will need 2400 mL or 2.4 L to correct her dehydration. Because Molly was in cardiovascular collapse and she became dehydrated quickly, the veterinarian decides her replacement phase will be given quickly over 12 hours. 2400 mL/12 hr = 200 mL/hr.
3. *Ongoing maintenance phase:* We also must consider that to keep up with Molly's maintenance needs she should receive fluids at a rate of 60 mL/kg/day. This is calculated as 60 mL $\times$ 20 kg = 1200 mL/day = 50 mL/hr. If we combine Molly's replacement and maintenance phases

together, she should receive the following: 200 mL/hr + 50 mL/hr = 250 mL/hr. Therefore, Molly should have her crystalloid run at 250 mL/hr for the first 12 hours. At that point, she should be reassessed and, if she is no longer vomiting, then her fluid rate should be lowered to 50 mL/hr.

4. *Ongoing fluid loss phase:* However, if Molly has ongoing losses, such as more vomiting or diarrhea, that amount should either be given back to Molly rapidly as part of a fluid bolus or added into her replacement phase needs.

Administering a Fluid Plan

To correctly determine the amount of IV fluids to deliver to a given patient per unit of time, the veterinary technician/anesthetist must know several things, including the weight of the patient (kg), the desired fluid rate (mL/kg/hr), and the "drops per mL" produced by the individual administration set size (gtt/mL). The drops per second are automatically calculated if using an IV pump or syringe pump/driver, which makes these devices helpful for the anesthetist (IV pumps and syringe pump/drivers are described later). For example, a 48 kg dog presents to the veterinary hospital for a routine neuter under general anesthesia. The attending veterinarian prescribes an IV fluid rate of 5 mL/kg/hr. The hospital stocks fluid administration sets that are 15 gtt/mL. The veterinary technician knows the following:

- Weight of patient: 48 kg
- Desired fluid rate: 5 mL/kg/hr
- Administration set size: 15 gtt/mL

Therefore,

$$48 \text{ kg} \times 5 \text{ mL/kg/hr} = 240 \text{ mL/hr}$$

$$240 \text{ mL/hr} \times 15 \text{ gtt/mL} = 3,600 \text{ gtt/hr}$$

$$3,600 \text{ gtt/hr} \div 60 \text{ min/hr} = 60 \text{ gtt/min}$$

$$60 \text{ gtt/min} \times 1 \text{ min/60 seconds} = 1 \text{ gt/sec}$$

A review of basic mathematics concepts is found in Appendix F.

> ### Tech Tip 🐾
>
> Postanesthetic fluid administration may be needed in patients with anesthetic complications and/or medical conditions such as renal disease or conditions resulting in ongoing fluid losses. In addition to proper IV catheter care, it is important to properly maintain the IV administration set (Figure 4-19). If an IV administration set needs to be disconnected (e.g., to take a dog for a walk), the connection site should be cleaned with single-use, isopropyl alcohol wipes and capped with injection caps (do not reuse injection caps). IV administration sets used for continuous crystalloid fluid administration should be changed every 72 hours to prevent intravascular device-related infections or immediately if the pathway of sterile fluid in the tube has been compromised.

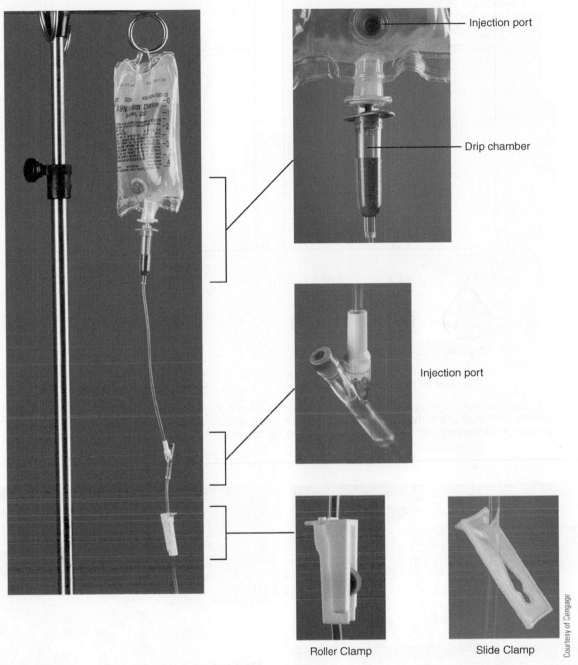

Injection port

Drip chamber

Injection port

Roller Clamp

Slide Clamp

Courtesy of Cengage

FIGURE 4-19 Supplies for fluid administration. Fluid bag and IV administration set. The 0.9% sodium chloride solution is yellow to allow visualization of the fluid in the bag and administration set.

Administration Sets

Administration sets deliver crystalloids and colloids from the fluid bag to the patient. They are available with filters (to use when administering blood products) or without filters. They also come in different types that deliver various amounts of drops/mL (gtt/mL). It is important to know how many gtt/mL an administration set delivers using gravity (i.e., not via an infusion pump), as this will alter the calculation used to determine drops to administer per second. Administration sets most commonly come in 15 gtt/mL and 60 gtt/mL (and occasionally 10 gtt/mL or 20 gtt/mL

that may be infusion pump specific) (Figure 4-20). Administration sets for smaller patients (generally less than 10 kg) deliver 60 gtt/mL to ensure precise delivery of the calculated rate. These administration sets are also referred to as "microdrip" sets and they slow down (or decrease) the size of the drop so that a smaller patient can have a "reasonable" or attainable number of drops per second. Attaching the administration set to a fluid bag is described in Figures 4-21a through 4-21k. In patients that weigh less than 5 kg, it is a good idea to use a burette to make sure the patient does not receive too much fluid (Figures 4-22a and 4-22b).

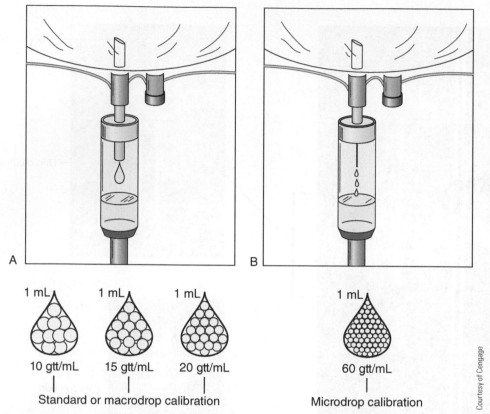

Standard or macrodrop calibration   Microdrop calibration

1 mL — 10 gtt/mL   1 mL — 15 gtt/mL   1 mL — 20 gtt/mL   1 mL — 60 gtt/mL

Courtesy of Cengage

FIGURE 4-20 Administration sets deliver a constant number of drops per mL, with the amount of drops differing based on whether the set contains a macrodrop or microdrop drip chamber. (a) Macrodrop drip chambers typically deliver 15 gtt/mL, but may also deliver 10 gtt/mL and 20 gtt/mL depending on the manufacturer. (b) Microdrop drip chambers typically deliver 60 gtt/mL.

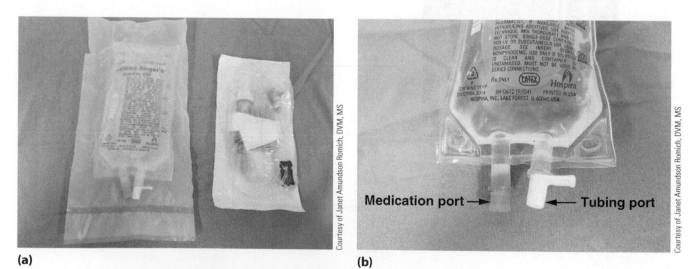

(a)   (b)

Medication port →   ← Tubing port

Courtesy of Janet Amundson Romich, DVM, MS

FIGURE 4-21 Attaching the administration set to the fluid bag is also called "spiking" a fluid bag. (a) Wash hands and gather supplies (fluid bag and administration set). Always check the fluid order to make sure the proper fluid bag and administration are selected and that expiration dates are not exceeded. (b) Tear the overwrap down side at slit and remove the fluid bag from the outer plastic covering. Some opacity of the plastic due to moisture absorption during the sterilization process may be observed. The fluid bag will typically have two "ports" at the bottom. One port has a beige stopper on the end and is called the medication port because it is used to inject medications into the fluid bag. This stopper does not come off. The other port is white (in this example) and is called the tubing port because it is where you insert the "spike" from the administration set. The cap of the tubing port pulls off. When the cap is pulled off, do not let the tip of the tubing port touch anything.

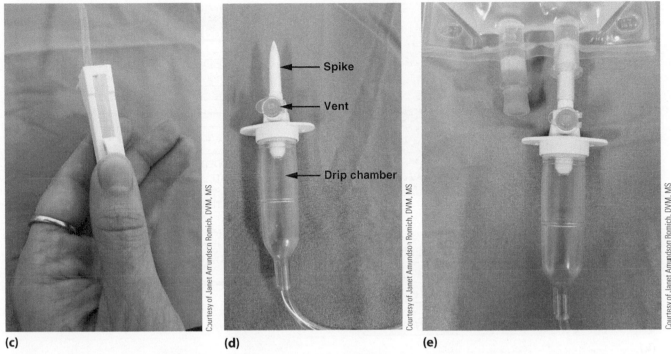

Courtesy of Janet Amundson Romich, DVM, MS

Courtesy of Janet Amundson Romich, DVM, MS

Courtesy of Janet Amundson Romich, DVM, MS

(c)   **(d)**   **(e)**

(c) Remove the administration set from the outer plastic covering. There is a roller clamp on the tubing that will be open. Slide the roller to the bottom until it pinches the tubing. This is the "closed" position and fluid will not flow in the tubing while it is closed. (d) Identify the end of the tubing that has the drip chamber. This end attaches to the IV bag. Remove the plastic cap to expose the plastic "spike." Do not let the "spike" touch anything once the cap has been removed. The spike should stay sterile, as it will go into the IV bag. (e) While holding the port on the fluid bag with the nondominant hand, use pressure to insert the spike until the shoulder of the spike is level with the port.

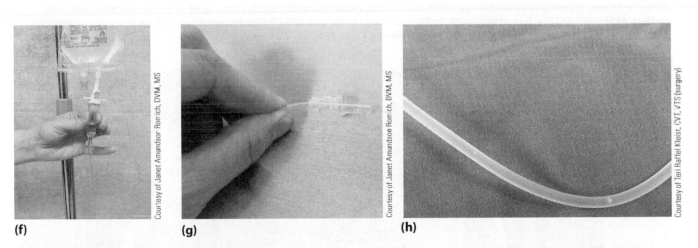

Courtesy of Janet Amundson Romich, DVM, MS

Courtesy of Janet Amundson Romich, DVM, MS

Courtesy of Teri Raffel Kleist, CVT, vTS (surgery)

(f)   **(g)**   **(h)**

FIGURE 4-21 (*Continued*) (f) Suspend the fluid container from a IV pole or hanger. Squeeze the drip chamber until it is ⅓–½ full of fluid. (g) Remove the cap on the other end of the tubing. Do not throw the cap away, as it may be used to temporarily cover the end of the tubing. It is important that the end of the tubing does not touch anything because it will be attached to the catheter and must remain sterile. Open the roller clamp on the tubing with one hand while holding the tip of the tubing with your other hand. Allow fluid to come out the end until the air is out of the line and then close the roller clamp. Removing the air from the line is called "priming" the tubing. (h) A few tiny bubbles in the line is fine; however, there should not be large bubbles in the tubing. Keep running the fluid through the line until the large air bubbles are gone. Attach set to the IV catheter. Regulate the rate of fluid administration with the roller clamp.

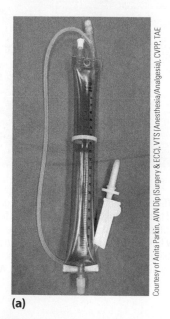

Courtesy of Teri Raffel Kleist, CVT, VTS (surgery)

(i)

WRAP AROUND IV TUBING	I.V. SET— _____ Hours Only
	Initial _____
	START—date/hr. _____
	DISCARD—date/hr. _____

(k)

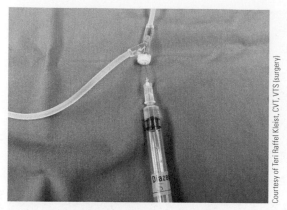

IV Label
Medication Added

MEDICATION ADDED

DRUG ADDITIVE   AMOUNT

_____ | _____
_____ | _____
_____ | _____
_____ | _____

ADDED BY _____
DATE _____ TIME _____ AM / PM
EXP. DATE _____
PATIENT _____
DATE OPENED _____
START TIME _____
START DATE _____
FLOW RATE _____

(j)

FIGURE 4–21 *(Continued)* (i) Medication can be added to the injection port. Close the roller clamp on the administration set to stop the flow of fluid to the patient. Use single-use or new vials when adding medication to fluids administered intravenously. Wipe all injection ports with alcohol before injection. Using a syringe with a needle, puncture resealable medication port and inject. Some medications will need a needle with a larger lumen if they are more viscous. Ideally, a smaller lumen needle should be used. After addition of the medication, it is important to mix the medication into the fluids by removing the fluid bag from the IV pole and inverting the bag at least four times. For high density medication (e.g., potassium chloride), squeeze ports while ports are upright and mix thoroughly. Return the fluid bag to the in-use position, open the roller clamp, and continue administration. (j) Commercially available labels can be used to provide all information needed on an IV fluid bag (patient name; date opened; date, name, and amount of any supplements/additives added; date and time fluid administration started; flow rate; expiration date; and initials of the person preparing the fluids). (k) Commercially available IV tubing labels are also available to identify when they need to be discarded.

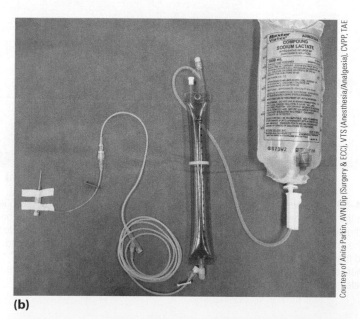

Courtesy of Anita Parkin, AVN Dip (Surgery & ECC), VTS (Anesthesia/Analgesia), CVPP, TAE

(a)   **(b)**

FIGURE 4–22 A burette is an inline volume calibrated chamber on IV tubing. These chambers can be filled with the exact amount of fluid to be delivered for a specific treatment and when the chamber is empty, the patient has received that specific volume. (a) The burette only holds 100 mL, so if the patient is to receive 20 mL of fluid per hour, the burette should be filled to 60 mL. In this case, the fluid in the burette should last 3 hours. When the patient is checked hourly, the blue ring will have moved down to the top of the fluid level, so the veterinary technician knows exactly how much fluid has been administered in that hour. It is important to fill the burette to the required level and to clamp off the roller clamp in between the burette and the fluid bag to stop the burette from filling. (b) An example of where the burette sits in relationship to the fluid administration set. The fluid bag is connected to the burette, then to the extension line, T-port, and finally to the catheter.

Be sure to check the package of each administration set to know exactly the number of drops delivered per mL. Sets can vary widely between different countries and with different manufacturers. Also be aware that some administration sets are specifically designed for use in certain infusion pumps.

Infusion Pumps and Syringe Pumps (Drivers)

Infusion pumps and syringe pumps (drivers) are mechanical pumps used to automatically dispense fluid at a given rate (Figures 4-23a and 4-23b); therefore, the user does not need to calculate drops/second. Basic infusion pumps infuse one type of fluid at a time, at a rate between 0.1 and 999 mL per hour. More complicated pumps can infuse two (and sometimes more) types of fluids at any given time (Figures 4-24a and 4-24b). Units called syringe pumps (drivers) can hold syringes that range from 1 mL to 60 mL (Figure 4-25). Some syringe pumps require a specific brand of syringe to correctly administer the prescribed fluid volumes. Some pumps are so advanced that they can be programmed with the microgram/milligram/gram (mcg/mg/g) dosages of drugs and the desired rate of administration.

Infusion pumps and syringe drivers can be useful tools that seemingly make the veterinary technician's life easier; but they can also malfunction and deliver incorrect volumes. They have alarms that beep when there is an issue (e.g., kinked tubing, air in line), which is helpful but also a source of frustration when multiple pumps are beeping at once.

Infusion pumps from individual manufacturers may require a specific type of administration set or proprietary fluid line that can be safely used with the pump. Some infusion pumps will tolerate any type of fluid line. When using an infusion pump that is able to use a generic type of administration set, it is important that the pump be calibrated to a particular amount of gtt/mL or a particular brand of administration set. If the incorrect set is used, the amount of fluid administered to the patient will be affected. When a multiple administration infusion pump with two or more fluid lines together (using three-way stopcocks as shown in Figure 4-26) is used, the fluid rates should be compatible to the pump. It is best not to use this method to administer fluids unless it is certain that the pumps are accurate; otherwise the patient may not receive the prescribed amount of medication because one of the infusions (the bag running at a higher rate) may push fluid into the second bag (via the three-way tap point) as the higher rate fluid has a greater pressure. This may cause the second infusion not to be administered or administered at a reduced rate. Always check with the manufacturer to make sure the proper administration set is being used. Examples of different fluid lines are shown in Figures 4-27a through 4-27d.

A quick reference guide for drip rates is provided in Appendix G.

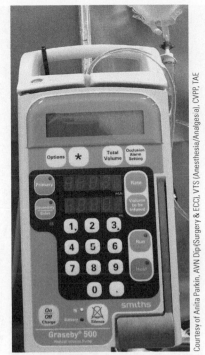

Courtesy of Anita Parkin, AVN Dip(Surgery & ECC), VTS (Anesthesia/Analgesia), CVPP, TAE

(a)

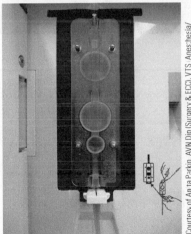

Courtesy of Anita Parkin, AVN Dip (Surgery & ECC), VTS Anesthesia/Analgesia), CVPP, TAE

(b)

FIGURE 4-23 Infusion pumps come in a variety of models used for different reasons. (a) A commercially available infusion pump in which the fluid rate per hour is set. A volume to be infused (VTBI) rate can also be set so an alarm will sound when that amount has been delivered. The infusion pump can then be reset for the next time the patient needs to be checked. Additional alarms can also be set for particular time frames. (b) The inside of the infusion pump. When putting an administration set into an infusion pump, all clamps (roller and slide clamps) must be fully opened for the pump to function properly. Occasionally an infusion pump door may need to be opened when administering fluids or changing fluid lines. Use caution when opening infusion pump doors as some pumps have an internal clamp inside the door that must be opened before fluids will run unchecked into the patient. If fluids are allowed to run full bore into a patient, it may cause fluid overload (especially in small patients) or a bolus of the fluids (such as KCl) will be administered to the patient. When closing the infusion pump door, take care not to pinch the tubing in the door.

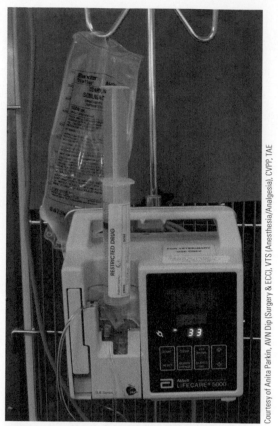

(a)

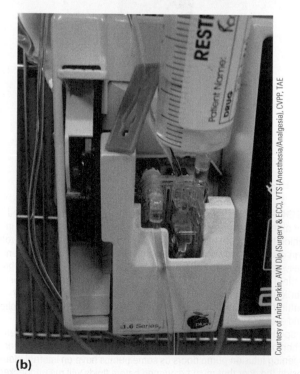

(b)

FIGURE 4-24 A dual infusion pump. (a) A dual infusion pump is able to administer two types of fluids using one administration set. (b) When the infusion pump is open, it is important to clamp off (green clamp) one of the fluids being administered to avoid the fluids from mixing when the pump is open.

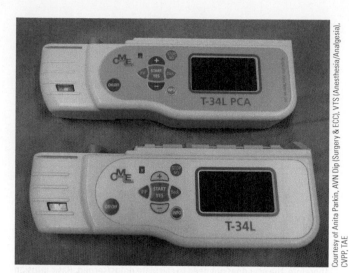

FIGURE 4-25 Example of syringe pumps (drivers) used for delivering small volumes of fluids. Syringe pumps can hold up to 60 mL syringes and deliver as little as 0.01 mL/hr.

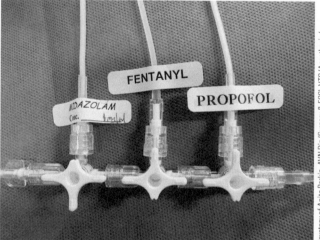

FIGURE 4-26 Example of four types of fluids being administered to one patient. The main fluid (isotonic solution), infusion 1: midazolam, infusion 2: fentanyl, infusion 3: propofol. With the three-way stopcocks (the white connectors), the long toggles indicate that particular inlet/outlet is open, therefore the fluid runs through it. In this example, the midazolam is turned off (arrow on the tap is pointing down), and the other two infusion are still running into the patient (arrows on the taps are pointing up).

SUMMARY

Fluid therapy and IV catheterization are important parts of any anesthesia plan and are the standard of care provided by today's well-trained veterinary anesthetist. Administering IV fluids to patients during the peri-anesthetic period ensures adequate intravascular volume and promotes tissue perfusion and oxygen delivery to the body's tissue (as well as providing immediate IV access in case of an emergency). Crystalloids, colloids, and blood products can

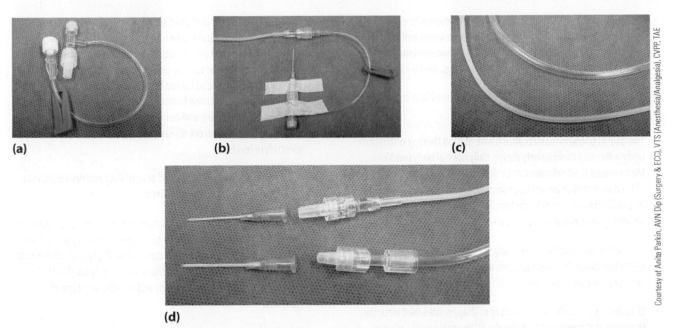

FIGURE 4-27 Different fluid lines. (a) A T-port is a shorter type of extension set that is attached to the catheter and connected to the terminal end of the IV line. T-ports have an injection port and the tubing is usually taped to the patient. (b) T-port connected to a catheter and extension set. (c) A high flow (the top line) and a low flow (bottom line) extension set. A low flow (small bore) extension set may be used for particularly small patients to prevent blood from flowing back into the IV line. (d) An example of a 22 gauge catheter (blue) ready to be connected to a low flow extension set, and a 18 gauge catheter (green) ready to be connected to a high flow extension set.

all be given IV and the choice of which to use depends on the patient's needs. Crystalloids are solutions that consist of a sodium or dextrose base dissolved in water and are most commonly used for maintenance and replacement of intravascular volume and interstitial hydration. Colloids contain large molecular weight particles that tend to remain in the intravascular space and help to retain fluid intravascularly. Colloids are used for intravascular volume expansion in treating hypovolemia and for managing disease in which hypoproteinemia is present and a low colloid osmotic pressure is suspected. Numerous types of blood products are available and the type selected for the patient is based largely on whether the anesthetist wishes to provide red blood cells, plasma volume, or both. All types of fluids and blood products have advantages, disadvantages, and potential adverse effects.

To deliver fluids effectively, veterinary technicians must understand the equipment used to place an IV catheter and methods used to deliver the fluids. Choosing the proper catheter type, size, and site as well as properly preparing and maintaining that site is a critical part of the veterinary technician's job on the anesthesia team. Accurately calculating the fluid rate to prevent too much or too little fluid from being delivered to the patient, as well as administering fluids using the proper administration set and mechanical pumps, ensures that veterinary technicians provide appropriate delivery of fluids to their patients. The success of fluid therapy depends on a veterinary professional who is well educated in fluid administration theory and techniques.

CRITICAL THINKING POINTS

- Intravenous access in anesthetized animals facilitates the delivery of IV anesthetic agents, avoids potentially harmful perivascular drug administration, provides access to rapid drug/fluid administration in emergency situations, and allows for continuous fluid administration.

- IV fluid therapy can help correct hypovolemia and replace ongoing fluid losses or dehydration in the surgical patient.

- IV fluids may be categorized as crystalloids, colloids, or blood products. Crystalloids are used for volume replacement within the intravascular space (e.g., to correct hypovolemia) and are beneficial to correct deficits within the interstitial space (e.g., dehydration). Colloids are used for intravascular volume expansion in treating hypovolemia and for treating disease in which hypoproteinemia is present and a low colloid osmotic pressure is suspected.

- Crystalloids may be balanced (have an electrolyte composition similar to plasma; also known as isotonic), hypertonic (have an electrolyte composition that exceeds plasma), or hypotonic (have an electrolyte composition less than plasma).

- Crystalloids are typically inexpensive, have a long shelf life, and are easy to obtain; however, they provide only a short duration of action in the intravascular space and in large volumes they may cause dilution of plasma proteins, red blood cells, and clotting factors.

- Colloids contain large molecules with higher molecular weight and contribute to oncotic pressure.

- Synthetic colloids include dextrans and hydroxyethyl starch, and can result in anaphylaxis, may increase the risk for acute kidney injury (especially in patients that are septic), can easily result in volume overload, and can cause coagulation abnormalities.

- Blood products such as whole blood, plasma, packed red blood cells, and albumin are natural colloids which can cause adverse effects such as fever, volume overload, and anaphylaxis.

- The goal of the resuscitation phase of fluid therapy is to expand intravascular volume by giving patients a large volume of fluid (e.g., crystalloid, colloid, blood products, or some combination thereof) rapidly over a short period of time (e.g., over 5–30 minutes).

- The goal of the replacement phase of fluid therapy is to correct dehydration and replace water previously lost.

- The goal of ongoing maintenance phase of fluid therapy **is** to replace body water lost continuously during daily normal body functions. Maintenance fluid volumes can be determined based on the amount of fluid expected to be lost from sensible (urine and feces) and insensible (result of normal metabolic processes such as sweating, breathing, and mucous membrane evaporation losses).

- The goal of ongoing fluid loss phase of fluid therapy is to replace body water lost through things like continued vomiting, diarrhea, oozing skin, and excessive drooling.

- IV catheters are available in a variety of shapes, sizes, and materials; therefore, it is important to know the different types of catheters available to best fit the needs of veterinary patients.

- The volume and rate of IV fluid (crystalloids, colloids, or blood products) given to a patient before, during, or after general anaesthesia depends on the patient's current level of hydration; need for intravascular volume replacement; potential for continued or future fluid loss; need for red blood cells and improved oxygen carrying capacity, clotting factor replacement, and colloidal oncotic pressure (COP) support; and numerous other patient specific factors such as concurrent disease states, acid-base balance, and electrolyte levels.

- Administration sets can be filtered or unfiltered and deliver varying number of drops of fluid per milliliter.

- Infusion pumps and syringe drivers are mechanical pumps used to automatically dispense fluid at a given rate. Infusion pumps from individual manufacturers may require a specific type of administration set or proprietary fluid line that can be safely used with the pump, while some infusion pumps will tolerate any type of fluid line.

REVIEW QUESTIONS

Multiple Choice

1. What are the types of extracellular fluids in animals?
   a. Intravascular, interstitial, and transcellular water
   b. Intracellular, synovial fluid, cerebral spinal fluid
   c. Urine, saliva, gastrointestinal tract secretions
   d. Intravascular, interstitial, and intracellular water

2. What is true in regards to types of fluid?
   a. Colloids contain microscopically dispersed suspended particles that are insoluble.
   b. Crystalloids are water solutions that consist of a dissolved sodium or dextrose base with the addition of plasma proteins or clotting factors.
   c. Blood products contain either whole blood or red blood cells.
   d. Synthetic colloids rarely result in volume overload and are the preferred fluid for patients undergoing surgeries that last longer than 2 hours.

3. Which fluid type would be recommended for a patient in heart failure?
   a. Balanced (isotonic) crystalloid solution
   b. Hypertonic crystalloid solution
   c. Colloid solution
   d. Fluid therapy is contraindicated in patients in heart failure

4. Which statement regarding electrolyte composition in the body is false?
   a. Sodium is the primary extracellular cation.
   b. Potassium is the primary intracellular cation.
   c. Chloride is the primary extracellular cation.
   d. Phosphate is the primary extracellular anion.

5. Which catheter material is less likely to initiate clot formation?
   a. Polyvinyl chloride
   b. Polyurethane
   c. Polyethylene
   d. Polypropylene

6. Approximately how much of the original volume of a crystalloid administered IV remains in the intravascular space after one hour?
   a. 25%
   b. 50%
   c. 75%
   d. 100%

7. What does the term gauge refer to when describing a catheter?

   a. A series of whole numbers that increases from zero in increments of 0.33 mm

   b. A unit used to measure catheter length and is expressed in inches

   c. A descending scale in which large numbers indicate a narrower catheter diameter

   d. A method of describing the degree of thrombogenicity of a catheter

8. What criteria is catheter selection based on?

   a. Patient size, type of fluid administered, and the outer diameter of the catheter

   b. Expected length of time the catheter will be in place, patient's tolerance to handling, and cost of maintaining the catheter

   c. Patient size, type of fluid administered, and expected length of time the catheter will be in place

   d. Skill of the veterinary technician to properly place the catheter, health status of the patient, and ability to secure the catheter in place

9. What is a disadvantage of over-the-needle catheters?

   a. They are made from rigid materials that may kink.

   b. They require long needles that may break during catheter placement.

   c. They are difficult to connect to the patient and are cumbersome.

   d. They require a flexible, sterile guide wire that is inserted in the lumen of the vein.

10. Which is true regarding transfusion reactions?

    a. Febrile non-hemolytic reactions are due mainly to an immune response; signs of this type of reaction include an increase in patient temperature and vomiting.

    b. Actual allergic reactions produce facial swelling, hives, itchiness, and difficulty breathing which can be safely treated once the entire transfusion is given.

    c. Transfusion associated circulatory overload (TACO) results in hives and edema from giving too much volume to the intravascular space.

    d. Transfusion related acute lung injury (TRALI) result in jaundice from giving too much volume causing acute liver injury.

11. Why should a wet bandage on an IV catheter be replaced?

    a. It indicates that the bandage is too tight.

    b. It affects fluid administration rate.

    c. It changes the pH of the catheter site.

    d. It wicks bacteria and can lead to infection.

12. When putting an administration set into an infusion pump, in which position should the roller and slide clamps be in once you are ready to administer the volume to the patient?

    a. Fully closed

    b. Partially closed

    c. Partially open

    d. Fully open

13. What are the commonly available administration sets for delivering calculated fluid rates?

    a. 5, 10, 15, and 20 gtt/mL

    b. 10, 20, 30, and 40 gtt/mL

    c. 10, 15, 20, and 60 gtt/mL

    d. 20, 40, 60, and 80 gtt/mL

14. Which type of fluid replaces daily volume and electrolyte loss from normal body functions?

    a. Ongoing maintenance

    b. Replacement

    c. Ongoing fluid loss

    d. Resuscitation

15. Why might plasma be administered to a patient?

    a. To provide oxygen to tissues as well as plasma volume to expand the intravascular space

    b. To cause water movement into the cells and to replace free water deficits in hypernatremic patients

    c. To expand the intravascular space and provide a source of albumin, clotting factors, and a source of colloidal osmotic pressure

    d. To expand plasma volume and to correct hyponatremia or metabolic alkalosis

16. Which is *not* a sign of fluid overload?

    a. Chemosis

    b. Weight gain

    c. Hypoxemia

    d. Bradypnea

17. What can cause increased resistance to flow during fluid administration?

    a. A wider IV catheter because it may make contact with the vessel wall causing more turbulent flow of fluid.

    b. A narrower IV catheter will dramatically increase resistance to flow and slow the rate of fluid passage though the catheter.

    c. A shorter catheter because it can cause vessel inflammation, which may make the catheter more likely to become dislodged.

    d. Administering fluid rates too quickly can result in fluid backup, which impedes the flow of fluid through a catheter.

18. What is recommended for flushing an IV catheter to maintain its patency?

    a. Heparin

    b. Saline/crystalloid fluids

    c. Sterile water

    d. EDTA

19. Which of these choices are considered blood products?

    a. 5% dextrose, LRS, and 0.9% NaCl

    b. 20% mannitol and 50% dextrose

    c. Ringers solution and Plasma-Lyte®

    d. Albumin and plasma

20. When is it helpful to use syringe pumps (syringe drivers) in the clinical setting?

    a. When administering small volumes of fluid

    b. When administering colloids through a filter

    c. When administering crystalloids using a long extension set

    d. When administering two fluid types at a time

Case Studies

Case Study 1: Max, a M/N, 3-year-old Border Collie canine weighing 24 kg, presents to the clinic with a laceration that needs to be sutured. While Max is in his cage awaiting premedication, a fluid administration rate needs to be calculated.

1. Using a fluid rate of 4 mL/kg/hr, how many mL/hr of LRS should be administered to Max?
2. If using a 15 gtt/mL administration set, what would be the drops per second?

Case Study 2: A 20-year-old Appaloosa mare is brought to the clinic with signs of colic. She is tachycardic, has a prolonged CRT, and has decreased GI sounds. The veterinarian is concerned that she might have an intestinal obstruction and asks you to place a jugular catheter.

1. Why is a catheter needed in this patient?
2. What concerns would you have when placing a catheter in this patient?

Case Study 3: A 1-year-old M/N DSH feline that was hit by a car is brought into the clinic. On physical examination the cat had altered mentation, pale pink mucous membranes, prolonged capillary refill time, weak femoral pulses, cold extremities, and tachypnea. It was determined that the cat was hemorrhaging into the abdominal cavity (via abdominal tap) and was going into hypovolemic shock. The decision is made to place an IV catheter and administer whole blood to this patient.

1. Why might this cat need whole blood?
2. Whole blood is mixed with an anticoagulant that affects its shelf-life. List four examples of anticoagulants used for whole blood.
3. Does this patient need cross-matching? Why or why not?

Case Study 4: A healthy 6-month-old male Welsh Springer Spaniel canine presents to the clinic to be neutered. An IV catheter is placed, a gravity administration set attached, and fluids are started. Since this is a young, healthy animal, you are tempted to work on the next case while the veterinarian is neutering this dog.

1. What complications with IV fluid administration could occur in this patient while you are not actively monitoring him?
2. Which clinical signs might help you determine if this patient received too much fluid?

Case Study 5: A 5-year-old F/S Labrador Retriever canine had surgery to remove an intestinal foreign body and is moved into the recovery area. Since the surgical procedure is done, is it necessary to keep the IV catheter in place and continue fluid therapy?

Critical Thinking Questions

1. Understanding the properties of blood products that can be administered to animals is important when providing fluid therapy for anesthetized patients. Complete the information in the chart below about various blood products, by placing an "x" in the box if the statement in the column on the left applies to the various blood products listed:

	Synthetic colloid	Whole blood	Packed RBC	Plasma	25% Albumin
Replaces intravascular volume					
Immediate increase in colloid osmotic pressure					
Carries oxygen					
Contains albumin					
Contains coagulation factors					

2. As a new veterinary technician, you want to be prepared when asked to place IV catheters. The three most commonly used sites for IV catheters in dogs in your practice are the cephalic, lateral saphenous, and jugular veins. Describe each of these sites and their advantages/disadvantages.

ENDNOTES

1. Bruce, J., Kriese-Anderson, L., Bruce, A., & Pittman, J. (2015). Effect of premedication and other factors on the occurrence of acute transfusion reactions in dogs. *Journal of Veterinary Emergency and Critical Care, 25*(5), 620–630.
2. Mitchell, A., et al. (1982). Reduced catheter sepsis and prolonged catheter life using a tunneled silicone rubber catheter for total parenteral nutrition. *British Journal of Surgery, 69*, 420.
3. Davis, H., Jensen, T., Johnson, A., et al. (2013). AAHA/AAFP fluid therapy guidelines for dogs and cats. *Journal of the American Animal Hospital Association, 49*(3), 149–159. doi:10.5326/JAAHA-MS-5868

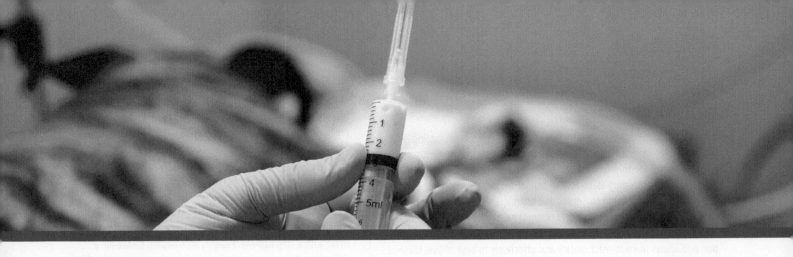

CHAPTER 5
Anesthetic and Analgesic Pharmacology

Janet Amundson Romich, *DVM, MS*

Emma Archer, *RVN, Dip AVN (Surgical), VTS (Anesthesia/Analgesia)*

Janel Holden, *CVT, VTS (Anesthesia/Analgesia)*

LEARNING OBJECTIVES

Upon completion of this chapter, it is expected that the reader should be able to:

5.1 Explain the meaning of pharmacological terms and their use in veterinary medicine

5.2 Describe the principles of anesthetic and analgesic drug absorption, distribution, metabolism, and elimination (ADME)

5.3 Explain factors that affect absorption, distribution, metabolism, and elimination of anesthetic and analgesic agents

5.4 Describe the mechanism of action of anesthetic agents

5.5 Describe the intended effects of anesthetic agents

5.6 List adverse effects of drugs used in anesthesia

5.7 Describe the mechanism of action of analgesic agents

5.8 Describe the intended effects of analgesic agents

5.9 List adverse effects of drugs used for analgesia

5.10 Describe the mechanism of action of neuromuscular blocking agents

5.11 Describe the intended effects of neuromuscular blocking agents

5.12 List adverse effects of drugs of neuromuscular blocking agents

KEY TERMS

Affinity

Agonist

Antagonist

Anxiolytic

Bioavailability (F)

Blood-brain barrier (BBB)

Blood-gas partition coefficient (BGPC)

Catalepsy

Ceiling effect

Competitive antagonist

Cumulative

Diffusion

Efficacy

Emergence delirium

First-pass metabolism

Full agonist

Hydrophilic

Ionization

IV bolus

Lipophilic

Minimum alveolar concentration (MAC)

Mixed agonist/ antagonist

Neuroleptanalgesia

Noncompetitive antagonist

Non-cumulative

Parenteral

Partial agonist

Partial pressure difference

Partial pressure

Pharmacodynamics

Pharmacokinetics

Plasma esterase

Potency

Protein binding

Surgical MAC

Therapeutic index (TI)

Toxic dosage

Vapor pressure

Volatile anesthetics

INTRODUCTION

Developing an anesthesia plan requires a thorough understanding of pharmacology in order to know how drugs are selected, administered, and monitored for their effects. Knowing what the animal does to the drug and what the drug does to the animal helps the anesthesia team make wise decisions about the drugs and techniques they choose for their patients. Recognizing adverse effects of drugs helps veterinary professionals anticipate possible disturbances to the patient's physiological condition and allows them to react quickly and effectively to best address these issues. For example, most anesthetic agents will cause significant changes in cardiovascular and pulmonary function. Reduced cardiac output, reduced blood pressure, and decreased respiratory rate and tidal volume are sequela of many anesthetic agents. Knowing what to expect when an animal is given a drug will help the anesthesia team manage their cases more effectively. All anesthetic agents produce a variety of effects; therefore, the goal for the anesthesia team is to exploit the beneficial effects of anesthetic agents while avoiding their adverse effects.

This chapter focuses on the principles of drugs and their movement throughout the body, as well as the characteristics of specific drugs used in the practice of anesthesia. In addition to knowledge of specific drugs and how they work, it is important when working with anesthetic and analgesic agents to know how to correctly calculate doses, draw up and label medications in syringes, and administer drugs that are described in multiple chapters in this text. Drugs specifically used for cardiopulmonary resuscitation (CPCR) are described in Chapter 13 (Cardiac Arrest and Cardiopulmonary Cerebral Resuscitation [CPCR]).

PHARMACOLOGY TERMINOLOGY

To understand how drugs work, familiarity with terms used to describe drug action or its properties is necessary. There terms include:

- *Affinity*: the measure of the strength with which a drug binds to its receptor.
  - *Agonist*: drug that binds to a cell receptor and stimulates a response characteristic of that receptor.
  - A *full agonist* is a drug that causes a maximal response because it has an exact fit at the receptor site.
  - A *partial agonist* is a drug that binds to the receptor to produce a submaximal response relative to the full agonist. The submaximal response is true even if the drug occupies all of a cell's receptors.
  - *Mixed agonist/antagonists* bind to more than one type of receptor and simultaneously stimulate one drug and block another.
- *Antagonist*: drug that binds to a cell receptor and prevents the cell from performing some function.
  - A **competitive antagonist** competes with the agonist for the same receptors and is reversible by giving high dosages of agonist (reversible or surmountable antagonism).
  - A **noncompetitive antagonist** binds to a different site than the agonist's binding site, which changes the agonist's receptor and is not reversible (irreversible or insurmountable antagonism).
- *Therapeutic index* (*TI*): comparison of the amount of a therapeutic agent that causes the therapeutic effect to the amount that causes toxicity; also called the *margin of safety*. A drug with a wide or greater therapeutic index can produce its desired effect without approaching toxicity. Many of the drugs used in anesthesia may have a narrow or lower therapeutic index; therefore, calculation or administration errors may have significant consequences for the patient.

- *Half-life*: time required for elimination of 50% of the drug. Drugs with a short half-life are administered more frequently than those with a long half-life. Patients with hepatic failure have difficulty with biotransformation of a drug that is metabolized by the liver, while patients with renal insufficiency and a decreased glomerular filtration rate have difficulty eliminating drugs that are removed from the body by the kidney. Because of their roles in drug elimination, patients with liver failure and/or kidney insufficiency can cause an increase in a drug's half-life, which can increase the risk of toxicity.
- *Onset of action*: when the drug achieves its therapeutic effect.
- *Duration of action*: the length of time the drug produces a therapeutic effect.
- *Synergism*: interaction between two drugs that produces a greater effect than the sum of their separate actions; for example, instead of $1 + 1 = 2$, $1 + 1 = 3$.

GENERAL PRINCIPLES OF PHARMACOLOGY

Anesthetic agents produce an effect in the patient based on the properties of the drug as well as patient factors such as species and health status. **Pharmacokinetics** is the study of physiological movement of a drug within the body after it is administered and includes its absorption, distribution, metabolism, and elimination (or excretion) (ADME). **Pharmacodynamics** is the study of the drug's mechanism of action and its effect on the animal (Figure 5-1). To produce a systemic effect, anesthetic agents must be absorbed and distributed to the site of action, which is usually the brain. Once the effect is no longer needed, anesthetic agents must be metabolized and eliminated. Knowing what the animal does to the drug (pharmacokinetics) and what the drug does to the animal (pharmacodynamics) is important when making decisions about which drugs and techniques to use in patients.

Drug Absorption

Absorption is the movement of a drug from the site of administration into the blood. The rate of drug absorption depends on a variety of factors that include:

- *Drug properties*: Unless given IV, drugs must cross cell membranes to be absorbed into the blood. Absorption occurs more readily with a drug that has a low degree of ionization, lower molecular weight, and high lipid solubility.[1]
  - *Ionization:* Many drugs are weak acids or bases that exist in both ionized (charged) and non-ionized (uncharged) forms in the body depending on the pH of the tissue in which they are given. Only a drug's non-ionized form is sufficiently permeable to the phospholipid bilayer of the cell membrane. Basic drugs such as

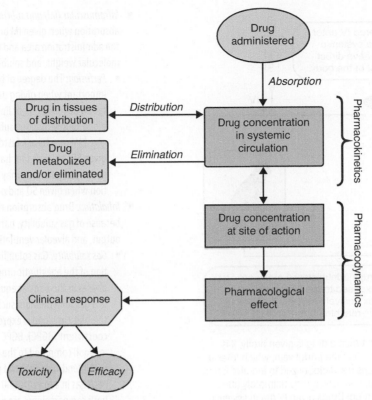

FIGURE 5–1 Pharmacokinetics versus pharmacodynamics. Pharmacokinetics: When a drug is administered to an animal, it is absorbed and the drug concentration increases in the systemic circulation. The drug can be distributed into tissues or metabolized and eliminated. Pharmacodynamics: When the drug is at the site of action, it can bind to receptors and produce a pharmacological effect, which may present as a clinical response in the patient.

lidocaine are ionized in acidic environments, while acidic drugs like thiopental are ionized in basic environments.

- *Molecular weight:* Molecular weight (the mass of a molecule) is a general measure of a molecule's size based on the number and types of atoms it contains. Lower molecular weight drugs are smaller and can cross the cell membrane more easily than higher molecular weight ones.
- *Solubility:* **Lipophilic** (lipid soluble) drugs dissolve in the phospholipid cell membrane because like molecules dissolve in like molecules. Lipophilic drugs tend to cross membranes more quickly than water soluble drugs. **Hydrophilic** (water soluble) drugs tend to stay within the blood and interstitial fluid that surrounds cells.

- *Diffusion properties:* Most drugs are absorbed by **diffusion** (particles move from an area of high concentration to an area where particles are in low concentration), and the rate of diffusion is proportional to the drug concentration, the amount of surface area available, and tissue thickness. The greater the concentration difference, the more rapid the drug absorption. The more surface area available for diffusion to occur, the greater the amount of drug absorbed and more rapid the drug effect. The greater the tissue thickness, the further the drug must travel and the slower the drug absorption.
- *Route of administration*: The site of administration determines drug absorption rate.

- *Oral:* Drugs given orally are typically absorbed more slowly than those given **parenterally** (not through the GI tract). Factors that affect absorption of oral drugs include:
  - **Bioavailability** (F) is the percent of drug administered that enters systemic circulation; basically, it is the degree to which a drug is absorbed and reaches circulation. A drug administered IV is in systemic circulation immediately, is 100% available, and has a bioavailability value of 1. Drugs given by routes other than IV generally have bioavailability values <1. The lower the drug's bioavailability, the less of it that enters the systemic circulation. Most drugs given orally are absorbed in the small intestine and must dissolve before they can cross the small intestinal mucosa to reach systemic circulation. Dissolving takes time and can also significantly reduce the final amount of drug in the blood stream. Properties of the drug (ionization, molecular weight, solubility, and mechanism of absorption), interaction with other drugs, feeding condition (fasted versus fed), patient age, gastrointestinal health, and animal species can all affect oral bioavailability.
  - *First-pass metabolism:* When a drug is given orally, it is absorbed from the small intestine into the portal vein, which takes it to the liver. The liver may metabolize some drugs to inactive forms before the drug enters systemic circulation. This is known as the **first-pass metabolism** (also known as

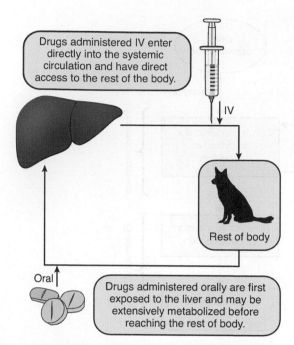

Drugs administered IV enter directly into the systemic circulation and have direct access to the rest of the body.

IV

Rest of body

Oral

Drugs administered orally are first exposed to the liver and may be extensively metabolized before reaching the rest of body.

FIGURE 5-2 First pass effect. When a drug is given orally, it is absorbed from the small intestine into the portal vein, which takes it to the liver. In the liver, some drugs are metabolized to inactive forms before they enter systemic circulation, which may ultimately affect the availability of active level of drug. Drugs given IV are in systemic circulation immediately.

first-pass effect or hepatic first pass) and may result in subtherapeutic levels of active drug in the blood stream (Figure 5-2). Drugs that have significant first-pass metabolism may not be effective when given orally and, despite administering higher dosages of the drug, therapeutic levels may never be reached. For example, lidocaine is not given orally because it has extensive first-pass metabolism and therefore never reaches therapeutic plasma levels when given orally.

- *Species:* Oral drug absorption varies between monogastric and ruminant species. Monogastric animals (such as dogs and cats) do not secrete hydrochloric acid continuously; therefore, stomach pH (and ultimately drug absorption) varies depending on whether food is in the stomach. Drug absorption in horses is also affected by feeding because, although they are monogastric animals, they can ferment feed in their colon, which affects intestinal pH and ultimately drug absorption. Dosing schedules for horses and ruminants typically take this into account; therefore, it is important to administer drugs as recommended.
- ○ *Parenteral routes:* Parenteral routes of administration vary in their rates of absorption.
  - *Intravenous (IV):* Drug action is fastest when a drug is given IV because no absorption is needed. Drugs given as an **IV bolus** (drug given rapidly via the IV route) achieve high blood levels to produce a rapid, immediate response. IV drugs can also be given via a continuous rate infusion (CRI) so drug levels in blood can be maintained at a certain level or given to effect.

- *Intramuscular (IM) and subcutaneous (SQ or SC):* The rate of drug absorption when given IM or SQ is determined by vascularity to the administration area and the drug's properties of ionization, molecular weight, and solubility.
  - *Perfusion:* The degree of blood flow to the absorption site is important when giving drugs IM or SQ— even more important than ionization or lipid solubility. Muscle has a greater blood supply than the subcutaneous space; therefore, drugs given IM may have a faster rate of absorption than those given SQ. Animals that have hypothermia or decreased peripheral perfusion may have a slower rate of drug absorption when given SQ and potentially IM.
- *Inhalation:* Drug absorption through the alveoli is effective because of gas solubility, partial pressure differences, cardiac output, and alveolar ventilation.
  - *Gas solubility:* Gas solubility, the measure of the distribution of the anesthetic drug between the blood and gas phases in the body at equilibrium, describes the affinity of a gas for a medium such as blood or adipose tissue. Gas solubility in blood is expressed as the **blood-gas partition coefficient** (BGPC). BGPC is the ratio of gas concentration in blood compared to the gas concentration in contact with blood when **partial pressures** (the pressure that each gas in a mixture will exert on a column of mercury) in both compartments are equal. The BGPC is used to predict the speed of anesthetic induction, recovery, and change of anesthetic depth. For example, isoflurane has a BGPC of 1.4. If the gas in the blood is in equilibrium with gas in the alveoli, the concentration of isoflurane in blood will be 1.4 times higher than the concentration in the gas in the alveoli. A higher BGPC means a higher uptake of the gas into the blood and therefore a slower induction time and longer recovery period since there is more anesthetic agent to be eliminated from blood. Inhalant anesthetic agents with lower BGPC such as sevoflurane produce a faster onset of anesthesia and faster emergence from anesthesia once the vaporizer is turned off, assuming no other drugs or conditions are present.
  - *Partial pressure difference:* Inhaled anesthetic agents must cross the membrane of the alveolus into blood. The difference in partial pressures between anesthetic gas in the alveoli versus that in the blood affects the rate of uptake (absorption) of the inhaled anesthetic agent (Figure 5-3). The greater the difference between the pressures of inhalant anesthetic agent in the alveoli to that in the blood, the greater the uptake of inhalant anesthetic agent and the faster the induction. When the partial pressure of anesthetic gas in the blood equals that in the alveoli, equilibrium is met and no further uptake occurs.[1]
  - *Cardiac output:* Blood flow through the lungs determines the amount of blood available to take up (absorb) inhalant anesthetic agents. Increased cardiac output exposes alveoli to more blood per unit of time. Greater blood volumes remove more inhalant anesthetic gas from the alveoli, which lowers alveolar partial pressure. The greater the

cardiac output, the lower the alveolar partial pressure, and the longer it will take for the inhalant anesthetic agent to reach equilibrium between the alveoli and brain, resulting in prolonged induction times.

- *Alveolar ventilation*: In general, the greater the minute ventilation, the more rapid the uptake of inhalant anesthetic agent. Minute ventilation is the amount of air an animal breathes in one minute (tidal volume multiplied by respiratory rate), which takes into account the depth and rate of breathing.

  ▪ *Topical application:* Drugs applied topically are put on the surface of the skin or mucous membranes (mouth, respiratory tract, esophagus, genitourinary tract, etc.) and are available in a variety of formulations (creams, lotions, pastes, and aerosol dose forms). Topical drugs must first dissolve and then penetrate the skin or mucous membrane. Systemic absorption of drugs administered topically is slow; however, at the application site the concentration of drug is high. Mucosal surfaces usually have a rich blood supply, which provides rapid drug transport to the systemic circulation and in most cases avoids degradation by first-pass hepatic metabolism.

  ▪ *Transdermal:* Drugs administered transdermally are absorbed through the skin. A common transdermal formulation is a patch that consists of a reservoir of drug covered with a membrane (which limits the rate of absorption) surrounded by an adhesive dressing that attaches to the patient's skin. Transdermal patches are designed to release a consistent amount of drug per hour as it is absorbed across skin.

Drug Distribution

After a drug is absorbed into the blood, it is transported to various tissues or target sites. Tissue concentration of a drug is primarily dependent on movement across capillary membranes, and tissue concentration must be sufficient for a long enough period of time to produce an effect. Distribution is a dynamic process, and drug concentrations in the blood and tissues are constantly changing. Factors affecting drug distribution include:

- *Drug properties:* Just as in absorption, drug properties affect distribution. In addition to lipid solubility (which greatly affects drug distribution to the brain, where the **blood-brain barrier** [partition formed by tight junctions between the capillary endothelial cells and also by the glial cells that surround the capillaries] restricts the penetration of polar and ionized molecules) and molecular weight (lower molecular weight drugs move more easily across cell membranes), drug distribution is also affected by the following:

  ○ *Protein binding*: Many drugs bind reversibly to plasma proteins to form drug-protein complexes. Protein-bound drugs are inactive while unbound drugs are active, can produce a therapeutic effect, and can be eliminated. Typically, equilibrium is established between the concentration of bound and unbound drug (Figure 5-4). As concentrations of unbound drug in blood decrease, bound drug is released from its binding sites. Protein binding decreases drug distribution because it limits transport of molecules across cell membranes. Propofol is a highly protein-bound injectable anesthetic agent with high affinity for its receptor and therefore tends to stay in circulation. Animals with liver or protein-losing diseases have less blood protein, which results in less protein binding of propofol. This means more available free drug for the target tissue, which makes accidental overdose a concern. Since anesthetic agents tend to be given to effect, protein binding is less of a concern in anesthetized patients compared to other drugs; however, it is still important to monitor

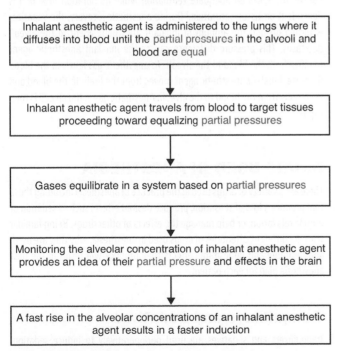

Inhalant anesthetic agent is administered to the lungs where it diffuses into blood until the partial pressures in the alveoli and blood are equal

↓

Inhalant anesthetic agent travels from blood to target tissues proceeding toward equalizing partial pressures

↓

Gases equilibrate in a system based on partial pressures

↓

Monitoring the alveolar concentration of inhalant anesthetic agent provides an idea of their partial pressure and effects in the brain

↓

A fast rise in the alveolar concentrations of an inhalant anesthetic agent results in a faster induction

FIGURE 5-3 Partial pressure difference. The difference in partial pressures affects the absorption rate of inhalant anesthetic agents (and ultimately affects distribution).

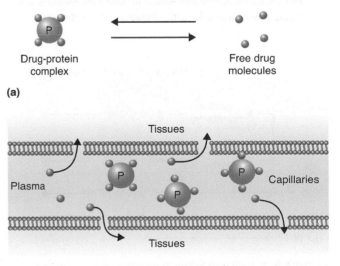

(a)

(b)

FIGURE 5-4 Protein binding. (a) An equilibrium is typically established between the concentration of bound and unbound drug. (b) Only unbound drug can leave systemic circulation to cause an effect.

patients for both desired and adverse effects when titrating the drug to effect.

- *Concentration gradient:* The greater the difference in concentration between the amount of drug in the blood and the site of action, the more rapid its distribution.
- *Perfusion:* The degree of blood flow to the target tissues influences drug distribution. Blood flow is high to tissues such as the brain, heart, kidney, and liver while it is moderate to muscle and low to fat. Because of the high blood flow to the CNS, anesthetic drugs are delivered rapidly to the brain, while low blood flow causes them to be delivered slowly to fat and subcutaneous tissue.
- *Affinity:* Drug diffusion to tissue is related to affinity of the drug for a particular tissue. Drugs with a high affinity for a tissue will diffuse more rapidly and bind more tightly than those with low affinity.

Drug Metabolism

Metabolism is the process by which the animal changes a drug into a metabolite that may be a more or less active form that can be eliminated. Metabolism ultimately increases a drug's water solubility and polarity because elimination of drugs occurs in body fluids. As a drug becomes more water soluble, there is less distribution of it to tissues and its water solubility facilitates removal of the drug from the body. The primary site of metabolism is the liver; however, some anesthetic agents are partially or totally metabolized by other tissues. The liver has an enzyme system called cytochrome P-450 to metabolize drugs. Some drugs up-regulate cytochrome P-450 (increase the amount of the enzyme), while other drugs down-regulate cytochrome P-450 to inhibit its action (decrease the amount of enzyme).

Drug metabolism occurs in two phases (Figure 5-5):

- Phase I: enzymes act on the drug and change it. Examples include oxidation, reduction, hydrolysis, and hydration, which add a functional group to the drug. Cytochrome P-450 is used for metabolism of most phase I reactions.
- Phase II: the metabolite joins another to make it more water soluble. An example is conjugation with compounds in the body.

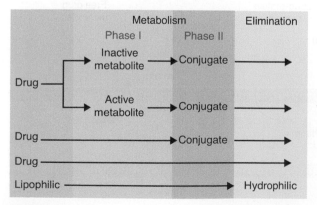

FIGURE 5–5 Drug metabolism. Drug metabolism can occur in two phases. The first phase involves enzymes acting on a drug and changing it. The second phase involves joining metabolites together to make a more water soluble (hydrophilic) molecule. As drugs become more water soluble, they are removed from the body.

Factors affecting metabolism include:

- *Drug interaction:* If more than one drug is metabolized by the same enzyme system, metabolism will be delayed for one of the drugs. Some drugs cause the inhibition of cytochrome P-450, therefore slowing the metabolism of drugs that use these enzymes.
- *Species:* Some species lack certain enzyme systems. For example, cats are deficient in the enzyme glucuronyl transferase, which conjugates acetaminophen to glucuronic acid to be excreted by the body. Cats should never be treated with acetaminophen, even in low dosages.
- *Liver dysfunction:* Animals with impaired liver function may metabolize drugs more slowly because they do not produce the level of enzymes needed for drug metabolism.

Drug Elimination

Recovery from anesthesia depends on removal of anesthetic agents from the site of action. Most drugs are eliminated from the body via urinary excretion or exhalation from the lungs. The kidneys eliminate most drugs via glomerular filtration after they have been metabolized in the liver into more polar (ionized), water-soluble forms. Drugs that are water-soluble, polar, lower molecular weight, and unbound to plasma proteins can be eliminated from the body. Since glomerular filtration depends on blood flow, conditions such as hypovolemia and hypotension that decrease the rate of blood flow will decrease the rate of drug elimination. Animals with kidney disease will have a reduced rate of drug elimination if they are excreted in the urine. Geriatric patients with altered kidney function or cardiac disease that affects blood flow will not eliminate drugs as rapidly as healthy patients.

For most inhalant anesthetic agents, elimination is the reverse of uptake and relies on adequate ventilation while its duration of action is dependent on tissue saturation.[1] During anesthetic recovery, the vaporizer is turned off and the amount of inhalant anesthetic agent in the alveoli decreases. This pressure difference causes the inhalant anesthetic agent to move from the blood to the alveoli. As anesthetic gas levels in the blood decrease, inhalant anesthetic agent moves from the brain to the blood and eventually from the blood to the alveoli. Anesthetic gas in the alveoli is then expired.

DRUGS USED IN ANESTHESIA

There are a variety of drugs used in the practice of anesthesia including those that produce a loss of sensation, produce desired effects such as sedation or muscle relaxation, or help manage the effects of other drugs. Being familiar with the different categories of drugs used in anesthesia, why they are used, and desirable and undesirable effects they produce will result in the best anesthesia plan for each patient.

Tranquilizers and Sedatives

Tranquilizers and sedatives are used perioperatively to induce sedation, cause muscle relaxation, provide restraint, and reduce the amount of injectable and inhalant anesthetic agents required to induce and maintain anesthesia.

Phenothiazines

Phenothiazines are tranquilizers that affect the central nervous system (CNS) by blocking dopamine receptors. Blocking dopamine receptors results in a calming and **anxiolytic** (anti-anxiety) effect because dopamine regulates behavior, fine motor control, and prolactin secretion. Phenothiazines are also used as antiemetics, usually to prevent motion sickness in animals, because they also depress parts of the brain that control vomiting. Phenothiazines do not provide analgesia. They can be recognized by their —azine suffix, and the main phenothiazine used in veterinary medicine is acepromazine; other phenothiazines include chlorpromazine and promazine.

- *Acepromazine* (Promace®, generic) is used in dogs and cats to provide reliable tranquilization to allow physical examination (restraint) or transport, for tranquilization prior to minor procedures, or as a preanesthetic agent because it reduces the amount of induction and inhalant anesthetic agents needed[2] (see Figure 8-2). In dogs, it is also used as an antiemetic. It is used in horses and cattle as a tranquilizer to aide in recovery from general anesthesia and to aide handling. A limitation of acepromazine is that it does not provide pain relief. Acepromazine can be given IV, IM, SQ, or PO. Its onset of action is fairly slow (it may take up to 15 minutes when given IV) and is 20 to 40 minutes after IM injection. Its duration of action is often long and unpredictable, especially if hepatic dysfunction is present. Acepromazine should be avoided in patients with hepatic disease because it is metabolized entirely by the liver and excreted in the urine. A benefit of acepromazine is that it has mild respiratory system effects, making it useful in patients with airway obstruction (such as in brachycephalic breeds) or those with airway disease. Acepromazine causes vasodilation because it also is an alpha-1 (α_1) adrenergic antagonist (in the peripheral nervous system [PNS] phenothiazines block α_1 adrenergic receptors found in the smooth muscle cells of peripheral blood vessels). This vasodilation may lead to hypotension when given with other vasodilating drugs or at increased dosages. Because vasodilation decreases vascular resistance, using low-dose acepromazine may be beneficial in patients with some types of cardiac disease (such as chronic valve disease). Vasodilation may lead to hypothermia, redistribution of blood to the periphery, a temporary decrease in packed cell volume levels, and transient splenomegaly. Acepromazine was once believed to lower the seizure threshold in animals with a seizure history; however, this is less problematic than once thought.[3] Acepromazine can rarely cause priapism (persistent penile erection) in horses, and both the degree and duration are dosage dependent. Low dosages of acepromazine can be used in geldings, but is sometimes avoided in breeding stallions. Dogs and cats given acepromazine may have protrusion of the nictitating membrane (third eyelid) that resolves as the drug is eliminated from the body. Acepromazine is long lasting and non-reversible; therefore, its use should be avoided in patients where hypotension or cardiovascular compromise is anticipated as a potential complication of anesthesia and surgery. Collies with a specific gene mutation (MDR-1) may have an exaggerated response to some drugs, such as acepromazine.[4]

Tech Tip

Using acepromazine for tranquilization is extra-label; therefore, the dosage listed on the label is not the dosage used for tranquilization. The dosage listed on the bottle label is 10 times higher than the tranquilization dosage. It is important to check veterinary drug formularies prior to calculating drug doses to ensure the proper dosage for the desired action is used.

Alpha-2 Adrenergic Agonists

Alpha-2 (α_2) adrenergic agonists inhibit release of the neurotransmitter norepinephrine, and because norepinephrine maintains alertness, its absence produces potent sedation. Alpha-2 adrenergic agonists are used perioperatively, alone or in combination with opioids, to produce sedation and analgesia for procedures and diagnostics, to significantly reduce the amount of induction and maintenance drugs, and to synergize with the effects of other analgesic drugs.[2] This group of drugs can be used as a CRI to provide analgesia as an adjunct to other methods in the intra- and post-operative period, especially when additional sedation is required. In dogs and cats, α_2 adrenergic agonists provide reliable dosage-dependent sedation, analgesia, and muscle relaxation, and decrease the animal's ability to respond to stimuli. In horses, α_2 adrenergic agonists produce deep and reliable sedation. Cardiovascular adverse effects, which include initial transient vasoconstriction (which causes hypertension, followed by a reflex bradycardia that causes decreased cardiac output over time), are profound with this drug class. Occasionally dosage-dependent bradyarrhythmias such as second degree AV block and escape beats may be observed. Because of their profound cardiovascular adverse effects, their use is reserved for animals with cardiovascular stability. Animals given α_2 adrenergic agonists are challenging to monitor as they will have significant bradycardia, increased blood pressure, and pale mucous membranes. An advantage of using α_2 adrenergic agonists is that they are reversible by administration of selective antagonists, and reversal may be indicated for some types of cardiovascular adverse effects.

Tech Tip

IV catheter placement in small patients who have received some types of α_2 adrenergic agonists, such as dexmedetomidine, is sometimes more challenging. The reason for this is because these drugs result in profound peripheral vasoconstriction, which can make finding peripheral blood vessels more challenging because the patient's actual blood pressure may be normal to even high.

Alpha-2 adrenergic agonists are metabolized by the liver and then excreted primarily into the urine. The use of α_2 adrenergic agonists should be avoided in animals with liver disease as their use reduces hepatic blood flow and the rate of metabolism of other drugs by the liver. They frequently cause vomiting due to activation of the central α_2 adrenergic receptors and are used as an emetic in cats. They should not be used in patients with an esophageal foreign body, gastrointestinal obstruction, intestinal torsion, and bloat because vomiting may cause further local damage. Vomiting in patients with head trauma and increased intracranial pressure and/or those with increased intraocular pressure will further increase these pressures; therefore, the use of α_2 adrenergic agonists in these patients should also be avoided. Alpha-2 adrenergic agonists reduce insulin secretion causing transient hyperglycemia, which should be kept in mind when interpreting blood values of animals given these sedatives.

Each α_2 adrenergic agonist drug has a different duration of action, causes slightly different adverse effects, and varies in cost. Most α_2 adrenergic agonists have a duration of action of less than two hours, but this depends on the species, individual drug, and dosage used (increasing the dosage prolongs the duration of activity).

- *Xylazine* (Rompun®, AnaSed®, Sedazine®, and Chanazine®) was the first α_2 adrenergic agonist used in veterinary medicine and is relatively non-selective in its binding to α_2 and α_1 receptors (see Figure 8-5a). Xylazine is commonly used in horses as a sedative and analgesic for colic, but although they may appear sedate, horses can still respond to stimuli by kicking. Xylazine is used extra-label in cattle for surgical procedures such as laceration repair and Cesarean (C-) sections. Cattle are extremely sensitive to xylazine requiring 1/10th the dosage required for horses to exhibit the same effect.[2] Xylazine may be combined with ketamine for short-term procedures such as castration in horses and surgical wound repair. Xylazine can be given IV, IM, or SQ. Its onset of action is 2 to 3 minutes when given IV, 5 to 15 minutes when given IM, and 10 to 20 minutes when given SQ. The duration of effect is dosage dependent and is typically 20 minutes in horses. Small-animal (20 mg/mL) and large-animal (100 mg/mL) concentrations are available, so care must be taken when determining drug doses so that the proper concentration is used. Xylazine can be reversed with yohimbine (Yobine®, Antagonil®), and in horses it can be reversed with tolazoline (Tolazine®).[5]

Tech Tip 🐾

When using xylazine, make sure you know the concentration per milliliter (100 mg/mL versus 20 mg/mL) to prevent overdosing or underdosing a patient. A 10% solution is 100 mg/mL and a 2% solution is 20 mg/mL.

- *Detomidine* (Dormosedan®) is a selective α_2 adrenergic agonist used mainly in horses for sedation, analgesia, and muscle relaxation (see Figure 8-5b). It is labeled for use in horses for sedation; however, it should be noted that the sedated animal can still respond to stimuli. Detomidine is also used for treating horses for pain associated with colic and allows for suturing of skin lacerations. Detomidine can be given IV (as a bolus or CRI), IM, or applied as a gel under the tongue for transmucosal absorption (Dormosedan gel®). The duration of its analgesic effect (30–45 minutes) is shorter than the duration of its sedative effect (30–90 minutes). When given IV it has an onset of

action of 2 to 5 minutes and a duration of action of 1 hour (depends on dosage).[2] It is not approved for use in cattle and has a withholding period of 7 days for meat and 72 hours for milk. Adverse effects include increased blood pressure followed by bradycardia that is more pronounced than with xylazine. Piloerection, sweating, ataxia, and salivation may also be seen with detomidine use. Detomidine can be reversed with yohimbine; it is only partially reversed with atipamezole and tolazoline.[5]

- *Dexmedetomidine* (Dexdomitor®) is a selective α_2 adrenergic agonist that produces fewer adverse effects than xylazine and has largely replaced xylazine use in small animal practice (Figure 5-6). Dexmedetomidine is labeled for use in dogs older than 6 months of age and cats older than 5 months of age as a preanesthetic agent for sedation, as an analgesic for minor surgical procedures and minor dental procedures, and as an aide in examinations and procedures. Two concentrations of dexmedetomidine are currently available (0.5 mg/mL and 0.1 mg/mL). Dexmedetomidine can be given IV (as a bolus or CRI), IM, or SQ. Its onset of action is 1 to 5 minutes when given IV, 5 to 15 minutes when given IM, and 10 to 20 minutes when given SQ. If given IM, it provides approximately 40 minutes of sedation in dogs and 60 minutes of sedation in cats. An eye lubricant should be applied to animals given dexmedetomidine to prevent corneal ulcers because it may cause decreased tear production. Following injection of dexmedetomidine, the animal should be allowed to rest quietly for 15 minutes; sedation and analgesia occur within 5 to 15 minutes of administration, with peak effects occurring 30 minutes after administration. It is reversed with atipamezole. If dexmedetomidine is given with an opioid such as butorphanol, the analgesic properties of the opioid are retained when reversed with atipamezole.

- *Romifidine* (Sedivet®) is a selective α_2 adrenergic agonist that causes potent, dosage-dependent sedation, muscle relaxation, and reduced

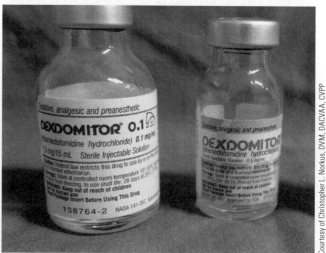

Courtesy of Christopher L. Norkus, DVM, DACVAA, CVPP

FIGURE 5–6 Dexmedetomidine is available in 0.1 mg/mL and 0.5 mg/mL concentrations and is the α-2 adrenergic agonist used most commonly in small animals. Other examples of α-2 adrenergic agonists include xylazine (Figure 8-5a), detomidine (Figure 8-5b), romifine (Figure 8-5c), and dexmedetomidine.

responsiveness to stimuli (see Figure 8-5c). It is approved for and mainly used in horses and is given IV, IM, or SQ. Its onset of action is 2 to 10 minutes when given IV and 10 minutes when given IM. Romifidine is preferred for procedures such as taking radiographs or lameness examinations because it produces a longer duration of sedative effects than detomidine and xylazine.[6] It is also used as a pre-anesthetic sedative in horses, and when combined with butorphanol, it is effectively used in pain management. Its duration of analgesia is shorter than the duration of sedation. Horses appear less ataxic when given romifidine in comparison to horses given xylazine or detomidine. It can be reversed with yohimbine.

- *Medetomidine* (Domitor®) is a selective alpha-2 adrenergic agonist (more selective than xylazine) that can be given IV (as a bolus or CRI), IM, or SQ. Its onset of action is 1 to 5 minutes when given IV, 5 to 15 minutes when given IM, and 10 to 20 minutes when given SQ. It is only available in the United States through compounding pharmacies and is used mainly in exotics and at higher concentrations in zoo animals. It still is available in Europe, but in the United States its use has been replaced by dexmedetomidine.

Alpha-2 Adrenergic Antagonists

Alpha-2 adrenergic antagonists reverse the effects of alpha-2 adrenergic agonists and include the following:

- *Atipamezole* (Antisedan®) is used to reverse medetomidine and dexmedetomidine and to partially reverse detomidine (Figure 5-7a). Adverse effects of atipamezole include occasional vomiting, diarrhea, hypersalivation, and tremors. The preferred route of administration is IM to avoid unwanted adverse effects.
- *Yohimbine* (Yobine®, Antagonil®) is approved for reversal of xylazine administration or overdose and is available in an injectable form for IV and IM use (Figure 5-7b). It is also used to reverse detomidine and romifidine. Adverse effects include transient CNS stimulation, muscle tremors, and salivation.
- *Tolazoline* (Tolazine®) is approved for reversal of effects associated with xylazine in horses; it is not approved for use in food producing animals (see Figure 5-7b). Adverse effects of tolazoline include transient tachycardia, peripheral vasodilatation (seen as sweating and injected mucous membranes of the gingiva), and piloerection, which has limited its clinical use. It is typically administered slowly IV.

Benzodiazepines

Benzodiazepines are a class of tranquilizers used in all species to provide mild to moderate skeletal muscle relaxation and amnesia. They are also highly-effective anticonvulsant drugs. They are less reliable anxiolytics in veterinary species.[7] Benzodiazepines are C-IV controlled substances and do not provide any analgesic effects. They work by increasing gamma aminobutyric acid (GABA), an inhibitory neurotransmitter in the brain. A major advantage of benzodiazepines is that they have minor effects on both the cardiovascular and respiratory systems, making them a good choice for patients with cardiovascular compromise. They also have minimal kidney effects which, along with the cardiovascular and respiratory advantage, make them good choices for critically ill, pediatric, and geriatric patients. They are sometimes

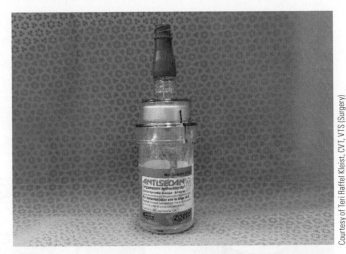

(a)

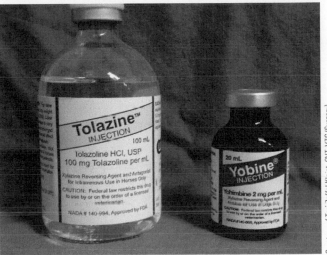

(b)

FIGURE 5-7 Examples of alpha-2 adrenergic antagonists include (a) atipamezole, (b) tolazoline, and yohimbine.

Veterinarians who wish to prescribe controlled substances must obtain a drug enforcement administration (DEA) license. Special documentation, storage, and prescribing rules surround their use.

used as appetite stimulants in cats and in combination with ketamine or tiletamine for short-term anesthesia. Despite their widespread use, benzodiazepines are not approved or marketed for use in animals.[2]

The most widely used benzodiazepines in veterinary medicine include diazepam and midazolam, which are both metabolized by the liver. Diazepam and midazolam have very similar clinical effects and can often be used interchangeably. Benzodiazepines are reversible with the agent

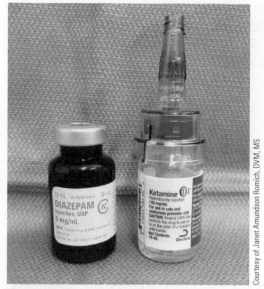
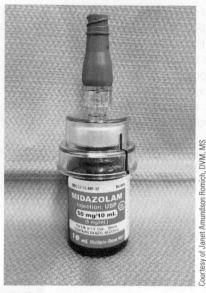
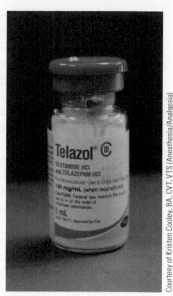

(a)   (b)   (c)

FIGURE 5-8 Benzodiazepines. (a) Diazepam, a benzodiazepine, is often mixed in the same syringe with ketamine, a dissociative, for anesthetic induction. (b) Midazolam is water soluble and can be injected and absorbed IV, IM, or SQ without causing pain or irritation. (c) Telazol® contains zolazepam, a benzodiazepine, and tiletamine, a dissociative.

flumazenil (Romazicon®, generic); however, reversal is seldom needed due to their limited adverse effects.

- *Diazepam,* which is commonly known by its human trade name Valium®, is lipid soluble (therefore, insoluble in water), which allows it to cross the blood-brain barrier. Diazepam is a C-IV controlled substance that has very unpredictable absorption when injected IM or SQ. Diazepam can cause pain and irritation when injected IV because the commercial product has propylene glycol as a carrier. Diazepam may be mixed in the same syringe with ketamine for anesthetic induction (Figure 5-8a); however, the propylene glycol in the commercial product limits its ability to be mixed in the same syringe with most other drugs.
- *Midazolam* (Versed®) is a water-soluble, C-IV controlled substance that can be injected and absorbed IV, IM, or SQ without causing pain or irritation (Figure 5-8b).[5] Midazolam is metabolized in the liver; however, its metabolites are inactive so it is shorter acting than diazepam with less risk of the **cumulative** effect (administration of the drug may produce effects that are more pronounced than those produced by the first dose).[5]
- *Zolazepam* (Telazol®, Tilzolan®) is a benzodiazepine that is only in the commercially available products called Telazol® and Tilzolan®, which are C-III controlled substances because of the presence of the tiletamine in these products (Figure 5-8c). (Telazol® and Tilzolan® are further described in the Dissociative Anesthetics section of this chapter.)

Guaifenesin

Guaifenesin (mostly made by compounding pharmacies, Guailazin®, Gecolate®), also known as glyceryl guaiacolate (GG), is a sedative and skeletal muscle relaxant that blocks nerve impulses in the CNS, which also makes it a good expectorant (drug that enables or increases coughing-up and swallowing of material from the lungs). Guaifenesin is not an anesthetic, analgesic, or controlled substance, but is used in large animals when combined with an anesthetic agent to facilitate endotracheal intubation and ease induction and recovery. Guaifenesin is available in a 5% or 10% solution; more concentrated solutions may cause hemolysis of red blood cells, and their use should be avoided. Guaifenesin's duration of action is short (approximately 10–20 minutes) in horses and ruminants and is typically given as a onetime IV bolus to induce general anesthesia or administered with other anesthetic agents via CRI to provide IV anesthesia.

Tranquilizers and sedatives are summarized in Table 5-1.

Anesthetic Induction Agents

Anesthesia induction agents are drugs used to induce or "bring about" anesthesia. Induction agents are typically used to facilitate the passing of an endotracheal tube so that inhalant anesthetic agents can be given to maintain general anesthesia.

Dissociative Anesthetics

Dissociative anesthetics cause **catalepsy** (condition characterized by a loss of sensation and consciousness accompanied by rigidity of the body), amnesia, and mild analgesia by altering neurotransmitter activity. They provide analgesia by inhibiting the excitatory NMDA (n-methyl-D-aspartate) receptor in the CNS that is responsible for central sensitization and produces anesthesia and analgesia by blocking the excitatory neurotransmitter glutamate;[8] they do not relieve deep pain. Dissociatives are C-III controlled substances used as induction agents when given with sedatives or tranquilizers that provide

TABLE 5-1 Tranquilizers and Sedatives Described in this Chapter

Drug Class	Example(s)	Mechanism of Action and Effects	Adverse Effects	Uses
Phenothiazine	acepromazine (Promace® and generic)	• Blocks dopamine receptors producing calming and anxiolytic effect • Antiemetic (depresses vomiting center) • Potential anti-arrhythmic • Causes vasodilation (via alpha-1 antagonist effect) • Provides no analgesia • Non-reversible	• Vasodilation may produce hypothermia and hypotension • Rare priapism (horses) • Protrusion of third eyelid (dogs/cats)	Dogs/cats: • Tranquilizer for restraint/transport • Sedation prior to procedures • Reduces amount of induction and inhalant anesthetic agents Horses/cattle: • Tranquilizer to help anesthetic recovery and aid in handling
Alpha-2 Adrenergic Agonists*	• xylazine (Rompun®, AnaSed®, Sedazine®, and Chanazine®) • detomidine (Dormosedan®) • dexmedtomidine (Dexdomitor®) • romifidine (Sedivet®) • medetomidine (Domitor®)	• Inhibits release of norepinephrine to produce sedation • Provides analgesia for minor procedures and diagnostics • Reduces amount of induction agent needed • Reversible	• Profound cardiovascular effects (vasoconstriction leads to ↑ blood pressure which results in a reflex bradycardia; over time, the cardiac output ↓ which leads to ↓ tissue oxygenation) • Not for use in animals with esophageal foreign body, intestinal torsion, bloat, head trauma, and/or intraocular pressure because they frequently cause vomiting • Cattle are extremely sensitive to xylazine • Decreased tear production (dexmedtomidine)	Dogs/cats: • Reliable dosage-dependent sedation, analgesia, muscle relaxation • Unexpected arousal to stimuli that can be very dangerous (e.g., dog reacts and bites someone on the face) • Emetic in cats Horses: • Produces deep and reliable sedation but can still respond to stimuli that can be very dangerous (e.g., horse suddenly kicks someone nearby) (xylazine & detomidine) • Produces analgesia in colic patients, allows skin suturing Cattle (used extra-label): • Allows surgical procedures such as laceration repair or C-section, but not invasive surgeries if used alone
Benzodiazepines**	• diazepam (Valium®; C-IV) • midazolam (Versed®; C-IV) • zolazepam (only in commercial products Telazol®, Tilzolan®); C-III)	• Increases the inhibitory brain neurotransmitter GABA • Minor cardiovascular and respiratory effects • Minimal renal effects • Controlled substance • No analgesia • Reversible	• Less reliable anxiolytic as paradoxical excitement is an adverse effect • Not approved for use in animals • Oral diazepam in cats has been linked to hepatic failure	• Provides mild to moderate skeletal muscle relaxation and amnesia • Effective anticonvulsant • Short term analgesia when combined with ketamine or tiletamine (cats)

(Continued)

TABLE 5–1 Tranquilizers and Sedatives Described in this Chapter (*Continued*)

Drug Class	Example(s)	Mechanism of Action and Effects	Adverse Effects	Uses
Guaifenesin	Also known as GG or glycerol guaiacolate (Guailazin®, Gecolate® may need to be compounded based on availability)	• Centrally acting muscle relaxant • Sedative and skeletal muscle relaxant • Expectorant • No analgesia	• More concentrated solutions (>5%) may cause RBC hemolysis • Can be challenging to source, pharmacy often has to compound it	Large animal • Combined with anesthetic agent to facilitate endotracheal intubation and ease of induction and recovery • Anesthesia induction agent (in combination with drugs such as ketamine) • Provides IV anesthesia when given via CRI (in combination with other drugs)

Note: Veterinary drug trade names are in purple and human drug trade names are in red.

*Alpha-2 adrenergic agonists can be reversed with atipamezole (Antisedan®), yohimbine (Yobine®, Antagonil®), and/or tolazoline (Tolazine®).

**Benzodiazepines can be reversed with flumazenil (Romazicon®, generic).

muscle relaxation, for restraint and sedation, and for analgesia via CRI. They stimulate the CNS; therefore, they are generally given to patients already well-sedated (especially in large animals to protect staff from injury). Patients under dissociative anesthesia appear awake (eyes open, palpebral and laryngeal reflexes intact) but cannot move and are oblivious to their surroundings (unconscious). Ocular lubricants are applied to animals receiving dissociatives to prevent corneal ulcers from forming if the eyelids remain open. During induction and recovery, tremors, spasticity, and convulsions may occur if high dosages are given or they are used without concurrent muscle relaxants or CNS depressants.

Tech Tip

Muscle rigidity, nystagmus (involuntary eye movement), and movement independent of surgical stimuli are commonly seen with dissociative anesthesia.

As a result of sympathetic nervous system stimulation, dissociative anesthetics increase heart rate and blood pressure which leads to an increase in cardiac work and myocardial oxygen demand. In critical patients and those with a depleted sympathetic reserve, dissociatives can decrease blood pressure and cardiac output. These drugs cause minimal respiratory depression, but may cause an apneustic breathing pattern (characterized by a prolonged inspiratory breath, a pause after inspiration and a short, insufficient expiration) especially in cats and horses. When used alone, dissociatives cause patients to maintain a central eye and keep many of their reflexes such as palpebral reflexes. These, however, are abolished in the face of other concurrent drugs such as isoflurane. Rough anesthetic recovery and **emergence delirium** (state in which the patient is agitated and inconsolable, uncooperative and typically thrashing, vocalizing and defecating) can occur when dissociatives are used alone. Emergence delirium is generally self-limiting in 5 to 15 minutes, but may require drug intervention to keep the patient from harming itself or others.

Dissociatives may increase intracranial pressure (ICP) and should not be used in high dosages as a sole agent in patients with a seizure history or history of head trauma or brain lesions/tumors. Dissociatives should also be avoided in patients with ocular disease such as a desmetocele (a corneal ulcer eroded through the stroma leaving only Descemet's membrane), which increases the risk of rupture or glaucoma, which increases the intraocular pressure.

The common dissociates used in veterinary medicine are ketamine and tiletamine. Ketamine (100 mg/mL) is often coupled with a sedative or tranquilizer. Ketamine and tiletamine/zolazepam (Telazol®, Tilzolan®) are metabolized in the liver of dogs and their use should be avoided in dogs with liver failure due to prolonged drug effects. In cats, dissociatives are excreted virtually unchanged in the urine and should be used cautiously in cats with renal disease or insufficiency due to prolonged drug effects.

Examples of dissociatives include the following:

- *Ketamine* (Ketaset®, VetaKet®, Ketaject®, Vetalar®) is a C-III controlled substance usually used in combination with acepromazine, xylazine, and/or diazepam to provide muscle relaxation and deepen anesthesia (see Figure 5-8a and Figure 9-10b). It is given IM, IV, and SQ. Pain at the injection site is frequently noted with ketamine due to its low pH. Ketamine is absorbed more slowly than other induction drugs given IV, but is rapidly absorbed (approximately 5 minutes) via IM and SQ routes. It is approved for cats and primates and is used extra-label in other species. Ketamine should not be given to cats with hypertrophic cardiomyopathy because it increases cardiac output, heart rate, mean aortic pressure, and pulmonary artery pressure. These cardiovascular parameter increases in turn increase myocardial work and oxygen consumption and potentially cause ventricular arrhythmias in cats. Ketamine provides somatic analgesia at subanesthetic dosages given via CRI.

- *Tiletamine* (in proprietary 50/50 mixtures with the benzodiazepine zolazepam in the trade name drugs Telazol® and Tilzolan®) is a C-III controlled substance manufactured as an injectable anesthetic approved for dogs and cats (see Figure 5-8c). Today it is most commonly used in the shelter medicine setting for high quality high volume (HQHV) spay/neuter programs and extra-label in zoo and wildlife medicine. Pain at the injection site is frequently noted with Telazol® and Tilzolan® use, due to their low pH.

- *Ketamine-diazepam mixtures* are prepared in clinics and used for IV induction (see Figure 9-10a). Induction with ketamine/diazepam is slower than induction with propofol, and patients will retain more muscle and jaw tone; therefore, the anesthetist needs to be patient to avoid over-dosing. Ketamine takes 2 minutes to reach peak effect and lasts approximately 20 minutes. It may precipitate if stored for more than one week; therefore, it should be mixed in a syringe as needed.

Tech Tip

Unlike other general anesthetics, dissociatives cause CNS stimulation.

Propofol

Propofol (Rapinovet®, PropoFlo™, PropoFlo™28, Propovan®) is a commonly used, short-acting IV hypnotic sedative that binds to and stimulates the inhibitory neurotransmitter GABA. Propofol is a lipophilic, **non-cumulative** drug (does not have more pronounced effects with serial doses than produced by the first dose) that provides a rapid, smooth induction and a short duration of action making its use ideal for outpatient anesthesia. It has a rapid onset (90 seconds to reach peak effect) because well-perfused tissues, like the brain, rapidly receive high concentrations of propofol when it is given by IV bolus. Its short duration of action (5 to 10 minutes) is due to its redistribution to muscle and fat and subsequent rapid metabolism. Redistribution results in a lower concentration of propofol in the blood than its concentration in the brain, which causes the drug to move from the brain to the bloodstream. The drug in the bloodstream then goes to poorly-perfused fat, which becomes a reservoir for propofol. As propofol continues to be redistributed from the brain to the bloodstream to the fat, the animal regains consciousness. Since propofol equilibrates between the CNS and plasma in approximately 2 minutes, an injection time of 2 minutes is recommended for induction doses to titrate to effect in order to avoid overdose, apnea, and cardiovascular effects. Propofol provides no analgesia while providing good muscle relaxation, but may cause pain on IV injection. At the time of publication, propofol is not a controlled substance.

Propofol is highly protein bound and unique in that it is a milky white emulsion of drug, soybean oil, egg lectin, and glycerol (Figure 5-9). Its composition makes it prone to bacterial growth and once opened should be handled aseptically to minimize bacterial contamination.[2] Propofol has a shelf life of only 6 to 8 hours; however, a similar preparation called Propfol-28 contains benzyl alcohol to retard bacterial growth, which prolongs its shelf life to 28 days. Propofol™ 28 is currently approved for use in dogs; in cats, high doses of benzyl alcohol can be toxic and may cause Heinz body anemia if repeated doses are given. In dogs, propofol can also be used for total intravenous anesthesia (TIVA), CRI (must be used within 6 hours if not using Propofol™ 28), and redosed via low-dose injections for procedures such as laceration repair, minor biopsies, and endoscopy. Coadministration of drugs such as lidocaine, diazepam, opioids, etc., can help facilitate endotracheal intubation and minimize adverse effects.

Tech Tip

Repeated administration of propofol over time could lead to high levels of drug in the fat, due to its redistribution from the brain, through the bloodstream, to body fat. As fat stores become saturated with propofol, subsequent doses of the drug may have a longer duration of action than initial doses, because the fat cannot quickly store more of the drug. This phenomenon may keep the animal, especially overweight ones given lipophilic drugs such as propofol, anesthetized too long. The concept of redistribution also helps explain propofol's adverse effect of apnea. One concern with administration of propofol is that the patient stops breathing if the drug is administered too rapidly due to its depressive effects on the respiratory centers in the brain stem. This propofol-induced apnea reverses in a few minutes after administration as propofol redistributes to poorly perfused tissues. One way to avoid apnea when administering propofol is to give it slowly intravenously.

Propofol causes short-lived, but profound adverse cardiovascular effects. A dosage-dependent decrease in systemic vascular resistance may cause vasodilation, and a decrease in cardiac contractility may lead to decreased cardiac output and hypotension. Heart rate generally is maintained and sensitization of the myocardium to arrhythmias does not occur with propofol. Propofol can cause significant respiratory depression (hypoventilation and apnea) and cyanosis that is more pronounced when

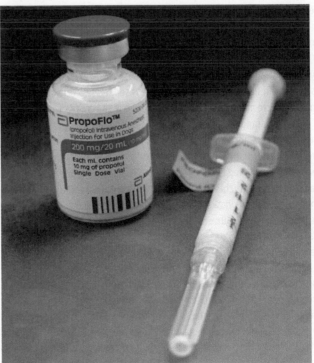

FIGURE 5-9 Propofol is an IV hypnotic sedative used as an induction agent.

Courtesy of Kristen Cooley, BA, CVT, VTS (Anesthesia/Analgesia)

the drug is administered rapidly or in large doses.[3] Being able to intubate and ventilate a patient following administration of propofol is critical due to its adverse respiratory effects. Animals given propofol must be monitored closely, supported (assist respirations to avoid hypoxemia), and concurrent oxygen administration is recommended. Muscle twitching and myoclonus (sudden, involuntary jerking of a muscle or group of muscles), which typically resolves once the drug is redistributed (around 5 minutes), may occur in patients that have not been premedicated.[9] Propofol should be avoided in patients with lipid metabolism disorders (e.g., hyperlipidemia or pancreatitis) because it could potentially be aggravated by the lipid emulsion formulation of propofol after long-term infusions.[10] Because it causes a decrease in intracranial pressure, it is a good choice for animals with head trauma or intracranial disease. It has minimal adverse effects on the fetus; therefore, it is considered safe for use in C-sections.

A mixture of propofol and ketamine has been used as a co-induction technique in dogs. "Ketofol" appears to improve the patient's blood pressure, cardiac output, and oxygen delivery to tissue while improving the quality of induction without altering the quality of recovery.[11] Despite cardiac advantages to "ketofol," this drug combination produces more profound respiratory depression.

Etomidate

Etomidate (Amidate®) is a short-acting injectable induction agent that works by modulating the action of the inhibitory neurotransmitter GABA. It is rapidly metabolized by the liver and **plasma esterases** (enzymes found in the plasma that cleave apart compounds and make them inactive), leading to a quick recovery. It is not currently a controlled substance, nor does it provide analgesia. Etomidate is prepared in propylene glycol, which may cause pain on injection (Figure 5-10). It penetrates the blood-brain barrier quickly and peak brain concentrations are reached within 1 minute.[12] Etomidate can produce 10 to 20 minutes of anesthesia, and doses should be calculated based on lean body weight with the entire calculated dose given as a bolus or slow IV push to well-sedated patients. Single injections produce relatively brief hypnosis, and the duration of hypnosis is dosage related. Etomidate is not used as a CRI for prolonged TIVA because it suppresses cortisol production and may cause hemolysis if propylene glycol is in the preparation.

An advantage to using etomidate is that it produces minimal cardiovascular effects and does not alter heart rate, cardiac contractility, blood pressure, or cardiac output. Etomidate causes mild respiratory depression that is not considered clinically significant. It is a great choice for the induction of high-risk patients including those with preexisting cardiac conditions and hepatic disease; however, there is a potential for acute hemolysis from its propylene glycol carrier if the drug is given in large volumes or is given via infusion.[13] Adrenal cortex suppression for 2 to 6 hours after induction has been seen in dogs given etomidate, which is why it is not used for CRI.[2] It is best to avoid the use of etomidate in patients with questionable adrenal corticosteroid levels or those undergoing an adrenalectomy because this drug can inhibit steroid production for up to 4 to 6 hours post-operatively. Etomidate is a good choice for patients with neurologic issues as it reduces cerebral metabolic oxygen requirements and cerebral blood flow, decreases intracranial and intraocular pressure, and has anticonvulsant properties. Retching, vomiting, excitement, and myclonus can occur during induction and is more common in patients that have not been adequately premedicated or those receiving an inadequate dose of etomidate.

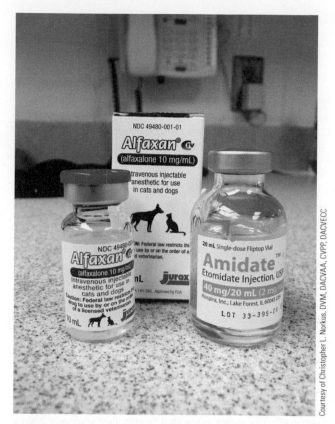

FIGURE 5-10 Etomidate and alfaxalone are induction agents.

Alfaxalone

Alfaxalone (Alfaxan®, Alfaxan®Multidose, Alfaxan®Multidose IDX) is a C-IV neuroactive steroid used to induce anesthesia by producing hypnosis and muscle relaxation through modulation of the inhibitory effect of GABA (see Figure 5-10). Alfaxalone provides rapid induction and hypnosis with reasonable muscle relaxation.[14] Alfaxalone is not an analgesic, is non-cumulative, and does not cause histamine release. Alfaxalone is rapidly metabolized by the liver resulting in a short duration of action, and hepatic disease will prolong anesthetic time.

Alfaxalone produces minimal cardiovascular and respiratory effects when given at clinical dosages; however, at very high dosages hypotension, decreased systemic vascular resistance, tachycardia, hypoventilation, and apnea can be seen.[14] Being able to intubate and ventilate patients given alfaxalone, as well as providing oxygen support, are necessary when administering this drug. The dose for alfaxalone should be calculated based on patient lean body weight and is given IV (bolus or CRI); in some countries it is labeled for IM use in cats (extra-label use in United States). Perivascular injection is not painful and does not produce tissue necrosis. In the United States, alfaxalone was only available in a formulation without preservatives and had a shelf life of only 6 hours; however, a multidose product containing preservatives is available and has a shelf life of up to 28 days after first use (see Figure 9-13). Alfaxan®Multidose IDX is labeled as an injectable sedative and anesthetic for non-food producing minor species such as reptiles (lizards, snakes, turtles, and tortoises), amphibians (toads), fish, birds, non-human primates (lemurs, marmosets, macaques), rodents (mice, rats, Guinea pigs), and ferrets (see Figure 9-13).

Barbiturates

Barbiturates are CNS depressants used mainly as anticonvulsants, anesthetics, and euthanasia solutions. They are easy and inexpensive to administer; however, they can cause potent cardiovascular and respiratory depression. They are highly protein-bound drugs, and plasma proteins can serve as reservoirs for these drugs. Acidotic animals (e.g., animals in shock) show less binding of barbiturates to plasma proteins. In animals with hypoproteinemia and acidosis, barbiturate dosages must be decreased to avoid adverse effects associated with overdosing.

Barbiturates are classified according to their duration of action: long-acting, short-acting, or ultrashort-acting. Thiobarbiturates are used for anesthesia and are not currently available in the United States, but may be used in other countries. Thiobarbiturates have an ultrashort duration of activity because they are very fat-soluble and move out of the CNS rapidly to fat stores within the body.

- *Thiopental* (Pentothal®) is an ultrashort-acting (duration of action is short 5 to 30 minutes) thiobarbiturate available in a vial as a sterile powder that must be reconstituted with sterile water for use. Thiopental is a C-III controlled substance that is used as an anesthetic induction agent and can only be given IV because it is very alkaline. If thiopental accidentally is injected peri-vascularly, severe inflammation, tissue swelling, and tissue necrosis may result. Care must be taken when administering thiobarbiturates to thin animals (such as sighthounds), because they have minimal fat to redistribute the drug to and they are deficient in enzymes needed to metabolize barbiturates. Following IV injection, thiobarbiturates rapidly enter the CNS (highly perfused) and then redistribute to fat (poorly perfused). When the drug redistributes to fat, the animal may not appear anesthetized and another dose may be given. For this reason, thiobarbiturates given IV are not re-dosed as this will prolong recovery times and are dosed based on lean body weight. Thiobarbiturates can also cause apnea if given too fast (resulting in the need for assisted ventilation) and CNS excitement if given too slowly. Typically animals are given one-third to one-half of the calculated dose rapidly and then the rest is given to effect. Adverse effects include cardiac arrhythmias, hypotension, and transient apnea.

Induction agents are summarized in Table 5-2.

Inhalant Anesthetic Agents

Inhalant anesthetic agents (also called **volatile anesthetics** or gas anesthetics) are usually liquids at room temperature, with the exception of desflurane, which boils at room temperature. The liquid form of the drug is vaporized in the presence of oxygen or a carrier gas and is delivered to a patient via the respiratory tract. Inhalant anesthetic agents are used to induce and maintain general anesthesia in a variety of species (Figure 5-11).

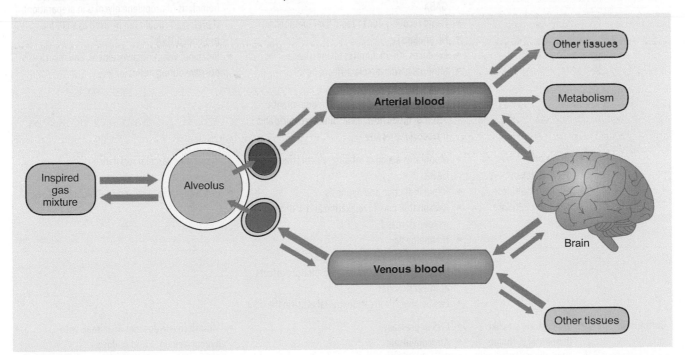

FIGURE 5–11 Inhalant anesthetic agents travel as a gas in the respiratory system until they reach the alveoli of the lung, where they diffuse across the alveolar membrane. The concentration of inhalant anesthetic agent is high in the alveoli and low in the blood during induction; therefore, there is rapid diffusion of inhalant anesthetic agent into the blood. The inhalant anesthetic agent travels from the blood to the areas of the body that are well perfused (like the brain). Diffusion occurs in the opposite direction when the vaporizer is shut off after a procedure. The concentration of inhalant anesthetic agent in the alveoli goes down as the vaporizer is turned off. Since the concentration of inhalant anesthetic agent is high in the blood, as the vaporizer is turned off, there is diffusion of particles from the blood into the alveoli. The inhalant anesthetic agent is expired from the alveoli. If an animal continues to receive pure oxygen when the vaporizer is turned off, diffusion will happen more rapidly because of the greater difference in inhalant anesthetic agent concentration in the blood versus the alveoli.

TABLE 5-2 Induction Agents Described in this Chapter

Drug Class	Example(s)	Mechanism of Action and Effect	Adverse Effects
Dissociatives	• ketamine (Ketaset®, VetaKet®, Ketaject®, Vetalar®; C-III) • tiletamine (Telazol®, Tilzolan®; C-III)	• NMDA receptor antagonist in the CNS • Causes muscle rigidity and analgesia • Usually given to sedated patients because dissociatives stimulate the CNS • Not recommended for cats with hypertrophic cardiomyopathy or patients with seizure history, head trauma, brain lesion (may increase intracranial pressure), some types of ocular diseases (e.g., desmetocole; may increase risk of rupture), or glaucoma (may increase intraocular pressure) • Only approved for use in cats and primates	• Corneal ulceration possible because eyes remain open (need to apply ocular lubricants) • Monitoring anesthetic depth is challenging because eyes remain open, palpebral and swallowing reflexes remain intact, and jaw and muscle tone is retained • Tremors, spasticity, and convulsions at high dosages during induction and recovery • Apneustic breathing • Rough anesthetic recovery and emergence delirium if used alone • Pain at injection site
Non-barbiturates	• propofol (Rapinovet®, PropoFlo™, Propovan®) • propofol with benzyl alcohol (PropoFlo™28)	• Stimulates the inhibitory brain neurotransmitter GABA • No analgesia • Rapid, smooth induction and short duration of action	• Titrate to effect to prevent overdose due to its redistribution to muscle and fat and rapid metabolism • Profound cardiovascular effects (hypotension, ↓ cardiac output) • Respiratory depression, especially transient apnea after induction • Lipid emulsion that should be avoided in patients with lipid metabolism disorders (e.g., pancreatitis, diabetic hyperelipidemia)
	• etomidate (Amidate®)	• Modulates action of inhibitory neurotransmitter GABA • Rapid recovery due to rapid liver metabolism • No analgesia • Produces 10–20 minutes of anesthesia • Minimal cardiovascular effects • Mild respiratory depression • ↓ cerebral metabolic oxygen requirements, cerebral blood flow, intracranial pressure, and intraocular pressure	• Suppresses cortisol production and may cause hemolysis (if propylene glycol is in preparation); therefore, it should not be used as a CRI for prolonged TIVA • Retching, vomiting, excitement, and myoclonus possible during induction
	• alfaxalone (C-IV) (Alfaxan®, Alfaxan®Multidose, Alfaxan®Multidose IDX)	• Modulates action of inhibitory neurotransmitter GABA • Rapid induction and hypnosis • Reasonable muscle relaxation and ↓ in cerebral oxygen demand • No analgesia • Rapidly metabolized by the liver • Minimal cardiovascular and respiratory effects at lower dosages • Can be given IM in cats (extra-label use in the U.S.)	• ↓ systemic vascular resistance • Tachycardia • Hypoventilation and apnea at higher dosages
Barbiturates	• thiobarbiturates like thiopental (C-III) are ultra-short acting; not currently available in the U.S.	• CNS depressant • Anticonvulsant • Anesthetic • Euthanasia agent • Highly protein bound (plasma proteins can serve as reservoirs) • Calculate based on lean body weight • No analgesia	• Must decrease dosages in animals with hypoproteinemia and acidosis • Apnea if given rapidly • Cardiac arrhythmia • Hypotension • Tissue sloughing if given perivascularly

They do not provide analgesia and should never be used as the sole drug for painful procedures. These drugs are halogenated hydrocarbons (carbon and hydrogen based molecules) that have fluorine, chlorine, bromine, and/or iodine attached to them. Halogenation of the molecule increases their potency and reduces flammability to a point that they are considered non-flammable. Isoflurane and sevoflurane are the halogenated anesthetics used most frequently in veterinary medicine.

Personnel using inhalant anesthetic agents are at risk for exposure to waste anesthetic gas (WAG), which can lead to health concerns such as feelings of light headiness, nausea, fatigue, headache, irritability, and depression. WAG can also lead to liver and kidney disease. Exposed workers can experience difficulty with cognitive, perception, judgment, and motor skills. The effects of WAG on female reproductive health is also a concern especially with the number of women who work in the veterinary profession.

Inhalant anesthetic agents can be administered via endotracheal tube (ETT), supraglottic airway device, fitted facemask, or induction chamber. The mechanism of action of inhalant anesthetic agents is not entirely understood; however, they may disrupt normal nerve transmission by interfering with the release or re-uptake of neurotransmitters. The various inhalant anesthetic agents are compared to each other based on their properties of vapor pressure, minimum alveolar concentration, and blood-gas partition coefficient (Table 5-3).

- *Vapor pressure* is the measure of the tendency for the molecules in the liquid state to enter the gaseous state. Vapor pressure determines how easily the anesthetic liquid evaporates and is both drug- and temperature-dependent. Drugs with high vapor pressures want to exist as a gas and are described as volatile. Because drugs with high vapor pressures evaporate readily, if they are not controlled, they will produce an extremely high concentration that could be fatal to the patient. Inhalant anesthetic agents with high vapor pressures (isoflurane, sevoflurane, desflurane) are usually delivered via a precision vaporizer to control the amount of inhalant anesthetic agent delivered to prevent overdosing the patient.
- *Minimum alveolar concentration* (MAC) is the lowest alveolar concentration of inhalant anesthetic agent required to prevent movement in response to noxious stimuli in 50% of patients and is used to compare the relative "potency" of inhalant anesthetic agents. Inhalant anesthetic agents with a lower MAC value are more potent, and those with a higher MAC value are less potent. Isoflurane has a lower MAC than sevoflurane; therefore, it is more potent than sevoflurane, making it necessary to provide the patient with a higher concentration or percent of sevoflurane to get the animal under anesthesia (lower MAC = more potent anesthetic = less anesthetic needed to produce the effect; higher MAC = less potent anesthetic = more anesthetic needed to produce the effect). **Surgical MAC** is the concentration of drug necessary to keep 95% of patients immobile during surgical stimulation and is calculated by multiplying 1.5 X the MAC value.
- *Gas solubility* in blood is expressed as the *blood-gas partition coefficient* (BGPC) and was described in the Drug Absorption section of this chapter.

All modern inhalant anesthetic agents produce dosage-dependent cardiovascular and respiratory depression including hypotension, hypoventilation, and potential apnea at higher concentrations. Cardiac output and heart rate are generally unaltered when using these agents at clinically relevant concentrations and in healthy adult patients; however, at higher dosages tachycardia may be observed as a reduction in cardiac output is partially compensated by an increase in heart rate. All inhalant anesthetic agents increase cerebral blood flow, raise intracranial pressure, reduce blood flow to the kidney, and reduce glomerular filtration rate. Since inhalant anesthetic agents are primarily excreted through the lungs, there is little liver or kidney metabolism, and they are a suitable selection for patients with hepatic or renal disease. Emergence delirium is seen with all inhalant anesthetic agents; therefore, using a tranquilizer or sedative prior to induction can smooth recovery.

Nonflammable, halogenated, inhalant anesthetic agents include the following:

- *Isoflurane* (IsoFlo®, Isosol®, Aerrane®) produces rapid induction of anesthesia and does not cause cardiac arrhythmias; however, it can cause respiratory depression and can trigger malignant hyperthermia (Figure 5-12a). Animals receiving isoflurane change planes of anesthesia rapidly and have very short recovery times. Mask induction of an animal with isoflurane and related inhalant anesthetic agents may be difficult because they irritate the respiratory system.

TABLE 5-3 Select Properties of Inhalant Anesthetic Agents

Inhalant Anesthetic Agent	Vapor Pressure (mmHg) at 20°C (68°F)	Blood-Gas Solubility Coefficient	MAC	Induction Setting for Dogs and Cats*	Maintenance Setting for Dogs and Cats*
isoflurane (IsoFlo®, Isosol®, Aerrane®)	240	1.41	1.3–1.5% (dog) 1.3–2.2% (cat)[15]	3%**	1–3%**
sevoflurane (SevoFlo®, Petrem®, Flurovess™ Ultane®)	160	0.69	2.1–2.3% (dog) 2.6–3.4% (cat)	4–5%**	2–4%**
desflurane (Suprane®)	664–700	0.42	7.2–10.3% (dog) 9.8–10.2% (cat)		

*Based on a fresh-gas flow rate of 1–2 L/min during the induction phase (i.e., the first several minutes following induction with an injectable drug) and a fresh-gas flow rate of at least 10 mL/kg/min during the maintenance phase. Vaporizer settings for closed breathing systems are typically 1–2% higher.

**Bednarski, R., Anesthesia and Analgesia for Domestic Species, In Kurt A. Grimm, Leigh A. Lamont, William J. Tranquilli, Stephen A. Greene, Sheilah A. Robertson, Veterinary Anesthesia and Analgesia, 5th edition of Lumb and Jones, Wiley & Sons, p. 824.

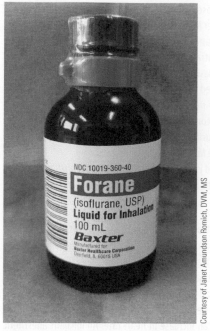

(a)

Courtesy of Janet Amundson Romich, DVM, MS

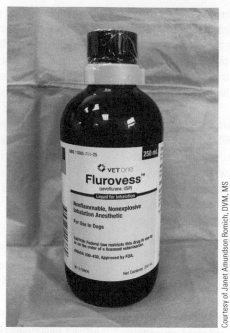

(b)

Courtesy of Janet Amundson Romich, DVM, MS

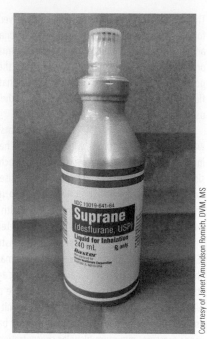

(c)

Courtesy of Janet Amundson Romich, DVM, MS

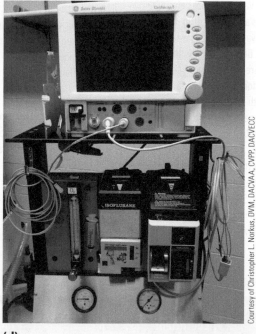

(d)

Courtesy of Christopher L. Norkus, DVM, DACVAA, CVPP, DACVECC

FIGURE 5–12 Nonflammable, halogenated, inhalant anesthetic agents include (a) isoflurane, (b) sevoflurane, and (c) desflurane. (d) An anesthesia machine with an isoflurane vaporizer. Note the purple stripe on the isoflurane vaporizer matches the purple on the isoflurane bottle neck and the blue stripe on the desflurane vaporizer matches the blue on the desflurane bottle label.

- *Sevoflurane* (SevoFlo®, Petrem®, Flurovess™, Ultane®), like isoflurane, produces rapid induction and rapid recoveries making diligent monitoring of animals receiving sevoflurane important (Figure 5-12b). It is also quickly eliminated making it a good choice for patients undergoing C-sections because any sevoflurane absorbed by the fetus is quickly eliminated. Sevoflurane has low tissue solubility resulting in rapid elimination of the drug by the body and rapid awakening making the judicious use of tranquilizers important. Sevoflurane quickly enters systemic circulation, rapidly distributes to the brain, may produce less injury to lung function than isoflurane, and is less irritating to mucous membranes, therefore, is the preferred inhalant anesthetic agent for mask and chamber inductions.
- *Desflurane* (Suprane®) has a very low blood-gas partition coefficient that produces very rapid induction and recovery times (twice as fast as isoflurane) (Figure 5-12c). Desflurane is the least potent inhalant anesthetic agent; therefore, higher vaporizer settings are needed

when administering this agent. Desflurane exists as a vapor at room temperature due to its low boiling point and requires a special heated vaporizer to keep it under pressure to prevent it from boiling. The cost of desflurane and its specialized vaporizer has limited its use in veterinary medicine.

The inhalant anesthetic agents are summarized in Table 5-4.

Safety Alert

Labels on bottles of inhalant anesthetic agents are color coded for safety. The color code on the label of the drug bottle matches the color on the vaporizer so that the anesthetic agent is not poured into the wrong vaporizer (Figure 5-12d).

TABLE 5-4 Summary of Inhalant Anesthetic Properties Described in this Chapter

Inhalant Anesthetic Agent	Effects	Advantages	Disadvantages
isoflurane (IsoFlo®, Isosol®, Aerrane®)	• CNS, respiratory, and cardiovascular depression • Decreased renal blood flow • Excellent muscle relaxant • Fetal depression (crosses placenta)	• Fast induction and recovery • Absorbed and eliminated virtually unchanged by alveoli • Potent anesthetic • Can be used for mask induction (has unpleasant odor that may irritate lung) • Does not cause cardiac arrhythmias	• Emergence delirium possible • Pungent odor • Can trigger malignant hyperthermia • No analgesia
sevoflurane (SevoFlo®, Petrem®, Flurovess™ Ultane®)	• CNS, respiratory, and cardiovascular depression • Decreased renal blood flow • Excellent muscle relaxant	• Faster induction than isoflurane • Fast recovery • Absorbed and eliminated virtually unchanged by alveoli • Good choice for C-section patients due to rapid elimination; however, still causes fetal depression because it crosses the placenta • Potent anesthetic • Can be used for mask and chamber induction because it quickly enters blood and travels to brain and is less irritating to mucous membranes than the other inhalant anesthetic agents	• Emergence delirium possible • Can trigger malignant hyperthermia • Requires higher vaporizer settings than isoflurane • No analgesia
desflurane (Suprane®)	• CNS, respiratory, and cardiovascular depression • Decreased renal blood flow • Excellent muscle relaxant	• Extremely rapid recovery; much more rapid than isoflurane and sevoflurane	• Emergence delirium possible • Can trigger malignant hyperthermia • May cause tachycardia and airway irritability • Needs special heated vaporizer with an electrical source because it has a low boiling point (can be in the vapor phase at room temperature) • Expensive • No analgesia • Extremely pungent odor and is irritating to airways

Local Anesthetic Drugs

Local anesthetics act directly on nervous tissue by blocking nerve conduction from the site of administration or application in the peripheral nervous system (PNS) and spinal cord. Local anesthetics inhibit sodium influx into the nerve cell membrane, which prevents action potentials and depolarization of the nerve cell. They are given by local infiltration of tissue, via epidural or intrathecal (into the subarachnoid space) injection, by application to the corneal or mucous membrane surface, or by regional nerve blocks (Table 5-5). Local anesthetics can be used to localize lesions in equine lameness exams, as nerve blocks in cattle to allow surgical and medical procedures, to aide endotracheal tube placement in cats and pigs by preventing laryngospasm, and to prevent or decrease pain both during and after surgery. They can also be used to perform surgery in a conscious patient, to relieve the pain caused by the original trauma, or to allow minimally invasive surgical procedures such as suturing minor wounds or taking a biopsy.

Tech Tip

Lidocaine, bupivacaine, and ropivacaine are amides that undergo extensive metabolism in the liver, making hepatic blood flow important in the clearance of these drugs. Careful selection of dosages of a local anesthetic should be part of a balanced anesthetic plan, especially for a patient with cardiovascular and liver disease.

There are many different local anesthetic agents available including lidocaine, bupivacaine, ropivacaine, mepivacaine, proparacaine, and prilocaine. Most local anesthetics have names that end in "caine" and are divided into two main classes, the amides and the esters.

TABLE 5–5 Local Anesthetic Uses

Type	Description	Benefit	Use	Drug Example
Infiltrative anesthesia	Small amounts of anesthetic solution are injected into the tissue surrounding the site to be worked on (surgical site, wound repair site, etc.)	Because small amounts of anesthetic are used, there is reduced danger of systemic adverse effects	• Wound suturing • Wound debriding • Skin biopsies	• lidocaine (Xylocaine®, Lidocaine® for injection 1% and 2%) • ropivacaine (Naropin®, LEA 103®) • bupivacaine (Marcaine®, Nocita®)
Topical anesthesia	Anesthetic agent is applied directly onto the surface of the skin or eye (tetracaine, proparacaine); also used to aid in diagnostic procedures and endotracheal intubation in cats (lidocaine)	Systemic absorption is limited from these sites	• Eye examination • Minor skin irritation • Larynx is sprayed or liquid is applied to prevent spasm during endotracheal intubation in cats	• tetracaine (Neo-Predef® with tetracaine, Altacaine®) • proparacaine (Ophthaine®, Ophthetic®, Alcaine® [in Europe this product's generic name is proxymetacaine]) • lidocaine
Nerve block anesthesia	Anesthetic solution is injected along the course of a nerve so that the area it innervates is desensitized	Ability to determine the source of pain Localization of pain relief	• Helps locate areas of injury • Provides local desensitization	• lidocaine • ropivacaine • bupivacaine • mepivacaine (Carbocaine®)
Line block anesthesia	A continuous line of local anesthetic is given SQ in the tissues proximal to the targeted area (such as an inverted L block in cattle) or where the incision is going to be made or to the side of the incision after closure in small animals	Allows for larger area of desensitization, but limits systemic effects	• Surgical procedures	• lidocaine • ropivacaine • bupivacaine
Regional (epidural)	Anesthetic agent is injected into a nerve plexus or area of the spinal cord	Provides adequate restraint and may prevent movement (can affect respiratory muscles if given in the cranial parts of the spinal cord)	• Surgeries like C-section, tail amputations, anal sac removal, and surgery of the rear limb	• lidocaine • ropivacaine • bupivacaine

Amides such as lidocaine, bupivacaine, ropivacaine, and mepivacaine are very stable, metabolized in the liver, and predominantly nonallergenic. Esters such as procaine and tetracaine have shorter durations of action and are highly allergenic, which limit their use as an anesthetic/analgesic. Keep in mind that onset and duration of action of local anesthetics depends on the volume administered, drug concentration, how close the placement of local anesthetic is in relation to the nerve, nerve involved, and species.

Toxic dosages (drug dosage at which a substance can damage an organism) of local anesthetics vary depending on the route of administration, the site of injection, and rate of absorption. Species variation also occurs; for example, cats are much more sensitive to the toxic effects of local anesthetics than dogs so their dosages should be reduced to ½ to ¾ of those for dogs.

With lidocaine, CNS toxicity is seen at much lower dosages than those required to produce cardiovascular toxicity. Signs of CNS toxicity include muscle twitching and seizures. Muscle twitching and seizures can be managed with diazepam. Bupivacaine will cause similar CNS signs as lidocaine and ropivacaine but these signs are much more closely followed by cardiac toxicity (arrhythmias and myocardial depression). Treatment of cardiac arrest is discussed in Chapter 13 (Cardiac Arrest and Cardiopulmonary Cerebral Resuscitation [CPCR]).

Safety Alert

To prevent the risk of toxicity, total doses of local anesthetic should be calculated carefully, based on lean body weight, and should not be exceeded.

- *Lidocaine* (Xylocaine®, Lidocaine® for injection 1% and 2%) has a rapid onset of action (10–15 minutes), but a short duration of action (1–2 hours). It is therefore useful for providing rapid intraoperative analgesia rather than postoperative analgesia. Some commercially available lidocaine products contain epinephrine to increase its duration of action. It is important that this product not be given IV as lidocaine is also used to treat cardiac arrhythmias. Cats are more sensitive to the adverse effects of lidocaine than dogs, so care should be taken to ensure dose rates are accurately calculated. Lidocaine is versatile and comes in a wide range of formulations (0.5–2% solution for injection, 2 to 4% for topical use, and a 5% transdermal patch) (Figures 5-13a and 5-13b). It can be used topically, infiltrated locally and perineurally, given IV for regional anesthesia, used as a transdermal patch, and administered for epidural and intrathecal blocks.

- *Bupivacaine* (Marcaine®) has a longer onset of action (20–30 minutes) than lidocaine and a long duration of action (4–6 hours). Bupivacaine's longer duration of action is due to its molecular weight, greater lipid solubility, and higher protein binding, which give it greater affinity for the sodium channel. The long duration of action makes it useful for intra- and post-operative analgesia in small animals; however, it has a much narrower therapeutic index (margin of safety) than lidocaine and ropivacaine, and the early warning signs of toxicity seen with other local anesthetics do not occur with bupivacaine before cardiovascular compromise. This makes correct dosing of bupivacaine crucial, and it should never be administered intravenously. Bupivacaine can be used for local, perineural, epidural, and intrathecal administration. It is available in different concentrations (0.25%, 0.5%, and 0.75%) and as an approved injectable suspension for dogs.

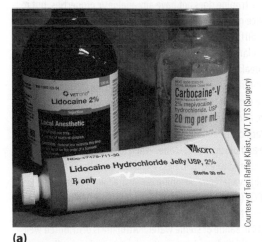  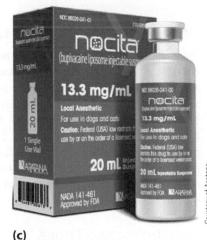

(a) **(b)** **(c)**

FIGURE 5-13 Local anesthetic agents. (a) Lidocaine and mepivacaine can be used as local anesthetics. (b) In cats and pigs, lidocaine can also sprayed bilaterally to the laryngeal cartilage to anesthetize the larynx and prevent laryngospasm. (c) Bupivacaine is available in a long-acting FDA-approved injection suspension for dogs.

Bupivacaine liposome (Nocita®) is an FDA-approved injectable suspension for use in dogs to provide local postoperative analgesia for cranial cruciate ligament surgery and in cats as a peripheral nerve block for regional postoperative analgesia following onychectomy (Figure 5-13c). It is used for postoperative analgesia for up to 72 hours after administration and is given as a single-dose infiltration dose into the tissues near the incisional site at the time of closure. The particles should be resuspended by inverting the vial and not shaking it. This product does not contain preservatives; therefore, it is in a single-use vial, and any remaining product should be discarded.

- *Ropivacaine* (Naropin®, LEA 103®) is structurally related to bupivacaine, has a similar onset of action (20 minutes) and duration of action (3–5 hours), and comparable selective sensory blockade. It also has similar indications to bupivacaine but a wider therapeutic index (margin of safety). It is available in different concentrations (0.25%, 0.5%, 0.75%, and 1.0%).

- *Proparacaine* (Ophthaine®, Ophthetic®, Alcaine®) drops can be used to desensitize the cornea for ocular procedures such as corneal foreign body removal or ocular ultrasound. It has an onset of action of 30 seconds, is applied in drop doses every 5 to 10 minutes, and provides 5 to 10 minutes of anesthesia to the cornea with limited drug penetration to the conjunctiva. Proxymetacaine is a similar product available in Europe.

- *Mepivacaine* (Carbocaine®) is available in a 1 to 2% injectable form (see Figure 5-13a) and without epinephrine. Its onset of action is 5 to 10 minutes and lasts 90 to 180 minutes. Mepivacaine causes little tissue reaction, making it a good choice for nerve blocks of the limbs in equine patients.

- *Tetracaine* (Neo-Predef® with tetracaine, Altacaine®) is available in a 0.1% injectable form used topically on the skin and a 0.2% solution for use as an ophthalmic or otic solution. It has an onset of action of 5 to 10 minutes and lasts 2 hours. It is typically used to desensitize the cornea and is not absorbed systemically.

Autonomic Nervous System Drugs

The autonomic nervous system (ANS) is the involuntary part of the peripheral nervous system (PNS) that innervates smooth muscle, cardiac muscle, and glands. There are two divisions of the ANS: the sympathetic nervous system and the parasympathetic nervous system (Figure 5-14). The sympathetic nervous system, also known as the "fight-or-flight" system, increases heart rate, respiratory rate, and blood flow to muscles while decreasing gastrointestinal function and causing pupillary dilation. The parasympathetic nervous system, also known as the homeostatic system, normalizes heart rate, respiratory rate, and blood flow while returning gastrointestinal function and pupil size to normal.

Anticholinergic Drugs

Acetylcholine is one type of neurotransmitter in the PNS and the only neurotransmitter in the parasympathetic nervous system. The PNS has two receptors that bind acetylcholine: muscarinic and nicotinic.

Muscarinic receptors stimulate smooth muscles and slow the heart rate, and nicotinic receptors affect skeletal muscles. Anticholingerics, also referred to as parasympatholytics, block the effects of the parasympathetic nervous system by competitively blocking the binding of acetylcholine at muscarinic receptors.

Anticholinergic drugs inhibit the actions of acetylcholine by occupying the acetylcholine receptors. The major body tissues affected by the anticholinergic drugs are the heart, respiratory and gastrointestinal tracts, urinary bladder, eyes, and exocrine glands. Anticholingeric drugs are used perioperatively to counteract parasympathetic effects such as bradycardia and atrioventricular (AV) block caused by surgical manipulation (stimulation of the vagus nerve) or other anesthetic agents such as opioids. They are also used to control salivation and respiratory secretions; however, anticholinergic drugs control secretions by decreasing the fluid component of the secretion, which may increase secretion viscosity leading to obstruction of small airways or endotracheal tubes. Anticholinergics should not be automatically given as a premedication to patients; the risks and benefits of their use should be assessed for each patient.

The most common anticholinergic drugs used in animals are atropine and glycopyrrolate. Both drugs are nonselective muscarinic antagonists. Administration of anticholinergics can cause sinus tachycardia, which can lead to an increase in myocardial workload and a decrease in myocardial perfusion in animals with cardiovascular disease. Perioperative administration of anticholinergics can reduce intestinal motility and lead to postoperative gastrointestinal complications, particularly concerning in large animal patients. Adverse effects of anticholinergics may include tachycardia, constipation, dry mouth, dry eye, and drowsiness. Examples of anticholinergics include:

- *Atropine* (Atrocare®, generic) is occasionally used as a premedicant because it increases heart rate, decreases secretions (including saliva), decreases gastrointestinal mobility, and causes bronchodilation and mydriasis (Figure 5-15). Ocular effects of atropine can last 1 to 2 days while its effect on other body systems subsides within a few hours. Atropine does not have any analgesic properties, but it may produce mild sedation and reduction of vomiting and nausea because it diffuses into the CNS. Atropine may be given SQ, IM, or IV. Cardiovascular effects occur within 5 minutes after IM administration with peak effects occurring within 10 to 20 minutes; they occur within 1 minute after IV administration with peak effects occurring within 3 to 4 minutes.[2] Atropine is also well absorbed after oral, inhalation, and endotracheal administration. Atropine is used in dogs for cardiopulmonary cerebral resuscitation (CPCR) to increase heart rate and block of vagal tone. When used for CPCR, atropine is indicated for asystole or pulseless electrical activity (PEA), especially if due to high vagal tone. Atropine is repeated with every other cycle of CPCR (every 4 minutes). If given intra-tracheally the dosage given is two- to tenfold of the standard dosage, and the drug is diluted 1:10 in saline or sterile water.[16] Atropine is distributed well throughout the body and crosses the placenta and in small quantities into milk. In food-producing animals, there is a 28-day meat and 6-day milk withdrawal time.[17] Atropine is not generally selected for use in rabbits because some rabbits produce the enzyme atropine

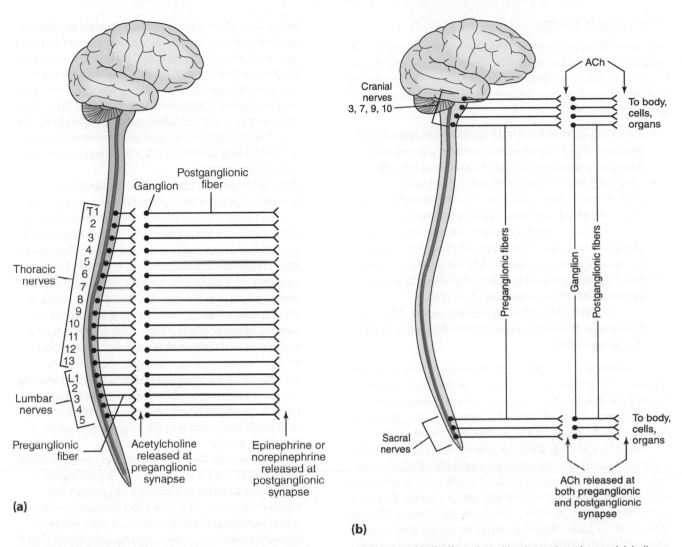

Postganglionic fiber

Ganglion

Postganglionic fiber

T1
2
3
4
5
6
7
8
9
10
11
12
13

Thoracic nerves

L1
2
3
4
5

Lumbar nerves

Preganglionic fiber

Acetylcholine released at preganglionic synapse

Epinephrine or norepinephrine released at postganglionic synapse

(a)

ACh

Cranial nerves 3, 7, 9, 10

To body, cells, organs

Preganglionic fibers

Ganglion

Postganglionic fibers

Sacral nerves

To body, cells, organs

ACh released at both preganglionic and postganglionic synapse

(b)

FIGURE 5-14 The autonomic nervous system. (a) The sympathetic nervous system, found in the thoracic and lumbar regions, has acetylcholine released at the preganglionic synapse and epinephrine or norepinephrine released at the postganglionic synapse. (b) The parasympathetic nervous system, found in the brain stem region and sacral segments, has acetylcholine released at both the preganglionic and postganglionic synapse.

esterase, which limits the drug's clinical effectiveness; glycopyrrolate is preferred for use in rabbits.

- *Glycopyrrolate* (Robinul®) is four times more potent than atropine, and its absorption, metabolism, and elimination is similar to atropine (see Figure 5-15). Cardiovascular effects occur within 5 minutes after IM administration with peak effects occurring within 30 minutes. Glycopyrrolate's onset of action after IV administration is slightly slower than the onset of atropine, which is why it is not selected for emergency situations. Glycopyrrolate has a longer duration of action than atropine, with its effects typically lasting approximately 90 minutes; however, reduced salivation may persist for up to 7 hours. Glycopyrrolate is more polar than atropine, which limits its diffusion into the CNS and across the placenta into fetal circulation. Due to its limited movement into the CNS, sedation is not observed. Glycopyrrolate does not cause mydriasis or alter intraocular pressure.

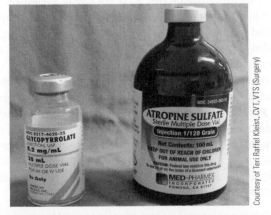

FIGURE 5-15 Atropine and glycopyrrolate are anticholinergic drugs used to increase heart rate, decrease secretions, and decrease gastrointestinal motility.

Adrenergic Drugs

The sympathetic nervous system has more than one type of neurotransmitter. Acetylcholine is the neurotransmitter released at the preganglionic synapse, and epinephrine or norepinephrine is the neurotransmitter released at the postganglionic synapse. The adrenergic receptors of the sympathetic nervous system include:

- alpha-1 (α_1), which are found in the smooth muscles of blood vessels. Stimulation of (α_1) receptors causes constriction of the arterioles (except in the gastrointestinal tract), increasing blood pressure.
- alpha-2 (α_2), which are found in the postganglionic sympathetic nerve endings. Stimulation of α_2 receptors causes inhibition of norepinephrine release in the brain, resulting in sedation and analgesia. Adverse effects of α_2 receptor stimulation are initial hypertension (due to vasoconstriction) followed by a reflex bradycardia that causes the central α_2 action of decreased blood pressure and cardiac output.
- beta-1 (β_1), which are in the heart (and fatty tissue). These cause increased heart rate, conduction, and contractility.
- beta-2 (β_2), which are found mainly in smooth muscles of the lung. Stimulation of these receptors causes bronchodilation and dilation of skeletal blood vessels.

Dopaminergic receptors are related to adrenergic receptors because dopamine is the precursor to norepinephrine. Dopaminergic receptors are located in the renal, mesenteric, and cerebral arteries. Stimulation of dopaminergic receptors may cause dilation of the coronary vessels, dilation of the blood vessels of the kidney, and dilation of mesenteric blood vessels.

Two groups of drugs affect the sympathetic nervous system: the adrenergic agonists and the adrenergic antagonists (blocking agents). Alpha adrenergic antagonists (α-blockers) usually promote vasodilation and a decrease in blood pressure; the only ones discussed in this chapter are the reversal agents yohimbine, atipamezole, and tolazoline (discussed previously in the Alpha-2 Adrenergic Antagonist section). Beta adrenergic antagonists (β-blockers) decrease heart rate and blood pressure and are not described in this textbook.

Catecholamines are chemicals that can cause a sympathetic (fight or flight) response, and drugs that mimic their effects are called adrenergic drugs or sympathomimetics. They act on one type of adrenergic receptors located on the cells of smooth muscles (either α or β receptors only; these are called selective drugs) or both types (both α and β receptors; these are called nonselective drugs). Adverse effects of adrenergic drugs include tachycardia, hypertension, and cardiac arrhythmias.

Xylazine is a nonselective adrenergic drug that was described in the Tranquilizers and Sedatives section. Other adrenergic drugs are used as inotropic agents to treat hypotension related to decreases in cardiac contractility and/or severe drops in systemic vascular resistance (seen in severe sepsis). Inotropy is cardiac contraction force, and inotropes work by stimulating beta-adrenergic receptors. Inotropes that increase the contractile force are called positive inotropes. Inotropic agents commonly used in anesthetized patients are as follows:

- *Dobutamine* (generic) is a beta-adrenergic agonist (predominantly β_1 and mild β_2 and α_1 activity) that also has positive inotropic properties (Figure 5-16a). It is used in horses and other large animal species to increase blood pressure by improving cardiac output and contractility. In these species, dobutamine does not increase peripheral vascular resistance or heart rate (except at very high dosage rates). In dogs, dobutamine produces dosage-related increases in myocardial contractility, cardiac output, stroke volume, and coronary blood flow, with no change in systemic arterial blood pressure. In cats dobutamine does not increase blood pressure,[18] and when given at higher dosages or in hypovolemic patients, tachycardia may occur. Dobutamine has a very short half-life and must be administered as a CRI and carefully monitored. It is available in a variety of concentrations and is usually administered in lactated ringer's solution, saline, or 5% dextrose in water.

- *Dopamine* (generic) is an adrenergic β_1 and α_1 agonist that has positive inotropic properties (Figure 5-16b). Dopamine is administered as a CRI due to its short half-life; however, clinical effects may not be seen for up to 5 to 10 minutes after initiating treatment and may last a few minutes after the drug is discontinued. Dopamine has different physiologic effects depending on the dosage at which it is administered. In many species it has dopaminergic effects when given at lower dosages (increases renal blood flow), β_1 adrenergic effects at moderate dosages (increases cardiac contractility), and mainly α_1 adrenergic effects at higher dosages (causing tachyarrhythmias, coronary vasoconstriction, and myocardial excitability). Blood pressure may be increased at the expense of peripheral perfusion, and the concurrent increase in contractility and heart rate may lead to a greater myocardial oxygen demand, which increases the risk of arrhythmias. It is available in a variety of concentrations and is usually administered in lactated ringer's solution, saline, or 5% dextrose in water.

- *Norepinephrine* (Levophed®) stimulates α_1, β_1, and β_2 adrenergic receptors and is used to increase systemic vascular resistance while improving cardiac contractility, blood pressure, and oxygen delivery to tissues (Figure 5-16c). Norepinephrine is used to treat hypotension in patients with sepsis and endotoxemia because it preferentially corrects inappropriately dilated areas.[19] Norepinephrine must also be administered as a CRI. Careful drug titration and monitoring of blood pressure and mucous membrane color is needed to prevent over-constriction, which would lead to a decrease in peripheral perfusion.

- *Ephedrine* (generic) is a short acting, noncatecholamine sympathomimetic drug that stimulates the release of the patient's endogenous stores of norepinephrine, leading to mild vasoconstriction (Figure 5-16d). Ephedrine has both β_1 and α_1 effects and is useful in many species to increase cardiac output and to treat perioperative hypotension. It is given as an IV bolus and is an ideal choice for a short-term (15–20 minute) boost in cardiac output and blood pressure. The dose can be repeated, but after 2 to 3 doses the endogenous stores of norepinephrine have been depleted and subsequent doses are no longer effective. Ephedrine does not cause arrhythmias but may cause tachycardia, reflex bradycardia, and hypertension. In some states, ephedrine is a controlled substance or has restrictions on its sale because it is used to make methamphetamine.

- *Vasopressin* (Vasostrict®), also known as anti-diuretic hormone, is a V1 (vasopressin 1) agonist that causes smooth muscle constriction within capillaries and small arterioles to increase systemic vascular resistance (Figure 5-16e). Vasopressin has no inotropic or chronotropic

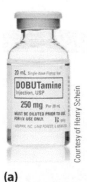

(a)

Courtesy of Henry Schein

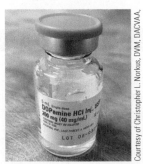

(b)

Courtesy of Christopher L. Norkus, DVM, DACVAA, CVPP, DACVECC

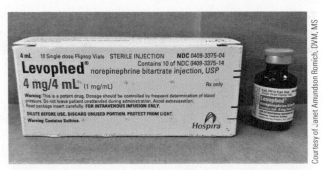

(c)

Courtesy of Janet Amundson Romich, DVM, MS

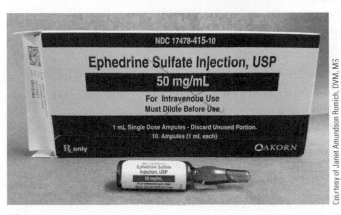

(d)

Courtesy of Janet Amundson Romich, DVM, MS

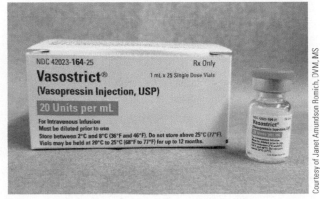

(e)

Courtesy of Janet Amundson Romich, DVM, MS

FIGURE 5–16 Adrenergic drugs used in veterinary medicine. (a) Dobutamine, a β adrenergic agonist that has positive inotropic properties. (b) Dopamine, an adrenergic inotropic β_1 and α_1 agonist that has positive inotropic properties. (c) Norepinephrine, a nonselective adrenergic drug that stimulates α_1, β_1, and β_2 adrenergic receptors that has vasoconstrictive properties. (d) Ephedrine, a selective adrenergic drug that produces both β_1 and α_1 effects that has positive inotropic and vasoconstrictive properties. (e) Vasopressin, a V1 (vasopressin 1) agonist that has vasoconstrictive properties.

effects; therefore, it does not worsen myocardial ischemia. It is used to increase vascular tone in patients with sepsis or endotoxemia because it works in the face of acidosis. Vasopressin can be administered IV or through the endotracheal tube. Its dosage is increased two- to tenfold when administered intra-tracheally and is diluted in 5- to 10-mL of sterile water (preferred) or saline.

- *Epinephrine* is used for cardiac arrest and is described in Chapter 13 (Cardiac Arrest and Cardiopulmonary Cerebral Resuscitation [CPCR]).

Autonomic nervous system drugs are described in Table 5-6.

DRUGS USED FOR ANALGESIA

Analgesics are administered as pre-medicants as part of a pre-emptive analgesic plan and may be continued in the post-operative stage to continue to provide pain relief for veterinary patients.

Opioids

Opioids are the most effective analgesics to treat acute pain in veterinary medicine and are a key component of "pre-emptive analgesia" (providing pain control before a painful stimulus occurs). When used alone they provide mild to moderate sedation (except in horses), but when given with a

sedative or tranquilizer they produce neuroleptanalgesia (described in the Neuroleptanalgesics section). Opioids have high efficacy (effectiveness of the drug to work once it is in the patient) in managing pain in a variety of animals including patients with acute trauma, painful medical conditions or disease processes, and those undergoing surgical procedures. They are also a key component in providing multimodal (balanced) analgesia. Opioids can be more challenging to utilize for management of chronic pain in veterinary patients.

Opioids provide analgesia by binding to the opioid receptors in the central and peripheral nervous system. There are three main receptor types of opioid receptors mu (μ), kappa (κ), and delta (δ), which are located mainly in the CNS but also develop at sites of inflammation in peripheral tissue. The analgesic effects are mainly mediated at μ receptors located in the spinal cord and to a lesser extent the brain. Kappa receptors play a much smaller role in analgesia and are responsible to a greater degree for sedation and dysphoria. Delta receptors provide some spinal cord analgesia, but to a much lesser degree than the μ receptors. Drugs acting on opioid receptors are classified as full agonists, partial agonists, mixed agonist/antagonists, or antagonists. Full mu (μ) agonist opioids typically produce the most profound analgesia, partial μ agonists produce moderate analgesia, mixed agonist/antagonists that are κ agonists and μ antagonists produce moderate analgesia usually with less respiratory depression and addictive potential, and μ antagonists reverse the effects of a μ agonist.

TABLE 5-6 Autonomic Nervous System Drugs Described in this Chapter

Drug Category	Drug	Effects
Anticholinergic	atropine (Atrocare®, generic)	↑ heart rate ↓ gastrointestinal motility Bronchodilation Mydriasis
	glycopyrrolate (Robinul®)	↑ heart rate ↓ gastrointestinal motility Bronchodilation Mydriasis Does not cross the blood-brain barrier or placenta
Adrenergic	dobutamine (generic) positive inotrope	↑ cardiac contractility ↑ blood pressure (horses) ↑ heart rate (when given to hypovolemic dogs and cats or at higher dosages)
	dopamine (generic) positive inotrope	↑ renal blood flow likely due to increased cardiac output (lower dosages) ↑ cardiac contractility (moderate dosages) Tachyarrhythmias, coronary vasoconstriction, and myocardial excitability (higher dosages)
	norepinephrine (Levophed®) vasoconstrictor	↑ systemic vascular resistance Improves cardiac contractility, blood pressure, and oxygen delivery to tissues
	ephedrine (generic) positive inotrope and vasoconstrictor	↑ cardiac output Treats perioperative hypotension
	vasopressin (Vasostrict®) vasoconstrictor	Smooth muscle constriction within capillaries ↑ systemic vascular resistance in small arterioles

When choosing an opioid, it is important to know the difference in the *potency* and the *efficacy* of an opioid. **Potency** is the measure of drug activity in terms of the amount of the drug required to produce an effect and refers to the dosage required to produce maximum effect. Oxymorphone has a higher degree of potency than morphine; therefore, a typical dosage for morphine may be 0.5 mg/kg while that for oxymorphone may be 0.05 mg/kg. **Efficacy** is the capacity to produce an effect and refers to the effectiveness of the drug to relieve pain once it is in the patient. For example, buprenorphine is approximately 30 times more potent than morphine, but its maximum effect is less than that of morphine.

Opioid receptors are widely distributed throughout the body; therefore, they have effects on multiple body systems and all effects, whether desirable or undesirable, are dosage dependent. Although adverse effects are rarely seen at the low dosages used clinically, higher dosages of opioids produce higher incidence of undesirable effects. Respiratory depression may be profound in humans, but is generally not a clinical concern in veterinary patients given standard dosages. Opioids may cause vagal-induced bradycardia and bradyarrhythmias; however, opioids generally have minimal effects on cardiac output, systemic vascular resistance, cardiac contractility, and blood pressure. They are metabolized by the liver and excreted by the kidneys. Adverse effects of opioids may be species related and include:

- *Small animals:* Gastrointestinal (vomiting and nausea), CNS (dysphoria), ocular (miosis in dogs and mydriasis in cats), and thermoregulatory (hypothermia in dogs and hyperthermia in cats; panting in dogs). At higher dosages cats may display excitement known as "morphine mania."
- *Horses:* Gastrointestinal (ileus with colic), CNS (excitement), and at higher or repeated dosages increased locomotion.
- *Cattle and small ruminants*: Hyperactivity, abnormal chewing behaviors, and hyperthermia. Opioids are lipid soluble and may reach milk, which could potentially produce drug residues in animal products. Opioid dosages for immobilization vary between ruminant species and give variable results.

Full Mu (μ) Agonists

- *Morphine* (Duramorph®, Astramorph®) is a full μ agonist but also has mild effects at κ and δ receptors (Figure 5-17a). Morphine is effective at relieving pain and despite the development of synthetic opioids continues to be the opioid gold standard. Morphine is a C-II controlled substance that may be given IV *slowly* as a bolus or CRI, IM, SQ, epidural, or intrarticular (IA). A preservative-free formulation is available for epidural administration that produces analgesic effects

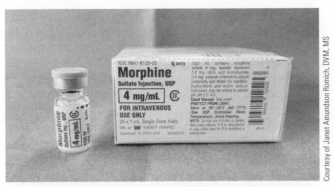

(a)

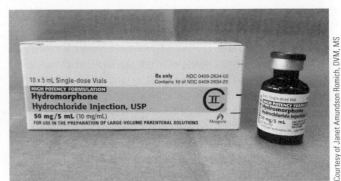

(b)

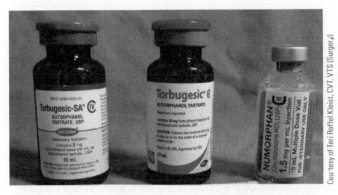

(c)

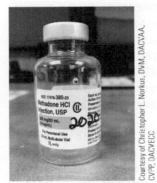

(d)

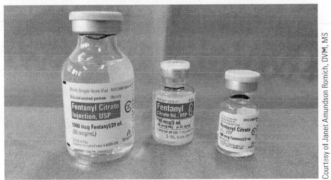

(e)

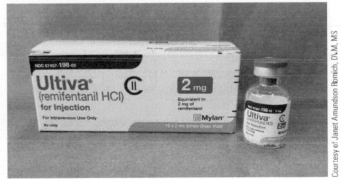

(f)

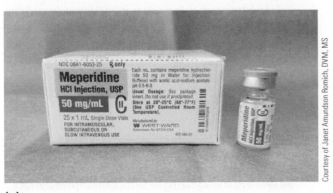

(g)

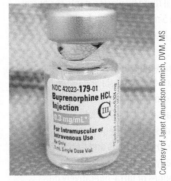

(h)

FIGURE 5–17 Types of opioid drugs include: (a) morphine; (b) hydromorphone; (c) butorphanol and oxymorphone; (d) methadone; (e) fentanyl; (f) remifentanil; (g) meperidine; and (h) buprenorphine.

that can last up to 24 hours. Oral tablets are also available. Its onset of action is 3 to 5 minutes when given IV and 5 to 5 minutes when given IM or SQ. Its duration of action is typically 4 to 6 hours. Morphine may cause histamine release when give quickly IV (which leads to vasodilation and hypotension), vomiting (avoid use in patients with an esophageal foreign body or upper airway dysfunction that may be prone to aspirate), and excitement in certain species such as cats, horses, etc. Vomiting can further increase intracranial pressure, and the use of morphine should be justified prior to administering it to patients with existing intracranial pressure. Vomiting caused by morphine is observed less frequently if the patient is given morphine when it is already painful.

- *Hydromorphone* (Dilaudid®) is a synthetic, full μ agonist that is a C-II controlled substance five to seven times more potent than morphine (Figure 5-17b). Hydromorphone does not cause histamine release; therefore, it can be given IV as a bolus. It is also given IM, SQ, CRI, or as an epidural. Its onset of action is 1 to 5 minutes when given IV and 10 to 20 minutes when given IM or SQ. Its duration of action is 2 to 4 hours. Hydromorphone may cause hyperthermia in cats. Hydromorphone, like morphine, should be avoided in animals in which vomiting is undesirable.

- *Oxymorphone* (Numorphan®) is a full μ agonist that is ten times more potent than morphine (Figure 5-17c). It is a C-II controlled substance used IV, IM, and SQ mainly in dogs and occasionally cats for sedation, as a restraining agent for diagnostic procedures, analgesic, and as a preanesthetic agent. It is also used in horses as an analgesic but may cause CNS excitement in them. Oxymorphone provides analgesia for approximately 3 to 6 hours. Adverse effects of vomiting, nausea, and panting in dogs commonly occur. Oxymorphone may be used alone or in combination with neuroleptic agents or barbiturates. Oxymorphone should not be mixed in the same syringe with barbiturates as this causes precipitates to form. Oxymorphone is currently more expensive than hydromorphone and difficult to source, which has limited its use.

- *Methadone* (Dolophine®) is a C-II controlled substance that is a full μ agonist and N-methyl-D-asparate (NMDA) receptor non-competitive antagonist, which may explain its ability to prevent and/or treat central sensitization (Figure 5-17d). Methadone rarely causes vomiting. Methadone is used to control moderate to severe pain and can be given IV as a bolus or CRI, IM, or SQ. In cats, methadone can be given orally via the transmucosal route by placing it into the buccal pouch. Its onset of action is 1 to 5 minutes when given IV and 10 to 20 minutes when give IM or SQ. Methadone's duration of action is 3 to 6 hours.

- *Fentanyl* ((Sublimaze®) is a full μ agonist that is 100 times more potent than morphine and is used for analgesia and sedation (Figure 5-17e). Fentanyl is a C-II controlled substance that can be given IV as a bolus or CRI and transdermally (it is labeled for IM and SQ administration but these routes are not frequently used). When given IV, its onset of action is 30 to 60 seconds and its duration of action is 20 to 30 minutes. Fentanyl is useful for short but painful procedures such as dressing changes or intra-operatively during times of intense surgical stimulation. It is also useful as an IV premedicant or co-induction agent. Fentanyl bolus IV produces bradycardia, but has minimal direct depressant effects on the myocardium or vasculature. Fentanyl transdermal patches (Duragesic®) are available and can be useful adjuncts to analgesia; however, the dose they provide may be uncertain due to unreliable absorption, and they may cause skin irritation. Fentanyl transdermal patches provide 3 to 5 days of analgesia; however, it takes 12 to 24 hours to reach effective plasma levels in dogs and cats, and it may be difficult to adhere to the patient's skin. A significant concern for veterinarians prescribing fentanyl transdermal patches is the abuse potential due to drug diversion. Fentanyl patches have been used in dogs (placed on the lateral thorax, base of tail, or metatarsal/carpal areas) and have also been used in other species including cats, rabbits, horses, goats, sheep, and pigs. Fentanyl may cause nausea and rarely causes vomiting and panting.

- *Remifentanil* (Ultiva®) is an ultrashort acting, full μ agonist that is a C-II controlled substance similar to fentanyl (Figure 5-17f). It is used during the perioperative or postoperative period as an analgesic.[20] Remifentanil is a unique opioid because it has an ester structure that is metabolized by nonspecific blood and tissue esterases;[21] therefore, it is not dependent on liver metabolism or renal excretion and is quickly cleared from the animal. Due to its short duration of action (<5 minutes), remifentanil is administered by CRI. The cost of remifentanil has limited its used in veterinary medicine, but it would be the drug of choice for patients with liver failure if available. Other ultrashort acting, full μ agonists that are structurally similar to fentanyl are alfentanil (Alfenta®) and sufentanil (Sufenta®).

- *Meperidine* (Demoral®) or *pethidine* are full μ agonists suitable for treating moderate to severe pain (Figure 5-17g). Meperidine is the generic name used in the United States, while pethidine is the generic name used in the United Kingdom. Meperidine is a C-II controlled substance that is administered IM or SQ. It should not be administered IV because it causes histamine release, which can produce hypotension and tachycardia. It is short acting (1 to 2 hours) and stings on injection. Meperidine causes light sedation and is unlikely to cause vomiting. It may cause less bradycardia than other opioids and may increase heart rate because it has a similar chemical structure to atropine.

- *Codeine* (generic) is a full μ agonist that is metabolized to morphine and therefore produces effects similar to morphine. It is a C-II controlled substance used orally mainly as an antitussive in cats and dogs rather than as an analgesic. It is however sometimes combined with acetaminophen and in dogs to relieve mild pain. Due to acetaminophen toxicity, codeine-acetaminophen products should not be used in cats.

Partial Mu Agonist

- *Buprenorphine* (Buprenex®) is a C-III controlled substance used for mild to moderate pain because as a partial μ agonist it does not have the same efficacy as full μ agonists and therefore does not provide the same level of analgesia (Figure 5-17h). Buprenorphine also has a **ceiling effect** in which increasing dosages will not produce a greater effect. Buprenorphine has a delayed onset of action (45–60 minutes) and because it has a high affinity for the opioid receptor site, it has a long duration of action of 4 to 6 hours depending on species and dosage. The duration of action may be variable in cats highlighting the importance of performing regular pain assessments rather than assuming a set duration of analgesic action. Buprenorphine's strong affinity for the μ receptor site may limit full μ agonists from binding to the μ receptor if given concurrently. This may alter the clinical effects of full mu agonists when giving different opioid drugs concurrently and varies significantly depending on the species, individual patient, and drug dosage and route of administration. Buprenorphine's strong affinity for the μ receptor also makes it difficult to reverse with the opioid antagonist naloxone. Buprenorphine can be given IV, IM, epidural, or transmucosally. Buprenorphine (Butrans®) is also available as a transdermal patch (see Chapter 17 [Analgesic Techniques]). It undergoes a high first-pass metabolism, which limits its oral use; however, it has been used for pain in laboratory animals because it can be formulated in feed and given orally to rodents. Buprenorphine has also been administered transmucosally to cats by putting it in the buccal pouch. Subcutaneous absorption of buprenorphine is questionable in cats; therefore, IV and IM are preferred injectable routes of administration. Buprenorphine provides minimal sedation. Simbadol™ is a specific high concentration formulation of buprenorphine (not an extended release formulation) that is FDA approved for control of postoperative pain in cats given SQ once a day (q 24hr) for up to 3 doses (or 72 hrs) (see Figure 8-11). The most frequent adverse effects with Simbadol™ are hypotension, tachycardia, hypothermia, hyperthermia, hypertension, anorexia, and hyperactivity.

Mixed Agonist/Antagonist

- *Butorphanol* (Torbugesic®, Torbutrol®) is a C-IV controlled substance that is a κ agonist and μ antagonist (see Figure 5-17c). As a synthetic mixed agonist/antagonist it is not as efficacious as full μ agonists or partial μ agonists; therefore, it is used to relieve mild pain when used alone. When used in combination with other analgesics such as NSAIDs, it can provide analgesia for moderately painful procedures such as ovariohysterectomies. Butorphanol provides mild to moderate sedation and rarely causes panting or vomiting. It can be used as an antagonist to partially reverse full mu agonists without reversing the κ receptor-mediated analgesia. Butorphanol is an effective antitussive, making it useful as a premedication in patients undergoing upper airway examination or bronchoscopy. Butorphanol can be given IV as a bolus or CRI, IM, or SQ. Butorphanol has a rapid onset of action (3–5 minutes when given IV and 5–15 minutes when given IM) and a short duration of action (20–60 minutes in dogs and 45–90 minutes in cats). In species with a higher distribution of κ receptors within their CNS (birds and horses), butorphanol provides equal or better analgesia than from a full mu agonist. It is used for moderate to marked pain relief in horses with a rapid onset of action (3 minutes when given IV and a peak analgesic effect of 15–30 minutes) and a duration of action that may last up to 4 hours. Butorphanol, as a mixed kappa agonist/ mu antagonist, has a ceiling effect in which increasing dosages will not produce a greater effect.

Butorphanol is useful, alone or in combination with other drugs, for diagnostic procedures due to its sedative effects in patients.

Atypical Opioids

- *Tramadol* (Ultram®) is a weak μ agonist in humans that is classified as an atypical opioid (see Figure 16-6). This appears to be the case in other species as well but not in dogs. Dogs appear to lack adequate quantities of the M1 (O-desmethyltramadol) metabolite, which is mostly responsible for the drug's analgesic effect. As a result, the drug is not believed to act as an opioid in dogs; however, it does provide analgesia in cats. It is a C-IV controlled substance due to its opioid receptor action in humans but also appears to inhibit serotonin and norepinephrine reuptake, which is thought to contribute to its analgesic effects. Tramadol is used as an oral analgesic in dogs and cats. Whether or not tramadol provides analgesia in dogs is up for debate; therefore, it may be prescribed along with other analgesic drugs in dogs. It is typically administered orally in tablet form to patients with chronic cancer pain or those with arthritis when nonsteroidal anti-inflammatory drugs (NSAIDs) alone are inadequate or are contraindicated. In the United States, tramadol is available in tablet form in addition to an extended-release version and in combination with acetaminophen. An injectable IV preparation is available in some countries. Adverse effects include sedation and dysphoria, especially in cats that also salivate when it is given orally as it appears to be unpalatable. Serotonin syndrome can develop in patients receiving a combination of drugs that increase serotonin; therefore, tramadol should not be used with other serotoninergic medications such as tricyclic antidepressants, serotonin-norepinephrine reuptake inhibitors, and amitraz-containing compounds. Serotonin syndrome is described in Chapter 16 (Chronic Pain Management). Potential human abuse should be kept in mind when prescribing and dispensing tramadol.
- *Tapentadol* (Nucytna®) is an atypical opioid whose parent compound (not metabolite) has a dual mechanism of action (μ opioid receptor agonist and norepinephrine reuptake inhibitor) that may make it superior to tramadol in dogs. After IV administration, tapentadol can be detectable in plasma for up to 6 hours, and it is rapidly absorbed in dogs after oral administration, but its bioavailability via this route is low. Dosage-related adverse effects such as salivation and sedation were observed following IV administration. It is a C-II controlled substance and is much more expensive than tramadol, which has limited its use in veterinary medicine.

Mu Antagonist

- *Naloxone* (Narcan®) is a full μ opioid antagonist that has a high affinity for opioid receptors and can displace opioid agonists from μ and κ receptors (Figure 5-18). Naloxone rapidly reverses all opioid-induced clinical effects including analgesia; therefore, its use should be reserved for emergency situations such as opioid overdose or profound respiratory depression. It increases alertness, responsiveness, coordination, and potentially increases perception of pain. Naloxone

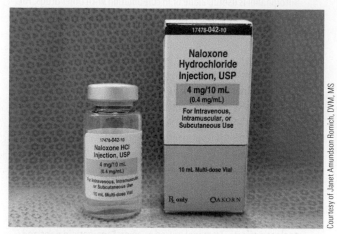

Courtesy of Janet Amundson Romich, DVM, MS

FIGURE 5-18 Naloxone, a full μ opioid antagonist, rapidly reverses all opioid-induced clinical effects including analgesia.

can be given IV or IM, although when reversing unwanted adverse effects such as respiratory depression, IV use is favored due to its rapid onset of action. It is also administered sublingually to neonates or injected into the umbilical vein to reverse respiratory depression following C-section in which the mother was given an opioid agonist. Naloxone has an onset of action of 1 to 2 minutes when given IV and sublingually; 10 to 20 minutes when given IM. Its duration of action is 30 to 60 minutes, which is shorter than the duration of action of many opioid agonists. When used to reverse long-acting μ agonist opioids, adverse effects may return, prompting the necessary administration of repeated doses of naloxone. Naloxone does not effectively reverse buprenorphine because this partial μ agonist has greater affinity for the μ receptor than pure antagonists. Excitement or anxiety may be seen after naloxone reversal of an opioid agonist.

Opioid types and their effects are summarized in Table 5-7.

Neuroleptanalgesics

Neuroleptanalgesia is the mixing of a sedative or tranquilizer with an analgesic drug (typically an opioid) to provide more profound sedation and potentially analgesia then if either drug was used alone. Neuroleptanalgesics cause CNS depression and analgesia and may or may not produce uncounsciousness. A benefit of neuroleptanalgesia is reducing the unpleasant adverse effects of the narcotic analgesic such as excitement reactions and vomiting. These drug combinations are used for premedication purposes or for chemical restraint for short procedures. Neuroleptanalgesics are prepared (compounded) by veterinarians and examples include xylazine and butorphanol, acepromazine and morphine, and acepromazine and oxymorphone. The opioid portion of the

TABLE 5-7 Types of Opioids and Their Effects*

Opioid Class	Drug	Pain Level Effectiveness
Full μ agonists	morphine (preservative-free recommended when given as an epidural/intrathecal) (Duramorph® and Astramorph® PF are injectables; Roxanol®, MS Contin®, and generic tablets are oral forms; low bioavailability when given orally)	Moderate to severe
	hydromorphone (Dilaudid®)	Moderate to severe
	oxymorphone (Numorphan®)	Moderate to severe
	methadone (Dolophine®)	Moderate to severe
	fentanyl (Sublimaze® [injectable], Duragesic® [transdermal fentanyl patch]), remifentanil (Ultiva®) sufentanil (Sufenta®) alfentanil (Alfenta®)	Moderate to severe
	meperidine/pethidine (Demoral®)	Mild to moderate
	codeine (generic)	Mild to moderate
	codeine 60 mg/acetaminophen 300 mg (generic)	Mild to moderate
Partial μ agonist	buprenorphine (Simbadol™, Buprenex®, Butrans® transdermal patch)	Mild to moderate
Mixed agonist-antagonist	butorphanol (Torbugesic®, Torbutrol®)	Mild (good visceral analgesia)
Atypical opioid	tramadol (Ultram®) tapentadol (Nucytna®)	Mild to moderate

*Naloxone (Narcan®) is a full μ opioid antagonist used to treat respiratory and CNS depression caused by opioids

neuroleptanalgesic can be reversed with an opioid anatagonist. Adverse effects of neuroleptanalgesics include panting, flatulence, bradycardia, and increased sensitivity to sound.

N-Methyl-D-Aspartate (NMDA) Receptor Antagonists

Chronic pain can be enhanced by a state of sensitization within the central nervous system (central sensitization) through numerous pathways including the N-methyl-D-aspartate (NMDA) receptor, which promotes excitatory neurotransmission through glutamate. The NMDA receptor in patients without chronic pain is blocked by magnesium; however, activation of the NMDA receptor causes sodium and calcium ion influx to occur within post-synaptic neurons, which activates second messenger signaling cascades and results in increases to the excitatory neurotransmitter glutamate. Activation of these signaling cascades produces up-regulation of all inputs to the spinal cord neuron, thus increasing the cell's response to pain stimuli and decreasing neuronal sensitivity to opioid receptor agonists. NMDA-receptor antagonists have a significant impact on the development of chronic pain and the development of tolerance to opioid analgesics. Consequently, NMDA-receptor antagonists may have potential as co-analgesics when used in combination with opioids.[7] The most commonly used commercially available NMDA-receptor antagonists include ketamine, tiletamine, amantidine, and methadone.

- *Ketamine* is a highly effective injectable NMDA antagonist, and subanesthetic dosages are commonly used as an adjunct to analgesia protocols. Ketamine is not given orally as it is not palatable and its potential adverse effects limit its regular use in the companion animals for chronic pain. Ketamine can be given as an analgesic IV via a CRI for patients in the hospitalized settings. Generally such low dosages of the drug have minimal adverse effects.
- *Tiletamine* is also a highly effective injectable NMDA antagonist that provides better analgesia than ketamine. It is available in injectable anesthetics approved for dogs and cats (Telazol® and Tilzolan®, proprietary 50/50 mixtures with the benzodiazepine zolazepam), which is its only use in veterinary medicine.
- *Amantadine* (Symmetrel®) is an NMDA antagonist that helps to "reset" a patient's nervous system in chronic pain. It is used orally in animals and has demonstrated usefulness in the treatment of osteoarthritis pain in companion animals when administered along with a primary non-steroidal anti-inflammatory drug (NSAID). Amantadine can be given orally for several weeks (e.g., pulse therapy) or for long term use, making it a popular NMDA antagonist choice. The most frequently reported adverse effects of amantadine include gastrointestinal upset. A cousin of amantadine is memantine, but less data is available about its use in companion animals. For more information on amantadine, refer to Chapter 16 (Chronic Pain Management).
- *Methadone*, a full μ agonist opioid, also has NMDA antagonist properties. The drug is commonly given via injection for acute pain and may be helpful in preventing the establishment of chronic pain states. The drug can also be administered transmucosally in cats to provide

a minimum of 4 hours of analgesia, making it appealing for home administration. Palatability is sometimes an issue.

Non-Steroidal Anti-Inflammatory Drugs (NSAIDs)

NSAIDs are probably the most commonly used analgesics in veterinary practice and are used to treat mild to moderate pain when used alone and moderate to severe pain when used in conjunction with other drugs such as opioids. They are effective, long acting analgesics that have anti-pyretic and anti-endotoxic effects with a wide range of indications. NSAIDs have a very low threshold for toxicity, and their potential for causing serious adverse effects may preclude their use in some patients under anesthesia. NSAIDs reduce prostaglandin production by inhibiting cyclooxygenase enzymes (COX) that are responsible for prostaglandin synthesis. Prostaglandins are mediators of inflammation and pain, so preventing their production reduces inflammation and pain at the site of injury. Spinal prostaglandins are involved in central sensitization as well as working peripherally at the site of inflammation; therefore, NSAIDs may also help prevent central sensitization and the development of chronic pain. Prostaglandins, however, also have many important roles in the body, including gastroprotection, preservation of renal perfusion in situations of reduced renal blood flow, and maintenance of normal coagulation. Reducing prostaglandin production impacts these systems and can result in gastro-duodenal ulceration, renal injury, and coagulopathy. Currently there are thought to be three COX types: COX-1, COX-2, and COX-3. Traditionally, it was thought that COX 1 was responsible for "house-keeping" roles within the body while COX-2 was induced during states of inflammation. This meant that inhibiting COX-2 would reduce pain and inflammation without affecting housekeeping systems. This theory resulted in a drive for production of "COX-2 selective" NSAIDs. Unfortunately, this is an oversimplification, and both COX-1 and COX-2 are important in gastric and renal protection and in coagulation. COX-3 is thought to be a variant of COX-1, which helps explain the anti-pyretic and analgesic effects of the COX-3 inhibitor acetaminophen and its ineffectiveness at reducing inflammation.

NSAIDs should never be administered concurrently with glucocorticoids nor should the recommended dosage ever be exceeded. Their use should be avoided in hypovolemic and hypotensive patients, because they may worsen existing hypovolemia and renal perfusion. Their use should be delayed at least until the patient is fluid resuscitated or until the anesthetic recovery period. Potential adverse effects of NSAIDs include gastrointestinal upset (vomiting, diarrhea, anorexia), gastrointestinal ulceration that may lead to perforation, acute renal injury, and rarely liver injury (in dogs and cats); gastrointestinal ulceration is the major adverse effect in horses.

Common NSAIDs used in dogs include carprofen (Rimadyl®), firocoxib (Previcox®), deracoxib (Deramaxx®), robenacoxib (Onsior®), meloxicam (Metacam®), and grapiprant (Galliprant®) (Figures 5-19a through 5-19e). Commonly used NSAIDs in cats include meloxicam and robenaxocib (see Figures 5-19c and 5-19d). Horses are commonly treated with flunixin meglumine (Banamine®), phenylbutazone (Butazolidin®, Equipalazone®), ketoprofen (Ketofen®), and firocoxib (Equioxx®) (Figures 5-19f and 5-19g). NSAIDs can be administered orally or via injection. Chapter 15 (Acute Pain Management) and Chapter 16 (Chronic Pain Management) describe some of the commonly used NSAIDs and their indications.

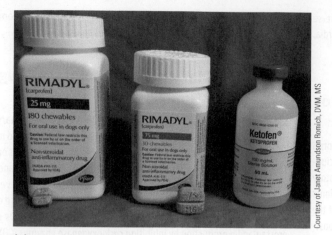

Courtesy of Janet Amundson Romich, DVM, MS

(a)

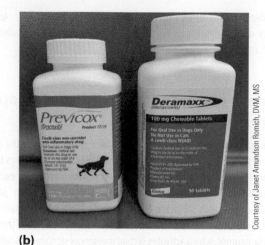

Courtesy of Janet Amundson Romich, DVM, MS

(b)

Courtesy of Janet Amundson Romich, DVM, MS

(c)

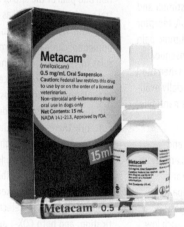

(d)

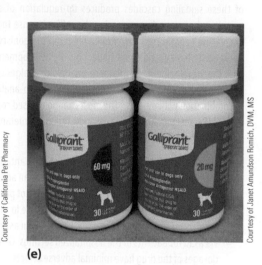

Courtesy of California Pet Pharmacy

Courtesy of Janet Amundson Romich, DVM, MS

(e)

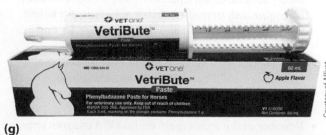

Courtesy of Livestock Concepts

(f)

Courtesy of Allivet

(g)

FIGURE 5–19 Examples of NSAIDs. (a) Carprofen and ketoprofen are propionic acid derivatives that inhibit COXs (ketoprofen also inhibits lipoxygenase). (b) Examples of COX-2 inhibitors include deracoxib and fibrocoxib, (c) robenicoxib, and (d) meloxicam. (e) Grapiprant is an EP4 receptor antagonist for Prostaglandin E$_2$ (PGE$_2$). (f) Flunixin meglumine is a COX inhibitor used in equines. (g) Phenylbutazone is a nonselective COX inhibitor used in equines.

Glucocorticoids

Glucocorticoids are potent anti-inflammatory agents used to control chronic pain in select companion animals because their adverse effects are considered greater than NSAIDs. Like NSAIDs, glucocorticoids reduce prostaglandin production by inhibiting cyclooxygenase enzymes (COX) that are responsible for prostaglandin synthesis; however, because they work at the level of phospholipase A_2, they also inhibit lipoxygenase (LOX), which blocks production of inflammatory mediators such as leukotrienes (Figure 5-20). Adverse effects of glucocorticoids are listed in Table 5-8. For these reasons glucocorticoids should be used cautiously and animals should be monitored frequently so that the lowest possible dosages can be given and the risk of developing adverse effects reduced. As stated above, glucocorticoids should never be administered concurrently with NSAIDs or other glucocorticoids due to the risk of developing gastrointestinal adverse effects, GI perforation, and potentially death.

Commonly used oral (tablet and liquid) glucocorticoids include prednisone (generic), prednisolone (PrednisTab®), and dexamethasone (Dexaject®, generic), and examples of injectable glucocorticoids include methylprednisolone (Depo-Medrol®) and triamcinolone (Vetalog®) (see Figure 16-5). Glucocorticoids can increase appetite and weight gain, which may be a beneficial adverse effect especially in underweight or anorexic patients.

Nerve Growth Factor Inhibitor

Nerve growth factor (NGF) is produced by a variety of inflammatory and immune cells and joint chondrocytes and is elevated in the joints of dogs with osteoarthritis. More information on nerve growth factor inhibitors is found in Chapter 16 (Chronic Pain Management).

TABLE 5-8 Adverse Effects of Glucocorticoids

Adverse Effects of Glucocorticoids
• Polyuria
• Polydipsia
• Polyphagia
• Immune suppression (which increases the risk of infection)
• Decreased wound healing
• Behavioral changes
• Panting
• Iatrogenic Cushing's disease
• Predisposition to diabetes mellitus
• Alopecia
• Similar adverse effects as NSAIDs (both block cycloxygenase):
○ Gastrointestinal upset
○ Gastrointestinal ulceration and potential perforation
○ Acute renal injury
○ Liver injury (rare)

Analgesic Adjunctives

Analgesic adjunctives are drugs that enhance the effect of another analgesic agent, have minimal pain relieving effects on their own, and may decrease the amount of analgesics needed to relieve clinical signs that compromise the patient's quality of life. One example of an analgesic adjunctive is gabapentin (Figure 5-21). Gabapentin (Neurontin®, generic) is an anticonvulsant;

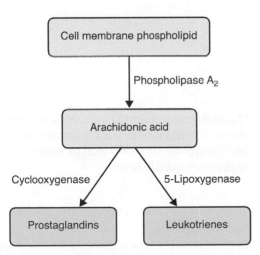

FIGURE 5-20 Site of glucocorticoid and NSAID action. The enzyme phospholipase A_2 releases arachidonic acid from phospholipids in the cell membrane. Arachidonic acid can be metabolized by cyclooxygenase to produce prostaglandins or by lipoxygenase to produce leukotrienes. Glucocorticoids inhibit phospholipase A_2 to decrease the release of both inflammatory mediators (prostaglandins and leukotrienes) by inhibiting the release of arachidonic acid while NSAIDs inhibit only cyclooxygenase to reduce prostaglandin production.

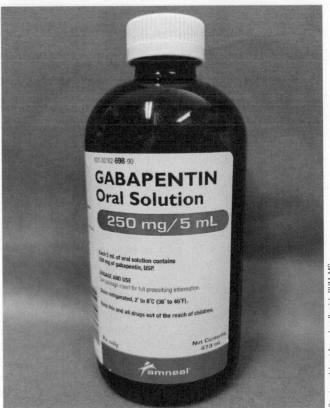

Courtesy of Janet Amundson Romich, DVM, MS

FIGURE 5-21 In small animals, gabapentin is sometimes used to control chronic pain in the peri-operative period.

however, it is also considered an analgesic adjunctive because it has also been shown to have analgesic effects in people with chronic pain, particularly neuropathic pain. In small animals, it is sometimes used to control chronic pain in the peri-operative period, especially in patients undergoing surgery for disc extrusions, particularly if there is nerve root involvement. Gabapentin's exact mechanism of action is unknown; however, it appears to be involved with the interaction of calcium channels in the central nervous system. It is given orally and often causes sedation. Gabapentin use is relatively new in small animals, but it is becoming more commonly used as an adjunct in chronic pain. There is increasing interest in its use in small animals both for chronic and acute pain. Gabapentin should be used in combination with other analgesics. Currently, Kentucky, West Virginia, Tennessee, Michigan, and Virginia have reclassified gabapentin as a Schedule V controlled substance.

Analgesic adjunctives tend to be used to treat chronic pain. Additional examples of analgesic adjunctives such as sodium channel blockers, alpha-2 adrenergic agonists, TrpV1 receptor agonists and antagonists, antidepressants, anticonvulsants, bisphosphonates, and

polysulfated glycosaminoglycan can be found in Chapter 16 (Chronic Pain Management).

Table 5-9 summarizes non-opioid analgesics.

DRUGS USED AS NEUROMUSCULAR BLOCKING AGENTS

Neuromuscular blocking agents are infrequently used in veterinary medicine and are mainly used for ophthalmic procedures that need the eye centrally oriented. Most neuromuscular blocking agents used in veterinary patients are competitive non-depolarizing agents that compete with acetylcholine for the same receptor sites, thus inhibiting the binding of acetylcholine. A decrease in acetylcholine binding decreases the ability of the muscle to generate an action potential, and muscle paralysis results. Atracurium (Tracrium®) is a commonly used neuromuscular blocking agent and is given as a single IV bolus (for 20 to 30 minutes of neuromuscular blockade) or via CRI (Figure 5-22). Atracurium will cause respiratory muscle paralysis

TABLE 5-9 Non-opioid Analgesics Described in this Chapter

Drug Category	Drug	Mechanism of Action
N-methyl-D-aspartate (NMDA) receptor antagonists	• ketamine (Ketaset®, VetaKet®, Ketaject®, Vetalar®) • amantadine (Symmetrel®) • methadone (Dolophine®)	Blocks the N-methyl-D-aspartate (NMDA) receptor, which promotes excitatory neurotransmission through glutamate. NMDA excitation produces up-regulation of all inputs to the spinal cord neuron thus increasing the cell's response to pain stimuli and leads to development of chronic pain.
Non-steroidal anti-inflammatory drugs (NSAIDs)	• carprofen (Rimadyl®) • firocoxib (Previcox®) • deracoxib (Deramaxx®) • robenacoxib (Onsior®) • meloxicam (Metacam®) • grapiprant (Galliprant®) • flunixin meglumine (Banamine®) • phenylbutazone (Butazolidin®, Equipalazone®) • ketoprofen(Ketofen®) • firocoxib (Equioxx®)	Reduces prostaglandin production by inhibiting cyclooxygenase enzymes (COX) that are responsible for prostaglandin synthesis (prostaglandins are mediators of inflammation and pain)
Glucocorticoids	• prednisone (many generic human products) • prednisolone (PrednisTab®) • dexamethasone (Dexaject®, generic) • methylprednisolone (Depo-Medrol®) • triamcinolone (Vetalog®)	Works at the level of phospholipase A_2 to inhibit cyclooxygenase enzymes (COX) responsible for prostaglandin synthesis and inhibit lipoxygenase (LOX), which blocks production of inflammatory mediators such as leukotrienes
Nerve growth factor inhibitor	• See Chapter 16 (Chronic Pain Management)	
Analgesic adjunctives	• gabapentin (Neurontin®) • See Chapter 16 (Chronic Pain Management) for information on sodium channel blockers, alpha-2 adrenergic agonists, TrpV1 receptor agonists and antagonists, antidepressants, anticonvulsants, bisphosphates, and polysulfated glycosaminoglycan	Used for neuropathic pain Exact mechanism of action is unknown; may be involved with the interaction of calcium channels in the CNS

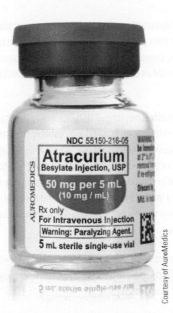

Courtesy of AureMedics

FIGURE 5-22 Atracurium is a competitive non-depolarizing neuromuscular blocking agent used as an adjunct to general anesthesia to inhibit muscle contraction. It can cause respiratory muscle paralysis resulting in hypoventilation and apnea; therefore, mechanical ventilation is required when atracurium is used.

resulting in hypoventilation and apnea; therefore, mechanical ventilation is required when neuromuscular blocking agents are used. Atracurium is not metabolized by the liver or kidneys, but its metabolite undergoes hepatic metabolism and can result in CNS toxicity and seizures. It must be refrigerated if not being used to prevent its decomposition.

Non-depolarizing neuromuscular blocking agents can be reversed with anticholinesterase inhibiting drugs including edrophonium (Enlon®, Tensilon®), neostigmine (Prostigmin®), and physostigmine (Mestinon®). Since cholinesterase breaks down acetylcholine, a drug that inhibits cholinesterase will result in more acetylcholine available to compete for binding sites. Anticholinesterase inhibiting drugs produce parasympathetic effects; therefore, their adverse effects include bradycardia, salivation, vomiting, nausea, diarrhea, and abdominal pain. Edrophonium has the least amount of adverse effects of the anticholinesterase inhibitors; adverse effects of this group of drugs can be treated with atropine or glycopyrrolate.

SUMMARY

There are a variety of drugs used in the practice of anesthesia including those that produce a loss of sensation, produce desired effects such as sedation or muscle relaxation, or help manage pain and the effects of other drugs. Each anesthetic protocol is different for every patient, and patient factors such as age, species, temperament, concurrent disease and medications, and previous response to anesthetic drugs should be considered before developing an anesthetic plan for each patient. The risks and benefits associated with each drug should be assessed and the safest drugs chosen for each patient. The

type of procedure, duration of the procedure, and level of pain the procedure causes should also be considered in drug selection. All effects, desirable and undesirable, are dosage dependent, making accurate drug calculation and diligent monitoring critical in veterinary patients.

CRITICAL THINKING POINTS

- Tranquilizers and sedatives are used perioperatively to induce sedation, cause muscle relaxation, provide restraint, and reduce the amount of injectable and inhalant anesthetic agents required to induce and maintain anesthesia.

- Induction agents are typically used to facilitate the passing of an endotracheal tube so that inhalant anesthetic agents can be given to maintain general anesthesia.

- Induction of general anesthesia with inhalant anesthetic agents is rapid but still much slower than induction with a rapid acting injectable agent such as propofol or ketamine/diazepam. This slower induction allows the patient to experience a prolonged excitement phase that can be minimized with the use of concurrent sedation.

- Inhalant inductions by mask or chamber induction increase patient stress, predispose them to cardiac arrhythmias, and expose staff to waste anesthetic gas; therefore, these methods of delivering inhalant anesthetic agents is not recommended.

- Inhalant anesthetic agents are usually liquids at room temperature (except desflurane) that are vaporized in the presence of oxygen or a carrier gas and are delivered to a patient via the respiratory tract. Inhalant anesthetic agents are used to induce and maintain general anesthesia in a variety of species.

- Isoflurane and sevoflurane undergo limited hepatic metabolism, and when the delivery of the drug is stopped, the patient begins to recover as the level of drug in the brain lowers.

- Local anesthetics block nerve conduction from the site of administration or application in the peripheral nervous system (PNS) and spinal cord. They are given by local infiltration of tissue, via epidural or intrathecal injection, by application to the corneal or mucous membrane surface, or by regional nerve blocks.

- Local anesthetics can be used to localize lesions in equine lameness exams, as nerve blocks in cattle to allow surgical and medical procedures, to aide endotracheal tube placement in cats and pigs by preventing laryngospasm, to prevent or decrease pain both during and after surgery, and can be used to perform surgery in a conscious patient, to relieve the pain caused by the original trauma, or to allow minor surgical procedures such as suturing minor wounds or taking a biopsy.

- Anticholingeric drugs are used perioperatively to counteract parasympathetic effects such as bradycardia and atrioventricular (AV) block caused by surgical manipulation (stimulation of the vagus nerve) or other anesthetic agents such as opioids.

- Alpha-adrenergic antagonists (α-blockers) usually promote vasodilation and a decrease in blood pressure and include the reversal agents yohimbine, atipamezole, and tolazoline.

- Beta-adrenergic drugs are used as positive inotropic agents to treat hypotension related to decreases in cardiac contractility and/or severe drops in systemic vascular resistance (seen in severe sepsis).

- Opioids are the most effective analgesics to treat acute pain in veterinary medicine and are a key component of pre-emptive analgesia.

- The opioid antagonist naloxone has a high affinity for opioid receptors and rapidly reverses all opioid-induced clinical effects including analgesia, therefore, its use should be reserved for emergency situations such as opioid overdose or profound respiratory depression.

- NMDA-receptor antagonists significantly impact the development of chronic pain and the development of tolerance to opioid analgesics. Consequently, NMDA-receptor antagonists may have potential as co-analgesics when used in combination with opioids.

- NSAIDs are effective, long-acting analgesics that have anti-pyretic and anti-endotoxic effects and are probably the most commonly used analgesics in veterinary practice. NSAIDs are used to treat mild to moderate pain when used alone and moderate to severe pain when used in conjunction with other drugs such as opioids. NSAIDs have a very low threshold for toxicity, and their potential for causing serious adverse effects may preclude their use in some patients under anesthesia.

- Glucocorticoids are potent anti-inflammatory agents that produce adverse effects greater than NSAIDs; therefore, they should be used cautiously in veterinary practice.

- Neuromuscular blocking agents are mainly used in veterinary medicine for ophthalmic procedures that need the eye centrally oriented.

REVIEW QUESTIONS

Multiple Choice

1. What is the study of the absorption, distribution, metabolism, and elimination of drugs?
   a. Metabolism
   b. Bioavailability
   c. Pharmacodynamics
   d. Pharmacokinetics

2. In which form do most drugs cross the cell membrane?
   a. Ionized
   b. Unionized
   c. Acidic
   d. Basic

3. What effects the bioavailability of a drug?
   a. Blood supply to the area, surface area of absorption, and properties of the drug
   b. Surface area of absorption, origin of the drug, and circadian rhythm of the animal
   c. Properties of the drug, origin of the animal species, and hydration status of the patient
   d. Origin of the drug, blood supply to the area, and patient weight

4. What is the process by which drug passes through the intestinal lumen to the liver, which reduces the amount of active drug in systemic circulation?
   a. Solubility
   b. Ionization
   c. First pass metabolism
   d. Hepatic bypass

5. A drug that has a bioavailability of 1 is most likely administered via which route?
   a. IV
   b. IM
   c. SQ
   d. Orally

6. What is a reason that juvenile or adolescent animals may require a higher dosage of induction agent than geriatric animals?
   a. They have a smaller body volume.
   b. They have a more acidic gastrointestinal environment.
   c. They have higher blood protein levels.
   d. They have a higher cardiac output.

7. What defines a drug with a narrow therapeutic index?

    a. Large margin of safety
    b. Small margin of safety
    c. Slow onset of action
    d. Rapid onset of action

8. What defines the duration of action of a hydrophilic opioid drug such as morphine?

    a. It is shorter than the duration of action of lipophilic opioids.
    b. It is longer than the duration of action of lipophilic opioids.
    c. It is equal to the duration of action of lipophilic opioids.
    d. Its duration of action is variable depending on patient's hydration status.

9. Administering opioids to dogs may cause _____, while giving them to cats may cause _____.

    a. hypothermia, hyperthermia
    b. bradycardia, tachycardia
    c. vomiting, salivation
    d. apnea, tachypnea

10. Which drug blocks dopamine receptors in the chemoreceptor trigger zone of the medulla, producing an antiemetic effect?

    a. Ketamine
    b. Xylazine
    c. Acepromazine
    d. Morphine

11. What effect does increasing the dosage of an alpha-2 adrenergic agonist cause in the body?

    a. Increases the intensity of sedation
    b. Prolongs the duration of activity
    c. Intensifies muscle rigidity
    d. Decreases muscle tremors

12. Which drug lacks a reversal agent?

    a. Diazepam
    b. Xylazine
    c. Morphine
    d. Acepromazine

13. Which drug should be avoided in patients with an esophageal foreign body?

    a. Xylazine because it may cause vomiting that will further damage the esophagus
    b. Acepromazine because it may lead to hypovolemia and cardiac disease
    c. Tramadol because it may cause dysphoria that will damage the vocal cords
    d. Butorphanol because it may cause respiratory and gastrointestinal depression

14. Which drug increases intracranial pressure?

    a. Ketamine
    b. Remifentanil
    c. Propofol
    d. Etomidate

15. Which drug does not provide analgesia?

    a. Xylazine
    b. Propofol
    c. Ketamine
    d. Butorphanol

16. Which drug is used in horses and other large animal species to increase blood pressure by improving cardiac output and contractility?

    a. Acepromazine
    b. Propofol
    c. Atropine
    d. Dobutamine

17. Why are inhalant anesthetic agents with high vapor pressures (isoflurane, sevoflurane, desflurane) usually delivered via a precision vaporizer?

    a. To prevent the anesthetic agents from producing cardiovascular adverse effects in animals
    b. To prevent malignant hyperthermia in large animal species
    c. To limit staff exposure to waste anesthetic gas
    d. To control the amount of anesthetic agent delivered to prevent overdosing the animal

18. Which anesthetic agent produces emergence delirium?

    a. Propofol
    b. Etomidate
    c. Ketamine
    d. Sevoflurane

19. Which class of analgesic drugs work by reducing prostaglandin production through inhibition of cyclooxygenase enzymes?

    a. Local anesthetics
    b. Opioids
    c. NSAIDs
    d. Anticholinergics

20. Which anticonvulsant is also considered an analgesic adjunctive?

    a. Methadone
    b. Diazepam
    c. Morphine
    d. Gabapentin

Case Studies

Case Study 1: You are working in a veterinary hospital where standard drug protocols are used. A 7-month-old F Collie puppy is scheduled for an ovariohysterectomy. The standard protocol is to give low-dosage acepromazine and a full μ opioid agonist as the preanesthetic medication. List some concerns regarding the use of this standard protocol for this patient. Suggest some preanesthetic options for this puppy.

Case Study 2: Scotty, a 6-month-old male Scottish Terrier canine, weighing 7.7 kg is admitted for a routine castration. What drug(s) would be a reasonable selection for induction of anesthesia following premedication with hydromorphone and acepromazine?

Case Study 3: A 6-month-old M DLH cat had a crushing injury to his left front paw and has been admitted for removal of the distal phalanx (P3) of digits IV and V. Which classes of drugs would be best to use for this patient? Start at premedication and go through the post-operative period.

Case Study 4: A 3-year-old Quarter Horse gelding presents with signs of colic. The veterinarian wants to give this horse xylazine for pain relief. What do you want to keep in mind when using xylazine in horses?

Case Study 5: A 5-year-old dog had a mass surgically removed and the veterinarian prescribes carprofen for analgesia in this patient. You notice in the medical record that the dog has been receiving glucocorticoids for atopy. What are your concerns, and how do you address these concerns?

Critical Thinking Questions

1. What would happen if isoflurane is accidently placed into a sevoflurane vaporizer?
2. What role does the ceiling effect have on the level of analgesia produced by a drug?

ENDNOTES

1. Roberts, F., & Freshwater-Turner, D. (2007). Pharmacokinetics and anaesthesia. Continuing education in anaesthesia. *Critical Care & Pain, 7*(1), 25–29.
2. Plumb, D. C. (2018). *Plumb's veterinary drug handbook* (9th ed., pp. 3–6; 94–96; 304–308; 403–404; 479–481; 690–693; 870–873; 1043–1046). Ames, IA: Blackwell Publishing.
3. Drynan, E. A., Gray P, & Raisis, A. L. (2012). Incidence of seizures associated with the use of acepromazine in dogs undergoing myelography. *J Vet Emerg Crit Care, 22*(2), 262–266.
4. Washington State University. (2014). Problem drugs, http://www.vetmed.wsu.edu/depts-vcpl/drugs.aspx
5. Hubbell, J. A., & Muir, W. W. (2006). Antagonism of detomidine sedation in the horse using intravenous tolazoline or atipamezole. *Equine Vet J., 38*(3), 238–241.
6. England, G. C., Clarke, K. W., & Goossens, L. (1992). A comparison of the sedative effects of three alpha 2-adrenoceptor agonists (romifidine, detomidine and xylazine) in the horse. *J Vet Pharmacol Ther., 15*, 194–201.
7. Simon, B. T., Scallan, E. M., Siracusa C., et al. (2014). Effects of acepromazine or methadone on midazolam-induced behavioral reactions in dogs. *Can Vet J., 55*(9), 875–885.
8. Sleigh, J., Harvey, M., Voss, L., Denny, B. (2014). Ketamine-More mechanisms of action than just NMDA blockage. *Trends in Anaesthesia and Critical Care, 4*(2–3), 76–81
9. Cattai A., Rabozzi R., Natale V., et al. (2014). The incidence of spontaneous movements (myoclonus) in dogs undergoing total intravenous anaesthesia with propofol. *Vet Anaesth Analg.*
10. Muniraj, T., & Aslanian, H. R. (2012). Hypertriglyceridemia independent propofol-induced pancreatitis. *Journal of Pancreas, 13*(4):451–453.
11. Kennedy M. J., & Smith L. J. (2014). A comparison of cardiopulmonary function, recovery quality, and total dosages required for induction and total intravenous anesthesia with propofol versus a propofol-ketamine combination in healthy Beagle dogs. *Vet Anaesth Analg.*
12. Dodam, J. R., Kruse-Elliott, K. T., Aucoin, D. P., & Swanson, C. R. (1990). Duration of etomidate-induced adrenocortical suppression during surgery in dogs. *Am J Vet Res., 51*(5), 786–788.
13. Ko, J., Thurmon, J., & Benson, G. (1993) Acute haemolysis associated with etomidate-propylene glycol infusion in dogs. *Veterinary Anaesthesia and Analgesia, 20*(2), 92–94.
14. Muir, W., Lerche, P., Wiese, A., et al. (2008). Cardiopulmonary and anesthetic effects of clinical and supraclinical doses of alfaxalone in dog. *Vet Anaesth Analg., 35*(6), 451–462.
15. Shaughnessy, M. R., & Hofmeister, E. H. (2014). A systematic review of sevoflurane and isoflurane minimum alveolar concentration in domestic cats. *Veterinary Anaesthesia and Analgesia, 41*(1), 1–13.
16. Fletcher, D., Boller, M., Brainard B., et al. (2012). Recover evidence and knowledge cap analysis on veterinary CPR. Part 7: Clinical guidelines. *JVECC, 22*(S1), S102–S131.

17. Haskell, S., Payne, M., Webb, A., Riviere, J., & Craigmill, A. (2005). Antidotes in food animal practice. *JAVMA, 226*(6), 884–887

18. Pascoe, P. J., Ilkiw, J. E., & Pypendop, B. H. (2006). Effects of increasing infusion rates of dopamine, dobutamine, epinephrine, and phenylephrine in healthy anesthetized cats. *Am J Vet Res., 67*(9), 1491–1499.

19. Mazzaferro, E., & Wagner, A. E. (2001). Hypotension during anesthesia in dogs and cats: Recognition, causes, and treatment. *Compendium, 23*(8), 728–737.

20. Egan, T. (2000). Pharmacokinetics and pharmacodynamics of remifentanil: An update in the year 2000. *Current Opinion in Anaesthesiology 13*, 449–455

21. Feldman, P., James, M. K., Brackeen, M. F., Bilotta, J. M., Schuster, S. V., Lahey, A. P., Lutz, M. W., Johnson, M. R. & Leighton, H. J. (1991). Design, synthesis, and pharmacological evaluation of ultra-short to long-acting opioid analgetics. *Journal of Medical Chemistry 34*, 2202–2208

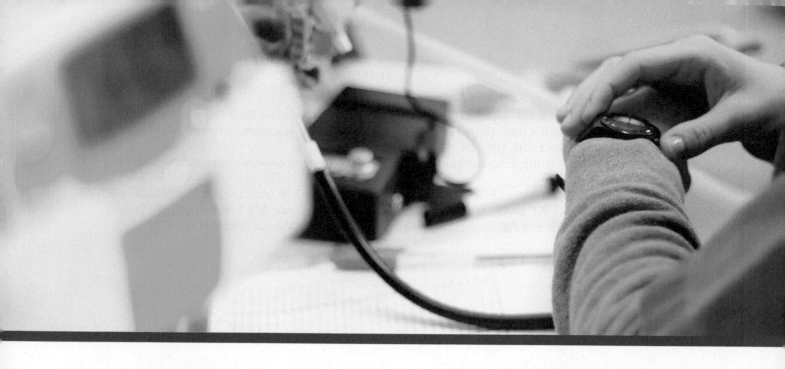

CHAPTER 6

Creating an Anesthetic and Analgesic Plan

Rebecca A. Johnson, *DVM, PhD, DACVAA*

LEARNING OBJECTIVES

Upon completion of this chapter, it is expected that the reader should be able to:

6.1 Explain the goals of the preanesthetic patient evaluation

6.2 Formulate an effective anesthetic plan beginning with the basic history and signalment

6.3 Create a "patient problem list" based on each individual patient

6.4 Describe the importance of preanesthetic patient preparation

6.5 Create a multimodal anesthetic protocol unique to each patient

6.6 Formulate an effective multimodal analgesic protocol complementary to the anesthetic protocol

6.7 Assess the adequacy and importance of a proper, complete anesthetic record

6.8 Complete anesthetic plan documents

INTRODUCTION

Formulation of an anesthetic plan is the most important step for a successful anesthetic event. When drug choices were limited and anesthetic monitoring was rudimentary at best, successful anesthesia was defined as one where the patient simply "woke up" afterward. With advances in anesthetic techniques and discovery of improved anesthetic agents, current anesthesia goals are to provide unconsciousness, muscle relaxation, and analgesia with minimal changes in normal physiology. Simply put, the goal is to proceed through anesthetic induction, maintenance, and recovery with minimal **morbidity** (diseased state, disability, or poor health) and NO **mortality** (state of death). Creating an individualized anesthetic plan which meets each patient's specialized needs is the first step in attaining this goal.

A significant amount of pre-anesthetic information must be obtained and evaluated prior to creation of a plan. This material includes patient **signalment** (description of the animal containing information such as species, breed, sex, age, and sexual status [intact or neutered]), prior anesthetic episodes, current physical exam findings, and diagnostic test results. Only after all data are gathered and assessed can the anesthetic plan be formulated choosing appropriate drugs and monitors. Appropriate documentation of the entire plan is essential as it is a legal record and provides a visual aid to assess patient condition in "real-time" during anesthesia and allows for quick intervention when problems occur.

PREANESTHETIC PATIENT EVALUATION

The primary goal of preanesthetic patient evaluation is to determine the animal's physical status[1] (the patient's overall condition). The physical status is not a means of assessing "risk" of the anesthetic procedure; however, it does aid in the selection of anesthetic/analgesic agents. Physical status may also alert the anesthetist to potential complications or adverse events that may arise during the procedure. Physical status evaluation consists of the: (1) history and signalment, (2) physical exam findings, and (3) diagnostic testing results[1] (see Appendix C). Although laboratory testing is commonly performed prior to anesthesia in veterinary patients, screening tests rarely improve patient outcome and seldom alter drug choice. Physical exam findings and history provide the most useful information and are the primary focus of the anesthetist when generating the physical status. The American Society of Anesthesiologists (ASA) Physical Status Classification System described in Chapter 3 (Cardiovascular and Respiratory Physiology Review) is the most commonly used system for assessing patient condition prior to anesthesia (see Appendix D).[2] ASA physical status is a subjective evaluation of the patient; however, increased mortality rates are somewhat associated with a higher ASA physical status.[3] Thus, the ASA physical status provides a useful tool for the anesthetist as the anesthetic plan is formulated.

During the preanesthetic evaluation, it is essential to focus on factors that may alter anesthetic drug activity and therefore patient management. This is relatively more important in less-than-healthy patients.[1] Important parts of the pre-anesthetic evaluation include:

- Gathering patient information including age, breed, temperament, history of previous anesthetic episodes, and type of procedure for which the patient is presenting.[4]
- Taking an in-depth history that includes previous and current disease status (noting both severity and duration), clinical signs of disease, and current drug administration (including over-the-counter medications).

- Performing a physical exam that concentrates on the cardiovascular and respiratory systems as they are significantly affected by many anesthetic agents.
  - ○ Recording body weight and vital signs (such as core body temperature, pulse rate, and respiratory rate) as part of every physical exam.

The history and physical exam will then guide the anesthetist in choosing further diagnostic tests. At a minimum, the "QATs" (quick assessment tests) are completed for future comparisons. The QATs include packed cell volume (PCV)/total protein (TP)/blood glucose/blood urea nitrogen (BUN) (from Azostick®). Additional analyses such as a blood chemistry and cell count, urinalysis, acid-base evaluation with blood gases, electrocardiography, or radiography may be warranted in particular situations. However, further tests are only completed if their results will change anesthetic management.[1] Only after the above information is assessed should an appropriate ASA physical status be assigned.

CREATING A PATIENT PROBLEM LIST

Once all pertinent patient information has been gathered, the anesthetist will formulate the anesthetic plan including a list of potential anesthetic problems. A patient "**SOAP**" form is recommended in each case and contains four parts: (1) *Subjective* component, (2) *Objective* component, (3) *Assessment*, and (4) *Plan*. **Subjective** information is based on a given person's experience, understanding, and feelings and includes the patient history, clinical signs, and behavioral/temperament assessment. **Objective** information is based on actual data and not influenced by emotions or personal prejudices and includes the body weight, vital parameters, physical exam findings, and additional results from diagnostic testing. The Assessment is then made, which includes the medical diagnosis and differential diagnoses for the problem. The Plan then follows, further detailing specific anesthetics and monitors (Figures 6-1a and 6-1b).

Using the information in the SOAP, a list of potential patient health concerns and how these may affect the anesthetic period is developed. This list should include the current pain status of the patient and the potential for the procedure to cause pain or stress. In addition to the specific health concerns each patient exhibits, specific anesthetic-associated problems must also be considered. Although each patient will have a unique list of problems associated specifically with the anesthetic agents and procedures chosen, some anesthetic-related adverse effects are more common in some species than others and are shown in Table 6-1.

TABLE 6-1 Common Anesthetic-Related Adverse Effects

Hypoxemia
Hypertension/Hypotension
Tachycardia/Bradycardia
Hyperventilation/Hypoventilation
Cardiac arrhythmias
Hyperthermia/Hypothermia
Pain
Regurgitation/Reflux
Hemorrhage (if applicable)

SOAP FORM

Procedure:
Date:
Anesthetist:

S.	Attitude & Temperament:		Body Score:

History:

Circle OR Non-rebreathing system	Oxygen flow rate: Induction:	Maint:	Recovery:
IV catheter size and location:	Breathing bag size (TV X 6):	ET Tube Sizes:	

O.	T:	P:	R:	Age:	Wt (kg):
PCV: (D: 35-55%, C: 30-50%)	TP: (D: 5-7.4 g/dL, C: 5.2-8.8 g/dL)	AZO: (5-15 g/dL)	GLU: (D: 70-140 mg/dL, C: 65-170 mg/dL)		

Cardio status & diagnostics: (CXR, Echo...)	Respiratory status: (CXR, etc.)	Neuro/musculoskeletal/GI/UG/ Abdominal exam & diagnostics: (Rads, US...)

Current/recent medications:

A.	Medical/Surgical problem list or Ddx:

ASA status		
I		
II	**Possible Anesthetic Complications:**	
III	1	6
IV	2	7
V	3	8
E	4	9
	5	

FIGURE 6-1 Example of a SOAP form.

Patient Name:		Species/Breed/Age/Sex:		Weight (kg):

P.		**Anesthetic and Pain Management Plan**		

Premedications	Route IM SC IV	Dosage	Dose (mg)/Volume (mL)

Induction Agent	Route	Dosage Range (mg/kg) AND Dosage volume (mL)	

Intraoperative dose if needed:

Inhalant: MAC:

IV Fluids	mL/kg/hr	mL/hr	gtt/sec (drops/sec)	Appropriate bolus vol

Other Techniques	Route	Drug	Dose	Intended effect

Total blood volume (est):		Shock dose crystlloids:

Emergency Drugs	Dose (in mg and mls)	
Glycopyrrolate (0.005-0.01 mg/kg)		[0.2 mg/mL]
Naloxone (0.04 mg/kg)		[0.4 mg/mL]
Atropine (0.02-0.04 mg/kg)		[0.4 mg/mL]
Epinephrine (0.01-0.02 mg/kg)		[1.0 mg/mL]
Lidocaine (2 mg/kg dog)		[20 mg/mL]

Post-op plan (pain meds, sedation etc)

Drug/Drug Type/Dosage (mg/kg)	Dose/Route/when to give/duration of action

Monitoring Plan:	Other considerations:
HR:	
RR:	
Blood pressure:	
EtCO$_2$:	
SpO$_2$:	
Temp:	
Other:	
	Plan Approval:

FIGURE 6-1 (Continued)

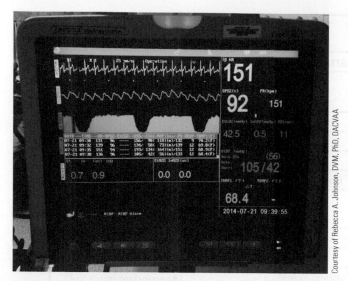

Courtesy of Rebecca A. Johnson, DVM, PhD, DACVAA

FIGURE 6-2 Multiparameter anesthetic monitor including pulse oximetry, non-invasive blood pressure assessment, inspired and expired gas concentrations, and electrocardiographic monitoring. Sinus tachycardia, commonly associated with pain or in patients at a light plane of anesthesia with pharmacologic agents such as ketamine, is present in this large breed dog.

It is important to understand how anesthetic agents can affect an animal's physiology. For example, inhalant anesthetic agents typically cause hypotension due to a decrease in systemic vascular resistance and vessel dilation.[5] In contrast, hypertension is associated with lack of proper anesthesia/analgesia. Bradycardia is common when sedatives such as alpha-2 adrenergic agonists or pure mu-opioids are administered. [6, 7] Hypoventilation is common with inhalant anesthetic agent and opioid use.[8] However, if pain is not controlled or if anesthetic depth is not sufficient, hyperventilation and sinus tachycardia may be seen (Figure 6-2). In addition, patients may experience hypothermia, cardiac arrhythmias (such as atrioventricular block or premature complexes), and regurgitation or reflux of gastric contents.

Unique situations present more specific problems. For example, hemorrhage would certainly be a potential problem in a dog undergoing liver lobectomy. Thus, each case is assessed on an individual basis and approached accordingly. To be prepared to intervene early and treat these abnormalities, the anesthetist documents them on the SOAP form. There should always be a well thought out treatment plan including pre-calculated emergency drug doses such as epinephrine, atropine, lidocaine, and the appropriate anesthetic reversal agents.

PREANESTHETIC PATIENT PREPARATION

When patients are not adequately prepared for anesthesia and/or surgery, the potential for complications increases greatly. Client communication is essential throughout all stages of the anesthetic protocol, but is especially important during patient preparation and formulation of the anesthetic plan. For example, appropriate fasting times should be discussed with clients. Dogs and cats are held off feed for approximately 6 to 12 hours

to minimize regurgitation. Some exotic animal species are fasted for significantly shorter time periods (less than 2 hours or in some species not at all) due to their high metabolic rates. Larger species such as ruminants usually require significantly longer fasts (calves, sheep, goats, camelids: 12–18 hours; adult cattle: 18–24 hours) to avoid hypoventilation, and hypoxemia.[1, 9] Previously, it was suggested that horses be fasted for 12 hours. However, recent techniques suggest shorter fasting periods (6 hours) or no fasting at all due to fasting-associated decreases in gastric pH and possible ileus although further studies concerning fasting in horses are still needed. Young animals require shorter fasting times (less than 4 hours) due to the risk of hypoglycemia; therefore, procedures on these patients should be scheduled earlier in the day.[4] Although water is usually not removed, inadvertent water deprivation can add to pre-existing dehydration.

Although all patients should have an extensive physical exam and assessment, emergency anesthetic procedures sometimes preclude these steps. Coexisting abnormalities are stabilized prior to anesthesia whenever possible to minimize future anesthetic complications. For example, patients with kidney disease are treated for any pre-existing dehydration in order to optimize kidney blood flow during the anesthetic procedure. Specific diseases and their unique anesthetic plans based on an individual patient's needs are described in Chapter 14 (Anesthetic Techniques for Special Cases).

Both acute/adaptive (physiologic pain that acts as a warning signal to notify the body of injury and/or tissue damage) and chronic (pathologic pain that serves no useful biologic purpose; also referred to as maladaptive) pain are treated prior to anesthesia if possible. In addition, the pre-emptive use of analgesics is recommended to reduce "**wind-up**" of pain pathways following a painful procedure and is a vital piece of the anesthetic plan (wind-up is concept that will be discussed in Chapter 16 (Chronic Pain Management) and is a heightened sensitivity that results in altered pain thresholds, both peripherally and centrally, so that pain is experienced in places unrelated to the original source). If pain is left untreated, the stress response is enhanced, which increases the release of endocrine and sympathetic factors.[10] The use of analgesic agents will decrease not only the pain response but also the stress response, resulting in less patient morbidity.

Although adequate patient preparation may not be possible in every situation (e.g., emergency procedures), the anesthetist should be organized and prepared to reduce inadvertent mistakes. All properly-working anesthetic equipment is assembled before anesthetic induction. Emergency supplies and protocols should be available and posted in a conspicuous place. The anesthetic record is started well before the patient is anesthetized (see Record Keeping below). The anesthetist should be extremely familiar with all aspects of the patient, anesthetic equipment, and reason for the anesthetic procedure. Communication between the anesthetist and surgeon (or other clinician) is key for the system of "checks and balances" to work appropriately and reduce potential complications.

Prior to the anesthetic procedure, the client should be made aware of any abnormal physical exam findings, diagnostic tests, etc. Clients are required to sign a **waiver** (a legal form that someone fills out with the intent of releasing someone else from liability) discussing the physical status of the patient as well as the potential risks associated with the plan. Although current anesthetic techniques are relatively safe due to improvements in drugs, monitoring devices, personnel training, etc., there is never a guarantee that an unanticipated event (such as death) will not occur. Thus, signed consent is a mandatory part of the anesthetic plan.

THE ANESTHETIC PROTOCOL

No single plan is appropriate for every patient due to differences in patient physiology and ASA physical status. The selection of anesthetic drugs is based on: (1) patient species, breed, and age, (2) ASA physical status, (3) procedure time, type, and severity, (4) the anesthetist's familiarity with the agents/anesthetic techniques, and (5) the anesthetic equipment available.[1] For example, anesthetic techniques appropriate for one species (such as birds) may not be effective in another (such as ruminants); species differences are discussed in later chapters. In addition, short-duration injectable techniques will not be suitable for longer procedures. For example, invasive soft tissue or orthopedic surgeries frequently require inhalant anesthetic agents. Inhalant anesthetic agent delivery requires the use of a vaporizer, anesthetic machine, and appropriate breathing system for the patient. If such equipment is not available, injectable anesthetics must then be used (each having their own advantages and disadvantages), as discussed in Chapter 5 (Anesthetic and Analgesic Pharmacology) and Chapter 9 (The Induction Period). However, in general, the anesthetic protocol combines different agents to minimize the adverse effects of each individual drug and ensures unconsciousness with sufficient muscle relaxation and analgesia.

Tech Tip

The anesthetic protocol must be formulated for each individual patient: "cocktails" or "protocols" for use in every patient are not recommended.

The anesthetic protocol is comprised of four individual parts: (1) premedication, (2) induction, (3) maintenance, and (4) post-anesthetic recovery. Premedication is frequently used, especially in fractious or overly-excitable animals, and has many advantages (Table 6-2). Premedication drug delivery is performed most often via the subcutaneous (SQ), intramuscular (IM), or intravenous (IV) route (although other routes such as oral or transmucosal are also possible).

TABLE 6-2 Advantages of Premedication

Aids in patient restraint
Reduces stress/anxiety
Decreases the quantity and associated adverse effects of concurrent drugs (e.g., induction and maintenance agents)
Evens the transition from consciousness into unconsciousness
Minimizes autonomic nervous system activity
May provide analgesia
May cause muscle relaxation

Tech Tip

Be organized—put together a "drug sheet" that lists all available premedication, induction, and analgesic agents your clinic has for use along with dosages and some potential adverse effects associated with each agent. This sheet can be posted conspicuously for all personnel to use.

Common classes of premedication agents include phenothiazines, benzodiazepines, alpha-2 adrenergic agonists, anticholinergics, opioids, and dissociatives. Specific drugs are chosen based on advantages and disadvantages of each agent and individual patient need as discussed in Chapter 5 (Anesthetic and Analgesic Pharmacology) and Chapter 8 (Premedication). The drug choice, dosage and dose, route, and time of administration are documented on the anesthesia record. Following administration (approximately 5–10 minutes if administered IV, 15–25 minutes following SQ or IM), the patient's response to premedication is also documented on the record. Examples may include graded, subjective statements such as "poor-, moderate-, or excess sedation." These descriptions provide information to the anesthetist for future procedures requiring sedation/anesthesia.

Following premedication, the patient should be relaxed and/or sedated but still conscious. An IV catheter is placed prior to anesthetic induction. An induction agent such as a dissociative/benzodiazepine mixture (e.g., ketamine [Ketaset®, VetaKet®, Ketaject®, Vetalar®]/diazepam [Valium®]), propofol (Rapinovet®, PropoFlo™, Propovan®, PropoFlo™28), etomidate (Amidate®), or alfaxalone (Alfaxan®, Alfaxan® Multidose, Alfaxan® Multidose IDX) is then administered IV. The IV route produces the quickest, smoothest transition period from consciousness to unconsciousness and should *always* be used if possible. Intravenous catheterization helps avoid drug leakage outside the vein, aids in fluid support, and provides a route for emergency drug administration. The specific agent, dosage, dose, and time of administration are documented on the record. Mask or chamber inductions are highly discouraged since they cause undue stress, environmental contamination, delayed airway control, and profound cardiorespiratory depression.[4]

Safety Alert

If you are presented with a fractious patient, the use of gloves, nets (for cats), muzzles (including "cone" muzzles for cats), and safety poles is warranted; the IM route is the most desirable in this situation. At-home "pre-sedation" with drugs such as gabapentin (Neurontin®, generic) or trazodone (Desyrel®) should also be considered to lessen overall patient stress prior to their arrival at the clinic. Mask or chamber ("box") inductions are not recommended and should only be used as the absolute last resort.

Following anesthetic induction, patients are frequently transitioned to inhalant anesthesia to provide anesthetic maintenance. Efficient delivery of inhalant anesthetic agents requires endotracheal intubation. The endotracheal tube (ETT) size is written on the record as well as the choice and level of inhalant anesthetic agent throughout the procedure (such as isoflurane [IsoFlo®, Isosol®, Aerrane®], sevoflurane [SevoFlo®, Petrem®, Flurovess™, Ultane®], or desflurane [Suprane®]). Endotracheal intubation offers the advantages of airway control with supplemental oxygen and the ability to breathe for the patient if required.

Similar to the choice of anesthetic agents, parameters to monitor during the maintenance period are chosen based on the specific needs of the patient. Monitoring always focuses on the cardiovascular and respiratory systems. According to the American College of Veterinary Anesthesia and Analgesia (ACVAA) guidelines, circulation, oxygenation, ventilation, and temperature should be monitored on every patient (Table 6-3).[11] Circulation ensures blood flow to the tissues; oxygenation ensures adequate oxygen levels in the blood; and ventilation supports adequate gas exchange at the lungs. The body temperature must be maintained near normal levels to maintain normal organ physiology and drug metabolism. Thus, direct visualization, heart and lung auscultation, and palpation are always performed. Incorporation of specific monitoring equipment such as invasive or non-invasive blood pressure monitors, pulse oximetry, core body temperature, and capnometry (measurement of CO_2) is an integral part of *every* anesthetic protocol (see Chapter 7 [Anesthesia Monitoring]). In addition, the ACVAA recommends that a dedicated anesthetist (with no additional responsibilities such as surgical preparation) be present throughout the entire procedure. All monitored parameters are also recorded during anesthetic maintenance (see Record Keeping below). Although constant vigilance and patient monitoring is required, parameters are written in the record every 5 to 10 minutes at a minimum. This time frame allows visualization of any trends in physiology that may occur so the anesthetist can intervene in a timely manner to reduce patient morbidity.

Tech Tip

It is helpful to keep all monitoring equipment in one localized area of the clinic where the batteries can be charged overnight and the monitors can be easily accessed for quick use on the following day. This is also a useful place to display an emergency drug chart (Figure 6-3).

TABLE 6-3 ACVAA Recommendations for Monitoring

Parameter	Common Monitoring Devices Used
Circulation	Ultrasonic Doppler device, invasive or non-invasive blood pressure monitor, pulse oximetry, pulse palpation
Oxygenation	Pulse oximetry, arterial blood gas analysis
Ventilation	Capnometry, spirometry, arterial blood gas analysis
Temperature	Esophageal or rectal temperature probe
Other	Electrocardiogram, muscle tone, anesthetic gas analyzer, visualization (mucous membranes, thoracic movement, etc.), palpation (pulses, ocular reflexes, etc.), auscultation (heart, lungs)

FIGURE 6-3 Organization and accumulation of anesthetic monitors in one location of the veterinary hospital. Emergency drug chart is clearly posted (white sheet).

Courtesy of Rebecca A. Johnson, DVM, PhD, DACVAA

During the recovery period, the time to extubation and body temperature are recorded. The patient is extubated at the return of the swallow (laryngeal) reflex and the patient should be monitored throughout recovery until stable and warm (greater than 98°F/37°C for dogs and cats). The patient's oxygenation is observed and recorded throughout the recovery period since many patients will still be expiring at least some levels of inhalant anesthetic agent (which produces respiratory depression) and will be transitioning from approximately 100% oxygen to room air (approximately 21% oxygen). Thus, hypoxemia is frequently seen during the recovery period; patients should remain on oxygen whenever necessary. The anesthetic protocol is not complete until the patient returns to near-normal physiology including consciousness.

THE ANALGESIC PLAN

The analgesic plan should be clearly thought through for each individual patient. Every animal does not require the same analgesic plan nor should every patient be administered "whatever is on the shelf" or "whatever doesn't require paperwork." Not all pain is created equal—it has different origins (e.g., mechanical, thermal, chemical), is transmitted through different pathways (e.g., Aβ fibers, Aδ fibers, C fibers), is modulated within the nervous system differently in individual patients, and is perceived differently based on species, concurrent drugs, etc. (see Chapter 15 [Acute Pain Management], Chapter 16 [Chronic Pain Management], and Chapter 17 [Analgesic Techniques]). However, pain does occur in every animal and every attempt to relieve pain should be made. Providing analgesia is not only humane, it is generally regarded as beneficial as patients return to normal physiology sooner and recover more quickly post-operatively when analgesics are given.[12] Unrelieved pain in veterinary species provides no benefit to animals,[13] although much variation still exists in when and how pain is treated in animals.[12, 14, 15,16] Specific analgesic agents, as well as their advantages and disadvantages are discussed in Chapter 5 (Anesthetic and Analgesic Pharmacology), Chapter 15 (Acute Pain Management), and Chapter 16 (Chronic Pain Management) of this book and in other references.[17] Although pain is assessed and managed differently in each species, one concept holds true—it is our obligation as veterinary professionals to alleviate pain in animals, and the focus must be on the individual patient whenever possible. The American Animal Hospital Association (AAHA) Pain Management Guidelines (https://www.aaha.org) provide in-depth descriptions of and suggestions for pain management in dogs and cats; algorithms on how to manage patient pain are an extremely useful part of the analgesic plan.[18]

The initial analgesic plan focuses on the type of pain since some analgesic agents may be better suited for acute/adaptive (e.g., opioids) versus chronic (e.g., gabapentin) pain. Pharmacologic intervention is therefore the mainstay of the analgesic plan. In addition, timing of analgesic administration is imperative. For example, administration of opioids or other analgesic agents (such as nonsteroidal anti-inflammatory drugs [NSAIDs]) *prior to* a painful procedure is more beneficial than administering the same dose of drug afterward.[18, 19]

The unique effects of each drug must be considered when formulating an effective analgesic plan. For example, administration of the short-acting, mild analgesic butorphanol (Torbugesic®, Torbutrol®) prior to surgical repair of a femur fracture will not adequately relieve pain and inflammation associated with the procedure. However, butorphanol may be appropriate for removal of a small dermal mass due to the short duration of the procedure and minimal inflammation/pain involved. Similarly, a constant rate infusion (CRI) of the potent opioid fentanyl (Sublimaze®) may be warranted in an invasive abdominal surgery that requires prolonged IV analgesia post-operatively. However, a fentanyl infusion would not be appropriate for a short dental procedure in which specific nerve blocks using local anesthetics would be an excellent choice (e.g., infraorbital nerve block with bupivacaine) (see Chapter 17 [Analgesic Techniques]).

Tech Tip 🐾

It is beneficial and efficient to have laminated pictures of commonly-performed local nerve blocks assembled near the anesthesia machine so veterinary personnel have visual aids to help with accurate needle placement near the targeted nerves.

In every case, the fundamental goal of the analgesic plan is to use a multimodal approach which uses a combination of different drug classes to target the pain pathway at multiple sites (translation, transmission, modulation, perception; see Chapter 15 [Acute Pain Management] and Chapter 16 [Chronic Pain Management]), thus reducing the dosage of each individual agent and their associated adverse effects. The concurrent use of drugs from different classes and specific analgesic techniques is common in the analgesic plan. See Tables 5-1 (alpha-2 adrenergic agonists), 5-5 (local anesthetic agents), 5-7 (opioids), and 5-9 (non-opioid analgesics) for specific information on these drugs.

The techniques and routes of drug administration vary greatly among patients and are again based on the agents themselves as well as specific patient requirements. Systemic administration (e.g., SQ, IM, IV) is common. However, other techniques include CRIs in which drug levels can be tightly controlled and changed quickly (especially suited for very short-acting agents such as fentanyl). Epidural or spinal administration is used to reduce adverse effects of various agents (e.g., profound respiratory depression associated with high blood levels of opioids). Oral, transdermal, or transmucosal administration of agents is commonly used when patients return home. Local infiltration of anesthetic is used to keep analgesia confined to a very specific location (e.g., dental nerves, sciatic nerve, femoral nerve). Each technique has its own advantages and disadvantages, but all are an integral part of the analgesic plan and should be incorporated when possible.

In addition to analgesic drugs, non-pharmacologic interventions are also employed. For example, cold therapy, acupuncture, physical therapy, chiropractic manipulation, and laser therapy are becoming increasingly popular in veterinary medicine (see Chapter 18 [Veterinary Physical Rehabilitation]). These interventions should be integrated into an overall analgesic plan for the patient in the immediate post-anesthetic period as well as during the entire recovery period, which can last many days, weeks, or even months.

An exceedingly important part of the analgesic plan is to evaluate the patient frequently and adjust the plan accordingly. The plan is not hard and fast, and deviations should and will occur. Pain scoring sheets have been validated in dogs[20] and cats[21] and have become an important part of clinical veterinary practice (see Chapter 15 [Acute Pain Management] and Chapter 16 [Chronic Pain Management]). In fact, AAHA Accreditation Standards include pain scoring as one of their criteria.[22] Thus, it is clear the analgesic plan, although initially formulated as an integral part of the anesthetic plan, continues well into the post-anesthetic period. Education of all personnel involved in the case—the veterinarians and veterinary staff as well as the client—is key, and everyone must work together to implement the overall anesthetic plan.

RECORD KEEPING AND ANESTHETIC MONITORING

Anesthetic-related morbidity and mortality are never acceptable in veterinary medicine. Perioperative risk for the patient depends on the interaction of anesthesia-, surgery-, and patient-specific factors.[23] Specific anesthesia-related factors linked to patient morbidity/mortality have been studied and reviewed in small animals[24] and horses.[25] Indeed, the overall risks of anesthetic- and sedation-related death in dogs and cats are relatively low (0.17%, and 0.24%, respectively) whereas rabbits and horses are somewhat higher (1.39% and approximately 1%, respectively).[25, 26] In an attempt to keep these figures low (or even further reduce them), anesthetic record keeping has been employed and is highly suggested by the ACVAA.[4, 11] Written documentation enables practitioners to obtain and record accurate, meaningful data throughout the procedure. It also provides a legal document for the veterinarian (and an additional record of DEA-controlled substances) and allows for recognition of trends in monitored parameters. A template can be found in Appendix H and on the AAHA website for use by the veterinary professional.[27] Although records may vary between clinics, specific information should be found on every document (Table 6-4).

TABLE 6-4 Information Included on the Anesthetic Record

Client's name and case number
Patient's name, species, breed, sex, age, and weight
ASA physical status
Procedure
Date of procedure
Pre-anesthetic vital signs
Pre-anesthetic diagnostic testing results
Endotracheal tube size and anesthetic system specifics (circle rebreathing vs. non-rebreathing)
All pre-anesthetic and anesthetic agents, route and time of administration with patient response
Monitored parameters
Additional drugs, problems encountered, and corrective measures
Extubation time and patient notes
Anesthetist's signature

Tech Tip

It is helpful to have a clipboard on which to write the anesthetic record and attach the completed SOAP form. On the reverse side, a sheet listing potential anesthetic drugs and dosages and/or the machine check off procedure could be attached so all pertinent information is within reach if quickly needed. The anesthetic record is a legal, written document and should be completed in blue or black ink

EXAMPLE OF "THE PLAN"

With an understanding of the rationale behind the anesthesia plan, the following example illustrates how an anesthesia plan is developed.

History and Signalment

You are presented with a 5-year-old female Golden Retriever, DiDi, whose owners finally agreed to have her spayed. She is scheduled for surgery in the clinic tomorrow. Her prior history is quite unremarkable except she has had two litters of puppies and the owner believes she is "in heat" at this time. She has no history of vomiting or diarrhea (except when getting in the garbage) and has no heart or lung disease. Infrequently, she has had seasonal allergies that are treated with the glucocorticoid prednisone. She has never had anesthesia before and the owner reports she has always been healthy and a good mother.

Physical Exam

The majority of her physical exam is unremarkable although her vulva appears red and swollen (consistent with signs of "heat"). Her heart and lungs sound clear and normal with no murmurs or arrhythmias. She is markedly obese (50 kg) with a body condition score of 8 out of 9. Her temperature is 101°F, pulse rate is 120 beats per minute, and respiratory rate is difficult to determine since she is panting.

Anesthetic Plan

DiDi's SOAP form (two pages) with all pertinent information, anesthetic/analgesic plan and further considerations is shown in Figure 6-4.

Note that the front page is dedicated to the "S", "O" and "A" portion while the "P" is on a separate page. Important items to note on her SOAP form include the following:

- History and signalment are documented at the top of the plan. Because of her obesity, her lean body weight is estimated and all drug doses are calculated based on this lean weight.
- The anticipated anesthetic equipment for the procedure is also included.
- All physical exam findings and diagnostic tests are listed followed by the medical/surgical problem list and the ASA physical status.
- Potential complications are well documented.
- The plan includes all premedications, induction, and inhalant anesthetic agents. Additional induction agent doses to use if the patient becomes "light" during the procedure are calculated. Anticipated inhalant anesthetic agent concentrations (minimum alveolar concentration [MAC] values) are noted.
- Fluid type and rate of administration are clearly calculated along with the patient's total blood volume (especially since hemorrhage is likely).
- All emergency drug doses are calculated in milligrams and milliliters.
- The analgesic plan is clearly detailed and encompasses the pre-anesthetic period as well as the "go home" plan.
- The plan is signed off by a veterinarian and can be implemented at any time. Since the plan is comprehensive, all personnel involved should be able to read the document and become effectively familiar with the case.

Tech Tip

It is helpful to have a clipboard on which to write the anesthetic record and attach the completed SOAP form. On the reverse side, a sheet listing potential anesthetic drugs and dosages and/or the machine check off procedure could be attached so all pertinent information is within reach if quickly needed.

Anesthetic Record

A sample anesthetic record for DiDi's procedure is shown in Figure 6-5.

Note that all pertinent information is detailed at the top of the document including an abbreviated history and signalment and all necessary laboratory values. The top part can be filled out ahead of the anesthetic procedure. All anesthetic/analgesic agents and physiologic responses are similarly documented. Monitoring and anesthetic equipment including endotracheal tube size are incorporated into the record. Although all parameters are monitored continuously, values are documented every 5 minutes using appropriate symbols. Trends are clearly seen as hemorrhage occurred, which resulted in quick interventions by the anesthetist (e.g., fluid bolus, hetastarch administration, decreased inhalant anesthetic agent level). Recovery notes and times are included. This record will provide a great deal of information for future anesthetic periods and provides a legal document for administrative use.

SOAP FORM

"DiDi" Johnson

5 year old, female

Golden Retriever

Procedure:	Ovariohysterectomy
Date: 25-July-2020	
Anesthetist: Becky Johnson	

S.	**Attitude & Temperament:**	**Body Score:**
	Very nice dog, energentic	8 out of 9

History:

Has had two previous litters	Seasonal allergies
Currently in estrus	No prior anesthetic procedures
Obese	

Circle OR Non-rebreathing system	**Oxygen flow rate: Induction:** 3 L/min	**Maint:** 1 L/min	**Recovery:** 3 L/min
Circle system			

IV catheter size and location:	**Breathing bag size (TV X 6):**	**ET Tube Sizes:**
20 gauge- cephalic	50 kg X 10 mL/kg X 6 = 3L	~ 10, 12, 14 mm

O.	**T:** 101°F	**P:** 120 bpm	**R:** pant	**Age:** 5 yrs	**Wt (kg):** 50 kg (lean weight 40 kg

PCV: 45%	**TP: 6.8 g/dL**	**AZO: 5-15 g/dL**	**GLU: N/A**
(D: 35-55%, C: 30-50%)	(D: 5-7.4 g/dL, C: 5.2-8.8 g/dL)	(5-15 g/dL)	(D: 70-140 mg/dL, C: 65-170 mg/dL)

Cardio status & diagnostics: (CXR, Echo...)	**Respiratory status:** (CXR, etc.)	**Neuro/musculoskeletal/GI/UG/ Abdominal exam & diagnostics:** (Rads, US...)
Heart auscults normal No radiographs done	Lungs auscult normal No radiographs done	Vulva red and swollen Markedly obese No further diagnostics done

Current/recent medications:
On heartworm and flea preventative currently and prednisone occasionally No other medications

A.	**Medical/Surgical problem list or Ddx:**
ASA	1. Obese
status	2. Intact female
I	3. In estrus
(II)	**Possible Anesthetic Complications:**
III	1. Hypoxia 6. Hyper/hypothermia
IV	2. Hyper/hypotension 7. Pain
V	3. Tachy/bradycardia 8. Regurgitation
E	4. Hyper/hypoventilation 9. Hemorrhage
	5. Arrhythmias

FIGURE 6-4 Example of a completed canine SOAP form.

Patient Name: DiDi Johnson		Species/Breed/Age/Sex: Canine/Golden Retriever/5yrs/F	Weight (kg): 50 kg
			Lean body weight = 40 kg

P.	**Anesthetic and Pain Management Plan**		
Premedications	**Route IM SC IV**	**Dosage**	**Dose (mg)/Volume (mL)**
Acepromazine	IM	40 kg X 0.02 mg/kg = 0.8 mg	0.8 mg/0.4 mL
Hydromorphone	IM	40 kg X 0.1 mg/kg = 4.0 mg	4.0 mg/1.0 mL
Carprofen	SC	40 kg X 4.4 mg/kg = ~175 mg	175 mg/3.5 mL
Induction Agent	**Route**	**Dosage Range (mg/kg) AND Dosage volume (mL)**	
Propofol	IV	40 kg X 6 mg/kg = 240 mg = 24 mL	
		Given "to effect"	

Intraoperative dose if needed: 1 mg/kg = 40 mg = 4 mL if patient "waking up"

Inhalant: isoflurane   Isoflurane MAC: 1.3%

IV Fluids	**mL/kg/hr**	**mL/hr**	**gtt/sec** (drops/sec)	**Appropriate bolus vol**
Plasmalyte-A	5	200 mL/hr	~ 0.5 drip per second	200-400 mL/15 minutes
			10 drip/mL set	

Other Techniques	**Route**	**Drug**	**Dose**	**Intended effect**
Incisional line block	SC	bupivicaine	< 1-2 mg/kg total	Block sensory and motor transmission
			~50 mg/10 mL	

Total blood volume (est): 3.6 liters		Shock dose crystalloids: 3.6 liters

Emergency Drugs	**Dose (in mg and mls)**	
Glycopyrrolate (0.005-0.01 mg/kg)	0.2-0.4 mg/1.0-2.0 mL	[0.2 mg/mL]
Naloxone (0.04 mg/kg)	1.6 mg/4 mL	[0.4 mg/mL]
Atropine (0.02-0.04 mg/kg)	0.8-1.6 mg/2.0-4.0 mL	[0.4 mg/mL]
Epinephrine (0.01-0.02 mg/kg)	0.4-0.8 mg/0.4-0.8 mL	[1.0 mg/mL]
Lidocaine (2 mg/kg dog)	80 mg/4 mL	[20 mg/mL]

Post-op plan (pain meds, sedation, etc.)	
Drug/Drug Type/Dosage (mg/kg)	Dose/Route/when to give/duration of action
Fentanyl patch 100 micrograms/hr	Placed just following surgery: 12 hours to effect, lasts 3 days
Hydromorphone 0.1 mg/kg	4.0 mg IM every 4 hours until patch takes effect
Carprofen 2.2 mg/kg PO BID	~75 mg PO every 12 hours for 3-5 days

Monitoring Plan:	**Other considerations:**
HR: Taken from pulse oximeter	Older, obese dog: increased risk for hemorrhage anticipated
RR: Taken from capnometer	Surgical procedure will be prolonged: increased pain expected
Blood pressure: Non-invasive monitor used	Doses based on lean body weight
EtCO$_2$: Taken from capnometer	Hypoventilation expected due to obesity and used of depressant drugs
SpO$_2$: Taken from pulse oximeter	
Temp: Esophageal temp probe	
Other: Electrocardiogram	
	Plan Approval: Becky Johnson, DVM

FIGURE 6-4 (Continued)

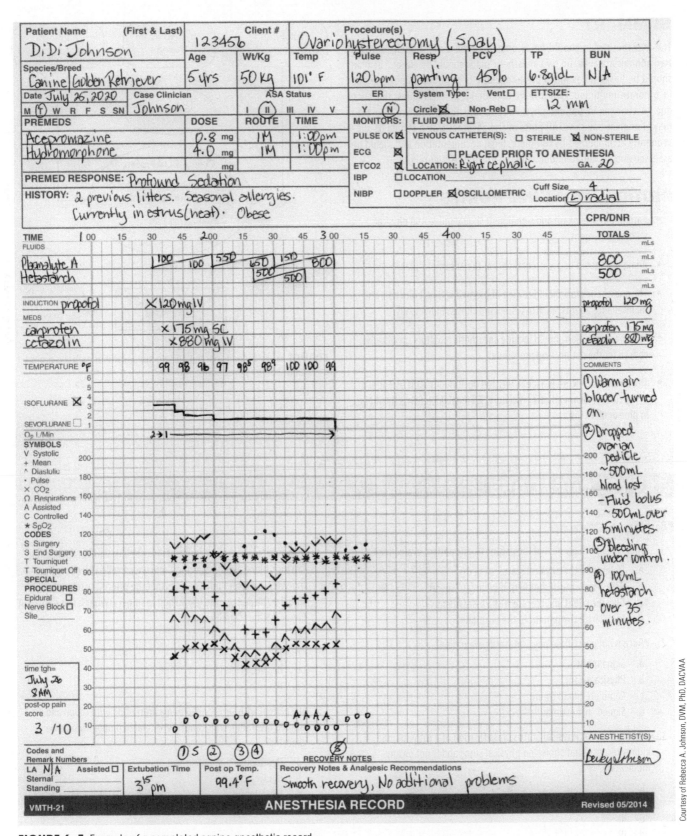

FIGURE 6-5 Example of a completed canine anesthetic record.

SUMMARY

The anesthetic plan should be as unique as the patient itself. Since no two animals are alike, no two anesthetic plans should be either. The plan should be based on as much pre-anesthetic information as possible and incorporate multimodal anesthetic and analgesic techniques. However, due to changes in patient physiology during the procedure, the plan may change at any time. All parts of the anesthetic plan should be subsequently accompanied by complete, legal documentation and should be assessed and reevaluated following the procedure (Figure 6-6).

CRITICAL THINKING POINTS

- Patient physical exam and signalment should guide ASA physical status assessment.

- Analgesia should be an integral part of the anesthetic plan.

- The anesthetic plan should consist of premedication, induction, and maintenance agents, each assessed and chosen individually.

- The anesthetic plan should include anesthetic monitors specifically chosen for the patient and procedure.

- The anesthetic record is a legal, written document that is essential to every successful anesthetic procedure. All legal documents should be in blue or black ink.

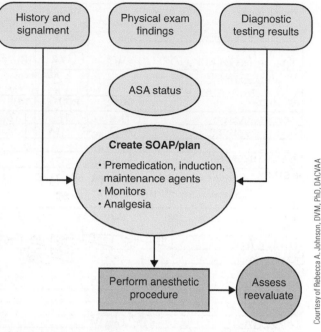

FIGURE 6-6 Steps for completing a successful anesthetic record and perianesthetic procedure.

Courtesy of Rebecca A. Johnson, DVM, PhD, DACVAA

REVIEW QUESTIONS

Multiple Choice

1. What is the primary goal of the preanesthetic patient evaluation?
   a. Determine the physical status
   b. Assess how much the client is willing to spend
   c. Give the client an accurate "risk" assessment
   d. Familiarize the patient to the hospital

2. Which is *not* part of the physical status evaluation?
   a. Signalment
   b. Physical exam
   c. Complete blood count
   d. Signed cost estimate

3. What is the ASA Physical Status Classification of a cat with mild kidney disease that shows no physical signs and is not on any medication?
   a. I
   b. II
   c. III
   d. IV

4. A dog that was hit-by-car presents to the clinic with an open femur fracture, blood around the lungs, and uncontrolled bleeding from the liver. The veterinarian decides that the veterinary team should stabilize the dog prior to taking him to surgery. What is this dog's ASA Physical Status Classification?
   a. I
   b. II
   c. III
   d. IV

5. In a patient with severe pneumonia presenting with cough and labored breathing, which anesthetic-related problem would be expected?
   a. Hypoxemia
   b. Bradycardia
   c. Hypertension
   d. Hemorrhage

6. Which emergency drug should be pre-calculated prior to the anesthetic procedure?

    a. Sodium bicarbonate
    b. Prednisone
    c. Epinephrine
    d. Penicillin

7. How long should dogs and cats be held off feed prior to anesthesia?

    a. 1 to 2 hours
    b. 6 to 12 hours
    c. 12 to 18 hours
    d. 18 to 24 hours

8. Which is not part of the selection process for anesthetic drugs in a patient?

    a. Patient age
    b. Requirement for DEA documentation
    c. Procedural time
    d. Anesthetic equipment available

9. Which is an advantage of premedication?

    a. Aids in patient restraint
    b. Causes the patient to vomit stomach contents
    c. Increases autonomic activity
    d. Speeds the entire anesthetic period duration

10. Which is an inappropriate drug selection in an anesthetic protocol designed for a healthy dog undergoing a neuter procedure (premedicant; induction agent; maintenance)?

    a. Acepromazine/butorphanol; propofol; isoflurane
    b. Dexmedetomidine/hydromorphone; propofol; sevoflurane
    c. Midazolam/butorphanol; mask with isoflurane; isoflurane
    d. Acepromazine/hydromorphone; ketamine/diazepam; sevoflurane

11. Which monitoring parameter recommended by the ACVAA as part of the anesthetic plan does a capnometer specifically measure?

    a. Oxygenation
    b. Temperature
    c. Cardiac arrhythmias
    d. Ventilation

12. How often should monitored parameters be documented on the anesthetic record?

    a. Every 3 minutes
    b. Every 5 minutes
    c. Every 15 minutes
    d. Every 30 minutes

13. Which is not part of the anesthetic record?

    a. Time to endotracheal extubation
    b. Time that premedications are administered
    c. Time to patient walking
    d. Total time of the anesthetic procedure

14. What is true concerning the analgesic plan when formulated as part of the overall anesthetic plan?

    a. Administration of analgesics before the painful procedure is beneficial to the patient
    b. Controlled analgesic agents (by the DEA) should only be used if absolutely necessary
    c. The analgesic plan should end at the conclusion of the anesthetic procedure
    d. A single analgesic agent (versus multiple agents) always works best

15. Which is not a common class of analgesic agent?

    a. Phenothiazines
    b. Alpha-2 adrenergic agonists
    c. Local anesthetics
    d. Non-steroidal anti-inflammatory agents

16. What anesthetic/analgesic agent is best suited for an intravenous constant rate infusion?

    a. One whose time to peak action is prolonged
    b. One that does not need to be changed frequently during the procedure
    c. One with a short duration of action
    d. One that is irritating when placed IV

17. Which is false concerning the anesthetic plan?

    a. Once set, the anesthetist should not deviate from the plan or protocol.
    b. A pain scoring system is recommended for use on every patient requiring anesthesia for painful procedures.
    c. Similar to anesthetic agents, anesthetic monitors should be chosen on an individual basis, based on specific patient requirements.
    d. Physical exam findings belong in the "O" portion of the SOAP.

18. Which species has the lowest overall risk for anesthetic and sedation-related death?

    a. Horses
    b. Rabbits
    c. Cats
    d. Dogs

19. Which should not be included on the patient's anesthetic record?

    a. Procedure type
    b. Patient's weight
    c. Client's phone number
    d. Endotracheal tube size

20. What is the appropriate action when a client leaves the hospital before signing the anesthesia consent form?

    a. Attempt to reach them on the phone as soon as possible to obtain verbal consent which is subsequently documented in the record.
    b. Continue to perform the anesthetic procedure without consent since current techniques are relatively safe.
    c. Postpone the procedure until the following day and have a discussion when the client returns to pick up the patient.
    d. Have the veterinarian sign the consent form and take full responsibility for the procedure.

Case Studies

Please fill out SOAP forms for the following cases presented to your hospital. SOAP forms are available in the online student resources for this product. Sign up or sign in at www.cengage.com to search for and access this product and its online resources. Note that there are multiple correct choices for pharmacologic agents and the answer key contains only some of the many acceptable protocols.

Case Study 1: An 8-year-old male neutered Siamese cat named Mike presents for removal of a fractured upper right first premolar tooth. He is a reserved cat with a "short fuse." You are able to obtain blood work, but he currently has lost his patience and is now hissing and striking at you. The heart rate is 180 beats per minute, and no cardiac murmurs are heard although the exam was short. The owner reports he has been losing weight. He is slightly thin at 3 kg with a body condition score of 3 out of 9. No temperature was taken due to his temperament, and his respiratory rate is 30 breaths per minute. No other physical exam was done, but he does appear painful near his mouth. Packed cell volume was 38% (normal 28–47%), total protein was 6.2 g/dL (normal 5.7–8.6 g/dL), and blood urea nitrogen was 32 mg/dL (normal 13–36 mg/dL) and creatinine was 2.0 mg/dL (normal 0.5–2.5 mg/dL), suggesting early-stage kidney disease; electrolytes are normal. Urine specific gravity was not done. He is scheduled for dentistry later this afternoon; the procedure is anticipated to be quite painful. The pharmacologic agents available for use in your clinic are bupivacaine, dexmedetomidine, fentanyl patches, hydromorphone, isoflurane, ketamine, midazolam, and propofol.

Fill in the SOAP form available in the online student resources for this product.

Case Study 2: A 2-year-old male intact Clydesdale horse named Bruce presents for castration. Since it is the first one your veterinarian has performed, he is scheduled for general anesthesia using inhalant anesthetic agents tomorrow. He weighs 900 kg with a body score of 4 out of 9. His heart and lungs auscult normal and he has no significant physical exam findings. His temperature is 101°F, heart rate is 30 beats per minute, and respiratory rate is 24 breaths per minute. A packed cell volume was 35% and total protein 5.6 g/dL (both within normal limits of your laboratory). Although young, he has a very nice disposition and is quite calm.

Fill in the SOAP form available in the online student resources for this product.

Case Study 3: A 5-year-old male neutered Greyhound dog named Speed presents for mid-tail amputation since he caught it in a door two days ago and continues to bleed from the nonhealing tip. His veterinarian administered carprofen and referred him for surgery. He is very nervous and is shaking during the physical exam. The heart rate is 150 beats per minute and a grade II/VI left systolic murmur is heard on physical exam. The owner reports no other significant history except he was rescued "off the track" 3 years ago. He is thin (as most Greyhounds are) at 30 kg with a body condition score of 2 out of 9. Body temperature is 103°F, and he is panting excessively. All other physical exam findings are normal except the distal tail tip is painful and dried blood is noted. On the blood work, the packed cell volume is 60% and total protein is 7.4 g/dL; all other values are within normal laboratory limits. He has fasted and is scheduled for surgery today, which is anticipated to be quite painful. The owner declines further work up for the heart murmur. The pharmacologic agents available for use in your clinic are acepromazine, bupivacaine, dexmedetomidine, fentanyl patches, hydromorphone, isoflurane, ketamine, midazolam, preservative-free morphine, and propofol.

Fill in the SOAP form available in the online student resources for this product.

Case Study 4: A 6-week-old male, medium-sized mixed breed puppy named Thor presents for front right dewclaw removal since he avulsed it playing yesterday following his first set of puppy vaccines. He is very excitable and is difficult to settle down for physical exam. The heart rate is 148 beats per minute and no cardiac murmurs are heard. The owner reports no other significant history except for intermittent diarrhea; he has been treated for intestinal parasites. He is small at 2 kg with a body condition score of 4 out of 9. Body temperature is 101°F and respiratory rate is 36 breaths per minute. All other physical exam findings are normal and both testicles are descended. Packed cell and total protein values are 36% and 4.6 g/dL, and glucose levels are 60 mg/dL; no additional diagnostic tests are performed. He is scheduled for surgery today and will go home tonight following the procedure. The pharmacologic agents available for use in your clinic are acepromazine, bupivacaine, dexmedetomidine, fentanyl patches, hydromorphone, isoflurane, ketamine, midazolam, preservative-free morphine, and propofol.

Fill in the SOAP form available in the online student resources for this product.

Case Study 5: A 14-year-old female, spayed Savannah cat named Binky presents for removal of a mass located between her scapulae that the owners recently found. She is very nice but is quite difficult to restrain. Her heart rate is 190 beats per minute and a III/VI left systolic murmur is ausculted, which has been present and stable for a few years. The owner does not want further diagnostics for the heart murmur and reports no other significant history except she may be drinking slightly more than usual. She weighs 5 kg with a body condition score of 5 out of 9. Body temperature is 102.5°F, and she is purring. All other physical exam findings are normal, and her claws are extremely sharp (and she is using them well!). Blood work could not be obtained since she is difficult to restrain, and her veins are small and friable. She is scheduled for surgery today, which is planned to be extensive. The pharmacologic agents available for use in your clinic are acepromazine, bupivacaine, dexmedetomidine, fentanyl patches, hydromorphone, isoflurane, ketamine, midazolam, preservative-free morphine, and propofol.

Fill in the SOAP form available in the online student resources for this product.

Critical Thinking Questions

1. When creating an anesthetic plan, it is important to be as prepared as possible and anticipate any problems that may occur during the anesthetic event. To prepare an anesthetic plan, it is important know the following information:

a. Name the four parts of the patient SOAP form:

S: _____

O: _____

A: _____

P: _____

b. List five common anesthetic-related problems:

a. _____

b. _____

c. _____

d. _____

e. _____

c. What are the four parts of general anesthesia that need to be addressed in an anesthetic protocol?

a. _____

b. _____

c. _____

d. _____

d. List four monitoring criteria recommended by the ACVAA as part of the anesthetic plan:

a. _____

b. _____

c. _____

d. _____

e. List three non-pharmacologic interventions that can be part of the overall anesthetic/analgesic plan:

a. _____

b. _____

c. _____

2. To think critically about a patient's anesthetic plan, it is important to have information committed to memory in order to make decisions about patient care and how to troubleshoot complications that may arise. Answer the following true/false questions as quickly and accurately as you can:

a. If present just prior to anesthesia, acute and chronic pain do not need to be treated until after the procedure.

b. It is appropriate to mix up premedications in one bottle and use the same mixture for multiple patients over one week.

c. If an analgesic is known to work for acute pain, it is acceptable to use it for all acute pain in every species.

d. Pain has been shown to provide no known benefit to animals when left untreated.

e. Acupuncture and physical therapy are frequently used as non-pharmacologic interventions within the analgesic plan.

f. The anesthetic record is a legal document.

g. Multimodal anesthetic techniques should be chosen as part of an adequate anesthetic plan.

h. Obtaining all diagnostic testing possible prior to formulating the anesthetic plan is more important that obtaining an adequate history and performing a thorough physical exam for physical status assessment.

i. All agents used on dogs and cats will also be effective in an anesthetic plan for a bird.

j. Hypotension is an uncommon occurrence when isoflurane is used as part of an anesthetic protocol.

ENDNOTES

1. Muir, W. W. (2007). Considerations for general anesthesia. In W. J. Tranquilli, J. C. Thurmon, & K. A. Grimm (Eds.) *Lumb & Jones' veterinary anesthesia and analgesia* (4th ed., pp. 7–30). Ames, IA: Blackwell.

2. ASA Physical Status Classification System. American Society of Anesthesiologists. www.asahq.org/Home/For-Members/Clinical-Information/ASA-Physical-Status-Classification-System

3. Vacanti, C. J., VanHouten, R. J., & Hill, R. C. (1970). A statistical analysis of the relationship of physical status to postoperative mortality in 68,388 cases. *Anesth Analg 49*, 564–566.

4. Bednarski, R., Grimm, K., Harvey, R., et al. (2011). AAHA anesthesia guidelines for dogs and cats. *J Am Anim Hosp Assoc 47*, 377–385.

5. Pagel, P. S., Farber, N. E., Pratt, P. F. Jr., et al. (2010). Cardiovascular physiology. In R. D. Miller (Ed.), *Miller's anesthesia* (7th ed., pp. 595–632). Philadelphia, PA: Churchill Livingstone.

6. Fukuda, K. (2010). Opioids. In R. D. Miller (Ed.), *Miller's anesthesia* (7th ed., pp. 769–824). Philadelphia, PA: Churchill Livingstone.

7. Reves, J. G., Glass, P. S. A., Lubarsky, D. A., et al. (2010). Intravenous anesthetics. In R. D. Miller (Ed.), *Miller's anesthesia* (7th ed., pp. 719–768). Philadelphia, PA: Churchill Livingstone.

8. Brunson, D. B., & Johnson, R. A. (2014). Respiratory disease. In Culp-Snyder LB, Johnson RA (Eds.) *Canine and feline anesthesia and co-existing disease*. Hoboken, NJ: Wiley.

9. Riebold, T. W. (2015). Ruminants. In W. J. Tranquilli, J. C. Thurmon, & K. A. Grimm (Eds.), *Lumb & Jones' veterinary anesthesia and analgesia* (5th ed., p. 748). Ames, IA: Blackwell.

10. Muir, W. W. (2002). Pain and stress. In J. S. Gaynor, & W. W. Muir (Eds.), *Handbook of veterinary pain management* (p. 46). St Louis: Mosby.

11. Small Animal Monitoring Guidelines. American College of Veterinary Anesthesia and Analgesia, www.acva.org, accessed May 28, 2014.

12. Crook, A. (2014). Pain: An issue of animal welfare. In C. M. Egger, L. Love, & T. Doherty (Eds.), *Pain management in veterinary practice* (pp. 3–8). Ames, IA: Blackwell.

13. ACVA. (1998). American College of Veterinary Anesthesiologists' position paper on the treatment of pain in animals. *J Am Vet Med Assoc 213*, 628–630.

14. Hellyer, P. W., Frederick, C., Lacy, M., et al. (1999). Attitudes of veterinary medical students, clinical faculty, and staff toward pain management in animals. *J Am Vet Med Assoc 214*, 238–244.

15. Hugonnard, M., Leblond, A., Keroack, S., et al. (2004). Attitudes and concerns of French veterinarians' use of analgesics in cattle, pigs, and horses in 2004 and 2005. *Can Vet J 48*, 155–164.

16. Whay, H. R., & Huxley, J. N. (2005). Pain relief in cattle: a practitioner's perspective. *Cattle Pract 13*, 81–85.

17. Egger, C. M., Love, L., &Doherty, T. (eds) (2014). *Pain management in veterinary practice*. Ames, IA: Blackwell.

18. Hellyer, P., Rodan, I., Brunt, J., et al. (2007). AAHA/AAFP Pain management guidelines for dogs and cats. *J Am Anim Hosp Assoc 43*, 235–248.

19. Kristiansson, M., Saraste, L., Soop, M., et al. (1999). Diminished interleukin-6 and C-reactive protein responses to laparoscopic versus open cholecystectomy. *Acta Anaesthesiol Scand 43*, 146–152.

20. Morton, C. M., Reid, J., Scott, M. E., et al. Application of a scaling model to establish and validate an interval level pain scale for assessment of acute pain in dogs. *Am J Vet Res 266*, 2154–2166.

21. Brondani, J. T., Mama, K. R., Luna, S. P., et al., Validation of the English version of the UNESP-Botucatu multidimensional composite pain scale for assessing postoperative pain in cats. *BMC Vet Res 9*, 143.

22. AAHA Accreditation General Standards. American Animal Hospital Association, www.aahanet.org/Accreditation/StandardsBySubCat, accessed May 28, 2014.

23. Fleisher, L. A. (2010). Risk of anesthesia. In R. D. Miller (Ed.), *Miller's anesthesia* (7th ed., pp. 969–999). Philadelphia: Churchill Livingstone.

24. Brodbelt, D. C., Pfeiffer, D. U., Young, L. E., et al. Results of the Confidential Enquiry into Perioperative Small Animal Fatalities regarding risk factors for anesthetic-related death in dogs. *J Am Vet Med Assoc 233*, 1096–1104.

25. Senior, J. M. (2013). Morbidity, mortality, and risk of general anesthesia in horses. *Vet Clin North Am Equine Pract 29*, 1–18.

26. Brodbelt, D. C., Blissitt, K. J., Hammond, R. A., et al. The risk of death: The confidential enquiry into perioperative small animal fatalities. *Vet Anaesth Analg 35*, 365–373.

27. Anesthesia Record. American Animal Hospital Association, www.aahanet.org/Store

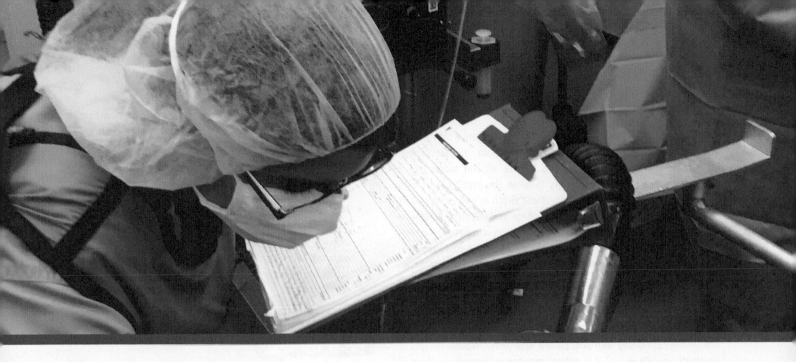

CHAPTER 7

Anesthesia Monitoring

Kristen Cooley, *BA, CVT, VTS (Anesthesia/Analgesia)*

Anita Parkin, *RVN, AVN Dip (Surgery & ECC), VTS (Anesthesia/Analgesia), CVPP, TAE*

Janet Amundson Romich, *DVM, MS*

LEARNING OBJECTIVES

Upon completion of this chapter, it is expected that the reader should be able to:

7.1 Explain how to assess anesthetic depth in veterinary patients

7.2 Describe invasive and non-invasive methods of monitoring circulation

7.3 Describe invasive and non-invasive methods of monitoring oxygenation

7.4 Describe invasive and non-invasive methods of monitoring ventilation

7.5 Describe invasive and non-invasive methods of monitoring temperature

7.6 Describe ways to combat the various types of heat loss

KEY TERMS

Arterial oxygen saturation (SaO_2)

Arterial oxygen tension (PaO_2)

Capnography

Capnogram

Cardiac contractility

Conductive heat loss

Convective heat loss

Diastolic pressure

End-tidal carbon dioxide (EtCO2)

Evaporative heat loss

Hypercapnia

Hyperthermia

Hypocapnia

Hypotension

Hypothermia

Hypoxemia

Hypoxia

Mean arterial pressure (MAP)

Minute volume (V_E)

Oscillometric

Oxygen consumption

Oxygen delivery (DO_2)

Oxygenation

Oxyhemoglobin dissociation curve

Perfusion

Peripheral capillary oxygen saturation (SpO_2)

Photoplethysmography

Pulse oximeter

Radiant heat loss

Systolic blood pressure

Transducer

Vasomotor tone

Weaning from a mechanical ventilator

INTRODUCTION

The goal of the anesthesia team is to balance the negative physiological effects of anesthesia with its benefits of analgesia, unconsciousness, and muscle relaxation. To achieve this balance, vigilant monitoring of the anesthetized patient is necessary to allow the anesthesia team to make adjustments to the patient's anesthetic depth as well as support its physiologic well-being. The American College of Veterinary Anesthesia and Analgesia (ACVAA) recommends monitoring circulation, oxygenation, ventilation, and body temperature frequently and continuously in anesthetized patients.[1] As discussed in previous chapters, this means constant vigilance and monitoring of the patient, being attentive, and recording parameters every 5 minutes. Hands-on monitoring in conjunction with physiologic monitors allows recognition and proper response to physiologic abnormalities that can occur during anesthesia. Despite statistics that state human error associated with poor patient monitoring accounts for 70% of anesthetic complications,[2] it is patient monitoring through all stages of anesthesia that provides early recognition and correction of problems that may arise secondary to an anesthetic event.

This chapter describes monitoring veterinary patients during anesthesia with a focus on dogs and cats. Monitoring other veterinary species is described in species-specific chapters. The care and maintenance of the monitoring equipment described in this chapter is found in Appendix B.

Tech Tip 🐾

What is the most expensive and best anesthetic monitor you can buy? YOU… nothing can replace the cost of your training, experience, and ability to interpret what your patient is doing.

PATIENT MONITORING

Anesthetic monitoring involves gathering physiologic data to keep the patient safe and to make adjustments to a patient's anesthetic depth. Monitoring cardiovascular and respiratory parameters, temperature, and muscle relaxation ensures appropriate anesthetic depth and analgesia for the patient as well as early recognition of anesthetic problems for the anesthesia team.

Unfortunately, there is no single reliable indicator of anesthetic depth in veterinary medicine; however, by evaluating a variety of parameters, veterinary technicians can monitor changes and evaluate trends in their patients to help them have the least physiological impact from anesthesia. Patient monitoring involves hands-on physical assessment of the patient through the evaluation of vital signs at regular intervals as well as judicious use of specialized monitors.

- The most basic vital signs include heart rate, respiratory rate, and body temperature, but assessment of mucous membrane color, capillary refill time (CRT), heart rhythm, respiratory character, muscle tone, muscle reflexes, and eye position are also important. All of this information can be easily gathered without the use of expensive equipment and provides the anesthesia team with a wealth of information regarding patient status and anesthetic depth.

- In addition to vital signs, physiologic monitors that measure blood pressure directly or indirectly, end-tidal carbon dioxide ($EtCO_2$), oxygen saturation (SpO_2), and heart rhythm and rate via electrocardiogram (ECG) can be used to assess patient status and may, in conjunction with hands-on monitoring, aid in assessment of anesthetic depth.

- More advanced monitoring may be used for specific cases and may include blood gas analysis, point of care blood testing (such as assessing blood glucose, electrolytes, lactate levels, etc.) as well as other modalities such as ultrasound and monitors that assess cardiac output and respiratory function parameters (pulmonary compliance, spirometry, etc.).

Tech Tip 🐾

There is a limit to the accuracy of any monitor; therefore, the trends they provide are often more meaningful than a single numerical value. Frequent and continuous monitoring by properly trained personnel improves the quality of care for veterinary patients.

Monitoring Anesthetic Depth

Determination of anesthetic depth should reflect information gathered from a variety of parameters including reflexes, eye position, and muscle tone (see Chapter 10 [The Maintenance Period]). Physiologic parameters (heart rate, respiratory rate, and blood pressure) provide information on cardiovascular status and homeostasis and may be influenced by the anesthetic agents given, surgery being performed, duration of the anesthetic event, and anesthetic/surgical complications such as **hypoxemia** (low partial pressure of dissolved oxygen in the arterial blood), hypovolemia, anemia, and **hypercapnia** (elevated carbon dioxide levels).

Reflexes

Reflexes are involuntary responses to stimuli and reflect nervous system activity (but not necessarily consciousness).[3] Reflexes are protective and predictable in conscious animals, but as a patient becomes more deeply anesthetized their reflexes will gradually decrease or slow and cease. Commonly, reflex status is communicated in terms like "present", "decreased", and "absent" to represent the speed at which the response occurs.

Reflexes assessed during anesthesia include the following:

- The *palpebral* or *blink reflex* is initiated with a light tickle of the medial or lateral canthus of the eye or by brushing the eyelashes (see Figure 10-2b). Using more than a light tickle may artificially cause blinking through stimulation of the corneal reflex and may mislead the anesthetist to think the palpebral reflex is present. This reflex protects the eye from injury in conscious animals, and most anesthetized animals will retain the palpebral reflex until the medium plane of anesthesia (the palpebral reflex will only exist in the light plane of anesthesia, Stage III Plane 1). The presence of a brisk palpebral reflex indicates a light plane of anesthesia, but the absence of one does not necessarily mean the patient is too deep. Sluggish palpebral reflexes are acceptable in some species, especially in equine and ruminant patients, but not dogs or cats.

- The *corneal reflex* occurs with stimulation of the cornea using a sterile object or eye drops. Stimulation of the cornea will cause the patient to blink or the eyeball to retract. This reflex will disappear during deep anesthesia and is most useful to determine large animal anesthetic depth as it is not always reliable in small animal patients.

- The *pupillary light reflex* (PLR) is characterized by constriction of the pupil in response to light (see Figure 10-2c). This reflex is lost during deep anesthesia, but should be present (although slow) in light to medium anesthetic planes.

- *Lacrimation* is tear production, and it may be totally absent in deep anesthetic planes. Lacrimation indicates a light level of anesthesia in all species; in horses lacrimation is a very sensitive indicator that a patient is getting lighter.

- The *pedal* or *digital withdrawal reflex* is flexion or withdrawal of a limb in response to a painful stimuli (typically, squeezing and twisting of a toe). The leg is extended and the toe pinched with the fingers or a hemostat. This reflex is present (the leg is withdrawn) during a light level of anesthesia and is absent when the patient is at an adequate plane of anesthesia.

- *Swallowing* and *laryngeal reflexes* protect the airway and occur in response to food/saliva/foreign material or stimulation of the pharynx (due to examples such as esophageal temperature probe, esophageal stethoscope, endotracheal tube (ETT) placement). Induction agents are meant to abolish the laryngeal reflex to facilitate endotracheal intubation. These reflexes are absent in all surgical anesthetic planes (all phases of Stage III), which is why the patient's airway should be protected with a properly fitted endotracheal tube. The swallowing reflex returns during recovery and help determine the proper time to remove the ETT (the ETT should be left in place to protect the airway until the patient can protect it on their own). Removing the ETT too early leaves the airway unprotected, while removing it too late may result in laryngospasm or the patient chewing on the ETT (which ruins the tube and has the potential to become a tracheal or gastrointestinal foreign body).

- *Reflex movement* is not a true reflex but is purposeful movement that occurs in response to a surgical stimulus. Reflex movement indicates a light level of anesthesia (Stage III Plane 1) and does not mean the patient is perceiving pain. Some drugs (such as propofol) may cause temporary involuntary myoclonus (sudden, involuntary jerking of a muscle or group of muscles) at medium planes of anesthesia and are generally not a cause for alarm.

Tech Tip

The pedal (digital withdrawal) reflex is a good test to perform if the anesthetist is unsure of the patient's anesthetic depth.

Ocular Signs

Ocular signs used to monitor anesthetic depth include eye position, nystagmus, and pupil diameter.

- The position of the eye within the orbit varies markedly depending on the stage of anesthesia, the anesthetic agent administered, and the species of animal (see Figure 10-1).
  - When administering inhalant anesthetic agents, eye positioning in small animals may change from a central position to a ventromedial one and then back to central as anesthetic depth increases (see Table 10-1). Ventromedial rotation typically indicates a medium plane of anesthesia (Stage III Plane 2).
  - Eye positioning is not as reliable in horses as they generally maintain a central eye regardless of anesthetic depth.

- Nystagmus, the involuntary, rapid oscillation of the eyeball, can vary during anesthesia.
  - Fast nystagmus occurs during the light plane of anesthesia and decreases as anesthetic depth increases.
  - Dogs and cats do not routinely get nystagmus, while nystagmus in anesthetized horses means the horse is at a very light plane of anesthesia and may begin to move.

- Pupil size can vary during the different planes of anesthesia and is affected by drugs such as opioids (dilated in cats, constricted in dogs), anticholinergics (dilated), and dissociative anesthetics (dilated).
  - Pupils are most often dilated during Stage II of anesthesia (the nonsurgical, excitement stage that is typically brief in properly premedicated and induced patients).
  - Pupil size is normal or constricted during lighter planes of anesthesia and progressively dilates as anesthetic depth deepens.

Muscle Tone

Muscle tone is an indicator of skeletal muscle relaxation and is also used to determine anesthetic depth. Some drugs such as ketamine increase muscle tone when not used in conjunction with other agents that relax muscle tone and thus interfere with this assessment. Methods to assess muscle tone include jaw tone (opening of the jaw to gauge "looseness") (see Figure 10-2a), anal tone (diameter of the anal orifice and tone of the anal sphincter), and general muscle tone. The deeper the plane of anesthesia, the less tone and more relaxed these muscles will be (resistance to jaw movement will decrease, the anal orifice will increase in size, and limbs will become more flaccid).

Indicators of anesthetic depth are summarized in Table 7-1.

Tech Tip

Jaw tone is difficult to assess in puppies (because they have minimal jaw tone) and in animals with large masseter muscles (such as large dog breeds and large animals).

TABLE 7-1 Indicators of Anesthetic Depth

Parameter	Level of Anesthetic Depth		
	Stage III Plane 1: Light	Stage III Plane 2: Medium	Stage III Plane 3: Deep
Heart rate	Normal or high	Normal to low	Low
Respiration rate	Normal or high	Normal to low	Low
Palpebral (blink) reflex	Present	Diminished to absent	Absent
Corneal reflex	Present	Present	Absent
Pupillary light reflex	Present (may be slow)	Present (may be slow)	Absent
Lacrimation	Normal to increased	Decreased	Absent
Pedal (digital withdrawal) reflex	Present	Absent	Absent
Swallowing reflex	Possible	Absent	Absent
Laryngeal reflex	Possible	Absent	Absent
Movement in the presence of surgical stimulation	Possible	Absent	Absent
Eyeball position (not reliable in horses)	Central to ventromedial	Ventromedial (dogs and cats); ventrally (ruminants)	Central
Nystagmus (dogs and cats do not get; typically used to assess horses)	Present	Slow to absent	Absent
Pupil size	Midrange	Midrange to dilated	Dilated
Muscle tone	Present	Relaxed	Relaxed

Adapted from S. Haskins (1992). General guidelines for judging anesthetic depth, *Vet Clin North America Small Animal Practice 22*, 432–434.

Monitoring Circulation

The goal of monitoring circulation in anesthetized patients is to ensure they have adequate circulatory function. Circulation can be monitored by many parameters, but the most basic are heart rate (HR) and rhythm, pulse strength, capillary refill time (CRT), mucous membrane color, and blood pressure (Table 7-2). Assessment of cardiac output and other modalities are less commonly used clinically due to cost, equipment limitations, and expertise.

Heart Rate and Rhythm Monitoring

One way to monitor anesthesia based on the circulatory system is via heart rate and rhythm.

- HR and rhythm are *subjectively* assessed by direct palpation of the apical pulse through the chest wall; by palpating a peripheral pulse (common pulse sites include the dorsal pedal artery, metatarsal/metacarpal artery, femoral artery, radial or brachial artery, coccygeal artery, auricular artery, and in anesthetized patients, the lingual artery); and by auscultation of the heart using a regular stethoscope, esophageal stethoscope, or audible heart monitor.
  - Palpation of a peripheral pulse provides information about the strength (strong or weak), presence or absence, and rhythm (regular or irregular) of the pulse.
  - Auscultation of the heart provides information about the presence or absence of the HR as well as it rhythm (regular or irregular) and the presence of murmurs.
- HR and rhythm are *objectively* determined by various physiologic monitors including electrocardiography (ECG) (Figure 7-1), non-invasive blood pressure monitors (NIBP) such as **oscillometric** (electronic blood pressure sensor with a numerical readout) monitors and Doppler blood flow detector (Doppler), or pulse oximeter (pulse-ox). The NIBP and Doppler will be described in the monitoring blood pressure section. The pulse oximeter will be described in the oxygen saturation monitoring section.

When assessing heart rate, it is important to know the normal rate for each species as well as what is normal for the breed, age, and the individual animal. It is common for HRs of anesthetized patients to be slower and quieter than their conscious counterparts because unconscious animals have low metabolic oxygen needs and less sympathetic nervous system stimulation. It is important for a patient's heart rate not to fluctuate from one extreme to the other of the normal range but rather to stay toward the median of the normal range to ensure appropriate cardiac output and to avoid excessive myocardial work and oxygen demand.

TABLE 7-2 Methods Used to Monitor Circulation, Oxygenation, Ventilation, and Temperature

Guideline	Monitoring Parameter	Normal Range for Dogs and Cats
Circulation		
	Heart rate	Dog: 60–140 bpm Cat: 120–200 bpm
	Mucous membrane color	Pale pink to pink
	Capillary refill time	1–2 seconds
	Palpation of pulse	Strong, regular beats
	Heart rhythm	Normal sinus rhythm
	Blood pressure	Systolic: 100–160 mm Hg Diastolic: 50–100 mm Hg MAP: 60–120 mm Hg
Oxygenation		
	Mucous membrane color	Pale pink to pink
	SpO_2	95–100%
Ventilation		
	Respiratory rate	8–15 breaths/min
	Respiratory character	Easy, regular breaths
	End-tidal CO_2	35–45 mm Hg
Body Temperature		
	Rectal or esophageal	Dog: 99.5–102.5°F (37.5–39.2°C) Cat: 100–102.5°F (36.7–39°C)

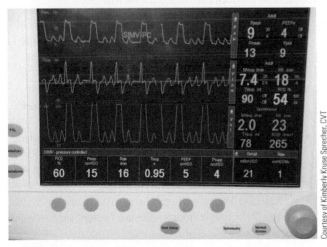

Courtesy of Kimberly Kruse Sprecher, CVT

FIGURE 7-1 An ECG is an objective measure of heart rate and rhythm. This is an example of an ECG monitor displaying the electrical activity of the heart.

Tech Tip

Take preoperative temperature, heart rate, and respiration rate in anesthesia patients to provide a baseline of the patient's "normal" values.

The Esophageal Stethoscope

The esophageal stethoscope is a soft, flexible blind-ended tube connected to the Y-piece of a standard stethoscope. The blind-ended tube is inserted into the esophagus to hear heart and lung sounds (Figure 7-2). The side wall at the distal blind end of the esophageal stethoscope has a number of holes that are covered in a thin plastic membrane to prevent fluid from entering the lumen of the stethoscope. Various diameters of esophageal stethoscopes are available to accommodate the variety of esophageal diameters of veterinary patients. The esophageal stethoscope in its simplest form is used to obtain information on heart rate and breathing in small animal patients. More advanced models can be found as part of a multiparameter monitor and may also provide temperature readings or ECG tracings.

Electrocardiography

In the anesthetized patient, electrocardiography is used to detect arrhythmias. Certain arrhythmias can be detected during auscultation or pulse palpation while others require the use of an electrocardiogram (ECG). A sinus rhythm is considered normal and sounds and feels like a strong, regular beat. If the heart rhythm is not regular and/or does not sound normal, it may indicate a more serious condition and should be brought to the attention of the veterinarian. Any patient with an abnormal heart rhythm will benefit from an ECG.

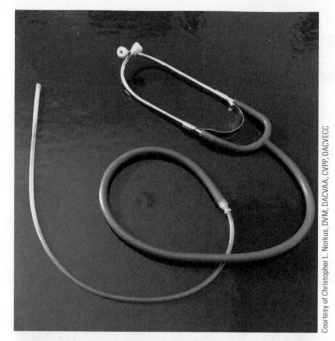

Courtesy of Christopher L. Norkus, DVM, DACVAA, CVPP, DACVECC

FIGURE 7-2 An esophageal stethoscope comes in a variety of diameters and is connected by a plastic adaptor to the Y-piece of a standard pair of stethoscope earphones.

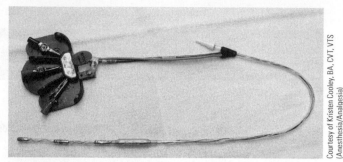

Courtesy of Kristen Cooley, BA, CVT, VTS (Anesthesia/Analgesia)

FIGURE 7-3 An esophageal electrocardiogram is performed by inserting a probe with a transducer down the esophagus rather than placing the transducer on the chest. The probe of the esophageal ECG must contact the mucosa of the esophagus, and the depth at which the probe is placed can result in variation of configuration and amplitude of the waves. Esophageal ECGs are best for assessing heart rhythm.

information on the electrical stability of the heart.[4] Ideally all anesthetized patients should have an ECG.

Blood Pressure Monitoring

Adequate blood pressure (BP) is essential for the circulatory system to deliver oxygen and nutrients to all tissues in the body. Blood pressure is the pressure within the systemic arterial circulation and is influenced by cardiac output (stroke volume × heart rate) and systemic vascular resistance. Stroke volume is influenced by blood volume, contractility, and tension in the left ventricle before the aortic valve opens. Systemic vascular resistance is influenced mainly by vessel diameter (Figure 7-4). Blood pressure varies throughout the cardiac cycle due to the relaxation and contraction of the ventricles.

In the conscious patient, an ECG can be used for other diagnostic purposes, but in anesthetized patients, it can really only be used to analyze the ECG waves and observe changes in those waves over time. It should never be used as the sole monitor of cardiac function because it only provides information about the electrical activity of the heart and nothing about its mechanical function. An ECG should always be used in conjunction with other cardiovascular parameters.

- The systolic reading reflects the peak pressure during contraction of the ventricle and is comprised of stroke volume and arterial compliance (how much the vessel distends and myocardial inotropy [the intrinsic strength of the heart's contraction]).
- **Diastolic pressure** is the lowest pressure during relaxation of the heart and is comprised of heart rate and systemic vascular resistance (vessel size) and volume.
- The **mean arterial pressure** (MAP) is the average pressure throughout the cardiac cycle (MAP = (Systolic Pressure − Diastolic Pressure)/3 + Diastolic Pressure). MAP is the driving pressure for organ perfusion and is an important value in the anesthetized patient.[5]
- Pulse pressure (PP) is the difference between the systolic and diastolic pressure (PP = Systolic Pressure − Diastolic Pressure) and is reflective of stroke volume.

Tech Tip

ECGs are a great way to monitor HR and rhythm; however, the ECG can appear normal when myocardial performance and blood pressure are poor.

The electrical impulse generated by the heart is detected throughout the body and is represented on the ECG strip. For this reason, electrodes are placed on the body's surface to detect these currents or an esophageal ECG is placed (Figure 7-3).

An ECG is beneficial in the diagnosis and prognosis of cardiac disease and serves as a baseline monitor for all patients. The ECG is becoming the standard of care in all cases and is essential for identification of arrhythmias for higher risk cases such as breeds predisposed to arrhythmias, animals whose arrhythmias were detected by auscultation, and patients with acute onset of conditions (such as dyspnea, shock, splenic disease, gastric dilatation volvulus, or syncope [fainting or passing out]).[4] It is also essential during the preoperative evaluation in high-risk or trauma patients, intra-operatively for all procedures, and postoperatively for those cases requiring follow up

Tech Tip

Monitoring and maintaining blood pressure is important during anesthesia since many patients will experience some degree of **hypotension** (decrease in blood pressure outside of the normal range) during an anesthetic event.

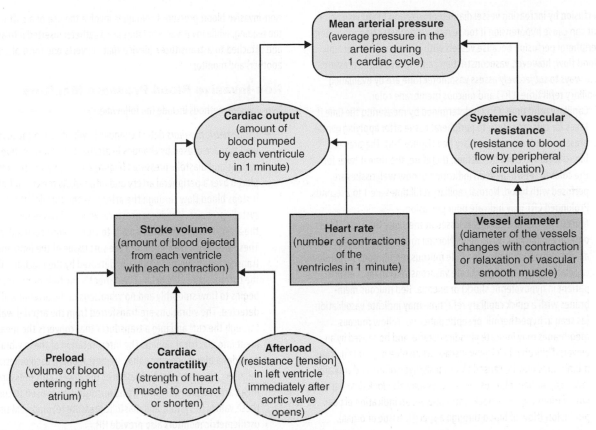

FIGURE 7-4 Mean arterial blood pressure is comprised of cardiac output and systemic vascular resistance. Cardiac output is the volume of blood ejected over 1 minute and is made up of stroke volume and heart rate. Preload, cardiac contractility, and afterload all contribute to stroke volume. Systemic vascular resistance is mainly influenced by vessel diameter (and to a lesser amount blood viscosity and vessel length).

Blood pressure can be assessed both subjectively and objectively. There are two ways to *subjectively* measure blood pressure: pulse palpation and assessment of pulse quality.

- Palpation of a patient's pulse is a *subjective* way of approximating blood pressure and is assessed for strength, rate, and regularity. Pulse strength feels different in larger arteries versus smaller arteries due to the effect of variation in blood vessel diameter on blood pressure.[4] Examples of different types of pulses include:
  - Bounding pulses occur due to large pulse pressure differences that feel like a beach ball in a tent. Bounding pulses are commonly seen not only with patent ductus arteriosis (PDA) but may also indicate vasodilation secondary to sepsis.
  - Weak or thready pulses occur from tighter blood vessel walls through which small waves of blood move and feel like a marble in a sock. Weak or thready pulses are a result of vasoconstriction that may be caused by certain drugs such as alpha-2 adrenergic agonists, pain, and hypovolemic shock.

Pulse palpation can be used to monitor trends or changes throughout an anesthetic event; however, it is important to palpate the patient's pulse before anesthetic induction to obtain baseline information. It is important to note that patients with strong peripheral pulses do not necessarily have adequate blood pressure and tissue perfusion. Pulse palpation is not a substitute for blood pressure monitoring because pulse quality is a subjective, not objective, measurement and does not necessarily correlate to a patient's actual blood pressure.[4]

Tech Tip

Palpate many pulses from different locations in different animals to make sure a normal pulse can be differentiated from an abnormal one. Checking pulses during various times of the day such as after exercise or during rest can help differentiate rapid versus slow pulse rates. It is important to remember when monitoring anesthesia that slow heart rates are often acceptable as long as blood pressure remains normal.

- Pulse quality *subjectively* measures blood pressure and is largely a reflection of stroke volume and **vasomotor tone** (the degree of vasodilation or vasoconstriction). Vasomotor tone regulates both peripheral and visceral perfusion. Vasodilation improves peripheral perfusion by increasing vessel diameter to expand blood volume but can cause hypotension if too severe. Vasoconstriction can impair peripheral perfusion because vessels with a narrow diameter limit blood flow; however, vasoconstriction can improve blood pressure. Two ways to *subjectively* assess vasomotor tone are by evaluating capillary refill time (CRT) and mucous membrane color.

- Capillary refill time (CRT) is determined by measuring the time it takes for color to return to peripheral tissue after applying gentle pressure to blanch the capillary bed (Figure 7-5). The pressure blanches blood from the tissues; therefore, the time it takes for the color to return gives an indication of how well tissues are perfused with blood. Normal capillary refill times are 1 to 2 seconds. Prolonged CRT may indicate poor perfusion.
- Mucous membrane color is assessed at the same time as CRT by describing gingival tissue color. Normal mucous membrane color is a light to medium pink color. Pale mucous membranes with a slow capillary refill time may indicate vasoconstriction (as seen in a patient in hypovolemic shock) or anemia. Red mucous membranes with a quick capillary refill time may indicate vasodilation (as seen in hyperthermic or septic patients). Yellow mucous membranes may indicate jaundice/icterus and be caused by an elevated bilirubin level/liver disease while blue or purple mucous membranes may be caused by cyanosis/hypoxemia and critically low oxygenation. Mucous membranes may also look dusky or grey with various hypoxic/shock states and are an indication of poor **perfusion** (flow of blood through a specific tissue or organ).

There are two ways to *objectively* measure blood pressure: indirectly using non-invasive methods and directly using invasive methods.

Tech Tip

Ambient light may alter the color of mucous membranes. Determining a baseline color prior to anesthesia induction will provide a normal reference for that patient under those lights.

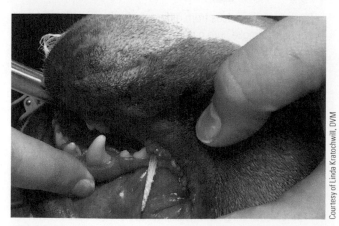

FIGURE 7-5 A subjective way to assess vasomotor tone is by performing a CRT.

Courtesy of Linda Kratochwill, DVM

Non-invasive blood pressure techniques involve the use of a cuff to obtain the reading, while invasive techniques use a catheter inserted into an artery and attached to a **transducer** (device that converts one form of energy to another) and monitor.

Non-Invasive Blood Pressure Monitors

Non-invasive methods include the following:

- *Oscillometric monitors* detect changes in vibrations in the arterial wall that are caused by changes in arterial blood flow between systolic and diastolic pressures (Figure 7-6). A pneumatic cuff is placed over a peripheral artery and inflated. As the cuff inflates, it stops blood flow through the artery. As the cuff deflates below systolic pressure, it reduces pressure, which allows blood to reenter the artery causing the artery walls to vibrate with each pulsation. These vibrations travel through the soft tissue of the limb and are transmitted through the cuff and detected by the machine. When the cuff pressure falls below the patient's diastolic pressure, blood begins to flow smoothly and no vibrations in the arterial wall are detected. The vibrations are transferred from the arterial wall through the cuff and into a transducer that converts the measurement into electrical signals. The interpretation of these vibrations provides a blood pressure value. In most modern oscillometric monitors, the strongest oscillation at the lowest cuff pressure is the mean arterial pressure (MAP), and an algorithm is used to convert these values to systolic and diastolic pressure readings.[5] Many oscillometric monitors also provide HR.
  - When taking an animal's blood pressure, the width of the blood pressure cuff should be 30% (cats) and 40% (dogs) of the circumference of the limb to obtain an accurate reading[6] (Figure 7-7a). Cuffs that are too large may lead to underestimated pressures, and cuffs that are too small may lead to overestimated pressures.[5]
  - If the correct cuff size is not available, a wider cuff should be chosen over a smaller one because the magnitude of error is greater with a small cuff because they do not occlude blood flow.[7]
  - Fit the cuff snugly around the limb either above the carpus, proximal to the hock (Figure 7-7b), distal to the hock, or around the base of the tail.
  - The cuff should be at the level of the heart on an unrestrained appendage for optimal results. If obtaining a blood pressure on an awake patient, place them into a comfortable position and bring their limb to the level of the heart (Figure 7-7c). It is important to minimize stress and anxiety to ensure an accurate reading.
  - The monitor should be set to read every 3 to 5 minutes for anesthetized patients (Figure 7-7d). Blood pressures read more frequently do not give the limb a chance to recover and re-perfuse with blood; pressures read less frequently may lead to missed events.
  - Non-invasive blood pressure monitors provide information on trends as opposed to absolute values.
- *Doppler flow monitors* use an ultrasonic probe with a crystal that has two piezoelectrical (ability to generate an electric charge in response to applied mechanical stress) elements and an electronic monitor to detect blood pressure. One piezoelectric element in the probe gives off ultrasound waves that are reflected back as echoes to the other piezoelectric element. The ultrasound waves bounce off red blood cells moving inside an artery and are sent back to an electronic monitor to be recorded.

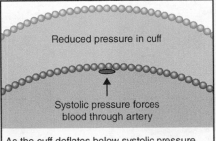

Inflated cuff

Upper arm   Artery

A pneumatic cuff is placed over a peripheral artery and is inflated to stop blood flow through the artery.

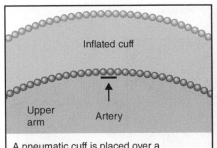

Reduced pressure in cuff

Systolic pressure forces blood through artery

As the cuff deflates below systolic pressure, blood reenters the occluded vessel causing vibrations of the artery walls with each pulsation. These vibrations are detected by sensors within the blood pressure cuff.

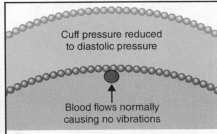

Cuff pressure reduced to diastolic pressure

Blood flows normally causing no vibrations

When the cuff pressure falls below the patient's diastolic pressure, blood flows smoothly and no vibrations in the arterial wall are detected.

(a)

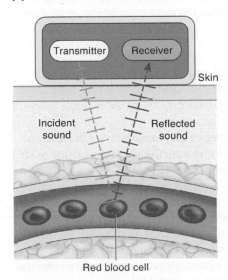

Transmitter   Receiver

Skin

Incident sound   Reflected sound

Red blood cell

(b)

FIGURE 7-6 Objective ways to assess blood pressure include non-invasive and invasive methods. Non-invasive methods include oscillometric blood pressure monitors and a Doppler blood pressure monitor. (a) Oscillometric flow monitors detect changes in blood flow through an artery between systolic and diastolic pressures. (b) A Doppler blood flow detector consists of an ultrasonic probe that is placed over a peripheral artery along with coupling gel to maximize contact of the probe with the patient. An audible "swooshing" sound is amplified so that each pulse wave within the vessel is heard.

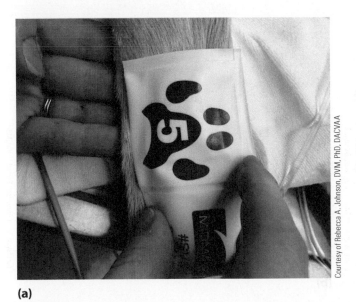

Courtesy of Rebecca A. Johnson, DVM, PhD, DACVAA

(a)

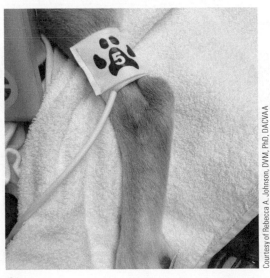

Courtesy of Rebecca A. Johnson, DVM, PhD, DACVAA

(b)

FIGURE 7-7 Procedure for blood pressure measurement using oscillometric flow detector. (a) A proper size blood pressure cuff is determined for each patient. For dogs, the blood pressure cuff is 40% of the circumference of the patient's limb (for cats, the blood pressure cuff is 30% of the circumference of the patient's limb). (b) The cuff fits snugly around the limb proximal to the hock.

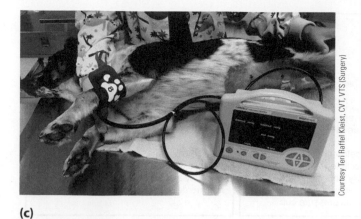

Courtesy Teri Raffel Kleist, CVT, VTS (Surgery)

(c)

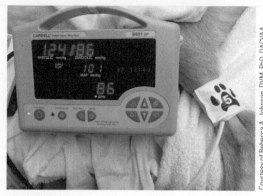

Courtesy of Rebecca A. Johnson, DVM, PhD, DACVAA

(d)

FIGURE 7-7 (*Continued*) (c) If obtaining a blood pressure on an awake patient, they should be in a comfortable position and their limb at the level of the heart. (d) This blood pressure monitor provides information on systolic and diastolic blood pressure, mean arterial pressure, and pulse rate.

The echoes make an audible swooshing sound that changes pitch with fluctuations in red blood cell velocity and changes intensity proportionally to the number of red blood cells detected. The Doppler only provides information about **systolic blood pressure** (peak pressure during contraction of the heart); however, it also gives a flow signal. As the cuff is deflated, pulses appear in cuff at systolic arterial pressure (SAP), pulses are maximal at MAP, and pulses disappear at diastolic arterial pressure (DAP). Cuff selection and placement, maximum inflation pressure, and deflation rate are determined by the operator;[8] therefore, results can vary greatly based on user experience or user error. An advantage to using a Doppler is that it gives constant audible feedback of blood flow within a tissue bed (see Figure 7-6).

○ Place the concave surface of the Doppler crystal and coupling gel over a peripheral artery to detect blood flow within that vessel

(metatarsal and metacarpal arteries are commonly used in small animals) (Figure 7-8a). Tape the crystal in place, being sure the tape is snug (you should not see any "air" or be able to move the Doppler crystal once taped).

○ Place a blood pressure cuff proximal to the crystal and connect it to a sphygomomanometer (Figure 7-8b). Turn the unit on and listen for an audible swoosh sound.

○ Inflate the cuff until the sound of blood flow stops.

○ Slowly deflate the cuff while watching the pressure gauge until the sound returns.

○ The pressure at which the sound returns reflects the systolic blood pressure. In dogs, the pressure reading reflects SAP. In cats, both the oscillometric monitor and Doppler pressure readings underestimate the actual SAP.[9]

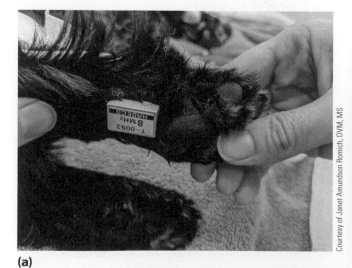

Courtesy of Janet Amundson Romich, DVM, MS

(a)

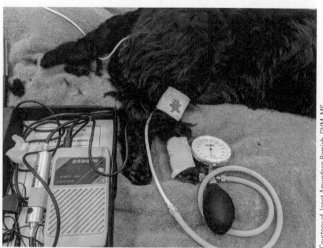

Courtesy of Janet Amundson Romich, DVM, MS

(b)

FIGURE 7-8 Procedure for blood pressure measurement using a Doppler. (a) The Doppler crystal and coupling gel are applied over a peripheral artery (in the hind limb it is the dorsal pedal artery). (b) A blood pressure cuff is placed proximal to the crystal, taped in place, and the cuff is inflated. The cuff is slowly deflated, and the value at which the sound returns reflects systolic blood pressure.

Invasive Blood Pressure Monitors

Invasive blood pressure (IBP) measurement is the gold standard of blood pressure monitoring. Invasive blood pressure measurement provides continuous, beat-by-beat assessment of blood pressure and requires the aseptic insertion of a catheter into an artery.

- *Arterial catheterization* is an invasive assessment of blood pressure used in a number of clinical settings including major surgery, trauma, and critical care. In dogs and cats, the primary artery used for arterial catheterization is the dorsal pedal artery. Other arteries that can be used include the coccygeal artery, lingual artery, and femoral artery in most species; the transverse facial and facial artery in the horse (Figure 7-9); and the auricular artery in cattle, sheep, goats, and dogs. The arterial catheter is connected via sterile tubing to a pressure manometer or transducer connected to a physiologic monitor (Figure 7-10). Most monitors capable of measuring invasive pressures

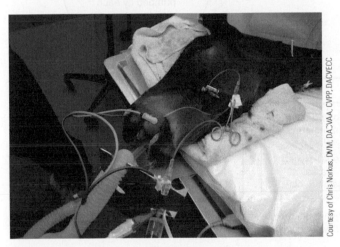

Courtesy of Chris Norkus, DVM, DACVAA, CVPP, DACVECC

FIGURE 7-9 Arterial catheterization is an invasive way to measure blood pressure. This arterial catheter is in the facial artery of the horse.

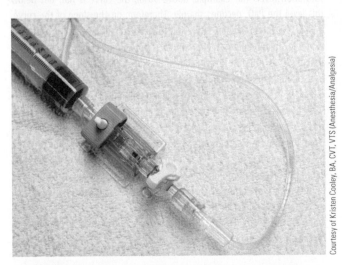

Courtesy of Kristen Cooley, BA, CVT, VTS (Anesthesia/Analgesia)

FIGURE 7-10 An arterial catheter is connected via sterile tubing to a transducer. The transducer transforms the measurement of pressures from a physical event into an electrical signal. The pressure transducer most commonly used consists of a mechanical element that is displaced as a result of changes in pressure. This pressure transducer is used for direct measurement of blood pressure.

display a continuous systolic, diastolic, and mean arterial blood pressure reading in addition to a waveform.

Tech Tip

Since veterinary patients vary in size, normal physiologic parameters, and temperature, veterinary-specific blood pressure equipment is essential for obtaining accurate information.

Monitoring Oxygenation

Oxygenation is the process of taking oxygen molecules from inspired air and delivering them to tissues for energy production (ATP) via aerobic cellular metabolism. **Oxygen delivery** (DO_2) is the rate of oxygen transport from the lungs to the peripheral tissues. DO_2 is dependent on cardiac output (CO) and oxygen content in arterial blood (CaO_2). CaO_2 is dependent on hemoglobin (Hb), arterial oxygen saturation (SaO_2), and physically dissolved oxygen in the plasma phase of an arterial sample (PaO_2) (Figure 7-11). Increased CO or CaO_2 increases DO_2, while decreased CO, anemia, and hypoxemia decreases DO_2. **Oxygen consumption** is the rate at which oxygen is removed from the blood for use by the tissues. Oxygen consumption increases as metabolism increases (e.g., fever, exercise) and decreases during anesthesia and as metabolism or body temperature decreases.

Oxygen Saturation Monitoring

Oxygenation can be *subjectively* measured by assessing mucous membrane color. Pink mucous membranes imply adequate oxygen saturation. It takes approximately 5 g/dL of unoxygenated Hb in the capillaries to generate the dark blue color seen clinically as cyanosis, so a patient may be severely hypoxemic by this point. Some patients, such as those with anemia, may not demonstrate cyanosis. Detecting hypoxemia by assessing mucous membrane color is subjective; therefore, more objective assessments of oxygenation should be performed along with mucous membrane color determination.

Tech Tip

Pigmentation of mucous membranes and anemia make assessment of mucous membrane color a less accurate method of assessing oxygenation.

Oxygenation can be *objectively* measured by invasively performing an arterial blood gas to obtain a PaO_2 and SaO_2 analysis or non-invasively using a pulse oximeter to obtain a SpO_2 analysis. The proportion of free to bound oxygen is used to assess oxygenation in veterinary patients based on the following principles:

- The amount of oxygen that binds to hemoglobin in the red blood cells is the **arterial oxygen saturation** or SaO_2 (think of *Sa* representing *sa*turation) and is determined invasively via an arterial blood gas analysis.

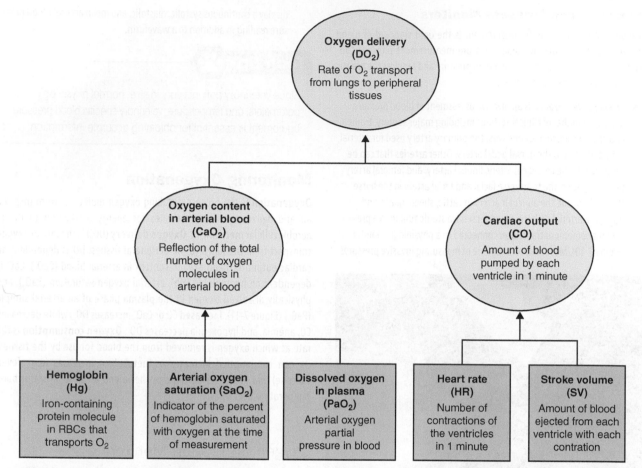

FIGURE 7–11 Oxygen delivery is dependent on cardiac output and oxygen content in arterial blood. Oxygen content in arterial blood is dependent on hemoglobin, arterial oxygen saturation, and dissolved oxygen in plasma, while cardiac output is dependent on heart rate and stroke volume.

- **Arterial oxygen tension** or the amount of oxygen molecules dissolved in the plasma phase of an arterial sample (i.e. not bound to Hb) is the PaO_2 (the little 'a' denotes that the sample is arterial) and is determined invasively via an arterial blood gas analysis.
- **Peripheral capillary oxygen saturation** (SpO_2) is an estimate of the oxygen saturation level and is determined non-invasively by a pulse oximeter (think of Sp as representing S-pulse ox).

There is a dynamic relationship between the free oxygen in the plasma and the bound oxygen in the hemoglobin. The more free oxygen molecules that are dissolved in plasma, the higher the PaO_2 value (more oxygen available to bind to hemoglobin) which in turn increases saturation and the SaO_2 value. SaO_2 always depends on and is proportional to the PaO_2; the higher the PaO_2, the higher the SaO_2. The SaO_2 value, however, represents the percentage of total binding sites on the arterial hemoglobin and can never be more than 100%.

The nonlinear relationship between the percent of peripheral capillary oxygen saturation of hemoglobin (SpO_2) to oxygen dissolved in plasma (PaO_2) is explained by the **oxyhemoglobin dissociation curve**[10] (Figure 7-12). This curve demonstrates how the values differ between the non-invasive (SpO_2) and invasive (PaO_2) methods of measuring oxygenation. Normal PaO_2 values depend among other things on the inspired concentration of oxygen, which directly influences oxygen saturation of hemoglobin (as PaO_2 increases, oxygen saturation increases), but this relationship is not linear. At higher

saturation levels (for example, above 90%), the curve is flat, but below this level the PaO_2 declines sharply. Therefore, on the slope of the curve, a large decline in oxygen saturation follows a small decrease in PaO_2. For example, an animal breathing room air (approximately 21%) should have a PaO_2 of 90 to 100 mm Hg while an animal on 100% oxygen should have a PaO_2 of approximately 500 mm Hg or more.[11] Despite this sizable difference in PaO_2 levels, the pulse oximeter reading will only change from 100% to 98% because it does not discriminate high PaO_2 values in areas where the oxyhemoglobin curve is flat (100–500 mm Hg). As PaO_2 decreases from 100 mm Hg to 80 mm Hg, oxygen saturation decreases from 98% to 95% and when PaO_2 decreases from 60 mm Hg to 40 mm Hg, oxygen saturation decreases rapidly from 90% to 75%. Based on this example, one of the limitations of the pulse oximeter is that it is a late indictor of problems in oxygenation especially in patients on 100% oxygen because PaO_2 can fall from 500 mm Hg to 100 mm Hg and show only a 2% change in oxygenation on the pulse oximeter.

Non-Invasive Oxygenation Monitor

The **pulse oximeter** indirectly and non-invasively measures oxygenation of blood hemoglobin by providing a continuous, automatic, and audible test of cardiopulmonary function. The pulse oximeter measures and displays the pulse rate and the oxygen saturation of hemoglobin by measuring pulsating signals

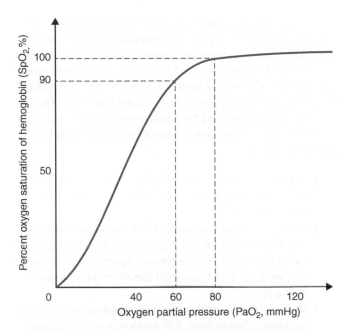

FIGURE 7-12 The oxyhemoglobin dissociation curve describes the relation between the partial pressure of oxygen (x axis) and the oxygen saturation of hemoglobin (y axis). Hemoglobin's affinity for oxygen increases as molecules of oxygen bind to it. Molecules continue to bind as the oxygen partial pressure increases until the maximum amount that can be bound is reached. As hemoglobin approaches saturation with oxygen, very little additional binding occurs and the curve levels out producing a sigmoidal or S-shape. At pressures above 95 mm Hg, the standard dissociation curve is relatively flat because the oxygen content of the blood does not change significantly even with large increases in the oxygen partial pressure. A SpO_2 of about 90% correlates to a PaO_2 of about 60 mm Hg. At higher saturation levels (e.g., above 90%), the curve is flat, but below this level the PaO_2 declines sharply. Therefore, on the slope of the curve, a large decline in oxygen saturation follows a small decrease in PaO_2.

from arteries through capillaries in a perfused tissue bed; therefore, patients need to have adequate perfusion and pulse for the pulse oximeter to properly function.

The pulse oximeter detects a patient's pulse from pulsing signals that pass a predetermined threshold. Anything that reduces the pulsing signal affects the accuracy of the reading. For example, it is difficult to obtain an accurate measurement from patients with poor peripheral perfusion secondary to drugs or disease states (such as hypotension, hypothermia, and vasoconstriction). Motion and electrical noise also interfere with the signal's detection, producing inaccurate readings. Newer units are improving with the use of proprietary technology to eliminate error and provide a more accurate value in the face of vasoconstriction, noise, or interference.

In addition to SpO_2 and heart rate, many pulse oximeters display a photoplethysmographic waveform (Figures 7-13a and 7-13b) associated with each heartbeat. **Photoplethysmography** (commonly called pleth) is a technique to detect blood volume changes in tissue beds. Analysis of pleth waveforms generated by pulse oximeters has been used to assess the strength of the monitor's signal and changes in peripheral perfusion.[4] Signals that are dampened by vasoconstriction will appear shorter than normal signals or those signals associated with vasodilation. More recently, pleth has been used to determine volume status and fluid responsiveness based on the effect mechanical ventilation has on cardiac output.

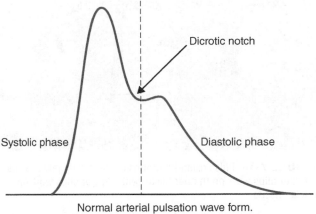

Normal arterial pulsation wave form.

(a)

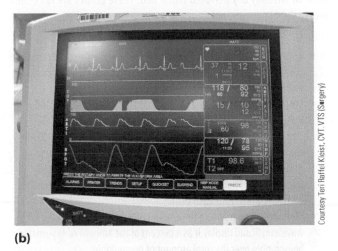

(b)

FIGURE 7-13 Pleth waveform. (a) The ascending part of the pulse wave indicates the systolic phase (left ventricle contractility), the dicrotic notch indicates closure of the aortic valve, and the descending phase indicates the diastolic phase (vascular resistance/arteriole diameter). (b) This multiparameter monitor shows the pleth waveform in the middle of the monitor.

Pulse oximeters with monitors that display a graph or utilize pleth are superior to those that do not. The lack of a waveform severely inhibits the ability to interpret the accuracy of the pulse oximeter reading. Checking the monitored pulse rate against the actual pulse rate can help determine accuracy as well.

Like all monitors, the pulse oximeter has its limitations:

- Pulse oximeters do not assess the adequacy of ventilation; they only estimate the amount of oxygen saturating the present hemoglobin.
- Anemic patients may have a SpO_2 reading of 100% because all available hemoglobin is saturated, yet their tissues are hypoxic. Since hemoglobin is the biggest contributor to the arterial oxygen content equation ($CaO_2 = (Hb \times 1.36 \times SaO_2/100) + (PaO_2 \times 0.003)$), its value greatly affects the SpO_2 value.[11]

There are many different pulse oximeter probes for use on veterinary patients. The most common is a clip-like probe supplied in different sizes to be used on tissues of varying thicknesses (cat or dog tongue or Achilles tendon). Smaller clips work on thinner tissue while larger clips work on thicker tissue (Figure 7-14). In addition to these clip-like

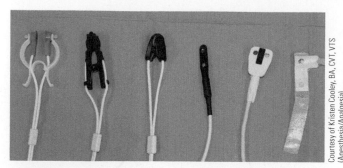

FIGURE 7-14 Pulse oximeter probes come in a variety of sizes. Smaller clips work on thinner tissue while larger clips work on thicker tissue.

probes, there are also a variety of flat reflectance probes that can be used on virtually any perfused tissue bed. These probes are more commonly used in people, not animals.

Proper use of the pulse oximeter is as follows (Figure 7-15):

- Apply the pulse oximeter to nonhaired skin over a perfused tissue bed (such as the tongue, pinna, digit web, base of tail, etc.) (Figures 7-15a and 7-15b).
- The pulse oximeter has a sensor that transmits both red and infrared light through the perfused tissue bed (Figure 7-15c). Oxygenated hemoglobin absorbs more infrared light (~950 nm) than deoxygenated hemoglobin, while deoxygenated hemoglobin absorbs more red light (~650 nm) than oxygentated hemoglobin. Therefore, these two types of hemoglobin can be differentiated from each other.
- Photodetectors measure the amount of light absorbed at each wavelength and display it as a percent of the amount of oxygenated hemoglobin over the total amount of hemoglobin.
- The pulse oximeter is designed to alert the anesthetist to hypoxemia. Hypoxemia is characterized as a PaO_2 below 80 mm Hg (which correlates to a SpO_2 of 94%). Keep in mind that a PaO_2 of 60 mm Hg correlates to a SpO_2 of 90%.[10] The rule of thumb is to keep the SpO_2 reading at 95% or above.

It is generally not recommended that anything be placed between the probe and the tissue. Placing items like damp gauze or using ultrasound gel to keep the tissues moist may result in skewed readings because the monitor functions by utilizing wavelengths of light (gauze is white and ultrasound gel is often blue). Abnormal readings may indicate poor perfusion and hypoxemia, but they may also be due to monitor malfunction. If a pulse oximeter reading is low, check the patient before altering anesthesia levels. To make sure the pulse oximeter probe is functioning properly, consider doing the following:

- Moving the probe because a probe that sits in one place for a period of time may cause pressure on the tissue, resulting in altered readings.
- Wetting the tongue because as the probe is taken off the tongue to wet it, reperfusion of the tongue occurs and a better signal is obtained. The pulse oximeter does not require moisture to function, and wetting the tongue itself does not help the equipment function better.
- Mechanically ventilating the patient with a few breaths may improve ventilation and gas exchange. If the breath improves the SpO_2 reading, then the anesthetist should continue to assist ventilation and/or decrease the depth of anesthesia, which decreases respiratory depression in an attempt to get the patient to adequately ventilate on their own.
- Checking blood pressure readings since a mismatch between ventilation and perfusion may be responsible for the low readings. If blood pressure is low, take steps to improve it.

Oxygenation Monitoring via Blood Gas Analysis

Blood gas analysis is an indicator of both oxygenation, ventilation, and acid-base status as it measures blood pH and dissolved oxygen and carbon dioxide molecules dissolved in the plasma phase of an arterial sample (PaO_2 and $PaCO_2$) or venous (PvO_2 and $PvCO_2$) blood. It is further described under the monitoring ventilation section of this chapter.

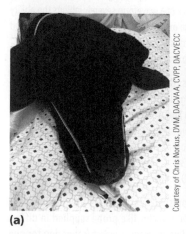

(a)

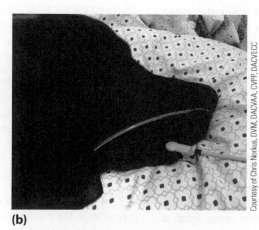

(b)

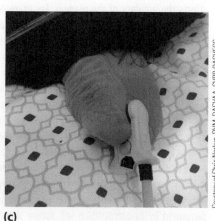

(c)

FIGURE 7-15 Procedure for placing a pulse oximeter. (a) The pulse oximeter is placed on nonhaired skin over a perfused tissue bed such as the tongue or (b) lip. (c) The sensor transmits red and infrared light through the perfused tissue bed.

Monitoring Ventilation

Many anesthetic drugs provide some degree of respiratory depression; therefore, monitoring a patient's ability to ventilate is crucial. Ventilation is comprised of three components

- Tidal volume (V_T) = the volume of each normal breath (air inhaled and exhaled in one breath) (10–15 mL/kg)
- Respiratory rate (*f*) = the number of breaths per minute (8–15 breaths/min in dogs and cats)
- **Minute volume** (V_E) = the volume of air breathed over one minute (200 mL/kg/min)

The volume of air breathed over one minute depends on the volume of each breath and the number of breaths per minute; $V_T \times f = V_E$.

Adequate ventilation, or how well the patient is moving air in and out of the lungs, provides an estimate of how well diffusion of oxygen and carbon dioxide is occurring in the patient (respiration).

Respiratory Rate and Character

One way to *subjectively* assess ventilation is to identify respiratory rate and describe respiratory character. Respiratory rate is the frequency of breaths taken over one minute and respiratory character describes the quality of the breaths. Breaths can be described as follows:

- *Shallow.* Shallow breaths are often associated with a small tidal volume because most anesthetic drugs decrease contraction of intercostal muscles. The patient increases respiratory rate in an attempt to compensate for the decrease in volume; however, these animals are predominately moving dead space, which results in poor alveolar ventilation. Mechanical ventilation usually benefits these patients.
- *Deep.* Deep breaths are characterized as having a large volume but slow rate (hypopnea) or having a large volume but rapid rate (hyperpnea).
- *Slow.* Bradypnea is a slow respiratory rate, and without an end-tidal carbon dioxide monitor or blood gas analyzer, it is difficult to determine if these patients are adequately ventilating. Hypothermic patients and those that are at a deep level of anesthesia will often exhibit bradypnea.
- *Fast.* Tachypnea is an increased respiratory rate with a number of possible causes including a light level of anesthesia, **hypoxia** (condition where the oxygen supply to the body or tissues is inadequate), hypercapnia, hyperthermia, and response to pain.
- *Irregular.* Irregular breathing may be due to the anesthetic agents used (e.g., ketamine may produce an apneustic breathing pattern).
- *Absent.* Apnea is the absence of spontaneous breathing and is common especially after induction of anesthesia with drugs such as propofol if they are given too rapidly or at too high of a dosage. It is important that the anesthetist support apneustic patients by breathing for them periodically until they can begin to breathe on their own.

Carbon Dioxide Monitoring

Carbon dioxide (CO_2) is a normal byproduct of aerobic cellular metabolism. As CO_2 within the cells increases, it diffuses into the capillaries and venous

circulation to be transported to the lungs. In the lungs, the CO_2 diffuses down a concentration gradient from the pulmonary capillaries into the alveoli. Once in the alveoli, CO_2 is eliminated through exhalation. In order for CO_2 to be effectively produced and eliminated from the body, the following must take place:

- Aerobic cellular metabolism in the tissues
- Adequate tissue perfusion to pick up CO_2 and transport it to the lungs
- Adequate pulmonary circulation to transport CO_2 from capillaries to alveoli
- Adequate gas exchange across alveolar membrane
- Sufficient ventilation[12, 13, 14]

Blood CO_2 levels are determined by three factors: the rate of production (cellular metabolism), the rate of transport to the lungs (CO_2 and perfusion), and the rate of elimination (respiratory function).

Non-Invasive Carbon Dioxide Monitors

CO_2 can be *objectively* measured via **capnography**, which is a graphical record of expired CO_2 concentrations during a respiratory cycle. It is a non-invasive, continuous method of sampling and analyzing inspired and expired CO_2 that helps identify situations that can lead to apnea, hypercapnia, or hypoxia. The amount of CO_2 is measured by infrared absorption as it is sampled near the ETT connector if it is *mainstream* or through a sampling at a peripheral sensor if it is *sidestream* and usually determines the respiratory rate and end-tidal carbon dioxide ($EtCO_2$) (some monitors will also provide inspired CO_2 levels). $EtCO_2$ reflects systemic arterial CO_2 and is commonly reported in millimeters of mercury (mm Hg). Capnography does not directly measure blood CO_2; however, in patients without significant cardiac or pulmonary pathology, the $EtCO_2$ correlates well with the CO_2 in blood ($PaCO_2$).[11,15] Since blood CO_2 levels are partially determined by the rate of transport to the lungs, $EtCO_2$ correlates well to cardiac output; therefore, a decrease in $EtCO_2$ may indicate a decrease in cardiac output and decreased oxygen delivery. While $EtCO_2$ levels in very ill patients should be interpreted with caution, trends in $EtCO_2$ correlate with changes in $PaCO_2$ and can provide an early warning of metabolic or cardiorespiratory problems.[16] Capnography allows for the observation of trends in the $EtCO_2$ value and a comparison of $EtCO_2$ with invasive arterial blood gas values.

A capnograph is a device that produces the actual waveform and consists of the following:

- A sensor that measures infrared light absorption, which is proportional to the CO_2 level. There are two types of sensors:
  - A mainstream capnograph has a sensor directly between the ETT and breathing circuit (Figure 7-16a). Mainstream sensors produce an immediate reading, are the more accurate type of sensor, and produce the best quality wave form. Although they are lightweight, they are larger and heavier than sidestream sensors, which increases mechanical dead space and poses a risk of the ETT putting torque on the airway. These sensors are also heated to prevent condensation, which may rarely pose a burn risk to patients. The CO_2 adapter is disposable but can get "fogged up" or be dirty and cause issues with getting an accurate reading.
  - A sidestream capnograph has a sensor located within a monitor away from the patient. A plastic adapter is placed between the ETT and breathing system, and this adapter is attached to a

suction line that aspirates expired gas into the monitor for reading (Figure 7-16b). Sidestream sensors can sample 50 to 150 mL per minute from the circuit, which may cause loss of gas from the breathing circuit and tidal volume. These sensors also have a few second delay in displaying CO_2 levels. The airway adapter is lightweight, so the risk of putting torque on the airway is less likely than with mainstream sensors. Sidestream sensors are also smaller than mainstream sensors; therefore, they theoretically produce less apparatus dead space. Sidestream adaptors use a filter within the line and can become easily clogged with airway sections. Adapters and lines are usually disposable and need to be changed on a regular basis. Sidestream capnographs need a waste gas scavenger.

- A computerized monitor with a digital readout. The monitor displays the respiratory rate, the $EtCO_2$ displays the CO_2 level at the end of expiration, the inspired carbon dioxide (iCO_2) displays the inspired carbon dioxide level, and a graphic waveform is produced of the CO_2 concentration during each respiratory cycle (capnogram).

The **capnogram** is a record of a graph that has time on the x-axis and CO_2 levels on the y-axis (Figure 7-17). The shape of the wave is determined by CO_2 levels as they pass through the ETT. Inspired CO_2 is 0 mm Hg (if the anesthetic machine is properly functioning and suitable oxygen flow rates are used), and this is known as the baseline. CO_2 levels increase rapidly at the beginning of expiration, increase slightly until the end of expiration, and then decrease rapidly to 0 mm Hg at the beginning of inspiration.

The capnogram has five important characteristics to monitor:

1. Height (normal 35–45 mmHg)
   a. Tall or elevated = High $EtCO_2$ (Hypercapnia)
   b. Small or lowered = Low $EtCO_2$ (Hypocapnia)
2. Rate
   a. Length depicts time
   b. Many waveforms = Tachypnea
   c. Few waveforms = Bradypnea
   d. Guidelines for normal rate in anesthetized dogs and cats is 8 to 15 breaths per minute
3. Rhythm
   a. Breathing pattern should be regular
   b. May increase or decrease over time in spontaneously breathing patients
4. Baseline
   a. Represents inspired CO_2
   b. Acceptable inspiratory CO_2 values should be near zero
   c. Elevation of baseline represents inspired CO_2 and rebreathing of CO_2 and should be investigated
5. Shape of the wave
   a. The wave should be relatively square and consistent. Readings that deviate from the normal waveform are indicative of many things including but not limited to airway obstruction, breathing circuit/airway leaks, spontaneous breathing during mechanical ventilation, etc. Waveforms that deviate from normal may alert the user of a problem or may not be accurate; therefore, the user should interpret the projected number cautiously. A full discussion on capnography and waveform analysis can be found at www.capnography.com.

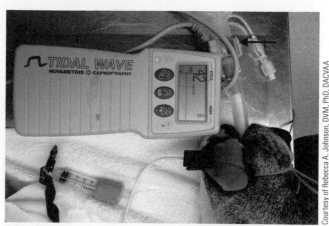
(a)

(b)

FIGURE 7-16 Types of capnographs. (a) A mainstream capnograph uses a sensor directly between the ETT and breathing circuit. (b) A sidestream capnograph has a sensor located in the computerized monitor that has air pulled into it through a tube attached to a fitting between the ETT and breathing circuit.

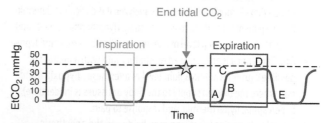

FIGURE 7-17 Typical capnograph waveform. The height of the waveform corresponds to the amount of exhaled CO_2, and the length of the waveform represents time. The taller the waveform, the higher the exhaled CO_2 value. The shorter the duration of the waveforms, the faster the respiratory rate. Note that inspiration is associated with a rapid drop in CO_2 concentration back to a zero baseline. A = Beginning of expiration; B = Replacement of dead space gas with alveolar gas; A-B = Baseline; B-C = Expiratory upstroke; C-D = Alveolar (expiratory) plateau; D = Beginning of inspiration (this is where $EtCO_2$ is measured); E = Alveolar gas diluted with inspired fresh gas. When looking at the capnograph, expiration initially continues along the zero baseline as dead space is expelled from the airway. This is followed by a rapid increase in exhaled CO_2 as alveolar air is exhaled. The peak (point immediately preceding inspiration) is the spot at which the end-tidal (or end of breath) CO_2 value is obtained. The minimal upward slope of the expiratory phase is the alveolar plateau, and it represents alveolar gases being exhaled.

Expired CO₂ The measurement of EtCO₂ via capnography is currently the optimal non-invasive method of continuously monitoring the adequacy of ventilation and subsequent circulation in veterinary patients because as stated previously, EtCO₂ will mimic cardiac output. Analysis of the capnogram can identify patient conditions such as **hypocapnia** (low carbon dioxide levels) and hypercapnia as well as identify leaks in the circuit, airway obstructions (Figure 7-18 and Table 7-3), and determine proper intubation. Because there is little to no carbon dioxide in the esophagus and stomach, an EtCO₂ reading on the capnometer of zero or near zero immediately following intubation may indicate that the endotracheal tube is not in the trachea (zero or near zero readings on a capnometer may also indicate apnea or cardiac arrest). Patients that are properly intubated should have a normal EtCO₂ reading. This method can also be used to check the placement of enteric feeding tubes to ensure they are not in the trachea.[5]

Capnography can also help the anesthetist determine appropriate settings when a patient is initially set up on a mechanical ventilator. Typically, tidal volumes are calculated based on a patient's lean body weight at 10 to 20 mL/kg.[11] This value gives the anesthetist a starting point for setting the ventilator. As the range of 10 to 20 mL/kg is wide and will vary with individual patients, monitoring EtCO₂ is essential to evaluating the adequacy of ventilation. The anesthetist can use that value to increase or decrease ventilator settings to achieve the best and most physiologic breaths possible.

End-tidal carbon dioxide values can also be used to help remove a patient from mechanical ventilation. **Weaning from a mechanical ventilator**, the process of getting the patient to breathe on its own so it no longer needs mechanical ventilation, ensures that the patient is clinically stable and able to maintain its expired CO₂ levels within a stable range. When weaning a patient from mechanical ventilation, it is important that the anesthetist allow the level of CO₂ to rise slightly (50–60 mm Hg) without allowing the patient to become apneic or excessively hypercapnic and to decrease the level of anesthesia.[17] If the patient's CO₂ is within this range and the patient does not begin to breathe spontaneously, it is likely that the patient is still too deeply anesthetized.

Capnography is a valuable tool during cardiopulmonary cerebral resuscitation (CPCR). Carbon dioxide levels fall abruptly when there is an absence of cardiac output (blood flow) and pulmonary blood flow. A patient with a decrease in cardiac output will not carry as much CO₂ per minute back to the lungs to be exhaled. This change will be reflected as a sudden drop in the EtCO₂ measurement. Studies have shown that resuscitation is most effective in terms of cardiac output when EtCO₂ levels are closer to normal.[18] This indicates that chest compressions are adequate and that blood has good forward flow. More information on CPCR is described in Chapter 13 (Cardiac Arrest & Cardiopulmonary Cerebral Resuscitation [CPCR]).

Ideally, all anesthetized patients will have ventilation quality assessed by determining end-tidal carbon dioxide levels via capnography. When this monitor is not available, the anesthetist should check body temperature and blood pressure, assess pain/analgesia and anesthetic depth if respiratory rate and/or character are abnormal, and make adjustments as needed.

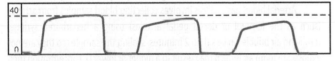

(a) Hypocapnia. Hyperventilation (increased tidal volume or respiratory rate or decrease in metabolic rate) will cause CO₂ to be exhaled more quickly than it is produced; therefore, the waveform will have a gradual decrease in EtCO₂ producing shorter rectangles.

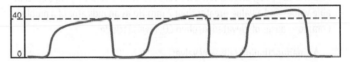

(b) Hypercapnia. Hypoventilation (decreased tidal volume or respiratory rate or increased metabolic rate) will cause CO₂ to remain in the airway and cause a gradual increase in EtCO₂, producing taller rectangles.

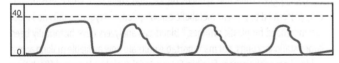

(c) Leak in circuit. A leaky endotracheal tube cuff will cause the downward slope of the plateau to blend in with the descending limb, producing sloppy up-strokes and down-strokes to the rectangles.

(d) Rebreathing CO₂. The capnogram of a patient that is rebreathing CO₂ will cause an elevation of the baseline above zero because the inspired air contains CO₂ (and thus levels do not reach zero). Rebreathing CO₂ may be caused by a faulty expiratory unidirectional valve, inadequate inspiratory flow, malfunction of the CO₂ absorber, or insufficient expiratory time.

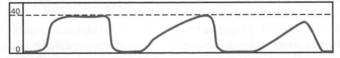

(e) Obstruction. Obstructed expiratory gas flow causes a change in the slope of the ascending limb of the capnogram (the expiratory plateau may be absent). The obstruction can be in the expiratory limb of the breathing circuit or an upper airway obstruction.

FIGURE 7–18 Examples of abnormal capnograms.

TABLE 7-3 Appearance of Abnormal Capnograms

Effect on Capnogram	Possible Problem
Sudden decrease to zero	• Blocked ETT • Disconnection of ETT from sensor • Esophageal intubation • Cardiac arrest
Rapid decline in plateau	• Sudden hyperventilation • Marked decrease in cardiac output
Low plateau value	• Hyperventilation • Hypothermia • Increased dead space
Low EtCO$_2$ measurement without a stable plateau	• Tachypnea
Increased baseline above zero	• Rebreathing CO$_2$
Increased EtCO$_2$ plateau	• Hypoventilation
Downward slope of plateau with sloppy up-stroke and down-strokes to the rectangles	• Leak in circuit • Kinked ETT
Change in ascending limb slope (the expiratory plateau may be absent)	• Obstructed expiratory gas flow

Inspired CO$_2$ Most capnographs also display inspired CO$_2$ levels. The inspired CO$_2$ (iCO$_2$) should remain near zero with 0 to 3 mm Hg as acceptable values and a value greater than 7 mm Hg as an indicator of a problem. Common causes of elevated inspired CO$_2$ levels include:

- malfunctioning one-way valve;
- excessive mechanical/apparatus dead space;
- exhausted carbon dioxide absorbent; and
- inadequate fresh gas flows on a non-rebreathing system.

Carbon Dioxide Monitoring via Blood Gas Analysis

CO$_2$ levels can be *objectively* measured through arterial and venous blood gas analysis. Arterial blood gas values provide the amount of CO$_2$ dissolved in arterial blood plasma (PaCO$_2$) as well as arterial oxygen (PaO$_2$), while venous blood gas values provide the amount of oxygen (PvO$_2$) and carbon dioxide (PvCO$_2$) dissolved in venous blood plasma. Both arterial and venous blood gas analysis provides blood pH measurements. Arterial blood gas analysis is an indicator of an animal's oxygenation, ventilation, and acid-base status, which are all affected by the patient's respiratory function. Venous blood gas analysis is helpful for assessing PvCO$_2$ as CO$_2$ levels are highest in venous blood; however, PvO$_2$ is not a useful parameter for judging O$_2$ status because O$_2$ levels are lowest in venous blood. Blood gas analysis is a superior method of determining a patient's respiratory function because although observing respiratory rate, depth, and character can provide an indication of respiratory function, it does not directly measure parameters of oxygenation and ventilation. Non-invasive modalities such as SpO$_2$ and EtCO$_2$ only provide estimates of arterial blood gas values; therefore, in critical patients, blood gases are preferred because they provide precise information about PaO$_2$ as

well as bicarbonate, pH, lactate, and base excess. These values all help identify underlying patient problems and guide therapy.

Arterial blood gas analysis involves collecting blood samples from arteries (such as the femoral or dorsal pedal arteries) using a heparinized syringe, removing air bubbles in less than 2 minutes, sealing the samples so they are not exposed to room air (which may result in a higher or lower O$_2$ concentration then that of the patient), and analyzing them immediately using a blood-gas analyzing machine. If samples cannot be analyzed immediately, transportation times should be minimal. Samples collected in plastic syringes should be transported at room temperature within 15 minutes if the partial pressure of oxygen (pO$_2$) or oxygen saturation are desired, and within 30 minutes otherwise. The practice of keeping glass collection syringes on ice after collection, which was historically performed to prevent metabolic consumption of oxygen, is thus no longer recommended for plastic syringes.[19] Blood-gas analyzers have historically been larger, stationary units, but many portable units are now available making access to blood gas analysis more feasible. Since arterial sampling is more difficult than collecting venous samples and evaluating repeated arterial samples is expensive, arterial blood gas analysis is impractical for most healthy animals and is reserved for use in sick patients when more precise data is needed.

Tech Tip

Elevated CO$_2$ values indicate hypercapnia, which is most commonly a result of hypoventilation, and low CO$_2$ values indicate hypocapnia, which is most likely due to hyperventilation. PaCO$_2$ has an inverse relationship with alveolar ventilation. PaCO$_2$ values < 35 mm Hg indicate hyperventilation and values > 45 mm Hg indicate hypoventilation.

Monitoring Temperature

Monitoring core body temperature is essential to proper anesthetic case management. Anesthetics, particularly inhalant anesthetic agents and opioids, such as hydromorphone, reset the thermoregulation threshold in the hypothalamus allowing tolerance of a wider range of temperatures without the patient attempting to regulate its body temperature. An animal's ability to conserve and generate body heat is affected by sedation and anesthesia. For example, hypothermia is common in anesthetized animals, which decreases anesthetic requirements and may delay recovery. If an animal shivers to counteract hypothermia, oxygen consumption will increase and may negatively affect the patient's oxygenation status.

Types of Heat Loss

During anesthesia, patients generally experience heat loss from four main causes:

- **Radiant heat loss** is loss of heat generated within the patient's body that is given off to the atmosphere. Radiant heat loss is proportional to the temperature difference between the patient and the environment.
- **Evaporative heat loss** is heat loss due to evaporation of liquids from the surface of the body or within an open body cavity.
- **Conductive heat loss** is heat loss through transfer of heat from the warmer patient body to cooler objects/surfaces in contact with the animal.
- **Convective heat loss** is heat loss to cooler objects not in direct contact with the patient's body (such as air current or ambient temperature).

During anesthesia maintenance it is inevitable that patients will have a decrease in body temperature for the following reasons:

- Some drugs that cause peripheral vasodilation such as acepromazine, isoflurane, and propofol may cause the patient's temperature to drop due to redistribution of heat from the patient's core to its periphery where it is lost more easily.
- The hypothalamus regulates the body's temperature (thermoregulation) and since anesthetic agents depress the hypothalamus, natural responses such as shivering are not initiated to increase the patient's temperature.
- The oxygen flow rate may play a minimal role in lowering the patient's temperature via evaporative heat loss when the patient breathes in the cold gas.
- Room temperature IV fluids are at a lower temperature than an animal's normal body temperature; therefore, when they are given to the patient they can lower their body temperature due to conductive heat loss.
- Transporting or prepping a patient on a cold, stainless steel gurney with no barrier between the patient and the table exposes the patient to lower temperatures that can lower their body temperature via conductive heat loss.
- Some aspects of the surgical prep (such as using cooler prep solutions that cause evaporative heat loss and shaving the patient) can chill the patient while preparing it for surgery.

- Ambient room temperature can contribute to convective and radiant heat loss; therefore, the operating room temperature should be kept at a reasonable level despite the desire of some surgeons to keep the operating room cool.
- Patients with open body cavities will lose heat due to radiant and evaporative heat loss.
- Irrigation of a wound or an open body can result in conductive heat loss.
- A patient's age may affect their ability to maintain normal body temperature during anesthesia (pediatric and geriatric patients are less able to regulate their temperature).
- The size of the patient is important because smaller patients lose heat faster than larger patients due to their proportionately higher body surface area.

Prevention of Hypothermia

Drugs are metabolized more rapidly at higher temperatures; therefore, keeping a patient as normothermic as possible will help them recover more quickly from anesthesia. **Hypothermia** (low body temperature) in veterinary patients is a concern because it can

- increase a patient's risk of infection as hypothermia impairs immune functions including the killing ability of neutrophils;[20]
- slow recovery times as metabolism is slowed and liver function altered, which slows metabolism of anesthetic agents;[20]
- increase oxygen consumption caused by the animal's shivering;
- alter cardiac function by causing bradycardia;
- decrease anesthetic requirements that can lead to overdosing of anesthetic agent(s) if the patient is not properly monitored;
- induce electrocardiographic changes;
- impair platelet function;
- increase fibrinolysis; and
- cause coagulopathy and increase the need for transfusion.[21]

Maintaining a patient's body temperature and preventing heat loss is easier than treating hypothermia. Heat loss begins during the preoperative period, which makes warming the patient a goal throughout the entire anesthetic event. It is best to start addressing heat loss by warming the patient early (at the premedication phase) and having as many techniques as possible available to prevent the body temperature from dropping further. The aggressiveness with re-warming efforts for a patient will depend upon the degree of hypothermia and the stability of the patient. Ways to prevent hypothermia in patients undergoing anesthesia include:

- Place premedicated patients in a warm area of the hospital (still in full view of veterinary staff) or turn on the cage heating and provide blankets/towels.
- Use forced air blankets and conductive fabric warming blankets (Figures 7-19a and 7-19b) on the surgical table that have been warmed before the patient gets into the surgical area. Rice socks, electric heating blankets, and warm water bags/bottles/gloves should be avoided because they have the tendency to cause burns.
- Use a covered heating pad (covering the pad with a towel avoids burns) on the prep bench for induction and prepping of the patient.

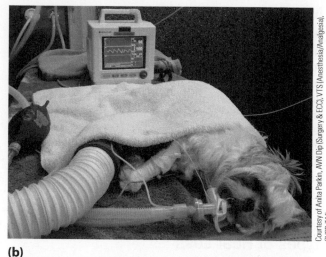

(a)

(b)

Courtesy of Kristen Cooley, BA, CVT, VTS (Anesthesia/Analgesia)

Courtesy of Anita Parkin, AVN Dip (Surgery & ECC), VTS (Anesthesia/Analgesia), CVPP, TAE

FIGURE 7–19 Forced air and conductive fabric warming blankets are commercially available products that are excellent ways to prevent hypothermia in anesthetized patients. Exogenous heat sources such as a Bair Hugger® forced air blanket or Hot Dog® conductive fabric warming blanket can prevent heat loss. (a) A Hot Dog® blanket consists of a control box and a flexible, vinyl conductive blanket connected by a cable to warm the patient by electrical resistance. The warming blanket may be secured around a patient with straps to maximize contact with the patient. The blanket only heats on one side, directing all the heat to the patient, not the surgery table. (b) This patient has a commercial forced air blanket (Bair Hugger®) that blows hot air through a large flexible pipe into a blanket (disposable) which can be covered with a light towel to prevent the blanket from falling off the patient. These blankets also come in a style that sits under the patient for use in the surgical suite (theatre).

- Warm IV fluids to 99.5°F (37.5°C) (not over 107.5°F [42°C] as this can cause hemolysis and damage the vascular endothelium and organs). Warming can be done with a water bath or with a commercial fluid warmer (Figure 7-20a). Heating fluids with the use of a microwave is not advisable as you can get an uneven distribution of temperature within the fluids and could potentially burn the patient (Figure 7-20b).
- Provide warm gases by warming the breathing hoses. Unnecessarily high oxygen flow rates can lower body temperature as cooler air enters the system. When using a rebreathing (circle) system or a

Universal F circuit, warm the breathing tubes by using an anesthetic circuit warmer (Darvall SWT Heated Circuit), which warms the inspired side of the circuit (Figure 7-21a, b, and c). The monitor displays two temperatures: the circuit warmer temperature and the patient's temperature. There are two circuit sizes available, 7.5–15 kg and 15–30 kg, so although they are good, a patient weighing under 7.5 kg cannot use this system. The efficacy of this system has largely been called into question by some anesthesiologists, mainly because loss of heat through the airway is a small contributing factor to overall heat loss.

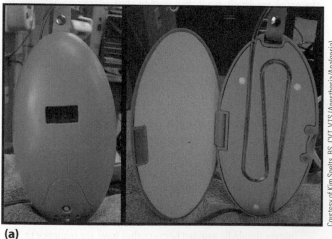

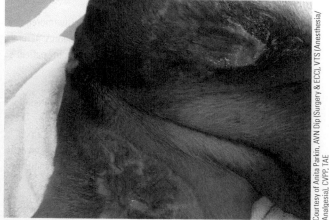

(a)

(b)

Courtesy of Kim Spelts, BS, CVT, VTS (Anesthesia/Analgesia)

Courtesy of Anita Parkin, AVN Dip (Surgery & ECC), VTS (Anesthesia/Analgesia), CVPP, TAE

FIGURE 7–20 Ways to warm a patient. (a) Commercial IV fluid warmer. Fluid warmers such as the iWarm™ (Midmark, Dayton, OH) come in a heated plate or heated water bath so as the fluid flows through the line, it is warmed up. These are best used close to the patient, so the fluid does not cool down again before reaching the patient. (b) Burns on a patient. Third degree burns obtained when using a fluid bag warmed in the microwave and put directly onto the patient's skin post-surgery.

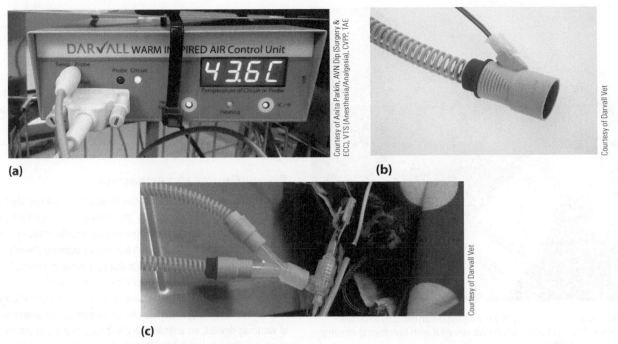

(a)

(b)

(c)

FIGURE 7-21 Anesthetic breathing circuit warmer. (a) The control panel of an anesthetic circuit warmer will alternate the temperature display between the patient's temperature and the circuit temperature. (b) The inspiratory hose has heating wires wound through it. The temperature of delivered gases is measured at Y-piece. (c) The circuit warmer has a sensor that is inserted into the patient's esophagus. This sensor feeds back to the unit and is automatically shut off if the patient is too warm.

- Provide warm gases by using heat-moisture exchange filters (Figures 7-22a and 7-22b). This device is connected between the anesthetic circuit and the endotracheal tube to filter expired gas and warm the air that goes to the patient. Disadvantages of these filters are that they increase mechanical dead space, their weight can be cumbersome in small patients, and they increase resistance to breathing in small patients. These disadvantages can lead to muscle fatigue, which will have to be monitored closely, particularly in small patients. Heat-moisture exchange filters are also expensive and can only be used once; therefore, they are not commonly used in veterinary practice.

- Transport the patient to the surgical area on a gurney that has a blanket on it. Do not place the patient directly onto the stainless steel table top, even if the patient is only on it for a short period of time because their body temperature will still decrease.
- Put baby socks or wrap any nonsurgical limbs in bubble wrap or reflective "space blankets" (the kind they give runners at races on a cold day) (Figure 7-23). These are economical ways of preventing heat loss. The head can be wrapped if needed, but only while the patient is intubated.
- Turn up the heat in the operating room.
- Ask the surgeon to lavage a surgical area with warm fluids if the patient is having a thoracotomy or laparotomy and having trouble maintaining body temperature.

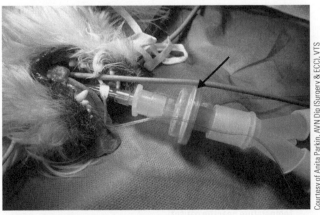

(a)

(b)

FIGURE 7-22 Heat moisture exchange filter. (a) The heat moisture exchange filter is connected between the anesthetic circuit and the ETT. (b) The heat moisture exchange filter filters expired gas and helps expired air remain humidified and theoretically keeps expired heat with the patient.

Courtesy of Anita Parkin, AVN Dip (Surgery & ECC), VTS (Anesthesia/Analgesia), CVPP, TAE

FIGURE 7–23 One way to prevent hypothermia is to wrap the patient's limbs in bubble wrap and secure it with bandaging material or children's socks.

Key points to remember when maintaining animals under anesthesia is to keep anesthetic time to a minimum, not oversaturate the patient in prep solutions (for most surgical cases, the initial surgical scrub can be warmed), and closely monitor the animal's body temperature. Know how to keep the patient warm and start warming the patient early.

Tech Tip

Any heating devices should be covered with a towel to help avoid burns to the patient. Thermostat controlled heating devices are preferred.

Treatment of Hypothermia

If a patient is hypothermic, passive rewarming is adequate to correct mild hypothermia in a stable patient. An example of passive rewarming is the use of blankets, which allows the patient's intrinsic mechanism of heat production (shivering) to provide heat. Intense monitoring is not needed. Active re-warming can involve external or core methods and should be used in moderate to severe cases of hypothermia. Active external rewarming involves the use of exogenous heat sources such as forced-air warming system (Bair Hugger®) or conductive fabric warming (Hot Dog®) blankets (see Figures 7-19a and 7-19b) and heat lamps. Active core rewarming involves the use of warm water enemas, warmed or humidified oxygen, and warmed IV fluids (see Figure 7-20a).

When warming a patient, heat is initially applied on the thorax in an attempt to raise the patient's core body temperature. The goal is to warm the animal by at least 1.8–3.6°F (1–2°C) per hour until a temperature in the normal range can be maintained.

Safety Alert

Exercise care with cords when setting up warming devices. Make sure they are placed out of harm's way or are safely covered so they do not pose tripping hazards. Cords that are out of the way also allow the cage door to be securely closed without crushing the cords.

Prevention of Hyperthermia

Hyperthermia, a core body temperature elevation that is above the accepted normal range for each species, can also develop in anesthetized patients. Malignant hyperthermia secondary to genetic predisposition may be exacerbated by the overuse of heating devices (such as forced air warming blankets left on a patient with thick fur and lots of surgical drapes), patients in a surgical area that is too warm, patients that have thick fur and multiple layers of surgical drapes, or patients under lights that are too hot, and the use of certain anesthetic agents.

Unintentional hyperthermia can be caused by the overzealous use of warming devices, an animal's thick hair coat, and warm ambient temperatures. Drug-induced hyperthermia is rare; however, there are two main types seen in animals: opioid-induced (seen more frequently with hydromorphone, morphine, and methadone but can also be seen with other opioids such as buprenorphine) and malignant hyperthermia.

- Opioid-induced hyperthermia can develop in cats and is associated with the use of mu opioid agonists such as hydromorphone and buprenorphine. Opioid-induced hyperthermia can cause high temperatures in some patients and can be distressing to both the patient and veterinary staff but in general is self-limiting. Cats whose temperature is greater than 104°F (40°C) are treated with gentle cooling (removing blankets, providing a fan, etc.) and those with extreme temperature elevations should be given naloxone to partially reverse the opioid. Naloxone can be given SQ, IM, or IV, depending on clinician preference, status of patient, and presence of an IV catheter, to partially reverse the opioid. In rare circumstances the dose of naloxone dose can be repeated. Keep in mind that any analgesia provided by the opioid will also be reversed with naloxone.
- Malignant hyperthermia is an inherited abnormality that occurs in pigs, dogs, cats, and horses. Malignant hyperthermia is most often associated with the use of halogenated inhalant anesthetic agents. Avoiding these agents in susceptible patients is warranted.

Treatment of Hyperthermia

The aggressiveness with cooling efforts for a patient will depend upon the degree of hyperthermia and the stability of the patient. Techniques for cooling a patient include:

- Removal of all blankets and allowing the patient to be on a cool surface
- Removal of all external heating devices
- Administration of room temperature IV fluids
- Increasing oxygen flow rates in animals on inhalant anesthetic agents
- Using a fan in the surgical/procedure room to lower the ambient temperature near the patient.
- Treating the cause of hyperthermia
- Wetting the patient with room temperature water

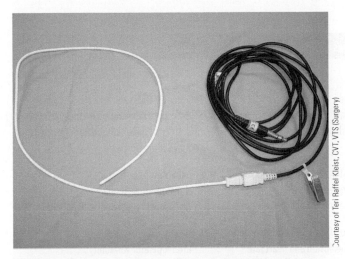

Courtesy of Teri Raffel Kleist, CVT, VTS (Surgery)

FIGURE 7-24 An esophageal temperature probe is placed in the lower esophagus as a way to measure body temperature.

Treatment continues until the temperature reaches slightly above the top of the normal range for the animal species since the patient will likely continue to cool on their own. Overzealous cooling can also lead to shivering (which ultimately produces more heat) and vasoconstriction (which slows heat loss, decreases cutaneous blood flow, and increases oxygen consumption).

Temperature Monitoring

The ACVAA recommends "the temperature should be measured periodically during anesthesia and recovery and if possible checked within a few hours after return to the wards.[2] It is wise to monitor the patient's temperature regularly and if it is low, provide thermal support until it is normal and take steps to proactively prevent heat loss. If the patient's temperature is high, limit or reduce thermal support. Continue to monitor the patient's body temperature for at least two readings following its return to normal before reducing the frequency of temperature assessment.

Temperature can be monitored intermittently with a rectal thermometer (ideally every 15–30 minutes) or continuously using an esophageal probe connected to a mechanical monitor. Properly placing a reusable probe in the esophagus provides an easy and accurate assessment of core body temperature (Figure 7-24). The distal end of the probe should be passed to the level of the base of the heart to get a core body temperature. To determine how far to insert the probe, it is recommended to "premeasure" the distance before feeding the probe into the esophagus.

SUMMARY

Monitoring anesthesia is one of the most important and challenging aspects of a veterinary technician's job. Anesthetic monitoring provides the veterinary technician the opportunity to perform at their highest capacity and therefore provide the greatest benefit to the veterinary care team when properly utilized. ACVAA has set forth guidelines on monitoring anesthetized patients that include both a hands-on approach as well as the use of mechanical monitors focusing on circulation, oxygenation, ventilation, and body temperature. Subjective measurements such as the patient's pulse

and respiratory rates as well as the character of the respiration and pulse have historically been used to monitor anesthesia patients. Over the last few decades, there has been a shift toward using electronic monitoring devices that provide objective measurements (summarized in Table 7-4). The enthusiasm for such devices has been driven partly by curiosity, partly by the need for objective measurements that gauge real scientific progress, and partly by the medical and legal climate, which necessitates accurate, objective medical records. There is no doubt that the quality of patient care has improved, but it must be remembered that the new monitoring devices act as adjuncts to, not replacements for, diligent, hands-on monitoring. A veterinary technician with very basic equipment can be a competent anesthetist.

CRITICAL THINKING POINTS

- Constant vigilance allows the anesthetist to be proactive rather than reactive in response to a patient's anesthetic depth and physical well-being.

- Anesthetic depth should be assessed regularly by evaluating reflexes, muscle tone, eye position, lacrimation, pupil dilation, response to surgical stimulus, and changes to physiologic parameters.

- Frequent, continuous monitoring of circulation, oxygenation, ventilation, and body temperature should be done by a designated anesthetist whose sole task is to monitor anesthesia.

- Anesthetic monitors are only as good as the person using them. Keep in mind that a single measurement of a monitoring parameter gives a "window" into a dynamic situation. Only repeated measurements give proper insight to the dynamic picture of an animal's anesthetic status.

- It is easy to do a "bad job" of monitoring with the very best of equipment. If a monitor fails to give an appropriate reading, the anesthetist's first priority must be to do a basic but thorough examination of the patient and the breathing system before attending to the monitor. Once the well-being of patient has been established, the monitor can be addressed.

- It is important to monitor and maintain blood pressure in all anesthetized patients since hypotension is common under anesthesia.

- Monitoring end-tidal carbon dioxide is recommended for all patients as it is the optimal way to non-invasively monitor the adequacy of ventilation and subsequent circulation on a continual basis.

- Hypothermia is common under anesthesia and can result in adverse effects on patient outcomes if left unrecognized and untreated.

- Insulating a patient from the cool environment with the use of blankets (forced air, conductive fabric warming, and circulating warm water) minimizes conductive heat loss; using small instead of excessive amounts of prep solutions can minimize evaporative heat loss; keeping the surgery or procedure room warm or using exogenous heat sources such as forced air warming blankets and heat lamps can prevent convective heat loss; and preventing or limiting vasodilation can minimize radiant heat loss.

TABLE 7-4 Summary of Monitoring Instruments

Monitor	Parameter	Advantages	Disadvantages
Esophageal stethoscope	Heart rate and lung sounds	Easy to use	Need different sizes for different sized patients
ECG	Electrical activity of the heart	Easy to place Provides heart rate Alerts to dyrhythmias	Alligator clips can cause tissue trauma Only gives information on the electrical activity of the heart and nothing about mechanical function
Oscillometric blood pressure monitor	Blood pressure trends	Provides systolic, diastolic, MAP, and heart rate values at regular intervals Easy to place and simple to use	Expensive Only monitors trends Not always reliable in small patients
Doppler blood pressure monitor	Blood pressure trends Blood flow to tissue bed	Audible blood flow allows anesthetist to hear changes in perfusion and heart rate Can be used in all species as a heart rate monitor Monitors blood pressure with proper equipment	Only monitors trends Inter-operator variability Requires some skill
Arterial catheterization	Real-time blood pressure measurement	Systolic, diastolic, and MAP value plus pressure waveform and heart rate Real-time value reflects changes as they happen	Expensive Requires technical skill Strict aseptic technique is essential Moderate risk of hemorrhage or hematoma
Pulse-oximeter	Peripheral saturation of hemoglobin with oxygen	Alerts to hypoxemia Easy to place Provides an SpO_2 value, heart rate, and pleth waveform	Prone to bias and variability Does not monitor the adequacy of ventilation
Capnograph	End-tidal carbon dioxide	Provides information about the adequacy of ventilation as well as circulation and metabolism Easy to use	Expensive
Blood gas analysis	Arterial sample: $PaCO_2$, PaO_2, and blood pH Venous sample: $PvCO_2$, PvO_2, and blood pH	Directly measures parameters of oxygenation and ventilation	Arterial sampling is difficult Repeated arterial samples are expensive to run
Rectal thermometer	Body temperature	Easy to perform Reliable results	May be difficult to do if patient is draped for surgery
Esophageal temperature probe	Body temperature	Easy to perform Reliable results	Need to make sure of proper placement

REVIEW QUESTIONS

Multiple Choice

1. What four parameters in the anesthetized patient does the American College of Veterinary Anesthesia and Analgesia recommend monitoring in the anesthetized patient according to their revised guidelines?

   a. Carbon dioxide, oxygen, pulse-oximetry, ECG
   b. Circulation, oxygenation, ventilation, temperature
   c. Temperature, pulse, respiration, blood pressure
   d. Perfusion, depth, reflexes, breathing

2. What two factors directly contribute to arterial blood pressure?

   a. Heart rate and vessel tone
   b. Preload and afterload
   c. Stroke volume and cardiac contractility
   d. Cardiac output and systemic vascular resistance

3. What are three reasons anesthetized patients could become hypothermic?

    a. Specific drugs can cause peripheral vasodilation; anesthetics depress the hypothalamus; patients with open body cavities will lose heat due to evaporative heat loss.
    b. Prep solutions can chill the patient; air conditioning is off; fasting leads to zero calories being burned.
    c. Pediatric and geriatric patients have impaired thermoregulation; oxygen is warmed before patient receives it; heat lamps are used in the operating room.
    d. Transporting patients on gurney with blanket on it; heat is turned up in operating room; fasting leads to zero calories being burned.

4. Which is true regarding constriction of the pupil in response to light?

    a. Pupillary constriction increases in deep levels of anesthesia.
    b. Pupillary constriction is lost during deep levels of anesthesia.
    c. Pupillary constriction is normal during deep levels of anesthesia.
    d. Pupillary constriction is not used to assess deep levels of anesthesia.

5. Which of the following indicates a light level of anesthesia?

    a. Lacrimation in horses
    b. Absence of the pedal reflex in dogs
    c. Ability to perform surgery in cats
    d. Slow palpebral reflex in dogs

6. Which of the following is an objective way to assess heart rate and rhythm in an anesthetized patient?

    a. Apical pulse
    b. Electrocardiography
    c. Auscultation of the heart with a stethoscope
    d. Capnography

7. Which piece of monitoring equipment uses an ultrasonic probe and electronic monitor to detect blood pressure?

    a. Oscillometric monitor
    b. Doppler flow monitor
    c. Arterial catheterization
    d. Capnography

8. What does MAP stand for?

    a. Monitoring anesthesia parameters
    b. Minimum arterial pressure
    c. Monitoring alveolar parameters
    d. Mean arterial pressure

9. What does a SaO_2 value represent?

    a. Oxygen dissolved in blood plasma
    b. Arterial oxygen saturation
    c. Peripheral capillary oxygen saturation
    d. Venous oxygen saturation

10. What greatly affects the SpO_2 reading on a pulse oximeter?

    a. Anemia
    b. Renal disease
    c. Pulmonary hypertension
    d. Hypotension

11. What complication is the pulse oximeter designed to alert the anesthetist to?

    a. Hypotension
    b. Cardiac arrhythmias
    c. Hypoventilation
    d. Hypoxemia

12. Cooling techniques should be performed on which of the following patients?

    a. A cat given hydromorphone who now has a temperature of 103°F (39.4°C)
    b. A pig that is prone to malignant hyperthermia and whose temperature has been steadily rising over the past few minutes and is now 104°F (40.0°C)
    c. A dog with a temperature of 102°F (38.9°C) who has been given propofol
    d. A 3-year-old horse in order to help speed recovery from anesthesia

13. What values are presented on the X and Y axis of a capnogram?

    a. Time on the X axis and CO_2 levels on the Y axis
    b. O_2 levels on the X axis and CO_2 levels on the Y axis
    c. CO_2 levels on the X axis and O_2 levels on the Y axis
    d. CO_2 levels on the X axis and time on the Y axis

14. Which term describes the volume in a single breath?

    a. Respiratory rate
    b. Minute ventilation
    c. Tidal volume
    d. Expiratory reserve volume

15. What often causes hypercapnia?

    a. Hypoxia
    b. Hypotension
    c. Hypothermia
    d. Hypoventilation

16. When cooler air or liquids not in direct contact with the warm surface of the body produce heat loss, which type is it?

    a. Convective
    b. Conductive
    c. Evaporative
    d. Radiant

17. What detrimental effect does hypothermia, leading to postoperative shivering, have in a patient?

    a. Atelectasis
    b. Increased metabolic oxygen demand
    c. Release of catecholamines
    d. Conductive heat loss

18. What is indicated by a tall wave on a patient's capnogram?

    a. Bradypnea
    b. Tachypnea
    c. Hypocapnia
    d. Hypercapnea

19. How could a mildly hyperthermic animal best be initially treated?

    a. Ice baths
    b. Take away bedding and apply fan
    c. Tepid water, IV fluids, and oxygen administration
    d. Put the animal in a crate outside

20. Which is *not* an appropriate warming device for a hypothermic patient?

    a. Space heater
    b. Forced air blanket
    c. Circulating water blanket
    d. All of the above are appropriate.

Case Studies

Case Study 1: Charlemagne, an 18-year-old M/N domestic short haired (DSH) cat weighing 4.2 kg presents for a dental cleaning. After premedication with midazolam and ketamine IM, an IV catheter is placed in a front limb. He is induced with propofol, intubated, maintained on sevoflurane, and connected to a capnograph and pulse oximeter. A Doppler monitor is also placed on his right front limb. During monitoring, his heart rate is 123 bpm, respiratory rate is 10 breaths/min, pulse oximeter reading is 95%, capnograph reading is 35 mm Hg, and the Doppler records his blood pressure as 65 mm Hg.

1. What are the normal ranges for these parameters in this species (see Table 7-2)?
2. Are these values normal? If they are not normal, what should you do?

Case Study 2: Bruce, a 6-month-old M/N Cattle Dog mix, presents to the clinic for a neuter. He is premedicated with dexmedetomidine and hydromorphone and induced with IV propofol, intubated, and maintained with isoflurane and oxygen. A pulse oximeter probe is placed on this tongue, an end-tidal CO_2 adapter is placed between the endotracheal tube and the breathing circuit and the non-invasive oscillometric blood pressure monitor is placed on his left hind leg above the hock. His heart rate is 62 bpm, his respiratory rate is 9 breaths/minute, the pulse oximeter is reading 93%, the $EtCO_2$ is 40 mm Hg, and the blood pressure is 120/80 with a MAP of 93 mm Hg.

1. What are the normal ranges for these monitors in this species (see Table 7-2)?
2. Are these values normal? If not, what might the cause be?

Case Study 3: Rhett, a M/N Neapolitan Mastiff canine, who is an overweight 62 kg, presents for a tibial plateau leveling osteotomy (TPLO). He is premedicated with hydromorphone and acepromazine, and an IV catheter is placed. He is induced with propofol, intubated, and connected to 3% sevoflurane and oxygen. His monitoring equipment includes a pulse oximeter, a capnograph, an ECG, and a blood pressure cuff around his forelimb above the carpus. He is positioned in dorsal recumbency as his leg is clipped and prepped for surgery. His first set of readings are as follows: SpO_2 94%, $EtCO_2$ 55 mm Hg, ECG normal sinus rhythm with a heart rate of 66 bpm, blood pressure 92/55, and MAP 67 mm Hg.

1. Are these values normal (see Table 7-2)? If not, which ones are abnormal and what might that mean?
2. What intervention can be provided for this patient?

Case Study 4: Chauncy is a 10-year-old M/N Poodle dog with grade 3 (out of 4) periodontal disease. He is premedicated with a combination of midazolam and morphine IM and an IV catheter are placed without issue. He is induced with propofol and connected to isoflurane and oxygen, and the endotracheal tube cuff is properly inflated. The pulse oximeter probe is placed over the pads of his feet and the capnograph is connected along with the ECG leads and blood pressure cuff. The $EtCO_2$ is 0.

1. What are the three main rule outs for an $EtCO_2$ be 0?
2. How can you determine what is the cause of the $EtCO_2$ to be 0?

Case Study 5: A tiny kitten found on the side of the road was brought to the clinic with a maggot-infested wound. The kitten weighs only 2 kg and needs general anesthesia to clean her wounds. She is mildly dehydrated (5%), her TPR is normal, and neurologic status is normal. She is premedicated with buprenorphine and dexmedetomidine SQ and an IV catheter is placed in a front limb. The kitten is induced with propofol, intubated, and maintained on isoflurane.

1. How should this patient be monitored?

Critical Thinking Questions

1. How can mechanical ventilation lead to hypotension?
2. The capnograph is recommended during cardiopulmonary cerebral resuscitation. What information does the capnograph provide in this situation and why?

ENDNOTES

1. American College of Veterinary Anesthesia and Analgesia position statement, 2009.

2. Brodbelt, D. C., Pfeiffer, D. U., Young, L. E., et al. (2007). Risk factors for anaesthetic-related death in cats: Results from the confidential enquiry into perioperative small animal fatalities (CEPSAF). *British Journal of Anaesthesia, 99,* 617–623.

3. Bleigenberg, E. H., van Oostroom, H., Akkerdaas, L. C., et al. (2011). A study into the relationship between the bispectral index and the clinically evaluated anaesthetic depth in dogs. *Vet Anaesth Analg, 38,* 536–543.

4. Haskins, S. C. (2007). Monitoring anesthetized patients. In W. J. Tranqulli, J. C. Thrumon, & K. A. Grimm (Eds.). *Lumb & Jones' Veterinary Anesthesia and Analgesia* (4th ed., pp. 534–559). Ames, IA: Blackwell Publishing.

5. Dorsch, J. A. & Dorsch, S. E. (2008). Gas monitoring. In *understanding anesthesia equipment* (pp. 687–720). Philadelphia, PA: Lippincott, Williams & Wilkins.

6. Haskins, S. C. (2007). Monitoring anesthetized patients. In W. J. Tranqulli, J. C. Thrumon, & K. A. Grimm (Eds.). *Lumb & Jones' Veterinary Anesthesia and Analgesia* (4th ed., p. 90). Ames, IA: Blackwell Publishing.

7. Love, L. & Harvey, R. (2006). Arterial blood pressure measurement; physiology, tools and techniques. *Compendium, 28*(6), 450–462.

8. Egner, B., Erhardt, W., Henke, J. & Carr, A. (2007). Indications for blood pressure measurement. In Egner, B, Carr, A & Brown, S. (Eds.), *Essential facts of blood pressure in dogs and cats* (4th ed., pp. 23–24). Germany: VBS.

9. Caulkett, N. A., Cantwell, S. L., & Houston, D. M. (1998). A comparison of indirect blood pressure monitoring techniques in the anesthetized cat. *Vet Surg., 27*(4), 370–377. doi:10.1111/j.1532-950x.1998.tb00143.x

10. Grubb, T. (2002). Veterinary anaesthesia. *Veterinary Anesthesia and Analgesia,* 10th ed., 29, 114.

11. Robertson, S. A. (2002). Oxygenation and ventilation. In S. Greene (Ed.), *Veterinary anesthesia and pain management secrets* (pp.15–20). Philadelphia, PA: Hanley & Belfus, Inc.

12. Ornato, J. P., Garnett, A. R., & Glauser, F. L. (1990). Relationship between cardiac output and the end-tidal carbon dioxide tension. *Annals of Emergency Medicine, 19*(10), 1104–1106.

13. Sanders, A. B. (1989). Capnometry in emergency medicine. *Ann Emerg Med 18,* 1287–1290

14. Paddleford R. R., &Greene S. A. (2007). Pulmonary disease. In W. J. Tranqulli, J. C. Thrumon, & K. A. Grimm (Eds.). *Lumb & Jones' Veterinary Anesthesia and Analgesia* (4th ed., pp. 899–900). Ames, IA: Blackwell Publishing.

15. Taskar, V., John, J., Larsson, A., Wetterberg, T. & Johnson, B. (1995). Dynamics of carbon dioxide elimination following ventilator resetting. *Chest, 8,* 196–202.

16. Weil, M. H, Bisera, J., Trevino, R. P., & Rackow, E. C. (1985). Cardiac output and ETCO2. *Crit Care Med, 13,* 907–909.

17. Hartsfield, S. M. (2007). Airway management and ventilation. In W. J. Tranqulli, J. C. Thrumon, & K. A. Grimm (Eds.). *Lumb & Jones' Veterinary Anesthesia and Analgesia* (4th ed., pp. 512–514). Ames, IA: Blackwell Publishing.

18. Touma, O., & Davies, M. (2013). The prognostic value of end tidal carbon dioxide during cardiac arrest: A systematic review. *Resuscitation, 84*(11), 1470–1479. doi: 10.1016/j.resuscitation.2013.07.011

19. Baird G. (2013). Preanalytical considerations in blood gas analysis. *Biochem Med (Zagreb), 23*(1), 19–27. doi:10.11613/bm.2013.005

20. Beal, M. W., Brown, D. C., Shofer, F. S. (2000). The effects of perioperative hypothermia and the duration of anesthesia on postoperative wound infection rate in clean wounds: A retrospective study. *Vet Surg, 29*(2), 123–127.

21. Bryant, S. (Ed.). (2010). *Anesthesia for veterinary technicians* (pp. 98–99). Wiley & Blackwell.

CHAPTER 8
Premedication

Christopher L. Norkus, *DVM, DACVAA, CVPP, DACVECC*

LEARNING OBJECTIVES

Upon completion of this chapter, it is expected that the reader should be able to:

8.1 Describe how veterinary patients benefit from premedication

8.2 Identify factors that influence premedication drug selection

8.3 Explain the general concept of allometric scaling

8.4 Identify premedication drug classes

8.5 Describe the advantages and disadvantages of premedication drug classes

8.6 Identify major adverse effects of premedication drug classes

8.7 Identify which premedication agents are reversible and their respective reversal agents

8.8 Recognize which premedication drugs have analgesic properties and which do not

8.9 Recognize the expected duration of action of different premedication agents

KEY TERMS

Allometric scaling

Analgesia

Antiemetic

Anxiolysis

Cataleptic

Controlled substance

Dose

Dosage

Dysphoria

Neuroleptanalgesia

Parasympatholytic

Perioperative period

Pre-emptive analgesia

Premedication

Sedative

Tranquilizer

INTRODUCTION

Agents that are given before anesthesia begins are known as pre-anesthetic drugs and are also referred to as **premedication**. These "pre-med" drugs commonly include sedatives, tranquilizers, analgesics, anticholinergics, and sometimes other miscellaneous adjunctive agents. Often several drugs are given together to maximize the benefits from each individual drug.

The importance of premedication is often overlooked. While premedication drugs are not always indicated, their use is generally recommended and adds to the safety and quality of the entire anesthesia experience. There are many reasons to utilize premedication in veterinary patients including anxiety and stress reduction in the patient (and therefore in the anesthetist!) as well as reduce nausea during the **perioperative period** (the entire duration of a patient's surgical/anesthetic experience including preoperative, intraoperative, and postoperative time periods). Premedication drugs may provide **analgesia** (pain relief) and cause muscle relaxation. Premedication may result in sedation and aid in patient restraint. They often reduce the amount of anesthetic induction and maintenance drugs that are needed later and, lastly, they can improve the smoothness of anesthesia induction and quality of the patient's recovery back to the unanesthetized state.

Appropriate selection of premedication drugs will have a major impact on the entire anesthetic experience. Therefore it is critical to know the effects of each premedication agent and select drugs specifically based upon the needs of the individual patient rather than using a single regimen or protocol for all animals. Predetermined premedication mixes, recipes, or "cocktails" are generally frowned upon as they usually do a poor job of considering individual patient needs.

Lastly, premedication and sedation are often used as an alternative to general anesthesia for minor procedures. There is, however, a misconception by owners and some veterinarians that heavy sedation is always a safer route than general anesthesia. As drugs used for premedication and sedation can have significant adverse effects, many high-risk patients may benefit from undergoing general anesthesia rather than heavy sedation. With general anesthesia an airway can be secured by way of an endotracheal tube (ETT), oxygen can be administered, ventilation supported, and more options for patient monitoring and physiological support are available.

Tech Tip 🐾

Not all premedication situations are the same. The approach to premedication prior to general anesthesia differs from that used for sedation adequate for a diagnostic or therapeutic procedure.

DRUG AND DOSAGE SELECTION

Selecting an individual drug protocol can seem overwhelming at first because of the many patient variables at play as well as the numerous drugs that are available. In many incidences, more than one drug option or **dosage** (amount of drug per animal species' body weight or measure) will meet a patient's needs. It is however the anesthetist's goal, working in conjunction with the attending veterinarian, to select drugs that are the "best fit" for the individual patient. Before selecting individual premedication drugs, the anesthetist should identify key points about a case that influence drug selection. This "patient problem list" is critical to successful drug selection and patient management. The anesthetist then determines which drug classes would be the worst choice to give to a particular patient. Once these choices are identified, selecting appropriate drug classes suddenly becomes much easier because one knows what *not* to select. Once the anesthetist has "ruled in" appropriate options, then in conjunction with and under the supervision of a licensed veterinarian, they can continue to narrow down the field by evaluating different options within an appropriate drug class and select which individual agent(s) would be the best fit for a given patient.

Key considerations the anesthetist should specifically contemplate when determining which premedication drugs are appropriate include:

- *Will the patient benefit from premedication in the first place?* In most incidences they do. Rare exceptions to this may include Cesarean section patients and patients who are already receiving potent analgesics or central nervous system depressing drugs in hospital. These agents have made them already sedate.
- *What is the species of the patient?* One species may respond differently than another species to particular drug class and individual drugs. For example, a middle-aged pig might sedate well with a benzodiazepine whereas a middle-aged horse or dog would not. You must know the species you are working with and their individual differences.
- *What is the age of the patient?* Patient age may influence drug selection as well as drug dosing. For example, older patients may be especially sensitive to the effects of sedative and tranquilizing agents, therefore requiring lower dosages of these drugs.
- *What is the breed and temperament of the patient?* Being knowledgeable about drug effects on specific animal breeds can help reduce adverse reactions in some patients. For example, some Boxer dogs have adverse effects when administered acepromazine. The anesthetist should question whether this drug should be avoided entirely or used sparingly in the breed. In addition, calm and friendly patients may require less drug and/or lighter sedation than an excited or fractious patient; therefore, drug dosing should be individualized in all patients.
- *Is the patient on other medications that could interact with my drug choices? Does the patient have any drug allergies?* Drug interactions and any known drug allergies should influence which medications are chosen for each patient. One such example might be a patient who is concurrently on an angiotensin converting enzyme inhibitor (ACEi) such as enalapril. The vasodilatory effects of the drug may increase the incidence of perioperative hypotension in the face of other vasodilating agents such as acepromazine. Although it is not contraindicated to utilize both drugs together, it is important for the anesthetist to consider such drug–drug interactions.
- *What is the duration and specifics of the procedure being performed?* If the procedure is short and the patient is scheduled to go home immediately afterwards, selecting a drug with a very long duration of action may not be ideal. If the procedure is painful, adequate analgesia must be provided for the proper amount of time.
- *Has the patient received premedication drugs before, what dosages were used, and what were the effects?* Always review a patient's medical records for information on which drugs have or have not produced the

desired effect for the animal. For example, if a drug failed to provide adequate sedation when used last time, a different dosage or different drug entirely should be considered in the future. If adverse effects were noted previously, changing drugs or dosages may prevent them from occurring.

• *Are there pre-existing conditions and how does this influence the patient's overall health status and anesthetic management of the case?* For example, a patient with heart disease, dogs with brachycephalic airway syndrome, horses with colic, or animals with kidney disease should be assessed on an individual basis to determine which premedication drugs would be best for them. For example, acepromazine might be a great choice or a detrimental choice for a patient with heart disease depending on the type of cardiac disease present. Alpha-2 adrenergic agonists may be a poor choice for an animal with renal disease. The anesthetist should consider any concurrent issues occurring in the animal and how these conditions influence perioperative management and drug selection in their patient.

Much of the above information is gathered through obtaining a detailed patient history, performing a thorough physical exam, and conducting the proper laboratory tests and diagnostic imaging. These factors help determine which drug(s) to use in a patient's individual premedication plan. Similarly all of these considerations also influence which drug dosage is selected.

Another critically important concept in drug dosing is metabolic or **allometric scaling.** In short, allometric scaling considers the influence of patient size in dosage selection. On a body weight scale alone, larger patients require a lower dosage on a mg/kg basis than small patients do. When given a drug dosage range, larger patients typically require dosages on the lower end of the scale and smaller patients require drug dosages on the upper end. For example, an effective dosage of acepromazine for a 5 kg cat might be 0.1 mg/kg while a 450 kg horse might need only 0.02 mg/kg. The clinical significance of not taking this important concept into account is that small patients may be inadvertently underdosed and larger patients may be grossly overdosed.

Lastly, patients who are overweight should have their drug doses adjusted to their lean "ideal" body weight or have their overall drug dose reduced. For example, if you wish to sedate an overweight cat that weighs 9 kg but should weigh only 5 kg, use the 5 kg weight when calculating its drug **dose** (amount of drug administered at one time to achieve the desired effect). Not adjusting drug doses to account for obesity will result in patients receiving more drug than they need.

ROUTE OF DRUG ADMINISTRATION

For most small animal patients, premedication drugs are usually administered via either the subcutaneous or intramuscular route. This is because minimal restraint is required for drug administration via these routes. In horses, the intravenous route is usually selected because the intravenous route is very easily accessible with minimal restraint. When an intravenous catheter is present, its use should be considered to avoid unnecessary injection site pain in the patient.

Intravenous drug administration will yield the fastest onset time for a drug and generally the most profound effect. Some drugs that commonly cause vomiting (e.g., hydromorphone) may have less of this effect when given intravenously. However, intravenous drug administration may have a slightly shorter duration of action than other routes such as subcutaneous or intramuscular. Drugs given via the subcutaneous or intramuscular routes can require at least 20 to 30 minutes to take clinical effect (an exception is dexmedetomidine with an onset of action 5–15 minutes after IM administration and 10–20 minutes after SQ injection).

Intramuscular drug administration may be more reliable in some patients compared to subcutaneous administration, especially those with poor peripheral perfusion or those with excessive subcutaneous fat. A disadvantage of the intramuscular route is that it may result in postinjection muscle soreness. Regardless of the route of administration, it is imperative that patients administered premedication agents be allowed to sedate in an environment that is calm, quiet, and non-stimulating. Failure to adhere to this guideline may result in a patient that becomes incompletely sedate. Furthermore, patients should be observed frequently following premedication. Because intervention may be needed rapidly if adverse events occur, those patients with critical illness or those with brachycephalic airway syndrome (Pugs, English Bulldogs, etc.) should never be left alone once given premedication drugs.

Table 8-1 summarizes the advantages and disadvantages of different routes of administration.

TABLE 8-1 Pros and Cons of Premedication Administration Routes

Route of Administration	Advantages	Disadvantages
Intravenous (IV)	• Immediate onset of action • Least painful if IV catheter in place • May result in less vomiting than via other routes	• May slightly lessen duration of action • Ideally requires IV catheter
Intramuscular (IM)	• Reliable • More rapid onset than SQ route • Preferred route for most premedication	• May result in muscle soreness • Slower onset than IV • In general, the IM dose is larger than an IV dose
Subcutaneous (SQ, SC, or SubQ)	• Easy and fast to administer • Potentially preferred in fractious patients • Can accommodate larger volumes than IM route • No muscle soreness	• Slower onset than other routes • Absorption can be questionable in some patients

TYPE OF PREMEDICANTS

Common premedicants or "premed" drugs include sedatives, tranquilizers, anticholinergics, analgesics, and other miscellaneous adjunctive agents. Often several drugs are given together to maximize the benefits from each individual drug.

Tranquilizers and Sedatives

The terms tranquilizer and sedative are often used synonymously in clinical practice. Technically, **tranquilizers** are specific drugs that calm animals and are used to reduce anxiety and aggression in animals. A **sedative** is a drug that may result in further central nervous system depression and sleepiness (Figures 8-1a and 8-1b). Both types of drugs may decrease irritability and excitement in animals and can be used to quiet excited animals. Even when given at high dosages, tranquilizers and sedatives do not produce reliable effects in veterinary patients. Even though patients may appear deeply sedated, patients often can still be aroused, and therefore technicians must always be careful, especially when working with fractious or dangerous patients. For example, patients sedated with alpha-2 adrenergic agonists may appear deeply sedate or even anesthetized but can unexpectedly be aroused.

Phenothiazines

Phenothiazines are tranquilizers that exert their effect on the central nervous system by blocking dopamine receptors. Historically they were used as anti-psychotic agents in humans and as a result have a calming and **anxiolytic** (anti-anxiety) effect. The most common phenothiazine drugs used in veterinary medicine today include acepromazine (Promace®, generic), chlorpromazine (Thorazine®), and promazine (Sparine®), with acepromazine being the mostly widely prescribed and therefore the focus of this section (Figure 8-2).

Acepromazine is probably the most commonly used premedication agent in veterinary medicine and for good reason. It provides reliable and predictable tranquilization across a wide variety of species. It is often used in conjunction with an opioid because it does not provide analgesia. Other advantages of the drug are that it has an **antiemetic** (anti-nausea and vomiting) effect, potential anti-arrhythmic effect, is inexpensive and readily available, has mild effects on ventilation and the respiratory system, and significantly reduces the amount of induction and maintenance anesthetic a patient requires.[1,2]

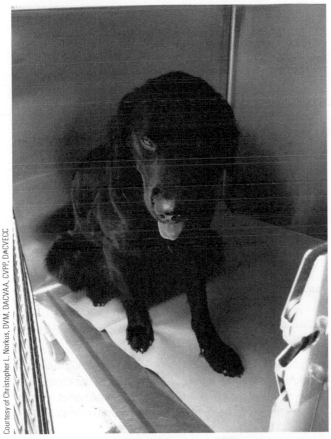
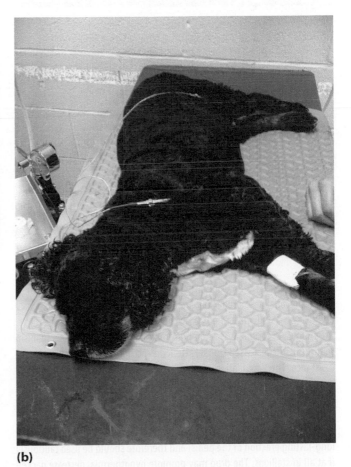

Courtesy of Christopher L. Norkus, DVM, DACVAA, CVPP, DACVECC

(a)　　　　　　　　　　　　　　　　　　**(b)**

FIGURE 8-1 Sedation in a dog. (a) This picture shows a dog under moderate sedation. Notice the low head and body carriage and the droopy eyes and ears of the patient. The patient is also drooling slightly and is considering lying down. This dog is appropriately premedicated for most purposes and would readily tolerate IV catheter placement. (b) This picture shows a dog under heavy sedation. Notice the lateral recumbency and disinterest in his surroundings; however, the patient is still arousable with stimulation. Electrocardiogram leads and an IV catheter have been placed prior to anesthesia induction.

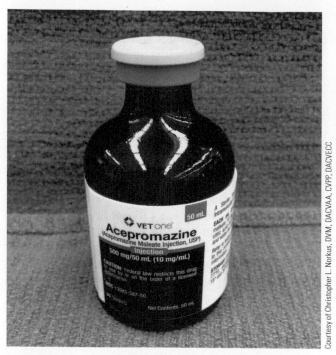

Courtesy of Christopher L. Norkus, DVM, DACVAA, CVPP, DACVECC

FIGURE 8-2 Acepromazine is yellow in color and commercially available in a 10 mg/mL concentration. Many small animal practices, however, will dilute their acepromazine with saline to make a 1 mg/mL or 2 mg/mL concentration, which makes dosing smaller patients easier. When calculating an acepromazine dose, double check the concentration of drug used to ensure the proper drug dose is determined.

Courtesy of Janet Amundson Romich, DVM, MS

FIGURE 8-3 Prolapse of the third eyelid (nictitating membrane) may be seen after acepromazine premedication. There is no negative effect of temporary third eyelid prolapse, but owners should recognize this condition so they are not alarmed if they see this phenomenon in their pet.

Historically acepromazine was thought to lower seizure threshold in patients susceptible to seizures. As a result, it was once recommended that the drug be avoided in any patient at risk or with a history of seizures. More recent evidence has shown this claim does not hold true.[3,4] Acepromazine was also previously thought to alter platelet function; however, newer research suggests that if this does occur it is minimal and not of clinical significance.[5] Acepromazine also has weak anti-histamine properties and therefore is generally avoided prior to intradermal allergy skin testing.

Some Boxer dogs appear to be uniquely sensitive to acepromazine, and reports of profound cardiovascular depression and syncope exist. It appears that certain lineages of Boxer dogs in Europe appear most susceptible to acepromazine. In the United States, acepromazine is still widely used in the breed without adverse effects although lower dosages are sometimes selected.

Tech Tip

For small animal use, acepromazine is typically diluted from its original concentration of 10 mg/mL to lower concentrations such as 2 mg/mL or 1 mg/mL.

Tech Tip

Administering acepromazine to extremely fearful/nervous or aggressive dogs is concerning since it is not thought to relieve anxiety or fear.

Disadvantages of acepromazine include long (4+ hours) duration of action, no analgesic properties, inability to easily reverse the drug once administered, and, most importantly, potent cardiovascular adverse effects. Acepromazine results in vasodilation from alpha-1 adrenergic antagonism that results in decreased cardiac afterload and decreased systemic vascular resistance. The vasodilation can lead to hypotension (a state of low blood pressure), especially when used with other vasodilating drugs (e.g., inhalant anesthetic agents). Temporary prolapse of the third eyelid may be seen in some species but has no negative effect on the patient (Figure 8-3).

Acepromazine may rarely cause priapism (persistent and painful long-lasting erection of the penis) and therefore should be used cautiously if at all in stallions. The drug may promote hypothermia, decrease packed cell volume levels, and results in splenomegaly. Acepromazine should be avoided in patients with liver dysfunction because the drug is heavily dependent on the liver for elimination from the body, so its duration of action will be prolonged in patients with liver disease and dogs with the MDR1 gene mutation.

Acepromazine is most commonly administered intravenously, intramuscularly, or subcutaneously. Cats and very small dogs may benefit from dosages a little higher than those for larger dogs. Generally, acepromazine is best reserved for patients who are younger and healthy with good cardiac reserves. In the very old, very young, or critical patient the drug should be used cautiously if at all.

Tech Tip

The recommend label dosage on the box of most injectable acepromazine products greatly exceeds what is needed clinically.

Phenothiazine Facts

Advantages:

- Antiemetic
- Anti-arrhythmic
- Mild respiratory effects
- Reduces amount of induction and inhalant anesthetic agents needed
- Reliable tranquilization
- Inexpensive and readily available
- Not a controlled substance

Disadvantages:

- No analgesia
- Potent cardiovascular adverse effects (may contribute to hypotension)
- Priapism in stallions
- Hypothermia
- Not reversible
- Long duration of action
- Not for use in patients with hepatic dysfunction

Alpha-2 Adrenergic Agonists

Alpha-2 (α_2) adrenergic agonists are potent sedative and effective analgesic drugs that work by binding to the α_2 adrenergic receptors on neurons that normally release the neurotransmitter norepinephrine. When α_2 adrenergic agonists bind, they inhibit the release of norepinephrine. Because norepinephrine maintains alertness, its absence produces sedation. Alpha-2 adrenergic agonists are widely used in many species with very frequent usage in horses (Figure 8-4). Alpha-2 adrenergic agonists used in equine veterinary practice today most commonly include xylazine (Rompun®, AnaSed®, Sedazine®, and Chanazine®) (Figure 8-5a), detomidine (Dormosedan®) (Figure 8-5b), and romifidine (Sedivet®) (Figure 8-5c). An α_2 adrenergic agonist used in small animal veterinary practice today is dexmedetomidine (Dexdomitor®) (Figure 8-6). Dexmedetomidine has largely replaced xylazine in small animal practice because of fewer adverse effects. Two concentrations of dexmedetomidine are currently available (0.5 mg/mL and 0.1 mg/mL) (see Figure 5-6). Medetomidine (Domitor®) was previously available for small animals; however, today it is only available through compounding pharmacies in the United States and is mostly used for zoo animals and wildlife. The main differences between each individual α_2 adrenergic agonist drug include different durations of actions, subtly different adverse effect profiles, and cost. On average, α_2 adrenergic agonists have a duration of action of less than two hours, but this is widely variable depending on the species, individual drug, and dosage selected.

Tech Tip

There are two concentrations of dexmedetomidine currently available (0.5 mg/mL and 0.1 mg/mL). Be careful to choose the proper concentration when administering dexmedetomidine to patients.

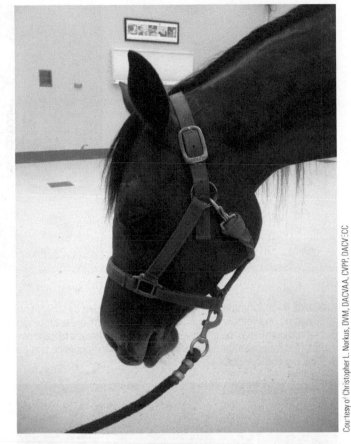

FIGURE 8-4 This horse is moderately sedated. Note the low hanging head carriage and drooping lips, eyes, and ears. He also is less responsive to his surroundings including sights and sounds and is mildly ataxic.

Courtesy of Christopher L. Norkus, DVM, DACVAA, CVPP, DACVECC

Safety Alert

Alpha-2 adrenergic agonists are profound sedatives that can be harmful to humans if accidental exposure occurs. Always be observant when administering them to patients to avoid accidental exposure to personnel!

A major advantage of this drug class is that it usually provides profound sedation and therefore can be effective in excited or fractious animals. It is best to allow an "excited" or stressed animal 10 to 15 minutes to calm down (e.g., if handling was just attempted and had to be aborted). Otherwise, the dosage needed to produce effective sedation increases and the effect becomes less reliable and can increase potential adverse effects. These drugs are also reversible with specific antagonist drugs such as atipamazole (Antisedan®) or yohimbine (Yobine®), making them useful for short outpatient procedures such as radiographs or bandage changes (see Figures 5-7a and 5-7b). Alpha-2 adrenergic agonists generally have mild effects on the respiratory system and ventilation at clinical dosages. At higher dosages and in some patients, respiratory depression can occur. Alpha-2 adrenergic agonists also dramatically reduce the dosage of other anesthetic induction and maintenance drugs.

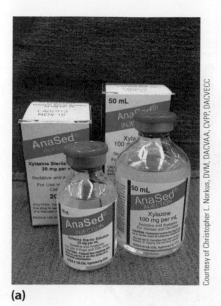

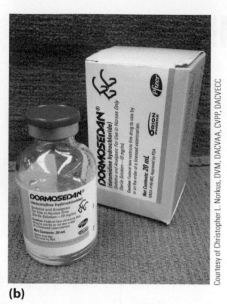

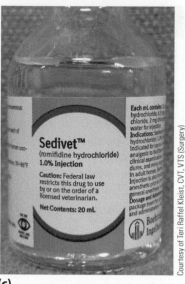

Courtesy of Christopher L. Norkus, DVM, DACVAA, CVPP, DACVECC

Courtesy of Christopher L. Norkus, DVM, DACVAA, CVPP, DACVECC

Courtesy of Teri-Raffel Kleist, CVT, VTS (Surgery)

(a)   **(b)**   **(c)**

FIGURE 8-5 Examples of α_2 adrenergic agonists used mainly in horses. (a) Xylazine is available in two concentrations (20 mg/mL and 100 mg/mL) and veterinary technicians should always ensure they are using the correct concentration to avoid drug overdose. (b) Detomidine has a long duration of action. It is used for treating horses for pain associated with colic and allows for suturing of skin lacerations. (c) Romifidine is preferred for procedures such as taking radiographs or lameness examinations because it produces a longer duration of sedative effects than detomidine and xylazine.

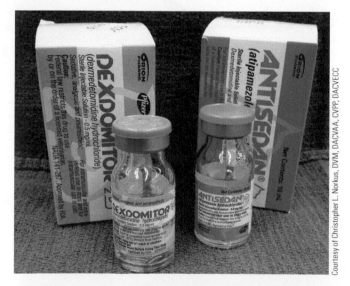

Courtesy of Christopher L. Norkus, DVM, DACVAA, CVPP, DACVECC

FIGURE 8-6 Dexmedetomidine is widely used in small animal practice for premedication and for outpatient procedures. The drug is shown here with its reversal agent atipamazole.

The major disadvantage of these drugs is that they result in profound cardiovascular adverse effects including initial transient vasoconstriction, which causes hypertension, followed by a reflex bradycardia that causes the central α_2 action of decreased blood pressure, cardiac output, and decreased tissue oxygenation. Anticholinergics are generally not recommended to treat or prevent the bradycardia associated with α_2 adrenergic agonists because tachycardia, arrhythmias, hypertension, and increased myocardial work occur.[6,7] Alpha-2 adrenergic agonists should be avoided in patients with hepatic dysfunction because this group of drugs compromise blood flow to organs such as the liver. Lastly, α_2 adrenergic agonists result in a transiently increased blood glucose as well as urine output.

Tech Tip 🐾

Monitoring patients who have received α_2 adrenergic agonists can be a challenge. Unlike most other drugs, significant bradycardia and increase in blood pressure are commonly observed. This is somewhat of a paradigm shift from what many are used to. The drug, however, can be reversed if the effects seen are detrimental to the patient.

Alpha-2 adrenergic agonists are often used alone or in conjunction with an opioid. In cats and dogs, α_2 adrenergic agonists and opioids are often used together to provide enhanced and longer duration analgesia and help produce neuroleptanalgesia, which is described later. In horses, α_2 adrenergic agonists may be used alone and are the cornerstone of safe equine sedation/premedication. These drugs work well when given via intravenous, intramuscular, and subcutaneous routes. As a result of the profound cardiovascular effects, α_2 adrenergic agonists should be reserved for patients who are young and healthy and have excellent cardiac reserves. In the old, very young, fragile, or critical patient this drug class should be used cautiously if not avoided altogether.

Safety Alert 🔔

Alpha-2 adrenergic agonists result in profound patient sedation; however, they do not cause anesthesia. Patients can sometimes be briefly aroused from their sedated state enough to bite. Technicians and staff should be careful when working with any fractious patients even if they have been sedated.

Pale mucous membranes and an inability to obtain consistent pulse oximetry readings can occur when using α_2 adrenergic agonists as a result of profound peripheral vasoconstriction. Because α_2 adrenergic agonists result in decreased cardiac output (due to decreases in heart rate and increases in peripheral vascular resistance), tissue oxygen delivery decreases; therefore, oxygen support should be provided to patients receiving them.

Alpha-2 Adrenergic Agonist Facts

Advantages:

- Profound sedation
- Muscle relaxation
- Analgesia
- **Anxiolysis** (anti-anxiety)
- Reversible

Disadvantages:

- Profound cardiovascular effects
- Transient hyperglycemia
- Transient increased urine output
- Avoid in animals with hepatic disease

Stand clear! Horses will often have to massively urinate after being administered an alpha-2 adrenergic agonist.

Benzodiazepines

Benzodiazepines are a class of tranquilizers widely used in both human and veterinary medicine. They work by enhancing the release of gamma aminobutyric acid (GABA), one of the body's main inhibitory neurotransmitters. The most widely used benzodiazepines for premedication in veterinary medicine include diazepam (Valium®) and midazolam (Versed®) (Figure 8-7). Diazepam and midazolam have very similar clinical effects and can often be used interchangeably. The main difference between the two agents is that midazolam is highly water soluble and diazepam is not. Diazepam also utilizes propylene glycol as a carrier. These properties both affect absorption and limit its ability to be mixed in the same syringe with other drugs. One exception to this rule is that diazepam is often mixed in the same syringe as ketamine for anesthetic induction; however, in general diazepam should not be mixed

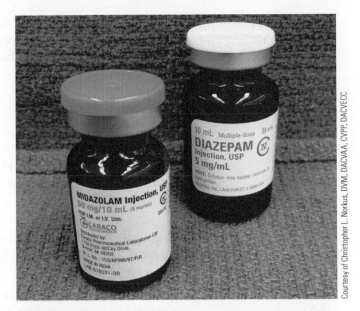

FIGURE 8-7 Benzodiazepines such as diazepam are a first-choice therapy for treating seizure activity. Both diazepam and midazolam are stored in brown vials because they are sensitive to light.

Courtesy of Christopher L. Norkus, DVM, DACVAA, CVPP, DACVECC

with other drugs. Midazolam is readily absorbed via the subcutaneous and intramuscular routes while diazepam is not. Both drugs can be given intravenously. Alprazolam (Xanax®) is another benzodiazepine that is sometimes prescribed in veterinary medicine for oral administration.

Advantages of this group of drugs are that they may provide anxiolysis, mild to moderate skeletal muscle relaxation, amnesia, and are effective anticonvulsant drugs. They are widely used as first choice drugs to stop active seizures. Benzodiazepines have minimal influence on the cardiovascular system and respiratory system that as a whole make them good choices for pediatric, geriatric, and critically ill patients. Additionally, the drugs are reversible with the agent flumazenil (Romazicon®).

A major disadvantage of this drug class is that these drugs do not provide analgesia and often do not provide profound sedation in healthy patients. For example, these drugs may provide unreliable sedation in patients who are neither pediatric, geriatric, nor critically ill. Some patients may even exhibit excitatory effects (i.e., "paradoxical excitement") including dysphoria or disinhibition that can make handling difficult. **Dysphoria** is recognized as the state of feeling unwell, unhappy, or emotional or mental discomfort. In pets this can include restlessness, anxiety, disorientation, vocalization, and even aggression or attempts to flee. Benzodiazepines should also be avoided or used cautiously in low dosages in patients with hepatic dysfunction.

The United States Drug Enforcement Agency (DEA) classifies benzodiazepines as controlled substances because there is a risk for human abuse and addiction. A **controlled substance** is a drug or chemical whose manufacturing, sale, possession, and use is strictly regulated by the federal government. As a result, veterinarians must obtain a DEA license to prescribe these drugs and then follow special procedures for storing and securing these agents as well as strict record keeping. Controlled substances are classified in the United States as Schedules numbered between I–V (1–5). Level one is considered to have the highest risk potential and five has the least (Table 8-2). Both diazepam and midazolam are classified as C-IV controlled substances.

TABLE 8-2 Controlled Substances Categories* and Anesthetic and Analgesic Agents Examples

Drug Schedule	Definition of Schedule	Examples
Schedule I (C-I)	Substance has high potential for abuse and has no currently accepted medical use; there is a lack of accepted safety for use (considered most dangerous, with virtually no medical benefit)	heroin, LSD, marijuana
Schedule II (C-II)	Substance has high potential for abuse but has currently accepted medical use, with severe restrictions	cocaine, morphine, methadone, hydromorphone, hydrocodone, pentobarbital**, etorphine, fentanyl, remifentanil, codeine, alfentanil, oxymorphone, merperidine, alfentanil, sufentanil, tapentadol
Schedule III (C-III)	Substance has potential for abuse less than schedule I and II drugs and has accepted medical uses	acetaminophen/codeine combinations, ketamine, thiamylal, thiopental, buprenorphine, tiletamine (found in the combination product tiletamine/zolazepam)
Schedule IV (C-IV)	Substance has low potential for abuse relative to drugs in schedule III and has accepted medical uses	diazepam, midazolam, phenobarbital, alfaxalone, butorphanol, tramadol
Schedule V (C-V)	Substance has low potential for abuse relative to drugs in schedule IV and has accepted medical uses	codeine cough syrups

*Check the DEA web site at https://www.deadiversion.usdoj.gov/schedules/ for the most current information regarding schedules of controlled substances.
**Euthanasia solutions may be C-II (if pentobarbital is the only narcotic ingredient) or C-III (if pentobarbital is combined with a least one other narcotic ingredient).

Tech Tip

Ensure that all controlled substances are properly stored in a locked and secured area and their use is immediately documented into a controlled substance log to ensure compliance with DEA standards. Failure to comply can result in massive fines and loss of the veterinary license for the veterinarian responsible for the controlled substances through their DEA registration.

Benzodiazepines are often combined with ketamine for anesthetic induction to reduce ketamine's adverse effects of muscle rigidity and central excitation. Benzodiazepines are typically combined with an opioid when used for premedication. In general this class of drugs is most useful in younger, older, and/or sicker patients.

Tech Tip

Controlled substances must be stored in a locked cabinet or, preferably, in a safe attached to a concrete floor. The registered veterinarian and other authorized handlers must keep records of orders, receipts, uses, discards, and thefts of controlled substances for two years following each transaction (commonly done with a controlled substance log). Records of controlled substance inventory are kept on file at the clinic, so they are available to the DEA if the facility is audited or has unannounced inspections by DEA agents.

Benzodiazepine Facts

Advantages:

- Anxiolysis
- Skeletal muscle relaxation
- Amnesia
- Anti-seizure effects
- Minimal cardiovascular and respiratory effects
- Reversible

Disadvantages:

- No analgesia
- Potential dysphoria
- Mild tranquilization

Butyrophenones

Historically butyrophenones were a fourth class of tranquilizers used in veterinary medicine for tranquilization and sometimes premedication. These drugs did not have analgesic effect nor were they reversible. Their drug profile was very similar to that of phenothiazines. Butyrophenones work through the mechanism of central dopamine and peripheral adrenergic blockage to produce a strong anticholinergic effect. These drugs most commonly included droperidol (Inapsine®), haloperidol (Haldol®), and azaperone (Stresnil®). Doperidol was available as a proprietary product mixed with fentanyl. Azaperone was widely used in pigs and is currently compounded for use in wildlife. This drug class is not used commonly for small animals; however, knowledge of these drugs is important for historical reasons and their occasional appearance on certification exams.

Anticholinergic Drugs

The vagus nerve provides *parasympathetic* innervation (the homeostatic branch of the autonomic nervous system; homeostasis is maintaining a stable internal environment). Parasympathetic effects include decreased heart rate, bronchoconstriction, increased gastric motility and secretions, and miosis (pupillary constriction). Drugs that block vagal tone are categorized as **parasympatholytics** because they "break down" the parasympathetic response. Parasympatholytic effects include increased heart rate, bronchodilation, decreased gastric motility and secretions, and mydriasis (pupillary dilation). Anticholinergic drugs cause parasympatholytic effects.

Anticholinergic drugs such as atropine (generic) and glycopyrrolate (Robinul-V®) are occasionally used as part of a premedication plan. These drugs work as parasympatholytic drugs meaning that they block the release of acetylcholine in the parasympathetic nervous system. Specifically, these drugs have their effect on muscarinic receptors and thereby block vagal tone. The results are increased heart rate, bronchodilation, decreased secretions (including saliva), decreased gastrointestinal mobility, and mydriasis. They do not have analgesic properties. Historically some veterinarians unnecessarily used these drugs as part of premedication plans with the goal of "raising the heart rate and drying up saliva." However, a decrease in secretions can result in increased secretion viscosity that could potentially result in the plugging of small airways and/or ETTs. Some anesthetists still utilize anticholinergics routinely in brachycephalic breeds, neonates, and in cases being anesthetized for ocular procedures. These drugs do not need to be routinely included in most premedication plans because they are simply not needed for most patients. Occasionally some patients with select pre-existing cardiac arrhythmias may benefit from having an anticholinergic drug included as part of premedication, but in general these drugs are used more on an "as needed" basis during the perioperative period.

Atropine and glycopyrrolate work rapidly via intravenous, intramuscular, and subcutaneous routes (see Figure 5-15). Atropine has a slightly faster onset time than glycopyrrolate and often yields a more robust change to heart rate. As a result, atropine is generally selected over glycopyrrolate in emergency situations. Glycopyrrolate has the advantage of a longer duration of action (approximately 90 minutes) when compared to atropine (approximately 60 minutes). Hypothermia can result in decreased patient response to these drugs.

Anticholinergics are infrequently used in horses as significant concern arises due to ileus (decreased gastrointestinal motility) and often leads to colic. Other adverse effects can include tachycardia, arrhythmias, and possible increases to intraocular pressure. As stated previously, anticholinergics are generally not recommended for the treatment of α_2 adrenergic agonist bradycardia due to adverse effects including excessive tachycardia, hypertension, arrhythmias, and increased myocardial work.

Opioids

Opioids are a very important group of drugs that are included in most small animal patient premedication plans. Opioids result in profound analgesia and are a key component to provide **pre-emptive analgesia** which means providing pain control before a painful stimulus occurs. Opioids alone produce mild to moderate sedation. When given in conjunction with a sedative or tranquilizer they produce **neuroleptanalgesia** which results in more profound sedation and potentially analgesia than when each drug class is given

Anticholinergic Facts

Advantages:

- Used for select patients to block vagal tone that results in:
  - increased heart rate,
  - bronchodilation,
  - decreased secretions (including saliva),
  - decreased gastrointestinal mobility, and
  - mydriasis
- Some patients with select pre-existing cardiac arrhythmias may benefit from administration of anticholingeric drug (see Chapter 12 [Anesthetic Complications]).

Disadvantages:

- No analgesia
- Decrease in secretions may increase secretion viscosity causing plugging of small airways and/or endotracheal tubes.
- Ileus and colic in horses
- Potential development of tachycardia and arrhythmias
- Possible increases to intraocular pressure

alone. An example of a neuroleptanalgesic combination that is commonly used in veterinary practice would be the opioid butorphanol combined with the sedative dexmedetomidine. Opioids are commonly given alone or in conjunction with sedatives or tranquilizers for premedication purposes or for chemical restraint for short procedures.

Opioids work predominantly by influencing mu (μ) and kappa (κ) opioid receptors within the central and peripheral nervous system. In humans they commonly cause euphoria and are addictive. In cats and dogs, common opioid unwanted adverse effects include vomiting, dysphoria, hypothermia in dogs, hyperthermia in cats, panting in dogs, and can occasionally result in excitement ("morphine mania") in cats when using higher dosages. In horses, the most common opioid adverse effects include ileus, colic, and increased locomotion with repeated or high dosages. Opioids may cause excitement in horses.

Opioids generally have mild effects on the cardiovascular system. They can increase vagal tone and therefore result in bradycardia; however, overall blood pressure, systemic vascular resistance, and cardiac output are largely unchanged. Their friendly cardiovascular profile makes them excellent choices for sick or geriatric patients. Opioids can result in respiratory depression but this generally is not a significant clinical issue in veterinary species unless patients are receiving very high dosages of opioids. Although opioids (with the exception of remifentanil) are metabolized by the liver, they are appropriate to administer in patients with hepatic disease because decreased metabolism only extends the analgesia. If undesirable effects do occur—for example, a prolonged recovery from anesthesia in a puppy with a portosystemic shunt—this class of drug is always reversible with opioid antagonists.

Opioids are a common source of abuse in people and can become addictive. As a result, opioids are classified in the United States by the DEA as controlled substances. Opioids can be administered by numerous routes including but not limited to intravenously, intramuscularly, subcutaneously, topically (transdermal patch), orally, and transmucosally. In most veterinary species, poor oral absorption limits their use by the oral route.

Opioid Facts

Advantages:
- Analgesia
- Mild cardiovascular effects
- Some sedative effects (species dependent)

Disadvantages:
- Bradycardia
- Hyperthermia (cats)
- Vomiting (dogs and cats)
- Potential dysphoria (dogs and cats)
- Hypothermia (dogs)
- Panting (dogs)
- Ileus, colic, excitement, and increased locomotion (horses)

Full Mu Agonists

Full μ agonists are opioid drugs that produce profound analgesia, and those most commonly used in veterinary medicine include morphine (Duramorph®, Astramorph®) (Figure 8-8), hydromorphone (Dilaudid®) (Figure 8-9), oxymorphone (Numorphan®) (see Figure 5-17c), methadone (Dolophine®) (see Figure 5-17d), fentanyl (Duragesic® transdermal patches, Sublimaze®) (see Figure 5-17e), and remifentanil (Ultiva®) (see Figure 5-17f). Agonists are

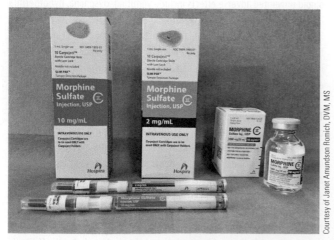

FIGURE 8-8 All μ agonist opioids have the same analgesic efficacy. Because each drug requires a different dosage to achieve this effect, different drugs are said to have different potencies. This image shows the full μ agonist morphine.

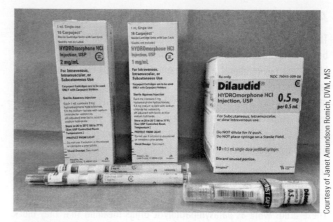

FIGURE 8-9 Humans are much more sensitive to the respiratory depressant effects of opioids than dogs, cats, and hoses. This image shows the full μ agonist hydromorphone.

substances that bind to a receptor (in this case the μ opioid receptor) and activate the receptor to produce a biologic response that mimics the release of an endogenous substance (in this case endorphins, enkephalins, etc.) in the body. As a result, these drugs have the highest *efficacy* or effectiveness and are the most effective analgesics known to date. They can be used for moderate to severe pain. These drugs have high abuse potential and are classified as C-II controlled substances in the United States.

All of these drugs can produce the same degree of pain control in a patient; however, each individual drug has a different *potency* meaning that different drug dosages are required to produce ultimately the same effect. For example, both morphine and fentanyl can provide the same degree of pain control in a dog. However, a typical dosage of morphine for a dog might be 0.5mg/kg while a typical dosage of fentanyl might be 0.005mg/kg. While both drugs produce the same pain control, fentanyl is said to be a more potent drug because less of the drug is needed to produce the desired effect.

Although each full μ agonist has slightly different traits, their properties are fairly similar with the exception of their duration of action. Remifentanil is administered intravenously and is an ultrashort acting drug that must be given by a constant rate infusion because of its short duration of action (<5 minutes). For this reason, is it rarely used as a premedication drug, but is sometimes used during the perioperative or postoperative period as an analgesic. Remifentanil is unique because it is not metabolized by the liver but rather by nonspecific tissue esterases; therefore, accumulation of this drug does not occur, and it is cleared quickly from the body. When available, this drug is likely the drug of choice for patients with liver failure; however, significant expense has limited the drug's use in veterinary medicine.

Fentanyl is a short acting opioid with duration of action of approximately 20 to 30 minutes. Because of this short duration of action, it is often poorly suited for routine use as a premedication drug. Clinically it appears to be less likely to cause vomiting than other opioids. Subjectively, cats often sedate well with fentanyl. Fentanyl can be continued during the perioperative and post operative period as a constant rate infusion for analgesia.

Hydromorphone, morphine, and oxymorphone all provide analgesia for approximately 3 to 6 hours (Figures 8-10a through 8-10c). Adverse effects

Courtesy of Janet Amundson Romich, DVM, MS

(b)

(c)

Courtesy of Christopher L. Norkus, DVM, DACVAA, CVPP, DAC/ECC.

(a)

FIGURE 8-10 Premedication in a dog. (a) Spritz, a healthy 7-month-old intact female Poodle, presents for a routine ovariohysterectomy. She is very friendly but excessively hyperactive. She is given hydromorphone SQ and acepromazine SQ for premedication. She is placed in a cage in a quiet environment and allowed to sedate. (b) After 10 minutes, Spritz is noticeably sedate, calm, and quiet. She is now willing to lay down, her demeanor is more relaxed, and her head carriage is low. She is still able to walk but would readily tolerate IV catheter placement at this time. (c) After 20 minutes, Spritz is markedly sedate. She is reluctant to walk. Although arousable, Spritz puts her head down and goes right to sleep when not disturbed.

of vomiting, nausea, and panting in dogs commonly occur. Morphine may result in histamine release that could result in hypotension if given rapidly by the intravenous route.

Methadone provides analgesia for approximately 4 to 6 hours and subjectively appears to cause less vomiting and panting than other opioids. Methadone has the added advantage of providing analgesia not just via opioid receptors but also by blocking the N-methyl-D-aspartate (NMDA) receptor and as a norepinephrine and serotonin reuptake inhibitor.

Out of all the opioids, full μ agonists tend to be the most likely opioids to cause bradycardia and respiratory depression. All full μ agonists are reversible and can be reversed by the opioid antagonist naloxone (Narcan®) if needed (see Figure 5-18). An antagonist binds to a receptor to block or dampen an agonist-mediated response. It is important to remember that reversal of an opioid may result in reversal of patient analgesia and an abruptly painful patient. The duration of action of naloxone is typically 20 to 30 minutes, so renarcotization may occur in a patient and therefore, the patient should be monitored carefully.

Partial Mu Agonists

Buprenorphine (Buprenex®) is the only partial μ agonist used in veterinary medicine today (see Figure 5-17h). A partial opioid agonist is a drug that binds to a receptor and activates the receptor to produce a biologic response that mimics the release of endogenous opioids (endorphins, enkephalins, etc.) in the body; however, the effect is not as robust as a full agonist. Buprenorphine, therefore, does not have the same efficacy as full μ agonists do, meaning it cannot provide the same level of analgesia. As a result, buprenorphine is best suited for pain of mild to moderate intensity. Buprenorphine, however, has a high affinity for the opioid receptor site, which means that it firmly secures itself to the receptor and results in a long duration of action. Depending on the dosage and species, buprenorphine can have a duration of action up to 6 to 8 hours and is typically considered the longest acting opioid available.

Buprenorphine has a slow onset time of up to 45 to 60 minutes, results in minimal sedation for most patients, and rarely results in panting or vomiting. Significant bradycardia with the drug is also uncommon. Buprenorphine is widely used in cats and is commonly administered into the buccal pouch and

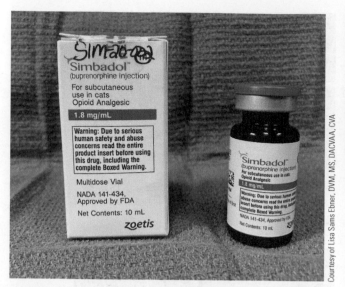

Courtesy of Lisa Sams Ebner, DVM, MS, DACVAA, CVA

FIGURE 8-11 Buprenorphine's absorption via the subcutaneous route in cats appears questionable, and it should not be given SQ in cats unless it is the Simbadol™ formulation of buprenorphine.

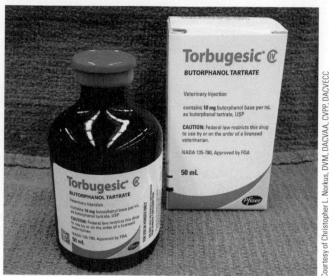

Courtesy of Christopher L. Norkus, DVM, DACVAA, CVPP, DACVECC

FIGURE 8-12 Butorphanol is a widely used drug in veterinary medicine; however, important limitations of the drug include a short duration of action (approximately1 hour) and limited analgesia in most species.

absorbed by the oral-transmucosal route. In cats, buprenorphine's absorption via the subcutaneous route appears questionable and for this reason the drug should not be given via this route unless the specific concentration formulation of buprenorphine (Simbadol™) is used.[8,9] Simbadol™ is FDA-approved for control of postoperative pain in cats and is given SQ once a day (q 24 hr) for up to 3 doses (or 72 hours) (Figure 8-11). The first dose should be given 1 hour prior to surgery. Because of the potential for adverse reactions associated with accidental injection, Simbadol™ should only be administered by veterinarians or veterinary technicians and not sent home for clients to administer to their pets. Its adverse effects include hypotension, tachycardia, hypothermia, hyperthermia, hypertension, anorexia, and hyperactivity.

Because of a strong affinity to the μ receptor site, reversing buprenorphine with naloxone can be difficult. Additionally, its strong affinity to the μ receptor site may temporally prevent full μ agonists from binding to the μ receptors and therefore may limit how much analgesia can be obtained. In other words, if you administer buprenorphine and then later administer a full μ agonist because you want better analgesia, you may find that the buprenorphine prevents the full μ agonist from reaching its maximum efficacy. Buprenorphine is currently classified as a C-III controlled substance.

Mixed Agonist-Antagonists

Butorphanol (Torbugesic®, Torbutrol®) is the most commonly used agonist-antagonist opioid used in veterinary medicine (Figure 8-12). Butorphanol works as a κ receptor agonist and a μ receptor antagonist. The drug has a rapid onset of action; however, its duration is typically short (approximately 1 hour). Additionally, the drug is not as efficacious as full μ agonists or partial μ agonists, meaning it cannot achieve as strong pain control properties and therefore is most appropriate for milder pain states. A unique exception to this rule may be in some species such as horses and birds (and perhaps ruminants). These species may have a higher distribution of κ receptors within their central nervous system and therefore may get as good or better analgesia from drugs like butorphanol than they would from a

full μ agonist. Butorphanol generally provides mild to moderate sedation and rarely causes panting or vomiting. Butorphanol can be used as an antagonist to partially reverse full μ agonists while still maintaining some analgesia via the κ receptor. It is currently classified as a C-IV controlled substance. Lower dosages given IV can be used for partial reversal of full mu agonists.

Tech Tip

In the ideal world, nonsteroidal anti-inflammatory drugs (NSAIDs) would be given before surgical pain and inflammation occur. Unfortunately, if a patient were to experience low blood pressure during anesthesia in the face of NSAID administration, this could result in acute kidney injury (AKI). Therefore, it is best to administer NSAIDs only after the conclusion of anesthesia and only if no other contraindications for the drug exist.

Miscellaneous Agents Used for Premedication

On a case by case basis, anesthetists sometimes include other agents for premedication than what has already been described in this chapter. Some additional agents that may be used in select cases may include antibiotics, antiemetics, anti-histamines, dissociative anesthetics, and glucocorticoids. Two of the more common miscellaneous drugs are described below.

Dissociative Drugs

The most frequently used dissociative drug in veterinary medicine is ketamine (Ketaset®, VetaKet®, Ketaject®, Vetalar®) (Figure 8-13). Ketamine is a C-III controlled substance and is an anesthetic agent commonly used

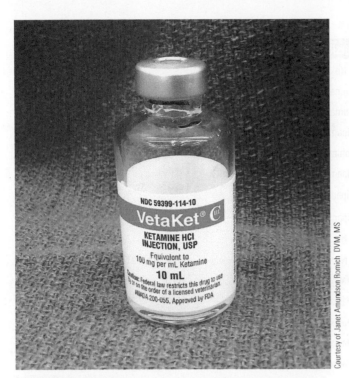

Courtesy of Janet Amundson Romich DVM, MS

FIGURE 8-13 Ketamine can be given via the oral route in very fractious or feral cats (it is typically squirted into the mouth via a red rubber catheter attached to a syringe). Based on the size of the cat, 100 to 200 mg can be used and its onset of action is 5 to 10 minutes. Oral ketamine is associated with transient excessive drooling due to the drug's bitter taste and low pH. Care should be taken so the drug is not sprayed in the cat's eye.

for starting and maintaining general anesthesia. Other uses for the drug can be as a sedative and analgesic drug. Ketamine can cause patients to enter a **cataleptic** or "dissociative-like" state where they become detached from and not responsive to their surroundings. The drug achieves its analgesic effect predominantly by N-Methyl-D-aspartate (NMDA) receptor antagonism.

When used as a premedication agent, ketamine is given in sub-anesthetic dosages in combination with other drugs (e.g., acepromazine and morphine) to provide enhanced sedation in difficult or fractious patients. It is generally very effective in cats that as a species can traditionally be difficult to sedate. Other benefits of the drug include being a reliable sedative, a bronchodilator, having brief analgesia properties, and having minimal cardiovascular and respiratory depressant effects. Ketamine results in tachycardia, increased cardiac output, and increased blood pressure. Unfortunately, the drug is not reversible, can sting upon injection, and can result in muscle rigidity. Ketamine can sometimes result in poor recoveries including dysphoria, hallucinations, and wild unpredictable behavior although this is largely dosage dependent. When combined with other drugs that have sedative and muscle relaxant properties, these adverse effects are greatly reduced. Ketamine should be avoided in dogs with hepatic dysfunction.

When administered, ketamine is best given by the intramuscular or intravenous route to avoid patient discomfort. Overall the agent is a very useful drug to have in one's premedication toolbox. Nonetheless it is probably not appropriate or necessary for everyday routine use.

Neurokinin-1 Antagonist

Maropitant (Cerenia®) is a neurokinin (NK)-1 antagonist that is licensed for use in veterinary medicine as an antiemetic drug (see Figure 11-8a). Maropitant works by inhibiting substance P, the key neurotransmitter involved in vomiting. In humans it is well recognized that nausea and vomiting frequently occur from the administration of anesthetics and analgesics in the post-operative period. In human medicine this phenomenon is referred to as post-operative nausea and vomiting (PONV). The exact incidence of this phenomenon in dogs and cats has not been explored; however, subjectively to the author many patients have decreased appetite, nausea, and vomiting following anesthesia that mimic PONV in humans. To improve patient comfort and mitigate these signs, maropitant is sometimes included as a premedication agent by anesthetists. Similarly, it is contra-indicated for some patients to vomit during the perioperative period (e.g., patients with fragile eyes, patients where rises in intraocular pressure or intracranial pressure may be detrimental) and the administration of maropitant preoperatively can be helpful to avoid this. As a premedication, maropitant is given SQ and an adverse effect is pain at the injection site, which can be decreased if the drug is kept refrigerated. When choosing to administer maropitant SQ, anesthetists should administer the drug at least 45 minutes before administering other drugs that are known to cause vomiting (e.g., opioids) so that adequate antiemetic effect is reached. Oral pretreatment with maropitant at least 2 hours prior to anesthesia is another option to control vomiting.

Table 8-3 summarizes the analgesic properties of premedicants described in this chapter as well as whether these drugs are reversible. Table 8-4 provides dosages for select premedicants described in this chapter.

TABLE 8–3 Characteristics of Premedication Agents

Drug	Reversible?	Analgesia Provided?
acepromazine	No	No
midazolam	Yes-flumazenil	No
diazepam	Yes-flumazenil	No
dexmedetomidine	Yes-atipamazole	Yes
xylazine	Yes-yohimbine	Yes
medetomidine	Yes-atipamazole	Yes
romidifine	Yes- atipamazole	Yes
detomidine	Yes-atipamazole	Yes
glycopyrrolate	No	No
atropine	No	No
hydromorphone	Yes-naloxone	Yes
morphine	Yes-naloxone	Yes
fentanyl	Yes-naloxone	Yes
methadone	Yes-naloxone	Yes
merperidine	Yes-naloxone	Yes
oxymorphone	Yes-naloxone	Yes
remifentanil	Yes-naloxone	Yes
buprenorphine	Yes-naloxone (but difficult)	Yes
butorphanol	Yes-naloxone	Yes
ketamine	No	Yes

SUMMARY

The importance of providing veterinary patients with appropriate premedication cannot be overlooked. Appropriate selection of premedication drugs will have a major impact on the entire anesthetic experience; therefore, it is critical for the anesthetist to be fully familiar with the effects of each premedication agent and select drugs specifically based upon the patient's individual needs. Table 8-5 contains a summary of the drugs described in the chapter.

Selecting premedication agents is just one aspect of providing excellent anesthetic management for your patients. Other chapters within this book discuss other important phases of the process including selection of anesthetic induction and maintenance agents, patient monitoring, and perioperative management techniques.

CRITICAL THINKING POINTS

- Having a thorough knowledge of each drug you use is necessary to be a competent anesthetist.

- For most patients, premedication agents are beneficial and should be used.

- Individual premedication plans should always be customized for each individual patient rather than simply using set protocols, recipes, "cocktails," or premade drug mixes.

- Most cats and dogs will benefit from having an opioid given as part of their premedication. Some, but not all patients, will then require additional drugs beyond this (e.g., a tranquilizer, anticholinergic, maropitant) to achieve individual patient goals.

- Opioids and benzodiazepines have mild cardiovascular adverse effects, which can make them useful in younger, older, and/or sicker patients.

- Avoid or restrict the use of acepromazine and alpha-2 adrenergic agonists in very young, very old, and/or sick patients.

- Contrary to what was once taught, most patients do not require premedication with an anticholinergic drug.

- If an analgesic is not included as part of the premedication for a painful procedure, one must be administered before the painful procedure begins.

TABLE 8-4 Dosages of Select Premedication Drugs Described in this Chapter*

Drug	Dog	Cat	Horse
acepromazine (Promace® and generic)	0.01–0.05 mg/kg IV, IM, SQ (Small dogs: 0.05–0.1 mg/kg IV, IM, SQ)	0.05–0.1 mg/kg IV, IM, SQ	0.01–0.04 mg/kg IV, IM, SQ
xylazine (Rompun®, AnaSed®, Sedazine®, and Chanazine®)	Not routinely used	Not routinely used	0.1–1 mg/kg IV, IM
romifidine (Sedivet®)	Not routinely used	Not routinely used	0.01–0.12 mg/kg IV, IM
detomidine (Dormosedan®)	Not routinely used	Not routinely used	0.02–0.04 mg/kg IV, IM
dexmedetomidine (Dexdomitor®)	0.001–0.01 mg/kg IV, IM lower dosages for IV and higher dosages for IM	0.001–0.01 mg/kg IV, IM lower dosages for IV and higher dosages for IM	0.001–0.005 mg/kg IV
midazolam (Versed®)	0.1–0.5 mg/kg IV, IM, SQ	0.1–0.5 mg/kg IV, IM, SQ	0.05–0.1 mg/kg IV
diazepam (Valium®)	0.1–0.5 mg/kg IV	0.1–0.5 mg/kg IV	0.05–0.1 mg/kg IV
ketamine (Ketaset®, VetaKet®, Ketaject®, Vetalar®)	2–5 mg/kg IV, IM	2–5 mg/kg IV, IM	Not routinely used
atropine (generic)	0.02–0.04 mg/kg IV, IM, SQ	0.02–0.04 mg/kg IV, IM, SQ	Not routinely used
glycopyrrolate (Robinul-V®)	0.005–0.2 mg/kg IV, IM, SQ	0.005–0.2 mg/kg IV, IM, SQ	Not routinely used
maropitant (Cerenia®)	1 mg/kg IV, SQ	1 mg/kg IV, SQ	Not used

*Opioid dosages can be found in Chapter 15 (Acute Pain Management)

TABLE 8-5 Summary of Premedication Drugs Described in this Chapter

Drug Category	Purpose	Examples	Mechanism of Action	Adverse Effects
Tranquilizer	• Calms animals, reduces anxiety and aggression in animals, may decrease irritability and excitement, can quiet excited animals	• Phenothiazines 1. acepromazine (Promace® and generic) 2. chlorpromazine (Thorazine®) 3. promazine (Sparine®)	• Blocks dopamine receptors in CNS (anxiolytic)	• No analgesia • No reversal agent • Potent cardiovascular effects • Priapism (stallions) • May promote hypothermia • Decrease PCV • Splenomegaly
		• Benzodiazepines 1. diazepam (Valium®) 2. midazolam (Versed®) 3. alprazolam (Xanax®)	• Enhances release of the main inhibitor neurotransmitter GABA (anxiolytic)	• No analgesia • Limited sedation • Dysphoria (disinhibition) displayed as restlessness, vocalization, and attempts to flee
		• Butyrophenones 1. droperidol (Inapsine®) 2. haloperidol (Haldol®) 3. azaperone (Stresnil®)	• Anti-dopamine and anti-adrenergic activity; may mimic the inhibitor neurotransmitter GABA	• No analgesia • No reversal agent

(Continued)

TABLE 8-5 Summary of Premedication Drugs Described in this Chapter (*Continued*)

Drug Category	Purpose	Examples	Mechanism of Action	Adverse Effects
Sedative	• Causes CNS depression, sleepiness, may decrease irritability and excitement, can be used to quiet excited animals	• Alpha-2 adrenergic agonists 1. xylazine (Rompun®, AnaSed®, Sedazine®, and Chanazine®) 2. romifidine (Sedivet®) 3. detomidine (Dormosedan®) 4. dexmedetomidine (Dexdomitor®) 5. medetomidine (Domitor®)	• Binds to alpha-2 adrenergic receptors on neurons and inhibits the release of the neurotransmitter norepinephrine.	• Profound cardiovascular effects (bradycardia, decreased cardiac output, dysrhythmia, increased systemic vascular resistance, decreased oxygen to tissues) • Hypertension • Transient hyperglycemia and increased urine output
Anticholinergic	• Causes increased heart rate, bronchodilation, decreased gastric motility and secretions, and mydriasis	1. atropine (generic) 2. glycopyrrolate (Robinul-V®)	• Blocks ACh release in parasympathetic nervous system	• No analgesia • Ileus and colic in large animals • Tachycardia • Arrhythmias • Increased intraocular pressure
Opioid	• Causes profound analgesia; mild to moderate sedation when used alone; produces neuroleptanalgesia when given in conjunction with a sedative or tranquilizer	• Full mu agonists 1. morphine (Duramorph®, Astramorph®) 2. hydromorphone (Dilaudid®) 3. oxymorphone (Numorphan®) 4. methadone (Dolophine®) 5. fentanyl (Duragesic® transdermal patches, Sublimaze®) 6. remifentamil (Ultiva®)	• Binds to activate mu opioid receptor; methadone also blocks NMDA	All full mu agonists: • Bradycardia and respiratory depression • Dysphoria • Hypothermia in dogs; hyperthermia in cats • Possible excitement in cats at higher dosages • Ileus, colic, increased locomotion, and possible excitement in horses (seen with repeated or higher dosages) Morphine, hydromorphone, oxymorphone: • Vomiting • Panting
		• Partial mu agonist 1. buprenorphine (Buprenex®); a specific formulation FDA-approved for control of postoperative pain in cats (Simbadol™)	• Binds to reactivate mu opioid receptors; however, less robustly than full mu agonists	• Milder bradycardia • Rare panting or vomiting in comparison to full mu agonists
		• Mixed agonist-antagonist 1. butorphanol (Torbugesic®, Torbutrol®)	• Works as a kappa opioid receptor agonist and mu receptor antagonist	• Occasional panting or rarely vomiting

(Continued)

TABLE 8-5 Summary of Premedication Drugs Described in this Chapter (*Continued*)

Drug Category	Purpose	Examples	Mechanism of Action	Adverse Effects
Dissociative	• Causes sedation and analgesia; used to induce and maintain general anesthesia	1. ketamine (Ketaset®, VetaKet®, Ketaject®, Vetalar®)	• Blocks NMDA receptor	• Tachycardia • Increased cardiac output • Hypertension • Dysphoria, hallucinations, and unpredictable behavior during recovery • May sting on injection • Muscle rigidity
Neurokinin-1 receptor antagonist	• Prevents vomiting	1. maropitant (Cerenia®)	• Inhibit substance P (the key neurotransmItter involved in vomiting)	• No analgesia • Pain at injection site

REVIEW QUESTIONS

Multiple Choice

1. Azaperone belongs to which category of drug?
   a. Alpha-2 adrenergic agonists
   b. Butyrophenone
   c. Phenothiazine
   d. Benzodiazepine

2. Xylazine and detomidine belong to which category of drug?
   a. Butyrophenone
   b. Phenothiazine
   c. Alpha-2 adrenergic agonist
   d. Benzodiazepine

3. Which statement about acepromazine is false?
   a. Acepromazine is an excellent analgesic.
   b. Acepromazine is not reversible.
   c. Acepromazine may cause vasodilation leading to hypotension.
   d. Acepromazine has antiemetic properties

4. Which agent could be used to reverse dexmedetomidine?
   a. Flumazenil
   b. Naloxone
   c. Butorphanol
   d. Atipamazole

5. Which opioid drug has the longest duration of action?
   a. Buprenorphine
   b. Fentanyl
   c. Butorphanol
   d. Morphine

6. Which opioid drug is an agonist-antagonist?
   a. Morphine
   b. Hydromorphone
   c. Butorphanol
   d. Oxymorphone

7. Which opioid drug also has NMDA antagonist properties?
   a. Oxymorphone
   b. Methadone
   c. Morphine
   d. Fentanyl

8. Which drug could be used to reverse fentanyl?
   a. Atipamazole
   b. Azaperone
   c. Naloxone
   d. Flumazenil

9. What kind of drugs are anticholinergics?
   a. Parasympathomimetics
   b. Sympathomimetics
   c. Sympatholytics
   d. Parasympatholytics

10. Which dog breed may have an increased sensitivity to acepromazine?
    a. Labrador Retrievers
    b. Boston Terriers
    c. Boxers
    d. Dalmatians

11. Which drug does *not* have analgesic properties?

    a. Midazolam
    b. Oxymorphone
    c. Dexmedetomidine
    d. Xylazine

12. Which is *not* an effect of atropine?

    a. Decreased salivation and secretions
    b. Decreased heart rate
    c. Pupil dilation
    d. Decreased intestinal motility (ileus)

13. Which is an example of allometric scaling?

    a. Calculating drug dosages based on the temperament of the patient so that excitable patients receive doses based on the higher end of the dosage range and mellow patients receive doses based on the lower end of the dosage range
    b. Determining drug dosages based on the age and health of the patient with geriatric and critically ill patients receiving doses based on the lower end of the drug dosage range and healthy patients receiving doses based on the higher end of the drug dosage range
    c. Choosing drug dosages based on the size of the patient so that smaller patients receive doses calculated from the higher end of the drug dosage range and larger patients receive doses calculated from the lower end of the drug dosage range
    d. Selecting drug dosages on the body condition of the patient with obese patients receiving doses based on the lower end of the drug dosage range and thin patients receiving drugs from the higher end of the drug dosage range

14. Which opioid drug has the shortest duration of action?

    a. Buprenorphine
    b. Morphine
    c. Fentanyl
    d. Hydromorphone

15. Which drug can be reversed with yohimbine?

    a. Xylazine
    b. Oxymorphone
    c. Diazepam
    d. Acepromazine

16. Which drug is most likely to cause vomiting after administration?

    a. Maropitant
    b. Hydromorphone
    c. Acepromazine
    d. Glycopyrrolate

17. Which drug does not have analgesic properties?

    a. Romifidine
    b. Butorphanol
    c. Glycopyrrolate
    d. Ketamine

18. Which is *not* a potential cardiovascular adverse effect of dexmedetomidine?

    a. Vasodilation
    b. Bradycardia
    c. Decreased cardiac output
    d. Arrythmias

19. Which premedication drug has been known to cause priapism in horses?

    a. Detomidine
    b. Ketamine
    c. Xylazine
    d. Acepromazine

20. Which sedative/tranquilizer drug would be expected to have the least profound effects on the cardiovascular system?

    a. Acepromazine
    b. Diazepam
    c. Xylazine
    d. Dexmedetomidine

Case Studies

Case Study 1: Chloe is a 6-month-old intact female yellow Labrador Retriever dog. Her weight is 32 kg. She presents to your practice for an elective ovariohysterectomy (OHE). Her physical exam findings are within normal limits (WNL), she has no significant past medical history, and she is not currently taking any daily medication. No preoperative laboratory testing has been performed. During your physical exam you find Chloe friendly but extremely hyperactive (not just rambunctious puppy behavior). You recognize that key issues to picking premedication agents for Chloe include ensuring adequate sedation because she is young and hyperactive as well as providing her with adequate pain control for her surgery. In conjunction with the attending veterinarian, you elect to administer acepromazine along with methadone subcutaneously. Within 20 minutes, Chloe is nicely sedated and she allows you to easily place an intravenous catheter. The rest of her anesthesia and recovery are uneventful.

1. Why was acepromazine chosen for this patient?
2. Why was methadone chosen for this patient?
3. List the adverse effects of acepromazine.
4. List the adverse effects of methadone.
5. Why did it take 20 minutes for this dog to become sedate?
6. Acepromazine is typically diluted from the 10 mg/mL concentration used for large animals to 2 mg/mL for use in small animals. How many mL of 10 mg/mL acepromazine and how many mL of sterile water would you need to make 10 mL of the 2 mg/mL concentration (enough to fill a 10 mL sterile vial)? See Appendix F if needed.

Case Study 2: Tinkerbell is a 12-month-old intact female Domestic Shorthair cat. Her weight is 4.5 kg. She presents to your practice for a routine OHE. Her physical exam findings reveal fleas, she has no significant past medical history, and she is not currently taking any daily medication. A packed cell volume (PCV), total protein (TP), and blood glucose are performed because of the history of fleas and all values are found to be WNL. During your physical exam you find Tinkerbell to be anxious. You recognize that key issues to picking premedication agents for Tinkerbell include ensuring adequate sedation because she is young and anxious, as well as providing her with adequate pain control for her surgery. In conjunction with the attending veterinarian, you elect to administer buprenorphine (regular injectable formulation [0.3 mg/mL]) along with dexmedetomidine subcutaneously. Within 20 minutes Tinkerbell is moderately sedated and she allows you to easily place an intravenous catheter with gentle restraint. The rest of her anesthesia and recovery are uneventful.

1. Why was buprenorphine used in this patient?
2. Why was dexmedetomidine used in this patient?
3. List the adverse effects of buprenorphine.
4. List the adverse effects of dexmedetomidine.
5. When using buprenorphine, what must you do because it is a controlled substance?

Case Study 3: Walter is a quiet 13-year-old castrated male Golden Retriever dog. His weight is 35 kg. He presents to your practice for a dental prophylaxis. You suspect he will require dental extractions which will be a source of pain. A physical exam reveals Walter to be in good general health. Walter has a history of several subcutaneous masses which cytology has confirmed to be lipomas (benign fat tumors). Walter also takes a glucosamine and chondroitin sulfate supplement for osteoarthritis as well as fish oil. He was placed under general anesthesia 2 years ago at another practice without event. A complete blood count, blood chemistry, and urinalysis are performed and are WNL. You recognize that the key issues to consider when picking premedication agents for Walter include providing adequate sedation in a quiet, older animal along with ample analgesia in case teeth must be removed during the dentistry. In conjunction with the attending veterinarian, you elect to administer 0.15mg/kg midazolam along with 0.4mg/kg morphine intramuscularly. Walter vomits once after administration of these drugs. An intravenous catheter is placed and the rest of his anesthesia and recovery are uneventful. Following dental radiographs two teeth were extracted.

1. Why was midazolam chosen for this patient?
2. Why was morphine chosen for this patient?
3. List the adverse effects of midazolam.
4. List the adverse effects of morphine.
5. Calculate a dose of midazolam and morphine for this patient.

Case Study 4: Taco is a 5-year-old male intact Rottweiler dog who presents to your hospital for an acute lameness of his left hind leg. His weight is 45 kg. The attending veterinarian suspects a rupture of the dog's cranial cruciate ligament and would like to obtain radiographs and palpate the limb further. Although Taco has always been friendly at the veterinarian, today he is clearly in pain and during the initial exam tries to snap at the attending veterinarian. For everybody's safety, a muzzle is placed and the rest of the rest of the physical exam is completed. No other issues are found on physical exam (PE) and a packed cell volume, total protein, and blood glucose are performed and are within normal limits. Taco does not take any daily medication and has no past anesthesia history. The owner of Taco plans to take him home after the radiographs are completed. You recognize that the key issues to consider when picking drug choices for Taco include providing adequate sedation for a dog that may be aggressive, providing adequate analgesia to facilitate manipulation of the leg and radiographs, and providing drug selection that allows the dog's owner to take Taco home after the procedure is completed. In conjunction with the attending veterinarian, you elect to administer 0.2 mg/kg butorphanol along with 0.01 mg/kg dexmedetomidine intramuscularly. Within 20 minutes Taco is heavily sedated. He is kept muzzled during the radiographs; however, oxygen supplementation is provided and his is carefully monitored by another technician while you obtain radiographs. The radiographs are consistent with a cranial cruciate ligament rupture, and Taco is referred to a board-certified surgeon to have surgery performed. You administered atipamazole intramuscularly at the completion of the procedure and Taco is awake and ambulatory to go home within the hour.

1. Why was butorphanol chosen for this patient?
2. Why was dexmedetomidine chosen for this patient?
3. List the adverse effects of butorphanol.
4. List the adverse effects of dexmedetomidine.
5. What is the function of atipamazole in this patient?

Case Study 5: Mamma's Boy is a 1-year-old intact male Quarter Horse. His weight is 420 kg. He presents to your mixed animal practice for an elective castration. Mamma's Boy has had limited handling and is flighty. A PE is performed, and the patient is found to be in good general health. He takes no daily medication and has no past medical history. You recognize that the key issues to consider when picking drug choices for Mamma's Boy include providing adequate sedation for a horse that is difficult to handle, providing adequate sedation that will allow for a smooth transition to general anesthesia, and providing adequate analgesia for the surgical procedure. The attending veterinarian administers IV acepromazine first, which allows Mamma's Boy to settle and cooperate. An intravenous catheter is placed. In conjunction with the attending veterinarian, you elect to administer xylazine and butorphanol intravenously. After 5 minutes the horse is profoundly sedate and is ready for anesthesia induction.

1. Why was acepromazine chosen for this patient?
2. Why was xylazine chosen for this patient?
3. Why was butorphanol chosen for this patient?
4. List the adverse effects of xylazine.
5. Is there a unique adverse effect of acepromazine use in horses?

Case Study 6: Earl is a 7-year-old castrated male Poodle dog. His weight is 37 kg. He presents to your hospital for a three hour duration of unproductive retching, anorexia, drooling, and depression. Upon physical exam he is tachycardic with a heart rate of 180 bpm, his respiratory rate is 40 breaths per minute, and his temperature is 100.1°F. His mucous membranes are pale and his capillary refill time is prolonged at 3 seconds. He is depressed and his abdomen is distended and painful. The attending veterinarian is concerned about gastric dilatation volvulus syndrome (GDV, aka "bloat" and torsion). An intravenous catheter is placed and an IV lactated ringers solution (LRS) bolus is administered twice. Radiographs confirm the diagnosis of GDV. Gastrocentesis is performed to further stabilize the patient. The patient's heart rate improves to 140 bpm. Earl is in a state of early decompensatory shock and requires emergency general anesthesia and surgery to correct his condition. Blood work is performed and reveals no major deviations from normal. Earl has a history of skin allergies and is being treated for a resolving skin infection but is otherwise healthy. He received cephalexin PO 6 hours ago. He had a dentistry performed at your practice 2 years ago during which time he was premedicated with acepromazine and hydromorphone subcutaneously. He sedated well from this combination. You recognize that the key issues to consider when picking drug choices for Earl today include: minimal sedation is needed for an already depressed and critically ill dog (e.g., administering acepromazine again is not indicated and would be contraindicated in this patient), Earl is still in a state of cardiovascular compromise and so drugs that have profound cardiovascular adverse effects should be avoided (such as acepromazine, alpha-2 adrenergic agonists, etc.), avoiding vomiting may be beneficial, achieving adequate analgesia for the surgical procedure is needed and may also reduce the amount of induction and general anesthesia required, and lastly, Earl's past medical history and current medication do not presently have an influence on his drug plan. In conjunction with the attending veterinarian, you elect to administer maropitant subcutaneously as an antiemetic after radiographs are taken and then administer hydromorphone intravenously for premedication. After 5 minutes Earl is profoundly sedate. His heart rate is now 120 bpm and a systolic blood pressure currently reads 120 mmHg. Now a better anesthetic candidate, Earl is placed under general anesthesia for surgery.

1. Why was acepromazine not chosen for this patient?
2. Why was maropitant chosen for this patient?
3. Why was hydromorphone chosen for this patient?
4. List the adverse effects of maropitant.
5. List the adverse effects of hydromorphone.
6. Are any of these drugs controlled substances? If so, which one(s)?

Critical Thinking Questions

1. A 20 kg intact male 2-year-old mixed breed dog presents to your practice for an elective castration. He is prescribed 0.05 mg/kg (1mg) acepromazine along with 0.4 mg/kg (8mg) morphine IM as his "premed." A new technician to the practice draws up and administered the drugs. Thirty minutes after drug administration, the dog is profoundly sedate, bradycardic, hypotensive, and has weak pulses. Upon speaking with the new technician, it becomes apparent that the individual accidently administered 1 mL of 10 mg/mL acepromazine rather than 1 mL of the 1 mg/mL concentration that your practice has available for easier dosing in smaller patients. The patient has received 10 mg rather than 1 mg, which reflects a tenfold drug overdose. How would the attending veterinarian and the staff handle this situation? Is there a reversal agent for acepromazine? What treatment options should be started to help treat this patient? Are there drugs that could be administered to specifically treat the cardiovascular effects of the acepromazine? How would you dilute 10 mg/mL acepromazine to a concentration of 1 mg/mL?

2. You are presented with a 6-year-old castrated male German Shepherd dog for euthanasia. He weighs 42 kg. The owner warns you that the dog is extremely aggressive and recently severely bit a neighbor, which, for safety reasons, is why the owner has elected to have the dog euthanized. The dog is apprehensive and growling but currently is muzzled and on a leash. What precautions would you take to ensure your safety and the safety of the staff you are working with? What drug options would you select to allow for safe placement of an intravenous catheter to facilitate the euthanasia? What dosages and route of drug administration would you select?

ENDNOTES

1. Koh, R. B., Isaza, N., Xie, H., et al. (2014). Effects of maropitant, acepromazine, and electroacupuncture on vomiting associated with administration of morphine in dogs. *J Am Vet Med Assoc., 244*(7), 820–829.
2. Valverde, A., Cantwell, S., Hernández, J., et al. (2004). Effects of acepromazine on the incidence of vomiting associated with opioid administration in dogs. *Vet Anaesth Analg., 31*(1), 40–45.
3. Drynan, E. A., Gray, P., & Raisis, A. L. (2012). Incidence of seizures associated with the use of acepromazine in dogs undergoing myelography. *J Vet Emerg Crit Care, 22*(2), 262–266.
4. Tobias, K. M., Marioni-Henry, K., Wagner, R. (2006). A retrospective study on the use of acepromazine maleate in dogs with seizures. *J Am Anim Hosp Assoc., 42*(4), 283–289.
5. Conner, B. J., Hanel, R. M., Hansen, B. D., et al. (2012). Effects of acepromazine maleate on platelet function assessed by use of adenosine diphosphate activated- and arachidonic acid- activated modified thromboelastography in healthy dogs. *Am J Vet Res., 73*(5), 595–601.
6. Congdon, J. M., Marquez, M., Niyom, S., et al. (2011). Evaluation of the sedative and cardiovascular effects of intramuscular administration of dexmedetomidine with and without concurrent atropine administration in dogs. *J Am Vet Med Assoc., 239*(1), 81–89.
7. Alvaides, R. K., Neto, F. J., Aguiar, A. J., et al. (2008). Sedative and cardiorespiratory effects of acepromazine or atropine given before dexmedetomidine in dogs. *Vet Rec., 162*(26), 852–856.
8. Giordano, T., Steagall, P. V., Ferreira, T. H., et al. (2010). Postoperative analgesic effects of intravenous, intramuscular, subcutaneous or oral transmucosal buprenorphine administered to cats undergoing ovariohysterectomy. *Vet Anaesth Analg., 37*(4), 357–366.
9. Steagall, P. V., Carnicelli, P., Taylor, P.M., et al. (2006). Effects of subcutaneous methadone, morphine, buprenorphine or saline on thermal and pressure thresholds in cats. *J Vet Pharmacol Ther., 29*(6), 531–537.

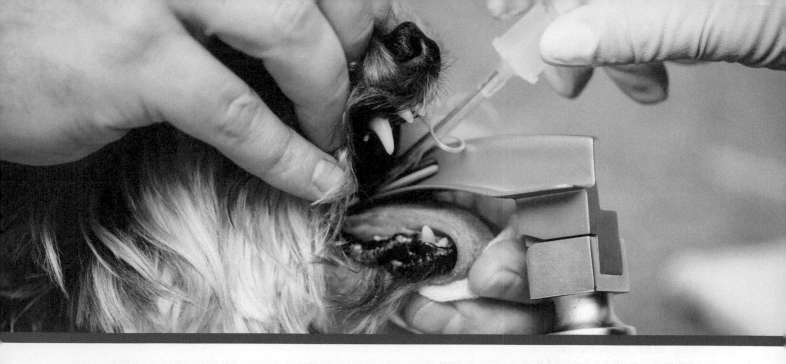

CHAPTER 9
The Induction Period

Kristen Cooley, *BA, CVT, VTS (Anesthesia/Analgesia)*

Teri Raffel Kleist, *CVT, VTS (Surgery)*

KEY TERMS

Anesthesia induction

Apneustic breathing

Bronchial intubation

Laryngospasm

Myoclonus

Patent airway

Titrated

To effect

Waste anesthetic gas (WAG)

LEARNING OBJECTIVES

Upon completion of this chapter, it is expected that the reader should be able to:

9.1 Describe different types of induction techniques

9.2 Explain how to select an appropriate endotracheal tube for a patient

9.3 Describe how to place an endotracheal tube

9.4 Outline the steps for proper endotracheal tube cuff inflation

9.5 Discuss properties of the most commonly used anesthesia induction agents

9.6 List situations in which the use of an inhalant induction agent is preferred over an injectable induction agent

9.7 Describe how to safely induce a patient using inhalant anesthetic agents

INTRODUCTION

Premedication, induction, maintenance, and recovery are the four phases of anesthesia. This chapter focuses on **anesthesia induction**, which includes the careful administration of appropriate drug(s) to induce unconsciousness or "bring about" anesthesia. Successful induction minimizes an animal's excitement, provides muscle relaxation, and sustains anesthesia long enough to allow completion of a procedure or transfer of the patient to the maintenance phase of anesthesia (often with an inhalant anesthetic agent). An ideal anesthetic induction involves giving the induction agent and passing an endotracheal tube (ETT or ET tube), which allows administration of inhalant anesthetic agents to maintain general anesthesia and provides the patient with a protected airway and a means to administer oxygen and ventilation.

A variety of equipment and drugs are available to use in the induction period; therefore, the anesthetist must decide which of these are best to use based on the patient's past history, health and temperament, as well as the procedure being performed. For example, a patient receiving a dental cleaning should be intubated with an ETT with an inflation cuff to protect against aspiration of water and debris. Another example is using injectable induction agents, such as propofol, due to their short duration of action for faster recovery for outpatient procedures. The anesthetist's role is to gather all pertinent information, and along with personal experience and consultation with a veterinarian, develop a suitable induction plan for their patient.

Tech Tip

Anesthetists are paid pessimists. They plan for the worst, which allows them to be prepared in case something goes wrong.

INDUCTION EQUIPMENT

The anesthetic equipment used to prepare a patient for anesthesia varies depending on the species, the patient's health status and temperament, as well as the procedure being performed and expected complications or problems that may develop. The anesthetist's role is to understand and maintain anesthetic equipment so it is ready to use for whatever type of procedure arises.

Methods and Equipment for Delivering Inhalant Anesthetic Agents

Inhalant anesthetic agents are most commonly delivered to the patient via a face mask, chamber, or ETT. A less common delivery method is via a supraglottic airway device.

- Mask induction involves placing a cone-shaped facemask with a tight-fitting diaphragm over the patient's face (Figure 9-1). Face masks are used to provide the patient with oxygen or inhalant anesthetic agent for induction of anesthesia. Face masks come in a variety of sizes, and the size chosen is based on the animal's body size; choosing one that is too large increases the chance of leaks and prolongs induction while

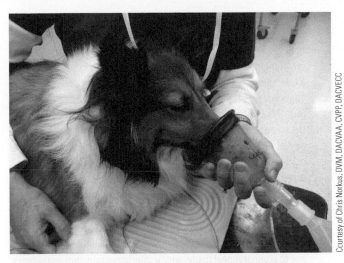

FIGURE 9-1 Dog fitted with a face mask for mask induction.

choosing one that is too small will not fit the patient's face. Face masks should be clear to visualize the animal's muzzle to detect regurgitation.

- Chamber inductions have been used for induction of aggressive cats and other small, exotic, or wild species that cannot be properly restrained; however, their use is declining due to safety concerns for staff and patients from exposure to waste anesthetic gas. Other disadvantages of delivering inhalant anesthetic agents to patients in induction chambers is that induction is slow, prolonged, and may result in excessive excitement of the patient. Induction chambers should be airtight, clear, and durable to allow for visualization of the animal and prevent the animal from breaking or shattering the chamber (Figure 9-2). Large chambers may have a partition so a small animal can be partitioned to one part of the chamber to lower the volume (and cost) of anesthesia agent used.

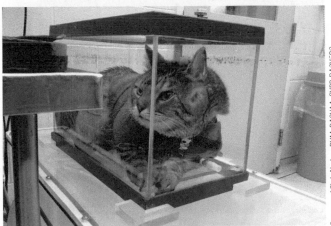

FIGURE 9-2 Cat in a chamber for chamber induction. This picture shows a cat inside an induction chamber moments before being induced. Anesthetic chamber inductions have significant downsides including slow anesthetic induction without a patient airway, large amounts of anesthetic waste gas, and rough excited inductions. The specific piece of equipment shown here, however, is novel (and not described in this book) because it uses a liquid injection technique that very rapidly induces anesthesia, preventing patient struggling, and has a built in scavenging system to prevent staff exposure. Such systems are usually handmade, as is this one, and are difficult to come by.

- Supraglottic airway devices, also referred to as laryngeal mask airways, are alternatives to endotracheal intubation in difficult to intubate species due to unique anatomical features (rabbits) and potential for laryngospasm (felines). Supraglottic airway devices consist of a tube connected to an elliptical mask that has an inflatable outer edge/cuff (see Figure 20-32). When placed and inflated correctly, they form a circumferential, low-pressure seal around the glottis allowing the safe delivery of inhalational anesthetic agents and supplemental oxygen. These devices can be quickly and efficiently placed without the use of a laryngoscope, do not enter the larynx or trachea, and, in some species, may be simpler and faster to place than ETTs. If properly placed and inflated, supraglottic airway devices can provide positive pressure ventilation and do not leak more anesthetic gas than with ETT placement (Figure 9-3; see Figure 20-33). Inappropriate use and/or patient selection can result in placement difficulties/failures, damage to oropharyngeal tissue, and/or improper protection and patency of the airway. In human anesthesia, supraglottic airway devices have been available since the mid-1980s; however, their use in veterinary anesthesia is relatively new. A commercially-available veterinary product (V-gel®) is designed for use in rabbits and felines. Its use in rabbits is described in Chapter 20 (Anesthesia and Analgesia for Pet Birds and Exotic Small Mammals).
- Endotracheal intubation, the preferred method of providing inhalant anesthetic agents and oxygen to patients, involves placing a semi-rigid polyvinylchloride (PVC) or rubber/silicone tube into the trachea of anesthetized or otherwise unconscious patients. In addition to maintaining a

patent airway (open passageway into and out of the lungs), endotracheal intubation prevents aspiration of stomach contents, fluid, blood, or other debris; delivers drugs (oxygen, inhalant anesthetic agents) via the respiratory tract; assists in ventilation; and reduces personnel exposure to inhalant anesthetic agents. In some settings, ETTs may be placed through the oropharynx while in other species (such as foals) the tube may be placed up the nasopharynx and into the trachea.

Endotracheal Tubes

The most rigid type of ETT is PVC (Figure 9-4), followed by silicone and red rubber. PVC ETTs can be warmed to soften them, which allows them to be molded to the shape of the trachea. Endotracheal tubes made of PVC also have a radiopague line in them making them visible on a radiograph. All ETTs are beveled at the patient end to allow visualization of and ease of placement in the patient's laryngeal opening.

There are three styles of ETTs (Figure 9-5).

- Murphy ETTs have an inflation cuff and escape hole at the end of the tube opposite the patient opening. The escape hole (also called an eye) allows gas to flow around any obstruction in the patient end of the ETT. The most commonly used type of ETT in both large and small animals is the Murphy ETT.
- Magill ETTs are similar to the Murphy but do not have the escape hole. They are available in a variety of sizes to accommodate animals ranging in size from cats to large horses. They are curved and may or may not have inflation cuffs. Plain tubes are used when the volume of the cuff would hinder insertion of a tube in very small animals (such as cats and birds).
- Cole ETTs have a tapered end (which seals the airway); therefore, they do not have inflation cuffs. Cole ETTs are designed so that the shoulder of the tube seals the airway at the laryngotracheal opening thus providing an air-tight seal; however, movement or intermittent positive pressure ventilation tends to dislodge the tube. Cole ETTs are most commonly used in birds because they have complete tracheal rings that could potentially be ruptured from pressure applied to the tracheal area when inflating a cuff. Cole ETTs are also used for intubating reptiles and pediatric patients whose tracheas would not tolerate cuff inflation.

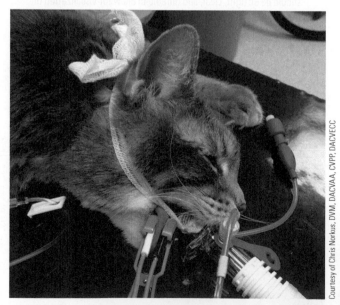

Courtesy of Chris Norkus, DVM, DACVAA, CVPP, DACVECC

FIGURE 9-3 A supraglottic airways device is designed to be passed blindly into the oropharynx beyond the larynx, then pulled out slightly to bring the epiglottis rostrally and position the cuffed lumen of the device over the larynx. These devices have anatomical features for specific species and are made of soft gel-like material to provide a pressure seal that avoids laryngeal and tracheal trauma. One example is the V-gel® ("v" for veterinary and "gel" describing its gel-like material), which provides a gas-tight seal and prevents the issues of laryngeal spasm, coughing, gagging, stridor, or tracheitis post-endotracheal intubation.

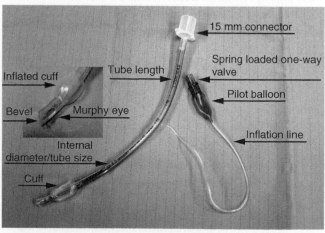

Courtesy of Kristen Cooley, BA, CVT, VTS (Anesthesia/Analgesia)

FIGURE 9-4 A cuffed ETT with its parts labeled.

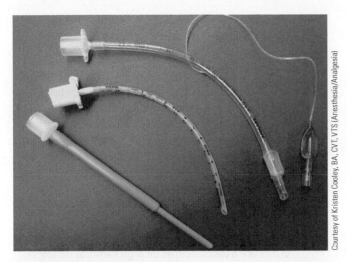

FIGURE 9-5 The different types of ETTs. The top tube is a cuffed PVC Murphy ETT, the middle tube is an uncuffed PVC Magill ETT, and the bottom tube is a Cole ETT, which do not have cuffs.

Courtesy of Kristen Cooley, BA, CVT, VTS (Anesthesia/Analgesia)

Tech Tip 🐾

Always check the inflatable cuff on the ETT to make sure it will inflate and hold air as expected. ETTs should also be kept clean as they are placed near the lung and could be a potential source of infection.

Choosing Proper Endotracheal Tube Size

There are three unvalidated methods for selecting ETT size. One method uses a chart of the animal's weight that applies the formula body weight (in kilograms) divided by 4 then plus 4 [(body weight in kg/4) + 4] to determine the recommended endotracheal tube size in millimeters (Table 9-1). This formula works better for dogs than cats. Another method uses the width of the nasal septum with the outer diameter of the ETT (Figures 9-6a and 9-6b). The last method bases the size of the ETT on palpation of the outer diameter of the animal's trachea in the region of the mid-neck (Figure 9-6c). It should be noted that some brachycephalic patients such as English bulldogs have a much smaller trachea than expected; therefore, in these patients is it advised to always use a smaller ETT than would be expected for another breed of dog of a similar size. The veterinary technician should always have an ETT one size larger and one size smaller than expected available when intubating animals. In time, veterinary technicians will become comfortable at estimating appropriately sized ETTs for their patients.

Laryngoscopes

A laryngoscope is an instrument that enhances visibility of the larynx to help pass an ETT. Laryngoscopes are used in small ruminants, calves, and swine and may be used in dogs and cats; horses are blindly intubated and cattle may be intubated blindly or by digital palpation. Other species such as exotic small mammals (e.g., rabbits, rodents, and ferrets) may have modified laryngoscopes to help with endotracheal intubation; laryngoscopes are not typically used in birds because their airway opening is easily visualized without one.

(a)

(b)

(c)

Courtesy of Teri Raffel Kleist, CVT, VTS (Surgery)

FIGURE 9-6 Methods of determining ETT size. In addition to using a chart based on the animal's weight, another method of determining ETT size includes measuring the width of the nasal septum. (a) This image shows an ETT that is the right size for this dog. (b) This images shows an ETT that is too big for this dog. (c) Another method uses palpation of the outer diameter of the animal's trachea to determine ETT size.

TABLE 9-1 Guidelines for Choosing an Endotracheal Tube (values provided serve as a starting guideline; keep in mind that individual patients will vary)*

Cat body weight	ETT size
1 kg	3.0 mm
2 kg	3.5 mm
3.5 kg	4.0 mm
4.0 kg	4.5 mm
5.0+ kg	4.5–5.0 mm

Dog body weight	ETT size
2.0 kg	4.5 mm
3.5 kg	5.0 mm
5.0 kg	5.5 mm
6.0 kg	5.5 mm
8.0 kg	6.0 mm
10 kg	6.5 mm
12 kg	7.0 mm
14 kg	7.5 mm
16 kg	8.0 mm
18 kg	8.5 mm
20 kg	9.0 mm
25 kg	10.5 mm
30 kg	11.5 mm
40–60+ kg	14–16 mm

* Most adult horses will take a 24–26 mm ETT.

Laryngoscopes consist of a handle and a curved or straight blade (Figures 9-7a and 9-7b). The tip of straight blades may be curved or straight (see Figure 9-7b).

- The laryngoscope blade consists of a spatula that passes over the lingual surface of the tongue and a flange that is used to direct the tongue, soft palate, or epiglottis to facilitate endotracheal intubation. Laryngoscope blades contain a fiber–lit or conventional bulb light source to aid in visualization of the laryngeal opening (Figure 9-7c). Laryngoscope blades vary in length and are available in sizes 0 (small) to 5 (large) (Figure 9-7d); longer custom blades for swine and small ruminants are available as are smaller custom blades for exotic animals.
  - Miller laryngoscope blades are straight and the side of the flange is reduced to minimize trauma to the larynx (see Figure 9-7b).
  - Macintosh laryngoscope blades are curved and when used the tip of the blade is compressed into the angle formed by the base of the tongue and the epiglottis (see Figure 9-7a).
  - Some laryngoscope blades are modifications of the straight and curved models. For example, Wisconsin laryngoscope blades are straight with a modest curve that expands slightly toward the distal portion of the blade to reduce trauma during endotracheal intubation.
- The handle contains a battery that powers the light source. The laryngoscope blade hooks onto the handle and is then locked into place. There are fiber-lit and conventional handles that fit the appropriately lit laryngoscope blades and are not interchangeable. A laryngoscope handle that uses fiber illumination has a green dot on the base and its laryngoscope blade has a green circle at the top of the handle. Conventional handles have an electrical connection and are not color coded. The light source should always be checked to ensure that it is working prior to use.

(a)

Courtesy of Teri Raffel Kleist, CVT, VTS (Surgery)

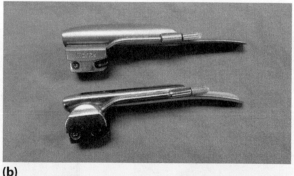

(b)

Courtesy of Teri Raffel Kleist, CVT, VTS (Surgery)

(c)

Courtesy of Teri Raffel Kleist, CVT, VTS (Surgery)

(d)

Courtesy of Teri Raffel Kleist, CVT, VTS (Surgery)

FIGURE 9-7 A laryngoscope is used to enhance visibility of the larynx. (a) A Macintosh laryngoscope has a curved blade. (b) The tip of some straight laryngoscope blades, such as these Miller laryngoscope blades, may be curved or straight. (c) Laryngoscope blades have a light to help visualize the laryngeal opening. (d) Laryngoscope blades vary in length as demonstrated by these Miller laryngoscope blades.

The laryngoscope blade is used to gently depress the base of the tongue just rostral to the epiglottis, which helps visualize the larynx for endotracheal intubation. The laryngoscope should generally not be placed over the epiglottis or into the glottis or trachea.

Preparing for Endotracheal Intubation

Prior to intubating a patient, all supplies should be gathered (Figure 9-8a) and confirmed to be clean and in working order (Appendix B). The supplies needed for endotracheal intubation include:

- Endotracheal tubes (three different sizes)
  ◦ ETTs should always be kept clean.
  ◦ Gather the proper size ETT plus one-half size smaller and one-half size larger. For brachycephalic breeds, choose the size you expect to use and at least two sizes smaller since these breeds tend to have narrow tracheas.
- Gauze tie or another type of tie to secure the ETT in place.
- Gauze sponge to hold the tongue.
- Syringe to inflate the cuff to check for leaks and to form a seal inside the trachea.
- Sterile water-based lubricant to put around the cuff to reduce friction from intubation and to help form a better seal within the trachea.
- Laryngoscope to allow visualization of the larynx.
- Lidocaine to desensitize the larynx to help avoid laryngospasm and coughing (common in cats and pigs).

The ETT length should be measured prior to intubation.

- Hold the ETT to the patient's neck (do not touch the fur). The ETT should measure from the tip of the nose to the thoracic inlet (Figure 9-8b).
- Mark the proximal end of the tube where it lines up with the incisors. If the ETT extends more than 1 inch past the incisors, trim the tube only if it is trimmed proximal to the inflation line/pilot balloon (may not be appropriate for silicone or wire reinforced tubes), or use a shorter tube. ETT that are too short may not pass through the larynx and could result in an unprotected airway or an ETT that is easily dislodged. ETT that are too long may lead to bronchial intubation (resulting in ventilation of one lung) and/or excessive mechanical dead space (resulting in rebreathing of exhaled carbon dioxide).

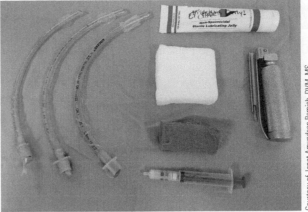

Courtesy of Janet Amundson Romich, DVM, MS

(a)

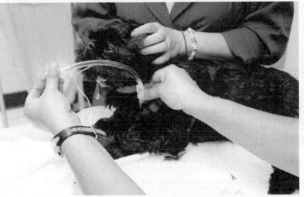

Courtesy of Kristen Cooley, BA, CVT, VTS (Anesthesia/Analgesia)

(b)

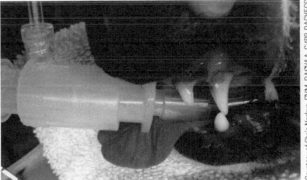

Courtesy of Chris Norkus, DVM, DACVAA, CVPP, DACVECC

(c)

FIGURE 9-8 Preparing for endotracheal intubation. (a) This image shows equipment/supplies gathered to perform endotracheal intubation in a dog including three different sizes of ETTs, sterile water-based lubricant, gauze sponges, material to tie ETT in place (tubing, roll gauze, or nonrestrictive material), Tru-cuff® ETT inflation syringe for cuff inflation, and laryngoscope. (b) The proper length ETT should extend from the tip of the nose to the thoracic inlet. This ETT is too long for this dog, which could result in the beveled end advancing into only one bronchus (supplying gas and air to only one lung) or the patient end extending too far from the mouth thus increasing mechanical dead space. Note that this ETT extends well beyond the tip of the nose and therefore should be cut before use. (c) ETTs that are too short may not pass through the trachea at all, resulting in an unprotected airway or an ETT that is easily dislodged.

Tech Tip 🐾

Overinflation of the ETT cuff can cause tracheal damage, including mucosal irritation, necrosis, stenosis, and tears. Ways to prevent tracheal damage include not overinflating the cuff (especially when performing dental prophylaxis), disconnecting the ETT before flipping/turning the patient, not allowing the stylet to stick out beyond the end of the ETT if using a metal stylet, and extubating a patient only after deflating the cuff. Ideally, the ETT cuff pressure should be high enough to seal the trachea while not impeding blood flow to the tracheal mucosa. There are several commercial devices specifically designed with pressure valves to monitor ETT cuff inflation and, if used, should be part of the equipment/supply list for endotracheal intubation (see Figure 9-8a).

Endotracheal Tube Leak Check

Every ETT should be checked for leaks prior to each use.

- Inflate the inflation cuff by attaching an air-filled syringe to the pilot balloon (Figure 9-9a).
- Allow it to sit for 3 to 5 minutes.
- Inflation cuff should remain inflated and not lose any air. To ensure that no air was lost due to a leak in the inflation cuff:
  - Use a syringe to withdraw all the air from the inflation cuff. The same volume of air that was placed into the inflation cuff should be removed and measured using the same syringe.
  - Alternatively, the inflation cuff can be inflated in a bowl of water (Figure 9-9b). If bubbles are seen during inflation, a leak exists and the ETT should be discarded. Do not overinflate the inflation cuff during checking as stretching it may decrease the life of the ETT.
- Each ETT should be inspected for damage or debris, and the adapter should be firmly seated within the tube.

DRUG SELECTION

Just like selecting a premedication drug, the anesthetist's goal for choosing an induction agent is to select drugs, working in conjunction with the attending veterinarian, that "best fit" the individual patient. The "patient problem list" described in Chapter 6 (Creating an Anesthetic Plan) and

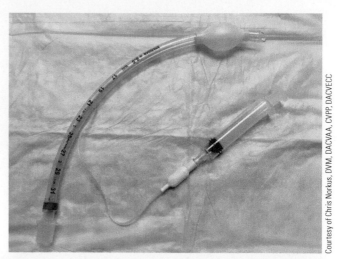

(a)

Courtesy of Chris Norkus, DVM, DACVAA, CVPP, DACVECC

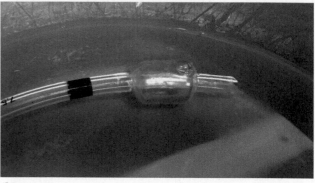

(b)

Courtesy of Teri Raffel Kleist, CVT, VTS (Surgery)

FIGURE 9–9 Inflation cuff leak check. (a) A leak check of an ETT involves keeping the inflation cuff inflated for 3 to 5 minutes to ensure air is not escaping from the cuff. (b) This inflation cuff is leaking air as is evident by the formation of air bubbles when it is held in a bowl of water.

Chapter 8 (Premedication) can also help the anesthetist choose the best induction agents for each patient based on evaluating choices within an appropriate drug class and selecting which induction agent(s) best fits a given patient.

Key considerations the anesthetist should specifically consider when choosing which induction drugs to use include:

- *Which premedication(s) are being given to the patient?* Premedication drugs and the patient's response to them influence the amount of induction agent needed. For example, premedicating an animal with acepromazine or an alpha-2 adrenergic agonist significantly reduces the amount of induction agent needed. In horses it is important to ensure that adequate premedication was given, and the patient is adequately sedate to prevent injury to the anesthetist and horse during induction. Being knowledgeable about the adverse effects of the premedication drugs helps the anesthesia team choose induction agents that do not exacerbate undesirable effects of premedication drugs and can also influence the anesthesia maintenance agent selected.

- *What is the species of the patient?* A drug's duration of action may vary from one species to another based on how each metabolizes the drug. Another consideration is drug cost. While alfaxalone might be an inexpensive option in a cat, it could be cost prohibitive to use in a horse.
- *What is the breed and temperament of the patient?* Even within a species, the action of a drug may vary from one breed to another. For example, sighthounds are sensitive to thiopental due to a liver enzyme deficiency responsible for metabolism of that drug. Likewise, excitable horses that are not adequately sedated prior to induction may be a danger to itself and veterinary personnel.
- *Does the patient have any drug allergies? Is the patient on other medications that could interact with my drug choices?* Drugs such as the anticonvulsant phenobarbital that activate hepatic enzymes can affect the duration of action of induction drugs such as ketamine that are metabolized by the liver.
- *What is the duration and specifics of the procedure being performed?* Induction agents that have short duration of action and leave patients with minimal residual effects (such as propofol) may be an ideal choice for outpatient procedures such as radiographs or CT scans.
- *Has the patient received induction drugs before, what dosages were used, and what were the effects?* If an induction drug did not produce a smooth induction the last time is was used, a different dosage or different drug entirely should be considered in the future. If adverse effects were noted previously, changing drugs or dosages may prevent them from occurring.
- *What pre-existing conditions exist and how does this influence the patient's overall health status and anesthetic management of the case?* An animal's overall health status and organ function will affect how quickly the drug is distributed, metabolized, and excreted, which ultimately affects when and how long those effects last.[1] Whether or not a patient has cardiovascular, respiratory, and neurologic stability is critical to know when determining which induction agent to use. For example, a patient with gastric dilatation volvulus who is being taken to surgery may not be able to handle the negative cardiovascular effects of propofol. The anesthetist should consider any concurrent issues an animal has and how these conditions influence management of the induction period and drug selection in their patient.

As with premedicants, patients who are overweight should have their induction drug doses adjusted to their lean "ideal" body weight or have their overall drug dose reduced. All induction doses should be administered "**to effect**," which means the drug is given until the desired effect (in this case unconsciousness) is obtained rather than giving the entire calculated dosage (an exception to this is typically in induction of zoo patients/wildlife and large animals such as horses).

ANESTHESIA INDUCTION AGENTS

Anesthesia induction agents can be injectable or inhalant anesthetic agents that produce unconsciousness in the patient (see Chapter 5 [Anesthetic and Analgesic Pharmacology] for complete drug profiles).

The ideal induction agent has a rapid onset of action, is rapidly cleared from the body, and does not negatively affect other body systems. Animals that are ill or debilitated in some way, including advanced age, will require less induction agent than their young, healthy counterparts.[2] Premedication drugs and the patient's response to them also influence the type and amount of induction agent needed. The administration of premedication drugs reduces the amount of induction drug needed and thus further reduces unwanted adverse effects of these agents.[3]

Safety Alert

Anesthesia induction can be done using inhalant anesthetic agents delivered via mask or chamber (often referred to as masking, gassing, or boxing down an animal). However, these methods should be reserved for situations in which IV induction is not possible. Mask or chamber induction is a stressful event that is potentially unsafe to patients and personnel due to the delay in obtaining airway control, staff exposure to waste anesthetic gas, slow induction, and the risk of drug overdose if the patient cannot be intubated quickly enough and the mask needs to be reapplied or the animal needs to be put back into the chamber.[4]

Anesthesia induction is best accomplished via injection in most animals. Depending on the type of injectable anesthetic induction agent, it may be able to be given via routes other than IV such as IM. The IM route may be helpful for fractious animals or in young, healthy patients being anesthetized for short routine procedures such as neuters in animal shelters. The disadvantage of IM induction is that titration of the dose is not possible, and the pain associated with IM administration. In contrast, the advantage of using an IV route to administer induction agents is that the anesthetic agent can be titrated to effect to provide a rapid, smooth transition from consciousness to unconsciousness. In most settings, the IV route of administration of induction drugs is preferred.

Tech Tip

Most IV anesthetic induction agents are **titrated** (drug is given in a series of IV bolus injections) "to effect," as overdosing is more likely when drugs are given too quickly and their effects are not appreciated as they happen.

Injectable anesthetic induction agents include dissociatives, propofol, etomidate, and alfaxalone.

Dissociative Anesthetics

Dissociatives are C-III controlled substances that stimulate the central nervous system and when used as an induction agent are given to patients with sedatives or tranquilizers that provide muscle relaxation. The common dissociates used in veterinary medicine include the following:

- *Ketamine* (Ketaset®, VetaKet®, Ketaject®, Vetalar®) is often coupled with a sedative or tranquilizer such as a benzodiazepine or propofol for IV induction of anesthesia (Figure 9-10a). For ketamine/diazepam induction, the dose is calculated based on the patient's lean body weight. Keep in mind that ketamine, in addition to the more commonly used 100 mg/mL concentration, is available in other formulations (including 50 mg/mL and 200 mg/mL) (Figure 9-10b). Patients that are induced with ketamine/diazepam will retain muscle and jaw tone; therefore, the anesthetist needs to be patient to avoid overdosing. For propofol/ketamine induction in dogs, "ketofol" is prepared by combining 2–4 mg/kg of each drug (ketamine 100 mg/mL + propofol 20mg/mL) in the same syringe and dosing at 2–4 mg/kg of the drug combination (see Appendix F for examples of these calculations). Ketofol produces profound respiratory depression despite its cardiac advantages of improving the patient's blood pressure, cardiac output, and oxygen delivery to tissue. Intravenous ketamine takes 2 minutes to reach peak effect and lasts approximately 20 minutes. Ketamine can provide somatic analgesia at subanesthetic dosages. Ketamine should be used cautiously in cats with renal disease or insufficiency because the drug is excreted unchanged in the urine, which may result in prolonged recovery if the drug is slow to exit the body.[5]
- *Tiletamine* is only available in the proprietary drugs Telazol® and Tilzolan®, which are available as a powder that is reconstituted according to bottle directions to yield a 100 mg/mL concentration (50 mg tiletamine, 50 mg zolazepam) (see Figure 5-8c). The dose is calculated based on the patient's lean body weight and ½ of the dose is given IV over 30 to 90 seconds and then titrated to effect. Patients that are induced with Telazol® or Tilzolan® also retain muscle and jaw tone; be patient to avoid overdosing. Telazol® and Tilzolan® should only be used as an induction agent in healthy dogs and cats because the amount of tiletamine and zolazepam cannot be individually dosed based on the needs of the patient, the drug is not reversible, and it has a relatively long duration of action. This drug is widely used in the shelter medicine and high quality/high volume (HQHV) spay/neuter clinics, has a shelf life of 56 days if kept in the refrigerator, and may cause long or poor-quality recoveries; otherwise its cardiovascular and respiratory effects and analgesic properties are similar to ketamine. Telazol® and Tilzolan® given IM can provide a surgical plane of anesthesia and may be the sole agent used for minor procedures (see Chapter 10 [The Maintenance Period]).

Tech Tip

Induction of anesthesia is a time when patients die; many from inattention. Patients with difficult to access veins or fractious/aggressive temperaments can be "induced" with IM injections of a dissociative cocktail (e.g., ketamine/diazepam) combined in the same syringe. Failure to monitor these patients as they are becoming unconscious is dangerous as the amount of drug has not been titrated to effect and, if not intubated, they have an unprotected airway.

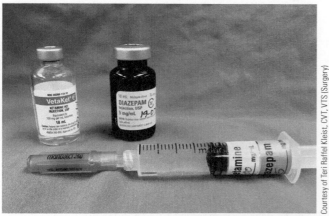

Courtesy of Teri Raffel Kleist, CVT, VTS (Surgery)

(a)

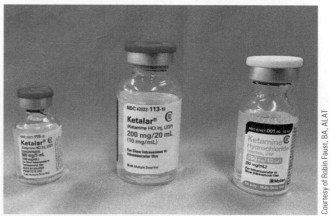

Courtesy of Robin Faust, BA, ALAT

(b)

FIGURE 9-10 Ketamine. (a) Ketamine is commonly mixed with diazepam for use as an induction agent. Ketamine-diazepam mixtures can be prepared in veterinary clinics by combining equal volumes of diazepam (5 mg/mL) and ketamine (100 mg/mL). It may begin to precipitate if stored for more than one week; therefore, it is typically mixed in a syringe as needed. Induction with ketamine-diazepam is slower than with other induction agents (e.g., propofol) and the animal retains more muscle and jaw tone; therefore, anesthetists need to use care to avoid overdosing the patient. (b) Ketamine is available in different concentrations; therefore, it is important to make sure the correct concentration is chosen when performing calculations and drawing up the patient's dose.

Dissociative Facts |||

Advantages:

- Short-term analgesia
- Wide therapeutic range
- Minimal respiratory depression
- Improves cardiac output and blood pressure

Disadvantages:

- Poor muscle relaxant
- Muscle twitching possible
- Tachycardia
- Potential for emergence delirium (state in which the patient may be agitated and inconsolable, uncooperative and typically thrashing, vocalizing, and defecating)
- Drooling or hypersalivation
- Increased intraocular pressure possible
- **Apneustic breathing** (prolonged inspirations with subsequent short exhalations) in cats and horses

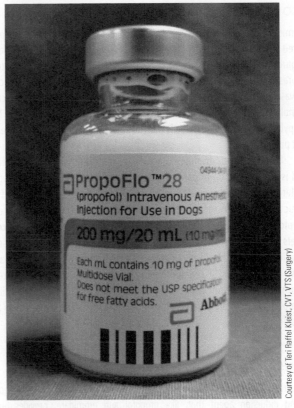

FIGURE 9-11 Propofol is a short-acting IV anesthetic that provides a rapid, smooth induction.

Courtesy of Teri Raffel Kleist, CVT, VTS (Surgery)

Propofol

Propfol (Rapinovet®, PropoFlo™) and propofol with benzyl alcohol (PropoFlo™28) provide a rapid, smooth induction and a short duration of action (Figure 9-11). Animals given propofol may benefit from pre-oxygenation prior to administration and must be monitored closely to avoid hypoxemia. Following administration of propofol, it is important to be able to intubate and ventilate a patient because the drug may cause significant respiratory depression, apnea, and cyanosis that is more pronounced when the drug is administered rapidly or in large doses.[5] Propofol causes short-lived but profound cardiovascular adverse effects. A dosage-dependent decrease in systemic vascular resistance may cause hypotension due to vasodilation, and a decrease in cardiac contractility may lead to decreased cardiac output and hypotension; however, titration to effect can decrease but not eliminate cardiac depression. Transient muscle twitching and **myoclonus** (involuntary jerking of a muscle or group of muscles) may occur more commonly in patients that have not been premedicated and/or those patients given large amounts of propofol.[6] Propofol's lipid emulsion formula can cause it to aggravate lipid metabolism disorders in patients with pancreatitis or diabetic hyperlipidemia. It provides no analgesia and like ketamine may cause pain on IV injection. Once opened, propofol should be handled aseptically to minimize bacterial contamination.

The dose of propofol should be given IV slowly over 60 to 90 seconds while monitoring for effect. Patients should be monitored and supplemented with oxygen. Coadministration of drugs such as lidocaine, diazepam, ketamine, and fentanyl may reduce the amount of propofol needed and allow for a smoother induction. Propofol takes approximately 90 seconds to reach its peak effect and another 5 to 10 minutes to redistribute to tissues and initiate recovery.

Propofol Facts |||

Advantages:

- Rapid onset and recovery
- Noncumulative
- Decreases intracranial pressure
- Minimal fetal affects, suitable choice for C-section
- Good for patients with liver dysfunction because it is metabolized via hepatic and extra hepatic sources
- Minimal residual effects making it appropriate for patients with liver dysfunction

Disadvantages:

- Dosage-dependent respiratory depression
- Profound cardiovascular effects such as hypotension and decreased cardiac contractility
- Does not provide analgesia
- High risk of bacterial contamination if bottle of regular propofol without preservatives is used more than 6 hours after opening
- Potentially toxic in cats with repeated consecutive day injections (due to benzyl alcohol in Propofol 28®)[7]

Etomidate

Etomidate (Amidate®) is a short-acting injectable induction agent prepared in propylene glycol that can produce 10 to 20 minutes of anesthesia (Figure 9-12). Etomidate produces minimal cardiovascular and respiratory depression, making it a great choice for the induction of patients with cardiovascular disease; however, its propylene glycol carrier can cause acute hemolysis and thus should be used only in select high risk patients. The propylene glycol carrier can cause pain on IV injection, making patients react and pull away.

Retching and myclonus can occur during etomidate induction and is more common in patients not adequately premedicated or those underdosed with etomidate. Adrenal suppression for 2 to 6 hours after induction has been seen in dogs given etomidate;[8] therefore, their use should be avoided in patients with questionable adrenal glucocorticoid levels (such as animals with hypoadrenocorticism [Addison's disease]). Etomidate doses should be calculated based on lean body weight and given to effect in well-sedated or very depressed patients. Etomidate is an appropriate option for patients with unstable cardiovascular function. Hemolysis is possible when giving etomidate and is the main reason for its selective use.

Alfaxalone

Alfaxalone (Alfaxan®, Alfaxan®Multidose, Alfaxan®Multidose IDX) is a short-acting induction agent that, like propofol and etomidate, provides rapid induction and hypnosis with reasonable muscle relaxation in approximately

Etomidate Facts

Advantages:

- Rapid onset and recovery
- Short duration
- Noncumulative
- Minimal cardiovascular and respiratory depression
- No alteration of heart rate, cardiac contractility, or blood pressure
- Minimal residual effects

Disadvantages:

- Retching and myoclonus if not adequately premedicated or underdosed with etomidate
- Pain possible on IV injection
- Acute hemolysis possible due to its propylene glycol carrier
- No analgesia
- Transient cortisol suppression

60 seconds[9] (Figure 9-13). Alfaxalone is rapidly metabolized by the liver; therefore, it has a short duration of effect. The cardiovascular and respiratory depressant effects of alfaxalone are generally mild (less than propofol) when given at recommended dosages in the healthy dog and cat. Respiratory depression and apnea may be exacerbated if higher dosages are used; therefore, it is important to be able to intubate and ventilate patients given alfaxalone, as well as providing oxygen support. Mild hypotension caused by vasodilation can occur at higher than recommended dosages, but cardiac output is generally well maintained. Tachycardia can occur immediately following induction, but the exact cause for this occurrence is not totally clear. Patients with pre-existing clinical disease may experience more profound cardiopulmonary depression with alfaxalone than a normal healthy patient, but in general, alfaxalone has mild effects on both body systems. The dose for alfaxalone should be calculated based on patient lean body weight. Alfaxalone is not an analgesic and is noncumulative.

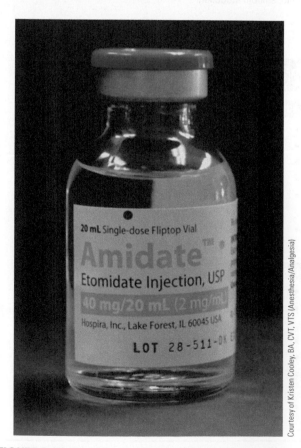

Courtesy of Kristen Cooley, BA, CVT, VTS (Anesthesia/Analgesia)

FIGURE 9-12 Etomidate is a short-acting injectable induction agent.

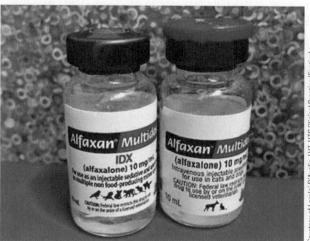

Courtesy of Lorelei D'Avolio, LVT, VTS Clinical Practice (Exotics)

FIGURE 9-13 Alfaxalone is available in single-dose vials (see Figure 5-10), in multidose vials, and in multidose IDX vials for nonfood producing minor species.

Alfaxalone Facts

Advantages:

- Rapid onset and recovery
- Noncumulative
- Excellent cardiovascular safety
- Less respiratory depression than propofol
- Reasonable muscle relaxation

Disadvantages:

- More expensive than ketamine and propofol
- Poor quality recovery possible

Table 9-2 summarizes dosages for IV induction agents described in this chapter.

Inhalant Anesthetic Agents

Inhalant anesthetic agents are used to induce and maintain general anesthesia in a variety of species by providing a relatively rapid induction and recovery while producing good muscle relaxation. The most commonly used inhalant anesthetic agents are isoflurane (Isoflo®, Isosl®, Forane®) and sevoflurane (SevoFlo®, Petrem®, Flurovess™, Ultane®) (Figure 9-14) and to a lesser extent desflurane (Suprane®). Inhalant anesthetic agents cause CNS, myocardial, and respiratory depression and dosage-dependent hypotension. They do not provide analgesia and should never be used as the sole drug for painful procedures.

Induction of anesthesia with inhalant anesthetic agents is sometimes performed in patients that may not be able to properly metabolize injectable agents. Induction of general anesthesia with inhalant anesthetic agents is rapid but still much slower than induction with a rapid acting injectable agent such as propofol or ketamine/diazepam. This

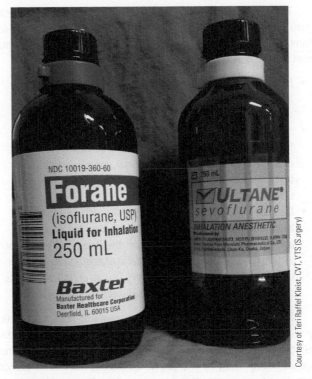

FIGURE 9-14 Isoflurane and sevoflurane are the most commonly used inhalant anesthetic agents in veterinary medicine.

slower induction allows the patient to experience a prolonged excitement phase if the patient is not adequately sedate or may induce a condition causing CNS depression.

Personnel using inhalant anesthetics are at risk for exposure to **waste anesthetic gas (WAG)**; therefore, proper scavenging equipment, properly inflated ETTs, and proper ventilation need to be used to reduce staff exposure to WAG. The properties of inhalant anesthetic agents and the effects of WAG exposure are described in Chapter 10 (The Maintenance Period).

TABLE 9-2 Dosages for Intravenous Induction Agents Described in this Chapter*

Induction Drug	Dog	Cat	Horse
ketamine (Ketaset®, VetaKet®, Ketaject®, Vetalar®)	5–10 mg/kg IV	5–10 mg/kg IV	2–2.5 mg/kg IV
tiletamine/zolazepam combination (Telazol®, Tilzolan®)	2–5 mg/kg IV	2–5 mg/kg IV	1–3 mg/kg IV
propofol (Rapinovet®, PropoFlo™; propofol with benzyl alcohol (PropoFlo™28)	2–8 mg/kg IV	2–8 mg/kg IV	Not routinely used
etomidate (Amidate®)	0.5–2 mg/kg IV	0.5–2 mg/kg IV	Not routinely used
alfaxalone (Alfaxan®, Alfaxan®Multidose, Alfaxan®Multidose IDX)	1–2 mg/kg IV	2–4 mg/kg IV	Not routinely used

* Please note that individual drug dosages required for anesthetic induction may dramatically differ based upon concurrent drug administration, level of central nervous system depression, age, and pre-existing disease.

The properties of nonflammable, halogenated inhalant anesthetic agents are described more extensively in Chapter 5 (Anesthetic and Analgesic Pharmacology) and Chapter 10 (The Maintenance Period); however, key points about their use as induction agents include the following:

- *Isoflurane:* Masking an animal with isoflurane may be difficult because it has an unpleasant odor that irritates the respiratory system. The vaporizer setting for induction with isoflurane is 3%.
- *Sevoflurane:* Sevoflurane is the preferred inhalant induction agent for mask and chamber inductions because it quickly enters systemic circulation, rapidly distributes to the brain, and is less irritating to mucous membranes. The vaporizer setting for induction with sevoflurane is 4 to 5%.
- *Desflurane:* Desflurane is the least potent inhalant induction agent hence higher vaporizer settings are needed when administering this agent. Desflurane has a low boiling point; therefore, it exists as a vapor at room temperature and requires a special heated vaporizer to keep it under pressure to prevent it from boiling. The cost of desflurane and its specialized vaporizer has limited its use in veterinary medicine.

The initial vaporizer settings for induction vary with the inhalant anesthetic agent and the health status of the patient, type of breathing circuit used, and the fresh-gas flow rate (see Table 5-3). The relatively high fresh-gas flow rate and vaporizer setting initially for induction are decreased to maintenance settings when the patient nears the desired anesthetic plane. The vaporizer setting is adjusted according to signs of anesthetic depth.

THE STEPS OF ANESTHESIA INDUCTION

Anesthesia induction occurs after the patient has been premedicated as described in Chapter 8 (Premedication). Once a patient has been given subcutaneous or intramuscular premedication(s), it takes time for the drugs to take effect. This "extra" time can be used to check equipment and set up the induction area with needed supplies. Tasks that can be performed during this time include:

1. Outfitting the anesthesia machine with appropriately sized breathing hoses and reservoir bag; pressure test the machine to ensure it is leak-free (see Chapter 2 [The Anesthesia Machine]).
2. Gathering, checking, and preparing needed supplies for use including:
   a. Intravenous catheter supplies
   b. Endotracheal tube supplies being sure to check for leaks
   c. Monitoring equipment
   d. Supplemental heating sources (e.g., turn on water blanket)
   e. Monitoring form; should be completed with patient information and procedure details
   f. Induction drugs; should be calculated, approved by supervising veterinarian, and drawn up
   g. IV fluids and appropriate administration equipment; calculate fluid administration rates for approval by supervising veterinarian
   h. An emergency plan including calculation of emergency drugs; the plan and drugs should be approved by the supervising veterinarian

Once the patient and anesthetist are ready, administration of the induction agent can begin.

Tech Tip

Auscultating the patient's heart and lungs, assessing mucous membrane color, determining capillary refill time, and palpating a pulse for strength and rhythm is essential to do while the patient is awake. This provides baseline information that is helpful when determining if a change in the patient's status has occurred during the anesthetic event.

Intravenous Induction and Endotracheal Tube Placement

For endotracheal intubation to be successful, the veterinary technician should be familiar with normal anatomy of the upper respiratory tract and laryngeal area in order to communicate any abnormalities to the veterinarian. The following is a set of guidelines for successful induction and endotracheal intubation in dogs and cats:

- Administer IV induction agent and check for readiness to be intubated (Figure 9-15a)
  - Jaw tone should be slack; patient should be unconscious and not moving.
  - Patient should not cough, gag, or swallow when ETT is placed. If they do, more induction drug is necessary and attempts at intubation should be stopped.
- Place and secure the ETT
  1. Patient should be in sternal recumbency; if not place the patient into sternal recumbency.
  2. Lubricate the inflation cuff of the ETT with sterile lubricant to facilitate intubation (Figures 9-15b and 9-15c).
  3. The restrainer should grasp the maxilla behind the canine teeth with one hand to raise and extend the head.
  4. The restrainer should grasp the tongue with the gauze square by the other hand to open the mouth by gently pulling the tongue rostrally and ventrally (Figure 9-15d). Be careful not to cut the tongue on the mandibular incisors.
  5. Place the laryngoscope blade in the mouth and press on the base of the tongue, which is directly underneath the epiglottis (Figure 9-15e). This will pull the epiglottis down and expose the glottis (Figure 9-15f). Do not touch the epiglottis with the laryngoscope blade as this structure is very delicate and, if damaged, swelling can inhibit breathing after extubation.
  6. Once the glottis is visible, lidocaine spray or one drop of lidocaine can be applied bilaterally on the laryngeal cartilage if intubating a cat to anesthetize the larynx and guard against laryngospasm (Figures 9-15g and 9-15h).
  7. Advance the ETT through the arytenoid cartilage aiming for the base of the opening (Figure 9-15i). The ETT should glide in easily; if it does not, the ETT might be too large. In some cases a gentle rotation of the ETT may be necessary to part the arytenoids.
  8. Once the ETT is in place, confirm its proper placement. Misplacement of the ETT in the esophagus is common, yet it may

appear to be in the trachea because air seems to move in and out of the tube and reservoir bag. Improper ETT tube placement results in inability to keep the patient anesthetized and may cause airway blockage and hypoxemia. There are several methods to confirm ETT placement described below. It is important to use more than one way to confirm proper placement.

a. Watch for fogging of the ETT with condensation during exhalation.

b. Feel for air movement from the tube connector as the patient exhales or when rapid, light pressure is applied to the thoracic wall.

c. Palpate the neck to determine how many firm tubes feel present. The trachea is the only firm structure in the neck; therefore, if the ETT tube is in the trachea, the trachea will be the only tube palpable. If the ETT is in the esophagus, both the trachea and the tube in the esophagus will be palpable.

d. Use the laryngoscope to revisualize the larynx, and confirm that the ETT is in the trachea.

e. Check the unidirectional valves for opening of the inhalation valve when the patient inhales and opening of the exhalation valve when the patient exhales.

f. Listen for vocalization from the patient. Vocal cords need to vibrate together to make a sound, which is impossible if the ETT is placed properly.

9. Once the ETT is in place, secure it by tightly tying a single half-hitch knot with a length of roll gauze or tubing around the ETT noting where it lines up with the canine teeth and then cinching the knot making sure it is behind the canine teeth (Figure 9-15j). Make sure the roll gauze or tubing is not tied around and obstructing the small tube to the pilot balloon. Then, snuggly tie the ETT to the patient. Tubes should be tied over the muzzle in dolicocephalic (long muzzled) and mesocephalic (medium sized muzzled) breeds, and behind the head in cats and brachycephalic (short, flat muzzled) breeds (Figures 9-15k and 9-15l). If the procedure involves the mouth, tie the ETT behind the head. ETTs can also be secured to the mandible when necessary, but care should be taken to monitor the tongue for entrapment and swelling.

10. Transfer the patient to the desired position for surgical/procedural preparation before connecting them to the anesthesia machine (unless there is concern about the patient being hypoxemic). The patient may be placed in lateral recumbency, or if the patient remains in sternal recumbency, place a rolled up towel to support the head and neck.

11. Connect the patient to the anesthesia machine and turn on the oxygen. If the patient needs to be repositioned or moved after being connected to the anesthesia machine, it is best to disconnect the patient prior to any turning/rolling over.

12. Inflate the ETT cuff with a small amount of air and then evaluate the patient by checking vital signs (taking a pulse, auscultating the heart, assessing muscle tone, and observing respirations and mucous membrane color). If not attached prior to induction, attach monitors (ideally monitors should be attached prior to induction whenever possible). If the patient is not stable, alert the attending veterinarian. Once the patient is stable, proceed to the next step.

13. Properly inflate the ETT cuff prior to turning on the vaporizer to minimize exposure of personnel to waste anesthetic gas.

   - The ETT cuff should be maximally inflated to 20 to 25 cm H_2O (adequate inflation of normal lungs can be achieved with inflation pressures of 15–20 cm H_2O). Feeling the balloon at the cuff valve is not an accurate gauge of inflation cuff pressure.
     - Overinflated cuffs/excessive cuff pressure can result in tracheal ischemia and necrosis that makes tearing of the trachea more likely and can lead to subcutaneous emphysema, pneumomediastinum, and pneumothorax. These conditions may heal on their own or may require surgery.
     - Underinflated cuffs do not protect the airway from aspiration, may cause the patient to not stay anesthetized, and do not minimize waste anesthetic gas which exposes staff to potentially unsafe levels of inhalant anesthetic agent.
   - To proficiently inflate the ETT cuff:
     - Straighten the airway by extending the patient's head.
     - Attach an appropriately sized syringe to the inflation balloon (3 mL for cats and 5 mL for dogs). Using a larger syringe may lead to inadvertent overinflation.
     - Close the pop-off valve (or occlusion valve with safety relief if present).
     - Pressurize the system by squeezing the reservoir bag while watching the pressure manometer and listening for leaking of gas around the tube (a "hiss" sound) (Figure 9-15m).
       - Add air to the inflation cuff until the sound of gas leaking around the cuff is no longer heard and the manometer is able to hold pressure at 15 to 20 cm H_2O. Lower pressures may be used if the patient is a bird (or other animal with complete tracheal rings) or a patient with tracheal pathology that would raise concern about potential damage to mucosa by ETT cuff.
     - The cuff is inflated so that the airway will leak pressure at 25 cm H_2O.
     - Patients that have been intubated with a larger ETT may not require cuff inflation, at least not initially.
     - Devices exist to help the anesthetist properly inflate the ETT cuff. The Posey Cufflator™ is a portable pressure gauge that can be connected to ETTs with low pressure, high volume cuffs to read the pressure within the cuff (Posey®~$300). The PressurEasy™ cuff pressure controller is a device that connects to the pilot balloon of most ETTs and reads the pressure within the cuff throughout a procedure (Smiths Medical®~$25). The AG Cuffill® syringe is a compact device that can detect and display the ETT cuff pressure value on its digital screen of the plunger (Mercury Medical®~$25)10 (Figure 9-15n). However, these devices may not be useful for all types of ETTs (e.g., silicone) because the elastic recoil of the cuff itself will alter measurements and make them invalid.
     - Open the pop-off valve.

14. Check vital signs by taking a pulse, auscultating the heart, assessing muscle tone, and observing respirations and mucous membrane color. Make a mental note of the vital sign values so they can be recorded on the anesthesia record. Ideally, a set of vitals should be recorded within 10 minutes of induction.

15. Reassess the ETT cuff and inflate if necessary to seal the system. Once the system is sealed, turn on the vaporizer to the induction setting appropriate for the specific inhalant anesthetic agent.

16. Connect the remaining monitors and take first reading. Record vital sign values on the anesthesia record (this includes the initial values taken before turning on the vaporizer and the ones from the first reading).

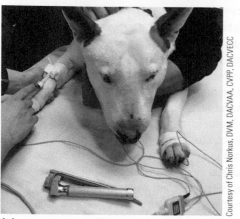

Courtesy of Chris Norkus, DVM, DACVAA, CVPP, DACVECC

(a)

▲ Patient receiving injectable induction agent in sternal recumbency.

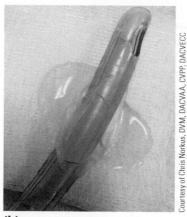

Courtesy of Chris Norkus, DVM, DACVAA, CVPP, DACVECC

(b)

▲ Lubricate the inflation cuff of the ETT with sterile lubricant by placing the lubricant on a gauze pad and rolling the ETT into the sterile lubricant.

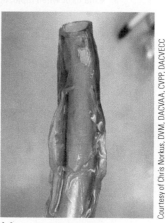

Courtesy of Chris Norkus, DVM, DACVAA, CVPP, DACVECC

(c)

▲ A properly lubricated ETT should have a small amount of sterile lubricant on the cuff.

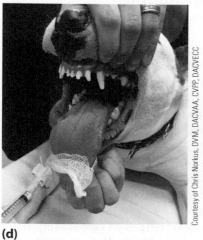

Courtesy of Chris Norkus, DVM, DACVAA, CVPP, DACVECC

(d)

▲ Restrainer grasping behind the canine teeth and extending the head. The restrainer is holding the tongue with the gauze square to open the mouth by gently pulling the tongue rostrally and ventrally.

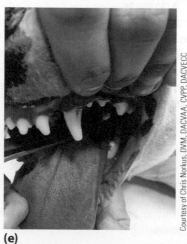

Courtesy of Chris Norkus, DVM, DACVAA, CVPP, DACVECC

(e)

▲ Place laryngoscope blade in mouth and press on the base of the tongue to expose the glottis (canine glottis).

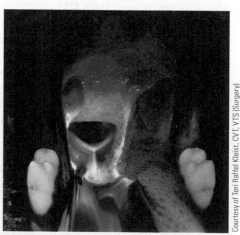

Courtesy of Teri Raffel Kleist, CVT, VTS (Surgery)

(f)

▲ Glottis in a dog.

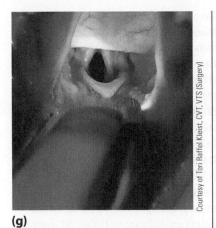

Courtesy of Teri Raffel Kleist, CVT, VTS (Surgery)

(g)

▲ Glottis in a cat.

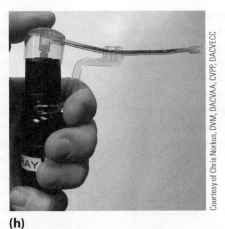

Courtesy of Chris Norkus, DVM, DACVAA, CVPP, DACVECC

(h)

▲ Lidocaine spray can be applied to prevent laryngospasm in cats.

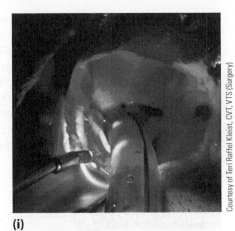

Courtesy of Teri Raffel Kleist, CVT, VTS (Surgery)

(i)

▲ The ETT should glide easily as it is passed through the arytenoid cartilage. Canine glottis with ETT in place. Always check for proper placement of the ETT before proceeding.

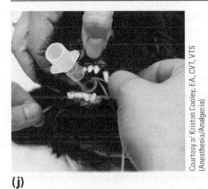

Courtesy of Kristan Cooley, EA, CVT, VTS (Anesthesia/Analgesia)

(j)

▲ The ETT is secured with roll gauze or tubing where it lines up with the canine teeth. Be sure to secure the roll gauze or tubing behind the canine teeth in dolichocephalic and mesocephalic dogs.

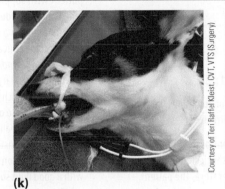

Courtesy of Teri Raffel Kleist, CVT, VTS (Surgery)

(k)

▲ ETTs should be tied over the muzzle in dolicocephalic (long muzzled) and mesocephalic (medium sized muzzled) breeds.

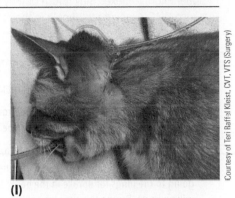

Courtesy of Teri Raffel Kleist, CVT, VTS (Surgery)

(l)

▲ The ETT is secured behind the head in brachycephalic (short, flat muzzled) dog breeds and cats or if the procedure involves the mouth.

Courtesy of Chris Norkus, DVM, DACVAA, CVPP, DACVECC

(m)

▲ The anesthetist closes the pop-off valve and squeezes the reservoir bag while watching the pressure manometer and listening for leaking gas around the ETT. The cuff is inflated with air until any sound of leaking gas is no longer heard and the pressure manometer is able to hold pressure at 15 to 20 cm H$_2$O. The pop-off valve is opened when the cuff check is complete.

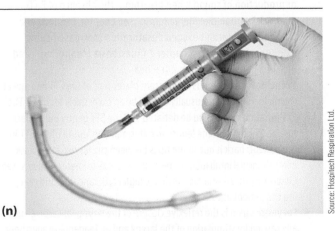

Source: Hospitech Respiration Ltd.

(n)

▲ Pressure sensors such as the AG Cuffill® syringe can detect and display the ETT cuff pressure value on the digital screen of the plunger. The pressure value can be reset by pressing the yellow button.

FIGURE 9–15 Placing an endotracheal tube.

Prior to connecting the patient to the anesthesia machine and starting the flow of oxygen, the breathing hoses and reservoir bag connected to the machine are filled with room air (which consists of 21% oxygen) or 100% oxygen in pre-oxygenated patients. To fill the breathing circuit with the anesthetic gas mixture (oxygen and inhalant anesthetic agent), the oxygen flow rate and vaporizer are set at a higher induction rate. For example, the oxygen flow rate may be set at 25 to 50 mL/kg/minute or greater and the vaporizer set at 2 to 2.5% for isoflurane or 3 to 4% for sevoflurane for 2 to 5 minutes immediately following induction, intubation, and connection of the patient to the breathing circuit to help facilitate machine filling.

Safety Alert

The first place to check for leaks within the system is the ETT cuff. Leaks around the ETT impact the delivered tidal volume in mechanically ventilated patients, may make the patient light (reflexes may return or may lift its head), and cause WAG exposure to people in the room.

Errors in Endotracheal Intubation

The veterinary technician should be aware of errors or conditions that can occur when passing an ETT, so that the error can be corrected.

- *Esophageal intubation* is inadvertent placement of the ETT into the esophagus resulting in the inhalant anesthetic agent not going to the lungs for uptake in the body; therefore, the patient will wake up. During this time, the airway is not secured and aspiration could occur. The ETT should be removed from the esophagus and a new ETT that was not in the esophagus should be placed into the trachea. Esophageal intubation can result in failure to keep the patient anesthetized, patient vocalization or production of sound when breathing, the absence of EtCO$_2$ (end tidal carbon dioxide) readings (see Chapter 7 [Anesthesia Monitoring]), patient oxygen desaturation (decreased pulse oximeter readings; see Chapter 7 [Anesthesia Monitoring]), and lack of a palpable ETT in the trachea.
- **Bronchial intubation** is inadvertent placement of the ETT into one of the mainstem bronchi usually caused by an excessively long ETT. The ETT inflation cuff should be deflated and the ETT gently pulled back (use an ETT of the same length to determine the length of the ETT in the patient; back it out so the tip is between the larynx and thoracic inlet). Bronchial intubation can result in low pulse-oximeter readings (see Chapter 7 [Anesthesia Monitoring]), high EtCO$_2$, and difficultly keeping the patient anesthetized.
- **Laryngospasm** is the reflexive closure of the laryngeal cartilage usually caused by stimulation of the larynx and an inadequate anesthetic depth. Laryngospasm is common in cats. Desensitizing the laryngeal cartilage by placing 2% lidocaine onto the area can help prevent/ treat laryngospasm. Deepening the level of anesthesia may suppress the reflex. Do not continue to push the ETT toward laryngeal cartilage without first desensitizing it or making the level of anesthesia deeper as these will only make the spasms worsen. Laryngospasm can result in closure of laryngeal cartilage, making it difficult if not impossible to place the endotracheal tube.
- *Forceful intubation* can result in damage to the delicate tissues at the back of the throat and lead to edema and potential airway obstruction. Endotracheal intubation should be gentle and easy. Although the largest sized ETT should be placed, do not forcefully attempt to place an ETT that is larger than necessary. Repeated attempts at intubation should be avoided as should touching the epiglottis with the laryngoscope blade. Forced intubation can result in reddened or swollen laryngeal tissues, difficulty breathing upon extubation, and/or complete airway obstruction. If a patient is having difficulty breathing post-extubation, it may be necessary to re-anesthetize him/her and place another ETT while determining the cause of respiratory difficulty. Provide the patient with supplemental oxygen and try to keep them calm, which may include providing sedation/tranquilization; be ready to intubate with a much smaller ETT. Glucocorticoids may be indicated to reduce swelling.
- *Brachycephalic syndrome* is seen in patients with elongated soft palates, everted laryngeal saccules, and redundant pharyngeal tissues, all of which make visualization of the laryngeal opening difficult. They also often have hypoplastic tracheas and require ETTs that are much smaller than expected based on body weight. Special considerations exist when intubating brachycephalic patients including preoxygenation of the patient (if tolerated), using a laryngoscope with a good light source and an appropriately long blade, and having a variety of pretested ETTs available including one rather small tube to use in an emergency. Brachycephalic syndrome may result in difficulty visualizing the laryngeal opening and difficulty intubating the patient. Having additional tools such as tongue depressors to elevate or hold back redundant tissue can be useful.

Mask and Chamber Induction

The time it takes to induce general anesthesia via mask or chamber induction is much longer in comparison to an IV anesthetic agent induction. During induction, the airway is unsecured for a prolonged period and the risk for airway compromise or obstruction is high. Inhalant inductions are costly as high concentrations of anesthetic agent and large volumes of anesthetic gases are required to achieve general anesthesia. These higher doses produce more cardiovascular and respiratory depression than seen with comparable doses of IV induction agents. Mask and chamber inductions are contraindicated in some brachycephalic breeds such as English and French Bulldogs, Pugs, and Boston Terriers that have stenotic nares, an elongated soft palate, hypoplastic trachea, and everted laryngeal saccules or any patient with respiratory compromise or for whom securing an airway rapidly is important.

Mask Induction

The procedure for mask induction is as follows:

- Administer preanesthetic sedative and analgesic. Ensure that the patient is deeply sedated.
- Once the patient is sedated, place a well-fitted clear mask over their face and administer 100% oxygen for 3 to 5 minutes (if patient tolerates the mask) (see Figure 9-1). A clear mask allows visualization of the animal's muzzle and detection of vomiting. To avoid ocular injuries, place ophthalmic ointment in the patient's eyes and do not let the mask cover the eyes.
- Start a low concentration of inhalant anesthetic agent (isoflurane 0.5%, sevoflurane 1%) at the beginning of induction and gradually increase it as the induction progresses (increase isoflurane by 0.5% every 30–60 seconds until 2% is reached; increase sevoflurane 1% every 30–60 seconds until 3% is reached).
- To complete induction, increase the isoflurane to 3.5 to 5% or the sevoflurane to 5 to 7% until the patient can be intubated. Sevoflurane has a more rapid induction and less obnoxious smell than isoflurane; therefore, it is the inhalant anesthetic agent of choice for mask induction.
- Once intubated, adjust the inhalant anesthetic concentration as needed, then monitor and treat the patient like any other. This method decreases struggling and helps to avoid relative overdosing of the inhalant anesthetic agent and subsequent hypotension, and cardiac and respiratory depression. It also minimizes personnel exposure to WAG.

Chamber Induction

The procedure for chamber induction is as follows:

- Chambers should be clear to allow patient monitoring during induction and large enough so the patient can lay in lateral recumbency without having to flex their neck, which can occlude the airway (Figure 9-16a). Excessive chamber volume slows the rate of induction and allows for more patient movement and potential for injury. An appropriate size chamber is just slightly larger than the patient.
- Oxygen from a non-rebreathing circuit should be piped into the chamber for 5 minutes while the patient acclimates to the chamber (Figure 9-16b).
- After acclimation, the concentration of inhalant anesthetic agent should be increased by 0.5% increments for every 10 seconds until 4 to 4.5% isoflurane or 5 to 5.5% sevoflurane is being administered.
- When the patient has lost its righting reflex, it can then be removed from the chamber and the induction completed by mask. Keep in mind that inhalant anesthetic agents wear off quickly, so veterinary technicians need to move quickly once the patient is removed from the chamber.
- Once the patient is removed from the chamber, the chamber should be moved from the populated work area and, if not hooked up to an active scavenging system, it needs to be placed somewhere it can scavenge (such as outside).

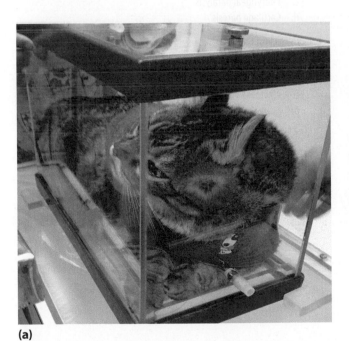

(a)

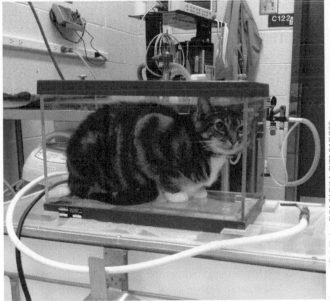

(b)

Courtesy of Chris Norkus, DVM, DACVAA, CVPP, DACVECC

FIGURE 9-16 Chamber induction. (a) Cat in an appropriate size chamber that is just slightly larger than the patient. (b) Oxygen from a non-breathing circuit is piped into the chamber and will eventually be replaced with inhalant anesthetic agent.

SUMMARY

The induction period is a busy time and requires the full attention of the anesthetist as numerous tasks need to occur while constantly monitoring the patient's status. The key to successful induction is to be prepared. Setting up supplies in advance, anticipating logistics of what may happen during the anesthetic event (such as possible extubation of a patient during the procedure or moving the patient to a special room and having to reconnect the patient to a new anesthesia machine so that a CT scan can be performed), and developing a plan on how to deal with a situation that may go wrong are all important components of an effective induction plan. Hoping for the best, yet preparing for the worst, is the motto of a competent anesthetist.

CRITICAL THINKING POINTS

- Mask and chamber inductions are stressful for patients, potentially unsafe for both patients and personnel, and rarely the first choice for induction, but they may be considered in select cases.

- All patients undergoing general anesthesia should be intubated to maintain a patent airway and deliver supplemental oxygen or inhalant anesthetics.

- Induction of anesthesia is best done with fast-acting IV medications that can be titrated to effect to provide rapid unconsciousness and airway control.

- Anesthesia induction agents have narrow therapeutic indexes (margins of safety); therefore, careful monitoring is important to avoid overdosing the patient.

- Patients given inhalant anesthetic agents can change anesthetic planes quickly, so they should be monitored vigilantly.

REVIEW QUESTIONS

Multiple Choice

1. Which statement is false?
   a. Laryngoscopes are typically used when intubating dogs and cats.
   b. Horses are typically intubated blindly.
   c. Cattle are typically intubated by digital palpation.
   d. Small ruminants and swine are typically intubated blindly.

2. Which drug provides analgesia?
   a. Alfaxalone
   b. Isoflurane
   c. Sevoflurane
   d. Ketamine

3. What is *not* considered a risk of inhalant inductions?
   a. Inhalant induction delays airway control.
   b. Inhalant induction increases risk of drug overdose.
   c. Inhalant induction increases exposure to waste anesthetic gas.
   d. Inhalant inductions are safer than injectable drug inductions.

4. Why should all patients undergoing general anesthesia be intubated?
   a. To limit hepatic metabolism of a drug by causing the patient to exhale waste products
   b. To trigger the gag reflex to prevent the patient from vomiting
   c. To maintain a patent airway and deliver supplemental oxygen or inhalant anesthetic agents
   d. To allow for smooth induction and recovery from general anesthesia

5. What term is used to describe the reflexive closure of the laryngeal cartilage from inadequate anesthetic depth and stimulation?
   a. Bronchospasm
   b. Laryngospasm
   c. Pharyngeal paralysis
   d. Cartilage paresis

6. Which signs indicate that an animal is ready to be intubated?
   a. Heavy head, minimal leg twitching, blinking
   b. Slack jaw tone, gag reflex, minimal leg twitching
   c. Absence of gag reflex, blinking, jaw tone
   d. Heavy head, slack jaw tone, absences of gag reflex

7. Inhalant inductions by mask or chamber induction cause which of the following?
   a. Increase patient stress, produce rapid induction, and expose staff to waste anesthetic gas
   b. Predispose patients to cardiac arrhythmias, expose staff to waste anesthetic gas, and increase patient stress
   c. Expose staff to waste anesthetic gas, produce rapid induction, and decrease patient risk of cardiac arrhythmias
   d. Produce prolonged induction, decrease patient stress, and reduce staff exposure to waste anesthetic gas

8. What are some signs that your patient may be bronchially intubated?

    a. Not staying anesthetized, low pulse oximeter reading, elevated $EtCO_2$

    b. Low pulse oximeter reading, patient vocalization, lack of a palpable ETT in the trachea

    c. Elevated $EtCO_2$, patient producing sound when breathing, patient staying anesthetized

    d. Absence of $EtCO_2$, patient producing sound when breathing, low pulse oximeter reading

9. During endotracheal intubation, press on the _____ to flip down the _____ to visualize the _____ .

    a. epiglottis, larynx, esophagus

    b. uvula, epiglottis, glossal

    c. base of the tongue, epiglottis, glottis

    d. glottis, epiglottis, base of the tongue

10. What can ETT cuff overinflation lead to?

    a. A safely sealed system and minimal staff exposure to WAGs

    b. Tracheal ischemia and necrosis

    c. Laryngospasm

    d. Bronchial intubation

11. What is the anesthetist's goal for selecting drugs for a patient's anesthetic induction?

    a. Choosing a drug that follows the clinic's standard anesthesia policy

    b. Choosing a drug that is rapidly eliminated from the animal's body

    c. Choosing a drug that has a long shelf life to lower overhead costs

    d. Choosing a drug that best fits the individual patient

12. Which induction agent contains a propylene glycol carrier that may cause pain on IV injection?

    a. Alfaxalone

    b. Acepromazine

    c. Etomidate

    d. Propofol

13. Which induction agent may produce adrenal suppression that inhibits glucocorticoid production?

    a. Propofol

    b. Ketamine

    c. Alfaxalone

    d. Etomidate

14. Which of the following indicates that an animal is not properly intubated?

    a. Fogging of the ETT with condensation during exhalation

    b. Vocalization from the patient

    c. Palpating only one firm tube in the neck after intubation

    d. Opening and closing of unidirectional valves when the patient inhales and exhales

15. Which IV drug is prone to bacterial growth if strict aseptic technique is not used?

    a. Propofol

    b. Ketamine

    c. Tiletamine

    d. Etomidate

16. Which induction agent may produce an apneustic breathing pattern?

    a. Ketamine

    b. Etomidate

    c. Alfaxalone

    d. Isoflurane

17. When determining the ETT length for proper placement, the distal end should measure to the _____ and the proximal end should measure to the _____ .

    a. larynx, canine tooth

    b. pharynx, incisor

    c. thoracic Inlet, incisor

    d. glottis, canine tooth

18. Which induction agent can cause retching and myoclonus if underdosed?

    a. Tiletamine

    b. Etomidate

    c. Alfaxalone

    d. Ketamine

19. Which induction agent produces minimal fetal affects, making it a suitable choice for C-sections?

    a. Tiletamine

    b. Ketamine

    c. Propofol

    d. Etomidate

20. What is the name given to the dissociated state of consciousness where the patient is agitated and inconsolable, uncooperative and typically thrashing, vocalizing, and defecating?

    a. Tetany

    b. Spasticity

    c. Emergence delirium

    d. Myoclonus

Case Studies

Case Study 1: During the intubation of a large dog, you have a hard time controlling the distal end of the 11 mm tube. After inserting the ETT and connecting the dog to oxygen, you close the pop-off valve and squeeze the bag to find there is a huge leak at the endotracheal tube cuff. You inflate the cuff until you appreciate some back pressure, close the pop-off valve, and try again. When you squeeze the bag there is still a giant leak and you hear gases escaping the airway. You are shocked at how much air you have already put into this ETT and still cannot get a seal.

1. Why might you not be able to get a good seal of the ETT?
2. What should be done to correct this?

Case Study 2: Barbie is a 6-month-old 3 kg Domestic Shorthair (DSH) cat in for an ovariohysterectomy. She is sedated and an IV catheter is placed into her cephalic vein. She is induced with propofol. When attempting to intubate her, you cannot find the opening to her trachea.

1. What might be causing the problem?
2. What should you do?

Case Study 3: It is Monday morning and you are gathering the supplies necessary for four surgeries that are planned for the day. Your patients include a 4-month-old mixed bred puppy weighing 7 kg for a spay, a 3-year-old 4 kg domestic short haired (DSH) for a neuter, a 5-year-old 14 lb overweight mini Dachshund for a dental, and a 2-year-old English Bulldog weighing 27 kg who needs to be anesthetized to explore and flush out a ruptured anal gland.

1. What type of ETTs do you select for each patient?

Case Study 4: Percy is a 3-year-old, 32 kg, M/N, Gordon Setter canine presenting to the clinic for an ear flush of the right ear (AD). He is premedicated and then induced with propofol IV and intubated. After induction and endotracheal intubation, you connect Percy to the anesthesia machine.

1. What gas/gases should be flowing at this point?
2. After inflating the cuff you notice that Percy is apneic. What does this mean and what might be the cause? What should you do to support Percy?

Case Study 5: Bernie, an 8-year-old M/N Chihuahua dog, presents for a dental cleaning. You calculate a dose of ketamine/diazepam for the patient. After giving ½ of the calculated dose IV, you check to see if Bernie is ready to be intubated only to find that he still has quite a bit of jaw tone.

1. Why does Bernie have jaw tone?
2. Should you give more induction drug?

Critical Thinking Questions

1. As a new hire at the veterinary clinic, the anesthesia team wants to make sure that you understand the intubation process prior to anesthetizing a patient. The lead veterinary technician asks what criteria you use to determine the best sized ETT to place in patients. She also wants you to explain why the endotracheal tube cuff just cannot be inflated with an unknown amount of air. How do you answer these questions?
2. As a veterinary technician, you are responsible for cleaning the equipment used for anesthesia. Describe how you would clean the ETT and laryngoscope for a healthy patient that had been neutered versus equipment that was used on a patient with a respiratory infection.

ENDNOTES

1. Lemke, K., Lin, H. C., Steffey, E. P., & Cullen, L. K. (1999). Pharmacology. In J. C. Thurman, W. J. Tranquilli, & G. J. Benson (Eds.), *Essentials of small animal anesthesia and analgesia* (p. 128). Baltimore, MD: Lippincott Williams and Wilkins.
2. Ko, J. C. H., & Galloway, D. S. (2002). Anesthesia of geriatric patients. In S. A. Greene (Ed.), *Veterinary anesthesia and pain management secrets* (pp. 215–216). Philadelphia, PA: Hanley & Belfus.
3. Bednarski, R. M. (2011). Anesthesia management of dogs and cats. In Grimm, et al. (Eds.), *Essentials of small animal anesthesia and analgesia* (2nd ed., p. 279). Ames, IA: Wiley-Blackwell.
4. Keegan, R. D. (2002). Inhalant anesthetics. In S. A. Greene (Ed.), *Veterinary anesthesia and pain management secrets* (p. 98). Philadelphia, PA: Hanley & Belfus.
5. Benson, G. J. (2002). Intravenous anesthetics. In S. A. Greene (Ed.), *Veterinary anesthesia and pain management secrets* (pp. 92–94). Philadelphia, PA: Hanley & Belfus.

6. Paddleford, R. R. (1999). Anesthetic agents. In R. R. Paddleford (Ed.), *Manual of small animal anesthesia* (pp. 38–66). Philadelphia, PA: W.B. Saunders Company.

7. Andress, J. L., Day, T. K., & Day, D. (1995). The effects of consecutive day propofol anesthesia on feline red blood cells. *Vet Surg., 24*(3), 277–282.

8. Dodam, J. R., Kruse-Elliott, K. T., Aucoin, D. P., & Swanson, C. R. (1990). Duration of etomidate-induced adrenocortical suppression during surgery in dogs. *Am J Vet Res., 51*(5), 786–788.

9. Muir, W., Lerche, P., Wiese, A., et al. (2008). Cardiopulmonary and anesthetic effects of clinical and supraclinical doses of alfaxalone in dog. *Vet Anaesth Analg., 35*(6):451–462.

10. Hung, W-C, Ko, J., Weil, A. B., & Weng, H-Y. (2020). Evaluation of endotracheal tube cuff pressure and the use of three cuff inflation syringe devices in dogs. *Frontiers in Veterinary Science, 7*(39). doi:10.3389/fvets.2020.00039. PMID: 32118062; PMCID: PMC7015870.

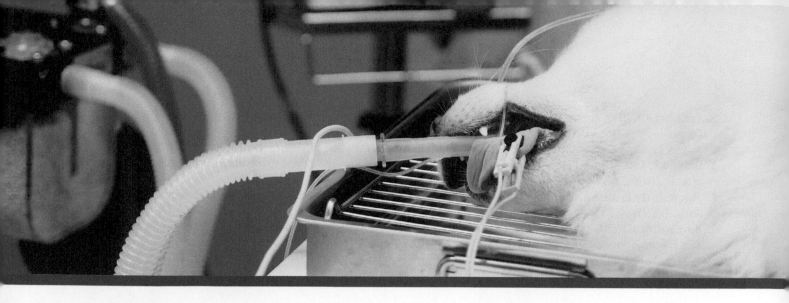

CHAPTER 10

The Maintenance Period

Anita Parkin, *RVN, AVN Dip (Surgery & ECC), VTS (Anesthesia/Analgesia), CVPP, TAE*

LEARNING OBJECTIVES

Upon completion of this chapter, it is expected that the reader should be able to:

10.1 Identify the different stages of anesthesia and planes of anesthetic depth in veterinary patients

10.2 Describe the characteristics of a patient that is in a surgical plane of anesthesia

10.3 Describe considerations for choosing various maintenance anesthetic agents for veterinary patients

10.4 Explain how the minimum alveolar concentration (MAC) is used to compare inhalant anesthetic agents

10.5 Identify factors that can affect the MAC of inhalant anesthetic agents

10.6 Explain how inhalant anesthetic agents are used to maintain the proper depth of anesthesia

10.7 Describe methods of maintaining anesthesia using injectable anesthetic agents

10.8 List expectations for filling out an anesthetic record correctly and accurately

10.9 Identify transporting techniques for anesthetized patients that will minimize staff injury

10.10 Describe ways to safely transition the patient from maintenance to recovery

KEY TERMS

Bagging

Minimum alveolar concentration (MAC)

Overpressurizing the vaporizer

Quick assessment tests (QATs)

Stage I of anesthesia

Stage II of anesthesia

Stage III of anesthesia

Stage IV of anesthesia

INTRODUCTION

Maintaining adequate analgesia, unconsciousness, and muscle relaxation are the goals of anesthetic maintenance. These goals must be met while maintaining cardiovascular and pulmonary stability in the patient and minimizing generalized morbidity. Maintenance of general anesthesia is usually accomplished using inhalant anesthetic agents following endotracheal intubation and connection to an anesthesia machine, but may also be maintained with repeated IV boluses, continuous rate infusion (CRI), or IM injection. No matter which administration route is used to maintain anesthesia, adjusting the animal's anesthetic depth to keep the patient at an appropriate plane of anesthesia is an art that involves excellent observational skills and successful interpretation of a variety of parameters. Successful maintenance of anesthesia is a team effort that also includes properly filling out an anesthetic record (see Chapter 6 [Creating an Anesthetic and Analgesic Plan]), monitoring vital signs and physiological parameters in order to intervene to maintain homeostasis in the patient, transporting animals safely into the surgery or procedure area, and monitoring patients as they move from an anesthetized to an awake state in the recovery area.

STAGES AND PLANES OF ANESTHESIA

Following premedication and induction, the maintenance period of anesthesia begins. Maintaining an animal at the proper anesthetic depth requires knowledge of the different stages and planes of anesthesia (Table 10-1 and Figures 10-1 and 10-2a through 10-2c). There are four stages of anesthesia with the third stage subdivided into three planes.

TABLE 10-1 Stages of Anesthesia and Expected Responses

	Stages of Anesthesia					
Parameter	**I** **Induction** **Stage**	**II** **Excitement** **Stage**	**III** **Light Surgical** **(Plane 1)**	**III** **Medium Surgical** **(Plane 2)**	**III** **Deep Surgical** **(Plane 3)**	**IV** **Anesthetic** **Overdose Stage**
Depth and behavior	Conscious but disorientated may struggle	Losing consciousness, may struggle, vocalize, paddle, chew, or yawn	Unconscious Lightly anesthetized	Unconscious Moderately anesthetized	Unconscious Deeply anesthetized	Unconscious Very deeply anesthetized and life threatening
Reflexes*	All present	Present and may be exaggerated	Present but may be diminished, swallowing poor or absent	Diminished pedal and palpebral reflexes Corneal reflex present	Absent	Absent
Muscle tone	Present	Present	Present	Relaxed	Relaxed	Relaxed
Pupil size	Normal	Normal to dilated	Normal	Medium to dilated	Dilated	Widely dilated
Pupillary light reflex	Normal response	Normal response	Normal response; may be slow	Sluggish response	Absent response	Absent response
Eyeball position**	Central	Central, nystagmus likely	Central to rotated ventromedially	Rotated ventromedially (dogs & cats) ventrally (ruminants)	Central	Central
Nystagmus (Horses; dogs or cats do not get with anesthesia)	Absent	Likely	Possible Fast	Slow to absent	Absent	Absent
HR	Normal	Normal to increased	Strong pulse with normal HR	Reduction in HR	Pulse weakening with decreased HR, prolonged CRT	Slow to absent HR, long CRT, pale MM
RR	Normal /breath holding /may be panting	Irregular	Normal or increased	Normal to decreased, shallow	Low, shallow	Jerky/Apnea
Response to surgery	Struggle and pull away	Struggle and pull away	May respond to stimulation by flicking skin or pulling limb back	May respond to stimulation with an increase in HR and/or RR	No response to surgery	No response to surgery

* Full list of reflexes can be found in Table 7-1.
**Eyeball position is not a reliable indicator of anesthetic depth in horses.

Eyeball center with
medium pupil
Light plane of
anesthesia

Eyeball ventromedial
with medium pupil
Surgical plane of
anesthesia

Eyeball center with dilated pupil
Deep plane of anesthesia
Opioids can cause mydriasis
in cats
Anticholinergic drugs can cause
mydriasis in cats and dogs

Eyeball center with
constricted pupil
Deep plane or
CNS disease
IV opioid administration
(dogs)

FIGURE 10-1 Eyeball position and anesthesia depth. In combination with other factors, the position of the eyeball and size of pupils may provide information regarding anesthetic depth. There is considerable variation among animal species (it is reliable in dogs, cats, and ruminants while unreliable in horses). In dogs and cats, the eyeballs rotate ventromedially as anesthesia deepens toward the surgical planes and becomes central if the animal is getting too deep. In ruminants, the eyeballs rotate ventrally as anesthesia deepens toward the surgical planes and becomes central if the animal is too deep. The eyeball position does not rotate when ketamine is administered by itself; therefore, this parameter is not reliable in assessing anesthetic depth in patients receiving solely ketamine.

- **Stage I:** *The Induction Stage.* The patient is gradually losing consciousness and properly premedicated patients given IV induction agents transition rapidly through Stage I. Most modern day induction agents induce anesthesia so rapidly that Stage I is not frequently observed and if observed is most commonly seen in poorly sedated patients. Breath holding and increased heart rate may be seen in Stage I. The Induction Stage is also referred to as the Stage of Voluntary Movement and is the most variable stage.
- **Stage II:** *The Excitement Stage.* In properly premedicated and induced patients, Stage II is brief and most patients do not visibly experience signs of excitement. Some patients may briefly exhibit involuntary reactions like vocalizing and paddling. An increase in heart and

respiratory rate, pupil dilation, and normal muscle tone and reflexes are all present during Stage II. Most modern day induction agents induce so rapidly that Stage II is not frequently observed and if observed is most commonly seen in poorly premedicated patients.
- **Stage III:** *The Surgical Anesthesia (Operative) Stage.* The patient is unconscious and has progressive muscle relaxation and loss of reflexes as the animal transitions from a light to a deep plane of anesthesia. Stage III is divided into three planes: light or plane 1, medium or plane 2, and deep or plane 3. The ideal surgical plane for each patient will depend on the type of procedure being performed, but in general, a medium plane of Stage III anesthesia is required for most invasive procedures and surgeries.

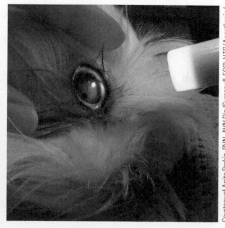

Courtesy of Anita Parkin, RVN, AVN Dip (Surgery & ECC), VTS (Anesthesia/Analgesia), CVPP, TAE

FIGURE 10-2 Eye position, jaw tone, and reflex signs such as palpebral reflex, pupillary light reflex, and corneal reflex are used to assess anesthetic depth. (a) Assessing jaw tone involves using the index finger and thumb to determine how much resistance there is when opening the mouth and releasing it. Jaw tone assesses the amount of skeletal muscle relaxation, which increases with depth of anesthesia. (b) Assessing the palpebral (blink) reflex involves gently tapping or tickling the medial or lateral canthus of the eye to see if the patient blinks or not. Absence of this blink reflex means the patient is in a deeper plane of anesthesia. When assessing the palpebral reflex, eye position can also be determined. In this image, the eyeball is rotated ventrally indicating a medium plane of anesthesia. (c) Assessing the pupillary light reflex (PLR) involves shining a light into the eye toward the retina. A normal PLR is pupil constriction with light and slight dilation with removal of the light. As anesthetic depth increases, the PLR diminishes and is eventually lost during deep anesthesia.

- **Stage IV:** *The Anesthetic Overdose (Danger) Stage.* The patient's body is extremely depressed, reflexes are absent, muscle tone is flaccid, pupils are dilated, and the patient experiences marked cardiopulmonary depression. This is a critical stage and if steps are not taken to lower the patient's anesthesia level, the heart will stop and cardiac arrest ensues.

DRUG SELECTION

Maintenance anesthetic agents are chosen for individual patients based on the "patient problem list" similar to that used for selection of premedication and induction agents and based upon what is the best choice for a given patient. The patient's health status, types of agents available, and desired route of administration all influence the choice of anesthetic agent used during the maintenance stage; however, there are limited options currently available for animals (partly due to cost and equipment availability in veterinary medicine).

Key considerations the anesthetist, working in conjunction with the attending veterinarian, should contemplate when choosing which maintenance drugs to use include:

- *What options are available for anesthesia maintenance?* To deliver inhalant anesthetic agents, the practice must have an anesthetic machine with a vaporizer to change the drug from a liquid state to the usable vapor state. The volatility of the inhalant anesthetic agent determines the vaporizer design for its use. Since most clinical practices only have one type of vaporizer (one that delivers either isoflurane or sevoflurane), the choice of which inhalant anesthetic agent to administer is predetermined by the practice. Other inhalant anesthetic agent options such as halothane and desflurane exist for human and veterinary anesthesia, however, due to cost, availability, and equipment limitations these agents are not currently used for veterinary patients in the United States. For example, halothane is no longer recommended for use in horses due to its effect on cardiac output and is currently not available in the United States. The use of injectable agents for anesthetic maintenance is less common than inhalant anesthetic agents. The most common injectable drug options are propofol, alfaxalone, or ketamine, often combined with other drugs such as muscle relaxants, sedative/tranquilizers, or analgesics. For example, propofol can be given continuously IV to maintain general anesthesia, while a combination of ketamine and diazepam IV may provide a short duration of anesthesia to facilitate a bandage placement in a fractious cat.
- *What is the anticipated duration of the procedure and what is the expected degree of analgesia/muscle relaxation needed?* The patient needs to be maintained at the proper plane of anesthesia for as long as needed to complete the procedure or surgery. For example, if the procedure is short such as lancing an abscess in a cat, premedication with an opioid and induction with ketamine/diazepam might be sufficient and allow an adequate duration of anesthesia. Conversely, if the procedure is long such as a dog undergoing a tibial plateau leveling osteotomy (TPLO) to repair a torn cruciate ligament, premedication and induction agents do not provide a long enough duration of anesthesia to complete the procedure nor do they dictate which anesthetic maintenance agents to use. In addition to needing a longer duration of anesthesia, a patient undergoing a TPLO needs more muscle relaxation than most IV anesthetic agents provide; therefore, an inhalant anesthetic agent would be used to provide an adequate duration of anesthesia and as a general rule provide better muscle relaxation than do IV agents.
- *What specific procedure is being performed?* In addition to the length of the procedure, it is also important to consider the type of procedure being performed. For example, anesthesia machines containing materials that are not magnetic resonance imaging (MRI) compatible cannot be brought into an MRI suite; therefore, use of total intravenous anesthesia (TIVA) may be more suitable for these cases.
- *What is the species of the patient?* Maintaining a large animal species such as a horse on inhalant anesthetic agents may not be an option for all practices based on availability of anesthetic equipment or the clinic's distance from a referral institution. Some animal species may respond differently than another species to a particular drug classes and individual drugs. For example, the development of malignant hyperthermia secondary to halogenated inhalant anesthetic agent use is more of an occurrence in pigs than some other species; therefore, using an injectable anesthetic agent might be a better choice when anesthetizing genetically predisposed pigs.
- *What pre-existing conditions exist and how does this influence the patient's overall health status and anesthetic management of the case?* An animal's overall health status and organ function will affect the rate of drug distribution, metabolism, and excretion. The anesthetist should consider an animal's concurrent health issues and how these conditions influence management and drug selection for the maintenance period. Modern inhalant anesthetic agents such as isoflurane and sevoflurane are minimally metabolized by the liver making them attractive options for patients with hepatic disease. Sevoflurane is less irritating to mucous membranes and has a lower blood-gas solubility than isoflurane, making it the preferred agent for mask and chamber inductions. Sevoflurane's lower blood-gas solubility allows for faster induction, depth changes, and potentially faster recovery also making it a good choice for patients undergoing Cesarean (C-) sections. Patients with hemodynamic compromise may be less tolerant of the effects of inhalant anesthetic agents such as vasodilation and negative inotropy (weakened cardiac muscle contractions) especially when given at high dosages; as a result, other drugs may need to be given concurrently to lower the inhalant anesthetic agent requirements or eliminate its use entirely.
- *Is pain relief or muscle relaxation desired in the patient?* Most maintenance agents do not produce analgesia (ketamine is the main exception). Inhalant anesthetic agents like isoflurane and sevoflurane do not provide pain relief; therefore, another drug(s) should be added to the anesthetic plan to provide analgesia for patients receiving these drugs.

Inhalant Anesthetic Agents

Inhalant anesthetic agents are commonly used in veterinary practice to maintain anesthesia. The benefits of using inhalant anesthetic agents include:

- They are directly inhaled into the respiratory system and are primarily eliminated by the lungs. This is beneficial because they do not undergo significant liver metabolism, making them well suited for patients with liver disease.
- Their uptake and elimination is rapid; therefore, anesthetic depth can be changed rapidly (thus requiring constant and vigilant monitoring).
- They are delivered via an endotracheal tube (ETT), face mask, or laryngeal airway mask; therefore, the anesthetist may be able to take over breathing for the animal if needed.
- They are administered in oxygen; therefore, the patient always receives concurrent O_2.

Despite the benefits of using an inhalant anesthetic agent, some disadvantages to their use include:

- Patients given inhalant anesthetic agents need to be monitored for alterations in breathing patterns, which can affect the anesthetic concentration they receive. If the patient is not breathing (apnea) or breathing rapidly and shallowly (polypnea), gases (oxygen and inhalant anesthetic agents) do not get to the alveoli and therefore cannot participate in gas exchange. Essentially, this "unused" inhalant anesthetic agent flows past the patient and out to the scavenging system. Assisting ventilation will help facilitate gas exchange (both oxygen and inhalant anesthetic agent).
- Using inhalant anesthetic agents results in concerns regarding waste anesthetic gases (WAGs) (Appendix A). The amount of WAG exposure is dependent on the proximity of the person to the WAG (veterinary technicians that recover patients from anesthesia may be close to the end of the ETT as the patient is disconnected from the anesthesia machine) and how WAGs are removed from the working environment. Any amount of inhalant anesthetic agent that leaks from the patient's anesthetic breathing circuit or is exhaled by patients recovering from anesthesia is considered a WAG.

Vapors of halogenated agents such as isoflurane, sevoflurane, and desflurane may produce a variety of health concerns in personnel including nausea, dizziness, headaches, fatigue, difficulties with judgment and coordination, sterility, miscarriages, birth defects, cancer, and liver and kidney disease. Exposure to WAGs can occur under the following conditions:

- When leaks occur in the anesthetic breathing circuit (inhalant anesthetic agent may leak if the connectors, tubing, and valves are not maintained and tightly connected).
  - Leak testing the anesthetic machine is one of the most influential ways that WAGs can be controlled. Leak tests (also known as pressure checks) should be performed before every anesthetic event (see Chapter 2 [The Anesthesia Machine]).
  - Since reservoir bags and Y-pieces are cleaned after each procedure, leak tests need to be done for each patient. New reservoir bags or Y-pieces could potentially be faulty; therefore, a new leak test could detect any concerns.
- When inhalant anesthetic agents escape during hookup and disconnection of the system. Once the system is connected, avoid disconnecting parts of the system to limit WAG exposure.
- When inhalant anesthetic agent escapes around a face mask. Due to the risk of WAG exposure, using a facemask is not ideal and is best avoided unless better options do not exist.
- When animals are recovering from anesthesia and have an ETT in place but are no longer connected to an anesthesia machine. With each breath the patient is still eliminating inhalant anesthetic gas and breathing it out into the room.
- When scavenging systems are not properly functioning (they need to be fully functional and tested regularly).

Safety Alert

Occupational Health and Safety Administration (OSHA) guidelines requires that exposure limit for all halogenated agents is not more than two parts per million. If you can smell an inhalant anesthetic agent, then you are over the OSHA level.

The pharmacology of inhalant anesthetic agents is described in Chapter 5 (Anesthetic and Analgesic Pharmacology). The most commonly used inhalant anesthetic agents are isoflurane (IsoFlo®, Isosol®, Aerrane®) and sevoflurane (SevoFlo®, Petrem®, Flurovess™, Ultane®) (desflurane [Suprane®] is rarely used outside of an academic setting mainly due to cost) (see Figures 5-12a through 5-12c). All modern day inhalant anesthetic agents are nonflammable and halogenated. Inhalant anesthetic agents depress the central nervous, respiratory, and cardiovascular systems. Inhalant anesthetic agents decrease systemic vascular resistance (SVR) and cause subsequent vasodilation; however, cardiac output and heart rate are generally unaltered when using these agents unless they are used at very high concentrations. All inhalant anesthetic agents increase cerebral blood flow, raise intracranial pressure, reduce renal blood flow, and reduce glomerular filtration rate. Inhalant anesthetic agents are primarily excreted through the lungs, and because there is little liver or kidney metabolism, they are good for patients with hepatic or renal disease. Emergence delirium can be seen with all inhalant anesthetic agents; therefore, using a tranquilizer or sedative prior to induction can smooth recovery. They do not provide analgesia and should never be used as the sole drug for painful procedures.

Isoflurane

Patients receiving isoflurane change planes of anesthesia rapidly and have very short recovery times. A benefit of isoflurane is that it does not generally promote arrhythmias; however, disadvantages are that it can cause respiratory and cardiac depression and can trigger malignant hyperthermia.[1]

Isoflurane Facts

Advantages:

- Rapid onset and recovery
- Potent anesthetic
- Excellent muscle relaxant

Disadvantages:

- Fetal depression (crosses placenta)
- Respiratory and cardiac depression
- Pungent odor
- Can trigger malignant hyperthermia
- No analgesia

Sevoflurane

Patients receiving sevoflurane also have rapid recoveries and ability to change anesthetic depth quickly, making diligent monitoring important in animals receiving sevoflurane. Sevoflurane is also rapidly eliminated from the patient due to its low tissue solubility; therefore, it is a good choice for patients undergoing C-sections because any sevoflurane absorbed by the fetus is quickly eliminated.[1] Sevoflurane requires higher vaporizer settings than isoflurane because it is less potent.

Sevoflurane Facts

Advantages:

- More rapid onset and recovery than isoflurane
- Potent anesthetic, but less potent than isoflurane
- Excellent muscle relaxant
- Pleasant odor (preferred inhalant anesthetic agent for mask and chamber inductions)
- Less negative respiratory effects (surfactant deterioration, lung injury, etc.) than isoflurane
- Slightly less effect on systemic vascular resistance than isoflurane

Disadvantages:

- Fetal depression (crosses placenta)
- Respiratory and cardiac depression
- Can trigger malignant hyperthermia
- No analgesia
- More expensive than isoflurane

Desflurane

Desflurane has a very low blood-gas solubility coefficient (desflurane = 0.42, sevoflurane = 0.69, and isoflurane = 1.41) that produces very rapid induction and recovery times (twice as fast as isoflurane).[1] Desflurane is the least potent inhalant anesthetic agent; therefore, higher vaporizer settings are needed when administering this agent. Desflurane has a very low boiling point causing it to exist as a vapor at room temperature; therefore, it cannot be used with typical anesthetic vaporizers. Special thermostat heated and pressurized vaporizers are needed to deliver desflurane and keep anesthetic

output stable, which require a power source and a cord to plug in. The cost of desflurane and its specialized vaporizer have limited its use in veterinary medicine; however, it is perhaps the most widely used inhalant anesthetic agent in human medicine in the United States at this time.

Desflurane Facts

Advantages:

- Extremely rapid recovery; much more rapid than isoflurane and sevoflurane
- Potent anesthetic, but least potent inhalant anesthetic agent
- Excellent muscle relaxant
- Good choice for C-section patients due to rapid elimination

Disadvantages:

- Fetal depression (crosses placenta)
- Respiratory and cardiac depression
- Can trigger malignant hyperthermia
- No analgesia
- Expensive
- Extremely pungent odor
- May cause tachycardia and airway irritability
- Needs special heated vaporizer because it has a low boiling point (can be in the vapor phase at room temperature)

Inhalant Anesthetic Agents and Minimum Alveolar Concentration

When using inhalant anesthetic agents, it is important to understand **minimum alveolar concentration** (MAC). MAC is the lowest concentration of drug in the lungs that is necessary to keep 50% of patients immobile during noxious stimuli. MAC is used to express potency of an anesthetic. MAC values are used clinically to compare the relative "potency" of inhalant anesthetic agents and serves as a starting concentration for an inhalant anesthetic agent (particularly when administering an agent for the first time). Remember, the higher the MAC value, the lower the potency. For example, in dogs the MAC value of isoflurane is approximately 1.3 to 1.5% while the MAC value of desflurane is 7.2 to 10.3%. An inhalant anesthetic agent that has a high MAC value (such as desflurane in dogs) needs to have more of it given to attain the desired minimum concentration in the alveoli than a highly potent anesthetic that has a lower MAC value (such as isoflurane in dogs).

Tech Tip 🐾

MAC values can be used to compare inhalant anesthetic agents and vary among species. MAC values are alveolar concentrations at equilibrium and are NOT vaporizer settings. A gas analyzer is needed to determine if the exhaled gas concentration from the patient is at MAC; therefore, the only way to know if the patient was at MAC is to have a gas analyzer set up with the monitoring system.

There are a variety of factors that can affect the MAC (potency) and must be taken into consideration when maintaining an anesthetized patient including:

1. Factors that may decrease the MAC and make it a more potent inhalant anesthetic agent.
   Patient factors: metabolic acidosis, severe hypoxia, anemia, blood loss, advancing age, pregnancy, and decreases in body temperature.
   Drug factors: concurrent use of opioids, sedatives, local anesthetics, acepromazine, alpha-2 adrenergic agonists, nonsteroidal anti-inflammatory drugs (NSAIDs), and maropitant.

2. Factors that may increase the MAC and make it a less potent inhalant anesthetic agent.
   Patient factors: fever,[2] increased metabolic rate, stress, and age (young animals have higher MAC requirements).
   Drug factors: concurrent use of ephedrine.

3. Factors that do *not* affect MAC:
   - The type of stimulation (or noxious stimulant)
   - Duration of anesthesia
   - Sex of the patient
   - Metabolic alkalosis
   - Potassium levels[2]

Table 5-3 shows the inhalant anesthetic agents and their MAC values for cats and dogs. These values are percents and are used to compare potencies of inhalant anesthetic agents. The MAC value can be used as a starting point for the vaporizer setting, but many patients will need more inhalant anesthetic agent than MAC for surgical procedures and some will need less, due to premedication drugs or decrease in body temperature.

Ways to Change Inspired Inhalant Anesthetic Agent Concentration

Recall from Chapter 5 (Anesthetic and Analgesic Pharmacology) that inhalant anesthetic agent uptake is influenced by the drug's gas solubility, the difference in alveolar to blood partial pressure (concentration) of the inhalant anesthetic agent, and the patient's cardiac output and alveolar ventilation status. There are ways to alter inspired inhalant anesthetic concentration depending on the type of breathing system used. In a circle rebreathing system, the concentration of inspired inhalant anesthetic agent can be influenced by adjusting both the vaporizer setting and oxygen flow rate (Figures 10-3a and 10-3b). In non-rebreathing systems such as the Bain, adjusting the vaporizer setting is the only way to influence the concentration of inspired inhalant anesthetic agent.

Tech Tip

Vaporizers convert liquid anesthetic agent to vapor and add a controlled amount of vapor to the carrier gas flowing through the machine. If the oxygen flow is turned off, no anesthetic agent is delivered to the patient.

Vaporizer Settings

When the vaporizer is set at a certain percent, the circuit of a circle rebreathing system will not reach that concentration of inhalant anesthetic agent for a period of time, which is largely dependent on the volume of the circuit and rate of new incoming gas. The percent that is on the vaporizer dial rarely

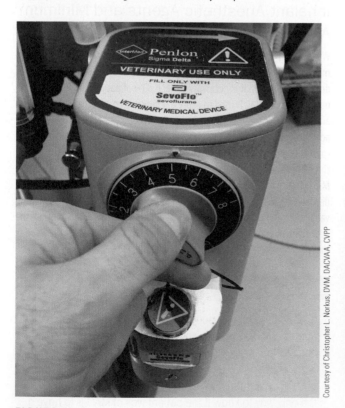

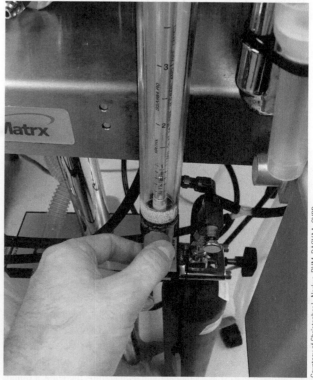

Courtesy of Christopher L. Norkus, DVM, DACVAA, CVPP

FIGURE 10-3 The concentration of inspired inhalant anesthetic agent in a circle rebreathing system can be influenced by (a) increasing the vaporizer setting and (b) increasing the oxygen flow rate.

reflects what the patient receives due to the dilution effect of the patient breathing out inhalant anesthetic agent that is at a lower concentration. One way to increase the anesthetic concentration in the circuit and therefore the patient is to **overpressurize the vaporizer**. Overpressurizing is done by setting the vaporizer at a higher than the desired concentration so the target concentration is met in a shorter period of time. After assessing the depth of anesthesia (using reflex signs), it can be determined if there is a need to titrate the inhalant anesthetic agent up or down (see Table 7-1 [Indicators of Anesthetic Depth]). Constant monitoring, reevaluation of anesthetic depth, and changing vaporizer settings to find an appropriate plane of anesthesia are essential during anesthesia maintenance. Overpressuring can be used in non-rebreathing systems but may not be needed because no dilution effect occurs (there is no rebreathing of inhalant anesthetic agent); therefore, depth changes and high inspired concentrations occur almost instantly.

Oxygen Flow Rate

After determining the vaporizer setting, it is important to realize that the oxygen flow rate can also change the amount of inhalant anesthetic agent delivered to the patient. In a circle rebreathing system, the oxygen flow rate determines how quickly the concentration that is set on the vaporizer is actually delivered to the patient. If the goal is to deliver 2% of an inhalant anesthetic agent to a dog, it will reach that concentration more rapidly if the oxygen flow rate is 4 L/min instead of 200 mL/min. Both oxygen flow rates will eventually deliver 2% to the patient; it will simply take much longer when using the lower oxygen flow rate.

When a patient is initially anesthetized using a rebreathing circuit (circle system), the oxygen flow rate will generally be higher (25–50 mL/kg/minute) to remove nitrogen from the system and increase the alveolar concentration of the inhalant anesthetic agent. When the patient reaches a desirable plane of anesthesia, the oxygen flow rate can then be changed to a maintenance rate required for the type of circuit being used. The oxygen flow rates for low-flow and semi-closed rebreathing circuits can be between 10 to 50 mL/kg/min (50 mL/kg/min for anesthesia induction and then the rate can be lowered to 10–20 mL/kg/min for low-flow and 20–40 mL/kg/min for semi-closed) because the CO_2 absorbent will remove carbon dioxide, thus leaving fresh gas to be rebreathed. When using these circuits at the end of the surgical procedure (after the vaporizer has been turned off), it is a good idea to turn the oxygen flow rate back up to high flow rates (50 mL/kg/min) to allow the inhalant anesthetic agent to be eliminated from the system and patient. This allows the patient to breathe pure oxygen before being transported to the recovery area.[3]

When using a non-rebreathing circuit (such as a Bain circuit), the oxygen flow rate needs to be higher to eliminate carbon dioxide produced from the patient so it is not rebreathed (as CO_2 absorbent is not used in these circuits). These rates are generally set at 200 to 300 mL/kg/minute depending on the type of non-rebreathing system used for the entire procedure.

The key point is that anesthetic depth changes can occur more rapidly not only by changing the vaporizer setting but also the oxygen flow rate. If an animal is under anesthesia using a low-flow technique, the low oxygen flow rate causes a significant lag period between the time the vaporizer setting is changed to when change in anesthetic depth is observed. To change anesthetic depth faster, increasing the oxygen flow rate substantially may be useful; therefore, high flow rates should be instituted transiently and only when a change of anesthetic depth is desired or immediately following induction and then is lowered 5 to 10 minutes later. When a patient is at an appropriate anesthetic depth, the oxygen flow rate can be lowered to conserve oxygen and body heat, and not waste money.

Injectable Maintenance Agents

In veterinary practice, injectable anesthetic agents are used far less frequently than inhalant anesthetic agents to maintain anesthesia. Occasionally, small doses of injectable anesthetic agents are given rapidly to deepen the anesthetic depth.

The most common routes for injectable anesthetic agents are IV bolus, constant rate infusion (CRI), and IM (intraperitoneal (IP) injections are used in rodents).

The properties of injectable maintenance agents are described in Chapter 5 (Anesthetic and Analgesic Pharmacology) and those used for small animals are emphasized in Chapter 9 (The Induction Period). Injectable maintenance agents used in large animals are further described in Chapter 19 (Anesthesia and Analgesia of Large Animal and Food Producing Species) and those used in pet birds and exotic small mammals in Chapter 20 (Anesthesia and Analgesia of Pet Birds and Exotic Small Mammals).

Examples of methods of administering injectable anesthetic agents for maintenance include:

- *IV bolus* administration of drug given as needed may be administered in volumes less than those given during induction to achieve an anesthetic plane. The duration of the induction dose may provide the amount of anesthesia necessary to complete a procedure of short duration. Propofol (Rapinovet®, PropoFlo™, PropoFlo™28, Propovan®), ketamine (Ketaset®, VetaKet®, Ketaject®, Vetalar®), etomidate (Amidate®), and alfaxalone (Alfaxan®, Alfaxan® Multidose, Alfaxan® Multidose IDX) can be given IV alone for procedures that can be completed in a short amount of time. For example, propofol may be administered IV for short procedures such as bronchoscopy and airway exams.
- *TIVA/CRI* infusion through a syringe placed in a syringe pump/driver is another way to perform injectable anesthesia maintenance. A calculated volume of injectable anesthetic agent is drawn into a syringe that can hold the anticipated total amount of drug needed to maintain surgical anesthesia for the desired duration of the procedure. The syringe pump/driver is then programmed with an infusion rate based on the calculated dose. Following induction, the syringe pump is started and the injectable anesthetic agent is delivered at a continuous rate. Propofol is the drug most commonly used for CRI as it does not accumulate in fat and is rapidly metabolized. Propofol TIVA is not usually done in cats for prolonged periods due to the concern that they may develop Heinz body anemia. Calculating CRI is covered in Chapter 17 (Analgesic Techniques). Propofol, alfaxalone, and ketamine mixtures (such as GKX [guaifenesin, ketamine, xylazine] in horses) are given IV to provide a short duration of anesthesia.
- *IM* injection can be used to induce and maintain anesthesia for short periods of time. Using a combination of injectable anesthetic agents intramuscularly can provide a surgical plane of anesthesia for approximately 30 minutes depending on the drugs administered. Telazol® and Tilzolan® (proprietary 50/50 mixtures of tiletamine and zolazepam), alfaxalone (Alfaxan®, Alfaxan® Multidose, Alfaxan® Multidose IDX), and ketamine combinations are typically the only anesthetic agents given IM. Ketamine combined in a syringe with an alpha-2 adrenergic agonist provides surgical anesthesia with good analgesia, and recovery can be hastened with the use of a reversal agent such as atipamezole (Antisedan®) or yohimbine (Yobine®, Antagonil®). Ketamine can also be mixed with a sedative/tranquilizer and opioid (e.g., ketamine/dexmedetomidine/

butorphanol) and administered IM to facilitate short-term procedures/ surgeries and provide analgesia in small animals. Although not ideal, giving injectable anesthetic agents IM is most commonly done for short procedures, fractious animals, zoo animals, wild life, spays and castrations in shelter animals (especially cats), or in field medicine where access to an anesthesia machine is not possible (Figure 10-4).

Tech Tip

Maintenance agents such as propofol may be administered via CRI for longer-duration procedures, a drug cocktail may be given IM for shorter procedures such as elective neuters on shelter animals, or a "triple drip" total intravenous anesthesia (TIVA) may be used in a cryptorchid horse for castration.

THE STEPS OF ANESTHESIA MAINTENANCE

Following anesthesia induction, patients enter the maintenance period in a light plane of anesthesia and must transition to a surgical plane of anesthesia. Assessing a patient's anesthetic depth consistently at regular intervals (at least every 5 minutes) is performed by looking at the whole patient including its behavior, reflexes, muscle tone, eye position, pupil size, and vital signs (Table 10-2; see Figures 10-1 and 10-2a through 10-2c, review Chapter 7 [Anesthesia Monitoring]). Looking at the entire patient to determine which level of anesthesia the majority of signs indicate allows smooth transitions from one phase of anesthesia to the next. Knowing which medications were administered or are currently being administered to the patient as well as what the surgeon is doing or about to do help gauge whether or not anesthetic depth may change.

Consider an anesthetized patient whose heart rate, respiratory rate, and blood pressure increase. First determine if the increase in physiologic parameters is significant or not. A mild increase in physiologic parameters (e.g., a 10% change) is acceptable while greater changes are not. If the increase is significant, ask yourself *"Are these changes due to the animal's response to a pain stimulus or is the patient at too light a plane of anesthesia?"* If reflex signs (palpebral reflex, pupillary light reflex, jaw tone, rotation of the eyes [species dependent], and anal tone) indicate that the patient is at a surgical plane of anesthesia, more than likely there is a physiologic response to pain and the patient requires some analgesics. However, if reflex signs are present (return of the palpebral reflex, pupillary light reflex, jaw and anal tone, and eye position) then the patient most likely requires more anesthetic agent.

Tech Tip

Because they may not blink and tear production decreases, animals under anesthesia need ophthalmic ointment applied approximately every 30 minutes to protect their eyes from corneal ulceration.

Courtesy of Christopher L. Norkus, DVM, DACVAA, CVPP

FIGURE 10-4 Injectable anesthetic agents are commonly used in zoo animals and wild life. This cheetah was given an IM injection of Telazol® and medetomidine remotely using a projectile syringe from a dart gun.

If an animal's depth of anesthesia is unknown, lowering the patient's depth of anesthesia can be accomplished by turning the vaporizer down/off. It is important to inform the veterinarian that the animal's anesthetic depth is unknown and have an injectable anesthetic agent ready to administer if needed. It is better for the patient to be in a light plane of anesthesia rather than one that is too deep, which could potentially be fatal. An animal should be monitored very closely until reflex signs indicate the stage of the patient's anesthetic depth.

Tech Tip

Since most anesthetized patients receive IV fluids and many animals will not urinate before they are anesthetized, manually expressing the urinary bladder while small animal patients are still under anesthesia may make them more comfortable during recovery.

The Anesthetic Record

The anesthetic record is a legal document that visually represents what happened during the anesthetic event and includes information about the patient's physiological parameters (heart and respiratory rate, temperature, etc.), anesthetic agents used, and any health concern of the patient (see Figure 6-5 in Chapter 6 [Creating an Anesthetic and Analgesic Plan]). Filling out an anesthetic record is important because it draws attention to any trends that may be occurring when monitoring patient parameters (such as blood pressure, CO_2 levels, or SpO_2 [peripheral oxygen saturation as

TABLE 10-2 Reflexes Monitored During Anesthesia

Reflex	Implementation	Assessment	Comments
Palpebral (blink)	Lightly tickle the medial or lateral canthus of the eye or brush the eyelashes	Most anesthetized animals retain reflex until medium plane of anesthesia (only consistently exists in the light plane of anesthesia, Stage III Plane 1)	Presence of brisk reflex indicates a light plane of anesthesia Absence of reflex does not necessarily mean the patient is too deep Sluggish reflexes are acceptable in some species (equine and ruminant patients, but not canines or felines)
Corneal	Stimulation of cornea using a sterile object or eye drops	Patient will blink or eyeball will retract	Disappears during deep anesthesia Most useful to determine large animal anesthetic depth (this reflex is not always reliable in small animal patients)
Pupillary light reflex	For depth of anesthesia assessment, determine pupil size after direction of focal light into the same eye (direct PLR).	Pupil constriction in response to light	Lost during deep anesthesia Should be present (although slow) in light to medium anesthetic planes
Lacrimation	Tear production	Indicates a light level of anesthesia in all species; in horses lacrimation is a very sensitive indicator that a patient is getting lighter	May be totally absent in deep anesthetic planes
Pedal (digital withdrawal)	Flexion or withdrawal of a limb in response to a painful stimuli (typically, squeezing and twisting of a toe) Leg is extended and toe pinched with fingers or hemostat	Present (the leg is withdrawn) during a light level of anesthesia Absent at an adequate plane of anesthesia	Good test to perform if the anesthetist is unsure of the patient's anesthetic depth
Swallowing reflex*	Occurs in response to stimulation of the pharynx (swallowing reflex)	Absent in all surgical anesthetic planes (all phases of Stage III) Swallowing returns during recovery	Used to determine the proper time to remove ETT
Laryngeal reflex*	Occurs in response to stimulation of the larynx usually observed during endotracheal intubation (laryngeal reflex)	Laryngeal reflex protects patient from tracheal aspiration and its absence allows passage of ETT	Induction agents should abolish this reflex to facilitate endotracheal intubation
Reflex movement	Purposeful movement in response to a surgical stimulus	Indicates light level of anesthesia (Stage III Plane 1) Does not mean the patient is perceiving pain	Not a true reflex Some drugs (such as propofol) may cause temporary involuntary myoclonus at medium planes of anesthesia and are generally not a cause for alarm
Eyeball position	Position of eyeball in the orbit	Ventromedial rotation typically indicates a medium plane of anesthesia (Stage III Plane 2) Small animals: may change from a central position to a ventromedial one and then back to central as anesthetic depth increases	Position of the eyeball within the orbit varies markedly depending on the stage of anesthesia, the anesthetic agent administered, and the species of animal Horses generally maintain a central eye regardless of anesthetic depth; therefore, is not a reliable indicator in this species

Reflex	Implementation	Assessment	Comments
Nystagmus	Involuntary, rapid oscillation of the eyeball	Fast nystagmus occurs during light plane of anesthesia and decreases as anesthetic depth increases	In anesthetized horses, indicates a very light plane of anesthesia and that patient may begin to move Can be seen in sheep and goats in response to anesthetic depth Dogs and cats do not routinely get nystagmus
Pupil size	Direct assessment of pupil size	Normal or constricted during lighter planes of anesthesia Progressively dilates as anesthetic depth deepens Can be dilated during Stage II of anesthesia	Can vary during the different planes of anesthesia Affected by drugs such as opioids (dilated in cats, constricted in dogs), anticholinergics (dilated), and dissociative anesthetics (dilated)
Muscle tone	Jaw tone: opening jaw to gauge "looseness" Anal tone: diameter of the anal orifice and tone of the anal sphincter General muscle tone: gauge looseness of muscles	Less tone and more relaxed muscles during deeper planes of anesthesia Resistance to jaw movement decreases Anal orifice increases in size Limbs become more flaccid	Some drugs increase muscle tone (e.g., ketamine) and thus interfere with this assessment Jaw tone difficult to assess in puppies (because they have minimal jaw tone) and in animals with large masseter muscles (e.g., large dog breeds and in large animals)

*The absence of the swallowing and laryngeal reflexes is why the patient's airway should be protected with properly fitted ETT.

determined by the pulse oximeter] levels) and allows them to be addressed before becoming problematic. For example, a gradual decrease in blood pressure may not be noticed if it happened over a 20 minute period, but by reviewing the anesthetic record while monitoring the patient it will be obvious by looking at the graph that blood pressure is declining. The sooner decreases in blood pressure can be noted, the more time there will be to make modifications before the patient becomes hypotensive.

An anesthetic record should have 5-minute graduations on the chart for recording monitoring parameters for every 5 minutes of anesthesia. On the side of the form should be a legend describing what the symbols mean (e.g., x = RR, o = HR, v = systolic blood pressure) (Appendix H). Each clinic has the option to create its own anesthetic record tailored to the clinic's needs; however, there are a variety of commercially available ones for use.

Tech Tip

Manufacturers of induction agents often supply anesthetic records to clients who order their products. The American Animal Hospital Association (AAHA) also created a form called the Anesthesia and Sedation Record that adheres to the 2020 AAHA Anesthesia and Monitoring Guidelines for Dogs and Cats and is available for purchase from their website (aaha.org).

Information that should be included on an anesthetic record (see Figure 6-5):

- Date, procedure, names of the attending veterinarian and anesthetist
- Patient details (names, surname, ID number, signalment)
- Abbreviated history including any pre-existing medical conditions

- Initial TPR (temperature, HR, RR), pulse quality, MM, CRT
- Body weight and BCS
- Laboratory test and results (**Quick Assessment Tests** [QATs], which include PCV/TP/blood glucose/BUN [from an Azostick*] as a minimum are often used)
- ASA physical status (I – VI +/-E) (see Appendix D)
- Premedications (drug, dosage, dose [both mg and mL administered to document exactly how much was given for a drug that comes in different concentrations], route of administration, and time given)
- Response to premedications (e. g., if a sedative is given, does it provide mild, moderate, or profound sedation)
- IV catheter (size and location)
- Induction agent (drug, dosage, dose, route of administration, and time given)
- ETT size
- Circuit size (rebreathing or non-rebreathing)
- Oxygen flow rate
- Inhalant anesthetic agent and vaporizer setting
- Fluids (type, rate, and cumulative amount given)
- Graduations that indicate every five minutes of anesthetic monitoring
- Surgery time (incision, completion of anesthesia, and extubation times)
- Physiological monitoring values such as HR, RR, CO_2, SpO_2, temperature, etc.
- Complications or Comments section where details of any changes made and why they were made can be written
- Any medication currently being administered (dosage, dose, route of administration, and time when last dose was given)
- Other laboratory results if they have been evaluated (biochemistry, complete blood count, blood gas, clotting times, etc.)
- Recovery notes (such as temperature, time to sternal and/or standing)
- Pain score and analgesic recommendations

Expectations for Completing an Anesthetic Record

The anesthetic record should be completed for every patient being anesthetized using blue or black ink and must be kept in real-time (staff cannot go back and change recordings after the fact). Staff training is important to make sure that everyone fills out the anesthetic record the same way, which makes it easier when assisting someone with anesthesia and reviewing the anesthetic events.

Tech Tip

Have a laminated copy of a sample anesthetic record on the anesthetic machines so all staff members know the correct procedure for filling out the charts.

When recording drug administration, always write the dose in mg (milligrams) or mcg (micrograms) as drugs come in different concentrations; therefore, it does not matter how many mL (milliliters) were given, but rather the exact amount of drug. This eliminates confusion and the question *"How much was actually given?"* For example, a drug may be diluted to avoid overdosing it in small patients. The concentration of methadone is 10 mg/mL, but using this concentration to administer an accurate dose to a 1 kg patient would be difficult due to the small volume needed; therefore, it is diluted to 1 mg/mL. If it was written on the anesthetic record that 0.1 mL of methadone was administered, either 1 mg (at 10 mg/mL) or 0.1 mg (at 1 mg/mL) could have been given. Recording all doses in weight (e.g., mg) and volume (e.g., mL) eliminates this confusion. Since many veterinary hospitals charge by the mL of drugs used, including mL volumes of each drug can help speed up the billing process (particularly if someone other than a veterinary technician is entering charges). Logging mL of drugs also helps with hospital inventory management. Keep in mind that the DEA will also pull anesthesia records to determine if controlled substances on records match controlled drug logs.

If an incident that requires intervention occurs during the anesthetic event, it needs to be documented as to what happened, why it happened, and what was done when it happened. For example, if a dose of atropine was administered to the patient, not only is it good to write in the comments section that atropine (dose in mg) IV was given but also the reason for its administration (*Was atropine given to correct bradycardia? Was it given to correct second degree heart block?*). This provides a very clear picture as to what happened to the patient and why this drug was administered.

In an emergency situation, readiness is the key and precalculation of emergency drugs (in mL and mg) can save time. This should not be written on the front of the anesthetic record as it could cause confusion as to whether or not these drugs were administered, but it is a good idea to write them on the back of the chart for easy access if needed.

Transporting the Patient

It is both the employer and employees' responsibility to safely transport animals in a manner that also protects personnel from injury. Employers must supply the correct staff training and equipment (e.g., gurneys [trolleys], back braces) so all employees understand the importance of and proper ways to transport patients. It is the employee's responsibility to follow the safety rules in order to prevent injury to themselves and their patients.

Prior to transporting a patient from the prep bench (where the patient is clipped and prepped for surgery) to the surgical or procedure area, the surgical area and patient should be properly prepared. Questions to answer prior to transporting a patient are listed in Table 10-3.

Once the patient is ready to be moved, the anesthesia team needs to work together to ensure everyone's safety. When lifting, pulling, or pushing animals, proper technique can help reduce the incidence of injury. Keeping the back straight when lifting can decrease back injuries. Pulling or pushing large animals during induction or recovery should be done with assistance, either with more personnel or with chain or electric hoists or other assistive devices.

Patients weighing more than 5 kg should be transported on a gurney into the surgical area to make it easier for staff and to keep the prepped surgical area as aseptic as possible. Ideally, all patients should be transported with the use of a gurney so that staff members do not get hurt when lifting and carrying patients from one area to another (Figure 10-5). Gurney transport also allows continuous monitoring of the patient during transport. If separate monitoring equipment is used in the prep area and the surgical area, it is still best to use some type of monitoring equipment during transfer, even if it is only the Doppler (audible pulse rate) or a portable pulse oximeter (visual and audible pulse rate) (monitoring methods are described in Chapter 7 [Anesthesia Monitoring]).

To be OSHA compliant, larger patients (>16 kg) should be transferred to the surgical area by at least two staff members and the use of a gurney to avoid injury to personnel. Proper communication making sure everyone lifts (or slides) the patient at the same time while engaging core abdominal muscles will decrease the amount of strain put on the back and reduce the risk of injury.

Safety Alert

Engage the core abdominal muscles when lifting to help prevent back injuries.

The patient is now ready to be transferred to the surgical area. Be sure to disconnect the machine from ceiling/wall drop down connectors and that the E tank is turned on prior to moving. If a portable anesthesia machine is used, the patient remains connected to the anesthesia circuit and the anesthesia machine is transported to the surgery site. After entering the surgical area, the patient is brought close to the surgery table and slid onto the surgery table in the correct position for surgery. The patient remains connected to the anesthesia circuit throughout the procedure/surgery.

If the patient is not connected to a portable anesthetic machine, it is important to:

- Disconnect the anesthetic circuit at the ETT and cover the end of the Y-piece to avoid WAGs.
- Turn off the inhalant anesthetic agent and oxygen.
- Flush oxygen through the system into the scavenging system.
- Move the patient by sliding the patient onto the gurney (that has a blanket in place) and move the patient into surgical area.

TABLE 10-3 Items to Consider Prior to Patient Transport

Surgical Area Preparation	Patient Preparation
• Is the surgical table in the correct position, with positioning aids in the proper location and with heating devices turned on? • Are all other heating devices warmed and ready to use (forced air blankets, conductive fabric warming blankets, fluid warmers, etc.)? • Is the anesthetic machine ready for the patient (circuit, reservoir bag, leak test, etc.)? • Is the scavenging system turned on (or not expired)? • Is there enough oxygen for the procedure? • Is there enough inhalant anesthetic agent for the procedure? • Are the correct monitoring equipment and ECG leads ready to be attached to the patient? • Is a working fluid pump +/– syringe pumps (syringe drivers) ready? • Are all medications in the surgical area (analgesia, antibiotics, etc.)? • Is the emergency cart (crash box) fully stocked with unexpired medications? • Is the oxygen turned on and ready to be connected when the patient gets into the surgical area?	• How will the patient be positioned? It is best to have the patient in the same position they will be in for the surgical procedure (e.g., a laparotomy patient would be prepped in dorsal recumbency and transported in dorsal recumbency) to keep the prepped area as aseptic as possible during transport. If the patient cannot breathe properly in the surgical position, it is best to ventilate for them, or leave them in a comfortable position (such as lateral recumbency), prep as good as possible, and when the surgeon is ready to drape, roll the patient into dorsal recumbency, assist ventilation, and have someone else do the final prep. The key is to make the patient as comfortable for as long as possible. • Is the patient at an adequate plane of anesthesia to allow enough time to transport it into the surgical area? If patients are too light, they will wake up during transport and may jump off the table with an ETT in place causing injury and contaminating the surgical prep area. • Are fluid lines ready (clamped, taken out of the prep area's infusion pump, on the gurney, ready to go into the surgical area with the patient)? • Are monitors that are staying outside of surgical area disconnected? • Are emergency doses of injectable anesthetic drawn up and ready to be administered if needed? • Is the proper number of staff dressed and ready to go into the surgical area to easily transport the patient?

Upon entering the surgical area, the patient should be brought close to the surgery table and slid onto the surgery table in the correct position for surgery. At this point it is important to:

• Connect the anesthetic circuit (oxygen is already on) and turn on the vaporizer.
• Connect all monitors.
• Insert fluid lines into the infusion pump.
• Stabilize the patient until the surgeon is ready.

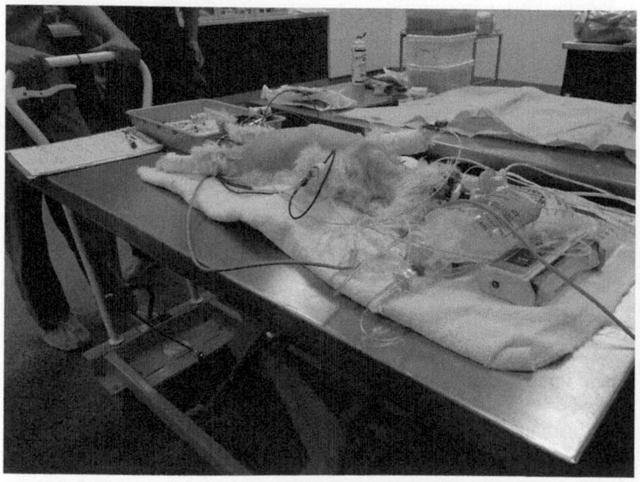

In some practices, the anesthesia machine is part of a station that rolls with the patient (Figure 10-6). In that situation, the machine will be brought, with the patient, into the surgical area from the prep room or whatever location where the patient was induced.

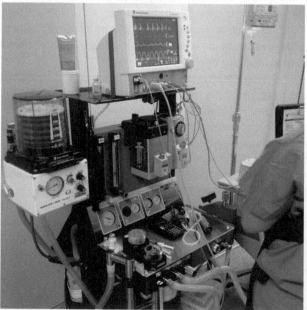

FIGURE 10-5 Patient on gurney with monitors attached and ready for transport into surgical area (theatre).

Courtesy of Anita Parkin, RVN, AVN Dip (Surgery & ECC), VTS (Anesthesia/Analgesia), CVPP, TAE

FIGURE 10-6 Example of an anesthesia machine that is part of a station that rolls with the patient.

Courtesy of Christopher L. Norkus, DVM, DACVAA, CVPP

Transitioning from Maintenance to Recovery

When the surgery is finished, it is time to prepare the patient for transport to the recovery area. Before transporting the patient:

- Check that a cage or bed is available and is suitable for the patient.
- Prior to moving the patient, double check recovery cages to make sure:
  - cage doors completely close.
  - water and food bowls are removed.
  - barriers or partitions in the cage do not move.
- Have postoperative analgesia and sedation calculated and ready for use in the patient if needed.
- Have any monitoring equipment that is needed available and/or connected to the patient.
- Have staff available to monitor the patient.

Depending on the practice and type of postoperative care that can be provided, the patient may be reconnected to the anesthetic circuit and allowed to breathe pure oxygen at a rate of 50 to 100 mL/kg/min (up to a maximum of 5 L/min) for 5 to 10 minutes to improve ventilation especially in hypoventilating patients. This is also a way to decrease the risk to staff of breathing inhalant anesthetic agent as it keeps anesthetic gas in the environment low.

After 5 to 10 minutes of breathing pure oxygen, prepare for recovery by doing the following:

- Disconnect the monitoring equipment from the patient (it is a good idea to keep some monitoring equipment attached to the patient, such as a Doppler or pulse oximeter).
- Have a cuff syringe ready for deflating the cuff if the patient becomes light and is ready to have the ETT withdrawn (Figure 10-7).
- Disconnect the anesthetic circuit from the patient and turn off the oxygen.

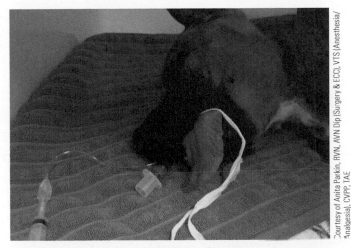

FIGURE 10-7 Cuff syringe (lower left) attached to ETT. The cuff of the ETT should be deflated prior to withdrawal to prevent injury to the trachea.

- Have adequate amount of staff available to transport the patient (with the use of a gurney) and with good communication, engage the core abdominal muscles, and slide the patient onto the gurney.
- Transport the patient into the recovery ward.

When lifting the patient into the cage, use correct lifting techniques (Figure 10-8a), engage the core abdominal muscles, and move the patient smoothly into the cage (Figure 10-8b). If this process is done with haste, it can cause the patient to have a rough recovery due to the stimulation of being put into the cage with force. Hence a nice, gentle, smooth, and quiet action is best. The patient should be placed in the cage or bed with the head facing outwards, so the ETT can be visualized, the appropriate monitoring equipment attached (Figure 10-9), and analgesia ready to be given. Constant supervision of the patient (either by

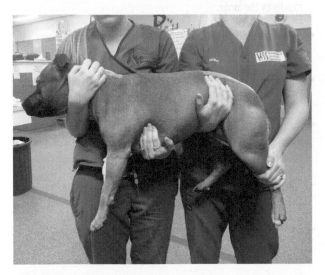

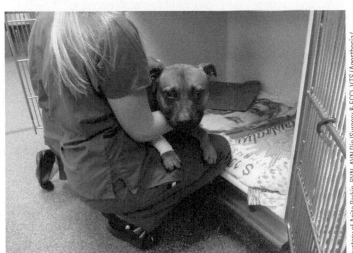

FIGURE 10-8 Correct lifting techniques for (a) taking the patient to the cage and (b) when placing a large patient into the cage.

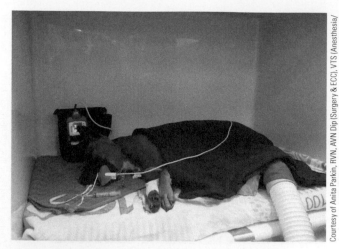

Courtesy of Anita Parkin, RVN, AVN Dip (Surgery & ECC), VTS (Anesthesia/Analgesia), CVPP, TAE

FIGURE 10–9 Patient in the cage in the recovery ward after an anesthetic event.

the anesthetist or other veterinary technician) continues as the patient enters the recovery period. A pulse oximeter is the standard of monitoring care in recovery and should be used on all small animal patients where it can be safely used. It is not necessary to record vital signs on the anesthesia record every 5 minutes in recovery, but the patient should be constantly monitored by some member of the veterinary staff. Oxygen should be available for support during recovery. Be sure that personnel are not placing their faces inside the cage or over the head of the animal to reduce exposure to WAGs. The goal is to complete all of the steps from maintenance to recovery before it is time to deflate the ETT cuff and extubate the patient. The guidelines for when to deflate the ETT cuff and extubate the patient are described in Chapter 11 (The Recovery Period and Post-Anesthetic Care).

Safety Alert

When transitioning the animal from maintenance to recovery, veterinary technicians need to closely monitor their patients for hypothermia, shivering (which increases oxygen consumption), regurgitation/vomiting, dysphoria, pain, and prolonged recovery.

SUMMARY

Maintaining a patient at a surgical plane of anesthesia (Stage III/Plane 2–3) while preserving cardiovascular and pulmonary stability is as much an art as it is a science. Inhalant anesthetic agents are commonly used for anesthesia maintenance and are absorbed through the alveoli based on the drug's blood-gas solubility and partial pressure differences and the patient's cardiac output and alveolar ventilation. Inhalant anesthetic agents are directly inhaled into the respiratory system

and are primarily eliminated by the lungs; therefore, their uptake and elimination may cause anesthetic depth to change rapidly. Ideally inhalant anesthetic agents are delivered via an ETT, which also provides patients with a protected airway during the maintenance period of anesthesia. Injectable anesthetic agents are less frequently used for anesthetic maintenance and are mainly used for shorter procedures or when anesthesia machines are not available or hinder access to the patient/large procedural equipment. No matter the administration route, balancing anesthetic depth with a drug's adverse effects are challenging parts of the anesthetist's job that can be met through proper training and hands-on practice. By knowing what can happen physiologically to the patient during maintenance of anesthesia and why it happens, the anesthetist can anticipate and prepare for changes during maintenance of anesthesia.

CRITICAL THINKING POINTS

- The goal of anesthesia maintenance is analgesia, unconsciousness, and muscle relaxation while maintaining cardiopulmonary stability and proper anesthetic depth.

- There are four stages of anesthesia with Stage III subdivided into three planes. Knowing which of these stages the patient is in will help ensure that the patient has the proper amount of muscle relaxation, analgesia, and unconsciousness for the procedure being performed.

- Reflexes are used to determine a patient's depth of anesthesia; therefore, the veterinary technician should be familiar with how they are assessed.

- Communication between the anesthetist and surgeon about a patient's depth of anesthesia or change in vital signs is key to successful patient outcomes.

- The advantage of administering inhalant anesthetic agents is that they are inhaled directly into the respiratory system and eliminated rapidly by the lungs.

- MAC values indicate the potency of an inhalant anesthetic agent and are not vaporizer settings.

- The oxygen flow rate can change the amount of inhalant anesthetic agent delivered to the patient on a circle rebreathing system.

- Maintaining accurate anesthetic records helps visualize trends that are occurring in an anesthetized patient and serves as a legal document.

- Precalculation of emergency drugs can save precious time during an anesthetic emergency.

- Safely transporting patients is critical for both patient and staff.

REVIEW QUESTIONS

Multiple Choice

1. During anesthesia, how often should assessment information ideally be recorded on the anesthetic record?

   a. Every 5 minutes
   b. Every 10 minutes
   c. Every 15 minutes
   d. When the veterinarian tells you to

2. What does placing ophthalmic ointment in anesthesia patients' eyes prevent?

   a. Keratoconjunctivitis
   b. Corneal ulceration
   c. Glaucoma
   d. Cataracts

3. What effect on MAC value can occur with any of the following conditions: metabolic acidosis, severe hypoxia, anemia, blood loss, advancing age, pregnancy, and decreases in body temperature?

   a. No change
   b. Increase by 20%
   c. Increase by 50%
   d. Decrease

4. What does MAC stand for?

   a. Minimum alveolar concentration
   b. Mean arterial concentration
   c. Minimum arterial concentration
   d. Mean alveoli concentration

5. Why are high oxygen flow rates required for non-rebreathing circuits?

   a. Because the patient breathes more
   b. To keep the patient warm
   c. No CO_2 absorber is used
   d. The tubing is smaller

6. What is the term for setting the vaporizer at a higher than desired concentration so the target concentration is met in a shorter period of time?

   a. Hyperconcentrating
   b. Vaporization
   c. Time constants
   d. Overpressurizing

7. What is a way to increase the concentration of inspired inhalant anesthetic agent in a circle rebreathing system?

   a. Increase oxygen flow rate
   b. Increase the drug's gas solubility
   c. Decrease the patient's cardiac output
   d. Decrease the patient's ventilation

8. What are benefits of inhalant anesthesia?

   a. The ability to rapidly change the level of anesthesia delivered to the patient
   b. The ability to take over breathing for the animal
   c. Both a and b
   d. None of the above

9. What can be said of an inhalant anesthetic agent that has a higher MAC value than another?

   a. An inhalant anesthetic agent that has a higher MAC value is more potent than one with a lower MAC value.
   b. An inhalant anesthetic agent that has a higher MAC value is equally as potent as one with a lower MAC value because MAC is a vaporizer setting.
   c. An inhalant anesthetic agent that has a higher MAC value is less potent than one with a lower MAC value.
   d. An inhalant anesthetic agent that has a higher MAC value is more flammable than one with a lower MAC value.

10. What is the range for oxygen flow rates on a circle rebreathing circuit?

    a. 5–15 mL/kg/min
    b. 25–50 mL/kg/min
    c. 60 mL/kg/min
    d. 100 mL/kg/min

11. In a non-rebreathing circuit, what should the minimum O_2 flow rate be set at?

    a. 200 mL/kg/min
    b. 100 mL/kg/min
    c. 50 mL/kg/min
    d. 25 mL/kg/min

12. How is the dial setting on the vaporizer determined so that the proper depth of anesthesia is maintained?

    a. 1.5 times the MAC value
    b. 2 times the MAC value
    c. It is set based on a chart provided by the drug manufacturer.
    d. It is adjusted based on patient needs.

13. What is the purpose of the anesthetic record?

    a. Allows the anesthetist to follow trends in the patient's vital signs
    b. Keeps the anesthetist's attention focused on what is happening with the patient
    c. Provides a legal record of what happened during anesthesia
    d. All of the above

14. Prior to anesthesia, emergency drugs should be

    a. locked away.
    b. precalculated for the patient.
    c. written down on the record.
    d. stored on ice until the premedication drugs are given.

15. Which inhalant anesthetic agent requires a specific vaporizer that is both heated and pressurized?

    a. Desflurane
    b. Sevoflurane
    c. Isoflurane
    d. Halothane

16. Which of the following affect MAC?

    a. Duration of anesthesia
    b. Fever
    c. Metabolic alkalosis
    d. Potassium levels

17. Which stages of anesthesia occur rapidly and if observed are most commonly seen in poorly sedated patients?

    a. I and II
    b. II and III
    c. III and IV
    d. I and III

18. Who is obligated to comply with OHSA rules?

    a. The employer
    b. The employee
    c. Both the employer and employee
    d. Legislation and the employer

19. A minimum of two people must be involved with lifting and/or moving any patient over

    a. 5 kg.
    b. 12 kg.
    c. 16 kg.
    d. 21 kg.

20. Why is propofol the drug most commonly used for CRI?

    a. Propofol can easily become contaminated and there is less risk of contamination of drugs administered as a CRI.
    b. Propofol is acidic and therefore is more rapidly distributed to the brain.
    c. Propofol does not accumulate in fat and is rapidly metabolized.
    d. Propofol is not a controlled substance; therefore, respiratory depression is not a concern.

Case Studies

Case Study 1: Brutus, a 12-month-old male Bull Mastiff canine (45 kg), presented to the clinic for fracture repair (plating) of the proximal tibia. His preanesthetic vital signs were as follows:

HR 90 bpm, RR 24 breaths per minute, temperature = 100.8°F (38.2° C), strong pulses, MM (mucous membranes) pink, CRT (capillary refill time) 1–2 seconds, PCV 40% (normal values 35–55%), TP 6.5 g/dL (normal values 6.5–8.0 gm/dL)

Brutus received the following anesthetic protocol:
Premedication: acepromazine (0.05 mg/kg SQ) and methadone (0.25 mg/kg SQ)
Induction agent: propofol titrating dose
Maintenance agent: isoflurane 1.5% and O_2 (1.5 L/min)
Thirty minutes after administration of premedication, Brutus is in surgery room and he is prepped for surgery. The incision has been made and the surgeon is dissecting tissues (in particular the periosteum) down to the tibia. The patient's vital signs change from/to:
HR 70 to 140 bpm, BP (SAP) 100 to 170 mm Hg, RR 12 to 28 breaths per minute, MM pink, CRT 1 second
Reflex signs are: eye ventromedial, no palprebral reflex, and small (to little) amount of jaw tone (consistent with the previous 10 minutes)

What would you do for the patient in this scenario and why?
    a. Turn off the isoflurane as the patient is in a deep plane of anesthesia.
    b. Turn down the isoflurane as the patient is in a deep plane of anesthesia.
    c. Administer more analgesics as the patient's pain appears poorly managed.
    d. Nothing is required to be done yet.

Case Study 2: Brutus (the 12-month-old male Bull Mastiff canine [45 kg] from Case Study 1) presented to the clinic for fracture repair (plating) of the proximal tibia. Preanesthetic vital signs were:
HR 90 bpm, RR 24 breaths per minute, Temp 100.8°F (38.2°C), strong pulses, MM pink, CRT 1–2 seconds, PCV 40%, TP 65g/dL

Premedication: acepromazine (0.05 mg/kg SQ) and methadone (0.25 mg/kg SQ)
Induction agent: propofol titrating dose
Maintenance agent: isoflurane 1.2% and O_2 (1.5 L/min)
The above patient is in the surgery room and the incision has been made.
The vitals change from HR 70 to 110 bpm, BP (SAP) 100 to 150 mm Hg, RR 12 to 28 breaths per minute, MM pink, CRT 1 second.
Reflex signs are: eye central with light reflex, palpebral reflex present, and small amount of jaw tone.

What would you do for the patient in this scenario and why?
   a.   Turn up the isoflurane as the patient is in a light plane of anesthesia.
   b.   Turn down the isoflurane as the patient is in a deep plane of anesthesia.
   c.   Administer more analgesic as the patient is in pain.
   d.   Nothing is required to be done yet.

Case Study 3: A 6-month-old male Golden Retriever puppy weighing 20 kg is scheduled for neutering. As the anesthesia technician, you must get the anesthesia machine ready for this surgery.
Based on the information in Chapter 2 (The Anesthesia Machine), what type of circuit and breathing system and preferred oxygen flow rate is most appropriate for this patient?
In a circle system, initially a higher oxygen flow rate is used (25–50 mL/kg/min) until the patient is at the desired plane of anesthesia (then lower to 10–25 mL/kg/min for maintenance). What is this patient's initial oxygen flow rate?
The patient has reached Stage III plane 2 of anesthesia and you decide to reduce the oxygen flow rate to maintenance oxygen flow rates (10–25 mL/kg/min). What is this patient's new oxygen flow rate?

The surgery is complete and the vaporizer is turned off. What oxygen flow rate would you use to help recover this patient?

Case Study 4: A 4.5 kg cat requires a mass removal. As the anesthesia technician, you must get the anesthesia machine ready for this surgery.
Based on the information in Chapter 2 (The Anesthesia Machine), what type of circuit and system is most appropriate for this patient?
What is this patient's oxygen flow rate?

Case Study 5: Brandy, a 2-year-old spayed female German Shepherd dog, presents to the emergency service for seizures. Her physical exam and initial diagnostic work up (CBC/Chemistry Panel/UA and blood pressure) are within normal limits. The attending clinician requests an MRI be performed with Brandy under general anesthesia. Your anesthetic machine is not compatible in the MRI suite. What options do you have to deliver anesthetic to this patient?

Critical Thinking Questions

1.   Frannie, an 8-month-old female 20 kg Boxer dog, is anesthetized for routine ovariohysterectomy (OHE). As the surgeon begins to manipulate the reproductive tract, the anesthetist notes that the patient is getting light based on jaw tone and palpebral reflex. What steps should be taken to bring the patient to a surgical plane of anesthesia?

2.   When a patient on a rebreathing system has a rapid, shallow breathing pattern, it can become "light" and no longer be at a surgical plane of anesthesia. Why does this occur? Why does manual ventilation ("**bagging**") of the patient help correct this situation?

ENDNOTES

1.   Plumb, D. C. (2018). *Plumb's veterinary drug handbook* (9th ed., pp. 333–334, 631–633, 1066–1068). Ames, IA: Blackwell Publishing.
2.   Steffey, E. P., Mama, K. R., & Brosnan, R. J. (2015). Inhalant anesthetics in Lumb and Jones 4th edition: Anesthetic risk and informed consent. In K. G. Grimm, L. Lamont, W. J. Tranquilli, S. Greene, & S. Robertson (Eds.), *Veterinary anesthesia and analgesia The fifth edition of Lumb and Jones*. (pp. 310–323). Ames, IA: Wiley Blackwell.
3.   McKelvey, D., & Hollingshead, K. W. (2003). *Veterinary anesthesia and analgesia* (3rd ed., pp. 149, 112, 204). St. Louis, Mo.: Mosby.

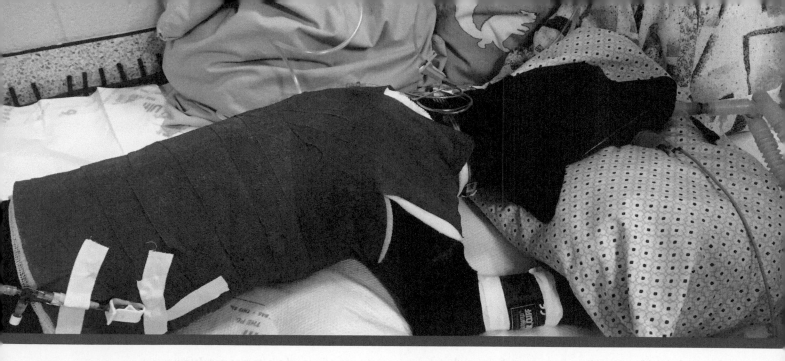

CHAPTER 11
The Recovery Period and Post-Anesthetic Care

Patricia R. Zehna, *RVT*

LEARNING OBJECTIVES

Upon completion of this chapter, it is expected that the reader should be able to:

11.1 Describe the veterinary technician's role in anesthetic recovery

11.2 Identify the best practices for post anesthetic monitoring

11.3 Explain how to prepare the patient recovery area

11.4 Recognize ways to ensure personnel and patient safety during recovery periods

11.5 Describe how to properly extubate patients

11.6 List special recovery concerns for ruminants and equine patients

INTRODUCTION

The anesthetic recovery period begins once delivery of the anesthetic agent is discontinued, and whether its endpoint is when the patient is sternal (typically used for small animals), standing (used for large animals), or takes it first meal, it is often erroneously considered a time when the anesthesia team can relax following the stress of anesthesia and surgery. Postanesthetic care may appear routine and straightforward; however, 47% of canine and 61% of feline anesthesia mortalities have been reported to occur in the postoperative period, making it the leading time during the anesthesia experience for patients to experience complications and death.[1] The residual effects of drugs on the cardiovascular, respiratory, and neurologic systems as well as complacent anesthesia team members who fail to continue monitoring and supporting the patient after anesthesia drug delivery is discontinued are main reasons complications increase during the recovery period. Animals with sub-optimal ASA status can be particularly challenging to the anesthesia team during the recovery period as can those with unique anatomy, such as brachycephalic Bulldogs (stenotic nares, elongated soft palate, everted saccules, and hypoplastic trachea which can lead to upper airway obstruction), or those that present unique concerns, such as horses (physical recovery involves getting them from asleep in dorsal or lateral to awake and standing without injury/fracture). Although anesthetists know in advance to pay special attention to patients at increased anesthetic risk during their recoveries, they should not let their guard down for healthy patients because *all* patients are at risk for anesthetic complications during recovery. Even a few minutes of unsupervised recovery may lead to major complications in patients.

THE RECOVERY PERIOD

Anesthesia recovery begins when delivery of anesthetic agents is discontinued. It can be challenging to manage patients in recovery as there is no distinction between anesthetic stages and there may be nothing obvious to tell you the patient is transitioning between Stage III to Stage II and then to Stage I. Optimally there is at least one recovery technician who specifically works with the veterinarian to ensure an efficient, safe, and effective recovery. The recovery technician closely monitors the patient's recovery, reports complications to the veterinarian in a timely fashion, ensures orders for ongoing supportive care (including adequate postoperative care) are completed, and makes sure the patient's record is properly maintained.

Signs of Recovery

Anesthesia is supposed to be a controlled situation that, when we stop it, the patient should awaken. How long it takes an animal to recover from anesthesia varies depending on procedural (e.g., anesthetic technique(s) used, type and duration of surgical procedure) and patient (e.g., patient health status, body temperature) factors. To avoid complications during recovery from anesthesia, every attempt should be made to develop a specific anesthesia plan for each patient.

The anesthetist should consider the following to anticipate how their patient's recovery will progress:

- *How long was the animal anesthetized? What depth of anesthesia was needed for the particular procedure and/or patient? Which premedication(s), induction agent(s), and maintenance agent(s) were given to the patient and how they are metabolized?* Patients who are under anesthesia for longer periods of time tend to take longer to recover. Consider the examples from Chapter 10 (The Maintenance Period) in which a cat needs to have an abscess lanced versus a dog that needs repair of a torn cruciate ligament who is undergoing a tibial plateau leveling osteotomy (TPLO). The cat's procedure is short and it is appropriate to select drugs with shorter durations of action, such as premedication with an opioid and induction with ketamine/diazepam without using a maintenance agent. The dog's procedure is longer and more painful; therefore, he will need a longer duration of anesthetic depth during the maintenance period to complete the surgery as well as more potent and sustained perioperative pain relief such as epidural or locoregional techniques and constant rate infusions (CRI) (see Chapter 17 [Analgesic Techniques]). A drug's mechanism of action, type of metabolism, and duration of action may also affect how a patient recovers from anesthesia. For example, cats given Telazol® IM may have a longer recovery time and be drowsy all day because zolazepam in Telazol® has a duration of action that provides tranquilization into the recovery period. In contrast, patients given repeated IV boluses of propofol, a drug that is more rapidly metabolized and thus has a shorter duration of action, will have shorter recovery times.

- *Which route of drug administration was used?* Whether an anesthetic agent is given via inhalation, repeated IV boluses, CRI, or IM injection may affect anesthetic recovery. For longer procedures, patients maintained on the inhalant anesthetic agent sevoflurane tend to recover faster than those who received an IM anesthetic combination such as Telazol®/dexmedetomidine/butorphanol.

- *What was the health status of the patient prior to anesthesia?* Patients may not recover normally because they are cold, hypotensive, or hypoglycemic or have hepatic failure and inability to metabolize drugs, major electrolyte abnormalities (e.g., hypernatremia that may cause them to be physiologically altered and appear "drunk"), or CNS disease, which may have worsened following anesthesia. For example, isoflurane and sevoflurane are minimally metabolized by the liver; therefore, they are good choices for patients with hepatic disease because they are less likely to lengthen recovery time. These patients, however, may be given drugs concurrently such as an opioid, which may need to undergo hepatic metabolism and thus prolong recovery. Animals who are hypoglycemic, especially small or neonatal patients, can be prone to prolonged recoveries. Hypoglycemia can be caused by immature hepatic glycogen storage in neonatal/juvenile patients, sepsis, paraneoplastic syndrome, hypoadrenocorticism (Addison's disease), and insulinoma and may not be clinically evident in anesthetized patients. In patients that are slow to recover from anesthesia, a blood glucose (BG) should be evaluated and in patients that have a BG < 60mg/dl, IV dextrose and a dextrose CRI should be started to address the neuroglycopenia. Additional diagnostic testing such as blood pressure evaluation, electrolyte evaluation, temperature monitoring, and neurologic examination should be done promptly in any patient not recovering as expected.

- *Is this patient a breed of animal that typically has difficulty with metabolism or elimination of drugs?* Some animal breeds have specific physiology or are prone to diseases that may impact anesthesia. Brachycephalic dog breeds prone to upper airway obstruction may have prolonged recoveries due to complications that develop during maintenance of anesthesia.[2] Large or giant-breed dogs are more commonly overdosed when milligram per kilogram dosing versus body surface area dosing is used, which can delay metabolism and elimination of anesthetic and analgesic drugs.[3] Sighthounds receiving thiobarbiturates have minimal fat to redistribute the drug to and are deficient in enzymes needed to metabolize barbiturates; therefore, they may experience prolonged recoveries associated with general anesthesia.[4,5] Dog breeds with the MDR1 mutation such as Collies have a more profound effect to some drugs (e.g., acepromazine and butorphanol), which could lead to more sedation and slower recovery.

- *Did the patient experience anesthetic complications?* Patients that experience anesthetic complications (e.g., hypothermia, blood loss/anemia, hypotension, severe hypercapnia, and hypoxemia) may have delayed drug metabolism, respiratory and cardiovascular compromise, impaired tissue perfusion, and cerebral depression.[6] These body system alterations can result in prolonged recovery. Anesthetic complications are described in Chapter 12 (Anesthetic Complications).

Tech Tip

For dogs and cats, an optimal recovery time is within 10 to 30 minutes of the end of anesthesia; however, the length of recovery depends on the patient health status and breed, any comorbidities the patient may have, type of anesthetic technique used (e.g., inhalant versus injectable), duration of anesthesia, and body temperature.[3]

The Recovery Area

The recovery area should be confined (so that patients are not able to escape or be injured by other animals), quiet and dimly lit (as to not simulate an awakening patient), and warm (to facilitate smooth recoveries). Many patients in recovery have an increased sensitivity to sound and light; therefore, loud/abrupt noises and bright lighting can startle or frighten a patient producing less desirable/smooth recoveries. An area with no or limited traffic and the ability to turn off lights is ideal for animals recovering from anesthesia. Limiting hypothermia by providing a warm environment and utilizing ways to preserve heat is also important during the recovery period.

Successful recovery from anesthesia relies on a recovery area stocked with the equipment necessary to assist in the monitoring and recovery of animals. Having the proper equipment and drugs available prepare the anesthesia team for an urgent/emergency event. The following list is an example of some of the supplies needed in a properly stocked recovery area for small animals (specific information on large and exotic animal recoveries is found in Chapter 19 [Anesthesia and Analgesia in Large Animal and Food Producing Species] and Chapter 20 [Anesthesia and Analgesia in Pet Birds and Exotic Small Mammals]):

- Emergency medications or "crash cart" should be in or at least near the recovery area and should contain sedatives/tranquilizers to treat dysphoric patient recoveries (such as acepromazine [Promace®, generic] and dexmedetomidine [Dexdomitor®]) and rapidly-acting induction agents (such as propofol [Rapinovet®, PropoFlo™, Propovan®, PropoFlo™28]) if re-intubation is needed. Calculators and/or emergency drug sheets for quick reference should be available.
- Analgesics (controlled substances need to be in a locked box that is often not in the recovery area; therefore, these drugs may need to be in an area close to recovery)
- Needles and syringes of various gauges and lengths
- IV catheters/extension sets to replace those that become dislodged or damaged
- Injection caps to cap off and protect an IV catheter while maintaining easy access to the vessel
- IV fluids of various types
- Saline flush
- Oxygen source
- Endotracheal tubes (ETTs) in various sizes/tube ties
- Oxygen masks
- Laryngoscope
- Suction/suction tips
- Bandage materials, including bandage scissors; a bandage cart that can be moved to the patient is handy
- Disposable absorbent cage pads ("chucks")
- Heat source(s)
- Monitors—blood pressure, pulse oximeter, ECG, etc.
- Elizabethan collars
- Animal handling gloves
- Leashes and muzzles
- Nasotracheal tubes of various sizes if performing equine anesthesia
- Face masks

Tech Tip

Set up basic patient recovery bed(s) in advance to better manage your time.

Ways to ensure environmental ambiance and patient comfort and safety include the following:

- A nicely padded recovery bed with clean linens/blankets provides an ideal environment for anesthesia recovery of small animals (Figure 11-1). Access to clean bedding is important should the bedding become soiled.
- Disposable pads that can be changed frequently help non-ambulatory patients avoid developing urine scald or contaminating wounds with bodily fluids. Be diligent about changing blankets/linens/pads as needed and cleaning the patient if soiled.
- A fully and continually staffed recovery area, in which patients are within plain view of the recovery technician(s), allows early detection of post-anesthetic complications.
- Removal of any objects that can cause discomfort/pain such as ECG clips, temperature probes, etc., at the conclusion of surgery and from the vicinity of the patient. Electrode pads for the ECG clips can be used if the ECG is necessary outside the surgical suite and during the recovery period.

Courtesy of Chris Norkus, DVM, DACVAA, CVPP, DACVECC

FIGURE 11-1 A nicely padded recovery bed, clean blankets, forced air warming blanket (Bair Hugger®), and head support provide an ideal environment for the patient to recover from anesthesia.

- The patient's urinary bladder size should be checked and expressed if necessary.
- Any new IV catheters, bandages, surgical incision coverings, etc., should be done *before* recovery. It is also good to ask if there are any radiographs that need to be taken or blood work that needs to be performed *before* recovery (e.g., Do I need to recheck the patient's packed cell volume [PCV], electrolytes, or blood glucose? Do I need to take a radiograph after an orthopedic surgery to ensure a fracture was properly repaired?).
- Play soothing music, such as Through a Dog's Ear, once the patients have recovered from the immediate post-op period.
- Avoid slamming doors or having loud conversations in or near the recovery area.
- Pediatric patients can be fed a small meal 2 to 3 hours postoperatively or as soon as they can safely eat since they are more prone to developing hypoglycemia when fasted.

Safety Alert

Ensure that recently cleaned areas are free from cleaning product fumes before using the area for patient recovery. Many products can cause respiratory irritation when used at higher concentrations and take time to diffuse.

Transport to the Recovery Area

Care, communication, and common sense must be used when moving a post-anesthetic patient from the surgical suite. For the safety of the patient and the members of the anesthetic team, many patients require co-lifting/moving or the use of a gurney (depending upon distance to the recovery area, size of patient, and type of surgery) to move them to the recovery area (see Chapter 10 [The Maintenance Period] and Figures 10-5 and 10-8). To prevent an animal from falling out of a cage, pulling out an IV catheter

that is connected to fluids, or knocking over an IV pole, make sure the cage doors completely close and that any barriers in the cage cannot move. The patient's surgical site and surgical procedure should *always* be a consideration when moving patients from one area to another. The wrong type of patient manipulation carries the risk of causing discomfort, pain, and/or injury and could lead to post-surgical complications. Prevention of patient injury is key to a safe recovery.

Tech Tip

Design and implement an area in the recovery room for patients who cannot be caged, for example, giant breeds of dogs.

During the recovery period, patients can expire a level of inhalant anesthetic agent above OSHA acceptable levels. The highest levels of anesthetic gas are expired immediately after a patient is removed from the anesthesia machine. Usually this is the time when the animal is either still on the operating table or in the recovery cage when personnel are performing procedures (such as nail trims or vaccinations) that put them in close proximity to the patient. It is extremely important that personnel keep themselves away from the end of the ETT or if the patient is in a cage, that personnel not put their heads into the cage with the patient. The veterinary practice must have proper ventilation in the building so that room air exchanges provide enough fresh air to keep waste anesthetic gases (WAGs) levels as low as possible.

MONITORING RECOMMENDATIONS FOR THE RECOVERY PERIOD

Veterinary clinics and hospitals have varying degrees of anesthetic sophistication, which impacts the approach to post-anesthetic monitoring. The veterinary technician's knowledge of the recovery process coupled with due diligence is the most important "monitor" since it is always available. All veterinary facilities operate with a minimum standard of care, which also applies to patients recovering from anesthesia.

Minimum Monitoring Recommendations

The minimum parameters to consider when safely recovering a patient include the following:

- Core body temperature
- Vital signs
  - Heart/pulse rate, rhythm, strength, and quality
  - Respiratory rate, rhythm, and quality
  - Mucous membrane color
  - Capillary refill time
- Oxygenation via pulse oximetry
- Maintenance of an IV catheter
- Pain assessment (using a pain scale for effective communication among personnel)
- Mentation assessment

Measuring and recording these parameters in the patient's anesthesia record alerts the anesthesia team of potential problems early so that prompt intervention can be initiated. If not already present, a post-anesthetic monitoring section of the anesthesia record should be added to the anesthetic record since monitoring during recovery starts as soon as the anesthetic event ends (the vaporizer is turned off or the IV anesthetic agent is no longer being administered to the patient). The post-anesthetic (recovery) monitoring section provides a readily available, visual representation of any complications or trends that have occurred. As in other parts of the anesthesia record, parameters are recorded at intervals as directed. A good example of a completed anesthetic record with a post-anesthetic section is illustrated at the bottom of Figure 6-5 (note that respiratory rate, pulse, and SpO$_2$ were monitored and documented until the patient was extubated). A simple recovery form can also be made for clinic use (Figure 11-2).

Core Body Temperature

It is important to monitor a patient's temperature regularly during recovery (minimally every 30–60 minutes) in order to reach and/or maintain it within the normal range. The normal temperature ranges for a variety of species are listed in Appendix I.

A rectal temperature needs to be taken; however, keep in mind that rectal thermometers overestimate the *core body* temperature. Be aware that taking a rectal temperature aggravates certain patients, especially cats. Esophageal thermometers can be inserted and used to determine a patient's temperature during anesthesia to provide a more accurate reading (except during thoracic surgeries) (Figure 11-3a). Esophageal thermometers should not be left in place during recovery from anesthesia because the animal could chew them in half and have an esophageal foreign body that needs to be retrieved. Many multi-parameter monitors, such as those that assess cardiopulmonary function, also include either a rectal or esophageal temperature probe (Figure 11-3b).

Patient Name: DiDi Johnson					
Procedure: Ovariohysterectomy					
Anesthesia Recovery Monitoring Record					
Time	3:00 pm	3:05 pm	3:10 pm	3:15 pm	3:20 pm
Mentation	Unconscious	Light	Light	Conscious	Alert
Temperature (°F)	99	99	99.2	99.4	Not taken
HR (bpm)	95	100	110	120	120
RR (breaths/min)	9	13	15	15	15
MM/CRT	pink/1.5 sec	pink/1.5 sec	pink/1 sec	pink/1 sec	pink/1 sec
SpO$_2$ (%)	98	98	96	98	Not taken
Urine	...	...	...	...	...
Feces	...	...	...	...	...
IV Fluids	Discontinued	...	...	...	...
Recumbency	Lt Lateral	Sternal	Sternal	Sternal	Sternal
Ambulatory	...	...	...	...	...
Pain	...	...	...	...	...
Notes				Extubated	Smooth recovery
Initials	baj				

FIGURE 11-2 An example of an anesthesia recovery monitoring record.

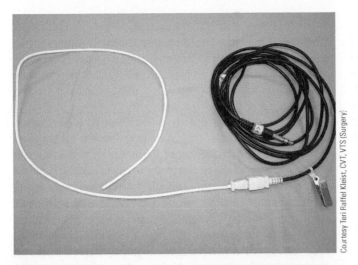

(a)

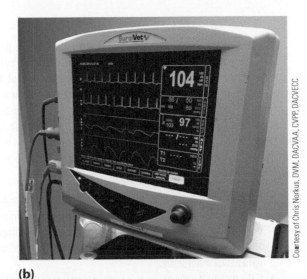

(b)

Courtesy Teri Raffel Kleist, CVT, VTS (Surgery)

Courtesy of Chris Norkus, DVM, DACVAA, CVPP, DACVECC

FIGURE 11-3 Methods of measure body temperature. (a) An esophageal thermometer can be placed into the patient's esophagus to provide accurate continuous measurement of core body temperature. Accidental placement of the esophageal thermometer in the stomach or mouth can reduce the accuracy of a core body temperature estimate. (b) A multiparameter monitor can provide a variety of monitoring data. This unit displays an electrocardiogram and heart rate (green tracing and numerical display), arterial blood pressure and wave form (purple tracing and numerical display), and SpO$_2$ (orange tracing and numerical display) on the monitor screen. It also can display numeric values such as non-invasive blood pressure (white) and body temperature (darker green) on the bottom of the monitor screen.

Ways to keep an anesthetized patient as normothermic as possible were discussed in Chapter 7 (Anesthesia Monitoring).

Vital Signs

Vital signs are important parameters to monitor regularly in the post-anesthetic/recovery period and include assessment of a variety of body systems including the respiratory, cardiovascular, and nervous systems. The most basic vital signs include heart rate, respiratory rate, and body temperature, but assessment of mucous membrane color, capillary refill time (CRT), heart rhythm, and respiratory character are also important. A thorough description of vital sign assessment was described in Chapter 7 (Anesthesia Monitoring) and Appendix I.

Oxygenation

The continued use of pulse oximetry is the standard of care in the post-anesthetic period (see Figure 7-15). It will aid in the early recognition and swift intervention of post-anesthetic hypoxemia. Hypoxemia as a sequela to anesthesia can be caused by residual drug effects leading to hypoventilation/hypercapnia, anesthesia- and positioning-induced atelectasis, and muscle relaxation resulting in less ability to protect the upper airway, which are all then made more significant when the patient is brought from 100% inspired oxygen during anesthesia to 21% once the ETT is removed and the patient is breathing room air. Shivering, which is a common postoperative occurrence, can also cause an increase in oxygen consumption and possibly lead to hypoxia. Supplemental oxygen is provided if SpO$_2$ < 94% and is stopped when the patient can keep these values when off oxygen for 5 minutes.

Tech Tip

In the recovery period shivering can increase the metabolic oxygen demand/consumption at a time when the oxygen supply has often been reduced (inspired oxygen going from 100% on the anesthetic machine to 21% in room air once the patient is extubated). It is a good idea to monitor the patient's postoperative oxygen saturation level, particularly in hypothermic patients.

IV Catheter Care

Patients recovering from routine surgery or procedures typically have their IV fluids stopped during recovery as they can maintain hydration status without supplementation and because animals often get tangled in the IV lines. For those patients who are hypovolemic and need fluid therapy, IV catheters should be checked regularly during the recovery period for patency, cleanliness, and whether they are a source of irritation to the patient. IV catheters should also be checked to make sure they are secure and should be re-bandaged if necessary. For the recovery process, many clinics prefer a bandage covering over the taped IV catheter consisting of a stretch gauze layer followed by a Vetrap® layer with a hole or opening fashioned so the injection port is readily available. IV catheters can easily become dislodged while transporting the patient from surgery to the recovery area, during post-anesthetic emergence delirium, and by patient self-trauma. A possible complication is the t-port (see Figures 4-27a and 4-27b) or cap coming off of the IV catheter, leading to blood loss from the patient.

Pain Assessment

Pain assessment is critical to recovery in the post-anesthetic period and beyond. Since pain is subjective, its assessment should be made by the same person each time (if possible) and in the same manner throughout the entire recovery process. Differentiating pain from **emergence delirium** (dissociated state in which the patient is agitated and inconsolable, uncooperative and typically thrashing, vocalizing, and defecating) and **dysphoria** (state of feeling unwell, unhappy, or emotional or mental discomfort) can be difficult in the recovery period. Animals that are experiencing emergence delirium, dysphoria, and/or pain will likely vocalize.

- If the patient is extubated and immediately seems to be vocal and begins thrashing, flailing, etc., this is likely emergence delirium and occurs as the patient regains consciousness to its surroundings. Often all that is needed is gentle restraint for a minute or two and the patient will often settle.
- If the patient is not settling or is getting worse, then the patient is assessed for pain versus dysphoria.
  - Patients who are in pain are generally not consolable. These patients react (move away from, tense up, try to bite, etc.) when the surgical site is gently touched. The patient might also be vocal, try to bite/chew its surgical site, etc. Parameters such as tachycardia, tachypnea, and elevated blood pressure are *poor* indicators of pain and can be altered in many other settings.
    - Patients who are in pain should receive more analgesia (e.g., rescue analgesia) with rapidly-acting opioids. Once the patient is more comfortable, the entire pain management plan should be reassessed and adjusted (e.g., a nonsteroidal anti-inflammatory drug [NSAID] should be given). Cold compressing a surgical site can also be done, but confirm with the veterinarian first.
  - Patients who have dysphoria can also be vocal or restless, but these cases are more commonly consolable and do not respond to gentle palpation of their surgical site.

If the patient seems restless or distressed, it is important to consider other differentials such as the patient needs to urinate or defecate, is hot, is hypoxemic and cannot breathe, or just wants to be held in someone's arms (very common in lap dogs/toy breeds). If drug therapy is needed, opioids are good for treating painful animals while acepromazine or dexmedetomidine are good for treating dysphoria or emergence delirium. Sedation/tranquilization should be provided when in doubt so that the patient does not hurt itself, its surgical site, or the recovery technician. Dexmedetomidine can be used if it is unclear whether the patient has dysphoria or pain because it will sedate the animal and provide pain control. Sedatives/tranquilizers should be given IV for quicker onset of action if the patient is painful or dysphoric. Veterinarians will use a much lower dosage of sedative/tranquilizer for patients in recovery than was used for pre-medication. In general, drugs such as benzodiazepines should *not* be used for tranquilization because they are less reliable at reducing anxiety in veterinary patients and they do not provide analgesia.

Veterinary clinics and hospitals employ a variety of pain assessment tools so they will differ depending upon preference (see Chapter 15 [Acute Pain Management] and Chapter 16 [Chronic Pain Management]). Pre-emptive management of pain is optimal. Providing ample pain control prior to recovery/extubation is important, but realize that it may delay recovery or extubation if, for example, a dose of hydromorphone is given IV at the end of anesthesia.

Mentation Assessment

Assessment of **mentation** (level of consciousness and the patient's reaction to its environment) should be performed at regular intervals. Significant changes or a trend toward less responsiveness should be reported to the veterinarian for further assessment. As with pain assessment, it is optimal for the same person to assess mentation.

In rare situations, patients who are not recovering as expected may remain extremely obtunded (dull, aware of environment but senses are blunted) in the postoperative setting. These patients should have their vital signs assessed as well as other parameters such as blood pressure and blood glucose and then treated accordingly. If these parameters are normal, sometimes a small amount of reversal drug (such as naloxone (Narcan®), atipamezole (Antisedan®), flumazenil [Romazicon®, generic]) can be diluted and titrated to effect to prevent rapid awakening, emergence delirium and rough recovery, and removal of analgesia. In general, reversal of drugs is not recommended unless the patient is having an adverse effect.

ANESTHETIC RECOVERY CONSIDERATIONS

Patient monitoring and supportive care must continue through all phases of anesthesia including recovery. The following are important procedures beyond monitoring basic vital signs to help ensure the smoothest recovery possible for the patient.

Oxygen Administration

Patients benefit from supplemental oxygen as they begin their recovery from anesthesia. Whenever possible, oxygen should be administered for 5 to 10 minutes before patients are moved into the recovery area. The patient will benefit from breathing oxygen since anesthetic drugs can depress the respiratory system.

The following are ways to supply oxygen to patients in the recovery period:

- The patient can either remain attached to the breathing circuit with the vaporizer turned off and O_2 left on (Figure 11-4a) or
- If the patient requires imminent extubation, the ETT should be deflated and removed. A face mask or the end of the breathing circuit can be used to deliver O_2.

If the administration of oxygen in an extubated patient during a routine recovery causes undo stress, it is best to discontinue providing O_2 or move the patient to an oxygen cage.

If post-anesthetic O_2 is a medical necessity (if $SpO_2 < 94\%$ via pulse oximetry or $PaO_2 < 80$ mm Hg via blood gas analysis) several methods could be chosen:

- The animal can be placed in an O_2 cage (Figure 11-4b) or be given oxygen via a face mask or nasal cannula. Oxygen cages and nasal cannulas can provide approximately 30 to 40% oxygen on average.
- A NT tube can be placed through the nares and into the trachea to deliver oxygen (Figure 11-4c) (placing a NT tube at this point is difficult). NT tubes are more commonly placed in horses than in small animals.

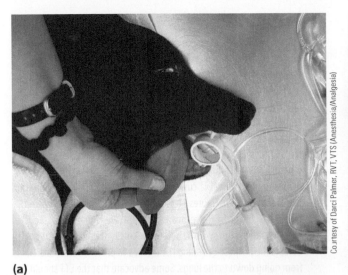

(a)

Courtesy of Darci Palmer, RVT, VTS (Anesthesia/Analgesia)

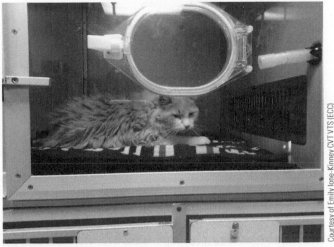

(b)

Courtesy of Emily Ione-Kinney CVT VTS (ECC)

(c)

Courtesy of Chris Norkus, DVM, DACVAA, CVPP, DACVECC

FIGURE 11-4 Post-anesthetic oxygen can be provided in a variety of ways. (a) This dog remains attached to the breathing circuit with the vaporizer turned off and O₂ left on. Post-anesthetic oxygen can also be administered via a face mask or end of the breathing circuit if the patient needs to be extubated. (b) This cat is in an oxygen cage, which provides a variable percent of oxygen (supplied through tubing on right of cage). (c) A NT tube can provide oxygen via nasal intubation. The nares should be clean and the ETT well lubricated so that it can be passed gently with nonrotational movement to limit trauma to this area. This horse was hand recovered and thus had a halter in place and a rope attached to the halter.

Patient Positioning

Keeping the airway open is always important when determining how to position a patient during recovery. In general, patients should be positioned however comfortable and in sternal recumbency whenever possible (horses are the exception and are allowed to lay in lateral recumbency). Sternal recumbency helps small animals ventilate better and more effectively in both lung fields. Keeping the patient's head elevated on a pillow or "propped up" with rolled towels or blankets under their head can be useful to prevent aspiration once the ETT is removed and for patient comfort. Foam wedges can be used to prop the patient's chin

but should be done carefully especially when the patient is still intubated to avoid kinking of the ETT and prevent airway obstruction. Sounds of stridor/stertor should prompt the recovery technician to reposition the patient's head, neck, tongue, and airway.

When positioning patients during recovery, be aware of the surgery site and areas of the body that might be painful postoperatively. In general it is best not to move the patient too much as this will stimulate them and may cause them to awake suddenly. The position of an ETT must be carefully watched if turning a patient to prevent tracheal damage due to the twisting motion or oxygen deprivation from a kinked ETT.

Clean blood, betadine, etc., off the skin ASAP with saline or hydrogen peroxide; it will be easier to remove if not left on the skin for a long period of time. Remember to avoid cleaning directly over the sutures.

Handling Fractious Patients

For those animals that need an Elizabethan collar and are considered frisky to fractious, placement of the Elizabethan collar should occur well before they completely awaken. The chance of placing the Elizabethan collar easily is drastically reduced when the patient has recovered sufficiently to bite or scratch at the collar. This also can help to reduce the level of fractiousness and degree of self-trauma to the surgery site. These animals must be watched diligently in case they develop difficulties with the Elizabethan collar, which can inhibit their movement during the early recovery period.

If the patient is too fractious to be handled safely, recovery in a crate may be the best option. The patient should be placed in the crate with its head at the door and the crate placed where the patient is visible and recovery can be monitored, although it may not be possible to assess any vital signs other than a respiratory rate. Placing a leash on patients who are dangerous or may dart out from a crate/cage may be warranted. Large dogs may need to have a basket muzzle placed once extubated.

Extubation

Extubation is the process of removing the ETT, and the main indicator for extubation is the swallow reflex. Once patients are able to swallow, they are able to protect their airway from aspiration should they vomit or regurgitate. Patients that begin to show voluntary movement of either their head or limbs, or even show subtle to spastic movement of the tongue should be extubated if they are close to consciousness even if the swallow reflex has not been observed. Swallowing, movement, or robust cranial nerve signs in response to touching or talking to the patient should prompt extubation. Failure to extubate the patient in a timely fashion may result in the patient biting the ETT, which can potentially lead to airway obstruction.

In the case of brachycephalic dogs, it is considered safest to recover these patients in sternal recumbency with their heads elevated, mouth open, and tongue pulled forward (Figure 11-5). This helps to counteract the anatomical differences in these breeds such as stenotic nares and elongated soft palates, which increase the risk of airway obstruction. A dedicated technician to monitor these recoveries is imperative for a patient's safe and smooth recovery. Oxygen therapy, pulse oximeter, clean ETT, and induction agent should be available to provide support to the patient. Brachycephalic dogs may benefit from adequate tranquilization to allow for a slow, regular, effort-free breathing pattern and to keep them calm as to not increase their oxygen requirements. Tranquilization should be considered on a case-by-case basis because if these dogs are overly sedate, especially with a drug such as an alpha-2 adrenergic agonist, they could experience respiratory depression and relaxation to the point of occluding the airway. Many brachycephalic dogs, as long as pain is controlled, are quite content to recover with the ETT in place for as long as possible because they ventilate so well. Often these dogs will eventually stand up and just "spit" the ETT out.

Guidelines for extubation:

- Prior to extubation it is optimal for the patient to continue breathing oxygen for 5 to 10 minutes after the vaporizer has been turned off. Delivery of 100% oxygen will help to remove the inhalant anesthetic agent, which causes dosage dependent respiratory depression. This also allows exhaled inhalant anesthetic agent to continue to be scavenged, which reduces personnel exposure to WAG.
- Before extubation visually inspect the oropharynx for foreign material and clear any accumulation of blood, gauze, lubricants, tartar, or other materials from the area.
- In most patients, the ETT cuff should be deflated immediately prior to removal of the ETT (Figures 11-6a and 11-6b) and the ETT protected from being bitten by the recovering animal (see Figure 11-4a). Be sure the patient can swallow before the cuff is deflated, otherwise, it defeats the purpose of sealing the trachea to prevent secretions/fluids from going down to the lungs. Some advocate that the ETT should be removed with the cuff partially inflated if fluid may have entered the trachea and has become sequestered proximal to the cuff. The cuff should be slightly deflated to avoid tearing the tracheal mucosa and the patient should be extubated with its head hanging over the edge of the table or facing down to avoid having fluid enter the trachea.
- The ETT is disconnected from the breathing hose (Figure 11-6c) and the oxygen turned off (Figure 11-6d).

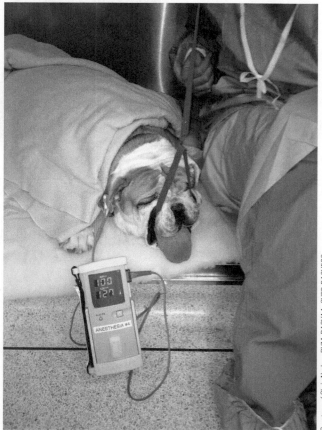

Courtesy of Chris Norkus, DVM, DACVAA, CVPP, DACVECC

FIGURE 11-5 A safe way to recover brachycephalic dogs is to have these patients in sternal recumbency with their heads elevated, mouth open, and tongue pulled forward.

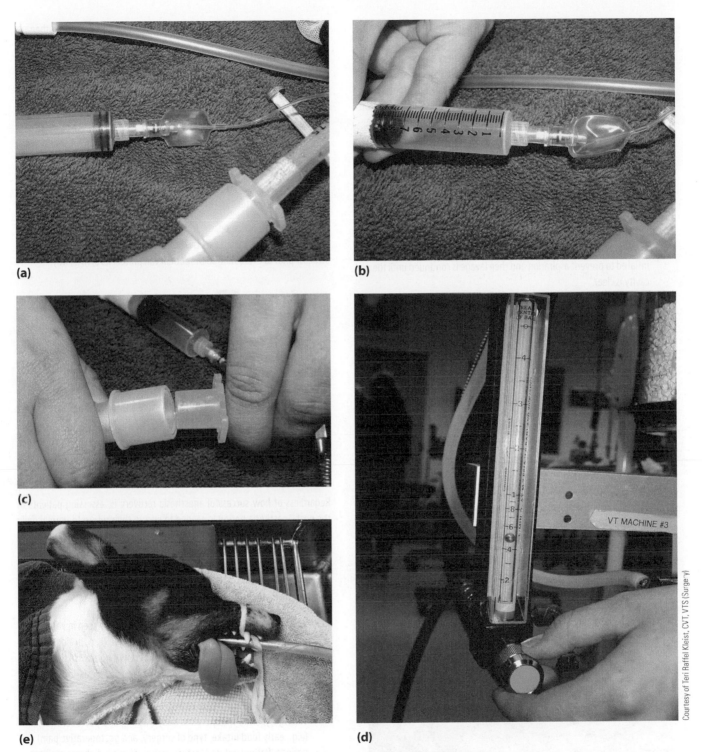

FIGURE 11–6 Extubation. (a) The ETT cuff should be deflated before it is removed. This ETT cuff is inflated. (b) This ETT cuff is deflated. (c) The ETT is disconnected from the breathing hose. (d)The oxygen is turned off. (e) The ETT is untied and then gently pulled to remove it from the patient.

Courtesy of Teri Raffel Kleist, CVT, VTS (Surgery)

- The ETT is untied (Figure 11-6e) and gently pulled to remove it from the patient.
- Animals anesthetized for gastrointestinal surgery such as dogs with gastric dilatation volvulus (GDV) and horses with colic are prone to passive movement of fluid into pharynx. Nasogastric or orogastric tubes commonly used during these surgical procedures may promote flow of gastric contents into the pharynx when the ETT is removed. With either species, the head should be positioned to allow drainage of fluid from the pharynx during surgery, and removing the ETT with the cuff inflated and head tilted down is advised (Figure 11-7). Evaluation and suctioning of the airway may be needed.
- If the patient regurgitates prior to extubation, the patient should be put back under anesthesia, the oropharynx and esophagus suctioned to remove the highly acidic material, and the area gently lavaged with saline and a bulb syringe or red rubber catheter. The ETT must be kept inflated to prevent aspiration and then lavage is continued until the oral cavity is clear.
- Extubation following the use of mechanical ventilation should follow the same procedures since the animal should be properly "weaned" off the ventilator, receiving supplemental oxygen until spontaneous breathing is relatively normal. To properly wean a patient off of a ventilator, the respiration rate is gradually reduced to a few breaths per minute as the patient is monitored for spontaneous breathing and end-tidal CO_2 is allowed to increase to 50 to 55 mg Hg. Lowering/decreasing anesthetic depth will also help wean a patient from a ventilator. The patient may still need to be given occasional manual breaths while regaining spontaneous respiration and tidal volume. The entire process usually takes about 5 minutes.
- Some veterinarians like to have their patient's **reticular activation center (RAC)** of the brain stimulated in order to facilitate recovery and extubation while others do not unless faced with a prolonged recovery (> 45 minutes off inhalant anesthetic agent). The RAC is the area of the brain responsible for maintaining consciousness in animals. Some stimulation might include, talking, moving limbs back and forth, opening mouth, etc. Aggressive stimulation in order to "trigger" initial swallowing should be avoided because it may cause the patient to be extubated too early which can result in the patient going back to sleep without a secured airway. Overly stimulated

animals can be startled and will awaken suddenly, panic, and then have a poor quality recovery.

- Continue to monitor the patient's respiratory function, tissue perfusion, and mentation level after extubation to ensure that recovery is progressing in a stable manner.
- In cats, the ETT may be removed as soon as the cat shows evidence of nearing arousal.
  - ○ Watch for blinking and ability to move voluntarily. Cats often do not swallow with an ETT in place, and delaying extubation may result in having a spastic, uncontrollable cat with an ETT in place that can quickly lead to the ETT being bitten.
  - ○ Delayed extubation may make cats more susceptible to laryngospasm.

Following extubation, the recovery technician must protect the patient's airway from foreign material. This is best achieved by waiting for the patient to swallow prior to extubation so the patient can protect its own airway.

Tech Tip 🐾

Signs that indicate recovery is complete are the patient's ability to swallow (indicates nerve impulse transmission is resuming in the brain), maintenance of normothermia (normal thermoregulatory mechanisms are functional), and return of normal respiratory and heart rate (anesthetic agents are clearing from the body).

Post-Anesthesia Concerns

Regardless of how successful anesthesia recovery is, assessing patients in the post-anesthesia period is needed to ensure their return to normal daily activities. Many times these assessments are done by clients at home making communication with them an important part of post-anesthetic care of patients. Veterinary staff should monitor the following if patients remain in the hospital/clinic or communicate the following information to clients if the patient is being discharged:

- Postoperative nausea and vomiting (PONV) can be seen in veterinary patients due to risk factors in the preoperative (e.g., prolonged preoperative fasting, full stomach if food was not restricted preoperatively, anxiety, and administration of emesis-inducing drugs), intraoperative (e.g., being in a deeper plan of anesthesia, administration drugs that known to cause vomiting, duration of surgery, and intraoperative dehydration), and postoperative (e.g., early food intake, type of surgery, and postoperative pain) periods. Patient-related factors such as obesity, diabetes mellitus, hypothyroidism, and abdominal conditions can also increase the risk of PONV. Sevoflurane (SevoFlo®, Petrem®, Flurovess™, Ultane®) administration has been known to cause nausea, while opioids and ketamine have been linked to vomiting. Antiemetic drugs such as the neurokinin receptor antagonist maropitant (Cerenia®) and/or the 5-HT3 receptor antagonist ondansetron (Zofran®) may be given to prevent or treat vomiting (Figures 11-8a and 11-8b).
- Normal bowel movements may not return for 48 to 72 hours after anesthesia. Fasting prior to anesthesia, anesthetics drugs including

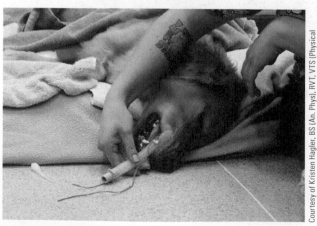

FIGURE 11-7 The patient's head is facing down to prevent fluid from entering the trachea.

Courtesy of Kristen Hagler, BS (An. Phys), RVT, VTS (Physical Rehabilitation), CCRP, CVPP, VCC, OACM, CBW

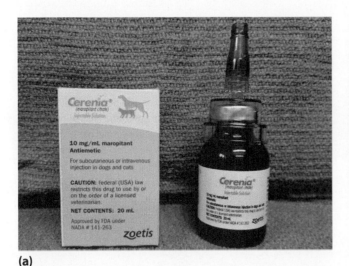

(a)

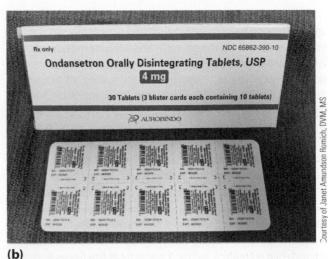

(b)

Courtesy of Janet Amundson Romich, DVM, MS

FIGURE 11-8 Antiemetic drugs such as (a) maropitant and/or (b) ondansetron can be used to manage PONV.

opioids and inhalants that slow GI motility, and consumption of small meals after anesthesia limits the amount of stool produced in the post-anesthesia period. Patients may have little interest in food or water for the first 12 to 24 hours after anesthesia; however, some are interested in food after an anesthetic procedure and may be fed a decreased amount of their normal meal. Once patients begin normal levels of eating and drinking, normal bowel movements should occur.

- An animal may exhibit changes in their behavior for several days after general anesthesia. They may seem unfamiliar with their surroundings including people or other animals. Young children should not be left unattended with an animal that has just recovered from general anesthesia no matter how trustworthy that animal normally is. Behavioral changes after general anesthesia typically resolve within a few days.

- An animal's ability to control its body temperature may be affected during the recovery period. Some anesthetic agents may alter the temperature set point in the hypothalamus or cause vasodilation of blood vessels in the skin to promote heat loss, while other animals may be unable to adequately respond to increases in environmental temperature due to altered natural cooling methods such as panting. For the first few days after general anesthesia, the patient should be kept in a warm, not overly hot area and assessed often to make sure they are comfortable.

- Animals who were intubated during anesthesia may have a slight cough for several days from irritation from the ETT used during anesthesia. If the cough becomes moist, progressive, or productive, the veterinarian may request that further diagnostic testing be performed.

- Restricted activity is important to allow healing from the surgery or initial injury. Keeping the patient quiet and not allowing running, jumping, bathing, or swimming is key to ensuring proper healing time and surgical success.

- Incisions should be assessed at least once daily for a minimum of 14 days or until any sutures are removed. Clients should be instructed not to touch, clean, or put topical medication on the incision. Clients should be reminded that not all sutures are external; therefore, restricting activity as described above is important for these incisions or injuries to heal. Any changes in the appearance of the incision or any redness, swelling discharge, pain, loss of sutures, or opening of the incision should be reported to the veterinarian.

Tech Tip

Anesthetic discharge instructions, in addition to a surgical discharge form, can describe potential complications that may arise and outline when the client should contact the veterinary team. These discharge instructions are important guides for postoperative care of patients. Examples of an online discharge instruction template can be found on the American Animal Hospital Association (AAHA) website at https://www.aaha.org/aaha-guidelines/2020-aaha -anesthesia-and-monitoring-guidelines-for-dogs-and-cats /anesthesia-discharge-template/ and one specifically designed for discharge of dental patients can be found at https://www.aaha.org/aaha-guidelines/dental-care /resource-center/oral-health-discharge-template/.

SPECIAL RECOVERY CONSIDERATIONS FOR RUMINANT AND EQUINE PATIENTS

Ruminants are generally docile while recovering from anesthesia and will stay recumbent; however, special measures are taken to accommodate for complications such as regurgitation and aspiration. Ruminants should be recovered in sternal recumbency with their heads propped up on something like a bale of hay or bucket to prevent aspiration. Because of the possibility of regurgitation, the ETT should remain inflated and in place until the time of extubation. Ruminants are extubated when they are able to support and move their own head. After extubation, the ruminant should remain in sternal recumbency with its head up and closely monitored for regurgitation until able to stand. Eructation is a good sign that the GI tract is returning to normal function. Equipment and drugs should be available to re-induce and re-intubate if needed. Ruminants usually begin to stand by lifting their hind end up in the air, following by the front limbs.

The recovery period for equine patients can be extremely unpredictable and therefore dangerous to both the patient and veterinary staff. Recovery strategies should be well thought out in advance. Some recovery techniques rely on head/tail ropes, inflatable mats, water pools, and slings to assist with transitioning the horse to a standing, recovered position and to reduce the risk of complications. A few complications that are commonly seen include re-injury of orthopedic repair, fractures, and lacerations. If intubated, the horse should continue to receive oxygen throughout recovery via the orotracheal tube. Alternatively, one or two NT tube(s) should be passed before extubation and standing, especially if the horse has any nasal edema or was under general anesthesia for more than an hour. The decision about when to pull the ETT should take into account the horse's behavior. Most horses should not be extubated until they are taking regular deep breaths and standing. Once extubated, observe the patient for signs of distress/restlessness, listen for stridor (a harsh vibrating noise during inspiration), watch for increased respiratory rate and effort of breathing, and check nostrils for air flow (horses are obligate nasal breathers) that may indicate that the patient has an airway obstruction. If an obstruction is suspected, a NT tube is passed or the horse may need to be re-induced and an orotracheal tube passed (very rare). It is advisable to check for airway obstruction by holding a hand over the nostrils to feel airflow while watching thoracic and abdominal movements. Horses are often given sedatives, such as small amounts of an alpha-2 adrenergic agonist and/or acepromazine, to allow for a smooth, slow recovery and kept in lateral recumbency for a long enough period of time so their inhalant anesthetic agent is eliminated from the body, which allows them to attempt standing when they are more clear-headed. This method allows the horse to recover on its own in a locked recovery stall with close monitoring in case intervention is required. Less monitoring is done when horses are allowed to recover in the field or in rare situations when a horse is hand recovered because it is too dangerous for someone to stay with the horse. Foals are generally hand recovered. Horses generally begin to stand by placing their front legs out first, followed by lifting in their hind end.

More information on ruminant and equine anesthesia and recovery is found in Chapter 19 (Anesthesia and Analgesia in Large Animal and Food Producing Species).

SUMMARY

Patient recovery starts when the delivery of anesthetic agents is discontinued and is not only a critical time for the prevention of patient morbidity and mortality but also for maximum patient comfort so that anesthetic recovery is as stress-free and rapid for the patient as possible.

Each recovery experience is unique; therefore, each patient should be treated as an individual. All patients should be assessed regularly during recovery by monitoring core body temperature, heart/pulse parameters (rate, rhythm, strength, and quality), respiratory parameters (rate, rhythm, and quality), mucous membrane color, CRT, oxygenation, IV catheter placement, pain status, and mentation status. In addition to monitoring vital signs, the use of pulse oximetry and close observation in addition to providing supportive care such as supplemental oxygen and transporting them appropriately to a recovery room that is warm, dark, and quiet benefits all patients recovering from anesthesia. Having a plan in place to handle both calm and fractious patients and to successfully extubate them helps ease patients through the recovery period. Careful planning and close monitoring combined with sound nursing practices contribute to the best possible outcome for the patient.

CRITICAL THINKING POINTS

- During recovery, the best "monitor" is a knowledgeable and diligent veterinary technician.

- The recovery period is the leading time during the anesthesia experience for patients to experience complications and death; therefore, preparation for recovery is a key part of the anesthesia plan.

- Diligent monitoring "early and often" is the key to a successful recovery.

- Parameters such as vital signs are a direct reflection of what is happening systemically in patients and serve as indicators of how to proceed with patient care.

- In the post-anesthetic period, patients benefit from supplemental oxygen, which should be administered for 5 to 10 minutes before patients are moved to the recovery area and then provided as needed based upon continuous monitoring with a pulse oximeter. Supplemental oxygen is provided if $SpO_2 < 94\%$ (pulse oximetry) or $PaO_2 < 80$ mm Hg (arterial blood gas) and is stopped when the patient can maintain normal values once oxygen has been discontinued for 5 minutes.

- The anesthetic monitoring record provides a visual representation of physiologic values and trends that play a key role in making adjustments during the recovery period and for planning future anesthesia in a patient.

REVIEW QUESTIONS

Multiple Choice

1. What anatomical differences in brachycephalic breeds contribute to complications in the peri-anesthetic period?

   a. Short legs and big necks
   b. An enlarged spleen
   c. Stenotic nares and elongated soft palates
   d. Round eyes and short tails

2. During the recovery period, when is the best time to extubate a dog?

   a. Immediately when the anesthetic gas is turned off
   b. About 10 minutes post anesthesia
   c. When the swallow reflex returns
   d. It does not make much difference

3. While in recovery, how should a patient's heart rate be taken?

    a. Via visual inspection of the thoracic wall.

    b. With a stethoscope while evaluating pulse rate.

    c. Only with monitoring equipment such as an ECG

    d. By auscultating the heart for 10 seconds and multiplying the number by 4

4. Which of the following is the best way for ruminants to be recovered?

    a. Lateral recumbency

    b. Sternal recumbency

    c. Standing

    d. In a water pool

5. Which of the following would be an appropriate esophageal lavage solution to use in case of regurgitation?

    a. Saline

    b. Dilute chlorhexidine scrub

    c. Dilute alcohol

    d. Betadine solution

6. Which of the following is a reliable indication that an animal is in pain?

    a. Biting the surgical site

    b. Tachycardia and restlessness

    c. Tachypnea and lethargy

    d. Elevated blood pressure

7. Why is it usually best to extubate a cat as soon as there are signs of arousal?

    a. To get it over with so that the next animal can be anesthetized

    b. In order to clean the recovery area faster to avoid delays with discharging patients

    c. In order to move them to a warmer area as soon as possible to prevent hypothermia

    d. The ETT can more swiftly lead to complications (e.g., laryngospasm, chewing the tube)

8. Which of the following is a good method of recovering a fractious animal?

    a. Placing the patient in a crate until mobile and then placing an Elizabethan collar

    b. Placing the patient in a crate without escape doors to prevent self-trauma

    c. Placing the patient in a crate with its head away from the door to prevent overstimulation

    d. Placing the patient in the crate with its head at the door and the crate placed where the patient is visible and recovery can be monitored

9. Which of the following is the best way to recover brachycephalic dogs?

    a. In lateral recumbency with their heads on a pillow and their tongue pulled laterally

    b. In sternal recumbency with their heads down and mouth gag in place to prevent them from biting the ETT

    c. In sternal recumbency with their heads elevated, mouth open, and tongue pulled forward

    d. In lateral recumbency with the ETT tied to their muzzle and their heads down to prevent aspiration of foreign material or secretions

10. During recovery, why does the rumen volume and length of time required for the rumen to empty put recovering cattle at risk?

    a. It can lead to regurgitation and aspiration.

    b. It can cause the animal to be unstable if a standing surgery is done.

    c. It may cause the mucous membranes to become icteric.

    d. It can lead to airway obstruction.

11. Which of the following is the area of the brain responsible for maintaining consciousness in animals?

    a. Cerebellum

    b. Hypothalamic pituitary pathway

    c. Hippocampus

    d. Reticular activation center

12. Which is the optimal way to measure body temperature in animals?

    a. Aurally

    b. Sublingually

    c. Rectally

    d. Lingually

13. Why might an anesthetized patient develop hypoxemia during the recovery period?

    a. Animals in recovery may still have the residual effects of drugs, which may cause hyperventilation/increased respiratory reserve volumes.

    b. Animals may shiver which decreases oxygen consumption.

    c. Animals in recovery have muscle relaxation that results in enhanced ability to protect the upper airway.

    d. Animals may develop anesthesia- and positioning-induced atelectasis during recovery.

14. SpO_2 values should not drop below

    a. 100%

    b. 94%

    c. 90%

    d. 85%

15. In what type of recumbency can a horse be best recovered?

    a. Lateral
    b. Sternal
    c. Dorsal
    d. Standing

16. Which of the following is the standard of care for post-anesthetic monitoring after extubation?

    a. Doppler blood pressure
    b. Capnography
    c. Arterial blood gas analysis
    d. Pulse oximetry

17. Depending on their behavior, when should most horses be extubated?

    a. When they are in sternal recumbency and swallowing
    b. When they are in lateral recumbency and swallowing
    c. When they are taking regular deep breaths and standing
    d. When they are taking regular deep breaths and in lateral recumbency

18. What is the best way to continue ECG monitoring in a recovery patient?

    a. Clips
    b. Electrode pads
    c. Wires
    d. A cuff

19. Which of the following are main reasons complications increase during the recovery period?

    a. Residual effects of drugs on the cardiovascular and respiratory systems
    b. Complacent anesthesia team members who stop monitoring and supporting the patient after anesthesia drug delivery is discontinued
    c. Residual effects of drugs on the neurologic system
    d. All of the above

20. During the recovery period, when is the greatest risk of exposure to WAGs?

    a. While the patient is on the operating table connected to the breathing circuit
    b. When the patient is being moved to the recovery cage
    c. When the patient is at a surgical plane of anesthesia
    d. When the patient is first disconnected from the anesthesia machine

Case Studies

Case Study 1: A 5-year-old female Dalmatian dog presented to the hospital for a routine spay (ovariohysterectomy). The recovery technician notices a trend of the dog's temperature consistently decreasing throughout the surgery. The first post-op temperature reading is 97.3°F (36.3°C). The dog has a recovery bed ready with sufficient padding to keep her from coming into contact with the cold surface of the cage. The dog is carefully moved to its recovery bed and a forced air blanket (Bair Hugger®) is placed on the dog.

1. The dog is not ready to be extubated, so what considerations should be followed for this patient?

Case Study 2: A 7-year-old M/N Great Dane dog presents to the hospital with gastric dilatation volvulus (GDV). Upon presentation the patient's blood lactate level is 8 mmol/L (elevated), hematocrit is 52%, total protein is 7.8 g/dL, and electrolytes and remaining blood chemistry parameters are within normal limits. The patient is stabilized prior to anesthesia and surgery with IV fluids, IV pain control, and gastric decompression. The patient is given ketamine and diazepam for induction and maintained on isoflurane. The patient experiences intraoperative hypotension, which is treated with an IV bolus of LRS. Occasional ventricular premature contractions (VPCs) are noted on the ECG. Once the surgery is complete, his care becomes the responsibility of the recovery technician.

1. As the anesthesia recovery technician, what considerations need to be addressed for this patient?

Case Study 3: A 12-year-old F/S Chesapeake Bay Retriever dog presents to the hospital in critical condition after collapsing. The patient is stabilized, then is diagnosed with a ruptured splenic mass and undergoes a splenectomy. While under inhalant anesthesia, the patient starts having premature ventricular contractions (VPCs). The recovery area was set up in advance with the usual comfortable bedding but now needs an ECG monitor and a fluid pump. Before the patient is extubated, the recovery technician clips small areas of fur and applies electrode pads in order for a continuous ECG to run while ensuring patient comfort. The continuous ECG will help the recovery team to continue closely monitoring the heart for sustained VPCs and intervene if they should escalate to more serious hemodynamic instability. The veterinarian asks for blood work, including PCV/total protein and electrolytes. The catheter is checked for patency and the patient is placed back on IV fluids to continue to support the patient after the stress of emergency surgery and maintain electrolyte balance. The patient continues to recover comfortably without incident. When she is stabilized, she continues to receive IV fluid therapy, and the ECG monitor is left on to continue monitoring her.

1. Although this patient had cardiac complications, she recovered well because the veterinary technician was prepared for this patient's recovery. List three measures/considerations that helped this patient in the recovery period.

Case Study 4: A 1-year-old male Siamese cat presents to the hospital for neutering. He was premedicated with buprenorphine and dexmedetomidine SQ and an IV catheter is placed in a front limb. The cat is induced with propofol, intubated, and maintained on isoflurane. The surgery is routine, the vaporizer is turned off, and the patient is extubated quickly. The postanesthetic temperature is 103.8°F (39.9°C).

1. List four measures/considerations that should be done about this patient's temperature.

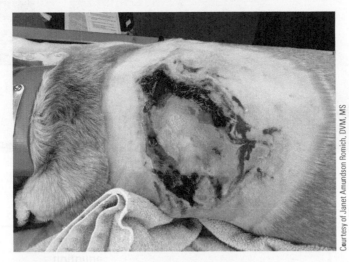

Courtesy of Janet Amundson Romich, DVM, MS

FIGURE 11-9 Large postoperative lesion on a dog.

Case Study 5: Max, a 5-year-old M/N yellow Labrador Retriever dog, presents to the clinic with a laceration that needs to be sutured. Max is premedicated with IV hydromorphone and midazolam and induced with IV propofol. An ETT is passed, isoflurane is administered via a precision vaporizer, and Max is placed in left lateral recumbency on a circulating warm water blanket so his laceration can be sutured. He recovers from surgery and is sent home. Two days later the owner calls and says Max has a big "sore" on his side. The owner brings Max to the clinic and a large lesion is seen on his left side (Figure 11-9). The lesion is not associated with the area that was sutured, but rather on the side he was lying on.

1. What is the likely cause of Max's lesion?
2. How could this lesion been prevented?

Critical Thinking Questions

1. A M/N German Shepherd dog is recovering from a tibial-plateau-leveling osteotomy (TPLO) surgery. The patient is noted to be taking more rapid and labored respirations while intubated. Once extubated, the dog begins howling and looking back at his incision. The patient is gently restrained for assessment. How would it be determined whether this dog is experiencing pain or emergence delirium?

2. A 6-month-old female Chihuahua puppy presents to the clinic for entropion surgery on her left eye. After surgery, she is placed in the recovery area by a member of the surgical staff. The puppy is scared and very quiet in the recovery cage, which is in the corner of the recovery area. As part of the anesthesia team, you discuss with the surgery technician the importance of monitoring patients during the recovery period. What advice do you give? What possible complications could arise if this patient was left in an unobservable part of the recovery area?

ENDNOTES

1. Brodbelt, D. C., Blisset, K. J., Hammond, R. A., et al. (2008). The risk of death: The confidential enquiry into perioperative small animal fatalities. *Veterinary Anesthesia and Analgesia, 35*(5), 365–373.
2. Gruenheid, M., Aarnes, T. K., McLoughlin. M. A., et al. (2018). Risk of anesthesia-related complications in brachycephalic dogs. *Journal of the American Veterinary Medical Association, 253*, 301–306.
3. Grubb, T., Sager, J., Gaynor, J. S., Montgomery, E., Parker, J. A., Shafford, H., & Tearney, J. (2020). 2020 AAHA anesthesia and monitoring guidelines for dogs and cats. *Journal of the American Animal Hospital Association, 56*(2), 59–82.
4. Jones, S. J., Mama, K. R., Brock, N. K., et al. (2019). Hyperkalemia during general anesthesia in two Greyhounds. *J Am Vet Med Assoc, 254*, 1329–1334.
5. Robinson, E. P., Sams, R. A., & Muir, W. W. (1986). Barbiturate anesthesia in greyhound and mixed-breed dogs: Comparative cardiopulmonary effects, anesthetic effects, and recovery rates. *Am J Vet Res, 47*, 2105–2112.
6. Haskins, S. (2007). Monitoring anesthetized patients. In W. J. Tranquilli, J. C. Thurmon, K. G. Grim (Eds.), *Lumb and Jones' veterinary anesthesia and analgesia* (4th ed., pp. 86–105). Ames, IA: Blackwell.

CHAPTER 12
Anesthetic Complications

Kim Spelts, *BS, CVT, VTS (Anesthesia/Analgesia)*

LEARNING OBJECTIVES

Upon completion of this chapter, it is expected that the reader should be able to:

12.1 Recognize the most common complications that can occur in anesthetized patients

12.2 Analyze abnormalities in temperature in anesthetized patients

12.3 Analyze abnormalities in blood pressure in anesthetized patients

12.4 Analyze abnormalities in heart rate and rhythm in anesthetized patients

12.5 Analyze abnormalities in oxygenation in anesthetized patients

12.6 Analyze abnormalities in ventilation in anesthetized patients

12.7 Determine the underlying cause of physiologic abnormalities associated with anesthesia

12.8 Describe how to take precautions to minimize the risk of anesthetic complications

12.9 Explain how to respond to anesthetic complications

KEY TERMS

Alveolar oxygen equation
Atrial fibrillation (A-fib)
Atrial premature contractions (APCs)
Atrioventricular (AV) block
Autoregulation of blood flow
Dead space ventilation
Diffusion impairment
1st-degree AV block
Hyperthermia
Hyperventilation
Hypocapnia
Hypotension
Hypothermia
Hypoventilation
Hypoxemia
Idioventricular rhythm
Inspired oxygen (F_iO_2)
Intermittent positive pressure ventilation
Malignant hyperthermia
Mobitz Type 1 (Wenckebach)
Mobitz Type 2
Oxygen delivery (DO_2)
Respiratory sinus arrhythmia
Right-to-left shunt
2nd-degree AV block
Sick sinus syndrome (SSS)
Sinus bradycardia
Sinus tachycardia
Supraventricular tachycardia
3rd-degree AV block
Ventilation-perfusion (V/Q) mismatch
Ventricular fibrillation (V-fib)
Ventricular premature contractions (VPCs or PVCs)
Ventricular tachycardia (V-tach)

INTRODUCTION

No anesthetic agent or procedure is 100% safe; therefore, every anesthetic event carries risk. The sicker the patient, the more likely he/she is to experience complications during the perianesthetic period. Physical observations of an astute anesthetist and the use of electronic monitoring devices help identify anesthesia-related complications. Managing anesthetic complications involves determining whether a patient is adequately anesthetized, if pain is appropriately managed, and if the autonomic nervous system is adequately subdued so that the patient is unconscious, amnesia is achieved, and physiologic responses to stimulation are blunted. Most importantly, appropriate anesthesia monitoring allows the observer to assess the current physiologic consequences of anesthesia to help improve the safety of the anesthetic procedure. Appropriate monitoring and interpretation of abnormalities will guide the anesthetist in making adjustments in order to mitigate the negative consequences of the anesthetic event.

Responding to these complications requires a clear understanding of the underlying problem, a calm and rational approach to addressing the problem, and an understanding of the short- and long-term consequences of *not* addressing the problem. Clear communication with the veterinarian is key when dealing with any anesthesia complication.

This chapter describes common complications encountered during the perianesthetic period, triggers that should prompt treatment, and treatment steps and corrective goals to rectify these complications.

TEMPERATURE

Maintaining proper core body temperature in anesthetized patients is important because hypothermia and hyperthermia have consequences as discussed in Chapter 7 (Anesthesia Monitoring). Monitoring body temperature throughout anesthesia and recovery will help identify temperature trends in patients and allow implementation of corrective measures. Body temperature is monitored via a rectal thermometer or rectal/esophageal probe connected to a monitor.

Hypothermia

Hypothermia, a condition in which heat loss is greater than heat production (a body temperature < 99.5°F in dogs and cats), occurs frequently in veterinary patients who undergo general anesthesia. Over 80% of dogs and nearly 98% of cats suffer from hypothermia after surgical procedures requiring anesthesia.[1, 2] Numerous factors contribute to heat loss and subsequent hypothermia as described in Chapter 7 (Anesthesia Monitoring). Thankfully, providing thermal support throughout the perianesthetic period is one of the easiest things an anesthetist can do to prevent hypothermia.

The majority of heat loss is from the body surface while a small amount is from the respiratory tract. Recall from Chapter 7 (Anesthesia Monitoring) that patients lose heat through convection, conduction, evaporation, and radiation.[3]

Anesthetic events can lead to hypothermia for a number of different reasons, including:

- Thermoregulatory mechanisms in the central nervous system are blunted.
- Body cavities exposed to room temperature air lose heat rapidly.

- Vasodilation, a common adverse effect of many anesthetic agents, predisposes patients to redistribute heat from their core to their periphery where it is more easily lost.
- Patients may become wet during aseptic preparation of the surgical site, abdominal cavity lavage, dental procedures, deep ear cleanings, and other procedures, greatly accelerating heat loss.
- Incoming fresh oxygen into the breathing circuit is cold and dry, which leads to body cooling.
- Patients have decreased cellular metabolic rates and muscle activity while anesthetized, which decreases heat production.
- Cool ambient room air and body contact with cold surfaces (e.g., stainless steel tables) hastens heat loss.

Some patients are more likely to experience hypothermia than others. Small patients have a lower weight:surface area ratio (i.e., more surface area compared to their body mass as described in Chapter 7 [Anesthesia Monitoring]); therefore, they lose heat more rapidly than larger patients[4] (Figure 12-1). Geriatric patients have decreased muscle mass and lower metabolic activity than younger patients, increasing their susceptibility to hypothermia. Debilitated and emaciated patients are also predisposed to hypothermia.[5]

The consequences of hypothermia are serious for a variety of reasons including:

- Hypothermia reduces minimum alveolar concentration (MAC); therefore, a lower percent of inhalant anesthetic agent is needed to maintain anesthetic depth. This situation increases the possibility for drug overdose.
- Hypothermic patients have a slower rate of drug metabolism that can lead to a prolonged recovery from anesthesia, increasing their risk of morbidity.[6]
- Severely hypothermic patients become bradycardic and unresponsive to anticholinergic drugs resulting in decreased cardiac output.[7] As hypothermia becomes severe, patients are at risk for developing arrhythmias such as atrioventricular block, premature ventricular contractions, and ventricular fibrillation, potentially leading to unresuscitatable cardiac arrest.[5]
- Other adverse effects associated with hypothermia include respiratory depression, coagulopathy, altered blood viscosity, an increased incidence of postoperative infections, and increased need for blood transfusions.[8]

FIGURE 12-1 Pediatric patients and those with a low weight:surface area ratio are particularly susceptible to hypothermia during the perianesthetic period.

In recovery, shivering increases metabolic oxygen demand by 200 to 600%, which can lead to hypoxia and acidosis.[5] As patients recover and their thermoregulatory mechanisms return, blood vessels constrict in an attempt to stop heat loss. Increased myocardial oxygen demand in the presence of this vasoconstriction can lead to increased work on the heart and cardiac arrhythmias. Shivering in recovery may also increase incisional pain.

Treatment of Hypothermia

The best treatment for hypothermia is *prevention*. Patients at risk of hypothermia should be kept warm as soon as they come into the hospital. Padded bedding and heated cages can reduce conductive heat loss, and blankets and jackets can reduce the heat lost through convection by producing a trapped layer of air between the blanket and the patient's body surface. Common methods used to maintain body temperature during anesthesia include the use of circulating warm water blankets (the efficacy of which is limited to the amount of body surface area that is covered), forced air blankets (Bair Hugger®), and conductive fabric patient warming blankets (HotDog®). Wrapping the patient's feet in bubble wrap, baby socks, or exam gloves helps prevent heat loss through the foot pads.

Tech Tip

Socks and bubble wrap are inexpensive and readily available items in the hospital for wrapping a patient's feet to help prevent heat loss (Figure 12-2).

Intravenous (IV) fluids should be warmed and/or run through a heating unit placed as close to the patient's IV catheter as possible (Figure 12-3; see Figure 7-20a). Keeping surgical prep solutions warm and avoiding the use of isopropyl alcohol will limit evaporative heat loss. Warming fluids for lavage prior to their administration into the thoracic or abdominal body cavities will also help prevent heat loss.

The use of electrical heating pads, microwavable heated disks, heated fluid bags or fluid-filled exam gloves, and heat lamps carry a high risk of causing thermal burns, and they should be used with extreme caution. Prewarming patients is the most effective strategy for preventing hypothermia. A neonatal ICU incubator can be purchased (typically a used/refurbished one) for prewarming patients.

Hyperthermia

Hyperthermia, a condition in which heat production is greater than heat loss (body temperature > 103°F in dogs and cats), is not common in anesthetized patients but is worthwhile to monitor. In healthy patients, the most common cause of hyperthermia in the perianesthetic period is iatrogenic overheating through aggressive attempts to prevent hypothermia. This type of hyperthermia usually occurs with the use of circulating warm water, conductive fabric warming or forced air blankets.[5] The use of opioids has been associated with hyperthermia in cats; however, this hyperthermia is usually self-limiting and not life threatening.[9]

Malignant hyperthermia is a rare genetic condition in which uncontrolled skeletal muscle activity and increased metabolism lead to rapid increases in body temperature during general anesthesia.[4,5] A sharp increase in end-tidal CO_2 ($EtCO_2$) and partial pressure of carbon dioxide dissolved in arterial

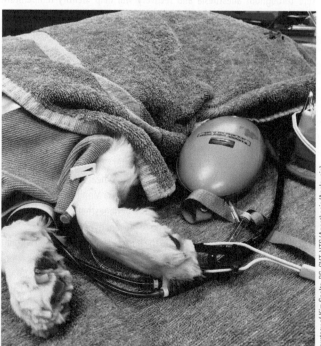
Courtesy of Kim Spelts, BS, CVT, VTS (Anesthesia/Analgesia)

FIGURE 12-2 Exam gloves on a patient's feet to prevent hypothermia.

FIGURE 12-3 Utilizing an IV fluid warmer such as the iWarm™ (Midmark, Dayton, OH) can help prevent hypothermia. The blue IV warmer unit can be seen next to the patient.

blood ($PaCO_2$) is also observed. Malignant hyperthermia has been reported most commonly in swine, although rare occurrences in dogs and horses have also been observed. The rapid and extreme increase in body temperature seen with malignant hyperthermia can lead to organ failure and death.

An increase in body temperature increases metabolic activity and energy consumption, which could result in cellular hypoxemia.[10] Increased metabolic oxygen demand increases the workload of the heart, brain, and skeletal muscles.[5] As hyperthermia progresses and the body temperature increases to approximately 107°F, permanent organ damage can occur, including brain damage, which can result in altered mentation and seizures. Other sequelae of severe hyperthermia include ventricular arrhythmias, organ failure, and potentially death.[5]

Treatment of Hyperthermia

Once an anesthetized dog's or cat's body temperature reaches approximately 101–102°F, supplemental heating sources should be removed. If using a circulating warm air blanket, turn down the temperature of the circulating air to ambient levels. It may be beneficial to increase the oxygen flow rate to increase heat loss through the respiratory tract. Opioid-induced hyperthermia can occur in cats (temperatures may reach 106°F or higher), and to restore temperature to a normal level, cats can be given naloxone (Narcan®) (1–5 mcg/kg IV) in addition to providing supportive care. Without intervention, the temperature of most cats with opioid-induced hyperthermia will return to normal within hours; therefore, reversing the opioid's effect, including its analgesic properties, should be carefully considered. Any cat receiving an opioid should have its body temperature closely monitored during anesthesia and in the perioperative period.

Active cooling measures must be taken if the patient's temperature continues to increase. Evaporative cooling can occur with the application of cool water or alcohol to the footpads and skin. The administration of room temperature IV fluids leads to an increase in peripheral perfusion and heat loss.[10] More aggressive cooling measures for life-threatening hyperthermia include cool water enemas, cool abdominal lavage, and even cool water submersion. IV acepromazine (Promace®, generic) can cause vasodilation that may also assist with cooling; however, administer IV acepromazine with caution because the resultant vasodilation may decrease blood pressure.[5]

The only definitive treatment for malignant hyperthermia is slowly administered IV dantrolene (Dantrium®), a drug not typically available in veterinary hospitals. If anesthesia is planned for a patient suspected of having malignant hyperthermia, a veterinarian may be able to obtain dantrolene from a human hospital prior to the procedure.[5] Consulting with and/or contracting a Diplomate of the American College of Veterinary Anesthesia and Analgesia (DACVAA) to anesthetize such a patient is ideal.

BLOOD PRESSURE

Adequate blood pressure is needed to effectively perfuse and deliver oxygen to organs such as the brain, kidney, and liver. Maintaining blood pressure in anesthetized patients can be challenging, but is possible with proper monitoring and intervention.

There are several ways to monitor blood pressure during an anesthetic procedure, the most common being ultrasonic Doppler or noninvasive oscillometric device. Each method has its limitations but are simple to use. The Doppler method of measuring blood pressure only reliably provides the systolic arterial pressure. Some studies have shown that Doppler tends to underestimate systolic blood pressure values in cats.[12] However, since this is not true of all cats, it is most prudent to treat patients whose measured blood pressures may be falsely low than to ignore the possibility of hypotension.[13] Noninvasive oscillometric blood pressure devices display all three pressure values (systolic, diastolic, and mean), but they are not as accurate as direct blood pressure measurement utilizing an arterial catheter.[14]

The gold standard of blood pressure measurement is placement of an arterial catheter, which is then connected by a fluid-filled tube (such as an IV extension set) to an electronic transducer. The transducer converts the pulsing fluid wave into a measurement of systolic, diastolic, and mean arterial pressures (Figure 12-4). This is an advanced technical skill, but it also provides an accurate, real-time measurement of a patient's blood pressure.

Manually palpating and assessing the quality of peripheral pulses only provides the anesthetist an impression of the difference between the systolic and diastolic arterial pressures of the patient; it is not necessarily indicative of the mean arterial blood pressure or perfusion pressure.[14] *If blood pressure is not measured, hypotension cannot be recognized and corrected.*

Blood Pressure Goals

Regardless of how blood pressure is measured, the goal is to maintain the following lower limits in systolic (SAP), diastolic (DAP), and mean (MAP) arterial pressures:

- SAP > 90 mm Hg. If mean and diastolic arterial pressures are not being monitored (e.g., if ultrasonic Doppler is utilized, it only reads the systolic arterial pressure), then the systolic pressure should be kept > 100 mm Hg to help ensure the MAP presumably stays higher than 60 mm Hg.
- MAP > 60 mm Hg. Renal and cerebral **autoregulation of blood flow** involves local and systemic feedback mechanisms that maintain constant blood flow and perfusion pressure to the kidneys and brain despite changes in systemic blood pressure. These autoregulatory mechanisms are most effective when MAP is 50 to 150 mm Hg for autoregulation of cerebral blood flow and 80 to 180 mmHg for autoregulation of renal blood flow.[15] Keeping the MAP above 60 mmHg helps to ensure autoregulation is not disrupted. Ideally, a MAP of 70 mmHg or greater can be achieved.
- DAP > 40 mm Hg. Perfusion of the myocardium (coronary blood flow) occurs during diastole and requires DAP > 40 mm Hg. A low diastolic pressure could result in decreased myocardial perfusion, myocardial hypoxemia, and cardiac arrhythmias.

Physiologic Considerations

One of the primary goals throughout the perianesthetic period is to provide adequate oxygen delivery to the tissues, which occurs at the capillary level. Even if ventilation is perfect and arterial blood is maximally oxygenated, oxygen delivery to the tissues is not guaranteed; if blood flow and perfusion to the capillary beds is inadequate, cellular hypoxia can develop.

Oxygen delivery (DO_2) to the tissues is a function of both cardiac output (CO) and arterial oxygen content (CaO_2), and it cannot be directly measured:[16]

$$DO_2 = CO \times CaO_2$$

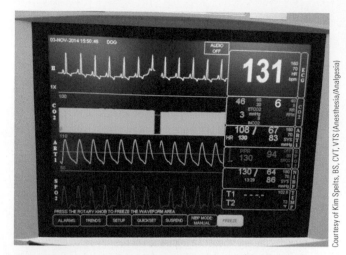

Courtesy of Kim Spelts, BS, CVT, VTS (Anesthesia/Analgesia)

FIGURE 12-4 Placement of an arterial catheter will allow the anesthetist to monitor systolic, diastolic, and mean arterial pressures in real time. This monitor displays simultaneous electrocardiography and direct arterial blood pressure readings.

To improve oxygen delivery, cardiac output and/or arterial oxygen content must increase.

Cardiac output is a direct contributor to tissue perfusion, and it is a function of heart rate (HR) and stroke volume (SV). Stroke volume in turn is a function of preload, afterload, and cardiac contractility[17]

$$CO = HR \times SV$$

preload   afterload   cardiac contractility

Cardiac output can be measured directly with the placement of a Swan-Ganz catheter into the right atrium and pulmonary artery (a thermodilution technique) or through other less invasive methods (such as transesophageal echocardiogram, lithium dilution, etc.). Placing a Swan-Ganz catheter is not common and beyond the capability of most non-university or specialty/referral veterinary hospitals.

In turn, blood pressure (specifically mean arterial pressure) is a function of CO and systemic vascular resistance (SVR) (the overall degree of vascular tone (vasodilation or vasoconstriction) that exists at that moment):[17]

$$MAP = CO \times SVR$$

Blood pressure alone is not an exact indicator of tissue perfusion and cardiac output. In fact, in some cases (such as with profound vasoconstriction), blood pressure may be high but cardiac output and tissue perfusion are actually low because of the increased afterload. Future technology may make direct measurement of cardiac output less invasive and more feasible, but until then monitoring blood pressure is the easiest, most reliable, and least invasive method we have to assess a patient's hemodynamic status.[18]

Hypotension

Hypotension, low blood pressure, is the most common complication seen during the perianesthetic period.[11] A number of different factors contribute to hypotension in anesthetized patients, including the effects of anesthesia drugs and the patient's underlying hemodynamic status prior to and during anesthesia. It is crucial that the anesthetist monitor blood pressure,

recognize abnormalities, identify the underlying cause, and treat hypotension quickly. Maintaining adequate cardiac output (reflected in blood pressure measurements) helps to ensure adequate tissue perfusion, oxygen delivery, and autoregulation of blood flow to the kidneys and brain.

Anesthetic drugs can negatively affect cardiac output and blood pressure for a number of reasons. Some drugs (e.g., opioids, α_2 adrenergic agonists) increase vagal tone and decrease heart rate. Others (e.g., inhalant anesthetic agents, propofol [Rapinovet®, PropoFlo™, Propovan®, PropoFlo™28]) decrease cardiac contractility and systemic vascular resistance. The underlying cause of a patient's hypotension varies depending on the combination of injectable and inhalant anesthetic agents utilized, along with the patient's underlying physical condition. To resolve hypotension quickly and efficiently, it is crucial for the anesthetist to first determine the most likely underlying cause.

Clinical hypotension (MAP < 60 mm Hg) results when the systolic and/or diastolic arterial pressures are low, due to low cardiac output and/or systemic vascular resistance (remember, MAP = CO × SVR). The anesthetist should assess all three pressure values whenever possible and understand the overall cardiopulmonary status of the patient in order to determine the appropriate course of treatment.

Cardiac output (and, hence MAP) may drop as a result of decreased stroke volume, decreased contractility (a common adverse effect of inhalant anesthetic agents), decreased afterload (e.g., vasodilation) and/or decreased preload (e.g., hypovolemia). Cardiac output can also decrease due to bradycardia.

Treatment of Hypotension

The first step to prevent low blood pressure readings and resultant complications begins with ensuring that the patient is as physiologically stable before anesthesia as possible. Selecting a multimodal (balanced) anesthesia plan is the next step. Multimodal anesthesia does not necessarily mean using a low number of anesthesia drugs; rather, it means using lower dosages of multiple drugs to achieve analgesia, immobilization, muscle relaxation, unconsciousness, and amnesia. Using lower dosages helps to prevent the dosage-dependent negative adverse effects that are commonly seen with many anesthetic agents.

The biggest contributor to hypotension in the perianesthetic period is the inhalant anesthetic agent. Inhalation anesthetic agents cause dosage-dependent cardiovascular depression primarily through a decrease in peripheral vascular resistance (vasodilation) and to a lesser degree through a decrease in cardiac contractility.[19] Hypotension can be exacerbated by the anesthesia-induced depression of the patient's sympathetic reflexes that normally provide the physiologic maintenance of blood pressure and tissue perfusion.[18]

The anesthetist should plan and strive for multimodal anesthesia and take steps to minimize the amount of inhalant anesthetic agent required. This can be achieved with appropriate preoperative sedation as well as the provision of pre- and intraoperative analgesia. Intraoperative infusions of opioids,[20, 21] lidocaine (Xylocaine®),[22] and ketamine (Ketaset®, VetaKet®, Ketaject®, Vetalar®)[23] have each been shown to reduce inhalant anesthetic agent requirements[24] (Figure 12-5). Local and regional anesthetic techniques and preoperative use of some NSAIDs can also reduce inhalant anesthetic agent requirements.[25]

Once multimodal anesthesia has been achieved, the first line of defense in responding to hypotension should always be to assess the

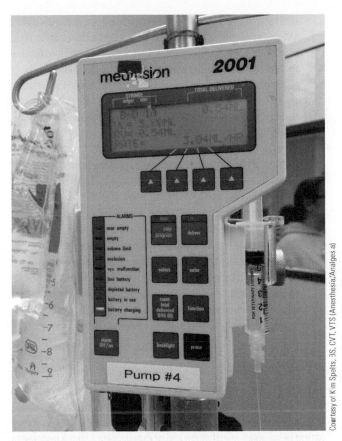

FIGURE 12-5 Continuous infusions of multimodal analgesic agents are part of a robust balanced anesthesia plan and can reduce the amount of inhalant anesthetic agent required to maintain an appropriate plane of anesthesia.

Courtesy of Kim Spelts, 3S, CVT, VTS (Anesthesia/Analgesia)

patient's depth of anesthesia and turn down the inhalant anesthetic agent if possible. Decreasing the depth of anesthesia can improve cardiac output and systemic vascular resistance, resulting in increased blood pressure.[18] The goal is to produce a patient that is still appropriately anesthetized while reducing the inhalant anesthetic agent. In some settings, this is not achievable without the patient awakening, in which case the anesthetist must address the hypotension with an alternative plan.

Bradycardia is common under anesthesia, especially during systemic opioid administration.[11] A hypotensive patient who is also bradycardic should be treated with an anticholinergic (such as atropine [Atrocare®, generic] or glycopyrrolate [Robinul®]) to increase the heart rate into a functional range and therefore increase cardiac output (see Figure 5-15). Depending on the severity and length of time remaining in the procedure, a partial reversal agent may be considered if the patient received an α_2 adrenergic agonist. Bradycardia is a common source of hypotension in puppies and kittens because they are more reliant upon heart rate to maintain their cardiac output than adults.[26]

Propofol, a common injectable anesthetic agent, can cause a dosage-dependent transient hypotension due to vasodilation and a decrease in cardiac contractility.[27] As discussed previously, inhalant anesthetic agents are also a common cause of decreased cardiac contractility and vasodilation. A patient who is vasodilated and/or hypovolemic will often display a low diastolic arterial pressure and may display a low systolic arterial pressure

as well. For these patients, a fluid bolus of crystalloid and/or colloid is warranted (see Chapter 4 [Fluid Therapy and IV Catheterization]).

Current (2013) American Animal Hospital Administration guidelines for fluid boluses in anesthetized patients are as follows:[28]

- Provide an IV bolus of an isotonic crystalloid such as LRS at 3 to 10 mL/kg delivered over 10 to 20 minutes. Repeat once if needed. This author recommends the lower end of bolus volume (3–5 mL/kg as well as longer time of administration [20–30 minutes]) for cats in order to reduce the risk of volume overload.
- If the response is inadequate, consider IV administration of a colloid such as hetastarch after confirming the patient does not have a coagulopathy or renal insufficiency. Administer 5 to 10 mL/kg for dogs and 1 to 5 mL/kg for cats.

Hypotension due to a decrease in cardiac contractility may be reflected in a low systolic arterial pressure. In this situation, administration of a positive inotropic agent is appropriate because they are effective at treating hypotension related to decreases in cardiac contractility. Hypotension may also arise from severe drops in systemic vascular resistance (such as that seen in septic shock), in which the diastolic pressure is usually very low. In this case, treatment with a vasopressor may be indicated. Unlike positive inotropes (which increase cardiac contractility), vasopressors induce vasoconstriction and thereby elevate mean arterial pressure. This vasoconstriction may decrease cardiac output (by increasing afterload) and blood flow to other organ systems (e.g., kidneys).[29] Most positive inotropes work by stimulating alpha- and/or beta-adrenergic receptors of the autonomic nervous system. Some drugs have both vasopressor and positive inotropic effects. Agents commonly used in anesthetized patients include the following and are summarized in Table 12-1:[18, 30]

- *Ephedrine* (generic) is a non-catecholamine sympathomimetic drug that stimulates the release of endogenous stores of norepinephrine, leading to mild vasoconstriction (see Figure 5-16d). Ephedrine also has a direct effect on β_1 and β_2 adrenergic receptors, improving myocardial contractility. Ephedrine is given IV and is an ideal choice for a short-term (15–20 minute) boost in cardiac output and blood pressure. The dose can be repeated, but after two to three doses the endogenous stores of norepinephrine have been depleted and subsequent doses are no longer effective. Ephedrine has an advantage over dopamine and dobutamine in that it can be administered as a bolus, eliminating the need for setting up a constant rate infusion (CRI). It is also inexpensive and is not arrhythmogenic. Patients may occasionally exhibit transient bradycardia, likely a reflex to a sharp increase in blood pressure.
- *Dobutamine* (generic) is a β_1 adrenergic agonist that provides an increase in cardiac contractility without an increase in peripheral vascular resistance or in heart rate (except at very high administration rates) (see Figure 5-16a). In dogs and cats, dobutamine may improve hemodynamics by improving cardiac output and oxygen delivery; however, dobutamine is unlikely to increase blood pressure unless the patient's issue is purely one of contractility.[31, 32] Because of its short half-life, dobutamine must be administered as a CRI. Dobutamine (as are all beta-agonists) is less effective in the presence of acidosis, since beta receptors become downregulated in the presence of acidosis.
- *Dopamine* (generic) is an adrenergic inotropic β_1 and α_1 agonist and has different physiologic effects depending on the rate at which it is

Tech Tip 🐾

Dilute ephedrine to a 1 mg/mL concentration by removing 0.6 mL diluent (e.g., 0.9% NaCl) from a 30 mL vial then adding 30 mg (0.6 mL) of 50 mg/mL ephedrine to the diluent (Figure 12-6). This makes a 30 mg/30 mL, or 1 mg/mL, ephedrine dilution. For a smaller volume (for single patient use), add 0.1 mL of 50 mg/mL ephedrine to 4.9 mL of diluent to make 1 mg/mL.

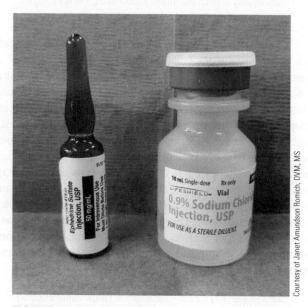

Courtesy of Janet Amundson Romich, DVM, MS

FIGURE 12-6 Dilute ephedrine to 1 mg/mL for ease of dosing.

administered (see Figure 5-16b). Like dobutamine, dopamine has a short half-life and must be administered as a CRI. At low administration rates, it may stimulate dopaminergic receptors to increase renal blood flow. Moderate administration rates result in the direct stimulation of β_1 adrenergic receptors, increasing cardiac contractility and, potentially, heart rate. At high administration rates in dogs and cats, stimulation of α_1 receptors can cause coronary, as well as systemic, vasoconstriction. Blood pressure may be increased at the expense of peripheral perfusion and oxygen delivery. The concurrent increase in cardiac contractility and heart rate often seen with dopamine may result in a greater myocardial oxygen demand, increasing the risk of arrhythmias. This is potentially dangerous in patients with underlying heart disease.

- *Norepinephrine* (Levophed®) stimulates α_1, β_1, and β_2 adrenergic receptors, increasing mean arterial pressure through an increase in systemic vascular resistance and cardiac contractility (see Figure 5-16c). Norepinephrine may be the drug of choice in septic/endotoxemic patients, since it preferentially corrects inappropriately dilated areas.[18] Norepinephrine must also be administered as a CRI. Careful monitoring of blood pressure and mucous membrane color should occur to guard against over-constriction and a subsequent decrease in peripheral perfusion.

Hypertension

Due to the vasodilatory and negative inotropic effects of many anesthetic agents, hypertension (MAP > 100 mm Hg) is rarely seen as an anesthetic complication. High blood pressure under anesthesia is most likely due to the patient being under-anesthetized and/or experiencing a strong response to a noxious stimulus (e.g., painful surgical stimulation). In both of these cases, the depth of anesthesia can be increased and additional pain medications administered.

Prolonged hypertension is detrimental to the brain and kidneys in particular, as autoregulatory mechanisms may not function properly.

TABLE 12-1 Drugs Used to Treat Hypotension in Dogs and Cats

Drug	Drug Class	Dosage and Route	Indications
atropine (Atrocare®, generic)	Anticholinergic	0.02–0.04 mg/kg IV, IM, SQ	Bradycardia with concurrent hypotension
glycopyrrolate (Robinul®)	Anticholinergic	0.005–0.2 mg/kg IV, IM, SQ	Bradycardia with concurrent hypotension
ephedrine (generic)	Beta agonist	0.1–0.2 mg/kg IV bolus*	Increase contractility (positive inotrope)
dobutamine (generic)	Beta agonist	1–10 mcg/kg/min IV CRI*	Increase contractility (positive inotrope)
dopamine (generic)	Dopaminergic agonist	0.5–3.0 mcg/kg/min IV CRI+	Dilate renal arterioles via dopamine 1 (DA1) receptors (dogs)**
dopamine (generic)	Beta agonist	4.0–10.0 mcg/kg/min IV CRI+	Increase contractility (positive inotrope)
dopamine (generic)	Alpha agonist	10–15 mcg/kg/min IV CRI+*	Increase SVR
norepinephrine (Levophed®)	Mixed alpha and beta agonist	0.05–0.3 mcg/kg/min IV CRI*	Increase SVR, increase contractility

* Grubb, T., Sager, J., Gaynor, J. S., Montgomery, E., Parker, J. A., Shafford, H., and Tearney, C. (2020). 2020 AAHA anesthesia and monitoring guidelines for dogs and cats. *Journal of the American Animal Hospital Association*, 56(2), 59–82.
+ Luisito. P. *Using inotropes and vasopressors in anesthesia (Proceedings)*, www.dvm360.com, July 13, 2011.
** Cats likely do not have DA1 receptors.

Some disease processes can predispose a patient to hypertension, including renal disease, thyroid disease, hyperadrenocorticism (Cushing's disease), intracranial disease, and pheochromocytoma. Prolonged, severe hypertension, if not responsive to an increase in inhalant anesthetic agent, may warrant the administration of a vasodilator such as acepromazine, nitroprusside (Nitropress®), or phentolamine (Regitine®) IV.

CARDIAC ARRHYTHMIAS

Heart rhythm abnormalities can impact stroke volume and cardiac output thus affecting oxygen delivery. Since some cardiac rhythm abnormalities may progress to more serious arrhythmias that could be fatal, it is important to recognize and document the potential significance of the abnormality. Quantification of cardiac rhythm is observed via the electrocardiogram (ECG). ECG abnormalities are a common occurrence in anesthetized patients. A number of factors contribute to the predisposition for arrhythmias, including the effects of some drugs, stress, pain, hypoxemia, electrolyte disturbances, and underlying cardiac or systemic disease. The decision to treat these arrhythmias or not is based in part on the hemodynamic consequence(s) of the abnormal rhythm, as well as the likelihood for further deterioration into a more life-threatening arrhythmia. This section will review the arrhythmias that an anesthetist is most likely to observe during the perianesthetic period.

ECG Evaluation

At times abnormal rhythms are obvious; at other times, they are not so clear. The anesthetist should be familiar with values for normal heart rate and rhythm for a particular patient, given its species, size, and age. When evaluating an ECG, it is beneficial to take a stepwise approach to determine which type of abnormality may be displayed on the screen or rhythm strip. Assessment of the questions in Table 12-2 can help guide the anesthetist in determining the arrhythmia.

Bradyarrhythmias

Sinus bradycardia, characterized by slow but normal complexes on the ECG, is most frequently caused by a drug-induced increase in vagal tone, such as that caused by the use of opioids and/or α_2 agonists.[17] Some patients (e.g., brachycephalic breeds) have high preexisting vagal tone and are predisposed to bradycardia under anesthesia. Hypothermia can also lead to bradycardia, as can the use of some cardiac medications (e.g., beta blockers).[5] Other pathophysiologies that can lead to bradyarrhythmias include central nervous system (CNS) disease (particularly if there is increased intracranial pressure) and hyperkalemia.[33]

Sinus bradycardia is a common adverse effect of mu-agonist opioid administration (Figure 12-7). Sinus bradycardia does not always require intervention; however, bradycardia can lead to decreased cardiac output (CO = HR × SV) and blood pressure (MAP = CO × SVR), resulting in decreased perfusion.[5] In this case, treatment with an anticholinergic drug is indicated. Caution should be taken when using anticholinergics in the presence of drugs that cause vasoconstriction (e.g., α_2 adrenergic agonists). Concurrent use may lead to hypertension, tachycardia, increased cardiac workload and oxygen consumption, and other arrhythmias.[34, 35] In these cases, full or partial reversal with an α_2 adrenergic antagonist may be preferred.

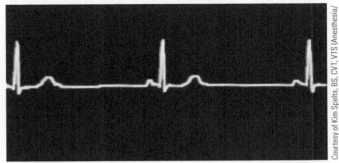

FIGURE 12-7 Sinus bradycardia. Rhythm is normal but slow.

Atrioventricular (AV) Blocks

Atrioventricular (AV) block results when there is a delay in conduction between the sinoatrial (SA) node and the AV node.

- **1st-degree AV block** results in a prolonged but consistent P-R interval on the ECG, but all impulses are conducted through the AV node to the ventricles and QRS complexes appear consistently on the ECG (Figure 12-8). The effect on cardiac output of this arrhythmia is minimal, and treatment is not generally warranted.
- **2nd-degree AV block** is characterized by a delay and intermittent lack of transmission of impulse through the AV node, resulting in intermittently absent QRS complexes and "dropped beats."[5] There are two subtypes of 2nd-degree AV block, with slightly different manifestations on the ECG:
  - *Mobitz Type 1 (Wenckebach):* This subtype is characterized by a gradually increasing P-R interval with an eventual complete block of the signal through the AV node, dropping the QRS completely (Figure 12-9).

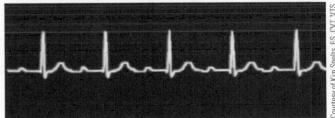

FIGURE 12-8 First-degree AV block. Note the prolonged P-R interval.

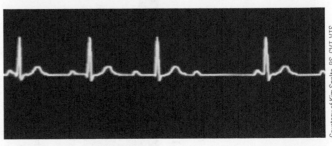

FIGURE 12-9 Second-degree AV block, Mobitz Type 1 (Wenckebach). Note the gradually increasing P-R interval with eventual complete AV block and drop of the QRS complex.

TABLE 12-2 Questions to Help Evaluate Arrhythmias

Question	Bradyarrhythmias				Tachyarrhythmias		Ventricular Arrhythmias		Atrial Arrhythmias	
	Sinus Bradycardia	1st degree AV Block	2nd degree AV Block	3rd degree AV Block	Sinus Tachycardia	Supraventricular Tachycardia	Ventricular Premature Contractions	Ventricular Tachycardia	Atrial Premature Contractions	Atrial Fibrillation
1. What is the heart rate?	Low	Low–Normal	Low–Normal	Low	High	High	Low, normal, or high	High	Low, normal, or high	High
2. Is the rhythm regular or irregular?	Regular	Regular	Irregular	Irregular	Regular	Regular	Regular	Regular	Irregular	Irregularly irregular
3. Is there a P wave for every QRS complex?	Yes	Yes	Yes	No	Yes	No	No	No	Yes	No
4. Is there a QRS complex for every P wave?	Yes	Yes	No	No	Yes	N/A	Yes	Yes	Yes	No P waves
5. How are the P waves related to the QRS complexes?	Normal	Prolonged P-R interval	Type 1: increasing P-R interval followed by dropped QRS Type 2: normal P-R interval with intermittent dropped QRS complexes	Unrelated; atria and ventricles are depolarizing at different rates	Normal	P waves are normal	P-QRS interval normal in normal complexes; VPC often do not follow a P wave (P waves not usually present)	P waves are not usually discernible	Normal	P waves are absent

○ *Mobitz Type 2*: This subtype shows a consistent and normal PR interval with occasional complete block of the signal through the AV node, again completely dropping QRS complexes (Figure 12-10). This arrhythmia has the potential for deteriorating into 3rd-degree AV block and should always be treated.

Anticholinergic drug administration can be used to treat both subtypes of 2nd-degree AV block. It should be noted, however, that IV administration of atropine will have a brief paradoxical effect of slowing the heart rate by acting on central muscarinic receptors and causing a transient increase in vagal tone. This effect is rapidly overcome by elimination of vagal tone peripherally, leading to increased heart rate. This usually resolves spontaneously without additional intervention, although a second dose can be given to resolve it as well. Generally the recommendation is to give the anticholinergic drug a few minutes longer to work rather than giving more drug.

- **3rd-degree AV block** is due to a defect in the cardiac conduction system in which there is no conduction through the AV node, therefore causing a complete dissociation between the SA and AV nodes.[5] Firing rate from the SA node is normal/regular and the P wave rate is greater than the QRS rate, but the two are independent of each other (Figure 12-11). This is a serious and potentially lethal arrhythmia, as it is unresponsive to anticholinergic drugs and has a potential for progressing to asystole. The ventricular beats are usually slow and ineffective, resulting in a marked decrease in cardiac output. Isoproterenol (Isuprel®) may be administered for its β_1 adrenergic effects to try to increase cardiac rate and contractility. However, isoproterenol's β_2 adrenergic effects may cause vasodilation and hypotension. For long-term management of 3rd-degree AV block, pacemaker implantation is typically the best option for the patient.

Some breeds of dogs (Miniature Schnauzers, Cocker Spaniels, Dachshunds, Pugs, West Highland White Terriers) are predisposed to a

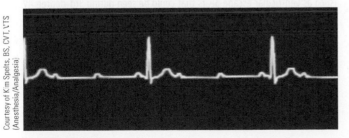

FIGURE 12-10 Second-degree AV block, Mobitz Type 2. P-R interval is normal, but there is an intermittent complete AV block and drop of the QRS complex.

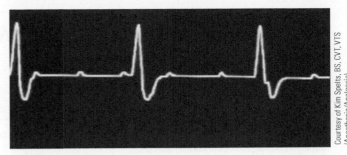

FIGURE 12-11 Third-degree AV block. The atria and ventricles are depolarizing independently of each other. The P waves and QRS complexes are asynchronous.

cardiac condition called **sick sinus syndrome** (SSS). These patients usually present with bradycardia and a history of intermittent weakness and collapse.[36] Anesthetic management of SSS can be a challenge, as long sinus pauses (up to 10 seconds) can be displayed, with functional pulse rates of < 30 beats per minute. Sick sinus syndrome is often unresponsive to anticholinergic drugs, although that should be the first treatment attempted. Dopamine administration has been described in anesthetized humans as a perianesthetic treatment for SSS.[37, 38] As with 3rd-degree AV block, pacemaker implantation may be needed.

Tachyarrhythmias

Sinus tachycardia is characterized by fast but normal-looking ECG complexes (Figure 12-12). Common causes include the use of drugs such as anticholinergics, ketamine, alfaxalone, etc., as well as sympathetic nervous system stimulation such as pain, a light plane of anesthesia, and certain physiologic conditions (e.g., hyperthyroidism, pheochromocytoma).[33] Hypotension, especially hypotension caused by hypovolemia, can also lead to sinus tachycardia, as an increase in heart rate is a compensatory mechanism driven by the baroreceptor reflex to maintain cardiac output in the face of decreased stroke volume. Note that under anesthesia, the baroreceptor reflex is often blunted or subdued, so it is possible for a patient to be hypovolemic without exhibiting tachycardia. Other important causes of tachycardia include hypoxemia, hypercapnia, a need to urinate, and anemia.

Supraventricular tachycardia (SVT) is a fast rhythm with an origin somewhere above the ventricles. Sinus tachycardia is a type of SVT with an origin from the SA node. Traditionally, SVT is defined as originating somewhere other than the SA node but above the AV junction. Consequently, an ECG strip of SVT will usually not show P waves (Figure 12-13). This is not often seen in anesthetized patients, but it should be addressed quickly. Prolonged (several weeks) untreated SVT can lead to progressive myocardial failure as well as

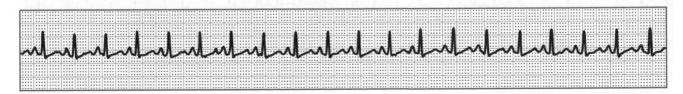

FIGURE 12-12 Sinus tachycardia. Rhythm is normal but fast.

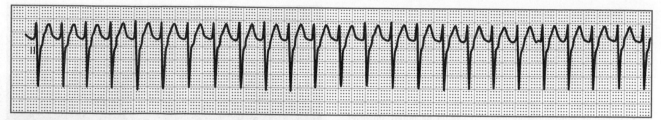

FIGURE 12-13 Supraventricular tachycardia. The rhythm is normal but fast, with no P waves.

Courtesy of Kim Spelts, BS, CVT, VTS (Anesthesia/Analgesia)

congestive heart failure. Short-term, perioperative SVT is unlikely to result in long-term cardiac changes but may cause decreased cardiac output and hypotension. A vagal maneuver, such as applying pressure to the eyes, may increase vagal tone, slowing the SA node discharge and the AV nodal conduction time, breaking the SVT. Esmolol (Brevibloc®, a selective β-$_1$ blocker with a rapid onset and short duration of action) or diltiazem (Cardizem®) administration is also a treatment option for SVT.[39]

Sustained periods of severe tachycardia can have a profound negative impact on cardiac output. The heart does not have time to adequately fill during diastole, and ejection time is decreased. Coronary perfusion time is also decreased, which can lead to myocardial hypoxia.[29] The underlying cause of the tachycardia should be identified and treated as soon as possible. For example, if sustained tachycardia develops as a result of surgical stimulation, the anesthetist should consider administering additional analgesics as well as potentially increasing the inhalant anesthetic agent level. IV fluid boluses (and perhaps blood products if severe hemorrhage is present) should be given to improve vascular filling and correct tachycardia-inducing hypovolemia.

Ventricular Arrhythmias

Ventricular premature contractions (VPCs or PVCs) are common occurrences. Some pathophysiologies (gastric dilatation-volvulus, splenic trauma and splenectomy, traumatic myocarditis) predispose patients to VPCs, as do some drugs (such as thiobarbituates). Hypoxemia and myocardial ischemia can also lead to VPCs.[33] Excitement, fear, and/or pain can also result in a catecholamine release that can sensitize the myocardium to arrhythmias.[40]

An irritable site (or sites) in the myocardium trigger VPCs. Impulses generated within the ventricles do not utilize the normal conduction pathway; rather, they are conducted from cell to cell. Consequently, these electrical impulses manifest on the ECG as QRS complexes that are wide, bizarre, premature in the underlying cadence, and often do not follow a P wave[5] (Figure 12-14). They also occur earlier than expected in the underlying cadence of the normal rhythm. VPCs are followed by a compensatory pause in the rhythm during which time no pulse wave is generated. This pulse deficit and compensatory pause are audibly discernable with ultrasonic Doppler.

VPCs that look alike are termed "uniform," while VPCs that look different from each other are termed "multiform." Multiform VPCs require additional scrutiny and are more likely to warrant treatment; they are more serious than uniform VPCs because they indicate that multiple sites within the myocardium are irritable. Treatment for rare or occasional uniform VPCs is often not required. The anesthetist should consider treatment with lidocaine for frequent (> 15 complexes per minute) VPCs, especially if the complexes are multiform and associated with a drop in blood pressure,[40] and/or if the patient has a high heart rate. Note that cats are more susceptible to lidocaine toxicity than are dogs, so lidocaine should be used judiciously in cats.

Care should be taken to differentiate an **idioventricular rhythm** (such as ventricular escape beats that could be seen during 3rd-degree AV block) from VPCs. In the absence of a signal from the SA and AV nodes, the ventricles have their own pacing capability, although the rate is typically much slower (20–40 bpm in dogs, 40–60 bpm in cats) than a normal sinus rate. Patients who are severely bradycardic may exhibit a persistent ventricular rhythm as the ventricular pacemaker is activated. The electrical

04 25 MM/SEC delay=9 (ECG) HR: 53 bpm C02 RR: 14 RPM SP02:---% NIBP (mmHg): 120 / 79 (93) 12:47

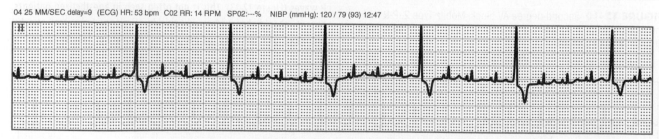

FIGURE 12-14 Uniform ventricular premature contractions. Note the wide and bizarre QRS complexes that come early in the rhythm.

Courtesy of Kim Spelts, BS, CVT, VTS (Anesthesia/Analgesia)

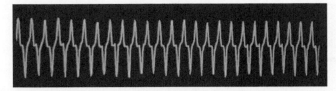

FIGURE 12-15 Ventricular tachycardia. Note the abnormally rapid heart rate and wide and bizarre QRS complexes without any associated P waves.

Courtesy of Kim Spelts, BS, CVT, VTS (Anesthesia/Analgesia)

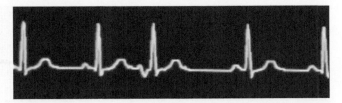

FIGURE 12-17 Atrial premature contraction. The third complex, normal in appearance, comes early in the underlying rhythm.

Courtesy of Kim Spelts, BS, CVT, VTS (Anesthesia/Analgesia)

complexes generated with this activation are wide and can be mistaken for VPCs. However, in this case administration of lidocaine is contraindicated, since it may block the ventricular rhythm completely and would result in asystole. Instead, anticholinergic therapy is warranted in order to increase rate at which the SA node fires.

A run of greater than three VPCs in a row, or a sustained ventricular rate above 150 to 180 beats per minute in dogs or 220 beats per minute in cats, is considered **ventricular tachycardia** (V-tach) (Figure 12-15).[5, 41] Ventricular tachycardia, as with sinus tachycardia, results in a decreased ability of the ventricles to fill during diastole. Stroke volume, cardiac output, and tissue perfusion decrease, leading to myocardial hypoxemia and additional arrhythmias. Sudden death can occur. Ventricular tachycardia may also progress to ventricular fibrillation. An IV lidocaine bolus followed by a lidocaine infusion should be administered to resolve ventricular tachycardia as rapidly as possible.

Ventricular fibrillation (V-fib) is chaotic electrical activity of the heart with no mechanical activity (Figure 12-16). It is a terminal rhythm if not treated immediately. If a defibrillator is not available, a strong blow using heel of the hand directly over the heart (precordial thump) may be attempted.[40] See Chapter 13 (Cardiac Arrest & Cardiopulmonary Cerebral Resuscitation [CPCR]) for additional information on ventricular fibrillation.

Atrial Arrhythmias

Atrial premature contractions (APCs) are normal-looking complexes that appear early in the normal cadence of the underlying rhythm (Figure 12-17). As with VPCs, they generate a compensatory pause, which is audibly discernible with ultrasonic Doppler. APCs rarely have a significant impact on the patient's hemodynamic status, and treatment is not warranted.

Atrial fibrillation (A-fib) may be seen in some giant breed dogs (e.g., Irish Wolfhounds, Giant Schnauzers), and/or in patients with congestive heart failure or dilated cardiomyopathy. Auscultation of A-fib has been described as "shoes in a dryer." In this arrhythmia, atrial depolarization and repolarization occur, but there is no coordinated atrial contraction. The AV node is stimulated frequently but also randomly. ECG complexes are characterized by a fast, "irregularly irregular" pattern of normal QRS complexes (the complexes are not regularly spaced apart, and this inconsistent interval between complexes also does not occur in a recognizable pattern) without discernible P waves (Figure 12-18). The baseline is not stable, and F waves (fine fibrillatory waves that are small and chaotic) are present. Cardiac output can be significantly decreased since atrial contractions are not coordinated to adequately fill the ventricles.[40] Ideally, a patient that presents with A-fib for an anesthetic workup should be medically managed prior to anesthesia. Treatment of A-fib during anesthesia is dependent upon the hemodynamic consequences of the rhythm. If significant tachycardia is exhibited and the patient does not have congestive heart failure, esmolol may be administered for heart rate control. In cases of chronic A-fib, atrial remodeling makes restoring and maintaining a sinus rhythm more challenging. A reasonable therapeutic goal is to optimize cardiac output by maintaining ventricular rate within a normal range. Longer-term management of A-fib includes electrocardioversion (procedure in which an arrhythmia is converted to a normal rhythm using electricity or drugs)[5] and/or oral cardiac medications such as diltiazem or digoxin.

Other ECG Abnormalities

Respiratory sinus arrhythmias are normal in dogs but not in cats. Due to fluctuations in vagal tone associated with respiration, the heart rate increases slightly during inspiration and decreases during exhalation (Figure 12-19). Brachycephalic dogs and dogs with respiratory disease may

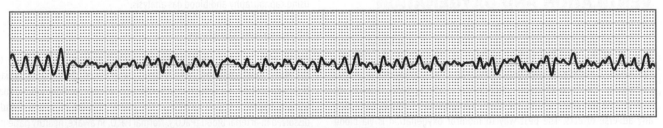

FIGURE 12-16 Ventricular fibrillation. Note the chaotic electrical activity of the heart.

Courtesy of Kim Spelts, BS, CVT, VTS (Anesthesia/Analgesia)

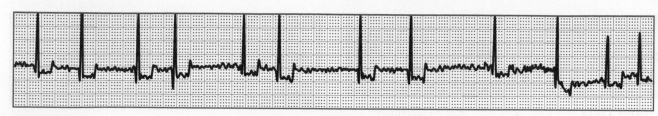

FIGURE 12-18 Atrial fibrillation. Note the unstable baseline, lack of discernible P waves, and irregularly irregular pattern of QRS complexes.

Courtesy of Kim Spelts, BS, CVT, VTS (Anesthesia/Analgesia)

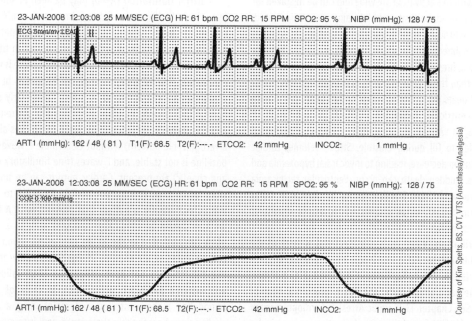

FIGURE 12-19 Respiratory sinus arrhythmia. Note the increase in heart rate that immediately follows inspiration and the decrease that follows expiration.

have dramatic fluctuations in their vagal tone, leading to a pronounced respiratory sinus arrhythmia.[36] Treatment is typically not warranted, but the administration of an anticholinergic drug may resolve the arrhythmia if clinical signs were to occur.

Table 12-3 summarizes treatment of cardiac arrhythmias.

VENTILATION AND OXYGENATION

Oxygen delivery involves the cardiovascular system (which pumps oxygenated blood to tissues) and the pulmonary system (which loads arterial blood with oxygen), while carbon dioxide removal involves ventilation by the lungs. A variety of anesthetic drugs can alter cardiovascular and/or pulmonary function; therefore, monitoring these systems in order to understand the cause of the anesthetic complication helps in correcting any adverse effects. The best way to monitor oxygenation is via blood gas analysis; however, collecting an arterial blood sample for blood gas analysis is technically more challenging and not every veterinary practice

has blood gas analyzing units. Although some general and emergency practices now have point of care blood gas analyzing units (such as the i-STAT®1), those without them still need to refer patients to a large referral/specialty or university setting. Repeated arterial blood gas testing can also become costly. Pulse oximetry therefore is more commonly used because it is a simple, noninvasive method to provide similar information about a patient's oxygenation. Ventilation is monitored via assessing CO_2 levels using capnography and blood gas analysis.

Hypoventilation

Along with hypothermia and hypotension, hypoventilation is one of the three most common complications experienced during anesthesia.[5] **Hypoventilation** is a reduction in ventilation rate and/or tidal volume that leads to hypercapnia, an increase in the partial pressure of carbon dioxide dissolved in arterial blood ($PaCO_2$). Normal $PaCO_2$ in awake, healthy patients at sea level is 35 to 45 mm Hg. Typical anesthetized patients have a $PaCO_2$ of 45 to 55 mm Hg.

TABLE 12-3 Cardiac Arrhythmia Treatments for Dogs and Cats

Drug/Treatment	Dosage and Route	Indication(s)
Atropine (Atrocare®, generic)	0.02 mg/kg IV++ 0.04 mg/kg IM++	Bradyarrhythmia
Glycopyrrolate (Robinul®)	0.005 mg/kg IV++ 0.01 mg/kg IM++	Bradyarrhythmia
Isoproterenol (Isuprel®)	0.1–2.0 mcg/kg/min IV CRI++	3rd degree AV block
Esmolol (Brevibloc®)	0.25–0.5 mg/kg IV slow bolus over 1–2 min followed by 50–200 mcg/kg/min IV CRI+++ 10–200 μg/kg/min IV CRI+++	SVT Atrial fibrillation
Lidocaine (Xylocaine®),	2 mg/kg of lidocaine IV given as a single bolus; this dose can be repeated up to a max of 8 mg/kg IV over a course of 5 to 10 minutes++ 25 to 80 mcg/kg/min IV CRI++	VPCs Ventricular tachycardia
Defibrillation		Ventricular fibrillation

++ Ko, Jeff, and Krimins, R. (2012). Anesthetic monitoring: Therapeutic actions. *Today's Veterinary Practice.*
+++ Riviere, J., and Papich, M. (2009). *Veterinary pharmacology and therapeutics,* 9th ed.

Mild increases of $PaCO_2$ are tolerated well by most patients. However, as $PaCO_2$ increases, respiratory acidosis develops. For every increase of 10 mm Hg in $PaCO_2$, the pH drops by 0.05. It is important to keep the pH > 7.2 to ensure proper enzymatic function. A mildly elevated $PaCO_2$ can lead to sympathetic nervous system stimulation, mildly increasing cardiac contractility, stroke volume, cardiac output, and blood pressure. An elevation in $PaCO_2$ should not be allowed in patients with a space-occupying brain lesion, such as a tumor or intracranial hemorrhage because an increase in $PaCO_2$ can further increase intracranial pressure (ICP), which could lead to brain stem herniation.[5]

Hypoventilation can contribute to hypoxemia when a low **inspired oxygen** (FiO_2) is also present, such as when the patient is breathing room air ($FiO_2 = 21\%$). The partial pressure of oxygen in the alveoli (PAO_2) is determined by the **alveolar oxygen equation:**[41]

$$PAO_2 = ((P_{bar} - P_{H_2O}) \times FiO_2) - (PaCO_2/0.8)$$

P_{bar} is the barometric pressure (at sea level, 760 mm Hg) and P_{H_2O} is the water vapor pressure in the atmosphere (47 mm Hg). A patient breathing 100% oxygen with mild hypercapnia ($PaCO_2 = 50$ mm Hg) would have a PAO_2 of approximately 650 mm Hg and a PaO_2 (partial pressure of

oxygen in arterial blood) of approximately 635 mm Hg (in normal patients, the difference between oxygen in the alveoli and arteries is usually < 15 mm Hg [650 mm Hg − 15 mm Hg = 635 mm Hg]).[42] However, that same patient breathing 21% oxygen with mild hypercapnia would have a PAO_2 of approximately 87 mm Hg and a PaO_2 of approximately 72 mm Hg. As $PaCO_2$ increases and/or P_{bar} decreases (such as occurs with increases in altitude), the patient is more likely to develop hypoxemia.

Patient breathing 100% oxygen:

$$100\% = 1.0$$
$$PAO_2 = ((P_{bar} - P_{H_2O}) \times FiO_2) - (PaCO_2/0.8)$$
$$PAO_2 = ((760 \text{ mm Hg} - 47 \text{ mm Hg}) \times 1) - (50 \text{ mm Hg}/0.8)$$
$$PAO_2 = (713 \text{ mm Hg}) - (63 \text{ mm Hg})$$
$$PAO_2 = 650 \text{ mm Hg}$$
$$PaO_2 = 650 \text{ mm Hg} - 15 \text{ mm Hg} = 635 \text{ mm Hg}$$

Patient breathing 21% oxygen:

$$21\% = 0.21$$
$$PAO_2 = ((P_{bar} - P_{H_2O}) \times FiO_2) - (PaCO_2/0.8)$$
$$PAO_2 = ((760 \text{ mm Hg} - 47 \text{ mm Hg}) \times 0.21) - (50 \text{ mm Hg}/0.8)$$
$$PAO_2 = (150 \text{ mm Hg}) - (63 \text{ mm Hg})$$
$$PAO_2 = 87 \text{ mm Hg}$$
$$PaO_2 = 87 \text{ mm Hg} - 15 \text{ mm Hg} = 72 \text{ mm Hg}$$

Hypoventilation and subsequent hypoxemia are concerns during induction, especially for patients who are in a hospital at an altitude higher than sea level. Most induction drugs have dosage-dependent respiratory depressive effects, and $PaCO_2$ can rise quickly when these drugs are administered. Oxygen supplementation via mask provides an FiO_2 of approximately 40% (Figure 12-20). Preoxygenation at 3 to 5 L/min via mask for 3 to 5 minutes prior to and during induction can greatly reduce the chance of the patient developing life-threatening hypoxemia before control of the airway is established.[42]

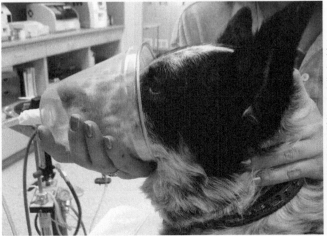

FIGURE 12-20 Preoxygenation reduces the risk of hypoxemia developing during induction.

$PaCO_2$ can only be determined by arterial blood gas analysis. The anesthetist can monitor end-tidal carbon dioxide ($EtCO_2$) as an estimate for $PaCO_2$. In normal patients, $EtCO_2$ is approximately 10% less than $PaCO_2$ due to **dead space ventilation** (areas where there is ventilated air but no gas exchange takes place). Recall from Chapter 3 (Cardiovascular and Respiratory Physiology Review) that anatomical dead space is the volume of air occupying the space in the nose, pharynx, larynx, trachea, bronchi, and bronchioles at inspiration, and alveolar dead space occurs when there are alveoli that are ventilated, but not perfused. Anatomical dead space and alveolar dead space together make up the physiological dead space. Mechanical dead space is the area in the breathing circuit or endotracheal tube (ETT) where bidirectional flow takes place yet no gas exchange occurs. Dead space should be minimized for every patient by selecting the appropriately sized breathing circuit for the patient as well as ensuring the ETT is not extending past the incisors.

The most common causes of hypoventilation and hypercapnia in anesthetized patients are the drugs utilized to induce and maintain anesthesia. Inhalant anesthetic agents, systemic opioids, and many induction drugs cause dosage-dependent respiratory depressive effects. These drugs depress the respiratory centers of the brain, blunting the normal CNS ventilatory response to hypercapnia, hypoxemia, and acidosis.[43] If possible, the anesthetist should first attempt to lighten the plane of anesthesia in order to improve patient ventilation and reduce CO_2 levels.

If the patient does not respond to decreases in anesthetic drugs, **intermittent positive pressure ventilation** (IPPV) may be required. IPPV can be accomplished manually with a dedicated person administering breaths or mechanically with a ventilator. The patient should be initially ventilated with a tidal volume of 10 to 15 mL/kg and a peak inspiratory pressure (PIP) of 10 to 12 cm H_2O (not to exceed 20 cm H_2O). Depending on the type of ventilator used, adjustments can be made to the ventilation rate, tidal volume, and/or PIP to bring $EtCO_2$ and $PaCO_2$ to normal or high normal levels. IPPV should not be initiated without ensuring the patient's cardiac output and blood pressure are maintained. Introducing positive pressure into the thorax can compress the vena cava, decreasing venous return to the heart[44] (Figure 12-21). This can decrease stroke volume, cardiac output, and blood pressure.

Tech Tip

When should the anesthetist intervene with IPPV? The answer varies among anesthesiologists, but one suggestion is to allow $EtCO_2$ to rise to about the mid-50s (some call this "permissive hypercapnia"), and then begin IPPV at no more than 2 to 4 breaths per minute to encourage the patient to continue spontaneously breathing. Once the patient consistently maintains $EtCO_2$ in the mid to upper 60s (and inhalant anesthetic agent has been decreased as much as possible, and other concerns such as hypothermia are being addressed), mechanical ventilation instead of manual ventilation can be initiated to free up the anesthetist.

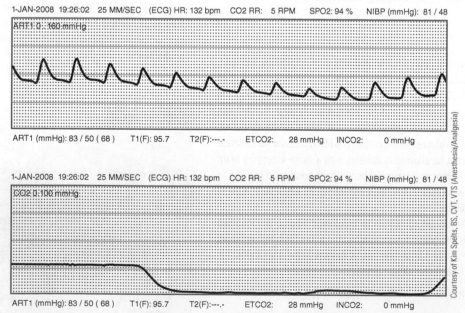

FIGURE 12-21 The use of intermittent positive pressure ventilation can reduce venous return to the heart, which in turn can decrease cardiac output. In this image, the amplitude of the arterial catheter pulse waveform decreases whenever the mechanical ventilator gives a breath (shown by the inspiratory phase of the capnogram).

Hyperventilation

Hyperventilation is an increase in respiratory rate and/or tidal volume that lead to **hypocapnia**, or decreased $PaCO_2$. The most common causes of hyperventilation in anesthetized patients are a light plane of anesthesia and overly aggressive IPPV.

If the hypocapnia is due to overventilation, then decrease the tidal volume, ventilation rate, and/or inspiratory pressure to allow $PaCO_2$ to approach normal levels. If the patient is hyperventilating due to a painful response to surgical stimulus, additional pain medications should be administered, and the inhalant anesthetic agent level potentially increased.

Hyperventilation may also be the patient's response to help compensate for hypoxemia (see below). In this case, rapid identification and resolution of the cause of hypoxemia is crucial.

Hypoxemia

Hypoxemia is a condition in which the blood is not sufficiently oxygenated to meet metabolic requirements. This occurs when the partial pressure of dissolved oxygen in arterial blood (PaO_2) drops to less than 80 mm Hg. PaO_2 is evaluated by blood gas analysis of an arterial blood sample.

There are five primary causes of hypoxemia as follows:[5,43]

- *Low inspired oxygen* (FiO_2). This is rare in anesthetized patients, since they are breathing 100% oxygen. Inadvertent disconnection of the patient from the breathing circuit, placement of the ETT into the esophagus instead of the trachea, a kinked ETT, or a depletion of the central oxygen source are all mistakes that can lead to life-threatening hypoxemia.
- ***Ventilation-perfusion (V/Q) mismatch***. This is the most common cause of hypoxemia in anesthetized patients. V/Q can be high (alveoli are ventilated but perfusion is low due to decreased cardiac output or, in the extreme case, pulmonary thromboembolism) or, more commonly, low (alveoli are not ventilated but perfusion exists, such as in extreme atelectasis). Obese patients, pregnant patients, patients with lung disease, and patients placed in lateral or dorsal recumbency are at high risk for experiencing V/Q mismatch.
- *Hypoventilation*. While hypoventilation is a common complication in anesthetized patients, it usually does not play a significant role in hypoxemia unless it occurs when the patient has a low FiO_2. See the "Hypoventilation" section above.
- ***Diffusion impairment***. Conditions such as chronic fibrosis of the lungs can impair gas exchange.
- ***Right-to-left shunt***. Physiologic shunts due to unusual pulmonary blood flow can cause deoxygenated blood to mix with oxygenated blood going back into systemic circulation through the left ventricle.

It is unusual for anesthetized dogs, cats, and other small animals breathing 100% oxygen to develop hypoxemia ($PaO_2 < 80$ mm Hg) in the absence of severe respiratory pathophysiology. However, a patient not oxygenating as expected at an FiO_2 of 100% may be susceptible to hypoxemia in recovery where the FiO_2 is only 21%. To assess this likelihood, the anesthetist can evaluate the P/F ratio.

The **PaO_2/FiO_2 (P/F) ratio** is used extensively in critical care medicine as an objective number that can be utilized to assess the adequacy of a patient's response to oxygen therapy. This number holds some value to the anesthetist as well to help predict the likelihood of a patient developing

hypoxemia once taken off 100% oxygen and moved to recovery. A healthy patient with a normal PaO_2 of 100 mm Hg breathing room air with an FiO_2 of 21% has a P/F ratio of approximately 500 (100/0.21). If that same healthy patient is anesthetized with a PaO_2 of 500mm Hg breathing 100% oxygen, the P/F ratio is still 500 (500/1).[45]

A P/F ratio > 400 indicates normal respiratory function,[45] and a ratio of < 300 indicates acute respiratory distress syndrome (ARDS).[46] If a patient is anesthetized on 100% oxygen and has a PaO_2 of 100 mm Hg, that patient is not technically hypoxemic (PaO_2 is not less than 80 mm Hg). The P/F ratio, however, is only 100. When that patient moves to recovery, the FiO_2 could drop to 21 mm Hg, a severe and immediately life-threatening hypoxemia.

Treatment of Hypoxemia

Treatment of hypoxemia is dependent on identifying the underlying cause. Anticipating a precipitous drop in PaO_2 when the patient is moved off 100% oxygen, the anesthetist should consider supplemental oxygen administration (via O_2 cage, nasal cannulas, or mask) in recovery for those patients with a low P/F ratio. V/Q mismatch exhibited while the patient is anesthetized may resolve with efforts to improve ventilation or improve perfusion. Hypoventilation during surgery is resolved with IPPV (Figure 12-22). Hypoventilation in recovery should resolve as the patient

Courtesy of Kim Spelts, BS, CVT, VTS (Anesthesia/Analgesia)

FIGURE 12-22 Intermittent positive pressure ventilation, either manually or via a mechanical ventilator, may be required to manage hypoventilation and/or V/Q mismatch in order to prevent hypoxemia.

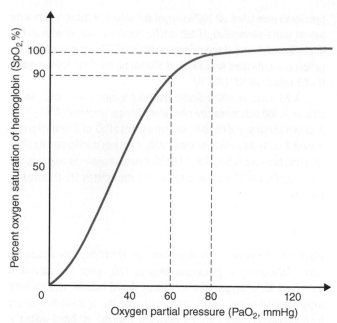

FIGURE 12-23 The oxyhemoglobin dissociation curve describes the relation between the partial pressure of oxygen (x axis) and the oxygen saturation of hemoglobin (y axis). Note that due to the sigmoid shape of this curve, PaO$_2$ can drop rapidly with very little change in SaO$_2$.

clears the effects of the inhalant anesthetic agent and other anesthetics; in the meantime, supplemental oxygen may be indicated to reduce the effect of lower inspired oxygen. SpO$_2$ should be monitored in recovery in order to watch the effect of residual hypercapnia and quickly identify hypoxemia as it starts to develop.

If it is not feasible to run arterial blood gases to measure PaO$_2$, the anesthetist can utilize a pulse oximeter and perform an oxygen challenge test to the patient prior to moving to recovery. The percentage of hemoglobin saturated with oxygen (SaO$_2$) is related to PaO$_2$ as shown in the oxygen-hemoglobin dissociation curve (Figure 12-23). It should be noted that, because of the sigmoid shape of this curve, PaO$_2$ can drop precipitously with very little change in SaO$_2$. A drop in SaO$_2$ in a patient breathing 100% oxygen is a late-stage sign of a drop in PaO$_2$ and should be immediately investigated and remedied by the anesthetist.

With a pulse oximetry measurement (SpO$_2$, which closely estimates SaO$_2$) of 95 to 100% the observer cannot know whether the PaO$_2$ is close to 80 mm Hg or > 400 mm Hg. A short deprivation of 100% oxygen and a close observation of the pulse oximetry reading will give the anesthetist some indication of where the patient's oxygenation may lie on the curve. This is a simple and inexpensive method of evaluation to ensure that a patient's hypoventilation resolves prior to being moved to a location with a lower FiO$_2$.

How to perform an oxygen challenge test:

1. At the end of the procedure, place the patient in lateral recumbency.
2. Allow the patient to spontaneously ventilate on 100% oxygen with the inhalant anesthetic agent turned off for 3 to 5 minutes (in theory it takes 3–5 minutes to desaturate).
3. Ensure that the pulse oximeter is displaying a good reading with a strong pulse signal.
4. Disconnect the breathing circuit and continue to watch the pulse oximeter for 1 to 2 minutes.
   a Some anesthetists prefer to keep the capnograph adaptor attached if the patient is still intubated to see how much of a contribution hypoventilation is playing in a low pulse oximeter reading. If the patient is hypoventilating (EtCO$_2$ > 45 mm Hg), then they are likely going to need oxygen a little longer before they are alert enough to spit out the endotracheal tube.
5. If the SpO$_2$ drops, reconnect the breathing circuit (with 100% oxygen) and allow the SpO$_2$ to increase.
6. Repeat this process until the patient can maintain the same SpO$_2$ reading on room air that is maintained by breathing 100% oxygen.

SUMMARY

Anesthetic agents compromise patient homeostasis; therefore, maintaining a balance between anesthetic depth and organ function is important for patient stability. In addition to avoiding preventable anesthetic complications, proper monitoring and early intervention minimize anesthetic complications (Table 12-4). Recognition of abnormalities ultimately allows identification of the underlying cause of the complication so that corrective measures and supportive care may be implemented if necessary. While no electronic monitoring equipment can replace the eyes, ears, hands, and instincts of a trained anesthetist, electronic monitoring devices may provide valuable insight to the patient's physiologic status thus alerting the attentive anesthetist to abnormalities with potentially life-threatening consequences. Being able to discern if an anesthetic complication needs to be treated based upon its etiology and likelihood to cause consequence if ignored is a learned clinical skill. Clearly communicating all information about anesthetic complications to the veterinarian is fundamental to resolving them effectively and efficiently.

TABLE 12-4 Preventable Anesthetic Complications

- Drug dose/calculation errors
- Fluid calculation errors
- Closed pop off valve
- Overinflated ETT cuff
- Tracheal tears from turning patient while connected
- Walking away from patient and it falls off the table
- Misassembled equipment and/or faulty equipment not identified
- Deflating the ETT cuff too early and patient aspirates regurgitated content/vomitus

CRITICAL THINKING POINTS

- The three most common anesthetic complications are *hypothermia*, *hypotension*, and *hypoventilation*. A fourth complication, *hypoxemia*, is less commonly seen; however, when it does occur it is often more life-threatening and needs to be immediately addressed. Remember the 4 H's.

- Prewarming patients is the most effective strategy for preventing hypothermia. Simple steps to help reduce hypothermia include the use of padded bedding and blankets (forced air warming, conductive fabric warming, and circulating warm water), wrapping the patient's feet, warming the IV fluids, and warming the surgical prep solutions.

- Oxygen delivery is a function of cardiac output and arterial oxygen content: $DO_2 = CO \times CaO_2$.

- Cardiac output is a function of heart rate and stroke volume (which in turn is a function of preload, afterload, and cardiac contractility): $CO = HR \times SV$.

- Mean arterial pressure is a function of cardiac output and systemic vascular resistance: $MAP = CO \times SVR$.

- Hypotension may result from decreased heart rate, decreased preload, decreased cardiac contractility, and/or vasodilation (which may be profound in some medical conditions). Identify the underlying cause in order to treat hypotension in the most effective and efficient manner possible.

- Cardiac arrhythmias are not uncommon, but not all arrhythmias require pharmaceutical intervention. Assess the impact on cardiac output as well as the likelihood for the arrhythmia to progress to a more serious abnormality prior to treatment.

- Preoxygenation is an inexpensive and valuable safety measure to prevent hypoxemia at induction.

- The five potential causes of hypoxemia are low FiO_2, V/Q mismatch, hypoventilation, diffusion impairment, and right-to-left shunt.

- A low P/F ratio while breathing 100% oxygen could lead to hypoxemia when the patient is breathing room air.

REVIEW QUESTIONS

Multiple Choice

1. Which of the following is *not* a mechanism of heat transfer?
   a. Radiation
   b. Conduction
   c. Convection
   d. Osmosis

2. Which of the following is *not* a cause of hypothermia in anesthetized patients?
   a. Vasoconstriction from the use of alpha-2 adrenergic agonists
   b. Exposure of tissues to room air
   c. Aseptic prep with room-temperature solutions
   d. CNS depression

3. Which drug is solely a beta-1 adrenergic agonist and is the least likely to cause tachycardia?
   a. Ephedrine
   b. Norepinephrine
   c. Dopamine
   d. Dobutamine

4. If a patient is both bradycardic and hypotensive, what is the first treatment choice?
   a. IV anticholinergic
   b. Dobutamine infusion
   c. IV fluid bolus
   d. IV dexmedetomidine

5. When does renal autoregulation occur?
   a. The systolic arterial pressure is > 60 mm Hg
   b. The mean arterial pressure is 50–150 mm Hg
   c. The mean arterial pressure is 80–180 mm Hg
   d. The diastolic arterial pressure is > 40 mm Hg

6. What instrument can measure oxygen delivery to the tissues (DO_2)?
   a. Respirometer
   b. Capnometer
   c. Pulse oximeter
   d. None of the above

7. What term is used for the overall degree of vasoconstriction or vasodilation that exists at any given moment?
   a. Preload
   b. Systemic vascular resistance
   c. Cardiac output
   d. Contractility

8. A norepinephrine CRI may be indicated in patients with which type of medical condition or injury because it preferentially corrects inappropriately dilated areas?
   a. Hypovolemic GDV
   b. Laceration
   c. Proptosed eye
   d. Sepsis

9. Which of the following are most commonly seen in anesthetized patients?
    a. Hyperventilation and hypoxemia
    b. Hypoventilation and hypercapnia
    c. Hypoventilation and hypoxemia
    d. Hypoventilation and hypocapnea

10. Why might an anesthetist use dobutamine in dogs and cats?
    a. It causes quick and reliable vasoconstriction and increases cardiac contractility and heart rate.
    b. It is inexpensive and currently is a C-V controlled substance.
    c. It combats the myocardial depression caused by the inhalant anesthetic agent by increasing cardiac contractility.
    d. It can be administered via IV bolus to quickly elevate heart rate.

11. Second-degree A V block (Mobitz Type 1) may be observed in normal small animal patients administered which of the following drugs?
    a. Dexmedetomidine
    b. Carprofen
    c. Acepromazine
    d. Diazepam

12. What condition causes multiform VPCs on the ECG?
    a. Sick sinus syndrome
    b. Multiple irritated sites within the myocardium
    c. Hypotension
    d. Rapid SA node firing

13. During a femoral fracture repair, your canine patient's heart rate approaches 160 beats per minute. Which of the following is *not* a potential cause of her sinus tachycardia?
    a. Pain
    b. Inadequate depth of anesthesia
    c. Premedication with hydromorphone
    d. Premedication with glycopyrrolate

14. You are anesthetizing a Great Dane with gastric dilatation-volvulus (GDV). The ECG displays a heart rate of 160 beats per minute and the complexes are wide and bizarre with no discernible P waves. What do you decide about this patient?
    a. The patient is painful, so administer additional fentanyl.
    b. The patient is at a light plane of anesthesia, so turn up the vaporizer.
    c. The patient is tachycardic, so give IV dexmedetomidine to slow the patient's heart rate.
    d. The patient is exhibiting ventricular tachycardia, so administer IV lidocaine.

15. Which arrhythmia is characterized by a normal P-R interval with an intermittent drop of the QRS complex?
    a. 1st degree AV block
    b. 2nd degree AV block, Mobitz Type 1
    c. 2nd degree AV block, Mobitz Type 2
    d. 3rd degree AV block

16. Which statement is false about preoxygenation?
    a. Preoxygenation reduces the patient's $PaCO_2$.
    b. Preoxygenation reduces the risk of the patient developing hypoxemia during anesthetic induction.
    c. Preoxygenation increases FiO_2 while the patient is breathing through the mask.
    d. Preoxygenation is especially important for patients who live at an altitude higher than sea level.

17. Your patient anesthetized for a laparotomy has an $EtCO_2$ of 60 mm Hg. What is causing this hypercapnia?
    a. Inhalant anesthetic agent
    b. Infusion of fentanyl
    c. CNS depression
    d. All of the above

18. What would be an appropriate treatment for the patient in #17?
    a. Turn up the inhalant anesthetic agent
    b. Reverse all analgesics
    c. Administer positive pressure ventilation
    d. Administer ephedrine

19. Which of the following is not a potential cause of hypoxemia in the perianesthetic period?
    a. V/Q mismatch
    b. High FiO_2
    c. Diffusion impairment
    d. Right-to-left shunt

20. What is a good idea to perform prior to moving your patient to recovery?
    a. Monitor the $EtCO_2$ with the breathing circuit disconnected to make sure the patient can stay oxygenated breathing room air.
    b. Monitor the SpO_2 with the breathing circuit disconnected to make sure the patient can stay oxygenated breathing room air.
    c. Monitor the $EtCO_2$ with the breathing circuit connected to make sure the patient can stay oxygenated breathing 100% oxygen.
    d. Monitor the SpO_2 with the breathing circuit connected to make sure the patient can stay oxygenated breathing 100% oxygen.

Case Studies

Case Study 1: A healthy, 30 kg, 2-year-old F/S Labrador Retriever dog presents to the hospital for a large but superficial laceration repair on her lateral thorax. Preoperative bloodwork is all within normal limits. After premedication with acepromazine and hydromorphone, you place an IV catheter, initiate IV fluids, and have your assistant start preoxygenation at 5 L/min via open mask. You administer 150 mg of propofol IV to induce anesthesia. Endotracheal intubation is successful. The ETT is secured, cuff sealed, and isoflurane is initiated at 2%. Your initial set of electronic monitoring parameters reveals a heart rate of 110 bpm,

10 respirations/minute, SpO_2 98%, $EtCO_2$ 45 mm Hg, and blood pressure (systolic/diastolic) 95/35 with a MAP of 55 mmHg. She has a brisk palpebral reflex and her respirations increase when her wound is being clipped. What do you think is the cause(s) of this patient's hypotension and what is the first step you will take to correct it?

Case Study 2: You are the anesthetist for a patient undergoing lung lobectomy. The patient was receiving intermittent positive pressure ventilation during the procedure, but now negative pressure has been restored in his chest and you have weaned him off the ventilator. He is spontaneously ventilating and breathing 100% oxygen. Your last set of monitoring parameters revealed a normal heart rhythm with a rate of 120 bpm, blood pressure 110/68 (MAP = 82), SpO_2 99%, and $EtCO_2$ 55 mm Hg. You decide to run an arterial blood gas before moving to ICU. The blood gas revealed a PaO_2 of 110 mm Hg and $PaCO_2$ of 60 mm Hg. Are you concerned about moving this patient to recovery? Why or why not? What steps would you take to help ensure the safety of your patient when you move to recovery?

Case Study 3: A 7-year-old M/N Domestic Shorthair cat presents to your hospital with urethral obstruction. Bloodwork at presentation shows a BUN of 63 mg/dL, creatinine 2.7 mg/dL, and serum potassium of 5.7 mEq/L. You are named the anesthetist for the cat. He is given SQ methadone as a preanesthetic. An IV catheter is placed and IV fluid therapy is initiated with LRS. Preoxygenation is initiated and you induce him with a combination of etomidate and midazolam. Endotracheal intubation is successful. When you hook up the ECG, you recognize that the cat is in ventricular tachycardia with a heart rate of 240 bpm. You alert the veterinarian immediately. What drug do you anticipate the veterinarian ordering for administration?

Case Study 4: You are anesthetizing a 5-year-old F/S Domestic Longhair cat for a dental prophylaxis and extraction of a fractured lower left canine tooth (304). Whenever the veterinarian tries to manipulate and elevate the tooth for extraction, the cat hyperventilates and becomes tachycardic. You try turning up the inhalant anesthetic agent to get the patient deeper, but when you do so she becomes hypotensive. What would you consider doing differently in order to smooth out this patient's plane of anesthesia and (hopefully) put an end to her anesthesia complications?

Case Study 5: A 5-year-old intact male German Shepherd dog is rushed to your hospital after being hit by a car. The dog is tachycardic, tachypneic, has a prolonged capillary refill time, weak peripheral pulses, and a depressed mentation. Further diagnostics by the veterinarian reveal a hemoabdomen as well as pulmonary contusions. Stabilizing measures are initiated and pain medications are administered. You have been selected to anesthetize this patient for laparotomy and possible splenectomy or liver lobectomy. What complications do you expect to encounter?

Critical Thinking Questions

1. You have been asked to anesthetize a 3 kg Yorkshire Terrier dog for a portosystemic shunt repair. This will be a long procedure, and the concern is how to keep this patient warm. What steps should be taken to prevent the development of hypothermia?

2. Describe ways to provide multimodal anesthesia to patients in order to minimize the amount of inhalant anesthetic agent used, thereby minimizing the dosage-dependent negative adverse effects of the inhalant anesthetic agent (decreased contractility, vasodilation).

ENDNOTES

1. Redondo, J. I., Suesta, P., Serra, I., et.al. (2012). Retrospective study of the prevalence of postanesthetic hypothermia in dogs. *Veterinary Record 171*(15), 374.
2. Ibid
3. Robertson, S. A. (2007). Hypothermia – "The Big Chill." In *Scientific proceedings: Companion animals programme* (pp. 51–52). European Veterinary Conference.
4. Muir, W. W., Hubbell, J. A. E., Bednarski, R. M., et.al. (2013). Temperature regulation during anesthesia: Anesthetic-associated hypothermia and hyperthermia. In W. W. Muir et.al. (Eds.), *Handbook of veterinary anesthesia* (5th ed., pp. 330–347). St. Louis, MO: Elsevier/Mosby.
5. Shelby, A. M., & McKune, C. M. (2014). Anesthetic complications. In A. M. Shelby, & C. M. McKune (Eds.), *Small animal anesthesia techniques* (pp. 191–231). Ames, IA: John Wiley and Sons Inc.
6. Pottie, R. G., Dart, C. M., Perkins, N. R., et al. (2007). Effect of hypothermia on recovery from general anaesthesia in the dog. *Australian Veterinary Journal, 85*(4), 158–162.
7. Ko, J. C. (1999). Other complications that result in anesthetic emergencies. In J. Ko (Ed.), *Small animal anesthesia and pain management* (p. 181). London: Manson Publishing Ltd.
8. Haskins, S. (1999). Perioperative monitoring. In R. Paddleford (Ed.), *Manual of small animal anesthesia* (2nd ed., pp. 123–146). Philadelphia, PA: WB Saunders Co.
9. Posner, L. P., Pavuk, A. A., Rokshar, J. L., et. al. (2010). Effects of opioids and anesthetic drugs on body temperature in cats. *Veterinary Anaesthesia and Analgesia, 37*(1), 35–43.
10. Levensaler, A. (2010). Monitoring: Pulse oximetry, temperature, and hands-on. In S. Bryant (Ed.), *Anesthesia for veterinary technicians* (1st ed., pp. 98–100). Ames, IA: Blackwell.

11. Gaynor, J. S., Dunlop, C. I., Wagner, A. E., et. al. (1999). Complications and mortality associated with anesthesia in dogs and cats. *Journal of the American Animal Hospital Association35*(1), 13–17.

12. Grandy, J. L., Dunlop, C. I., Hodgson, D. S., et al. (1992). Evaluation of the Doppler ultrasonic method of measuring systolic arterial blood pressure in cats. *American Journal of Veterinary Research, 53,* 1166–1169.

13. Gordon, A. M., & Wagner, A. E. (2006). Anesthesia-related hypotension in a small-animal practice. *Veterinary Medicine,* 22–26.

14. Wagner, A. E., & Brodbelt, D. C. (1997). Arterial blood pressure monitoring in anesthetized animals. *Journal of the American Veterinary Medical Association, 210,* 1279–1285.

15. Muir, W. W., & Mason, D. (1996). Cardiovascular system. In J. C. Thurmon, W. J. Tranquilli, & G. J. Benson (Eds.), *Lumb and Jones' veterinary anesthesia* (3rd ed., pp. 62–114). Baltimore, MD: Williams & Wilkins.

16. Peterson, N. W., & Moses, L. (2011). Oxygen delivery. *Compendium of Continuing Education Veterinary, 33*(1), E5.

17. Thurmon, J. C., Tranquilli, W. J., & Benson, G. J. (1999). Anesthesia and the cardiovascular, respiratory, and central nervous systems. In J. C. Thurmon, W. J. Tranquilli, & G. J. Benson (Eds.) *Essentials of small animal anesthesia and analgesia* (1st ed., pp. 61–125). Baltimore: Lipincott Williams & Wilkins.

18. Mazzaferro, E., & Wagner, A. E. (2001). Hypotension during anesthesia in dogs and cats: Recognition, causes, and treatment. *Compendium, 23*(8), 728–737.

19. Mutoh, T., Nishimura, R., Kim, H. Y., et al. (1997). Cardiopulmonary effects of sevoflurane, compared with halothane, enflurane, and isoflurane, in dogs. *American Journal of Veterinary Research, 38,* 885–890.

20. Ferreira, T. H., Steffey, E. P., Mama, K. R., et al. (2001). Determination of the sevoflurane sparing effects of methadone in cats. *Veterinary Anaesthesia and Analgesia, 38*(4), 310–319.

21. Reilly, S., Seddighi, R., Egger, C. M., et al. (2013). The effect of fentanyl on the end-tidal sevoflurane concentration needed to prevent motor movement in dogs. *Veterinary Anaesthesia and Analgesia, 40*(3), 290–296.

22. Ortega, M., & Cruz, I. (2011). Evaluation of a constant rate infusion of lidocaine for balanced anesthesia in dogs undergoing surgery. *Canadian Veterinary Journal, 52,* 856–860.

23. Solano, A. M., Pypendop, B. H., Boscan, P. L., et.al. (2006). Effect of intravenous administration of ketamine on the minimum alveolar concentration of isoflurane in anesthetized dogs. *American Journal of Veterinary Research, 67*(1), 21–25.

24. Muir, W. W., Wiese, A. J., & March, P. A. (2003). Effects of morphine, lidocaine, ketamine, and morphine-lidocaine-ketamine drug combination on minimum alveolar concentration in dogs anesthetized with isoflurane. *American Journal of Veterinary Research, 64*(9), 1155–1160.

25. Yamashita, K., Okano, Y., Yamashita, M., et.al. (2008). Effects of carprofen and meloxicam with or without butorphanol on the minimum alveolar concentration of sevoflurane in dogs. *Journal of Veterinary Medical Science, 70*(1), 29–35.

26. Farry, T. (2010). Anesthesia for pediatric patients. In S. Bryant (Ed.), *Anesthesia for veterinary technicians* (1st ed., pp. 267–274). Ames, IA: Blackwell.

27. Fuehrer, L. (2010). Induction drugs. In S. Bryant (Ed.), *Anesthesia for veterinary technicians* (1st ed., pp. 143–151). Ames, IA: Blackwell.

28. Davis, H., Jensen, T., Johnson, A., et.al. (2013). AAHA/AAFP fluid therapy guidelines for dogs and cats. *Journal of the American Animal Hospital Association, 49*(3), 149–59

29. Gross, M. E., Giuliano, E. A., Raffe, M. R., et. al. (2011). Anesthetic considerations for special procedures. In K. A. Grimm, W. J. Tranquilli, & L. A. Lamont (Eds.), *Essentials of small animal anesthesia and analgesia* (2nd ed., p. 483). Ames, IA: John Wiley & Sons.

30. Ko, J. C. (2013). Anesthesia monitoring and management. In J. C. Ko (Ed.), *Small animal anesthesia and pain management* (pp. 123–162). London: Manson Publishing Ltd.

31. Pascoe, P. J., Ilkiw, J. E., & Pypendop, B. H. (2006). Effects of increasing infusion rates of dopamine, dobutamine, epinephrine, and phenylephrine in healthy anesthetized cats. *American Journal of Veterinary Research, 67*(9), 1491–1499.

32. Rosati, M., Dyson, D. H., Sinclair, M. D., et al. (2007). Response of hypotensive dogs to dopamine hydrochloride and dobutamine hydrochloride during deep isoflurane anesthesia. *American Journal of Veterinary Research, 68*(5), 483–494.

33. Egger, C. (2011). Anaesthetic complications, accidents, and emergencies. In S. Seymour, & K. Duke-Novakovski (Eds.), *BSAVA manual of canine and feline anaesthesia and analgesia* (2nd ed., pp. 310–332). Gloucester, England: British Small Animal Veterinary Association.

34. Congdon, J. M., Marquez, M., Niyom, S., et. al. (2011). Evaluation of the sedative and cardiovascular effects of intramuscular administration of dexmedetomidine with and without concurrent atropine administration in dogs. *Journal of the American Animal Hospital Association, 239*(1), 81–89.

35. Monteiro, E. R., Campagnol, D., Parrilha, L. R., et. al. (2009). Evaluation of cardiorespiratory effects of combinations of dexmedetomidine and atropine in cats. *J Fel Med & Surg, 11*(10), 783–792.

36. Tilley, L. P., & Smith, F. W. K. (2008). Electrocardiography. In L. P. Tilley, F. W. K. Smith, M. A. Oyama, & M. M. Sleeper (Eds.), *Manual of canine and feline cardiology* (4th ed., pp. 49–77). St. Louis, MO: Saunders.

37. Watson, K. (2012). Abnormalities of cardiac conduction and cardiac rhythm. In R. L. Hines, & K. E. Marschall (Eds.), *Stoelting's anesthesia and co-existing disease* (6th ed., pp. 73–103). Philadelphia, PA: Saunders.

38. Shirasaka, W., Ideshita, K., Toriyama, S., et.al. (2014). Intraoperative asystole in a patient with concealed sick sinus syndrome: a case report. *Masui., 63*(3), 338–341.

39. Curtis-Uhle, W., & Waddell, K. W. (2010). Anesthesia for patients with cardiac disease. In S. Bryant (Ed.) *Anesthesia for veterinary technicians* (1st ed., pp. 205–218). Ames, IA: Blackwell.

40. Keefe, J. (2010). Introduction to monitoring: Monitoring the ECG and blood gases. In S. Bryant (Ed.), *Anesthesia for veterinary technicians* (1st ed., pp. 85–94). Ames, IA: Blackwell.

41. Muir, W. W., Hubbell, J. A. E., Bednarski, R. M., et.al. (2013). Acid-base balance and blood gases. In W. W. Muir (Eds.) *Handbook of veterinary anesthesia* (5th ed., pp. 285–305). St. Louis, MO: Elsevier/Mosby.

42. McNally, E. M., Robertson, S. A., & Pablo, L. S. (2009). Comparison of time to desaturation between preoxygenated and nonpreoxygenated dogs following sedation with acepromazine maleate and morphine and induction of anesthesia with propofol. *American Journal of Veterinary Research, 70*(11), 1333–1338.

43. McMillan, S. (2010). Anesthetic complications and emergencies. In S. Bryant (Ed.), *Anesthesia for veterinary technicians* (1st ed., pp. 167–185). Ames, IA: Blackwell.

44. Dyson, D. H. (2012). Positive pressure ventilation during anesthesia in dogs: Assessment of surface area derived tidal volume. *Canadian Veterinary Journal, 53*(1), 63–66.

45. Irizzary, R., & Reiss, A. J. (2009). Beyond blood gases: making use of additional oxygenation parameters and plasma electrolytes in the emergency room. *Compendium on Continuing Education for the Practicing Veterinarian, 31*(10), E1–E5.

46. Snyder, K. (2009). Acute lung injury and acute respiratory distress syndrome: Two challenging respiratory disorders. Available at http://veterinarymedicine.dvm360.com/vetmed/Medicine/Acute-lung-injury-and-acute-respiratory-distress-s/ArticleStandard/Article/detail/640122, accessed October 15, 2014.

CHAPTER 13

Cardiac Arrest & Cardiopulmonary Cerebral Resuscitation (CPCR)

CHRISTOPHER L. NORKUS, *DVM, DACVAA, CVPP, DACVECC*

LEARNING OBJECTIVES

Upon completion of this chapter, it is expected that the reader should be able to:

13.1 Identify the common causes of cardiopulmonary arrest (CPA) in dogs and cats

13.2 Identify the impending signs of CPA in dogs and cats

13.3 Compare and contrast the realistic prognosis for dogs and cats experiencing CPA during anesthesia and when in the general hospitalized population

13.4 Recognize the importance of standing resuscitation orders and "do not resuscitate (DNR)" codes

13.5 Describe the current clinical airway management guidelines for performing basic life support in dogs and cats

13.6 Describe the current ventilation clinical guidelines for performing basic life support in dogs and cats

13.7 Describe the current chest compression clinical guidelines for performing basic life support in dogs and cats

13.8 Describe the current clinical guidelines for performing advanced life support in dogs and cats

13.9 Discuss the mechanism of action of epinephrine, atropine, and vasopressin in relation to cardiac arrest treatment

13.10 Identify all cardiac arrest ECG rhythms and their treatment options Explain treatment goals for post cardiac arrest care once return of spontaneous circulation is achieved

KEY TERMS

Agonal breathing

Arterial oxygen saturation

Cardiac pump theory

Cardiopulmonary arrest (CPA)

Central venous oxygen saturation

Defibrillation

Impedance threshold device (ITD)

Interposed abdominal compressions

Precordial thump

Reassessment Campaign on Veterinary Resuscitation (RECOVER)

Thoracic pump theory

INTRODUCTION

Cardiopulmonary arrest (CPA) is a sudden total failure of both the circulatory and respiratory systems. The absence of appropriate cardiac output and delivery of oxygen to tissue quickly results in unconsciousness and body-wide cellular death. Without aggressive intervention, cerebral hypoxia leads to complete biological brain death within 4 to 6 minutes of CPA.[1,2] Prompt and effective cardiopulmonary cerebral resuscitation (CPCR) including basic life support (BLS) and advanced life support (ALS) techniques are imperative to maximize the chances of sustained return of spontaneous circulation (**ROSC**) and long-term patient survival. Veterinary technicians in both the anesthesia and critical care arena play a key role in ensuring that patients receive this treatment within that period.

CAUSES AND FREQUENCY OF CPA DURING GENERAL ANESTHESIA

In anesthetized and hospitalized companion animals, leading reasons for CPA include

- Anesthetic overdose
- Anesthetic equipment failure
- Anesthetist accident (e.g., closed pop-off valve)
- Severe hypoxemia
- Vagal stimulation
- Profound blood loss (e.g., hypovolemia)
- Unstable cardiac arrhythmias (e.g., unstable ventricular tachycardia)
- Severe electrolyte disturbances (e.g., hyperkalemia)
- Anaphylaxis, cardiorespiratory disorders (e.g., congestive heart failure, pericardial tamponade, pneumothorax)
- Debilitating or end-stage diseases (e.g., sepsis, cancer)

Whenever possible, rapid pursuit and treatment of the underlining CPA causes is essential to long-term survival.

Based upon current literature, it is estimated that the rate of healthy dogs and cats (American Society of Anesthesiology Status [ASA] 1 and 2) experiencing CPA during anesthesia is approximately 0.12%. For sick animals this number is higher with approximate mortality rates of 2.9% for ASA three cases, 7.58% for ASA four cases, and 17.33% for ASA five cases.[3] Based on several studies, horses and exotic species may have higher risk for anesthetic-related morbidity and mortality. Some risk factors for anesthetic-related deaths in dogs include an increase in ASA physical status, urgency of the procedure, increased patient age, duration of the procedure, and decreased body weight.[4]

SIGNS OF IMPENDING CPA

Being able to identify when CPA may occur can help the veterinary team prepare for aggressive intervention. Signs that CPA may be imminent during anesthesia are listed in Table 13-1.

In the non-anesthetized patient, many of these signs can also be observed along with patient collapse. Some patients will also demonstrate unusual behavior or distressed vocalizations before CPA.

TABLE 13-1 Signs of Imminent Cardiopulmonary Arrest During Anesthesia

- Absence of surgical bleeding
- Dramatic changes in the characteristics of breathing (e.g., **agonal breathing** [abnormal pre-death breathing pattern characterized by short, gasping breaths], marked increased effort, sudden decreased rate or apnea, abnormal pattern)
- Absence of a palpable pulse
- Severe hypotension
- Diminishing or inaudible Doppler pulse wave sounds
- Dramatic changes in heart rate or electrocardiographic rhythm
- Changes to mucous membrane color (e.g., white or cyanotic)
- Fixed and dilated pupils
- An accompanying rapid fall in end-tidal carbon dioxide (Figure 13-1)

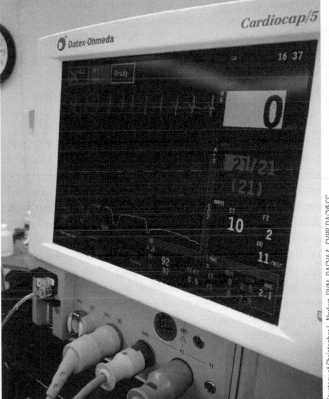

Courtesy of Christopher L. Norkus, DVM, DACVAA, CVPP, DACVECC

FIGURE 13-1 This figure amazingly documents a patient under general anesthesia just seconds after experiencing cardiac arrest. The red line on the multi-parameter monitor should read an arterial blood pressure wave form but now shows no activity because the heart is no longer generating any systemic blood pressure. Additionally on the bottom of the screen is a white capnograph tracing, which is rapidly decreasing with each expired breath. The white end-tidal CO_2 value reads 10 mm Hg, telling the anesthetist that there is a massive decrease in cardiac output and delivery of blood to the lungs. Lastly, the green electrocardiogram at the top of the screen first appears to be normal but, after considering the context of other parameters, we realize the patient has the cardiac arrest rhythm known as pulseless electrical activity (PEA).

PROGNOSIS FOLLOWING CPA

Approximately 50% of dogs and cats experiencing CPA during general anesthesia have return of spontaneous circulation (ROSC) and ultimately survive following CPCR.[5] This number is vastly better than the overall long term prognosis for companion animals experiencing CPA than those in the general hospital population as noted in several other studies.[6,7,8] For example, survival statistics of 4% or less have been documented at one week for both cats and dogs that received CPCR following CPA in the hospital.[6] Another more recent publication documented that while 35% of dogs and 44% of cats initially had ROSC, only 6% of dogs and 7% of cats went on to ultimately survive to hospital discharge because these patients either had additional CPAs or were euthanized.[5] It is therefore important for anesthetists to know that if they can learn to provide effective CPCR efforts to patients experiencing CPA under their anesthesia care, their patients often have a favorable prognosis. Interestingly, for those pets that do survive CPA, functional recovery has been reported to be good for most animals.[9] Other factors that appear to influence survival from CPA include patients that are treated promptly, are resuscitated by multiple rescuers, have a reversible underlying CPA cause (e.g., anesthetic overdose, upper airway obstruction, hemorrhage, or electrolyte abnormalities), and who only experience respiratory arrest.[5,6,8,9]

STANDING RESUSCITATION CODES AND DO NOT RESUSCITATE ORDERS

Each patient, before being admitted to the hospital, must have a designated resuscitation code assigned. This important responsibility falls to the attending veterinarian and includes having a practical discussion with each pet's owners to explain realistic prognosis, cost, and ethical considerations surrounding CPA and CPCR effort. Similarly, the attending veterinarian must have the same conversation with each owner whose pet will undergo sedation or general anesthesia. During this conversation the risk of related morbidity and mortality must be discussed. Once owners are informed and educated, they may elect a "do not resuscitate" (DNR) order for their pet and request that no resuscitation efforts be made if their pet were to experience CPA. These patients are typically assigned the color code of *RED* which is then carefully noted on their cage card, hospital chart, and anesthesia record. Owners that wish the veterinary team to perform full resuscitation efforts for their pet, including open chest CPCR, are assigned the color code of *GREEN*. Some facilities also allow for limited resuscitation efforts, often noted by the color *YELLOW*, which may include CPCR but generally excludes open chest compression. If for whatever reason a patient has not received a resuscitation code and proceeds to experience CPA while under anesthesia or in hospital, the patient is treated aggressively with full resuscitation efforts while the attending veterinarian consults with the pet's owner.

Tech Tip 🐾

Whenever the resuscitation status of a hospitalized or anesthetized patient is unknown, it should be assumed a pet owner wants all efforts made in an emergency until directed otherwise.

REASSESSMENT CAMPAIGN ON VETERINARY RESUSCITATION (RECOVER)

In an effort to critically evaluate the evidence-based literature on CPA in companion animals, publish consensus guidelines on veterinary CPCR, and ultimately improve patient survival, over 100 specialists in emergency/critical care and anesthesiology from around the world met during the 2011 International Veterinary Emergency & Critical Care Symposium (IVECCS). This project, known as the *Reassessment Campaign On Veterinary Resuscitation* (**RECOVER**) initiative, was ultimately published in the Journal of Veterinary Emergency and Critical Care in 2012 and has subsequently become the standard of care for providing veterinary CPCR.[10] While these guidelines do not specifically focus on CPCR during anesthesia, they are the first evidence-based guidelines on veterinary CPCR and have become the standard of care for veterinary anesthesia providers as well. The guidelines in their full capacity are available free to the public at https://www.veccs.org/recover-cpr/. The remainder of this chapter focuses heavily on the clinical guidelines of the RECOVER initiative.

CPCR PREPAREDNESS

Evidence from the RECOVER initiative would suggest that having a well led, organized, cohesive, and knowledgeable team that adheres to evidence-based CPCR guidelines should improve overall survival from CPA. To this end, a well-stocked and consistently audited crash cart should be readily available[10] (Figure 13-2 and Table 13-2). In addition, the presence of cognitive aids

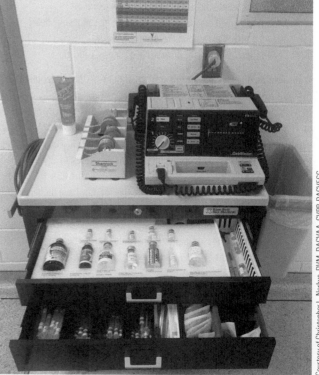

FIGURE 13-2 This is an example of an excellent crash cart. Note that the cart is well organized, well stocked, clean, contains only necessary supplies, and is in an accessible location. Checklists, algorithm charts, and dosing charts should also be available to improve compliance with veterinary CPCR guidelines.

Courtesy of Christopher L. Norkus, DVM, DACVAA, CVPP, DACVECC

TABLE 13-2 Recommended Contents for Crash Carts

General Equipment	• ECG or multiparameter monitor • ECG jelly • Defibrillator • Suction unit • Clippers • Stethoscope • Minor surgical pack, scalpel blades, surgical gloves
Airway and Breathing Equipment	• Clear endotracheal tubes (ETTs) in various sizes • Laryngoscope and blades • ETT ties • Doyen intestinal clamp (or similar tool for grasping objects obstructing the airway) • Tracheostomy tubes • Thoracocentesis setup (60 cc syringe, butterfly catheters, extension sets, three-way stop cocks) • AMBU bags
Circulation Supplies	• 3 or 5 French red rubber feeding tube for tracheal drug administration • IV catheters of various sizes • Tape • Crystalloid fluid bags with IV line attached • Colloid fluid bags • 7.2% hypertonic saline • Pressurized "slam" bags • 3 cc saline flushes
Emergency Drugs	• Syringes of various sizes with large bore hypodermic needs attached • Epinephrine • Atropine • Vasopressin • Lidocaine (without epinephrine) • Sodium bicarbonate • Calcium gluconate • 50% dextrose • Naloxone • Propofol

such as checklists, algorithm charts, and dosing charts is likely to improve compliance with veterinary CPCR guidelines.[10] Availability and clear visibility of these resources in areas in which CPA may occur, such as intensive care units, procedure rooms, anesthesia induction rooms, and surgery suites are recommended.[10]

Tech Tip

Syringes in a crash cart can be taken out of their protective casing and have an appropriately sized sterile needle attached to save time in an emergency.

Adherence to CPCR guidelines can only be accomplished if personnel received effective, standardized training and have regular opportunities to refresh their skills. Refresher training at least every 6 months is recommended to reduce the risk of the decay of skills.[10] Simple mock codes run every 3 to 6 months using manikins or stuffed animals will likely improve

team awareness of CPCR guidelines and help establish improved team cohesiveness and organization.

Tech Tip

It is becoming a standard in many teaching hospitals and veterinary clinics to take certification courses for CPCR, such as the online certification and recertification courses through Cornell University, https://www.ecornell.com /about-ecornell/recover/.

Following real resuscitation efforts or simulated CPCR, it is beneficial to allow for team members to review and critique their performance and the performance of the team as a whole. An open, honest discussion about opportunities for improvement immediately after a CPCR attempt can lead to significant enhancement of overall team performance.[10]

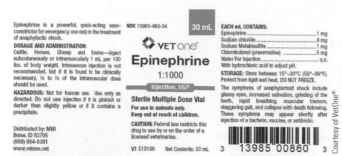

FIGURE 13-3 Epinephrine 1:1000 is a 1 mg/mL concentration.

As research in humans has found no advantage on survival from CPA when a physician directs CPCR efforts, either veterinarians or veterinary technicians may be considered effective leaders of a CPCR team. Crucial roles of the team leader include assigning tasks for other team members and ensuring overall quality of care and consistency of procedures. Overall team performance can be enhanced by using focused, clear communication directed at individuals when tasks are assigned, and utilization of closed loop communication. Closed loop communication is performed by a clear, directed order being given to one team member by another (Dr. Smith: Jane, please give Fluffy 0.2 mL of epinephrine 1:1000 IV now), after which the receiving team member repeats the order back to the requestor to verify the accuracy of the receiver's perception (Jane: I am giving 0.2 mL epinephrine 1:1000 IV to Fluffy now). This simple technique can dramatically reduce medical errors, especially in emergency and CPCR situations.

> **Tech Tip** 🐾
>
> Epinephrine 1:1000 is just another way of writing that the drug is a 1 mg/mL concentration (Figure 13-3).

INITIAL RESUSCITATION INTERVENTIONS & GENERAL CONSIDERATIONS

Rapid assessment and prompt treatment of the patient is crucial if CPA is suspected. Assessment of the patient with initial attention to the patency of the airway (A), presence and effectiveness of breathing (B), and presence and effectiveness of circulation (C) is referred to as "checking the ABCs." This initial assessment should take no more than 5 to 10 seconds. Several human studies have documented poor sensitivity of pulse palpation for the diagnosis of CPA.[11, 12] There is also strong evidence in humans that less than 2% of patients not in CPA experience serious harm when basic life support techniques are started when not needed.[13] Therefore, aggressive administration of CPCR in patients suspected of being in CPA is recommended as the risk of injury due to CPCR in patients not in CPA is low.

Initially the rescuer should summon for help and quickly note the time. The response by all of the veterinary team to a potential arrest must be immediate. If there is any doubt as to whether or not the patient has experienced CPA, chest compressions should be initiated immediately while further assessment to support the diagnosis of CPA is started by other personnel or after an initial cycle (2 minutes) of CPCR.[10] Those patients in CPA will be unresponsive to verbal or physical stimulus (assuming they are not already anesthetized), apneic, and without a pulse and heart beat on auscultation. Additional signs such as absence

of a Doppler pulse wave, electrocardiographic evidence of CPA, and decreasing to absent end-tidal CO_2 may aid in the diagnosis in the anesthetized patient.

> **Tech Tip** 🐾
>
> Summoning for help is one of the most important things a veterinary technician can do in an emergency. If nobody knows help is needed, nobody will come.

Adapted from human emergency medicine, veterinary CPCR is initiated in three tiers: basic life support (BLS), advanced life support (ALS), and ongoing post-resuscitative care. The BLS phase involves

- Establishing an open and clear airway
- Providing assisted ventilation
- Performing effective chest compressions

These steps mimic the ABCs as previously discussed—airway, breathing, and circulation. The ALS stage includes such advanced care as

- Initiation of monitoring ($EtCO_2$ and ECG)
- Venous access
- Drug administration (including reversal of anesthetics)
- Defibrillation

Often many ALS techniques (e.g., ECG monitoring, IV catheter usage, ETT placement, and oxygen supplementation) are already underway in patients who experience CPA under anesthesia. Post-resuscitative care after ROSC includes ongoing intensive care monitoring as well as aggressive cardiovascular and ventilation support.

Patients experiencing CPA will need to be transported to an accessible location that is adjacent to an oxygen source if not already near one. In some instances this may require moving a patient to a central hospital location or treatment room (e.g., a patient is found in its cage in CPA); however, during anesthesia and in the surgical suite it may not be necessary at all (e.g., the patient is already intubated and on oxygen). A transportable crash cart should have treatment and diagnostic supplies at the ready and be moved to the patient bedside if available. If treatment and diagnostic supplies are not mobile, the patient may need to be transported to a centralized treatment area. Typically patients are placed in right or left lateral recumbency for CPCR, but dorsal recumbency may be most appropriate for patients during surgical settings (e.g., when they are already in this position) or in barrel-chest dogs (e.g., English bulldogs).[10] It is important to turn off inhalant anesthetic agents and any constant rate infusions (CRIs) that would contribute to unsuccessful CPCR.

BASIC LIFE SUPPORT

The first tier of veterinary CPCR is BLS, which focuses on the airway, breathing, and circulation.

Airway Management

Initial airway management for patients not previously intubated involves extending the patient's neck to straighten the airway and pulling the patient's tongue forward. The veterinary staff quickly examines the upper airway and initiates suctioning, as needed, to clear any foreign material or vomit observed in the patient's oropharynx.

If the patient's airway is fully obstructed, sharp abdominal thrusts and digital finger sweeps of the pharynx can be used to dislodge the obstruction. An emergency tracheotomy may be necessary if the airway obstruction is not promptly resolved. Percutaneous insertion of a needle or intravenous (IV) catheter directly into the trachea distal to the obstruction, along with oxygen administration, can be lifesaving while foreign material is removed or a tracheotomy is being performed. A tracheotomy is a surgical procedure where an opening is made directly into the trachea and should only be performed by a veterinarian. In some cases, material fully obstructing the airway (e.g., a bone or a tennis ball) can be successfully manually removed with long hemostats or Doyen intestinal clamps.

If the patient is already intubated and has proceeded into CPA from an unknown cause, the anesthetist should always confirm that the ETT has not become kinked or occluded. Capnography can be a useful tool to determine if there is an obstruction. If an obstruction is found, the ETT may need to be removed and replaced. Otherwise the rescuer proceeds on the breathing management.

Breathing

After the patient's airway has been evaluated and cleared as necessary, an appropriately sized ETT should be inserted if not already done. It is critical that correct ETT placement (e.g., making sure it is not accidently placed into the esophagus) be confirmed and the ETT secured and the pilot balloon inflated. The patient in CPA is then ventilated with 100% oxygen. If the patient is under general anesthesia, the maintenance inhalant anesthetic agent (e.g., isoflurane) should be discontinued immediately and the breathing circuit should be flushed with 100% oxygen to ensure no residual anesthetic is delivered to the patient. From there, a ventilation rate of 10 breaths per minute with an approximate tidal volume of 10 mL/kg and a short inspiratory time of 1 second are recommended.[10] Hyperventilation has been shown to be detrimental to outcome in humans because it results in decreased heart and brain perfusion.[10, 14, 15]

Patients who experience CPA can be effectively ventilated by hand via an anesthetic machine, by way of a mechanical ventilator, or by an AMBU® bag (bag-valve-mask [BVM]) (Figure 13-4). Difficultly in achieving or a complete lack of appropriate chest wall motion during ventilation should prompt the rescuer to immediately search for an incorrectly positioned tube (such as accidental one lung intubation, esophageal intubation) or a life-threatening pleural space disorder (such as pneumothorax). If the ETT is correctly placed and a pleural space disorder is identified, emergency thoracocentesis or thoracotomy may be indicated and would be performed by the attending veterinarian.

Tech Tip

For an AMBU® bag to deliver oxygen, a reservoir bag must be attached to the device (see Figure 13-4). Sometimes these reservoir bags are accidently removed and discarded, making the device ineffective.

Occasionally, an impedance threshold device (ITD) is placed on the proximal end of an ETT during resuscitation to create an increase in negative pressure during the chest-recoil phase of chest compression (Figure 13-5). With an increase in negative pressure in the chest cavity, a vacuum is created that increases venous return and more blood being delivered to the heart. During the

FIGURE 13-4 Three different sized AMBU® bags are shown here with an accompanying reservoir bag. Users must remember that an attached reservoir bag is necessary to deliver high concentrations of inspired oxygen to the patient.

Courtesy of Christopher L. Norkus, DVM, DACVAA, CVPP, DACVECC

next chest compression, increased cardiac output is achieved. Some ITDs also include a timing light that helps the rescuer know when to ventilate a patient to avoid accidental hyperventilation. The device has shown some application in both humans and animal models and may be used if available.[16, 17]

Circulation

The goal of maximizing myocardial and cerebral perfusion during CPCR is achieved initially by effective chest compressions and later aided by advanced life support modalities such as drug therapy. Chest compressions should be started immediately in any patient suspected of being in CPA, and if multiple rescuers are not present, airway and ventilation management should not delay the start of chest compressions. In other words, if you are the only rescuer then you should start chest compressions, interposing every 30 compressions with two ventilations, while you await help to arrive.

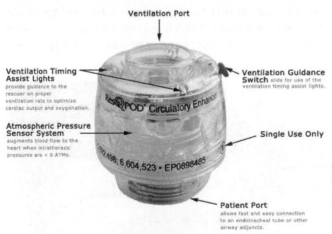

FIGURE 13-5 An **impedance threshold device** (ITD) is a valve preset to open at a specific pressure (cracking pressure). The ITD limits the entry of air into the lungs during chest recoil between chest compressions (reduces intrathoracic pressure) to improve venous return to the heart. It is attached between the ETT and breathing circuit, limiting inflow of air from the ETT during chest recoil until the preset cracking pressure is reached.

Courtesy of Combat Medical.

There are two theories describing the mechanisms by which external chest compressions lead to blood flow during CPCR. The **cardiac pump theory** proposes that the ventricles of the heart are directly compressed between the ribs. The **thoracic pump theory** suggests that chest compressions increase overall intrathoracic pressure, thereby compressing the aorta and collapsing the vena cava leading to blood flow out of the thorax.[10] Upon elastic recoil of the chest, subatmospheric intrathoracic pressure draws blood volume from the periphery back into the thorax and the cycle continues with each compression. Present guidelines recommend that small patients such as cats and small dogs should receive compressions directly over the heart (cardiac pump theory), whereas larger patients should receive more caudal compressions that are directed over the widest part of the chest (thoracic pump theory).[10]

Cats and small dogs can have effective external chest compressions delivered with a one-hand technique with the compressor's fingers wrapped around the sternum at the level of the heart. For bigger dogs, a two-hand technique is used either over the widest portion of the chest (thoracic pump theory) or over the heart (cardiac pump theory).

If CPA occurs during a body cavity surgery, the surgeon should enter the chest cavity rapidly and perform internal chest compressions by directly squeezing the heart. Similarly in patients who are not under general anesthesia but have significant intrathoracic disease, such as a tension pneumothorax or pericardial effusion, opening the chest cavity, directly visualizing the heart, and performing internal chest compressions may be most effective. A compression rate of 100 to 120/minute is recommended for all types of compressions regardless of whether the chest cavity is open or not.[10] For music fans, chest compression to the beat of the classic disco song "Stayin' Alive" by the Bee Gees will achieve these rates.

Based on the American Heart Association's 2010 guidelines for CPCR, the rescuer performing chest compressions should push hard, push fast, allow full chest recoil after each compression, and minimize interruptions in chest compressions.[18] There is a linear relationship between chest compression depth and mean arterial pressure, thus deep chest compressions of ⅓–½ the width of the thorax are recommended. Full elastic recoil between chest compressions allows for improved venous return and improved coronary and cerebral perfusion. Rescuers who do not allow for adequate chest recoil are said to be "leaning" into their patient, and this mistake should be avoided (Figure 13-6). It is also critical that chest compressions be continuous and not stopped once they are started (Figure 13-7). Due to

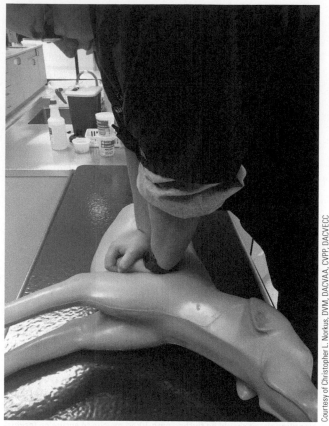

FIGURE 13-7 Here the same registered veterinary technician demonstrates *effective* chest compressions. Note that her arms form a 90-degree angle with the table, that her hand position is secure, and that she has gotten a stool to be above her patient. She is also allowing for adequate chest recoil between compressions. Remember to push hard, push fast, and do not stop until your cycle is finished!

rapid rescuer fatigue during CPCR, chest compressions should be performed by a given rescuer for no more than a 2-minute cycle before being handed over to another rescuer to ensure effective compressions are always being performed. Switching from one rescuer to another rescuer should be done as quickly as possible to avoid pauses in chest compressions.

To facilitate venous return from the abdomen and improve cardiac output, abdominal compressions over the cranial abdomen, known as **interposed abdominal compressions**, may also be considered by a second rescuer when an adequate number of rescuers are available to perform CPCR.[10, 19, 20] A rate of 70 to 90 abdominal compressions per minute has been proposed.

ADVANCED LIFE SUPPORT

As previously discussed, the ALS stage of CPCR includes advanced care such as establishing venous access, interpretation of an electrocardiogram (ECG), specific drug administration, and electrical **defibrillation** (stopping of fibrillation of the heart by administering a controlled electric shock in order to allow restoration of the normal rhythm). Often many of these ALS techniques (e.g., ECG monitoring, IV catheter usage, ETT placement, and oxygen supplementation) are already underway in patients who experience CPA under anesthesia. In patients who are not anesthetized but

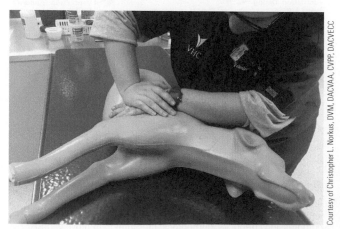

FIGURE 13-6 Here a registered veterinary technician demonstrates *ineffective* chest compressions. Note that she is leaning into the patient from the side and has her hands poorly positioned.

Courtesy of Christopher L. Norkus, DVM, DACVAA, CVPP, DACVECC

experience CPA, efforts must be made to establish venous access, attach an ECG, establish the underlining cardiac arrest rhythm, and treat the cardiac rhythm disturbance with drug administration and/or defibrillation promptly.

Tech Tip

Alcohol should never be put on electrocardiogram leads because if a patient needs to be defibrillated at a later time, the alcohol will cause sparks and potentially fire leading to patient burns. Specific electrode jelly should be used to ensure conductivity while preventing patient burns and fire.

Establishing venous access is important, but it should never interfere with chest compressions or the delivery of defibrillation. Once venous access has been achieved, aggressive fluid administration (e.g., hypertonic saline, crystalloids, colloids, or blood products) should be performed only if hypovolemia existed/was suspected before the CPA event or if the patient is experiencing ongoing blood loss. Fluid administration in euvolemic (adequate circulating plasma volume) patients in CPA may be detrimental and decrease coronary perfusion pressure, and should be avoided.[10, 21]

Safety Alert

Before defibrillation of a patient, it should be ensured that no rescuer is in contact with the patient or the surface that the patient is on. This includes the team member performing the defibrillation yelling "CLEAR" to inform staff of an impending defibrillation and confirming that all rescuers are safe before administering a shock.

If venous access cannot be quickly obtained, emergency drugs, such as *Naloxone, Atropine, Vasopressin, Epinephrine,* and *Lidocaine* (remember the pneumonic "NAVEL") can be administered through the ETT into the lungs at two to ten times the typical intravenous dosage.[10] A small saline chaser is often is administered immediately after the drug to ensure delivery to the lung. Typically, a long red rubber catheter is inserted down the ETT and drugs are administered through the catheter deep into the lung. Some older texts describe directly injecting medication into the heart. This technique has largely fallen out of favor, especially during closed-chest CPCR, as vessel or cardiac laceration and hemorrhage, along with inaccurate location administration, are common.[10]

Emergency Drug Pharmacology

A description of CPCR drugs is provided below. Dosages are provided in Table 13-3.

Epinephrine

Epinephrine 1:1000 (Epinject®, generic) is the standard drug prescribed for use in cardiac arrest for both people and animals (see Figure 13-3). The drug acts as a catecholamine and sympathomimetic agent and has nonspecific adrenergic agonist effects. Its use in cardiac arrest is predominantly for its vasoconstrictive effects due to alpha-1 adrenergic agonist (α_1) activity. Epinephrine also has beta-1 (β_1) and beta-2 (β_2) adrenergic agonist activity depending on the dosage; however, the positive inotropic (cardiac contractility) and chronotropic (heart rate) effects from the beta agonism are less crucial in cardiac arrest and may actually be harmful because of increased myocardial oxygen demand, exacerbating myocardial ischemia, and potentially predisposing patients to arrhythmias if ROSC is achieved. Dosages between 0.01 to 0.2 mg/kg IV have been proposed for use in cats and dogs. This dosage is increased two- to tenfold when it is administered intra-tracheally. The drug is administered every other 2-minute cycle of BLS with dosages of 0.01 mg/kg IV/IO (0.02–0.1 mg/kg IT) being used early in CPCR efforts and higher dosages such as 0.1 mg/kg IV/IO (0.2 mg/kg IT) being considered only after prolonged (> 10 minutes) CPA.[10] If administered IT, epinephrine should be diluted (1:1) with sterile water or saline and administered via a catheter longer than the ETT.

Atropine

Atropine (Atrocare®, generic) is a parasympatholytic anticholinergic agent that has been used widely for CPCR (see Figure 5-15). The main effect is an increase in heart rate and blockade of vagal tone. In humans, many studies have evaluated the use of atropine for treatment of CPA, and have largely shown no benefit or a detrimental effect following its use at standard dosages and higher dosages. As a result, atropine has recently been removed from human CPCR guidelines. However, because some data in dogs suggest a potential benefit under some forms of CPA and because CPA in dogs and cats is generally of different etiology than in humans, atropine remains a treatment consideration for dogs and cats with ventricular asystole, pulseless electrical activity, and in patients with high vagal tone. A dosage of 0.04 mg/kg IV may be considered every other 2-minute cycle of BLS (therefore, repeating every 4 minutes).[10] This dosage is increased two- to tenfold if it is administered intra-tracheally and the drug is diluted 1:10 in saline or sterile water and administered via a catheter longer than the ETT. An exception to atropine's routine use is in the rabbit because this species produces atropine esterase, which limits the drug's clinical effectiveness. Therefore if an anticholinergic drug is selected, the agent glycopyrrolate (Robinul®) is preferred in this species.

Vasopressin

Vasopressin (Vasostrict®) is a V1 agonist that results in vasoconstriction during CPA by targeting specific receptors on vascular smooth muscle (see Figure 5-16e). This mechanism of action is completely independent of alpha-1 (α_1) adrenergic agonist effects, which may be beneficial because the V1 receptor remains responsive in the face of acidemia unlike adrenergic receptors. Additionally, the drug has no inotropic or chronotropic effects that could potentially worsen myocardial ischemia. Current data on the effectiveness of vasopressin compared to epinephrine is mixed although further study is needed in cats and dogs and with special emphasis looking at simultaneous use of epinephrine and vasopressin. At this time, vasopressin at 0.8 U/kg IV may be considered as a substitute or in combination with epinephrine every other 2-minute BLS cycle.[10] This dosage is increased two- to tenfold when administered via the trachea and the drug is diluted (1:1) with sterile water or saline and administered via a catheter longer than the ETT.

TABLE 13-3 CPCR Emergency Drug Dosing Chart

Drug	Dose	Weight (Kg) 2.5	5	10	15	20	25	30	35	40	45	50
		Weight (lbs) 5	10	20	30	40	50	60	70	80	90	100
		mL	mL	mL	mL	mL	mL	mL	mL	mL	mL	mL
epinephrine 1:1000 Low Dose	0.01 mg/kg*	0.03	0.05	0.1	0.15	0.2	0.25	0.3	0.35	0.4	0.45	0.5
epinephrine 1:1000 High Dose	0.1 mg/kg*	0.3	0.5	1	1.5	2	2.5	3	3.5	4	4.5	5
atropine 0.54 mg/mL	0.04 mg/kg	0.2	0.4	0.8	1.1	1.5	1.9	2.2	2.6	3	3.3	3.7
vasopressin 20 U/mL	0.8 U/kg	0.1	0.2	0.4	0.6	0.8	1	1.2	1.4	1.6	1.8	2
lidocaine 20 mg/mL	2 mg/kg	0.25	0.5	1	1.5	2	2.5	3	3.5	4	4.5	5
External Defibrillation (Monophasic)	4–6 J/kg	10	20	40	60	80	100	120	140	160	180	200
External Defibrillation (Biphasic)	2–4 J/kg	5	10	20	30	40	50	60	70	80	90	100

*The dosage range for epinephrine is 0.01–0.2 mg/kg ("low dosage" is 0.01 mg/kg and 0.02 mg/kg, while "high dosage" at 0.1 mg/kg and 0.2 mg/kg. The 0.01 mg/kg and 0.1 mg/kg are often used and represent a 10 × difference in low versus high dosage of epinephrine). Dosages vary depending on the route of administration (IV/IO or IT).

This chart is just one example of numerous emergency drug dosing charts that are commercially available. The author recommends those that can be purchased online at www.veccs.org.

Specific CPA Arrhythmias and Their Treatment

The individual treatment for each CPA depends on the underlining cause of the CPA and the type of cardiac arrhythmia present. To determine the type of cardiac arrhythmia present, an electrocardiogram (ECG) evaluation is necessary. For cats and dogs, the two most frequent CPA arrhythmias are ventricular asystole and pulseless electrical activity (PEA). Ventricular fibrillation (VF) is the most common CPA arrhythmia in humans, but it appears to be far less common in dogs and cats than in people.[22]

Ventricular Asystole

Ventricular asystole is the complete absence of all significant electrical and mechanical activity in the ventricles. During this rhythm, an occasional P wave may be observed on the ECG, reflecting atrial activity, but no QRS complexes and ventricular activity are seen (Figure 13-8). The rhythm is sometimes crudely referred to as "flatline" on television.

Treatment includes chest compressions and IV epinephrine administered at 0.01 to 0.2 mg/kg every 3 to 5 minutes, IV atropine administered at 0.04 mg/kg every 3 to 5 minutes, and potentially IV vasopressin at 0.8 U/kg every 3 to 5 minutes.[10] Patients in ventricular asystole generally have the worst prognosis for ROSC. As a result, it is important that the veterinary team also attempt to identify and treat the underlining cause of CPA, if possible.

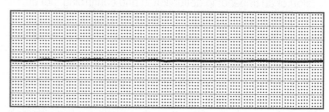

FIGURE 13-8 An electrocardiogram tracing showing a patient in ventricular asystole. CPCR must be started immediately.

Pulseless Electrical Activity (PEA)

Pulseless electrical activity, previously known as electromechanical dissociation (EMD), occurs when electrical activity remains in the heart; however, the heart is unable to develop adequate cardiac output to deliver oxygen to the body to sustain life. During PEA, the ECG may appear relatively normal, which highlights the importance of never using electrocardiography as the sole source of monitoring patients under general anesthesia. Consequently the presence of normal appearing P waves, QRS complexes, and T waves may be seen (Figure 13-9).

The treatment for PEA includes chest compressions and administration of IV epinephrine at 0.01 to 0.2 mg/kg every 3 to 5 minutes, IV atropine at 0.04 mg/kg every 3 to 5 minutes, and vasopressin 0.8 U/kg IV every 3 to 5 minutes.[10] Historically the use of different adjunctive drug therapies,

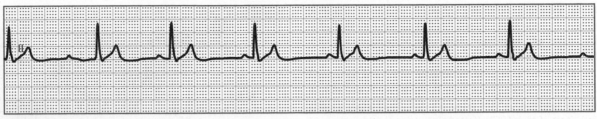

FIGURE 13-9 An electrocardiogram tracing showing a patient in pulseless electrical activity. This rhythm can and often does look identical to a normal sinus rhythm.

including naloxone, dexamethasone, and calcium, have all been investigated for the treatment of PEA with limited success. The key to treatment of PEA is rapid identification and correction of the underlining offending problem with special emphasis and evaluation of the 6 "H's" (hypovolemia, hypoxia, hydrogen ion imbalance/acidemia, hyperkalemia or hypokalemia, hypothermia, and hypoglycemia) and 5 "T's" (toxins, cardiac tamponade, tension pneumothorax, thrombosis, and trauma). For example, severe hypoglycemia can lead to death because low levels of glucose in the brain can result in a sympathetic response that can lead to cardiac arrhythmias and failure followed by respiratory arrest; therefore, it is important to monitor blood glucose and correct any abnormalities in these patients.

Ventricular Fibrillation

Patients who experience ventricular fibrillation (VF) have electrical activity of their heart that is disorderly, erratic, and fails to produce adequate cardiac output to sustain life. Dogs and cats do not commonly experience VF in the clinical setting, unlike humans who experience it most commonly. In patients with VF, the electrocardiographic waveform is chaotic with no clear P waves, QRS complexes, or T waves (Figure 13-10). Left untreated, the presence of electrical activity quickly abolishes and the patient proceeds into asystole.

The treatment for VF for patients whose VT is immediately identified on ECG involves immediate electrical defibrillation with an initial shock dose of 2 to 4 joules/kg for biphasic defibrillators and 4 to 6 joules/kg for monophasic defibrillators.[10, 23] If the duration of VF is less than 4 minutes, chest compressions should be continued for 2 minutes to help replenish oxygen to tissue, and then the patient should be defibrillated immediately. If the duration of VF is more than 4 minutes, one full cycle of CPCR should be performed before defibrillating to allow the myocardial cells to generate enough energy to restore a normal membrane potential, thereby increasing the likelihood of success.[23]

Following a defibrillation attempt, CPCR should then be continued for a two minute BLS cycle before another defibrillation attempt is made at a higher joule setting. Following unsuccessful defibrillation low-dosage epinephrine at 0.01 mg/kg IV and vasopressin at 0.8 U/kg IV may be administered and may make successful defibrillation more likely. Drugs such as amiodarone, magnesium chloride, esmolol, and lidocaine have all been utilized as drug therapy for refractory VF.

Without a defibrillator, it is very difficult to successfully treat VF. A hard **precordial thump**, delivering a strong blow using the heel of the hand directly over the heart, may be administered to the chest wall and may occasionally convert VF to a sinus rhythm in very small patients when a defibrillator is not available. Old textbooks also discuss the use of pharmacologic defibrillation techniques using drugs such as potassium chloride, insulin and dextrose, and acetylcholine. Because these drugs were never effective at treating VF, they have subsequently been removed from the current veterinary literature.

Additional ALS Considerations

Countless other drugs have been investigated for the use in the treatment of CPA. Some older textbooks discuss the use of the respiratory stimulant drug doxapram (Dopram-V®, Respiram®) for respiratory arrest; however, this agent is not recommended at this time. Despite what might seem like a promising treatment, doxapram increases cerebral and myocardial oxygen demand while reducing cerebral blood flow.[24, 25, 26, 27] The most effective therapy for respiratory arrest remains endotracheal intubation, oxygen supplementation, and ventilation as needed.

Sodium bicarbonate at 1 mEq/kg IV should be considered during prolonged CPCR (e.g., after 10–15 minutes) to manage the metabolic acidosis induced by CPA. Sodium bicarbonate also should be administered if metabolic acidemia was diagnosed before the arrest or if the patient had preexisting

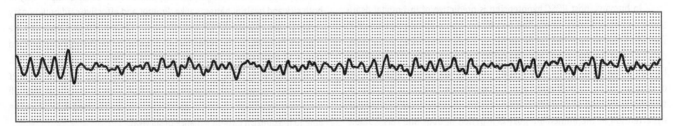

FIGURE 13-10 An electrocardiogram tracing showing a patient in ventricular fibrillation. Immediate defibrillation is required.

hyperkalemia. Historically calcium was routinely administered during CPCR; however, this is no longer recommended.[10] Its selective use, however, may be justified in patients with severe hyperkalemia or hypocalcemia.

The use of capnography and end-tidal carbon dioxide ($EtCO_2$) monitoring may be useful to guide resuscitation efforts and offer insight into prognosis. $EtCO_2$ values greater than 15 mm Hg are associated with higher survival rates in dogs.[10] The veterinary staff should regularly assess the effectiveness of their CPCR efforts by palpating for the presence of pulses as compressors are being quickly switched out and by using a Doppler ultrasound transducer. However, if this comes at the expense of stopping chest compressions then these steps should be avoided. With sufficient water-based lubricant, a Doppler transducer can be placed in direct contact with the cornea (not through the eyelid). The presence of a "swooshing" wave sound from the Doppler unit during concurrent chest compressions may provide a crude estimate of whether forward blood flow is being delivered to the brain. If chest compressions are determined not to be generating adequate cardiac output, the patient should be repositioned or the resuscitation technique changed.

Patients that present to the hospital suffering from severe hypothermia (e.g., patient is found in a snow bank) should be fully rewarmed before resuscitation efforts are deemed unsuccessful. In humans, the remarkable ability to resuscitate patients with hypothermia even hours after their cardiac arrest has promoted the bold statement that you "aren't dead until you're warm and dead!"

RETURN OF SPONTANEOUS CIRCULATION (ROSC) AND POST-CPA CARE

When to stop unsuccessful resuscitation efforts is directed by the attending veterinarian after consultation with the patient's owner. Animals that have returned to spontaneous circulation (ROSC) commonly experience CPA again often within minutes to several hours, especially when the original cause of the CPA event has not been identified and/or resolved. As a result, patients who achieve ROSC usually require significant cardiovascular and pulmonary support during the period following CPA. Additional emphasis is also placed on cerebral support to help protect neurologic function.

Normalizing **central venous oxygen saturation** (percentage of hemoglobin bound to oxygen in a central vein such as the cranial vena cava), lactate, blood pressure, central venous pressure, and **arterial oxygen saturation** (the amount of oxygen actually bound to hemoglobin compared with the oxygen capacity) are all goals of post-CPA care. For dogs and cats who achieve ROSC, transfer to a specialty center with 24-hour care, higher healthcare provider to patient ratios, and advanced critical care capabilities should be considered.[10]

In humans, post ROSC hypotension is predictive of in-hospital death and is associated with diminished functional status among survivors.[27, 28] Therapies may include the use of agents such as norepinephrine, vasopressin, dopamine, or dobutamine to improve cardiac output and blood pressure (see Figures 5-16a 5-16b, 5-16d, and 5-16e).[10] The routine use of large volumes of intravenous fluids post-arrest is not recommended except in the cases of strongly suspected or confirmed hypovolemia. It is reasonable to tolerate hypertension in the immediate post-CPA period in dogs and cats.[10] Stable

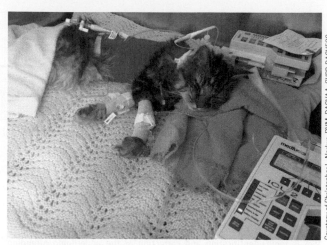

Courtesy of Christopher L. Norkus, DVM, DACVAA, CVPP, DACVECC

FIGURE 13-11 Patients who achieve return of spontaneous circulation are extremely critical and require aggressive and attentive monitoring and life support to prevent repeat cardiac arrest. Note that this patient has two intravenous catheters, bilateral nasal oxygen catheters, has its head elevated 30 degrees to mitigate cerebral edema, and is on several constant rate infusions.

and perfusing ventricular escape rhythms should never be suppressed with drug therapy (e.g., lidocaine).

Agents such as mannitol and 7.2% hypertonic saline may be used to treat cerebral edema. Patients may benefit from having their head elevated 30 degrees to mitigate cerebral edema (Figure 13-11). Glucocorticoids should not be routinely used following ROSC. Diazepam IV may be needed to treat seizure formation, and barbiturates such as phenobarbital may be considered for seizure prophylaxis being mindful of cardiovascular adverse effects.[10] Mild (90°F; 32°C) hypothermia after ROSC decreases cerebral oxygen, improves outcomes in dogs and humans, and may be considered.[10, 29, 30] Patients who are hypothermic post-arrest should be slowly rewarmed at a rate of 0.45 to 0.9°F (0.25–0.5°C)/hr.[10] Hyperglycemia has been shown to be associated with worse neurologic outcomes in human patients and should be avoided after ROSC.[31]

As the entire body was starved of adequate oxygen delivery during CPA, subsequent brain injury, coagulation disorders, gastrointestinal tract injury, and kidney injury may all occur. Intensive and ongoing monitoring and aggressive supportive care are required to ensure patients who achieve ROSC have the greatest chance of hospital discharge.

SUMMARY

Cardiopulmonary arrest (CPA) is the most serious medical condition a patient can have and without rapid and aggressive treatment, often proves fatal. General anesthesia is not without risk, and CPA can occur unexpectedly in both sick and healthy patients alike. Prompt and effective cardiopulmonary cerebral resuscitation (CPCR) including basic life support and advanced life support techniques are imperative to maximize the chances of sustained return of spontaneous circulation (ROSC) and long-term patient survival. As approximately 50% of cats and dogs experiencing CPA while under anesthesia survive with appropriate medical care, it is critical that all veterinary technicians acting as anesthetists be well trained in CPCR and be able to maximize a patient's chance of survival.

CRITICAL THINKING POINTS

- The evidence supports that in anesthetized patients with CPA, prompt CPCR should be attempted considering that these patients have a better prognosis for survival (47%) and discharge from a hospital than the overall CPCR survival rate (4–9.6%).

- It is likely that early intubation and ventilation in veterinary CPCR is highly valuable, with a ventilation rate of approximately 10 breaths per minute, a tidal volume of 10 mL/kg, and an inspiratory time of 1 second delivered simultaneously with compressions.

- High-quality chest compressions should be delivered in uninterrupted cycles of 2 minutes with most patients in lateral recumbency, at a compression rate of 100 to 120/min and a compression depth of 1/3–1/2 the width of the chest while allowing for full elastic recoil of the chest between individual compressions.

- After each 2-minute cycle of BLS, the compressor should be rotated to prevent fatigue, which may decrease the quality of chest compressions.

- IV epinephrine administered at 0.01 to 0.2 mg/kg every other cycle of CPCR (every 4 min) (0.01 mg/kg IV/IO; 0.02–0.1 mg/kg IT) early in CPCR and higher dosages such as 0.1 mg/kg IV/IO (0.2 mg/kg IT) being considered only after prolonged (>10 minutes) CPA, IV atropine administered at 0.04 mg/kg every other BLS cycle, and potentially IV vasopressin at 0.8 U/kg every other BLS cycle should be used to treated ventricular asystole and pulseless electrical activity. Drugs administered IT are typically given at a dosage two- to tenfold of the IV/IO dosages and are diluted prior to administration with a catheter longer than the endotracheal tube.

- The treatment for ventricular fibrillation involves immediate electrical defibrillation with an initial shock dose of 2 to 4 joules/kg for biphasic defibrillators and 4 to 6 joules/kg for monophasic defibrillators.

- Rescuers should aggressively search for and treat the underlining cause of the CPA whenever possible to help prevent the patient from having CPA reoccur.

REVIEW QUESTIONS

Multiple Choice

1. Epinephrine's mechanism of action in treating cardiac arrest is believed to be prominently due to vasoconstriction. Which adrenergic receptor is responsible for this?

   a. V1
   b. Alpha-1
   c. Beta-1
   d. Beta-2

2. Which of the following rates is most appropriate when performing chest compressions on a cat in CPA?

   a. 60 bpm
   b. 80 bpm
   c. 110 bpm
   d. 190 bpm

3. Which of the following rates is most appropriate when providing assisted ventilation in a dog experiencing CPA?

   a. 5 bpm
   b. 10 bpm
   c. 15 bpm
   d. 20 bpm

4. Which of the following is most appropriate for the initial treatment of ventricular fibrillation?

   a. Atropine
   b. Vasopressin
   c. Epinephrine
   d. Defibrillation

5. Which of the following pieces of anesthetic monitoring equipment would be most useful in evaluating effectiveness of CPCR during CPA?

   a. Pulse oximeter
   b. Doppler blood pressure
   c. Apnea monitor
   d. End-tidal carbon dioxide

6. A dog experiences CPA during a routine general anesthesia. Based upon current evidenced-based literature, what is the approximate rate of success for ROSC and survival?

   a. 5%
   b. 10%
   c. 25%
   d. 50%

7. A dog experiences CPA while in the intensive care unit. Based upon current evidence-based literature, what is the approximate success rate for ROSC and eventual discharge from the hospital?

   a. <10%
   b. 15 to 20%
   c. 30 to 50%
   d. >50%

8. Based on Table 13-3, which of the following dosages of epinephrine 1:1000 are considered "low dose" and initially appropriate for the treatment of ventricular asystole and pulseless electrical activity?

   a. 0.001 mg/kg
   b. 0.01 mg/kg
   c. 0.1 mg/kg
   d. 1.0 mg/kg

9. Based on Table 13-3, which of the following doses of atropine may be considered for the treatment of ventricular asystole and pulseless electrical activity in the dog and cat?

   a. 0.0004 mg/kg
   b. 0.004 mg/kg
   c. 0.04 mg/kg
   d. 0.4 mg/kg

10. Which of the following etiologies commonly cause CPA in dogs and cats?

    a. Severe blood loss and hypovolemia, severe hypocalcemia, and severe hypokalemia
    b. Severe hypoxemia, severe dehydration, and severe hypokalemia
    c. Severe hyperkalemia, severe blood loss and hypervolemia, and severe hypercalcemia
    d. Severe blood loss and hypovolemia, severe hypoxia, and severe hyperkalemia

11. A patient in the intensive care unit collapses and is unresponsive. Which of the following should *not* be done?

    a. Call for help.
    b. Note the time.
    c. Wait to start chest compression until it is confirmed the patient does not have a pulse and is in fact in CPA.
    d. Confirm the patient's resuscitation code and whether or not do not resuscitate (DNR) orders are in place.

12. Which of the following species produces atropinase (an atropine esterase), which may make the use of atropine ineffective?

    a. Rabbit
    b. Dog
    c. Cat
    d. Chinchilla

13. Which of the following patient positions is inappropriate for CPCR?

    a. Right lateral recumbency
    b. Left lateral recumbency
    c. Sternal recumbency
    d. Dorsal recumbency

14. After this duration of time, the rescuer performing chest compressions should be rotated to prevent fatigue, which may decrease the quality of chest compressions?

    a. 1 minute
    b. 2 minutes
    c. 5 minutes
    d. 10 minutes

15. Which of the following statements regarding vasopressin is false?

    a. It is effective in the presence of metabolic acidemia.
    b. It is a potent vasoconstrictor due to V1 agonism.
    c. It has alpha-1 antagonist properties.
    d. It does not have beta adrenergic agonist properties.

16. Chest compressions performed to the rhythm of which classic disco song achieves a target compression rate of 100–120bpm?

    a. "Stayin' Alive" by the Bee Gees
    b. "What Am I Going to Do with You?" by Barry White
    c. "Last Dance" by Donna Summer
    d. "Love is in the Air" by Jay Black

17. Which of the following drugs should be administered during prolonged CPA (> 10–15 minutes) to treat metabolic acidosis?

    a. Doxapram
    b. Calcium gluconate
    c. Sodium bicarbonate
    d. Dexamethasone

18. Which of the following is an inappropriate treatment for respiratory arrest?

    a. Endotracheal intubation
    b. Oxygen supplementation
    c. Ventilation as needed
    d. Doxapram administration

19. What is the concentration of epinephrine 1:1000?

    a. 0.001 mg/mL
    b. 0.01 mg/mL
    c. 0.1 mg/mL
    d. 1 mg/mL

20. Which of the following treatments is inappropriate during post-arrest care for routine treatment of cerebral edema following ROSC?

    a. Mannitol
    b. Glucocorticoids
    c. Elevating the patient's head by 30 degrees
    d. 7.2% hypertonic saline

Case Studies

Case Study 1: Marco is a 5-month-old intact male Labrador retriever under general anesthesia for a routine castration. He was premedicated with acepromazine IM and morphine IM. General anesthesia was induced with IV propofol, and he has been maintained with isoflurane at 1.75% in 100% oxygen via a circle breathing system. His anesthesia has been routine with normal vital signs, and he is assessed to be appropriately anesthetized. However upon traction to his left spermatic cord, Marco's heart rate drops suddenly from 75 bpm to 15 bpm and then he is noted to be in ventricular asystole. The anesthetic vaporizer is immediately turned off and the breathing circuit is flushed with 100% oxygen. The time is noted and you call for help, simultaneously alerting the surgeon of the CPA. The surgeon begins closed chest compression at a rate of 100 bpm and you provide ventilation at a rate of 10 breaths per minute. A single dosage of

0.04 mg/kg IV atropine is administered along with 0.0 1mg/kg IV epinephrine by a third rescuer. Marco regains a sinus rhythm at a rate of 75 bpm immediately and is noted to have a blood pressure of 120 mm Hg via Doppler. The procedure is aborted and Marco is transferred to the intensive care unit where ongoing monitoring will occur. A profound vagal event is believed to have triggered his CPA.

1. Why was Marco's vaporizer turned off?
2. What was the purpose of flushing the breathing circuit with 100% oxygen?
3. What is a "vagal" event?

Case Study 2: You are busy working an extra shift one weekend when you hear over the loudspeaker, "Triage STAT to the emergency room." Paco a 15-year-old male neutered Chihuahua weighing 12 pounds presents laterally recumbent, unresponsive, cyanotic, and with agonal respirations. Several team members are available to assist you and the time is noted. While a veterinary technician goes to speak with the owner about Paco's medical history and the owner's resuscitation wishes, Paco is intubated and placed on 100% oxygen. His ETT is secured and the tube's cuff inflated. Chest compressions are started at a rate of 120 bpm, and he is ventilated at 10 breaths per minute. An electrocardiogram and IV access is obtained. The ECG reveals Paco is in ventricular fibrillation. The defibrillator is charged and all rescuers clear themselves from the patient. An initial shock of 20 joules is administered via a biphasic defibrillator. Paco is noted to still be in ventricular fibrillation and a 2-minute cycle of CPCR is continued along with the administration of 0.01 mg/kg IV epinephrine and 0.8 U/kg vasopressin IV. After the 2-minute cycle, the ECG is evaluated, and ventricular fibrillation is still present and no pulse detected. The defibrillator is charged to 30 joules and a second shock is administered. The ECG is checked and ventricular asystole is noted. At this time the veterinarian who was talking with the owner returns to the treatment room and updates the team that Mrs. Fletcher, Paco's owner, wishes to stop all resuscitation efforts. CPCR efforts are halted and Paco is still noted to be in ventricular asystole at which time he is pronounced dead.

1. What some possible causes of Paco's cardiac arrest?
2. Why were epinephrine and vasopressin administered?
3. What is the prognosis for ROSC once Paco was noted to be in ventricular asystole?

Case Study 3: Murphy a 4-year-old castrated male Golden Retriever dog presents to the emergency room after his owners noted him to be choking on a play toy. Murphy is collapsed, unresponsive, and cyanotic on presentation. You immediately call for help and note the time. Chest compressions are started by a second rescuer at 100 bpm. Upon evaluation of the airway, Murphy is noted to have a toy lodged at the most proximal part of his tracheal opening. Several attempts to remove the toy with Doyen intestinal clamps are unsuccessful. Murphy is positioned on his back and the attending emergency room doctor performs an emergency tracheostomy. Once established, a temporary tracheostomy tube is placed and Murphy is placed on 100% oxygen and ventilated at 10 breaths per minute. Murphy is noted to have sinus bradycardia on his ECG. A dosage of 0.4 mg/kg (0.04 mg/kg × a factor of 10) of atropine is administered via a red rubber catheter down his tracheostomy tube into the lungs. Murphy's mucous membrane color improved to a healthy pink, and he is noted to be in a sinus rhythm with a heart rate of 70 bpm. A blood pressure is obtained and is 80 mm Hg via Doppler. Murphy is started on a dobutamine CRI at 5 mcg/kg/min, and he is transferred to the critical care unit for further care.

1. Why were chest compressions started in Murphy?
2. Was Murphy ever in cardiac arrest?
3. Why was dobutamine started?

Case Study 4: Bobert is a 16-year-old castrated male old Sphynx cat. He is hospitalized in the intensive care unit for congestive heart failure secondary to severe hypertrophic cardiomyopathy. He is presently in an oxygen cage when you witness Bobert cry out and collapse.

1. What interventions should be started immediately?

Case Study 5: You are a member of a team currently performing CPCR for 4 minutes on a 20 kg dog. End-tidal CO_2 readings are 5 mm Hg. After applying a water-based lubricant, a Doppler blood pressure crystal is placed directly on the cornea and no flow signal is detected.

1. What things could you as rescuers do to improve effectiveness of the resuscitation?

Critical Thinking Questions

1. One of the most frequent and dangerous accidents that can occur during general anesthesia is with the adjustable pressure limiting (APL) or "pop-off" valve. Under what circumstances is a pop-off valve likely to be accidently left closed? How serious is a closed pop-off valve and how can it result in patient demise? What treatment options are available if a patient is compromised because of a closed pop-off valve?

2. Compared to humans, the rate of mortality for dogs, cats, and horses undergoing general anesthesia is much higher. Similarly, the rates of return of spontaneous circulation and patient survival are higher in humans who experience CPA than in dogs and cats. What are some possible reasons for these statistics?

ENDNOTES

1. Safar P. (1986). Cerebral resuscitation after cardiac arrest: A review. *Circulation, 74*(6 Pt 2), 138–153.
2. Safar P. (1988). Resuscitation from clinical death: Pathophysiologic limits and therapeutic potentials. *Crit Care Med, 16*(10), 923–941.
3. Bille, C., Auvigne, V., Libermann, S., et al. (2012). Risk of anaesthetic mortality in dogs and cats: An observational cohort study of 3546 cases. *Vet Anaesth Analg., 39*(1), 59–68.
4. Brodbelt, D. C., Pfeiffer, D. U., Young, L. E., et al. (2008). Results of the confidential enquiry into perioperative small animal fatalities regarding risk factors for anesthetic-related death in dogs. *J Am Vet Med Assoc., 233*(7), 1096–1104.
5. Hofmeister, E. H., Brainard, B. M., Egger, C. M., et al. (2009). Prognostic indicators for dogs and cats with cardiopulmonary arrest treated by cardiopulmonary cerebral resuscitation at a university teaching hospital. *J Am Vet Med Assoc., 235*(1), 50–57.
6. Kass, P. H., & Haskins, S. C. (1992). Survival following cardiopulmonary resuscitation in dogs and cats. *J Vet Emerg Crit Care, 2*(2), 57–65.
7. Wingfield, W. E., & Van Pelt, D. R. (1992). Respiratory and cardiopulmonary arrest in dogs and cats: 265 cases (1986–1991). *JAVMA, 200*(12), 1993–1996.
8. de Vos, R., Koster, R. W., de Haan, R. J., et al. (1999). In-hospital cardiopulmonary resuscitation: Prearrest morbidity and outcome. *Arch Intern Med, 159*(8), 845–850.
9. Waldrop, J. E., Rozanski, E. A., Swanke, E. D., et al. (2004). Causes of cardiopulmonary arrest, resuscitation management, and functional outcome in dogs and cats surviving cardiopulmonary arrest. *J Vet Emerg Crit Care, 14*(1), 22–29.
10. Fletcher, D., Boller, M., Brainard, B., et al. (2012). Recover evidence and knowledge cap analysis on veterinary CPR. Part 7: Clinical guidelines. *JVECC, 22*(S1), S102–S131.
11. Dick, W. F., Eberle, B., Wisser, G., et al. (2000). The carotid pulse check revisited: What if there is no pulse? *Crit Care Med., 28*(11 Suppl), N183–N185.
12. Eberle, B., Dick, W. F., Schneider, T., et al. (1996). Checking the carotid pulse check: Diagnostic accuracy of first responders in patients with and without a pulse. *Resuscitation, 33*(2), 107–116.
13. White, L., Rogers, J., Bloomingdale, M., et al. (2010). Dispatcher-assisted cardiopulmonary resuscitation: Risks for patients not in cardiac arrest. *Circulation, 121*(1), 91–97
14. Aufderheide, T. P., Sigurdsson, G., Pirrallo, R. G., et al. (2004). Hyperventilation-induced hypotension during cardiopulmonary resuscitation. *Circulation, 109*, 1960–1965
15. Plunkett, S. J., & McMichael, M. (2008). Cardiopulmonary resuscitation in small animal medicine: An update. *J Vet Intern Med, 22*(1), 9–25.
16. Buckley, G., Shih, A., Garcia-Pereira, F., et al. (2012). The effect of using an impedance threshold device on hemodyamic parameters during cardiopulmonary resuscitation in dogs. *JVECC, 22*(4), 435–440.
17. Yannopoulos, D., & Aufderheide, T. P. (2007). Use of the impedance threshold device (ITD). *Resuscitation, 75*(1), 192–193.
18. American Heart Association. (2010). CPR & ECC guidelines, www.heart.org/cpr
19. Voorhees, W. D., Ralston, S. H., & Babbs, C. F. (1984). Regional blood flow during cardiopulmonary resuscitation with abdominal counterpulsation in dogs. *Am J Emerg Med, 2*(2), 123–128.
20. Ralston, S. H., Babbs, C. F., & Niebauer, M. J. (1982). Cardiopulmonary resuscitation with interposed abdominal compression in dogs. *Anesth Analg, 61*(8), 645–651.
21. Ditchey, R. V., & Lindenfeld, J. (1984). Potential adverse effects of volume loading on perfusion of vital organs during closed-chest resuscitation. *Circulation, 69*(1), 181–189.
22. Rush, J. E., & Wingfield, W. E. (1992). Recognition and frequency of dysrhythmias during cardiopulmonary arrest. *JAVMA, 200*(12), 1932–1937.
23. Fletcher, D., & Boller, M. (2013). Small animal pulmonary resuscitation. *Vet Clin Small Anim, 43*, 971–987, http://dx.doi.org/10.1016/j.cvsm.2013.03.006
24. Soma, L. R., & Kenny, R. (1967). Respiratory, cardiovascular, metabolic, and electroencephalographic effects of of doxapram hydrochloride in the dog. *Am J Vet Res, 28*(12), 191–198.
25. Bruckner, J. B., Hess, W., Schneider, E., et al. (1977). Doxapram-induced changes in circulation and myocardial efficiency. *Anaesthesist, 26*(4), 156–164.
26. Miletich, D. J., Ivankovich, A. D., Albrecht, R. F., et al. (1976). The effects of doxapram on cerebral blood flow and peripheral hemodynamics in the anesthetized and unanesthetized goat. *Anesth Analg, 55*(2), 279–285.
27. Dani, C., Bertini, G., Pezzati, M., et al. (2006). Brain hemodynamic effects of doxapram in preterm infants. *Biol Neonate, 89*(2), 69–74.
28. Trzeciak, S., Jones, A. E., Kilgannon, J. H., et al. (2009). Significance of arterial hypotension after resuscitation from cardiac arrest. *Crit Care Med., 37*(11), 2895–2903.
29. Nozari, A., Safar, P., Stezoski, S. W., et al. (2006). Critical time window for intra-arrest cooling with cold saline flush in a dog model of cardiopulmonary resuscitation. *Circulation, 113*, 2690–2696.
30. Nozari, A., Safar, P., Stezoski, S. W., et al. (2004). Mild hypothermia during prolonged cardiopulmonary cerebral resuscitation increases conscious survival in dogs. *Crit Care Med, 32*(10), 2110–2116.
31. Steingrub, J. S., & Mundt, D. J. (1996). Blood glucose and neurological outcome with global brain ischemia. *Crit Care Med, 24*(5), 802–806.

CHAPTER 14
Anesthetic Techniques for Special Cases

Kate Lafferty, *BFA, CVT, VTS (Anesthesia/Analgesia)*

Rebecca A. Johnson, *DVM, PhD, DACVAA*

KEY TERMS

Chemoreceptor trigger zone (CRTZ)

Cushing response

Diabetes mellitus (DM)

Hyperadrenocorticism (Cushing's disease)

Hyperthyroidism

Hypoadrenocorticism (Addison's disease)

Hypothyroidism

LEARNING OBJECTIVES

Upon completion of this chapter, it is expected that the reader should be able to:

14.1 Identify unique factors of anesthetic and analgesic drugs that may affect their use in select patients

14.2 Formulate an effective anesthetic plan for patients with unique systemic disease (e.g., renal, hepatic, neurologic, cardiopulmonary impairment)

14.3 Create an appropriate anesthetic plan for the unique physiology of neonatal patients

14.4 Create an appropriate anesthetic plan for the unique physiology of geriatric patients

14.5 Describe the advantages and disadvantages of using total intravenous anesthesia (TIVA)

14.6 Choose cases where TIVA or constant rate infusions (CRIs) are appropriate

14.7 Choose appropriate drugs for use in TIVA or CRI techniques

INTRODUCTION

As preventative, medical, and surgical care continually improve in veterinary medicine, our companion animals are living longer. Consequently, it is becoming more common for the veterinary professional to anesthetize patients that have at least some systemic disturbance or disease (e.g., aged cat with chronic renal disease that needs a dental procedure). Formulation of a unique anesthetic plan specifically based on an individual patient's needs is the most important step toward a successful anesthetic period (see Chapter 6 [Creating an Anesthetic and Analgesic Plan]). Since many anesthetic agents reduce organ blood flow and alter normal physiology, the goal is to formulate an anesthetic protocol that minimizes harmful consequences on the affected organs yet still provides adequate anesthesia and analgesia for the patient. Although there are many "correct" anesthetic protocols for a specific case, some suggested techniques associated with specific states are described here along with specific anesthetic procedures such as TIVA and CRIs.

CARDIOPULMONARY DISEASE

Patients with cardiopulmonary disease present unique challenges to the anesthetist. Since many anesthetic agents (including inhalant anesthetic agents [volatile gases]) depress cardiovascular and/or respiratory function, it is difficult to maintain acceptable cardiopulmonary parameters during anesthesia in *normal* patients. In fact, these effects are more pronounced in patients with cardiopulmonary disease. Additionally, inhalant anesthetic agents are uniquely absorbed through the lungs and distributed to the body. If the pulmonary system is impaired, it may be difficult to stabilize patients during anesthesia due to inefficient gas exchange. In comparison to other organ systems (such as ocular or hepatic), impairments in the cardiac or pulmonary systems can directly lead to patient mortality and should be the primary focus of the anesthetist. Although management and pathogenesis of both cardiac and pulmonary disease vary widely based on etiology and have been reviewed in detail elsewhere, there are some commonalities in anesthetic management across patients.[1,2]

Preanesthetic Considerations

Performing a preanesthetic evaluation is crucial when anesthetizing patients with cardiopulmonary disease. A thorough history and physical exam centering on the heart and lungs is key. Patient medications should be documented, and the anesthetist needs to be fully aware of the effects these drugs may have on cardiovascular physiology (e.g., reduce blood pressure, slow heart rate). In-depth preanesthetic assessment including electrocardiography, thoracic radiographs, and echocardiography are required to provide the anesthetist with all pertinent information needed to choose drug and monitoring protocols. Patients should be stabilized as much as possible prior to anesthesia since hypoxemia, fluid overload, reduced cardiovascular function, and depressed ventilation are commonly associated with anesthesia in compromised patients; all of these may directly lead to patient demise. Elective procedures should be postponed in patients showing signs of congestive heart failure. Preoxygenation via mask (Figure 14-1) can be used to increase the arterial oxygen concentration in patients that may have cardiopulmonary disease (see Chapter 9 [The Induction Period]). However, it must be provided continually over at least 3 minutes to increase oxygen levels substantially and should not be performed if the patient is struggling or stressed by the technique.[2,3]

FIGURE 14-1 Preoxygenation of a dog with valvular pulmonic stenosis being anesthetized for a balloon valvuloplasty using a tight-fitting face mask. Note the condensation within the mask resulting from a tight seal around the muzzle; 100% oxygen is being delivered at 5 L/min.

Anesthetic Management

Preanesthetic medication is directed at reducing stress and maintaining normal heart rate and rhythm with minimal depression of cardiovascular function. Premedication is essential to smooth the induction period and to reduce the amount of induction and inhalant anesthetic agents required. The use of "organ-friendly" drugs such as the combination of a benzodiazepine and opioid is recommended since these patients are frequently compromised and this combination has minimal cardiovascular effects. For example, hydromorphone (Dilaudid®) and midazolam (Versed®) administered intravenous (IV) or intramuscular (IM) are commonly used for sedation, muscle relaxation, and analgesia prior to anesthesia of a dog. Premedication with long-acting drugs such as acepromazine (Promace®, generic) is not usually recommended due to the possibility of prolonged hypotension and non-reversibility. However, acepromazine can be useful to reduce stress in patients with inspiratory dysfunction that are struggling to breathe (such as laryngeal paralysis) and may be helpful in low dosages in cases that will benefit from afterload reduction (such as patients with chronic heart valve disease).[2] Similarly, alpha-2 adrenergic agonists may be useful in patients with select respiratory diseases to reduce stress[2] but are not usually recommended for patients with cardiac disease due to reductions in cardiac output and increases in systemic vascular resistance.[4]

Anesthetic induction can be accomplished with commonly used agents administered IV. Propofol (Rapinovet®, PropoFlo™, PropoFlo™28, Propovan®), although associated with apnea and desaturation, can be used

TABLE 14-1 Approximate FiO$_2$ Values from Various Oxygen Supplementation Techniques*

Technique	Approximate FiO$_2$ Obtainable	Oxygen Flow Rate
Face mask (varies depending on how tightly fitting the face mask is and oxygen flow rates)	30–60% (dogs/cats) 35–60% (horses)	2–8 L/min (dogs/cats) 10–15 L/min (horses)
Nasal oxygen catheter (oxygen delivered via nasal oxygen catheter [see Figure 7-15a] or an appropriately sized French red rubber catheter used as a nasal tube)	30–50% (one nasal line) 30–70% (two nasal lines)	100–150 mL/kg/min (patient discomfort noted at oxygen flow rates > 100 mL/kg/min)
Oxygen cage	25–50%	Variable; can be as high as 15 L/min

*Supplemental O$_2$ can be administered to anesthetic patients to ↑PaO$_2$ and promote O$_2$ delivery to tissue. This table summarizes a few methods of oxygen supplementation techniques. Adapted from *Lumb and Jones Veterinary Anesthesia and Analgesia*, 5th edition.

if administered slowly "to effect;" oxygen should be supplemented throughout induction.[5] Although controversial, ketamine (Ketaset®, VetaKet®, Ketaject®, Vetalar®) can be used as an induction agent in patients with respiratory disease and may be used in specific patients with cardiac disease. Ketamine has been associated with increased cardiac contractility and cardiac work so is not recommended in cases with hypertrophic disease (e.g., cat with hypertrophic cardiomyopathy). However, it can be useful in certain cardiac conditions such as dilated cardiomyopathy.[1] Alfaxalone (Alfaxan®) or etomidate (Amidate®) can also be used successfully in patients with cardiopulmonary disease due to their minimal cardiovascular effects, but similarly, oxygen should be supplemented throughout the induction period to reduce hypoxemia.[6]

Maintenance of general anesthesia is usually accomplished with inhalant anesthetic agents. However, both sevoflurane (Petrem®, SevoFlo® Flurovess™, Ultane®) and isoflurane (Isoflo®, Isosl®, Aerrane®) are profound respiratory and cardiovascular depressants, and the levels should be minimized.[1,2] This is frequently accomplished by the administration of other agents in a multimodal (balanced) analgesia plan such as fentanyl (Sublimaze®, Duragesic®) or ketamine CRI (see PIVA in Canine and Feline Patients later in this chapter). Fluid administration is necessary to maintain cardiac output and circulating volume. However, if the patient has cardiac dysfunction, care must be taken to avoid fluid overload, especially if contractility is impaired. Thus, fluid rates are often reduced to approximately 2 to 5 mL/kg/hr of balanced crystalloid administration, and colloids should be administered cautiously. Bradycardia should be avoided since cardiac output relies, at least in part, on a normal heart rate; anticholinergics such as atropine (Atrocare®, generic) or glycopyrrolate (Robinal®) can be used to reduce vagal-mediated bradycardia. On the other hand, tachycardia should also be avoided since very high heart rates may increase cardiac work and oxygen consumption and worsen the underlying disease. If hypotension is present and the heart rate is stabilized, agents that increase cardiac contractility may also be warranted (e.g., dobutamine (generic).[1]

Patients should be closely monitored throughout anesthetic maintenance and well into anesthetic recovery. Pulse oximetry, capnometry, electrocardiography, and arterial blood pressures should be recorded every 5 minutes to quickly detect any alterations in physiology. Mean arterial pressure should be kept above 60 mm Hg and ventilation should be assisted if end-tidal carbon dioxide levels rise above approximately 45 to 50 mm Hg. Patients should be supplied with oxygen even after extubation since residual respiratory depressant effects of inhalant anesthetic agents may be present well into recovery, resulting in hypercapnia and/or hypoxemia.

The anesthetist should be ready to re-intubate the patient or to provide supplemental oxygen (via nasal cannula, oxygen cage, mask, etc.) during recovery if needed (Table 14-1). With vigilant monitoring throughout the procedure, patients with cardiopulmonary disease can be anesthetized safely. However, if the disease is severe and the patient is not successfully managed preprocedure, anesthesia should not be attempted or the patient should be referred to a board-certified anesthesiologist.

NEUROLOGIC DISEASE

Although not as common as cardiopulmonary disease, patients with neurologic disease do present for anesthesia related to their disease process or even unrelated to their disease. In any case, it is imperative that patients do not recover from general anesthesia with neurologic dysfunction due to the procedure. The anesthetic plan should therefore be directed at optimizing cerebral blood flow (CBF) and perfusion and to prevent increases in intracranial pressure (ICP). CBF depends on many factors including, but not limited to, perfusion pressure, and arterial oxygen and carbon dioxide levels (Figure 14-2). CBF is normally autoregulated between mean arterial pressures of approximately 50 to 150 mm Hg. However, disease processes such as intracranial masses, hypertension, and traumatic brain injury as well as inhalant anesthetic agents impair autoregulation.[7]

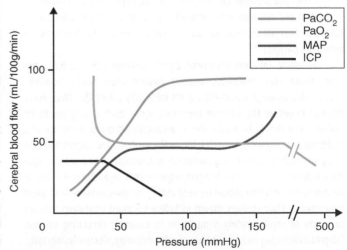

FIGURE 14-2 The relationship between intracranial pressure (ICP), partial pressure of oxygen (PaO$_2$), partial pressure of carbon dioxide (PaCO$_2$), mean arterial pressure (MAP), and cerebral blood flow (CBF).

Preanesthetic Considerations

Patients with neurologic disease should be stabilized prior to anesthesia if possible. Blood pressure and cardiac output should be normalized with volume replacement therapy if necessary. Hyperglycemia or hypoglycemia should also be treated in order to lower mortality rates and improve neurologic outcomes.[8] Ventilation and oxygenation should also be optimized prior to anesthesia to maintain normal arterial oxygen and carbon dioxide values and reduce their effects on increasing ICP. Clinical signs of increased ICP include abnormal pupillary reflexes, mentation, and ventilatory patterns. The **Cushing response**, also known as Cushing reflex vasopressor response, can also be present and results from increased sympathetic discharge to the peripheral vasculature (sometimes resulting in hypertension) with a reflex increase in parasympathetic discharge to the heart (resulting in bradycardia) in an attempt to maintain cerebral perfusion.[9] If increased ICP is suspected or clinical signs are present, mild hyperventilation (arterial carbon dioxide approximately 30–35 mm Hg) is frequently used to reduce ICP by decreasing carbon dioxide levels and cerebral blood flow (see Figure 14-2).[10] Hyperosmotic agents (e.g., mannitol or hypertonic saline) may also be administered to reduce ICP prior to or even during anesthesia.[11]

Anesthetic Management

Sedatives and analgesics can be quite beneficial in patients with neurologic disease. However, acepromazine should be used with caution since administration in large dosages can produce hypotension.[7] Although previously thought to reduce the seizure threshold in patients with a seizure history, current evidence does not support this, and acepromazine can be used safely in this patient population.[12] Both dexmedetomidine (Dexdomitor®) and midazolam can be used in patients with intracranial disease since they both decrease ICP.[7] Opioids also have minimal direct effects on ICP. However, they can indirectly increase ICP secondary to hypoventilation/hypercapnia and can cause vomiting, which temporarily raises ICP. Alternatively, pain also can increase ICP, and withholding opioids in painful patients to reduce adverse effects is not recommended.

Propofol and etomidate can both be used for anesthetic induction since they decrease CBF and ICP.[7] The use of ketamine remains controversial since it does not decrease CBF or ICP.[13] However, in patients with controlled ventilation and in combination with other sedative agents (not used alone), ketamine does not increase ICP and so may be a reasonable alternative in veterinary patients.[14]

Generally, inhalant anesthetic agents increase CBF and ICP when used above their minimum alveolar concentration (MAC). However, below MAC, inhalant anesthetic agents minimally affect CBF. Thus, every attempt to reduce the inhalant anesthetic agent levels during anesthetic maintenance should be made. This is accomplished through the use of a multimodal analgesic approach with local anesthetic techniques and CRIs of agents such as opioids (e.g., fentanyl) or ketamine, which substantially reduce the inhalant anesthetic agent requirements. In addition, assisted or mechanical ventilation should be used during inhalant anesthesia to avoid hypercapnia. Fluid therapy should include a balanced crystalloid solution at a rate of approximately 5 mL/kg/hr to maintain circulating volume. Diligent monitoring includes the use of pulse oximetry, electrocardiography, capnometry, and blood pressure measurement since arterial carbon dioxide, oxygen, and systemic blood pressure levels play an integral role in cerebral perfusion (see Figure 14-2).

Avoid the use of spring-loaded mouth gags when doing procedures that require opening of the mouth, such as oral or dental procedures (Figure 14-3). The use of spring-loaded mouth gags is associated with reductions in systemic (maxillary) and cerebral blood flow and subsequent central neurologic injury and blindness, especially in cats.[15] If the mouth must be held open, reduce the amount of pressure used and the duration of the procedure to minimize the development of neurologic complications.

Neurologic patients should recover from anesthesia with the head slightly elevated to encourage venous outflow and prevent further increases in ICP.[7] The end-tidal carbon dioxide and glucose levels and MAP should be closely monitored until fully recovered to prevent large deviations in these variables. The neurologic exam is difficult to interpret immediately following anesthesia, and the patient should be allowed to recover completely for many hours before fully assessing the neurologic status. When appropriately monitored, patients with neurologic disease can be safely anesthetized although every precaution should be taken to avoid increases in ICP.

FIGURE 14-3 Spring-loaded mouth gags are NOT recommended for use in any patient (especially cats) since they can reduce cerebral and maxillary blood flow and result in neurologic injury, such as postanesthetic cortical blindness.

Courtesy of Rebecca A. Johnson, DVM, PhD, DACVAA

GASTROINTESTINAL DISEASE

Patients with gastrointestinal (GI) disease commonly present for anesthesia. Although the primary insult can affect the anesthetic management (e.g., gastric foreign body), the secondary effects such as anorexia, dehydration, hypovolemia, acid-base and electrolyte abnormalities, hypoproteinemia, abdominal pain, and emaciation can significantly alter the anesthetic protocol.[16] GI diseases encompass everything from oropharyngeal and esophageal disease (e.g., neoplasia, obstruction, megaesophagus) to gastric and small intestinal disease (e.g., foreign bodies, gastric dilatation-volvulus) to disease of the large intestine, rectum, and perineum (e.g., obstruction, megacolon, perineal hernia) (Figure 14-4). Each specific disease process has its own unique pathophysiologic changes that affect the anesthetic protocol, and each case should be evaluated individually. However, many patients with GI disease share common considerations for anesthetic management such as those discussed below.

Preanesthetic Considerations

Patients with GI disease should have an extensive history taken with close attention paid to any current medications. If the patient with GI disease presents with pain prior to anesthesia, selection of analgesic drugs that will not exacerbate vomiting should be considered. Since patients frequently present with secondary issues such as electrolyte abnormalities and hypovolemia, every attempt to stabilize them prior to anesthesia should be made. A complete blood count and blood chemistry panel should be performed prior to anesthesia; abdominal radiographs and ultrasound should also be completed if warranted. In some emergent situations (e.g., decompensating gastric dilatation volvulus), a complete work-up and stabilization may not be possible; however, in many cases, correction of the underlying abnormalities and complete diagnostic testing beforehand will greatly

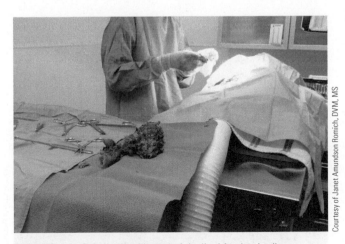

FIGURE 14-4 Patients with gastrointestinal foreign bodies can present with a variety of clinical signs including dehydration, vomiting, anorexia, weight loss, lethargy, and sometimes abdominal pain and diarrhea. This dog ingested Gorilla Glue® and presented with electrolyte abnormalities and hypovolemia, which needed to be addressed prior to surgery. Surgical removal of the entire amount of hardened Gorilla Glue® took multiple gastrotomy and enterotomy incisions which increased the length of time this dog was under anesthesia.

Courtesy of Janet Amundson Romich, DVM, MS

improve patient outcome. In general, small animal patients with GI disease should be fasted for 8 to 12 hours before anesthesia, during which any electrolyte, acid-base, or volume abnormalities should be addressed. Patients with GI disease are frequently at risk for vomiting, gastro-esophageal reflux or regurgitation, and aspiration pneumonia. Thus, every attempt to reduce these effects should be made.

Anesthetic Management

In patients with GI disease, it is important to minimize vomiting and gastric reflux/regurgitation associated with anesthesia. The act of vomition is complex, resulting from stimulation of the **chemoreceptor trigger zone** (CRTZ) that lies outside of the blood-brain barrier near the fourth ventricle of the brain. The CRTZ sends information to the vomiting/emetic center located in the medulla, which integrates many inputs that result in vomiting.[17] Opioids can have emetic or antiemetic effects depending on the specific drug's lipid solubility, the dosage, and route of administration. The emetic effects of opioids are believed to result from stimulation of delta opioid receptors in the CRTZ, and the antiemetic effects are due to stimulation of mu receptors at the vomiting center. To reduce vomiting associated with peri-anesthetic opioids, the use of opioids that are not always associated with vomiting such as butorphanol (Torbugesic®, Torbutrol®), buprenorphine (Buprenex®), or methadone (Dolophine®) has been recommended.[18] If the procedure is expected to produce severe pain, a full mu opioid agonist is required. Although full mu opioid agonists such as hydromorphone have been associated with vomiting, when administered IV vomiting is less frequent. In addition, administration of the neurokinin-1 antagonist, maropitant (Cerenia®) 1 mg/kg SQ, 30 minutes prior to premedication with an opioid such as hydromorphone significantly reduces opioid-induced vomiting (see Figure 11-8a).[19] Thus, opioids should not be withheld from the painful patient with GI disease, and alterations to the anesthetic plan should be made to minimize the chance of vomiting.

Sedatives such as acepromazine, midazolam, and dexmedetomidine can also be used in these patients. Although associated with decreases in gastro-esophageal sphincter pressure during general anesthesia, which may predispose to gastro-esophageal reflux,[20] they are useful as premedications to sedate the patient and to reduce the amount of inhalant anesthetic agents required for the procedure; most commonly-used inhalant anesthetic agents also increase the risk of gastro-esophageal reflux.[21] Attempts to reduce gastro-esophageal reflux using drugs such as maropitant,[22] metoclopramide (dopamine antagonist [Reglan®]),[23, 24] and omeprazole (proton pump inhibitor [Prilosec®])[25] have not shown consistent results, and reflux can still be seen with administration of these agents. However, the H_2-antagonist famotidine (Pepcid®), as well as omeprazole are effective in increasing the pH of gastric secretions and may aid in reducing damage to the esophageal mucosa when administered during the peri-anesthetic period.[16] Thus, these agents are commonly administered to patients undergoing anesthesia although their use remains controversial.

Because of the risk of vomiting and gastro-esophageal reflux (with potential for aspiration of stomach contents into the lungs), anesthetic induction should be rapid and smooth with immediate endotracheal intubation. Agents such as intravenous propofol or alfaxalone are best suited for this purpose. Intubation should occur with the patient in sternal recumbency and the head elevated. The endotracheal tube (ETT) cuff should be promptly inflated to hold an inspiratory pressure of approximately 20 cm H_2O within the breathing system in order to minimize the possibility that stomach contents will enter the lungs during anesthesia. A suction system should be available during this time to remove any contents seen in the oral cavity or esophagus. Patients should experience minimal movement throughout the anesthetic procedure to avoid unnecessary increases in intraabdominal/intragastric pressures. Anesthesia is maintained with inhalant anesthetic agents, and patients should be administered a balanced crystalloid solution at 5 mL/kg/hr based on the patient's individual requirements. If the GI disease is associated with decreased protein levels, the oncotic pressure within the vascular space may also be reduced, predisposing the patient to fluid shifts outside of the vasculature. These patients may also benefit from IV colloid therapy (such as VetStarch®) to increase or maintain intravascular oncotic pressure. Patient monitoring should include pulse oximetry, capnometry, electrocardiography, and blood pressure measurement, similar to most anesthetized patients. In addition, electrolyte or blood glucose monitoring may be warranted to assess the effectiveness of therapeutic interventions.

During anesthetic recovery, care must be taken to protect the airway from gastric secretions. Extubation should not take place until the patient is sufficiently awake to swallow and ideally maintaining sternal recumbency. If the patient cannot maintain in sternal recumbency on their own and is trying to chew the ETT, consider propping them in sternal recumbency with towels/pillows. The ETT cuff should remain inflated (or partially inflated) throughout extubation to remove any contents proximal to the cuff within the trachea. If vomiting or regurgitation occurs during extubation, the patient's head should be lowered and suctioning should be used if possible. Reversal agents (e.g., atipamezole [Antisedan®], flumazenil [Romazicon®], etc.) should be administered if the patient is too sedate to remain sternal. However, appropriate analgesics should not be withheld, even if some sedation is expected. An exception is the use of nonsteroidal anti-inflammatory agents (NSAIDs). Because of the association of NSAIDs with GI ulceration and delayed GI healing, vomiting, and diarrhea, these agents are not recommended for patients with GI disease.[26]

Safety Alert

Always be aware of the risk of being bitten or equipment being damaged if the animal is awake and you are trying to reach in to their mouth (e.g., when performing suctioning).

HEPATIC/RENAL DISEASE

Patients with severe dysfunction of the hepatic or renal systems are infrequently anesthetized in small animal private practice since these organs are required for drug disposition. However, patients with mild-to-moderate changes in hepatic or renal physiology are quite common (e.g., feline chronic renal disease, dogs with slight elevation of liver enzymes) and may require anesthesia. Since many anesthetic agents are removed from the body via hepatic metabolism and/or renal excretion, any decrease in function of these organs may prolong anesthetic effects. An exception would be the inhalant anesthetic agents that are mainly removed via the lungs with adequate alveolar ventilation and do not necessarily rely on hepatic and/or renal

function for removal. However, patients with hepatic or renal disease can be safely anesthetized with appropriate choice of agents and dosages.

Preanesthetic Considerations

Preanesthetic bloodwork may alert the veterinary professional to the presence of hepatic or renal dysfunction. However, it is important to remember that a significant amount of kidney dysfunction must be present before elevations in blood urea nitrogen (BUN) and creatinine levels are seen; therefore, a SDMA (symmetric dimethylarginine) should be considered because it detects renal dysfunction sooner than changes in BUN and creatinine. In addition, elevations in hepatic enzymes (such as ALT, AST, etc.) do not necessarily represent liver dysfunction; more sensitive tests such as bile acid evaluation may be required. Thus, these patients frequently require a significant work up including blood work, radiographs, and abdominal ultrasound to quantify their disease process prior to anesthesia. Secondary consequences of hepatic or renal disease are commonly seen. For example, hepatic dysfunction is associated with decreased protein levels including albumin and clotting factors and thus should always be evaluated prior to anesthesia in patients with liver dysfunction (Figure 14-5). In addition, alterations in electrolytes may also be associated with hepatic and renal disease. A potentially life-threatening example is hyperkalemia associated with acute renal injury or renal obstruction; very high levels result in abnormal cardiac conduction of electrical impulses and should be addressed prior to anesthesia if possible.[27]

Anesthetic Management

Many times, the choice of specific anesthetic agent is not as important as the overall anesthetic management itself. The overall goal during anesthesia in patients with hepatic or renal dysfunction is to maintain adequate perfusion of these organs to reduce further organ compromise. Patients

Courtesy of Getty Images

FIGURE 14–5 Patients with hepatic disease, such as this cat with jaundice, may present with pre-existing fluid abnormalities (e.g., dehydration) that ideally should be addressed prior to anesthesia. Hepatic dysfunction may also result in hypoproteinemia that may require additional oncotic support. Propofol is commonly used as an induction agent in animals with liver disease because it has significant extra-hepatic metabolism and does not accumulate quickly in the patient.

should have blood pressures measured prior to anesthesia, and their blood pressure during anesthesia should be kept within 20% of the awake values to maximize renal and hepatic blood flow. Thus, agents that are associated with hypotension such as acepromazine should be avoided or their dosages substantially reduced, especially in hepatic disease where drug duration may be extremely prolonged due to reduced metabolism. Premedications such as midazolam combined with a full mu opioid agonist such as hydromorphone are often recommended at dosages previously mentioned in this chapter.[27, 28] Availability of reversal agents for these premedications should be confirmed prior to administration. These combinations have minimal effects on blood pressure and cardiac output and can reduce the amount of inhalant anesthetic agent needed during the procedure. Because of variation in blood pressure and reduction in cardiac output associated with alpha-2 adrenergic agonists, these should be avoided in patients with substantial organ disease.[27] However, since they are easily reversed, they may be used in patients with hepatic disease if heavy sedation is required for the procedure; they should be reversed with atipamezole as soon as possible at the completion of the procedure.[28]

Anesthetic induction can be accomplished with commonly used induction agents such as propofol or alfaxalone as previously described. The minimum dosage required for intubation should be used and the dose should be administered "to effect" while avoiding re-dosing during the procedure if possible. Since propofol has significant extra-hepatic metabolism,[29] it is commonly used in patients with hepatic disease as it does not accumulate quickly. Ketamine may also be used in patients with organ dysfunction. However, although it is metabolized by the liver in most species, elimination of ketamine relies on renal excretion, especially in cats, and should be avoided or used with caution due to the potential for prolonged pharmacologic activity.[27, 30]

Animals with hepatic or renal disease should be intensely monitored during the entire anesthetic period. Pulse oximetry, capnometry, and electrocardiography are critical, especially in patients with concurrent electrolyte disturbances since bradyarrhythmias, including atrial standstill, can be seen with extreme cases of hyperkalemia.[27] Intravenous fluid therapy should be considered on an individual basis since these patients may present with pre-existing fluid abnormalities such as dehydration or even over hydration. In addition, patients with hypoproteinemia associated with hepatic dysfunction may require additional oncotic support; overhydration resulting from aggressive fluid therapy is possible, so attentive patient assessment (e.g., thoracic auscultation) during and following the procedure is mandatory. Patients with chronic renal disease may also be anemic and if warranted blood products should be made available. For these reasons, blood pressure evaluation is required throughout the anesthetic procedure and if hypotension is detected, quick and aggressive interventions to improve organ perfusion are necessary. During recovery, patients should continue to be monitored until they are warm and in sternal recumbency. Core body temperature should be maintained close to normal to facilitate drug metabolism and excretion, and reversal agents should be administered when appropriate.

The use of analgesic agents such as opioids should never be withheld from painful patients with hepatic or renal disease, even though their duration of action may be prolonged. Fentanyl IV is commonly used due to its quick onset and short duration of action. Its dose can be easily titrated to patients' needs. Remifentanil (Ultiva®), an ultrashort acting full mu opioid, can be given IV for pain relief. NSAIDs are not usually recommended in patients with severe hepatic or renal disease. NSAIDs can alter renal blood flow and cause dosage-dependent detrimental functional effects to the kidneys, and their use during anesthesia should be avoided if pre-existing renal disease is present[26, 27] although they are generally considered safe in healthy patients. Because NSAIDs have also been implicated in causing detectable alterations in platelet function, it may also be prudent to avoid their use in patients with hepatic disease associated with coagulopathy. However, if the patient simply has mild elevations in liver enzymes with no evidence of liver dysfunction, NSAIDs may be used cautiously although they have been associated with idiosyncratic liver failure in rare instances.[26] Although patients with hepatic and renal dysfunction can have highly variable conditions and presentations, they can be effectively and safely anesthetized with careful anesthetic agent choices and diligent monitoring throughout the procedure.

ENDOCRINE DISEASE

Patients presenting with endocrine disease are common in veterinary medicine. These patients rarely have severe complications during anesthesia and usually perform well during the procedure. However, every effort to stabilize these patients prior to anesthesia should be made to result in a successful outcome, especially for patients with thyroid disease, adrenal disease, and diabetes mellitus. A thorough history and complete physical exam with appropriate bloodwork and diagnostic testing specific to the patient should be performed prior to anesthesia.

Preanesthetic Considerations

Thyroid Disease

Hypothyroidism is a common canine endocrinopathy. Because untreated patients may have reduced cardiac output, decreased circulating blood volume, and the potential for decreased hepatic metabolism and renal excretion of drugs, patients should be treated with oral levothyroxine for at least 4 weeks prior to an elective procedure if possible.[31] The untreated patient may show greater hemodynamic instability during anesthesia and a propensity for hypothermia, so short-acting and/or reversible anesthetic agents should be considered.

In cats, **hyperthyroidism** is more common. Excessive thyroid hormones result in increases in metabolic rate and oxygen consumption, weight loss, hyperactivity, tachycardia, and tachypnea (Figure 14-6).[31] Increased packed cell volume (PCV), ALT, and ALKP (which is *not* steroid-induced in cats) are seen due to increased erythropoiesis and bone turnover. Electrolyte disturbances such as hypernatremia and hypokalemia are common. If possible, animals should be treated with methimazole (Felimazole®, Tapazole®), radioiodine therapy, or thyroidectomy to reduce levels of thyroid hormones prior to any anesthetic procedure.[31]

Adrenal Disease

Hyperadrenocorticism (Cushing's disease) is commonly encountered in dogs presented for anesthesia and results from either over secretion of cortisol from the adrenal gland due to a pituitary gland tumor (most common) or adrenal gland tumor or excessive cortisol from overuse of glucocorticoids. Dogs with Cushing's disease often present with a pendulous

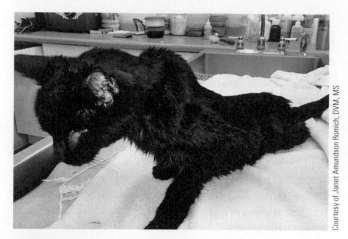

Courtesy of Janet Amundson Romich, DVM, MS

FIGURE 14-6 Cats with hyperthyroidism have increases in metabolic rate and oxygen consumption, weight loss, hyperactivity, tachycardia, and tachypnea. Preanesthetic blood work may reveal increased PCV, ALT, and ALKP as well as electrolyte disturbances such as hypernatremia and hypokalemia.

abdomen, alopecia, and muscle weakness. Hypertension is present in many cases. Neutrophilia, monocytosis, lymphopenia, and eosinophilia (the "stress leukogram") is often seen. In addition, increased ALT, ALKP, and cholesterol are frequently present. Treatment usually includes an adrenocorticolytic drug such as mitotane (Lysodren®) or trilostane (Vetoryl®), which can rapidly destroy adrenal tissue resulting in unwanted effects such as vomiting, diarrhea, lethargy, and Addisonian crisis; anesthesia should not be performed until this phase of treatment is complete.[31]

Although rare, dogs with **hypoadrenocorticism** (Addison's disease) resulting from decreased production of glucocorticoids and mineralocorticoids from the adrenal gland present with weakness, dehydration, hypotension, vomiting, and weight loss associated with electrolyte abnormalities such as hyperkalemia, hyponatremia, and hypochloremia. These animals are physiologically unstable and should have aggressive stabilization prior to any anesthetic episode. If anesthesia is required, supplementation with steroid hormones should be provided since stress associated with anesthesia and surgery can decompensate these patients; these patients are frequently referred to an anesthesiologist.

Diabetes Mellitus

Diabetes mellitus (DM) results from deficient serum insulin levels (hypoinsulinemia; Type I) or decreased insulin secretion and sensitivity of peripheral tissues to its effects (Type II). Dogs and cats present with polyuria, polydipsia, polyphagia, and weight loss.[31] Persistent hyperglycemia is the most common bloodwork finding. Treatment includes administration of various insulin preparations. Anesthesia and surgery in the diabetic patient are associated with higher complication and mortality rates and longer hospital stays.[32, 33, 34] Thus, stabilization of blood glucose levels should be done prior to anesthesia since anesthetic agents and the stress of anesthesia itself affect blood glucose levels.[35] Prior to anesthesia, patients should be fasted overnight and owners instructed NOT to give insulin to a fasted patient because fasting will affect insulin requirements. Blood glucose levels should be analyzed upon patient presentation. If the glucose is < 100 mg/dL, no insulin should be administered and a

2.5 to 5% dextrose solution may be infused to stabilize blood glucose levels. If the glucose is > 300 mg/dL, the usual full insulin dose may be given. Between 100 to 300 mg/dL, one-half of the dose is usually administered. Blood glucose should then be monitored every 30 to 60 minutes throughout the procedure, especially into recovery when increased glucose demands associated with thermoregulation may alter its levels. The ultimate goal during anesthesia is to avoid hypoglycemia and to minimize hyperglycemia.

Anesthetic Management

Similar to anesthetic management of patients with renal or hepatic dysfunction, agents that are reversible, short-acting, or have minimal cardiovascular and respiratory effects are preferred for patients with endocrine disease. For example, a benzodiazepine-opioid combination is a good choice for premedication to sedate the patient and reduce the inhalant anesthetic agent requirements throughout the procedure. Acepromazine is usually avoided due to its long duration of action because patients with endocrine disorders need to return to normal function as soon as possible following anesthesia, and prolonged sedation is undesirable. Alpha-2 adrenergic agonists are also usually avoided due to their pronounced cardiovascular depressant effects. In addition, administration of alpha-2 adrenergic agonists to diabetic patients may inhibit insulin secretion and worsen hyperglycemia.[36] However, this effect may be balanced by a reduction in sympathetic tone, and the use of alpha-2 adrenergic agonists remains controversial in diabetic patients.

Anesthetic induction with propofol or alfaxalone to rapidly achieve tracheal intubation is recommended. Because cortisol can potentiate the actions of norepinephrine and exacerbate tachycardia, hypertension, and hyperglycemia, ketamine is not recommended in patients with pre-existing hypertension such as that seen in hyperadrenocorticism or hyperthyroidism[37] or in hyperglycemic diabetic patients.[31] Patients should be monitored throughout the anesthetic procedure using pulse oximetry, capnometry, and electrocardiography. Blood pressure should also be intensely monitored and kept within 20% of awake values since blood flow autoregulation to organs may be shifted to the right in the chronically hypertensive patient (Figure 14-7).

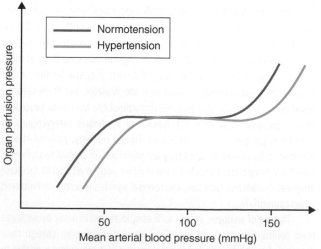

FIGURE 14-7 The relationship between organ perfusion pressure and a rightward shift in the autoregulatory curve in a patient that may have chronic hypertension. As a result of these shifts, these patients may require higher blood pressures to maintain organ blood flow.

Body temperature should be maintained near normal to aid with drug metabolism, especially in the hypothyroid patient. Fluids should be administered as needed, usually between 5 mL/kg/hr IV; dextrose should be added to diabetic patients if hypoglycemia is present. Blood glucose should be monitored every 30 to 60 minutes in the diabetic patient. Patients may benefit from having a central venous catheter placed to facilitate frequent blood draws.

Tech Tip 🐾

Frequently, it is difficult to draw blood during a surgical procedure due to the presence of surgical drapes, patient positioning, etc. A 25 gauge needle can usually be placed in either a small lingual vein or directly into an accessible paw pads, and blood samples directly placed into a hematocrit tube or a very small drop of blood can be used in most handheld glucometers for blood glucose analyses.

Anesthetic recovery is similar to any other anesthetized patient with routine monitoring throughout until the patient is warm and remains in sternal recumbency unassisted. Normal feeding schedules and drug administration routines should be started as soon as possible following the procedure. Although various endocrinopathies affect whole-body homeostasis, which can thereby impact the anesthetic period, with proper stabilization beforehand, these patients can undergo anesthesia safely and effectively. Severe complications are uncommon if "organ-friendly" agents are chosen and patients are properly monitored throughout.

OPHTHALMIC DISEASE

Patients with ophthalmic disease present special challenges to the anesthetist. For example, intraocular pressures and tear production are affected by many anesthetic agents and large variations in them should be avoided. In addition, concurrent medications and co-existing diseases can alter normal physiology. However, with care, these patients can be safely anesthetized.

Preanesthetic Considerations

Intraocular pressure (IOP) is an important factor in maintaining blood flow to the retina and optic nerve. Thus, any increase in IOP can greatly reduce optic nerve function and loss of vision, and it is important to minimize changes in IOP in anesthetized patients with ophthalmic disease.[38] In addition, increases in IOP of patients with potential for loss of globe integrity (e.g., descemetocele, trauma; Figures 14-8a and 14-8b) should be avoided to prevent complete globe rupture. Any pressure on the jugular veins or eyes such as that seen with physical restraint or placement of an oxygen mask is not recommended. Vomiting, retching, coughing, and the use of neck leads also increase IOP along with hypercapnia and hypoxemia and thus should be avoided.[39]

In general, most injectable anesthetic agents decrease tear production in veterinary patients, possibly for up to 24 hours;[40,41,42] therefore, topical artificial tear ointment or solution should be applied to both eyes liberally throughout the procedure and even afterwards. However, patient

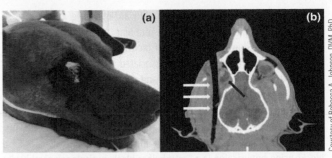

FIGURE 14-8 Ophthalmic Injury. (a) Ophthalmic trauma in a dog with a large (approximately 5 inch) stick protruding from the right orbital tissues. (b) Computed tomography (CT) scan showing the length of the stick (black linear area with white arrow) within the peri-orbital tissues. Premedication included IV fentanyl to reduce vomiting, provide analgesia, and smooth the induction process. The globe was intact (black arrow); the stick was removed and the eye remained with the dog following surgery.

Courtesy of Rebecca A. Johnson, DVM, PhD, DACVAA

positioning may reduce accessibility to the eyes during the procedure (e.g., head is away from the anesthetist and draped in), making it difficult to consistently apply ointment. In addition, patient positioning can make patient evaluation during the procedure challenging since eye reflexes and jaw tone cannot usually be assessed. IV catheters, blood pressure monitors, and pulse oximeters should be placed away from the patient's head to be accessible.

Many patients who present for ophthalmic procedures are being treated with ophthalmic drugs such as topical glucocorticoids, atropine, beta-adrenergic receptor antagonists (e.g., timolol [Timoptic®]), or other antiglaucoma agents. Antiglaucoma agents such as cholinergic drugs (e.g., pilocarpine [Isopto Carpine®]) or carbonic anhydrase inhibitors (e.g., acetazolamide [Diamox®]) can lead to adverse effects such as a bradyarrhythmia and bronchoconstriction (pilocarpine) or may predispose to metabolic acidosis and hypokalemia (acetazolamide).[38] Thus, a thorough physical examination centering on the heart and lungs with complete blood work evaluation should be performed prior to anesthesia. Additional diagnostic imaging techniques such as radiographs or ultrasound should be performed on an individual basis. A detailed patient evaluation will also aid in detecting any potential co-existing diseases that may affect anesthetic management (e.g., diabetes mellitus, renal disease, cardiovascular disease).

Anesthetic Management

The eye has extensive sensory innervation, and treatment of ocular pain should be a priority in any ophthalmic surgery or condition. Thus, it is common to include opioids or NSAIDs in the anesthetic protocol to manage pain. However, if increased IOP or potential globe rupture due to vomiting is a concern, opioids should be administered IV or withheld until after the patient is anesthetized. In addition, maropitant may be administered prior to the procedure to reduce opioid-associated vomiting. The patient should be gently restrained with minimal pressure on the neck and eyes for premedication and IV catheter placement. Typically a benzodiazepine-opioid combination is administered IV prior to anesthetic induction (e.g., midazolam-hydromorphone or midazolam-fentanyl). However, the use of acepromazine, alpha-2 adrenergic agonists, or other opioids can also be acceptable based on the individual patient's physiologic status.

Induction can be performed IV with an injectable agents such as propofol. Although opinions differ, ketamine and etomidate are not frequently used due to the chance of increased IOP associated with the drug itself (ketamine)[43] or with associated myoclonus (etomidate).[38] Tracheal intubation should be smooth and rapid and only attempted when a sufficient level of anesthesia is reached to avoid retching or coughing. Minimal pressure should be applied to the ocular area and the neck.

Maintenance is usually achieved with inhalant anesthetic agents; additional techniques such as opioid or lidocaine (Xylocaine®) CRIs or local blocks using bupivacaine (e.g., retrobulbar block, auriculopalpebral block) are also commonly used to reduce sensation and provide immobility to the affected structures but may also make patient monitoring more challenging. Monitoring should include pulse oximetry, capnometry, electrocardiography, and blood pressure measurement throughout the entire procedure. Specific attention should be paid to heart rate and rhythm since the oculocardiac reflex can occur with manipulation of any ocular or extraocular structure, especially if the patient is hypercapnic.[44] It arises from stimulation of the trigeminal and vagal nerves, and the most frequent arrhythmia noted is bradycardia; however, ventricular arrhythmias and asystole may also occur. Treatment includes stopping any periorbital tissue manipulation with immediate IV administration of an anticholinergic drug, such as atropine.

Anesthetic recovery should be smooth and the patient should be extubated without repeated attempts to swallow or cough. Pain should be treated and sedatives used prior to recovery if there is a possibility of the patient waking up excited or agitated. Excessive restraint should be kept to a minimum and Elizabethan collars are frequently necessary to keep the patient from scratching the eye.

Some ophthalmic procedures (such as cataract removal) require that the globe be in a central position with no movement during the procedure. In these instances, complete muscular paralysis with neuromuscular blocking agents should be used (e.g., atracurium [Tracrium®], vecuronium [Norcuron®]). However, these procedures are usually referred to a specialty practice or teaching hospital for advanced ventilation techniques and physiologic monitoring since the diaphragm becomes paralyzed as well and patients frequently have associated multisystemic disease (such as diabetes mellitus that caused the cataracts to form).

PREGNANCY AND CESAREAN SECTION

Anesthesia involving pregnant patients includes concerns and complications beyond the safety of a single patient. A comprehensive anesthetic plan must include considerations for the gravid patient as well as the fetuses. Most anesthetic agents cross the blood-brain barrier, thus allowing the drug to exert its effect. The mechanism that allows passage across the blood-brain barrier also enables anesthetic agents to enter the placenta and thereby affect the fetus.[45] Cardiac output, total blood volume, oxygen requirement, and intra-gastric pressure are all increased during pregnancy. Respiratory tidal volume is decreased due to the effects of the gravid uterus pushing on the diaphragm, and the patient compensates by increased respiratory frequency. These physiologic changes impact how anesthetic agents and the anesthetic event itself affect the patient and fetus.[46] In addition, procedural timing should be taken into consideration since maternal and

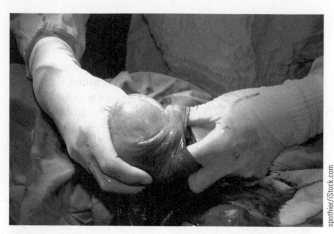

FIGURE 14-9 Anesthetic agents and the anesthetic event impact both the pregnant patient and fetus. Most anesthetic agents cross the blood-brain barrier; therefore, anesthetic agents and techniques selected for Cesarean section should minimize respiratory, central nervous system, and cardiovascular depressive effects of anesthetic drugs in the fetus so live, vigorous offspring are delivered. Pregnant animals have increases in cardiac output, total blood volume, oxygen requirement, respiratory tidal volume, and intra-gastric pressure; therefore, choosing drugs that minimize these effects in the mother are also important.

fetal fatality rates are significantly higher following emergency Cesarean sections when compared to planned Cesarean cases (Figure 14-9).[47,48] Thus, significant thought should be put into creation of the anesthetic plan, and the anesthetist should be well-prepared from anesthetic induction of the patient throughout delivery and recovery.

Preanesthetic Considerations

Many compensatory measures and physiologic changes occur during pregnancy (Table 14-2). In women, blood volume increases as much as 35%. While there are few studies quantifying the changes in canine and feline blood volume during pregnancy, it is reasonable to assume there is a similar increase.[49] Despite the increase in blood volume, pregnancy-associated relative anemia results from a greater increase in plasma volume than red blood cell concentration.[50,51] Oxygen requirements are elevated during pregnancy to meet fetal requirements. To compensate, cardiac output, heart rate, stroke volume, and blood pressure subsequently increase.

TABLE 14-2 Examples of Pregnancy-Related Physiologic Changes*

Increased Parameters	Decreased Parameters
Heart rate	Red blood cell volume
Cardiac output	Functional lung capacity
Total blood volume	Gastrointestinal motility
Minute ventilation	
Oxygen requirements	

*Paddleford, R. R. (1992). Anesthesia for Cesarean section in the dog. *Vet Clin North Am Small Anim Pract, 22*, 481–484; Raffe, M. R., & Carpenter, R. E. (2007). Anesthetic Management of Cesarean Section Patients. In W. J. Tranquilli, J. C. Thurmon, & K. A. Grimm (Eds.), *Lumb & Jones' veterinary anesthesia and analgesia* (4th ed., pp. 955–967). Ames, IA: Blackwell Publishing: Ames.

Ventilation is significantly impacted during pregnancy.[49] The enlarged uterus compresses and displaces the diaphragm and abdominal organs. The decrease in pulmonary function puts pregnant patients at increased risk for hypoventilation, hypoxemia, and possibly hypercapnia.[52] Thus, preoxygenation is an essential part of preanesthetic management in pregnant patients. Preoxygenation increases arterial oxygen concentrations; therefore, anemic patients and those at risk of hypoxemia greatly benefit from even a few minutes of preoxygenation. The only exception is for patients who become unduly stressed by preoxygenation, potentially worsening the situation.[46] Pregnancy causes decreased gastric motility, delayed gastric emptying, and enlargement of the uterus leading to displacement of the stomach. As a result, patients in late stage pregnancy are at increased risk of regurgitation and aspiration.[53] Many Cesarean section patients present as an emergency, and these patients should be assumed to have a full stomach. Metoclopramide in conjunction with cimetidine or famotidine (histamine-2 antagonist drugs) may be used as part of the preanesthetic plan.[54]

Anesthetic Management

The ideal anesthetic plan for Cesarean patients includes appropriate analgesic and anesthetic techniques without physiologic depression of the patient or fetus that may lead to decreased survival rates.[46] Most parturient patients do not usually require preanesthetic sedation, and an IV catheter is easily placed. However, for highly anxious or excited patients, preanesthetic sedation may be warranted. Sedation will not only reduce stress, it also will decrease the amounts of induction drug and inhalant anesthetic agent required. Acepromazine administration results in detectable fetal levels and may cause prolonged maternal and fetal depression.[47] In addition, acepromazine can cause vasodilation and should not be used in dehydrated or critical patients. Acepromazine is not reversible in either mother or neonate and, if administered, is given IV or IM (or less as needed). Benzodiazepines, such as midazolam, can also cause depression of mother and fetus, but may be comparatively less than with acepromazine.[55] Midazolam can be given IV or IM and can be reversed in both mother and neonate with flumazenil.[46]

Administration of any opioid will affect mother and fetus, although the extent to which opioids cause depression varies. Full mu opioid agonists may result in profound respiratory depression in the fetus, and levels can be found in the fetal/neonatal circulation days after clearance from the mother.[56] However, analgesia is best if full mu opioid agonists are used. For example, hydromorphone, oxymorphone (Numorphan®), and methadone can be given IM or IV. As with all drugs, if the drug is administered IV, the lower end of the dosage range should be used.[57] Opioids can be reversed, both in the mother and neonate, using naloxone (Narcan®). In the mother, this can be given IM or *slowly* IV to effect. In neonates, naloxone can be diluted and administered sublingually.[46] In addition, opioids may be withheld until after the babies are removed to reduce the depressant effects in the neonates. Another option is the use of an opioid, such as preservative-free morphine, in the lumbosacral epidural space. Generally, the dose administered in the space is lowered by 25% in pregnant patients to account for decreased space due to increased collateral circulation.

As previously mentioned, pregnant patients should be preoxygenated before anesthesia. Even as few as five minutes of preoxygenation has been shown to have beneficial effects on mother and fetuses. Anesthetic induction is most commonly achieved using propofol IV to effect. Propofol is readily available, has a rapid onset, short duration, and is quickly metabolized via hepatic and extra-hepatic pathways.[58] Propofol does cause transient apnea and hypotension, but these effects can be minimized with prompt endotracheal intubation, manual ventilation, and appropriate blood pressure support.[59] Alternatively, alfaxalone IV may be used for induction and, although survivability is similar to propofol, it is associated with greater neonatal vitality post-birth.[60] Alfaxalone also produces apnea similar to propofol. The anesthetist should be prepared to intubate immediately after induction. Pregnant patients are at increased risk of hypoxia and regurgitation of stomach contents, requiring the airway to be quickly secured. Thus, mask inductions using inhalant anesthetic agents are not recommended since they are relatively slow with delayed tracheal intubation. Patients can be maintained on either isoflurane or sevoflurane; however, both inhalant anesthetic agents cause significant cardiorespiratory depression. Physiologic changes during pregnancy can lead to a reduction in MAC; as a result any inhalant anesthetic agent used should be maintained at the lowest level possible.[61]

Regional analgesia in the form of an epidural or line block is commonly used for Cesarean patients (see Chapter 17 [Analgesic Techniques]). Regional anesthesia has little effect on the fetus and is successfully used to provide analgesia and/or muscle relaxation in pregnant patients.[62] Preservative-free local anesthetic agents such as lidocaine or bupivacaine can be used in the epidural space (Figure 14-10). Another option is the use of an opioid, such as preservative-free morphine, in the lumbosacral epidural space. However, the size of the epidural space in pregnant patients may be smaller, which can result in farther cranial migration of local anesthetic agents, and therefore it is advised to decrease overall volume of the epidural drug by 25%.[63]

As with any type of anesthesia, the patient should be instrumented with all available monitoring equipment as quickly as possible so as not to delay entry in to the operating room. Pulse oximetry, capnometry, electrocardiography, noninvasive blood pressure, and temperature should be monitored. Recording parameters in the anesthetic record should occur every 5 minutes, allowing for assessment of physiologic trends and quick detection (and correction) of issues is necessary for optimal survival of mother and neonates. Hemoglobin saturation should be maintained above 95% and end-tidal carbon dioxide levels should be maintained between 35 to 45 mm Hg. Assisted ventilation should be instituted if needed to maintain end-tidal carbon dioxide levels. Mean arterial pressure should be kept above 70 mm Hg to maintain cardiac output and therefore uterine blood flow and oxygen

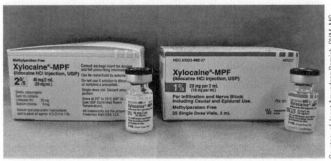

FIGURE 14-10 Regional anesthesia is commonly used for Cesarean patients because it has minimal effect on the fetus and can be used to provide analgesia and/or muscle relaxation in pregnant patients. Preservative-free local anesthetic agents such as lidocaine can be used in the epidural space.

delivery to fetuses. Blood pressure support should also be provided. The patient should be closely monitored into recovery, providing supplemental oxygen and thermal support as needed. Analgesia will have to take into consideration the transfer of drug residues in the milk.

Care of Neonates

Staff should be available for resuscitation of neonates immediately after delivery. The neonates should be cleared of placental membranes and the mouth and nose cautiously suctioned free of fluid and membrane. Gentle, but not overly vigorous, rubbing of patients can help warm and dry the neonates, as well as stimulate breathing. "Swinging" or "slinging" of neonates is no longer recommended to loosen secretions and/or stimulate the neonate due to the risk for injury (e.g., cerebral hemorrhage from the centrifugal forces, aspiration of stomach contents, dropping of the puppy). Oxygen masks should be available. A kit with neonate resuscitation supplies such as a laryngoscope and small ETTs, small intravenous catheters, supplies for intraosseous catheters, and emergency/reversal drugs and dosages should be readily available.[46] An incubator with oxygen and thermal support should be ready for all viable neonates.

Tech Tip

IV catheters (20 gauge to 16 gauge), with the stylet removed, can be used to intubate small puppies or kittens. A size 2.0 mm ETT adapter can be fitted into the hub of the IV catheter and connected to non-rebreathing anesthetic hoses to provide oxygen and respiratory support.

NEONATAL AND PEDIATRIC PATIENTS

Puppies and kittens may present for anesthesia ranging from elective (ovariohysterectomy, castration, dental extractions) to emergent (hit-by-car, intestinal obstruction, fractured limb, trauma) to critical (liver shunt, patent ductus arteriosus, persistent right aortic arch) procedures. When formulating an anesthetic plan, neonatal or pediatric patients cannot be treated as "little" cats and dogs. Neonatal and pediatric patients exhibit drug responses that drastically differ from adult patients. For example, they have unique anatomic and physiologic differences that make anesthesia more challenging. Defining what constitutes "neonate" versus "pediatric" can be challenging and often varies by breed. As a general rule, puppies and kittens are termed "neonate" until they are approximately 4 to 6 weeks old. Young patients are considered "pediatric" until approximately 12 weeks old. These classifications are not a hard and fast rule, but may provide guidelines from which to create an anesthetic plan.[64, 65, 66]

Very young neonatal patients may require lower dosages of anesthetic agents. They generally have lower albumin levels compared to an adult, which leads to highly protein-bound drugs having a greater effect due to more free drug in circulation; therefore, protein-bound drugs such as ketamine, propofol, etomidate, and NSAIDs may have a greater effect.[62] Neonatal and pediatric patients have a high metabolic rate and increased oxygen demands.[67] They may also have incomplete or immature hepatic,

renal, cardiovascular, and respiratory functions. As hepatic and renal systems are not fully functional until 8 weeks of age, avoid administering drugs with extensive metabolism or reduce the dose given. Young patients are at increased risk of hypothermia due to their small size (they have a large body surface to body mass ratio), limited amount of body fat, high metabolic rate, and immature thermoregulatory ability. These physical and physiologic differences require careful consideration when formulating the anesthetic plan.[64]

Preanesthetic Considerations

Much of organ development in mammals occurs while *in utero*. However, maturation of organs and organ systems does not reach completion for weeks to months after birth. In puppies and kittens, the cardiovascular system is still quite immature and they may not yet have mature "normal" blood pressure. Neonates and pediatric patients have minimal ability to increase their cardiac contractility and have little cardiac reserve.[68] Young patients rely on heart rate to drive cardiac output, thus bradycardia in a pediatric patient can have a greater detrimental effect on blood pressure than in adults.[69, 70] Cardiac defects such as patent ductus arteriosus, valvular dysplasia, and septal defects are more common in neonatal patients. The physical exam should center on the cardiopulmonary systems, and cardiac auscultation is extremely important in young patients. All cardiac murmurs and/or arrhythmias should be further evaluated.[71]

Neonatal and pediatric patients have a higher oxygen consumption rate relative to their small size. To compensate for this increased oxygen demand, young patients have a higher respiratory rate and minute volume. They also have a more compliant ribcage and weaker thoracic musculature, which fatigue more easily than adult patients.[72] In addition, neonatal and pediatric patients have smaller airways and may be at a greater risk for respiratory obstruction.[70] Whenever possible, it is recommended to provide at least 3 minutes of supplemental oxygen therapy before anesthetic induction to counteract potential hypoxia and hypoxemia.[3]

The underdevelopment of renal, hepatic, and microsomal systems that metabolize and eliminate anesthetic agents in neonates can lead to slow metabolism and elimination of anesthetic agents and prolonged recovery in patients[73,74]. Hypoglycemia is also a risk for young patients. Neonates and pediatric patients are generally able to regulate their glucose while awake and able to feed regularly. However, they have minimal glycogen reserves, and when food is withheld or a patient is put under stress, they can easily become hypoglycemic.[62,73]

Anesthetic Management

Sedation is often not required for neonatal and pediatric patients. Benzodiazepines are the sedative of choice for very young patients that are excited or stressed. They provide some sedation (especially if combined with opioids and/or low dosage ketamine), have little negative respiratory or cardiovascular effects, and can be reversed (Figure 14-11).[57] Midazolam can be given IM or IV. Phenothiazines such as acepromazine are not recommended as they may cause hypotension, have a long duration of effect, are not reversible, and may affect thermoregulation.[75] Alpha-2 adrenergic agonists are also not recommended as they are associated with bradycardia and vasoconstriction and reduced cardiac output.[64]

Courtesy of Rebecca A. Johnson, DVM, PhD, DACVAA

FIGURE 14-11 Twelve-week-old kitten sedated with midazolam, ketamine, and oxymorphone. Note the dilated pupils and relative tachycardia (242 beats per minute, as shown on the pulse oximeter).

Analgesics should be included in every anesthetic plan where pain is anticipated. Opioids provide excellent analgesia and some sedation and are reversible if a critical situation arises. For example, oxymorphone or hydromorphone IM or IV can be used for both canine and feline patients for moderate to severe pain; buprenorphine IM or IV can be used for mild to moderately painful patients.[64] Butorphanol can be utilized for its sedative effects, allowing for ease of IV catheter placement and followed by more effective opioids for pain management. Other opioids, such as fentanyl, can be used for young patients, but may cause more profound respiratory depression.[76]

As previously mentioned, pediatric and neonatal patients should be pre-oxygenated before anesthesia. Propofol, alfaxalone, and etomidate are good choices for anesthetic induction IV. These drugs are easily eliminated, which is advantageous to patients with immature or decreased hepatic function. Patients can be maintained on either isoflurane or sevoflurane; however, both inhalant anesthetic agents cause significant cardiovascular and respiratory depression, particularly in neonatal and pediatric patients, and should be maintained at the lowest level possible.[64]

As with any type of anesthesia, pulse oximetry, capnometry, electrocardiography, noninvasive blood pressure, and temperature should be monitored and recorded every 5 minutes to detect trends in physiology. Normal heart rate in puppies and kittens can near 200 beats per minute (depending on age), and respiratory rate ranges from 15 to 35 breaths per minute. Assisted ventilation should be instituted if needed to maintain

normal end-tidal carbon dioxide levels. Mean arterial pressure in neonatal and pediatric patients is lower than that of adult patients. The goal should be to keep mean arterial pressure at or above 60 mm Hg but is again based on age and body weight, and lower pressures may be acceptable for some patients.[69]

Neonatal and pediatric patients have higher fluid requirements, but are at increased risk of fluid overload due to their small size and potential for inadvertent operator error. Young patients may have immature renal function and have difficulty coping with over-hydration.[64] Thus, IV fluid rates at 5 mL/kg/hr are recommended with frequent monitoring of hydration status.

Neonatal and pediatric patients should be closely monitored into recovery, providing supplemental oxygen and thermal support as needed; oxygen masks or flow-by oxygen should be available. For very young and/or very small patients, an incubator with oxygen and thermal support should again be ready and waiting.

GERIATRIC PATIENTS

The term "geriatric" is generally defined by both the patient size and species. Dogs weighing less than 9 kg are considered geriatric at 9 years of age, whereas dogs weighing 9.5 to 41 kg are considered geriatric between 9 and 10 years of age. Large breed dogs weighing greater than 41 kg are defined as geriatric by 6 years of age. Cats are considered geriatric at 8 to 10 years of age.[77] With this all said, age is not a disease and it is important to address specific issues a patient has as those can lead to a poor anesthesia experience, not the age itself.

Tech Tip

Defining the term geriatric can be difficult in the wide variety of veterinary patients. Instead of stating a year that identifies a patient as geriatric, some veterinary professionals prefer to define geriatric as when the patient has reached 75% of its life expectancy.

As any mammalian species ages, optimal function of normal physiologic systems eventually declines. Additionally, co-morbidities develop in geriatric patients that affect how they respond to various anesthetic agents and techniques. Before formulating an anesthetic plan, geriatric patients should undergo a thorough physical examination, blood work, and additional diagnostic testing as needed. Thoracic radiographs are recommended to evaluate respiratory and cardiovascular status, or check for neoplasia if warranted; additional cardiovascular assessment is needed to evaluate any murmur, arrhythmia, or signs of cardiac enlargement. Ultrasound may be required for further evaluation of abdominal structures. Preanesthetic blood pressure measurements can be useful to evaluate cardiovascular function. Complete records from the owner regarding current conditions and medications are also important. Older patients may have endocrine disturbances, age-related degeneration of heart valves, a degenerative neural condition, loss of kidney function, decrease in liver mass and function, and decreased joint flexibility and mobility. Geriatric patients generally have a decreased metabolic rate and potentially decreased muscle mass. They are at an increased risk of hypothermia during the premedication, anesthetic maintenance, and recovery periods. These physiologic changes require careful consideration and modification of the anesthetic plan for geriatric patients.[64]

> **Tech Tip** 🐾
>
> Many clients are reluctant to have their geriatric pets anesthetized. As a veterinary technician, you can help clients overcome their reluctance by knowing the altered physiology of geriatric patients and explaining how these alternations affect anesthesia and recovery.

Preanesthetic Considerations

Geriatric patients often have both age-related cardiovascular changes as well as cardiac disease. Older patients may have valvular insufficiencies, thickening of cardiac walls, fibrosis of myocardial tissue, and conduction impairments.[78] These cardiovascular changes can decrease the overall function of the heart, and geriatric patients may suffer hyper- or hypotension and decreased cardiac output.[79] Geriatric patients rely heavily on preload to maintain appropriate blood pressure and do not tolerate hypovolemia.

Lung tissue in geriatric patients has decreased elasticity and expansion ability.[80] Older patients often have decreased respiratory rate and tidal volume. The ribcage may be less compliant, and thoracic muscles are often weaker and more prone to fatigue than younger patients. Geriatric patients may lose protective airway reflexes and are at increased risk for pulmonary pathology such as fibrosis, neoplasia, or pneumonia.[64, 81] Whenever possible, it is recommended to pre-oxygenate older patients before anesthetic induction to counteract potential hypoxia and hypoxemia.[3]

Geriatric patients have decreased liver mass as well as decreased hepatic blood flow. The renal, hepatic, and microsomal systems that metabolize and eliminate anesthetic agents are diminished, which can lead to slow elimination of anesthetic agents and extended recovery times.[82] Geriatric patients are at increased risk for hepatic pathology (e.g., hepatic neoplasia) that may further affect anesthetic drug metabolism and clotting factor production. In addition, geriatric patients with hepatic dysfunction are at increased risk for hypoproteinemia and hypoglycemia.[82]

Older patients usually have decreased renal function.[83] The renal system is unable to tolerate hypotension, hypo-perfusion of kidney tissue, and hypovolemia. Caution should be used when considering drugs excreted by the kidney (such as ketamine) or drugs that could impair renal function (such as NSAIDs). Careful monitoring of tissue perfusion and blood pressure is required in all geriatric patients.[77]

Geriatric patients often have decreased brain mass, fewer functioning neurons, and lower levels of available neurotransmitters. Cognitive function and perception may be impaired. As a result they may have a more exaggerated response to anesthetic agents.[84]

Anesthetic Management

Sedation may or may not be required for geriatric patients. Quiet, depressed, or debilitated patients may not need any sedation whereas excited, nervous, or stressed patients will benefit from sedation. Benzodiazepines, such as midazolam IM or IV, provide mild to moderate sedation, have little negative respiratory or cardiovascular effects and can be reversed. They are the "go-to" drug for seizure cases and are safe to use in most neurologic patients. However, benzodiazepines may have a prolonged effect in patients with hepatic dysfunction.[64,77] Acepromazine should be used cautiously in geriatric patients. It can relieve stress and anxiety in very excited or nervous patients. However, because acepromazine causes vasodilation and hypotension, is not reversible, and may affect thermoregulation, it is not frequently used in geriatric patients. Alpha-2 adrenergic agonists are not usually recommended in older patients as they may have profound negative effects on the cardiovascular and circulatory systems.[77] Opioids should never be withheld from geriatric patients when warranted, especially since they are reversible. For example, oxymorphone, hydromorphone, buprenorphine, or fentanyl can all be used to provide analgesia in dogs and cats.[64]

As previously mentioned, geriatric patients should be preoxygenated before anesthetic induction. Propofol, etomidate, alfaxalone, and ketamine-benzodiazepine (midazolam or diazepam [Valium®]) are all possible choices for anesthetic induction. Etomidate has minimal negative cardiorespiratory effects and is an excellent choice for elderly patients with significant cardiovascular compromise. Ketamine-benzodiazepine (midazolam or diazepam) can be used in healthier geriatric patients. Ketamine has minimal respiratory effects and may increase cardiac output. However, ketamine may not be appropriate for patients with certain cardiac or neurologic diseases. Ketamine and benzodiazepine (midazolam or diazepam) are metabolized by the liver and ketamine is excreted by the kidneys, so patients with hepatic or renal disease may not be good candidates for ketamine-midazolam inductions. Patients can be maintained on either isoflurane or sevoflurane at the lowest levels required keeping in mind that MAC is reduced with increased age.[64]

Pulse oximetry, capnometry, electrocardiography, non-invasive blood pressure, and temperature should be monitored and recorded every 5 minutes, similar to all anesthetized patients. Maintenance of normal arterial pressures in geriatric patients is essential. It is necessary to keep mean arterial pressures at or about 70 mm Hg in order to maintain appropriate perfusion to renal tissue and other vital organs. Geriatric patients do not tolerate over-hydration well. They may have cardiac dysfunction that is worsened by overly aggressive fluid therapy. The goal should be to provide adequate fluid volume to maintain appropriate perfusion and volume; 2.5 to 5 mL/kg/hr of crystalloid administration are recommended with an attempt to decrease the hourly fluid rate as the anesthetic period goes beyond an hour (e.g. a 5-hour-long dental procedure). Geriatric patients should be monitored closely in recovery. Pulse oximetry monitoring, supplemental oxygen, and thermal support should be provided as in any case.

TOTAL INTRAVENOUS ANESTHESIA (TIVA) AND CONSTANT RATE INFUSIONS (CRIs) IN CANINE, FELINE, AND EQUINE PATIENTS

The use of CRIs for induction and maintenance of anesthesia and as adjunct analgesic/anesthetic agents is dramatically increasing in veterinary medicine.[85,86] When considering general anesthesia for veterinary patients, most plans assume inhalant anesthetic agents to be part of the process. For longer procedures—those lasting longer than an hour—using an inhalant anesthetic agent for anesthetic maintenance is a more appropriate choice. However, some cases may benefit from using TIVA. For example, TIVA may be indicated for short procedures (bandage changes, small mass removals, and equine castrations). It may also be needed in situations where a patient cannot be intubated (tracheal procedures, laser surgeries in or near the mouth and airway), where patients need to remain intubated and unconscious for extended periods (cases requiring long term ventilation assistance), or in patients with autosomal disorders where inhalant anesthetic agents cannot be used (malignant hyperthermia). In any case, there are many advantages to injectable anesthesia including potential to avoid inhalant-associated cardiovascular effects, no exposure of patient or personnel to waste anesthetic gases, and smoother induction and recovery periods.[85] There is also incomplete data to suggest that avoiding inhalant anesthetic agents leads to less cardiopulmonary depression. However, there are also disadvantages. While recovery from TIVA tends to be smoother than from procedures where inhalant anesthetic agents are used, there is the risk for prolonged recovery, especially in very young neonates, geriatrics, and those with significant hepatic dysfunction.[87] Use of TIVA still requires a source for oxygen administration, and standard anesthetic monitoring equipment is needed. It may be harder to control anesthetic depth when using TIVA, so diligent monitoring is essential.[86]

In addition to using injectable agents for general anesthesia, many drugs can be used as adjunctive analgesic/anesthetic agents. These infusions are referred to as intravenous analgesic adjuncts (IVAA) or partial intravenous anesthesia (PIVA). Agents such as opioids, ketamine, lidocaine, or alpha-2 adrenergic agonists can be added to a TIVA protocol to decrease the amount of injectable anesthetic needed. Intravenous analgesic adjuncts can be added to an anesthetic plan where additional pain management is needed (e.g., ketamine CRI for a fracture repair, lidocaine CRI for an equine colic surgery, fentanyl/lidocaine/ketamine CRI for an abdominal surgery).

When using TIVA, a CRI technique should be chosen versus an intermittent bolus injection technique. When giving intermittent boluses of anesthetic or analgesic agents, it is nearly impossible to achieve a stable level of drug. By administering a constant, steady infusion, stable tissue concentrations are reached. CRI also prevents peaks and valleys of comfort versus discomfort or consciousness versus unconsciousness that occur when bolus doses of anesthetic or analgesic agents are given.[88] When drugs are administered intermittently, the drugs reach a peak, providing anesthesia and/or analgesia, then depending on how the drug is dosed can rapidly fall below therapeutic levels allowing for breakthrough pain or awakening. CRI administration eliminates these fluctuations. CRIs can be adjusted to meet the needs of each individual patient. The ability to make minor adjustments allows for lower total amounts of drug administration. Lower dosages also decrease the incidence and severity of adverse effects.[89] Patients receive better analgesia, safer anesthesia, and less money is spent on drugs and supplies (syringes, needles, etc.) although syringe or fluid pumps are usually required to deliver precise volumes (Figure 14-12).[87]

TIVA in Canine and Feline Patients

Because of their pharmacokinetic properties, short-acting drugs that minimally accumulate are best suited for TIVA techniques. Thus, propofol continues to be the mainstay for TIVA in canine and feline patients. Propofol has the benefits of quick onset and rapid redistribution. TIVA using propofol

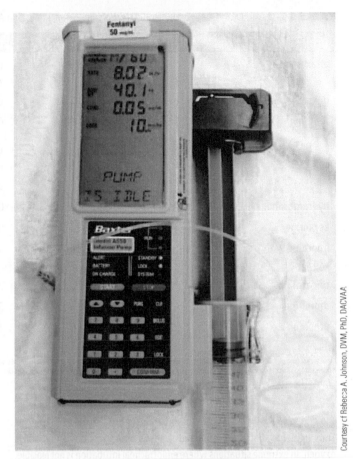

Courtesy of Rebecca A. Johnson, DVM, PhD, DACVAA.

FIGURE 14-12 A syringe pump primed and set to deliver a fentanyl CRI.

generally results in smooth recoveries. However, it has no analgesic properties and causes dosage dependent cardiovascular and respiratory depression.[90] Recently, alfaxalone has also been successfully used for TIVA in dogs, cats, and ruminants.[90, 91, 92] Benzodiazepines such as midazolam can be added to the TIVA protocol to decrease the total amount of propofol or alfaxalone used; however, they are not potent enough sedatives to produce an unconscious state when given singly in dogs and cats.[85] Adding an opioid such as fentanyl CRI to the TIVA protocol can further decrease the overall amount of drug needed and will provide analgesia for the procedure. Decreasing the overall amount of propofol or alfaxalone used produces less respiratory depression and maintains better arterial blood pressure.[90] Table 14-3 summarize loading dosages and CRI rates for the most useful pharmacologic agents used in dogs and cats.

Anesthetic Management in Canine and Feline Patients During TIVA

Premedication and/or the use of loading dosages before TIVA are recommended to rapidly achieve initial therapeutic blood levels. Sedative and analgesic choices will depend on the patient and type of procedure. Premedication will allow for smoother induction and lower infusion amounts.[85] Respiratory depression is a risk associated with any type of TIVA, and a patient breathing room air may become hypoxemic over time.

TABLE 14-3 CRI Drug Dosages for Canine and Feline Anesthesia and Analgesia*

Drug	IV Loading Dosage (mg/kg)	CRI Dosage (mg/kg/hr)	Notes
alfaxalone	1.0–2.0	6.0–7.0 (premedicated dogs) 7.0–8.0 (premedicated cats)	Can use lower dosages if combined with opioid
dexmedetomidine	0.0005–0.002	0.0005–0.002	Bradycardia, vasoconstriction
fentanyl	0.001–0.005	0.001–0.01 May see higher dosages (0.02) for short time periods during intense surgical stimulation	Lower end dosage for post-operative management
hydromorphone	0.05–0.2	0.01–0.015	Lower end of dose for feline patients
ketamine	0.5–1.0	0.3–0.6	Often used with an opioid
lidocaine	1.0–2.0 (given slowly)	1.8–3.0	NOT RECOMMENDED FOR USE IN CATS
midazolam	0.2–0.4	0.01–0.05	No analgesia
morphine	0.1–0.3	0.1–0.2	Give IV loading dose SLOWLY
propofol	2.0–6.0	12–24	Can use lower dosages if combined with opioid

These are just examples; all drug dosages may be adjusted, depending on individual patient needs, procedural requirements, and whether the patient has been premedicated.
* Duke, T. (2013). Partial intravenous anesthesia in cats and dogs. *Can Vet J, 54,* 276–282; Pablo, L. S. (2011). Total IV anesthesia. In *ACVS 2011 Symposium proceedings* (452–456). Paper presented at ACVS 2011 Symposium: The Surgical Summit, Chicago.

Thus, patients should be intubated if possible and maintained on 100% oxygen. In extreme situations where endotracheal intubation is not feasible, a tight fitting anesthetic mask should be placed over the mouth and nose and an oxygen flow rate of approximately 2 L/min maintained.[90] Routine monitoring should be applied to these patients including pulse oximetry, capnometry, electrocardiography, non-invasive blood pressure, and core body temperature. Fluids should also be administered during the anesthetic period; however, consider the volume of the TIVA infusion when calculating fluid rates. Small patients, geriatric patients, and those with cardiac dysfunction can be at increased risk of fluid overload.[64] Recoveries from TIVA can be prolonged due to drug accumulation, especially when long-acting drugs are used.[93] All patients should be monitored carefully throughout recovery until the patient is awake, aware, and can maintain thermoregulation.

PIVA in Canine and Feline Patients

Adding CRIs to an anesthetic plan can produce an anesthetic sparing effect, lowering the total amount of inhalant anesthetic agent needed to maintain an appropriate surgical plane. Adjunctive analgesic and supplementary drugs can produce a more hemodynamically stable anesthetic plane and decrease the negative effects associated with inhalant anesthetic agents, primarily hypotension.[88] Patients receiving a CRI have more constant levels of analgesia during the surgery and are potentially more comfortable post-operatively. Many of the drugs used in a CRI have sedative effects, which can lead to a smoother emergence from general anesthesia.[94]

Many classes of analgesic drugs can be given as a CRI including opioids, local anesthetic agents (specifically lidocaine), NMDA receptor antagonists (ketamine), and alpha-2 adrenergic agonists (dexmedetomidine). Loading dosages and CRI rates are shown in Table 14-3. Opioids such as morphine,

fentanyl, hydromorphone, and butorphanol are commonly used in CRIs. Morphine, fentanyl, and hydromorphone can be used for moderate to severe pain, while butorphanol is most appropriate for use in mildly painful cases. CRIs are delivered at a more consistent level than bolus administration, which may decrease some opioid adverse effects such as vomiting, dysphoria, and respiratory depression. Patients should still be monitored closely for any sign of distress.[95]

Lidocaine works by blocking sodium ion channels and causing membrane stabilization. It can reduce the amount of opioid analgesic and inhalant anesthetic agents required to maintain anesthesia. Lidocaine is relatively inexpensive and has anti-arrhythmic and anti-inflammatory properties. It may be useful in cases where gastrointestinal pain is involved (GDV, laparotomies, etc.). Cats have an increased sensitivity to local anesthetic agents, and it is currently not recommended to use lidocaine infusions on feline cases.[96] Bupivacaine is quite cardiotoxic when delivered intravenously and cannot be substituted for lidocaine in an infusion.[97]

Ketamine is a dissociative agent that works by antagonizing the NMDA receptors, which may be responsible for central sensitization, hypersensitization, and "wind-up" pain. Ketamine, especially when administered in combination with opioid analgesics, can significantly lower anesthetic agent requirements[98] and is particularly useful in chronic pain states.

Dexmedetomidine produces analgesia and has anxiolytic and sedative properties. Dexmedetomidine CRIs are most commonly used in the postoperative phase, for patients that are anxious and/or vocal despite an appropriate anesthetic regimen.[95] All alpha-2 adrenergic agonists have the potential to have significant cardiovascular effects such as bradycardia, vasoconstriction, and hypertension or even hypotension.[99] Thus, they are not recommended for prolonged sedation; however, they are useful to reduce anesthetic agent requirements.[100]

Combinations of the previously described drugs can provide excellent multimodal analgesia. In painful cases, combinations of opioid-lidocaine-ketamine infusions can provide a significant reduction in inhalant anesthetic agent requirements. Most commonly used combinations are fentanyl-lidocaine-ketamine (FLK) and morphine-lidocaine-ketamine (MLK) infusions, but hydromorphone or butorphanol can also be substituted.[95]

TIVA in Equine Patients

TIVA is the most common method of field anesthesia in equine patients and has many benefits over inhalant anesthesia. TIVA is relatively easy to use, is often less expensive as costly anesthetic equipment is not needed, can be used away from a hospital setting (e.g., field castrations, bandage changes), and typically results in a smoother recovery.[101] There are similar disadvantages as with small animal TIVA. There is the risk for prolonged recovery with extended procedures and drug accumulation, especially in neonates, geriatrics, and those with hepatic dysfunction. In addition, it may be difficult to provide supplemental oxygen when out in the field. TIVA may not achieve adequate muscle relaxation for all procedures. It is also more difficult to control anesthetic depth when using TIVA and appropriate monitoring is critical.

TIVA in horses can be divided into two groups: procedures lasting less than 20 minutes and those lasting 20 to 60 minutes. In either case, the induction protocol will remain the same. Horses should be sedated with an alpha-2 adrenergic agonist. For example, xylazine, detomidine, or romifidine IV can be used. Horses are properly sedated when the head is hanging, lips are drooping, and there is a wide stance with the front limbs. An appropriately sedated horse should not be responsive to external stimuli such as surrounding noises, movement of personnel, flicking of the horse's ears, or pulling on the animal's tongue/lips. The safety of personnel and patient are put at risk with improperly sedated animals; *never* give induction drugs to an unsedated horse.[102]

Equine induction is most commonly achieved using either a ketamine/benzodiazepine bolus or guaifenesin/ketamine bolus.[101] Induction techniques for horses are further discussed in Chapter 19 (Anesthesia and Analgesia for Large Animal and Food Producing Species). For procedures lasting less than 20 minutes, the induction protocol may provide adequate anesthetic duration. However, in some horses, the duration can be very short (5–15 minutes), and the anesthetist should be prepared to re-dose the patient if it reaches an inadequate anesthetic plane. One-third to one-half the original doses of ketamine and alpha-2 adrenergic agonist are generally enough to provide longer anesthetic duration, but with repeated re-dosing there is increased risk for prolonged recovery.

If the anesthetic period needs to be extended beyond 15 or 20 minutes, the patient can be put on a "triple drip" TIVA infusion, which can be used for procedures lasting up to 60 minutes.[86,103] Guaifenesin is used as a base with ketamine and a sedative (e.g., an alpha-2 adrenergic agonist) added to the solution. Guaifenesin is a centrally-acting agent that provides excellent muscle relaxation with minimal effects on respiratory rate and cardiac output.[104] The most common combinations are guaifenesin-ketamine-xylazine (GKX), guaifenesin-ketamine-detomidine (GKD), and guaifenesin-ketamine-romifidine (GKR).[103] Other triple drip combinations involve a base of 0.9% saline with midazolam-ketamine-xylazine (MKX).[104] TIVA drug combinations, amounts, and rates are listed in Table 14-4.

TABLE 14-4 Examples of TIVA Drug Dosages for Equine Anesthesia and Analgesia*

Drug	Amount Added	Concentration (mg/mL)	Rate (mL/kg/hr)
guaifenesin (5%)	500 mL	50	1.0–2.0
ketamine	500 mg	1	
xylazine	250 mg	0.5	
guaifenesin (5%)	500 mL	50	1.2–1.6
ketamine	500 mg	1	
detomidine	10 mg	0.020	
guaifenesin (5%)	500 mL	50	1.0
ketamine	3.3 g	6.6	
romifidine	40 mg	0.08	
0.5% saline	500 mL	5	1.6
midazolam	25 mg	0.5	
ketamine	650 mg	1.3	
xylazine	325 mg	0.65	

These are just commonly-used examples; all drug doses and additive amounts may be adjusted, depending on individual patient needs and procedural requirements.

* Lerche, P. (2013). Total intravenous anesthesia in horses. *Vet Clin Equine, 29*, 123–129; Muir, W. W., & Hubbell, J. A. (2009). *Equine anesthesia* (2nd ed., pp. 260–276). St Louis, MO: Saunders Elsevier; Hubbell, J. A., Aarnes, T. K., Lerche, P., et al. (2012). Evaluation of a midazolam-ketamine-xylazine infusion for total intravenous anesthesia in horses. *Am J Vet Res, 73*(4), 470–475.

Anesthetic Management in Equine Patients During TIVA

Many TIVA procedures in equine patients will occur in the field. In this situation, monitoring will be primarily hands-on but is nonetheless extremely important. Heart rate, respiratory rate and character, and mucous membrane color should be closely observed. Reflexes are usually maintained in protocols involving ketamine, so patients maintain the palpebral response, tearing, and blinking in response to lid palpation. Very light anesthetic planes include spontaneous blinking, nystagmus, increased respiratory rate, trembling, increased lacrimation (eye is wet), or tensing of the neck. Signs of a deep stage of anesthesia include loss of a palpebral reflex, decreased lacrimation (eye is dry), decrease in respiratory rate or apnea, and weakened pulses.[86] Recovery from TIVA procedures tends to be stable and smooth. However, in cases where the procedure lasts longer than 60 minutes, recovery may be slow due to drug accumulation. Horses should continue to be monitored for as long as it is safe for the patient and anesthetist.[86]

PIVA in Equine Patients

As with small animal patients, horses can require anesthesia for prolonged periods. Surgical colic cases, fracture repair, airway cases, etc., require extended anesthesia. As in small animal cases, the inhalant anesthetic agent will cause significant cardiovascular depression, more greatly seen in critical anesthetic cases. Adding analgesic PIVA protocols to the anesthetic plan can provide better analgesic management, decrease anesthetic agent requirements, and provide more stable anesthesia. Ketamine, lidocaine, alpha-2 adrenergic agonists, and opioids are anesthetic agents commonly added to anesthetic and analgesic plans.[86,101,105] Table 14-5 contains drugs, loading dosages, and CRI rates for these protocols.

TABLE 14-5 PIVA Drugs, Loading Dosages and CRI Ranges for Analgesia in Equine Patients*

Drug	IV Loading Dosage (mg/kg)	CRI Dosage (mg/kg/hr)	Notes
ketamine	0.3–3	1.0–3.6	Ketamine induction agent can serve as loading dosage if CRI is begun immediately following
lidocaine	1.5–5	1.5–6.0	Discontinue 30 minutes before turning off inhalant anesthetic agent
butorphanol	0.025	0.025	Well sedated or anesthetized patient
morphine	0.05–0.15	0.03–0.1	Well sedated or anesthetized patient

These are just commonly-used examples; all drug doses may be adjusted, depending on individual patient needs and procedural requirements.

** Lerche, P. (2013). Total intravenous anesthesia in horses. *Vet Clin Equine, 29*, 123–129; Muir, W. W., & Hubbell, J. A. (2009). *Equine anesthesia* (2nd ed., pp. 260–276). St Louis, MO: Saunders Elsevier; Hubbell, J. A., Aarnes, T. K., Lerche, P., et al. (2012). Evaluation of a midazolam-ketamine-xylazine infusion for total intravenous anesthesia in horses. *Am J Vet Res, 73*(4), 470–475.

Low-dosage ketamine CRIs produce moderate analgesia and lower required inhalant anesthetic agent levels. They preserve cardiac function and generally do not produce negative adverse effects that can be associated with larger anesthetic induction agent dosages such as tachycardia/hypertension and muscle rigidity.[106, 107] However, it is not recommended to give large boluses immediately before discontinuing inhalant anesthesia and moving to recovery. If a ketamine bolus must be given shortly before anticipated recovery, an alpha-2 adrenergic agonist should be co-administered to provide some sedation.[101]

Lidocaine is an inexpensive and effective adjunctive analgesic in equine protocols. Lidocaine has analgesic and anti-inflammatory effects, with little to no negative effect on respiratory or cardiovascular function and may have positive effects in cases of equine ileus.[108,109] However, lidocaine infusions may result in ataxia and tremors and should be turned off approximately 30 minutes before the inhalant anesthetic agent prior to recovery.[110,111]

Morphine and butorphanol are the opioids most commonly used in equine anesthesia; however, they are used less often than other adjunctive groups and significantly less than in small animal anesthesia. Horses may have an excitatory reaction to mu-opioid analgesics, thus it is important to administer opioids slowly to either well-sedated or anesthetized patients.[101, 104, 112]

Tech Tip 🐾

Animals with chronic disease typically receive long-term medication to manage their condition. It is important to discuss with clients whether the patient's medication should be discontinued or given on the day of anesthesia. Table 14–6 summarizes select disease conditions and whether medication should be given to the patient on the day of anesthesia.

TABLE 14-6 Recommendations for Chronic Medications and Anesthesia*

Disease Condition/Medication	Recommendation
Hypothyroidism/thyroid supplement	Continue as scheduled
Hyperthyroidism/methimazole	Continue as scheduled
Behavioral conditions/behavior modifying drugs	Continue as scheduled
Painful conditions/analgesics	Continue as scheduled
Cardiac conditions/pimobendan, furosemide	Continue as scheduled
Bacterial infections/antibiotics	Continue as scheduled
Pruritus, autoimmune conditions/glucocorticoids	Continue as scheduled; should not discontinue abruptly
Hypertension/antihypertensive medications (especially ACE inhibitors such as enalapril and benazepril)	Discontinue day of anesthesia
Bleeding conditions/anticoagulants	May need to be discontinued 2 weeks before anesthesia to decrease bleeding risk
Diabetes mellitus/insulin	Because of fasting, do not give full dose on day of anesthesia due to risk of hypoglycemia

*Adapted from Tamara Grubb, Jennifer Sager, James S. Gaynor, Elizabeth Montgomery, Judith A. Parker, Heidi Shafford, and Caitlin Tearney. (2020). 2020 AAHA anesthesia and monitoring guidelines for dogs and cats. *Journal of the American Animal Hospital Association, 56*(2), pp. 59–82.

SUMMARY

Many veterinary patients present for anesthesia with unique and specific physiologic alterations. With proper preanesthetic evaluation, a safe and effective anesthetic plan can be formulated based on each patient's individual needs. Oftentimes, the analgesia or anesthetic plan incorporates CRIs or even TIVA to smooth the anesthesia period, provide steady levels of drugs, and reduce the amount of other more depressant agents. Once the precise anesthetic agents are chosen for premedication, induction, and maintenance, proper anesthetic monitoring techniques throughout the procedure will enable the anesthetist to assess whether any interventions should be made to ensure proper patient physiology. If diligent patient monitoring is present during the entire procedure, anesthesia can be performed with minimal risks to the patient.

CRITICAL THINKING POINTS

- Each anesthetic plan should be tailored to the individual patient, and concurrent disease states should be carefully evaluated.

- When in doubt, use "organ-friendly" drugs that have minimal cardiovascular depressant effects and tend to maintain adequate blood pressure and organ blood flow; multimodal anesthetic techniques work best to accomplish this goal.

- There are many advantages to injectable anesthetic techniques, including potential for less cardiovascular depression than with inhalant anesthetic agents (e.g., hypotension), no exposure of patient or personnel to waste anesthetic gases, and smoother induction and recovery periods.

- Total intravenous anesthesia (TIVA) may be beneficial for cases where inhalant anesthesia is not an option (tracheal procedures,

- laser surgeries in or near the mouth and airway) or for long-term mechanical ventilation.

- Constant rate infusions (CRIs) provide a steady infusion of a drug to provide stable levels of analgesia.

- CRIs decrease the amount of overall drugs used, decrease severity of adverse effects, and provide better analgesia and safer anesthesia.

REVIEW QUESTIONS

Multiple Choice

1. Which anesthetic/analgesic agent is unique since it is mainly eliminated through the lung?

   a. Propofol
   b. Isoflurane
   c. Acepromazine
   d. Hydromorphone

2. Which group of drugs is *not* recommended for use in patients with cardiovascular disease due to their depressant effects?

   a. Alpha-2 adrenergic agonists
   b. Benzodiazepines
   c. Phenothiazines
   d. Opioids

3. Which of the following is *not* associated with the use of ketamine?

   a. Increased cardiac contractility
   b. Decreased intracranial pressure
   c. Increased intraocular pressure
   d. Increased cardiac work

4. What is the term for increased sympathetic discharge to the peripheral vasculature resulting in a reflex bradycardia commonly seen in patients with increased intracranial pressure?

   a. Addison's symptom
   b. Cushings response
   c. Brahman's sign
   d. Oculocardiac reflex

5. Which of the following is correct regarding anesthetic agents in patients with neurologic disease?

   a. Below MAC, inhalants minimally affect cerebral blood flow.
   b. Acepromazine should never be administered to a patient with seizures.
   c. Patients should be allowed to become hypercapnic to maintain cerebral blood flow.
   d. Opioids should be withheld from neurologic patients due to increases in intracranial pressures.

6. Which of the following increases gastric pH and should be administered to patients with GI disease to reduce damage due to and prevent the incidence of gastric reflux?

   a. Omeprazole
   b. Maropitant
   c. Midazolam
   d. Methadone

7. Which class of anesthetic/analgesic agents should not be administered to patients with GI disease?

   a. Opioids
   b. NSAIDs
   c. Benzodiazepines
   d. Inhalants

8. Which drug is highly excreted unchanged by the kidneys, especially in cats, and may have prolonged activity in patients with renal disease?

   a. Propofol
   b. Acepromazine
   c. Butorphanol
   d. Ketamine

9. Which induction agent is not recommended for use in patients with intraocular disease since it can produce myoclonus?

   a. Propofol
   b. Ketamine
   c. Alfaxalone
   d. Etomidate

10. Which metabolic derangement is associated with the use of carbonic anhydrase inhibitors in ophthalmic patients?

    a. Metabolic acidosis
    b. Metabolic alkalosis
    c. Respiratory acidosis
    d. Respiratory alkalosis

11. Which of the following is a physiologic change seen in pregnant patients?

    a. Decreased blood volume
    b. Decreased oxygen requirements
    c. Decreased packed cell volume
    d. Decreased cardiac output

12. Cardiac output in neonatal and pediatric patients is largely driven by which factor?

    a. Breed
    b. Increased oxygen demand
    c. Hepatic function
    d. Heart rate

13. Neonatal and pediatric patients are at increased risk for hypoglycemia due to the immaturity of which organ system?

    a. Cardiovascular
    b. Respiratory
    c. Hepatic
    d. Urogenital

14. Neonatal and pediatric patients are more likely to suffer respiratory fatigue due to which factor?

    a. Weak thoracic muscles
    b. Low respiratory rate
    c. Noncompliant ribcage
    d. Low oxygen consumption rate

15. Geriatric patients rely on which of the following to maintain blood pressure?

    a. Afterload
    b. Preload
    c. Hypovolemia
    d. Myocardial fibrosis

16. Which drug is usually avoided due to its long duration of action in patients with endocrine disorders?

    a. Midazolam
    b. Propofol
    c. Oxymorphone
    d. Acepromazine

17. Which of the following is *not* a true statement regarding the respiratory system of a geriatric patient?

    a. Lung tissue has decreased elasticity.
    b. The ribcage may be less compliant than in younger patients.
    c. Patients are likely to have pulmonary fibrosis.
    d. Patients often have increased tidal volume.

18. Recovery from total intravenous anesthesia (TIVA) may be prolonged due to which factor?

    a. Drug accumulation
    b. Decreased cardiovascular depression
    c. Increased exposure to waste gases
    d. Decreased oxygen consumption

19. Which of the following is *not* a situation where TIVA would be considered appropriate?

    a. Canine tracheal surgery
    b. Canine long-term ventilator management
    c. Equine colic surgery
    d. Equine bandage change

20. Which of the follow drugs should *not* be used as a CRI in feline patients?

    a. Fentanyl
    b. Lidocaine
    c. Ketamine
    d. Butorphanol

Case Studies

Case Study 1: A 14-year-old, female spayed Labrador Retriever dog named Beau presents for removal of a large lipoma on the right thorax. She is slightly overweight at 40 kg and on warm summer days, the owner reports that she has some difficulty with breathing as she makes loud inspiratory noises, pants, and uses her abdominal muscles to inspire. She has been previously diagnosed with laryngeal paralysis. The owner also reports that the mass is large enough that movement of her front right limb is impaired. Her physical exam reveals a heart rate of 160 beats per minute and she is panting with inspiratory stridor. Her body temperature is 103°F but no other physical exam findings were abnormal. The surgery is assumed to be quite painful since the mass is large.

1. What diagnostic testing, if any, would be warranted at this time?

2. What special considerations should you give this patient with respiratory disease that requires general anesthesia for a mass removal prior to administering general anesthesia?

3. What premedications would you choose prior to surgery (drug, route, effect)?

4. How would you induce this patient to unconsciousness and provide maintenance anesthesia to the patient (drugs, analgesia, monitors)?

5. What specific techniques will be required upon anesthetic recovery?

Case Study 2: A 7-year-old, male neutered Schnauzer dog named Moses presents for consultation prior to routine dentistry. One year ago, he was diagnosed with diabetes mellitus. He has been doing well at home with NPH insulin injections twice a day (5 units, SQ). The clients report that he does urinate in the house at times and has infrequent diarrhea. He is 10 kg and is in good body condition. His physical exam is mostly unremarkable with a heart rate of 130 beats per minute, respiratory rate of 30 breaths per minute, and thoracic auscultation within normal limits. His hair coat is slightly thin. He is scheduled for a dentistry tomorrow morning.

1. What directions will you give to the client regarding preanesthetic fasting and insulin administration?

2. What additional diagnostic testing will you perform prior to anesthesia?

3. What premedications, induction, and maintenance anesthetic agents will you choose (drug, route, effects) and why?

4. What special anesthetic techniques will you employ throughout the anesthetic period for this diabetic patient?

5. What are some specific considerations you have for this patient in recovery and while at home following the procedure?

Case Study 3: A 3-year-old, female intact Boston Terrier dog named Babe presents for emergency Cesarean section. She is 11 kg and 68 days pregnant. She delivered one stillborn pup and continued to labor for 18 hours without producing any additional pups. The owner reports that there were at least four pups seen on radiograph on day 65, performed at the referring veterinary hospital. Her physical exam reveals a heart rate of 180 beats per minute, and she is panting heavily and with some increased effort. Her rectal temperature is 104.5°F, and she is pacing, unable to settle, and seems very uncomfortable. It is deemed that a Cesarean section is needed to save the bitch and determine if any other pups are viable. It is unknown if the patient has been fasted.

1. What diagnostic testing and blood work, if any, would be recommended at this time?

2. What premedications would you choose prior to surgery (drug, route, effect)?

3. What drugs would you use postdelivery (drug, route, effect)?

4. What special considerations should you consider regarding the pups (effects of drugs, postdelivery care)?

5. What considerations would you have for this patient regarding recovery (extubation time, brachycephalic syndrome, analgesia)?

Case Study 4: A 12-year-old, male neutered German Shepherd dog named Hank presents in severe respiratory distress. He is extremely over conditioned at 52 kg. Hank presents with muddy mucous membrane color, heart rate of 170 beats per minute, panting heavily, and a rectal body temperature of 106°F. Before further examination can be completed, the patient collapses and is unconscious. A catheter is quickly placed, propofol given, and the patient intubated with a size 14 mm ETT. Hank was placed on a mechanical ventilator in the critical care unit. The patient struggles against the ventilator but requires intubation and ventilation to keep his end-tidal carbon dioxide levels within normal limits and his hemoglobin saturation above 90%.

1. What drugs/techniques are available to manage this patient on a mechanical ventilator (drugs, route, effects)?

2. What special considerations should you give this patient given the nature of his cardiovascular status, hyperthermia, and extreme obesity?

3. What concerns do you have regarding maintaining an extended anesthetic period?

4. What monitors would you choose for this patient? Why?

5. What considerations and monitoring techniques will be required upon recovery?

Case Study 5: A 3-year-old, male intact quarter horse named Lil Mockingbird presents for routine castration and removal of two wolf teeth. He is 347 kg and has a normal physical exam with heart rate of 37 beats per minute, respiratory rate of 12 breaths per minute, and a rectal temperature of 99.2°F. Mockingbird is feisty but friendly. The patient has been fasted for 4 to 6 hours and surgery is expected to last 30 to 45 minutes.

1. What diagnostic testing and blood work, if any, would be recommended at this time?

2. What premedications would you choose prior to surgery (drug, route, effect)?

3. What type of TIVA protocol would you choose (drug combinations, expected duration)?

4. What special preparations would you have in the event the surgery takes longer than 45 minutes?

5. What considerations would you have for this patient recovery (monitoring, sedation)?

Critical Thinking Questions

1. When presented with a patient with decreased cardiac contractility (e.g., dilated cardiomyopathy) that is predisposed to fluid overload, what type, route, and rate of fluid administration would you choose?
   a. Fluid type:
   b. Route of administration:
   c. Rate of administration:

2. List four monitors you would use during anesthesia of a cat with chronic renal failure for removal of a gastric foreign body. What information would you expect to get from each monitor?

Monitor	Information
a. _____	_____
b. _____	_____
c. _____	_____
d. _____	_____

ENDNOTES

1. Congdon, J. M. (2014). Cardiovascular disease. In L. B. C. Snyder, & R. A. Johnson (Eds.), *Canine and feline anesthesia and co-existing disease* (pp. 1–54). Hoboken, NJ: Wiley.

2. Brunson, D. B., & Johnson, R. A. (2014). Respiratory disease. In L. B. C. Snyder, & R. A. Johnson (Eds.), *Canine and feline anesthesia and co-existing disease* (pp. 55–70). Hoboken, NJ: Wiley.

3. McNally, E. M., Robertson, S. A., & Pablo, L. S. (2009). Comparison of time to desaturation between preoxygenated and non-preoxygenated dogs following sedation with acepromazine maleate and morphine and induction of anesthesia with propofol. *American Journal of Veterinary Research, 70*(11), 1333–1338.

4. Congdon, J. M., Marquez, M., Niyom, S., et al. (2011). Evaluation of the sedative and cardiovascular effects of intramuscular administration of dexmedetomidine with and without concurrent atropine administration in dogs. *Journal of the American Veterinary Medical Association, 239*(1), 81–9.

5. Murison, P. J. (2001). Effect of propofol at two injection rates or thiopentone on post-intubation apnoea in the dog. *Journal of Small Animal Practice, 42*(2), 71–74.

6. Rodríguez, J. M., Muñoz-Rascón, P., Navarrete-Calvo, R., et al. (2012). Comparison of the cardiopulmonary parameters after induction of anaesthesia with alphaxalone or etomidate in dogs. *Veterinary Anaesthesia and Analgesia, 39*(4), 357–365.

7. Wendt-Hornickle, E. (2014). Neurologic disease. In L. B. C. Snyder, & R. A. Johnson (Eds.), *Canine and feline anesthesia and co-existing disease* (pp. 71–81). Hoboken, NJ: Wiley.

8. Syring, R. S., Otto, C. M., & Drobatz, K. J. (2001). Hyperglycemia in dogs and cats with head trauma: 122 cases (1997–1999). *Journal of the American Veterinary Medical Association, 218*(7), 1124–1129.

9. Fodstad, H., Kelly, P. J., & Buchfelder, M. (2006). History of the cushing reflex. *Neurosurgery, 59*(5), 1132–1137.

10. Curley, G., Kavanagh, B. P., & Laffey, J. G. (2010). Hypocapnia and the injured brain: More harm than benefit. *Crit Care Med, 38*(5), 1348–1359.

11. Cottenceau, V., Masson, F., Mahamid, E., et al. (2011). Comparison of effects of equiosmolar doses of mannitol and hypertonic saline on cerebral blood flow and metabolism in traumatic brain injury. *J Neurotrauma, 28*(10), 2003–2012.

12. Tobias, K. M., Marioni-Henry, K., & Wagner, R. (2006). A retrospective study on the use of acepromazine maleate in dogs with seizures. *J Am Anim Hosp Assoc, 42*(4), 283–289.

13. Harvey, R., Greene, S., & Thomas, W. (2007), Neurologic disease. In W. J. Tranquilli, J. C. Thurmon, & K. A. Grimm (Eds.), *Lumb & Jones' veterinary anesthesia and analgesia* (4th ed., pp. 903–913). Ames, IA: Blackwell Publishing: Ames.

14. Chang, L. C., Raty, S. R., Ortiz, J., et al. (2013). The emerging use of ketamine for anesthesia and sedation in traumatic brain injuries. *CNS Neurosci Ther, 19*(6), 390–395.

15. Martin-Flores, M., Scrivani, P. V., Loew, E., et al. (2014). Maximal and submaximal mouth opening with mouth gags in cats: implications for maxillary artery blood flow. *Vet J, 200*(1), 60–64.

16. Figueiredo, J. P., & Green, T. A. (2014). Gastrointestinal disease. In L. B. C. Snyder, & R. A. Johnson (Eds.), *Canine and feline anesthesia and co-existing disease* (pp. 93–115). Hoboken, NJ: Wiley.

17. Hay Krause, B. L. (2013). Efficacy of maropitant in preventing vomiting in dogs premedicated with hydromorphone. *Vet Anaesth Analg, 40*(1), 28–34.

18. Bennett, R. (2007). Gastrointestinal and hepatic disease. In C. Seymour, & T. Duke-Novakovski (Eds.), *BSAVA manual of canine and feline anaesthesia and analgesia* (2nd ed., pp. 244–256). Gloucester: British Small Animal Veterinary Association.

19. Hay Krause, B. L. (2014). Effect of dosing interval on efficacy of maropitant for prevention of hydromorphone-induced vomiting and signs of nausea in dogs. *J Am Vet Med Assoc, 245*(9), 1015–1020.

20. Strombeck, D. R., & Harrold, D. (1985). Effects of atropine, acepromazine, meperidine, and xylazine on gastroesophageal sphincter pressure in the dog. *Am J Vet Res, 46*(4), 963–965.

21. Wilson, D. V., Boruta, D. T., & Evans, A. T. (2006). Influence of halothane, isoflurane, and sevoflurane on gastroesophageal reflux during anesthesia in dogs. *Am J Vet Res, 67*(11) 1821–1825.

22. Johnson, R. A. (2014). Maropitant prevented vomiting but not gastroesophageal reflux in anesthetized dogs premedicated with acepromazine-hydromorphone. *Vet Anaesth Analg, 41*(4), 406–410.

23. Wilson, D. V., Evans, A. T., Mauer, W. A. (2006). Influence of metoclopramide on gastroesophageal reflux in anesthetized dogs. *Am J Vet Res, 67*(1), 26–31.

24. Favarato, E. S., Souza, M. V., Costa, P. R., et al. (2012). Evaluation of metoclopramide and ranitidine on the prevention of gastroesophageal reflux episodes in anesthetized dogs. *Res Vet Sci, 93*(1), 466–467.

25. Panti, A., Bennett, R. C., Corletto, F., et al. (2009). The effect of omeprazole on oesophageal pH in dogs during anaesthesia *Small Anim Pract, 50*(10), 540–544.

26. Monteiro-Steagall, B. P., Steagall, P. V., & Lascelles, B. D. (2013). Systematic review of nonsteroidal anti-inflammatory drug-induced adverse effects in dogs. *J Vet Intern Med, 27*(5), 1011–1019.

27. Schroeder, C. A. (2014). Renal disease. In L. B. C. Snyder, & R. A. Johnson (Eds.), *Canine and feline anesthesia and co-existing disease* (pp. 116–128). Hoboken, NJ: Wiley.

28. Schroeder, C. A. (2014). Hepatobiliary disease. In L. B. C. Snyder, & R. A. Johnson (Eds.), *Canine and feline anesthesia and co-existing disease* (pp. 82–92). Hoboken, NJ: Wiley.

29. Stoelting, R. K., & Hillier, S. C. (2006). Liver and gastrointestinal tract. In *Pharmacology & physiology in anesthetic practice* (4th ed., pp. 831–843). Philadelphia, PA: Lippincott Williams & Wilkins.

30. Hanna, R. M., Borchard, R. E., & Schmidt, S. L. (1988). Pharmacokinetics of ketamine HCl and metabolite I in the cat: A comparison of I.V., I.M., and rectal administration. *J Vet Pharmacol Ther, 11*(1), 84–93.

31. Fischer, B. L. (2014). Endocrine disease. In L. B. C. Snyder, & R. A. Johnson (Eds.), *Canine and feline anesthesia and co-existing disease* (pp. 151–174). Hoboken, NJ: Wiley.

32. Rand, J. S., & Marshall, R. D. (2005). Diabetes mellitus in cats. *Vet Clin North Am Small Anim Pract, 35*(1), 211–224.

33. Kohl, B. A., & Schwartz, S. (2009). Surgery in the patient with endocrine dysfunction. *Anesthesiol Clin, 27*(4), 687–703.

34. Kadoi, Y. (2010). Anesthetic considerations in diabetic patients. Part II: intraoperative and postoperative management of patients with diabetes mellitus. *J Anesth, 24*(5), 748–756.

35. Robertshaw, H. J., & Hall, G. M. (2006). Diabetes mellitus: Anesthetic management. *Anaesthesia, 61*(12), 1187–1190.

36. Restitutti, F., Raekallio, M., Vainionpää, M., et al. (2012). Plasma glucose, insulin, free fatty acids, lactate and cortisol concentrations in dexmedetomidine-sedated dogs with or without MK-467: A peripheral α-2 adrenoceptor antagonist. *Vet J, 193*(2), 481–485.

37. Reusch, C. E., Schellenberg, S., & Wenger, M. (2010). Endocrine hypertension in small animals. *Vet Clin North Am Small Anim Pract, 40*(2), 335–323.

38. Lerche, P. (2014). Ophthalmic disease. In L. B. C. Snyder, & R. A. Johnson (Eds.), *Canine and feline anesthesia and co-existing disease* (pp. 179–186). Hoboken, NJ: Wiley.

39. Feldman, A. F., & Patel A. (2009). Anesthesia for eye, ear, nose and throat surgery. In R. D. Miller (Ed.), *Miller's anesthesia* (7th ed., pp. 2378–2385). Ames, IA: Blackwell.

40. Herring, I. P., Pickett, J. R., Champagne, E. S., et al. (2000). Evaluation of aqueous tear production in dogs following general anesthesia. *J Am Anim Hosp Assoc, 36*(5), 427–430.

41. Ghaffari, M. S., Malmasi, A., & Bokaie, S. (2010). Effect of acepromazine or xylazine on tear production as measured by Schirmer tear test in normal cats. *Vet Ophthalmol, 13*(1), 1–3.

42. Mouney, M. C., Accola, P. J., Cremer, J., et al. (2011). Effects of acepromazine maleate or morphine on tear production before, during, and after sevoflurane anesthesia in dogs. *Am J Vet Res, 72*(11), 1427–1430.

43. Gross, M. E., & Giuliano, E. A. (2007). Ocular patients. In W. J. Tranquilli, J. C. Thurmon, & K. A. Grimm (Eds.), *Lumb & Jones' veterinary anesthesia and analgesia* (4th ed., pp. 943–954). Ames, IA: Blackwell Publishing: Ames.

44. Blanc, V. F., Hardy, J. F., Millot, J., et al. (1983). The oculocardiac reflex, a graphic and statistical analysis in infants and children. *Can Anaesth Soc J, 30*, 360–369.

45. Ward, R. (1989). Maternal placental-fetal unit: Unique problems of pharmacologic study. *Pediatr Clin North Am, 36*, 1075–1088.

46. Raffe, M. R., & Carpenter, R. E. (2007). Anesthetic Management of Cesarean Section Patients. In W. J. Tranquilli, J. C. Thurmon, & K. A. Grimm (Eds.), *Lumb & Jones' veterinary anesthesia and analgesia* (4th ed., pp. 955–967). Ames, IA: Blackwell Publishing: Ames.

47. Moon, P. F., Erb, H. N., Ludders, J. W., et al. (1998). Perioperative management and mortality rates of dogs undergoing Cesarean section in the United States and Canada. *J Am Vet Med Assoc, 213*, 365–369.

48. Moon, P. F., Erb, H. N., Ludders, J. W., et al. (2000). Perioperative risk factors for puppies delivered by Cesarean section in the United States and Canada. *J Am Anim Hosp Assoc, 36*, 359–368.

49. Camann, W., & Ostheimer, G. (1990). Physiological adaptations during pregnancy. *Intern Anesthesiol Clin, 28*, 2–10.

50. Kaneko, M., Nakayama, H., Igarashi, N., et al. (1993). Relationship between the number of fetuses and the blood constituents of beagles in late pregnancy. *J Vet Med Sci, 55*, 681–682.

51. Brooks, V., & Keil, L. (1994). Hemorrhage decreases arterial pressure sooner in pregnant compared with nonpregnant dogs: role of baroreflex. *Am J Physiol, 266*, H1610–1619.

52. Hollinshead, F., Hanlon, D., Gilbert, R., et al. (2010). Calcium, parathyroid hormone, oxytocin, and pH profiles in the whelping bitch. *Theriogenology, 73*, 1276–1283.

53. Linden, A., Erikson, M., Carlquist, M., et al. (1987). Plasma levels of gastrin, somatostatin, and cholecystokinin immunoreactivity during pregnancy and lactation in dogs. *Gastroenterology, 92*, 578–584.

54. Paddleford, R. R. (1992). Anesthesia for Cesarean section in the dog. *Vet Clin North Am Small Anim Pract, 22*, 481–484.

55. Luna, S. P. L., Cassu, R. D., Castro, G. B., et al. (2004). Effects of four anaesthetic protocols on the neurological and cardiorespiratory variables of puppies born by Cesarean section. *Vet Rec, 154*, 387–389.

56. Loftus, J. R., Hill, H., & Cohen, S. (1995). Placental transfer and neonatal effects of epidural sufentanil and fentanyl administered with bupivacaine during labor. *Anesthesiology, 83*, 300–308.

57. Matthews, K. A. (2008). Pain management for the pregnant, lactating, and neonatal to pediatric cat and dog. *Vet Clin North Am Small Anim Pract, 38*, 1291–308.

58. Funkquist, P. M. E., Nyman, G. C., Lofgren, et al. (1997). Use of propofol-isoflurane as an anesthetic regimen for Cesarean section in dogs. *J Am Vet Med Assoc, 211*, 313–317.

59. Andaluz, A., Tusell, J., Trasserres, O., et al. (2003). Transplacental transfer of propofol in pregnant ewes. *Vet J, 166*, 198–204.

60. Doebeli, A., Michel, E., Bettschart, R., et al. (2013). Apgar score after induction of anesthesia for canine Cesarean section with alfaxalone versus propofol. *Theriogenology, 80*, 850–854.

61. Gin, T., & Tan, M. (1994). Decreased minimum alveolar concentration of isoflurane in pregnant humans. *Anesthesiology, 81*, 829–832.

62. Pascoe, P. J., & Moon, P. F. (2001). Periparturient and neonatal anesthesia. *Vet Clin North Am Small Anim Pract, 31*, 315–340.

63. Egger, C., & Love, L. (2009). Local and regional anesthesia techniques, Part 4: Epidural anesthesia and analgesia. *Vet Med, 104*(10), 460–466.

64. Pettifer, G., & Grubb, T. (2007). Neonatal and geriatric patients. In W. J. Tranquilli, J. C. Thurmon, & K. A. Grimm (Eds.), *Lumb & Jones' veterinary anesthesia and analgesia* (4th ed., pp. 985–991). Ames, IA: Blackwell Publishing: Ames.

65. Vogt, A. H., Rodan, I., Brown, M., et al. (2010). AAFP-AAHA: Feline life stage guidelines. *J Am Anim Hosp Assoc, 46*, 70–85.

66. Bartges, J., Boynton, B., Vogt, A. H., et al. (2012). AAHA: Canine life stage guidelines. *J Am Anim Hosp Assoc, 48*, 1–11.

67. Zoetis, T., & Hurtt, M. E. (2003). Species comparison of lung development. *Birth Defects Res B Dev Reprod Toxicol, 68*, 121–124.

68. Adelman, R. D., & Wright J. (1985). Systolic blood pressure and heart rate in the growing beagle puppy. *Dev Pharmacol Ther, 8*, 396–401.

69. Magrini, F. (1978). Hemodynamic determinants of the arterial blood pressure rise during growth in conscious puppies. *Cardiovascular Res, 12*, 422–428.

70. Rankin, D. C. (2002). Neonatal anesthesia. In S. A. Greene (Ed.), *Veterinary anesthesia and pain management secrets* (1st ed., pp. 211–214). Philadelphia, PA: Hanley and Belfus.

71. MacDonald, K. A. (2006). Congenital heart diseases of puppies and kittens. *Vet Clin North Am Small Anim Pract, 36*, 503–531.

72. Haddad, G. G., & Mellins, R. B. (1984). Hypoxia and respiratory control in early life. *Annu Rev Physiol, 46*, 629–643.

73. Tavoloni, N. (1985). Postnatal changes in hepatic microsomal enzyme activities in the puppy. *Neonatology, 47*, 305–316.

74. Zoetis, T., & Hurtt, M. E. (2003). Species comparison of anatomical and functional renal development. *Birth Defects Res B Dev Reprod Toxicol, 68*, 111–120.

75. Hosgood, G. (1992). Surgical and anesthetic management of puppies and kittens. *Compend Contin Educ Pract Vet, 14*, 345–357.

76. Luks, A. M., Zwass, M. S., Brown, R. C., et al. (1998). Opioid-induced analgesia in neonatal dogs: pharmacodynamic differences between morphine and fentanyl. *J Pharmacol Exp Ther, 284*, 136–141.

77. Ko, J. C. H., & Galloway, D. S. (2002). Anesthesia of geriatric patients. In S. A. Greene (Ed.), *Veterinary anesthesia and pain management secrets* (1st ed., pp. 215–227). Philadelphia, PA: Hanley and Belfus.

78. Saunders, A. B. (2012). The diagnosis and management of age-related veterinary cardiovascular disease. *Vet Clin North Am Small Anim Pract, 42*, 655–668.

79. Wei, J. Y. (1992). Age and the cardiovascular system. *N Engl J Med, 327*, 1735–1739.

80. MacDougall, D. F., & Barker, J. (1984). An approach to canine geriatrics. *Br Vet J, 140*, 115–123.

81. Robinson, N. E., & Gillespie, J. R. (1975). Pulmonary diffusing capacity and capillary blood volume in aging dogs. *J Appl Physiol, 38*, 647–650.

82. Mosier, J. E. (1989). Effect of aging on body systems of the dog. *Vet Clin North Am Small Anim Pract, 19*, 1–12.

83. Bartges, J. W. (2012). Chronic kidney disease in dogs and cats. *Vet Clin North Am Small Anim Pract, 42*, 669–692.

84. Carpenter, R. E., Pettifer, G. R., & Tranquilli, W. J. (2005). Anesthesia for geriatric patients. *Vet Clin North Am Small Anim Pract, 35*, 571–580.

85. Pablo, L. S. (2011). Total IV anesthesia. In *ACVS 2011 Symposium proceedings* (452–456). Paper presented at ACVS 2011 Symposium: The Surgical Summit, Chicago.

86. Lerche, P. (2013). Total intravenous anesthesia in horses. *Vet Clin Equine, 29*, 123–129.

87. Tsai, Y. C., Wang, L. Y., & Yeh, L. S. (2007). Clinical comparison of recovery from total intravenous anesthesia with propofol and inhalation anesthesia with isoflurane in dogs. *J Vet Med Sci, 69*(11), 1179–1182.

88. Ilkiw, J. E. (1999). Balanced anesthetic techniques in dogs and cats. *Clin Tech Small Anim Pract, 14*, 27–37.

89. Lucas, A. N., Firth, A. M., Anderson, G. A., et al. (2001). Comparison of the effects of morphine administered by constant-rate intravenous infusion or intermittent intramuscular injection in dogs. *J Am Vet Med Assoc, 218*(6), 884–891.

90. Mendes, G. M., & Selm, A. L. (2003). Use of a combination of propofol and fentanyl, alfentanil, or sufentanil for total intravenous anesthesia in cat. *J Am Vet Med Assoc, 223*, 1608–1613.

91. Granados, M. M., Dominguez, J. M., Fernandez-Sarmiento, A., et al. (2012). Anaesthetic and cardiorespiratory effects of a constant-rate infusion of alfaxalone desflurane-anaesthetised sheep. *Vet Rec, 171*(5), 125.

92. Herbert, G. L., Bowlt, K. L., Ford-Fennah, V., et al. (2013). Alfaxalone for total intravenous anaesthesia in dogs undergoing ovariohysterectomy: a comparison of premedication with acepromazine or dexmedetomidine. *Vet Anaesth Analg, 40*(2), 124–133.

93. Mama, K. R., Wagner, A. E., Steffey, E. P., et al. (2005). Evaluation of xylazine and ketamine for total intravenous anesthesia in horses. *Am J Vet Res, 66*(6), 1002–1007.

94. Liehmann, L., Mosing, M., & Auer, U. (2006). A comparison of cardiorespiratory variables during isoflurane-fentanyl and propofol-fentanyl anaesthesia for surgery in injured cats. *Vet Anaesth Analg, 33*, 158–168.

95. Duke, T. (2013). Partial intravenous anesthesia in cats and dogs. *Can Vet J, 54*, 276–282.

96. Pypendop, B. H., & Ilkiw, J. E. (2005). Assessment of the hemodynamic effects of lidocaine administered IV in isoflurane-anesthetized cats. *Am J Vet Res, 66*(4), 661–668.

97. Ortega, M., & Cruz, I. (1999). Evaluation of a constant rate infusion of lidocaine for balanced anesthesia in dogs undergoing surgery. *Can Vet J 52*: 856–860, 2011.

98. Wagner, A. E., Walton, J. A., Hellyer, P. W., et al. Use of low doses of ketamine administered by constant rate infusion as an adjunct for postoperative analgesia in dogs. *J Am Vet Med Assoc, 221*, 72–75.

99. Moraz-Muñoz, R., Ibancovichi, J. A., Gutierrez-Blanco, E., et al. (2014). Effects of lidocaine, dexmedetomidine or their combination on the minimum alveolar concentration of sevoflurane in dogs. *J Vet Med Sci, 76*(6), 847–853.

100. Ansah, O. B., Raekillo, M., & Vainio, O. (2000). Correlation between serum concentrations following continuous intravenous infusion of dexmedetomidine or medetomidine in cats and their sedative and analgesic effects. *J Vet Pharmacol Therap, 23*, 1–8.

101. Muir, W. W., & Hubbell, J. A. (2009). *Equine anesthesia* (2nd ed., pp. 260–276). St Louis, MO: Saunders Elsevier.

102. Kerr, C. L., McDonnell, W. N., & Young, S. S. (1996). A comparison of romifidine and xylazine when used with diazepam/ketamine for short-duration anesthesia in the horse. *Can Vet J, 37*, 601–609.

103. Muir, W. W., Lerche, P., Robertson, J. T., et al. (2000). Comparison of four drug combinations for total intravenous anesthesia of horses undergoing surgical removal of an abdominal testis, *J Am Vet Med Assoc, 217*, 869–873.

104. Hubbell, J. A., Aarnes, T. K., Lerche, P., et al. (2012). Evaluation of a midazolam-ketamine-xylazine infusion for total intravenous anesthesia in horses. *Am J Vet Res, 73*(4), 470–475.

105. Valverde, A. (2013). Balanced anesthesia and constant-rate infusions in horses. *Vet Clin Equine, 29*, 89–122.

106. Muir, W. W., & Sams, R. (1992). Effects of ketamine infusion on halothane minimal alveolar concentration in horses. *Am J Vet Res, 53*, 802–806.

107. Fielding, C. L., Brumbaugh, G. W., Matthews, N. S., et al. (2006). Pharmacokinetics and clinical effects of a subanesthetic continuous rate infusion of ketamine in awake horses. *Am J Vet Res, 67*(9), 1484–1490.

108. Malone, E., Ensink, J., Turner, T., et al. (2006). Intravenous continuous infusion of lidocaine for treatment of equine ileus. *Vet Surg, 35*, 60–66.

109. Doherty, T. J., & Seddighi, M. R. (2010). Local anesthetics as pain therapy in horses. *Vet Clin North Am Equine Pract, 26*(3), 533–549.

110. Valverde, A., Gunkel, C., Doherty, T. J., et al. (2005). Effect of a constant rate infusion of lidocaine on the quality of recovery from sevoflurane or isoflurane general anaesthesia in horses. *Equine Vet J, 37*(6), 559–564.

111. Valverde, A., Rickey, E., Sinclair, M., et al. (2010). Comparison of cardiovascular function and quality of recovery in isoflurane-anaesthetised horses administered a constant rate infusion of lidocaine or lidocaine and medetomidine during elective surgery. *Equine Vet J, 42*(3), 192–199.

112. Clutton, R. E. (2010). Opioid analgesia in horses. *Vet Clin North Am Equine Pract, 26*(3), 493–514.

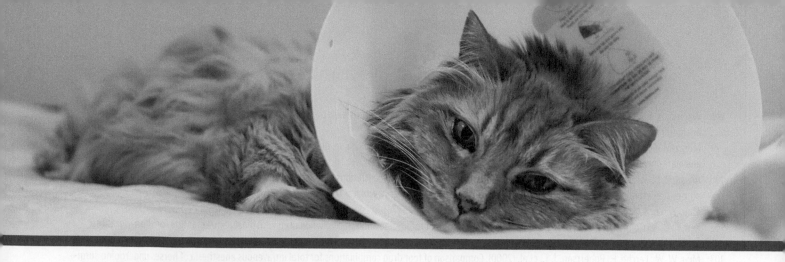

CHAPTER 15
Acute Pain Management

Christopher L. Norkus, *DVM, DACVAA, CVPP*

LEARNING OBJECTIVES

Upon completion of this chapter, it is expected that the reader should be able to:

15.1 Recognize events that cause pain in companion animal patients

15.2 Define terms related to acute pain

15.3 Describe the basic mechanism and physiology of the pain pathway

15.4 Explain the concepts of transmission, transduction, modulation, and perception of pain

15.5 Differentiate central and peripheral sensitization

15.6 Identify common signs of pain in cats and dogs

15.7 Describe different companion animal pain scales

15.8 Explain how to perform a pain assessment in companion animals

15.9 Identify several different therapeutic options (pharmacological and non-pharmacological methods) for the treatment of acute pain in companion animals

KEY TERMS

Adaptive pain

Allodynia

Central sensitization

Hyperalgesia

Hyperesthesia

Modulation

Neuropathic pain

Nociception

Nociceptive pain

Nociceptor

Pathologic pain

Perception

Peripheral sensitization

Pre-emptive analgesia

Referred pain

Somatic pain

Threshold

Transduction

Transmission

Transmucosal

Visceral pain

INTRODUCTION

According to the International Association for the Study of Pain (IASP), pain is defined as "an unpleasant sensory and emotional experience associated with actual or potential tissue damage."[1] Identifying and addressing pain in companion animals is challenging for two main reasons. First, veterinary patients are nonverbal and cannot tell when and where they hurt. Second, many companion animal species are stoic when painful and hide their pain well. This instinctive behavior once served as a protective mechanism for them in the wild to avoid being seen as prey by other animals. Historically, the absence of behavioral displays of pain when an animal has significant trauma or illness has been a major factor in the inadequate treatment of companion animal pain.[2]

Despite these unique limitations, the veterinary profession has over the past 20 years made great strides in improving the recognition of pain in companion animals and developing effective therapeutic modalities. Today pain recognition and management has developed into an entire specialty and is recognized in the veterinary profession as ethical, a standard of care, and a moral obligation that our clients and colleagues expect, and that our patients benefit the most from.

GENERAL TYPES AND CAUSES OF PAIN

Pain in companion animals is broadly classified into two main categories: acute pain and chronic pain. Acute pain most commonly results from trauma (e.g., a dog fight, being hit by a car, intervertebral disc disease) (Figure 15-1), inflammatory conditions (e.g., pancreatitis, otitis, cystitis), or surgical pain (e.g., a spay, fracture repair, mastectomy). Additionally, many procedures performed in veterinary medicine may be overlooked as painful such as IV catheter placement, ear cleaning, or anal gland expression. For the purpose of this chapter, however, we will focus on pain originating from surgery or specifically, acute postoperative pain. Pain from other sources such as trauma or inflammatory conditions is managed in the same fashion, while also addressing an underlying cause of the pain.

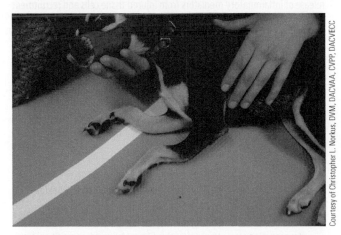

Courtesy of Christopher L. Norkus, DVM, DACVAA, CVPP, DACVECC

FIGURE 15-1 This patient presented to the emergency room in severe pain after breaking both front limbs. Her pain was addressed by intravenous oxymorphone and dexmedetomidine which calmed the patient and allowed for radiographs to confirm the diagnosis.

Tech Tip

The American Veterinary Medical Association (AVMA) describes the human–animal bond as "a mutually beneficial and dynamic relationship between people and animals that is influenced by behaviors considered essential to the health and well-being of both. The bond includes, but is not limited to the emotional, psychological, and physical interactions of people, animals, and the environment." An animal in pain can put the human–animal bond in great jeopardy. The role of veterinary professionals is to minimize pain in their patients in order to maximize the human–animal bond and to promote the health and well-being of both people and animals.

In general veterinary professionals should anthropomorphize with patients and recognize that if something causes pain in a human, it is likely to cause similar pain in a companion animal. Biologically acute pain serves as **adaptive** or **physiologic pain** meaning it acts as a warning signal to the body to notify it of injury and/or tissue damage. Chronic pain, however, is **pathologic pain** or **maladaptive** in nature meaning that it serves no useful biologic purpose. In veterinary species, chronic pain may be better defined as pain extending beyond the expected period of healing that possesses the ability to cause neurophysiological and psychological changes in the patient. Although the subject of chronic pain is peripherally discussed within this chapter, the reader is referred to Chapter 16 (Chronic Pain Management) for further discussion of chronic pain and its unique management. It is important at this point for the reader to understand that the treatment of chronic pain is quite different than the treatment of acute pain.

Acute pain can include **nociceptive pain** (pain caused by direct activation of special sensory pain neurons) and **neuropathic pain** (pain caused by direct damage to or indirect alteration of the nervous system) (Table 15-1). Nociceptive pain includes somatic and visceral types. **Somatic pain** includes pain originating from ligaments, tendons, bones, blood vessels, and muscles and in people is described as cramping, aching, throbbing, and sharp. An example of somatic pain might include a dog with pain originating from a femur fracture. **Visceral pain** originates from internal organs and in people is often described as dull, aching, and can be difficult to specifically localize. An example of visceral pain might include a horse with colic or a cat with pancreatitis. Visceral pain can often produce **referred pain** or pain that is perceived to originate from a location other than its true source (e.g., it is common in humans having myocardial infarction or "heart attack" to perceive pain in the left arm when really the problem is in the heart).

Neuropathic pain is divided into pain originating from the peripheral nervous system (peripheral nerves) or central nervous system (brain or spinal cord) and is often described as burning, electrical, tingling, stabbing, or as having a "pins and needles" quality. An example of neuropathic pain might include a dog with a brachial plexus injury.

TABLE 15-1 Nociceptive Pain vs. Neuropathic Pain

Nociceptive Pain		Neuropathic Pain	
Normal processing of stimuli that damages normal tissues or has the potential to do so		Abnormal processing of sensory input by the peripheral or central nervous system	
Somatic Pain	**Visceral Pain**	**Centrally Generated Pain**	**Peripherally Generated Pain**
• Arises from muscle, blood vessels, and connective tissue such as bone, tendons, and ligaments • Aching, sharp, or throbbing in quality • Well localized	• Arises from visceral organs • May be aching in quality and fairly well localized (tumor involvement) • May cause intermittent cramping and poorly localized (obstruction to hollow organs)	• May be due to central nervous system injury • May be associated with dysregulation of the autonomic nervous system	• Pain is felt along the distribution of peripheral nerves (polyneuropathies such as diabetic neuropathy) • Pain is associated with a known peripheral nerve injury (mononeuropathies such as nerve root compression)

EFFECTS OF ACUTE AND CHRONIC PAIN

Pain can be mild, moderate, or severe and when uncontrolled may increase patient morbidity and mortality. Transmission of pain stimuli can result in a neuroendocrine stress response and a myriad of local inflammatory substances (e.g., cytokines, prostaglandins). The predominant physiologic response to pain is a neuroendocrine response meaning there is involvement of the hypothalamic-pituitary-adrenal (HPA) axis, the sympathetic nervous system, and the adrenal gland itself (Figure 15-2). The effects of the neuroendocrine response are widespread across the body and can include, but are not limited to:

- Increases to heart rate and blood pressure
- Increased myocardial oxygen demand
- Increased catabolic hormone release (e.g., cortisol, adrenocorticotropin hormone, antidiuretic hormone, renin)
- Elevated blood glucose
- Water and sodium retention
- Increased oxygen consumption
- Increased risk for hypercoagulation and thromboembolic disease
- Immunosuppression and poor wound healing
- Postoperative ileus and decreased gastrointestinal function

Pain can be totally incapacitating and emotionally draining to the patient and additionally can result in fatigue, sleeplessness, decreased activity and play behavior, decreased appetite, decreased grooming behavior, disability (e.g., inability to go up stairs or jump into the car), decreased learned behavior (e.g., failure to use the litter box), anxiety, fear, and even aggression. Perhaps most important, poorly addressed acute pain can quickly lead to chronic pain states that can be ongoing and difficult to treat. Chronic postsurgical pain has been documented to occur in up to 65% of patients following some surgical procedures in humans.[3] Although data does not exist in companion animals, it is reasonable to conclude that chronic pain occurs with a similar or greater frequency in veterinary patients. One classic example is poorly managed pain from a feline declaw procedure leading to chronic postsurgical pain. Adequately addressing acute pain is critical for both the short- and long-term welfare of our patients.

PAIN PATHWAYS

How does tissue injury from a dog being hit by a car or a patient undergoing elective surgery for castration result in the animal perceiving it's in pain? Ultimately the pain pathway must take what was initially a mechanical event and convert it into an electrical and chemical response. This is not a simple process and therefore the answer to the original question is complex, often confusing, and only beginning to be understood within the medical field.

In the most basic terms, following an initial tissue injury there is a rapid release of inflammatory mediators from injured tissue cells and recruitment of leukocytes to the site of inflammation. Examples of inflammatory mediators include bradykinin, prostaglandins, histamine, calcitonin gene related polypeptide, substance P, leukotrienes, endocannabinoids, cytokines (e.g., interleukin (IL)-1, IL-6, tumor necrosis factor (TNF)-alpha), chemokines, proteases, serotonin, glutamate, adenosine, ATP, and nerve growth factor. Along with the release of these inflammatory mediators, there is activation of special sensory neurons known as peripheral **nociceptors**, which initiate transduction and transmission of the pain signal to the central nervous system (CNS).

- The term **transduction** specifically refers to the action where nociceptors activate and initially respond to and receive a noxious stimulus.
- **Transmission** occurs as the pain signal travels from the site of transduction along the nociceptor fibers (A δ and C nerve fibers from peripheral visceral and somatic sites specifically) to the dorsal horn of the spinal cord (Table 15-2).

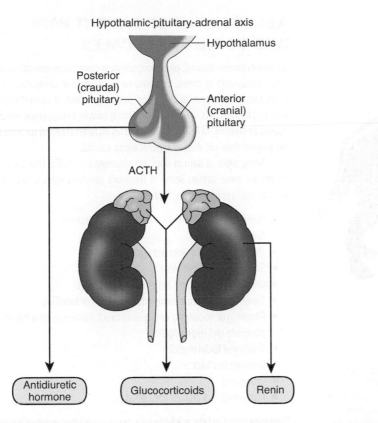

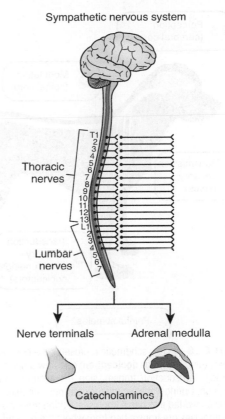

FIGURE 15-2 The neuroendocrine response to pain. Both the hypothalamic-pituitary-adrenocortical and sympathoadrenal systems are involved in the neuroendocrine response to pain producing widespread effects throughout the body.

- Within the spinal cord, modulation of the traveling pain signal occurs. Just like it sounds, **modulation** alters the traveling pain by a complex balance of stimulatory and inhibitory substances so that the pain signal is either enhanced or decreased. Some of the stimulatory substances include glutamate, N-methyl-D-aspartate (NMDA), α-Amino-3-hydroxy-5-methyl-4-isoxazolepropionic acid (AMPA), kainite, nerve growth factor, substance P, calcitonin gene related polypeptide, and transient receptor potential cation channel subfamily V member 1 (TRPV-1). Inhibitory substances include serotonin, norepinephrine, γ-aminobutyric acid (GABA), glycine, endocannabinoids, and endogenous opioids (enkephalin, endorphins, etc.).
- The pain signal then travels up through the spinothalamic and spinoreticular tracts into the brain where multiple areas such as the prefrontal cortex (PFC), somatosensory cortex (S1 and S2), anterior cingulated cortex (AC), insular cortex (IC), cerebellum, amygdala, and thalamus are activated to ultimately produce the end result of conscious **perception** or pain realization in the patient (Figure 15-3). All of this happens at lightning speed, similar to turning on an electrical switch!

Because the initial tissue injury does not immediately go away, this process is exacerbated because of the continuous release of inflammatory mediators in the periphery. The persistent release of inflammatory mediators continues the input process and results in **peripheral sensitization**, which

TABLE 15-2 Pain Fiber Types

Fiber Type	Characteristics	What They Transmit
Small fibers: • Aδ • C	• Thin, myelinated, slow • Thin, unmyelinated, slow	• Sharp, prickly pain • Dull, aching pain
Large fibers: • Aβ	• Thick, myelinated, fast	• Become recruited to transmit noxious stimuli resulting in pain during hyperalgesic/chronic pain states

means that the peripheral nociceptors outside the CNS become sensitized or overreactive. A perfect example of peripheral sensitization is inflammation. If a veterinary technician is scratched by a cat, the area around the incision quickly becomes red, swollen, and painful. Not only does the actual incision hurt (primary hyperalgesia), but the area around it does too (secondary hyperalgesia).

In addition to peripheral sensitization, dormant nociceptors can activate (Aβ fibers) resulting in new routes for the pain to travel. These processes result in a marked decreased **threshold** for initial pain activation

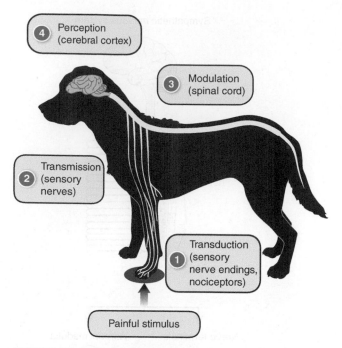

FIGURE 15-3 This schematic represents the nociceptive pain pathway (also known as **nociception**) and how a noxious stimulus undergoes transduction, transmission, modulation, and ultimately perception. During transduction, noxious pain stimulus (such as surgical pain) is converted into nerve impulses. During transmission, these nerve impulses are transmitted from where the pain originated to the spinal cord. During modulation, the nerve impulses are either inhibited or amplified within the dorsal horn of the spinal cord. Perception occurs when pain is recognized by the brain.

ASSESSMENT OF PATIENT PAIN AND USING PAIN SCALES

As previously mentioned, pain recognition in companion animals is difficult if not impossible at times. Learning to assess pain in companion animals takes careful observation skills and past experience. If something causes pain in a human, it is likely to cause similar pain in a companion animal and should be treated. In many incidences this includes treating for pain even if the patient does not display signs of being painful.

Some signs of pain in dogs are apparent and easily detectable, while others are more subtle. Some of the more common signs of pain in dogs include changes in behavior such as

- Being scared or having a submissive appearance
- Being aggressive
- Unwilling to eat or interact with people
- Showing hiding behavior
- Unwilling to lay down or general restlessness
- Trembling constantly without stimulation or handling
- Flinching or vocalizing with gentle touch alone may be a sign of allodynia and hyperalgesia
- Tensing of facial muscles
- Drawing lips back
- Grimacing
- Flattening ears against their head

Dogs may guard or bite at a body part, tense the abdomen when palpation is attempted, or have abnormal posture or gait (Figure 15-4).

Crying, whining, and whimpering can also be associated with pain although these signs may also be related to communication or behavioral causes and should not be over interpreted. Short shallow breathing, increased blood pressure, pupil dilation, nausea, and increased heart rate

and an increased rate of discharge with activation. Increased pain from a stimulus that should normally provoke pain is **hyperalgesia**. Clinically the result is a noxious stimuli causing more pain than it should (e.g., hitting one's thumb with a hammer hurts but it hurts even more if the person has hyperalgesia).

Intense pain input from the periphery may also result in **central sensitization** or persistent post-injury changes in the central nervous system that result in pain hypersensitivity. Repeated pain input may lead to **allodynia** or exaggerated and prolonged responsiveness of neurons to even normal input after tissue damage. In simple terms, allodynia is defined as pain due to a stimulus that does not normally provoke pain (e.g., you gently touch your arm and you feel pain). **Hyperesthesia**, an abnormal increase in sensitivity to sensory stimuli, can also occur. Ultimately repeated pain input could lead to functional changes in the dorsal horn of the spinal cord that can have long-term effects. Hence acute pain can quickly transition into chronic pain, especially if the acute pain is severe. Certain receptors such as the N-methyl-D-aspartate (NMDA) receptor appear to play an important role in the development of central sensitization and chronic pain after an acute injury and therefore have become just one target for drug therapy. This process is further described in Chapter 16 (Chronic Pain Management).

Courtesy of Mary Ellen Goldberg, BS, CVT, SRA, CCRA

FIGURE 15-4 This dog is displaying classic signs of abdominal pain. Note the clear body language including unwillingness to lie down, tense abdominal muscles, arched back, and low head carriage. This dog also resented abdominal palpation.

may be observed in a painful patient; however, physiologic parameters can be influenced by numerous non-painful causes and therefore are typically considered unreliable at best. Cats in pain show many of the same findings as dogs. As a species they are very prone to hide, stop grooming, become aggressive, or be reluctant to move. Some cats in pain may even purr, making detection of pain difficult.

Tech Tip 🐾

In people, physiologic parameters are unreliable for determining pain and are likely unreliable in animals too. When a patient has normal "vital signs" such as heart rate and blood pressure that does not mean they are comfortable or pain-free.

Tech Tip 🐾

Cats in pain may purr and therefore the notion that a purring cat is comfortable is likely a fallacy.

Unfortunately, many of the signs of pain mentioned in dogs and cats can occur from other non-pain related phenomena such as behavioral or communication causes, anxiety, or even drug-induced dysphoria. Therefore it is important to not over interpret any single sign, to frequently reassess patients, and to critically evaluate a patient's response to analgesic treatment. Dysphoria most commonly occurs in patients during emergence from general anesthesia or in patients receiving high or repeated doses of opioids or other drugs. Patients who are dysphoric often are consolable and will calm with time, reassurance, and gentle restraint. Sometimes it can be necessary to provide dysphoric patients with sedation or tranquilization and occasionally reduction or reversal of some of a patient's opioids or other drugs. An advantage of using multimodal or balanced analgesia (described later in the chapter) is that it allows for a reduction in individual analgesic drugs and therefore theoretically fewer adverse effects. Changes in patient behavior are perhaps one of the biggest clues to a patient being in pain (e.g., the normally docile Golden Retriever who tries to bite when you look in its ear). Also, gentle palpation of an incision site or injured area can be highly clinically useful (Figure 15-5).

In an effort to more objectively identify companion animals in acute pain, several pain assessments or "scales" have been developed and are available for clinical and research use. Each method has limitations, and some experts have questioned the true validity and usefulness of current pain scales in companion animals. When choosing which pain scale to use, it is important to use ones that are validated and identify the ones that are best to use in a given scenario. No pain scale is perfect, but they do encourage repeat assessment of patients and allow for documentation. Performing these tasks puts patient needs first and provides a method to improve patient outcomes.

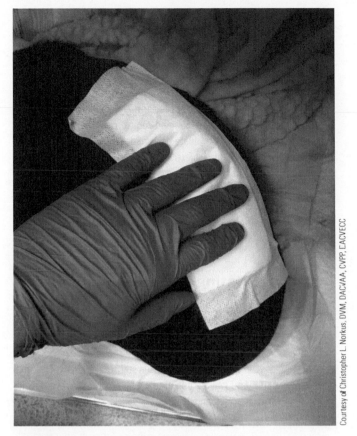

Courtesy of Christopher L. Norkus, DVM, DACVAA, CVPP, CACVECC

FIGURE 15–5 Very gentle palpation of an incision site or injured area can be useful in detecting pain clinically. Does the patient cry out? Try to bite? Look at the incision? Or move away from your hand? All of these reactions indicate patient discomfort.

Tech Tip 🐾

Pain detection and therapeutic intervention should be based on the use of objective, validated pain scoring systems. Ideally, pain assessment scales should be easy to use, reliable, repeatable (agreement in scoring by the same rater), reproducible (agreement in scoring between raters), unbiased, and efficient for people with varying levels of education to use. With appropriate training and experience, assigning pain scores tend to become more repeatable and reliable.

There are two categories of pain scales: unidimensional and multi-dimensional (also known as composite). Unidimensional pain scales are a simple way to rate the intensity of pain and trend scores over time. They are subjective and use words, images, or descriptors to measure pain or pain relief and have traditionally been used to assess intensity of acute pain. Common unidimensional pain scales include standardized pain scoring systems such as Numerical Rating Scales, Visual Analog Scales, and Simple Descriptive Scales that can be used as a subjective approach to assessing pain in animals. While these subjective assessments are imprecise and can often vary between observers, these judgments tend to become more repeatable and reliable with appropriate experience and training.

Multidimensional (or composite) scales incorporate quantitative measurements of pain behaviors (both spontaneous and evoked responses) and categorize, group, and weigh observer interpretations of a patient's pain. Individual categories are added to form an immediate pain score to produce more objective, repeatable, and robust assessments of pain that can be tracked over time to evaluate pain management efficacy. Although multidimensional scales are objective, they do have limitations, which include using different definitions or descriptors developed by different authors and variability in how the scale is applied clinically. Examples of multidimensional pain scales include the Glasgow Composite Measure Pain Score, the University of Melbourne Pain Score, and UNESP-Botucatu Multidimensional Composite Pain Scale.

Numerical Rating Scales

Numerical rating scales (NRS) use a number, instead of a descriptor, to indicate the severity of pain (e.g., 0–10 where 0 is no pain and 10 is the worst pain imaginable) (Figure 15-6). This unidimensional type of scale is discontinuous with unequal weighting between the categories, which can sometimes cause confusion or lead to errors in statistical analysis of research data. NRS are also used when scoring individual behavioral characteristics in multidimensional scales and have been developed for use in horses following surgery.

Visual Analogue Scales

Visual analogue scales (VAS) are unidimensional scales that were initially adapted from human psychometric questionnaires to subjectively assign a degree of pain to a given patient. They require an observer to assign a numerical position along a continuous line between two endpoints (Figure 15-7). Often the two endpoints are the numbers 0 and 100 mm or 0 and 10 cm.

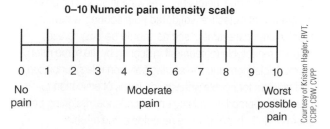

FIGURE 15–6 Numerical rating scale (NRS). A number line with individual numerical markings (1–10) that are chosen as the score.

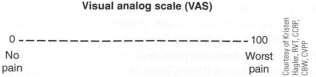

FIGURE 15–7 Visual analog scale (VAS). A line with no markings is used and, in this example, numbers are at each end with 0 being no pain and 100 being worst.

The 0 side of the continuous line represents a patient with no demonstrated pain and the 100 or 10 side of the scale indicates a patient who is assessed to be in the most severe pain possible. The observer subjectively assigns a number along this continuous line scale. For example, an observer might look at a dog and say, "I think this patient is mildly painful and therefore I will assign him a score of 27." The number the observer generates is entirely random and based upon intuition. Limitations of this assessment modality are that reliance on this evaluation technique can have wide variability among observers, it is subjective allowing for observer bias, and it has not been validated in companion animals. Nonetheless, VAS are easy and rapid to perform and have been used as a modality for assessing pain in companion animals in numerous research publications. Dynamic interactive visual analogue scales (DIVAS) are a subset of visual analogue scales that require the observer to touch the patient (including palpation of wounds or other known painful areas) and observe the patient during movement.

Simple Descriptive Scales

Another unidimensional scale similar to the VAS is the simple descriptive scale (SDS). This scale requests the evaluator assign a subjective pain intensity such as "no pain," "mild pain," "moderate pain," or "severe pain" to a patient. This type of scale suffers from the same limitations of VAS including the scale can have wide variability among observers, does not use objective criteria, does not overcome observer bias, and is very difficult to validate in companion animals. For example, one observer might look at a patient and say, "This patient is in mild pain" while a colleague might look at the same patient and say, "This patient is in moderate pain." Who is right? How does this influence treatment decisions?

Colorado State University Veterinary Medical Center has developed an expanded SDS for dogs, cats, and horses with acute pain that has become widely used in clinical practice (Figure 15-8). These scales ultimately ask the observer to rank pain based upon severity but also gives the observer some specific examples and criteria of what the patient might look like in pain including psychological and behavior traits, response to patient palpation, and observed levels of body tension. The scales are available at no charge at the links below:

- http://csu-cvmbs.colostate.edu/Documents/anesthesia-pain-management-pain-score-feline.pdf
- http://csu-cvmbs.colostate.edu/Documents/anesthesia-pain-management-pain-score-canine.pdf
- http://csu-cvmbs.colostate.edu/Documents/anesthesia-pain-management-pain-score-equine.pdf

Glasgow Composite Measure Pain Score

The Glasgow Composite Measure Pain Score (Glasgow CMPS) is a multidimensional scoring system developed at the University of Glasgow to aid in the clinical decision making for dogs in acute pain (Figure 15-9a). The Glasgow CMPS is validated for dogs in acute pain (surgical [orthopedic and soft tissue], medical, inflammatory, or traumatic). A short form (SF) has been developed to aid in the scoring system's clinical practicality. It includes 30 descriptor options within 6 behavior categories, including mobility. Within each category, the descriptors are ranked numerically according to

Your Clinic Name Here

Date _____

Time _____

Canine Acute Pain Scale

Rescore when awake	☐ Animal is sleeping, but can be aroused - Not evaluated for pain ☐ Animal can't be aroused, check vital signs, assess therapy

Pain Score	Example	Psychological & Behavioral	Response to Palpation	Body Tension
0		☐ **Comfortable** when resting ☐ **Happy, content** ☐ Not bothering wound or surgery site ☐ Interested in or curious about surroundings	☐ **Nontender** to palpation of wound or surgery site, or to palpation elsewhere	Minimal
1		☐ **Content to slightly unsettled** or restless ☐ **Distracted easily** by surroundings	☐ **Reacts to palpation** of wound, surgery site, or other body part by **looking around, flinching, or whimpering**	Mild
2		☐ Looks **uncomfortable** when resting ☐ May **whimper** or cry and may **lick or rub wound** or surgery site when unattended ☐ Droopy ears, **worried facial expression** (arched eye brows, darting eyes) ☐ **Reluctant to respond** when beckoned ☐ **Not eager to interact** with people or surroundings but will look around to see what is going on	☐ Flinches, whimpers cries, or guards/pulls away	Mild to Moderate **Reassess analgesic plan**
3		☐ **Unsettled, crying, groaning, biting or chewing** wound when unattended ☐ **Guards or protects** wound or surgery site by altering weight distribution (i.e., limping, shifting body position) ☐ **May be unwilling to move** all or part of body	☐ May be **subtle** (shifting eyes or increased respiratory rate) if dog is too painful to move or is stoic ☐ May be **dramatic**, such as a sharp cry, growl, bite or bite threat, and/or pulling away	Moderate **Reassess analgesic plan**
4	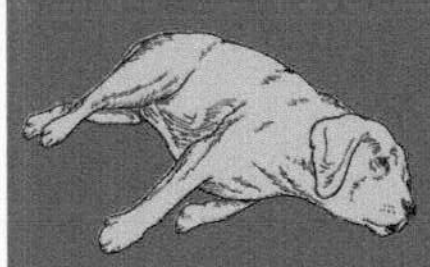	☐ **Constantly groaning or screaming** when unattended ☐ May bite or chew at wound, but unlikely to move ☐ **Potentially unresponsive** to surroundings ☐ **Difficult to distract** from pain	☐ **Cries at non-painful palpation** (may be experiencing allodynia, wind-up, or fearful that pain could be made worse) ☐ May react aggressively to palpation	Moderate to Severe May be rigid to avoid painful movement **Reassess analgesic plan**

RIGHT   LEFT

○ Tender to palpation
✕ Warm
■ Tense

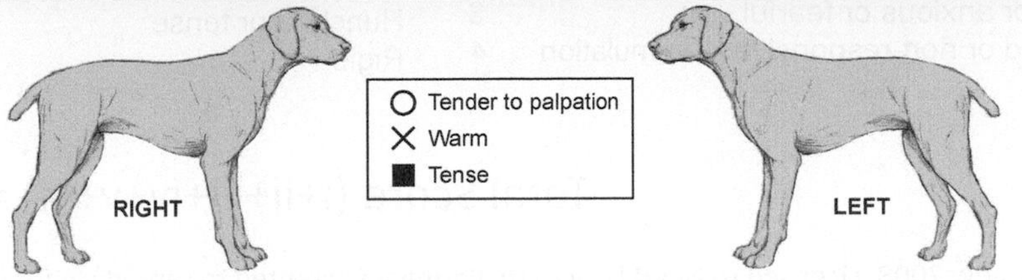

Comments _____

© 2006/PW Hellyer, SR Uhrig, NG Robinson

Colorado State University
VETERINARY TEACHING HOSPITAL

Courtesy of Colorado State University

FIGURE 15-8 Simple descriptive scale (SDS). Numbers are assigned to descriptions that categorize different levels of pain intensity. The Colorado State University Canine Acute Pain Scale is an example of an SDS.

SHORT FORM OF THE GLASGOW COMPOSITE MEASURE PAIN SCALE

Dog's name _____   Date   /   /   Time

Hospital Number _____

Procedure or Condition_____

_____

In the sections below please circle the appropriate score in each list and sum these to give the total score

A. Look at dog in Kennel
Is the dog

(i)

Quiet	0
Crying or whimpering	1
Groaning	2
Screaming	3

(ii)

Ignoring any wound or painful area	0
Looking at wound or painful area	1
Licking wound or painful area	2
Rubbing wound or painful area	3
Chewing wound or painful area.	4

> In the case of spinal, pelvic or multiple limb fractures, or where assistance is required to aid locomotion do not carry out section **B** and proceed to **C**
> *Please tick if this is the case* ☐ then proceed to C

B. Put lead on dog and lead out of the kennel
When the dog rises/walks is it?

(iii)

Normal	0
Lame	1
Slow or reluctant	2
Stiff	3
It refuses to move	4

C. If it has a wound or painful area including abdomen, apply gentle pressure 2 inches round the site
Does it?

(iv)

Do nothing	0
Look round	1
Flinch	2
Growl or guard area	3
Snap	4
Cry	5

D. Overall
Is the dog?

(v)

Happy and content or happy and bouncy	0
Quiet	1
Indifferent or non-responsive to surroundings	2
Nervous or anxious or fearful	3
Depressed or non-responsive to stimulation	4

Is the dog?

(vi)

Comfortable	0
Unsettled	1
Restless	2
Hunched or tense	3
Rigid	4

Total Score (i+ii+iii+iv+v+vi) = _____

Courtesy of Jacky Reid BVMS, PhD, DVA, DipECVAA, MRCA, MRCVS

FIGURE 15-9 The Glasgow Composite Measure Pain Score (Glasgow CMPS) is likely the most widely used acute pain scoring system for dogs today. It was developed and validated for dogs in acute pain (surgical [orthopedic and soft tissue], medical, inflammatory, or traumatic). It includes 30 descriptor options within 6 behavioral categories, including mobility. The descriptors are ranked numerically based on the evaluator's choice of descriptor within each category. The total score has been shown to be a useful indicator of analgesic requirement, and the recommended level for analgesic intervention in dogs is 6/24 or 5/20. It is important to carry out the assessment procedure as described on the questionnaire. It is important to carry out the assessment procedure as described on the questionnaire and to use only validated translations of the CMPS available from NewMetrica at www.newmetrica.com. An English version of the tool should be used if a specific language translation is not provided on the NewMetrica website; however, the English version of the scale is only validated for use by people whose first language is English.

Glasgow Feline Composite Measure Pain Scale: CMPS - Feline

Choose the most appropriate expression from each section and total the scores to calculate the pain score for the cat. If more than one expression applies choose the higher score

LOOK AT THE CAT IN ITS CAGE:

Is it?
Question 1
Silent / purring / meowing	0
Crying/growling / groaning	1

Question 2
Relaxed	0
Licking lips	1
Restless/cowering at back of cage	2
Tense/crouched	3
Rigid/hunched	4

Question 3
Ignoring any wound or painful area	0
Attention to wound	1

Question 4
(a) Look at the following caricatures. Circle the drawing which best depicts the cat's ear position?

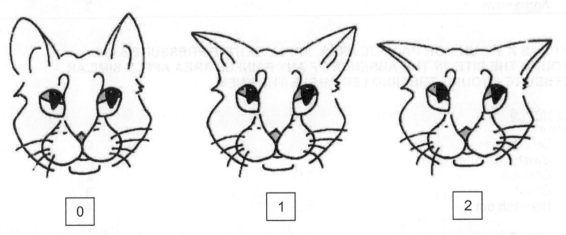

0   1   2

FIGURE 15-9 The Glasgow Feline Composite Measure Pain Scale (Glasgow CMPS-F) is a validated clinical decision-making tool for use in cats in acute pain. It includes 28 descriptor options within 7 behavioral categories. The descriptors are ranked numerically based on the evaluator's choice of descriptor within each category. The pain score is the sum of the rank scores with 20 being the maximum score possible. The total score has been shown to be a useful indicator of analgesic requirement, and the recommended analgesic intervention level is 5/20. It is important to carry out the assessment procedure as described on the questionnaire and to use only validated translations of the CMPS available from NewMetrica at www.newmetrica.com. An English version of the tool should be used if a specific language translation is not provided on the NewMetrica website; however, the English version of the scale is only validated for use by people whose first language is English.

(b) Look at the shape of the muzzle in the following caricatures. Circle the drawing which appears most like that of the cat?

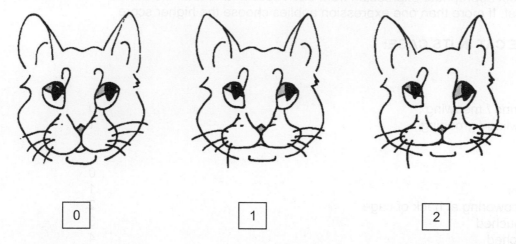

| 0 | 1 | 2 |

APPROACH THE CAGE, CALL THE CAT BY NAME & STROKE ALONG ITS BACK FROM HEAD TO TAIL

Question 5
Does it?

Respond to stroking	0

Is it?

Unresponsive	1
Aggressive	2

IF IT HAS A WOUND OR PAINFUL AREA, APPLY GENTLE PRESSURE 5 CM AROUND THE SITE. IN THE ABSENCE OF ANY PAINFUL AREA APPLY SIMILAR PRESSURE AROUND THE HIND LEG ABOVE THE KNEE

Question 6
Does it?

Do nothing	0
Swish tail/flatten ears	1
Cry/hiss	2
Growl	3
Bite/lash out	4

Question 7
General impression
Is the cat?

Happy and content	0
Disinterested/quiet	1
Anxious/fearful	2
Dull	3
Depressed/grumpy	4

Pain Score ... /20

FIGURE 15-9 (Continued)

their associated pain severity, and the person carrying out the assessment chooses the descriptor within each category that best fits the dog's status. The pain score is the sum of each individual rank scores. The maximum score for the six categories is 24, or 20 if mobility cannot be assessed due to spinal injury, pelvic or multiple limb fractures, or where assistance is required to aid locomotion. Maximum scores would represent the worst possible pain. A score of 0 would be a patient not experiencing any pain. The recommended level for analgesic intervention in dogs is 6/24 or 5/20. Keep in mind that some patients who do not trigger the descriptors in the scale may not be identified as painful. As a result, patients with low scores may still be painful.

The Glasgow Feline Composite Measure Pain Scale (Glasgow CMPS-F) is a validated multidimensional scoring system for assessing medical-, surgical-, and trauma-associated acute pain cats (Figure 15-9b). It includes 28 descriptor options within 7 behavioral categories. The descriptors are ranked numerically according to their associated pain severity based on the evaluator's choice of descriptor within each category. The pain score is the sum of the rank scores. The maximum score for the 7 categories is 20. The total score has been shown to be a useful indicator of analgesic requirement, and the recommended analgesic intervention level is 5/20.

University of Melbourne Pain Score

Similar to the Glasgow CMPS, the University of Melbourne has established a validated scoring system to aid in the decision to treat dogs in acute pain (Figure 15-10). The multidimensional scoring system asks an observer to assign weighted numbers based upon traits seen in a given patient. For example, the category of "mental status" contains four levels: submissive, overly friendly, wary, and aggressive. These characteristics are awarded scores of 0, 1, 2, and 3 respectively. The sum of individual scores from the six categories is totaled to produce a maximum score of 27 (most intense pain possible), and a score of 5 or greater should prompt the observer to treat for pain. This scale also has limitations as a dog two days post limb amputation and without analgesia may lay quietly, be unwilling to move, fail to eat, appear very depressed, and yet only score a 4/27 and not receive any analgesia.

UNESP-Botucatu Multidimensional Composite Pain Scale

The UNESP-Botucatu Multidimensional Composite Pain Scale (UNESP-Botucatu MCPS) is a validated pain scale for the assessment of acute postoperative pain in cats. This multidimensional scale assesses pain expression (nonspecific behaviors, response to palpation of the surgical wound, reaction to palpation of the abdomen/flank, and vocalization), psychomotor changes (posture, comfort, activity, and attitude), and physiological variables (arterial blood pressure and appetite). The scores may be classified as mild pain (0–8 points), moderate pain (9–21 points) and severe pain (22–30 points). The maximum value on this scale is 30 points and analgesic intervention is recommended when the score is ≥ 8. The UNESP-Botucatu MCPS was developed and validated in multiple languages and cultures.[4] Printable versions of this pain scale that contain directions about how to use the pain scale and videos relating to each score criterion are available from the Animal Pain website: http://www.animalpain.com.br/en-us/avaliacao-da-dor-em-gatos.php. A UNESP-Botucatu MCPS has been developed for cattle and is described in Chapter 19 (Anesthesia and Analgesia of Large Animal and Food Producing Species).

Category	Description	Scale
Biological variables		
	Dilated pupil	2
	Normal pupil	0
	Percentage of increase in Cardiac frequency	
	<20%	0
	>20%	1
	>50%	2
	>100%	3
	Salivation	2
	No salivation	0
Behavior variables		
Response to palpation	No changes	0
	Reaction to touch	2
	Reaction before being touched	3
Motor activity	Resting, sleeping	0
	Semiconscious	0
	Awake	1
	Restless, moving around	3
	Eating	0
Mental state	Submissive	0
	Sociable	1
	Cautious	2
	Agressive	3
Posture	Protects the affected area (fetal position)	2
	Lateral position	0
	Prone position	1
	Sitting or standing	1
	Moving	1
	Abnormal posture	2
Vocalization	Does not vocalize	0
	Vocalizes when touched	2
	Intermitent vocalization	2
	Continuous vocalization	3

FIGURE 15-10 The University of Melbourne Pain Scale (UMPS) combines physiologic data and behavioral response to help in the decision to treat dogs in acute pain.

Courtesy of the University of Melbourne

Grimace Scales

Grimace scales are scoring systems that assess changes in facial expressions following a painful procedure or condition. These scales were developed for mice, rats, rabbits, horses, sheep, lambs, piglets, cats, and ferrets. Most of these scales consist of four to five action units (AU) that are rated and scored as absent (zero), partially present (one), or markedly present (two). AU such as orbital tightening and ear position are listed across all species; however, other facial features and some specific changes are different (i.e., flattening of the nose and cheek regions are assessed in rats, while bulging is assessed in mice). The Feline Grimace Scale is a validated method that assesses five AU to assesses several unique facial expressions cats display if they are in pain including changes in ear position, orbital tightening, muzzle tension, whisker position, and head position (e.g., a painful cat may exhibit squinted eyes and an increased distance between the ear tips compared with a non-painful cat) (Figure 15-11).[5] A total score of 4 or greater indicates the cat is in pain and needs rescue analgesia. Species-specific grimace scales are described in Chapter 19 (Anesthesia and Analgesia of Large Animal and Food Producing Species), Chapter 20 (Anesthesia and Analgesia of Pet Birds and Exotic Small Mammals), and Chapter 21 (Pain Management in Laboratory Rats and Mice).

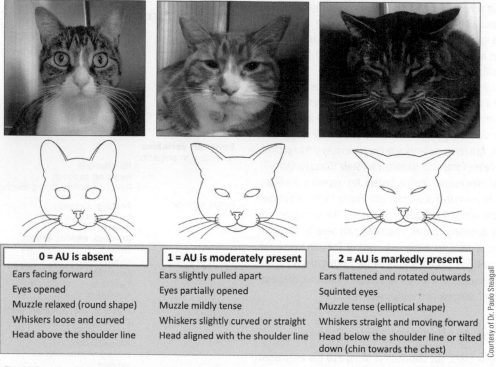

0 = AU is absent	1 = AU is moderately present	2 = AU is markedly present
Ears facing forward	Ears slightly pulled apart	Ears flattened and rotated outwards
Eyes opened	Eyes partially opened	Squinted eyes
Muzzle relaxed (round shape)	Muzzle mildly tense	Muzzle tense (elliptical shape)
Whiskers loose and curved	Whiskers slightly curved or straight	Whiskers straight and moving forward
Head above the shoulder line	Head aligned with the shoulder line	Head below the shoulder line or tilted down (chin towards the chest)

Courtesy of Dr. Paulo Steagall

FIGURE 15-11 The University of Montreal Feline Grimace Scale assesses five action units to determine a cat's pain level. The action units are ear position, orbital tightening, muzzle tension, whisker position, and head position. Descriptions of the scores within each action unit are provided below each score (e.g., scores of 0 are in the left column and include ears facing forward, eyes opened, muzzle relaxed, whiskers loose and curved, and head above the should line; scores of 1 are described in the middle column and scores of 2 are described in the right column).

Tech Tip

Grimace scales are non-invasive and require short observation of the animal, are easily applicable, and rapidly detect pain.

Obel Grading System

The Obel grading system is an SDS commonly used for equine lameness and, more specifically, evaluation of acute laminitis. This subjective grading system has been validated in horses as being reproducible between observers.[6, 7] The Obel grading system categorizes clinical signs of lameness or laminitis-related pain into the following four categories:

- *Grade I:* Horses shift weight from one foot to the other or incessantly lift feet. Lameness is not evident at a walk, but at the trot horses will have a shortened stride.
- *Grade II:* Horses move willingly at a walk and trot but with a noticeably shortened and stabbing stride. A foot can be lifted off the ground without difficulty.
- *Grade III:* Horses move reluctantly and resist attempts to lift affected or contralateral feet.
- *Grade IV:* Horses express marked reluctance or absolute refusal to move.

The Obel grading system is a widely accepted method for assessing the severity of laminitis; however, its use requires the horse to walk and trot, which can cause further pain to the patient.

American Association of Equine Practitioner's Lameness Grading System

Another scoring system used to evaluate lameness in horses is the American Association of Equine Practitioners' (AAEP) lameness grading system. This system has a 0 to 5 range of lameness scores in which 0 indicates no perceptible lameness and five indicates minimal weight bearing or non-weight bearing.[8] This system describes where the lameness can be seen (e.g., at walk, trot, circles); however, it does not imply severity of the condition. The severity of the lameness (mild, moderate, or severe) is assigned at the discretion of the veterinarian; therefore, it is subjective and highly variable.

- *Grade 0:* No detectable lameness under any circumstances.
- *Grade 1:* Lameness that is difficult to observe and is inconsistently apparent regardless of the circumstances (e.g., in hand or under saddle, hard surface, incline, or circling).
- *Grade 2:* Lameness is difficult to detect at a walk or trot in a straight line, but is consistently apparent under particular circumstances (e.g., under saddle, hard surface, incline).
- *Grade 3:* Lameness is consistently observed at a trot in all circumstances.
- *Grade 4:* Lameness is obvious with a marked head nod, hip hike, and/ or shortened stride.
- *Grade 5:* Lameness is obvious with minimal weight bearing either during motion or at rest. The horse might be unable to move.

Limitations of the AAEP lameness grading system are that it does not permit grading of the lameness independently at a walk and trot nor under different circumstances (e.g., circles and straight lines), and a horse with sporadic severe lameness cannot be classified.

TREATMENT OF ACUTE PAIN

In light of the limitations and shortcomings of companion animal pain scales, it is likely that some patients are evaluated incorrectly for their true degree of pain and treated inappropriately. In the end, pain assessment in companion animals is an art that takes time to develop and is often influenced by careful observation, intuition, anthropomorphizing, and past experience. In many incidences, the owner or caretaker may be the best person to assess and identify pain in their pet. When in doubt, the most ethical answer is to treat for pain.

Tech Tip

The value of using pain scales is they force the observer to take a step-by-step approach to evaluating a patient, provide an objective means to communicate, and once the user becomes comfortable with i their use, do not add any significant time to the physical examination process.

Preventive/Pre-emptive Analgesia

Preventive or **pre-emptive analgesia** was once described as the concept of providing a patient with appropriate analgesia before a painful stimulus occurs. Providing analgesia before a painful stimulus occurs, of course, is only feasible when the onset of the pain is known and controlled (e.g., an animal undergoing surgery is expected to be painful). Today the concept of pre-emptive analgesia has evolved to have the more specific goal of preventing the establishment of central sensitization and hyperexcitability with the ultimate end goal of decreasing pain in the immediate postoperative period and reducing the incidence of chronic pain, thereby improving the patient's long term quality of life. An example of pre-emptive analgesia in companion animals historically might include administering an opioid and an alpha-2 adrenergic agonist as a patient's preanesthetic medication in an effort to provide sedation but also presurgical analgesia. An example of pre-emptive analgesia in companion animals today might include the use of a ketamine constant rate infusion (CRI) during surgery with the specific goal of reducing opioid consumption postoperatively and reducing the establishment of central sensitization and hyperexcitability.

Multimodal Analgesia

Multimodal analgesia (also known as balanced analgesia) is the concept of using a combination of different analgesic drug classes to target different sites along the nociceptive pain pathway in an effort to maximize patient comfort and recovery. An additional benefit of multimodal analgesia is that it often allows for the reduction in dosage and frequency of individual drugs, thereby reducing adverse effects. For example, a dog recovering from a thoracotomy may have better analgesia and fewer adverse effects if several different analgesic drugs and techniques are used (e.g., a CRI of fentanyl and ketamine, intra-pleural administration of the local anesthetic bupivacaine, and then meloxicam as a nonsteroidal anti-inflammatory drug [NSAID]), rather than if only one type of medication were used in high quantities (e.g., just a CRI of fentanyl and nothing else). A continuation of this multimodal analgesia philosophy includes the goal of controlling acute pain to allow for early mobilization, early return to eating, and less time spent in hospital.

Systemic Analgesic Techniques

Various drugs are used for multimodal analgesia during surgery and painful procedures because systemically administered analgesic drugs help relieve pain through a variety of mechanisms.

Opioids

Opioids are one of the most effective and widely used classes of drugs for acute pain in both humans and companion animals. They are generally seen as the standard for treating moderate to severe pain. Opioids exert their analgesia effect by blocking opioid receptors within the central nervous system and work along the modulation and perception parts of the nociceptive pain pathway. Common adverse effects depend on the dosage administered and species.

- In dogs, signs may include panting, vomiting, nausea, hypothermia, and bradycardia.
- In cats, commonly observed adverse effects can include vomiting, nausea, bradycardia, myosis, and hyperthermia. Occasionally some patients may become dysphoric following opioid administration.
- In horses, common adverse effects include increased locomotion and excitement (depending on the drug and dosage) and decreased gastric motility leading to colic.
- In all companion animal species, opioids can result in respiratory depression; however, this is not usually of great magnitude or of large clinical significance unless very high dosages are used. Humans can experience profound respiratory depression following opioid use.

Some research questions the effectiveness of opioid use in horses in general, and subjectively it would appear opioids do not provide as reliable analgesia in horses as other classes of drugs might (e.g., NSAIDs, alpha-2 (α_2) adrenergic agonists). Further research into this subject is also ongoing. Opioids are controlled substances.

Full mu (μ) agonist opioids including morphine (Duramorph®, Astramorph®), methadone (Dolophine®), hydromorphone (Dilaudid®), oxymorphone (Numorphan®), and fentanyl (Sublimaze®) are generally reserved for patients experiencing moderate to severe pain states (see Figures 5-17a through 5-17e). When less painful states occur, such as mild or moderate pain, partial μ agonists such as buprenorphine (Buprenex®) and mixed agonist/antagonist drugs such as butorphanol (Torbugesic®, Torbutrol®) or nalbuphine (Nubain®) can be selected or other non-opioid options considered (see Figures 5-17c and 5-17h). An exception to this rule may be in some species such as horses and birds. Some evidence suggests that because of different distribution of opioid receptors in the central nervous system, horses and birds may receive more analgesia from drugs that have kappa agonist properties (butorphanol); however, research into this topic remains ongoing. Full μ agonists can be reversed with opioid antagonists such as naloxone (Narcan®) or naltrexone (ReVia®) (see Figure 5-18).

Opioids are commonly administered parenterally via the intravenous, intramuscular, or subcutaneous route. They can be administered continuously as a CRI (Table 15-3). Additionally, opioids can be utilized in various preparations including those given via the oral transmucosal route, applied topically, injected into the epidural or intrathecal space, or administered orally. Buprenorphine is commonly selected for oral transmucosal (abbreviated OTM and also known as buccal) administration in cats although methadone has also been shown to be effective via this route.[9, 10] Simbadol™, a specific high concentration formulation of buprenorphine, is

TABLE 15-3 Canine and Feline Opioid Dosages

Drugs	Canine Dosage	Feline Dosage	Comments
morphine	0.1–0.5 mg/kg IV, IM, SQ q3–6h 0.1–0.3 mg/kg/hr CRI	Same	May cause hyperthermia in cats Cautious use with IV administration due to histamine release
methadone	0.1–0.5 mg/kg IV, IM, SQ q4–6h	Same (can also be given by the oral transmucosal route)	Has NMDA antagonist properties
fentanyl	1–10 mcg/kg/hr CRI May see higher dosages for short time periods during intense surgical stimulation (20 mcg/kg/hr)	Same	Due to short duration of action, a CRI is preferred
oxymorphone	0.05–0.2 mg/kg q4–6h, IV, IM, SQ	Same	Difficult to source
hydromorphone	0.05–0.2 mg/kg q4–6h, IV, IM, SQ	Same	May result in hyperthermia in cats
buprenorphine	0.01–0.03 mg/kg q4–8h, IV, IM, SQ, OTM 0.04 mg/kg/day CRI	0.01–0.03 mg/kg q4–8h, IV, IM, OTM 0.24 mg/kg SQ q24h for up to 3 doses (or 72 hrs)	Slow onset but long duration of action, unreliable SQ absorption in cats Simbadol™ is a specific high concentration formulation of buprenorphine that is FDA-approved for SQ administration in cats
butorphanol	0.2–0.4 mg/kg SQ, IM, IV 0.1–0.4 mg/kg/hr CRI	Same	Often insufficient for moderate to severe pain, short duration of action (about 1 hour)

Food and Drug Administration (FDA) approved for control of postoperative pain in cats given SQ once a day (q 24hr) for up to 3 doses (or 72 hrs) (see Figure 8-11). The reader is referred to Chapter 5 (Anesthetic and Analgesic Pharmacology) for further discussion on the pharmacology of opioids and additional information about their use.

Tech Tip

All full μ agonist opioids have the same efficacy as analgesics, meaning they can provide the same degree of pain control. Different drugs have different potencies, however, meaning different drugs require different dosages on a milligram per kilogram of body weight scale to achieve analgesia.

Tech Tip

Constant rate infusions do not necessarily provide "better analgesia" than intermittent boluses of a drug. The only advantage to using a CRI of a drug is that it maintains plasma concentration at a constant level as opposed to the plasma concentration rising and falling ("peaks and troughs"). When administered appropriately, intermittent boluses can also maintain plasma levels within effective ranges.

Non-opioid Analgesics

Depending on the degree of pain, non-opioid analgesics can be used in lieu of or in conjunction with opioids to treat acute pain. Examples of commonly used non-opioid drug classes for treating acute pain include NSAIDs, α_2 adrenergic receptor agonists, intravenous "systemic" lidocaine, ketamine, and local anesthetic techniques (Figure 15-12).

Nonsteroidal Anti-Inflammatory Drugs (NSAIDs)

NSAIDs work as anti-inflammatory agents to reduce peripheral sensitization along the transduction part of the nociceptive pain pathway. They are well suited for mild to moderate pain as solo agents or in conjunction with other drugs such as opioids for multimodal analgesia in moderate to severe pain states. NSAIDs are most commonly given via injection or via the oral route (Tables 15-4 through 15-6).

- Commonly used NSAIDs in dogs include carprofen (Rimadyl®), meloxicam (Metacam®), deracoxib (Deramaxx®), and firocoxib (Previcox®) (see Figures 5-19a, 5-19b, and 5-19d).
- Commonly used NSAIDS in cats include meloxicam and robenaxocib (Onsior®) (see Figures 5-19c and 5-19d).
- Horses are commonly treated with flunixin meglumine (Banamine®), phenylbutazone (Butazolidin®, Equipalazone®), ketoprofen (Ketofen®), and firocoxib (Equioxx®) (see Figures 5-19a, 5-19f, and 5-19g); NSAIDs (specifically flunixin) are highly effective in treating colic pain in horses and are often selected over opioid drugs.

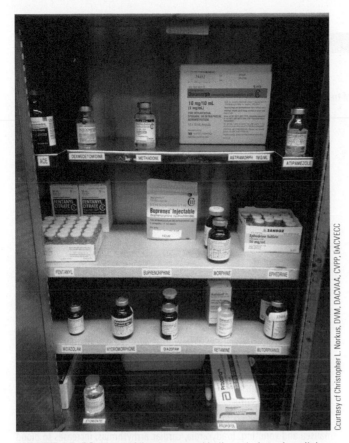

Courtesy cf Christopher L. Norkus, DVM, DACVAA, CVPP, DACVECC

FIGURE 15–12 Many of the drugs used in the veterinary hospital to provide excellent anesthesia care are useful in treating acute pain in companion animals. This well-stocked drug lock box includes many examples including opioids and ketamine.

Tech Tip

In 2010, the U.S. FDA issued a black box warning for repeated use of meloxicam in cats due to its association with acute renal failure and death. These adverse effects may be caused by the combination of the higher perioperative dose and then follow-up dosing of meloxicam.

Older drugs such as aspirin, naproxen (Equiproxen®), piroxicam (Feldene®), and ibuprofen (Motrin®) are not routinely used or recommended in any of the above species due to high risk for adverse effects and toxicity. Due to wide variation and controversy surrounding NSAID dosing, the reader is referred to an outside drug formulary for drug dosing information.

In dogs and cats, potential adverse effects of NSAID administration include

- Gastrointestinal upset including vomiting, diarrhea, and anorexia
- Gastrointestinal ulceration potentially leading to gastrointestinal perforation
- Acute renal injury and hepatic injury

In horses, the most common adverse effect from their administration is

- Gastrointestinal ulceration

NSAIDs should not be administered to patients in shock or those who are dehydrated, in hepatic or renal failure, have or are at risk for low blood pressure, or are on or recently received other NSAIDs or glucocorticoids. Some of these precautions may limit at times the use of NSAIDs in

TABLE 15-4 Commonly Used NSAIDS in Dogs for Acute Pain

Drug	Dosage and Route and Duration	Comment
carprofen	2.2 mg/kg PO q12h 4.4 mg/kg SQ preoperatively q24h	Licensed for preoperative use pre-emptively
meloxicam	0.2 mg/kg PO, IV, SQ once then 0.1 mg/kg PO, IV, SQ q24h	Licensed for treatment of osteoarthritis and inflammatory conditions
deracoxib	1–2 mg/kg PO q24h postop 3–4 mg/kg PO q24h	Licensed for osteoarthritis pain and inflammation; post-operative dental pain and pain associated with orthopedic surgery
firocoxib	5 mg/kg PO q24h	Licensed for postoperative pain and inflammation associated with soft tissue, orthopedic, and dental surgery and for chronic pain associated with osteoarthritis
robenacoxib	2 mg/kg SQ once perioperatively 1 mg/kg PO once daily for a maximum of 3 days	Licensed for postoperative pain and inflammation associated with soft tissue surgery Administered 30 minutes prior to surgery

TABLE 15-5 Commonly Used NSAIDs in Cats for Acute Pain*

Drug	Dosage and Route and Duration	Comment
meloxicam	0.05–0.3 mg/kg PO, SQ, IV q24h	Licensed for acute inflammatory pain FDA warning: repeated use in cats can cause acute renal failure and death. Safe for one time use in hydrated, normotensive, normovolemic cats with no signs of renal insufficiency
robenacoxib	1–2.4 mg/kg PO q24h	Licensed for acute inflammatory pain from orthopedic surgery, ovariohysterectomy, and castration

* Contraindications: Clotting disorders, volume depletion, gastric mucosal compromise

TABLE 15-6 Commonly Used NSAIDs in Horses for Acute Pain

Drug	Dosage and Route and Duration	Comment
flunixin meglumine	0.3 mg/kg q8h, 0.5 mg/kg q12h, or 1.1 mg/kg q24h IV not to exceed 5 successive days of treatment	Licensed for treatment of visceral and musculoskeletal pain; better for visceral pain Oral dose is extra label
phenylbutazone	2.2 mg/kg IV slowly, PO; q12h or 4.4 mg/kg IV slowly, PO; q24h	Licensed for treatment of inflammatory pain; better for treatment of musculoskeletal pain
ketoprofen	2.0–2.5 mg/kg IV, IM; q24h	Licensed for treatment of inflammation and musculoskeletal pain
firocoxib	0.1 mg/kg IV q24h	Licensed for treatment of pain and inflammation associated with osteoarthritis
carprofen	0.7 mg/kg IV once then 0.7 mg/kg PO q24h for up to 4–9 days	Extra label use in U.S. for treatment of pain

the emergency room or surgical setting. Patients should be carefully evaluated before receiving this class of drug. However, for most patients, NSAID use is safe and the benefits of their use far outweigh their potential risks.

Tech Tip

It is strictly contraindicated to administer NSAIDs and glucocorticoids at the same time! Concurrent use of these drugs can lead to profound gastrointestinal adverse effects such as ulceration, GI perforation, and potentially death.

Tech Tip

As a general rule, NSAIDs should be withheld until after general anesthesia because anesthesia can be unpredictable and their use during anesthesia-induced hypotension and hypoperfusion states may result in acute kidney injury.

Alpha-2 Adrenergic Agonists

Alpha-2 adrenergic agonists are commonly used as sedatives but are often overlooked as a class of analgesic drugs. Alpha-2 adrenergic agonists can enhance the analgesia of other drugs or provide short lived but excellent analgesia on their own. The class of drug works along the modulation and perception part of the nociceptive pain pathway. In dogs and cats, dexmedetomidine (Dexdomitor®) is the most commonly used α_2 adrenergic agonist (see Figure 8-6). Microdoses of dexmedetomidine IV prn can be very useful in complimenting other drugs in very painful patients and will often provide concurrent sedation and improve restfulness. It can also be useful to facilitate bandage or split changes or diagnostic procedures like radiographs

or ultrasound. For prolonged administration, a CRI of dexmedetomidine IV can be used. Other α_2 adrenergic agonists were used in the past; however, due to adverse effects they are no longer recommended.

In horses, α_2 adrenergic agonists such as xylazine (Rompun®), romifidine (Sedivet®), detomidine (Dormosedan®), and dexmedetomidine (Dexdomitor®) are utilized (see Figures 8-5a through 8-5c and Figure 8-6). Commonly observed adverse effects in dogs and cats include sedation, nausea and vomiting, bradycardia, cardiac arrhythmias, increases to blood pressure and systemic vascular resistance, decreased cardiac output, increased blood glucose, and increased urine output. Horses also get many of the same adverse effects as cats and dogs (except for nausea and vomiting) but also commonly experience significant sweating and potentially clinically significant decreases to gastrointestinal motility. Alpha-2 adrenergic agonists are most commonly given via the injectable route (including via CRI); however, in horses they are occasionally administered under the tongue (detomidine [Dormosedan®gel]).

The profound cardiovascular effects of α_2 adrenergic agonists limit their use to patients who are stable with excellent cardiovascular reserves. Even low and microdoses of α_2 adrenergic agonists can have profound adverse effects. Before their use, it is critical to consider whether or not a patient can tolerate the adverse effects. These limitations mean that for some sick patients, their selection is inappropriate. For example, an unstable dog in shock with gastric dilatation-volvulus syndrome (GDV) should not be administered an α_2 adrenergic agonist for analgesia because α_2 adrenergic agonists have profound cardiovascular adverse effects including dramatic decreases to cardiac output and delivery of oxygen to tissue. Alpha-2 adrenergic agonists can be reversed with drugs such as atipamazole (Antisedan®) or yohimbine (Yobine®).

Lidocaine

Intravenous or "systemic" lidocaine (Xylocaine®) has well known anti-arrhythmic properties but is also commonly used in humans as an adjunctive analgesic alongside other analgesic drugs (see Figure 5-13a). Systemic lidocaine is most commonly used in acute pain settings to treat severe or refractory pain and, most specifically, hyperalgesia. In recent years systemic lidocaine has also been used in companion animals. The exact mechanism by which systemic lidocaine provides analgesia is unclear but it appears to work as an anti-hyperalgesic drug and impact the modulation and perception parts of the nociceptive pain pathway. It is most useful to address the potential for or to treat existing chronic pain. It should not be used as a sole agent to treat acute pain.

Currently, literature validating the effectiveness of systemic lidocaine alone for treatment of acute pain in companion animals is limited; however, it is generally well tolerated and generally a low risk treatment. The most serious adverse effects of systemic lidocaine include neurologic and cardiovascular toxicity. These adverse effects are seen most commonly with high dosages or prolonged use and can be serious and potentially life threatening. Less serious adverse effects include nausea and vomiting. Cats may be especially sensitive to the effects of lidocaine, and systemic lidocaine should be used sparingly, if at all, in this species. Care should be taken when administering other local anesthetics concurrently to avoid local anesthetic toxicity. In dogs, a loading dose of IV lidocaine given slowly followed by a CRI has been used.

Ketamine

Ketamine (Ketaset®, VetaKet®, Ketaject®, Vetalar®) is a dissociative anesthetic that has known N-methyl-D-aspartate (NMDA) antagonist properties (see Figure 8-13 and Figure 9-10b). As described previously under the pain pathway section of this chapter, the NMDA receptor and its impact on glutamate act as a main source for hyperalgesia and the establishment of chronic pain. Clinically, ketamine therefore can be given in small sub-anesthetic dosages to work along the modulation and perception parts of the nociceptive pain pathway as an adjunctive analgesic alongside other analgesic drugs. Ketamine has become widely utilized in both humans and companion animals to address the potential for or treat existing chronic pain. Noteworthy is the increased interest in its use for the treatment of human depression. Ketamine should not be administered as a sole analgesic to treat acute pain as it is an adjunctive agent. Ketamine is most commonly selected for animals experiencing severe or refractory pain. In dogs and cats, ketamine commonly is given as a CRI following an IV loading dose. Generally, the author prefers to begin most CRI of ketamine in awake companion animals and titrates from there.

Because of the potentially profound effects of anesthetic dose ketamine, clinicians are often anxious about using ketamine as an analgesic in the unanesthetized hospitalized patient. Clinically, however, the dosage of ketamine used for analgesia is sub-anesthetic levels and generally has minimal cardiovascular, respiratory, and neurological adverse effects. The most commonly observed adverse effects in the author's experience include mild sedation, ataxia, and agitation or dysphoria. In susceptible patients, these effects are generally abolished with titration of the drug.

Regional and Local Anesthetic Techniques

Many regional and local anesthetic techniques can be used with local anesthetics to complement the treatment of acute pain. The most commonly used local anesthetic drugs in companion animals include lidocaine, bupivacaine (Marcaine®), mepivacaine (Carbocaine®), and ropivacaine (Naropin®) (see Figure 5-13a). The reader should refer to Chapter 5 (Anesthetic and Analgesic Pharmacology) and Chapter 17 (Analgesic Techniques) for more specific information about their pharmacology and their use. In short, some commonly used local anesthetic techniques in cats and dogs include epidural or intrathecal administration, brachial plexus nerve blocks, radial ulnar median and musculocutaneous nerve (RUMM) blocks, distal limb blocks for feline declaws, femoral sciatic nerve blocks, dental nerve blocks, topical application (e.g., transdermal lidocaine patches, topically onto the cornea), and local infusion. Local anesthetics can also be placed into body cavities such as the pleural cavity following thoracotomy, peritoneal cavity (e.g., for treatment of refractory pancreatitis pain), or into joint spaces for perioperative analgesia (risk of chrondrotoxicity via this route) (Figure 15-13).

Tech Tip

Bupivicaine must never be administered IV because it will cause profound cardiovascular collapse and potentially death when given by this route.

Tech Tip

In dogs and cats, the author generally limits the lidocaine dosage to no more than 5 mg/kg and bupivacaine dosage to less than 2 mg/kg for local blocks. Typically, lower dosages of lidocaine can accomplish the desired level of blockade (e.g., 2 mg/kg).

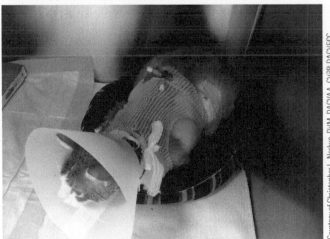

Courtesy of Christopher L. Norkus, DVM, DACVAA, CVPP DACVECC

FIGURE 15–13 This patient is hospitalized in the intensive care unit following a median sternotomy (thoracotomy) and chest tube placement. Analgesia for this patient has included a multimodal/balanced approach with both opioids and non-opioids. This cat is doing the "loaf of bread" position which can mean they are comfortable or could mean they are worried about something (whatever is happening in the next cage). This cat is relaxed, bright, and resting comfortably.

To prolong the duration of action of local anesthetics, drugs that have vasoconstrictive effects (e.g., epinephrine, α_2 adrenergic agonists) are occasionally added; however, consideration must be made to avoid impacting blood flow to local tissue. Neurologic and cardiovascular toxicity remain the major concern when administering local anesthetics. Careful consideration should be made with drug calculations and frequency of administration to prevent toxicity.

Tech Tip

Lidocaine initially stings upon administration. To prevent this sensation in the awake patient, consider mixing 1 part 8.4% sodium bicarbonate solution to 9 parts 2% lidocaine by volume for administration. Sodium bicarbonate raises the pH of lidocaine, which helps ease the pain on injection but could affect the onset time and duration of action for the drug.

Other Techniques

Numerous nonpharmacologic techniques have been explored to control acute pain. Some modalities that have gained popularity in recent years include acupuncture, transcutaneous electrical nerve stimulation (TENS), electromagnetic field therapy, chiropractic adjustment, physical therapy, massage, and low level laser therapy (see Chapter 18 [Veterinary Physical Rehabilitation]). The mechanisms by which these modalities achieve analgesia often remain unclear, and the lack of strong evidence to support their use in humans and companion animals has made their use controversial. Although the analgesic efficacy of these techniques may be open for debate, in general these approaches to address acute pain are safe, noninvasive, and devoid of the systemic adverse effects seen with other pharmacologic based treatment options. The precise role of these treatments for companion animal acute pain remains unclear; however, it appears these techniques, if used, are most appropriate as a compliment to pharmacologic treatments and not as the sole treatment modality.

Other practical non-pharmacologic techniques include the application of cryotherapy (e.g., ice or ice packs applied directly to a surgical incision site for 10 minutes) and splinting of fractured limbs and supportive bandaging when appropriate.[11, 12] Hospitalized post-surgical patients or those that are sick or injured should have excellent nursing care at all times, which includes, but is not limited to, being kept clean, gentle handing, a quiet minimally stressful environment, easy access to food and water, adequate time for sleep, ample comfortable bedding, and gentle, tender love and care.

SUMMARY

The American Animal Hospital Association (AAHA) now includes pain management as a specific core standard for small animal veterinary practices across the United States and requires pain assessment to be performed for all patients seen in practice. Ultimately the entire veterinary team plays a critical role in educating clients about pain management, assessing patients for pain, and ensuring adequate pain management is delivered to the patients in our care. Several options exist to aid in identifying acute pain in companion animals, and numerous pharmacological and non-pharmacological modalities are available to effectively treat it. Finally, the reader is reminded to stay abreast of new developments in the science and art of pain management and encouraged to review additional sources for further information on this important subject.

CRITICAL THINKING POINTS

- The presence of pain should be evaluated in every patient seen and treatment for pain should be provided accordingly.

- Anything that can cause pain in a human is likely to result in pain to an animal. Although we often remember that big events like trauma, inflammatory conditions, and surgery are all sources of pain, we sometimes overlook everyday procedures such as ear cleaning or IV catheter placement as painful to the patient.

- Recognizing pain in companion animals is challenging and takes time to develop as a skill. Numerous components should be evaluated including behavior, psychological well-being, physiologic parameters, body position/posture, response to palpation, and gait/response to movement.

- Pain scales are available to help assess companion animal pain; however, some of these scales often have limitations and may not always identify patients in pain.

- The nociceptive pain pathway turns a mechanical event into an electrical and chemical event for the body to recognize via transduction, transmission, modulation, and perception.

- Pain is easier to treat before it gets established. Pre-emptive/preventative analgesia and multimodal analgesia are two key strategies for the effective treatment of acute pain.

- Opioids are the mainstay treatment for acute pain in most settings, but non-opioid analgesics such as NSAIDs, alpha-2 adrenergic agonists, local anesthetics, ketamine, and systemic lidocaine can also be used as either primary modalities or adjunctive therapies depending on the setting.

- Non-pharmacologic treatments for pain should not be forgotten and commonly may include cryotherapy, splinting an injured limb, gentle handing, adequate patient comfort, and gentle, tender love and care.

REVIEW QUESTIONS

Multiple Choice

1. Which term describes the phenomena of pain that is caused by a non-painful stimulus?
   a. Allodynia
   b. Hyperalgesia
   c. Hyperesthesia
   d. Transduction

2. Which part of the nociceptive pain pathway specifically refers to the action where nociceptors activate and initially respond to a noxious stimulus?
   a. Transmission
   b. Transduction
   c. Modulation
   d. Perception

3. Which drug is a full mu agonist opioid?
   a. Buprenorphine
   b. Butorphanol
   c. Hydromorphone
   d. Nalbuphine

4. Which statement is false regarding visceral pain?
   a. It originates from structures such as muscle, tendon, ligament, blood vessels, or bone.
   b. It is often described in humans as being dull or aching in quality.
   c. It can often produce referred pain.
   d. An example of visceral pain would be a cat with pancreatitis.

5. Which term describes a phenomenon of pain that originates following a painful stimulus but is increased in intensity?
   a. Allodynia
   b. Hyperesthesia
   c. Hyperalgesia
   d. Perception

6. Which statement is false regarding full mu agonist opioid administration in dogs?
   a. They can result in nausea and vomiting.
   b. They are severe respiratory depressants and commonly can cause apnea.
   c. They often result in panting.
   d. They are reversible with naloxone.

7. Ketamine is useful in the prevention and treatment of central sensitization through what main mechanism?
   a. Stimulating kappa opioid receptors
   b. Blocking sodium channels
   c. Blocking the N-methyl-D-aspartate receptor (NMDA)
   d. Enhancing α-Amino-3-hydroxy-5-methyl-4-isoxazolepropionic acid (AMPA) receptor

8. Which nonsteroidal anti-inflammatory drug (NSAID) is *not* approved for use in dogs?
   a. Carprofen
   b. Meloxicam
   c. Firocoxib
   d. Ibuprofen

9. Which drug class should never be administered concurrently with nonsteroidal anti-inflammatory drugs (NSAIDs)?
   a. Angiotensin converting enzyme inhibitors (ACEi)
   b. Corticosteroids
   c. Cephalosporin antibiotics
   d. Calcium channel blockers

10. Which analgesic drug would be the most appropriate choice to administer to treat pain associated with a fractured femur in a cardio-vascularly unstable dog who was just hit by a car?
    a. Hydromorphone
    b. Ketamine
    c. Lidocaine
    d. Carprofen

11. Which drug would have the highest efficacy in treating severe acute pain?
    a. Butorphanol
    b. Buprenorphine
    c. Morphine
    d. Nalbuphine

12. Which opioid is commonly administered to cats via the oral transmucosal (aka buccal) route?
    a. Buprenorphine
    b. Morphine
    c. Fentanyl
    d. Butorphanol

13. Administering the analgesic morphine along with the tranquilizer acepromazine to a dog as premedication before an elective castration is an example of what analgesic principle?
    a. Balanced or multimodal analgesia
    b. Hyperalgesia
    c. Pre-emptive or preventative analgesia
    d. Physiologic pain

14. What two major body systems are impacted during local anesthetic overdose and toxicity (e.g., lidocaine, bupivacaine)?
    a. Respiratory and neurologic
    b. Neurologic and hepatic
    c. Cardiovascular and respiratory
    d. Cardiovascular and neurologic

15. Which nonsteroidal anti-inflammatory drug (NSAID) has been shown to be highly effective in treating colic pain in horses and is commonly used for this purpose?

    a. Phenylbutazone
    b. Flunixin meglumine
    c. Firocoxib
    d. Robenacoxib

16. Which drug is *not* a local anesthetic?

    a. Lidocaine
    b. Ropivacaine
    c. Mepivacaine
    d. Deracoxib

17. Which phase of the nociceptive pain pathway is where entering pain signals are adjusted by the body and either enhanced or diminished?

    a. Transduction
    b. Transmission
    c. Modulation
    d. Perception

18. Which drug could be used to reverse the adverse cardiovascular effects and central nervous system depression (such as sedation) of an alpha-2 adrenergic agonist?

    a. Naloxone
    b. Butorphanol
    c. Naltrexone
    d. Atipamezole

19. Regarding the assessment and treatment of companion animal acute pain, which statement is most correct?

    a. Physiologic parameters such as heart rate, respiratory rate, and blood pressure are the most reliable way to determine if a patient is in pain.
    b. It is generally advisable to anthropomorphize with patients and acknowledge that if something causes pain in a human, it is likely to cause similar pain in a companion animal.
    c. Dogs and cats in pain always show dramatic signs of pain that are obvious to observers.
    d. It is best for patients to feel pain following surgery because this makes them less likely to move and disrupt the surgery site and more likely to heal.

20. Which of the following are *not* potential effects of acute pain in companion animals?

    a. Immunosuppression and poor wound healing
    b. Decreased appetite, ileus, and decreased gastrointestinal function
    c. Aggression and/or changes in behavior
    d. Decreased blood glucose, heart rate, and blood pressure

Case Studies

Case Study 1: Agnes is a 6-month-old, intact female German Shepherd dog weighing 30 kg who presents to your practice for an elective ovariohysterectomy (OHE). The procedure and postoperative pain are expected to be moderate in severity. For premedication and to provide pre-emptive analgesia, you administer hydromorphone and dexmedetomidine IM. These drugs have different mechanisms of action and therefore help to provide a multimodal analgesic plan. A newly graduated veterinarian performs the surgery, which results in the surgery taking longer than normal. Two hours after the surgery, hydromorphone is administered SQ along with meloxicam SQ. Buprenorphine is then continued SQ overnight q6h while Agnes stays in the hospital, and she is discharged to her owner with meloxicam q24h PO for 5 days.

1. Why was hydromorphone chosen for this patient?

2. Why was dexmedetomidine included for this patient?

3. Why was more hydromorphone and meloxicam given to this patient postoperatively?

4. Hydromorphone's concentration is 2 mg/mL. How many mL would you administer to this patient if you were to administer 0.1 mg/kg?

Case Study 2: Timber is a 5-year-old gelding American Paint horse weighing 420 kg that is seen on the farm for a 2-hour history of colic. The horse's heart rate is 80 bpm with otherwise normal vital signs. Timber is actively showing signs of colic and is noted to be moderately to severely painful. The attending veterinarian evaluates the horse and diagnoses a large colon displacement that requires surgical correction. An intravenous catheter is placed, and the horse is administered intravenous fluids for cardiovascular stabilization. For analgesia, the horse is administered flunixin IV, butorphanol IV, and xylazine IV. These drugs are selected based upon the degree of the horse's pain and act as multimodal analgesia. Timber is immediately more comfortable and loaded into a horse trailer for transport to an equine hospital for surgery. Care should be taken using alpha-2 adrenergic agonists in patients with a compromised cardiovascular system.

1. Why should care be taken using alpha-2 adrenergic agonists in patients with compromised cardiovascular systems?

2. Why was butorphanol used in this case and not a full mu agonist?

Case Study 3: Princess is a 7-year-old intact female Chihuahua canine weighting 3 kg. She presents to the emergency room after she jumped out of her owner's arms. She was lame on her right front limb but is now non-weight bearing on it. The emergency room veterinarian evaluates the limb and suspects a distal radius and ulna fracture. Oxymorphone IV is administered prior to taking radiographs because Princess is assessed to be in moderate to severe pain. The radiographs confirm the fracture. The limb is splinted, and nerve blocks are performed with bupivacaine at the radial, ulnar, median, and musculocutaneous nerves. Meloxicam is administered IV every 24 hours and buprenorphine is continued q6h while the patient awaits surgery the next day.

1. By performing a radial, ulnar, median, and musculocutaneous nerve block, what parts of nociceptive pain pathway are targeted?

2. What parts of the pain pathway do oxymorphone, buprenorphine, and meloxicam work on?

Case Study 4: Gio is a 2-year-old intact male Domestic Shorthair cat weighing 4.6 kg. Gio presents for an elective castration, which is anticipated to be mild to moderately painful. Gio is given acepromazine SQ and morphine SQ as premedication, which also acts as pre-emptive analgesia. Gio is given ketamine IV and diazepam IV for anesthetic induction. The ketamine may have some short-lived analgesia properties for this patient. Lidocaine 2% is administered directly into each testicle as a local anesthetic block. Following recovery from anesthesia, Gio is given buprenorphine via the oral transmucosal (OTM) route and started on the oral NSAID robenacoxib. Gio is discharged from the hospital with three additional days of robenacoxib.

1. Describe the technique of administering a drug via the oral transmucosal (OTM) route in cats.

2. Are NSAIDs safe to administer to cats?

Case Study 5: Murphy is a 12-year-old castrated male Golden Retriever dog weighing 29 kg who is recovering in the intensive care unit following a left lateral thoracotomy to remove a primary lung tumor. Murphy has no other co-morbidities. For premedication he was given fentanyl IV and midazolam IV. Anesthesia was induced with IV ketamine and IV propofol. He was maintained on sevoflurane in 100% oxygen and received fentanyl and ketamine CRIs during surgery. The surgeon performed intercostal nerve blocks with 2% lidocaine and bupivacaine instilled through his chest tube prior to recovery. Once in ICU, Murphy received a fentanyl CRI along with a ketamine CRI. He was also started on carprofen SQ q12h and gabapentin at q8h PO. Bupivacaine was continued q6h through the chest tube for the first 24 hours.

1. Would you describe the level of postoperative pain in this patient as mild, moderate, or severe?

2. How would this patient's pain be managed after the first 24 hours?

Critical Thinking Questions

1. After preforming a brachial plexus block with bupivacaine on a 5 kg Boston Terrier dog, you realize that the concentration of the product was 0.75% and not the 0.25% your hospital normally carries. While you had intended to administer 2 mg/kg, you have inadvertently administered 6 mg/kg. The patient's Doppler reading is 40 mm Hg, indicating severe hypotension and arrhythmias are noted on the electrocardiogram (ECG). What is the cause of this patient's cardiovascular collapse, and what could be done to treat it?

2. I'm interested in learning more about pain management in companion animals and helping to improve its use at my veterinary practice. Where can I go to get more resources, knowledge, and help?

ENDNOTES

1. Part III: Pain terms, a current list with definitions and notes on usage. In H. Merskey and N. Bogduk (Eds.), *Classification of chronic pain* (2nd edition, pp. 209–214). Seattle, WA: IASP Press.
2. Hansen, B., & Hardie, E. (1993). Prescription and use of analgesics in dogs and cats in a veterinary teaching hospital: 258 cases (1983–1989). *Journal of American Veterinary Medical Association, 202,*1485–1494.
3. Kehlet, H., Jensen, T. S., & Woolf, C. J. (2006). Persistent postsurgical pain: Risk factors and prevention. *Lancet, 3367,* 1618.
4. Brondani, J. T., Mama, K. R., Luna, S. P. L. et al. (2013). Validation of the English version of the UNESP-Botucatu multidimensional composite pain scale for assessing postoperative pain in cats. *BMC Veterinary Research, 9,* 143. https://doi.org/10.1186/1746-6148-9-143

5. Evangelista, Marina C., Watanabe, R., Leung, V. S. Y., Monteiro, B. P., O'Toole, E., Pang, D. S. J., Steagall, P. V. (2019). Facial expressions of pain in cats: the development and validation of a Feline Grimace Scale. *Scientific Reports, 9*(1), 1–11. doi:10.1038/s41598-019-55693-8.

6. Menzies-Gow, N. J., Stevens, K. B., Sepulveda, M. F., et al. (2010). Repeatability and reproducibility of the Obel grading system for equine laminitis. *Veterinary Record, 167*(2), 52–55.

7. Viñuela-Fernández, I., Jones, E., Chase-Topping, M. E., & Price, J. (2011). Comparison of subjective scoring systems used to evaluate equine laminitis. *Veterinary Journal, 188*(2), 171–177. doi:10.1016/j.tvjl.2010.05.011

8. American Association of Equine Practitioners. LAMENESS EXAMS: Evaluating the lame horse, https://aaep.org/horsehealth/lameness-exams-evaluating-lame-horse, accessed August 13, 2020.

9. Ferreira, T. H., Rezende, M. L., & Mama, K. R. (2001). Plasma concentrations and behavioral, antinociceptive, and physiologic effects of methadone after intravenous and oral transmucosal administration in cats. *American Journal of Veterinary Research, 72*(6), 764–771.

10. Robertson, S. A., Taylor, P. M., & Sear, J. W. (2003). Systemic uptake of buprenorphine by cats after oral mucosal administration. *Veterinary Record, 152*(22), 675–678.

11. Rexing, J., Dunning, D., Siegel, A. M., et al. (2010). Effects of cold compression, bandaging, and microcurrent electrical therapy after cranial cruciate ligament repair in dogs. *Veterinary Surgery, 39*(1), 54–58.

12. Drygas, K. A., McClure, S. R., Goring, R. L., et al. (2011). Effect of cold compression therapy on postoperative pain, swelling, range of motion, and lameness after tibial plateau leveling osteotomy in dogs. *Journal of the American Veterinary Medical Association, 238*(10), 1284–1291.

CHAPTER 16
Chronic Pain Management

Mary Ellen Goldberg, BS, CVT, SRA, CCRA, CCRVN, CVPP, VTS (Lab Animal Medicine, Research/Anesthesia, Physical Rehabilitation)

LEARNING OBJECTIVES

Upon completion of this chapter, it is expected that the reader should be able to:

16.1 Differentiate between acute and chronic pain

16.2 Describe how pain may progress from acute to chronic pain

16.3 Explain the physiologic changes that can occur due to chronic pain

16.4 Determine a patient's pain level using a variety of pain scales

16.5 Identify which drugs are effective in managing chronic pain in various parts of the pain pathway

16.6 Explain alternative therapies and life style modifications that can help ease chronic pain

16.7 Describe treatment strategies for osteoarthritis pain

16.8 Describe treatment strategies for cancer pain

16.9 Describe treatment strategies for various non-osteoarthritis, non-malignant chronic pain conditions

KEY TERMS

Acute pain

Afferent

Alternative medicine

Analgesic adjunctives

Central sensitization

Chronic pain

Cyclooxygenase

Endorphins

First-order neurons

Focal erosive lesions

Lancinating

Modulation

Neuroplasticity

Noxious

Nutraceuticals

Palliative radiation therapy (PRT)

Paroxysmal

Perception

Peripheral sensitization

Pulse therapy

Second-order neurons

Subchondral sclerosis

Third-order neurons

Transduction

Transmission

TrpV1 receptor

Wind up

INTRODUCTION

Chronic pain is an often misunderstood, ill-defined condition in veterinary medicine that can develop from a variety of conditions and can present in patients as mild to excruciating, periodic to constant, or uncomfortable to fully debilitating and emotionally draining. Some chronic pain conditions are due to past injuries, poorly addressed acute pain (including surgical pain), or conditions such as osteoarthritis (OA), cancer pain, pancreatitis, intervertebral disc disease (IVDD), oral/dental pain, and otitis, while other types of chronic pain have no obvious physical cause (the type most frequently recognized in people). For example, if a dog suffers from a fracture of his femur and then the fracture is surgically repaired, we would expect the pain to resolve once the injury has fully healed. If, however, the dog's fracture fully heals following surgery yet the patient remains with some degree of pain, we would consider the patient has developed chronic pain.

Although numerous definitions for chronic pain have been proposed, any pain that persists for longer than the expected time frame (typically defined as lasting 3–4 weeks beyond the initial injury) for normal healing should be considered chronic. Recall from Chapter 15 (Acute Pain Management) that **acute pain** is adaptive (physiologic) pain that serves as a warning signal to notify the body of injury or tissue damage. **Chronic pain** is different as it is pathologic (maladaptive) pain that serves no useful biological purpose. In contrast to acute pain that arises suddenly in response to a specific injury and is usually treatable, chronic pain persists over time, can cause neurophysiological and psychological changes in the patient, and is often resistant to medical treatments, making its management challenging.[1]

Once chronic pain becomes established, its treatment is quite different than the treatment of acute pain. While treatment of acute pain aims to address its underlying cause and interrupt the nociceptive signals throughout the nervous system, chronic pain therapy relies on a multidisciplinary approach to manage the patient's neurophysiological and psychological changes and ultimately quality of life. It is therefore important to differentiate between acute and chronic pain and understand how their treatments are unique so that the patient's comfort level can be improved, their pain lessened, and overall quality of life improved.

Tech Tip 🐾

It is important to understand that treatments for acute and chronic pain differ dramatically and that drug therapy used to treat one is often not particularly effective at treating the other.

REVIEW OF THE PHYSIOLOGY OF PAIN

Recall from Chapter 15 (Acute Pain Management) that nociception is the processing of converting a **noxious** (harmful or destructive) mechanical stimulus into an electrochemical signal that ultimately reaches the brain where perception occurs (e.g., you step on a nail and those signals travel from your foot all the way to your brain so that you feel pain and go "Ouch!"). The initial noxious stimulus may include thermal (such as burning yourself on the stove), mechanical (such as slamming your finger in the car door), or chemical injury (such as spilling acid on your skin in the lab). Nociceptive pain can be superficial or deep, with deep pain originating from somatic (ligaments, tendons, bones, blood vessels, and muscles) and visceral (internal organs) sources. The process of nociception can be further broken down into individual pathway components that include transduction, transmission, modulation, and perception. Understanding these components is important because modifying them offers the best way to address acute pain with different medications.

- **Transduction** is the process of initially receiving the noxious stimulus be it mechanical, chemical, or thermal and rapidly transforming it into an electrochemical signal that can travel in the body. Peripheral nociceptors are free nerve endings located throughout the body and can receive this incoming information. This first step in the pain process is best inhibited by local anesthetics but can also be influenced by nonsteroidal anti-inflammatory drugs (NSAIDs) and glucocorticoids.

- **Transmission** describes the next step in the nociceptive pathway where the electrochemical signal travels through the peripheral nervous system up to the spinal cord. This occurs via **first-order neurons** or primary afferent (receiving) fibers. Nerve fibers primarily involved include A-delta (Aδ) (fast) fibers responsible for initial sharp, prickly pain and C (slow) fibers that cause secondary dull, throbbing or aching pain. A-beta (Aβ) (tactile) fibers do not normally transmit pain signals; however, this can change in the face of chronic pain (refer to Table 15-2). Transmission can be inhibited by local anesthetics such as lidocaine used in a nerve block and can be influenced by medication such as alpha-2 (α_2) adrenergic agonists.

- **Modulation** occurs when first-order neurons connect or synapse with **second-order neurons** in the dorsal horn cells of the spinal cord (Figure 16-1). Here in the spinal cord there is a constant give and take between pain signals trying to reach the brain and the body attempting to subdue the potential bombardment of stimuli to the nervous system. Numerous excitatory substances such as glutamate, N-methyl-D-aspartate (NMDA), α-amino-3-hydroxy-5-methyl-4-isoxazolepropionic acid (AMPA), nerve growth factor (NGF), substance P, and transient receptor potential cation channel subfamily V member 1 (TrpV-1) promote and amplify incoming pain signals to reach the brain. At the same time, endogenous substances such as serotonin, norepinephrine, γ-aminoutyric acid (GABA), cannabinoids, and **endorphins** (hormones secreted within the brain and nervous system that activate the body's opiate receptors causing an analgesic effect) attempt to dampen the nociceptive response. Numerous different drug classes work on the modulation part of the pain pathway to either blunt excitatory input (e.g., ketamine is an N-methyl-D-aspartate antagonist ultimately inhibiting glutamate), promote the inhibitory response (such as opioids, tricyclic antidepressants [TCAs] anticonvulsants, etc.), or both. Alpha-2 adrenergic agonists and NSAIDs can also have an effect on the modulation pathway.

- Lastly, pain signals travel from the spinal cord to the brain via **third-order neurons** where **perception** occurs (see Figure 15-3). *In other words, perception is the result of the first three steps and is the final phase where you consciously realize you've stepped on a nail and go "Ouch!"* This phase can be altered by opioids, (α_2) adrenergic agonists, ketamine, sedatives, and NSAIDs. Although they do not provide analgesia, inhalant anesthetic agents, benzodiazepines, and phenothiazines can block the perception of pain.

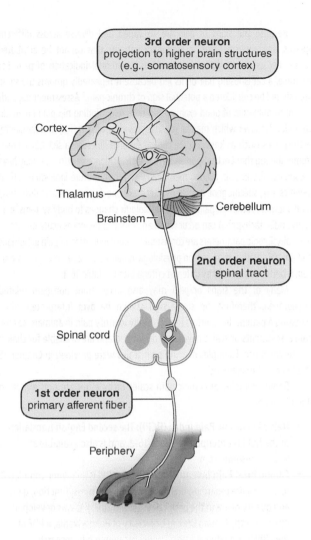

FIGURE 16–1 The afferent pain pathway. The afferent pain pathway is a nociceptive pathway that consists of three neurons: the first order neuron originates in the periphery and projects to the spinal cord; the second order neuron ascends the spinal cord; and the third order neuron projects into the cerebral cortex.

TRANSITION OF ACUTE PAIN TO CHRONIC PAIN

The establishment of chronic pain is a complex topic. In most simple terms, acute pain transitions to chronic pain through three broad phases:

- *Phase 1* – An initial noxious stimulus results in acute (physiological) pain. In some situations, this initial noxious stimulus may not be identified or in some situations in humans may never exist (in other words phase 2 may begin with little or no phase 1).
- *Phase 2* – Prolonged noxious stimulation occurs (either because of untreated or inadequate pain management). Rapid modification of the nervous system begins to occur as a result of an overworked pain pathway. **Neuroplasticity** or the *central nervous system's* ability to reorganize itself by forming new neural connections begins.

Peripheral and central sensitization occur leading to phenomena such as hyperalgesia and allodynia (see Chapter 15 [Acute Pain Management]). This phase is essentially the body undergoing changes in expectation for further pain to come.
- *Phase 3* – Prolonged noxious stimulation continues and modification of the neurologic system is fully engaged. Pain now becomes chronic (pathologic or maladaptive) and continues as ongoing persistent pain even when the original stimulus is removed or resolves. In this phase, the body has undergone potentially permanent changes in expectation to pain such that even if the initial noxious stimulation is removed the body is still is hardwired differently and may continue to feel pain.

> **Tech Tip** 🐾
>
> Pet owners often assume behavior changes and decreased mobility are inevitable with their animal's advancing age and often do not realize that therapies do exist for treatment of chronic pain.

Mechanisms for Developing Chronic Pain

Some of the main clinical phenomena that occur in chronic pain are as follows:

- A mixture of numerous local inflammatory mediators occurs, which results in **peripheral sensitization** or an increased sensitivity of peripheral nociceptors at or around the point of the original noxious stimulation.
- Sensory fibers that are not normally used for pain transmission (e.g., Aβ fibers normally used for tactile sensation) learn to transmit pain as part of genetic remodeling and **neuroplasticity.**
- "**Wind up,**" in response to a barrage of afferent nociceptive impulses, occurs causing *peripheral* nociceptor sensitivity and widening or expansion of the peripheral receptive fields. One way to think of wind up is a bad fishing analogy: that the body is using a larger net to catch incoming pain (fish).
- "**Central sensitization**" occurs as up regulation of the *central* nervous system causing increased excitability of neuron populations and enhancement of incoming pain signals. In other words, using the same bad fishing analogy, central sensitization is after the body has caught fish, modifying the size of the caught fish to make them bigger.
- Concurrent decreased activity of the body's natural endogenous inhibitory mechanisms lead to the reduced suppression of excitability within nociceptive pathways.

> **Tech Tip** 🐾
>
> The terms "wind up" and "central sensitization" are commonly used incorrectly or are misquoted as interchangeable terms. Wind up refers to "the progressive increase in the magnitude of C-fiber evoked responses of dorsal horn neurons." Central sensitization refers to "enhanced excitability of dorsal horn neurons." Wind up actually leads to central sensitization.

EFFECTS OF CHRONIC PAIN

Animals experiencing chronic pain have many of the same clinical signs as with acute pain. Chronic pain can significantly influence immunologic, endocrine (hypothalamic-pituitary-adrenal axis), and behavioral functions. Humans with chronic pain can also experience depression, sleep disturbance, fatigue, anxiety, and decreased overall physical functioning[2] and although we do not know if all of these events occur in animals with chronic pain, it is reasonable to suspect that many of them do. Chronic pain can also lead to patient suffering, which is stressful for pets, their owners, and their caregivers alike.

ASSESSING CHRONIC PAIN

Veterinary technicians are advocates for the patients under their care, which includes the identification and alleviation of pain and documentation of these findings in the medical record. Being able to recognize and score patient pain is a valuable skill for those working in veterinary medicine. Regardless of the presenting complaint or condition, pain assessment is considered an essential part of every patient evaluation; therefore, patients should never be required to "prove that they are in pain" but rather we should be on the lookout for it. By directly assessing patients including the prediction of noxious stimuli and identification of already established pain, the veterinary technician can request and advocate for appropriate analgesia for patients. A sound approach to pain management favors anticipation of the severity and duration of pain that is likely to occur with any procedure, condition, or surgery.

To assess a patient's level of pain, it is important to obtain a thorough and accurate history for all patients. When obtaining a patient history, open-ended questions may help uncover subtle changes in the pet's behavior that are indicative of pain. Such questions might include:

- *What differences have you noticed in your pet's disposition or personality?*
- *What differences have you noticed in your pet's sociability (both with people and other pets)?*
- *Could you describe how your pet gets up and down after resting?*
- *You said your pet is "just not himself." Could you tell me what concerns you?*
- *How does your pet's activity level compare with when he/she was younger?*
- *What other differences have you seen in your pet lately?*
- *What can you tell me about the changes in your pet's sleeping habits?*

Tech Tip

Owners often mistake vocalization as a sign of pain. While it is true that patients in pain may vocalize, vocalization is a nonspecific finding and often one that animals use for communication, to express dysphoria or anxiety, or simply as part of normal behavior and thus should not be over interpreted.

Because the signs of pain are so varied and diverse across different species, any abnormal sign in veterinary patients that cannot be attributed to another cause should be considered as a potential indication of pain. For example, a cat is losing hair on its leg because it repeatedly grooms the same area should be considered a potential sign of chronic pain.[3] Assessment depends on the combination of good examination skills (e.g., finding hip pain on an old Labrador Retriever which may be consistent with osteoarthritis) and familiarity with species (such as knowing the clinical signs of illness in a sick bird), breed (remembering that Cocker Spaniels have a high incidence of otitis and may have secondary chronic pain around their head), and in the future individual patient genetics (e.g., specific gene mutations may make one dog process pain differently than another). Although pain-related physiologic changes to body systems (e.g., tachycardia, tachypnea) can occur when an animal experiences acute or chronic pain, physiologic parameters are consistently a poor indication of pain in humans and animals. Rather than rely on physiologic parameters, assessment of chronic pain is better determined by different criteria listed in Table 16-1.

Many of the signs of pain may also occur from non-pain related phenomena; therefore, no single sign should be over interpreted when assessing a patient. In an effort to objectively identify pain in animals, several pain assessments or scales have been developed and are available for clinical and research use. Examples of acute pain scales were provided in Chapter 15 (Acute Pain Management).

Some examples of chronic pain scales include but are not limited to the following:

- **Helsinki Chronic Pain Index (HCPI):** The second English translation of the HCPI is described in Figure 16-2, and is also available at www.pawsomer.com
- **Canine Brief Pain Inventory (CBPI):** The CBPI is an owner-completed questionnaire regarding their dog's level of stiffness, function, gait, and quality of life in the past 7 days (Figure 16-3). It was developed by Dr. Dorothy Cimino Brown, University of Pennsylvania; a PDF of the CBPI is available at https://www.vet.upenn.edu/research/clinical-trials-vcic/our-services/pennchart/cbpi-tool
- **Liverpool Osteoarthritis in Dog (LOAD) instrument:** LOAD is an owner-completed clinical tool that can be recommended for the measurement of canine osteoarthritis. The dog's mobility is assessed in 13 areas (5 General domains and 8 At Exercise domains) on a 0 to 4 Simple Descriptive Scale. It was developed by Dr. John Innes, University of Liverpool; refer to evaluation of Construct and Criterion Validity for the Liverpool Osteoarthritis in Dogs' (LOAD) Clinical Metrology Instrument and Comparison to Two Other Instruments at http://www.brayvet.com/wp-content/uploads/2014/02/LOAD-Initial-Visit-Questionnaire.pdf
- **The Cincinnati Orthopedic Disability Index (CODI):** The CODI is a questionnaire that is a validated index different than the other scales because it consists of an owner-generated list of five activities that are impaired in their dog. The owner notes if the impaired activity is a Little Bit of a Problem, Quite a Bit of a Problem, a Severe Problem, or Impossible for the dog to do. Answers are assigned a score of 1 to 4, respectively; then transformed into a 0 to 100 score, with 100 denoting a perfectly normal dog. The score decreases with increasing degrees of disability. The CODI is available on the IVAPM's website at ivapm.org (for members), and is also available at: https://www.fourleg.com/media/Cincinnati%20Orthopedic%20Disability%20Index.pdf

TABLE 16-1 Criteria for Assessing Chronic Pain*

Chronic Pain Assessment Criteria
• Decreased activity and disability (e.g., inability to go up the stairs or jump into the car or get in or out of the litter box)
• Decreased socialization and lack of interest in play with people and other pets
• Decrease in learned behavior (e.g., failure to use the litter box)
• Abnormal postures, hunched back, muscle flaccidity or rigidity (refer to Figure 15-4)
• Poor grooming
• Constipation
• Decreased food and/or water consumption
• Weight loss (generally 20–25% of baseline), failure to grow, or loss of body condition (cachexia); it is important to also consider that weight loss can occur from numerous sources of pathology, and systemic disease should also be evaluated for and ruled out as a potential cause.
• Physical response to touch (withdrawal, lameness, abnormal aggression, vocalizing, abdominal splinting, increase in pulse or respiration)
• Teeth grinding (seen in rabbits, large animals, and food producing species)
• Self-aggression (such as excessive grooming or chewing) or aggression to others
• Decreased fecal or urine output
• Dehydration
• Decrease or increase in body temperature
• Decrease or increase in pulse or respiratory rate
• Inflammation
• Photophobia
• Vomiting or diarrhea
• Objective criteria of organ failure demonstrated by hematological or blood chemistry values, imaging, biopsy, or gross dysfunction

*Goldberg, M. E. The fourth vital sign in all creatures great and small. *The NAVTA Journal*, 31–54.

A form with pain descriptors (such as inability to chew food, decreased enthusiasm for walks) designed to detect chronic pain in cats and dogs may be a useful tool for both veterinarian and client. This form can be given to the client to help identify and localize pain (such as dental or abdominal) or to detect pain when reduced activity is presumed to be associated with simple aging.

Chronic pain scales use images and categories such as behavior, demeanor, and ease of movement as criteria for assessment (Figures 16-2 through 16-4; also refer to scale descriptions and illustrations in Chapter 15 [Acute Pain Management], Figures 15-6 through 15-11).

Tech Tip

No single pain scoring system is right for all practices or patients. In fact, it is not as important which system you choose as it is to simply choose one system that is used consistently by the entire team. Once a pain scoring system is chosen, set a goal to look at each patient and assign it a pain score.

PHARMACOLOGIC OPTIONS FOR CHRONIC PAIN

There are a variety of analgesic drug classes used in veterinary medicine to manage acute and chronic pain. Ideally, multimodal (balanced) analgesia is administered prior to a painful event (such as surgery) or before chronic pain sets in, which may decrease the incidence of neuroplasticity because it can lead to central hypersensitivity and chronic pain states. Multimodal

analgesia provides additive effects of multiple analgesic drugs that work through different mechanisms on different parts of the pain pathway to theoretically decrease the potential for any one drug to induce adverse effects. This section will focus on more specific modalities for the treatment of chronic pain; however, keep in mind that excellent acute pain management is the first step to addressing and preventing chronic pain establishment. The reader is directed to Chapter 5 (Anesthetic and Analgesic Pharmacology) for additional information on drugs used to treat pain.

Nonsteroidal Anti-Inflammatory Drugs (NSAIDs)

NSAIDs are analgesic, anti-inflammatory, and antipyretic (refer to Figure 5-19a through 5-19g) agents that are perhaps the most commonly used drugs for chronic pain in companion animals. Examples of common NSAIDs used in the United States include carprofen (Rimadyl®), meloxicam (Metacam®), deracoxib (Deramaxx®), firocoxib (Previcox®), and robenacoxib (Onsior®) in dogs; robenacoxib and meloxicam in cats; and flunixin meglumine (Banamine®), phenylbutazone (Butazolidin®, Equipalazone®), ketoprofen (Ketofen®), and firocoxib (Equioxx®) in horses. Another NSAID commonly used internationally is mavacoxib (Trocoxil®), a very long-acting COX-2 inhibitor used for the treatment of pain and inflammation associated with osteoarthritis in dogs. Mavacoxib is given orally after a meal to dogs greater than 1 year of age and weighing over 5 kg. The treatment is repeated in 14 days and then dosing is at 1-month intervals, not to exceed seven consecutive doses. All NSAIDs work by inhibiting **cyclooxygenase** (COX) enzymes leading to the blocking of the inflammatory mediator prostaglandin. There are numerous subtypes of prostaglandin that are responsible for many biological functions (both good and bad) within the body including inflammation and

HELSINKI CHRONIC PAIN INDEX (HCPI), FOR VETERINARY USE

Name of Dog _____   Diagnosis_____

Owner _____   Signature ____ (is important so we know that **the same owner** answer!)

Date _____   Questionnaire no. ____(remember to mark date and if it is pre or post treatment!)

"Tick only one answer – the one that best describes your dog during the preceding week"

Points	0	1	2	3	4

1. Rate your dog's mood:

Very alert	alert	neither alert, nor indifferent	indifferent	very indifferent	Points
☐	☐	☐	☐	☐	_____

2. Rate your dog's willingness to participate in play:

Very willingly	willingly	reluctantly	very reluctantly	does not play at all
☐	☐	☐	☐	☐
_____

3. Rate your dog's vocalization (audible complaining, such as whining or crying out):

Never	hardly ever	sometimes	often	very often
☐	☐	☐	☐	☐
_____

4. Rate your dog's willingness to walk:

Very willingly	willingly	reluctantly	very reluctantly	does not walk at all
☐	☐	☐	☐	☐
_____

5. Rate your dog's willingness to trot:

Very willingly	willingly	relu ctantly	very reluctantly	does not trot at all
☐	☐	☐	☐	☐
_____

6 Rate your dog's willingness to gallop:

Very willingly	willingly	reluctantly	very reluctantly	does not gallop at all
☐	☐	☐	☐	☐

7 Rate your dog's willingness to jump (eg. into car, onto sofa...)

Very willingly	willingly	relu ctantly	very reluctantly	does not jump at all
☐	☐	☐	☐	☐
_____

Kliinisen hevos- ja pieneläinlääketieteen laitos   PL 57 (Viikintie 49), 00014 Helsingin yliopisto
Eläinlääketieteellinen tiedekunta   Puhelin (09) 1911 (vaihde), faksi (09) 191 57298, www.vetmed.helsinki.fi

Institutionen för klinisk häst- och smådjursmedicin   PB 57 (Viksvägen 49), FIN-00014 Helsingfors universitet
Veterinärmedicinska fakulteten   Telefon +358 9 1911 (växel), fax +358 9 191 57298, www.vetmed.helsinki.fi/svenska/

Department of Equine and Small Animal Medicine   P.O. Box 57 (Viikintie 49), FIN-00014 University of Helsinki
Faculty of Veterinary Medicine   Telephone +358 9 1911, fax +358 9 191 57298, www.vetmed.helsinki.fi/english/

FIGURE 16-2 The Helsinki Chronic Pain Index (HCPI), developed in 2003 and validated in 2009, is a reliable and responsive method of assessing treatment response in osteoarthritic dogs. The HCPI includes 11 descriptor options that are scored 0 to 4 based on Simple Descriptive Scale for demeanor, behavior, and locomotion (mood, participation in play, vocalization, and willingness to walk, trot, gallop, and jump) and Visual Analog Scale for pain and locomotion (ease in lying down and rising from a lying position, and movements after long rest and after major activity or heavy exercise). A score of 0 and 1 is assumed to indicate normal canine behavior, whereas 2, 3 and 4 indicate increasingly severe pain-related behavior. Dogs with chronic pain will have a score of 12–44. The HCPI is the most widely used chronic pain index in the world and is available in a paper and digital form. The digital form and information on other aspects of pain management are available at www.pawesomer.com.

8. Rate your dog's ease in lying down:

With great ease	easily	neither easily, nor difficultly	with difficulty	with great difficulty	
☐	☐	☐	☐	☐	_____

9. Rate your dog's ease in rising from a lying position:

With great ease	easily	neither easily, nor difficultly	with difficulty	with great difficulty	
☐	☐	☐	☐	☐	_____

10. Rate your dog's ease of movement after a long rest:

Never difficult	hardly ever difficult	sometimes difficult	often difficult	very often/always difficult	
☐	☐	☐	☐	☐	_____

11. Rate your dog's ease of movement after major activity or heavy exercise:

Never difficult	hardly ever difficult	sometimes difficult	often difficult	very often/always difficult	
☐	☐	☐	☐	☐	_____

Points	0	1	2	3	4

Total up the answers to all 11 questions. Total chronic pain index score: _____
(The points and how to calculate the score has not been shown to the owners)

_____

Veterinarians note: _____

_____

_____

_____

_____

_____

For information about the HCPI, please contact Anna Hielm-Björkman, DVM, PhD
at anna.hielm-bjorkman@helsinki.fi

FIGURE 16-2 (Continued)

Canine Brief Pain Inventory

Description of pain:

Rate your dog's pain:

1. Fill in the oval next to the one number that best describes the pain at its **worst** in the last 7days.

  ○0   ○1   ○2   ○3   ○4   ○5   ○6   ○7   ○8   ○9   ○10

No pain   Extreme pain

2. Fill in the oval next to the one number that best describes the pain at its **least** in the last 7 days

  ○0   ○1   ○2   ○3   ○4   ○5   ○6   ○7   ○8   ○9   ○10

No pain   Extreme pain

3. Fill in the oval next to the one number that best describes the pain at its **average** in the last 7 days.

  ○0   ○1   ○2   ○3   ○4   ○5   ○6   ○7   ○8   ○9   ○10

No pain   Extreme pain

4. Fill in the oval next to the one number that best describes the pain as it is **right now**.

  ○0   ○1   ○2   ○3   ○4   ○5   ○6   ○7   ○8   ○9   ○10

No pain   Extreme pain

Description of function:

Fill in the oval next to the one number that best describes how during the last 7 days **pain has interfered** with your dog's:

5. General Activity

  ○0   ○1   ○2   ○3   ○4   ○5   ○6   ○7   ○8   ○9   ○10

Does not interfere   Completely interferes

6. Enjoyment of Life

  ○0   ○1   ○2   ○3   ○4   ○5   ○6   ○7   ○8   ○9   ○10

Does not interfere   Completely interferes

7. Ability to Rise to Standing From Lying Down

  ○0   ○1   ○2   ○3   ○4   ○5   ○6   ○7   ○8   ○9   ○10

Does not interfere   Completely interferes

Source: University Of Pennsylvania, Penn Vet Veterinary Clinical Investigations Center (VCIC)

FIGURE 16-3 The Canine Brief Pain Inventory (CBPI) has been validated for assessment of chronic pain associated with osteoarthritis and bone cancer. Dog owners are asked four questions about the severity of pain evident in a dog (the pain severity score) and six questions about how pain interferes with a dog's activity level (the pain interference score). The questionnaire also has one overall impression score, and although not used in the final calculation, it is useful along with the calculated score in order to assess whether a form of treatment (e.g., medication, surgery, supplementation) has had a measurable effect. Owners describe their dog's pain and function on a scale of 0 to 10 with 0 being the best and 10 being the worst score. The highest possible score is 100; the higher the final score the greater the degree of pain.

peripheral sensitization. Broadly speaking, COX enzymes are subdivided into COX-1, COX-2, and COX-3. COX-1 enzymes historically were believed to affect normal physiological "housekeeping" function, and COX-2 enzymes were thought to be more associated with pain and inflammation. This concept is no longer considered to be entirely true. COX-3 is a variant of COX-1, thus some prefer the name *COX-1b* or *COX-1 variant (COX-1v)*.[4] COX-3 is associated with the relief of pain via the central nervous system (CNS) but does not contribute to reducing inflammation. Examples of specific COX-3 inhibiting drugs are acetaminophen and dipyrone (dipyrone was discovered to cause leukopenia in people and although widely used in developing countries is not available in the United States except through compounding pharmacies).

Acetaminophen is well tolerated in dogs although liver damage is possible with its use at supratherapeutic doses. Cats have low activity of the enzyme glucuronyl transferase, which is used to change acetaminophen to glucuronic acid for excretion from the body. Cats given acetaminophen can develop severe methemoglobinemia that can cause hemolysis and death; therefore, it is contraindicated for use in cats. Some texts do not refer to COX-3 inhibitors as true NSAIDs as they do not have anti-inflammatory properties and can be given with NSAID drugs safely. For example, some headache medications for people have aspirin, acetaminophen, and caffeine as a proprietary product (Excedrin® Migrane). Traditional NSAIDs act to reduce local inflammation and peripheral sensitization and may also have activity within the CNS.

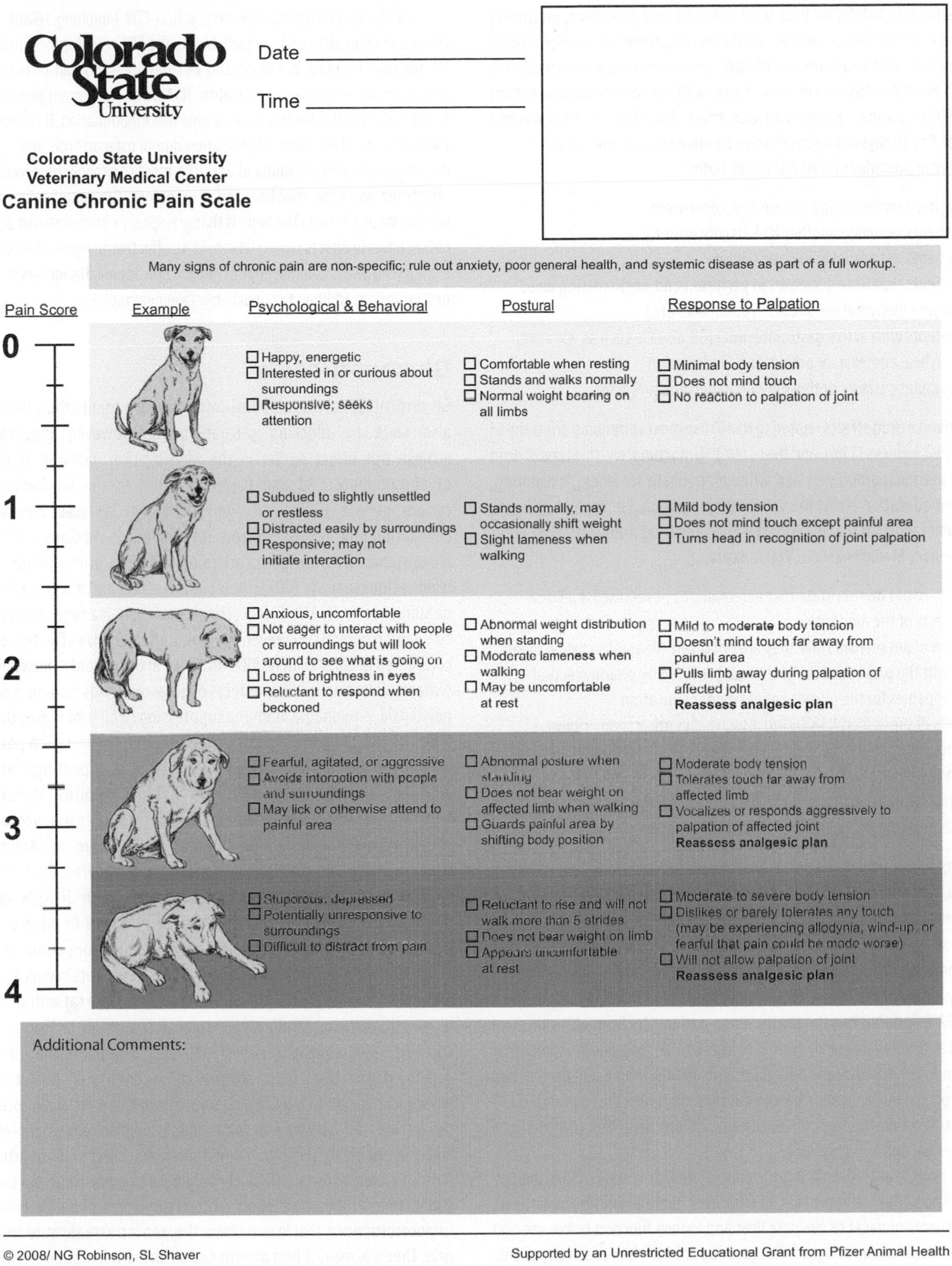

Colorado State University
Veterinary Medical Center
Canine Chronic Pain Scale

Date _____

Time _____

Many signs of chronic pain are non-specific; rule out anxiety, poor general health, and systemic disease as part of a full workup.

Pain Score	Example	Psychological & Behavioral	Postural	Response to Palpation
0		☐ Happy, energetic ☐ Interested in or curious about surroundings ☐ Responsive; seeks attention	☐ Comfortable when resting ☐ Stands and walks normally ☐ Normal weight bearing on all limbs	☐ Minimal body tension ☐ Does not mind touch ☐ No reaction to palpation of joint
1		☐ Subdued to slightly unsettled or restless ☐ Distracted easily by surroundings ☐ Responsive; may not initiate interaction	☐ Stands normally, may occasionally shift weight ☐ Slight lameness when walking	☐ Mild body tension ☐ Does not mind touch except painful area ☐ Turns head in recognition of joint palpation
2		☐ Anxious, uncomfortable ☐ Not eager to interact with people or surroundings but will look around to see what is going on ☐ Loss of brightness in eyes ☐ Reluctant to respond when beckoned	☐ Abnormal weight distribution when standing ☐ Moderate lameness when walking ☐ May be uncomfortable at rest	☐ Mild to moderate body tension ☐ Doesn't mind touch far away from painful area ☐ Pulls limb away during palpation of affected joint **Reassess analgesic plan**
3		☐ Fearful, agitated, or aggressive ☐ Avoids interaction with people and surroundings ☐ May lick or otherwise attend to painful area	☐ Abnormal posture when standing ☐ Does not bear weight on affected limb when walking ☐ Guards painful area by shifting body position	☐ Moderate body tension ☐ Tolerates touch far away from affected limb ☐ Vocalizes or responds aggressively to palpation of affected joint **Reassess analgesic plan**
4		☐ Stuporous, depressed ☐ Potentially unresponsive to surroundings ☐ Difficult to distract from pain	☐ Reluctant to rise and will not walk more than 5 strides ☐ Does not bear weight on limb ☐ Appears uncomfortable at rest	☐ Moderate to severe body tension ☐ Dislikes or barely tolerates any touch (may be experiencing allodynia, wind-up, or fearful that pain could be made worse) ☐ Will not allow palpation of joint **Reassess analgesic plan**

Additional Comments:

     Supported by an Unrestricted Educational Grant from Pfizer Animal Health

FIGURE 16-4 CSU Canine Chronic Pain Scale The Colorado State University Canine Chronic Pain Scale is a tool used for observational assessment of patients from a distance (psychological and behavioral evaluation and postural interpretation) and a hands-on evaluation period to assess response to gentle palpation. The scale uses a 0 to 4 scale with quarter marks along with a color scale and renderings of animals at various levels of pain as visual cues for progression along the 5-point scale. Specific descriptors for individual behaviors are provided which decreases inter-observer variability. A comments section allows documentation of specific areas of concern in the medical record. Although still unvalidated, this pain scale represents a useful and freely available tool for chronic pain assessment in dog.

While extremely helpful for long-term use in several conditions, prescribing NSAIDs are not without some risk. Monitoring of patients is therefore critical for both short- and long-term use. NSAIDS should only be used one product at a time to avoid development of adverse effects. When switching a patient from one NSAID to another (when no adverse effects have been seen), a washout period of 7 to 10 days minimizes chances for adverse drug interactions.

Contraindications for NSAID use include:[5]

- Patients receiving any systemic glucocorticoid
- Patients receiving another NSAIDs concurrently
- Patients with renal or hepatic failure
- Patients that have decreases in perfusion and/or circulating blood volume (dehydration, shock, hypotension, etc.)
- Patients with active gastrointestinal (GI) disease such as vomiting, diarrhea, anorexia, or preexisting GI ulceration
- Pregnant patients or those trying to become pregnant

Adverse drug effects related to NSAID use most commonly affect the GI tract (64%), kidneys (21%), and liver (14%).[6] Unfortunately, this class of drug is often used inappropriately and without adequate screening, monitoring, and client education across the veterinary profession. For example, adverse drug event reports at the United States Food and Drug Administration Center for Veterinary Medicine (FDA-CVM) indicate:

- 23% of pet owners state that veterinarians never discuss adverse effects of the medication
- 22% of pet owners state they are not given client information sheets about the prescribed drugs that are provided by pharmaceutical companies for the purpose of pet owner education
- 14% of prescribed NSAIDs are dispensed in other than original packaging, thereby denying pet owners drug information provided on the label
- Only 4% of pet patients prescribed drugs are given pre-administration blood analyses or biochemistry screening tests

Gastrointestinal problems associated with NSAIDs can include vomiting, diarrhea, anorexia, melena, regurgitation, gastric ulceration, GI perforation, and potential death. Vomiting has been identified as the most frequent clinical sign associated with gastric perforation. Pet owners should be instructed that if a pet begins vomiting while on an NSAID, the drug should be stopped and the patient promptly examined.

NSAID-induced hepatotoxicity is rare and results from either massive accidental drug overdose or from uncommon idiosyncratic reactions that occur randomly. NSAIDs can also induce acute kidney injury, and the greatest risk factors for this appear to be excessively high NSAID dosages and NSAID administration during time of concurrent sodium depletion, dehydration/hypovolemia, and hypotension.

In dogs, most adverse drug events related to chronic NSAID therapy occurs 14 to 30 days after the start of treatment; therefore, screening patients with a chemistry panel to evaluate liver and kidney function before use and then 2 to 4 weeks later is logical. The question of how often to screen patients after this will depend on factors such as the overall patient's health, associated cost, and client convenience. Many veterinarians for practicality reasons screen patients every 6 months thereafter. In general, the key to preventing adverse drug events is to catch problems early and err on the side of caution. All adverse drug events should be reported to the relevant pharmaceutical company or regulatory board.

Lastly, grapiprant (Galliprant®), a non-COX inhibiting NSAID, received FDA approval in 2016 and uniquely blocks the EP4 receptor for prostaglandin E2, the primary mediator of OA and inflammation. Instead of reducing the production of prostaglandins, as other NSAIDs do, grapiprant interferes with the ability of prostaglandins to cause pain and inflammation. It is approved for the treatment of pain and inflammation due to osteoarthritis in dogs greater than 8 pounds and 9 months of age. The most commonly observed adverse effects are vomiting, diarrhea, and inappetence. Grapiprant is currently not labeled for use in cats. The hope is that grapiprant's mechanism of action will reduce adverse effects while still providing effective analgesia. Initial research in dogs (and also cats in the hope of getting FDA approval) suggests the drug to be a promising addition to our pain management tool box.

Glucocorticoids

Glucocorticoids are potent anti-inflammatory agents that, like NSAIDs, also work by blocking prostaglandins. However, glucocorticoids achieve this effect earlier in the physiological pathway at the level of phospholipase A2 and therefore result in the blockade of both cyclooxygenase (COX) and lipoxygenase (LOX) (see Figure 5-20). LOX ultimately results in the inflammatory mediators known as leukotrienes. Because glucocorticoids can block prostaglandins at the level cyclooxygenase (COX), their clinical effect and adverse effects can be similar to an NSAID. Glucocorticoids can help manage chronic pain in companion animals; however, their potential adverse effects are largely considered greater than NSAIDs and thus are often only chosen in select patients. Potential adverse effects of glucocorticoids including polyuria, polydipsia, polyphagia, immune suppression (which increases the risk of infection), decreased wound healing, GI upset, GI ulceration potentially leading to GI perforation, behavioral changes, panting, iatrogenic Cushing's disease, predisposition to diabetes mellitus, alopecia, and laminitis are significant risks, thus the decision to use glucocorticoids should be made with caution and reassessed frequently. Additionally, the lowest possible effective dose should be considered. Glucocorticoids should never be administered concurrently with NSAIDs or other glucocorticoids because of a high risk of profound GI adverse effects including ulceration, GI perforation, and potential for patient death.

Glucocorticoids can be used in chronic pain patients in which profound anti-inflammatory effects can be beneficial such as for a cat with an inflamed ulcerated oral tumor. Commonly used oral glucocorticoids include prednisone (generic), prednisolone (PrednisTab®), and dexamethasone (Dexaject®, generic) (Figure 16-5). Glucocorticoids can be given orally (tablet or liquid) or via injection (which provide longer yet irreversible effects). Glucocorticoids can be injected directly into joints (methylprednisolone [Depo-Medrol®] and triamcinolone [Vetalog®] are sometimes injected intra-articular in horses to manage osteoarthritis) or epidurally (triamcinolone can be used in dogs to treat lumbosacral stenosis) to treat chronic pain. An additional benefit to glucocorticoids is that in some cases they can increase appetite and weight gain. Their use really is best determined on an individual patient basis.

Tech Tip

Glucocorticoids can be recognized by their generic names by their –one ending.

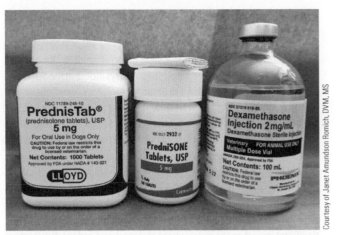

FIGURE 16-5 Glucocorticoids, such as prednisolone, prednisone, and dexamethasone, can help manage chronic pain in companion animals; however, their potential adverse effects limit their use except in select patients.

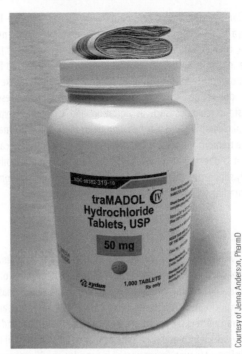

FIGURE 16-6 Tramadol, a mu agonist opioid as well as a serotonin and norepinephrine reuptake inhibitor, is considered an adjunctive agent for treatment of chronic pain in dogs. Tramadol may provide analgesia through opioid-mediated routes in species such as cats and horses.

Opioids

While opioids are effective in treating moderate to severe acute pain, they may not be as desirable of a choice for treating chronic pain as they are often less effective in established chronic pain states and patients may develop opioid tolerance and opioid dependency. Human patients taking opioids long term may develop opioid-induced hyperalgesia (increased sensitivity to pain), and whether or not this condition exists in animals remains controversial. The concern over clients abusing controlled substances along with the poor bioavailability of oral opioids in some companion animal species (e.g., dogs), their potential adverse effects (such as decreased GI motility and colic in horses), and questionable efficacy in treating chronic pain have limited their use. Opioids are not first line drugs for long-term management of companion animal chronic pain.

The main opioids used to treat chronic pain in companion animals are codeine (generic), hydrocodone (Vicodin®), oxycodone (Percocet®, OxyContin®), buprenorphine (Buprenex®), and morphine (Duramorph®, Astramorph®). When using an oral opioid to treat chronic pain in companion animals, keep in mind that limited data is available to confirm efficacy and thus each patient should be evaluated individually as to dosing and response to treatment.

Tech Tip

Although advantageous in dogs, any opioid preparation that also contains acetaminophen must not be given to felines as it could be fatal. It might be labeled as Tylenol® with Codeine (#3, #4).

Atypical Opioids

Atypical opioids used in companion animals include tramadol (Ultram®) and the newer agent tapentadol (Nucytna®) (Figure 16-6). These drugs have been used for both acute and chronic pain management in dogs,

cats, horses, and other species. In humans, tramadol's mechanism of action is as a mu agonist opioid as well as a serotonin and norepinephrine reuptake inhibitor. Tramadol effectively increases serotonin and norepinephrine for use as inhibitory substances within the modulation part of the nociceptive pain pathway. Tramadol initially became a popular oral analgesic choice in dogs over 10 years ago because it was not an NSAID and was not classified as a controlled substance. Unfortunately, we now know that dogs appear to lack adequate quantities of the M1 (O-desmethyltramadol) metabolite following tramadol administration, which is responsible for the drug's opioid effect in humans. This has led many to now question tramadol's efficacy as an opioid analgesic in dogs. Tramadol may provide some analgesia to dogs via serotonin and norepinephrine routes; however, newer research has also suggested that tramadol's bioavailability becomes very low in dogs with repeated administration after approximately one week. As a final nail in tramadol's coffin, it recently become a C-IV controlled substance in the United States. If tramadol is to be administered to dogs orally, it should never be used as a sole form of analgesia and should be considered as an adjunctive agent only. Tramadol may provide analgesia through opioid-mediated routes in other species such as cats and horses. Adverse effects of tramadol include poor palatability, vocalization, increased panting, dysphoria, and sedation. Tapentadol, a C-II controlled substance, is an atypical opioid whose parent compound has a dual mechanism of action (mu opioid receptor agonist and norepinephrine reuptake inhibitor). Initial data in dogs suggests the drug may be more promising as an analgesic than tramadol in dogs. The agent is significantly costlier than tramadol and further research is indicated to determine its clinical usefulness.

A condition known as serotonin syndrome can develop in patients receiving a combination of drugs that increase serotonin (such as methadone, tramadol, tricyclic antidepressants such as amitriptyline [Elavil®], serotonin reuptake inhibitors such as trazodone [Oleptro® which is used for anxiety and to promote sleep], or serotonin-norephinephrine reuptake inhibitors such as duloxetine [Cymbalta®]). Serotonin syndrome presents with mental status changes such as agitation and dysphoria, hyperactivity, tachycardia, vomiting and diarrhea, and neuromuscular abnormalities such as muscle tremors and seizures. In severe forms the syndrome can be fatal. Treatment is directed at removal of the offending drug combination and providing supportive therapy with agents that decrease serotonin such as cyproheptadine (Periactin®) and/or dexmedetomidine.

Nerve Growth Factor Inhibitor

Nerve growth factor (NGF) is produced by a variety of inflammatory and immune cells and joint chondrocytes and is elevated in the joints of dogs with osteoarthritis. When NGF binds to the tropomyosin receptor kinase A (trkA) receptor on nociceptors, it activates ion channels, TrpV1 receptors, and secondary neurotransmitters such as substance P to cause immediate and long-term excitability, which increases pain sensation and the sprouting of new nerve fibers in inflamed tissues. Ranevetmab® is a species-specific monoclonal antibody that binds to nerve growth factor in order to inhibit it from binding to its receptor thus decreasing the sensation of pain.[7,8] The monoclonal antibodies are made so that amino acid sequences are recognized as "self" by the dog's immune system to reduce undesirable adverse effects. In dogs, Ranevetmab® is approved to control pain associated with osteoarthritis and increase mobility for up to 4 weeks with a single IV injection.[8] Research continues to produce a monoclonal antibody for treatment of cats with osteoarthritis.

Analgesic Adjunctives

Analgesic adjunctives are drugs that on their own have minimal pain relieving effects, but are included to enhance the effect of another analgesic agent. Analgesic adjunctives may decrease the amount of analgesics needed to relieve clinical signs that compromise the patient's quality of life and act more specifically on chronic pain.

Sodium Channel Blockers

Sodium channel blockers are agents that impair conduction of sodium ions (Na+) through sodium channels, which blocks depolarization of the nerve and development of an action potential. Sodium channel blockers such as lidocaine (Xylocaine®), bupivacaine (Marcaine®), and ropivacaine (Naropin®, LEA 103®) traditionally act as local anesthetics. Lidocaine has been specifically explored in humans and companion animals via the intravenous route (systemically) as an adjunctive analgesic for various acute and chronic pain states. The verdict is still out as to the efficacy of intravenous lidocaine as an analgesic in humans, let alone in companion animals. Very limited data is available in the veterinary literature documenting efficacy, and the exact mechanism of action via the intravenous route is not understood. Lidocaine may also be applied as a transdermal patch for pain relief. Other local anesthetics such as bupivacaine and ropivacaine should never be administered intravenous due to toxicity. These drugs are described in Chapter 17 (Analgesic Techniques).

N-Methyl-D-Aspartate (NMDA) Receptor Antagonists

Chronic pain can be enhanced by a state of sensitization within the CNS (central sensitization) through numerous pathways including the N-methyl-D-aspartate (NMDA) receptor, which promotes excitatory neurotransmission through glutamate. The NMDA receptor in patients without chronic pain is blocked by magnesium; however, activation of the NMDA receptor causes sodium and calcium ion influx to occur within post-synaptic neurons, which activates second messenger signaling cascades and results in increases to the excitatory neurotransmitter glutamate. Activation of these signaling cascades produces up-regulation of all inputs to the spinal cord neuron thus increasing the cell's response to pain stimuli and decreasing neuronal sensitivity to opioid receptor agonists. NMDA-receptor antagonists have a significant impact on the development of chronic pain and the development of tolerance to opioid analgesics. Consequently, NMDA-receptor antagonists may have potential as co-analgesics when used in combination with opioids.[9] The most commercially available NMDA-receptor antagonists include amantidine, ketamine, and methadone.

- Ketamine (Ketaset®, VetaKet®, Ketalar®, Vetalar®) is a highly effective injectable NMDA antagonist, and subanesthetic microdoses are commonly used as an adjunct to analgesia protocols. Ketamine is not given orally as it is not palatable, and its potential adverse effects limit its regular use for chronic pain in companion animals. Ketamine can be given as an analgesic IV via a constant rate infusion (CRI) for patients in hospital settings. Generally when ketamine is given at such low dosages, it produces minimal adverse effects.

- Amantadine (Symmetrel®) is used in animals as an oral NMDA antagonist. When administered along with the primary NSAID, amantadine has demonstrated usefulness in companion animals with osteoarthritis that is unresponsive to conventional NSAID treatment. The drug helps reset a patient's nervous system in chronic pain. Amantadine can be given orally for several weeks (e.g., **pulse therapy** in which the patient is given the drug for several days then is off of the drug for several days in order to limit potential adverse effects) or for long term use, making it a popular NMDA antagonist choice. The most frequently reported adverse effects of amantadine include GI upset. A cousin of amantadine is memantine (Memenda®, which is used to treat Alzheimer's disease in people), but less data is available about its use in companion animals.

- Methadone (Dolophine®), a full mu agonist opioid, also has NMDA antagonist properties (see Figure 5-17d). The drug is commonly given via injection for acute pain and may be helpful in preventing the establishment of chronic pain states. The drug can also be administered transmucosally in cats to provide a minimum of 4 hours of analgesia, making it appealing for home administration. Palatability is sometimes an issue with transmucosal administration of methadone.

Alpha-2 Adrenergic Agonists

Alpha-2 adrenergic agonists are potent sedative/analgesic drugs that are not considered first line analgesics for management of chronic pain, but are commonly used as analgesic adjunctives. These drugs produce centrally-mediated effects as well as inhibition of pre-synaptic norepinephrine release. Parenteral administration of (α_2) adrenergic agonists is important in managing acute pain in veterinary patients, and their use in the treatment of chronic pain (especially the drug clonidine [Catapres®, an antihypertensive drug in humans]) is being explored for oral and intrathecal administration as an adjunctive agent for varied pain states in people.[10, 11]

Antidepressants

Antidepressants used in humans have been effective adjunctive therapies for a variety of chronic pain states in animals. Antidepressants primarily work by increasing the neurotransmitter levels (serotonin and norepinephrine), which are believed to play a role in pain modulation in the CNS and may also have anti-inflammatory effects on microglia cells. Antidepressants have the added benefit of potentially improving mood, reducing depression, and reducing anxiety in patients in which chronic pain is psychologically and emotionally draining.

- Amitriptyline (Elavil®) is a tricyclic antidepressant that has the best documented efficacy in humans for the treatment of chronic pain syndromes.[12,13] Amitriptyline's analgesic mechanism of action is based on decreasing reuptake of serotonin and norepinephrine at either spinal terminals or the brain stem, which ultimately results in higher levels of these neurotransmitters for inhibitory modulation in the spinal cord; however, other mechanisms may also be involved. Amitriptyline has been used for pain associated with feline interstitial cystitis (FIC) and for other chronic pain syndromes.[14] Experience with amitriptyline use is limited to dogs and cats.
- Duloxetine (Cymbalta®) is a serotonin-norepinephrine reuptake inhibitor (SNRI) that is used in humans for depression and anxiety disorders as well as chronic pain states such as diabetic peripheral neuropathy, fibromyalgia, interstitial cystitis, and musculoskeletal pain. The use of duloxetine in companion animals is currently minimal, but its use may increase in the next few years.

Anticonvulsants

Anticonvulsants have been used in people to reduce chronic pain by suppressing the spontaneous neuronal discharges and neuronal hyperexcitability. The following anticonvulsants are used in animals:

- Gabapentin (Neurontin®), an anticonvulsant drug, has also been used widely as an analgesic adjunctive for chronic pain, particularly neuropathic pain. The mechanism of action of gabapentin, for either its anticonvulsant or analgesic actions, is not understood; however, it appears to be involved with the interaction of calcium channels in the CNS. When serotonin activity is increased, which has been linked to facilitation of chronic pain states, gabapentin works well. When serotonin is *not* increased, gabapentin efficacy is more questionable. Gabapentin is currently being investigated for the treatment of chronic pain states in companion animals; however, confirmed efficacy remains unclear.[15,16,17,18] Gabapentin is typically given long term in conjunction with other chronic pain drugs such as NSAIDs, NMDA antagonists, and opioids and is well tolerated in dogs and cats. The main adverse effect is transient sedation, and geriatric animals with weakness and ataxia from their disease condition may temporarily worsen. Starting the dose at the lower end of the dosage range and increasing it over several days may alleviate this effect.[19] Gabapentin

has also been explored in numerous other species including horses. Currently, Kentucky, West Virginia, Tennessee, Michigan, and Virginia have reclassified gabapentin as a Schedule V controlled substance.
- Pregabalin (Lyrica®) is another gabapentanoid that has an anticonvulsant effect and has been used for chronic pain states. It has been shown to be effective against diabetic neuropathy and posttherapeutic neuralgia in people. It is currently a C-V controlled substance and is expensive, which has limited its use in veterinary medicine.

TrpV1 Receptor Agonists and Antagonists

TrpV1 receptor agonists and antagonists alter nociceptive pain pathways. The TrpV1 receptor (transient receptor potential cation channel subfamily V member 1; previously named the capsaicin receptor and the vanilloid receptor 1) is mainly found in nociceptive neurons of the peripheral nervous system and in lesser amounts in other tissues such as the CNS. The TrpV1 receptor detects and regulates body temperature and provides pain and scalding sensation. Endogenous and exogenous stimuli such as chemical and physical injury and temperatures greater than 98.6 to 113°F will activate this channel, which can lead to a painful, burning sensation. TrpV1 receptor activation also plays a role in the transmission and modulation of inflammatory and chronic pain stimuli. Capsaicin and resiniferatoxin (RTX) are TrpV1 agonists that activate TrpV1, which results in receptor desensitization, potential reduction of substance-P, and alleviation of pain.

- Capsaicin (Capzasin®, Salonpas-Hot®, generic) is an over-the-counter (OTC) example of a TrpV1 agonist that has been used topically in ointments and patches in concentrations of 0.025 to 0.15% to relieve pain caused by osteoarthritis, neck and back injuries, and sprains in people. Capsaicin products have been used to treat osteoarthritis pain in dogs and horses. A large issue with this technique is preventing patients from licking the capsaicin.
- Resiniferatoxin (RTX) is a TrpV1 agonist that aims to cause C-fiber neurolysis. RTX is an extremely potent, naturally occurring capsaicin analog that has been injected experimentally into the intrathecal space of dogs with bone cancer pain producing profound and prolonged analgesia.[20] RTX treatment requires general anesthesia and after administrating it to dogs may increase heart rate and blood pressure as well as cause short term panting and hypothermia.

Bisphosphonates

Bisphosphonates such as alendronate (Fosamax®) and pamidronate (Aredia®) work by blocking osteoclasts (cells that breakdown bone) to prevent bone loss and are used to prevent osteoporosis, alleviate osteosarcoma-related pain, reduce pathologic fractures, and treat hypercalcemia in people. In companion animals they are used as part of a treatment protocol for destructive bone diseases such as osteosarcoma in dogs and navicular disease in horses. They should not be used in patients with severe kidney disease, and a veterinary oncologist should be consulted prior to using these drugs in cancer patients.

Polysulfated Glycosaminoglycan

Polysulfated glycosaminoglycan (PSGAG) (Adequan®) is an FDA-approved drug for IM injection in dogs and horses to reduce inflammation, restore synovial joint lubrication, repair joint cartilage, and reverse the disease cycle.

The commercial product is prepared from bovine tracheal cartilage and is made of a chain of repeating disaccharide units. The primary glycosaminoglycan in PSGAG is chondroitin sulfate. PSGAG is safe and easy to administer, and its mechanism of action likely involves blocking inflammatory mediators, reducing synovial fluid protein levels, and increasing synovial fluid hyaluronic acid concentration in damaged joints. The commercial product is well tolerated by veterinary patients and has been shown to be effective in improving lameness, range of motion, and pain on manipulation of joints for treatment of osteoarthritis.[21] Adequan® is administered every 3 to 5 days for a total of eight injections in dogs and every 4 days for 28 days or seven treatments in horses.[22] The product is then administered as needed. Adequan® can be administered extra label subcutaneously and in cats.

NON-PHARMACOLOGIC OPTIONS FOR CHRONIC PAIN

Managing chronic pain not only involves the use of medications to ease a patient's pain, but also assessment of the animal's nutrition, weight, and lifestyle. The following are some considerations when treating a patient with chronic pain.

Nutraceuticals

Nutraceuticals are foods or food products that theoretically provide health and medical benefits, including the prevention and treatment of disease (the term nutraceutical is made by combining the words "nutrition" and "pharmaceutical"). Such products may range from isolated nutrients, dietary supplements, and specific diets to genetically engineered foods, herbal products, and processed foods such as cereals, soups, and beverages. With recent breakthroughs in cellular-level nutraceutical agents, researchers and medical practitioners have integrated nutraceuticals into medical practice with questionable efficacy. Unfortunately these products do not undergo FDA evaluation.

- A common group of supplements that have been used for chronic conditions include "joint supplements" containing glucosamine, chondroitin sulfate, methylsulfonylmethane (MSM), avocado/soybean unsaponifiables, hyaluronic acid, and other proprietary substances. The use of these nutraceuticals for treatment of osteoarthritis is popular despite limited clinical trial evidence to support their use.[23] The hope of a clinical benefit together with limited adverse effects have allowed glucosamine and chondroitin sulfate products to continue to have a role in the treatment of chronic musculoskeletal conditions. Keep in mind that not all glucosamine and chondroitin sulfate products truly contain the ingredients they report to contain; therefore, products from more reputable manufacturers such as Nutramax Laboratories, Inc. and have the National Animal Supplement Council (NASC) Quality Seal are recommended if these products are to be used. Other nutraceuticals used in the treatment of joint disease/osteoarthritis include phycocyanin (PhyCox®), omega-3 fatty acids (eicosapentaenoic acid [EPA] and docosahexaenoic acid [DHA]), microlactin (Duralactin®), S-adenosylmethionine (SAMe), *Perna canaliculus* (green-lipped mussel preparation), and elk antler.
- Fish oil omega-3 fatty acids (eicosapentaenoic acid [EPA] and docosahexaenoic acid [DHA]) are believed to have anti-inflammatory effects and have been used in the treatment of osteoarthritis in dogs despite

unclear dosing recommendations.[24, 25] It is believed that fish oil may be beneficial in managing pruritus (and therefore its associated pain) in dogs. Fish oil may produce diarrhea and should be added gradually to avoid this adverse effect.

Tech Tip

Efficacy and safety data may be lacking in "natural" and herbal products, and some may produce undesirable effects when given with prescription drugs. These products are not approved by the FDA and can often have limited assurance of efficacy or may not contain what they are reported to contain. Only veterinarians trained in use of herbal products should recommend their use.

Palliative Radiation Therapy

Palliative radiation therapy (PRT) is a type of radiation therapy administered to cancer patients with the goal to alleviate pain associated with the condition. During the PRT procedure, radiation is delivered to the cancer site over an extended period of time in hopes of slowing tumor growth and eliminating existing cancer cells. It is not clear how radiation therapy provides pain relief, but it may be due to reduction of inflammatory cells and prostaglandin production. PRT can give rapid and prolonged pain relief to some veterinary patients depending on the tumor type. Palliative radiation therapy is a specialized technique offered by veterinary radiation oncologists and is typically used in cases in which surgery is not possible. Several treatment protocols have been described based on the patient and type of cancer and should be considered in companion animals with chronic pain due to cancer. Animals must remain perfectly still during PRT; therefore, pets will be anesthetized during the procedure. Palliative radiation therapy usually only causes adverse effects in the areas where the treatment is localized (such as mild to moderate skin irritation or hair loss), but will occasionally cause vomiting and diarrhea.

Alternative Medicine

Alternative medicine consists of a many types of health care practices not typically taught in western medicine and may be combined with conventional western medical treatments (when used together it is known as complementary or integrative medicine) to help alleviate pain in animals. Examples of alternative medicine include acupuncture and body therapies including massage and chiropractic manipulation. Companion animal pet owners often ask about these therapies and how they can be effectively included in managing a patient's chronic pain. The reader is directed to Chapter 18 (Veterinary Physical Rehabilitation) for more detailed information on these interesting subjects.

Maintaining Proper Weight

Maintaining proper weight can improve chronic pain conditions especially those related to osteoarthritis. Decreasing calorie intake and increasing exercise gradually will help overweight patients achieve their ideal weight and avoid developing conditions such as hepatic lipidosis in cats. Commercial weight reducing diets are available to help patients lose weight and then maintain their ideal weight. Avoiding pet treats and human food will also help veterinary patients maintain a proper weight. Animals that are too thin need to increase calorie intake through the use of highly palatable diets, enteral feeding tubes (e.g., esophagostomy and PEG tubes), and perhaps the use of appetite stimulants (e.g., mirtazapine (Mirataz®, Remeron®), glucocorticoids, or cyproheptadine). Some chronic diseases or their treatment may cause GI upset; therefore, antiemetics may be prescribed for these patients. Ensuring proper fluid intake (orally or through the use of fluid therapy) will also help the overall patient health and quality of life.

Physical Therapy and Exercise

Physical therapy and exercise can help alleviate pain in animals with chronic musculoskeletal conditions. Physical therapy techniques are described in Chapter 18 (Veterinary Physical Rehabilitation).

CHRONIC PAIN CONDITIONS

Chronic pain management in animals should aim to lessen the patient's pain, make the animal more comfortable, and enhance its overall quality of life. The optimal treatment plan for chronic pain includes a multimodal approach using drug therapy, alternative/complementary therapy such as physical therapy, and lifestyle changes such as weight loss. Chronic pain conditions can be divided into osteoarthritis, cancer, and non-osteoarthritis, non-malignant pain. Select chronic pain conditions and their treatments are described below.

Osteoarthritis

Osteoarthritis (OA) is any disorder of articular joints characterized by deterioration of articular cartilage; osteophyte formation and bone remodeling; pathology of periarticular tissues including synovium, subchondral bone, muscle, tendon, and ligament; and non-purulent inflammation of variable degree[26] (Figure 16-7). Osteoarthritis is the most common orthopedic problem in dogs and cats, affecting about 20% of the population across all age ranges, and is perhaps the leading cause of chronic pain in companion animals. It has been shown that 90% of cats over the age of 12 have radiographic evidence of osteoarthritis.[27]

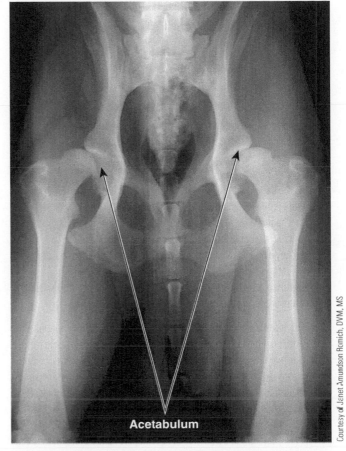

Acetabulum

Courtesy of Jenet Amundson Remich, DVM, MS

FIGURE 16-7 Radiograph of a dog with osteoarthritis. Note the improper alignment of the femoral head and the acetabulum.

Osteoarthritis is either primary or secondary to underlying joint pathology and is characterized clinically by pain, deformity, and limited range of motion (Figure 16-8a through 16-8d). Pathological lesions of osteoarthritis include **focal erosive lesions** (local destruction that occurs at the joint margins and in

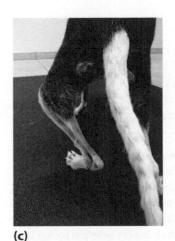

(a) (b) (c) (d)

Courtesy of Shari Sprague MSPT, CCRT

FIGURE 16-8 Dog with osteoarthritis pain secondary to lupus erythematosus. (a) This dog has been in physical rehabilitation for 5 years, which has allowed the patient to enjoy a high quality of life. (b) An anterior image of deformed left hind leg. (c) A posterior image of deformed left hind leg. (d) The left front elbow deformity.

subchondral bone of patients with arthritis), cartilage destruction, **subchondral sclerosis** (increased bone density or thickening in the subchondral layer), and formation of large osteophytes (bone spurs) at the joint margins.

The clinical signs of osteoarthritis are similar regardless of whether the disorder is primary or secondary and regardless of the species in question. The onset is often insidious but progressive.

- Early in the course of the disease, the pet may sporadically be reluctant to perform previous tasks or activities, such as jumping into the car.
- In the next stage, a lameness or stiffness occurs following periods of excess activity or overexertion. These signs often disappear after several days of rest.
- As the degeneration progresses, the stiffness and lameness may be most pronounced following periods of rest.
  - The pet typically "warms up" with subsequent reduction of clinical signs.
  - Any cold or damp weather will increase the severity and duration of the symptoms.
- Continuous stiffness, lameness, and chronic pain encompass the final stage producing an irritable, reclusive, and restless pet.

The most common locations of osteoarthritis in the dog include the pelvis, stifles, elbows, and vertebrae. Common locations for osteoarthritis in cats include the vertebrae followed by the elbow and the pelvis. Common feline signs include grooming difficulties, inappropriate eliminations, reduced jumping, aggressiveness when being handled, and lameness.

Most often treatment is initiated once a patient begins exhibiting clinical signs. Unfortunately, many pets never receive treatment because owners perceive clinical signs as a patient "just getting old" and never make the connection that a medical problem exists. Effective management of chronic pain from osteoarthritis requires a multimodal treatment approach. Multimodal therapy encompasses not only drug therapy, which acts at various levels of the nociceptive pathways, but also multiple non-drug therapies. The objectives of treatment for osteoarthritis should include reducing patient pain and discomfort, decreasing clinical signs visible to the owner, slowing the progression of the disease, promoting the repair of damaged tissue when possible, and improving the patient's overall quality of life.

Tech Tip 🐾

A study from the Journal of the American Veterinary Medical Association (JAVMA) showed that in overweight dogs with hind limb lameness secondary to hip osteoarthritis, weight reduction alone may result in a substantial improvement in clinical lameness.[28]

A thorough orthopedic and neurologic examination should be performed by a veterinarian before beginning therapy for osteoarthritis. Another important part of the evaluation should include information about the pet's living conditions and activities of daily life so that caregivers can modify these to improve patient quality of life. For example, questioning the owner about the type of flooring in the house can be helpful because hardwood floors can be slippery for many older pets and result in increased risk for falls and soft tissue strains. Simply placing more skid-free rugs around the house to give the pet more traction may be helpful.

A wide variety of treatment strategies for patients with osteoarthritis exist and vary in their documented efficacy. Some of the more common therapies include:

- Gradual weight loss. Patients with osteoarthritis should be kept lean, and if they are overweight, gradual weight loss should be encouraged.
- Regular low intensity exercise. Low intensity exercise programs improve quality and quantity of movement. Exercise programs should be developed for patients individually, addressing functional limitations and needs (see Chapter 18 [Veterinary Physical Rehabilitation]).
- Analgesic drugs. NSAIDs are often selected as the first line therapy in patients with osteoarthritis. Numerous adjunctive medications can also be added as needed including amantadine, gabapentin, amitriptyline, and opioids.
- Oral and injectable chondroprotective drugs: This category includes many agents such as glucosamine, chondroitin sulfate, MSM, avocado/soybean unsaponifiables, omega-3 fatty acids, hyaluronic acid, and polysulfated glycosaminoglycan.
- Alternative medicine including massage, acupuncture, therapeutic ultrasound, and low-level laser therapy. Alternative medicine techniques provide a non-invasive, holistic approach to managing chronic pain (see Chapter 18 [Veterinary Physical Rehabilitation]).
- Physical rehabilitation including therapies such as cryotherapy, moist heat, passive range of motion exercises, stretching exercises, balance and proprioception exercises, electrical stimulation, and active exercise (e.g., aquatic, cardiovascular, muscular strength, endurance) (see Chapter 18 [Veterinary Physical Rehabilitation]).
- Nutritional therapy including food rich in omega-3 fatty acids and commercial products such as Hills J/D® and Royal Canin® Mobility Support JS.
- Surgical intervention, regenerative stem cell therapy, shock wave therapy, and platelet rich plasma therapy. Researching new therapies for osteoarthritis treatment, such as radiosynoviorthesis (injection of a radioisotope into the synovial space to treat joint inflammation and mitigate chondromalacia), as they become available will expand the treatment choices that can be provided for your patients.
- Cannabidiol (CBD). This is one of many substances collectively called cannabinoids that occur naturally in the *Cannabis sativa* plant and has been studied for use in treating canine osteoarthritis. A 2018 clinical trial evaluating the effect of CBD on osteoarthritis in dogs showed reduction in pain scores when treated with CBD at 2 mg/kg q12h versus placebo with no negative side effects.[29] CBD has neuroprotective and anti-inflammatory effects; however, the amount of available scientific evidence pertaining to its use in animals is currently limited. Each state has its own regulations concerning cannabis; therefore, it is important to contact state veterinary medical associations regarding its use to maintain licensure in good standing. One cannabis-derived product (Epidiolex®) has been approved by the FDA for use in people to manage two rare forms of seizures (Lennox-Gastaut syndrome and Dravet syndrome). That drug can be used in an extra label manner by veterinarians in accord with the Animal Medicinal Drug Use Clarification Act (AMDUCA).

Lastly, osteoarthritic animals may benefit from household and life style modifications. For example, cats with osteoarthritis may find it difficult to jump and reach their favorite high points where they feel safe and content, be it a bed, sofa, or window ledge. Moving furniture to provide "stepped access" will help an osteoarthritic cat reach their favorite resting places. A variety of purpose-built steps and ramps are also commercially available. Harnesses may be useful to help improve mobility of dogs. Litter pans can be adapted so that it is easy for an arthritic cat to enter and exit from them and to comfortably position itself when urinating and defecating. The list for household and life style modifications is endless, and it is up to both the veterinary professional and pet owner to help come up with creative solutions for the patient.

Cancer Pain

Cancer pain typically progresses slowly and is often undetected for some length of time before owners notice a change in their animal's demeanor, function, or routine. Cancer pain in its early stages involves nerve fibers and nociceptors; however, as the cancer progresses the amount of nerve involvement and pain mechanisms in abnormal tissue varies from those observed in other types of pain. Cancer pain can be due to the tumor itself (caused by pressure and tissue invasion or destruction that can destroy normal tissue as well as trigger the patient's immune response, etc.), metastases (e.g., pleural pain in metastases that spread to the lung), paraneoplastic conditions (such as peripheral neuropathies), diagnostic procedures (biopsies), treatment (radiation treatment, chemotherapy, etc.), or secondary to unrelated aging conditions (such as osteoarthritis in older patients). For example, a dog with appendicular osteosarcoma can present in excruciating limb pain due to the osteolytic nature of this cancer, a rabbit with an uterine adenocarcinoma may have intense cancer pain attributed to stretching the capsule of visceral organs, a dog with a very large soft tissue sarcoma can have pain because the mass pushes inward toward the abdomen, or a cat with gastrointestinal lymphoma may have pain caused by obstructing flow in the affected area (GI tract In this case, but could also include biliary or urinary obstruction). Additionally, the host's immune response to the cancer itself may generate pain as a result of the liberation of inflammatory cytokines and chemicals.[30] All of these reasons make cancer pain challenging to manage.

Cancer pain due to the mass itself can be divided into two categories: pain that is caused by chemical irritation and pain that is caused by pressure and inflammation from direct tumor invasion.[28]

- Chemical irritation can occur for a variety of reasons such as
  - Release of growth factors, for example, nerve growth factor that regulates the growth and sensitivity of neurons. An increase of peripheral sensory neurons in or near the tumor can increase pain sensitivity, decrease the pain threshold, and produce long-lasting pain. This type of pain may be controlled by the use of anti-nerve growth factor antibodies.[31]
  - Tissue hypoxia. As cancer cells replicate, their rapid growth results in tissue expansion and increased innervation and vasculature to the tumor, which results in tissue hypoxia and nutrient deficiency. As the tumor is deprived of oxygen and nutrients, it releases harmful chemicals that can lead to peripheral and central pain sensitization.[31] This type of pain can be helped by surgical removal of the mass or palliative therapy to reduce the size of the mass.
- Direct tumor invasion causes pressure and compression, which lead to mechanical and pressure-induced pain due to
  - Pain signal cycling. As tumors enlarge, they stretch mechanoreceptors in the skin and muscles and compress nerves in surrounding tissues. If this pain is not treated, the tumor produces a cycle of pain signaling that results in chronic pain.[31] This type of pain is controlled by surgical removal of the mass or palliative therapy to reduce the size of the mass.
  - Inflammation. Uncontrolled proliferation of cells into normal tissue causes death of normal tissue and triggers the patient's inflammatory response.[31] This type of pain is similar to the inflammation of osteoarthritis and typically responds well to anti-inflammatory drugs.

The goal of managing animals with cancer is to control pain and improve the patient's overall quality of life by using traditional anticancer therapies,

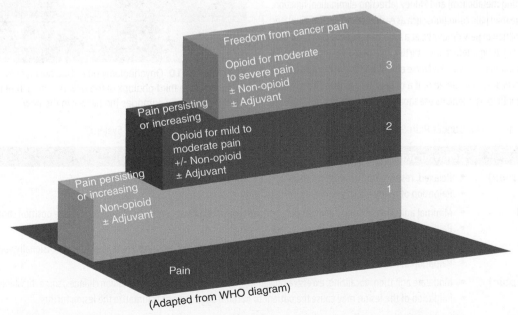

(Adapted from WHO diagram)

FIGURE 16-9 WHO pain ladder. The three-step analgesic ladder was designed for controlling mild, moderate, and severe pain in people. Any type of adjuvant therapy can be added at any level.

various analgesic drugs and techniques, and supportive care.[30] In some cancer patients medical care concentrates on palliative care that reduces the severity of disease signs, rather than striving to halt, delay, or reverse progression of the disease itself or provide a cure. For example, a treatment protocol may involve palliative surgeries to remove painful tumors even though the surgeries may not positively affect the overall prognosis for patient survival.[32] Consider a dog with osteosarcoma of the humerus in which metastatic disease ultimately determines the survival time. In this example, amputating the limb provides pain relief to the patient, and adjunctive chemotherapy is administered to prolong survival and delay the growth of metastases.[33]

Management of pain in cancer patients involves its early recognition and frequent assessment to determine if the patient's level of pain is improving or worsening. The World Health Organization (WHO) developed a three-step analgesic ladder used in humans to control mild, moderate, and severe pain (Figure 16-9)[34] and a modification of this scale can be used in veterinary patients. For example, one modification of the WHO three-step analgesic ladder may consist of using NSAIDs for the first step of the ladder for mild pain, then adding analgesic adjunctives on top of the NSAID for persisting pain/moderate pain in step two, and then considering opioids for persisting pain/severe pain in the third step along with an NSAID and adjunctive pain medication. Note that opioids are introduced as a last/late stage intervention. This ladder can be useful when considering the approach to treating a degree of pain and for understanding how to introduce additional drugs if pain management is inadequate. The Purdue Integrated Pain Score System can be used to adjust analgesic use in cancer patients (Table 16-2).

Contemporary thinking also suggests using more aggressive analgesic therapy early in fully established severe pain states. For example, if you are presented with a dog with severe pain from cancer, beginning with just an NSAID is likely inadequate. Jumping directly to a multimodal approach including NSAIDs, opioids, local anesthetics, NMDA antagonists, anticonvulsants, tricyclic antidepressants, bisphosphonates, and complementary therapies like acupuncture (may be contraindicated in patients with neoplasia) and rehabilitation medicine are likely beneficial. Caution should be used when treating geriatric patients with a variety of agents due to their potential to have altered liver (affecting metabolism) and kidney (affecting elimination) function. When treating more mild pain conditions such as early osteoarthritis in a geriatric patient, the introduction of new drugs one at a time, in a sequential manner may avoid and identify any drug interactions. Starting said dog on a dozen therapies at one time may result in undesirable adverse effects and overwhelm an owner. When adverse effects are encountered or if a drug is used at its maximum safe dosage without benefit to a patient, its use should be discontinued.[35]

Owners frequently ask veterinary professionals, "How will I know when is the right time to euthanize my pet?" Every animal has basic needs that should be met at a satisfactory level to justify preserving the life of ill patients. Pain scales and quality of life scales (see Figure 16-17) serve as guidelines to ensure pet owners that they can maintain a rewarding relationship with their pet that nurtures the human-animal bond. Using these scales helps the veterinary team actively support owners in the care and end-of-life decision making and alleviates owners' feelings of guilt when they decide to euthanize their pet.

Non-Osteoarthritis, Non-Malignant Chronic Pain

Non-osteoarthritis, non-malignant chronic pain conditions include post-amputation including onychectomy (Figure 16-10), degloving injury (Figure 16-11a through 16-11c), neck/back pain (such as discospondylosis and intervertebral disc disease), and periodontal disease (including stomatitis).

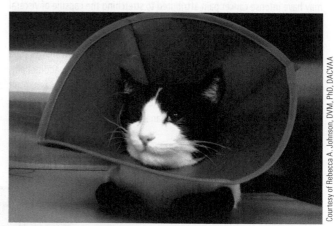

Courtesy of Rebecca A. Johnson, DVM, PhD, DACVAA

FIGURE 16-10 Onychectomy pain. Declawing cats involves amputating the third phalanx of the paw resulting in neuropathic pain because it directly damages the nerves in the paw.

TABLE 16-2 In-Hospital Cancer Pain Score System (part of the Purdue Integrated Cancer Pain Score System)*

Pain Score	Behavior Signs
1 (minimum pain)	• Relaxed, resting comfortably, not vocalizing, moving freely, calm or asleep • Palpation of lesion elicits no reaction from patient
2 (faint pain)	• Minimal agitation, resting calmly, barely noticeable alteration from signs of minimal pain, some position changes • Palpation of lesion elicits minimal response
3 (mild pain)	• Mild agitation, some position changes, responds to calm voice and stroking, some salivation, occasionally vocalizing • Palpation of lesion may cause patient to turn head, lick and/or scratch the lesion
4 (moderate pain)	• Moderate agitation, vocalizing, excessive salivation, muscle trembling, frequent position changes, some thrashing movements • Palpation of the lesion may cause the patient to become aggressive or traumatize the lesion further
5 (severe pain)	• Severe agitation, vomiting, vocalizing, excessive salivation, extremely depressed, inactive • Palpation of the lesion increases the level of agitation

* Impellizeri, J. A., Tetrick, M. A., & Muir, P. (2000). Effect of weight reduction on clinical signs of lameness in dogs with hip osteoarthritis. *JAVMA*, 216(7), 1089–1091.

Source: Nicholas Rancillo, Jean Poulson and Jeff Ko of Purdue University.

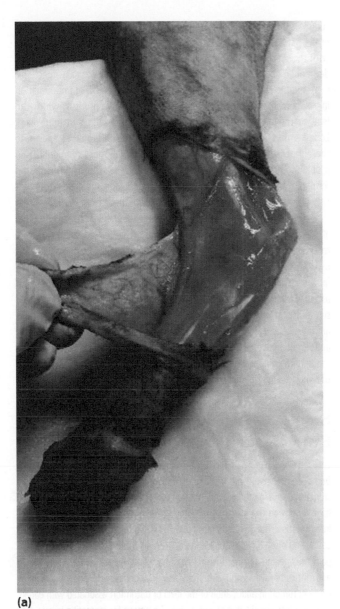

(a)

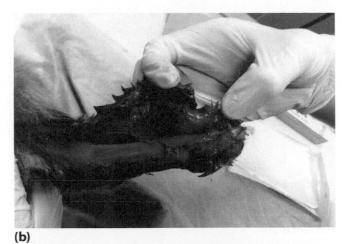

(b)

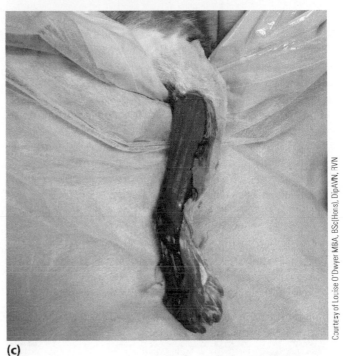

(c)

Courtesy of Louise O'Dwyer MBA, BSc(Hons), DipAVN, RVN

FIGURE 16-11 Feline degloving injury. A degloving injury, named because its process resembles removing of a glove, is an avulsion injury in which an extensive section of skin is completely torn off the underlying tissue, severing its blood supply and damaging nerves. Typically, degloving injuries affect the tail and limbs and produce neuropathic pain.

Select non-osteoarthritis, non-malignant pain conditions are described below.

Neck and Back Pain

Many medical conditions cause neck and back pain in dogs and cats including interverbral disc disease (IVDD) and discospondylosis. Neck and back pain can be caused by a variety of structures including the intervertebral discs, vertebral ligaments, vertebral periosteum, and the meninges and may present in animals as reluctance to walk, jump, or go up the stairs; inability to lift or lower the head to eat or drink; vocalization; or tenseness of the muscles. Depending on the type of disease process and its location within or surrounding the spinal cord, it can directly or indirectly stimulate pain sensors (nociceptors). Inflammatory diseases such as meningitis may hypersensitize nociceptive pathways due to the release of inflammatory substances such as serotonin and histamine, while diseases that cause mechanical compression of nociceptive pathways such as IVDD put pressure on the spinal cord or spinal nerve roots resulting in chronic neck or back pain. Rupture or protrusion (herniation) of an intervertebral disc puts pressure on the spinal cord or spinal nerve roots (Figure 16-12) and can occur in any area of the spinal cord. The clinical signs associated with IVDD are due to direct mechanical pressure on nerve tissue and hypoxic changes to the spinal cord due to pressure on the blood vessels supplying this area. Common clinical signs include back pain, ataxia, lameness, and/or inability to walk in the hind limbs or all four limbs depending on the area of the spinal cord affected. Dachshunds appear to be predisposed to developing IVDD. Young to middle-aged dogs are most commonly affected with dogs less than 1 year of age rarely being affected with IVDD. Geriatric dogs can also be affected with IVDD. Clinical signs of IVDD occur in cats as well.

IVDD degeneration occurs in all breeds of dogs; however, it is observed most frequently in the chondrodystrophoid breeds such as Dachshunds, Pekingese, and French Bulldogs. IVDD usually presents between 5 and 12 years of age in nonchondrodystrophoid dogs (Labrador Retrievers,

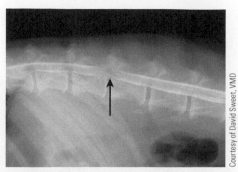

Courtesy of David Sweet, VMD

FIGURE 16-12 A myelogram showing IVDD. Dye injected into the spinal column shows compression of the spinal cord due to a ruptured intervertebral disc. Compression of the spinal cord results in neuropathic pain.

German Shepherd dogs, etc.). IVDD is diagnosed based on clinical signs and advanced diagnostic imaging such as computed tomography (CT) or magnetic resonance imaging (MRI) and sometimes myelogram. Radiographs of the spine are no longer considered useful at diagnosing IVDD in most cases. Medical treatment of IVDD when clinical signs are mild includes strict rest and analgesic drugs such as NSAIDs (such as carprofen, meloxicam, deracoxib, and firocoxib), possibly a muscle relaxant such as methocarbamol (Robaxin®), and an oral opioid (e.g., codeine) plus an analgesic adjunctive such as gabapentin. Dogs with severe pain or neurological defects often require surgery and aggressive multimodal intravenous analgesia. Alternative therapies such as acupuncture, chiropractic treatment, and lifestyle modifications such as weight loss may also be recommended.

Chronic Otitis

Otitis is inflammation of the inner, middle, or external part of ear and can be caused by a variety of factors such as bacteria, yeast, parasites, allergies, and immune-mediated disorders (Figure 16-13). Pain from chronic ear disease is

Courtesy of Janet Amundson Romich, DVM, MS

FIGURE 16-13 Chronic otitis externa in a dog. The shape of the patient's ear can lead to retention of glandular secretions, the presence of irritants, and moisture, which can all contribute to the development of inflammation. Inflammation may cause pruritus, which can lead to trauma and become painful.

common and is related to the inflammatory nature of the disease. The shape of the patient's ear can lead to retention of glandular secretions and along with the presence of irritants (bites of parasites, immune mediators from degranulation of mast cells in allergic reactions, chronic yeast infections, etc.) and moisture can all contribute to the development of inflammation in an animal's ear. Inflammation in turn may cause pruritus, which can lead to trauma and further damage to the ear resulting in more pain. Long-term treatment of the ears with cleaners, irritating drugs applied to the epithelium, and the use of cotton-tipped applicators compounds the problem. Chronic inflammation of the ears may lead to hyperplasia and dysplasia. Infections, improper treatments, and repeated treatments that may produce resistant microbes only exacerbate the matter.

Clinical signs of chronic otitis may include shaking the head, pawing at the affected ear, tilting the head, leaning to the side of the affected ear, and in severe cases an altered sense of balance. Treatment of chronic pain caused by otitis usually involves the use of an anti-inflammatory agent such as an NSAID to reduce inflammation, swelling, and potentially scar tissue and analgesic adjunctives such as gabapentin or NMDA-receptor antagonists such as amantadine to treat central sensitization. Treating the underlying cause of otitis (antibiotics for bacteria, antifungal agents for yeast, determination of allergens and immunotherapy or antipruritic drugs for allergies, etc.) is a critical component to resolving the disease condition. Sometimes a patient needs to be placed under general anesthesia for a thorough cleaning (or flushing) of the ear canal.

Stomatitis and Feline Orofacial Pain Syndrome (FOPS)

Cats can develop oral pain for a variety of reasons such as dental disease, thermal burns from chewing electrical cords, malignant tumors, tooth reabsorption, stomatitis, and orofacial pain syndrome. For example, cats with stomatitis have lesions on the mouth, gums, and upper throat due to long term, severe inflammation. The affected areas typically have a bright red appearance, bleed easily, and often have a cobblestone appearance. These cats will exhibit severe pain on opening the mouth and may have bad breath, excessive drooling, and difficulty swallowing. Cats may avoid food despite being hungry, may hiss and run off in anticipation of the level of discomfort associated with eating, and may lose weight due to not eating. Treatment for stomatitis in cats may include dietary changes (such as switching to semi-solid food or inserting a feeding tube), antibiotics, antiseptic mouthwashes, dental prophylaxis, and tooth extraction along with the use of analgesics (an anti-inflammatory such as a glucocorticoid and perhaps a short course of an opioid such as buprenorphine and an analgesic adjunctive such as gabapentin or amitriptyline).

Orofacial pain disorders are well described in humans, and a syndrome in cats known as feline orofacial pain syndrome (FOPS) has been recognized especially in Burmese cats, although occasional cases have been seen in the Domestic Shorthair, Burmilla, and Siamese cats.[36] FOPS is believed to cause chronic pain that may involve the processing of information from the trigeminal nerve, which produces **paroxysmal** (a sudden attack or increase in clinical signs) bouts of pain that are typically unilateral. The pain is mainly precipitated by facial movement (such as chewing), and cats present with exaggerated licking and chewing movements and pawing at the mouth. Trigeminal nerve pain is caused by a combination of peripheral damage (such as dental disease, fractured tooth, etc.) and decreased cerebral brainstem inhibition of the trigeminal nerve that results in a paroxysmal discharge of pain impulses. In many of the affected cats, tongue discomfort seems to be the primary problem, and it is not unusual for the primary presentation to be to an emergency clinic with acute onset of severe tongue mutilation. Preventing mutilation is important while investigating the cause of the patient's pain. Analgesic treatment may be similar to that described for stomatitis.

Post-Amputation

Phantom limb pain that occurs after amputation is a recognized syndrome in humans, but it may also occur in veterinary patients including post-onychectomy in cats (see Figure 16-10). Post-amputation pain may be attributed to peripheral sensitization either as a result of spontaneous activity from sprouting regenerating nerve endings or from neuroma formation that gives rise to secondary changes in otherwise silenced small dorsal root ganglia cells, central sensitization, or cortical reorganization.[32] In humans there has been a correlation between the severity of pre-amputation pain with post-amputation phantom pain. Prevention of phantom pain by preoperative epidural analgesia and postoperative local anesthesia, however, resulted in variable responses.[32] There seems to be no consistent effective treatment for post-amputation pain. Many therapies that are effective in other chronic conditions, including surgical exploration, tricyclic antidepressants (TCAs), sodium channel blockers, topical capsaicin, and gabapentin, may be ineffective or unproven in controlled studies of phantom limb pain.[32] Interestingly, chronic pain occurs in 60% of human patients after limb amputation but usually not until 1 year after surgery. This highlights the ongoing changes occurring in the peripheral nervous system (PNS) and CNS that are established by a precipitating event before chronic pain is experienced in some patients and with some conditions. Phantom limb pain is rarely reported in veterinary medicine; however, this does not mean it does not exist.[25] In cats, post-amputation pain is believed to occur post-onychectomy, which can result in a cat not walking normally and using its front feet effectively. Potential clinical signs of post-amputation pain other than those occurring from nerve entrapment may include chewing at the stump, intermittent unprovoked crying, or "jumping up or away" indicating **lancinating** (stabbing) pain attributable to ectopic nerve firing. Frequently, amputation is performed because of motor nerve injury; however, it is important to identify the exact level of the lesion to ensure that the nerve injury is relieved so as to prevent ongoing or potential future development of chronic pain.

Ocular Pain and Glaucoma

Ocular pain can be quite intense because keratitis, iritis, uveitis, and glaucoma can produce a deep, dull pain with inflammation. Corneal lesions and foreign bodies can produce sharp, acute pain due to nerve exposure.[37] Ocular surgery, trauma, or infection of the corneal epithelium can cause acute, severe pain that is difficult to control (Figure 16-14). Patients with glaucoma have increased intraocular pressure, which is painful and causes degeneration of retinal ganglion cells and their axons (Figure 16-15). For veterinary patients, enucleation in combination with analgesic drugs may be the best way to alleviate the animal's ocular pain. If bright lights cause pain, then reduced lighting may be necessary to help alleviate pain.

Tech Tip

The cornea has the highest concentration of free nerve endings in the body and can therefore be particularly painful.[38]

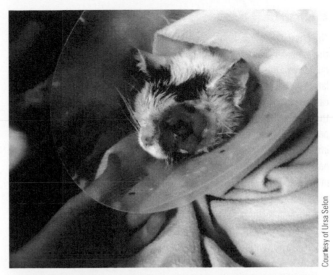
Courtesy of Ursa Selon

FIGURE 16-14 Ocular pain. Damage to the cornea, which has many free nerve endings, and inflammation associated with many ocular conditions can lead to pain.

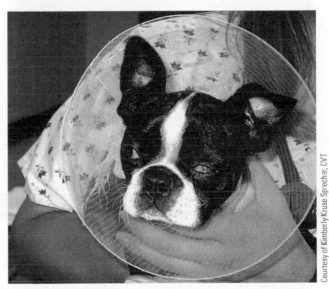
Courtesy of Kimberly Kruse Sprecher, CVT

FIGURE 16-15 Glaucoma in a dog. Increased intraocular pressure can lead to pain in patients with glaucoma.

Feline Interstitial Cystitis

Feline interstitial cystitis (FIC) is a well-recognized condition in cats and humans and is an example of visceral inflammation resulting in pain.[30] The term feline interstitial cystitis is reserved for those cats that have frequent recurrences of clinical signs or persistence of chronic clinical signs. Idiopathic cystitis can be acute or chronic, but interstitial cystitis by definition is a chronic process. Idiopathic cystitis is classified as a noninfectious inflammatory lower urinary tract disease. Since the underlying cause of FIC is unknown, current treatment recommendations are multifaceted and tentative. Recommendations center on provision of pain relief (e.g., buprenorphine), changes in diet (e.g., feeding urinary diets enriched with omega-3 fatty acids and antioxidants) and feeding practices, an increase in water intake (including transition to moist diet), optimal litter box management, nutraceuticals (e.g., glucosamine/chondroitin sulfate, which has been reported to improve the intercellular barriers in the urinary bladder epithelium), and drug therapy for underlying conditions.

SUMMARY

The physiological consequences of chronic pain have serious repercussions for the patient's ability to heal and affect immediate and long-term quality and quantity of life. Pain can lead to aggression, self-mutilation, reduced ambulation, or decreased interest in normal activities in veterinary patients, which can negatively impact not only the patient but other animals and the people they contact as well. The veterinary profession has an ethical duty to anticipate, identify, and provide pain relief for their patients. To meet this duty, it is important to assess pain using consistent parameters, understand how acute pain can transition into chronic pain, identify the mechanism of action of drugs in the pain pathway in order to optimize their use, and recommend adjunctive therapies that can be used to ease an animal's pain. Improving patient comfort through the use of a multimodal analgesia plan that is administered preemptively to prevent acute pain from becoming chronic pain is the first line of defense in preventing chronic pain. Once chronic pain is established, it is important to develop a pharmaceutical and life-style management plan to minimize a patient's level of pain. Having a working knowledge of a variety of analgesics and pain management techniques will allow the veterinary team to work together on a holistic approach to managing chronic pain in their patients.

CRITICAL THINKING POINTS

- Differentiating between acute and chronic pain can help determine treatment modalities for veterinary patients.

- The key to preventing chronic pain is addressing acute pain so that sensitization of the peripheral and central nervous systems does not occur.

- The pain pathway consists of transduction, transmission, modulation, and perception. Analgesics work at different parts of the pain pathway; therefore, using more than one drug to treat pain is often beneficial.

- Veterinary patients with chronic pain exhibit many clinical signs including decreased activity, anxiety, exhaustion, decreased behaviors such as grooming and using the litter box, fear, and aggression, which make identifying and controlling this type of pain challenging.

- Physiologic changes from pain affect many body systems including the cardiovascular, respiratory, and ocular systems.

- Learning how to assess pain using various pain scales will help monitor a patient's level of pain and the success or failure of treatment.

- Chronic pain scales differ from acute pain scales; both are valuable for grading different types of pain.

- Hindrances to assessment and alleviation of pain include:
  - Lack of appreciation that many chronic disease processes are associated with significant pain
  - Inability to accurately and consistently assess chronic pain in animals
  - Deficient knowledge of drugs, drug therapy, and other pain relieving techniques
  - Miscommunication with clients in the assessment and treatment of chronic pain
  - Underuse of veterinary staff for assessment and reevaluation of pain in hospitalized patients

REVIEW QUESTIONS

Multiple Choice

1. Which term describes the process of converting a noxious mechanical stimulus into an electrochemical signal that reaches the brain?

    a. Neuroplasticity
    b. Allodynia
    c. Nociception
    d. Wind up

2. Which is false regarding chronic pain?

    a. It is pain that persists beyond the expected time period for normal healing.
    b. It is pathologic (maladaptive) because it serves no biologic purpose.
    c. It is often resistant to medical treatments.
    d. It only occurs in geriatric patients.

3. What can prolonged noxious stimulation that is not treated or inadequately managed lead to?

    a. Rapid modification of the nervous system due to an overworked pain pathway
    b. Numbing of nerve fibers that can lead to neuron death
    c. Re-establishment of acute pain due to repeated nerve stimulation
    d. Adaptation of nerve stimuli to remind the body of tissue injury or damage

4. Which is the correct order of the individual pathway components of nociception?

    a. Perception, transduction, transmission, modulation
    b. Modulation, transmission, transduction, perception
    c. Transduction, transmission, modulation, perception
    d. Transmission, modulation, transduction, perception

5. What is the term for the central nervous system's ability to reorganize itself by forming new neural connections?

   a. Hyperalgesia
   b. Neuroplasticity
   c. Allodynia
   d. Adaptation

6. Chronic pain that is not relieved or controlled can affect which of the following?

   a. Urinary system (development of kidney stones and polyuria)
   b. Cardiovascular system (worsening of arrhythmias and hypotension)
   c. Central nervous system (increased incidence of ataxia and seizure activity)
   d. Endocrine system (modification of the hypothalamic-pituitary-adrenal axis)

7. As chronic pain develops, which of the following mechanisms occur?

   a. Wind up, which causes expansion of the central receptive fields in the brain.
   b. Sensory fibers disengage from transmitting pain, yet they continue to be damaged from local inflammatory mediators.
   c. Large amounts of local inflammatory mediators produce increased sensitivity of central nociceptors.
   d. Increased excitability of neuron populations within central nociceptive pathways lead to central sensitization.

8. What are examples of changes that could occur with chronic pain?

   a. Decreased activity, socialization, and learned behavior
   b. Increased grooming, appetite, and interaction with other animals
   c. Normal posture and appetite yet significant weight loss
   d. Increased interaction with owner and desire to be touched (petted, groomed, etc.)

9. Which enzyme is associated with pain and inflammation?

   a. Cyclooxygenase 1 (COX-1)
   b. Cyclooxygenase 2 (COX-2)
   c. Cyclooxygenase 3 (COX-3)
   d. Glucuronyl transferase

10. What is a benefit of using systemic glucocorticoids versus nonsteroidal anti-inflammatory drugs when treating select cases of chronic pain?

    a. Speeding of wound healing postoperatively
    b. Profound anti-inflammatory effects
    c. Decreased appetite, which can help reduce weight in overweight patients
    d. Protection of mucus secreting cells of the gastrointestinal tract to prevent gastric ulceration

11. Which statement is true regarding the use of tramadol for chronic pain?

    a. It should never be used as a sole form of analgesia.
    b. It is categorized as a subtype of NSAID and cannot be used concurrently with other NSAIDs.
    c. It is palatable to dogs and cats, which makes oral administration easier.
    d. It is a C-IV controlled substance, which is a low enough abuse rating so that it does not need to be documented in a controlled substance log.

12. Capsaicin belongs to which drug category?

    a. Antidepressant
    b. TrpV1 agonist
    c. Anticonvulsant
    d. NMDA receptor antagonist

13. Why may dogs with destructive bone diseases such as osteosarcoma benefit from the administration of bisphosphonate drugs?

    a. Prevent formation of cyclooxygenase enzymes
    b. Block osteoclast activity to prevent bone loss and pathologic fractures
    c. Provide anti-inflammatory effects
    d. Prevent serotonin and norepinephrine reuptake

14. What is the goal of palliative radiation therapy?

    a. Slowing tumor growth and possibly eliminating existing cancer cells
    b. Decreasing development of metastases associated with some types of cancers
    c. Blocking inflammatory mediators that can trigger production of harmful substances
    d. Destroying tissue in an area ravaged by cancer cells

15. Which group of analgesic adjunctives suppresses spontaneous neuronal discharges and neuronal hyperexcitability?

    a. Opioids
    b. Antidepressants
    c. Anticonvulsants
    d. Bisphosphonates

16. What are opioids effective for?

    a. Treating chronic pain
    b. Managing behavioral changes due to chronic pain
    c. Treating acute pain and as part of pre-emptive analgesia
    d. Managing inappetence associated with chronic pain

17. Which analgesic is an oral NMDA antagonist that is used along with NSAIDs in companion animals for problematic osteoarthritis?

    a. Tramadol
    b. Amantadine
    c. Buprenorphine
    d. Clonidine

18. What are the most common organs in which adverse drug reactions associated with NSAIDs occur?

    a. Gastrointestinal tract, kidney, and liver
    b. Heart, lungs, and spleen
    c. Urinary bladder, brain, and bones
    d. Reproductive, muscular, and skin

19. What is a contraindication for NSAID use in a patient?

    a. Also receiving systemic glucocorticoids or other NSAID
    b. Less than 1 year of age as their liver and kidneys may not be fully developed
    c. With a past history of diarrhea due to parasitic or inflammatory causes
    d. That have been rehydrated with lactated ringers solution (LRS)

20. Which group of drugs increases serotonin and norepinephrine levels that are believed to affect pain modulations in the central nervous system?

    a. NMDA receptor antagonists
    b. Anticonvulsants
    c. Tricyclic antidepressants
    d. TrpV1 receptor agonists

Case Studies

Case Study 1: Grayson, a 9-year-old, M/N, Greyhound dog weighing 75 lb (34.0 kg), is diagnosed with osteosarcoma (OSA) of the distal radius. The standard therapy for canine OSA includes radical surgical resection, adjunctive chemotherapy, and palliative radiation therapy. While some veterinary cancer patients' pain may be relieved adequately by treating the underlying malignancy, most will require symptomatic therapy to achieve improved quality of life, the primary objective of veterinary oncology. The median survival time of appendicular OSA treated with amputation followed with adjunctive systemic chemotherapy (doxorubicin, platinum compounds) is approximately one year. The attending veterinarian has elected to treat the pain at the time of and immediately following amputation, but is not pursuing chemotherapy due to the owner's requests.

Because the goal in this case is to attempt to "reset" the CNS in hopes of improving response to oral analgesics, it is important to be aggressive with calculating a dose of analgesic for this dog to manage his pain at the time of amputation. A CRI of morphine, lidocaine, and ketamine (MLK) was started knowing that if adverse effects are seen, the drug dosage rate will be lowered while still realizing reasonable anti-hyperalgesic/analgesic effects. It is ideal to run the drug combination for 24 hours, though 12 hours can be helpful if overnight care is not available.

1. Look at the pain pathway and describe where the three medications are effective in the nociceptive pathway. Identify if each drug is effective in transduction, transmission, modulation, or perception.

2. Would using a pain scale in this case be helpful? If so, why?

Case Study 2: Phideaux is an 11-year-old, M/N Beagle canine and weighs 35 lbs (15.9 kg). His owners bring him in because "he isn't moving around much. He has a hard time getting up when he is lying down or sitting. Sometimes he cries out. He hasn't been spending as much time with the family." When the owners see the number on the scale combined with the knowledge they already have about their pet slowing down, they are concerned that their pet has gained so much weight compared to the last wellness visit. Your veterinarian suggests the owners complete the Helsinki Chronic Pain Index (HCPI) form before he sees the patient (Figure 16-2).

1. How do you explain to the owners the importance of answering the questions on the HCPI scale?

2. How do you delicately find out what could be the reasons for the dog's increase in weight?

Case Study 3: Gertrude is a 4-year-old, F/S Cocker Spaniel canine that presents for chronic ear problems. She shakes her head constantly and has a foul smelling discharge from her right ear. She is shy and hides behind her owner, but when the ear is examined she attempts to bite. The owner says she has always had ear trouble. She's "been on every medication you can think of." After obtaining Gertrude's temperature, pulse, and respiration (TPR), the veterinarian wants to use the Psychological & Behavioral section of the Colorado State University Chronic Pain Scale for Dogs to find out what her score is upon presentation (the other sections of the Colorado State University Chronic Pain Scale are typically used for dogs with musculoskeletal concerns) (Figure 16-4).

Read the CSU Scale to determine the following:

1. What is Gertrude's pain score?

2. Why did you assign that particular score?

Case Study 4: Molly is an 11-year-old, F/S Golden Retriever dog that was diagnosed with T-cell lymphoma. Originally, she had Stage III (multiple nodes enlarged on both front and back halves of body). She was placed on chemotherapy and after several months of treatment went into remission. Molly's owners have brought her in because she has recently been weak and has pale mucous membranes, loss of appetite, and is lethargic. Molly's owners have asked the attending veterinarian about how they will know when it is time to consider euthanasia. The veterinarian has spoken to them and gives you (attending veterinary technician) the Quality of Life Scale to explain to the owners (Figure 16-16).

1. How do you accomplish this task? Are there any special considerations for the owners? For the patient?

Quality of Life Scale

HURT

SCORE

Adequate pain control, including breathing ability, is first and foremost on the scale. Is the pet's pain successfully managed? Is oxygen neccesary?

HUNGER

SCORE

Is the pet eating enough? Does hand feeding help? Does the patient require a feeding tube?

HYDRATION

SCORE

Is the patient hydrated? For patients not drinking enough, use subcutaneous fluids once or twice daily to supplement fluid intake.

HYGIENE

SCORE

The patient should be kept brushed and cleaned, particularly after elimination. Avoid pressure sores and keep all wounds clean.

HAPPINESS

SCORE

Does the pet express joy and interest? Is he responsive to things around him (family, toys, etc.)? Is the pet depressed, lonely, anxious, bored or afraid? Can the pet's bed be close to the family activities and not be isolated?

MOBILITY

SCORE

Can the patient get up without assistance? Does the pet need human or mechanical help (e.g., a cart)? Does he feel like going for a walk? Is he having seizures or stumbling? (Some caregivers feel euthanasia is preferable to amputation, but an animal who has limited mobility but is still alert and responsive can have a good quality of life as long as his caregivers arecommitted to helping him.)

MORE GOOD DAYS THAN BAD

SCORE

When bad days outnumber good days, quality of life might be too compromised. When a healthy human-animal bond is no longer possible, the caregiver must be made aware that the end is near. The decision needs to be made if the pet is suffering. If death comes peacefully and painlessly, that is OK.

TOTAL

SCORE

A total over 35 points represents acceptable life quality.

Adapted by Villalobos, A.E., Quality of Life Scale Helps Make Final Call, Veterinary Practice News, Sept. 2004; for Canine and Feline Geriatric Oncology Honoring the Human-Animal Bond, by Blackwell Publishing, Table 10.1, released 2005.

FIGURE 16–16 Alice Villalobos' Quality of Life Scale is a list of basic desires or factors (the 5 H's and 2 M's) presumed to be essential for quality of life. Each factor should be carefully monitored by both the veterinary professional and caretaker. A score of 1 is poor and a score of 10 is the best. A score above 5 on most of these issues is acceptable/average. A total score of >35 is acceptable Quality of Life.

Case Study 5: Cleo is a 5-year-old, black F/S DSH cat current on all vaccinations. She was brought into the emergency room with a degloving injury of her tail. At the Emergency and Critical Care Hospital, Cleo had radiographs taken, was given IV fluids, and had a tail amputation performed. She was left with only 2 caudal vertebrae, a "stump" of a tail. To close the defect left after all the injured tissue was removed, skin was stretched down from the dorsal area of her back. Small sutures were placed under the stretched skin to hold it in its new position (called walking sutures), a drain was placed for fluid to escape as the wound healed, and the skin came together perfectly over her new short tail.

Cleo was released after 1 day of hospitalization. She had been on a closely monitored on a fentanyl-lidocaine-ketamine (FLK) CRI. She was sent home with tramadol and amantadine.

The following day Cleo returns to her primary care veterinarian for continued evaluations. The veterinarian asks that you perform a basic physical exam on Cleo. He gives you the CSU Acute Pain Scale for Cats and asks you to rate Cleo's pain (Figure 16-17).

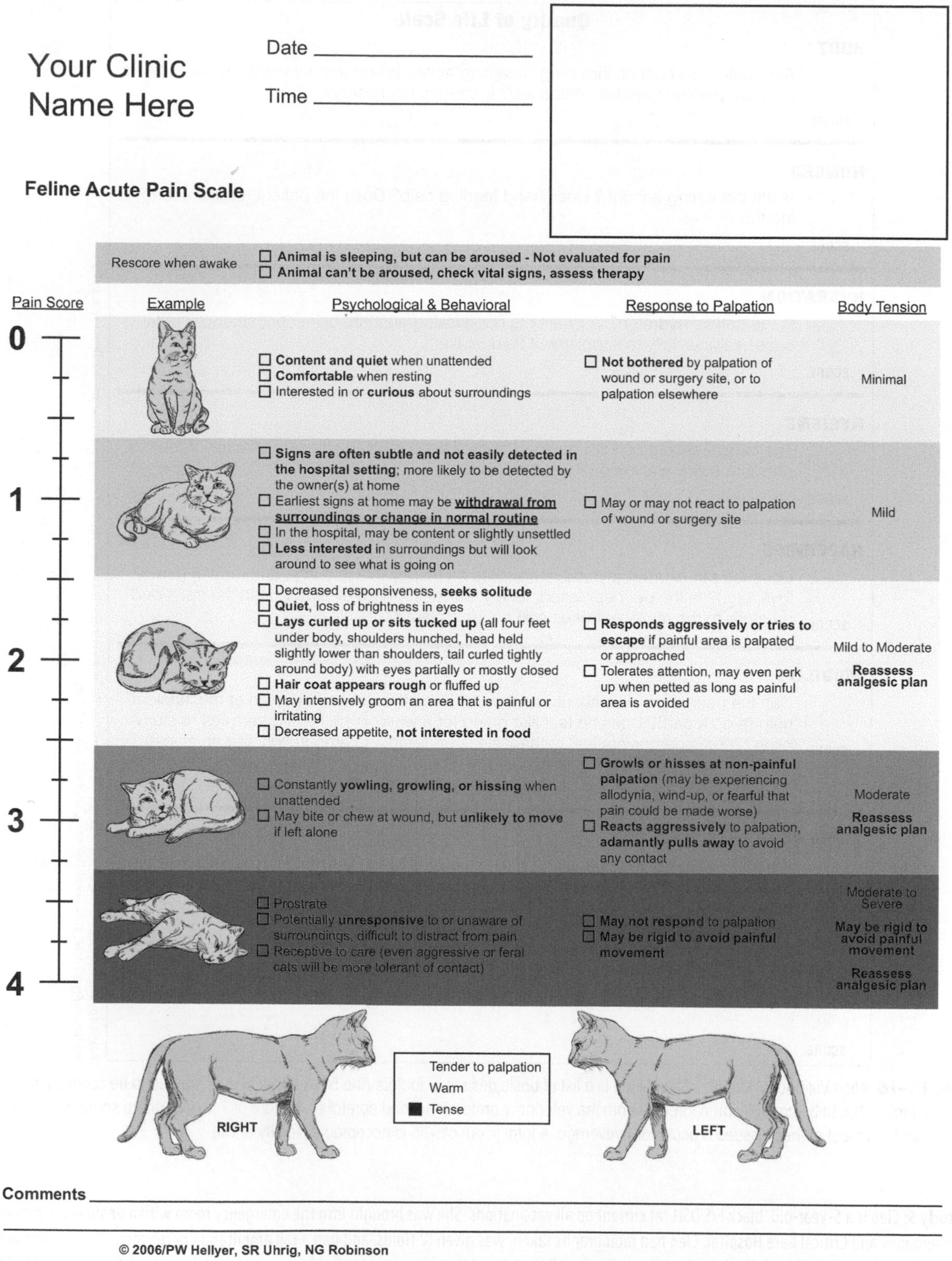

Your Clinic
Name Here

Date _____

Time _____

Feline Acute Pain Scale

Rescore when awake	☐ Animal is sleeping, but can be aroused - Not evaluated for pain ☐ Animal can't be aroused, check vital signs, assess therapy

Pain Score	Example	Psychological & Behavioral	Response to Palpation	Body Tension
0		☐ **Content and quiet** when unattended ☐ **Comfortable** when resting ☐ Interested in or **curious** about surroundings	☐ **Not bothered** by palpation of wound or surgery site, or to palpation elsewhere	Minimal
1		☐ **Signs are often subtle and not easily detected in the hospital setting**; more likely to be detected by the owner(s) at home ☐ Earliest signs at home may be <u>withdrawal from surroundings or change in normal routine</u> ☐ In the hospital, may be content or slightly unsettled ☐ **Less interested** in surroundings but will look around to see what is going on	☐ May or may not react to palpation of wound or surgery site	Mild
2		☐ Decreased responsiveness, **seeks solitude** ☐ **Quiet**, loss of brightness in eyes ☐ **Lays curled up or sits tucked up** (all four feet under body, shoulders hunched, head held slightly lower than shoulders, tail curled tightly around body) with eyes partially or mostly closed ☐ **Hair coat appears rough** or fluffed up ☐ May intensively groom an area that is painful or irritating ☐ Decreased appetite, **not interested in food**	☐ **Responds aggressively or tries to escape** if painful area is palpated or approached ☐ Tolerates attention, may even perk up when petted as long as painful area is avoided	Mild to Moderate **Reassess analgesic plan**
3		☐ Constantly **yowling, growling, or hissing** when unattended ☐ May bite or chew at wound, but **unlikely to move** if left alone	☐ **Growls or hisses at non-painful palpation** (may be experiencing allodynia, wind-up, or fearful that pain could be made worse) ☐ **Reacts aggressively** to palpation, **adamantly pulls away** to avoid any contact	Moderate **Reassess analgesic plan**
4		☐ Prostrate ☐ Potentially **unresponsive** to or unaware of surroundings, difficult to distract from pain ☐ Receptive to care (even aggressive or feral cats will be more tolerant of contact)	☐ May not respond to palpation ☐ May be rigid to avoid painful movement	Moderate to Severe May be rigid to avoid painful movement **Reassess analgesic plan**

RIGHT   LEFT

Tender to palpation
Warm
■ Tense

Comments _____

FIGURE 16-17 The CSU Acute Pain Scale for Cats uses both a quiet observation of the patient from a distance and a gentle hands-on evaluation to assess reaction to gentle palpation (wound as well as the entire body), indicators of muscle tension and heat, response to interaction, etc. The scale uses a 0 to 4 scale with quarter marks along with a color scale and renderings of animals at various levels of pain as visual cues for progression along the 5-point scale. Specific descriptors for individual behaviors are provided which decreases inter-observer variability. A comments section allows documentation of specific areas of concern in the medical record. There is also provision for non-assessment in the resting/sleeping patient. The CSU Acute Pain Scale has not been validated and is intended primarily as a teaching tool and to guide observations of clinical patients.

On examination, Cleo has a temperature of 102.6°F, HR = 225 bpm, RR = 28 breaths per minute. She is quiet, but does not want the surgical site palpated. She likes attention as long as the surgical site is left alone. The owners state that she has not been eating normally and has been hiding at home. Rate Cleo's pain with the CSU Pain Scale.

1. Which picture on the scale best resembles Cleo's posture?
2. What would you rate Cleo's Psychological & Behavioral response?
3. What is her response to palpation?
4. What is the result for Body Tension?
5. What is suggested on the scale?

After your veterinarian makes the assessment from your use of the pain scale, he is concerned that if this patient's pain does not resolve it could progress to chronic pain. He asks you to be sure to ask the owner if they can get the cat to take the tramadol or if the cat is spitting out the tablets because they are bitter tasting. You remind him that the smallest size tablet of tramadol is 50 mg, and the dosage of tramadol is 1 to 2 mg/kg (total dose for a 6 kg [13 lb cat] would be 12 mg or ¼ tablet). Compounding tramadol to a liquid form can allow for more accurate dosing (manufacturers only guarantee doses in tablets that are scored to be split into ¼ tabs and most tablets are not) and potentially help with palatability issues because the cat would taste the bitterness of the pill more when given as a split tablet. Ultimately the pain management plan is revised to include transmucosal buprenorphine, oral meloxicam, and the addition of gabapentin along with the amantadine. A follow up 1 week later reports a dramatic improvement in this patient.

Case Study 6: Greta is a 12-year-old, F/S, Domestic Longhair (DLH) cat. She has been in perfect health and is up to date on all her vaccinations. Her owners report that she is not jumping up on the bed anymore. For that matter, she is not jumping and she yowls for someone to pick her up. She's eating her food, but is slowing down (not chasing toys or playing as much). The veterinarian examines her and thinks she may have osteoarthritis. While blood tests are being run, she asks you to explain how the behavioral chart below can be used for this cat (see Figure 16-16).

Think about how this cat used to be and compare this with how she is now, using the type of activities listed in the figure as a guide and grade the severity of change 1 to 10. After starting treatment, think about any changes that have occurred since the start of treatment, but again grade severity of any behavioral change by comparing with when she was a young adult.

Critical Thinking Questions

1. You arrive at work in the morning for another typical day at the veterinary practice. Several dogs have been dropped off for their annual vaccinations, fecal examinations, and heartworm checks. The practice owner is interviewing new prospective candidates for a specialty position and has left strict instructions that they do not want to be disturbed during this time period. You and several veterinary assistants begin examining the dogs (one by one) to draw the necessary blood and fecal samples for microscopic examination. The fourth patient, Sophie Jones, a 10-year-old black Labrador Retriever is currently on carprofen for limping and has been on this medication for 33 days. When you get her out to draw samples, you see that Sophie has a bloody diarrhea and her temperature is 104°F. You call the owner and she tells you that Sophie has not eaten well for 4 days and that she vomited prior to bringing her in. Sophie has a depressed attitude and is sensitive if you touch her belly.

    a. Do you draw the samples and put Sophie back in her cage?

    b. Do you interrupt the veterinarian and present him with Sophie's record explaining what you have found?

    c. What is your reasoning for this?

    d. What has this situation required you to assess?

2. You have developed an interest in pain management while going through veterinary technician school. After graduation, you accepted a position at a practice where the owner is a veterinarian that graduated over 20 years ago who is not really keen on your interests. When he graduated from veterinary college, pain management was not emphasized the way it is today. Through your continued study, you have learned specific reasons why pain management is very important. The veterinarians in the practice give you an opportunity to discuss with them why pain management should be something to incorporate into daily planning and treatments.

    a. What important information should you include in your presentation?

ENDNOTES

1. Turk, D. C., & Okifuji, A. (2001). Pain terms and taxonomies. In D. Loeser, S. H. Butler, J. J. Chapman, et al. (Eds.), *Bonica's management of pain* (3rd ed., pp. 16–25). Lippincott Williams & Wilkins.

2. Ashburn, M. A., & Staats, P. S. (1999). Management of chronic pain. *The Lancet, 353,* 1865–1869.

3. Goldberg, M. E. (2010). The fourth vital sign in all creatures great and small. *The NAVTA Journal,* 31–54.

4. Chandrasekharan, N. V., Dai, H., Roos, K. L., Evanson, N. K., Tomsik, J., Elton, T. S., & Simmons, D. L. (2002). COX-3, a cyclooxygenase-1 variant inhibited by acetaminophen and other analgesic/antipyretic drugs: cloning, structure, and expression. *Proc. Natl. Acad. Sci. U.S.A., 99*(21), 13926–13931.

5. Budsberg, S. (2009). Nonsteroidal anti-inflammatory drugs. In J. S. Gaynor & W. W. Muir (Eds.), *Handbook of veterinary pain management* (p. 189). St. Louis, MO: Mosby/Elsevier.

6. Hampshire, V. A., Doddy, F. M., Post, L. O., et al. (2004). Adverse drug event reports at the United States Food and Drug Administration Center for Veterinary Medicine. *Journal of the American Veterinary Medical Association, 225,* 533–536

7. Gearing, D., Virtue, E., Gearing, R., & Drew, A. (2013). A fully caninised anti-NGF monoclonal antibody for pain relief in dogs. *BMC Veterinary Research, 9,* 226.

8. Lascelles, B. D., Knazovicky, D., Case, B., Freire, M., Innes, J., Drew, A., & Gearing, D. (2015). A canine-specific anti-nerve growth factor antibody alleviates pain and improves mobility and function in dogs with degenerative joint disease-associated pain. *BMC Veterinary Research, 11,* 101.

9. Hewitt, D. J. (2000) The use of NMDA-receptor antagonists in the treatment of chronic pain *Clinical Journal of Pain* 16(2-supplement), 73–79.

10. Walters, J. L., Lonergan, D. F., Todd, R. D., & Jackson, T. P. (2012). Idiopathic peripheral neuropathy responsive to sympathetic nerve blockade and oral clonidine. *Case Reports in Anesthesiology,* 407539.

11. Singh, S., & Arora, K. (2011). Effect of oral clonidine premedication on perioperative haemodynamic response and postoperative analgesic requirement for patients undergoing laparoscopic cholecystectomy. *Indian Journal of Anaesthesia, 55*(1), 26–30.

12. Bryson, H. M., & Wilde, M. I. (1996). Amitriptyline A review of its pharmacological properties and therapeutic use in chronic pain states. *Drugs Aging, 8*(6), 459–476

13. Saarto, T., & Wiffen, P. J. (2005). Antidepressants for neuropathic pain. *Cochrane Database System Reviews* (3), CD005454

14. DeMarco, G. J. (2008). Management of chronic pain. In R. E. Fish, M. J. Brown, P. J. Danneman, & A. Z. Karas (Eds.), *Anesthesia and analgesia in laboratory animals* (2nd ed., p 585). San Diego, CA: Academic Press/Elsevier.

15. Aghighi, S. A., Tipold, A., Piechotta, M., Lewczuk, P., & Kästner, S. B. (2012). Assessment of the effects of adjunctive gabapentin on postoperative pain after intervertebral disc surgery in dogs. *Veterinary Anaesthesia and Analgesia, 39*(6), 636–646.

16. Wagner, A. E., Mich, P. M., Uhrig, S. R., & Hellyer, P. W. (2010). Clinical evaluation of perioperative administration of gabapentin as an adjunct for postoperative analgesia in dogs undergoing amputation of a forelimb. *Journal of the American Veterinary Medical Association, 236*(7), 751–756.

17. Vettorato, E., & Corletto, F. (2011). Gabapentin as part of multi-modal analgesia in two cats suffering multiple injuries. *Veterinary Anaesthesia and Analgesia, 38*(5), 518–520.

18. Pypendop, B. H., Siao, K. T., & Ilkiw, J. E. (2010). Thermal antinociceptive effect of orally administered gabapentin in healthy cats. *American Journal Veterinary Research, 71*(9), 1027–1032.

19. Bonnie Wright DVM, DACVA, CVPP, CVA personal communication.

20. Brown, D. C., Iadarola, M. J., Perkowski, S. Z., et al. (2005). Physiologic and antinocioceptive effects of intrathecal resiniferatoxin in a canine bone cancer model. *Anesthesiology, 103*(5), 1052–1059.

21. Aragon, C. L., Hofmeister, E. H., & Budsberg, S. C. (2007). Systematic review of clinical trials of treatments for osteoarthritis in dogs. *Journal of the American Veterinary Medical Association, 230*(4), 514–521.

22. Plumb, D. C. (2011). Plumb's veterinary drug handbook, 7th Edition, pp. 746–747. Ames, IA: Wiley-Blackwell.

23. McKenzie, B. A. (2010). What is the evidence? There is only very weak clinical trial evidence to support the use of glucosamine and chondroitin supplements for osteoarthritis in dogs. *Journal of the American Veterinary Medical Association, 237*(12), 1382–1383.

24. Hielm-Björkman, A., Roine, J., Elo, K., Lappalainen, A., Junnila, J., & Laitinen-Vapaavuori, O. (2012). An un-commissioned randomized, placebo-controlled double-blind study to test the effect of deep sea fish oil as a pain reliever for dogs suffering from canine OA. *BMC Veterinary Research, 8,* 157.

25. Roush, J. K., Dodd, C. E., Fritsch, D. A., Allen, T. A., et al. (2010). Multicenter veterinary practice assessment of the effects of omega-3 fatty acids on osteoarthritis in dogs. *Journal of the American Veterinary Medical Association, 236*(1), 59–66.

26. Koopman, W. J. (Eds.) (1993). Arthritis and allied conditions: A textbook of rheumatology 12th ed, pp. 1761–1769. Philadelphia, PA: Lea & Febiger.

27. Hardie, E. M., Roe, S. C., & Martin F. R. (2002). Radiographic evidence of degenerative joint disease in geriatric cats: 100 cases (1994–1997). *Journal of the American Veterinary Medical Association, 220*(5), 628–632.

28. Impellizeri, J. A., Tetrick, M. A., & Muir, P. (2000). Effect of weight reduction on clinical signs of lameness in dogs with hip osteoarthritis. *Journal of the American Veterinary Medical Association, 216*(7), 1089–1091.

29. Gamble, L. J., Boesch, J. M., Frye, C. W., et al. (2018). Pharmacokinetics, safety, and clinical efficacy of cannabidiol treatment in osteoarthritic dogs. *Frontiers in Veterinary Science, 5,* 165.

30. Lucroy, M. D. (2013). Cancer pain management. In J. Ko (Ed.), *A color handbook small animal anesthesia and pain management,* p. 306. London, UK: Manson Publishing, Ltd.

31. Rancilio, N., Poulson, J., & Ko, J. (2015). Strategies for managing cancer pain in dogs & cats, Part 1: Pathophysiology & assessment of cancer pain, *Today's Veterinary Practice*, 60–67.
32. Mathews, K. A. (2008). Neuropathic pain in dogs and cats: If only they could tell us if they hurt. *Veterinary Clinics of North America: Small Animal Practice-Update on Pain Management, 38*(6), 1385–1389.
33. Fan, T. M., & de Lorimier, L. P. (2005). Cancer pain: Real cases. American College of Veterinary Internal Medicine Meeting Proceedings.
34. Fox, S. M. (2010). Pathophysiology of cancer pain. In *Chronic pain in small animal medicine*, pp. 97–100. London, UK: Manson Publishing Ltd.
35. De Lorimier, et al. (2005). Treating cancer pain in dogs and cats. *Veterinary Medicine.*
36. Heath, S., et. al. (2010). Feline orofacial pain syndrome. Australian Veterinary Association Proceedings, VIN Proceedings Library.
37. Fox, S. M. (2010). Chronic pain in selected physiological systems: Ophthalmic, aural, and dental. In *Chronic pain in small animal medicine* (p. 206). London, UK: Manson Publishing Ltd.
38. Marfurt, C. F., Murphy, C. J., & Florczak, J. L. (2001). Morphology and neurochemistry of canine corneal innervation. *Investigative Ophthalmology and Visual Science, 42*, 2242–2251.

CHAPTER 17

Analgesic Techniques

Emma Archer, *RVN, Dip AVN (Surgical), VTS (Anesthesia/Analgesia)*

KEY TERMS

Contralateral

Epidural analgesia

Epidural anesthesia

Ipsilateral

Locoregional anesthesia

Subanesthetic dosage

LEARNING OBJECTIVES

Upon completion of this chapter, it is expected that the reader should be able to:

17.1 Explain the rationale for selecting different types of analgesics for pain relief in dogs and cats

17.2 Explain the rationale for selecting different routes of analgesic administration for pain relief in dogs and cats

17.3 Calculate drug doses for constant rate infusions (CRIs)

17.4 Calculate administration rates for CRI

17.5 Describe how to successfully prepare and administer CRI

17.6 Explain the rationale for using locoregional anesthesia in animals

17.7 Describe techniques for commonly used nerve blocks in animals

17.8 Identify indications for specialized analgesic techniques such as wound catheters, epidural catheters, epidural analgesia and anesthesia, and spinal (intrathecal) anesthesia/analgesia

INTRODUCTION

Prevention and alleviation of pain are important medical and ethical components of a veterinary anesthetist's job. Despite analgesia's literal definition being "without sensation," the clinical goal of providing analgesia is to reduce the patient's pain to a tolerable level through the use of a wide variety of drugs, administration routes, and techniques. Oral and parenteral pain-modifying drugs are common and familiar ways to provide analgesia for veterinary patients; however, techniques such as constant rate infusion (CRI) and locoregional anesthesia can also provide intra- and postoperative pain control and can be performed in many clinical practices by veterinary technicians depending on state veterinary board practice acts. Other advanced techniques such as wound catheters, epidural catheters, and intrathecal anesthesia/analgesia are now being performed by those with specialized training to provide pain relief in patients undergoing more invasive and painful surgical procedures. With the increased emphasis on improving pain management for veterinary patients, understanding these analgesic techniques will allow veterinary technicians to be advocates for their patients by suggesting a wide range of pain relief options in order to provide their patients the best care possible. This chapter focuses on analgesic techniques for dogs and cats; large animal techniques are described in Chapter 19 (Anesthesia and Analgesia of Large Animal and Food Producing Species), pet birds and exotic mammal techniques are described in Chapter 20 (Anesthesia and Analgesia of Pet Birds and Exotic Small Mammals), and laboratory mice and rats are described in Chapter 21 (Pain Management in Laboratory Mice and Rats).

Tech Tip

Veterinary technicians should familiarize themselves with their state's practice acts to ensure they can legally perform the techniques described in this chapter.

ANALGESIC SELECTION

It is the anesthetist's goal, working in conjunction with the attending veterinarian, to select analgesic drugs and administration routes that best fit the individual patient. Addressing the following questions will help the anesthetist achieve that goal:

- *What level of pain is anticipated for the patient?* Understanding the procedure to be performed on a patient and predicting the level of pain it will produce helps in the selection of an analgesic and administration route. For example, a procedure such as a routine dental cleaning without extractions produces mild or no discomfort while a thoracotomy has the potential to produce severe and potentially chronic pain. Acute severe pancreatitis may be more painful than mild osteoarthritis; however, some patients with chronic severe debilitating osteoarthritis can have more severe pain than a patient with mild pancreatitis. It is therefore important that each patient and every procedure or condition be considered on an individual basis when making analgesic choices. To apply this clinically, butorphanol might be acceptable for the mild discomfort associated with a routine dental cleaning while an aggressive multimodal analgesic approach using full mu agonist opioids, nonsteroidal anti-inflammatory drugs (NSAIDs),

intra-pleural infusion of local anesthetics, and analgesic adjunctive drugs such as NMDA-receptor antagonists, the anticonvulsant gabapentin, and other agents would be more appropriate for a thoracotomy.

- *What species is the patient?* Veterinary patients vary in their anatomy and physiology; therefore, different species may respond differently to a particular drug class and even individual drugs within that class. For example, cats and dogs can be given codeine for pain relief; however, dogs can be given the codeine/acetaminophen combination product while cats should not because they lack the enzyme that alters acetaminophen for metabolism and administering acetaminophen may lead to toxicity and death. Another simple example is that butorphanol may be an inadequate analgesic for a moderately painful procedure in a dog; however, it may be appropriate for another species such as birds or horses because of different opioid receptor distribution in these species.

- *Can the desired drug be given by the desired route?* Always consider whether a drug is licensed for use in the species and can be given by the route the veterinarian wishes. For example, in the United Kingdom buprenorphine is licensed for use in cats and dogs by intramuscular but not intravenous injection whereas methadone is licensed for IV and IM administration in dogs. In the United States, the Food and Drug Administration (FDA) allows for extra-label drug use in certain situations, but a drug's formulation may prevent its administration by particular routes. When choosing an administration route, also consider:

  ○ *Does the patient already have an IV catheter in place?* If a catheter is already in place, it is often easier and less stressful to the patient to use the catheter and give the drugs IV.

  ○ *What is the least stressful way for my patient to receive a positive effect?* Consider the following when trying to minimize stress for painful patients:

    ■ Although intramuscular (IM) injection provides rapid absorption with therapeutic plasma levels typically seen in approximately 30 minutes, it can also cause painful muscle soreness especially if large volumes or repeated IM injections are used. Switching to a drug whose dose results in lower volumes (e.g., switching to morphine instead of hydromorphone) or starting with a small dose of a drug such as dexmedetomidine and waiting 10 to 15 minutes until that drug takes effect before giving the other drugs (with larger volumes) are options to reduce IM injection pain if the patient is a good candidate for it. Most opioids (except short-acting potent ones such as fentanyl that are most commonly given IV) can be administered effectively by IM injection. IM injections should be avoided in patients with coagulopathies or severe thrombocytopenia to prevent hematoma formation.

    ■ If a friendly canine patient without an intravenous catheter needs premedication for a routine splenectomy, perhaps its premedication drugs can be given subcutaneously rather than intramuscularly to minimize patient discomfort. Subcutaneous (SQ) injection provides slower and less predictable absorption than drugs given IM because the vascularity of the SQ space is less than that of muscle. SQ administration results in lower and much less predictable plasma levels than IV administration;[1] however, depending on the drug these lower plasma levels may still be effective. Irritating solutions should not be given SQ and the amount of solution given SQ should not be so large as to cause patient discomfort. Despite these concerns, some preanesthetic drugs are given SQ if they can be in order to avoid the pain associated with IM injection or if the volume of an IM injection is too large.

- If a hospitalized cat without an IV catheter is to receive buprenorphine, giving the drug via the oral transmucosal (OTM) route may be preferable to repeated IM injections.

- *How quickly is analgesia required?* Analgesia administered IV has a much more rapid onset of action and reaches a higher plasma concentration than if the same drug is administered by the IM or SQ route. If analgesia is required quickly and profoundly, IV administration is the preferred route. Remember that because the drug rapidly enters circulation, it should be given slowly (over 1–2 minutes) to avoid adverse effects. Depending on the drug and intended goal, IV injections can be administered as a single or intermittent bolus or via constant rate infusion (CRI) (described later in this chapter). Keep in mind that IV administration is not recommended with some opioid drugs such as meperidine because it can release histamine, which may lead to vasodilation and hypotension. Morphine can also have this effect; therefore, if morphine is administered IV it is often diluted and given slowly.

- *What is the duration of the procedure being performed or duration of analgesia needed?* Adequate analgesia must be provided for the proper amount of time. For example, in dogs fentanyl IV has a short clinical duration of action (approximately 30 minutes), while morphine IM has a longer duration of action (approximately 3–5 hours). Fentanyl may be a good analgesic choice for a procedure that needs a shorter duration of analgesia, such as endoscopic removal of an esophageal foreign body, while morphine is a good choice for a procedure that requires longer duration of analgesia such as an abdominal exploratory. Choosing a drug that provides sustained analgesia is simpler than redosing a shorter duration drug more often. Another way to utilize shorter duration drugs (such as fentanyl) is to administer them by a CRI as described in more detail later in this chapter. Similarly if the expected duration of pain is very short, then providing shorter acting drugs may reduce patient adverse effects in the long term.

- *Does the patient have pre-existing conditions and how does this influence the patient's overall health status and analgesic management of the case?* Concurrent health issues could influence a patient's pain management needs and should be considered when selecting an analgesic for the patient. For example, NSAIDs block prostaglandins and alter a pathway that affects renal vasodilation and thus blood flow to the kidney; therefore, their use should be avoided in hypovolemic, dehydrated, and/or hypotensive patients who may already have compromised renal blood flow. Another example is a dog with severe chronic valvular heart disease such as mitral valve insufficiency should not be given dexmedetomidine as a CRI as its vasoconstrictive effects increase afterload and worsen the workload on their heart. It is important to consider a drug's mechanism of action as well as its physiologic effect in relationship to the patient's concurrent health issues. These considerations are a key responsibility of the attending veterinarian, and the reader is referred to textbooks and resources that describe this topic in more detail.

- *Has the patient received analgesic drugs before, what dosages were used, and what were the effects?* If an analgesic drug failed to provide pain relief or produced an adverse effect last time it was used, a different dosage or different drug should be considered.

- *Is the patient on other medications that could interact with my drug choices? Does the patient have any drug allergies?* For example, patients should not be given NSAIDs and glucocorticoids concurrently as this may lead to gastric ulceration and perforation in dogs and cats. Although uncommon, drug allergies have been reported with nearly every drug available.

ANALGESIC ROUTE SELECTION

Both the route of administration and the actual analgesic drug are most likely chosen in tandem with each other in order to provide the adequate level of pain relief at the proper time for the appropriate length of time.

Parenteral Injection

Parenteral routes of administration are commonly used for pain management and vary in their rates of absorption, and the reader is referred to Chapter 5 (Anesthetic and Analgesic Pharmacology) for a review of this material. Analgesic drugs administered via parenteral injection and their dosages are provided in Chapter 15 (Acute Pain Management).

Topical Application

Topical analgesics are applied to the surface of the skin or mucous membranes and are available in a variety of formulations. It is important to understand that some topical analgesics only work on mucous membranes and not intact skin. Consider the topical application of the local anesthetic lidocaine (Xylocaine®). Lidocaine spray (which is actually injectable lidocaine put into a spray bottle) or gel is applied topically to desensitize oral, nasal, or laryngeal *mucous membranes* to facilitate placement of nasoesophageal feeding tubes, nasal oxygen catheters, or for endotracheal intubation in cats. Injectable lidocaine does not penetrate the epidermis of intact skin where the peripheral nerves are located, which is why applying it on its own to intact skin (such as a "splash block" performed by dribbling local anesthetic over a surgical site) may not be efficacious. In contrast, a commercially available local anesthetic cream (EMLA®, generic) containing a mix of lidocaine and prilocaine penetrates intact skin to provide analgesia of the skin to a maximal depth of 3 mm after a 60-minute application, and 5 mm after a 120-minute application in humans and can be used in small animals prior to placement of venous or arterial catheters. The skin is shaved so the medication is applied directly to the skin and not the hair. The cream should be covered with an occlusive dressing for 30 to 60 minutes to reach maximum effect.

Other local anesthetics applied topically include proparacaine (Alcaine®) and tetracaine (TetCaine®) drops, which are used to desensitize the cornea in order to perform ocular procedures such as intraocular pressure monitoring, corneal foreign body removal, or ocular ultrasound. Proparacaine and tetracaine provide anesthesia to the cornea for 5 to 10 minutes. Proparacaine and tetracaine also desensitize mucous membranes. Proparacaine is kept in the refrigerator to prevent a decrease in drug efficacy.

Tech Tip

Systemic toxicity is a concern with local anesthetics, and care should be taken to avoid using excessive amounts of topical agent, especially in cats.

Transmucosal administration is a very effective way of administering analgesics such as buprenorphine to cats by putting it in the buccal pouch (the pH of dog saliva reduces the amount of drug available for transmucosal absorption in that species but still may be useful in smaller dogs). The preservative in some multidose vials of buprenorphine (multidose vials are not available in the United States but are available in other countries) has anecdotally been reported to be unpalatable.

Transdermal

Drugs administered transdermally are absorbed through the skin. A common transdermal formulation is a patch that is designed to release a consistent amount of drug per hour as it is absorbed across skin (Figure 17-1). Human preparations of fentanyl (Duragesic®), buprenorphine (Butrans®), and lidocaine (Lidoderm®) are available as transdermal patches. Some transdermal drugs are taken up systemically (such as fentanyl and buprenorphine), while others do not reach significant systemic circulation levels (such as lidocaine). Patches are typically applied to the lateral thorax or dorsal neck. The hair is clipped from the skin prior to applying the patch and a dressing is used to cover the patch and hold it in place. The patch should be covered to prevent ingestion and clearly labeled with date, time, and dose.

Drugs commonly given in cats and dogs via transdermal patches are fentanyl and lidocaine; buprenorphine patches are used less frequently.

Transdermal fentanyl patches provide long-lasting analgesia but take 12 to 24 hours after application to reach therapeutic plasma concentrations in small animals. They last approximately 72 hours after application, but because their effectiveness varies from patient to patient, they should never be used as the sole form of analgesia. They are available in different strengths (12, 25, 50, 75, and 100 mcg/hr). Fentanyl is a C-II controlled substance whose use needs to be regulated. Fentanyl patches should be used with caution in patients that will be discharged as children or even adults can become seriously ill or even die from ingesting these patches. Remind clients about the risk of other pets in the household ingesting the patch (e.g., after it is thrown in the trash or when they groom the pet with the fentanyl patch). Patients with fentanyl patches should return to the practice for removal and to ensure that the patch is properly disposed. Changes in an animal's body temperature during anesthesia can affect transdermal drug absorption. Hypothermia during an anesthetic event can cause decreased drug absorption, whereas the use of heating devices such as forced air blankets will increase absorption, which may potentially lead to respiratory depression and bradycardia.

Transdermal lidocaine patches, originally developed for chronic pain in humans, provide local analgesia in dogs and cats (Figures 17-2a and 17-2b). A constant rate of absorption occurs 12 to 72 hours after patch application.[2,3] Systemic absorption of lidocaine in dogs and cats is minimal; therefore, the local tissue concentration of lidocaine is higher than plasma concentrations. Lidocaine's low rate of absorption ensures its plasma concentration remains significantly below toxic levels. Lidocaine patches preserve motor function while producing analgesia for up to 72 hours. These patches should be applied close to the painful site for maximum effectiveness. Toxicity is a concern if the patch is ingested.

Buprenorphine patches were also originally licensed for use in humans and have been used in cats and dogs to provide analgesia for approximately 7 days. The onset of action in dogs is slow (approximately 17 hours), and in

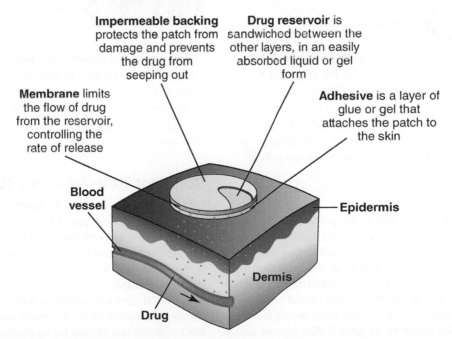

FIGURE 17-1 Transdermal patch. A transdermal patch slowly releases the drug through the skin to the bloodstream. The patch works by diffusion, and the rate of drug release is controlled by an intervening membrane or by suspending the drug in another material called a matrix, which lowers its initial concentration.

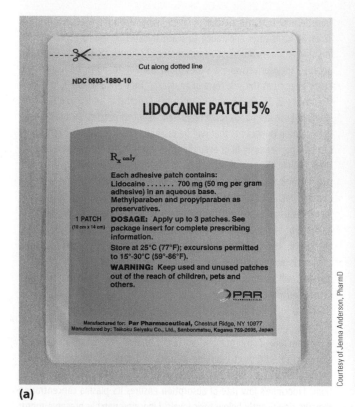

Courtesy of Jenna Anderson, PharmD

(a)

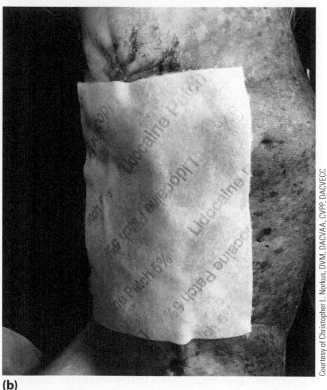

Courtesy of Christopher L. Norkus, DVM, DACVAA, CVPP, DACVECC

(b)

FIGURE 17-2 Transdermal lidocaine patch. (a) Example of a lidocaine patch. (b) Lidocaine patches should be applied close to the painful site for maximum effectiveness as shown in this image.

cats plasma levels are very low even up to 36 hours after application. The inability of buprenorphine patches to provide adequate plasma levels has limited its use in companion animals.

Tech Tip

A manufacturer may voluntarily implement a RiskMAP for drugs that pose unusual risks and limit the drug's use to those that are RiskMAP certified. RiskMAPs may include the use of tools such as provider training and certification programs, informed owner consent, and enhanced monitoring programs to minimize the risk when administering the drug and assess whether or not the RiskMAP is achieving its goals.

Capsaicin, the active component in chili peppers, has been used topically in humans to treat pain associated with diabetic neuropathy and osteoarthritis. Capsaicin is available as a 0.025% cream (Capzasin®, Salonpas-Hot®, generic) and has been used in dogs with atopic dermatitis to control pruritus and in dogs and horses with osteoarthritis. Capsaicin binds to the transient receptor potential cation channel subfamily V member 1 (TrpV-1) receptor on afferent pain fibers causing increased nerve firing through the release of substance P. With repeated application of capsaicin, neurons are depleted of neurotransmitters, which causes desensitization due to temporary loss of membrane potential and blockage of neurogenic inflammation. If capsaicin use is discontinued,

the neurons recover. The benefits of capsaicin include specific loss of pain sensation without the concurrent loss of motor and nonpain sensation.

Tech Tip

Capsaicin is listed in the 2017 Equine Prohibited Substances List created by the Federation Equestrian International (FEI). Prior to application or administration of any substance to competitive horses, consult the FEI website (https://inside.fei.org/fei/cleansport/) for a complete list of prohibited substances in equestrian sport.

Constant Rate Infusions (CRIs)

Constant rate infusion (CRI) provides a consistent plasma level of a given drug to effectively provide patient analgesia. Intermittent drug administration can also result in adequate patient analgesia, but that technique can produce more fluctuations in plasma levels that may at times result in adverse effects at plasma level peaks and inadequate analgesia at plasma level troughs. If intermittent drug administration is selected instead of CRI, conscious redosing of analgesics is needed to achieve plasma concentrations in the therapeutic range to minimize adverse effects. Constant rate infusions are mainly administered by the IV route; however, drugs can also be given continuously via an epidural catheter or a wound catheter (a sterile catheter place directed into a wound, which is described later in this chapter).

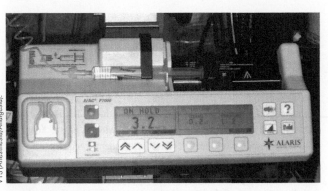

FIGURE 17-3 Syringe pump (syringe driver) with a constant infusion of fentanyl.

TABLE 17-1 MLK Recipe

Drug	Effect	Dosage rate of Individual Drug
morphine (15 mg/mL)*	• Opioid analgesic • Spares amount of inhalant anesthetic agent needed • May cause sedation and nausea/vomiting	100–500 mcg/kg/hr
lidocaine (20 mg/mL)	• Anti-hyperalgesic • Anti-arrhythmic • Spares amount of inhalant anesthetic agent needed • May cause sedation and nausea/vomiting	25–50 mcg/kg/min
ketamine (100 mg/mL)	• NMDA (n-methyl-D-aspartate) receptor antagonist • Spares amount of inhalant anesthetic agent needed	100–600 mcg/kg/hr

*Remember to check concentration of morphine as many different concentrations are available.

Prior to starting a CRI, a bolus or "loading dose" of the drug is often administered to achieve a rapid and effective dose in the plasma before starting a continuous rate, which simply maintains the plasma levels within a target range. In other words, just enough of drug is added to the body to balance the amount of drug leaving the body and therefore the plasma levels remain relatively constant. CRIs of analgesic drugs are preferred for animals expected to require high levels of analgesia during the intra-operative period as a part of multimodal anesthesia or postoperatively. CRIs can also be used in the intensive care unit (ICU) with painful medical conditions such as pancreatitis or trauma. Dose rates should be carefully calculated and clearly written in the medical record. To ensure accurate drug delivery, properly calibrated syringe pumps (syringe drivers) (Figure 17-3) or infusion pumps should be used whenever possible. The syringe pump and fluid bags should be clearly labeled with the drug name and its concentration. Diluting the drug is recommended when the volume of drug to be administered is very low because some syringe pumps/drivers cannot run accurately at small volumes (e.g., 0.1 mL). For example, if a patient is to receive 0.1 mL of a drug per hour, diluting the drug so that a larger volume such as 1 mL per hour is given makes for easier and more accurate administration. Sterile saline 0.9% or 5% dextrose in water (D_5W) should be used to dilute drugs for CRI and the dilution recorded on the bag and in the medical record. Regular servicing and proper maintenance of equipment is important to ensure accuracy and proper functioning of the infusion pump or syringe pump/driver.

Tech Tip

The accuracy of infusion pumps can be checked by running a bag of fluids through the pump. For example, a 500 mL bag running at 100 mL/hr should take 5 hours to finish.

Opioids via CRI

Full mu agonist opioids are commonly administered via CRI in small animals to provide high levels of analgesia in the perioperative period as well as in hospitalized animals. Currently in the United States, morphine (Duramorph®, Astramorph®) and fentanyl (Sublimaze®) are the most commonly used opioids via the CRI route; however, essentially any opioid including hydromorphone (Dilaudid®), methadone (Dolophine®), butorphanol (Torbugesic®, Torbutrol®), and buprenorphine (Buprenex®) can be selected. Often times fentanyl (F) or morphine (M) are combined with lidocaine (L) and/or ketamine (K) to produce a combinations known as "FL," "FK," "ML," "MK," "MLK," or "FLK" respectively for more painful procedures (Table 17-1).

Other short acting opioids such as alfentanil (Alfenta®), sufentanil (Sufenta®), and remifentanil (Ultiva®) have been used during small animal anesthesia with the specific goal of reducing the amount of inhalant anesthetic agent required (MAC reduction). When given at higher or repeated dosages in addition to inhalant anesthetic agents, opioids can cause a cumulative level of respiratory depression that may result in hypoventilation and require assisted ventilation. Opioids given intermittently or via CRI can result in a vagally induced reduction in heart rate that may progress to bradycardia. The veterinary technician should monitor blood pressure and can reduce the dosage rate and/or administer anticholinergic drugs if the bradycardia is impacting blood pressure and cardiac output. Other common adverse effects seen with opioid CRIs can include sedation, nausea, vomiting, panting in dogs, hyperthermia in cats, hypothermia in dogs, and dysphoria.

Alpha-2 Adrenergic Agonists via CRI

The alpha-2 (α_2) adrenergic agonist dexmedetomidine (Dexdomitor®) can be administered in very low dosages via CRI (in addition to other administration routes) for pain relief in the intra and postoperative period as well as to hospitalized patients. CRI of other α_2 adrenergic agonists are commonly administered in large animal species during anesthesia to improve analgesia and reduce inhalant anesthetic agent requirements (see Chapter 19 [Anesthesia and Analgesia of Large Animal and Food Producing Species]). Alpha-2 adrenergic agonists increase peripheral vascular resistance, reduce cardiac output, and produce bradycardia even at low dosages; therefore, they should only be used on hemodynamically stable patients. Sedation occurs even at low dosages and depending on the patient and treatment goals, this may be seen as a beneficial or an adverse effect. Dexmedetomidine CRIs can be very useful adjunctive agents when given with an opioid to enhance analgesia and provide a level of antianxiety and/or sedation effect. Alpha -2 adrenergic agonists can also increase blood glucose levels, increase urine output, and result in nausea and vomiting in cats. Basic nursing care such as frequent turning of recumbent patients may be required to minimize the risk of complications such as atelectasis and decubital ulcers and frequent walks to allow for urination/defecation to avoid urine scaled and contamination of wounds. Depending on the dosage used, it is recommended to check the patient for a proper swallowing reflex before offering food and water and to offer water regularly in smaller volumes rather than leave water bowls near heavily sedated animals. Dexmedetomidine can result in respiratory depression when given concurrently with other drugs such as opioids. Fortunately, the dosage of dexmedetomidine can be adjusted as required to ensure adequate analgesia without excessive sedation and adverse effects as needed. Dexmedetomidine CRIs should not be relied upon as a patient's sole form of analgesia but rather be selected to enhance the effect of other drugs such as opioids.

Lidocaine via CRI

Lidocaine is a sodium channel blocking agent with local anesthetic properties that may be administered intravenously (systemically) by CRI as an analgesic agent. It may be used in treatment or prevention of chronic pain states but may have a role for other types of pain as well. Metabolism of lidocaine relies on adequate hepatic blood flow, so care should be taken in patients with reduced blood flow to the liver (such as hypotensive patients) because accumulation of the drug and subsequent toxicity may theoretically occur. The anesthetist should monitor the patient for signs of central nervous system (CNS) toxicity such as twitching, ataxia, or seizures and if observed discontinue the infusion immediately, alert the veterinarian, and treat supportively. If lidocaine toxicity worsens, respiratory and cardiac arrest can occur. In the case of cardiopulmonary depression, specific treatment includes therapies such as IV lipid emulsion and insulin/dextrose/potassium therapy.[4] Lidocaine CRIs reduce the amount of inhalant anesthetic agent required under inhalant anesthesia, but this is not synonymous with analgesia.[5] Lidocaine also is used to prevent postoperative ileus. It may also have anti-inflammatory and free radical scavenging effects, so it is sometimes considered in cases such as gastric dilatation volvulus or sepsis for these reasons. Lidocaine CRIs are also widely used to control ventricular arrhythmia. Extra care should be taken in anesthetized patients because CNS toxicity signs are not apparent in unconscious animals. Intravenous lidocaine is not typically used in cats due to adverse cardiovascular adverse effects, toxicity concerns, and no clear documentation that it provides analgesia in this species.[6] Lidocaine CRIs' true efficacy as an analgesic in humans (let alone animals) is controversial.

Ketamine via CRI

Ketamine, an N-methyl-D-aspartate (NMDA) receptor antagonist, inhibits the excitatory glutamate receptor in the CNS, which may help with the prevention or treatment of central sensitization and chronic pain. Additionally, ketamine may be useful to reduce opioid consumption and

provide analgesia in more acute pain scenarios. Ketamine can be used as a CRI at **subanesthetic dosages** to produce profound analgesia with minimal adverse effects on the cardiovascular system or patient behavior. Using ketamine at such low dosages as an analgesic has been shown to improve the demeanor and well-being of dogs following limb amputation as assessed at home by their owners.[7] Adverse effects are unusual at low dosages. As dosages increase, especially in the awake patient, more behavior and cardiovascular adverse effects will be noted. Ketamine infusions may reduce inhalant anesthetic agent requirements (and therefore potentially their adverse effects) during anesthesia, especially when given in higher dosages.[5] Anesthetists should monitor the patient and discontinue the infusion and alert the veterinarian if any adverse effects are observed.

Safety Alert

In the United States, opioids and ketamine are regulated under the Controlled Substances Act, which has rules regarding purchase, handling, and dispensing these drugs. Their use requires compliance with detailed record keeping due to the potential for misuse. A trusted person should be assigned to monitor and periodically check to ensure all controlled drugs are properly stored and accounted for.

Calculating CRIs

When administering drugs via CRI, the constant dose is delivered over a specified period of time and needs to be calculated. For example, the drug dosage for CRI is often expressed in micrograms (mcg or μg) per kilogram per minute (*mcg/kg/min*), micrograms per kilogram per hour (*mcg/kg/hr*), milligrams per kilogram per minute (*mg/kg/min*), or milligrams per kilogram per hour (*mg/kg/hr*). The drug concentration is usually expressed as milligrams per milliliter (*mg/mL*) or micrograms per milliliter (*mcg/mL*) and then ultimately the drug is delivered at a rate expressed in volume (usually milliliters per hour (*mL/hr*). All of the different units make calculating CRIs appear confusing to new veterinary technicians, although with organization and practice the process rapidly becomes second nature.

To perform CRI calculations, the following information is needed:

- The drug dosage (e.g., 2 mcg/kg/min)
- The drug's concentration (e.g., ketamine is 100 mg/mL)
- The patient's weight in kg (e.g., the dog weighs 20 kg)
- Fluid rate in mL per hour (e.g., 0.024 mL/hr)

Remember that all of the above information is not always provided; sometimes you will need to look it up in a drug formulary or solve for an answer (e.g., *How many mL per hour rate will I administer?*). To determine the final rate, consider the following example that applies all of the information above:

A 20 kg dog is prescribed a ketamine CRI at a dosage of 2 mcg/kg/min. The concentration of ketamine is 100 mg/mL. What is the rate (mL/hr) at which the syringe pump should be set?

- First make sure the information is in the same units. In this example, the dosage is in mcg/kg/min while the concentration of ketamine is in mg/mL and the desired rate is in mL/hr. These units will need to be converted to a common unit (basic metric conversions are summarized in Appendix F).
  - Convert 2 mcg/kg/min to a dosage in mg.

    2 mcg/kg/min × 1 mg/1000 mcg = 0.002 mg/kg/min

  - Convert 0.002 mg/kg/min to a dosage in hr.

    0.002 mg/kg/min × 60 min/hr = 0.12 mg/kg/hr

  - Determine the amount of drug to administer based on the patient's weight

    0.12 mg/kg/hr × 20 kg = 2.4 mg/hr

  - Use the drug concentration to determine the administration rate in mL/hr

    2.4 mg/hr × 1 mL/100 mg = 0.024 mL/hr

The syringe pump/driver should be set to administer ketamine at 0.024 mL/hr. It may be possible to set this rate when using some but not other types of syringe pumps/drivers because it is such a small volume. Hence the 0.024 mL of ketamine could be added to 0.976 mL of saline to make a total volume of 1 mL and run at 1 mL/hr. Typically, enough concentration for a 6, 8, 12, or 24 hour period is made.

Another option would be to add ketamine to a bag of fluids to make a dilution and administer it by gravity flow or an IV fluid administration pump at a greater rate. Consider this new information to determine the following:

> The veterinarian would like to administer ketamine in a 250 mL bag of fluids (0.9% NaCl) at 10mL/hr.

It was previously determined that we would like to administer 0.024 mL/hr ketamine to this dog. We now need to figure out how much ketamine to add to an entire 250 mL bag of 0.9% NaCl. Before performing this calculation, it is important to know that if we administer 10 mL/hr of the NaCl to this patient, then the entire 250 mL bag of fluid will last exactly 25 hours (10 mL/hr × 25 hours = 250 mL). The veterinarian selected the 10 mL/hr rate at random but it seems to be a reasonable rate (not too much, not too little) to give to a 20 kg dog. A simple way to figure out how much ketamine to add to the bag is to know there are 25 hours of NaCl available (250 mL ÷ 10 mL/hr = 25 hours). We know we want to administer ketamine at a rate of 0.024 mL/hr. If we take 25 hours × 0.024 mL/hr of ketamine, this equals 0.6 mL to be added to the 250 mL bag. To ensure we do not have a volume greater than 250 mL in the bag, we first remove 0.6 mL of 0.9% NaCl from the bag and then add in 0.6 mL of ketamine (100 mg/mL). If we administer this bag of 0.9% NaCl at 10 mL/hr, the ketamine CRI will run at 0.024 mL/hr of ketamine for 25 hours.

Another way to look at this problem is as follows:

- To determine how much ketamine to add to the 250 mL bag, use the following equation:

$$M = \frac{D \times W \times V}{R}$$

  M = amount of drug (in units used to solve the equation so everything cancels out) to be added to the solution = unknown (?)
  D = dosage = 2 mcg/kg/min
  W = weight of the patient in kg = 20 kg
  V = volume of the solution = 250 mL
  R = infusion rate = 10 mL/hr

- First convert 2 mcg/kg/min to hours

  2 mcg/kg/min × 60 min/1hr = 120 mcg/kg/hr

- Then convert 120 mcg/kg/hr to mg/kg/hr

  120 mcg/kg/hr × 1 mg/1000 mcg = 0.12 mg/kg/hr

- Be sure to "cancel out" units when you are first learning how to do these calculations. In this example, the equation now looks like this:

$$M = \frac{(0.12 \text{ mg/kg/hr}) (20 \text{ kg})(250 \text{ mL})}{(10 \text{ mL/hr})}$$

- Perform the calculation and unit calculation:

$$M = \frac{600 \text{ mg}}{10}$$
$$M = 60 \text{ mg}$$

- Use the concentration of the drug (ketamine = 100 mg/mL) to determine the amount of mL to add to the bag

  60 mg × 1 mL/100 mg = 0.6 mL

- In this example, 0.6 mL of ketamine (100 mg/mL) needs to be added to the 250 mL bag. Again, remember to remove 0.6 mL of 0.9% NaCl from the bag before adding the ketamine to not exceed 250 mL of total volume, which will alter the concentration.

Veterinarians sometimes mix several analgesic drugs for very painful procedures. Each drug is calculated separately and is based on the dosages provided by the veterinarian. For example, a veterinarian requests a CRI of morphine, lidocaine, and ketamine (MLK). One set of example dosages is 200 mcg/kg/hr for morphine; 30 mcg/kg/min for lidocaine, and 600 mcg/kg/hr for ketamine. The concentration of morphine is 15 mg/mL, lidocaine is 20 mg/mL, and ketamine is 100 mg/mL. The veterinarian determines that the administration rate is 5 mL/kg/hr. Based on the administration rate of 5 mL/kg/hr in the 15 kg dog, determine the amount of each drug needed for a 500 mL bag of fluids.

Fluid rate:

- 5 mL/kg/hr × 15 kg = 75 mL/hr

For morphine:

- First make sure the information is in the same units. In this example the dosage is in mcg/kg/min while the concentration of morphine is in mg/mL and the desired rate is in mL/hr.
  - 200 mcg/kg/hr × 15 kg = 3000 mcg/hr
  - 3000 mcg/hr × 1 mg/1000 mcg = 3 mg/hr
  - 3 mg/hr ÷ 75 mL/hr = 0.04 mg/mL
  - 0.04 mg/mL × 500 mL bag = 20 mg (20 mg is $\frac{D \times W \times V}{R}$)
  - 20 mg ÷ 15 mg/mL (concentration of morphine) = 1.3 mL
- In other words,

$$M = \frac{D \times W \times V}{R} = 20 \text{ mg}$$

  M ÷ concentration of morphine = volume of morphine
  20 mg ÷ 15 mg/mL = 1.3 mL

Safety Alert

There are several concentrations of morphine available, and with current opioid shortages dictating availability it is important to double check concentrations of each drug, even if you just did the same calculations for a patient the previous week or month.

For lidocaine:

- First make sure the information is in the same units. In this example the dosage is in mcg/kg/min while the concentration of lidocaine is in mg/mL and the desired rate is in mL/hr.
  - ○ 30 mcg/kg/min × 15 kg = 450 mcg/min
  - ○ 450 mcg/min × 1 mg/1000 mcg = 0.450 mg/min
  - ○ 0.450 mg/min × 60 min/hr = 27 mg/hr
  - ○ 27 mg/hr ÷ 75 mL/hr = 0.36 mg/mL
  - ○ 0.36 mg/mL × 500 mL bag = 180 mg (180 mg = $\frac{D \times W \times V}{R}$)
  - ○ 180 mg ÷ 20 mg/mL (concentration of lidocaine) = 9 mL
- In other words,

$$M = \frac{D \times W \times V}{R} = 180 \text{ mg}$$

M ÷ concentration of lidocaine = volume of lidocaine
180 mg ÷ 20 mg/mL = 9 mL

For ketamine:

- First make sure the information is in the same units. In this example the dosage is in mcg/kg/hr while the concentration of ketamine is in mg/mL and the desired rate is in mL/hr.
  - ○ 600 mcg/kg/hr × 15 kg = 9000 mcg/hr
  - ○ 9000 mcg/hr × 1 mg/1000 mcg = 9 mg/hr
  - ○ 9 mg/hr ÷ 75 mL/hr = 0.12 mg/mL
  - ○ 0.12 mg/mL × 500 mL bag = 60 mg (60 mg = $\frac{D \times W \times V}{R}$)
  - ○ 60 mg ÷ 100 mg/mL (concentration of ketamine) = 0.6 mL
  - ○ In other words,

$$M = \frac{D \times W \times V}{R} = 60 \text{ mg}$$

M ÷ concentration of ketamine = volume of ketamine
60 mg ÷ 100 mg/mL = 0.6 mL

In this example, 1.3 mL morphine (15 mg/mL), 9 mL lidocaine (20 mg/mL), and 0.6 mL ketamine (100 mg/mL) needs to be added to the 500 mL bag. Remember to remove 10.9 mL of fluid (1.3 mL + 9 mL + 0.6 mL) before adding the drugs to not exceed a total volume of 500 mL, which will alter the concentration.

Based on the administration rate of 5 mL/kg/hr in the 15 kg patient, determine how long a 500 mL fluid bag will last.

- Calculate the rate of fluids needed per hour for the weight of the patient.

5 mL/kg/hr × 15 kg = 75 mL/hr

(This is the delivery rate set on the infusion pump or syringe pump.)

- To determine how long the bag will last, take the fluid bag size and divide it by the hourly rate.
  - ○ 500 mL ÷ 75 mL/hr = 500 mL × 1 hr/75 mL = 6.67 hr

Most CRI bags are discarded after 24 hours. All bags of IV fluids should not be used after 72 hours regardless of their content because of the potential for bacterial growth.

LOCAL AND REGIONAL ANESTHESIA/ NERVE BLOCKS

Local and regional anesthesia (also known as **locoregional anesthesia**) techniques desensitize a localized area of the body so that surgical procedures can be performed in conscious animals or to decrease the level of general anesthetic needed in anesthetized animals, which is especially helpful in patients with altered cardiopulmonary status. Locoregional techniques as part of a multimodal analgesia and anesthesia plan can improve analgesia and help reduce the need for other drugs that have greater adverse effects and can add to muscle relaxation for a wide variety of surgical procedures. The use of long-acting local anesthetics can also smooth recovery in some patients.

Local anesthetics block sodium ion channels in the cell membrane of peripheral neurons, which stops nerve impulse conduction along the nerve until the drug is absorbed into the local circulation. The local anesthetic agent must cover a sufficient length of nerve to block nerve conduction (either by injecting the proper volume or concentration of drug). Points to remember about local anesthetics include:

- Depending on the drug dosage and affected region, sensory and motor neurons may both become "blocked" resulting in a temporary loss of both sensation *and* motor function. Although sensory fibers tend to be blocked before motor fibers, patients should be monitored so they do not injure themselves due to altered motor function.
- Depending on where they are administered, local anesthetics can affect the autonomic nervous system causing sympathetic blockade, which can lead to vasodilation and hypotension. For example, dehydrated, hypovolemic, or shocky patients are not good candidates for epidural techniques with local anesthetics.
- As local anesthetics are concentration dependent drugs, avoid diluting the drug or mixing it with other drugs whenever possible. Anecdotally some anesthetists used to combine lidocaine with bupivacaine to get "a rapid onset and longer duration of action" of local block. Clinically, however, ample literature has shown that this concept is not effective and it is better to administer one drug in a larger quantity rather than two drugs in smaller quantities.
- Local anesthetic preparations have also been historically combined with very small amounts of vasoconstrictors such as epinephrine to slow systemic absorption of the local anesthetic thereby prolonging its duration of action. This practice is rarely performed clinically anymore because local anesthetics such as lidocaine and bupivacaine already provide adequate duration of effect and combining local anesthetics with epinephrine may delay wound healing and cause tissue necrosis due to intense vasoconstriction and secondary ischemia. Recently in human medicine, dexmedetomidine and/or opioids have been added to peripheral nerve blocks with the goal of prolonging the duration of the block. The exact mechanism of action is not fully known. In some

patients, extending the duration of local anesthesia may be beneficial and in other cases it may not. For example, if you go to the dentist to have dental work done, having your mouth numb during the procedure is beneficial, but having your mouth numb for 18 hours is excessive.

Peripheral Nerve Stimulator and Ultrasound Guided Nerve Blocks

Performing local and regional anesthesia involves injecting local anesthetic as close to the nerve as possible in order to successfully block nerve conduction (sensation) for pain management but as not to enter into the nerve as to potentially damage it. Unfortunately, as nerves can be located deep among muscle and other soft tissue structures, it can be technically challenging to inject local anesthetic close to a nerve accurately and there is always the risk of injuring the nerve itself by getting local anesthetic into a nerve. One way to ensure a more accurate location for a locoregional block and avoid intraneural injection is to use a peripheral nerve stimulator, a hand-held instrument designed to accurately locate peripheral nerves. Using a peripheral nerve stimulator may reduce the amount of local anesthetic needed for some nerve blocks because the nerve's location is precisely determined, which eliminates the need to inject more local anesthetic in hopes of getting the drug to the area needed (Figure 17-4). The peripheral nerve stimulator has one electrode that is attached to the skin (the positively charged lead wire, which has a red plug with an alligator clip) and another electrode that is attached to an insulated needle, which is typically uninsulated at the tip (the negatively charged lead wire, which has a black plug with an alligator clip). Insulated needles are coated with nonconducting material such as Teflon® over the length of the needle except at the tip. The nonconducting

material prevents electric current from leaking along the needle shaft into the surrounding tissue. A syringe with local anesthetic is attached to an extension set and the insulated needle. After anatomical landmarks are identified for the desired nerve block, the insulated needle is inserted and directed toward the nerve to be blocked. An electrical current (such as 1 mA) is used to stimulate the nerve and cause muscle contraction of the muscles innervated by the nerve. Once the target nerve is located using the stimulating needle, the current is reduced incrementally until a twitch is located at approximately 0.5 mA. If motor response of the nerve can be evoked at values as low as 0.2 mA, this may be consistent with the needle being in the nerve/nerve sheath and the needle should be repositioned. The lowest current above 0.2 mA that will induce any degree of muscle contraction when the needle is in position should be obtained prior to injection of the local anesthetic to ensure that the needle is close to but not in the nerve. The needle is then held in position while an assistant can aspirate and then inject through the pre-attached extension tubing connected to the needle. Resistance to injection of the local anesthetic may indicate that the needle is within a nerve sheath or that some amount of tissue or fat is obstructing flow. The needle should be slightly repositioned if this occurs.

Another technique used widely in human anesthesiology to improve accuracy is performing nerve blocks with the aid of ultrasound guidance. This technique is becoming more popular in veterinary medicine as well. The user employs an ultrasound machine to visualize the nerve and then while watching the screen can adjust placement of a needle so that local anesthetic can be deposited in close proximity to the nerve. Ultrasound guided nerve blocks are rapidly becoming the new "gold standard;" however, performing them requires both advanced equipment and advanced training. Presently this technique is most commonly performed by board certified veterinary anesthesiologists in academic and private referral hospital settings.

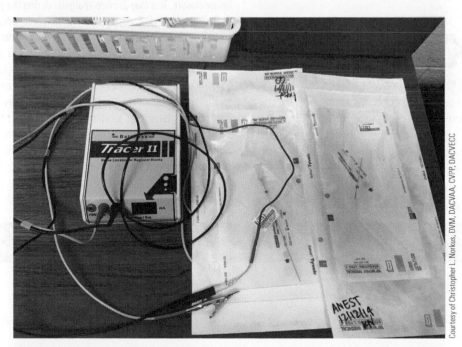

Courtesy of Christopher L. Norkus, DVM, DACVAA, CVPP, DACVECC

FIGURE 17-4 A peripheral nerve stimulator can be used to locate peripheral nerves for local nerve blocks. The unit can be set at different Hertz and milliampere (mA) levels to deliver current; the positive lead wire (red plug) is attached to skin and the negative lead wire (black plug) is connected to the insulated needle. The insulated needle is kept sterile until use and is then attached to an extension set and syringe.

Local Anesthetic Toxicity

Veterinary anesthetists using lidocaine, bupivacaine, and ropivacaine must be aware of the risk for local anesthetic toxicity and should use proper dosages to avoid toxicity in their patients. When looking for dosages of local anesthetics in reference books, most recommended dosages are listed as "do not exceed" dosages and are well below what is needed to reach toxic levels in animals. Maximum dosages of local anesthetics vary depending on the route of administration, the injection site, species, and rate of uptake. Cats are more sensitive to the toxic effects of local anesthetics than dogs;[8] therefore, the maximum dosages of local anesthetics for cats will be lower than those for dogs.

Lidocaine, bupivacaine, and ropivacaine all cause CNS and cardiovascular adverse effects. Lidocaine produces CNS toxicity (muscle twitching and seizures) at much lower dosages than those required to produce cardiovascular toxicity (Table 17-2). Bupivacaine will cause similar CNS signs to lidocaine, but these signs are much more closely followed by cardiac toxicity (arrhythmias and myocardial depression leading to cardiac arrest). Muscle twitching and seizures can be managed by stopping further administration of the local anesthetic and treating supportively including administering medications such as diazepam (Valium®). In the event of cardiovascular collapse, hypotension can be managed with IV fluids and cardiac arrest should be treated with cardiopulmonary cerebral resuscitation (refer to Chapter 13 [Cardiac Arrest & Cardiopulmonary Cerebral Resuscitation (CPCR)] and Critical Thinking Question 1 in Chapter 15 [Acute Pain Management]); however, cardiac arrest secondary to local anesthetic toxicity can be very difficult to treat with standard efforts. Newer and more specific therapies including intravenous lipid emulsion (20%) is now recommended as the treatment of choice for all local anesthetic toxicity, and insulin/dextrose/potassium therapy has also shown promise in experimental models.[11]

Safety Alert

To prevent the risk of toxicity, total doses of local anesthetic should be calculated carefully, based on lean body weight, and should not be exceeded. Be sure to also double check the concentration of the local anesthetic as it can vary (e.g., ropivacaine).

Routes of Local Anesthetic Administration

Local anesthetics can be administered by a variety of methods, and potentially multiple methods and sites can be used on the same patient. Guidelines to follow when preparing to give local anesthetics include:

- When determining the amount of local anesthetic to inject, the maximum "do not exceed" dose should be calculated to ensure it is not exceeded when administering the drug dose. For example, if Mary wishes to administer bupivacaine for an infraorbital dental nerve block in a 10 kg patient, based on the dosage of 2 mg/kg she would initially calculate that the "do not exceed" dose is 20 mg (2 mg/kg × 10 kg = 20 mg). Mary now knows that if the concentration of bupivacaine is 0.5% (5 mg/mL) she can administer up to 4 mL (20 mg) of bupivacaine in this patient (20 mg ÷ 5 mg/mL = 4 mL). She only ends up needing 5 mg (1 mL) for the block, and she has additional drug she can safely give if further blocks are needed. This "do not exceed" dose (in this example 20 mg) can be divided by the number of nerve block sites to be injected to ensure that the total maximum dose is not exceeded. In other words, Mary can perform four 5 mg (1 mL) nerve blocks in this patient.
- If the volume to be administered at each site is very small, sterile saline can be used to dilute (up to 1:1) the drug to increase the volume to be given. This should be avoided whenever possible as it reduces the concentration and likely the efficacy of the drug.

"Splash Blocks"

Local anesthetics such as lidocaine or bupivacaine can be directly applied to surgery sites by "splashing" or dripping them onto open surgical sites before closure. This may provide analgesia during the procedure and some degree of pain relief postoperatively; however, additional analgesia is always needed postoperatively. Currently, limited data in the veterinary literature support this practice. For this technique to be potentially effective, the local anesthetic would need to remain in contact with the tissue for a significant amount of time. When this technique is attempted in clinical practice, it appears to work best when the local anesthetic is squirted on relatively dry tissue (e.g., an abdominal incision after the body wall [linea alba] is closed) and the surgeon then continues with the subcutaneous layer closure. Whenever possible, to ensure efficacy, local infiltration via injection into tissue or specific perineural injection is preferred.

TABLE 17-2 Local Anesthetics*

Agent	Dosage (infiltration)	Onset time (in minutes)	Duration (hours)	Toxicity
lidocaine 2% 20 mg/mL	4–6 mg/kg (dog) 2–4 mg/kg (cat)	< 5	1–2	CNS, Cardiac
bupivacaine 0.5% 5 mg/mL	1–2 mg/kg (dog) 1 mg/kg (cat)	< 20	3–6	CNS, Cardiac
ropivacaine 0.75% 7.5 mg/mL	1–3 mg/kg (dog) 1–2 mg/kg (cat)	< 20**	3–6	CNS, Cardiac (Less cardiotoxic than bupivacaine)

*Grubb, T., & Lobprise, H. (2020). Local and regional anaesthesia in dogs and cats: Overview of concepts and drugs (Part 1). *Veterinary Medicine and Science*, 6(2), 209–217. doi:10.1002/vms3.219.
**Fanelli, G., Casati, A., Beccaria, P., et al. (1998). A double-blind comparison of ropivacaine, bupivacaine, and mepivacaine during sciatic and femoral nerve blockade. *Anesthesia & Analgesia*, 87(3), 597–600. doi:10.1097/00000539-199809000-00019.
Onset of ropivacaine can be faster than bupivacaine depending on concentration used, area blocked, and other factors.

Local Infiltration

Local anesthetics can be injected intradermally or subcutaneously to desensitize the skin and subcutaneous tissues for minor surgery such as removal of small masses (such as a 2 mm skin tag). After the skin is clipped and aseptically prepared over the surgical site, the local anesthetic is injected intradermally or subcutaneously to desensitize a small area. This can be repeated in a line ("line block"), rectangular, or triangular pattern in order to desensitize a larger area. As with all nerve blocks, always aspirate before injecting to avoid injecting the drug into a blood vessel.

Intrapleural Blocks

Local anesthetics can be applied into the pleural cavity via a chest drain to provide analgesia postoperatively in patients who have undergone a thoracotomy by either a lateral thoracotomy or median sternotomy approach. Bupivacaine 0.5% up to a total dosage of 2 mg/kg is the most commonly used local anesthetic for intrapleural blocks. When administering the local anesthetic, the chest tube should be flushed with 3 to 5 mL of sterile saline and the patient placed incision down for 10 minutes after administration to ensure the drug gravitates to the affected region (lateral thoracotomy). The bupivacaine diffuses across the pleura and blocks intercostal nerves near the administration site producing a rapid onset of action that persists for 4 to 6 hours.[12] If the patient had a median sternotomy, the animal is placed in sternal recumbency for 10 minutes to allow the local anesthetic to pool near the incision site. This technique is easy to perform once a chest drain is in place (Figures 17-5a and 17-5b); however, the risk of systemic toxicity is higher than with other techniques and the quality of analgesia variable due to the rapid uptake of local anesthetics from the pleural space into the systemic circulation.[13] Drugs given intrapleurally must be calculated carefully to avoid systemic toxicity. Some literature in humans supports including opioids such as morphine with bupivacaine for intrapleural analgesia.

Peritoneal Infiltration

Peritoneal infiltration of local anesthetic may be useful in patients with pancreatitis or other intra-abdominal disease whose pain is difficult to manage. Bupivacaine can be injected into the peritoneal cavity via abdominocentesis at the umbilicus. Following drug administration, the patient should be placed in dorsal recumbency for 10 minutes, then ventral/sternal recumbency with the hindquarters elevated for 10 minutes to promote cranial migration of the drug.[11] Strict asepsis should be observed and the dose calculated carefully to avoid toxicity. The bupivacaine is typically diluted with sterile saline and supplemented with sodium bicarbonate to reduce pain on injection in conscious patients. Local anesthetic can also be directly deposited into the abdominal cavity if it is open during surgery. Local anesthetics are commonly sprayed into the abdomen during human laparoscopic surgery. The efficacy of said techniques lacks evidence-based research regarding their use in animals.

Intravenous Regional Analgesia (IVRA)

Intravenous regional anesthesia (IVRA) can be used to produce analgesia of distal limbs in dogs and cats for procedures such as toe amputations. Lidocaine is injected IV distal to a tourniquet (an Esmarch bandage is typically applied to the limb as the tourniquet) (Figure 17-6). Anesthesia begins distally and moves proximally; therefore, the anesthetic should be injected as distally as possible. Onset of analgesia is rapid (around 10 minutes) and persists for the length of time the tourniquet is in place. The Esmarch bandage can be left in place for a maximum of 90 minutes, but should be removed as soon as the procedure is done to prevent vascular injury. Once the bandage is removed, analgesia remains for up to 30 minutes. A relatively high dosage of lidocaine is typically required and the risk of systemic toxicity is increased; therefore, some anesthetists avoid the technique entirely in cats. Lidocaine is the only local anesthetic used for IVRA. Bupivacaine has a much lower therapeutic index (margin of safety) than lidocaine and should NEVER be administered intravenously due to the potential for life threatening cardiac toxicity.

Intra-Articular Blocks

Local anesthetics can be given intra-articularly (IA) to provide analgesia when performing joint surgery. Bupivacaine or ropivacaine are commonly given IA to dogs after surgical repair of ruptured cruciate ligaments in the stifle and into elbows following arthroscopy, but can be administered into almost any joint space depending on the case. Buprenorphine or morphine

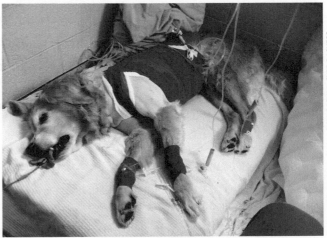

(a)

(b)

Courtesy of Christopher L. Norkus, DVM, DACVAA, CVPP, DACVECC

FIGURE 17-5 A previously placed chest drain is a method through which local anesthetic can be given to perform an intrapleural block. (a) The patient is prepped for placement of the chest drain. (b) The patient receiving analgesics through the chest drain.

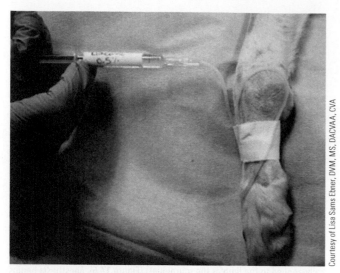

Courtesy of Lisa Sams Ebner, DVM, MS, DACVAA, CVA

FIGURE 17-6 Intravenous regional anesthesia (IVRA) in a distal limb of a dog. Lidocaine is injected IV distal to a tourniquet, and anesthesia begins distally and moves proximally.

have also been used alone or as an adjunct with a local anesthetic for intra-articular analgesia.[14, 15] When administering drugs IA, strict asepsis should be followed to reduce the risk of infection. Some surgeons may be concerned about the effects drugs administered IA have on chondrocytes.

Perineural Infiltration and Specific Nerve Blocks

Perineural infiltration is a technique involving injection of local anesthetics directly around the nerve. When performing nerve blocks by perineural infiltration, key points to adhere to regardless of the site include the following:

- The anesthetist should have a working knowledge of the anatomy of the area being blocked as well as key anatomical landmarks such as foramen, bony prominences, and blood vessels.
- An aseptic technique should be observed. Washing hands with an antiseptic scrub and wearing gloves is recommended.
- The total doses of local anesthetic should be calculated carefully based on lean body weight. Care should be taken not to administer more than the "do not exceed" dose.
- As veins and arteries tend to run along similar paths to nerves, the syringe should always be aspirated prior to injecting local anesthetic to avoid inadvertent injection into blood vessels. If blood is observed, the needle should be withdrawn and both the needle and syringe discarded before beginning another attempt. If the volume of drug being administered is large, the syringe should be re-aspirated several times during injection to ensure the position of the needle has not changed.
- If resistance is felt during injection, the needle should be withdrawn and repositioned as this may indicate the needle is directly in a nerve, fat, or unintended tissue, which may cause adverse effects.
- Inject the local anesthetic slowly to observe for adverse reaction.

There are many different nerves that can be blocked by perineural infiltration to provide perioperative analgesia in dogs and cats. This chapter will describe just some of these different "nerve blocks" and the reader is directed to other texts that specifically focus on the topic for more information. The nerve blocks described in this chapter are summarized in Table 17-3.

TABLE 17-3 Summary of Nerve Blocks

Nerve Block	Area Affected
Infraorbital	• Bone, soft tissue (upper lips and nares), and teeth rostral to upper fourth premolar (including canine and incisors) on side of the maxilla injected
Maxillary	• Entire upper dental arcade, soft and hard palates, muzzle to midline on the side injected
Mental	• Lower bone, soft tissue (lower lip), and teeth rostral to lower second premolar on side of the mandible injected
Mandibular	• Lower dental arcade and associated soft tissue including the tongue on the side injected
Intercostal	• Tissues of thorax on side injected • Pain management for lateral thoracotomy and fractured ribs
Brachial plexus	• Forelimb distal to elbow
Radial, Ulnar, Median, Musculoskeletal (RUMM)	• Forelimb distal to elbow
Radial, Ulnar, Median (RUM); also called the "ring block"	• Forelimb distal to carpus
Femoral	• Distal femur, stifle, and skin of the dorsomedial tarsus and first digit
Sciatic	• Hock and foot
Retrobulbar	• Structures of the eye including the conjunctiva, cornea, and uvea

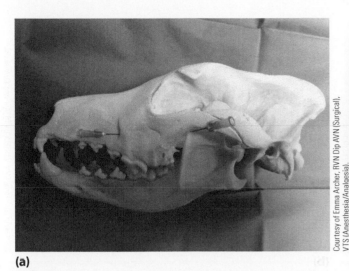

Courtesy of Emma Archer, RVN Dip AVN (Surgical), VTS (Anesthesia/Analgesia).

(a)

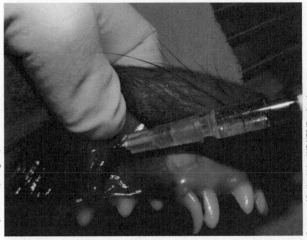

Courtesy of Benita Altier, LVT, VTS (Dentistry)

(b)

FIGURE 17-7 Examples of dental nerve blocks. (a) Canine skull showing the site for infraorbital and maxillary nerve blocks. The infraorbital nerve exits the skull through the infraorbital foramen (needle on the left); the maxillary nerve enters the skull through the maxillary foramen (needle on the right). (b) Infraorbital nerve block in a dog. Note that the dog is in left lateral recumbency.

Dental Nerve Blocks

There are four nerves commonly blocked to provide excellent analgesia in dogs and cats undergoing tooth extractions and dental work, maxillectomies, and mandibulectomies. These nerves are the infraorbital, maxillary, mental, and mandibular (also referred to as the inferior alveolar). When performing nerve blocks multiple sites may be injected; therefore, the total volume can approach or exceed the "do not exceed" dose and enter the toxic dosage range despite administering the recommended volumes at each site. For example, the recommended volumes per site when injecting 2% lidocaine might be 0.1 to 0.3 mL for cats, 0.25 to 0.5 mL for small dogs, and 1 to 1.5 mL for large dogs. If a 2 kg cat gets 4 dental blocks of lidocaine at 0.3 mL each, the cat will have received 24 mg of lidocaine total (4 injection sites × 0.3 mL each = 1.2 mL total × 20 mg/mL = 24 mg) which for the 2 kg cat is approximately 10 mg/kg (double the "do not exceed" dosage) and could result in toxicity.

Infraorbital Blocking the infraorbital nerve provides analgesia to the rostral portion of the maxilla on the **ipsilateral** (same) side, including the incisors up until the midline, upper canine and premolar teeth rostral to the fourth premolar, and the associated soft tissues (upper lips and nares) (Figure 17-7a) .

1. Retract the upper lip dorsally and palpate the infraorbital foramen (easily palpated in most dogs on the maxilla, rostral to the orbit).
2. The infraorbital foramen cannot be palpated well in cats, although a ridge can often be felt. In cats and brachycephalic dogs, the infraorbital canal is short, and care must be taken to avoid inserting the needle beyond the medial canthus.
3. With clean hands and wearing gloves, block the nerve by inserting a 22 to 25 gauge needle and advancing it caudally into the infraorbital foramen either through the inside of the mouth or through the skin on the muzzle (Figure 17-7b).

4. The needle appears to lock into place once the foramen is entered.
5. Aspirate before injecting to ensure a blood vessel has not been entered; withdraw and reposition if blood is seen.
6. Inject the local anesthetic slowly to observe for adverse reaction. Only a small amount of local anesthetic is required and is typically based on clinical judgement; for example, 0.25 to 0.5 mL lidocaine (2%) may be used for a cat whereas 1.0 to 1.5 mL of lidocaine (2%) may be used for a large dog. The key is to give less than the "do not exceed" dose. Bupivacaine or ropivacaine can also be used instead of lidocaine.

Maxillary The maxillary nerve and its branches provide sensory innervation to the entire upper dental arcade, soft and hard palates, and muzzle;[8] therefore, blocking the maxillary nerve desensitizes these areas (Figure 17-8a). A variety of techniques have been described for performing the maxillary nerve block. The intraoral approach is described as follows:

1. With clean hands and wearing gloves, open the patient's mouth and pull the lip commissure caudally.
2. Advance a 22 gauge needle dorsally and perpendicular to the palate into the mucosa caudal to the second molar of the maxilla (Figures 17-8b and 17-8c). The needle should only be advanced approximately 3 to 5 mm.
3. Refer to steps 5 and 6 from the intraorbital nerve block described above.

Mental The mental foreman and nerve can be tricky to successfully locate and block and therefore this nerve block is often replaced with the mandibular (inferior alveolar) nerve block for mandibular teeth. When attempted, the mental nerve can be blocked rostrally at the mental foramen, which is located ventral to the lower first or second premolar and is useful for lower canine and incisor tooth extractions, or procedures

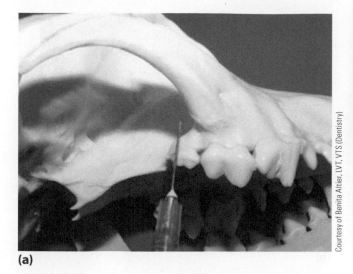

(a)

Courtesy of Benita Altier, LVT, VTS (Dentistry)

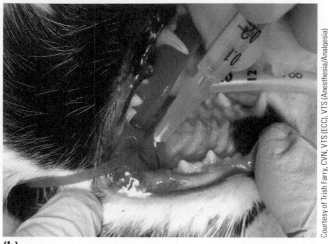

(b)

Courtesy of Trish Farry, CVN, VTS (ECC), VTS (Anesthesia/Analgesia)

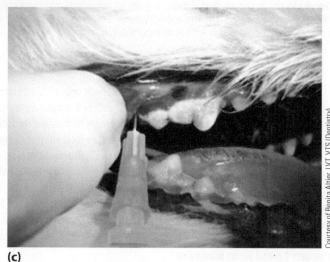

(c)

Courtesy of Benita Altier, LVT, VTS (Dentistry)

FIGURE 17-8 Maxillary nerve blocks. (a) Canine skull showing the site for the maxillary nerve block. (b) The maxillary nerve block being performed on a cat. Note the needle is bent to make it perpendicular to the mucosa caudal to the second molar of the maxilla. (c) The maxillary nerve block being performed in a dog.

performed on the lower lip (Figure 17-9a). The mental foramen is difficult to palpate in cats and some dogs, making the mandibular nerve block a better option in these animals.

1. Retract the mandibular labial frenulum ventrally with one hand and palpate the mental foramen on the ipsilateral side ventral to the first or second premolar.
2. With clean hands and wearing gloves, advance a 22 to 25 gauge needle caudally and slightly ventrally through the gingiva (Figure 17-9b).
3. Refer to steps 5 and 6 from the intraorbital nerve block described above.

Mandibular (Inferior Alveolar) The mandibular or inferior alveolar nerve block desensitizes the lower dental arcade and the associated soft tissue, including the tongue, on the ipsilateral side. Self-trauma to the tongue may occur if a bilateral mandibular block is performed. The mandibular foramen is located on the medial side of the mandibular ramus, just rostral to the angular process and can easily be palpated in most dogs from inside of the mouth (Figure 17-10a). The aim is to block the mandibular nerve before it enters the foramen as the foramen cannot be directly entered.

1. This block is performed extraorally by first palpating the mandibular foramen on the ipsilateral side from inside the mouth. Then palpate the indentation on the ventral border of the caudal mandible just rostral to the angular process of the mandible (this is at the same level as the lateral canthus of the eye).
2. With clean hands and wearing gloves, block the nerve by inserting a 22 to 25 gauge needle through the skin on the medial side of the mandible below the rostral part of the angular process and directing the needle dorsally toward the mandibular foramen (Figure 17-10b).
3. Ensure the bevel of the needle is facing the foramen.
4. Refer to steps 5 and 6 from the intraorbital nerve block described above.

Other Nerve Blocks

There are many different nerve blocks that can be used to provide complete analgesia to various parts of the body. The type of block performed depends on the surgical site, anatomy of the animal, skills and experience of the

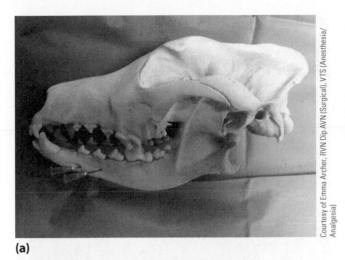
(a)

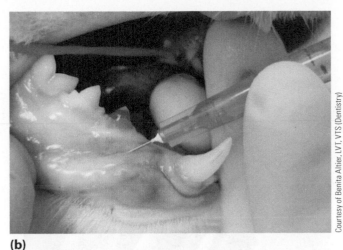
(b)

FIGURE 17–9 Mental nerve block. (a) Canine skull showing site of injection for mental nerve block. The mandibular nerve exits the mental foramen as shown by the needle on the left. The needle is placed into the mandibular canal via the mental foramen. (b) Mental nerve block in a cat.

(a)

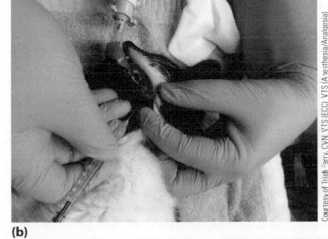
(b)

FIGURE 17-10 Mandibular (inferior alveolar) nerve block. (a) Canine skull showing site of injection for the mandibular nerve block on the medial side of the mandible. (b) The mandibular nerve block being performed on the medial side of the mandible in a cat. For the extraoral technique, the patient should be in dorsal or lateral recumbency, while for the intraoral technique, the patient can be in dorsal or lateral recumbency with the mouth completely open.

veterinarian and veterinary technician performing the blocks, and the type of surgery being performed. Some commonly performed nerve blocks are described below.

Intercostal Nerve Blocks Intercostal nerve blocks can be used to help manage perioperative pain from lateral thoracotomies or to manage pain from fractured ribs. These blocks may also improve ventilation in these patients as well as those recovering from lateral thoracotomies. Pneumothorax is a possible adverse effect of the intercostal nerve block.

The intercostal nerves (along with blood vessels) run alongside the caudal border of each rib. Two to three intercostal nerves should be blocked both cranially and caudally to the affected site. The intercostal nerve block is typically performed with 0.5% bupivacaine 0.25 to 1 mL per nerve up to a total dosage of 2 mg/kg.[16]

1. The patient is placed in lateral recumbency with the injured side up.
2. Using aseptic technique and wearing gloves, a 22 to 25 gauge needle is placed on the caudal border of the rib near the intervertebral foramen (Figure 17-11a).
3. Before the injection is performed, the needle should be "walked off" the caudal edge of the rib and as with all local injections, care should be taken to aspirate before injecting the drug to ensure you are not injecting into a blood vessel (Figure 17-11b). This is particularly important with intercostal nerve blocks as there are blood vessels in close proximity to the nerves; therefore, the risk is greater of injecting drug into a blood vessel, resulting in a higher blood concentration. It is also important to make sure you are not aspirating air, which would indicate you have entered the thorax. Inject the local anesthetic slowly to observe for adverse reaction.

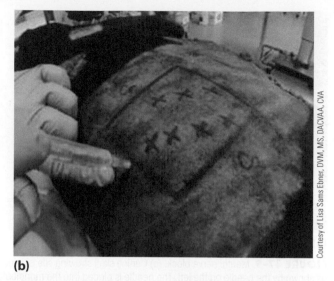

Courtesy of Heidi Reuss-Lamky, LVT, VTS (Anesth and Analgesia) (Surgery)

Courtesy of Lisa Sams Ebner, DVM, MS, DACVAA, CVA

(a)

(b)

FIGURE 17-11 Intercostal nerve block. (a) Location of needle placement on the caudal border of the rib near the intervertebral foramen on a dog skeleton. (b) The needle is placed, "walked off" the caudal aspect of the rib, aspirated with a syringe to confirm the needle is not in a blood vessel or thorax, and then each intercostal nerve is blocked by slowly injecting local anesthetic. On this cadaver dog, the red X's represent where a rib fracture would be.

Brachial Plexus Nerve Blocks The brachial plexus is a bundle of five spinal nerves (C5–T2) located in the axillary space slightly cranial to the first rib. The brachial plexus block involves the radial, median, ulnar, musculocutaneous, and axillary nerves and desensitizes the forelimb distal to the elbow to provide analgesia for ulnar and radial fractures or other soft tissue surgery distal to the elbow. Local anesthetics such as lidocaine or bupivacaine may be used alone or in combination with α_2 adrenergic agonists, opioids, or both. Bupivacaine 0.5% (or ropivacaine 0.75%) combined with 0.5 mcg/mL dexmedetomidine can last 12 to 28 hours.[16] The brachial plexus nerve block is difficult to perform effectively, and complications of this nerve block include pneumothorax, accidental IV injection, motor blockade, and toxicity.

1. Using aseptic technique and wearing gloves, locate the landmarks for the brachial plexus. A peripheral nerve stimulator is recommended to avoid complications such as entering the pleural cavity.
2. A long (7.5 cm) 20 to 22 gauge needle is required, and a spinal needle preferred (the length of the needle should be about the length of the humerus). An insulated needle is needed if using a nerve stimulator.
3. The needle is inserted in the axillary space (proximal to glenohumeral joint) just medial to the scapula (between the scapula and chest wall), directing it caudally toward the elbow (Figure 7-12). The needle is advanced caudally using a nerve stimulator. There is a risk of entering the pleural cavity and causing a pneumothorax if the needle is directed too far toward the midline.
4. When the stylet is removed, listen for any "whoosh" of air, which will be heard if the pleural cavity is entered. The needle should be withdrawn if this occurs.
5. Aspirate before injecting to ensure a blood vessel has not been entered; withdraw and reposition if blood is seen.

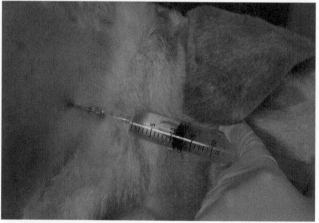

Courtesy of Heidi Reuss-Lamky, LVT, VTS (Anesth and Analgesia) (Surgery)

FIGURE 17-12 The brachial plexus block being performed in a dog. The needle is inserted in the axillary space just medial to the scapula, directing it caudally toward the elbow.

6. Inject the local anesthetic slowly to observe for adverse reaction. Re-aspirate during the injection to ensure the needle position has not changed.
7. Unilateral blockade of the phrenic nerve, which results in hemiparalysis of the diaphragm, is extremely rare but has been reported.[17]

Radial, Ulnar, Median, Musculocutaneous (RUMM) Nerve Block The radial, ulnar, median, musculocutaneous (RUMM) nerve block is an alternative to the brachial plexus block for providing analgesia to the forelimb distal to the elbow and is easier and safer to do than a brachial plexus block (Figures 17-13a and 17-13b). In the editor's

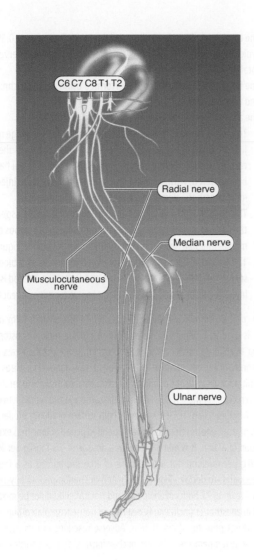

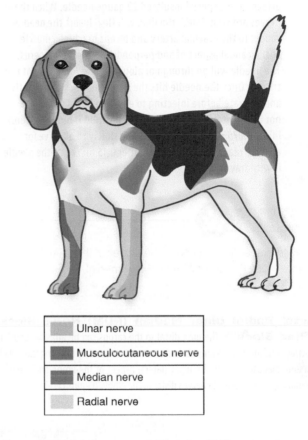

(a)   **(b)**

FIGURE 17-13 RUMM nerve block. (a) The location of the radial, ulnar, median, and musculocutaneous nerves in a dog. (b) The areas desensitized with the RUMM nerve block.

opinion, this block has largely replaced the brachial plexus nerve block in clinical practice due to ease of administration and increased safety. As the name implies, the RUMM block desensitizes the radial, ulnar, median, and musculocutaneous nerves.

1. Bupivacaine 0.5% is typically used for the RUMM block with one dose administered for the radial nerve and another dose administered for the ulnar, median, and musculocutaneous nerves.[16] There are only two injection sites.
2. The RUMM nerve block should be performed using aseptic technique and wearing gloves. A peripheral nerve stimulator is recommended to locate these nerves.
3. The patient is placed in lateral recumbency with the elbow flexed (leg to be blocked is uppermost) for the radial nerve block.[18]

○ Locate the radial nerve (*caudolateral* aspect of the mid-humerus between the medial and lateral head of the triceps and brachialis muscles) using the peripheral nerve stimulator, which is attached to a special insulated 22 gauge needle. When the radial nerve is stimulated, the carpus will extend. The injection site is on the caudal surface (back) of the humerus at the junction of the proximal 2/3 and distal 1/3 length of the humerus (i.e., 2/3 distal to the greater tubercle or 1/3 proximal to the lateral epicondyle). The needle is inserted at the caudal aspect of, and perpendicular to, the humerus. The needle will go through the long head of the triceps muscle or between the long and lateral heads of the triceps. Once it hits the humerus, back out slightly. Aspirate before injecting to ensure a blood vessel has not been entered; withdraw and reposition if blood is seen. Inject 0.1 mL/kg of local anesthetic slowly to observe for adverse reaction.[18]

4. The patient should be in lateral recumbency, lying on the limb that is being blocked for the ulnar, median, and musculocutaneous nerve block. Pull the limb caudally so that it lies behind the upper limb and palpate the brachial artery at mid-humerus.[18]
    ○ Locate the ulnar, median, and musculocutaneous nerves (these three nerves are in close proximity to each other on the *medial* aspect of the mid-humerus, cranial and caudal to the brachial artery) using the peripheral nerve stimulator attached to a special insulated 22 gauge needle. When these nerves are stimulated, the limb will flex. Insert the needle caudal to the brachial artery and biceps brachialis muscle at the caudal aspect of and perpendicular to the humerus. The needle will go through or alongside the long head of the triceps. Once the needle hits the humerus, back out slightly and aspirate before injecting to ensure a blood vessel has not been entered; withdraw and reposition if blood is seen. Inject 0.075 mL/kg local anesthetic slowly to observe for adverse reaction and inject another 0.15 mL/kg as the needle is withdrawn.[18]

Tech Tip

To remember the location of the nerves in the RUMM block, think R = late**r**al and UMM = **m**edial.

Distal Radial Ulnar Median (RUM) Nerve Block ("Ring Block")

The limb distal to the carpus can be anesthetized for onychectomy in cats, toe amputations, or mass removal by blocking the radial (dorsomedial aspect of limb), ulnar (lateral aspect of the limb), and median (palmar aspect of the limb) nerves distal to the elbow (Figure 17-14a).

1. The total dose of local anesthetic is calculated. Bupivacaine 0.5% is typically used for this block.[16] When injecting these three nerves, the total dose is divided by 3.
2. The RUM nerve block should be performed using aseptic technique and wearing gloves.
3. The animal is positioned in lateral or dorsal recumbency.
    ○ The radial nerve is blocked by inserting the 22 to 25 gauge needle just proximal to the carpus on the dorsomedial aspect of the metacarpus. Aspirate before injecting to ensure a blood vessel has not been entered; withdraw and reposition if blood is seen. Inject the local anesthetic slowly to observe for adverse reaction.
    ○ The median and ulnar nerves are blocked on the caudal aspect of the limb by inserting the needle just proximal to the carpus on both medial (median nerve) and lateral (ulnar nerve) points (Figures 17-14b and 17-14c). Aspirate before injecting to ensure a blood vessel has not been entered; withdraw and reposition if blood is seen. Inject the local anesthetic slowly to observe for adverse reaction.

Nocita®, a commercially-available liposome-encapsulated form of bupivacaine, is used as a peripheral nerve block to provide regional postoperative analgesia following onychectomy in nonpregnant, non-lactating cats greater than 5 months of age not intended for breeding (it is also labeled for use as a single tissue infiltration treatment during surgical closure to control postoperative pain in cranial cruciate ligament [CCL] surgery in the same age and reproductive status of dogs). After injection, Nocita®'s exterior lipid surface gradually erodes, which allows bupivacaine to be released over an extended period of time thus extending its duration of action. It is administered at a dosage of 5.3 mg/kg per forelimb (0.4 mL/kg per forelimb, for a total dose of 10.6 mg/kg/cat) as a four-point nerve block prior to declaw (Figure 17-15). Administration prior to onychectomy may provide up to 72 hours of analgesia, and Nocita® should not be mixed with other local anesthetics (including lidocaine and non-encapsulated bupivacaine) or other drugs prior to administration. Adverse reactions in cats may include elevated body temperature, infection, or chewing/licking at the surgical site.

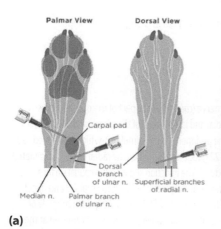

(a)

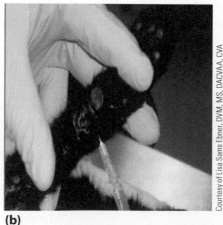

(b)

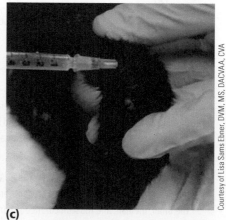

(c)

FIGURE 17-14 Ring block of cats. (a) This image illustrates the nerves (radial, ulnar, and median) that are desensitized in the RUM block. (b) Local anesthetic is injected subcutaneously on the caudal aspect of the limb by inserting the needle just proximal to the carpus on both medial (median nerve) and (c) lateral (ulnar nerve) points. A third injection of local anesthetic for the radial nerve is injected on the dorsomedial aspect of the aspect of the metacarpus.

Administration in cats

Legend

Needle insertion point ⊃

Drug injection point ○

Needle withdrawal + drug injection →

Needle redirection to a 90° angle to the palmar plane

Abbreviations

SpU - Styloid process of the ulna

ACb - Accessory carpal bone

A.

0.14 mL/kg (35%)
Superficial Branch of the Radial Nerve

At the center of the limb, on the dorsal aspect at the level of the antebrachio-carpal joint, insert the needle subcutaneously with the bevel up (·). Advance the needle subcutaneously and inject (○) adjacent to the confluence of the accessory cephalic and cephalic veins.

Dorsal

B.

0.08 mL/kg (20%)
Dorsal Branch of the Ulnar Nerve

Palpate a groove between the accessory carpal bone (ACb, in the base of the carpal pad) and the styloid process of the ulna (SpU). Distal to this groove, insert the needle subcutaneously with the bevel up and advance the needle proximally. Inject once the tip reaches the midpoint of the groove.

Lateral

C.

0.16 mL/kg (40%)
Median Nerve and Superficial Branch of the Palmar Branch of the Ulnar Nerve

Insert the needle subcutaneously with the bevel up lateral to the distal tip of the accessory carpal pad and advance the needle medially 2/3 the width of the limb, until the tip is located near the base of the first digit. Inject 2/3 of the volume at this point and the remaining volume while withdrawing the needle (solid teal arrow). Gently massage for 5 seconds.

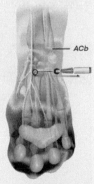

Palmar

D.

0.02 mL/kg (5%)
Deep Branch of the Palmar Branch of the Ulnar Nerve

Orient the needle perpendicular to the long axis of the limb at the level of the ACb. Insert the needle subcutaneously and advance the needle laterally until it contacts the medial aspect of the ACb. Redirect the needle dorsally by rotating the needle 90°, advance it along the medial side of the ACb 2-3 mm until it penetrates the flexor retinaculum, and inject.

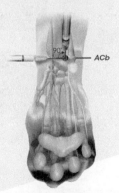

Palmar

INDICATION: For use as a peripheral nerve block to provide regional postoperative analgesia following onychectomy in cats.

IMPORTANT SAFETY INFORMATION FOR CATS: NOCITA® (bupivacaine liposome injectable suspension) is for use as a peripheral nerve block in cats only. Do not use in cats younger than 5 months of age, that are pregnant, lactating, or intended for breeding. Do not administer by intravenous or intra-arterial injection. Adverse reactions in cats may include elevated body temperature, infection or chewing/licking at the surgical site. Avoid concurrent use with bupivacaine HCl, lidocaine or other amide local anesthetics. Please see the full Prescribing Information for more detail.

11400 Tomahawk Creek Parkway, Suite 340, Leawood, KS 66211 1-844-ARATANA
©2019 Aratana Therapeutics, Inc. NOCITA is a registered trademark of Aratana Therapeutics, Inc.

NOC-0233

Courtesy of Aratana (Elanco)

FIGURE 17-15 Four-point nerve block using Nocita®, liposomal bupivacaine.

Femoral and Sciatic Nerve Blocks The femoral and sciatic nerve blocks are typically done together and are described together. To successfully perform these nerve blocks, a peripheral nerve stimulator or ultrasound is needed.

The femoral nerve in dogs and cats is located from spinal cord segments L3 to L5 and can be blocked to provide anesthesia of the distal femur, stifle, and skin of the dorsomedial tarsus and first digit. The femoral nerve block provides adequate postoperative analgesia following cruciate surgery and has few complications associated with its use.

The sciatic nerve in dogs and cats is from spinal cord segments L6 to S1 and can be blocked to provide anesthesia to the hock and foot. The sciatic nerve block can be combined with the femoral nerve block to provide anesthesia to the stifle and tibia.

1. Bupivacaine is commonly used for femoral and sciatic nerve blocks. Bupivacaine 0.5% (or ropivacaine 0.75%) combined with dexamethasone can last 14 hours.[16]
2. For the femoral nerve block using the medial approach, the patient is in lateral recumbency with the leg to be blocked in a natural extended position.
3. Using aseptic technique and wearing gloves, locate the femoral nerve (it lies in the "femoral triangle" on the medial surface of the thigh, bordered dorsally by the rectus femoris muscle, cranially by the caudal portion of the sartorial muscle, and deep by the iliopsoas muscle) using the peripheral nerve stimulator and 22 gauge insulated needle or ultrasound.
4. Insert the insulated needle at an approximately 25 degree angle to the skin. As the needle reaches the femoral nerve, the stifle will extend due to contractions of the quadriceps muscle.
   ○ Aspirate before injecting to ensure a blood vessel has not been entered; withdraw and reposition if blood is seen. Inject the local anesthetic slowly to observe for adverse reaction.
5. The femoral nerve block using the pre-iliac approach is described in Figure 17-16a.
6. The sciatic nerve block should be performed using aseptic technique and wearing sterile gloves. Locate the sciatic nerve (between the greater trochanter and ischial tuberosity) using the peripheral nerve stimulator or ultrasound (Figure 17-16b).
7. Insert the 22 gauge insulated needle perpendicular to the skin. As the needle reaches the sciatic nerve, dorsoflexion of the paw will occur.
   ○ Aspirate before injecting to ensure a blood vessel has not been entered; withdraw and reposition if blood is seen. Inject the local anesthetic slowly to observe for adverse reaction.

Retrobulbar Nerve Block

Retrobulbar nerve blocks can provide excellent analgesia while keeping the eye in a central position for a variety of ophthalmic procedures

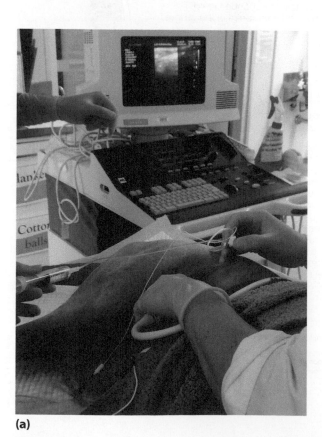
(a)

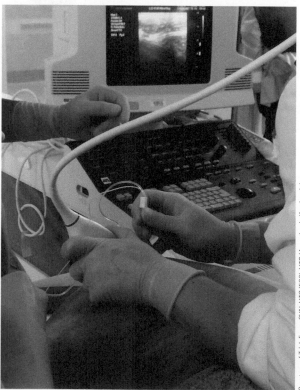
(b)

Courtesy of Trish Farry, CVN, VTS (ECC), VTS (Anesthesia/Analgesia)

FIGURE 17–16 The femoral and sciatic nerve blocks. (a) The femoral nerve block using a pre-iliac approach and ultrasound. The intersection of two lines serves as the injection site. One imaginary line is made starting at the transverse process of L6 and perpendicular to the spine. The second imaginary line is at the most cranial prominence of the wing of the ilium and parallel to the spine. Direct the needle 35 degrees to 45 degrees caudally. This approach avoids the femoral artery. (b) An ultrasound-guided sciatic nerve block.

such as enucleation (Figure 17-17a). Retrobulbar blocks are more commonly performed in large animals to avoid the use of general anesthesia. These blocks are useful in providing analgesia and reducing the anesthetic agent requirements for dogs undergoing enucleation or evisceration surgery. The retrobulbar nerve block can put pressure on the optic nerve of the non-blocked eye in cats and may result in **contralateral** (opposite side) blindness. This nerve block must be performed by a properly trained professional who has a thorough understanding of the anatomy in the area and the risks involved. Complications such as retrobulbar hemorrhage, perforation of the globe, and injection of local anesthetic into the subarachnoid space, which may lead to respiratory arrest, cardiac arrest, and death, are possible with this nerve block. Before injecting the local anesthetic agents, close attention should be paid to the hub of the needle to ensure there is no blood present and the syringe should be aspirated several times during drug administration to ensure the needle has not entered a vessel. Specifically designed curved, blunt needles are available to aid injection into the retrobulbar space, which greatly facilitates this procedure but does not necessarily add to the procedure's safety (Figure 17-17b).

1. Typically 0.5 to 2 mL of 2% lidocaine or 0.5 to 2 mL of 0.5% bupivacaine depending on the size of the patient is used for the retrobular nerve block.[16] Calculate the maximum dose of the local anesthetic as usual; however, a maximum volume of 2 mL of 2% lidocaine or 0.5% bupivacaine is recommended.[12]
2. Routine ophthalmic skin/ocular preparation is performed.

3. Using aseptic technique and wearing sterile gloves, a retrobulbar needle (or a 1.5 inch 22 gauge needle bent to approximately a 20 degree angle) is advanced either through the skin at the level of the orbital rim (Figure 17-17 c) or transconjunctivally at a 6 o' clock position if the orbit is a clock face.
4. The needle is directed along the floor of the orbit until it gets to the back of the eye. When the needle penetrates the orbital fascia, a popping sensation may be detected.
5. Aspirate prior to injection to observe for blood or cerebrospinal fluid (CSF).
6. Inject the local anesthetic slowly to observe for adverse reaction. If resistance occurs, then withdraw the needle slightly and reposition.
7. The pupil will dilate and the eye will rotate to a central position if the block is successful.

Tech Tip

An alternative to the retrobulbar block is performing a splash block once the surgeon has the eye removed and before the area is sutured back up. The splash block is useful for patients with ocular or orbital neoplasia because of the risk of a retrobulbar block seeding the tumor to other tissues.

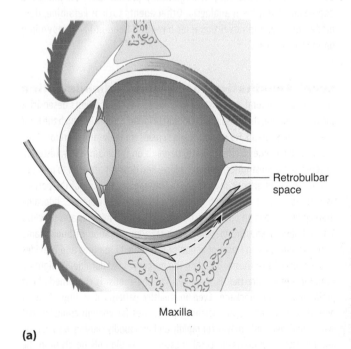

Retrobulbar space

Maxilla

(a)

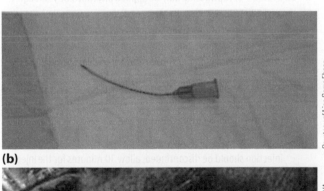

Courtesy of Lisa Sams Etner, DVM, MS, CACVAA, CVA

(b)

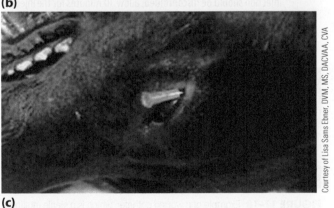

Courtesy of Lisa Sams Ebner, DVM, MS, DACVAA, CVA

(c)

FIGURE 17-17 Retrobulbar block. (a) A retrobulbar needle is advanced transconjunctivally at a 6 o'clock position and is directed along the floor of the orbit then redirected dorsally to perform the retrobulbar nerve block. (b) Special curved, blunt needles are available to aid injection into the retrobulbar space. (c) On this cadaver dog, a retrobulbar needle was advanced through the skin at the level of the orbital rim.

Specialized Options for Local Anesthetic Administration

Local anesthetic agents are incredibly versatile, and there are several more advanced options for administering local anesthetic agents that require specialized knowledge and skills. These advanced techniques include wound catheters, epidural anesthesia and analgesia, epidural catheters, and spinal analgesia; they are summarized below for completeness to provide an understanding of how the procedure is performed and how to assist with the procedure.

Wound Catheters (Continuous Wound Infiltration)

Continuous infiltration of a wound may provide postoperative analgesia for a variety of procedures such as mammary chain removal, limb amputations, and extensive tumor resections. Wound catheters are sterile multi-pore catheters that can be placed into the surgical site before the wound is closed to provide analgesia in the postoperative period. Specific wound infusion catheters are commercially available through a number of companies (Figure 17-18).

1. The catheter should be pre-flushed with local anesthetic agent before placement.
2. Wound catheters should be handled aseptically and sterile gloves worn. These catheters are placed during surgery by the surgeon under sterile conditions (similar to the placement of surgical drains).
3. To minimize the risk of wound infection, the catheter should exit the skin at a site separate from the wound as bacteria can migrate retrograde on the surface of the catheter.
4. Bupivacaine or ropivacaine may be applied intermittently through the catheter (every 4–6 hours as needed), or lidocaine may be used as a continuous infusion using a syringe pump/driver or a commercial reservoir infusion pump. Bupivacaine and ropivacaine have longer durations of action with ropivacaine having a wider therapeutic index (margin of safety). Newer research in the veterinary literature suggests the intermittent bolus administration of drug rather than continuous infusion results in better efficacy.[19]
   - Care should be taken not to produce local anesthetic toxicity.
   - Bupivacaine may cause discomfort on injection. If this occurs, the injection should be discontinued, allow 10 minutes for the injected amount of local anesthetic agent to desensitize the area, then the injection can often be continued without discomfort.

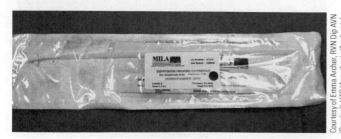

FIGURE 17-18 Example of a wound catheter, which is a sterile multipore catheter that can be placed in a surgical site before wound closure.

5. The patient's pain level should be regularly accessed. The catheter can be left in place for as long as the patient requires frequent doses of local anesthetic (normally 1–3 days). Other analgesia should always be provided.

Epidural Analgesia and Anesthesia

Epidural anesthesia refers to the complete sensory, motor, and possible autonomic blockade produced by local anesthetic agents injected epidurally (autonomic blockade occurs only if local anesthetics are used and the drug migrates cranially enough), whereas **epidural analgesia** produces segmental analgesia by providing a drug at its target (e.g., when an opioid is given epidurally, it migrates to the spinal cord's dorsal horn and the mu receptors at that location). The terms epidural and extradural are used interchangeably to describe the space superficial to the dura mater (the tough, fibrous, outermost layer of the meninges surrounding the spinal cord) (Figure 17-19). Epidural administration of local anesthetic agents and opioids is frequently used in small animals in combination with general anesthesia as part of a multimodal anesthetic technique to provide analgesia, to reduce the amount of anesthetic agents needed, and to provide muscle relaxation. They provide profound perioperative pain relief of surgical procedures involving the hind limbs, pelvis, and caudal abdomen.

Epidural Drugs

Numerous agents can be administered epidurally. Local anesthetics such as bupivacaine are most commonly combined with an opioid such as morphine because they synergistically provide both an improved degree and duration of analgesia. Other agents such as ketamine, dexmedetomidine, and novel therapies have also been utilized depending on the specific case.

Local Anesthetics for Epidural Administration

Lumbosacral epidural administration of local anesthetic agents desensitize the nerves exiting the spinal cord and the surrounding area in which the local anesthetic has spread. How far the drug spreads cranially (and therefore how cranially the analgesic effects reach) depends on the volume administered. Different local anesthetics, concentrations, and dosages are used to produce a wide variety of effects and depend on the species and size of the patient as well as desired onset and duration of effect. For example, lidocaine used epidurally has an onset of action of less than 10 minutes and provides 1 to 2 hours of anesthesia, while epidural administration of bupivacaine or ropivacaine has an onset of action of less than 15 minutes and provides 4 to 6 hours of anesthesia. Based on the properties of lidocaine, it might be used for a procedure that is short and the anesthetist is concerned about prolonged motor blockade. Even in healthy patients it is important to note that if enough local anesthetic progresses far enough cranially that sympathetic blockade may occur rapidly and profoundly causing vasodilation and refractory hypotension. Local anesthetics should only be given in the epidural space to normotensive patients, and adverse cardiovascular effects should be treated aggressively.

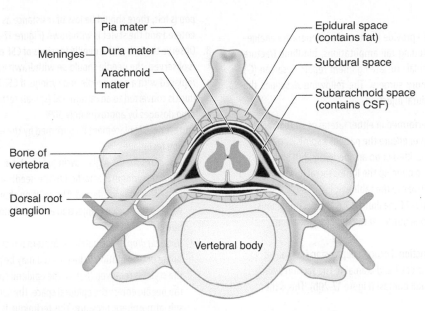

FIGURE 17–19 Cross section of spinal cord showing the epidural space, which is where the drug is injected for epidural technique.

Opioids for Epidural Administration Preservative-free morphine is the opioid most commonly used for epidural injection as it produces longer lasting analgesia with fewer adverse effects than more lipid soluble opioids such as fentanyl or hydromorphone. Morphine may be combined with local anesthetics such as bupivacaine. A single epidural injection of morphine has an onset of action of 30 to 60 minutes and provides 12 to 24 hours of analgesia. Hydrophilic agents like morphine spread cranially over time; therefore, lumbosacral morphine injections may provide reasonable analgesia for upper abdominal and thoracic procedures as well as for surgery performed caudal to the injection site. Epidural injections of opioids avoid the risk of motor blockade and the sympathetic blockade induced hypotension that may occur with local anesthetics; however, they may rarely cause urinary retention, nausea and vomiting, and pruritus.

Safety Alert

Preservative-free preparations of morphine and local anesthetic should be used to prevent the risk of damage to the spinal cord caused by preservatives. All drugs administered epidurally should come from previously unopened sterile vials using aseptic technique.

Alpha-2 Adrenergic Agonists for Epidural Administration The α_2 adrenergic agonist dexmedetomidine is occasionally used epidurally to provide analgesia because there are many alpha receptors in the CNS and periphery; however, if systemic absorption occurs, adverse effects such as sedation, marked bradycardia, decreases to cardiac output, and an increase in systemic vascular resistance may be seen.

These adverse effects make the use of α_2 adrenergic agonists unsuitable for use in animals with some cardiovascular diseases. Research into the use of α_2 adrenergic agonists given via the epidural space in animals is ongoing.

Patient Selection

To avoid complications, the anesthetist should select suitable patients for epidural injection. Contraindications of epidural injections include patients with thrombocytopenia or coagulopathies, skin infection/infection around the lumbosacral junction, or sepsis. Patients that are obese or have conditions such as anatomical abnormalities of the pelvis or lumbar spine (either congenital or acquired through trauma) may obscure landmarks making the procedure difficult and are best left to experienced anesthetists or avoided altogether.

Epidural injection of local anesthetic agents may cause vasodilation and hypotension via sympathetic blockade; therefore, they should not be given epidurally in hypovolemic or cardiovascularly unstable patients. Treatment for these adverse effects is specifically targeted at vasoconstriction by IV fluid loading and administering a drug with strong vasoconstriction properties such as phenylephrine, vasopressin, epinephrine, or norepinephrine. Depending on the dosage, volume, and type of local anesthetic used, motor nerves as well as the sensory nerves may be blocked. Postoperatively, motor nerve blockage may result in hind limb ataxia and weakness, and even leave a patient unable to walk until the effects of the injection wears off. Be prepared to turn the patient if prolonged recumbency occurs (to prevent decubital ulcers) and to assist the patient with walking using a sling under the abdomen if necessary. Urinary retention may rarely occur in patients receiving a morphine epidural; therefore, all patients should have their urinary bladder expressed at the end of surgery and have it checked periodically in the postoperative period. Delayed hair growth over the lumbosacral junction may occur so the owner should be suitably informed. In other cases, hair may grow back a different color.

Epidural Technique

Epidural analgesia may be used to provide intra- and postoperative analgesia for a variety of surgeries including tail amputations, hindlimb fracture repair, total hip replacement, cranial cruciate ligament repair, Cesarean (C-) section, and abdominal exploratory surgery. The following describes the procedure for performing an epidural injection.

1. Epidural injection can be performed in either lateral or sternal recumbency. Care should be taken to ensure the patient's spine is straight, parallel to the table, and not tilted or on any bedding. Pull both the hind limbs evenly cranially to open up the lumbosacral junction (the sacrum should be perpendicular to the table).

2. Palpate the dorsal most wings of the ilium (L7 is the dorsal spinous process in line with the cranial border of the wings of the ilium) (Figure 17-20a).

3. Palpate the lumbosacral junction. Locate a depression just caudal to the dorsal spinous process of L7 and cranial to the sacrum (which can be palpated as three small bumps) (Figure 17-20b). This is the lumbosacral space (L7–S1).

4. Clip a small patch of the fur over the lumbosacral junction and prepare the skin aseptically (Figure 17-20c). Place a sterile drape to prevent contamination from surrounding hair or skin. Full sterile technique is needed when performing an epidural to avoid the formation of an epidural abscess.

5. Wear sterile gloves after a thorough surgical hand scrub (Figure 17-20d).

6. Advance a spinal needle (normally 22 gauge 1.5 or 2 inch needle for dogs) (Figure 17-20e) through the skin over the midline at a 90 degree angle to the skin at the lumbosacral junction (Figure 17-20f).

7. The needle is advanced gently until a change in resistance is felt. This is sometimes described in textbooks as a "pop" as the needle advances through the ligamentum flavum. Others describe the feeling as some initial resistance and then sudden smoothness. Once the pop is felt, there should be loss of resistance as the epidural space is entered and the stylet is withdrawn (Figure 17-20g).

8. Observe the hub of the needle for signs of CSF or blood. If blood is observed, the needle should be withdrawn and repositioned or replaced with a new needle and syringe. If CSF is present, the procedure is converted to an intrathecal (spinal) technique by reducing the drug dosages by approximately 50%.

9. Correct needle placement is confirmed by the use of one of the following techniques:
   - *Loss of resistance.* A glass syringe (which has less resistance than a plastic syringe) is attached to the needle and a small amount of sterile saline is injected. The solution should be injected easily with no resistance to injection as if "pushing through soft melted butter."
   - *Hanging drop.* If the patient is in sternal recumbency, a drop of saline or the solution to be injected may be placed in the hub once the stylet is removed (before the epidural space is entered). As the needle enters the epidural space, the solution is "sucked in" by sub-atmospheric pressure. This technique is less reliable in small patients and will only work if the patient is in sternal recumbency.

10. Once the needle is in the proper location, hold the needle firmly supporting your hand on the patient. Attach the syringe containing the solution to be injected securely to the hub, taking care not to move the needle (Figure 17-20h).

11. Inject the drug slowly over about 1 to 2 minutes (Figure 17-20i). The addition of a small amount (0.1–0.5 mL) of air to the syringe helps confirm correct needle placement; compression/collapse of the air bubble during injection means there is increase of resistance and the needle is no longer in the epidural space. The air should not be injected into the epidural space as this may result in a less effective block and in people has been reported to cause nerve root compression.

12. After injection, position the patient as required for surgery. Rotating the patient may facilitate movement of the local anesthetic to the nerve roots supplying the surgery site.

▼ Palpate the dorsal most wings of the ilium.

▼ Palpate the lumbosacral space.

▼ After clipping a small patch of the fur over the lumbosacral junction, aseptically prepare the skin. A sterile drape would then be placed.

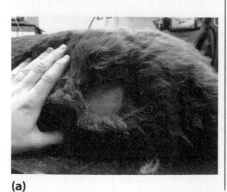

(a)

(b)

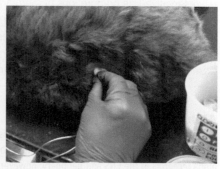

(c)

▼ Don sterile gloves after a thorough surgical hand scrub.

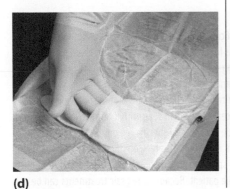

(d)

▼ Example of a spinal needle.

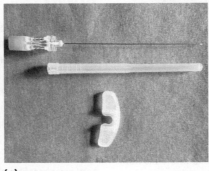

(e)

▼ Advance a spinal needle through the skin over the midline at a 90 degree angle to the skin at the lumbosacral junction.

Courtesy of Janet Amundson Romich, DVM, MS

(f)

▼ The needle is advanced gently until a change in resistance is felt which is described as a "pop". Once the pop is felt, there should be loss of resistance as the epidural space is entered and the stylet is withdrawn.

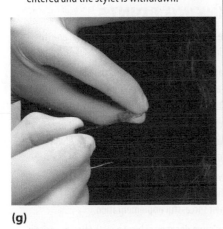

(g)

▼ Attach the syringe containing the solution to be injected securely to the hub, taking care not to move the needle.

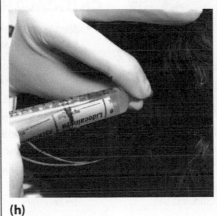

(h)

▼ Inject the drug slowly over about 1–2 minutes.

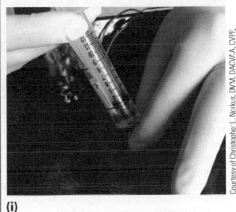

Courtesy of Christopher L. Norkus, DVM, DACVAA, CVPP, DACVECC

(i)

FIGURE 17–20 A cadaver dog receiving an epidural injection. Epidural injections are prepped and draped prior to insertion of the needle; the drape is removed from these photos to better demonstrate the landmarks.

Tech Tip

Be mindful of how you shave patients for epidurals, surgery sites, and IV catheters (Figure 17-21). Pet owners notice these things!

Epidural Catheters

Commercially available epidural catheter kits that are suitable for dogs can be used to provide long term postoperative analgesia of up to and beyond 7 days.

1. Following aseptic technique and wearing sterile gloves, an appropriately sized Tuohy needle is placed the same as for standard epidural injection (Figure 17-22).

2. A specifically designed catheter is then threaded through the needle prior to removing the needle. The catheter is sutured in place and the location of the tip of the catheter can be confirmed using radiography.

3. A filter and an injection port are placed on the end of the catheter and an adhesive dressing applied to keep the site covered. Make sure that the insertion site is kept clean, dry, and covered and that the animal cannot self-traumatize the area (typically an Elizabethan collar will be required).

Spinal (Intrathecal) Anesthesia/Analgesia

Spinal or intrathecal anesthesia/analgesia is accomplished by injection of local anesthetic agents and/or opioids into the subarachnoid space. This procedure is less frequently performed in small animal patients than epidural injections because it is difficult to perform and may need fluoroscopy or

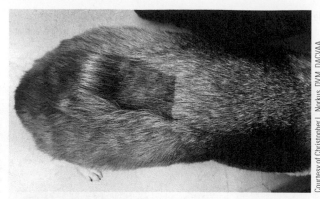

FIGURE 17-21 Note the poorly shaved epidural location. Veterinary technicians must remember to be careful of where and how they shave because shaved hair will be visible for the pet's owner to see (correct or incorrect) for months.

Courtesy of Christopher L. Norkus, DVM, DACVAA, CVPP, DACVECC

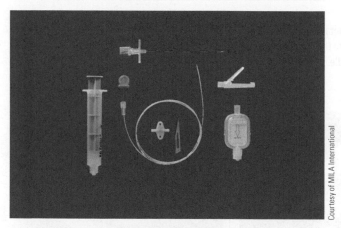

FIGURE 17-22 An epidural catheter kit. The epidural catheter kit has a drape; a catheter with a wire stylet, suture wing, and blade (middle of image); and from the top of the image clockwise: Tuohy introducer needle, luer lock catheter clamp, filter, and loss of resistance syringe. The needle provided in the epidural catheter kit varies in size depending on the kit (e.g., larger needles for bigger catheters and smaller needles for smaller catheters).

Courtesy of MILA International

other imaging technique to perform accurately. A more consistent analgesic effect may be observed with spinal (intrathecal) injections because the nerve roots are not protected by the dura mater. The dosages of drugs used intrathecally need to be decreased by one third to one half compared to epidural administration and administered more slowly.

SUMMARY

Being knowledgeable of commonly used analgesic drugs, methods of administration, and indications for their use help the veterinary team make optimal decisions when providing pain relief for their patients. IV, IM, SQ, topical, and transdermal routes of analgesic drug administration provide pain relief in animals; however, other techniques such as CRIs that may provide more consistent plasma drug concentrations than seen with intermittent drug boluses; nerve blocks that provide analgesia to specific regions of the body; and specialized techniques such as wound catheters, epidural anesthesia/analgesia, epidural catheters, and spinal analgesia should also be considered and can be of great benefit to the patient. Because analgesic treatments can be delegated to the veterinary technician for patients needing ongoing analgesia, a thorough knowledge of the potential adverse effects and complications when administering analgesic and anesthetic drugs is important. Specialized techniques often require special training and experience; however, understanding the rationale for their recommendation and how they are performed can help veterinary technicians explain these procedures to clients.

CRITICAL THINKING POINTS

- Always consider pre-emptive analgesia and multimodal anesthesia for pain relief in animals.

- When considering analgesia, veterinary technicians should be aware of how drugs work and their administration routes, potential adverse effects, and onset and duration of action.

- Local anesthetic techniques are often under-utilized but are an incredibly useful way of providing excellent analgesia. Always consider their use for painful procedures.

- Although most nerve blocks are relatively simple to perform, an understanding of the anatomy, potential adverse effects, and drugs are necessary for their successful implementation.

- Be aware of the maximum dosages of local anesthetics, especially when injecting into multiple sites.

- Understanding how specialized analgesic techniques are performed can help veterinary technicians counsel clients about the benefits and risks of these procedures.

REVIEW QUESTIONS

Multiple Choice

1. Which of the following routes of drug administration provides rapid drug transport due to its rich blood supply?
   a. Transmucosal
   b. Transdermal
   c. Subcutaneous
   d. Topical

2. A 25 kg dog requires a lidocaine CRI at 50 mcg/kg/min. Lidocaine's concentration is 20 mg/mL. At which rate (mL/hr) should the syringe pump/driver be set?
   a. 3.75
   b. 37 mL
   c. 75 mL
   d. 2.75

3. A 10 kg dog requires ketamine at 5 mcg/kg/min. How much ketamine should be added to a 500 mL bag of LRS if the fluid rate is 2 mL/kg/hr and the ketamine is 100mg/mL?

   a. 1.5 mL
   b. 0.75 mL
   c. 3 mL
   d. 7.5 mL

4. When performing intercostal nerve blocks, it is important to remember that intercostal nerves run alongside which border of each rib?

   a. Dorsal
   b. Ventral
   c. Caudal
   d. Cranial

5. IV regional anesthesia can be used for which type of procedure since the onset of analgesia is rapid and persists for the length of time the tourniquet is in place?

   a. Nephrectomy
   b. Abdominal closure
   c. Tooth extraction
   d. Digit amputation

6. Which of the following nerve blocks would be most useful to provide analgesia for an upper canine tooth extraction in a dog?

   a. Infraorbital
   b. Mental
   c. Mandibular
   d. IVRA

7. Which statement is true of providing analgesia by CRI?

   a. Provides fluctuating levels of anesthesia depending on the patient's level of pain
   b. Is usually achieved via the IM route
   c. Is useful in animals that need minimal monitoring because dose rates do not have to be calculated accurately since the drug is given to effect
   d. Is typically started by first administering as a bolus or loading dose of drug to achieve effective plasma levels

8. Why should local anesthetics not be given epidurally in hypovolemic or cardiovascularly unstable patients?

   a. Epidural injection may lead to spinal cord damage that can lead to paresis and lack of urinary tone.
   b. Local anesthetics affect blood vessels that are in close proximity to nerves and lead to perivascular leakage of plasma.
   c. Local anesthetics are often combined with glucocorticoids that produce immunosuppression that can lower the patient's ability to fight infection.
   d. Epidural injection of local anesthetic agents can cause vasodilation and refractory hypotension via sympathetic blockade.

9. What is a consideration for performing intrapleural blocks?

   a. Are more effective for patients with lateral thoracotomies than those with median sternotomies
   b. Are easy to perform and risk-free if a chest drain is already in place
   c. Provide variable quality of analgesia because of rapid uptake of local anesthetic from the pleural space
   d. Are used to desensitize the vagus nerve, which can result in bradycardia and hypotension

10. How long after application does a transdermal fentanyl patch last in dogs?

    a. 72 hours
    b. 24 hours
    c. 48 hours
    d. 120 hours

11. Which of the following nerve blocks would provide anesthesia to the stifle for a cruciate repair surgery?

    a. Femoral and sciatic
    b. Intercostal
    c. RUMM
    d. Brachial plexus

12. What is one advantage of using a retrobulbar nerve block for an enucleation?

    a. The use of a longer needle allows the anesthetic to penetrate the retina
    b. It keeps the eye in a central location
    c. It provides a stable plane of anesthesia for chronic pain
    d. The anesthetic agent is quickly metabolized in the body

13. What are sterile multi-pore catheters that can be placed into the surgical site before closure to provide analgesia in the postoperative period?

    a. Closure catheter
    b. Wound catheter
    c. Epidural analgesia
    d. Intrathecal analgesia

14. Which of the following is false regarding nerve blocks (perineural infiltration)?

    a. An aseptic technique should be observed.
    b. You should always aspirate prior to injection and observe to ensure no blood is visible.
    c. A peripheral nerve stimulator may be useful to locate nerve for some of the nerve blocks.
    d. It is recommended to inject the local anesthetic quickly to reduce the risk of adverse reaction.

15. What does the RUMM nerve block provide analgesia to?

    a. Proximal hindlimb

    b. Distal hindlimb

    c. Distal forelimb

    d. Proximal forelimb

16. Which of the following would be an appropriate way to manage pain caused by pancreatitis?

    a. Peritoneal infiltration

    b. IV regional anesthesia

    c. RUMM

    d. Intraplural block

17. Where is intrathecal analgesia administered?

    a. Peritoneum

    b. Thorax

    c. Space above the dura mater

    d. Subarachnoid space

18. What statement is true concerning IV regional anesthesia?

    a. Provide excellent long last intra- and postoperative analgesia

    b. Only provide analgesia intra-operatively for the duration of the tourniquet placement

    c. Are a good choice of analgesia for cats

    d. Are the local anesthetic technique of choice for total ear canal ablations

19. What is a main concern with intra-articular blocks?

    a. Weakening of ligaments secondary to injection into the joint

    b. Infection when strict asepsis is not followed

    c. Dilution of the drug with synovial fluid

    d. The volume of drug that is needed to provide analgesia

20. What choice correctly describes ketamine and its use?

    a. An NMDA receptor antagonist, which helps inhibit central sensitization

    b. Used to increase inhalant anesthetic agent requirements to help provide maximal pain relief

    c. A drug that is used via IV, IM, and SQ routes, but not via CRI

    d. Not recommended for use in dogs undergoing limb amputation

Case Studies

Case Study 1: A 20 kg dog undergoing spinal surgery requires a ketamine CRI at 10 mcg/kg/min. The ketamine is diluted to 10 mg/mL.

1. How many mL/hr of ketamine does the dog require?

2. Postoperatively, the ketamine needs to be added to the bag of intravenous fluids instead of through a syringe pump/driver. The same 20 kg dog is on LRS at 2 mL/kg/hr, but the rate needs to be reduced to 5 mcg/kg/min. How much undiluted ketamine (100 mg/ mL) would you need to add to a 1 liter fluid bag?

Case Study 2: Beuford, a 3 year-old M mixed breed dog, is admitted for a lateral thoracotomy for lung lobectomy. What would be an appropriate analgesia protocol and why?

Case Study 3: Rufus, a 10-year-old M Springer Spaniel dog with chronically painful ears, is admitted for total ear canal ablation (TECA) surgery. What would be an appropriate analgesic protocol and why?

Case Study 4: Terry, a 9-year-old F/S Yorkshire Terrier canine, is admitted for a dental prophylaxis; the right upper canine and incisors are expected to require extractions along with a left lower canine and possibly some lower premolars/molars on the same side. Discuss which types of local anesthetic techniques are appropriate for her.

Case Study 5: Skylar, a 16-year-old M/N Domestic Longhair (DLH) cat, is aggressive and requires surgery and anesthesia for enucleation due to end stage glaucoma. Preanesthetic blood work revealed that he is azotemic and he has been hospitalized on IV fluid therapy overnight in the clinic. Suggest an appropriate analgesia plan for surgery.

Critical Thinking Questions

1. What are the adverse effects seen with local anesthetic toxicity, and what is done to treat them?

2. A dog is recumbent while on a dexmedetomidine infusion. Discuss the nursing considerations for caring for this dog.

ENDNOTES

1. Ingvast-Larsson, C., Holgersson, et al. (2010). Clinical pharmacology of methadone in dogs. *Veterinary Anaesthesia and Analgesia, 37*(1), 48–56.

2. Ko, J. C., Maxwell, L. K., Abbo, L. A., & Weil, A. B. (2008). Pharmacokinetics of lidocaine following the application of 5% lidocaine patches to cats. *Journal of Veterinary Pharmacology and Therapeutics, 31*(4), 359–567.

3. Weiland, L., Croubels, S., Baert, K., Polis, I., De Backer, P. & Gasthuys, F. (2006). Pharmacokinetics of a lidocaine patch 5% in dogs. *Journal of Veterinary Medicine Series A, 53*(1), 34–39.

4. Kim, J. T., Jung, C. W., & Lee, K. H. (2004). The effect of insulin on the resuscitation of bupivacaine-induced severe cardiovascular toxicity in dogs. *Anesthesia & Analgesia, 99*(3). doi:10.1213/01.ANE.0000132691.84814.4E

5. Muir, W. W., Wiese, A. J., & March, P. A. (2003). Effects of morphine, lidocaine, ketamine, and morphine-lidocaine-ketamine drug combination on minimum alveolar concentration in dogs anesthetized with isoflurane. *American Journal of Veterinary Research, 64*, 1155–1160.

6. Pypendop, B., & Ilkiw, J. (2005). Assessment of the hemodynamic effects of lidocaine administered IV in isoflurane-anesthetized cats. *American Journal of Veterinary Research, 66*, 661–668.

7. Wagner, A. E., Walton, J. A., Hellyer, P. W., Gaynor, J. S., & Mama, K. R. (2002). Use of low doses of ketamine administered by constant rate infusion as an adjunct for post-operative analgesia in dogs. *Journal of the American Veterinary Medical Association, 221*, 72–75.

8. Webb, A. I. & Pablo, L. S. (2009). Local anesthetics. In J. E. Riviere & M. G. Papich (Eds.), *Veterinary pharmacology and therapeutics* (p. 381–399). Ames, IA: Blackwell Publishing.

9. Grubb, T., & Lobprise, H. (2020). Local and regional anaesthesia in dogs and cats: Overview of concepts and drugs (Part 1). *Veterinary Medicine and Science, 6*(2), 209–217. doi:10.1002/vms3.219

10. Fanelli, G., Casati, A., Beccaria, P., et al. (1998). A double-blind comparison of ropivacaine, bupivacaine, and mepivacaine during sciatic and femoral nerve blockade. *Anesthesia & Analgesia, 87*(3), 597–600. doi:10.1097/00000539-199809000-00019

11. O'Brien, T., Clark-Price, S., Evans, E., Di Fazio, R., & McMichael, M. (2010). Infusion of a lipid emulsion to treat lidocaine intoxication in a cat. *Journal of the American Veterinary Medical Association, 237*(12), 1455–1458.

12. Hansen B. (2008). *Analgesia for the critically ill dog or cat: An update, 38*(6), 1357.

13. Lemke, K. A., & Dawson, S. D. (2000). Local and regional anesthesia. *Veterinary Clinics of North America: Small Animal Practice, 30*(4), 839.

14. Lemke, K. A. (2007). Pain management II: Local and regional anaesthetic techniques. In C. Seymour & T. Duke-Novakovski (Eds.), *BSAVA manual of canine and feline anaesthesia and analgesia* (2nd ed., p. 108). Gloucester: BSAVA.

15. Kapitzke, D., Vetter, I., & Cabot, P. J. (2005). Endogenous opioid analgesia in peripheral tissues and the clinical implications for pain control. *Therapeutics and Clinical Risk Management, 1*(4), 279–297.

16. Mama, K. R. (2002). Local anesthetics. In J. S. Gaynor & W. W. Muir (Eds.), *Handbook of veterinary pain management* (p. 232). St. Louis, MO: Mosby.

17. Gaynor, J. & Muir, W. (2014). *Handbook of veterinary pain management*, 3rd ed., pp. 231–245. Elsevier.

18. Grubb, T., & Lobprise, H. (2020). Local and regional anaesthesia in dogs and cats: Descriptions of specific local and regional techniques (Part 2). *Veterinary Medicine and Science, 6*(2), doi:10.1002/vms3.218

19. Hansen, B., Lascelles, B. D., Thomson, A., & DePuy, V. (2013). Variability of performance of wound infusion catheters. *Veterinary Anaesthesia and Analgesia, 40*(3), 308–315. doi: 10.1111/vaa.12016

CHAPTER 18
Veterinary Physical Rehabilitation

Kristen L Hagler *BS (An.Phys), RVT, VTS (Physical Rehabilitation), CCRP, CVPP, OACM, CBW*

LEARNING OBJECTIVES

Upon completion of this chapter, it is expected that the reader should be able to:

18.1 Identify how credentialed veterinary technicians can provide physical rehabilitation therapy

18.2 List medical conditions that may benefit from or be eligible for physical rehabilitation therapy

18.3 Describe basic therapeutic exercise principles and modalities when applied to animals

18.4 Identify the role of rehabilitation and physical therapy/exercise as legitimate pain management controls in addition to pharmacologic interventions for acute and chronic pain conditions

18.5 Identify what veterinary technicians certified in veterinary physical rehabilitation techniques can perform under the supervision of a veterinarian

KEY TERMS

Active assisted range of motion

Active range of motion (AROM)

Active restricted range of motion

Acupuncture

Cluster care

Compensation

Conduction

Convection

Cryotherapy

Evaporation

Extracorporeal shockwave therapy

Functional Independence Measure (FIM)

Goniometer

Goniometry

Linear tracking

Low level laser therapy

Neuromuscular electrical stimulation

Passive range of motion

Photobiostimulation

Physical rehabilitation

Pulsed electromagnetic field therapy

Range of motion

Therapeutic ultrasound

INTRODUCTION

Adequately managed pain is essential for proper recovery as well as an animal's overall health. As an adjunct to traditional pharmacologic pain management techniques, patient care can be improved through the application of human physical therapy techniques and concepts (described in Appendix J). Veterinary physical rehabilitation techniques such as therapeutic exercises, thermal therapy, laser treatments, and acupuncture help animals lead more active lives, and veterinary technicians aware of these procedures can be advocates for their use in these patients. By assisting with individual patient evaluation, performing prescribed therapeutic modalities, educating and training pet owners about these techniques, and discussing potential rehabilitative strategies with the supervising veterinarian or veterinary rehabilitation therapist are all ways veterinary technicians can promote the use of veterinary physical rehabilitation techniques in clinical practice. Regulatory considerations for veterinary physical rehabilitation providers and veterinary physical rehabilitation credentialing are described in Appendix J.

Despite the fact that much of the veterinary literature focuses on canine and equine rehabilitation care, all animal species are equal candidates for physical rehabilitation. Felines, small mammals, birds, and other species have often been excluded from these procedures in part due to challenges of therapeutic treatments, including patient compliance and reliability of objective measurement tests; however, their inclusion in these treatments is increasing. This chapter will focus on rehabilitative strategies for the canine and reserves discussion of other species to other chapters within this textbook as well as other resources.

VETERINARY PHYSICAL REHABILITATION

Veterinary **physical rehabilitation** is the use of noninvasive techniques, excluding veterinary chiropractic, for the rehabilitation of injuries, disease processes, and congenital deformities in nonhuman animals. Using a multimodal approach including the use of veterinary physical rehabilitation alleviates pain better than providing drugs as the only therapeutic option for treating acute and chronic pain conditions.

The Credentialed Veterinary Technician's Role in the Physical Rehabilitation Team

The veterinary technician plays a key role in veterinary physical rehabilitation by assisting with treatment, handling patients, providing specialized nursing and rehabilitative care, collecting feedback from owners, and helping assess pain status, changes in limb use, joint stability and overall performance under the direction of a veterinarian or physical therapist. Veterinary technicians are responsible for carrying out patient care prescribed by veterinarians and may not initiate therapy without approval. Depending on each state's Veterinary Practice Act, veterinary technicians may be allowed to slightly alter therapy (e.g., repetitions of a therapeutic exercise, treadmill speed, or length of exercise sessions) based on the current patient performance and/or tolerance. Deviations from the treatment plan laid out by the veterinarian or physical therapist must be addressed immediately, especially if the patient appears in pain.

With knowledge of the patient condition and types of veterinary physical rehabilitation techniques available, the credentialed veterinary technician becomes highly skilled, serves as a source of essential information, and communicates effectively between veterinarians and other colleagues. With appropriate training in all aspects of pain management, including preemptive and multimodal modalities, veterinary technicians become a vital force in providing optimum care for surgical, medical, and chronically painful patients.[1,2]

ASSESSMENT AND PAIN RECOGNITION FOR PHYSICAL REHABILITATION PATIENTS

There are a significant number of households (68%) in the United States with a pet (56.7 million of them being dogs and 45.3 million being cats)[3], and as these pets age, the management of their injuries and chronic pain conditions continues to rise. In addition to routine veterinary care, specialized services such as physical rehabilitation for dogs and cats may include weight loss programs; recovery from orthopedic, neurological, or certain soft tissue procedures; conditioning programs for sporting and working dogs; maintenance of chronic osteoarthritis; and risk assessment for development of chronic orthopedic disease or progressive conditions. This chapter focuses on rehabilitation of dogs although many of the same principles may be applied to cats.

Patient rehabilitation evaluations occur with or after a veterinary orthopedic or neurologic examination. The veterinarian evaluates the patient to obtain an accurate diagnosis, prescribes medical or surgical treatment plans as needed, and works with the rehabilitation specialist who collects information through physical fitness testing to create an individualized rehabilitation therapy plan. Veterinary physical rehabilitation performed by non-veterinarians should be limited to the use of stretching; massage therapy; stimulation by use of low-level lasers, neuromuscular electrical stimulation devices, magnetic fields, and ultrasound; rehabilitative exercises; hydrotherapy; and applications of heat and cold.[4]

Rehabilitation evaluations include unique assessment tools for the measurement of muscles, joint motion, joint stability, pain,[5] functional abilities/capabilities at home, gait or movement capabilities, overall conformation, and establishment of baseline orthopedic or neurologic health values. When gathering information during an evaluation, the rehabilitation therapist must also consider the overall body type of the dog and intended physical activities desired. The quadruped nature of the canine body provides for both the efficient movement through the positioning of the thoracic limbs, pelvic limbs, and spinal column and allows the dog to easily compensate for injury to a limb. The dog can shift its center of gravity located at mid chest level just behind the scapulae (Figure 18-1) to other less affected areas, causing varying degrees of lameness.[6] Lameness and abnormal gait patterns are closely linked to functional physical capabilities with pathological changes resulting from conformation abnormalities. When conformation abnormalities are present, the dog is then predisposed to musculoskeletal disease later in life.[7,8]

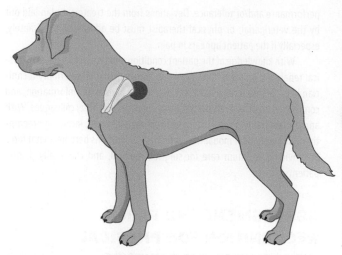

FIGURE 18-1 Center of gravity. The solid circle located at mid chest level just behind the scapulae is the center of gravity in the canine.

Animal Pain Communication and Behavior

The benefits of rehabilitation therapy are achieved with a willing and relaxed patient because a physical rehabilitation session can subject an animal to potentially painful or frightening procedures that may result in undesirable behaviors.[9] The individual performing treatments should continuously monitor the patient for body language signs that indicate pain or discomfort. The International Association for the Study of Pain (IASP) Subcommittee on Taxonomy states that "pain is always subjective and each individual learns the application of the word through experiences related to injury in early life."[10] Pain is a subjective sensation, yet animals experiencing pain cannot directly communicate they are feeling pain. Other methods for describing pain, including communication and behavior cues, provide information regarding an animal's level of pain.[11]

An understanding of different criteria used for the assessment of pain or distress is important and should be recorded in the medical record after therapy sessions (see Table 16-1). Standardized pain scoring systems can be used to assess pain in animals (see Chapter 15 [Acute Pain Management] and Chapter 16 [Chronic Pain Management]). Pain scales cannot be used alone and should be combined with other assessment tools such as physical examination, patient history, and client impression of home pain levels.

Knowing how animals communicate when they are in pain will help avoid injury in veterinary patients as well as staff members. In most cases, the painful animal will volunteer a series of escalating behaviors indicating discomfort. The first clue to a dog's emotional state is usually body posture and facial expressions. Very early signs of stress can be detected by carefully watching an animal for yawning, panting, or lip smacking during treatment and evaluation periods. Narrowing of the eyes, pulling back ears, drawing lips back tightly, pausing while panting, and pacing or hiding can indicate escalation of discomfort and/or

anxiety. If the patient continues to feel threatened, they may begin to verbalize their stress through growling, barking, and showing teeth. As a last resort, physical contact is made through biting. In rare instances an animal may immediately resort to biting.[12]

Tech Tip

Interaction with the pet owner is crucial when assessing animal pain because the owner helps the veterinary technician determine how "polite" an animal will be with their stress behavior cues. This politeness, or comfort space, varies between each dog and must be evaluated individually.

The veterinary technician can educate the pet owner about animal pain behaviors when discussing the home environment. Subtle behavioral changes may occur in the home and the owner may easily overlook them. Pet owners are typically unaware of behavioral changes linked to pain signs, often attributing them to being grumpy or part of the natural course of aging. For example, the painful animal often chooses to reside in a quiet room alone instead of amidst the family to achieve rest without disruption. Other examples may include changes in eating and urinating/defecating patterns or becoming clingier. When discussing physical challenges at home with the owner, asking about walk length and quality, interaction level, feeding schedules, and relieving capabilities can provide insight regarding current pain levels. It is common for owners to report that their pet is continuing to take regular walks as they have been for the last several years despite the fact that the pet is slowing down near the end of the walk or taking more rest stops. It is rare for the pet to actually resist going on walks, and if that is the case, they are usually in a great deal of pain. Another common and more obvious functional change owners notice is the pet not wanting to jump onto the bed or into the car. This is usually a result of progressive physical weakness or deterioration that could potentially have been delayed if signs were recognized earlier (Table 18-1).

Tech Tip

Thunder shirts® or compressive body wraps can help keep patients calm during an extended hospital stay.

A tool veterinary technicians can use to assess a patient's ability to carry out activities of daily living safely and autonomously is a **Functional Independence Measure** (FIM) system. This is a seven-point scale rating system to specifically address the level of dependence a patient requires for daily activities ranging from "fully dependent" (1) to "independence with no aids" (7) (Table 18-2). These objective scores can more accurately track patient progress in the medical record and help define the roles of the caretaker.

TABLE 18-1 Criteria for Early Home Pain Assessment*

Ability to ascend and descend stairs
Ability to enter and exit vehicles
Ability to cope with difficult surfaces such as wooden or tiled floors
Ability to remain standing while eating
Willingness to exercise and exercise tolerance
Ability to remain squatting while defecating
Ability to posture for urination
Inappropriate elimination
Willingness to play
Change in demeanor
Response to grooming
Response or lack thereof to medication
Effect of exercise on the lameness/pain
Effect of rest on the lameness/pain
Duration and intensity of the lameness/pain
Changes in sleep patterns

*Davies, L. Chapter 11. Canine rehabilitation. In C. M. Egger, L. Love, & T. Doherty (Eds.), *Pain management in veterinary practice*, (1st ed., p. 135, Ames, IA: John Wiley and Sons.

TABLE 18-2 Functional Independence Measure (FIM) Seven-Point Scale Rating System*

Number	Level of dependence	Description
7	Complete Independence	Fully independent
6	Modified Independence	Requires the use of a device but no physical assistance
5	Supervision	Requiring only standby assistance or verbal prompting for safe use of device
4	Minimal Assistance	Requiring incidental physical assistance; patient performs over 75% of the task
3	Moderate Assistance	Patient performs 50–75% of the task
2	Maximal Assistance	Patient contributes 25–49% of the effort
1	Total Assistance	Patient contributes < 25% of the effort or is unable to perform the task

*Uniform Data System for Medical Rehabilitation (UDSMR). (1993). *Guide for the uniform data set for medical rehabilitation*. Buffalo, NY: UB Foundation Activities.
Borghese, I. Fair, L. et.al. (2013). Assistive devices, orthotics, prosthetics and bandaging. In C. M. Zink & J. B. Van Dyk (Eds.), *Canine sports medicine and rehabilitation* (pp. 208–210). Wiley-Blackwell.

Tech Tip 🐾

When assessing an animal's response to pain, consider the following classifications:[13]

- Those that modify the animal's behavior through learning so the animal avoids recurrence of the experience (e.g., positive reinforcement training techniques for professionals working with animals in the clinical setting)
- Those that are often automatic and protect parts or the whole of the animal (e.g., withdrawal response)
- Those that minimize pain and assist healing (e.g., preemptive analgesia, and physical rehabilitation therapies and modalities)
- Those that are designed to modify infliction of pain on animal (e.g., low stress environment, bandages to protect incisions, multimodal pain management)

In the clinical environment, animals can disguise pain responses making assessment difficult. High pain thresholds allow animals to avoid showing signs of weakness, such as limping, until the threshold has been exceeded. Pet owners often report limping on/off for several weeks prior to an examination when really the animal has been in pain for several months, hiding the disability through **compensation** (an abnormal movement or posture from abnormal skeletal alignment or associated movement).[14] Common examples of compensation include sitting in an un-square position in the pelvic limbs (Figure 18-2), "flopping" to lie down, defecating in several areas on

Courtesy of Kristen Hagler, BS (An. Phys), RVT, VTS (Physical Rehabilitation), CCRP, CVPP, VCC, OACM, CBW

FIGURE 18-2 Compensation. The dog depicted in the image is sitting asymmetrically as a result of pain in the rear end.

the lawn, preferentially landing on one limb when jumping, "pulling" themselves up with the thoracic limbs after lying down, refusing to lie down, taking breaks during mealtime, and leaning against furniture or walls. All of these seemingly harmless characteristics or "behaviors" can be indicators of an underlying pain state. The veterinary technician can help pet owners better understand pain behaviors and cues when managing acute and chronic conditions by reviewing potential problem areas.

CANINE STRUCTURE AND FUNCTION

Understanding basic canine structural anatomy and its function help guide treatment of injuries, and educates pet owners about preventing injuries during sport specific activities. Recognizing abnormal structural components during rehabilitation therapy may help reduce risk of injury based on structural predispositions.

The Thoracic Limbs

The forelimbs bear 60% of the total body weight in the dog (Figure 18-3) and are constructed to permit significant **range of motion** (movement around a joint), provide lift when jumping, and allow directional changes when running. Dogs with abnormalities of the spine or pelvic limbs compensate by increasing weight bearing on joints of the thoracic limb. Over time abnormal postural changes give rise to secondary conditions[15] such as arthritis in the joints or repetitive overuse injury such as tendonitis. Structural variation in the thoracic limbs occurs between breeds, with differences in the degree of shoulder layback, length of the humerus, and ratio of humerus to scapula length affecting forelimb structure and ultimate functional capabilities. The perfectly built dog should have an ulna and scapula

of equal length and 30 degrees of shoulder layback. Shoulder layback is measured as seen in Figure 18-4.

The Pelvic Limbs

The pelvic limbs bear 40% of the total body weight in dogs and are designed to provide propulsion and directional changes while on level surfaces and "drive" the cranial portions of the body forward during activity.[16] Variations in pelvic limb angulation can be measured as seen in Figure 18-5. Ideally the dog has an evenly balanced distance between the two perpendicular lines, allowing for large bony areas available for muscle attachment. In reality, most dogs are not evenly balanced and have either too much or too little rear angulation. There are advantages and disadvantages to having abundant rear angulation. Those breeds with a lot of rear angulation such as the German Shepherd are able to take longer strides and expend less energy, but often are less stable in the rear as it takes tremendous muscular strength and coordination to stabilize the very angulated rear. As a result, dogs with straighter pelvic limbs such as the Border Collie tend to be able to turn more sharply than dogs with very angulated pelvic limbs.[17]

The Vertebral Spine

The canine vertebral spine is parallel to the ground and allows a significant amount of movement while supporting the head and abdominal contents. It also absorbs and distributes forces from the four limbs while protecting the spinal cord. Each area of the spine allows

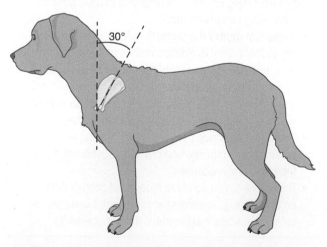

FIGURE 18-4 Shoulder layback. Shoulder layback is measured from a line drawn perpendicular to the ground through the point of the shoulder and another line drawn to follow the scapular spine. Dogs with less than 30 degrees of shoulder layback are more prone to shoulder and elbow conditions, have less bony regions for muscles to attach, have less power when jumping, less cushion when landing, and lower stability when changing direction while running.

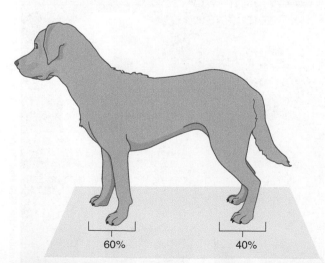

FIGURE 18-3 Body weight distribution. In the canine, the forelimbs bear 60% of the total body weight and the hind limbs bear 40% of the total body weight.

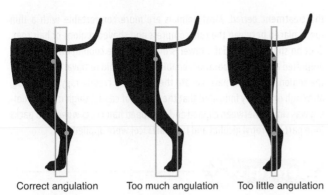

Correct angulation   Too much angulation   Too little angulation

FIGURE 18-5 Pelvic limb angulation. Variations in pelvic limb angulation can be measured between animals by drawing two perpendicular lines from the ground—one from the ischeal tuberosity of the pelvis and the other from the tarsi perpendicular to the ground—then drawing a line along the caudal aspect of the tarsi. The distance between the two lines is then compared.

for varying amounts of motion between two vertebral segments that interlock with its adjacent vertebra.

- The cervical region (C1–C7) permits just enough movement so the dog is able to touch any portion of its body except between the scapulae.
- The thoracolumbar (T13–L1) and lumbosacral (L7–S1) vertebral regions allow the highest degree of motion between vertebrae and are often areas of significant injury.
- Vertebral segment 11 (T11) is unique in that the dorsal spine acutely reverses direction from a caudal sweep (T10) to a dorsal sweep (T11). This vertebra (T11) is also known as the "anticlinal vertebra" and is nearly vertical in position. This change results in potential increased mechanical stress on the fibrous annulus, a strong cross fiber firmly attaching one vertebra to another, which contains supportive structures to the spinal cord.
- The sacral segments one through three (S1–S3) are fused and do not allow for much motion, if any. The coccygeal (Cg) vertebrae vary in number depending on the breed.[18,19,20,21,22,23]

VETERINARY PHYSICAL REHABILITATION MODALITIES AND TECHNIQUES

Careful selection of modalities and techniques for an animal's rehabilitation program should be done with clear goals in mind and in conjunction with the veterinarian. Patient condition, stage of recovery or illness, and client compliance are all taken into consideration when selecting which modality or technique is most appropriate, especially if the client will perform them at home. Physical modalities address pain, inflammation, or joint tension and are often used in conjunction with other treatment techniques such as individualized therapeutic exercise programs to improve treatment plan outcome. Physical modalities may include superficial thermal agents, massage therapy, range of motion and therapeutic exercise, aquatic

therapy, low level laser therapy, electrical stimulation, and therapeutic ultrasound.

Superficial Thermal Agents

Superficial thermal agents such as cold and heat therapy are frequently used to treat hospitalized patients and often continue through the time of discharge. Both cold and heat therapy are effective in altering cellular metabolism and providing pain management through the physical laws of heat transfer and conduction. The use of superficial thermal agents, their properties, and their physical effects on the patient are frequently misunderstood, resulting in their misuse.

Cryotherapy

Cryotherapy (cold therapy) is the application of cold that is used during the acute inflammatory phase of tissue healing and after exercise to minimize the metabolic rate of reactions involved in tissue injury and healing.[24,25] In human physical therapy, it is common to use **RICE** therapy during treatment, which combines compression and elevation of the affected part to reduce or prevent pain, bleeding, and swelling.[26] RICE is an acronym for *R*est (halting further injury), *I*ce (minimizing cell death), *C*ompression (decreasing edema), and *E*levation (decreasing edema). When applying RICE therapy to animals, compression and elevation of the affected body region to decrease edema may be difficult. Patients with edema in a single limb should have this limb elevated to prevent swelling, which may be facilitated by lying the patient on the contralateral side and using a rolled towel or pillow to elevate the edematous limb (if tolerated).[27]

There are several methods of cryotherapy, and the type selected depends on the desired effects, depth of penetration desired, stage of tissue healing, treatment area, and physiologic goals. Heat transfers from body tissues to the cold modality via **conduction** (direct contact between the body and cold pack/cold bath), **convection** (motion of immersed body part or agitation of water in cold baths), or **evaporation** (vapocoolant sprays). When compared to heat, it is a much longer acting treatment (upwards of 30 minutes depending on location and method chosen).[28] Long-term effects from cryotherapy are due to decreases in local circulation, edema formation, hemorrhage, histamine release, cellular metabolism, muscle spindle activity, rate of sensory and motor nerve conduction, muscle spasm, and pain. Topical cold treatments decrease the temperature of the skin and underlying tissues to a depth of 2 to 4 cm.[29] These changes cause an increase in connective tissue stiffness and muscle viscosity that creates a decreased ability to perform rapid movements, especially in older animals. Although less commonly used in veterinary medicine, cryotherapy may also be applied over the whole body or in large regions, causing changes that should be monitored carefully. Physiological effects of whole body or large region cryotherapy include the following:[30]

- Decreased respiratory and heart rates
- Generalized vasoconstriction (the posterior hypothalamus responds to cold by causing vasoconstriction to keep core body temperature normal)
- Increased muscle spindle bias, which can increase spasticity
- Increased muscle tone, which can be accompanied by shivering

Cryotherapy is extensively used for both acute and chronic pain management in humans; however, animals tend to disapprove of the cold. During cold application, human patients experience sensations such as intense cold, burning, aching, and numbness,[31] for varying amounts of time. It is assumed animals also feel these sensations based on reactions observed during treatment; therefore, they should be monitored for intolerances of cryotherapy. Following a successful treatment period, several local effects occur at the dorsal horn of the spinal cord aiding pain relief and relieving pain transmission signals from the affected area.[32]

The neurologic response to cryotherapy is mediated by peripheral thermal receptors and is categorized based on myelination, size, and nerve conduction velocity. Detection of sensory information or pain sensation is carried out by specialized nerve endings such as A δ and C nerve fibers that terminate as free nerve endings in the skin, subcutaneous tissue, periosteum, joints, muscles, and viscera. The terminals of these nerve fibers are essential for detection for all pain sensations. Many of the A δ nociceptors respond only to specific stimuli and discharge at higher rates than C fibers providing more discriminatory information and are responsible for prickling, sharp pain, and cold. C fiber receptors are found in large numbers in the skin, skeletal muscle, and joints and are responsible for aching pain and temperature detection.[33] Pain resulting from different tissue injuries can vary in characteristics and mechanisms due to the fiber receptors involved. Postoperative incisional pain is a unique and common form of acute pain that is most likely attributed to the conversion of mechanically insensitive silent nociceptors to mechanically responsive fibers.[34]

Cryotherapy reduces pain via the following mechanisms:

- Temporary numbness in the affected area from local vasoconstriction
- Raised activation threshold of tissue nociceptors
- Cooling of the nerves, which increases the refractory period duration for stimulation of a second impulse
- Reduction in nerve conduction velocity of pain nerves
- Over stimulation of cold receptors preventing pain transmission to higher centers via the gate theory and local anesthetic effects such as cold induced neuropraxia[35]

Application of cryotherapy to animals must be appropriately managed in order to avoid frostbite and tissue damage. Checking skin color is often difficult because of pigmentation and hair coat. As a general rule, a timer should be used and the patient evaluated several times during the treatment period. Most animals are more comfortable with a thin towel placed between the cold compress and shaved regions of hair coat. Caution should be used if a towel is moist because cooling effects can be amplified through evaporation, enabling more rapid removal of heat from the region. Hair coat may insulate the effects of cryotherapy in the dog, although one study indicated that the extent of caudal thigh muscle cooling was similar between clipped and unclipped hair coats when cold packs (one part isopropyl alcohol and two parts ice) were applied.[36]

Tech Tip

A thin wet towel will make a cold compress colder; therefore, setting a timer based on the animal size when applying cold compresses will help prevent tissue damage.

Cryotherapy can be applied to nearly any region of the body; however, the following should be considered when using cryotherapy:

- Superficial regions around peripheral nerves such as the peroneal and ulnar nerves should receive treatment with caution because they have been known to cause cases of cold-induced nerve palsy in humans.
- Patients with vascular compromise or who possess an impaired ability to thermoregulate should receive cryotherapy conservatively.
- Those patients with regions of poor sensitization or the very young or very old may receive cryotherapy, but with caution.
- Extreme precaution should be taken in anesthetized or sedated patients because of their inability to communicate adverse effects.
- Cryotherapy should not be used in cases of known nerve damage, open wounds 48 to 72 hours after injury, or history of cold urticaria (which causes wheals and swelling on the skin from histamine release).[37, 38]

Duration of cryotherapy treatment in animals will vary depending on location, patient body size, and client compliance. In humans, a skin surface temperature drop occurs immediately once a cold pack is applied and has been reported to reach a maximum effect in approximately 30 minutes regardless of the cold pack type. Once the cold pack is removed, skin surface temperature rises rapidly yet may not reach normal temperatures for up to several hours after removal.[39] Cryotherapy application recommendations for dogs have been extrapolated from human rehabilitation as follows:

- Cryotherapy should be applied for the first 24 to 72 hours following acute injury when the acute signs of inflammation are present (swelling, redness, heat, and pain).
- Cold packs can be applied for 10 to 20 minutes depending on the muscle group and patient size with a realistic frequency of treatment occurring every 3 to 4 hours.
- For patient comfort, small breed dogs and cats should not exceed more than 10 minutes of cold application for all joints including the extremities and 15 to 20 minutes for vertebral injuries.
- Medium to large breed dogs have a larger surface area and muscle mass compared to smaller animals and should receive at least 20 minutes of treatment to the stifle or shoulder joints and vertebral injuries. Cryotherapy may be applied for 10 to 15 minutes to the extremities and carpal, tarsal, and elbow joints.

Tech Tip 🐾

Cryotherapy for small animal rehabilitation is recommended if

- range of motion is decreased because of pain and inflammation,
- the patient needs stimulation of muscle function, and
- further pain and swelling may occur following an exercise session.

FIGURE 18-7 Thermal therapy-Cryotherapy. Ice cup preparation using rubber molding and Vetrap® on the molded ice block.

Courtesy of Kristen Hagler, BS (An. Phys), RVT, VTS (Physical Rehabilitation), CCRP, CVPP, VCC, OACM, CBW

There are a variety of ways to deliver cold therapy to a patient, and some are more suitable in the clinical veterinary hospital than others. Commercially prepared cold packs (Figure 18-6) are convenient and retain cold temperatures for a long period of time, but are not as effective as ice and add cost to treatment. The ideal method for cryotherapy application is crushed ice placed in a double wrapped plastic bag because it provides desired temperatures better than commercially prepared cold packs and easily molds to smaller joints or awkward locations. A simple and cost effective alternative to crushed ice is making a mixture of 1/3 part 70% isopropyl alcohol and 2/3 part water in a bag and placing it in the freezer for several hours to produce a moldable slush. For application of either crushed ice or isopropyl alcohol/water slush, use clear plastic food wrap to affix the cold bag into place achieving mild compression. Other forms of cryotherapy include ice cups (Figure 18-7), cold compression systems, cold immersion baths, ice towels, vapocoolant sprays, and contrast baths. With the exception of ice cups and cold compression systems, these forms of cryotherapy are often reserved for specialty centers. Ice cups are ideal for small joints and applying a mild amount of localized compression. Veterinary specific cold compression systems are commercially available with preprogrammed settings for duration, temperature, and intensity of compression applied.

Heat Therapy

Heat therapy is the application of warmth to an area, and its properties are opposite of and last for a shorter duration than cold therapy. With cryotherapy, energy moves from the body to the cold object; with heat therapy, it moves from the warm object to the body. Heat is retained by the tissues through energy transfer as heat moves from the warmer object (the thermal agent) to the cooler object (the body).[40] Energy transfer is then carried away from tissue by local circulation. Heat therapy sources are classified as radiant, conductive, or convective and for most veterinary applications conductive heat therapy (warm packs) is used. In some conditions, convective methods may be indicated (warm water hosing/bathing)[41] to assist with treatment.

Tech Tip 🐾

Factors affecting the rate of heat transfer to cooler objects include the following:[42]

- Temperature differences between the thermal agent and the body
- Regeneration of body heat and thermal agent cooling
- Heat storage capacity of the cold thermal agent
- The physical size of the thermal agent
- The area of the body in contact with the thermal agent
- Application duration of the thermal agent
- Individual variability in body composition

Tech Tip 🐾

Homemade cold packs can be double bagged in zipper sealed bags such as Ziplocs® so they do not leak.

Physiological effects from local application of heat therapy include decreases in blood pressure, muscle spasm, and pain while increases occur in body temperature, respiratory rate, capillary pressure and permeability, muscle relaxation, tissue elasticity, local circulation, and leukocyte migration. Heat penetrates into tissue deeper than cold, and heat therapy is classified as superficial or deep. Superficial heat therapy, such as warm packs and whirlpool baths, penetrate about 2 cm into tissues. Deep heat therapy, such as therapeutic ultrasound, elevates tissue temperatures to a depth of 3 cm or more. Heat therapy works by increasing the cutaneous thermal receptor pain threshold that inhibits pain transmission at the dorsal horn of the spinal cord. With nociceptive impulses altered, painful conditions

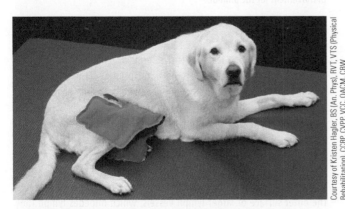

FIGURE 18-6 Thermal therapy—Cryotherapy. Commercially prepared cold pack placed on the stifle joint of a patient

Courtesy of Kristen Hagler, BS (An. Phys), RVT, VTS (Physical Rehabilitation), CCRP, CVPP, VCC, OACM, CBW

improve. Superficial heat therapy is most appropriately applied *after* the acute inflammatory phase of tissue healing has resolved and the reparative or remodeling phases of tissue healing have started (typically 24–72 hours after injury or surgery). If thermal therapy is applied too early during tissue healing or injury, any swelling, pain, cellular metabolism, or heat present is exacerbated.

Three mechanisms occur during local heat therapy application, with the greatest effects at cutaneous blood vessels and ischemic tissues (decreasing pain receptor activity). The first mechanism results from superficial temperature increases within 1 cm of tissue depth, causing the release of histamine and prostaglandins. The second mechanism involves the activation and synapse of cutaneous thermoreceptors on cutaneous blood vessels releasing the peptide bradykinin, which causes relaxation of smooth muscle walls. The third mechanism results in decreased smooth muscle contraction from a reduction in sympathetic activation via spinal dorsal root ganglia.[43]

Tech Tip

When treating painful conditions, it is often unclear whether heat or cold therapy is indicated. In people, the type of superficial thermal therapy is selected based on what "feels" the best, mainly because both heat and cold relieve pain and muscle spasm.[44] If there is any doubt when treating animals, it is best to apply cold and monitor for an adverse reaction (i.e. the patient reacts or owner reports worsening signs).

Neuromuscular effects resulting from heat therapy are opposite those of cold. Cold slows nerve conduction velocity, whereas heat increases firing rates by 2 m/sec for every 1.8°F (1°C) increase in temperature. The threshold for muscle spindle activation is lowered along with muscle tone, which decreases mechanically induced nociceptive pain during physical activity. Improvements in connective tissue extensibility and decreased muscular splinting from heat therapy enable quality muscle contractions during therapy exercises and less pain during passive manual therapy treatments.[45]

To avoid undesirable adverse effects, regular assessment of the original injury or condition and the patient's level of consciousness and overall physiologic health including age should be considered during application of all heat therapy. For example, a patient with an acute cranial cruciate ligament injury should avoid heat therapy, which could contribute to osteoarthritis by increasing destructive enzyme activity and catabolic rates within tissues. In the anesthetized patient, adverse thermoregulatory events can occur with heat therapy, causing burns (see Figure 7-20b) or hyperthermia. General precautions should be followed while placing an animal in a whirlpool to avoid submersion of the patient and hyperthermia. Patient evaluation should also include tolerance of the heat pack because the weight over a local application site can be painful due to stimulation of mechanical pain receptors. Heat therapy is contraindicated in regions of active bleeding, acute inflammation, cardiac insufficiency, fever, malignancy, decreased or impaired circulation, or known poor thermoregulatory capabilities.[46] During all heat therapy applications, patient skin should be assessed whenever possible for any white or mottled regions. If this occurs remove the heat source immediately.[47] As with all treatment modalities, the patient should never be left unattended.

Heat therapy is for patients experiencing pain secondary to stiffness and/or decreased range of motion.[48] With the application of heat to tissues, blood circulation carries surface heat energy away quickly. When using heat therapy for conditions related to animal physical rehabilitation, a treatment time of 15 to 30 minutes (more time for well muscled body regions) will promote the desired warming effects. If a warming pack falls off at any time during a treatment period, it is recommended to restart the timer. In doing so, adequate penetration and warming of tissues is ensured.

Heat therapy is achieved many ways with the most common applications being commercially available fabric bags found in most human pharmacies. Homemade heat therapy bags can be created using rice and dried lavender (for calming effects) in a sock or other cloth bag that can be closed and are more effective than commercially prepared heat packs because custom sizes can be made. Heat packs are generally warmed in the microwave (time in the microwave depends on power settings, but generally 1 minute per rice sock), and upon removal the contents should be manually mixed to avoid any localized hot spots. The heat pack should also be placed on human skin prior to treating a veterinary patient, and if blanching or pain occurs it is too hot for the animal. Adding a thin towel between the heat pack and the patient's skin is advised. Other heat therapy applications include circulating warm water blankets, forced air blankets, and warm water hosing. For veterinary technicians working at animal physical rehabilitation centers, warm whirlpools, swimming pools, or underwater treadmills are also considered applications of heat therapy. Hot water bottles, heating pads, and infrared lamps are considered to have a higher risk of burning animals and should not be used.[49]

Manual Therapy Techniques

Superficial thermal agents when used alone are often not sufficient to address mechanical restriction or dysfunction caused by chronic conditions. Joint motion or pain levels improve by combining techniques such as massage, passive range of motion, or stretching to assist with overall quality of movement for the animal.

Massage Therapy

One of the easiest and least expensive forms of rehabilitation therapy is massage. Therapeutic massage is one of the primary forms of "hands-on healing," with most cultures having developed specific techniques for both therapeutic and pleasurable purposes.[50] Massage has positive influences on the physical and psychological well-being of dogs of all ages and conditions[51] and based on human evidence serves to prevent injury in competitive athletes and aides with pain relief and improved mobility when incorporated with medical management.[52]

During inflammation, tissues and fluid are compressed against the outer surface of larger lymph vessels, which impedes lymph flow. Massage helps to move fluid back into the lymphatic system and restore normal function similar to the effects of using a compression bandage (although compression bandages also reduce swelling). Increased manual pressure within the tissues during a massage creates fluctuating pressure differences between one area of the tissue and another, which promotes fluid movement within tissues from areas of higher to lower pressure. Massage also replenishes fluid in tissue spaces, producing a flushing effect and bringing additional nutrients to the area. Evidence in humans suggests that chemical irritants in the tissues (substance P, prostaglandins, and waste products of metabolism) may decrease the pain threshold by sensitizing free nerve endings. By replenishing tissue fluid through massage therapy, inflammatory products are removed and sensitization may be reduced, preventing or reducing some types of chronic pain.

When treating medical conditions, massage therapy produces immediate effects to promote well-being, improve healing, initiate nervous system inhibition, reduce tension, and strengthen the bond between the patient and owner.[53] In sick animals in the ICU, massage therapy relieves stress caused by prolonged separation from their owners, continuous high-intensity noise and bright light, disrupted sleep cycles, and constant monitoring. Stress reduction is important for patient comfort, and massage is one of the most effective means of relaxing an animal.[54] Keep in mind there are some animals that become more stressed out by touch, so you must pay attention to the patient's behavior cues.

Tech Tip

Massage can stimulate the bowel and urinary bladder; therefore, the patient should be given the opportunity to eliminate after a massage session.

Massage therapy produces both mechanical and reflexive effects. Mechanical effects result from physical contact caused by pressure applied on the body and is directly proportional to pressure applied onto tissues.[55] Mechanical effects include improved lymphatic and venous drainage, removal of edema and metabolic waste, increased arterial circulation,[56]

breakdown of adhesions in soft tissues,[57] muscle relaxation, and reduction of stress hormones (which may lower blood pressure, slow breathing, improve digestion, and release endorphins).

Reflexive effects stimulate peripheral receptors producing general relaxation. A pure nervous reflex effect is a class of massage movements that solely influence the nervous system and is achieved with a very light touch that stimulates cutaneous sensory nerve endings. It is used primarily to soothe and relax an animal in a state of general tension, anxiety, shock, or pain. It does not increase gland secretion, cause chemical effects, or have a mechanical impact on circulatory fluids.[58] The two types of peripheral nerve endings involved in a nervous reflex effect are the cell's Golgi apparatus (proprioceptive input) and muscle spindle (prevents overstretching of muscle).[59] Interaction between the person applying the massage and tissue causes tissue movement that stimulates nerve endings to induce changes in the nervous system,[60] thus providing pain relief from an increased pain threshold and release of endorphins (gate theory of pain relief) once relaxation occurs.[61]

Tech Tip

Mechanical benefits of massage are produced directly by pressure, force, or range of motion and include increased blood circulation, reduced swelling, and reduced scar tissue formation. Reflexive benefits of massage are produced indirectly by stimulation of the nervous system and include decreased nervous system arousal producing general relaxation, triggering of stretch receptors, increased blood vessel diameter, and decreased blood pressure.

The veterinary technician will likely be the patient advocate for massage therapy as most veterinarians are not going to be the ones doing the massage; therefore, it is the perfect opportunity for the trained veterinary technician to provide this service at a cost that is both reasonable to the client and allows the technician to be adequately compensated for their time and skill set. Knowing which conditions benefit from massage therapy and how to effectively communicate with the veterinarian will promote its use as part of the treatment plan (Table 18-3).

TABLE 18-3 Indications, Precautions, and Contraindications for Massage Therapy

Indications for Massage Therapy	Precautions and Contraindications for Massage Therapy
• Reduce or prevent venous stasis and lymphostasis	• Open wounds
• Mobilize adhesions	• Unstable fractures
• Regulate muscle tone	• Severe pain
• Prepare muscles for physical training, preventing injury	• Coagulation disorders
• Accelerate muscle recovery after training, decrease fatigue and soreness	• Certain types of neoplasia
• After surgery to maintain flexibility and prevent further loss of function, decrease edema/inflammation	• Shock
• Chronic musculoskeletal problems (such as osteoarthritis), which develop into postural and gait compensations from muscle tension	• Fever
• Improve joint and muscle function	• Acute inflammation
• Maintain owner and animal bond	• Skin problems, infectious disease, or acute stages of viral disease
	• Pregnancy

Massage is most helpful in restoring proper movement to injured limbs and joints. When injured muscle tissue is held in a fixed, unnatural position it develops a new "neurologic pathway," which translates into the unconscious way a portion of the body is held in a given physiological or emotional situation. Animals develop new neurologic pathways as a result of any trauma that injures a bone or the alignment of the spine and can also result from poor conformation,[62] athletic performance, or training. The term "muscle memory" is sometimes used to describe this phenomena; however, "muscle memory" is an incorrect term as muscles have no memory. During a massage therapy session, the person performing the massage must be relaxed. If the therapist is not relaxed, the animal will sense tension and the massage will not be beneficial. Massage therapy may be performed every 24 to 48 hours with sessions lasting 10 to 20 minutes[63] and should follow a sequence or pattern from lighter to deeper strokes, with a return to lighter strokes in the end.[64,65,66] The hand should move in a distal to proximal direction to facilitate fluid movement from an extremity toward the central body core.[67] However, if the massage therapy is intended to prevent or remove inflammation from a body region, hand movements should start at the affected joint and move to more distal joints, keeping in mind that massaging an acutely inflamed area could cause pain to a patient. If hand movements are initiated at the distal extremity first, lymphatic flow will "back up" in the smaller spaces compared to the larger regions in the proximal portions of the limb. When no inflammation is present, initial hand movement starting location is not crucial for a successful therapy session. The speed of hand movements also affects the outcome of a therapy session. Faster movements will stimulate while slower, gentler movements will sedate or calm.

The basic massage therapy techniques include hold, stroking, effleurage, and petrissage (Figures 18-8a through 18-8d).

Other massage therapy techniques include trigger point therapy, acupressure, percussion, friction, deep transverse friction, and Tui-na (Chinese massage and physical rehabilitation). These techniques are often used during a massage therapy session, but should only be utilized with adequate skill and knowledge to prevent harm to the patient.

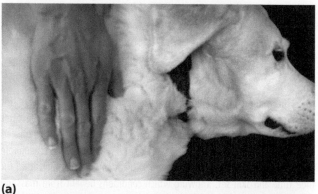
(a)

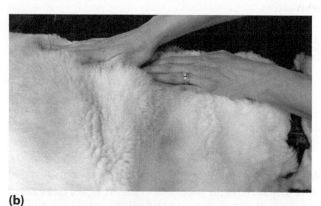
(b)

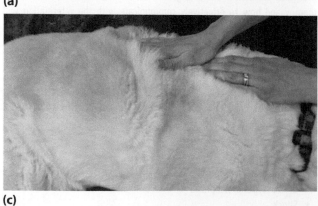
(c)

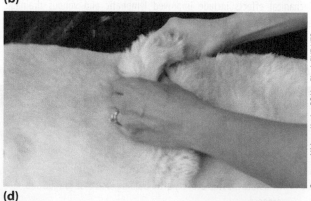
(d)

Courtesy of Kristen Hagler, BS (An. Phys), RVT, VTS (Physical Rehabilitation), CCRP, CVPP, VCC, OACM, CBW

FIGURE 18–8 Massage therapy. (a) Hold is the static placement of the hands on the animal's body and is used to introduce touch to the animal. Use at the beginning and completion of a massage therapy session. (b) Stroking is a medium to light pressure, similar to "petting." Stoking aids in relaxation and allows assessment of tissues, muscle tone, swelling, masses, and temperature differences. Slow, long movements will sedate the patient while faster shorter movements will stimulate. Stroking is used at the beginning and end of the massage after a hold. (c) Effleurage is "to skim lightly" and is a medium pressure, more than stroking, producing small "rolls" or "waves" of skin folding in front of the fingertips in the direction of the muscle fibers. The skin is allowed to glide gently over the underlying fascia, reducing adhesions. Effleurage is often used in between techniques as a transition. Effleurage functions to move fluid to lymph nodes, aide drainage, improve mobility between tissues, and decrease muscle tone. (d) Petrissage is "to knead" and is the rolling, lifting, kneading, or compression of the muscle bellies. Petrissage promotes muscle relaxation, decreased stiffness, increased blood flow, improved length of fibrous tissue, and increased scar tissue mobility.

Range of Motion and Stretching

Maintenance of normal joint range of motion (ROM) is one of the most important aspects of any physical rehabilitation program aside from therapeutic walks. In order to maintain normal ROM, joints and muscle must periodically be moved throughout their available ranges. ROM exercise is a therapeutic movement about a joint to maintain integrity of the tendon, ligament, articular cartilage, and muscle. ROM exercises may described as follows:

- **Passive ROM** where *no voluntary movement* is required by the patient because the motion is performed by an external force such as the therapist's hands. Referred to as PROM.
- **Active assisted ROM** where external forces are used to guide a joint through motion *voluntarily generated* by the patient. For example, the therapist/veterinary technician helps to move the patient's limb at appropriate phases of the gait cycle while the animal is walking on a treadmill.
- **Active restricted ROM** where a device or equipment such as a weighted vest is used to *increase voluntary* effort by the patient.
- **Active ROM** *without external intervention*; therefore, the patient causes the movement. For example, when an animal is guided through a series of obstacles,[68] Referred to as AROM.

Active range of motion (AROM) and passive range of motion (PROM) exercises are useful in diminishing the effects of disuse and immobilization[69] after surgical procedures or in the treatment of chronic musculoskeletal or neurological disease.

Depending on the facility, the veterinary technician's role in carrying out ROM therapy varies. In a general veterinary practice, treatment plans may be limited to only PROM therapy while in a facility dedicated to animal physical rehabilitation treatment plans can include all phases of ROM therapy. Regardless of the ROM therapy type, the veterinary technician should perform joint risk assessment for each patient under his/her care and discuss rehabilitation therapy strategies with the veterinarian.

Tech Tip

PROM therapy can provide unique feedback to veterinarians about a patient's joint health during a therapy session. This information can be used to develop a more effective instruction plan for clients to continue after the animal is discharged.

PROM is motion of a joint performed without muscle contraction and remains within the patient's available range of motion. PROM uses an external force (e.g., the hands of a veterinary technician or client) to move the joint in a proper plane of movement. It should be used whenever a patient is unable to move joints on its own or active motion is contraindicated, such as in an articular fracture[70] or difficult soft tissue procedure. PROM may also be used to prevent joint contracture during healing and recovery in the paralyzed patient, in the management of chronic osteoarthritis, when inflammation is present, or when active assistive ROM is painful. PROM therapy *cannot* prevent muscle atrophy, increase strength and endurance, or assist with circulation to the extent that voluntary muscle contraction does.[71] Daily treatment plans for recumbent or activity restricted hospitalized patients should include PROM and ideally be initiated early in the course of hospitalization and continuing through the time of discharge. Critical care patients often get overlooked for PROM therapy because of the perceived risk for decompensation despite their high risk for tendon, ligament, articular cartilage, and muscle health deterioration. In humans during controlled experimental settings, range of motion did not seem to adversely affect cardiopulmonary parameters, and in those who were critically or systemically ill, limb movements performed passively actually showed a statistically significant increase in oxygen consumption, heart rate, and blood pressure over baseline values.[72]

Tech Tip

Exercise pens (ex-pens) can be used to house nonambulatory patients in high traffic areas of the hospital for close supervision and easier access.

The goal of PROM therapy is to gently extend and flex individual joints through a comfortable range of motion so as to not injure joint structures. The patient should remain pain free during therapy and not react negatively to any movements. Over-aggressive PROM exercise will result in pain, reflex inhibition, delayed use of the limb, and ultimately more fibrosis of the tissues around the joint.[73] The limb should not be placed in varus/valgus or torsional forces that can be avoided with proper support during therapy. Select a quiet, calm location for PROM therapy; then place the patient in lateral recumbency with the affected limb up. Proper body mechanics should be maintained as much as possible, especially in patients that cannot be easily moved or access is limited by bedding or supportive care equipment. In the hospitalized patient, treatment times should be coordinated with other procedures (known as "**cluster care**") to minimize patient stress and allow rest in between necessary medical evaluations. Ideally all limbs should have PROM due to potential abnormal forces being placed on them from injury or dysfunction. If it is medically inappropriate to move the patient in order to access a desired limb, focus on the limbs available. Use caution in patients with acute or chronic inflammation because local edema and joint effusion may mechanically limit joint motion. In all cases, communication must occur between the veterinary technician and surgeon/supervising veterinarian to avoid inappropriate movements and ensure range of motion is performed within ranges that would not create damage.[74]

Tech Tip

The only times PROM therapy is contraindicated is when motion may result in further injury or instability such as with unstable fractures near joints and unstable ligament or tendon injuries.

Once the patient is relaxed, the individual performing therapy should place one hand on the affected limb above the joint and the other hand below the joint so that the limb is in a neutral position. Figures 18-9a through 18-9h describe PROM of the rear leg.

In addition to the previously described sagittal movements (around an axis of rotation that is directed mediolaterally) of pelvic limb, the digits should individually be placed through gentle flexion and extension. Pay special attention to those patients who have existing toe handling sensitivities. For these patients, the stress of handling far outweighs manipulation and the supervising rehabilitation specialists should be alerted. One final PROM exercise often overlooked that is important for neurological or weakly ambulatory patients is limb abduction (limb moves laterally away from the midline of the body). While supporting the entire limb, gentle abduction of the entire limb about 5 degrees is recommended[75] to prevent shortening of the medial musculature (Figure 18-9h). For most routine conditions, each joint

should go through 15 to 20 repetitions and be performed two to four times daily spaced evenly apart. Depending on patient condition, PROM may be prescribed as bilateral, unilateral, contralateral, or all four limbs. Figures 18-10a through 18-10g describe PROM of the thoracic limb.

As range of motion returns to normal and the tissues regain normal movement, the frequency can be decreased and AROM exercise therapies can be integrated into the treatment plan (see description of AROM below).[76] Directions for at home PROM can be provided by the veterinary technician along with other home instructions from the veterinarian. Progress can be monitored during follow-up assessments through objective joint and muscle measurements.

When the patient has been cleared to perform more active movements, range of motion therapy can become more specific to target muscle groups and functional patterns. Continuous passive range of motion (CPROM), in which all joints of the limb are moved simultaneously and appear similar to a bicycling movement while the animal is lying down,[77,78] may be recommended for patients to prevent contractures and restore joint function. PROM can be considered a form of active assisted ROM if the animal is being encouraged to walk on therapeutic equipment such as an underwater treadmill where a therapist is moving the animal's limbs for them. This form of range of motion exercise is more appropriate as an animal nears active use of a limb and can be a form of neuromuscular re-education. In all forms of range of motion therapy, the limb should remain in a natural plane of motion as when the dog is normally moving in a forward direction. Abduction or adduction of joints occur primarily during dynamic joint states or movement and should be avoided during range of motion therapy. **Linear tracking** of the limbs (maintaining a plane of motion consistent with the sagittal plane) occurs during ambulation and should be preserved when providing neuromuscular re-education to

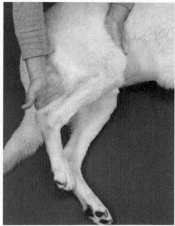

(a)

▲ Neutral starting position of the pelvic limb. One hand should be placed mid-femur at the cranial-medial portion of the rear leg and the other hand placed medial and caudal to the hock.

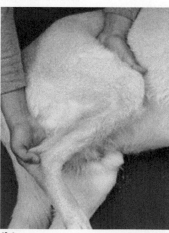

(b)

▲ Flexion of hip. The entire limb should be supported to avoid unnecessary stress on involved joints. Begin by slowly flexing the hip joint to the patients end tolerance range.

(c)

▲ Hip extension. The entire limb should be supported to avoid unnecessary stress on involved joints. After flexion of the hip, slowly move the joint into extension in a slow, rhythmic fashion maintaining neutral positioning of the remaining joints of the limb to the patients end tolerance range.

(d)

▲ Stifle flexion. Hand positioning and pelvic limb support during stifle joint flexion, to patient tolerance, after hip joint range of motion. Upon completion of the most proximal joint PROM, hand positioning should remain the same and the individual should focus on the next joint distal to the one in the beginning of the sequence, in this case the stifle.

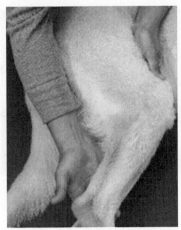

(e)

▲ Stifle extension. Hand positioning and pelvic limb support during stifle joint extension after to patient tolerance. It is important to continue to stabilize the joint below the segment being exercised to separate certain joint motions in other joint.

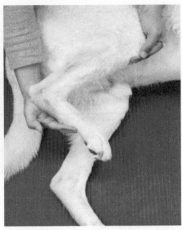

(f)

▲ Hock flexion. The hock is addressed after hip and stifle PROM. Hand positioning and pelvic limb support during hock flexion to patient tolerance.

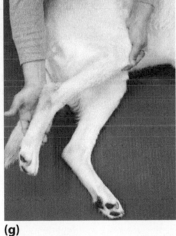

(g)

▲ Hand positioning and pelvic limb support during hock extension to patient tolerance.

(h)

▲ Pelvic limb abduction.

Courtesy of Kristen Hagler, BS (An. Physl, RVT, VTS (Physical Rehabilitation), CCRP, CVPP, VCC, DACM, CBW

FIGURE 18-9 Passive range of motion of the rear leg

the joints. Limbs should not circumduct, abduct, or rotate excessively off the sagittal plane during movement. Maintaining linear tracking is important so that reeducation of joint motion in unnatural planes of motion does not occur, which may be detrimental for long term appropriate movement and joint motion as the patient regains mobility and strength.

The final phase of ROM therapy is **active range of motion** (AROM). This can include active assisted, active restricted, or active range of motion therapy. Patients with functional stability and strength may begin participating in active assisted ROM therapy such as ambulating on a ground treadmill or during hydrotherapy (such as walking on an underwater treadmill) with the therapist/veterinary technician moving

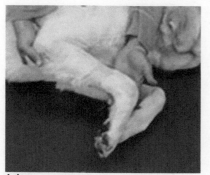

(a)

▲ Neutral starting position of the thoracic limb.

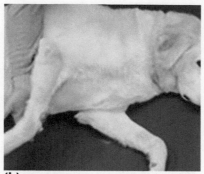

(b)

▲ Shoulder flexion. The entire limb should be supported to avoid unnecessary stress on involved joints. Begin by slowly flexing the shoulder joint to the patient's end tolerance range.

(c)

▲ Shoulder extension. Keeping hand position the same, the entire limb should be carried to extend the shoulder joint to patient tolerance.

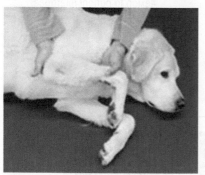

(d)

▲ Elbow flexion. Keeping hand position relatively the same, the elbow is flexed to patient tolerance.

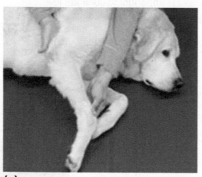

(e)

▲ Elbow extension. Keeping hand position the same, the elbow is extended to patient tolerance.

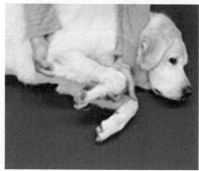

(f)

▲ Carpal flexion. Keeping hand position the same, the carpus is flexed to patient tolerance.

(g)

▲ Carpal extension. Keeping hand position the same, the carpus is extended to patient tolerance.

Courtesy of Kristen Hagler, BS (An. Phys), RVT, VTS (Physical Rehabilitation), CCRP, CVPP, VCC, OACM, CBW

FIGURE 18-10 Passive range of motion of the thoracic limb

the limb at appropriate phases of the gait cycle and the patient's muscles assist to some degree during active assisted ROM. Active restricted ROM encourages active movement of the joints as well as proprioceptive stability with minimal to no assistance. An example of active restricted ROM is an animal exercising with a weighted vest or leg weights. Active ROM (AROM) occurs without assistance such as when the animal is guided through a series of obstacles (e.g., on leash or following a food lure). During normal walking and trotting, joints do not go through a complete range of motion [79] with each joint having a large amount of reserve range of motion. AROM exercises facilitate increased use of the joints. AROM therapy exercises may include walking in snow, sand, tall grass, and crawling through a crawl tunnel, stairs, or over rails laid on the ground.

Tech Tip

The goal of the therapist is to imprint normal gait patterns in patients with abnormal gait movement (e.g., those who have suffered neurological injury, proprioceptive deficits, or disruption in the neuromuscular system such as in progressive non-surgical myelopathies). During imprinting of normal gait pattern, therapists facilitate movement in a patient's limbs using their hands or assistive devices to encourage proper footfall (gait) patterns. Walking on a ground or underwater treadmill does not adequately promote normal gait pattern imprinting; however, treadmill work is important for cardiovascular and muscular strengthening when walking on the ground is not possible or available.

A well-designed treatment plan should include stretching as well as ROM therapy. The theory of stretching has evolved throughout the years and continues to be debated by some experts in the field. The goal of stretching is to take tissues beyond the available ROM and realign soft tissues, increase flexibility and restore joint motion in normal and abnormal tissues.[80,81,82] A stretch should be maintained at each joint for at least 15 to 30 seconds, allowing the tissues to return to a neutral position before re-applying, and repeated at least three to five times in a session before other activities begin. Stretching is ideally performed after massage and PROM therapy or after tissues have been warmed so blood supply within muscles and tissue extensibility is improved, lessening the risk of tissue damage. Two to four stretching sessions should occur per day or 3 to 5 times per week. This is the ideal schedule to improve flexibility in patients with stiffness. It may take up to 2 to 3 weeks to see noticeable improvements, with more affected animals taking longer to improve.[83,84] Caution should be taken in patients with recently repaired fractures and after long periods of immobilization. Injured ligaments or tendons may take up to 3 months of rest/rehabilitation to recover; therefore, stretching should not be performed until the tissue can withstand the stress.[85]

Individualized Therapeutic Exercise Programs

Aside from aquatic therapy, therapeutic exercise programs are the most commonly prescribed form of physical rehabilitation therapy and should be considered for successful postoperative outcomes and management of chronic conditions. Therapeutic exercise programs improve recovery rates and quality and quantity of movement. Patients as well as owners benefit psychologically during therapeutic exercise sessions while strengthening the human-animal bond. In addition, owners pay closer attention to their pet's condition during therapeutic exercise sessions and seek medical care sooner for their animals.[86] Exercise programs are developed for individual patients, which addresses each animal's functional limitations and needs. Initial assessments made by a rehabilitation specialist determine baseline values and reasonable goals of therapy.[87] Goals often include improving active pain-free range of motion and flexibility, limb use, muscle mass, daily function, and prevention of further injury. As the patient and owner continue the program, exercise duration and intensity are adjusted based on regular assessments.[88]

Tech Tip

The well-designed therapeutic exercise program will include ROM and stretching, aerobic conditioning, muscle strength and endurance training, and correction of gait abnormalities if possible.

Therapeutic exercise programs generally focus on endurance, cardiovascular fitness, or obesity management and are initially addressed through endurance activities. Activity level of the patient is modified by first increasing the frequency of the activity with adequate rest periods between sessions. After appropriate adaptation and conditioning have occurred, the length of the activities may be increased to provide further challenges. Increasing speed or adding strengthening exercises may achieve additional strength and conditioning. Ideally patients should participate in daily exercises once or twice per week to avoid overuse injuries. Two exercise periods per day are typically adequate for most conditions and owner's level of time commitment. On occasion, when an owner is able to commit to therapy sessions more frequently, the patient may exercise up to four times daily spaced evenly apart.

In the case of chronic pain patients, continual reevaluation must occur to assess increases in soreness, lameness, or inability to perform an exercise properly. Osteoarthritis waxes and wanes, and the patient will experience "bad days" regardless of treatments received. Typically the goal for osteoarthritis management is to minimize the number of "bad days" the patient experiences and improve recovery times from these unfortunate periods of pain. When increasing the length of an activity, a general rule of thumb is to increase exercise length by 10 to 15% each week and if soreness or lameness occur from an increase in exercise, return to the previous activity level or less (near 50% decrease) for 3 to 4 days and gradually work toward the previous goal. A patient should never return from an exercise period sore or lame; if this occurs any physical benefit is lost. Without adequate recovery time, exercise periods become poor in quality as the patient compensates from pain. It is critical to educate owners about overuse exercise principles when reviewing a home exercise program.[89]

The role of the veterinary technician in carrying out a therapeutic exercise program is understanding how therapeutic exercises affect a patient and how these exercises will prevent injury, ensuring the patient is performing the exercise properly, and educating the client about the exercise. The veterinary technician who can demonstrate knowledge and skill when carrying out a prescribed exercise program will gain owner confidence as they relay questions between client and veterinarian or give feedback on their pet's progress.

A common misconception regarding therapeutic exercise is the need for specialized equipment. This often becomes a barrier for clinicians and leaves owners without tools to help their pet's condition. When designing an exercise program, be creative. Common household items and the natural environment can provide the "equipment" to perform the selected exercises. The list of therapeutic exercise types is endless, and this is important because not every exercise suits every patient (or owner). Points to consider when establishing a therapeutic exercise program include the following:

- Educating owners about therapeutic exercise should always include a verbal and visual demonstration by the rehabilitation practitioner
or veterinary technician followed by the owner repeating the exercise.
- If the owner displays hesitancy or the patient will not perform a selected exercise, the rehabilitation practitioner should be informed so a different exercise can be selected to achieve desired results.
- The sequence should always vary to prevent patient boredom and muscle accommodation.
- A designated area in the home should be set up with exercise equipment, preferably near a high traffic area, to eliminate additional set-up or break-down time and act as a visual reminder for the owner.
- The owner should be using a written log to document patient progress or setbacks. This log should be brought to follow-up appointments. This type of communication will help ensure optimal outcomes for each patient and the veterinary rehabilitation care team.

Regular patient assessment should use tangible measurements such as goniometry and muscle girth measurements. Veterinary technicians often perform patient progress assessments and should be proficient and able to communicate findings to the rehabilitation practitioner and veterinary care team. In addition to measurable outcomes, subjective outcome assessments may include functional and body composition scoring systems, pain assessments, owner impressions, and gait analysis. These findings are used by the rehabilitation practitioner to adjust functional goals and therapeutic exercise repetition, duration, and frequency.

Joint angle measurement is one method to monitor patient progress. **Goniometry** is the measurement of a joint's range of motion. To perform goniometry, the therapist places the two arms of the goniometer (Figure 18-11) along the bones immediately proximal and distal to the joint being examined. The two arms of the goniometer are lined up with specific anatomical landmarks for each joint[90] with movement measured in both flexion and extension. Measurements can be affected when patients exhibit a flexor response or muscle guarding; therefore, maximum angles should be cautiously obtained. If the patient is exhibiting apprehension, document the angle at which the animal flexes or extends a joint comfortably. Joints should be slowly extended, then flexed until the patient begins to show an indication of discomfort. The mean of three independent measurements is used to document reliable, reproducible measurements. As measurements are taken, joints should simultaneously be assessed for crepitus or pain during manipulation.[91]

FIGURE 18–11 A **goniometer** is an objective measurement tool to measure a joint's range of motion.

Courtesy of Kristen Hagler, BS (An. Phys), RVT, VTS (Physical Rehabilitation), CCRP, CVPP, VCC, OACM, CBW

Carpal flexion and extension are demonstrated and described in Figure 18-12.

Elbow flexion and extension are demonstrated and described in Figure 18-13.

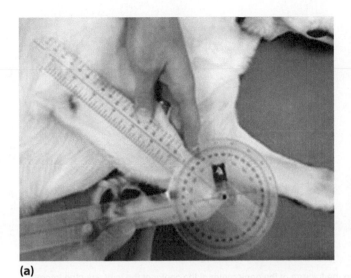

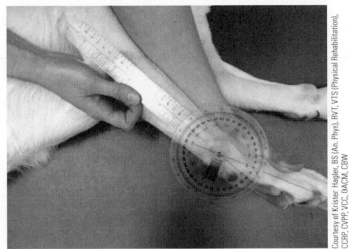

FIGURE 18-12 Goniometry. (a) Carpal flexion is measured by placing the fulcrum of the goniometer in the center of the joint and the arms along the axis of the long bones flexing the joint to end comfortable range. (b) Carpal extension is measured by placing the fulcrum of the goniometer in the center of the joint and the arms along the axis of the long bones. The joint is extended to an end comfortable range.

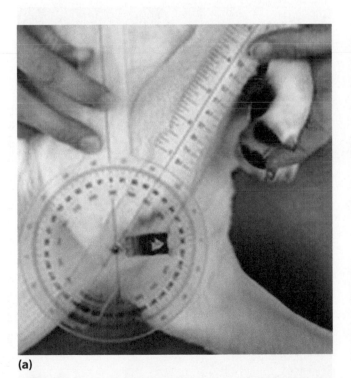

(a)

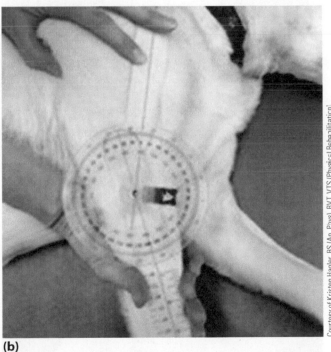

(b)

FIGURE 18-13 Goniometry. (a) Elbow flexion is measured by placing the fulcrum of the goniometer at the center of the elbow joint on the lateral aspect at the epicondyle of the humerus. One arm follows the line of the antebrachium to the ulnar styloid process, and the other arm follows a line to the lateral humeral epicondyle on the greater tubercle of the humerus. The joint is flexed to an end comfortable range. (b) Elbow extension is measured by placing the fulcrum of the goniometer at the center of the elbow joint on the lateral aspect at the epicondyle of the humerus. One arm follows the line of the antebrachium to the ulnar styloid process, and the other arm follows a line to the lateral humeral epicondyle on the greater tubercle of the humerus. The joint is extended to an end comfortable range.

Shoulder flexion and extension are demonstrated and described in Figure 18-14.

Tarsal flexion and extension are demonstrated and described in Figure 18-15.

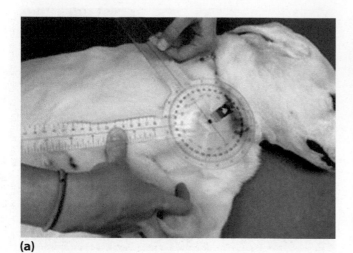
(a)

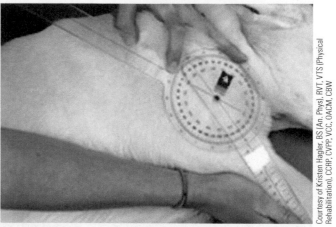
(b)

FIGURE 18-14 Goniometry. (a) Shoulder flexion is measured by placing the fulcrum of the goniometer at the center of the joint, over the lateral humeral epicondyle on the greater tubercle of the humerus. One arm of the goniometer follows the spine of the scapula and the other along the humeral shaft to the lateral epicondyle of the distal humerus. The joint is flexed to an end comfortable range. (b) Shoulder extension is measured by placing the fulcrum of the goniometer at the center of the joint, over the lateral humeral epicondyle on the greater tubercle of the humerus. One arm of the goniometer follows the spine of the scapula and the other along the humeral shaft to the lateral epicondyle of the distal humerus. The joint is extended to an end comfortable range.

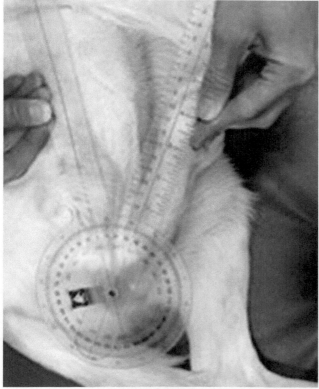
(a)

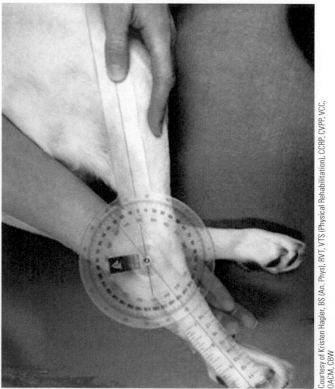
(b)

FIGURE 18-15 Goniometry. (a) Tarsal flexion is measured by placing the fulcrum of the goniometer at the center of the joint at the lateral malleolus. One arm of the goniometer follows the long axis of the metatarsal bones and the other along the axis of the tibial shaft. The joint is flexed to an end comfortable range. (b) Tarsal extension is measured by placing the fulcrum of the goniometer at the center of the joint at the lateral malleolus. One arm of the goniometer follows the long axis of the metatarsal bones and the other along the axis of the tibial shaft. The joint is extended to an end comfortable range.

Stifle flexion and extension are demonstrated and described in Figure 18-16.

Hip flexion and extension are demonstrated and described in Figure 18-17.

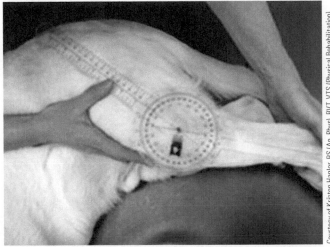

(a)

(b)

FIGURE 18–16 Goniometry. (a) Stifle flexion is measured by placing the fulcrum of the goniometer at the center of the joint near the lateral femoral condyle, one arm following the long axis of the tibial shaft to the lateral malleolus and the other following a line to the greater trochanter along the femur. The joint is flexed to an end comfortable range. (b) Stifle extension is measured by placing the fulcrum of the goniometer at the center of the joint near the lateral femoral condyle, one arm following the long axis of the tibial shaft to the lateral malleolus and the other following a line to the greater trochanter along the femur. The joint is flexed to an end comfortable range.

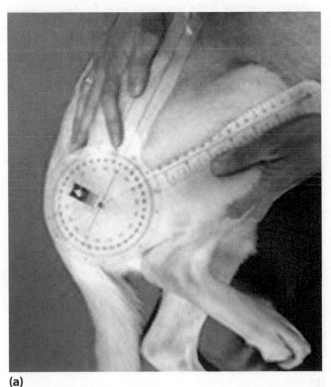

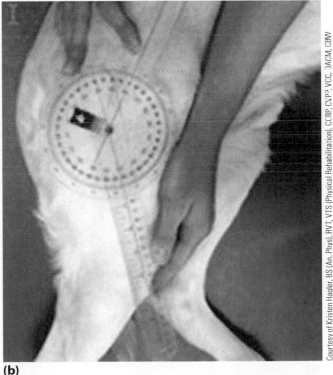

(a)

(b)

FIGURE 18–17 Goniometry. (a) Hip flexion is probably the most technically difficult to obtain and is measured by placing the center of the goniometer over the hip joint, which is found just proximal to the greater trochanter. One arm of the goniometer forms a line to join the lateral femoral epicondyle of the femur and greater trochanter, and the other arm forms a line joining the tuber sacrale and tuber ischiadicum. The joint is flexed to an end comfortable range. (b) Hip extension is also technically difficult to obtain and is measured by placing the center of the goniometer over the hip joint, which is found just proximal to the greater trochanter. One arm of the goniometer forms a line to join the lateral femoral epicondyle of the femur and greater trochanter, and the other arm forms a line joining the tuber sacrale and tuber ischiadicum. The joint is extended to an end comfortable range.

TABLE 18-4 Normal Range of Motion (Degrees) for Appendicular Joints in Healthy Labrador Retrievers

Joint	Flexion (degrees)	Extension (degrees)
Shoulder	57	165
Elbow		166
Carpus	32	196
Hip	50	162
Stifle	41	162
Hock	38	165
Joint	**Varus (degrees)**	**Valgus (degrees)**
Carpus**	7	12

*Jaegger, G., Marcellin-Little, D. J., Levine, D. (2002). Reliability of goniometry in Labrador retrievers, *Am J Vet Res, 63*, 979–986; Millis, D. L., Levine, D., Taylor, R. A (Eds.). 2004. *Canine rehabilitation & physical therapy*, pp. 441–446. St Louis, MO: Saunders.
**Carpal valgus and varus are not depicted—the technical difficulty of this measurement is recommended to be performed by an experienced rehabilitation professional.

Normal joint angles of the shoulder, elbow, carpus, hip, stifle, and hock have been documented in the Labrador Retriever and are used as guidelines for normal angles in other breeds (Table 18-4).

Limb circumference measurements are an indirect method of assessing changes in muscle mass. A measuring tape with a spring tension (Gulick tape measure) is used to improve consistent tension on the tape when taking measurements (Figure 18-18). Following guidelines used in one research study[92] thigh length and circumference is determined as described in Figures 18-19a and 18-19b.

Additional muscle circumference measurements may be taken in the forelimbs (Figures 18-20a and 18-20b) and the thoracic region (Figure 18-21).

Active Therapeutic Exercises

Patients with sufficient strength, proprioceptive awareness, and endurance may benefit from active therapeutic exercises[93] to encourage

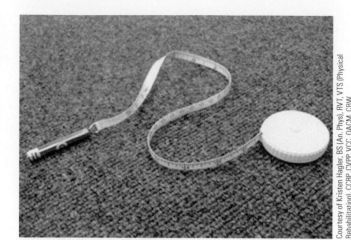

FIGURE 18-18 Image depicting a tension tape measuring device, also known as a Gulick girthometer.

voluntary joint motion through a wider range than that achieved with only walking or trotting.[94] Variations of active therapeutic exercises are nearly limitless and cannot be covered within the text of this chapter. The most commonly used active therapeutic exercises have been highlighted to assist the veterinary technician working in the field of veterinary physical rehabilitation.

Perhaps the most important and most incorrectly performed therapeutic exercise is the slow leash walk. Emphasis is placed on the *quality* of the walk, not *quantity*. The patient should never return from a walk sore or tired. They should be constantly walking and not stopping to relieve themselves or greet other animals. Owners often walk at a pace too brisk for the patient's functional capabilities, reinforcing compensation efforts and abnormal gait kinematics (properties of motion). Slow leash walks should consist of a continuous, even pace allowing the patient to use the limb correctly or encourage limb use. If the patient is having difficulty

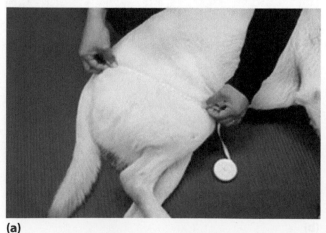

(a)

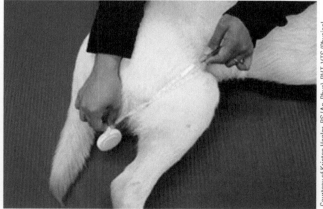

(b)

FIGURE 18-19 Muscle girth. (a) Following guidelines used in one study, thigh length is determined by measuring from the tip of the greater trochanter to the distal aspect of the lateral fabella (sesamoid bone). (b) Limb circumference is determined at points equal to 70% of the thigh length. Measurements are technically easier to obtain at 70% of length compared to any other distance because the skin on the flank does not impede measurements. Limb position should be at a fairly neutral angle as in a functional standing position or in full extension. Muscle circumference should be the mean of three measurements to ensure repeatability.[106]

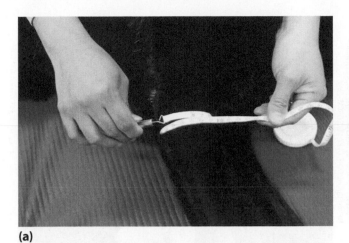

(a)

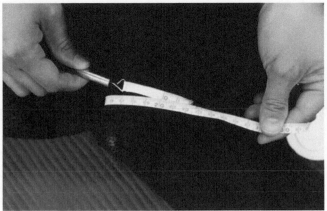

(b)

Courtesy of Kristen Hagler, BS (An. Phys), RVT, VTS (Physical Rehabilitation), CCRP, CVPP, VCC, OACM, CBW

FIGURE 18–20 Muscle girth. (a) In the forelimb, circumference of the antebrachial flexor and extensor muscles may be measured by placing the tension tape measure approximately 1 inch below the lateral epicondyle of the humerus. (b) Triceps muscle group measurements may be taken at approximately 0.5 inch above the lateral epicondyle of the humerus where the triceps muscle group transitions into the tendon inserting onto the olecranon process of the ulna.

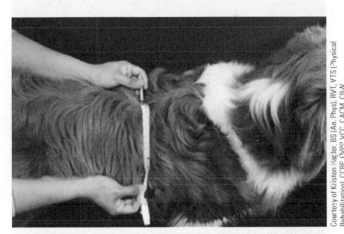

Courtesy of Kristen Hagler, BS (An. Phys), RVT, VTS (Physical Rehabilitation), CCRP, CVPP, VCC, CACM, CBW

FIGURE 18–21 Muscle girth. Thoracic circumference measurements may be taken by wrapping the tape measure around the thorax at T11 and the xyphoid process. Changes in muscle mass are related to increases or decreases in strength and correlate with improved overall body conditioning and redistribution of body weight.

using a limb, a sling under the abdomen or thoracic harness may be beneficial to alleviate pressure on a sore joint and promote proper gait patterning. If the walk is too long or strenuous, the patient will not have enough energy to participate in active range of motion (AROM) exercises or properly recover before the next therapy session. Those owners feeling guilty about a shorter walk length need to be reminded the prescribed exercises are important for a successful outcome. When starting an exercise program for improving strength or conditioning, walking length starts at a time with a 50 to 60% initial reduction based on the "regular" walking length described by the owner. Once the patient has demonstrated mastery of the leash walk coupled with therapeutic exercises, the walk length may be

increased each week until the patient is performing at a level equal to their activity prior to injury or debilitation. Walks should begin on flat, even surfaces, and once proficient, the patient may begin to incorporate more challenging terrain such as hills, ramps, inclines, and declines.

> **Safety Alert** 🚨
>
> Whenever possible and medically appropriate assistive equipment such as body harnesses, slings, and/or mobility carts or stands should be used for patient transportation. Retractable leashes should be avoided. The risk of injury to the handler and patient is reduced when equipment is utilized appropriately.

Patients regularly attending outpatient appointments may further improve joint motion and limb use with land treadmill walking (Figure 18-22). A land treadmill reduces the stress and pain of limb movement in some conditions because the belt provides assistance to extend the hip and stifle by pulling the rear limbs caudally. Subtle lameness may also be detected through gait evaluation on a land treadmill because the animal is forced to maintain linear movement and consistent speeds. Land treadmills provide active assistance for movement of painful joints, provide some cushion on impact, improve stifle extension, and increase the stance time of the limbs. The patient should wear a nonrestrictive harness to provide mild support and prevent falls and if equipped, sidewalls can prevent stepping off the treadmill. Many human treadmills are suitable for canine patients but depending on the breed, achieving a full stride can be affected by belt length. Points

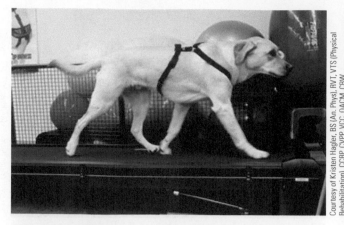

FIGURE 18-22 Active exercise. A land treadmill reduces stress and pain of limb movement in some conditions because the belt provides assistance to extend the hip and stifle by pulling the rear limbs back. Subtle lameness may also be detected through gait evaluation on a land treadmill because the animal is forced to maintain linear movement and consistent speeds.

to consider when using a treadmill for canine patients include the following:

- Equipment should be facing the middle of the room or hallway, not the wall.
- Treats or a favorite toy should be utilized with a "cheerleader" at the head of the dog to maintain motivation.
- The treadmill operator should have easy access to the control panel to adjust speed and desired incline/decline settings. An "emergency stop" button/cord should be within reach.
- When first introducing the patient to the treadmill, do so carefully. Unless the patient has received training, they will be unfamiliar with the equipment. It is recommended to begin with the patient walking onto and off the treadmill on leash a few times, following a treat or toy, and then initiate the belt rate slowly.
- Some patients respond to land treadmill walking better than others, with those who are more conditioned and vertically stable adapting quicker.

Tech Tip

Yoga mats placed on the outside of kennels with slippery flooring can help provide traction for the patient.

Sit-to-stand exercises (Figures 18-23a and 18-23b) help strengthen the hip and stifle extensors without causing extension of the hip, stifle, and hock and improve range of motion. Performing this movement requires strength of the quadriceps, hamstring (biceps femoris, semitendinosus, and semimembranosus), and gastrocnemus muscle groups. It is important to perform this exercise properly because it is considered an essential activity of daily living. Different techniques can be utilized depending on the level of patient obedience training and condition. For those patients with difficulty sitting, they may benefit from sitting on a stack of cushions or stair step (Figure 18-24). This reduces the amount of power needed to stand up and amount of control needed to return to the sitting position. Both pelvic limbs should be squarely positioned underneath the patient. If the patient is weak on one limb, a wall or foam block may be used against the affected limb to

(a)

(b)

FIGURE 18-23 Active exercise. (a) Sit positioning for a healthy dog. Note square positioning of the rear limbs. (b) Proper standing position for a healthy dog transitioning from a sitting position. Note even positioning of all four limbs.

FIGURE 18-24 Active exercise. Positioning of a patient onto a stair step to reduce the amount of power needed to get up from a sitting position.

encourage proper placement. The patient should not be allowed to stand if one limb is splayed to the side.

Ramp or stair walking is used to improve power in pelvic limb extensors and may be used if the patient is consistently using the limb at a walk. The patient should always be using the limb when going up and down a ramp or stairs—they should never hop. It is best to begin with ramp walking because of the gradual and smooth transition, progressing to using the stairs. If a patient has access to stairs and a ramp at the same time, have the patient walk up the stairs, then down the ramp. Going down stairs requires proprioception of the limbs as well as increased extension, and depending on the condition, the patient may benefit from walking down a ramp first before progressing to stairs (Figure 18-25).

Caveletti rails (Figure 18-26) are a series of obstacles set at certain distances apart and at certain heights that look like a ladder laid flat on the ground. When constructed from home equipment, the rails should be evenly spaced apart. Caveletti rails are useful with increasing stride length, stance time, active flexion of the limbs, and proprioceptive awareness. Again, most patients are unaccustomed to walking over objects and will often hop over a series of objects. It is recommended to begin with 1 to 2 rails laid flat on the ground, and assess the body's awareness. If the patient does not hop over these rails, add a few more. Once the patient has demonstrated mastery of rails flat on the ground, then the height can be raised. Begin with 1 to 2 rails and assess patient body awareness and hopping tendencies.

Safety Alert

Equipment cleaning protocols using chemicals such as chlorine, bromine, or bleach should follow recommendations for personal protections and chemical safety from the Safety Data Sheets (SDS). Many of these sanitizing chemicals cannot be mixed and can create noxious fumes or cause superficial skin irritation if personal protective equipment is not used. It is also recommended to have adequate ventilation when using chemicals for cleaning therapy equipment to avoid respiratory problems.

FIGURE 18–26 Active exercise. Cavaletti rails are useful with increasing stride length, stance time, active flexion of the limbs, and proprioceptive awareness.

FIGURE 18–27 Active exercise. Weaving around obstacles is a useful exercise for encouraging active lateral flexion of the spine, proprioceptive training, weight shifting, flexion/extension of the limbs, and conditioning of adductor and abductor muscles depending on pole spacing.

Weaving around obstacles (Figure 18-27) is a useful exercise for encouraging active lateral flexion of the spine, proprioceptive training, weight shifting, flexion/extension of the limbs, and conditioning of adductor and abductor muscles depending on equipment spacing. The patient is led through a series of vertical poles placed in the ground spaced approximately half a body length apart to achieve spinal flexion. If vertical poles are unavailable, objects may be placed in a line to mimic the equipment.

Other active assisted therapeutic exercises can include commando crawl, wheelbarrowing, dancing, pulling or carrying weights, elastic bands, controlled ball playing, and serpentine walks up and down a street

FIGURE 18–25 Active exercise: Stairs and ramps. Combining stairs with a ramp may improve overall success when re-introducing environmental obstacles.

curb. Additional exercises to assist pelvic limb strength include walking backwards (retro-walking), high-fives ("shake" with front paws) while standing, side stepping, toe tickling, playing tug, digging, or jogging. Exercises specific for anticipated patient activity can be designed based on long-term performance goals. Advanced active therapeutic exercises require consistent assessments by the rehabilitation practitioner and adequate owner knowledge to heighten patient safety.

Assisted Therapeutic Exercises

Assisted therapy exercises are useful for patients with mild to moderate deficiencies in ability to stand or ambulate well on their own,[95,96,97] Therapy exercises are designed to improve strength and endurance, enhance proprioceptive training and neuromuscular awareness, facilitate balance, and prepare the patient for more active exercise. Animals with bilateral pelvic injuries, neurological injuries, and severe deconditioning benefit the most from assisted therapy exercises. Standing exercises may be more appropriate in the early stages of recovery, utilizing devices such as body slings or mobility carts. Proprioceptive exercises are available to help these animals regain their ability to appropriately use and place their limbs. These can include weight-shifting, balance board, and PhysioRolls.

During each exercise, the animal should always maintain square foot positioning, and the person assisting the animal should maintain continuous contact with the dog to ensure proper positioning. Duration of assisted therapeutic exercises begins with very short time periods because patients are often severely deconditioned and they fatigue quickly. As the patient improves, the amount of assistance should be decreased.

Weight-shifting exercises are useful for patients recovering from neuromuscular injury or preserving neuromuscular feedback. With the patient standing, they are gently moved in all directions through manual pressure at the hips or shoulders. Manual pressure should be enough to activate muscle groups, but not cause the patient to fall over. The patient should be manually moved in all directions or encouraged to move slightly with low calorie treats. Assistive devices such as a pelvic harness or a sling under the abdomen should be used to facilitate proper placement of the limbs and aid in patient handling.

Radiography positioning blocks or pieces of foam may be used to position recumbent patients. An elevated platform can assist a patient during therapy exercises (e.g., sit-stand, weight shifting, square positioning).

Balance boards vary in size with traditional boards accommodating only the front or rear limbs, or are specialized boards to accommodate the size of the dog. Depending on outcome goals, either size of balance board may be used for assisted therapeutic exercise. The patient should always be supported during the introductory period to prevent falls, with the level of assistance decreased over time as stability improves. The patient should remain in a square functional position. Board movements should be slight and manually controlled by the therapist or veterinary technician. If the balance board is not manually stabilized, the patient will become fearful (Figure 18-28a) of the unstable surface and muscle contractions will not be coordinated nor therapeutic. In this case the therapist must provide assistance to the patient (Figure 18-28b). Treats may be used to assist the patient into weight-shifting, and the patient may also benefit from beginning the exercise lying down and working into a standing position.

(a)

(b)

Courtesy of Kristen Hagler, BS (An. Phys), RVT, VTS (Physical Rehabilitation), CCRP, CVPP, VCC, OACM, CBW

FIGURE 18-28 Balance board. (a) Patient resisting therapy exercise resulting from improper technique or fear. (b) Therapist providing assistance to a fearful or unsteady patient during balancing exercise.

PhysioRolls, PhysioBalls (Figure 18-29), and Swiss balls are used for dogs with severe proprioceptive deficits to help provide stability and support during assisted weight-bearing exercises. PhysioRolls must be properly inflated depending on the degree of support desired. The patient's limbs must be placed or maintained with manual assistance in a proper functional standing position (Figures 18-30a and 18-30b).

Tech Tip 🐾

When planning therapy exercises, it is better to perform PhysioBall or balancing exercises prior to aquatic therapy to minimize slipping.

Patients may be asked to perform weight shifting with treats moved slightly out of reach, or the therapist may perform joint approximations (joint surfaces are pressed together with the patient in a weight-bearing posture) to improve neuromuscular feedback while the patient is in an assisted standing position. As strength and stability return, the front half of the dog may be placed on the PhysioRoll, increasing the amount of weight placed on the rear limbs. Some patients may further benefit by standing with assistance on top of a properly sized PhysioBall. Neuromuscular activities such as rhythmic stabilization may be performed by gently "bouncing" the patient while maintaining support to prevent falls and provide security. This exercise tends to be very challenging for the patient and the person providing stability. Proper equipment choice is based on patient size, desired direction of movement, patient functional ability, and number of persons available to assist during the exercise. Swiss balls tend to be better for smaller patients and provide movement in all directions. PhysioBalls tend to be better for larger patients and provide movement in the cranial-caudal direction.

(a)

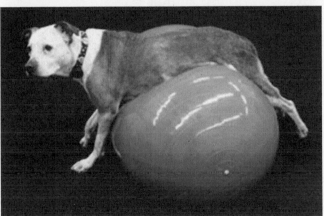

(b)

FIGURE 18-30 Weight-shifting exercises. (a) A patient standing in a non-functional position without assistance. (b) A patient utilizing a PhysioBall to improve functional standing positioning of the limbs.

Tech Tip 🐾

Baby crib sheets can be used to cover patient bedding or exercise equipment to maintain cleanliness and reduce damage from the patient.

Aquatic Therapy

Aquatic therapy can provide a postoperative patient with early weight bearing and psychological benefits; assistance with standing, conditioning, and endurance training; improvement of strength, range of motion, and agility; and reduction of pain.[98] Equipment can include whirlpool tanks, therapeutic swimming pools, underwater treadmills, and natural locations like lakes and streams. In most physical rehabilitation facilities, underwater treadmills or swimming pools are used for patient therapy. The rehabilitation practitioner designs aquatic therapy treatment plans; however, during therapy

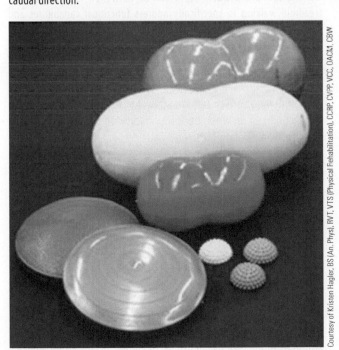

FIGURE 18-29 Image depicting various sized and shaped exercise balls and balance discs.

subtle changes in duration, rest periods, or walking speed may be necessary as the patient exercises (Figure 18-31).

Safety Alert

When it is necessary for the therapist to enter the aquatic chamber and submerge a portion of his/her own body to assist with patient treatment, it is recommended to wear protective shoes or waist waders to protect the skin from accidental injury or acquiring a contaminant from the water.

Therapeutic benefits from aquatic therapy derive from the physical properties of water: thermal effects, buoyancy, resistance or cohesion, turbulence, and hydrostatic pressure.

- Thermal effects are similar to those of superficial heat therapy and affect immersed body parts. Water has high specific heat and thermal conductivity (able to heat or cool the body rapidly) allowing heat to penetrate deeply. Most aquatic therapy equipment maintain temperatures between 82°F and 86°F (27–30°C) for various patient conditions or body type. Precautions should be taken in patients with cardiovascular disease or thermoregulatory difficulties.
- Buoyancy is the upward thrust of water on the body, and patients experience decreases in body weight from buoyancy. Buoyancy helps in the rehabilitation of weak muscles and painful joints, allowing patients to exercise with minimal weight bearing on joints.[99] In the underwater treadmill, water depth adjustments change total body weight and affect the amount of strength required to move a limb through the water (Figure 18-32).
- Water molecules tend to adhere to one another by cohesive forces and require energy to split molecules apart. These forces contribute to the resistance encountered while moving a body through water and increase as water depth decreases.
- Movement through water creates turbulence and occurs as a void behind the body. When moving through water briskly, turbulence increases and the patient needs to use more muscular strength to continue moving at a constant speed. In the underwater treadmill, for instance, doubling the walking speed for a patient increases the amount of resistance and requires 16 times more effort to move through the water.[100]
- Hydrostatic pressure occurs when a fluid exerts an equal amount of pressure on all surfaces of the body at a given depth of water and provides benefits to the patient with circulatory problems or edema in the extremities.[101]

Tech Tip

Compressive wraps or fabric body wraps can help wick away water from a patient's hair and prevent them from getting cold after aquatic therapy.

FIGURE 18–31 Aquatic therapy. Image depicting a patient walking in an underwater treadmill.

Animal physical rehabilitation facilities utilize various forms of aquatic therapy with swimming and water walking in an underwater treadmill being the most common. Aquatic therapy type is selected based on patient outcome goals, functional capabilities, and access to equipment. Patients participating in swimming as a therapeutic exercise experience different joint motion when compared to ground activity or when walking in an underwater treadmill. During swimming, total range of motion increases the overall flexion of joints when compared to walking on land and can be beneficial for patients experiencing pain during joint extension or when joint extension is contraindicated. Swimming can also provide cardiovascular conditioning as the patient guides activity intensity and duration. Unlike swimming, factors like walking speed and water depth can be controlled with underwater treadmill walking to specifically address functional capabilities and joint extension.[102]

Tech Tip

A gardening seat or human shower stool is excellent for a person to sit on when performing facilitated walking in the underwater treadmill.

Low Level Laser Therapy

Low level laser therapy (LLLT), "cold laser" therapy, or phototherapy has been used extensively in animal physical rehabilitation facilities for over a decade and is now commonly found in general veterinary practice. The most common application of LLLT is for adjunctive therapy in the management of chronic pain conditions like osteoarthritis. LLLT may also be used for musculoskeletal, tendon, or ligament injury, reduction of scar tissue formation, treatment of myofascial trigger points, stimulation or sedation of acupuncture points, and pain management. It is also used to improve healing times of chronic or acute skin wounds, reduce localized bacterial counts, improve vascular and lymphatic flow, and stimulate nerve regeneration.[103,104,105,106] Advantages of LLLT include short treatment times, non-invasive application techniques, wide

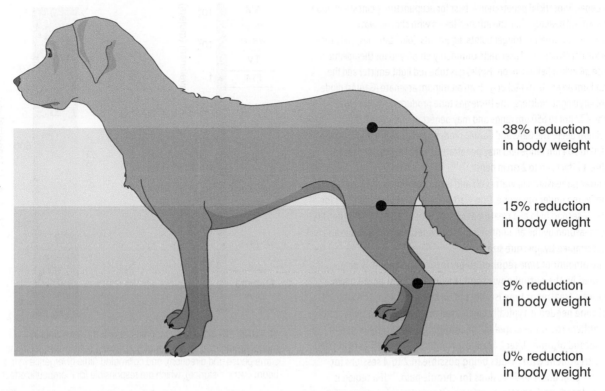

38% reduction
in body weight

15% reduction
in body weight

9% reduction
in body weight

0% reduction
in body weight

FIGURE 18-32 Aquatic therapy. In the underwater treadmill, water depth adjustments change overall weight bearing on joints and affect the amount of strength required to move a limb through the water. Percentages listed reflect overall reduction in total body weight compared to land.

safety margins, and reasonable prices depending on the laser type. Using a therapy laser is easy to do once the rehabilitation practitioner or veterinarian establishes parameters for dosing and treatment.

If a condition ends with an –itis, it may benefit from LLLT.

Laser is an acronym for *l*ight *a*mplification by *s*timulated *e*mission of *r*adiation, and LLLT light consists of a pure light source of a single wavelength. Many different types of lasers are available for medical or therapeutic use and differ in maximum power output. For example, surgical lasers are capable of thermal destruction to cells and tissues due to their very high output power (typically 30–100 watts), which is well above the power output of the therapeutic lasers (ranging from 1–15 watts).[107] Laser apertures of hand-held probes can range in size from 1 mm to 10 cm² or larger depending on the laser manufacturer. Smaller probe tips allow the laser beam to pinpoint smaller surface areas needing treatment and may be better suited for smaller animals while larger probe tips may be better suited for larger animals with more skin (surface area) to treat. Using a smaller aperture probe will increase the treatment time since the laser beam from a probe with a smaller opening covers less surface area than one with a larger opening.

The Food and Drug Administration (FDA) organizes lasers into four categories with classification based on maximum power produced.

- Class 1 lasers include CD and DVD players and are very mild.
- Class 2 lasers are always visible and include bar code scanners.
- Class 3 lasers are broken down into two groups; 3a produces visible light and includes laser pointers while 3b produces non-visible light and includes laser light show projectors and low level lasers.
- Class 4 lasers are also broken down into two subcategories with 4a including the therapeutic lasers and 4b the surgical lasers.

One additional type of laser found in physical rehabilitation and pain management are the LED lasers, which are classified by the FDA as heat lamps to treat pain.[108] Each laser utilizes different wavelengths of light with varying depths of tissue penetration. Low powered lasers with longer wavelengths penetrate superficially and high-powered shorter wavelength lasers penetrate into deeper structures.

Low Level Laser Treatment Guidelines

Three variables affect LLLT application and include the wavelength, power, and time (in seconds) needed to deliver the light energy by the therapy laser.

- Penetration depth to tissues is determined by *wavelength*, with longer wavelengths penetrating superficially and shorter wavelengths much deeper. Superficial penetration is best for acupuncture point stimulation or wound healing while the infrared lasers with shorter wavelengths are best for treating trigger points, ligaments, joint capsules, and intra-articular structures. Laser units commonly employed for therapeutic use include a helium-neon (HeNe) gas tube red light emitter and the gallium arsenate (GaAa) or gallium aluminum arsenate (GaAlAs) diode infrared light emitters. The HeNe gas tube produces wavelengths in the 632 nm to 650 nm range and may penetrate between 0.8 to 15 mm deep. The GaAs and GaAlAs diodes produce wavelengths in the 820 to 904 nm range and may penetrate 10 mm to 5 cm deep, with direct effects up to 2 cm in depth.[109,110]

- *Power* is measured in watts (W) and is the amount of light energy delivered under the probe. One watt of power delivers 1 joule (J) for every second held at a treatment spot (1 W = 1 J/sec). Energy density is measured by the amount of joules delivered per cm^2 of tissue and is determined by aperture size.

- The amount of *time* required to perform a treatment is determined by the output power and the desired total dose known (once a treatment type is selected on the laser unit, the amount of time needed is typically calculated and set by the equipment). Light energy is measured in the number Joules (J) delivered in 1 second of time (J/sec). For most conditions, 1 to 15 J are needed for treatment with results being possible in 2 to 4 sessions for acute pain and 1 to 7 sessions for chronic pain.[111] If a region is overdosed, desired effects may be retarded making determination of accurate time essential prior to initiating treatment. In other words, more power does not equate to a better treatment outcome.

Light energy properties are important in differentiating LLLT from normal sources of light like the sun (Figure 18-33). Lasers used in rehabilitation and pain management are low power and modulate cellular processes through **photobiostimulation** (light energy modulating or stimulating cellular function). Photobiostimulation occurs when light meets the criteria for laser light classification. Photons, the packets of light energy, are absorbed within the mitochondria and cell membranes. Absorption of light in the red and near infrared range (600–1000 nm) causes intracellular photochemical changes, which can stimulate the production of ATP (adenosine triphosphate) and DNA (deoxyribonucleic acid) as well as increasing cell membrane permeability.[112,113]

Analgesic properties from LLLT most likely occur through inhibition of the gate theory of pain by blocking pain transmission to the brain at the level of the dorsal horn of the spinal cord. Other factors such as release of endorphins, increased nerve cell action potentials, and nerve cell regeneration further any analgesic benefits.[114,115] Clinical analgesic effects, the result of nerve conduction inhibition at small and medium diameter peripheral nerve fibers, resolved inflammation, and reduced muscle fatigue, can persist nearly 24 hours.[116] The analgesic effects of LLLT are nearly equal to the analgesic of nonsteroidal anti-inflammatory drugs (NSAIDs) without their associated adverse effects.

Veterinary technicians performing low level laser therapy under the direction of a veterinarian must understand the importance of technique. When treating a certain area, different techniques can be used depending on the individual's knowledge of anatomy, pressure applied

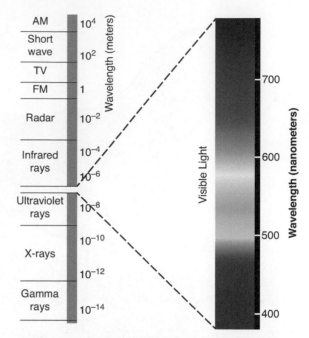

FIGURE 18-33 Electromagnetic spectrum. Laser light is monochromatic (one wavelength), coherent (photons travel in the same phase and direction), and collimated (little divergence in the beam over a distance), which are responsible for therapeutic effects.

to the area, angle of the laser probe head, and length of time (depending on the laser type) the probe head is held in an area. When treating a known anatomical point located just beneath the skin, the amount of energy or "dose" per point depends on the physical size of the probe head (aperture) in addition to the laser type. Some laser types require the probe to be held in one position for the duration of dosage delivery while others recommend a scanning technique where the probe head is continuously moved over the desired treatment area. For example, when using a laser type where the probe is kept in one location for the duration of the dosage delivery, a probe head may be 10 cm^2 in surface area and is the area where light energy is emitted. When the same probe head is placed on the animal's skin for a desired period of time, the treatment area receiving laser light energy is then 10 cm^2. The duration of time the laser probe is kept in that one location is determined by the veterinarian's prescription (Joules per point) and laser type. Often animals will need a large area treated, which is illustrated in (Figures 18-34a through 18-34c).

When applying LLLT to animals, the hair can absorb 50 to 99% of laser light and clipping should be considered. In cases where clipping the coat is undesirable, hair should be parted to expose the skin. In most cases, the probe is best held in direct contact with the skin for the calculated period of time, moving in a grid fashion over the selected treatment area (see Figures 18-34b and 18-34c).

The animal, owner, and persons performing or assisting in LLLT treatment should be in a comfortable position in a relatively quiet area. Restraint of the animal is usually not required, but may be necessary depending on the treatment area and level of pain. LLLT is contraindicated over a pregnant uterus, open fontanels, tattoos, growth plates of skeletally immature animals, photosensitive areas

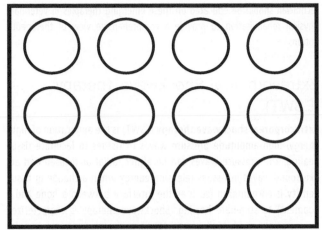

(a)

(b)

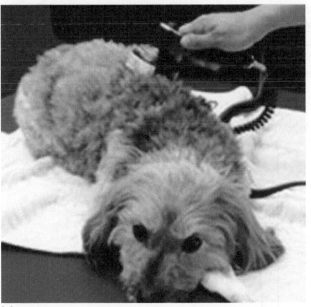

(c)

Courtesy of Kristen Hagler, BS (An. Phys), RVT, VTS (Physical Rehabilitation), CCRP, CVPP, VCC, OACM, CBW

FIGURE 18–34 Laser treatment. (a) Delivery of the laser treatment may be done through several techniques, and the author recommends using a grid technique for veterinary technicians. If the large square represents the desired treatment area (120 cm²) and the circles represent the size of the laser probe head (10 cm²), then the veterinary technician would be holding the laser probe in one spot to deliver the desired dosages prescribed in 12 different locations. (b) Low level laser therapy. Patient receiving laser therapy to the stifle—site 1 of treatment grid. (c) Low level laser therapy. Patient receiving laser therapy to the gluteal muscles—site 2 of treatment grid.

of the skin, malignant areas, directly into the cornea, and in areas of hemorrhage. Patients receiving LLLT postoperatively must have any remaining iodine or povidone washed off because laser light absorption may be intensified and become painful. Topical medications, especially glucocorticoids, may have negative effects on healing and must also be washed off treatment areas. Depending on the therapy laser power output, patients with dark skin or fur may need the laser probe moved more frequently as the melanin in pigmented areas absorbs laser light faster, causing heating effects. [117, 118]

> **Safety Alert**
>
> When performing LLLT, the therapist and any other persons in proximity should wear personal protective eye equipment (supplied by the laser manufacturer). Patients should either have a towel placed over their eyes during treatment or be restrained appropriately away from the laser probe.

Very few risks or adverse effects exist with LLLT. The most common "side effect" from LLLT is a false sense of health for the patient as pain can disappear very quickly after treatment. The owner must be reminded to faithfully follow exercise restriction guidelines during healing and reparative phases to avoid overloading damaged tissues. Patients will often overexert and reinjure themselves after treatment, giving owners the impression that LLLT was ineffective. On occasion patients may experience extreme tiredness from the systemic release of metabolites and easing of long-standing pain. There are examples of trotting horses that were completely "knocked out" for two days after laser therapy but went back to competing as usual after treatment. In rare instances patients may experience a pain reaction, which is usually temporary. [119,120]

Neuromuscular Electrical Stimulation

Neuromuscular electrical stimulation (NMES) devices are used in veterinary physical rehabilitation to address neurologic or orthopedic dysfunction and chronic pain. NMES by definition is the administration of an electrical current generated by a stimulator that travels through leads to electrodes placed over the affected area via contact gel (Figure 18-35). The electrical current depolarizes motor nerves and produces skeletal muscle contraction or constant nociceptive input at low frequencies. Current, intensity, and wave form vary during NMES therapy to recruit skeletal muscle fibers, increase joint mobility, decrease joint contracture and edema, enhance circulation, minimize disuse atrophy, improve muscle strength and sensory awareness, decrease spasticity, and correct gait abnormalities. [121] NMES is effective in re-educating muscles after prolonged disuse. Transcutaneous electrical nerve stimulation (TENS), a form of NMES, is most often used for adjunctive pain control in the management of chronic conditions like osteoarthritis and should be combined with other pain-management techniques rather than a sole therapy. During TENS supplication, nerves are stimulated with a constant low intensity current, sending information to the spinal cord to block pain pathways (gate theory).

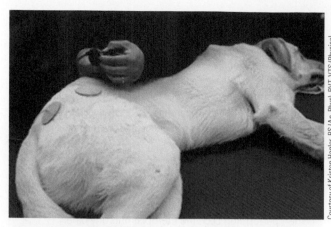

FIGURE 18-35 Neuromuscular electrical stimulation (NMES) administration of an electrical current generated by a stimulator that travels through leads to electrodes placed over an area with contact gel. Patient receiving NMES to the hip joint.

With parameters established by the rehabilitation practitioner, electrical stimulation is easily applied. The role of a veterinary technician during electrical stimulation therapy includes monitoring the patient during treatment and owner education when NMES is dispensed for home use, which is often the case for chronic pain management. Precaution should be taken in patients with impaired sensation or an inability to react to nociceptive input and skin irritation or damage. NMES is contraindicated over the heart and carotid sinus, over the trunk during pregnancy, over neoplastic areas, and in patients with pacemakers and seizure disorders.[122,123]

Therapeutic Ultrasound

Found more often in specialty veterinary rehabilitation centers than in the general veterinary practice, **therapeutic ultrasound** (US) delivers energy in the form of acoustic vibrations to produce thermal and non-thermal effects. The thermal effects of therapeutic US include deep heating to treat pain; increased blood flow, metabolic rates, or tissue extensibility; facilitating wound healing; and rehabilitating musculoskeletal conditions such as active restricted ROM resulting from joint contracture and muscle spasm. Non-thermal effects from therapeutic US include increased cell membrane permeability, calcium transport across cell membranes, and nutrient exchange. With tissue temperatures elevated at depths of 3 cm or more, the following precautions and contraindications are advised:

- Ultrasound exposure should be avoided and is contraindicated in patients with cardiac pacemakers, injured areas immediately after exercise, the testes, carotid sinus, heart, gravid uterus, malignancy, spinal cord after a laminectomy, and recent incision sites.
- Precaution should be taken with acute injury or inflammatory joint disease, bony prominences, fractures, physeal growth plates of young animals, cold therapy before ultrasound, decreased local blood circulation or pain sensation, and in patients that are sedated or under anesthesia.[124,125]

Emerging Therapies and Modalities

Additional modalities and treatment therapies may be encountered when working in the field of veterinary physical rehabilitation, and owners may want to discuss them as therapeutic options for their pet. The veterinary technician should be educated on these emerging therapies in veterinary physical rehabilitation to provide a comprehensive view of therapeutic options.

Extracorporeal Shockwave Therapy (ESWT)

Extracorporeal shockwave therapy (ESWT) is the application of high-energy, high-amplitude pressure waves to tissues to facilitate tissue healing. The pressure waves can be either radial or focused and act like sound waves in tissues releasing energy when a change in tissue density is encountered (such as the interface between a bone and a ligament) to stimulate healing. Shockwave therapy is characterized by extremely rapid, high-energy sound waves producing an impulse or small controlled, focused "explosion." An extremely rapid rise time or pressure front occurs with a slight negative pressure dip causing cavitation (Figure 18-36). Clinical effects include reduced inflammation or swelling, short term analgesia (about 4 days), improved vascularity, formation of bone, realignment of tendon fibers, and enhanced wound healing. The physical energy generated during a "shock" creates a biological response and the desired clinical effects. ESWT has been used since 1980 for the treatment of renal calculi in humans and is a standard of care for equine conditions. The most common application of ESWT in small animals is the adjunctive treatment of osteoarthritis of the major joints and chronic non-union fractures.

Shock waves can penetrate to depths of 110 mm, allowing a focused shock wave to stimulate healing and pain relief at affected joints. Patients must often be under heavy sedation or general anesthesia during treatment application because the release of energy is very uncomfortable and also produces a loud popping sound that could startle an animal; however, some patients are able to tolerate the treatment with minimal medications (Figure 18-37).

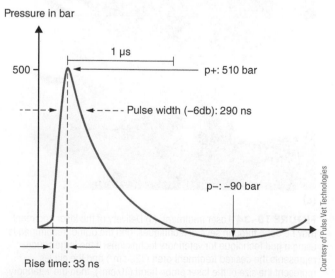

FIGURE 18-36 Extracorporeal shock wave. Graph depicting the physics surrounding shockwave therapy, characterized by extremely rapid, high energy sound waves producing an impulse or a small controlled, focused "explosion." An extremely rapid rise time or pressure front occurs with a slight negative pressure dip causing cavitation.

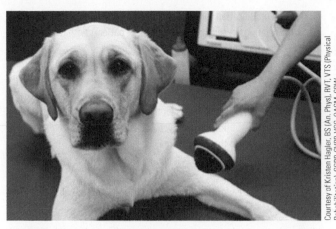

FIGURE 18-37 Extracorporeal shock wave. Image depicting a patient receiving shockwave treatment to the elbow with minimal sedation.

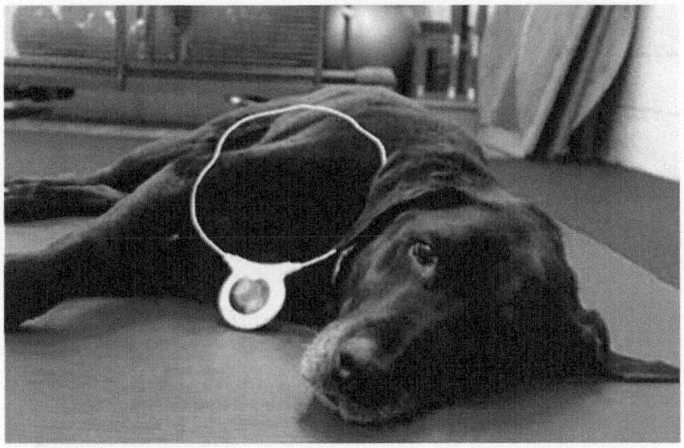

Courtesy of Kristen Hagler, BS (An. Phys), RVT, VTS (Physical Rehabilitation), CCRP, CVPP, VCC, OACM, CBW

FIGURE 18-38 Pulsed electromagnetic field (PEMF) loop. Image depicting a patient wearing a PEMF loop around their shoulder for cervical pain and inflammation.

Courtesy of Kristen Hagler, BS (An. Phys), RVT, VTS (Physical Rehabilitation), CCRP, CVPP, VCC, OACM, CBW

The method of action for pain modulation is unclear but may occur from a reduction in inflammation, production and release of growth factors, or stimulation of nociceptors affecting pain signals to higher centers of the brain. Patients with immune-mediated joint disease, neurologic deficits, neoplastic joint disease, infectious arthritis, diskospondylitis, acute unstable fractures, or orthopedic hardware should take precautionary measures when undergoing treatment. ESWT is contraindicated in patients with concurrent use of NSAIDs that affect platelet function or over the lungs, brain, heart, major blood vessels, nerves, neoplasms, or a gravid uterus. [126,127,128]

Magnetic Fields

A popular treatment modality for chronic pain conditions is the use of magnetic fields. Magnetic fields may either be static or pulsed, with many pet owners choosing static magnets as a complimentary modality for osteoarthritis therapy. *Static magnets* may be embedded in collars, wraps, and bedding making them cheap, easy to use, and commercially available. Although there is little evidence supporting the effectiveness of magnet therapy, it is thought static magnets release endorphins, produce anti-inflammatory effects, and improve local blood flow. Magnetic fields decrease significantly with distance, and for static magnet therapy to be effective, the magnet must be placed in direct contact with the treatment area. A magnet may be rated highly in strength (measured in gauss) nearing the hundreds, but if the treatment area moves less than an inch away strength can diminish into the low double digits. Static magnets have not been shown to cause any adverse effects; however, they should be used with caution if a patient has a cardiac pacemaker. [129,130]

The other form of magnet field is pulsed and is used in **pulsed electromagnetic field (PEMF) therapy**. This form of magnet therapy can involve expensive equipment, but consumer availability is increasing. Patients must remain still for an extended period of time (often 15–30 minutes at a time) or wear the equipment continuously, which makes this modality less than ideal for some patients depending on the treatment location and patient temperament (Figure 18-38). PEMF creates a magnetic field as pulsed electrical current flows through flexible coated metal wires that can be molded and worn around a joint in a light compression wrap. The pulsed

electrical current acts as a stimulus for new bone growth and can affect blood flow in treatment areas or provide analgesia through opioid-mediated effects. Although no controlled studies exist for dogs, PEMF research is available with documented effects on chondrocytes, fibroblasts, and bone growth. There have been no recognized adverse effects with short term or intermittent PEMF use. [131,132]

Acupuncture

Acupuncture is one of the five components of Traditional Chinese Medicine (TCM), the others being herbal medicine, massage (Tui na), gentle exercise (qigong), and food therapy. Traditional acupuncture is based on Qi (patterns of energy flow) throughout the body. Disruptions in energy flow along hypothesized lines called meridians are thought to be the basis of disease in the body. Meridians are believed to follow nerve pathways or fascial planes. The neuroanatomical approach to acupuncture (taught by CuraCore and Canine Rehabilitation Institute [CRI]) is very different than TCM, but with similar goals in mind. Veterinary acupuncture is often used as an adjunctive therapy or alternative treatment for musculoskeletal disorders including osteoarthritis, hip dysplasia, intervertebral disc disease, rheumatoid arthritis, and neurological disorders. Medications and surgery are frequently effective in treating disorders and conditions, but in some cases medication or surgery is not possible or desirable. The non-invasive holistic approach offered through veterinary acupuncture is a popular adjunctive approach to traditional pharmaceutical management of chronic conditions, and veterinary technicians may find themselves educating owners about techniques and mechanisms used during therapy. Treatment involves inserting fine needles into specific points of the body called acupoints, often corresponding with motor points in the muscle (where the motor nerve meets a muscle), to affect bodily systems and healing processes. Most acupoints lie in cutaneous areas, supplied by relatively high concentrations of free nerve endings, nerve bundles, and nerve plexi, which specialize in pain perception. When an acupuncture needle is placed into a point, microinjury occurs providing chronic stimulation of peripheral nerves, and therefore, needles should

Courtesy of Kristen Hagler, BS (An. Phys), RVT, VTS (Physical Rehabilitation), CCRP, CVPP, VCC, OACM, CBW

FIGURE 18-39 Acupuncture. Image depicting a patient receiving acupuncture for back pain.

never be placed in a tumor or area where a tumor is suspected. Pain modulation mechanisms from acupuncture cannot be explained by a single system but are biochemically similar to electrical stimulation, which provides segmental analgesia by altering segmental and super-segmental spinal pathways[133,134] (Figure 18-39). Acupuncture also increases plasma endorphin concentrations, which inhibits transmission of pain impulses.

Acupuncture needles range in size from 1.3 to 12.7 cm in length and 26 to 36 gauge in diameter. They have blunted tips, so bleeding from an acupuncture site is rare. However, according to TCM bleeding allows release of heat in certain patients and areas, which indicates decreasing inflammation. Acupuncture, when performed using TCM, can allow more rapid healing and treat many disease processes to prevent further injury while healing. TCM also believes in the balance of *yin* (cold, moisture, dimness, inward movement, quietness, and slowing) and *yang* (heat, dryness, brightness, outward movement, forceful action, and speed). Acupuncture can also treat conditions that are not recognized by conventional medicine such as an animal being yin deficient (which can present as the patient being too warm at night), which may keep the animal and owners awake and restless. To enhance the therapeutic effects of acupuncture, the needle can be stimulated through the addition of manual rotation, heat, or a mild electrical current (electroacupuncture). Chinese herbal medicines, when prescribed correctly according to Chinese diagnosis, determined by tongue, pulses, diagnostic point, and physical exam, can also complement the effects of acupuncture by supporting healing and providing pain relief.

ENVIRONMENTAL MODIFICATION

A major component of veterinary physical rehabilitation success is the home environment. Patients are often sent home with little attention paid to the environment. Things to review with clients before they take their rehabilitation animal home include the following:

- Discuss the route a patient needs to travel in order to move to the outside of the home. This may reveal problematic functional challenges for the postoperative recovery period or for the otherwise debilitated patient.

- Home modification suggestions may include:
  - Elevating the food and water bowl for patients with forelimb conditions or cervical pain
  - Bedding that can easily be stepped into for patients suffering from range of motion deficiencies
  - Multiple beds and water bowls throughout the home to provide alternative sleeping areas for the restless or deconditioned patient, additional rugs or yoga mats in slippery areas of the home (especially in the feeding area)
  - Baby gates to prevent falls or unwanted stair use, ramps in and out of the house or car
  - A quiet area for the pet to retreat especially if the household is noisy

In some instances the owner may need to utilize assistive devices such as a sling, mobility cart, harness, or non-slick boots/socks to assist with mobility throughout the home or vehicle. Without adequate modification tailored toward functional needs, frustration and injury occur, slowing progress.

Baby gates can be used to block access points to patients while allowing doors to remain open.

CONDITIONS BENEFITTING FROM PHYSICAL REHABILITATION

Veterinary physical rehabilitation can be applied to most conditions seen in the general veterinary practice with orthopedic and neurologic conditions being the most common. The veterinary technician should also be aware of less common disease conditions to effectively communicate with the veterinary care team. The role of the veterinary technician in animal physical rehabilitation is to first recognize candidates for treatment, and then act as an advocate for the patient when discussing medical therapy plans with the veterinarian. Similar to nursing and analgesic plan development, the veterinary technician needs a plan of action when making the case for utilizing veterinary physical rehabilitation therapies and be prepared to present treatment scenarios and anticipated outcomes (Table 18-5).

NURSING CARE FOR THE HOSPITALIZED AND CRITICAL PATIENT

Veterinary patients experiencing prolonged hospitalization periods are excellent candidates for veterinary physical rehabilitation techniques as an adjunctive therapy to supportive care. Veterinary technicians working closely with these animals are well equipped to assess and carry out detailed treatment plans. The veterinary technician should work closely with a rehabilitation practitioner or veterinarian to make proper adjustments depending on the patient's status and response to treatment. Hospitalized patients should be assessed daily to appropriately modify the regimen and identify additional problems if needed. Patients can improve or

TABLE 18-5 Conditions Suitable for Physical Rehabilitation Therapy

Orthopedic Injuries and Conditions	Soft Tissue Injuries and Conditions	Neurological Injuries and Conditions	Other Conditions
• Arthrodesis • Joint replacement • Fracture management • Angular deformities • Hip dysplasia • Cranial cruciate ligament injury • Patellar luxation • Osteochondritis dissecans • Elbow dysplasia • Spondylosis • Amputation • Osteoarthritis	• Myopathies • Hyperextension injuries • Shoulder instability • Contracture • Scar tissue • Tendonopathies • Muscular strains • Generalized weakness	• Fibrocartilaginous embolism • Intervertebral disc disease • Degenerative myelopathy • Spinal stenosis • Cauda equina/Lumbosacral disease • Peripheral nerve injury • Neuromuscular disease	• Obesity • Lymphatic improvements or drainage • Edema • Supportive care (hospitalized patients) • Hospice care support • Chronic pain support • Vestibular disorders

Zink, C. M., & VanDyke, J. (2013). *Canine sports medicine and rehabilitation*. Wiley-Blackwell.

decline very quickly, and attention to detail is critical for detecting stress or decompensation associated with the increased activity or intervention. Monitoring parameters that may be used include edema or bruising scores, disability assessment with timed walking or standing, subjective mentation scoring, pain assessment scoring (especially during massage or passive range of motion therapy), heart rate and rhythm, respiratory rate, and overall demeanor. In addition, limb girth measurements and goniometric measurements may be obtained, especially if the patient is expected to have a prolonged recovery.

Physical rehabilitation plans should not disrupt other treatments such as chemotherapy, fluid therapy, or the administration of medications. Timing of treatments should be coordinated to minimize patient contact time. "Cluster care" is often utilized in human neonatal intensive care units to prevent patient overstimulation and allow appropriate rest between treatment and assessment times. On occasion, modifications to a therapy plan may be necessary, especially if the patient is encouraged to stand or is needed to move into another recumbent position, to accommodate for the location of intravenous, epidural, and urinary catheters; feeding or thoracotomy tubes; and oxygen supplementation. Those patients confined to cage rest are at risk for a rapid reduction of muscle mass and exercise intolerance. Attempts to combat deterioration in function and improve circulation or lymphatic drainage can be achieved through facilitated standing, assisted walking with a sling or cart, therapy ball exercise, and passive range of motion therapy. Facilitated standing can be especially important in compromised patients because the simple act of standing can be a complicated activity, requiring neuromuscular coordination. Ensuring specialized therapy plans occur for the hospitalized or critical patient can help patients retain postural balance, mobility, and functional capacity as well as support them mentally during difficult or stressful hospitalizations.[135]

SUMMARY

The role of the credentialed veterinary technician in physical rehabilitation is unique and offers the opportunity to improve overall patient care and pain management strategies in the hospital setting and in the home environment. Care often continues until the time of patient discharge and the credentialed veterinary technician plays a key role in client education in the care of their pet.

Discharge summary discussion should not be limited to reviewing prescribed medications or activity restriction but should include home pain assessment and physical functional level scoring systems to help empower the caregiver and improve the overall success for patient recovery.

Physical rehabilitation veterinary technicians are intimately involved in carrying out physical medicine treatments prescribed by the rehabilitation veterinarian. The veterinary technician is leading or physically assisting a patient safely through a set of exercises or performing therapeutic modalities to help alleviate pain. The veterinary technician observes and collects information that is invaluable to the rehabilitation veterinarian to determine the success or lack of progress for a particular treatment plan. This can include many types of assessments such as pain scores, behavior, or willingness to perform an exercise, successful outcomes or trends during therapy exercises, and the client's impressions of home care activities. In addition, the strategies developed through physical rehabilitation practice enhance those of traditional veterinary medicine. Utilizing assistive equipment and making appropriate environmental modifications can help improve overall safety in the hospital setting, prevent injury, and promote patient psychological well-being while ensuring the veterinary hospital maintains a high standard of care for veterinary rehabilitation patients.

Be prepared for the client to ask a lot of questions during a therapy session. It is often therapeutic for them as well.

CRITICAL THINKING POINTS

• Veterinary physical rehabilitation begins with the patient's first examination by the veterinarian. Each patient should receive an early risk assessment as part of routine preventative care.

• Pain in veterinary patients can be relieved and their hospital/clinic experiences improved by veterinary technicians who understand the appropriate use of therapeutic modalities such as thermal and manual therapies.

- Veterinary technicians who understand animal pain communication and behavior can detect signs of pain earlier and act as an advocate for the patient.

- Veterinary technicians can perform therapeutic modalities such as aquatic therapy, low level laser therapy, neuromuscular stimulation, extracorporeal shockwave therapy, massage, and guiding patients through therapeutic exercise routines under the direction of a licensed veterinarian.

- Understanding the multi-disciplinary team approach associated with veterinary physical rehabilitation (e.g., integration of the physical therapist) allows the veterinary technician to confidently educate clients and provide accurate information about these therapies.

- Recognizing conditions improved with veterinary physical rehabilitation techniques or assessments beyond those traditionally targeted (e.g., osteoarthritis and spinal cord injury) will promote open mindedness among colleagues about these procedures.

REVIEW QUESTIONS

Multiple Choice

1. What is the primary focus when providing physical therapy?
    a. Restore, maintain, and promote function, fitness, wellness, and quality of life as they relate to movement disorders and health
    b. Restore, maintain, and promote function, fitness, wellness, and quality of life as they relate to moving around the environment
    c. Change old habits, encourage new movement methods, and promote lifestyle modifications as they relate to the health of others
    d. Letting the physical therapist develop an exercise program to promote strength and function regardless of health and movement ability

2. When describing therapeutic modalities applied to animals, it is recommended to use which term?
    a. Animal Physical Therapy
    b. Veterinary Physical Therapy
    c. Physical Therapy
    d. Veterinary Physical Rehabilitation

3. Where is the center of gravity in the dog located?
    a. Mid pelvic level just behind the iliac crest
    b. Mid neck level just behind the fifth cervical vertebrae
    c. Mid chest level just behind the scapulae
    d. Mid chest level just in front of the scapulae

4. In the normal canine, the thoracic limbs bear _____% of the total body weight and the pelvic limbs bear _____% of the total body weight.
    a. 90, 10
    b. 60, 40
    c. 40, 60
    d. 30, 70

5. What does the term compensation refer to?
    a. An abnormal movement or posture in dogs from abnormal skeletal alignment or associated movement
    b. An abnormal amount of weight placed on abnormal skeletal alignment or associated movement
    c. An abnormal movement or posture in dogs from normal skeletal alignment or associated movement
    d. A normal movement or posture in dogs from normal skeletal alignment or associated movement

6. What sensations may animals experience during cold application?
    a. Intense aching, burning, anesthesia, numbness
    b. Intense cramping, burning, aching, numbness
    c. Intense cold, burning, anesthesia, numbness
    d. Intense cold, burning, aching, numbness

7. Which of the following areas should be avoided during cryotherapy?
    a. Peroneal and ulnar nerves
    b. Femoral and tibial nerves
    c. Pudenal and ulnar nerves
    d. Peroneal and sciatic nerves

8. What are three vasodilatory mechanisms that occur during local heat application?
    a. Superficial temperature increase, activation and synapse of cutaneous thermoreceptors, decreased smooth muscle contraction
    b. Superficial temperature decrease, activation and synapse of cutaneous thermoreceptors, increased smooth muscle contraction
    c. Superficial temperature increase, activation and synapse of cutaneous thermoreceptors, increased smooth muscle contraction
    d. Superficial temperature increase, deactivation and sedation of cutaneous thermoreceptors, decreased smooth muscle contraction

9. When using heat therapy for conditions related to animal physical rehabilitation, it is for patients experiencing pain secondary to what condition(s)?

    a. To an old injury and/or decreased range of motion

    b. To stiffness and/or decreased range of motion

    c. To stiffness and/or increased range of motion

    d. From the use of cold applications

10. What are the effects of massage therapy?

    a. Prevention against injury, promote well-being, muscle relaxation, improved digestion

    b. Increase blood pressure, slow breathing, improve digestion, and release endorphins

    c. Removal of edema and metabolic waste, decrease arterial circulation, breakdown adhesions, muscle relaxation and increase mobility

    d. Psychological benefits only

11. What effects can passive range of motion therapy have on a patient?

    a. Cause muscle atrophy, reduce strength and endurance, or reduce circulation to the extent that voluntary muscle contraction does

    b. Prevent muscle atrophy, increase strength and endurance, or assist with circulation to the extent that voluntary muscle contraction does

    c. Prevent muscle atrophy, decrease strength and endurance, or assist with circulation to the extent that voluntary muscle contraction does

    d. Prevent muscle adhesions, increase stamina, or assist with removing waste products from muscle

12. What components will the well-designed therapeutic exercise program include?

    a. Range of motion and stretching, anaerobic conditioning, muscle strength and endurance training, correction of gait abnormalities

    b. Range of motion and counter stretching, aerobic conditioning, muscle strength and endurance training, correction of gait abnormalities

    c. Range of motion and stretching, aerobic conditioning, muscle lengthening and endurance training, correction of gait abnormalities

    d. Range of motion and stretching, aerobic conditioning, muscle strength and endurance training, and correction of gait abnormalities if possible

13. When increasing the length of an activity, a good general rule of thumb is to increase exercise length by _____ to _____% each week.

    a. 10, 20

    b. 15, 20

    c. 10, 15

    d. 12, 16

14. When is LLLT contraindicated?

    a. In animals prior to breeding if the probe will move over the uterus because it can lead to birth defects

    b. In mature animals after their growth plates have closed as it can lead to deformity of the long bones

    c. Postoperatively if iodine is still present on the surgical site because laser light absorption may be intensified and become painful

    d. In animals with benign tumors as the photons emitted from the laser can make the masses malignant

15. What are included as measurable objective outcome assessments?

    a. Strength and endurance

    b. Goniometry and muscle girth measurements

    c. Daily activity level function scores

    d. Pain assessment scores

16. What is the correct name for this objective measurement tool?

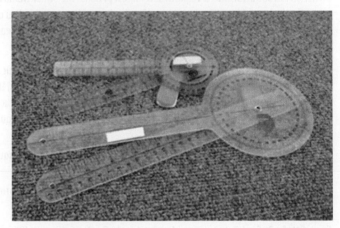

    a. Goniometer

    b. Tension tape measurer

    c. Shoulder layback measurement tool

    d. Geomometer

17. Which of the following is the most important therapeutic exercise that is often performed incorrectly?

    a. Sit-to-stand

    b. Slow leash walks

    c. Land treadmill

    d. Caveletti rails

18. What are the five therapeutic properties of water?

    a. Thermal effects, buoyancy, resistance or cohesion, turbulence, hydrostatic pressure

    b. Thermal effects, buoyancy, fluidity, turbulence, hydrostatic pressure

    c. Walking speed, water depth, thermal effects, turbulence, vasodilatation

    d. Thermal effects, gait training, turbulence, psychological, physiological

19. Which of the following is true regarding acupuncture?

    a. Most acupuncture points lie in muscles; therefore, acupuncture needles must be placed deep in the muscle belly.

    b. When acupuncture needles are placed, microinjury occurs providing chronic stimulation of peripheral nerves.

    c. Acupuncture involves inserting multiple thick needles into different points of the animal's body to stimulate peripheral nerves.

    d. Bleeding from acupuncture sites is common because the needle tips are sharp.

20. What is the mechanism by which LLLT modulates cellular processes?

    a. Photobiostimulation

    b. Photoneostimulation

    c. Gate control theory

    d. Proton pump inhibition

Case Studies

Case Study 1: Falcon, a 5-year-old, male neutered Beagle dog, weighs 25 kg and is coming to rehabilitation services for a weight-loss program (Body Condition Score 8/9 [ribs palpable with difficulty; heavy fat cover]). Previous health history has been uneventful other than his current obesity. Falcon has a Functional Independence Measure (FIM) score of 7/7. The physical rehabilitation plan outlined by the rehabilitation veterinarian includes three times weekly aquatic therapy in the underwater treadmill and core strengthening exercises. Falcon currently takes 10-minute walks twice a day on a loose leash with his owner on flat surfaces, and he requires several rest periods. The owner would like to start running with him because she feels this would be the best strategy to ensure she keeps him active long term. Her running goal is two miles. You only have a total of 30 minutes each session to exercise Falcon and also need to squeeze in one core strengthening exercise. Limitations regarding exercise equipment were not given to you, and the rehabilitation veterinarian is allowing you to choose the best exercise for Falcon.

1. The rehabilitation veterinarian has asked you to monitor Falcon's aquatic therapy program for 6 weeks. The water level height should also be varied (with gradual increases of intensity and duration during exercise times) through the 6-week period as Falcon improves. Initial guidelines for Falcon's hydrotherapy include beginning with the water level at a height to reduce approximately 38% of his total body weight, then gradually reducing to a 9% total body weight reduction. During Falcon's first session where should the water high begin (see Figure 18-32)? At 6 weeks where should the water level be?

2. Each week how would you progress the conditioning program during aquatic therapy time, and what guidelines would you give the client for home walking?

3. What assisted active exercises could you choose to help Falcon improve his overall core strength and center of gravity awareness while at the rehabilitation center?

Case Study 2: Sam is a 2-year-old spayed male Border Collie dog (BCS 6/9) who weighs 25 kg and started coming to your physical rehabilitation facility for injury prevention. Sam regularly competed in agility events and does not have any physical or structural abnormalities noted by the rehabilitation veterinarian. Sam has been prescribed therapeutic massage and stretching for injury prevention. The client has scheduled therapeutic massage sessions 1 week before their next competitive performance event and a session 2 days after the competition event. The client does not feel comfortable performing any manual therapy techniques at home. In addition, Sam will not focus or lie down when at home because he wants to play. The client has told you Sam is a little body sensitive (doesn't like being brushed on his ribs or back) and has never received a therapeutic massage. Sam's favorite piece of equipment is the weave poles. You have already received massage therapy and stretching training from your rehabilitation veterinarian and understand muscular anatomy well.

1. How could Sam benefit from therapeutic massage?

2. How would you approach Sam's first therapeutic massage session? Where would you decide to perform the massage in your rehabilitation facility?

3. Since Sam is a little body sensitive, what massage therapy techniques would you focus on the most during your session (see Figure 18-8)?

4. Knowing the weave poles are Sam's favorite thing to do, what areas of his body are most likely to be sore, and where should you place the focus during a massage therapy session?

5. Other than using manual stretching techniques, what type of active stretching could Sam do before and after performance events that you could show to the owner? Remember Sam probably will not stay still for too long.

Case Study 3: Junior, a 3-year-old male neutered Labrador Retriever dog who weighs 30 kg, presented to rehabilitation services 14 days after tibial plateau leveling osteotomy (TPLO) corrective surgery for a torn cranial cruciate ligament. He is an excitable, young, friendly dog and was having trouble standing on the flooring of the examination room. You are taking a brief history from the owner, and she tells you they have a flight of stairs leading to the upstairs bedroom, wood flooring throughout the home, and another dog in the household. Junior is allowed to have free access to the home including the stairs, but he has not used them since the surgery. The client would like to continue agility training with Junior long term.

1. What recommendations would you suggest to the rehabilitation veterinarian regarding environmental modifications?

2. Would this dog be considered a good candidate for agility classes after recovery from his surgical procedure?

Case Study 4: You are working with Jake, a male neutered 10-year-old German Shepherd dog who weighs 40 kg has been coming to rehabilitation services for 5 months. Jake's rehabilitation goals are to strengthen the gluteal and hamstring muscles in the pelvic limbs and manage symptoms associated with chronic osteoarthritis using aquatic therapy and low level laser therapy. Jake has been doing well but today when discussing pain assessment scores with the client, she mentioned that over the weekend Jake was incontinent. This is slightly alarming to you, and you begin to discuss altering the day's treatment plan until you can speak with the rehabilitation veterinarian but you are interrupted by the client. She is upset that Jake might not be able to do his aquatic therapy today and places a major emphasis on how long Jake is able to walk in the aquatic treadmill. You also notice a couple of Jake's rear toenails, numbers three and four on the right side, are bleeding and worn down.

1. Based on these recent reports from the client and exam findings, should Jake continue to participate in hydrotherapy today?
2. Should you still perform Jake's prescribed low level laser therapy program?
3. What other information would be important for you to obtain from the client to relay to the rehabilitation veterinarian?

Case Study 5: A patient named Zoey, a 2-year-old mixed female spayed terrier mix dog who weighs 12 kg, presents for hydrotherapy. It is her first day of treatment. The rehabilitation veterinarian saw Zoey 1 week ago and a treatment plan was developed to help Zoey build muscles in her quadriceps, which have atrophied from congenital hip dysplasia. The owners adopted Zoey from a rescue organization 1 year earlier and have three other dogs in the house. When Zoey walked into the treatment room with the exercise equipment, she stayed behind her owner. The owner felt it would be best for Zoey to put on the safety vest before getting into the underwater treadmill. Zoey did not want to cooperate much with this and tried to run away. After getting the safety vest on, you reached for the safety vest handle to get Zoey into the underwater treadmill and Zoey resisted. Since so much time was spent getting Zoey dressed for aquatic therapy, you decided to pick Zoey up and place her into the hydrotherapy exerciser. After filling the water level to the appropriate depth, you started the treadmill belt and Zoey would not walk.

1. What could have been done to improve Zoey's success on her first day of hydrotherapy?
2. Should Zoey have been given more time to acclimate to the new environment and equipment?
3. Could positive reinforcement behavior training techniques been utilized to increase success?
4. What parameters could have been changed to help improve Zoey's success walking in the underwater treadmill?

Critical Thinking Questions

1. Daisy, a female spayed 5-year-old Dachshund dog who weighs 19 kg, has been receiving LLLT to aid in the postoperative recovery of a T11–T13 hemilaminectomy performed three days ago. Daisy is unable to get up on her feet own and has a Functional Independence Measure (FIM) score of 1/7. She does have control of her bowel and bladder functions and is very perky. The rehabilitation veterinarian has prescribed laser therapy for pain management and healing of the surgical areas. You have been asked to target the laser two vertebrae above and two vertebrae below the surgical site. The client would like to hold Daisy during the laser therapy because she is anxious about somebody else touching her. How would you respond to this request?

2. You are setting up your equipment for the day in the rehabilitation treatment room. There are a lot of patients being dropped off for the day and one of your support staff just called in sick. You usually begin your day by checking the equipment, but none of the kennels are ready for your patients. Instead of checking the water levels and chemicals, you get accommodations ready for the patients. As each of them check in, the clients request for them to be done an hour sooner than planned. You agree to this for three of the five dogs for the morning group. You get the first patient ready for hydrotherapy and place them into the aquatic exercises. When you press the fill button on the exerciser, no water enters it. You leave the patient in the exerciser alone and check the reservoir, but there is no water in the tank. You turn on the hose and begin filling the water reservoir and return to your patient. They have defecated in the exerciser while unattended. Frustrated, you remove the patient from the exerciser and begin to clean it. Next thing you know a staff member comes back to tell you the water reservoir is spilling water all over the floor because the hose had come out while filling. Now therapy treatments will be delayed for at least 2 hours.
   a. What could have been changed at the beginning of the day?
   b. Should the veterinary technician have altered the regular morning routine equipment checks?
   c. Should the clients have been allowed to change the pick-up times for their pets even though the best intention was to accommodate the client?
   d. When the veterinary technician left the patient unattended, the patient soiled the exercise equipment and then the water hose created an even bigger mess. What sequence of events could have occurred to alter the final outcome (no water, equipment needing cleaning, further delay of patient treatment)?

ENDNOTES

1. Shaffran, N. (2008). Pain management: The veterinary technician's perspective. *Vet Clinics of North America: Small Animal Practice*, 1415.
2. Marcellin-Little, D. J., Danoff, K., Taylor, R. A., & Adamson, C. (2005). Logistics of companion animal rehabilitation. *Vet Clinics of North America: Small Animal Practice*, 1476.

3. American Pet Products Association (APPA). National Pet Owners Survey of 2011–2012. http://www.americanpetproducts.org/press_industrytrends.asp
4. Levine, D., & Millis D. (2014). Regulatory and practice issues for veterinary and physical therapy professions. In D. Millis & D. Levine (Ed.), *Canine rehabilitation and physical therapy*. Elseveir-Saunders.
5. Levine, D., Millis, D. L., & Marcellin-Little, D. J. (2005). Introduction to veterinary physical rehabilitation. *Vet Clinics of North America: Small An Practice*, 1247–1248.
6. Gross, D. M. (2002). Introduction to small animal physical therapy. In R. M. Woodman (Ed.), *Canine physical therapy – orthopedic physical therapy* (pp. 7–11). Connecticut, CI: Wizard of Paws.
7. Gillette, R. (2004). Gait analysis. In D. L. Millis, D. Levine, & R. A. Taylor (Eds.), *Canine rehabilitation & physical therapy* (p. 206). St Louis: Saunders.
8. Riegger-Kruch, C., Millis, D. L. & Weigel, J. P. (2004). Canine anatomy. In D. L. Millis, D. Levine, & R. A. Taylor (Eds.), *Canine rehabilitation & physical therapy* (p. 206). St Louis, MO: Saunders.
9. Albright, J. (2014). Canine behavior. In D. Millis & D. Levine (Eds.), *Canine rehabilitation and physical therapy* (p. 31). Elsevier-Saunders.
10. Mersky H. (1991). The definition of pain. *European Journal of Psychiatry, 6*, 153–159.
11. Hielm-Björkman, A. K., Kuusela, E., Liman, A., Markkola, A. et.al. (2003). Evaluation of methods for assessment of pain associated with chronic osteoarthritis in dogs. *Journal of the American Veterinary Medical Association, 222*(11), 1552.
12. Albright, J. (2014). Canine behavior. In D. Millis & D. Levine (Eds.), *Canine rehabilitation and physical therapy* (p. 31). Elsevier-Saunders.
13. Davies, L. Canine rehabilitation. In C. M. Egger, L. Love & T. Doherty (Eds.), *Pain management in veterinary practice*, (1st ed., p. 135). Ames, IA: John Wiley and Sons, Inc.
14. Riegger-Krugh, C., Millis, D. L., & Weigel, J. P. (2004). Canine anatomy. In D. L. Millis, D. Levine, & R. A. Taylor (Eds.), *Canine rehabilitation & physical therapy* (p. 57). St Louis, MO: Saunders.
15. Saunders, D. G., Walker, J. R., & Levine D. (2005). Joint mobilization. *Vet Clinics of North America: Small An Practice*, 1295.
16. Gross, D. M. (2002). Introduction to small animal physical therapy. In R. M. Woodman (Ed.), *Canine physical therapy—Orthopedic physical therapy* (p. 8). Connecticut: Wizard of Paws.
17. Zink, M. C. (2013). What is a canine athlete? In C. M. Zink & J. B. Van Dyk (Eds.), *Canine sports medicine and rehabilitation* (p. 11). Wiley-Blackwell.
18. Steinberg, HS. Coates JR. Diagnosis and Treatment Options for Disorders of the Spine. In C. M. Zink & J. B. Van Dyk (Eds.), *Canine sports medicine and rehabilitation* (p. 311). Wiley-Blackwell.
19. Saunders, D. G., Walker, J. R., & Levine D. (2005). Joint mobilization. *Vet Clinics of North America: Small An Practice*, 1312.
20. Riegger-Kruch, C., Millis, D. L., & Weigel J. P. (2004). Canine anatomy. In D. L. Millis, D. Levine, & R. A. Taylor (Eds.), *Canine rehabilitation & physical therapy* (pp. 39–41). St Louis, MO: Saunders.
21. Gross, D. M. (2002). Introduction to small animal physical therapy. In R. M. Woodman (Ed.), *Canine physical therapy—Orthopedic physical therapy* (p. 7). Connecticut: Wizard of Paws.
22. Gilette R. (2004). Gait analysis. In D. L. Millis, D. Levine, & R. A. Taylor (Eds.), *Canine rehabilitation & physical therapy* (p. 205). St Louis: Saunders.
23. Saunders, D. G., Walker, J. R., & Levine D. (2005). Joint mobilization. *Vet Clinics of North America: Small An Practice*, 1311–1312.
24. Millis, D. L. (2009). Physical therapy and rehabilitation in dogs. In J. S. Gaynor, & W. W. Muir (Eds.), *Handbook of veterinary pain management* (2nd ed., pp. 507–511). St. Louis, MO: Mosby.
25. Ho, S. S. W. et al., (1994). The effect of ice on blood flow and bone metabolism in knees. *American Journal of Sports Medicine, 22*, 537–540.
26. Rothstein, J. M., Roy, S. H., & Wolf, S. L. (2005). The rehabilitation specialists handbook (3rd ed., p. 801). F.A. Davis Company.
27. Drum, M., & Werbe, B. et al. (2014). Nursing care of the rehabilitation patient. In D. L. Millis & D. Levine, D. (Eds.), *Canine rehabilitation and physical therapy* (2nd ed., p. 281). Elsevier-Saunders.
28. Dragone, L., Heinrichs, K. et.al. (2014). Superficial thermal modalities. In D. L. Millis & D. Levine, D. (Eds.), *Canine rehabilitation and physical therapy* (2nd ed., p. 317). Elsevier-Saunders.
29. Millis, D. L. (2009). Physical therapy and rehabilitation in dogs. In J. S. Gaynor, & W. W. Muir (Eds.), *Handbook of veterinary pain management* (2nd ed., p. 509). St. Louis, MO: Mosby.
30. Steiss, J. E., & Levine, D. (2005). Physical agent modalities. *Vet Clinics of North America: Small An Practice*, 1317–1321.
31. Ibid.
32. Millis, D. L. (2009). Physical therapy and rehabilitation in dogs. In J. S. Gaynor, & W. W. Muir (Eds.), *Handbook of veterinary pain management* (2nd ed., p. 508). St. Louis, MO: Mosby.
33. Iggo A. (1969). Cutaneous thermoreceptors in primates and sub-primates, *Journal of Physiology, 151*, 332–341.
34. Meyer, A. M., Ringkamo M., et. al. (2006). Peripheral mechanisms of cutaneous nociception. In S. B. McMahon & M. Koltzenburg (Eds.), *Wall and Melzack's textbook of pain* (5th ed., p. 21). Elsevier: Churchill.
35. Heinrichs, K. (2004). Superficial thermal modalities. In D. L. Millis, D. Levine, & R. A. Taylor (Eds.), *Canine rehabilitation & physical therapy* (p. 280). St Louis, MO: Saunders.
36. Vannatta, M. L., Millis, D. L., Adair, S., et al. (2004). Effects of cryotherapy on temperature change in caudal thigh muscles of dogs. In D. J. Mercellin (Ed.), *Proceedings of the Third International Symposium on Rehabilitation and Physical Therapy in Veterinary Medicine* [abstract] (p. 205). Raleigh, NC: Department of Continuing Education, North Carolina State College of Veterinary Medicine.
37. Heinrichs, K. (2004). Superficial thermal modalities. In D. L. Millis, D. Levine, & R. A. Taylor (Eds.), *Canine rehabilitation & physical therapy* (p. 285). St Louis, MO: Saunders.

38. Millis, D. L. (2009). Physical therapy and rehabilitation in dogs. In J. S. Gaynor, & W. W. Muir (Eds.), *Handbook of veterinary pain management* (2nd ed., p. 508). St. Louis, MO: Mosby.

39. Heinrichs, K. (2004). Superficial thermal modalities. In D. L. Millis, D. Levine, & R. A. Taylor (Eds.), *Canine rehabilitation & physical therapy* (p. 279). St Louis, MO: Saunders.

40. Knight, K. L. (1995). Temperature changes resulting from cold application. In *Cryotherapy in sport injury management*. Champaign, IL: Human Kinetics Publishers.

41. Steiss, J. E., & Levine, D. (2005). Physical agent modalities. *Vet Clinics of North America: Small An Practice,* 1321.

42. Heinrichs, K. (2004). Superficial thermal modalities. In D. L. Millis, D. Levine, & R. A. Taylor (Eds.), *Canine rehabilitation & physical therapy* (p. 278). St Louis, MO: Saunders.

43. Travell, J. G., & Simons, D. G. (1992). *Myofascial pain and dysfunction: The trigger point manual,* vols I and II. Baltimore, MD: Williams and Wilkens.

44. Hayes, K. (1993). Cryotherapy. In *Physical agents* (4th ed., pp. 49–59). Norwalk, CT: Appleton & Lange.

45. Heinrichs, K. (2004). Superficial thermal modalities. In D. L. Millis, D. Levine, & R. A. Taylor (Eds.), *Canine rehabilitation & physical therapy* (p. 286). St Louis, MO: Saunders.

46. Steiss, J. E., & Levine, D. (2005). Physical agent modalities. *Vet Clinics of North America: Small An Practice,* 1321.

47. Heinrichs, K. (2004). Superficial thermal modalities. In D. L. Millis, D. Levine, & R. A. Taylor (Eds.), *Canine rehabilitation & physical therapy* (p. 287). St Louis, MO. Saunders.

48. Steiss, J. E., & Levine, D. (2005). Physical agent modalities. *Vet Clinics of North America: Small An Practice,* 1323.

49. Ibid (1321).

50. Schwatrz, C. (1996). Introduction to acupressure and massage techniques. *Four Paws Five Directions: A guide to Chinese medicine for dogs and cats* (p. 117). Berkeley, CA: Celestial Arts.

51. Hourdebaigt, J. P. (2004). *Canine massage: A complete reference manual,* 2nd ed. Dogwise Publishing div. of Direct Book Service, Inc.

52. Sutton, A. (2004). Massage. In D. L. Millis, D. Levine, & R. A. Taylor (Eds.), *Canine rehabilitation & physical therapy* (p. 303). St Louis, MO: Saunders.

53. Ibid (p. 308).

54. Dunning, D., Halling, K. B., & Ehrhart, N. (2005). Rehabilitation of medical and acute care patients. *Vet Clinics of North America: Small Animal Practice,* 1420–1421.

55. Hourdebaigt, J. P. (2004). *Canine massage: A complete reference manual,* 2nd ed. Dogwise Publishing div. of Direct Book Service, Inc.

56. Dunning, D., Halling, K. B., & Ehrhart, N. (2005). Rehabilitation of medical and acute care patients. *Vet Clinics of North America: Small Animal Practice,* 1420–1421.

57. Geiringer, S. R. (1988). Traction, manipulation and massage. In J. A. DeLisa (Ed.), *Rehabilitation medicine: Principles and practice* (pp. 276–294). Philadelphia, PA: J.B. Lippincott Co.

58. Hourdebaigt, J. P. (2004). *Canine massage: A complete reference manual,* 2nd ed. Dogwise Publishing div. of Direct Book Service, Inc.

59. Ibid.

60. Sutton, A. (2004). Massage. In D. L. Millis, D. Levine, & R. A. Taylor (Eds.), *Canine rehabilitation & physical therapy* (p. 304–308). St Louis, MO: Saunders.

61. Millis, D. L. (2009). Physical therapy and rehabilitation in dogs. In J. S. Gaynor, & W. W. Muir (Eds.), *Handbook of veterinary pain management* (2nd ed., p. 531). St. Louis, MO: Mosby.

62. Schwatrz, C. (1996). Introduction to acupressure and massage techniques. *Four paws five directions: A guide to Chinese medicine for dogs and cats* (pp. 119–120). Berkeley, CA: Celestial Arts.

63. Hanks, J., & Spodnick, G. (2005). Wound healing in the veterinary rehabilitation patient. *Vet Clinics of North America: Small An Practice,* 1461.

64. Manning, A. M. (1997). Physical therapy for critically ill veterinary patients. Part II: The musculoskeletal system. *Compend Contin Educ Pract Vet,* 803–809.

65. Millis, D. L. (2009). Physical therapy and rehabilitation in dogs. In J. S. Gaynor, & W. W. Muir (Eds.), *Handbook of veterinary pain management* (2nd ed., p. 532). St. Louis, MO: Mosby.

66. Sutton, A. (2004). Massage. In D. L. Millis, D. Levine, & R. A. Taylor (Eds.), *Canine rehabilitation & physical therapy* (p. 317). St Louis: Saunders.

67. Ibid (p. 304).

68. Dunning, D., Halling, K. B., & Ehrhart, N. (2005). Rehabilitation of medical and acute care patients. *Vet Clinics of North America: Small An Practice,* 1418.

69. Brody, L. T. (1999). Mobility and impairment. In C. M. Hall & L. T. Brody (Eds.), *Therapeutic exercise: Moving towards function* (1st ed.). Philadelphia, PA: Williams & Wilkins.

70. Millis, D. L., Lewelling, A., & Hamilton S. (2004). Range-of-motion and stretching exercises. In D. L. Millis, D. Levine, & R. A. Taylor (Eds.), *Canine rehabilitation & physical therapy* (p. 229). St Louis, MO: Saunders.

71. Millis, D. L. (2012). Module V: Exercise and manual therapy of patients with osteoarthritis (Unit 1: Manual therapy techniques). *Canine osteoarthritis case manager certificate series.* The University of Tennessee: University Outreach and Continuing Education.

72. Dunning, D., Halling, K. B., & Ehrhart, N. (2005). Rehabilitation of medical and acute care patients. *Vet Clinics of North America: Small An Practice,* 1418.

73. Bockstahler, B., Millis, D. L., Levine, D., & Mlacnik, E. (2004). Physiotherapy: What and how. In D. Bockstahler, B. Levine, & D. Millis (Eds.), *Essential facts of physiotherapy in dogs and cats rehabilitation and pain management* (p. 57). Babenhausen, Germany: BE Vet Verlag.

74. Millis, D. L., Lewelling, A., & Hamilton S. (2004). Range-of-motion and stretching exercises. In D. L. Millis, D. Levine, & R. A. Taylor (Eds.), *Canine rehabilitation & physical therapy* (p. 230). St Louis, MO: Saunders.

75. Millis, D. L. (2009). Physical therapy and rehabilitation in dogs. In J. S. Gaynor, & W. W. Muir (Eds.), *Handbook of veterinary pain management* (2nd ed., p. 515). St. Louis, MO: Mosby.

76. Millis, D. L. (2009). Physical therapy and rehabilitation in dogs. In J. S. Gaynor, & W. W. Muir (Eds.), *Handbook of veterinary pain management* (2nd ed., p. 515). St. Louis, MO: Mosby.

77. Millis, D. L., Lewelling, A., & Hamilton, S. (2004). Range-of-motion and stretching exercises. In D. L. Millis, D. Levine, & R. A. Taylor (Eds.), *Canine rehabilitation & physical therapy* (p. 231). St Louis, MO: Saunders.

78. Bockstahler, B., Millis, D. L., Levine, D., & Mlacnik, E. (2004). Physiotherapy: What and how. In D. Bockstahler, B. Levine, & D. Millis (Eds.), *Essential facts of physiotherapy in dogs and cats rehabilitation and pain management* (p. 58). Babenhausen, Germany: BE Vet Verlag.

79. Millis, D. L. (2012). Module 5B: Proprioception exercises for the arthritic patient. (Unit 2 proprioceptive exercises). In *Canine osteoarthritis case manager certificate series*. The University of Tennessee: University Outreach and Continuing Education.

80. Bockstahler, B., Millis, D. L., Levine, D., & Mlacnik, E. (2004). Physiotherapy: What and how. In D. Bockstahler, B. Levine, & D. Millis (Eds.), *Essential facts of physiotherapy in dogs and cats rehabilitation and pain management* (p. 58). Babenhausen, Germany: BE Vet Verlag.

81. Millis, D. L., Lewelling, A., & Hamilton, S. (2004). Range-of-motion and stretching exercises. In D. L. Millis, D. Levine, & R. A. Taylor (Eds.), *Canine rehabilitation & physical therapy* (p. 239). St Louis, MO: Saunders.

82. Millis, D. L. (2012). Module 5A: Range of motion, stretching, massage and joint mobilization (Unit 3 stretching). In *Canine osteoarthritis case manager certificate series*. The University of Tennessee: University Outreach and Continuing Education.

83. Millis, D. L. (2009). Physical therapy and rehabilitation in dogs. In J. S. Gaynor, & W. W. Muir (Eds.), *Handbook of veterinary pain management* (2nd ed., p. 516). St. Louis, MO: Mosby.

84. Millis, D. L. (2012). Module VA: Range of motion, stretching, massage and joint mobilization (Unit 3). In *Canine osteoarthritis case manager certificate series*. The University of Tennessee: University Outreach and Continuing Education.

85. Millis, D. L., Lewelling, A., & Hamilton, S. (2004). Range-of-motion and stretching exercises. In D. L. Millis, D. Levine, & R. A. Taylor (Eds.), *Canine rehabilitation & physical therapy* (p. 239). St Louis, MO: Saunders.

86. Bockstahler, B., Millis, D. L., Levine, D., & Mlacnik, E. (2004). Physiotherapy: What and how. In D. Bockstahler, B. Levine, & D. Millis (Eds.), *Essential facts of physiotherapy in dogs and cats rehabilitation and pain management* (p. 56). Babenhausen, Germany: BE Vet Verlag.

87. Millis, D. L. (2009). Physical therapy and rehabilitation in dogs. In J. S. Gaynor, & W. W. Muir (Eds.), *Handbook of veterinary pain management* (2nd ed., p. 514). St. Louis, MO: Mosby.

88. Bockstahler, B., Millis, D. L., Levine, D., & Mlacnik, E. (2004). Physiotherapy: What and how. In D. Bockstahler, B. Levine, & D. Millis (Eds.), *Essential facts of physiotherapy in dogs and cats rehabilitation and pain management* (p. 56). Babenhausen, Germany: BE Vet Verlag.

89. Millis, D. L. (2009). Physical therapy and rehabilitation in dogs. In J. S. Gaynor, & W. W. Muir (Eds.), *Handbook of veterinary pain management* (2nd ed., p. 517). St. Louis, MO: Mosby.

90. Norkin, C. C., & White, D. J. (1995). *Measurement of joint motion: A guide to goniometry* (2nd ed). Philadelphia, PA: FA Davis.

91. Millis, D. L. (2004). Assessing and measuring outcomes. In D. L. Millis, D. Levine, & R. A. Taylor (Eds.), *Canine rehabilitation & physical therapy* (p. 215). St Louis, MO: Saunders.

92. Millis, D. L., Scroggs, L., & Levine, D. (1999). Variables affecting thigh circumference measurements in dogs. *Proc 1st Int Symp Rehab Physical Ther Vet Med*, 157.

93. Bockstahler, B., Millis, D. L., Levine, D., & Mlacnik, E. (2004). Physiotherapy: What and how. In D. Bockstahler, B. Levine, & D. Millis (Eds.), *Essential facts of physiotherapy in dogs and cats rehabilitation and pain management* (p. 64–68). Babenhausen, Germany: BE Vet Verlag.

94. Millis, D. L. (2009). Physical therapy and rehabilitation in dogs. In J. S. Gaynor, & W. W. Muir (Eds.), *Handbook of veterinary pain management* (2nd ed., p. 517). St. Louis, MO: Mosby.

95. Bockstahler, B., Millis, D. L., Levine, D., & Mlacnik, E. (2004). Physiotherapy: What and how. In D. Bockstahler, B. Levine, & D. Millis (Eds.), *Essential facts of physiotherapy in dogs and cats rehabilitation and pain management* (p. 62–63). Babenhausen, Germany: BE Vet Verlag.

96. Hamilton, S., Millis, D. L., Taylor, R. A., & Levine, D. (2004). Therapeutic exercise. In D. L. Millis, D. Levine, & R. A. Taylor (Eds.), *Canine rehabilitation & physical therapy* (p. 244–263). St Louis, MO: Saunders.

97. Millis, D. L. (2009). Physical therapy and rehabilitation in dogs. In J. S. Gaynor, & W. W. Muir (Eds.), *Handbook of veterinary pain management* (2nd ed., pp. 517–524). St. Louis, MO: Mosby.

98. Levine, D., Rittenberry, L., Millis, D. L. (2004). Aquatic therapy. In D. L. Millis, D. Levine, & R. A. Taylor (Eds.), *Canine rehabilitation & physical therapy* (p. 264–275). St Louis, MO: Saunders.

99. Miller, L. (2000). Breakthroughs. *Veterinary Forum*, 11.

100. Millis, D. L. (2012). Module 5C: Aquatic exercises. In *Canine osteoarthritis case manager certificate series*. The University of Tennessee: University Outreach and Continuing Education.

101. Levine, D., Rittenberry, L., & Millis, D. L. (2004). Aquatic therapy. In D. L. Millis, D. Levine, & R. A. Taylor (Eds.), *Canine rehabilitation & physical therapy* (p. 264–275). St Louis, MO: Saunders.

102. Millis, D. L. (2012). Module 5C: Aquatic exercises (Unit 3: Aquatic kinematics). In *Canine osteoarthritis case manager certificate series*. The University of Tennessee: University Outreach and Continuing Education.

103. Millis, D. L. (2009). Physical therapy and rehabilitation in dogs. In J. S. Gaynor, & W. W. Muir (Eds.), *Handbook of veterinary pain management* (2nd ed., pp. 528–530). St. Louis, MO: Mosby.

104. Hode, J. & Lars, H. (2010). Therapeutic instruments. *The new laser therapy handbook* (pp. 53–54). Grangesberg, Sweden: Prima Books, 2010.

105. THOR Photomedicine Ltd. (2011). Training course booklet. *LLLT/Photobiomodualation for Musculoskeletal Pain Training Manual*. Stuarts Draft.

106. Millis, D. L. (2012). Module 6B: Low level laser for osteoarthritis (Unit 1: Introduction). In *Canine osteoarthritis case manager certificate series*. The University of Tennessee: University Outreach and Continuing Education.

107. Millis, D. L., & Saunders, D. G. (2014). Laser therapy in canine rehabilitation. In D. L. Millis & D. Levine (Eds.) *Canine rehabilitation and physical therapy* (p. 363). St Louis, MO: Saunders.

108. Millis, D. L. (2012). Module 6B: Low level laser for osteoarthritis (Unit 1: Introduction). In *Canine osteoarthritis case manager certificate series*. The University of Tennessee: University Outreach and Continuing Education.

109. McCauley, L., Glinski. (2004). Acupuncture for veterinary rehabilitation. In D. L. Millis, D. Levine, & R. A. Taylor (Eds.), *Canine rehabilitation & physical therapy* (p. 345). St Louis, MO: Saunders.

110. Millis, D. L. (2012). Module 6B: Low level laser for osteoarthritis (Unit 3: Types of lasers). In *Canine osteoarthritis case manager certificate series*. The University of Tennessee: University Outreach and Continuing Education.

111. Millis, D. L. (2012). Module 6B: Low level laser for osteoarthritis (Unit 3: Types of lasers and Unit 5: Doses and application). In *Canine osteoarthritis case manager certificate series*. The University of Tennessee: University Outreach and Continuing Education.

112. Millis, D. L., Francis, D., & Adamsom, C. (2005). Emerging modalities in veterinary rehabilitation. *Vet Clinics of North America: Small An Practice,* 1338.

113. THOR Photomedicine Ltd. (2011). Training course booklet. *LLLT/Photobiomodualation for Musculoskeletal Pain Training Manual*.

114. Millis, D. L. (2009). Physical therapy and rehabilitation in dogs. In J. S. Gaynor, & W. W. Muir (Eds.), *Handbook of veterinary pain management* (2nd ed., pp. 528–530). St. Louis, MO: Mosby.

115. Millis, D. L. (2012). Module 6B: Low level laser for osteoarthritis (Unit 2: Research). In *Canine osteoarthritis case manager certificate series*. The University of Tennessee: University Outreach and Continuing Education.

116. THOR Photomedicine Ltd. (2011). Training course booklet. *LLLT/Photobiomodualation for Musculoskeletal Pain Training Manual*. Stuarts Draft.

117. Millis, D. L., Francis, D., & Adamspm, C. (2005). Emerging modalities in veterinary rehabilitation. *Vet Clinics of North America: Small An Practice,* 1335–1345.

118. Millis, D. L. (2009). Physical therapy and rehabilitation in dogs. In J. S. Gaynor, & W. W. Muir (Eds.), *Handbook of veterinary pain management* (2nd ed., pp. 528–530). St. Louis, MO: Mosby.

119. Tuner, J., & Hode, L. (2010). *The new laser therapy handbook* (pp. 131–132). Grangesberg, Sweden: Prima Books.

120. THOR Photomedicine Ltd. (2011). Training course booklet. *LLLT/Photobiomodualation for Musculoskeletal Pain Training Manual*. Stuarts Draft.

121. Millis, D. L. (2012). Module 6D: Transcutaneous electrical nerve stimulation (Unit 1: Introduction). In *Canine osteoarthritis case manager certificate series*. The University of Tennessee: University Outreach and Continuing Education.

122. Millis, D. L. (2009). Physical therapy and rehabilitation in dogs. In J. S. Gaynor, & W. W. Muir (Eds.), *Handbook of veterinary pain management* (2nd ed., pp. 526–528). St. Louis, MO: Mosby.

123. Johnson, J., & Levine, D. (2004). Electrical stimulation. In D. L. Millis, D. Levine, & R. A. Taylor (Eds.), *Canine rehabilitation & physical therapy* (p. 289–302). St Louis, MO: Saunders.

124. Millis, D. L. (2012). Module 6E: Therapeutic ultrasound. In *Canine osteoarthritis case manager certificate series*. The University of Tennessee: University Outreach and Continuing Education.

125. Steiss, J. E., & McCauley, L. (2004). Therapeutic US. In D. L. Millis, D. Levine, & R. A. Taylor (Eds.), *Canine rehabilitation & physical therapy* (p. 324–325). St Louis, MO: Saunders.

126. Millis, D. L., Francis, D., & Adamsom, C. (2005). Emerging modalities in veterinary rehabilitation. *Vet Clinics of North America: Small Animal Practice,* 1345–1350.

127. Millis, D. L. (2009). Physical therapy and rehabilitation in dogs. In J. S. Gaynor, & W. W. Muir (Eds.), *Handbook of veterinary pain management* (2nd ed., pp. 534–536). St. Louis, MO: Mosby.

128. Millis, D. L. (2012). Module 6C: Extracorporeal shock wave therapy for the treatment of OA. In. *Canine osteoarthritis case manager certificate series*. The University of Tennessee: University Outreach and Continuing Education.

129. Millis, D. L. (2009). Physical therapy and rehabilitation in dogs. In J. S. Gaynor, & W. W. Muir (Eds.), *Handbook of veterinary pain management* (2nd ed., p. 533). St. Louis, MO: Mosby.

130. Millis, D. L. (2012). Module 6F: Pulsed electromagnetic field and static magnets (Unit 2 Static magnets). *Canine osteoarthritis case manager certificate series*. The University of Tennessee: University Outreach and Continuing Education.

131. Millis, D. L. (2009). Physical therapy and rehabilitation in dogs. In J. S. Gaynor, & W. W. Muir (Eds.), *Handbook of veterinary pain management* (2nd ed., p. 320). St. Louis, MO: Mosby.

132. Millis, D. L. (2012). Module 6F: Pulsed electromagnetic field and static magnets (Unit 1 Pulsed electromagnetic fields). In *Canine osteoarthritis case manager certificate series*. The University of Tennessee: University Outreach and Continuing Education.

133. Millis, DL. Module 7C: Acupuncture *Canine Osteoarthritis Case Manager Certificate Series*. The University of Tennessee – University Outreach and Continuing Education; 2012.

134. Mccauley, L., & Glinski, M. H. (2004). Acupuncture for veterinary rehabilitation. In D. L. Millis, D. Levine, & R. A. Taylor (Eds.), *Canine rehabilitation & physical therapy* (p. 337). St Louis, MO: Saunders.

135. Dunning, D., Halling, K. B., & Ehrhart, N. (2005). Rehabilitation of medical and acute care patients. *Vet Clinics of North America: Small An Practice,* 1412–1419.

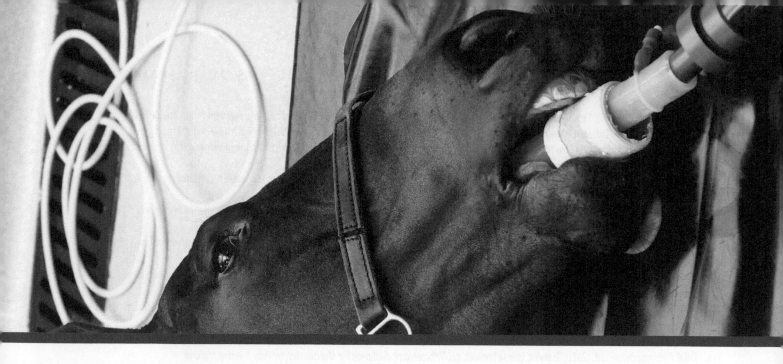

CHAPTER 19

Anesthesia and Analgesia of Large Animal and Food Producing Species

Mary Ellen Goldberg, *BS, CVT, SRA, CCRA, CCRVN, CVPP, VTS (Lab Animal Medicine, Research/Anesthesia, Physical Rehabilitation)*

Trisha Roehling, *CVT, VTS (Anesthesia/Analgesia)*

LEARNING OBJECTIVES

Upon completion of this chapter, it is expected that the reader should be able to:

19.1 Explain the importance of and where to access drug withdrawal times in food producing species

19.2 Identify preanesthetic preparations for ruminants (cattle, sheep, and goats)

19.3 Explain strategies of premedication for ruminants (cattle, sheep, and goats)

19.4 Explain strategies of anesthesia induction for ruminants (cattle, sheep, and goats)

19.5 Explain strategies of anesthesia maintenance for ruminants (cattle, sheep, and goats)

19.6 Describe how to safely monitor ruminants (cattle, sheep, and goats) when using local, regional, and general anesthesia

19.7 Describe how to respond to and/or prevent anesthetic complications specific to cattle

19.8 Explain strategies of anesthesia recovery for ruminants (cattle, sheep, and goats)

19.9 Identify methods of assessing pain in ruminants (cattle, sheep, and goats)

19.10 Identify available options for pain management in ruminants (cattle, sheep, and goats)

19.11 Identify preanesthetic preparations for monogastric large animals (horses and pigs)

19.12 Explain strategies of premedication for monogastric large animals (horses and pigs)

19.13 Explain strategies of anesthesia induction for monogastric large animals (horses and pigs)

19.14 Explain strategies of anesthesia maintenance for monogastric large animals (horses and pigs)

19.15 Describe how to safely monitor monogastric large animals (horses and pigs) when using local, regional, and general anesthesia

19.16 Describe how to respond to and/or prevent anesthetic complications specific to monogastric large animals (horses and pigs)

19.17 Explain strategies of anesthesia recovery for monogastric large animals (horses and pigs)

19.18 Identify methods of assessing pain in monogastric large animals (horses and pigs)

19.19 Identify available options for pain management in monogastric large animals (horses and pigs)

INTRODUCTION

Large animal and food producing species (equids, large and small ruminants, and swine) are often described together in textbooks; however, it is critical for the reader to appreciate that each species presents distinct clinical anesthetic and analgesic concerns because of their widely varied anatomy, individual response to anesthetic and analgesic agents, and unique physiology. Consider for example the anatomical variation between a horse and pig and image how these differences can dramatically influence the approach to endotracheal intubation. Similarly, an adult goat's physiological response to intravenous administration of an opioid and benzodiazepine may be the exact opposite than when it is administered to an adult horse (effective sedation verses ineffective sedation and potential excitement respectively). Recognizing pain in these species is also challenging and they are often under treated for their discomfort because of fear of adverse effects and difficulty in the detection of pain in these species. For instance, some veterinarians unjustly avoid administering opioids to horses because they are concerned about decreased gastrointestinal (GI) motility and inducing colic. While this is a potential clinical concern with repeated or high dosages, when used appropriately opioids can be a safe and beneficial part of equine pain management.

Despite many nuances to providing effective anesthesia and analgesia to individual large animal and food producing species, fortunately there are numerous similarities that help veterinary professionals identify and address comparable concerns. For example, some animals discussed in this chapter may be used for human consumption; therefore, the anesthetist must always consider the influence of drug administration on the patient's potential meat and milk production. Another similarity among these species is their sheer physical size that poses an inherent safety risk to all personnel involved in providing care to these species. Ultimately it is the anesthetist's knowledge of large animal and food producing species anatomy, physiology, and pharmacology that allows the anesthesia team to provide safe and effective anesthetic and analgesic management to those patients entrusted to their care.

This chapter strives to provide the reader with basic tools for providing anesthesia and analgesic care to ruminants (cattle, sheep, and goats), horses, and pigs (these specific large animal and food producing species will be referred to simply as large animal species throughout this textbook). Whenever in doubt, the reader is advised to remember and return to the fundamental concepts and principles of anesthesia and analgesia introduced in earlier chapters and apply these concepts to large animal species.

GENERAL CONSIDERATIONS FOR ANESTHESIA IN LARGE ANIMALS

Many of the questions to consider when providing anesthesia management to large animal species are similar if not identical to those asked during small animal cases. The reader should review said questions presented in earlier chapters and consider their application to the large animal setting. Some potentially unique questions for large animals include the following:[1, 2]

- *Based on the procedure and available facilities and instrumentation, will standing sedation, locoregional anesthesia, or general anesthesia be used?* Many diagnostic or surgical procedures in large animals can be safely and effectively performed by applying basic restraint principles combined with sedation, analgesia, and locoregional anesthesia techniques. For example, a laceration may be cleaned and sutured in a horse under sedation and local anesthetic. This may be faster, safer, and less stressful to all parties than placing the horse under full general anesthesia. When more complex, lengthy, or painful procedures that may require enhanced muscle relaxation arise, general anesthesia is then selected. Additionally in some cases general anesthesia may be preferred for the safety of personnel.

- *For the safety of all involved, how will the patient be restrained?* It is important for anyone working with the animals in this chapter to remember that these patients are large, which makes the safety of all personnel of great importance. Tools for patient restraint may include devices such as halters, lead ropes, hobbles, twitches, pens and chutes, swinging doors, and numerous others. It is important to have a discussion before starting a case as to what each person's role will be in restraining a patient and to ensure that all equipment that might be needed is available, functional, and ready to use.

- *If the patient is to be made recumbent, how will the patient be positioned and padded?* Because of the large size of horses and cattle, once a patient is made recumbent impaired tissue perfusion, tissue compression, and poor patient positioning can quickly lead to nerve and muscle injury, which can lead to long-term and potentially life-threatening postanesthetic complications. Great care is therefore taken in patient positioning, and it is not uncommon for material such as soft thick compressible foam padding to be used to disperse a patient's weight over a larger area.

- In food producing animals such as ruminants and pigs, what is the animal's use (e.g., is the patient to be used for its meat or milk, is it a family pet)? **Drug withdrawal times** (an estimate of the amount of time it takes 99.9% of the drug to be eliminated from the plasma) dictate which drugs can or cannot be used in food producing animals and how long an animal's meat or milk needs to be withheld from the human food chain to ensure that said product does not contain drug residues that could be eaten by a human. Most anesthetic drugs have short half-lives; however, those administering drugs to food producing animals must identify and follow the legal withdrawal times as determined by the Food Animal Residue Avoidance Databank (FARAD) at www.farad.org.

- Are there species differences or considerations in response to a given drug? Some medications may be frequently used in small animals but not for large animals and vice versa. For example, the nonsteroidal anti-inflammatory drugs (NSAIDs) phenylbutazone (Phenylbute®, Equipalazone®) and flunixin meglumine (Banamine®) are commonly used in ruminants and horses but are typically considered poorly tolerated and unsafe for use in dogs and cats. Similarly, we do not think twice about administering an anticholinergic drug to a dog with bradycardia; however, administering a large dose of an anticholinergic drug to a horse might have significant effects on that horse's GI mobility, which may lead to colic. Whenever in doubt about the effects of a given drug, turning to a drug textbook and formulary may be helpful to identify inter-species differences.

- If general anesthesia is to be performed, will an inhalant anesthetic agent (and thus necessitating endotracheal intubation) or total intravenous anesthesia (TIVA) be used? Anesthesia can often be maintained for short procedures on healthy large animals using TIVA just as in small animals, although different drugs (other than propofol and alfaxalone) are often used. Inhalant anesthesia is often selected instead for procedures that are longer in duration, more complicated or painful, or require more muscle relaxation. Sometimes providing inhalant anesthetic agents is preferred because passing an endotracheal tube (ETT) protects the airway, which is desirable for the patient. For example, endotracheal intubation with a cuffed ETT is important in ruminants if they become recumbent as it prevents aspiration of saliva and ruminal contents into the trachea and lungs if regurgitation were to occur. If general anesthesia is to be provided, consider whether an anesthesia machine or mechanical ventilator will be needed. In general it is safest to have an anesthesia machine and mechanical ventilator available even if you do not plan to use one.

- Are there species-specific recovery considerations of the patient? Because of the size, anatomy, and temperament of large animals, complications in anesthetic recovery occur with an increased rate of frequency compared to small animals. For example, placing ruminants in a sternal position during recovery reduces pressure on their rumen and may lower the risk of the patient regurgitating or aspirating. Another example is the importance of providing a safe environment for equine recovery. When horses awaken from anesthesia they commonly attempt to stand before they are physically able and coordinated. This can result in a limb fracture or soft tissue injury that could end in catastrophic damage and death.

- What is the patient's physical status (assigning physical status is the same as for assessing small animal patients; see Appendix D)? What co-morbidities are present? What medications is the patient on? What have past anesthesia experiences been? High risk patients such as a dehydrated horse with colic who has severe pain, hypovolemia, and electrolyte abnormalities that has recently received flunixin and numerous large doses of xylazine (ASA physical status IV or V) has different considerations and risks than a young stallion being anesthetized for a routine castration (ASA physical status I). Assigning an ASA physical status to patients and viewing each case as a unique individual with specific concerns is essential and requires that the anesthesia team evaluate a patient's specific case and physical health to help determine the most suitable anesthetic agents, monitoring, and support to use for each patient.

PAIN IN LARGE ANIMALS

The pain pathway for large animals mimics that of small animals. As in other animal species, pain in large animals can alter behavioral, biochemical, and physiological parameters. Assessing pain in large animal species is broadly classified into subjective or objective methods just as in other species.

- Subjective methods are value judgments made by a human observer and involve the assessment of criteria such as behavior, posture, movement, and other cues (e.g., palpation of a surgical site). Simple descriptive scales (SDS), visual analogue scales (VAS), and numerical rating scale (NRS) were described in Chapter 15 (Acute Pain Management). While these subjective assessments are imprecise and can often vary between observers, these judgments tend to become more repeatable and reliable with appropriate experience and training. Several pain scales used for large animal species are described in more detail throughout this chapter.

- Objective methods quantitate physiological stress responses (e.g., plasma cortisol levels), changes in levels of biochemical markers (e.g., elevations in acute phase proteins, cortisol, norepinephrine, and epinephrine and a decrease in insulin), or the incidence of clearly defined patterns of behavior (e.g., vocalization) and attempt to correlate these factors to a degree of pain or a response to treatment. Physiological factors such as vital signs (respiration rate, heart rate, rectal temperature) are also sometimes used as objective methods and have long been recognized as mostly unreliable in identifying pain in humans or animals. Unfortunately both subjective and objective methods of assessing pain in large animals have significant limitations and may fail to correctly identify patients as painful or in other incidences suggest patients are painful when they are not. It is helpful to have a consistent approach to pain assessment and realize that it is usually most ethical to treat patients for pain whenever in doubt.

Tech Tip

Large animals in pain can have elevations in cortisol, norepinephrine, and epinephrine levels, which may lead to vasoconstriction and increased myocardial work and myocardial oxygen consumption. The long-term outcome of these changes can precipitate to a catabolic state, increasing morbidity, suffering, and mortality; therefore, there are health consequences of not treating pain in large animal species.[3]

TABLE 19-1 Species-Specific Behavioral Signs of Pain*

Species	Vocalizing	Posture	Locomotion	Temperament
Cows, calves, goats	• Grunting • Bruxism (teeth grinding)	• Rigid • Head lowered • Back humped	• Limp • Reluctant to move the painful area	• Dull • Obtunded (depressed) • Acts violent when handled
Sheep	• Grunting • Bruxism	• Rigid • Head down	• Limp • Reluctant to move the painful area	• Disinterested in surroundings • Dull • Obtunded (depressed)
Horses[2]	• Grunting • Typically no vocalizing • Bruxism	• Looking at location of pain • Lowered head • Sweating	• Reluctant to move • May paw or roll • Lameness	• Obtunded (depressed) • Disinterested in surroundings • May be aggressive or display avoidance
Pigs	• From excessive squealing to no sound at all	• All four feet close together under body	• Unwilling to move • Unable to stand	• From passive to aggressive depending on severity of pain

This chart is meant to display some of the different signs species may exhibit if in pain. Individuals may not show any of these signs or may show signs not listed. This is meant as a general guide. Chart by M. E. Goldberg.
*Riebold, T. W., Geiser, D. R., & Goble, D. O. (1995). Clinical techniques for food animal anesthesia. In *Large animal anesthesia: Principles and techniques* (2nd ed.). Ames, IA: Iowa State University Press.

Methods of Pain Management

To diligently manage pain in large animal species, it is important to consistently assess the patient, which can be accomplished by being familiar with normal species behavior, understanding how these species commonly display signs of pain, and using established pain scales regularly whenever available. Basic lists of signs and behaviors that can indicate pain in the species discussed in this chapter are found in Table 19-1. Some large animal species (particularly small ruminants and swine) are used in research, and their pain management is covered by the Animal Welfare Act. When in doubt whether an animal is in pain or not, it should always be assumed that anything that would result in pain in humans will cause pain in large animal species and therefore should be treated accordingly.

Tech Tip

Despite limitations of pain scales (some are subjective and scores may differ widely between observers), proper training and their consistent use will help identify pain in animals.

The overall strategies and methods of pain management in large animal species are similar to those used in small companion animals. For example, the concept of preemptive and multimodal analgesia described in earlier chapters is employed in large animals as well. Many but not all of the drugs used in small animal practice are utilized when treating large animal patients. Broadly speaking the same classes of drugs are used for analgesia although some individual drugs within each class may differ. An example of this is flunixin meglumine and phenylbutazone, two NSAIDs that are widely used in large animal species that are rarely if ever used in small animals due to excessive adverse effects. Ultimately adequate pain management for large animal species can involve medications that can often be combined with complementary or alternative

therapies such cold therapy, physical therapy, massage, and acupuncture. The reader is directed to Chapter 5 (Anesthetic and Analgesic Pharmacology) for further description of specific drugs and Chapter 18 (Veterinary Physical Rehabilitation) for information on complementary or alternative therapies.

GENERAL CONSIDERATIONS FOR ANALGESIA IN LARGE ANIMALS

Choosing how to approach appropriate pain management for large animal species is based on many factors and must always be done in conjunction with and under the supervision of a licensed veterinarian. Most of the factors described in previous chapters that addressed small animal analgesia and pharmacology can be applied to large animals as well. When determining which analgesic options should be used, some of the common factors that need to be considered include:

- *What is the likely severity of pain? What is its anticipated duration of pain? Is the pain of short duration or has it become ongoing and potentially progressed to chronic pain?* For example, is the pain mild such as the brief placement of a subpalpebral lavage catheter to administer eye medication or is the pain severe such as a horse with bilateral front limb laminitis?
- *What drugs are available?* If care is provided outside a hospital setting, not all drug options may exist due to the client's inability to give a drug by a particular route such as by constant rate infusion (CRI).
- *Are there any special patient factors that will influence the choice of analgesic?* For example, does a patient have a known allergy or contraindication to a given medication?
- *Is there a pre-existing co-morbidity?* For example, large repeated doses of a NSAID would be a poor choice in a dehydrated goat with renal insufficiency.
- *Are there species specific factors that need to be taken into consideration?* For example, providing a 520 kg horse with a large volume

epidural of a local anesthetic might result in motor blockade making the horse ataxic or unable to stand and cause anxiety and potential injury to patient or care giver.

- *Is the patient a food producing animal and, if so, does the therapy selected have a meat or milk withdrawal time that must be adhered to?* For example, a single 2.2 mg/kg IM dose of flunixin meglumine has a meat withdrawal time of up to 30 days in cattle, which is why it is only approved for IV use in this species. As with anesthetic agents, information about withdrawal times of analgesic drugs can be accessed via the Food Animal Residue Avoidance Database.
- *How will the analgesic plan be implemented?* For example, the size of the animal may affect which treatment options are able to be given. Consider the following:
  - *What facilities are available for management of the animal?*
  - *What level of nursing care and monitoring of the animal is available?*
  - *Can staff attend to the patient throughout a 24-hour period to monitor for adverse effects?*
  - *Is there equipment such as infusion pumps for constant infusion of analgesics?*

REGULATORY CONCERNS FOR ANESTHETIC AND ANALGESIC DRUGS IN FOOD PRODUCING ANIMALS

The Food and Drug Administration's (FDA) Center for Veterinary Medicine (CVM) ensures that safe and effective animal drugs are approved and available for use in the United States. Despite the fact that not all analgesics are approved for use in large animal species, the CVM supports the ethical treatment and management of these animals and aims to improve the availability of safe and effective drugs for anesthesia and pain control. CVM encourages drug companies to use innovative approaches to demonstrate the effectiveness of drugs for use in large animals.

RUMINANT ANESTHESIA

The gastrointestinal anatomy of ruminants provides unique challenges when providing their anesthesia care. The risk of a ruminant regurgitating ingesta and inhaling it into the respiratory tract during any stage of anesthesia could result in hypoxemia, chemical pneumonitis/aspiration pneumonia, and death. Anesthesia does not slow the large amounts of saliva ruminants normally produce, making them prone to aspiration if the airway is not protected. Anesthesia only slightly decreases normal fermentation processes and because eructation is impaired, patients are predisposed to gas build up (sometimes called bloat), which results in the patient's inability to ventilate and can cause cardiovascular impairment, especially when they are in dorsal or lateral recumbency. Despite these risks, general anesthesia can be performed safely in ruminants if certain precautions are taken.

Cattle Anesthesia

Under normal circumstances cattle accept physical restraint well. Many procedures on ruminants can be performed without general anesthesia using standing restraint together with sedation, analgesia, and locoregional anesthetic techniques and should always be considered as a method to accomplish surgical procedures. When diagnostic or surgical procedures are more complex, painful, require deep muscle relaxation, require additional time to complete, or patient status directs otherwise, general anesthesia may be required or preferred over standing restraint.[4]

Preanesthetic Preparation

Preanesthetic preparation for cattle includes physical examination, a thorough history, preanesthetic fasting, blood work, venous catheterization, and actual or estimation of body weight. A thorough history includes the patient signalment (species, breed, age, sex, and sexual status [intact vs. altered; pregnancy status]), recent health concerns, current medications, past anesthesia history, and information about the presenting complaint/procedure/surgery. Perhaps the most important food animal question to ask is the animal's use (e.g., meat, dairy, pet), which can influence drug residues and may dictate which drugs can or cannot be used.

Domestic ruminants have a multicompartmental (more than one) stomach with a large rumen that will not empty completely despite fasting. A gas filled rumen can apply sufficient pressure on the diaphragm and large blood vessels in the abdomen leading to respiratory and circulatory compromise. To improve oxygenation and ventilation, calves greater than 3 months of age should be fasted 12 to 18 hours and deprived of water for 8 to 12 hours prior to anesthesia[5] (Table 19-2). Calves less than 3 months of age don't have a functional rumen and are typically not fasted. Adult cattle should be fasted 18 to 24 hours and deprived of water for 12 to 18 hours.[4] Neonates and young nursing ruminants are typically anesthetized without fasting to reduce the risk of hypoglycemia, and blood glucose levels should be checked periodically during anesthesia and supplemented accordingly.[6]

The benefits of fasting include:

- Reduced GI tract weight on the heart and lungs which may be beneficial during abdominal surgical procedures
- Improved respiratory and cardiovascular function
- Possible prevention of bloat
- Possible reduced severity of aspiration pneumonia

Preanesthetic blood work should be performed on an individual case basis. In some large animals the price of diagnostic testing can be cost prohibitive to many owners. However, tests such as complete blood count (CBC) and serum chemistry with electrolytes can provide useful information about the general health status of the patient and help influence patient management during anesthesia. For example, a patient that has identified electrolyte abnormalities should have these values corrected prior to anesthesia. Similarly, a patient who is noted to be anemic before anesthesia may be at

TABLE 19-2 Fasting/Water Withdrawal Times for Cattle

Age	Food Withdrawal Times	Water Withdrawal Times
Neonate/Young nursing	No fasting	No withholding
Calves > 3 months of age	Should be fasted 12–18 hours prior to anesthesia[5]	Should be deprived of water for 8–12 hours prior to anesthesia[5]
Adults	Should be fasted 18–24 hours prior to anesthesia[4]	Should be deprived of water for 12–18 hours prior to anesthesia[6]

risk for compromised oxygen delivery and prompt the anesthetist to provide blood products if surgical blood loss occurs. Whenever possible, treatments to address abnormal bloodwork results should be done prior to anesthesia.

Tech Tip

For accurate drug administration, large animals should be weighed; however, in a field-type situation, estimation of weight may be needed. The average weight range for mature cows is 1000 to 1500 pounds, and mature bulls is 1500 to 2000 pounds. The average weights of some common breeds include beef cattle such as the Hereford and Angus (cows approximately 1200 pounds and bulls up to 1800 pounds) and dairy cattle such as the larger breed Holstein (cows approximately 1500 pounds and bulls up to 2000 pounds) and smaller breed Jersey (cows approximately 1000 pounds and bulls 1500 pounds).

Venous Catheterization

The jugular vein is the most commonly used site for intravenous (IV) catheter placement in cattle.[6] A 12 to 14 gauge over-the-needle catheter may be used for adults and 16 gauge over-the-needle catheters are typically used for calves under approximately 250 pounds; however, some individual variation may exist depending upon the placer's preference. Infiltration of local anesthetic such as 2% lidocaine to desensitize skin structures before catheterization is recommended. Frequently an access hole is used to reduce peel back, drag, or kinking of the catheter during catheter threading, which is achieved by the use of a small incision with a scalpel blade. A veterinary technician's use of this technique, however, may be restricted by a state's veterinary practice act and may be limited solely to veterinarians depending on the state. IV catheters are then secured with suture passed through subcutaneous tissue and tied in place snugly around the catheter hub. The carotid artery lies close to the jugular vein in cattle; therefore, make sure that the catheterized vessel is not the carotid artery, which could allow accidental drug administration into the arterial circulation and could result in the patient's death. The jugular vein is superficial to the carotid artery and has much lower pressure than the artery. If the IV catheter is in the carotid artery, the higher pressure will cause blood to squirt out faster, farther, and with a pulsating effect. If the IV catheter is in the vein, its lower pressure causes blood to flow freely without pulsating. As arterial blood is more oxygenated, it may be a brighter red in color than venous blood; however, it is not reliable to differentiate venous blood from arterial blood solely on color as it can easily be subjective and misinterpreted.

A general outline for placing an over-the-needle jugular vein catheter in any large animal species is described in Appendix E.

Premedication, Sedation, and Restraint

Many diagnostic or surgical procedures can be done on cattle using basic restraint along with locoregional anesthetic/analgesic techniques and sedation, which are described under the Cattle Analgesia section. In some circumstances, restraining "crushes" or "chutes" (Figure 19-1) may be safer to use for appropriately sedated patients.[2,7] Dependent on facilities, ropes can be useful additions for restraining sedated or unsedated cattle. Specialized equipment required for restraint or positioning anesthetized patients, such as head gates, transporters, and tilt tables, should be checked for proper functioning.[8]

FIGURE 19-1 Restraint for ruminant sedation. For the safety of the patient and veterinary staff, this Brahman bull is restrained in a chute.

Anticholinergic drugs are not routinely administered to small animal patients before general anesthesia and the same is true for large animals such as cattle. Anticholinergic drugs such as atropine (Atrocare®, generic) or glycopyrrolate (Robinul®) may be used in cattle to treat life-threatening bradyarrhythmias or during cardiopulmonary resuscitation.

Sedatives/tranquilizers are used prior to general anesthesia in cattle, and common examples include the following: [8,9]

- Acepromazine (Promace®, generic) IM or IV: Acepromazine will typically produce mild to moderate tranquilization in cattle but can be given as part of a premedication plan along with other drugs.
- Xylazine (Rompum®, AnaSed®, Sedazine®, Chanazine®) IV or IM: Xylazine will produce profound sedation in cattle. It is important for the anesthetist to understand that cattle are very sensitive to xylazine on a mg/kg basis, requiring approximately 1/10 the dosage that would be used in a comparable sized equine. For example, a xylazine dosage in a horse could be 1 mg/kg; however, a dosage of 0.1 mg/kg would be on the very high end of the dosage range for cattle. In horses a dosage of 0.5mg/kg IV xylazine might produce moderate sedation while in cattle a dosage of 0.05mg/kg IV xylazine might cause heavy sedation or even recumbency (which is often not desired from premedication alone). Other alpha-2 (α_2) adrenergic agonists such as detomidine (Dormosedan®) and romifidine (Sedivet®) have also been used in cattle.
- Midazolam (Versed®) or diazepam (Valium®) IV or IM: Benzodiazepines generally are not used alone for tranquilization in adult cattle as their effects are mild and could produce excitement. When administered, they are most commonly given with other drugs for anesthetic induction (e.g., ketamine/diazepam). An exception to this may be in calves, however, as they will often tranquilize well with an opioid and benzodiazepine (e.g., butorphanol and diazepam).
- Opioids Many opioids have been utilized in cattle. The most commonly used opioids historically are butorphanol (Torbugesic®, Torbutrol®) and morphine (Duramorph®, Astramorph®). Concern may arise over using large and frequent doses of opioids in cattle as this may interfere with normal rumen function and cause GI stasis. Opioids, however, should be used as part of a multimodal analgesia plan in cattle.

Most anesthetic and analgesic drugs are not approved for use in food animal species, and these drugs are administered to large animals extra label. If animals are shipped to market and a residue is found, the FDA is authorized to take legal action against the individual who administered the drug. Clients should be counseled on withdrawal times (if they are known) and the consequences of drug residues in animal products intended for human consumption.

Induction Procedures and Techniques

General anesthesia can be achieved in cattle by administering a variety of medications with the overall aim of ensuring analgesia, muscle relaxation, and immobility. These goals are the same as for small animals. When consulting drug formularies, keep in mind that each patient is unique and one should not attempt a "cookie cutter" approach to all animals. For example, a drug dosage provided for a calf may be appropriate for a 4-month-old calf but not a 3-day-old calf. An example of this might be a dosage of xylazine for a calf. Some anesthesiologists would not recommend xylazine use in a 3-day-old calf over concerns of bradycardia and decreased cardiac output but may use it in a 4-month-old calf who has more cardiovascular maturity (see Neonatal and Pediatric Patients section of Chapter 14 [Anesthetic Techniques for Special Cases]).

Another example of selecting drugs and dosages based on the individual is the use of xylazine and ketamine for adult cattle. Sometimes xylazine and ketamine are listed together in books under induction agents; however, the xylazine is actually being given as the premedication and the ketamine is being given as the induction agent. An inexperienced veterinary technician could become confused and administer xylazine as a premedication and then more xylazine along with ketamine for induction when this could be potentially unnecessary. Working closely with the veterinarian when selecting medication and calculating doses will help veterinary technicians become skilled at understanding and delivering anesthesia to their patients.

Tech Tip

Dairy breeds (Holstein, Jersey, Brown Swiss, Ayrshire, and Guernsey) are relatively tolerant of the effects of anesthesia. Beef breeds such as the Beefmaster, Santa Gertrudis, and Brahman often require a much lower dose of anesthetic agents.[2]

Induction Medications

Guaifenesin (glyceryl guaiacolate or GG) (Guailazin®, Gecolate®, and compounded) is a centrally acting IV skeletal muscle relaxant and sedative used commonly in large animal species immediately before anesthetic induction to cause central nervous system depression and to smooth the anesthetic induction. This agent produces little respiratory and cardiovascular depressive effects while producing optimal muscle relaxation. Guaifenesin, when given with an anesthetic induction agent, often produces a very smooth induction and recovery. It is available in various concentrations; however, a 5% solution is always used in ruminant species because more concentrated

solutions may cause hemolysis of red blood cells. Guaifenesin is becoming increasingly difficult to source by manufacturers; therefore, most practices are having it compounded. When using guaifenesin, a pressure bag is commonly placed over the fluid bag to infuse a large volume at a very fast rate. This agent is given until signs of muscle relaxation are evident, which include head ptosis, droopy lips, partially shut eyelids, swaying, and buckling at the knees. Once the patient is unaware and disinterested in surroundings, the induction agent (e.g., ketamine) should be administered as a bolus to effect. It is also possible to mix an induction agent (e.g., ketamine or if available thiopental) with the "GG" (this is often referred to as a "double-drip" induction) and then administer this mixture to well sedated cattle to effect until the animal becomes recumbent and is able to be intubated. Benzodiazepines can also be substituted for GG if it is not available. Ketamine is the most commonly used induction agent currently for adult cattle. Thiopental when available was also frequently used. Other induction agents commonly used in small animals (e.g., propofol [Rapinovet®, PropoFlo™, Propovan®, PropoFlo™28], alfaxalone [Alfaxan®, Alfaxan®Multidose]) are generally too cost prohibitive to use in such large patients; however, they can be selected in calves. In general, cardiovascular and respiratory effects of induction agents are very similar to those seen in dogs and cats. An example of a common general anesthesia plan for a bovine might be sedation with xylazine +/− administration of GG to further sedation, and then ketamine or ketamine/benzodiazepine for induction.

Intubation of Cattle

Endotracheal intubation with a cuffed ETT provides airway access for oxygen and inhalant anesthetic agent delivery, allows for assisted ventilation, and prevents aspiration of saliva and ruminal contents if regurgitation occurs. Cattle should always be endotracheally intubated if they undergo general anesthesia as they are always at high risk for catastrophic regurgitation and aspiration. There are common ways to perform intubation in cattle including:[10]

- Direct palpation of the larynx with the patient positioned so the neck is extended. A guide tube may be utilized.
- Blind intubation through the pharynx (Figure 19-2).
- Introduction of the ETT using a laryngoscope (performed only in calves) (Figure 19-3).

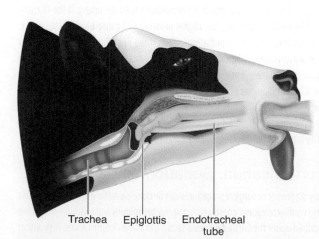

Trachea   Epiglottis   Endotracheal tube

Source: adapted from Riebold T. W., Geiser D. R. and Goble D. O. " Chapter 5 Clinical Techniques for Food Animal Anesthesia" in Large Animal Anesthesia: Principles and Techniques 2nd Edition. Iowa State University Press, Ames, IA. 1995 pg 148

FIGURE 19-2 Blind intubation through the pharynx method of ETT placement. Blind passage of an ETT can be done by extending the patient's head, passing the ETT through the mouth to the pharynx, elevating the larynx by external manipulation, and guiding the ETT into the larynx while holding an arm over the ETT with the hand cupped over the tube's end.

TABLE 19-3 Endotracheal Tube Sizes for Ruminants (mm ID)*

Body Weight (kg)	Oral	Nasal
< 30	4–7	4–6
30–60	8–10	6–8
60–100	10–12	8–10
100–200	12–14	10–12
200–300	14–16	
300–400	16–22	
400–600	22–26	
> 600	26	

ID = internal diameter

*Reiblod, T. W. (2015). Ruminants. In K. A. Grimm, L. A. Lamont, W. J. Tranquilli, S. A. Greene, & S. A. Robertson (Eds.), *Veterinary anesthesia and analgesia: The 5th edition of Lumb and Jones* (pp. 912–927). Ames, IA, John Wiley & Sons.

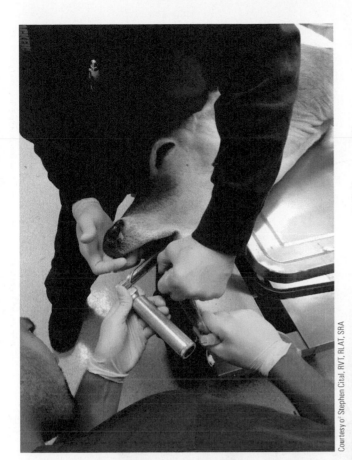

Courtesy of Stephen Cital, RVT, RLAT, SRA

FIGURE 19-3 Introduction of the ETT using a laryngoscope. Passage of an ETT can be done by visually observing the larynx with a laryngoscope and advancing the tube toward the trachea. A 350 mm Miller laryngoscope blade and a size 11 ETT are used for this calf.

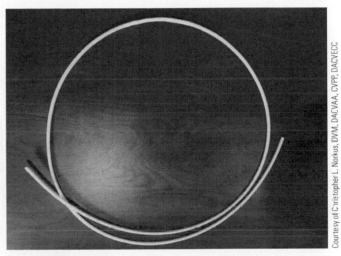

Courtesy of Christopher L. Norkus, DVM, DACVAA, CVPP, DACVECC

FIGURE 19-4 Example of a guide tube. Guide tubes are much longer and more flexible than stylets and are recommended for use when intubating large animals.

To ensure a smooth and safe intubation for the patient, make sure all equipment for endotracheal intubation is available and in working order prior to starting a case. Equipment commonly utilized for endotracheal intubation in cattle include:

- Three sizes of ETT (Table 19-3). Choose a size that is considered the correct size; add two additional ETTs, one size larger and one size smaller so that three different size ETTs are within reach during the sometimes hectic intubation period.
  - Check the ETT for patency and that it is free of dirt, blood, or debris. Check whether the cuff can be inflated and will hold pressure before use. The ETT should not have holes, damage, or deterioration. The ETT may be lubricated with a sterile water based lubricant.
- A guide tube (soft flexible piece of plastic tubing that is at least two to three times the length of the selected ETT) or stylet is recommended (Figure 19-4).
- Mouth gag to hold jaws apart (only use if absolutely necessary)
- Laryngoscope with a long laryngoscope blade (250–350 mm) will be needed in a calf
- Gauze sponges to grab the tongue
- 60 mL syringe to inflate the cuff

Endotracheal Intubation of Cattle Direct palpation of the larynx is the most commonly used method of endotracheal intubation of cattle and is described below (Figure 19-5a):[9]

- Ruminants are intubated in sternal recumbency to keep pressure off the rumen and prevent regurgitation immediately following induction.
- Place the mouth speculum or gag (only use if absolutely necessary) (Figure 19-5b).
- Grasp the tongue and pull it forward.
- Hold the ETT at the patient end, covering the cuff with the nondominant hand. Extend the nondominant arm into the mouth and advance to the larynx. Use the dominant hand to assist with advancement of the machine end of the ETT.
- While still holding the ETT, palpate the epiglottis and laryngeal opening with one or two fingers of the nondominant hand.
- Pass the ETT into the larynx, using the nondominant hand to guide the end of the ETT into the airway, and the dominant hand to advance

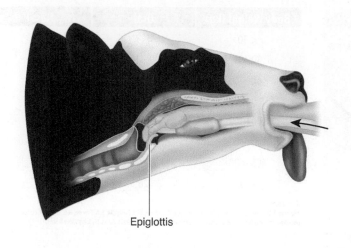

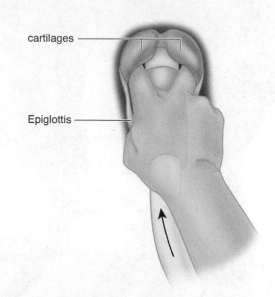

cartilages

Epiglottis

Epiglottis

(a)

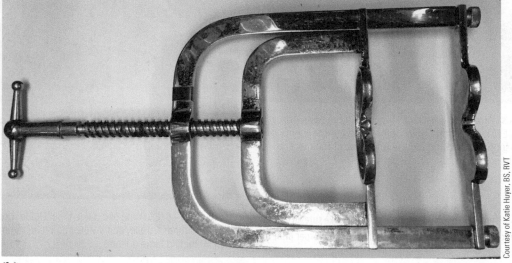

(b)

FIGURE 19-5 Direct palpation of larynx method of ETT placement. (a) Endotracheal intubation of adult cattle may be done by direct palpation of the laryngeal entrance. With a mouth speculum or gag firmly in place only if needed, the tongue is grasped and pulled forward. The ETT cuff is covered with the nondominant hand and the tube is advanced into the larynx. The epiglottis and laryngeal opening are palpated, and the ETT is passed into the larynx using the nondominant hand to guide the end of the tube into the airway and the dominant hand to advance the tube. (b) An example of a metal mouth gag. Mouth gags are typically not needed for intubation; however, using one for a minute or two if absolutely necessary will not typically affect maxillary blood flow in large animals.

the ETT. A guide tube can be utilized to aid in this process by digitally placing the guide tube into the trachea and then feeding an ETT over the guide tube.

- Check to ensure that the ETT is in the trachea by feeling air pass during exhalation or when an assistant presses down on the patient's chest. An end tidal carbon dioxide ($EtCO_2$) reading is the gold standard to confirm placement.
- Place air in the cuff (20–60 mL) and secure the ETT to the halter or muzzle before the patient is hoisted or positioned.

In calves the space is too tight to reach the arm in to the pharynx; therefore, it is recommended to use a laryngoscope and direct visualization of the larynx to swiftly secure the airway. Utilization of a guide tube can also be helpful as it is often easier to pass this into the trachea initially than an ETT itself. Nasotracheal intubation is an advanced skill technique that can be used in calves; however, it is not commonly used and beyond the scope of introductory information for veterinary technician students.

Maintenance of Anesthesia

Following anesthetic induction and endotracheal intubation, cattle are often in a light plane of anesthesia and likely need to be deepened to reach a desired anesthetic depth appropriate for surgical procedures. The appropriate level of anesthesia can be maintained in cattle with inhalant anesthetic agents such as isoflurane (IsoFlo®, Isosol®, Aerrane ®) (MAC value 1.3%), sevoflurane (Ultane®, SevoFlo®, Flurovess®, Petrem®), or desflurane (Suprane®) (MAC values not currently determined for sevoflurane and desflurane)[10] using a circle rebreathing system and large animal anesthesia machine in patients > 150 kg (small animal anesthesia machines can be used in patients < 150 kg because the size ETT placed in these patients can be connect to the anesthesia machine). Anesthesia can also be maintained in cattle using total intravenous anesthesia (TIVA), which is generally reserved for short procedures (≤ 20 minutes) on healthy patients. Unlike in small animals, TIVA in cattle is not routinely performed with propofol or alfaxalone due to the large volume of drug needed and expected cost. An example of a cattle TIVA technique is a "Triple Drip"[9] in which 1 g of ketamine and 100 mg xylazine are added to a 1 liter bag of 5% GG, making it a triple-drip of drugs (guaifenesin/ketamine/xylazine or GKX). The final concentration of each drug in the "Triple Drip" is 1 mg/mL ketamine, 0.1 mg/mL xylazine, and 50 mg/mL GG. A loading dosage can be given as an IV drip "to effect" for intubation and then continued as a slow drip until the inhalant anesthetic agent has fully taken effect (usually 5–10 minutes) or the GKX solution can be started after induction with other agents. Patients should have an ETT placed regardless of how anesthesia is maintained. TIVA may be selected more commonly when anesthesia is preformed out of hospital settings where access to a large animal anesthesia machine may be limited. General anesthesia with an inhalant anesthetic agent is often performed for major procedures as it facilitates oxygen delivery, allows for assisted ventilation, and can provide deeper muscle relaxation then can be achieved in most TIVA regimens.

Tech Tip

Remember that cattle are more sensitive to xylazine than horses; therefore, the dosage of xylazine described for cattle is much lower than what is used for "Triple Drip" in equine patients.

Patient Positioning

Once cattle are made recumbent, risk for regurgitation and concern over aspiration is high. Until the airway is protected with an ETT, cattle should be kept sternal. It is then recommended to change the patient's positioning to left lateral recumbency if possible (or keeping them in sternal recumbency). These positions alter intra-abdominal/intragastric pressure to minimize the risk of passive regurgitation as abdominal or gastric pressures overcome lower esophageal sphincter pressure. Once cattle are endotracheally intubated, many anesthetists breathe a little sigh of relief. Some anesthetist's then pass a "stomach tube" down the bovine's esophagus to allow excess gas and rumen contents to drain into a bucket. Other anesthetists skip this step as they feel it is likely unnecessary. It is, however, highly advisable to have suction equipment available in case of regurgitation. Suction equipment requires either an in-line suction system or electricity. Likewise, padding or a pillow may be used to slope the patient's head downward to allow saliva and any regurgitated ruminal content to drain out of the mouth; however, other anesthetists feel this step is not needed as long as the airway is protected.

Adult cattle will benefit from either water beds, air (dunnage) bags, or 10 cm high-density foam pads to prevent musculoskeletal injury.[4] In dorsal recumbency, the patient should be balanced equally on their back with the distribution of weight equal on the gluteal areas. The legs should be flexed and relaxed.[4] The dependent foreleg is drawn forward so the weight of the chest rests on the triceps rather than on the humerus to minimize pressure on the radial nerve.[4] The underlying principle is that all bony protuberances must be padded (Figures 19-6a and 19-6b).

When ruminants are placed in lateral recumbency, care should be taken to ensure the lids of the down eye are closed and the eye is lubricated and protected. A towel or thin pad can be placed under the down eye to further protect it when patients are anesthetized on the floor or ground. Ophthalmic ointment should be placed in the eyes to protect them during anesthesia. If the down eye has been bathed in saliva or regurgitation, it should be rinsed out during recovery and evaluated by a veterinarian.[11]

Monitoring

Once the patient is intubated and positioned it will typically be connected to a large animal anesthesia machine (assuming an inhalant anesthetic agent is being used) and is brought to a surgical plane of anesthesia (Figures 19-7a and 19-7b). Monitoring is done just as in small animal cases and is directed predominantly to the cardiovascular, respiratory, and level of central nervous systems depression (anesthetic depth). Recording values on an anesthesia record provides a legal document for the veterinarian (and an additional record of DEA-controlled substances) and allows for recognition of trends in monitored parameters.

Depth of anesthesia is monitored by eye position in the orbit (Figures 19-8a through 19-8e). In cattle, the eye begins to rotate ventrally during induction. The eye continues rotating ventrally so that in a light plane of anesthesia the edge of the iris and pupil are below the lower lid.[1,4] When the patient is in a surgical plane of anesthesia (Stage III, plane 2), the pupil becomes hidden by the lower eyelid. As the patient's depth of anesthesia increases, the eye rotates dorsally and returns to the central position as deep surgical anesthesia and profound muscle relaxation is achieved. At this point there is concern the patient is in too deep an anesthetic plane and should be lightened. As the depth of anesthesia is decreased, eye movements occur in a reverse order.[4]

Monitoring direct arterial blood pressure with an arterial catheter (e.g., in the auricular artery) remains the gold standard in large animals and is the best defense to poor tissue perfusion during anesthesia. A Doppler blood pressure probe may be used on the medial aspect of the carpus or the ventral midline of the tail artery; however, this and other oscillometric techniques do not provide as accurate readings in large animals as direct blood pressure.[12] Oxygen flow rates should be kept at approximately 5–10 L/min for adult cattle with flow rates in the lower part of this range selected once a patient has reached an appropriate anesthetic depth.

Conditions that compromise oxygen delivery status (e.g., hypovolemia, anemia, excessive anesthetic depth) require corrective measures perioperatively. IV fluid administration can be used to reduce volume deficits. Patients should receive 3 to 10 mL/kg/hr of a balanced crystalloid solution during routine general anesthesia. Higher fluid delivery rates are used to restore blood volume in hypovolemic patients. Depending on the case, packed cell volume (PCV) and total protein (TP) can be periodically evaluated to guide fluid delivery in hypovolemic patients. Blood transfusions can be

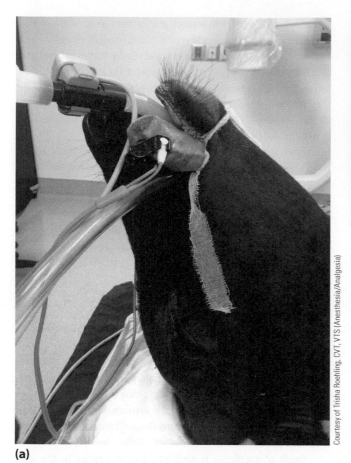

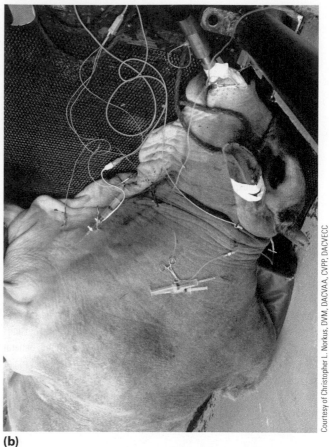

(a)

(b)

FIGURE 19-6 Padding is important when positioning cattle. (a) Positioning of a bovine's head while in dorsal recumbency. (b) A Brahman bull on blue padding to minimize pressure on protuberances.

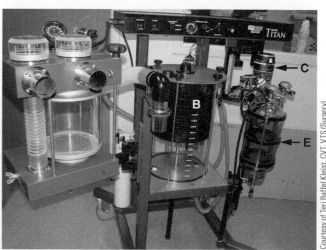

(a)

(b)

FIGURE 19-7 Large animal anesthesia machines. Circle rebreathing systems used for large animals have breathing hoses with an internal diameter of 50 mm and are recommended for patients weighing greater than 135 kg. Special adapters are available to connect endotracheal tubes to the large diameter y-piece. (a) Large animal anesthesia machine with breathing hoses. A = reservoir bag, B = ventilator bellows and housing, C = vaporizer, D = large animal breathing hoses, E = CO_2 absorber canister. (b) Large animal anesthesia machine without breathing hoses and reservoir bag. B = ventilator bellows and housing, C = vaporizer, E = CO_2 absorber canister.

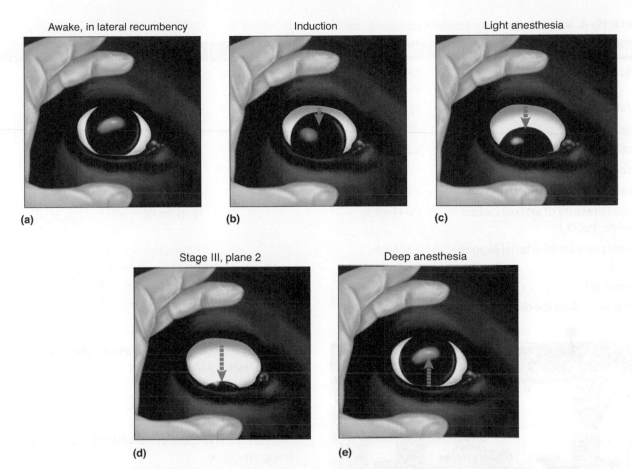

Awake, in lateral recumbency **(a)** Induction **(b)** Light anesthesia **(c)**

Stage III, plane 2 **(d)** Deep anesthesia **(e)**

FIGURE 19–8 Ocular rotation in cattle under anesthesia. (a) An unanesthetized bovine in lateral recumbency has a centrally located eye. (b) During induction, the eye begins to rotate ventrally. (c) The eye continues to rotate ventrally while the bovine is in a light plane of anesthesia. (d) The cornea is hidden by the lower eyelid as the bovine's depth of anesthesia increases (Stage III, plane 2). (e) The eye starts to rotate dorsally to the central position as the bovine is in deep surgical anesthesia. As the depth of anesthesia is decreased, eye movements occur in a reverse order.

used to treat moderate to severe anemia.[11] Dependent on status, ruminants younger than 2 months of age can have their blood glucose levels monitored utilizing **point-of-care instrumentation** (medical diagnostic instruments that can be used at the time and place of patient care) and supplemented if needed with dextrose during anesthesia and recovery. A less expensive alternative is to add 2.5% dextrose to the IV fluid bag for calves less than 2 months of age requiring a lengthy anesthetic procedure (greater than 30 minutes). Normal values for anesthetized calves and adult cattle are summarized in Table 19-4.

Cattle are usually ventilated because on their own they will often take rapid shallow breaths and can develop poor alveolar ventilation and hypercapnia. When mechanically ventilated, cattle often have a peak inspiratory pressure as high as 20 to 40 cm H_2O. Because domestic ruminants have a respiratory pattern characterized by rapid respiratory rate and small tidal volume (with more dead-space ventilation), higher vaporizer settings may be required to better maintain anesthesia in spontaneously breathing patients; however, as previously said, cattle will often benefit from mechanical ventilation.[4] Based on the patient's end-tidal carbon dioxide ($EtCO_2$) and breathing pattern, consideration can be given to instituting and adjusting mechanical ventilation (Figure 19-9). Common ventilator settings are a respiratory rate of 6 to 10 breaths/minute with a tidal volume of 10 to 15 mL/kg. Final ventilator settings are dependent on results of blood gas analysis and $EtCO_2$.

Recovery

Ruminant patients must be placed in sternal recumbency for recovery as soon as they are in the recovery area. Placing patients in a sternal position assists venting of fermentation gas trapped in the rumen during recumbency and anesthesia, keeps pressure off the rumen, and allows for the patient to ventilate well with both left and right lungs.[11] Sternal recumbency also allows saliva and any regurgitated content to drain from the oral cavity. Patients may need to be propped up to maintain sternal recumbency early in the recovery period (Figure 19-10a). Hay or straw bales and/or propping a patient against walls or fences can be used as positioning aids. Patients should also be placed on sufficient floor bedding to reduce scraping the carpus or tarsus during recumbency and provide proper footing when patients attempt to rise (Figure 19-10b).

Atipamezole (Antisedan®) and yohimbine (Yobine®, Antagonil®) are (α_2) adrenergic antagonist agents that can be given to cattle to reverse (α_2) adrenergic agonists; however, unlike in small animals, they are not frequently administered and typically only given in emergency situations. Atipamezole given IV reverses sedation in cattle with restoration of ruminal motility to normal.[13]

During recovery ruminants need to be monitored for bloating. If bloating occurs, a stomach tube should be passed to decompress the rumen. The ETT cuff should be left inflated during extubation to prevent any material

TABLE 19-4 Normal Values for Anesthetized Cattle, Sheep, and Goats

Parameter	Value for Cattle[1, 13]	Value for Sheep and Goats[26, 28]
Respiratory rate	20–30 breaths/min in adult cattle 20–40 breaths/min in calves	20–40 breaths/min
Tidal volume	10–15 mL/kg	10–15 mL/kg
Heart rate	60–90 bpm	80–150 bpm
Capillary refill time	< 2 seconds	< 2 seconds
Arterial pressure (systolic)	> 100 mm Hg and < 160 mm Hg	> 100 mm Hg and <160 mm Hg
Mean arterial pressure (MAP)	> 60 mm Hg	>60 mm Hg
Partial pressure of arterial carbon dioxide ($PaCO_2$)	35–45 mm Hg	35–45 mm Hg
Partial pressure of arterial oxygen (PaO_2)	> 80 mm Hg	> 80 mm Hg
Arterial pH	7.35–7.45	7.35–7.45
End tidal carbon dioxide ($EtCO_2$)	35–45 mm Hg	35–45 mm Hg

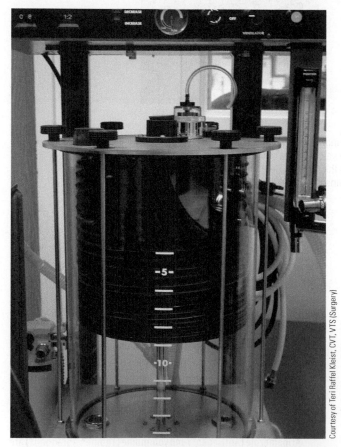

Courtesy of Teri Raffel Kleist, CVT, VTS (Surgery)

FIGURE 19-9 An example of a large animal mechanical ventilator.

that was lodged in the pharynx during anesthesia from being aspirated (some use it as a "squeegee" for any material in the pharynx). Oxygen should be supplemented when available. Extubation should not take place until the patient is able to support its head. After extubation, check for air flow from the nostrils or mouth of patient. Most cattle will lie quietly in a sternal position during recovery and unlike horses will have a quiet, smooth recovery. It is not necessary to withhold food or water after anesthesia unless otherwise indicated by health concerns of the patient.

Tech Tip

Cattle are the only species in which the ETT is left inflated during extubation.

Cattle Analgesia

Cattle are stoic animals by nature because, as a species, they have been subject to a strong evolutionary pressure to mask pain as it may be perceived as weakness by predators. As a result, they frequently do not demonstrate appreciable signs of pain until it is severe. Unwillingness to move may be a key indicator of discomfort, particularly in adult cattle. Aside from pain induced by surgery, numerous medical conditions can commonly lead to pain in cattle and should be treated.

Cattle that have chronic pain can exhibit all the signs of acute pain plus the following:[14, 15]

- Altered movement, posture, appearance, or behavior
  - Decreased movement/locomotion
  - Changes relevant to the source of the pain being experienced (e.g., altered locomotion, flank watching or kicking, or ear twitching)
  - Changes in normal postures associated with pain (e.g., lateral recumbency, standing motionless, or drooping of the ears)
  - Bruxism (teeth grinding)
  - Poor coat condition (e.g., rough, dusty, or unkempt) caused by decreased grooming
  - Standing with the abdomen "tucked"
  - Weight gain or loss
  - Decreased interaction with other animals in the group (avoidance)
  - Level of mental activity/responsiveness (animals in severe pain often show reduced responsiveness to stimuli)
  - Vocalizing

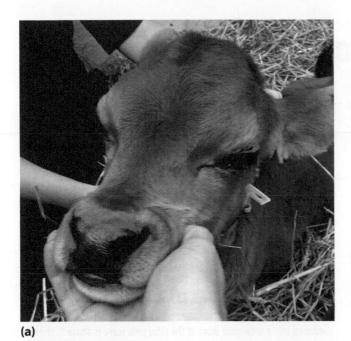

(a)

(b)

Courtesy of Christopher L. Norkus, DVM, DACVAA, CVPP, DACVECC

FIGURE 19-10 Anesthetic recovery of bovines. (a) This calf is propped up in sternal recumbency with hay. Sternal recumbency helps vent fermented gas trapped in the rumen and allows saliva and any regurgitated content to drain from the oral cavity. This calf's head is elevated to help saliva drain from the oral cavity. (b) This cow was recovered in a stall with sufficient floor bedding that provided proper footing when she attempted to rise.

- Changes in eating habits
  - Decreased feed intake (e.g., "hollow" left flank caused by a near-empty rumen)
- Variation of vital signs
  - Easily measurable indicators of physiological stress (e.g., increased heart rate, increased pupil size, altered rate and depth of respiration, or trembling)

Mastitis is painful in cows as evidenced by animals with moderate clinical mastitis having significantly higher heart rates, rectal temperatures, respiratory rates, and cortisol levels compared to normal cows and those with mild clinical mastitis.[16, 17] Cows with both mild and moderate cases of mastitis have significantly altered stances with larger hock-to-hock distances compared to normal cows.[18] Affected animals also exhibit an increased sensitivity to a mechanical pressure stimulus on the leg closest to the affected mammary quarter, suggesting a change in pain information processing as a result of inflammation.[16]

Lameness in cattle can originate from countless musculoskeletal conditions and is often associated with tissue damage, pain, and discomfort and manifests as an inability to walk normally. Usually, the affected animal attempts to reduce weight borne by a particular limb. Patients may exhibit unwillingness to put weight on a given limb or may shift their weight from one leg to another. In economic terms, lameness is currently ranked as the third most important disease affecting the United Kingdom dairy industry, after reduced fertility and mastitis.[19] Chronic pain associated with lameness is considered one of the most significant welfare concerns in dairy cows in the United States.[15] The most crucial step toward improving welfare and reducing the pain experienced by lame cattle is through prompt, effective treatment along with investigating husbandry such as nutrition and flooring.[19]

Table 19-5 summarizes painful conditions in cattle. Table 19-6 summarizes parameters used to identify and assess pain in ruminants.[20]

TABLE 19-5 Painful Conditions in Cattle*

Common painful conditions in cattle	Procedures that cause pain in cattle
• Foot problems	• Abomasopexy – Moderate Pain
• Claw problems	• Castration – Moderate Pain
• Joint problems	• Claw removal – Moderate to Severe Pain
• GI pain	• Dehorning – Moderate to Severe Pain
• Colic	• Teat surgery – Moderate to Severe Pain
• Mastitis	• Perineal surgery or injuries (obstetrical, rectal, vaginal, bladder) (Figure 19-11) – Moderate to
• Tissue damage and inflammation	Severe Pain
• Uveitis	• GI or colic surgery – Severe Pain
• Fracture of the tuber coxae	
• Acute metritis	

*Coetzee, J. F. (2013). A review of analgesic compounds used in food animals in the United States. *Veterinary Clinics of North America: Food Animal Clinics, 29*(1), 11–28.

TABLE 19-6 Parameters Used to Identify and Assess Pain in Ruminants*

Physiological	Biochemical	Behavior
• Heart rate variability • Respiratory rate and depth • Rectal temperature • Blood pressure • Eye temperature infrared thermography • EEG activity	• Cortisol concentrations • ACTH response • Adrenaline and noradrenaline concentrations • Endorphin levels	• Avoidance behavior • Pain threshold • Weight gain or loss • Wound healing

*Stafford, K. J. (2014). Recognition and assessment of pain in ruminants. In C. Egger, L. Love, & T. Doherty (Eds.), Pain management in veterinary practice (pp. 349–356). Ames, IA: Wiley/Blackwell.

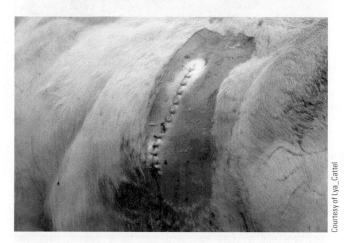

Courtesy of Lya_Cattel

FIGURE 19-11 Postsurgical pain in a cow due to a Cesarean section.

Tech Tip

The best type of restraint when administering injections to cattle is a chute or restraint pen.

Pain Scoring Systems in Cattle

Pain scales used specifically to assess pain in cattle are the UNESP-Botucatu Unidimensional Pain Scale for Acute Postoperative Pain Assessment in Cattle and the Cow Pain Scale. The UNESP-Botucatu Unidimensional Pain Scale is a valid and reliable tool used to assess acute postoperative pain in cattle. It assesses locomotion, interactive behavior, activity, appetite, and miscellaneous behaviors (Table 19-7). A series of videos illustrating the various behaviors and how they are scored can be found at https://animal-pain.com/dor-em-bovinos-2/. The scores within each category range from 0 to 2 and total scores ranging from 0 (no pain) to 10 points (maximum pain). Total scores greater than 4 correspond to cattle in pain and is the cut-off point for rescue analgesia.

The Cow Pain Scale is a rapid, easy to use tool that focuses on seven less obvious pain behaviors of cows. The seven behaviors are attention toward the surroundings (if a cow is in pain, she tends to be less focused on the environment), head position (pain often results in lower head carriage), ear position (cows in pain keep their ears straight backward or very low like lamb's ears), facial expression (the cow has a changed facial expression when in pain, a so-called pain face), response to approach (a cow in pain is less interested in social interaction and will therefore try to avoid an approaching person), back position (pain in legs or abdomen may result in an arched

back), and lameness (lameness is a result of pain in one or several limbs) (Figure 19-12).[21] The first four behaviors (attention toward the surroundings, head position, ear position, and facial expression) are evaluated from a distance so that the cow is not aware they are being observed.[21] Then the cow is approached, and the response to approach is evaluated.[21] When the cow is standing up, the back position may be evaluated, and finally the lameness is evaluated.[21] The cow pain scale scores each behavior from 0 to 2 and is then combined into a total pain score. If the total pain score is above 5, the cow may be in pain and administration of analgesic drugs should be considered.

Locomotion scoring scales for pain assessment are also available for cattle. One locomotion scoring is based on the observation of cattle walking (gait), with emphasis on head bob and stride length. The system uses a simple 0 to 3 scale to

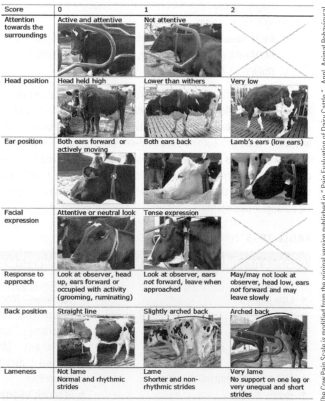

The Cow Pain Scale is modified from the original version published in "Pain Evaluation of Dairy Cattle", Appl. Animal Behavioural Science, 2015, Open Access Pain evaluation in dairy cattle

FIGURE 19-12 The Cow Pain Scale is the sum of the scores of six behaviors including attention toward the surroundings, head position, ears position, facial expressions, response to approach, and back position. The seventh assessment is based on lameness.

TABLE 19-7 UNESP-Botucatu Unidimensional Pain Scale for Acute Postoperative Pain Assessment in Cattle*

Item	Score/Criterion
Locomotion	• (0) Walking with no obviously abnormal gait. • (1) Walking with restriction, may be with hunched back and/or short steps. • (2) Reluctant to stand up, standing up with difficulty or not walking.
Interactive behavior	• (0) Active; attention to tactile and/or visual and/or audible environmental stimuli; when near other animals, can interact with and/or accompany the group. • (1) Apathetic: may remain close to other animals, but interacts little when stimulated. • (2) Apathetic: may be isolated or may not accompany the other animals; does not react to tactile, visual, and/or audible environmental stimuli.
Activity	• (0) Moves normally. • (1) Restless, moves more than normal or lies down and stands up with frequency. • (2) Moves less frequently in the pasture or only when stimulated.
Appetite	• (0) Normorexia and/or rumination. • (1) Hyporexia. • (2) Anorexia.
Miscellaneous behaviors	• Wagging the tail abruptly and repeatedly. • Licking the surgical wound. • Moves and arches the back when in standing posture. • Kicking/foot stamping. • Hind limbs extended caudally when in standing posture. • Head below the line of spinal column. • Lying down in ventral recumbency with full or partial extension of one or both hind limbs. • Lying down with the head on/close to the ground. • Extends the neck and body forward when lying in ventral recumbency. (0) All of the above described behaviors are absent. (1) Presence of 1 of the behaviors described above. (2) Presence of 2 or more of the behaviors described above.

*de Oliveira, F. A., Luna, S. P. L., do Amaral, J. B. et al. (2014). Validation of the UNESP-Botucatu unidimensional composite pain scale for assessing postoperative pain in cattle. *BMC Vet Res, 10*(200). https://doi.org/10.1186/s12917-014-0200-0

assess the severity of lameness in beef cattle (0 = normal; 1 = mild lameness; 2 = moderate lameness; 3 = severe lameness). This pain assessment can be accessed at http://www.angusbeefbulletin.com/extra/2014/04apr14/0414mg_locomotion.html#.YGN-87CSnid. Locomotion scoring scales are also available for dairy cattle that assess lameness while the animal is standing and walking. The system uses a 1 to 5 scale to assess the severity of lameness in dairy cattle (1 = normal; 2 = mild lameness; 3 = moderate lameness; 4 = lame; 5 = severe lameness). This pain assessment can be accessed at http://cdrf.org/wp-content/uploads/2012/06/11_1_Zinpro_locomotion_scoring.pdf.

Treating Cattle Pain

When treating cattle for pain, keep in mind that they may be raised for meat or for their milk; therefore, knowledge of drug withdrawal times is important if cattle or their products are being used for human consumption (see Regulatory Concerns for Anesthetic and Analgesic Drugs in Food Producing Animals in this chapter). Few drugs are currently FDA-approved for analgesic use in cattle; however, flunixin meglumine (Banamine®) is one NSAID approved for use in lactating dairy cattle for controlling pyrexia associated with acute mastitis, pneumonia, and endotoxemia, as well as controlling the inflammation in endotoxemia such as that caused by coliform mastitis. Banamine® has a milk withdrawal during treatment and for 36 hours after the last treatment. Other analgesic drugs more commonly used extra label include local anesthetics and NSAIDs and, less frequently, opioids, (α_2) adrenergic agonists, and ketamine. Although a multimodal analgesic technique is always desirable, it is significantly underused in ruminant practice and the duration of treatment with analgesic drugs is often shorter than would be ideal (e.g., one administration at the time of surgery and then nothing given after surgery).

NSAIDs are commonly used in ruminants for their anti-inflammatory, anti-pyretic, and analgesic effects. Gastrointestinal ulceration is a concern in all cattle treated with NSAIDs especially phenylbutazone. The NSAIDs most commonly used in cattle are flunixin meglumine, phenylbutazone, ketoprofen (Ketofen®), salicylic acid derivatives, carprofen (Rimadyl®), meloxicam (Metacam®).[8]

Opioid usage in ruminants is infrequent, yet this class of drug has experimentally been proven to provide analgesia for cutaneous thermal and mechanical pain. Most opioids are administered epidurally, intraarticularly, systemically, or intrathecally in cattle. There are currently no opioid analgesics approved for use in cattle in the United States.[15] The most common opioid used in cattle is butorphanol, but morphine, methadone (Dolophine®), buprenorphine (Buprenex®), and fentanyl (Sublimaze®) are also used.[22] Cattle generally become lightly sedated with opioids but may exhibit behavioral changes, such as restlessness and vocalization with higher or repeated dose. With prolonged use, opioids can cause GI stasis. Opioids should therefore be used at the lowest effective dosage and for the shortest duration possible to avoid adverse effects. If adverse effects occur, dosage or administration frequency can be titrated to balance meeting the patients' needs with minimizing adverse effects.

Commonly used (α_2) adrenergic agonists in ruminants include xylazine (Rompum®, AnaSed®, Sedazine®, and Chanazine®), detomidine (Dormosedan®), medetomidine (Domitor®), and romifidine (Sedivet®). Epidural administration of (α_2) adrenergic agonists at the sacrococcygeal or first intercoccygeal space in cattle produces analgesia that is more profound in the hind limbs, perineum, and abdomen. Systemic absorption may also result in generalized analgesia and sedation. A combination of romifidine and morphine administered at the first coccygeal space provides analgesia in cattle for up to 12 hours against a noxious electrical stimulus applied to the flank.[23] Paresis (weakness) is common with epidural xylazine and therefore should be avoided. This latter effect is not documented for other (α_2) adrenergic agonists. In general, cattle are more sensitive than horses to the sedative effects of (α_2) adrenergic agonists.

The dissociative drug ketamine (Ketaset®, Ketaject®, VetaKet®, Vetalar®) has been administered in sub-anesthetic dosages to cattle via a CRI in conjunction with other classes of analgesics such as opioids, (α_2) adrenergic agonists, and NSAIDs.[24] However limited published data is available on the efficacy or safety of this practice.

Locoregional anesthetic techniques are commonly used in large animals to avoid the risks associated with general anesthesia, reduce cost, and minimize the issues surround drug withdrawal times in food animals. Local anesthetic techniques usually are relatively straightforward to perform and have relatively few adverse effects. Lidocaine (Xylocaine®), bupivacaine (Marcaine®), and mepivacaine (Carbocaine®) are the most commonly used local anesthetics in cattle. Numerous locoregional techniques have been described in the literature including techniques to desensitize regions for abdominal surgery (e.g., paravertebral block, inverted "L" block), urogenital surgery (e.g., epidural), and head and limb surgery (e.g., retrobulbar block). Nerve blocks commonly used in cattle are summarized in Table 19-8.

Small Ruminant Anesthesia (Sheep and Goats)

Sheep and goats compared to cattle are less frequently used as food animals and often present as a family pet or in the laboratory research setting. Many of the concerns and techniques for small ruminants are the same as those for cattle; therefore, the reader should refer to the Cattle Anesthesia

section for further reading. Just as in large ruminants, the practitioner can often complete many diagnostic or surgical procedures using basic restraint principles along with locoregional techniques, analgesia, and sedation. Local or regional anesthetic or analgesic techniques are summarized in the Sheep and Goat Analgesia section. However, general anesthesia is performed for procedures that require full immobilization because of the nature of the procedure, anticipated pain or procedure complexity, requirement for deeper muscle relaxation, or need for assisted ventilation.

Preanesthetic Preparation

Preanesthetic preparation begins with an appropriate physical exam and history (see Preanesthetic Preparation under Cattle Anesthesia). Bloodwork and diagnostic testing are selected based upon the individual needs of the patient and may be limited or extensive. Prior to anesthesia, sheep and goats should be fasted for 12 to 18 hours and deprived of water for 8 to 12 hours.[25]

Venous Catheterization

An IV catheter should be placed prior to induction in sheep and goats undergoing general anesthesia. The technique is similar to that previously described for cattle; however, a 16-gauge over-the-needle IV catheter can be placed in the jugular vein of most sheep or larger goats while 18 or 20 gauge over-the-needle IV catheters are used for smaller sheep and goats (Figure 19-13).

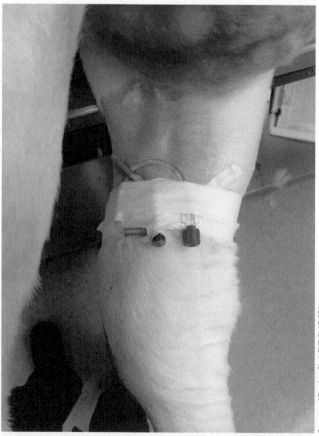

FIGURE 19–13 An IV catheter in a sheep. This cephalic IV catheter illustrates one way to stabilize a short-term catheter on an anesthetized sheep.

Courtesy of Stephen Cital, RVT, RLAT, SRA

TABLE 19–8 Nerve Blocks Commonly Used in Cattle*

Nerve Block	Area of Analgesia	Comments
Distal paravertebral	Flank on side anesthetic is given	• Spinal nerves T13, L1, and L2 are desensitized at the distal ends of L-1, L-2, and L-4. • Quick and simple way to provide anesthesia for standing flank surgery such as C-section, rumenotomy, cecotomy, correction of gastrointestinal displacement, intestinal obstruction, and volvulus.
Proximal paravertebral	Flank on side anesthetic is given	• Spinal nerves T-13, L1, and L2 are desensitized as they emerge from the intervertebral foramina. • Provides anesthesia for standing flank surgery such as C-section, rumenotomy, cecotomy, correction of gastrointestinal displacement, intestinal obstruction, and volvulus. • Technically more challenging than the distal paravertebral block.
Line	Skin and muscles of flank Parietal peritoneum	• Multiple SQ injections of anesthetic given; then anesthetic is infiltrated into the muscle layers and parietal peritoneum. • Easy technique to desensitize an area.
Inverted L	Flank (caudal and ventral to injection site)	• Injection of anesthetic into tissues bordering the dorsocaudal aspect of the last rib and ventrolateral aspect of the lumbar transverse processes. • Similar to line block but does not have anesthetic agent on the incision line.
Cornual	Horn and surrounding skin in calves	• Desensitizes the corneal nerve which is a branch of the trigeminal nerve (cranial nerve V). • Less effective in adults. • Used for dehorning.
Retrobulbar	Eye and adnexa	• Desensitizes cranial nerves II, III, IV, VI, and the ophthalmic and maxillary branches of V. • May result in damage to adnexal structures; usually reserved for enucleation.
Peterson	Eye and adnexa except eyelids	• Desensitizes the ophthalmic and maxillary branches of the trigeminal nerve (cranial nerve V). • Less destructive than retrobulbar block; need to anesthetize eyelids separately for enucleation.
Auriculopalpebral	Eyelids (motor function only)	• Desensitizes the auriculopalpebral branch of the facial nerve (cranial nerve VII). • Provides paralysis but not desensitization of the eyelids.
Bier (IV regional anesthesia)	Limb (depending on where tourniquet is placed)	• A restrictive bandage (such as an Esmarch) is put on the limb for exsanguination and then a tourniquet is placed around the limb and local anesthetic is injected into the vein. • After approximately 15 minutes the area distal to the tourniquet is anesthetized until the tourniquet is removed. • Commonly used for amputation of a digit.
Common peroneal and tibial	Hindlimb distal to tarsus	• Desensitizes two branches of the sciatic nerve (common peroneal and tibial nerves) • Good alternative to IV regional anesthesia, although technically more difficult.
Intratesticular	Testicle and spermatic cord	• Local anesthetic is injected into the neck of the scrotum, both spermatic cords, both testes, or combinations of these • Used for castration.
Caudal epidural	Hind limbs, perineum, abdomen	• Given at sacrococcygeal or first intercoccygeal space. • Used for obstetrical procedures (e.g., dystocia, uterine/vaginal/rectal prolapse) and tail surgery.

*Hudson, C. et al. (2008). Recognition and management of pain in cattle. *Practice, 30*, 126–134.

Premedication, Sedation, and Restraint

Sheep and goats are smaller and often easier to handle than adult cattle. Additionally, numerous drug options are available to provide sedation, tranquilization, and premedication in sheep and goats. Benzodiazepines such as midazolam and diazepam that can be combined with opioids such as butorphanol, buprenorphine, methadone, morphine, hydromorphone, or fentanyl are popular choices for many anesthetists. Depending on the dosage and drugs selected, these agents may make sheep and goats lie down; therefore, it is important to keep the patient in sternal recumbency with its head elevated and protected until induction and ETT placement occurs. Acepromazine can also be used as a sedative instead of a benzodiazepine. Alpha-2 adrenergic agonists are typically not recommended, especially in sheep, because while they do provide profound sedation, xylazine and potentially other (α_2) adrenergic agonists can cause acute lung injury resulting in pulmonary edema, hypoexemia, and death. Therefore, as other options are available and work well, (α_2) adrenergic agonists are often avoided. Anticholinergic drugs are not administered routinely to sheep and goats prior to induction of anesthesia, but may be used in these species to combat life-threatening bradyarrhythmias or during cardiopulmonary resuscitation.

Induction Procedures and Techniques

As in cattle, general anesthesia can be achieved in sheep and goats by administering a variety of medications with the overall aim of ensuring analgesia, muscle relaxation, and immobility.

Induction Medications

All of the induction agents used in small animals (ketamine, propofol, etomidate, and alfaxalone) have been used successfully in sheep and goats. Ketamine when administered may be combined with a benzodiazepine, especially if an induction agent was not given as part of the premedication. An example of a typical anesthesia plan for a small ruminant might be IV butorphanol and IV diazepam for premedication followed by IV propofol or ketamine or a propfol/ketamine combination with or without additional benzodiazepine for induction.

Intubation of Small Ruminants

Endotracheal intubation should be performed in small ruminants during general anesthesia. Similar to other species, intubation will prevent the aspiration of ruminal contents or saliva into the trachea and lungs, permit unobstructed administration of oxygen, plus allow for assisted or controlled ventilation if apnea occurs. Refer to the endotracheal intubation equipment list for cattle to ensure supplies are in reach for small ruminant intubation. ETTs of 5 to 14 mm internal diameter should be used in small ruminants. ETTs required for goats are 1 to 2 sizes smaller than those required for sheep of the same body weight.[26]

Endotracheal intubation of sheep and goats is similar to cattle, although the preferred technique is to visualize the glottis with a laryngoscope and is described below (Figure 19-14).

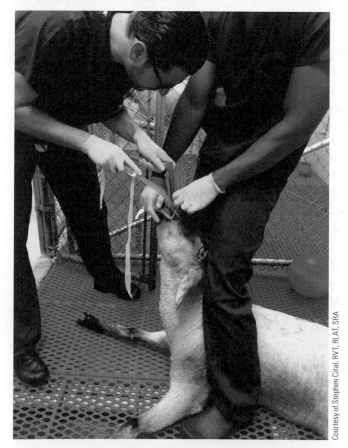

FIGURE 19-14 Intubation of a sheep. This sheep is in ventral recumbency with its head and neck fully extended. The mouth and tongue is held by the restrainer and body position is controlled by the legs and feet of the restrainer. The restrainer can feel the respirations of the sheep with their legs. A 350 mm Miller laryngoscope blade and a size 11 ETT is used for this sheep.

Courtesy of Stephen Cital, RVT, RLAT, SRA

- Maintain the animal in sternal recumbency and keep the patient's neck as straight as possible.
- Extend the head and neck to keep the airway protected and allow for visualization of the airway.
- Open the mouth as wide as possible using either bandage material (preferred) or a gag (only use if necessary).
- Pull the tongue forward or laterally to visualize the epiglottis using a laryngoscope with a long blade or normal laryngoscope with a separate tongue depressor.
- Spray lidocaine 2% solution onto the larynx and arytenoid cartilages to desensitize the arytenoids and ease endotracheal placement. The use of the benzocaine, butamben, and tetracaine hydrochloride combination product Cetacaine® is *not* recommended because overdosing can easily occur and benzocaine-based local anesthetics can cause methemoglobinemia.[27]
- Advance a guide tube with a protected tip (as to not injure the trachea) into the trachea during abduction of the arytenoid cartilages on inspiration. Once the guide tube is several inches within the trachea, an ETT is then fed over the guide tube and into the trachea and the guide tube is then removed. If a laryngoscope is not available, intubation can be attempted with the head and neck extended and the ETT is placed in the oropharynx. The larynx is located with the ETT, and it is then advanced on inspiration when the epiglottis is in an open position.
- Inflate the cuff.

Maintenance of Anesthesia

Conventional small animal anesthesia machines can be used to anesthetize ruminants weighing less than 150 kg. Larger ruminants need a modified small animal or large animal anesthetic machine. Inhalant anesthesia begins most commonly with isoflurane or sevoflurane and with oxygen flow rates of 25 to 50 mL/kg/min. Once an appropriate anesthetic depth has been achieved, anesthesia can then be maintained with isoflurane (MAC value 1.6% for sheep and 1.2% for goats) or sevoflurane (MAC value unknown for sheep and 2.3% for goats) on a circle rebreathing system with oxygen flow rates at 10 mL/kg/min.[4, 28] Isoflurane and sevoflurane have similar cardiovascular and pulmonary effects, which include vasodilation and dosage-dependent decreases in arterial blood pressures, respiratory rate, tidal volume, and minute ventilation.[28]

Patient Positioning

Sheep and goats may be maintained under anesthesia in dorsal, lateral, or sternal recumbency ensuring that there are no "kinks" or obstruction of the ETT (Figure 19-15). When sheep and goats are placed in lateral recumbency, care should be taken to ensure the lids of the down eye are closed. A towel or thin pad can be placed under the down eye to further protect it when patients are anesthetized on the floor or ground (Figure 19-16). Ophthalmic ointment should be placed in the eyes to protect them during anesthesia.

Monitoring

Sheep and goats are monitored in the same way as cattle and small animals. Vital signs, anesthetic depth parameters, capnometry, pulse oximetry, and indirect or ideally direct blood pressure are used to monitor small ruminants under anesthesia. Rotation of the eye does not occur in response to anesthetic depth in sheep and goats under anesthesia, but nystagmus can be seen.[4]

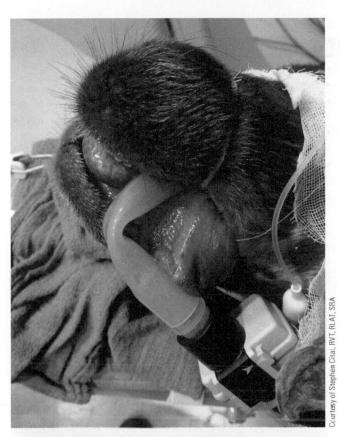

FIGURE 19-15 The patient should be positioned so there are no "kinks" of the ETT.

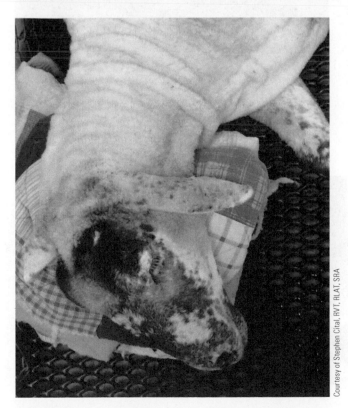

FIGURE 19-16 In ruminants that are anesthetized on the floor or ground, placing a towel under the down eye will protect it.

A balanced crystalloid solution (5 mL/kg/hr) should be administered IV to maintain perfusion and replace ongoing losses during anesthesia. Circulating warm water, forced air (Bair Hugger®), and conductive fabric patient warming (HotDog®) blankets can be used to maintain normal body temperature during anesthesia and recovery. Adequate ventilation is confirmed by the use of a capnograph, and end tidal carbon dioxide ($EtCO_2$) values should be kept in the range of 35 to 45 mm Hg[28] (see Table 19-4). In recovery, ventilation support is maintained until the animal starts to breathe spontaneously after appropriate weaning from a mechanical ventilator (see Chapter 7 [Anesthesia Monitoring]).

Recovery

Small ruminants should be placed in a sternal position at the end of anesthesia. Recovery in sheep and goats is usually smooth and similar to cattle. Points to remember when recovering small ruminants include the following:

- Examine the oral cavity and remove solid rumen material from the pharynx before the end of anesthesia and prior to extubating.
- Regurgitation may occur during recovery; therefore, the ETT must not be removed until the animal is able to protect its own airway, which may include chewing, swallowing, and ability to withdraw its tongue back into its mouth. The ability of sheep and goats to protect their airway may take considerable time and may be well after the animal is able to support its own head.
- The patient should be monitored for a period of time after extubation to make sure it is continuing to recover normally and physiologic parameters are within normal limits.
- Hay or grass and water may be allowed several hours after anesthesia

Sheep and Goat Analgesia

Sheep and goats are fairly shy prey species that usually display subtler signs of pain than some other species. Sheep in particular tolerate severe injury without showing signs of pain. Rubber ring castration and tail docking is painful in lambs as they will stand up and lie down frequently. Abnormal standing or inability to stand still can be seen after castration.[20] Lameness due to footrot is a painful condition in sheep, and there may be a close correlation between the observed lameness score, foot pathology, pain severity, and nociceptive thresholds.[20] Sheep that have chronic pain can exhibit all the signs of acute pain plus the following:[20]

- Altered movement, posture, appearance, or behavior:
  - Immobility or walking less
  - Abnormal standing (hunched posture, kicking, stamping, walking backwards, and walking on knees)
  - Abnormal gait may be the only signs observed in sheep[29]
  - Standing up and lying down frequently
  - Bruxism
  - Shivering may be a sign of anxiety[29]
- Modified eating habits
  - Inappetence
- Variation of vital signs
  - Easily measurable indicators of physiological stress (e.g., increased heart rate, increased pupil size, altered rate and depth of respiration, or trembling)

Goats are more vocal than sheep and may scream when restrained. After disbudding or castration, kids will lie down, vocalize, struggle, and kick

TABLE 19-9 Painful Conditions in Sheep and Goats

Common Painful Conditions in Sheep and Goats	Procedures That Cause Pain in Sheep and Goats
• Footrot and foot abscesses • External myiasis (flystrike) • Some dental problems • Urethral obstruction	• Castration • Tail docking (Figures 19-17a and 19-17b) • Mulesing (removal of skin from the perineal area in lambs to reduce the incidence of flystrike) [10] • Ear tagging or marking[10] (Figure 19-17c) • Cuts that occur during shearing and crutching • Disbudding of goats

out.[20] Goats that have chronic pain can exhibit all the signs of acute pain plus the following:[20]

- Altered movement, posture, appearance, or behavior:
  ○ Increased vocalization (screaming)
  ○ Lying down
  ○ Lameness
  ○ Struggling
  ○ Kicking and stamping of feet
  ○ Bruxism
  ○ Head pressing
  ○ Lip curling
- Modified eating habits
  ○ Inappetence
- Variation of vital signs
  ○ Easily measurable indicators of physiological stress (e.g., tachypnea and shallow breathing)

Painful conditions in sheep and goats are summarized in Table 19-9. Some parameters used to identify and assess pain in ruminants are found in the cattle section of this chapter.

(a)

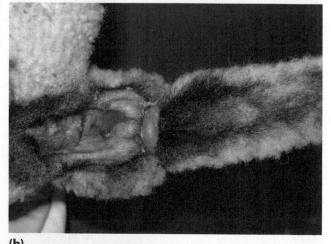

(b)

(c)

FIGURE 19-17 Painful conditions in sheep. (a) Tail docking in a lamb using the rubber band technique. (b) Close-up view of the rubber band applied to a lamb's tail in order to dock it. (c) Application of ear tags can be painful in lambs.

Pain Scoring Systems in Sheep and Goats

In 2009, a numerical rating scale was developed for use in goats postoperatively (Table 19-10).[29] To date it is the only pain scale available for goats. The Sheep Pain Facial Expression Scale (SPFES) was developed in 2016 by observing and interpreting facial expressions in sheep with naturally occurring footrot and mastitis.[30] The SPFES includes five traits: orbital tightening, cheek tightening, abnormal ear posture, abnormal lip and jaw profile, and abnormal nostril and philtrum (vertical indentation in the middle area of the upper lip) shape.[30] The five traits are each scored as 0 (not present), 1 (partially present), or 2 (present) with a maximum score of 10. Non-painful sheep do not show changes in facial expression, but those with a total score of 5 or greater are assumed to be in pain and administration of analgesic drugs should be considered.[30] The SPFES can be accessed at https://www.ablamb.ca/images/documents/newesletters/SPFES-insert.pdf. In 2017, the Sheep Grimace Scale (SGS) was developed by observing and interpreting facial expressions following unilateral tibia osteotomy in laboratory sheep.[31] The SGS includes three traits: orbital tightening (scored 0–2), ear and head position (scored 0–2), and flehming (scored 0–3) (Figure 19-18). A score of 0 indicates the action was not present. In orbital tightening, sheep with half-closed eyes are identified as moderate pain (score 1) and those with completely closed eyes are identified as severe pain (score 2).[31] For the ear and head position, sheep with flattened ears and a slanted head are identified as moderate pain (score 1) and those with hanging ears and head are identified as severe pain (score 2).[31] For flehming (curled upper lip), sheep with puckered lips are identified as moderate pain (score 1) and those exhibiting flehming are identified as severe pain (score 3).[31] The maximum total score is 7.

Treating Sheep and Goat Pain

Like other species, there are a variety of ways to treat pain in sheep and goats including NSAIDs, opioids, (α_2) adrenergic agonists, ketamine, and local anesthetics. Those caring for sheep and goats must keep in mind drug withdrawal times if their patient or their patient's products are being used for human consumption.

The most commonly used NSAIDs in ruminants are phenylbutazone, flunixin, ketoprofen, meloxicam, and carprofen. Clinically phenylbutazone seems to provide the best analgesia for musculoskeletal pain relief but less relief for the treatment of visceral pain. Unfortunately, GI ulceration is a concern in all ruminants treated with NSAIDs, especially phenylbutazone.

Most if not all opioids that are used in dogs and cats, including fentanyl patches, have been successfully used in sheep and goats.[28] In

TABLE 19-10 Numeric Rating Scale for Assessment of Postoperative Pain for Goats*

Criteria	Assessment	Score
Comfort	Awake, interested in surroundings, patient recumbent or standing, chewing cud, eating	0
	Awake, standing or recumbent, not interested in surroundings, not chewing cud, reduced appetite	1
	Lethargic, depressed appearance, ears drooped, not chewing cud, not eating	2
	Head down, very lethargic, ears stay drooped when aroused, not chewing cud, bruxism	3
	Recumbent, no response when approached, fixed look and staring or eyes half closed, little response when gently prodded, bruxism	4
Movement	Normal ambulation, full weight-bearing, no lameness	0
	Slight lameness on operated limb, toe touching on all steps	1
	Lameness on operated limb, toe touching on some, but not all steps	2
	Lameness on operated limb, not toe touching on all steps when walking voluntarily, but will toe touch when herded	3
	Lameness on operated limb, not toe touching on all steps when walking voluntarily and when herded	4
Flock Behavior	Normal, moves with the rest of the flock	0
	Mild changes, lethargic or lags behind rest of flock when flock is moved, but eventually joins them voluntarily	1
	Moderate changes, lags behind rest of flock when flock is moved, but eventually joins them if encouraged to do so	2
	Severe changes, no interest in rest of flock, always separated from flock	3
Total		**0–11****

*Hall, L. W., Clarke, K. W., & Trim, C. M. (2001). Anaesthesia of sheep, goats and other herbivores. In *Veterinary anaesthesia* (10th ed., pp. 341–366). London: WB Saunders.
**The higher the pain score, the greater the need for re-assessment and additional pain medication

Häger, C., Biernot, S., Buettner, M., Glage, S., Keubler, L. M., Held, N., Bleich, E. M., Otto, K., Müller, C. W., Decker, S., Talbot, S. R., & Bleich, A. (2017). The Sheep Grimace Scale as an indicator of post-operative distress and pain in laboratory sheep. PloS one, 12(4), e0175839. https://doi.org/10.1371/journal.pone.0175839

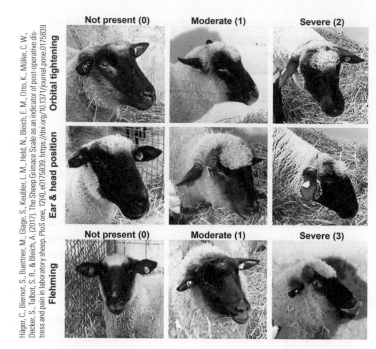

FIGURE 19–18 The Sheep Grimace Scale (SGS) scores three traits: orbital tightening (scored 0–2), ear and head position (scored 0–2), and flehming (scored 0–3) to determine if the animal is experiencing pain. The maximum possible score is 7. It is important to note that other factors, such as fear and stress, can affect facial expression and care should be taken to assess animals when they are calm.

goats, intramuscular buprenorphine every 6 hours after orthopedic surgery provided satisfactory analgesia.[28] Propulsive walking, rapid and frequent head movements, chewing, and hypersensitivity to auditory and visual stimuli have been observed in sheep receiving buprenorphine.[28] The atypical opioid tramadol (Ultram®) has been studied in the goat and has been shown to have a bioavailability of 36.9% following a single oral dosage of 2 mg/kg.[32] Interestingly, goats did not make significant quantities of the M1 metabolite (the metabolite responsive for tramadol's opioid analgesic effect in humans); therefore, if a dosage of tramadol provides analgesia in goats it likely does not relieve pain via an opioid mediated effect.[32]

Alpha-2 adrenergic agonists can provide short-term analgesia in small ruminants. The severity of their potential adverse effects (e.g., xylazine can cause fulminant pulmonary edema, hypoxemia, and respiratory failure leading to peracute death in sheep) limit their use in sheep and goats.

Similar to other species, the dissociative ketamine may offer analgesia to sheep and goats through NMDA (N-methyl-D-aspartate) antagonism. The drug has been widely used as an anesthetic induction agent and it has been administered together with other agents in the intrathecal and epidural space. Limited data, however, has been published in small ruminants describing its use IV in sub-anesthetic dosages for analgesia.

Local anesthetics have been used in a similar fashion in sheep and goats as in other species. Lidocaine specifically has been used in sheep and goats intravenously, topically through wound catheters and transdermal patches, and as part of a variety of locoregional techniques including nerve blocks and epidurals.[33] Some examples of common nerve blocks in sheep and goats include dehorning (corneal) nerve blocks, nerve blocks for limb surgeries, and the distal and proximal paravertebral nerve block as well as the "inverted L" block for abdominal procedures. In goats, IV lidocaine with or without ketamine reduced isoflurane requirements although this does not necessarily imply analgesic efficacy.[34] Infusions of lidocaine and ketamine have had similar results in sheep.

Local nerve blocks routinely performed on sheep and goats are listed in Table 19-11. The details of these blocks is beyond the scope of this chapter but can be found in several references.[4, 29, 34, 35]

MONOGASTRIC ANESTHESIA

Monogastric large animals include horses (which are also hindgut fermenters) and swine and although they are grouped together as a pair in this chapter, each species has very unique anatomical and physiological differences that make anesthetic comparisons between the two species difficult. For example, equine are obligate nasal breathers, which makes anything causing nasal obstruction during the anesthetic event or recovery (including pressing nostrils against a wall or support, swelling of the nose, collapse of the upper airways) a potential respiratory crisis. Species-specific anatomy can also affect how anesthetic procedures such as endotracheal intubation are performed. For example, intubation in swine can be challenging because they have a narrow dental arcade and a ventral laryngeal diverticulum, while horses typically can be blindly intubated with ease due to their oral and laryngeal anatomy. Being aware of anatomical and physiological differences in these monogastric species will allow for adaptation of anesthetic procedures and the knowledge to make the best anesthetic choices for these patients.

Equine Anesthesia

Just as in other large animal species, many procedures can be safely and appropriately performed on standing horses using physical and chemical restraint. Although there are several ways to facilitate physical restraint in horses, they do not always welcome this restraint due to their natural flight response, which, coupled with their overall size, can present significant safety concerns for veterinary personnel. Horses are large, at times flighty, and potentially unpredictable. For these reasons, chemical restraint may be necessary although it is critical that those working around sedated or anesthetized horses maintain a constant level of vigilance as to what is going on with their surroundings (continually ask yourself "Is the patient becoming lightly anesthetized?", "Is the patient panicking and trying to escape?", etc.) as well as having ways to escape from the horse's vicinity if needed.

A horse's temperament can also affect anesthesia management. For example, a Thoroughbred just off the race track may be difficult to handle, requiring additional drugs or higher dosages compared to an aged mellow Quarter Horse that is used for pleasure riding. Horses that are anxious will have an increased level of circulating catecholamines (norepinephrine, epinephrine), which may increase heart rate, blood pressure, and cardiac output in anesthetized horses.

When performing equine general anesthesia, the patient must be totally immobile since any movement in a very large horse under anesthesia

TABLE 19-11 Nerve Blocks Commonly Used for Sheep and Goats

Nerve Block	Area of Analgesia	Use
Cornual	Horn and surrounding skin on injected side	• Desensitizes the zygomaticotemperal and infratrochlear branches of the trigeminal nerve • In goats, local anesthetic is infiltrated between the lateral canthus of the eye and lateral horn base (different than in cattle since both corneal branches supply sensation to the horns in goats) • In sheep, local anesthetic is infiltrated as in cattle • Dehorning, disbudding
Inverted L-block (left) or reverse 7 (right)	Flank	• Flank is infiltrated with local anesthetic in inverted L (left side) or reverse 7 (right side) • Flank laparotomy
Paravertebral	Flank	• Desensitizes T-13, L1, and L2 (possibly L3 and L4) • Cesarean section
Caudal epidural	Hind quarters	• Local anesthetic is injected into the intervertebral space between the fifth sacral and first coccygeal vertebrae • Reproductive and obstetric procedures
Bier (IV Regional Anesthesia)	Limb (depending on where tourniquet is placed)	• Tourniquet is applied and local anesthetic injected usually against the direction of blood flow • Common sites are proximal or distal to the carpus or tarsus • Used for limb perfusion
Intratesticular	Testicle and spermatic cord	• Local anesthetic is injected into the neck of the scrotum, both spermatic cords, both testes, or combinations of these • Castration

*Reiblod, T. W. (2015). Ruminants. In K. A. Grimm, L. A. Lamont, W. J. Tranquilli, S. A. Greene, & S. A. Robertson (Eds.), *Veterinary anesthesia and analgesia: The 5th edition of Lumb and Jones* (pp. 912–927). Ames, IA, John Wiley & Sons; Lin, H. C., Caldwell, F., & Pugh, D. G. (2012). Anesthetic management. In D. G. Pugh & A. N. Baird (Eds.), *Sheep and goat medicine* (2nd ed., pp. 517–538). Maryland Heights, MO: Elsevier; Hall, L. W., Clarke, K. W., & Trim, C. M. (2001). Anaesthesia of sheep and other herbivores. In *Veterinary anaesthesia* (10th ed., pp. 341–366). London: WB Saunders; Staffleri, F., Driessen, B., Lacitignola, L., & Crovace A. (2009). A comparison of subarachnoid buprenorphine or xylazine as an adjunct to lidocaine for analgesia in goats. *Veterinary Anaesthesia and Analgesia, 36,* 502–511; Valverde, A., & Doherty, T. J. (2008). Anesthesia and analgesia of ruminants. In R. E. Fish, M. J. Brown, P. J. Danneman, & A. Z. Karas (Eds.), *Anesthesia and analgesia in laboratory animals* (2nd ed., pp. 385–411). London, UK: Academic Press.

can put both the horse as well as veterinary personnel in grave danger. It should also be noted that achieving immobility in the patient still demands maintaining acceptable physiology including cardiopulmonary function and muscle perfusion.

Preanesthetic Preparation

Preanesthetic preparation for an equine includes physical examination (during rest and exercise because exercise may reveal diseases such as recurrent airway obstruction [RAO] and cardiac diseases that are not apparent at rest), a thorough history, blood work, venous catheterization, and accurate body weight.[36] During the physical examination it is important to pay particular attention to auscultation of the respiratory tract because elective surgeries should be postponed one month following resolution of clinical signs of respiratory viral infection. Veterinary technicians can obtain an ECG reading from equine patients prior to elective procedures and any abnormalities detected should be addressed prior to anesthesia. Preanesthetic bloodwork is recommended based upon the horse's age and health status and may include a CBC with fibrinogen, chemistry panel with electrolytes, blood gas analysis, and other tests (e.g., radiographs, electrocardiogram) based on the individual patient. It is debatable whether or not to fast equine species prior to anesthesia. Horses can be fasted from 6 to 12 hours prior to anesthesia, but many believe that fasting is not needed because horses do not vomit or regurgitate, it stresses them, and fasting decreases postsurgical GI motility, which may lead to ileus. Some veterinarians prefer to withhold food for about 4 hours (grain should be withheld longer than hay/grass) to provide enough time for "wads" of plant material to move out of the mouth prior to induction and possible intubation. Water can be made available until just prior to anesthetic induction.[37]

Along with these general considerations, the breed of the horse and the patient's overall temperament are also important considerations when preparing them for anesthesia. For example, some horse breeds may be predisposed to particular diseases that affect their response to anesthesia. A classic example of this is Quarter Horses with the genetic disease hyperkalemic periodic paralysis (HYPP). This disease can produce attacks that are triggered by stress or diet. During the anesthesia period, a horse experiencing a HYPP crisis may develop life threatening hyperkalemia, dysrhythmias, bradycardia, muscle weakness, collapse, and difficulty breathing. Horses with known or suspected HYPP should be properly sedated (to avoid stress, which can exacerbate clinical signs), monitored with an ECG (to detect any cardiac effects of hyperkalemia), and have their electrolytes assessed both before and during anesthesia. Preoperative care may also include diet changes and medication such as oral administration of the carbonic anhydrase inhibitor acetazolamide (Diamox®) (2.2 mg/kg BID-TID) for 2 days prior to elective procedures. In horses with known or suspected HYPP, avoid administering drugs with the potential to exacerbate the condition (depolarizing neuromuscular blockers, potassium-penicillin, potassium sparing diuretics, NSAIDs, ketamine, halothane, isoflurane, and IV fluids containing potassium). More aggressive therapy is indicated if a hyperkalemic crisis occurs and largely involves intravenous fluids, treating the hyperkalemia, and ensuring adequate oxygenation and ventilation.

Some additional preparations for horses before anesthesia include brushing debris and extra hair from the patient's coat, removing horse shoes from feet, and cleaning or covering hooves (to reduce contamination of the surgical area). Rinsing feed out of the patient's mouth is also important to prevent airway obstruction and food from being pushed into the trachea during intubation. The horse's mouth should be rinsed with a dose syringe placed between the cheek and the teeth on each side of the mouth to flush out any feed material.

Tech Tip

It is important to tell clients why some preanesthetic procedures are necessary. For example, explaining that removing a horse's shoes prevents self-injury/injury to staff and/or damage to recovery stall helps clients understand its rationale because it can be stressful to the patient and creates an extra farrier cost to re-shod the horse. Removed shoes should not be discarded and should be properly labeled and returned to the client when the horse is discharged.

Venous Catheterization

An IV catheter should always be placed in horses prior to anesthesia so that drugs can be given quickly in an emergency situation such as if a horse accidently becomes lightly anesthetized, begins to move, and puts veterinary staff safety at risk.[36] Placing a sterile IV catheter in a horse is very similar to cattle. Useful catheter sizes for horses include 14 gauge for adult horses to prevent jugular thrombophlebitis and 14 to 20 gauge for foals. The jugular vein is most commonly used in horses for IV catheters (see Figure 19-19 and Appendix E). If the jugular veins are not patent, the median, cephalic, saphenous, or lateral thoracic veins can be used.[38]

Premedication, Sedation, and Restraint

Compared to cattle, horses are not as tolerant of standing procedures without sedation. Various standing procedures can be performed on horses with sedation and locoregional anesthesia techniques described in the Horse Analgesia section. Regional nerve blocks are also commonly performed on patient's limbs for diagnostic purposes to help identify the location(s) of equine lameness. Standing chemical restraint holds a lower risk of patient complications and in many situations reduces procedure costs and time.[39] Physical restraints such as stockades, ropes, halters, and twitches can also aid in standing procedures (Figures 19-20a and 19-20b).[40] All equipment should be prepared and organized including testing the equipment to make sure it is in proper working order prior to sedating the animal.

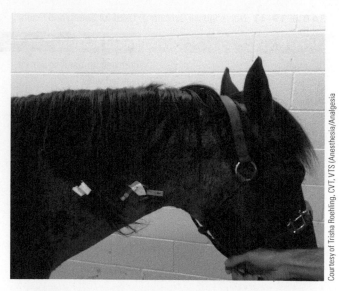

Courtesy of Trisha Roehling, CVT, VTS (Anesthesia/Analgesia)

FIGURE 19-19 A jugular catheter in a horse.

Sedatives/tranquilizers should be used prior to general anesthesia in equine. They are most commonly given IV but can also be given IM in some situations. Common examples include the following:

- Alpha-2 adrenergic agonists. The most commonly used sedative drugs in horses are (α_2) adrenergic agonists, which may be used in combination with opioids or phenothiazines or both. Alpha-2 adrenergic agonists including detomidine, xylazine, romifidine, and to a lesser extent medetomidine and dexmedetomidine (Dexdomitor®) are widely used in equine patients because they provide consistent and reliable sedation while other drug classes do not. The difference in choosing between each individual (α_2) adrenergic agonist is largely due to desired duration of effect and cost. For example, the duration of sedation from shortest to longest is medetomidine/dexmedetomidine < xylazine < romifidine < detomidine. Unlike in small animals, the (α_2) adrenergic agonists medetomidine or dexmedetomidine are not commonly used for IV bolus because their duration of action is exceedingly short, but they sometimes are administered via CRI. When administered IM, most (α_2) adrenergic agonists have an onset of action of 10 to 20 minutes.[38, 39]
- Phenothiazines. Acepromazine can be used to produce mild to moderate tranquilization in horses. When administered IV, its onset of action may take up to 15 minutes and is 20 to 40 minutes after IM injection.[38, 39] Use of acepromazine can rarely and generally temporarily cause priapism (persistent penile erection) in male horses, but it is reported and does have devastating effects for those patients (and owners). In breeding stallions, acepromazine has sometimes been avoided because if the erection continues and cannot be reduced it could result in penile injury and potential inability to breed in the future.
- Opioids. An opioid can be added to any of the above drugs (e.g., butorphanol, buprenorphine, morphine, or methadone IV); however, opioids in horses are more useful for analgesia then to increase sedation. Their effects are typically mild, and some horses will

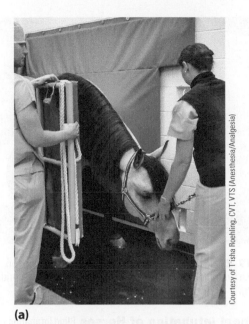

Courtesy of Tisha Roehling, CVT, VTS (Anesthesia/Analgesia)

(a)

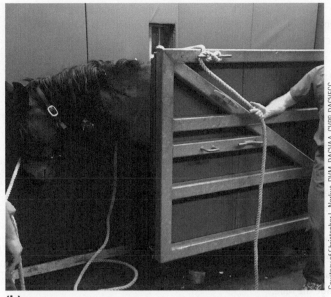

Courtesy of Christopher L. Norkus, DVM, DACVAA, CVPP, DACVECC

(b)

FIGURE 19-20 An adequately sedated horse ready for induction of anesthesia. (a) Note the physical restraints (stockades, rope, and halter) and that the head should be at the level of the carpus. Also note that though it may appear that the person restraining the horse is standing directly in front of the horse, which is a safety issue, they are actually standing slightly to the horse's right side. (b) This sedated horse is also restrained with ropes and its head lowered to the level of the carpus. The horse is further restrained by a padded induction squeeze gate and padded wall that helps support the animal's weight as it becomes more sedate.

actually become more aroused following opioid administration and may even have increased locomotion (they become restless and try to ambulate). Because of this, some anesthesiologists avoid opioids for standing sedation unless the patient requires additional analgesia or will wait until after anesthetic induction and the horse to be immobile before administering them.

Anticholinergic drugs are not routinely administered as preanesthetic medication to horses because horses do not excessively salivate, they infrequently develop clinically relevant bradycardia during induction or maintenance of anesthesia, and the risk for decreasing GI motility and causing colic is high. Due to their intestinal hypomotility effects, the administration of anticholinergic drugs should be reserved for the treatment of life-threatening bradyarrhythmias or cardiac arrest.

Although rarely done in clinical equine practice, antagonist drugs can be used to reverse (α_2) adrenergic agonists (atipamezole), opioids (naloxone [Narcan®]), and benzodiazepines (flumazenil [Romazicon®, generic]). These drugs should be reserved for use in resuscitation efforts or in cases where recovery is taking longer than 90 minutes and other options have been exhausted. Occasionally they are used in outpatient standing sedation cases or for prolonged recovery in foals as well. Generally speaking, it is uncommon to have to reverse any of these drugs in equine practice.

Induction Procedures and Techniques

For many surgical (such as orthopedic or abdominal exploratory surgery) or diagnostic techniques (for such as a computed tomography [CT] or magnetic resonance imaging [MRI] scan) performed in horses, the animal needs to be recumbent and fully unconscious and immobilized. Anesthesia induction may occur in the field or in a veterinary hospital in an induction stall with padded walls and floor. Horses are restrained, appropriately sedated, and then given induction agents and allowed to fall as the drugs quickly take the horse from a standing position to unconsciousness and recumbency.

Induction Medications

A number of induction agents can be used in horses. All induction agents require deep sedation prior to use to ensure smooth induction and safety of both the horse and personnel. To help achieve this, the muscle relaxant and sedative guaifenesin (GG) is sometimes administered IV just before or in conjunction with an induction agent to improve muscle relaxation and sedation and to smooth induction. The goal prior to anesthetic induction of a horse is that the patient is sedate enough so it is not responsive to external stimulus. One common test anesthetists will use is to clap or make a noise and to carefully observe whether or not the horse acts indifferent. The appropriately sedated horse is indifferent to the stimulus. At this point, an induction agent can be administered as a bolus. Ketamine, with or without a benzodiazepine, is currently the most commonly used equine induction agent in the United States. Thiopental when available is another commonly used induction agent. Induction agents, unlike in cats and dogs, are NOT titrated to effect in horses for the safety of personnel and patient. A common anesthetic plan for a horse might be premedication with xylazine, administration of GG for further sedation, and then anesthetic induction with ketamine. The opioid butorphanol might also be included for analgesia more so than sedation. Drugs such as alfaxalone, propofol, and etomidate are infrequently used in horses due to cost and the potential for poor quality induction.

Intubation of Equine

Following induction, horses are frequently intubated and transitioned to maintenance anesthesia techniques. Intubation of horses given injectable anesthetic agents for a short duration procedure (< 15 minutes such as a route castration) is at the discretion of the anesthetist; however, in general, horses should at least be nasotracheally if not orotracheally intubated for procedures lasting longer than 5 to 10 minutes. Ideally, oxygen should be administered to every horse under anesthesia even if the procedure is simply a field castration. Oxygen flow rates of 15 L/min should be provided to adult horses. It is important for the anesthetist to note that equine ETTs are relatively expensive, often costing over $100. Therefore, the anesthetist should be especially careful of the equine dental arcade during both during intubation and extubation as to not damage the tube.

Tech Tip

At an oxygen flow rate of 15 L/min, a full E tank would last approximately 40 to 45 minutes. The veterinary technician should be aware of that when packing the truck with supplies for the day.

TABLE 19-12 Endotracheal Tube Sizes for Horses (mm ID)*

Body Weight (kg)	Oral	Nasal
< 100	10–14	7–11
100–250	16–22	12–14
200–400	22–24	14–16
450 or greater	24–30	18–22

ID = internal diameter
*Bednarski, R. M. (2009). Tracheal and nasal intubation. In W. W. Muir & J. A. E. Hubbell (Eds.), *Equine anesthesia* (p. 282). Elsevier.

Equipment for orotracheal intubation in horses:

- Three sizes of ETT (Table 19-12). Choose a size that is considered appropriate and then select two additional ETTs, one that is a size larger and one a size smaller so that three different size ETTs are within reach.
  - Check the ETT for patency and that it is clean and kept free of dirt, blood, or debris. Check whether the cuff can be inflated and will hold pressure prior to attempting to use the tube. The ETT may be lubricated with a sterile, water-based lubricant.
- A mouth gag (typically polyvinyl chloride [PVC] piping or similar material is used) to hold jaws apart (Figure 19-21)
- 60 mL syringe to inflate the cuff

FIGURE 19-21 Mouth gags are commonly used for intubation in horses. An example of a mouth gag made from polyvinyl chloride (PVC) piping.

Orotracheal Intubation of Horses Blind intubation is the most commonly used method of orotracheal intubation of horses and is described below.[38, 41, 42]

- Keep the horse in lateral or sternal recumbency following anesthetic induction.
- Place a taped PVC tube (used as a mouth gag) between the horse's incisors. They will have a lot of jaw tone, so be aware that you may need to use considerable force to open the mouth enough to fit the gag in place.
- Extend the horse's neck and head to align the oral cavity with the larynx and trachea.
- Extend the tongue laterally out of the mouth, through the "bars of the mouth" (space between molars and canines).
- Smoothly insert ETT through mouth gag (Figure 19-22). Special care is needed so the tube stays centered within the mouth gag for smooth intubation and avoids damaging the ETT as it passes the molars.
- Smoothly advance the ETT toward the larynx. Make sure the ETT's concave surface is directed toward the palate.
- Once past the larynx, the ETT should slide smoothly into the trachea. If resistance is met, the tube is likely in the esophagus and it should be withdrawn approximately 6 to 9 inches, rotated 45 degrees, and then advanced again. If resistance is still met, then the ETT is again withdrawn, rotated, and advanced. This may have to be attempted several times. Often if you wait to pass during inspiration (by watching the movement of the thoracic wall), this will assist in widening the space between the arytenoids.
- Ensure that the ETT is in the trachea by detecting air flow at the end of the tube via spontaneous ventilation or compressing the chest manually.
- In rare circumstances (usually adult horses with pathology), going down a size of ETT may be needed (e.g., you originally selected a 26 mm ETT for your patient but after several attempts went down to the 24 mm).
- Once the ETT is confirmed to be in the trachea, the cuff should be inflated. Commonly in horses, between 60 and 20 mL of air are used to fill the cuff and inflation is stopped when positive pressure in the syringe is noted. Risk for tracheal injury or tear is uncommon in adult horses but has been reported in foals, so use caution with this age group.

Courtesy of Katie Huyer, BS, RVT

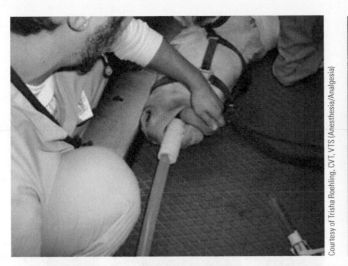

FIGURE 19-22 Placement of an orotracheal tube in a horse.

Courtesy of Trisha Roehling, CVT, VTS (Anesthesia/Analgesia)

FIGURE 19-23 Nasotracheal intubation in a horse.

Courtesy of Christopher L. Norkus, DVM, DACVAA, CVPP, DACVECC

Tech Tip

Orotracheal intubation is easier in horses by keeping the patient's head and neck in a straight line; hyperextending the neck may make it even easier.

Nasotracheal Intubation of Horses Nasotracheal intubation may be performed on awake or anesthetized horses and is most commonly selected in very young foals to induce general anesthesia and in adult horses following orotracheal tube removal for recovery from anesthesia. Nasotracheal intubation and inhalant anesthetic agent induction in full-sized adult horses is not performed in clinical settings (it is sometimes performed under special research settings) as it results in excitement and poor quality induction leading to a dangerous induction for both staff and patient. In anesthetized horses, a nasotracheal tube can be used for the recovery period and is placed after removal of the orotracheal tube so that the airway is protected and oxygen can be administered. Another option is to just leave the ETT (taped) in place for recovery. The following technique is used on an anesthetized foal or horse:[42]

- Keep the horse in lateral recumbency.
- Choose a smaller diameter ETT than orotracheal intubation (e.g., an 18–20 mm tube for adult sized horse and between an 8–12 mm tube for a foal).
- Insert a *well lubricated* (sterile water-based lubricant) ETT into the nostril guiding the tube in a ventral medial direction (Figure 19-23).
- Advance the tip of the ETT gently into the pharynx and from this point the procedure is the same as orotracheal intubation. If the tube meets resistance or increased airway noise is heard, then the ETT is not correctly placed in the trachea. The ETT should be withdrawn and then readvanced. Never force intubation; the ETT should always move smoothly because traumatic placement of a

nasotracheal tube can lead to severe epistaxsis in horses and should be avoided. Gentle placement and using a well lubricated tube helps reduce this risk. Intranasal phenylephrine, a vasoconstrictor, could also be considered.
- Secure the ETT to the patient (this is often done with rolled tape).
- An oxygen line can then be run down the nasotracheal tube to provide oxygen to the patient in recovery.

Maintenance of Anesthesia

Once the equine patient is induced and intubated, a maintenance anesthetic protocol is needed to continue anesthesia. For shorter procedures this can be accomplished with total intravenous anesthesia (TIVA) such as "Triple Drip," which is a combination of three drugs most often given as a CRI. One "Triple Drip" formulation for horses is guaifenesin/ketamine/xylazine (GKX), which typically contains 1 liter of 5% guaifenesin with 1 to 2 g of ketamine and 500 mg of xylazine. Variations from this combination exist depending on the individual anesthetist. It is important to note that numerous other drug combinations have and can be used instead of GKX such as guaifenesin/ketamine/detomidine (GKD), guaifenesin/ketamine/romifidine (GKR), and ketamine/xylazine/midazolam (KXM), and these also may be referred to as "Triple Drip." The anesthetist should be clear to use generic drug names to avoid confusion of the "Triple Drip" combination. Propofol combinations have also been reported in the equine literature for TIVA but are not commonly utilized in clinical practice.

Tech Tip

There are many GKX recipes that vary slightly from each other. The important thing to remember is that the ketamine is added at about twice the amount of xylazine.

TIVA for equine patients requires minimal equipment and is easy to use in the field. A major advantage of TIVA such as GKX is that cardiopulmonary function may be better preserved compared to inhalant anesthesia and that most patients who receive GKX have exceedingly smooth and good quality anesthetic recoveries.[43] TIVA also has several disadvantages. One limitation is that when using TIVA in horses, it is not easy to quickly alter the depth of anesthesia. Once the injectable drug has been administered, it also then must undergo metabolism often by the liver and eliminated by the kidneys. This can be a limitation in horses with altered cardiac output or altered hepatic/renal function. Another limitation with equine TIVA is much less muscle relaxation then when using an inhalant anesthetic agent. There is also risk for drug "build up," especially with combinations that have guaifenesin, resulting in prolonged recoveries and adverse effects when given much beyond 60 minutes. It is critical for the anesthetist to remember that horses maintained on GKX will have a fairly brisk palpebral reflex at an appropriate anesthetic plane, making the patient appear light to the inexperienced anesthetist thus causing them to administer more drug than necessary. A patient that does not have a strong palpebral reflex under GKX is deeply anesthetized. Spontaneous movement, changes in ventilation characteristics (increased depth or rate of breathing), spontaneous blinking, or increased tearing should immediately prompt the anesthetist that the horse is becoming inadequately anesthetized and light.

A typical example of the use of TIVA in a horse might be premedication with xylazine and butorphanol, induction with ketamine, and maintenance with guaifenesin, ketamine, and xylazine (GKX). The maintenance dose of "GKX Triple Drip" is given "to effect" based on anesthetic depth. Because guaifenesin can be difficult to purchase and is expensive, other drug combinations are becoming more commonly used.

For procedures that last longer than an hour or when increased muscle relaxation is needed, an inhalant anesthetic agent is typically selected for anesthetic maintenance. Isoflurane (MAC value 1.3–1.6%), sevoflurane (MAC value 2.3–2.8%), and desflurane (MAC value 7–8%) are all used in equine practice. Maintenance of anesthesia using an inhalant anesthetic agent requires expensive equipment such as a large animal anesthesia machine, which is not feasible to use in the field; however, it is beneficial as it delivers oxygen, can provide intermittent positive pressure ventilation (IPPV), and may allow easier alteration of anesthetic depth. Since the patient and anesthesia machine are large, it is important to remember that a horse's response to changes in inhalant anesthetic agent level and oxygen flow rates occur slowly. For this reason, a syringe of an injectable anesthetic such as ketamine is always kept near the patient or attached to a three-way stopcock in the fluid line so it can be administered if the patient becomes inadequately anesthetized in order to prevent movement and injury to personnel (Figure 19-24). Adverse effects and limitations for the use of isoflurane, sevoflurane, or desflurane in horses are similar to those seen in small animals.

Patient Positioning

Adult horses are at significant risk for developing neuropathy and myopathy when recumbent. Thick foam padding (10–12 inches) or an inflatable air mattress disperses the weight of the horse over a larger area (Figure 19-25a). As is true of ruminants, when a horse is in lateral recumbency the lower thoracic leg should be pulled forward to avoid radial nerve damage.[37] All extremities should lie parallel to the floor on leg supports attached to the table top (Figure 19-25b). The head should be positioned naturally with the ears not bent under the poll. The eye should

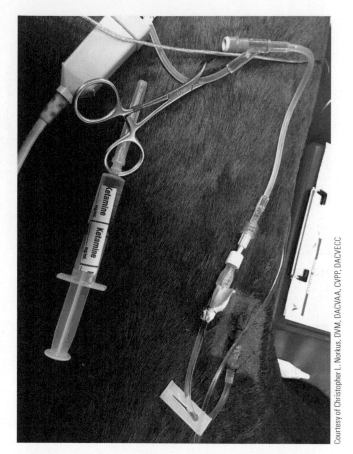

Courtesy of Christopher L. Norkus, DVM, DACVAA, CVPP, DACVECC

FIGURE 19–24 Preparing for horses that may become light during anesthesia. If a horse becomes inadequately anesthetized and moves during general anesthesia, there is a significant risk for human safety as a horse kick to the head is easily fatal. For the safety of everyone working with anesthetized horses, it is strongly recommended that a dose of injectable anesthetic such as ketamine be kept easily accessible near the horse at all times. If a horse begins to move or is showing signs of being lightly anesthetized, this dose can be given in an emergency because it will rapidly allow the anesthesia team to regain control of the patient. Note in this picture a syringe of ketamine located near the IV port for this purpose.

be well lubricated, closed, and pressure reduced on the orbit or eyeball to avoid injury (Figure 19-25c). Any halter should be removed to prevent facial nerve injury. In dorsal recumbency, the legs should be supported in a natural position using extra pads around the shoulders and hobbles around the thoracic limbs to hold the patient centered over the mat. A Tom fool knot can be used to hobble/tie the thoracic limbs to the surgery table (Figure 19-26 a through Figure 19-26c). Padding may need to be adjusted for horses with high withers. The pelvic limbs can be allowed to relax in a natural position and the head padded around the poll (Figure 19-25d). Avoid overextension of the head by placing wedge-shaped pads under the head (Figure 19-25e). This may also reduce the incidence of nasal edema and risk for postanesthetic airway obstruction. When positioning the patient, make sure are no "kinks" or obstruction of the ETT. In general large animals (especially horses) should not be anesthetized for over 3 hours as risk for complications increase significantly. Careful monitoring and management of blood pressure is very important to ensure adequate blood flow and oxygen delivery to muscles to avoid postanesthetic myopathy.

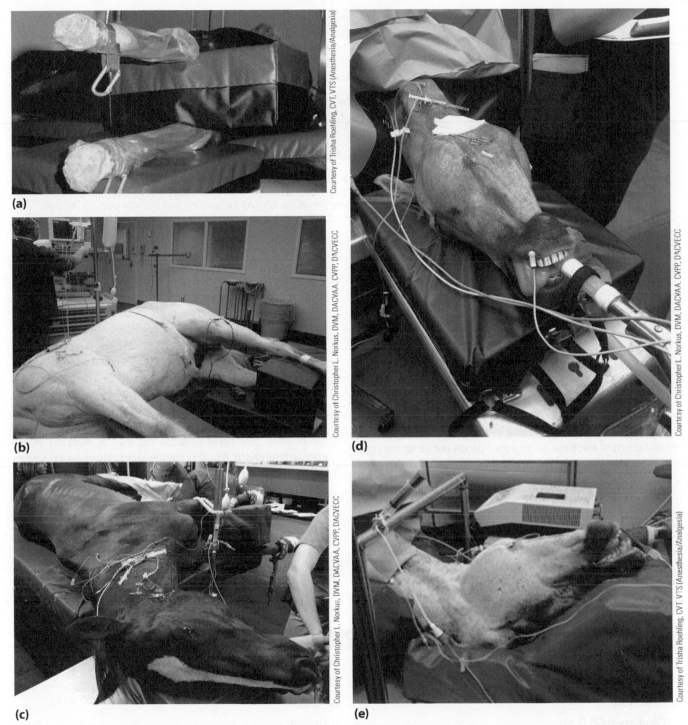

Courtesy of Trisha Roehling, CVT, VTS (Anesthesia/Analgesia)

Courtesy of Christopher L. Norkus, DVM, DACVAA, CVPP, DACVECC

Courtesy of Christopher L. Norkus, DVM, DACVAA, CVPP, DACVECC

Courtesy of Christopher L. Norkus, DVM, DACVAA, CVPP, DACVECC

Courtesy of Trisha Roehling, CVT, VTS (Anesthesia/Analgesia)

FIGURE 19-25 Positioning a horse during anesthesia. (a) Use thick foam padding or air cushions when positioning horse's rear legs in lateral recumbency. Thick foam padding disperses the weight of a horse over a large area, and padding between the legs allows the extremities to lie parallel to the floor. These steps are important to prevent neuropathy. (b) Horses in lateral recumbency should have their extremities maintained parallel to the floor on leg supports, while their head should be positioned naturally with unbent ears and well lubricated and protected eyes. (c) Horses in lateral recumbency should have their head positioned naturally with the ears not bent under the poll, eyes well lubricated, closed, and pressure reduced on the orbit or eyeball to avoid injury, and any halter removed to prevent facial nerve injury. (d) Positioning a horse's head while in dorsal recumbency. Note that the halter is removed to avoid facial nerve damage (it is hanging from the ETT) and the head padded around the poll. (e) Positioning of a horse's head while in dorsal recumbency. Note the patient in picture has an arterial catheter placed to allow for direct blood pressure monitoring in the facial artery. The syringe secured to the surgery table is attached to a three-way-stopcock and contains saline for flushing the catheter.

FIGURE 19-26 A Tom fool knot can be used to hobble/tie the thoracic limbs to the surgery table. (a) Grasp center of rope with both hands; rotate both hands to form two loops. (b) Bring loops together so that the right loop overlaps the left loop. Pass the two loops through each other by pulling the side of the right loop through the left loop and the near side of the left loop through and over the right loop. (c) Pull both hands apart to create two loops knotted together in the middle (see arrows). Place each loop over a limb and pull each end to snug the loops around the legs.

Monitoring

Monitoring short-term anesthesia in a horse is accomplished with a physical or "hands on" approach similar to a physical exam. When assessing anesthetic depth, physical signs are the most reliable. At an adequate anesthetic plane, the horse's eyes should have a dull palpebral reflex and a strong corneal reflex. Rapid nystagmus, lacrimation, and blinking are signs of very light levels of anesthesia in horses. Heart rate is not a dependable detector of anesthetic depth and changes very little during anesthesia.

Longer anesthetic events require in-depth monitoring due to the profound effects of anesthesia on the equine patient that commonly include hypotension, hypoxemia, and hypercapnia. Monitoring devices should include pulse oximetry, capnography, blood pressure (ideally direct but can be indirect) (Figure 19-27a), electrocardiogram, and a method to access temperature (Figure 19-27b). Direct arterial blood pressure is the preferred method for longer anesthesia periods (see Figure 19-25e). Maintenance of MAP higher than 70 mm Hg is a critical factor in preventing postanesthetic myopathies in horses (Table 19-13). Blood pressure also indicates level of anesthesia. An ideal MAP for horses is 70 to 90 mm Hg. If this value is above

90 mm Hg, the patient may be in a light plane of anesthesia, painful, have a drug-induced increase in blood pressure (from drugs such as (α_2) adrenergic agonist, inotropes, etc.), or have underlying disease.[36] A sudden change of blood pressure can also indicate the patient is in a light plane of anesthesia (e.g., if a horse's MAP is 65 mm Hg for the last hour and suddenly increases to 85 mm Hg).

Hypotension during anesthesia is commonly observed due to the cardiovascular depressant effect of inhalant anesthetic agents. Hypotension can lead to decreased skeletal muscle blood flow and the resulting poor tissue perfusion can lead to myopathy in the postoperative period. Providing maintenance fluids during a procedure can help to improve peripheral/muscle perfusion. Maintenance fluids in horses during these procedures should range from 5 to 10 mL/kg/hr. If hypotension occurs in the face of an appropriate level of anesthesia and maintenance fluid administration, numerous treatments can be considered. Fluids such as crystalloids and colloids can be given as a bolus if hypovolemia is suspected. However, most fluid pumps do not have the capability to deliver fluids rapidly; therefore, a pressure

TABLE 19-13 Normal Values for Anesthetized Horses*

Parameter	Value
Respiratory rate	10–30 breaths/minute
Tidal volume	10–15 mL/kg
Heart rate	28–40 bpm
Capillary refill time	< 2 seconds
Arterial pressure (systolic)	> 100 mm Hg and <160 mm Hg
Mean arterial pressure (MAP)	> 70 mm Hg to prevent myopathy
Partial pressure of arterial carbon dioxide ($PaCO_2$)	35–45 mm Hg
Partial pressure of arterial oxygen (PaO_2)	> 80 mm Hg
Arterial pH	7.35–7.45
End tidal carbon dioxide ($EtCO_2$)	35–45 mm Hg

*Robertson, J. T., & Scicluna, C. (2009). Preoperative evaluation: General consideration. In W. W. Muir & J. A. E. Hubbell (Eds.), *Equine anesthesia* (p. 123). Elsevier.

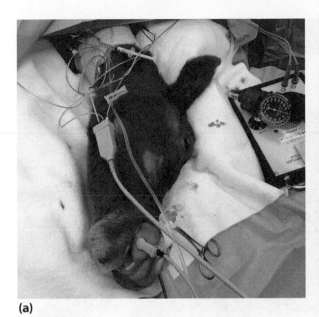

(a)

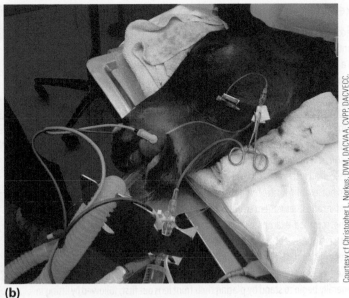

(b)

Courtesy of Christopher L. Norkus, DVM, DACVAA, CVPP, DACVECC.

FIGURE 19–27 Monitoring horses during anesthesia. (a) A horse with a pulse oximeter and indirect blood pressure monitoring device. (b) A horse that has multiple wires and hoses contained in a metal plate in order to avoid tangling them.

infusor or specialized pressure cuff may be needed. A pressure infusor is a device used to put pressure on a collapsible IV bag so fluid can be rapidly infused into the patient (Figure 19-28). Alternatively, specialized pressure cuffs can be wrapped around a fluid bag, secured with Velcro® closures, and air pumped into the cuff to deliver fluids rapidly to patients. Since most IV fluids for equine patients are manufactured in 5 to 10 liter bags, a 5 liter pressure cuff is needed. In cases of hypotension without hypovolemia or if the hypotension is multifactorial, then positive inotropes such as dobutamine or vasoconstricting medications

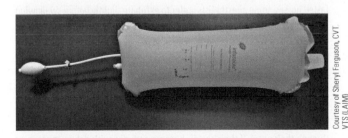

Courtesy of Sheryl Ferguson, CVT VTS (LAIM)

FIGURE 19–28 A pressure infusor helps rapidly infuse fluid into the patient. The IV bag is inserted into the cuff and the fluid bag is hung on a bag hook of an IV stand. The longer tube on the pressure infusor has a stopcock and inflation bulb. To fill the pressure infusor, turn the "OFF" mark on the stopcock handle to point toward the open stopcock vent (as seen in the photo) and then pump the inflation bulb until the pressure gauge indicates the desired pressure (not to exceed 300 mm Hg). To maintain a leak proof seal, turn the "OFF" mark of the stopcock handle toward the pressure infusor bag. To deflate the device, turn the "OFF" mark toward the inflation bulb. The inflate, maintain pressure, and deflate stopcock directions are also illustrated on the pressure infusor. The short tube on the pressure infusor measures the pressure. Pressure infusors are available in a variety of sizes including 1000 mL, 3000 mL, and 5000 mL.

such as dopamine, ephedrine, or norepinephrine can be helpful.[37] Titration of the anesthetic agent is always a consideration.

Animals that are poorly ventilating, no matter how long the procedure will last, need mechanical ventilation. The length of the procedure is not the main factor in determining whether or not to provide mechanical ventilation but rather it is the hypercapnia that results from the use of inhalant anesthetic agents that dictates the optimal type of ventilation (spontaneous versus mechanical) for veterinary patients. Animals are typically given inhalant anesthetic agents for procedures lasting longer than 60 minutes and should be mechanically ventilated if they are ventilating/oxygenating poorly. Mechanical ventilation is indicated if PaO_2 is lower than expected for a patient on 100% oxygen or patient has a $PaO_2 < 80$ mm Hg on room air, in hypercapnic patients ($EtCO_2 > 60$ mm Hg), those given neuromuscular blocking agents, neurologic patients at risk for increases to intracranial pressure secondary to inhalant anesthetic agent use, those not ventilating well despite having a normal $EtCO_2$, and for anesthetic management to improve alveolar ventilation of lightly anesthetized patients (e.g., in a horse just following anesthetic induction). Arterial blood gas analysis will be helpful to assess proper oxygenation and ventilation and is commonly performed in equine anesthesia cases to monitor case progress.

Patient positioning and size influence the PaO_2. Alveolar-arterial gradients measure the difference between the alveolar concentration (A) of oxygen and the arterial (a) concentration of oxygen. A-a gradients occur because there is a difference between the oxygen tension in the alveoli and that found in the arteries (see Chapter 12 [Anesthetic Complications]). Anesthetized horses can develop large alveolar-arterial gradients (A-a gradients) resulting in inappropriately low oxygen tensions and hypoxemia even if they are on 100% oxygen. Large A-a gradients are the result of alveolar shunting, an extreme form of V/Q (ventilation/perfusion) mismatch.

Recovery

Recovery is a critical phase of equine anesthesia yet in many ways is the least controllable. A fracture or injury that occurs during recovery could potentially lead to a patient's death; therefore, providing a safe environment for equine recovery is an important part of successful anesthesia. Horses are often recovered in most equine hospitals in a well-designed, designated recovery room; however, during field anesthesia patients can also be recovered outdoors in a large open area. Some horses when they awaken from anesthesia may attempt to stand before they are physically able resulting in incoordination and potential injury. For this reason, it is not uncommon to provided additional tranquilization or sedation to horses in an attempt to facilitate a smooth recovery. Sedation is often provided to keep horses calm and recumbent in order to give them adequate time to eliminate inhalant anesthetic agent so when they attempt to stand, they are "clear headed." Some common examples include acepromazine, xylazine, or romifidine. Note that sedation is not needed if the patient was on TIVA that included an (α_2) adrenergic agonist. Providing sedation allows the horse to recover on its own in a locked recovery stall with close monitoring in case intervention is required. Horses generally begin to stand by placing their front legs out first, followed by lifting in their hind end. Foals are usually hand recovered.

Patient positioning is also important for the horse recovering from anesthesia as they may remain recumbent for some time (e.g., it is not uncommon for a horse to remain in recovery for up to an hour). Horses are routinely recovered in lateral recumbency. It is important for the anesthetist to pull the dependent (down) thoracic leg forward to take pressure off the radial nerve to prevent nerve injury. Additionally the patient's eyes should also be protected. Halters are typically removed from a horse in recovery to prevent facial nerve paralysis. An exception to this, however, is when a horse is "hand recovered" with rope (such as with a lead rope and tail rope). If the halter is left in place, it should be padded to prevent facial neuropathy. A special recovery halter is available with padding placement where the webbing or leather intersect (Figure 19-29).

Postanesthetic myopathy has historically been a major concern for horses recovering from general anesthesia and has been a major source of morbidity and mortality. Fortunately postanesthetic myopathy has dramatically decreased over the past decade as older general inhalant anesthetic agents such as halothane have been phased out and replaced with newer agents (isoflurane, sevoflurane, desflurane) and more attention has been paid to anesthetic monitoring and aggressive blood pressure management. When there is poor blood flow to a horse's large muscle groups, the horse can develop ischemia resulting in severe muscle damage. In recovery, horses with postanesthetic myopathy cannot stand up after awakening from anesthesia or they have marked weakness. In other cases that initially appear to recover normally, lameness can develop several hours later. Steps to reduce the incidence of myopathy are accomplished by ensuring appropriate muscle perfusion through administering IV fluids, monitoring blood pressure and aggressively treating hypotension, avoiding the inhalant anesthetic agent halothane, and minimizing anesthetic duration (ideally under 3 hours). Additionally, providing proper patient padding and positioning may be of benefit (see the Patient Positioning section of this chapter).

The horse's behavior and the anesthetist's preference should be considered when deciding when to extubate. Many times horses are not extubated until they are taking regular deep breaths and standing; however, there are incidences that make waiting until the horse is standing dangerous (e.g., a

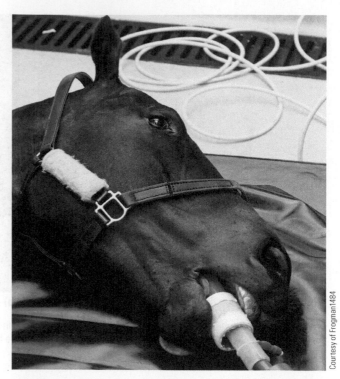

FIGURE 19–29 A special recovery halter has padding where the webbing or leather intersect to prevent facial nerve damage.

Courtesy of Frogman1484

horse smashing against a recovery stall in an attempt to stand could end kinking the ETT and developing pulmonary edema). Most horses do not tolerate the ETT remaining in place once they regain consciousness, and the benefit of maintaining an airway during recovery should be weighed against preventing injury to the horse. Placing one or two nasal tubes before extubation and when they are standing, especially if the horse has any nasal edema or was under general anesthesia for more than an hour, has been used to maintain airway patency and allow oxygen administration during the recovery phase. In fasted horses, some anesthetists prefer to extubate slightly before swallowing is expected (especially in young, difficult horses).[4] Once extubated, observe the patient for signs of distress/restlessness and airway obstruction (listen for stridor, watch for increased respiratory rate and effort of breathing, hold a hand over the nostrils to feel airflow while watching thoracic and abdominal movements). If an obstruction is suspected, a nasotracheal tube is passed or very rarely the horse may need to be re-induced and an orotracheal tube passed.

Horses are obligate nasal breathers (meaning they can only breathe through their nose). If their nostrils become occluded with debris, mucous, or blood or become edematous, then the patient can quickly obstruct its airway and suffer respiratory arrest. Nasal edema and congestion is common in horses that have had surgical procedures in which the patient's head is at or below the level of the rest of the body. This edema may be reduced by intranasal administration of phenylephrine but it is not eliminated. For this reason, ensuring an open airway is critical for horses during anesthetic recovery. If an orotracheal tube was placed for surgery, it may be kept in place until the horse is standing and recovered. Alternatively, the orotracheal tube can be removed and a nasotracheal tube can be placed. Oxygen (15 L/min) should be supplemented through either tube until the horse is awake and the patient is standing (see Figure 19-23). The orotracheal or nasotracheal tube must be secured to the patient to prevent it being dislodged or aspirated. A **Hudson demand valve** is a portable valve that when attached to an oxygen source can provide a large jet of oxygen (such as 75 L/min) at the push of a button. This device is useful for horses in recovery to both provide them with oxygen but also to provide intermittent positive pressure breaths as needed (Figures 19-30a and 19-30b).

The horse recovering from anesthesia should be continuously monitored (often from a distance by observation only for staff safety). A quiet and darkened environment with ample space will also help a horse to have a smoother recovery. A towel can be placed over the patient's upper eye and lights can be shut off in a recovery stall. Dry footing is important when the patient attempts to stand, as they need to plant their feet on solid, non-slippery surfaces. The ideal way to recover a horse from anesthesia is up for debate and is as much of an art as it is a science. Some anesthetists elect to perform different variations on a "hang recovery." This type of recovery often uses ropes or leads attached to the patient's halter and tail to help assist the horse with balance for recovery. There are many other different methods for surgical recovery that include air mattress pillows (Figure 19-31a), slings, ropes (Figures 19-31b and 19-31c), and water pools. The most important consistency for recovering horses is to maintain safety of personnel and the animal to the extent possible at all times.

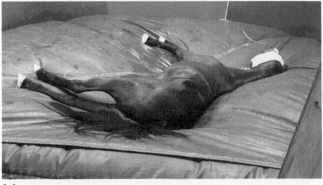

(a)

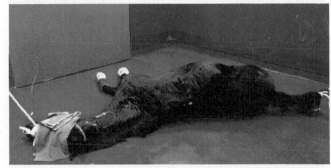

(b)

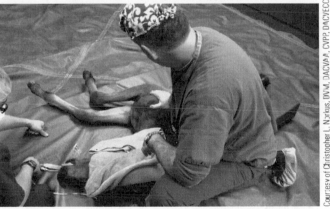

(c)

Courtesy of Christopher L. Norkus, DVM, DACVAA, CVPP, DACVECC

(a)

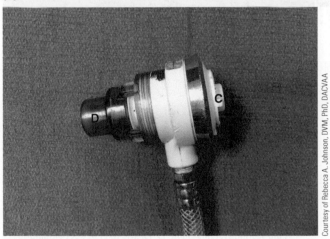

(b)

Courtesy of Rebecca A. Johnson, DVM, PhD, DACVAA

FIGURE 19-30 Oxygen demand valve. (a) The oxygen demand valve (A) and the connector to the oxygen supply hose (B). (b) The oxygen demand valve showing the manual button (C) and the connector to the ETT (D).

FIGURE 19-31 Types of recovery. Numerous forms of assistance are available to help promote a smooth and safe equine anesthetic recovery. (a) Large inflatable air mattresses shown here cover the entire floor to assist with equine recovery. The horse is placed on the deflated air pad and the pad is then inflated with a fan system. The thick air cushion keeps the horse in lateral recumbency for a longer period of time, allowing the horse to eliminate more inhalant anesthetic agent so that they are more coordinated when they do attempt to stand. The pad rapidly deflates within 30 to 60 seconds, allowing the horse to stand on a smooth surface once a coordinated attempt to stand is made. (b) Rope recovery in horses typically consists of one rope attached to the halter and another rope attached to the tail. Head and tail ropes should be placed through rings high on the wall (> 6 feet). The ropes are not meant to lift the horse when they try to stand but rather provide some support and a sense of balance to the ataxic horse once it is attempting to stand. (c) Foals can be hand recovered using ropes. This foal is on a mat to minimize pressure on protuberances. A towel is placed over its eyes to darken the recovery area.

Horse Analgesia

Like other large animals, horses are a species undermanaged for their pain. This underutilization of pain management in horses stems from a combination of reasons including the fact that horses are a prey species that hide their pain and discomfort, difficulty in assessing and quantifying equine pain, mixed efficacy of some analgesics, and a fear of adverse effects from analgesics such as opioids (decreased GI motility and colic as well concern for increased locomotion and excitement).[44] Many of these reasons are likely overstated and when given appropriately, multimodal analgesia can be effective and safe in horses as it is in small animal species.

A horse's behavioral response to pain depends on breed, age, and disease or surgical process and its duration. Keep in mind that pain is an individual experience so each patient reacts differently; therefore, it is important to pay attention to the overall picture of the animal in order to properly assess pain levels in each patient. When possible, a horse should be observed either from a distance or without the horse seeing the observer to avoid them masking pain behavior.[45] Horses that have chronic pain can exhibit all the signs of acute pain plus the following:

- Altered movement, posture, appearance, or behavior:
  - Restless, agitated, and/or anxious
  - Dullness and depression
  - Rigid stance and reluctance to move
  - Detached, unsocial behavior
  - Fixed stare and dilated nostrils
  - Aggression toward own foal, other horses, and handlers
  - Vocalization (groaning or grunting)
  - Lowered head carriage

- Rolling
  - Kicking at abdomen
  - Flank glancing or watching
  - Stretching
  - Weight-shifting between limbs
  - Limb guarding
  - Abnormal weight distribution
  - Abnormal movement
  - Arched back
  - Headshaking
  - Abnormal bit behavior
  - Bruxism
- Modified eating habits
  - Inappetence
- Variation of vital signs
  - Easily measurable indicators of physiological stress (e.g., tachypnea and changes in breathing pattern or rate)

Table 19-14 summarizes signs and causes of pain in horses as well as parameters used to identify and assess pain in horses.[3]

Tech Tip

Horses may mask pain-associated behaviors in an unfamiliar setting such as a clinic or in the presence of strangers; therefore, a thorough patient history is important because the owner/trainer sees the patient in its normal environment.[3]

TABLE 19–14 Causes and Assessment of Pain in Horses*

Common Painful Conditions in Horses	Procedures That Cause Pain in Horses	Parameters Used to Identify and Assess Pain in Horses[27]

• Colic (Figure 19-32)
• Arthritis
• Trauma
• Fractures
• Hoof problems
• Laminitis (Figure 19-33)
• Joint problems
• Dental problems
• Uveitis
	• Colic/Abdominal surgery
	• Castration
	• Arthroscopy
	• Perineal procedures (e.g., a perineal laceration due to foaling)
	• Umbilical hernia repair
	• Fracture repairs
	• Laceration repairs
		Objective assessment
		  • Heart rate
		  • Beta-endorphins
		  • Catecholamines and glucocorticoids
		  • Ground reaction force
		  • Response to pressure
		  • Gait analysis
		  • Thermographic imaging
		  • Electroencephalography
		  • Behavioral signs
		  • Response to analgesics
		
		Subjective assessment
		  • Behavior
		  • Pain scoring systems (Table 19-15; also refer to Obel Grading System and the American Association of Equine Practitioner's [AAEP] lameness grading system in Chapter 15 [Acute Pain Management])

*Lerche, P., & Muir, W. W. (2009). Perioperative pain management. In W. W. Muir & J. A. E. Hubbell, Equine Anesthesia (pp. 371–377). Elsevier.

FIGURE 19-32 Colic is a painful condition in horses. Rolling is a common sign with equine colic. The list of colic signs is lengthy, and the signs are not displayed consistently between patients. Signs include frequent weight shifting, decreased GI sounds, increased heart rate, flank watching or "glancing," restlessness, pawing, rolling, sweating, and frequent lying down and rising. The causes of colic are extensive but can include abrupt changes in diet, systemic illness, dental abnormalities, GI blockage, gastric ulceration, and pregnancy complications.

Courtesy of Shutterstock

FIGURE 19-33 Laminitis is a painful condition in horses. Laminitis can be caused by many disease processes including retained placenta, systemic illness, high concentrate feeds, and turnout on lush pasture. This horse is displaying a common sign of laminitis, leaning weight into the back limbs to reduce pain. Other signs can include increased digital pulses, reluctance to move or stand, increased recumbency, and frequent weight shifting. Supportive care can include supportive leg wraps and deep, soft bedding of straw, shaving, and peat moss.

Courtesy of Rita Tigert, CVT

Pain Scoring Systems in Horses

Pain assessment of horses is beneficial to their welfare and ensures that each patient is being assessed as an individual. The use of pain scales was described in Chapter 15 (Acute Pain Management), and Table 19-15 shows a simple visual analog scale for horses. All pain scales have limitations; however, composite pain scales and facial expression-based pain scales appear to be the most promising tools for pain assessment in horses. The Horse Grimace Scale (HGS) was developed in 2014 to assess facial expression changes to score pain following surgical castration.[46] Since then the HGS has also been used in assessing pain in horses due to acute laminitis. The HGS has six facial action units that are scored based on intensity of each facial expression as not present (score of 0), moderately present (score of 1), and obliviously present (score of 2) (Figure 19-34).[46] The maximum score is 2, and the higher the score, the greater the level of discomfort experienced by the horse. The HGS is available as a phone app, which includes a video and training session, at https://play.google.com/store/apps/details?id=info.awinhub.HorseGrimacePainScale&hl=en_US. The Colorado State University Veterinary Medical Center's Equine Comfort Assessment Scale was described, and a link to the scale and a list of behavioral descriptors of pain can be found, in Chapter 15 (Acute Pain Management). The Obel Grading System and the American Association of Equine Practitioners (AAEP) pain scales for assessing equine lameness are also provided in Chapter 15 (Acute Pain Management).

> **Tech Tip** 🐾
>
> Grimace scales are especially helpful in large animal pain assessment because they are non-invasive and require short observation of the animal, are easily applicable, and rapidly detect pain.

Treating Horse Pain

Due to concerns associated with opioid use in horses, there is an overutilization of NSAIDs to treat pain in the horse. NSAIDs remain a mainstay of equine analgesia and can be administered in a variety of applications including pre- or postoperatively for surgical pain, repeatedly for chronic pain such as osteoarthritis or to treat inflammatory medical conditions. NSAIDs are also used occasionally for anti-endotoxemia effect in patients with colic. Although it would be ideal to administer these drugs before inflammation occurs, the safest practice is to administer NSAIDs after anesthesia to avoid potential renal injury if a patient were to experience perioperative hypotension. NSAIDs are best avoided in patients with renal insufficiency, in patients who have or are at risk for gastric ulceration, are receiving concurrent glucocorticoids, or are dehydrated, hypovolemic, or hypotensive. Horses given flunixin meglumine IM can develop bacterial myositis (condition in which endospores of the bacteria *Clostridium* can lie dormant in healthy muscle, begin to proliferate if muscle is damaged, and can result in a serious and sometimes fatal condition).

> **Tech Tip** 🐾
>
> The NSAID flunixin meglumine is commonly administered to horses for analgesia; however, it should never be administered IM due to risk of severe, life-threatening bacterial myositis. A safer route is to administer flunixin meglumine IV or orally.

The Horse Grimace Scale (HGS) *Dalla Costa et al. (2014)*

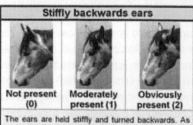

Stiffly backwards ears

Not present (0) | Moderately present (1) | Obviously present (2)

The ears are held stiffly and turned backwards. As a result, the space between the ears may appear wider relative to baseline.

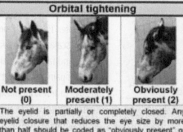

Orbital tightening

Not present (0) | Moderately present (1) | Obviously present (2)

The eyelid is partially or completely closed. Any eyelid closure that reduces the eye size by more than half should be coded as "obviously present" or "2".

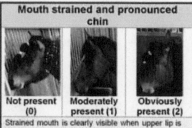

Mouth strained and pronounced chin

Not present (0) | Moderately present (1) | Obviously present (2)

Strained mouth is clearly visible when upper lip is drawn back and lower lip causes a pronounced "chin".

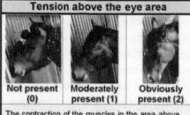

Tension above the eye area

Not present (0) | Moderately present (1) | Obviously present (2)

The contraction of the muscles in the area above the eye causes the increased visibility of the underlying bone surfaces. If temporal crest bone is clearly visible should be coded as "obviously present" or "2".

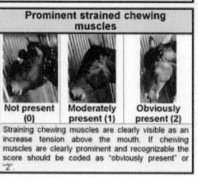

Prominent strained chewing muscles

Not present (0) | Moderately present (1) | Obviously present (2)

Straining chewing muscles are clearly visible as an increase tension above the mouth. If chewing muscles are clearly prominent and recognizable the score should be coded as "obviously present" or "2".

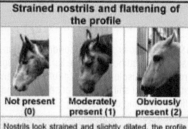

Strained nostrils and flattening of the profile

Not present (0) | Moderately present (1) | Obviously present (2)

Nostrils look strained and slightly dilated, the profile of the nose flattens and lips elongate.

PLOS | ONE

Facial Coding Unit	Score
Ears stiffly backwards	1
Orbital tightening	0
Tension above eye area	0
Prominent strained chewing muscles	0
Mouth strained and pronounced chin	0
Strained nostrils and flattening of the profile	0
Total pain score	1

(a)

Facial Coding Unit	Score
Ears stiffly backwards	2
Orbital tightening	2
Tension above eye area	0
Prominent strained chewing muscles	2
Mouth strained and pronounced chin	1
Strained nostrils and flattening of the profile	1
Total pain score	8

(b)

Facial Coding Unit	Score
Ears stiffly backwards	0
Orbital tightening	0
Tension above eye area	1
Prominent strained chewing muscles	0
Mouth strained and pronounced chin	0
Strained nostrils and flattening of the profile	0
Total pain score	1

(c)

Facial Coding Unit	Score
Ears stiffly backwards	0
Orbital tightening	0
Tension above eye area	1
Prominent strained chewing muscles	0
Mouth strained and pronounced chin	0
Strained nostrils and flattening of the profile	0
Total pain score	1

(d)

FIGURE 19–34 Horse Grimace Scale. The six action units of the Horse Grimace Scale include stiffly backward ears, orbital tightening, tension above the eye area, prominent strained chewing muscles, mouth strained and pronounced chin, and strained nostrils and flattening of the profile. Examples of how the Horse Grimace Scale is scored in four horses (a through d).

TABLE 19-15 Simple Visual Analog Scale for Horses

0		10

No Pain	Moderate Pain	Worst Pain
Alert	Anxious/uneasy	Focused-glazed stare
Active	Less attentive	Depressed
Playful	Uncomfortable	Tense/trembling
Interactive	Reduced activity	Frozen
Performs normally	Mouths food	Hypersensitive
Eat, drinks, defecates normally	Plays in water	Restless
	Lying down	Grunting
	Reluctant to perform work	Kicking, pawing, stomping
	Looking at flank	Abnormal posture
	Lame	Thrashing/rolling
	Elevated respiratory rate	Tachycardia
		Tachypnea
		Inappetence

Full mu agonist opioids such as morphine and methadone and the mixed kappa agonist/mu antagonist butorphanol are commonly used opioids in horses. An approved buprenorphine product (Vetergesic® Multidose) is available in the United Kingdom in horses, and research suggests that buprenorphine may provide better analgesia than butorphanol for up to 6 hours or longer.[47] In the UK it is labeled for IV use in horses 5 minutes after administration of an IV sedative, and the dose may be repeated (with sedation) if necessary, once, after not less than 1 to 2 hours. In the United States, buprenorphine use in horses is currently extra label.

The sedation and analgesic effects of (α_2) adrenergic agonists are reliable and profound in horses, which make this drug class widely used. A perfect example of this in clinical practice is in severely painful colic cases in which (α_2) adrenergic agonists will do an excellent job of quieting a distressed horse and providing it with pain relief while an opioid alone such as morphine would not. Unwanted sedation and significant cardiovascular effects remain the largest adverse effects seen by this drug class. In horses, (α_2) adrenergic agonists have been used extensively as a single dose to provide short term analgesia for diagnostic procedures, via CRIs to aid in analgesia in the hospital or during anesthesia, and as part of an epidural technique.

Ketamine is commonly used as an anesthetic induction agent, but it can also be used in sub-anesthetic dosages for analgesia in awake or anesthetized horses. Ketamine's main proposed benefit is that it has an anti-hyperalgesic effect which is beneficial in treating or preventing central sensitization and chronic pain states in the horse. Ketamine in sub-anesthetic dosages may also be useful to reduce the amount of other pain drugs needed (such as opioids) and should be considered whenever treating refractory or very painful patients. Ketamine CRI during anesthesia should be discontinued prior to recovery to avoid ataxic and rough recoveries.[48] Ketamine also has minimal respiratory depression and minimal effect on gastrointestinal motility.[49]

Many diagnostic and surgical procedures can be performed safely and humanely in a standing horse by using a combination of sedation, physical restraint, and regional anesthesia. Regional analgesia can also be used as an alternative or adjunct to systemic analgesic therapy for long-term pain control with minimal adverse effects.[46]

Local anesthetics can be administered locally, topically, regionally, intra-articularly, epidurally, and as a CRI (Figure 19-35 and Table 19-16). Lidocaine specifically has been used systemically via the IV route in horses for analgesia, as an anti-inflammatory, and to resolve postoperative ileus and return normal bowel function (which can ultimately shorten the length of hospitalization). Data on each of these applications is limited and/or mixed in efficacy. Fortunately, IV lidocaine when used at appropriate dosages is inexpensive and appears to be largely safe if administered at non-toxic levels. As a result,

Courtesy of Kate Lafferty, CVT, VTS (Anesthesia/Analgesia)

FIGURE 19-35 Intratesticular block for castration in a stallion. The scrotum is prepped with an antiseptic solution. When performing this technique, clean gloves should be worn as well as a new needle (18–20 gauge 1.5 inch) placed on the syringe of 2% lidocaine. The lidocaine is injected as the skin is tensed over the testicle. The needle should be inserted through the skin below the tail of the epididymis and pushed into the center of the testicle at an angle of approximately 30 degrees. Aspirate to verify the needle is not in a blood vessel, then inject enough lidocaine into the testicle so that it palpates firm. Remove the syringe and needle to complete the procedure.

TABLE 19-16 Nerve Blocks Commonly Used for Horses

Nerve Block	Area of Analgesia	Use
Infraorbital	Nose, muzzle, incisors, and rostral maxilla on injected side	• Desensitizes the infraorbital nerve (part of maxillary nerve as it exits the maxillary sinus) • Horses do not tolerate this nerve block well and tend to move the head as the nerve is touched by the needle • Use for dentistry and procedures of the maxilla
Maxillary	Upper teeth, maxilla, mucosa inside the nose and maxillary sinuses, muzzle, and nose	• Desensitizes the maxillary branch of trigeminal nerve • Used for dentistry and procedures of the maxilla
Mandibular	Lower teeth, mandible, and lower lip on injected side	• Desensitizes mandibular branch of trigeminal nerve • Used for dentistry and procedures of the mandible
Paravertebral	Flank (paralumbar fossa)	• Desensitizes the dorsal and ventral branches of T18, L1, and L2 • Used for surgery in the flank area of a standing horse
Limb	Area of limb depending on how distally the local anesthetic is given	• Performed distally in the limb first then progresses to more proximal blocks to determine area of lameness • Used diagnostically
Bier (IV regional anesthesia)	Limb (depending on where tourniquet is placed)	• Tourniquet is applied and local anesthetic injected usually against the direction of blood flow. • Common sites are proximal or distal to the carpus or tarsus • Used for limb perfusion
Caudal epidural	Pelvic organs, hind quarters, genitalia, tail, and skin	• Desensitizes caudal to coccygeal space1(local anesthetic is injected into the epidural space between Co1 and Co2) • Used to procedures on the tail, anus, vulva, perineum, rectum, vagina, urethra, and urinary bladder
Intratesticular	Scrotum and testicles	• Desensitizes scrotum and testicles by injecting local anesthetic directly into center of testicle • Used for equine castration

IV lidocaine can be considered a potential therapeutic option in moderately to severely painful horses when given in conjunction with other analgesics. Other local anesthetics are never used IV due to their low safety margins. When using a lidocaine CRI, it is important to discontinue its use 15 to 30 minutes prior to anesthesia recovery to reduce the risk of ataxia and injury during the recovery period.[50] Overall, adverse effects of lidocaine are minimal at proper dosages. With higher plasma concentrations, lidocaine can produce adverse effects, which include seizures, cardiac arrhythmias, and methemoglobinemia.[51]

Epidural or spinal administration of high dosages of local anesthetics in horses is avoided or done with care as patients may lose motor control of the hind limbs, which would be undesirable and unsafe.[3] Small volume epidurals with local anesthetics can be administered via the first and second coccygeal space to provide regional anesthesia to the caudal portion of the horse including the vagina, perineum, and anus. Epidural catheters can also be placed allowing for ongoing administration of local anesthetics for up to 14 days.[52]

Other drugs including tramadol and adjunctive drugs such as gabapentin have also been used in horses. Complementary therapies such as acupuncture, physical therapy, and chiropractic therapy are used in horses to alleviate pain. Many of these alternate treatments are described in Chapter 18 (Veterinary Physical Rehabilitation); the reader should research equine specific therapies and refer patients to an equine specialist if the use of these therapies is desired.

Swine Anesthesia

Swine can be challenging to anesthetize compared to other large animals because they are tricky to restrain and intubate. Swine come in a variety of sizes and can range in size from being piglets (0.5 to 3 kg) or miniatures (10–30 kg) to mature sows and large boars weighing in excess of 400 kg. Pigs have an excessive amount of jowl, tongue, and cheek, a difficult to open mouth, a relatively narrow glottis, pharyngeal diverticulum (outpouching), and a semi-V shaped trachea, which makes passage of an ETT more challenging than other species for inexperienced anesthetists. Taking the time to learn about these unique characteristics of swine will make understanding and performing swine anesthesia more successful.

General Considerations for Anesthesia in Swine

Just as in other large animal species, some procedures can be safely and appropriately performed on standing pigs using physical and chemical restraint. Locoregional anesthetic techniques and analgesic medications for swine are described in the Swine Analgesia section.

Intramuscular injection in pigs may necessitate the use of longer needles (up to 2.75 inches [7 cm] in length) because of the amount of subcutaneous fat present especially on large pigs with excess adipose tissue. Using longer needles requires adequate restraint, and the patient's excess fat can make it difficult to assess whether medication intended to be administered deep into the muscle was inadvertently administered more shallowly into the poorly vascularized subcutaneous tissue. Medication deposited into the subcutaneous tissue is not well absorbed. Therefore, if the patient inadequately responds to a drug, consideration should be made as to whether the drug was inadvertently deposited into the fat layer. For most pigs the preferred site for IM injection is in neck muscle because it is a region well tolerated for injection by pigs, has a low amount of subcutaneous fat, causes the least amount of muscle damage to the carcass if the animal may later be used for food, and has few superficial veins and arteries, which reduces the risk of accidental injection of drug into the blood stream (Figure 19-36). Typically, giving an IM injection to a pig is similar to giving an IM injection in a dog. Keep in mind to aspirate before administration and that many pigs will markedly vocalize. One IM injection technique that has worked well for difficult pigs is described as follows:

- Gather supplies needed which include:
  - Standard 21 gauge or 19 gauge butterfly catheter with 12 inch (30.5 cm) tubing for injection with 1 ½" needle length
  - Syringe(s) with premedication drug(s) in it
  - Sterile saline in a 3 to 5 mL syringe
  - Restraint assistance
- Next, approach the pig slowly if it is loose in a pen. If possible try to get it heading into a corner or use squeeze boards (Figures 19-37a and 19-37b)

in pens, which assist with immobilization and confine them into a stationary corner.

- With the unattached butterfly catheter, approach the pig and insert the needle into the neck musculature.
- Once the pig has settled, attach the syringe to the end of the butterfly catheter and inject the medication. Occasionally the pig will run around during this procedure, and having the 12-inch extension tubing allows you to follow the pig and still inject the medications simultaneously.
- Remove the medication syringe and attach the sterile saline syringe and inject it to ensure that the medication is injected completely into the muscle and does not remain in the plastic tubing.
- Remove the butterfly catheter and leave the pig alone in a dark quiet environment for at least 15 to 20 minutes to allow medications to take effect.

(a)

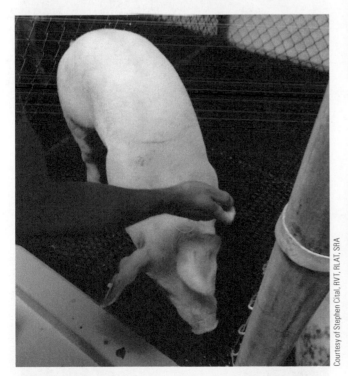

FIGURE 19-36 Preanesthetic IM injection into neck of pig. The neck is the preferred site for IM injections because it has the least amount of subcutaneous fat and causes the least amount of muscle damage to the carcass

Courtesy of Stephen Cital, RVT, RLAT, SRA

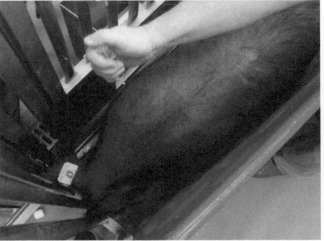

(b)

Courtesy of Caroline Mead CVT, RLATg

FIGURE 19-37 Pigs should be handled minimally prior to anesthesia. (a) A restraint board. (b) The use of a restraint board guides pigs where they should move.

Preanesthetic Preparation

Preanesthetic preparation for swine is similar to that of the bovine patients. Swine first should be observed from a distance before initiating any examination or procedures. Some pigs are docile and friendly (often those used as pets) while others may be excessively vocal or aggressive. A pig's temperament may be related to its age or sex with older boars or sows protecting their litters and potentially becoming aggressive. Observation of the patient at a distance can provide clues as to the patient's behavior, how they ambulate, how they breathe, and what conditions make them upset.

Once this observation phase is over, a physical exam should be performed on all swine prior to anesthesia. Having adequate equipment and staff to help restrain the pig is important. A more limited physical exam may need to be performed based on the patient's temperament. Preanesthetic blood work including a CBC, serum chemistry, serum electrolytes, and other diagnostic testing may be useful prior to anesthetizing a particular pig and is best determined by assessing the patient's individual needs and risks. Gaining IV access prior to sedation can be quite challenging in some patients, which may limit the amount of blood work performed. Following appropriate assessment of the pig, swine are generally kept off food for 6 to 12 hours, similar to a dog. Water is permitted until the beginning of the procedure. Pigs can vomit and therefore aspirate, so it is important to hold them off food.

Premedication, Sedation, and Restraint

Pigs can become vocal and loud when restrained; therefore, it is often advisable to sedate or tranquilize pigs via the IM route prior to restraining them or placing an IV catheter (one method of IM injection is described in the General Considerations for Anesthesia in Swine section). Pigs should be handled as gently and as little as possible; therefore, directing the pig into a corner or using restraint boards in pens to assist with confining the patient can be useful (see Figures 19-37a and 19-37b). Pigs should be left in a quiet dark area after drug administration for at least 15 to 20 minutes to allow the medication to take maximum effect; repeated stimulation will cause less than desirable tranquilization, and sedation in pigs and may lead to increased stress and/or the development of malignant hyperthermia in certain heavily muscled or lean breeds. Malignant hyperthermia is described later in this chapter.

Often several drugs are administered together for premedication in pigs. While all drug classes can be utilized in pigs, some drug classes provide more reliable sedation and tranquilization than others. Remember different species have unique responses to the various drug classes. In general, pigs sedate well with (α_2) adrenergic agonists and benzodiazepines and less reliably with acepromazine and ketamine combinations. In smaller, quieter, or sick pigs premedication with a benzodiazepine and an opioid may provide adequate sedation. A common example is midazolam and butorphanol. Diazepam can be utilized in pigs, but its poor absorption when given by the IM route usually makes midazolam a better choice. Other opioids such as buprenorphine or a full mu agonist opioid can be selected if improved analgesia is indicated; however, the level of sedation provided may be altered (e.g., buprenorphine may provide less sedation than butorphanol). If this combination fails to be effective or if the pig is larger, healthier, aggressive, or more agitated to begin with, the addition of an (α_2) adrenergic agonist greatly affects the level of sedation. A common premedication drug combination in swine is dexmedetomidine, midazolam, and butorphanol or buprenorphine. Dexmedetomidine is often selected over older (α_2) adrenergic agonists such as xylazine although these drugs may still have a place in some settings or when other drugs are not available. Alfaxalone given IM

can also be utilized in pigs, often in conjunction with a benzodiazepine and opioid. This drug combination can work well in small pigs, but the drug cost and drug volume of alfaxalone in large swine make the choice prohibitive. Another drug that has been used predominantly in swine is azaperone (Stresnil®). Azaperone is a butyrophenone neuroleptic drugs with tranquilization and antiemetic effects. The drug is an effective tranquilizer in pigs; however, the drug is not currently available in United States markets outside of special compounding pharmacies. This has made the drug's use fall dramatically as it is often unavailable. Other drug options such as ketamine, tiletamine/zolazepam, and acepromazine can be used in swine but are often not particularly useful in improving sedation and tranquilization. Some advocate giving pigs metoclopramide (Reglan®), maropitant (Cerenia®), or another antiemetic as part of their premedication plan to reduce postoperative nausea and vomiting. Just as in other species, anticholinergic drugs are not routinely administered as preanesthetic medication to swine. Although they may salivate more than other large animal species, they infrequently develop clinically relevant bradycardia during induction or the maintenance of anesthesia. As a result, the administration of anticholinergic drugs should be reserved for the treatment of life-threatening bradyarrhythmias or cardiac arrest.

Venous Catheterization

A pig's ear (auricular) vein is one of the most accessible and easy to visualize for IV catheter placement and is often the anesthetist's vessel of choice (Figure 19-38). Applying a tourniquet around the base of the ear results in venous distention and improved vessel visualization and may help facilitate catheter placement. Other common but more technically challenging venous

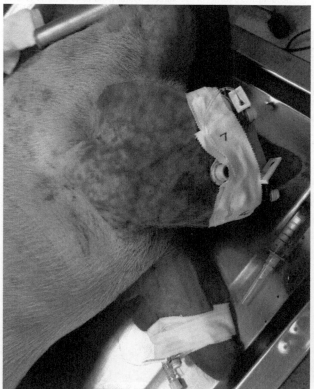

FIGURE 19-38 The auricular vein is the most accessible and easy to visualize in pigs for IV drug administration. Applying a tourniquet around the base of the ear causes vessel distention and easier visualization.

access sites include the cephalic, saphenous, and jugular vein. Venipuncture of the jugular vein is often a blind technique, and the use of ultrasound may make this technique markedly more successful. An appropriately gauged IV catheter is selected (often a 20–22 gauge over-the-needle IV catheter). The modified Seldinger technique can also be used to secure the catheter for long term usage (see Chapter 4 [Fluid Therapy and Intravenous Catheterization]). The catheter should then be secured using suture, tape, or some combination thereof.

Induction Procedures and Techniques

As with other large animal species, general anesthesia in swine can be achieved by administering a variety of medications with the overall aim of ensuring analgesia, muscle relaxation, and immobility. Depending on the individual animal, induction may be followed by administration of inhalant anesthetic agent or total intravenous anesthesia (TIVA); therefore, every patient needs an individualized anesthetic plan and drugs should be selected based on this plan.

Induction Medications

Induction agents in swine are selected based upon the same factors that you would use to select a drug in a small animal patient. Propofol, alfaxalone, ketamine with or without a benzodiazepine, and etomidate have all been used successfully in swine. An induction agent can be used to maintain TIVA for short procedures or they can allow for intubation and maintenance using an inhalant anesthetic agent. In some situations where endotracheal intubation is not performed, a pig may be sedated and then maintained on a face mask with an inhalant anesthetic agent but not actually intubated. Whether this technique will be acceptable will depend on the individual patient and its needs. Induction can also be performed by administering inhalant anesthetic agents with a face mask. But this should only be done if the patient is adequately sedated from drugs given via the IM route.

Intubation of Swine

Endotracheal intubation can be performed in swine in lateral, dorsal, or ventral recumbency; however, this author prefers dorsal recumbency as it allows the excessive pharyngeal tissue to "fall away" from the glottis, thus providing a more open airway (Figure 19-39). The sizes of ETT suitable for pigs are unexpectedly small when compared with those used in dogs of a similar body weight. For example, a 6 mm ETT may be the largest that can be passed in a pig weighing about 25 kg, a 9 to 10 mm ETT might be suitable for a 50 kg animal, and large boars and sows may accommodate an ETT of 16 to 18 mm diameter. A local anesthetic should be applied to the larynx once visualized, usually 2% lidocaine, as pigs are highly prone to laryngospasm. The use of a laryngoscope is advised and a straight Miller laryngoscope blade appears well suited for pigs. Standard laryngoscope blades, 195 mm or longer, are often sufficient for swine less than 50 kg. For larger pigs, modified custom blades with 3 to 5 cm extensions may be needed. The use of a guide tube can also help intubation. Once the larynx is visualized with the aid of the laryngoscope, the blunt ended guide tube can be directed into the rostral larynx and the ETT slid over the guide tube into the trachea. This process should be gentle, as being forceful may easily damage the pig's larynx.

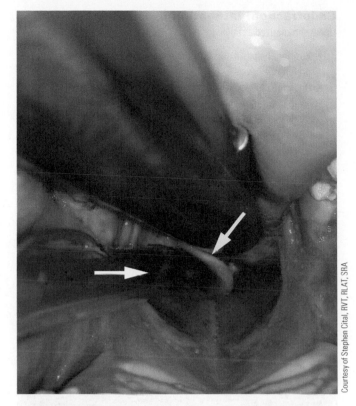

Courtesy of Stephen Cital, RVT, RLAT, SRA

FIGURE 19-39 Laryngeal structures of a pig. The view of the larynx of a pig with ETT (white arrow) in place and pig in dorsal recumbency. The epiglottis is labeled with the yellow arrow.

Equipment for endotracheal intubation in pigs:

- Three sizes of ETT (Table 19-17). Choose a size that is considered the best fit and then one size larger and one size smaller so that three different size ETTs are within quick reach during intubation.
  - The ETTs should be checked for patency prior to use and should be clean and free from debris. The ETT may be lubricated with sterile water based lubricant to help facilitate smooth passage into the trachea.
- Mouth gag to hold jaws apart (only use if necessary)
- Laryngoscope (straight Miller laryngoscope is suggested)
- Topical anesthetic (2% lidocaine) spray
- Flexible plastic guide tube
- Gauze sponges to grab the tongue
- 10 mL syringe to inflate the cuff

TABLE 19-17 Swine Endotracheal Tube Sizes for Swine*

Swine Size	Tube Size
Piglets	3–5 mm
10–15 kg	5–7 mm
20–50	8–10 mm
50–70	10–12 mm
100–200	12–14 mm
Larger pigs	16–18 mm

*Muir, W. W., Hubbell, J. A., Bednarski, R. M., & Lerche, P. (2013). Anesthetic procedures and techniques in pigs. In *Handbook of veterinary anesthesia* (5th ed.). St. Louis, MO: Elsevier/Mosby.

Endotracheal Intubation in Swine
Endotracheal intubation is summarized as follows:[53]

- Position the pig in dorsal recumbency or anesthetist's recumbency of preference.
- Open the pig's mouth and grasp the tongue with a gauze sponge.
  - If performing endotracheal intubation alone, insert mouth gag (Figure 19-40a)
- Insert the laryngoscope to identify the arytenoid cartilages and visualize the larynx.
- Spray 2% topical lidocaine onto the arytenoids and wait a minimum of 30 seconds for the drug to take effect.
- Advance a small-diameter guide tube approximately 2 cm into the trachea.
- Place the ETT over the guide tube and slide it down into the trachea. If resistance is met upon entering the larynx, gently rotate the tube 90 degrees to redirect the bevel. (Figure 19-40b).
- Once the ETT has advanced successfully into the trachea, remove the guide tube.

- Confirm placement into the trachea with $EtCO_2$ readings and auscultating bilateral lung sounds during ventilation.
- Inflate the ETT cuff and secure the ETT in place.

Maintenance of Anesthesia

Injectable anesthetic agents used for induction can also be given "to effect" for maintenance of general anesthesia or as a CRI for TIVA as previously discussed. This may be desired in patients that are not intubated. Another option for those patients in which endotracheal intubation is not performed is to maintain anesthesia with an inhalant anesthetic agent via a face mask. Oxygen flow rates of 1–3 L/min should be provided to pigs in either scenario.

For longer or more invasive procedures, inhalant anesthetic agents are selected. For smaller pigs (< 150 kg), a typical small animal circle system can be used while larger pigs (>150 kg) will require a large animal anesthesia machine and circle breathing system. Inhalant anesthesia can be maintained in swine with isoflurane (MAC value 1.45%), sevoflurane (MAC value 2.4%), or desflurane (MAC value 9.2%). The use of inhalation anesthetic agents in swine may trigger malignant hyperthermia. Patients who have a history of malignant hyperthermia should not have inhalation anesthetic agents used and TIVA (e.g., propofol, alfaxalone) is more appropriate.

Monitoring

For typical swine anesthesia, monitoring of the cardiovascular and pulmonary systems involves assessment of anesthetic depth, monitoring of vital signs, and the use of pulse oximetry, capnography, blood pressure (ideally direct), and ECG.[54] For more complex cases or when working with swine in research, it is not uncommon to use additional monitoring modalities such as central venous pressure (CVP) (Figure 19-41), blood gas analysis, pulmonary artery catheterization and pulmonary artery occlusion pressure monitoring, cardiac output monitoring, spirometry, **agent gas analysis** (an instrument that determines the amount of inhalant anesthetic agent present at inspiration [coming from machine] and expiration [having just left the alveoli]), and other

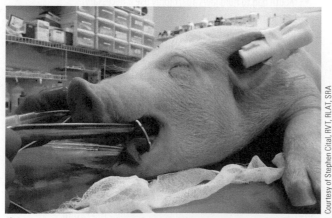

(a)

(b)

Courtesy of Stephen Cital, RVT, RLAT, SRA

FIGURE 19-40 Endotracheal intubation of a pig. (a) Placement of the mouth gag (lock nut in this example) behind the top and lower incisors. (b) Using a 350 mm Miller laryngoscope blade, the glottis is visualized and intubated with a 7 mm ETT.

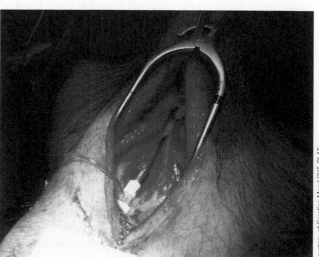

Courtesy of Caroline Mead CVT, RLATg

FIGURE 19-41 A jugular cut back in a pig to allow for central venous pressure determination.

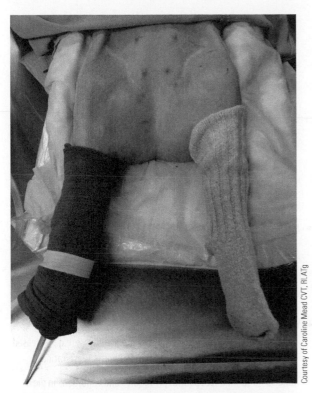

Courtesy of Caroline Mead CVT, RLATg

FIGURE 19-42 The extremities of young patients should be kept warm during anesthesia to prevent hypothermia.

techniques. Intraoperative fluids should be administered during swine anesthesia, and a balanced crystalloid solution such as lactated ringer's solution should be administered at 5 mL/kg/hr to hydrated, normovolemic pigs.

Thermoregulation is an extremely important consideration during swine anesthesia as pigs have little hair but increased amounts of subcutaneous fat. Prevention of hypothermia will often require circulating warm water, forced air (Bair Hugger®), and conductive fabric patient warming (HotDog®) blankets. The extremities of young patients should be kept warm to prevent hypothermia and the use of socks or "bubble wrap" may be useful (Figure 19-42). Table 19-18 summarizes normal values of anesthetized swine.

Malignant Hyperthermia in Swine

Malignant hyperthermia, also referred to as porcine stress syndrome, is a genetic condition in certain heavily muscled or lean breeds of domestic swine, such as the Landrace, Poland China, and Duroc. The condition has also rarely been reported in other species such as humans, dogs, cats, and horses. The condition is inherited as an autosomal dominant gene though influence of this gene in pigs has been drastically reduced in recent years by selective genetics. The condition in swine may be induced by stress (e.g., transport), the use of inhalant anesthetic agents (halothane, isoflurane, sevoflurane, desflurane), or by the use of the neuromuscular blocking agent succinylcholine. Although fairly rare, the anesthetist never knows which patient could suddenly develop malignant hyperthermia. The signs of malignant hyperthermia in pigs and other species include:

- Sudden onset of signs during anesthesia
- Rapidly increasing $EtCO_2$ level
- Rapidly increasing temperature
- Tachycardia and arrhythmias
- Hyperkalemia
- Dramatic muscle rigidity, often leading to toe separation
- Increased blood lactate levels
- Very hot, pink blotchy skin
- Very red mucous membranes
- Retraction of the eyeballs far into the sockets due to increased extraocular muscle tone
- Death

Once clinical signs are observed, patients often deteriorate rapidly, and treatment may be unsuccessful. Treatment for malignant hyperthermia must be immediate, is potentially lifesaving, and includes:

- Immediately removing the patient from an inhalant anesthetic agent. This likely involves beginning TIVA if the patient must remain anesthetized (e.g., if the patient is in the middle of a surgery). The patient should ideally not remain connected to an anesthesia machine to avoid breathing residual inhalant anesthetic agent.
- Oxygen supplementation.
- Aggressive mechanical ventilation in an effort to combat life-threatening hypercapnia.

TABLE 19-18 Normal Values for Anesthetized Swine*

Parameter	Value
Respiratory rate	10–25 breaths/minute
Tidal volume	10–15 mL/kg
Heart rate	80–130 bpm
Capillary refill time	< 2 seconds
Arterial pressure (systolic)	> 100 mm Hg and < 160 mm Hg
Mean arterial pressure (MAP)	> 60 mm Hg
Partial pressure of arterial carbon dioxide ($PaCO_2$)	35–45 mm Hg
Partial pressure of arterial oxygen (PaO_2)	> 60 mm Hg
Arterial pH	7.35–7.45
End tidal carbon dioxide ($EtCO_2$)	35–45 mm Hg

*Bollen, P. J. A., Hansen, A. K., & Rasmussen, H. J. (2000). *The laboratory swine: A volume in the Laboratory Animal Pocket Reference Series.* M. A. Suckow (Ed.), pp. 80–81. Boca Raton, FL: CRC Press.

- Administering dantrolene (Dantrium®) intravenously.
- Cooling the patient with cool IV fluids, wet towels, fans, etc.
- Administering sodium bicarbonate to correct acid/base balance.
- Treatment of hyperkalemia with dextrose and insulin to promote movement of potassium into the cells.

In patients with a known history of malignant hyperthermia or with familial history, adequate premedication to reduce stress and TIVA should be used. Inhalation anesthetic agents must be avoided and the drug dantrolene used prophylactically.

Recovery

Anesthetic recovery of a pig should occur in a quiet, warm, controlled environment and not in a loud pen with other pigs. The pig should be placed in sternal position and oxygen supplementation provided. Vital signs, temperature, oxygenation, ventilation, and pain should be carefully monitored. When the pig is able to protect its own airway, the ETT is removed. The general principles applicable to extubation and recovery for small animals apply to pigs, including detection and treatment of hypothermia. Typically the IV catheter is removed before full awakening, although if a pig is to be hospitalized it may be prudent to secure the IV catheter for administration of IV medications. Once the pig is able to support its head, the monitoring process can become more "hands off," which generally includes watching the pig from afar.

Swine Analgesia

Pigs attempt to hide pain, which makes its identification difficult. The scarcity of agents licensed for use in swine, drug withdrawal times, and the cost and labor involved in administering analgesics are additional challenges of controlling pain in swine. Despite these concerns, pain control can improve productivity and as effective products become available, analgesia will be a part of good animal husbandry in this species.

Normal behaviors observed in swine include interest in the surroundings (including staff), willingness to move around, explorative behavior, tail wagging, reaction to handling, vocalization when presented with feed, and willingness to eat. When swine react differently from this pattern, pain and distress might be the cause; for example, when swine lie unresponsive in sternal recumbency or are reluctant to eat, pain or distress should always be considered.[55]

Pigs that have chronic pain can exhibit all the signs of acute pain plus the following:[55]

- Altered movement, posture, appearance, or behavior:
    - Changes in ambulation (slower pace, occasional limping)
    - Carpal walking (walking on knees)
    - Standing with a hunched posture
    - Persistent sitting and reluctance to lie down
    - Frequent shifting of position when recumbent
    - Difficulty rising and/or lying down
    - Limping
    - Decreased roaming, exploring, and grazing
    - Changes in urination and defecation habits (toileting close to the sleeping area, urinating or defecating less frequently)
    - Absence of bed making

    - Apathy
    - Unwillingness to move
    - Hiding in bedding
    - Decreased activity levels
    - Remaining isolated from others
    - Kicking at the abdomen
    - Squealing when painful areas are palpated
    - Trembling
    - Lip smacking, teeth grinding, "teeth champing"
    - Rapid tail wagging or tail "flicking"
    - Increased irritability and/or aggression
    - High pitched vocalizations, squealing, screaming, staccato grunting
- Modified eating habits
    - Decreased food intake
    - Reluctance to eat
    - Desire to eat foods other than the regular feed
    - Slow ingestion of a meal
    - Leaving small amounts of feed in the bowl.
- Variation of vital signs
    - Easily measurable indicators of physiological stress (e.g., increased heart rate or increased rate and altered depth of respiration)

Table 19-19 summarizes painful procedures and conditions in pigs.

Overweight pigs are extremely common, which results in increased rates of arthritis. There is no normal weight for a pet pig, but animals should be maintained at a body condition score of 2 to 2.5 from nine months to eight years of age, and at 2 to 3 after that (the scale ranges from 1 to 5) (Figure 19-45).[56] Obesity can cause sores on pressure points and can result in self-inflicted scratches and skin-fold infections. Overweight pigs have difficulty moving, which may be due to severe arthritis, and increased snoring when sleeping.[56]

Arthritis is the typical disease seen in geriatric pigs, yet it can be seen on radiographs as early as 2 years old. The elbow, distal limb joints (metacarpals, metatarsals, phalanges), and spine are typically affected, and larger joints such as the shoulder and coxofemoral joints are often spared. Arthritic pigs may initially demonstrate difficulty rising in the morning, and a stiffness that decreases with movement throughout the day. Lameness is typically associated with a forelimb and is usually unilateral, but may be shifting. Decreased roaming, exploring, and grazing are common signs. Arthritic pigs may frequently seek out warm, sunny locations for resting. More severe signs include ambulation in a kneeling position ("carpal walking"). Other signs that are indicative of severe arthritic pain include sleeping for most of the day, spending the majority of time in one area, such as the shelter or sleeping area, and a hunched posture that is often mistaken for constipation. The range of motion of elbow joints is often markedly decreased, and even carpal walking may be impossible. Valgus deformity and enlargement of the elbow joint are commonly observed in advanced cases.

Pigs in pain might show changes in their overall demeanor, social behavior, gait, and posture. For example, pigs normally squeal and attempt to escape when handled; however, those in pain have exaggerated squealing, will squeal when painful areas are palpated, or adults may become aggressive. Pigs with moderate pain may simply reduce activity levels, be less responsive to familiar handlers, and reluctant to feed or drink.

TABLE 19-19 Painful Procedures and Conditions in Pigs*

Painful procedures and conditions in pigs
• Castration—baby pigs, pet pigs, and adult or larger pigs
• Aural hematoma
• Cesarean section
• Amputation—digital
• Atresia ani
• Entropion—frequent in pot-bellied pigs
• Exploratory laparotomy
• Fractures
• Hernia—inguinal and umbilical
• Hoof trimming
• Joint lavage
• Mastectomy
• Ovariohysterectomy
• Prolapse—rectal, vaginal, uterine (Figure 19-43)
• Tail docking (Figure 19-44)
• Tooth clipping
• Tusk removal
• Cryptorchidism
• Vasectomy

*Bollen, P. J. A., Hansen, A. K., & Rasmussen, H. J. (2000). *The laboratory swine: A volume in the Laboratory Animal Pocket Reference Series*. M. A. Suckow (Ed.), pp. 80–81. Boca Raton, FL: CRC Press.

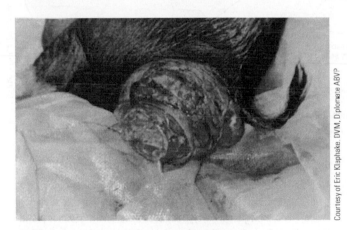

Courtesy of Eric Klaphake. DVM, D plomate ABVP

FIGURE 19-43 Rectal prolapse in a pig.

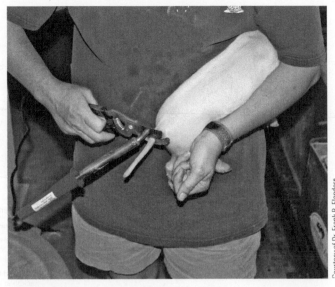

Courtesy of Dr. Frank B. Flanders

FIGURE 19-44 Tail docking in a pig.

Pain Scoring Systems in Swine

There are pain scales designed for use in swine. One example is the Pain Assessment for Surgical Cases in Pigs that uses number designations to reference a lack of pain perception or a painful reaction[57] (Table 19-20). Another example is the Piglet Grimace Scale (PGS) (University of Guelph), which assesses three facial expressions: ear position, cheek tightening/nose bulge, and orbital tightening (Figure 19-46).[58] Each expression is scored based on whether it is absent (score of 0), moderately present (score of 1), or obviously present (score of 2), with the exception of orbital tightening, which is scored on a 2-point scale of absent (score of 0) and present (score of 1).[58] The maximum total score is 5 with total scores of 0 to 1 correlating to a piglet experienced "no-to-low pain" and scores of 3 to 5 representing "moderate-to-high pain."[58] A higher grimace score indicates a greater pain state. A second grimace scale for piglets (Newcastle University) assesses ten facial expressions, which include temporal tension, forehead profile, orbital tightening, tension above the eyes, cheek tension, snout angle, snout plate change, upper lip contraction, lower jaw profile, and nostril dilation.[59] Each expression is scored as not present (0), moderately present (1), and obviously present (2) and the ability to provide a "don't know" score. This PGS can be accessed at https://www.frontiersin.org/articles/10.3389/fvets.2016.00100/full.

TABLE 19-20 Pain Scoring System for Swine*

Level of Pain	Description
1	Deep palpation of the surgical site and immediate surrounding tissue does not provoke a response. Remember that freshly opened tissue is susceptible to infection, and palpation should be done with a gloved hand.
2	Deep palpation of the surgical site and immediate surrounding tissue provokes a response, but a similar response can be seen on the contralateral side or limb, suggesting a hyperesthetic or hyperreflexive state.
3	Deep palpation of the surgical site and immediate surrounding tissue that provokes a response much greater than a similar stimulus on a nonsurgical part of the body. Probably indicative of some pain, and appropriate analgesic should be administered.
4	Deep palpation of the surgical site and immediate surrounding tissue that provokes a response much greater than a similar stimulus on a nonsurgical part of the body and accompanied by vocalization in an otherwise quiet patient. Requires analgesia.

*Swindle, M. M. (2000). Anesthesia, analgesia, and perioperative care. In *Swine in the laboratory surgery, anesthesia, imaging, and experimental techniques* (2nd ed., pp. 49, 55, 65–66). Boca Raton, FL: CRC Press.

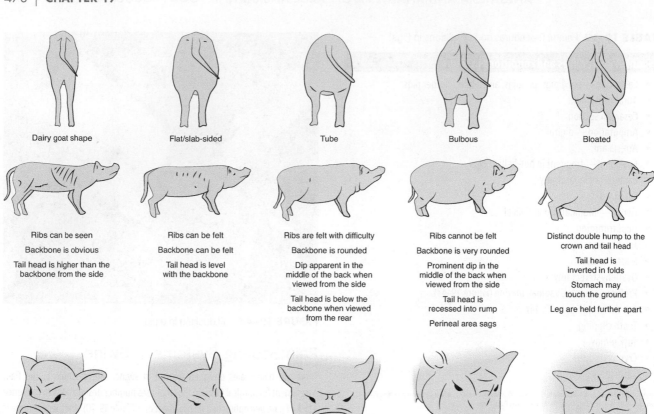

Dairy goat shape   Flat/slab-sided   Tube   Bulbous   Bloated

Ribs can be seen	Ribs can be felt	Ribs are felt with difficulty	Ribs cannot be felt	Distinct double hump to the crown and tail head
Backbone is obvious	Backbone can be felt	Backbone is rounded	Backbone is very rounded	Tail head is inverted in folds
Tail head is higher than the backbone from the side	Tail head is level with the backbone	Dip apparent in the middle of the back when viewed from the side	Prominent dip in the middle of the back when viewed from the side	Stomach may touch the ground
		Tail head is below the backbone when viewed from the rear	Tail head is recessed into rump	Leg are held further apart
			Perineal area sags	

Eyes can be clearly seen	Eyes can be clearly seen	Eyes can be clearly seen	Head is rounded	Folds of flesh hang from the head
			Ears separating and recessed into the head	Folds of fat on the head push the ears forward
			Eyes seen with difficulty	Pig may be fat blind

Courtesy of https://cochonsminiatures.com/en/final-size/

FIGURE 19–45 Body condition scores (BCS) in pigs can be used to maintain productivity in a swine herd by guiding choices for better nutrition and helping manage feed costs, improving reproductive performance sows, and preventing excessive weight in pigs.

Treating Swine Pain

When treating pigs for pain, keep in mind that they may be raised for food; therefore, knowledge of drug withdrawal times are important if swine or their products are being used for human consumption.

Tech Tip 🐾

When administering injectable medications to pigs, it is often best to inject IM into the neck muscles using a Butterfly needle (see Figure 4-9a) and extension set. Another option is to restrain the pig with restraint board (instead of chasing them around) and inject the drug IM. The neck muscle is often used to avoid damage to the ham muscle in the event that the pig is a food producing animal.

An NSAID such as carprofen can be combined with opioids such as buprenorphine to control pain during muscular or orthopedic surgery in swine. Flunixin meglumine is typically used alone and is good for providing analgesia for perioperative pain and chronic musculoskeletal conditions. Many NSAIDs are given to reduce fever due to respiratory disease. Most adverse effects are seen with long-term use of NSAIDs rather than for short-term postoperative pain. Not all NSAIDs (e.g., meloxicam and ketoprofen)[60] are effective in controlling pain in pigs, and because some NSAIDs are used extra label and may have longer withdrawal times, it is important to stay current of analgesic recommendations for use in pigs.

Opioids have been used in swine as adjuncts with other drugs to induce and maintain surgical anesthesia. They have also been used as CRI for maintenance.[61] Full mu agonist opioids, such as fentanyl, and partial mu agonist opioids, such as buprenorphine, are effective in pigs. Morphine is given epidurally at the lumbosacral space to produce analgesia caudal to the umbilicus.[62] Fentanyl can be administered as a CRI in the perioperative period or for moderate to severe pain. Transdermal fentanyl patches (12.5 or 25 mcg/hr)

The Piglet Grimace Scale

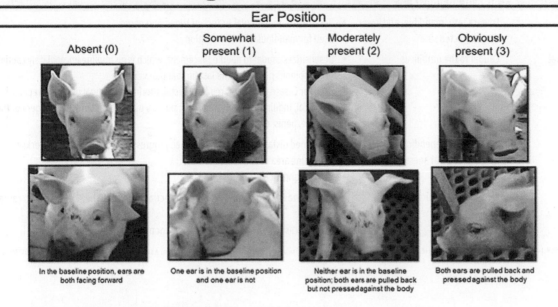

Ear Position

Absent (0)	Somewhat present (1)	Moderately present (2)	Obviously present (3)

In the baseline position, ears are both facing forward

One ear is in the baseline position and one ear is not

Neither ear is in the baseline position; both ears are pulled back but not pressed against the body

Both ears are pulled back and pressed against the body

- Determine ear position by looking at the base of the ear (point that connects to the piglet's head)

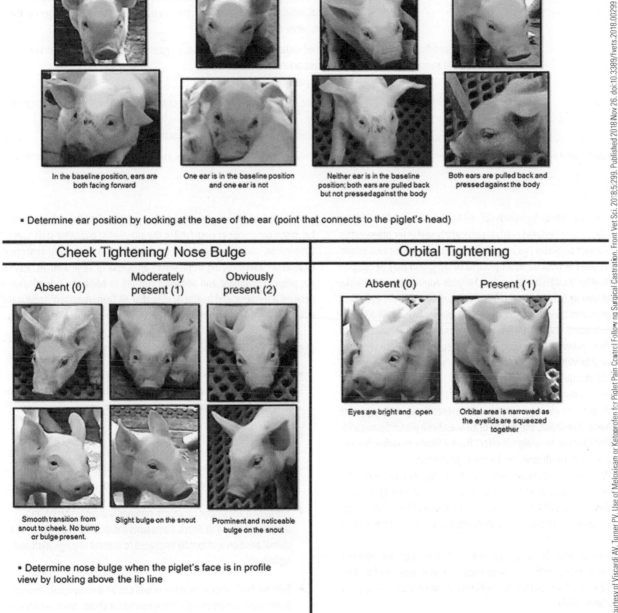

Cheek Tightening/ Nose Bulge

Absent (0)	Moderately present (1)	Obviously present (2)

Smooth transition from snout to cheek. No bump or bulge present.

Slight bulge on the snout

Prominent and noticeable bulge on the snout

- Determine nose bulge when the piglet's face is in profile view by looking above the lip line

Orbital Tightening

Absent (0)	Present (1)

Eyes are bright and open

Orbital area is narrowed as the eyelids are squeezed together

Courtesy of Viscardi AV, Turner PV. Use of Meloxicam or Ketoprofen for Piglet Pain Control Following Surgical Castration. Front Vet Sci. 2018;5:299. Published 2018 Nov 26. doi:10.3389/fvets.2018.00299

FIGURE 19–46 The Piglet Grimace Scale assesses three facial expressions: ear position, cheek tightening/nose bulge, and orbital tightening. When the pig is in pain, the ears are drawn back from baseline (forward) position and a bulge of skin is apparent on the snout in response to check tightening. Ear position and check tightening/nose bulge is scored according to whether it is not present (0), moderately present (1), and obliviously present (2). When the pig is in pain, the orbital area is narrowed as the eyelids are squeezed together. Orbital tightening is scored on a 2-point scale: Not present (0), present (1).

TABLE 19-21 Nerve Blocks Commonly Used for Pigs*

Nerve Block	Area of Analgesia	Use
Mandibular	Lower teeth, mandible, and lower lip on injected side	• Desensitizes mandibular branch of trigeminal nerve • Used for mandibular tusk extraction
Lumbosacral epidural	Caudal to the umbilicus	• Desensitizes caudal to needle placement, which is on midline immediately caudal to the spinous process of the last lumbar (L6) vertebra • Used for Cesarean section; repair of rectal, uterine, or vaginal prolapse; repair of umbilical, inguinal, or scrotal hernias; surgery of scirrhous cord; and surgery of the prepuce, penis, or rear limbs
Limb	Area of limb depending on how distally the local anesthetic is given	• Performed distally in the limb first then progresses to more proximal blocks to determine area of lameness • Used diagnostically
Intratesticular	Scrotum and testicles	• Desensitizes scrotum and testicles by injecting local anesthetic directly into center of testicle • Used for castration in pigs up to about 6 months of age

*Viscardi, A. V., Hunniford, M., Lawlis, P., Leach, M., & Turner, P. V. (2017). Development of a piglet grimace scale to evaluate piglet pain using facial expressions following castration and tail docking: A pilot study. *Front Vet Sci., 4*, 51. doi:10.3389/fvets.2017.00051

have been used successfully in pet pigs (27–82 kg), and can be efficacious for as long as 3 days.[63] The fentanyl patch is particularly useful for preoperative analgesia for major surgeries (such as uterine tumor removal or tusk extraction) when applied at least 12 hours prior to surgery, and may be complemented with other analgesics as needed. Suitable application sites are the interscapular area or behind the ear.[63] Oral tramadol is reported to be very effective in pigs, and they seem to do well with once to twice daily administration.[61] In advanced severe arthritis, tramadol and an NSAID may be given in combination. Individual pigs have variable responses to tramadol, thus the owner must be vigilant in observing for efficacy versus sedation, and the dosage should be adjusted accordingly.

Alpha-2 adrenergic agonists have been used successfully in swine alone or with other drugs (e.g., ketamine and benzodiazepines) to provide analgesia, chemical restraint, and as part of an anesthesia plan.[62] Some earlier evidence failed to show an analgesic effect from xylazine in swine, but this has not been observed with other (α_2) adrenergic agonists.

IV or IM ketamine in conjunction with other agents have been used to provide pain relief and short-term analgesia for a wide variety of procedures. Low dosage rates of ketamine administered as a CRI with or without other agents can be continued peri- and postoperatively for major surgical procedures.

The most commonly used local anesthetics in pigs are lidocaine and bupivacaine. The agents have been used in a wide variety of modalities as described in other species. Nerve blocks commonly used in pigs are described in Table 19-21.

SUMMARY

Anesthesia and analgesia for large animal and food producing species are similar to and utilizes many of the core principles utilized in small animal anesthesia and pain management; however, the anesthetist working with large animal and food producing species must also be familiar with species-specific differences in anatomy, physiology, and pharmacology. Veterinary

professionals must always remember that because of their size, large animals can prove dangerous or even fatal to the anesthetist and other staff. Careful observation, common sense, and quick thinking is paramount regardless of whether anesthetizing or providing analgesia to large animals. Having the proper equipment and adequate staffing in addition to developing a detailed anesthetic and analgesic plan that all team members understand will help veterinary personal provide excellent care to large animal species.

CRITICAL THINKING POINTS

• Providing anesthesia care to and appropriately controlling pain in large animal species while protecting the safety of veterinary personnel poses unique challenges for the veterinary team.

• Drug residue in food producing animals is a significant public health concern. Knowing specific drug withdrawal times for food animals is mandatory and can be found at the Food Animal Residue Avoidance Databank (FARAD) website at www.farad.org.

• The largest anesthetic concern in ruminants is regurgitation and aspiration leading to death. Ruminants should always be placed in a sternal position and rapidly intubated to prevent regurgitation and aspiration.

• Endotracheal intubation in ruminants can be accomplished through direct visualization using a laryngoscope for sheep, goats, and calves while most adult cattle are intubated manually with direct palpation of the larynx.

• The gold standard for assessing arterial blood pressure in large animals is direct blood pressure.

• Immobility is essential when properly anesthetizing horses. Anesthesia induction should never be performed in a horse unless it is deeply sedated and uninterested in external stimuli.

- Horses are typically intubated using a blind technique.

- Proper positioning of the equine patient under general anesthesia will keep the horse comfortable during and after the procedure.

- Maintaining normal blood pressure during anesthesia and ensuring the shortest duration of anesthesia possible is necessary to reduce the risk of postoperative myopathy in horses.

- There is no way to completely control the recovery phase with equine patients, but proper preparation and sedation can help increase the likelihood of a smooth and safe recovery.

- Pig intubation can be challenging because it is difficult to visualize the larynx and because of their anatomy. A laryngoscope, guide tube, and 2% lidocaine spray are crucial to success.

- Malignant hyperthermia is a rare but fatal condition in pigs that must be treated immediately and aggressively if it occurs.

- Every person in the veterinary practice has an important role to play in pain management of large animal species. Knowing the signs of pain in various large animal species and consistently using pain scoring scales will give veterinary professionals the confidence needed to identify, monitor, and manage pain in horses, cattle, goats, sheep, and pigs.

- The nociceptive pain pathway and general approach to the treatment of pain in large animals is similar to that of small animals.

- With many different analgesics and routes of drug administration available, providing pain relief for large animals is possible for a variety of situations and should be done in a similar fashion as in small animals.

- Large animal species are often unnecessarily undertreated for pain. Reasons given include lack of veterinarian education on the subject, excessive fear over adverse effects, cost, and concern about drug residues. These factors must be balanced with providing acceptable and ethical patient care.

- Numerous pain scales have been published for large animal species; however, all pain scales developed to date have limitations and may fail to identify pain in large animal patients. Therefore, when in doubt it is always best practice to treat patients for pain when suspicion arises and to assume that anything that would cause pain to a person would cause pain in a large animal species.

REVIEW QUESTIONS

Multiple Choice

1. How long should adult cattle be fasted prior to anesthesia?

    a. 18 to 24 hours
    b. 12 to 15 hours
    c. 5 to 8 hours
    d. 24 to 36 hours

2. In comparison to horses, cattle require a dramatically lower mg/kg dosage of which drug?

    a. Acepromazine
    b. Ketamine
    c. Xylazine
    d. Propofol

3. What signs of muscle relaxation could you expect to see when guaifenesin is given to a properly premedicated horse before induction?

    a. Falling down, swaying, compromised respiration
    b. Eye closure, slow heart rate, snoring
    c. Lack of interest in external stimulation and surroundings
    d. Droopy lips, rapid respiration and heart rate

4. When intubating cattle using direct palpation of the larynx, what is unique about this in comparison to species like dogs and cats?

    a. Only women can intubate cattle because a man's arm is too big.
    b. The person intubating the patient uses the non-dominant hand to feel the larynx and epiglottis while the dominant hand advances the tube into the trachea.
    c. If the person intubating the patient only has one hand then they cannot do it.
    d. Cattle get laryngospasm so you cannot touch the epiglottis.

5. You are monitoring a cow during general anesthesia. You notice that the eye is starting to rotate dorsally as it returns to the central position. What does this indicate and what should your response be?

    a. The cow is becoming too deep and the vaporizer should be turned down.
    b. The cow is becoming too light and you need to deepen the anesthetic level.
    c. It means nothing so you ignore it.
    d. The cow is about to die and you need to perform cardiopulmonary cerebral resuscitation (CPCR).

6. As with cattle, what is the position that sheep and goats are placed in for recovery?

   a. Dorsal

   b. Ventral

   c. Lateral

   d. Sternal

7. What is the preferred intramuscular injection site in pigs?

   a. Rump or gluteal muscles

   b. Shoulder muscles

   c. Intramuscular agents cannot be given to pigs

   d. Neck region muscles

8. What is malignant hyperthermia?

   a. A form of cancer that causes tissue necrosis under anesthesia and can be treated with chemotherapy

   b. A genetic condition that occurs in anesthetized swine that can rapidly lead to death

   c. A condition that affects the ability for an animal to keep warm resulting in severe hypothermia

   d. A condition that is triggered by propofol in swine but rarely leads to death

9. What are drug withdrawal times based upon?

   a. Time it takes for 99.9% of the drug to be eliminated from the plasma

   b. Length of time it takes for 100% of animals to complete the drug treatment

   c. Amount of drug given per 100 pounds of animal

   d. Variation in drug elimination based on animal weight

10. Which leg on adult horses should always be pulled forward to avoid radial nerve damage while lying in lateral recumbency?

    a. The lower (dependent) pelvic limb

    b. The lower (dependent) thoracic limb

    c. The upper (nondependent) thoracic limp

    d. The upper (nondependent) pelvic limb

11. Above what mean arterial blood pressure is a critical factor in preventing postanesthetic myopathies in horses?

    a. 70 mm Hg

    b. 90 mm Hg

    c. 30 mm Hg

    d. Blood pressure really does not matter

12. When receiving inhalant anesthetic agents, ruminants weighing less than 150 kg can be anesthetized using

    a. a small animal anesthesia machine because the size of the ETT used fits this machine.

    b. a large animal anesthesia machine because the size of the ETT used fits this machine.

    c. a face mask attached to any size anesthesia machine.

    d. an induction chamber attached to a large animal anesthesia machine.

13. Which of the following is true regarding monitoring large animals under anesthesia?

    a. They should be monitored from a distance to ensure staff safety.

    b. They should be monitored by assessing parameters such as HR and RR as large animal monitoring devices are limited in their accuracy.

    c. They should not be monitored using with instruments such as Dopplers or pulse oximeters since they can break when an animal recovers from anesthesia.

    d. They should be monitored similar to small animals using equipment modified or specially made for large animals.

14. Which drug can be given to pigs to prevent vomiting in the recovery period?

    a. Diazepam

    b. Midazolam

    c. Metoclopramide

    d. Acepromazine

15. Which class of drugs is most likely to cause decreased GI motility and colic in horses after repeated doses?

    a. Alpha-2 adrenergic agonists

    b. Local anesthetics

    c. NSAIDs

    d. Opioids

16. Which of the following NSAIDs provides excellent musculoskeletal pain relief, but little for the treatment of visceral pain?

    a. Flunixin meglumine

    b. Phenylbutazone

    c. Ketoprofen

    d. Carprofen

17. Which drug should be used with severe caution in sheep because it can cause pulmonary edema, hypoxemia, respiratory failure, and peracute death?

    a. Ketamine

    b. Lidocaine

    c. Xylazine

    d. Morphine

18. Which of the following is the best source of information about drug withdrawal times?

    a. Food Animal Residue Avoidance Database

    b. United States Department of Agriculture

    c. Health and Human Services Department

    d. Food Animal Producers

19. Which analgesic has been proposed to have pro-kinetic effects (increase gastrointestinal motility) in horses?

    a. Xylazine
    b. Morphine
    c. Lidocaine
    d. Methadone

20. Which of the following can result in bacterial myositis if given IM to a horse?

    a. Butorphanol
    b. Flunixin meglumine
    c. Lidocaine
    d. Detomidine

Case Studies

Case Study 1: A 7-year-old, female Holstein Cow weighing 681 kg needs to have a surgical procedure for a left displaced abomasum. The cow is admitted to the veterinary college and will be placed under general anesthesia for student learning. You are asked to get together the equipment for the anesthesia resident to intubate this cow. Please describe what supplies are needed.

1. You are asked to watch the cow and monitor vital signs during recovery. Please explain how you would recover this cow and why your choices are important.

2. The veterinarian decides to give the cow flunixin for pain following the surgery. What concern is there when administering this drug to the cow?

Case Study 2: An exploratory laparotomy is scheduled for a 5-year-old female Mouflon Sheep weighing 35 kg. After an uneventful surgery, you are asked to recover this patient. What should you consider when recovering a small ruminant?

Case Study 3: A 5-year-old castrated male Nubian goat weighing 175 pounds (79 kg) with a history of recurrent urethral obstruction presented to the clinic with decreased urine flow. On physical exam the goat was overweight and all vital signs were within normal limits except for mild tachycardia. Blood work (CBC and serum chemistry panel) was normal. Abdominal radiographs were taken and two uroliths were seen in the urethra. Due to this goat's repeated episodes of urethral obstruction, the veterinarian recommended tube cystotomy surgery for this patient.

1. What preanesthetic preparation should be done in this goat?

2. Which preanesthetic agents could be used in this patient?

3. Which induction agents could be used in this patient?

4. The veterinarian would like to maintain this goat on inhalant anesthesia. Which type of anesthetic machine should be used in this patient?

5. Which maintenance agents could be used in this patient?

6. What are some adverse effects of isoflurane and sevoflurane?

Case Study 4: A piglet was castrated by a veterinarian from another clinic 6 weeks ago (at 7 weeks of age). Upon examination it is noted that the piglet is not eating well, his abdomen is tucked up, his eyes are dull, and he is vocal and is pacing back and forth. The piglet is fairly inactive and not very alert. He is acting painful and does not want to be touched near the surgical site. The veterinarian gives you the pain scale below and asks you to evaluate this piglet's pain. Even though you are unfamiliar with this particular pain scale, you feel confident that you can assign this piglet a pain score.

Score	Variable 1: Abdominal Palpation Following Surgery or at Time of Illness
0	Normal, abdomen soft, piglet tolerates palpation well
1	Abdomen soft, piglet actively withdraws from palpation
2	Abdomen tense, piglet withdraws from palpation
3	Abdomen rigid, piglet immobile

Score	Variable 2: Physical Appearance
0	Normal (normal coat, eyes clear and bright, head up when standing)
1	Slight change in coat, i.e., piglet starting to look a little hairy, eyes bright, normal position, head up when standing
2	Piglet's coat looks very hairy, nasal/ocular discharge, dull eyes, eyes not centrally positioned, but will follow movements of people or objects when prompted to do so, piglet holds head down, presses nose on ground occasionally, moderate mouth breathing (panting), piglet is pacing back and forth and vocal
3	Very rough, hairy coat, abnormal posture (hunches up), eyes look dry, cloudy, and do not follow movements of people or objects, presses nose to ground, will not hold head up when standing, severe mouth breathing, panting with labored breathing

Score	Variable 4: Unprovoked Behavior
0	Normal behavior pattern
1	Minor changes; is awake less often, plays with toys less often, plays when encouraged
2	Abnormal behavior, reduced mobility, decreased alertness, inactive
3	Unsolicited vocalizations, either very restless or immobile, expiratory grunts

Score	Variable 5: Behavioral Responses to External Stimuli (hypo/hyperesthesia)
0	Normal (behavioral responses normal for the expected conditions)
1	Minor depression/exaggeration of responses (pain or touch response)
2	Moderately abnormal responses, moderate change of behavior (pain or touch response)
3	Violent reactions to stimuli, or very weak muscular responses as in a pre-comatose state (pain or touch response)

How to use scoring sheet:

A. The overall score will be evaluated based on hours after surgery and the following day. It can also be used for any chronically painful pig.

B. If the piglet has a score of 6–7 at any time, the piglet is to be closely monitored for any decline in condition and analgesics are to be administered.

C. If the piglet has a score of 8, 1–24 hours after surgery, additional pain medication is to be administered.

D. If the piglet has a score of 9, 48 hours from surgery and on, additional pain management, antibiotic treatment, blood sampling, and/or fluid balance correction must all be considered.

E. If the piglet has a score of 10–11, the piglet is to be euthanized.

1. What score would you give this piglet for Abdominal Palpation Following Surgery or at Time of Illness?
2. What score would you give this piglet for Physical Appearance?
3. What score would you give this piglet for Unprovoked Behavior?
4. What score would you give this piglet for Behavioral Responses to External Stimuli (hypo/hyperesthesia)?
5. What is the Total Score?
6. What should you expect the veterinarian to do?

Case Study 5: The owner of a 12-year-old Morgan gelding weighing 450 kg calls the clinic to say his horse has "colic like" signs. To be prepared when the patient arrives, name three pain signs the horse may display.

Upon arrival you take a history of this horse from the owner who states he came home from work to find his horse rolling with numerous scuffs and scraps around its head. Upon presentation the patient is very depressed, grunting, and kicking at the flank area. The gelding continues to have the urge to roll but constant walking deters the patient from doing so. Physical exam reveals that the patient has tachycardia, tachypnea, and no gut motility sounds.

Based on what history we have obtained so far, what level of pain do you believe this horse is in? (Use the simple VAS found in Table 19-15.)

The veterinarian would like to start by giving the patient an NSAID pain reliever as well as a sedative to facilitate further diagnostics. With the veterinarian's approval, what would be the best choices for this patient?

Despite giving the horse analgesics, the veterinarian determines that the gelding needs an abdominal exploratory due to his continued clinical signs of colic. You are asked to provide anesthesia care for this horse. To prepare him for surgery, this patient is being positioned on the surgery table using a hoist. How should this patient be positioned on the surgery table, and what precautions should be taken with positioning to prevent development of neuropathy or myopathy? Thirty minutes into the surgery his mean arterial blood pressure has dropped to 50 mm Hg. You notify the veterinarian in charge of this case. What are some recommendations for the anesthetic team to consider for this patient?

Critical Thinking Questions

1. Species-specific anatomy and physiology are important to understand when anesthetizing the variety of animals encountered in veterinary practice. For example, horses are obligate nasal breathers that only breathe through their nose. What are the implications for the anesthesia team to consider when performing anesthesia in obligate nasal breathing patients?

2. Maintaining normal body temperatures is a concern in all animals being anesthetized. Pigs are unique because they could experience hypothermia or hyperthermia. List things that can be done to prevent hypothermia in anesthetized pigs.

   Pigs that undergo anesthesia also can develop malignant hyperthermia. What are clinical signs of malignant hyperthermia?

   How can the risk of malignant hyperthermia be reduced in swine?

ENDNOTES

1. Riebold, T. W., Geiser, D. R., & Goble, D. O. (1995). Clinical techniques for food animal anesthesia. In *Large animal anesthesia: Principles and techniques* (2nd ed.). Ames, IA: Iowa State University Press.

2. Hall, L.W, Clarke, K.W., & Trim, C.M. (2001). Anaesthesia of cattle. In *Veterinary anaesthesia* (10th ed.). London, UK: W. B. Saunders, Harcourt Publishers LTD.

3. Lerche, P., & Muir, W. W. (2009). Perioperative pain management. In W. W. Muir & J. A. E. Hubbell, Equine Anesthesia (pp. 371–377). Elsevier.

4. Reiblod, T. W. (2015). Ruminants. In K. A. Grimm, L. A. Lamont, W. J. Tranquilli, S. A. Greene, & S. A. Robertson (Eds.), *Veterinary anesthesia and analgesia: The 5th edition of Lumb and Jones* (pp. 912–927). Ames, IA, John Wiley & Sons.

5. Lin, H. C. (2014). *Farm animal anesthesia: Cattle, small ruminants, camelids, and pigs* (1st ed., p. 8). Ames, IA: John Wiley & Sons, Inc.

6. Abrahamsen, E. J. (2009). Ruminant field anesthesia. In D. E. Anderson, & D. M. Rings (Eds.), *Food animal practice* (5th ed.). St. Louis, MO: Saunders/Elsevier.

7. University of Missouri, Standard Operating Procedures Animal Sciences Research Center Large Animal Housing Facilities, 2013, http://animalsciences.missouri.edu/resources/Large%20Animal%20SOP%20ASRC%20amended%207_22_13.pdf

8. Thomas, J. A., & Lerche P. (2010). Veterinary anesthesia. In J. M. Bassert & D. M. McCurnin (Eds.), *Clinical textbook for veterinary technicians* (7th ed.). St. Louis, MO: Saunders/Elsevier.

9. Ibid

10. Trim C. M. (2010). Special anesthesia considerations in the ruminant. In C. E. Short (Ed.), Principles and practice of veterinary anesthesia (p. 295). Baltimore, MD: William and Wilkins.

11. Abrahamsen, E. J. (2009). Inhalation anesthesia in ruminants. In D. E. Anderson, & D. M. Rings (Eds.), *Food animal practice* (5th ed.). St. Louis, MO: Saunders/Elsevier.

12. Aarnes, T. K., Hubbell, J. A., Lerche, P., & Bednarski, R. M. (2014). Comparison of invasive and oscillometric blood pressure measurement techniques in anesthetized sheep, goats, and cattle. *Vet Anaesth Analg., 41*(2), 174–185. doi: 10.1111/vaa.12101

13. Re, M., Blanco Murcia, F. J., San Miguel, J. M., & Gómez de Segura, I. A. (2013). Reversible chemical restraint of free-range cattle with a concentrated combination of tiletamine–zolazepam, ketamine, and detomidine. *Canadian Journal of Veterinary Research, 77*, 288–292.

14. Hudson, C. et al. (2008). Recognition and management of pain in cattle. *Practice, 30*, 126–134.

15. Coetzee, J. F. (2013). A review of analgesic compounds used in food animals in the United States. *Veterinary Clinics of North America: Food Animal Clinics, 29*(1), 11–28.

16. Leslie, K. E., & Petersson-Wolfe, C. S. (2012). Assessment and management of pain in dairy cows with clinical mastitis. *Veterinary Clinics of North America: Food Animal Clinics, 28*, 289–305.

17. Fitzpatrick, J. L., Nolan, A. M., Young, F. J., et al. (2000). Objective measurement of pain and inflammation in dairy cows with clinical mastitis. *Proceedings of the International Symposium on Veterinary Epidemiology and Economics*, Breckenridge, CO), 73.

18. Milne, M. H., Nolan, A.M., Cripps, P. J., et al. (2003). Preliminary results of a study on pain assessment in clinical mastitis in dairy cows. *Proceedings of the British Mastitis Conference, Stoneleigh, Lancashire*, North West England, UK, 117–119.

19. O'Callaghan, K. (2002). Lameness and associated pain in cattle: Challenging traditional perceptions. *Practice, 24*, 212–219.

20. Stafford, K. J. (2014). Recognition and assessment of pain in ruminants. In C. Egger, L. Love, & T. Doherty (Eds.), Pain management in veterinary practice (pp. 349–356). Ames, IA: Wiley/Blackwell.

21. Gleerup, K. (2017). Identifying pain behaviors in dairy cattle. *WCDS Advances in Dairy Technology, 29*, 231–239.

22. Shaffran, N., & Grubb T. (2014) Pain management. In J. M. Bassert & D. M. McCurnin (Eds.), *McCurnin's clinical textbook for veterinary technicians* (8th ed.). St. Louis, MO: Saunders/Elsevier.

23. Valverde A. (2014). Treatment of acute and chronic pain in ruminants; In C. Egger, L. Love, & T. Doherty (Eds.), Pain management in veterinary practice (pp. 359–371). Ames, IA: Wiley/Blackwell.

24. Lin Hui, C. (2014). Pain management for farm animals. In H. C. Lin & P. Walz (Eds.), *Farm animal anesthesia: Cattle, small ruminants, camelids and pigs* (pp. 174–214). Ames, IA: John Wiley & Sons, Inc.

25. Caulkett, N. (2003). Anesthesia in ruminants. *Large Animal Veterinary Rounds, Department of Large Animal Clinical Sciences Western College of Veterinary Medicine, 3*(2).

26. Galatos, A.D. (2011). Anesthesia and analgesia in sheep and goats. *Veterinary Clinics of North America: Food Animal Practice, 27*, 47–59.

27. Lagutchik, M. S., Mundie, T. G., & Martin, D. G. (1992). Methemoglobinemia induced by a benzocaine-based topically administered anesthetic in eight sheep. *Journal American Veterinary Medical Association, 201*, 1407–1410.

28. Lin, H. C., Caldwell, F., & Pugh, D. G. (2012). Anesthetic management. In D. G. Pugh & A. N. Baird (Eds.), *Sheep and goat medicine* (2nd ed., pp. 517–538). Maryland Heights, MO: Elsevier.

29. Hall, L. W., Clarke, K. W., & Trim, C. M. (2001). Anaesthesia of sheep, goats and other herbivores. In *Veterinary anaesthesia* (10th ed., pp. 341–366). London: WB Saunders.

30. McLennan, K., et al. (2016). Development of a facial expression scale using footrot and mastitis as models of pain in sheep. *Applied Animal Behaviour Science, 176,* 19–26. DOI: 10.1016/j.applanim.2016.01.007

31. Häger, C., Biernot, S., Buettner, M., Glage, S., Keubler, L. M., Held. N., et al. (2017). The Sheep Grimace Scale as an indicator of post-operative distress and pain in laboratory sheep. *PLoS ONE, 12*(4), e0175839. https://doi.org/10.1371/journal.pone.0175839

32. De Sousa, A. B., et al. (2008). Pharmacokinetics of tramadol and o-desmethyltramadol in goats after intravenous and oral administration. *Journal of Veterinary Pharmacology and Therapeutics.*

33. Doherty, T., et al. (2007). Effect of intravenous lidocaine and ketamine on the minimum alveolar concentration of isoflurane in goats. *Vet Anaesth Analg.*

34. Staffieri, F., Driessen, B., Lacitignola, L., & Crovace A. (2009). A comparison of subarachnoid buprenorphine or xylazine as an adjunct to lidocaine for analgesia in goats. *Veterinary Anaesthesia and Analgesia, 36,* 502–511.

35. Valverde, A., & Doherty, T. J. (2008). Anesthesia and analgesia of ruminants. In R. E. Fish, M. J. Brown, P. J. Danneman, & A. Z. Karas (Eds.), *Anesthesia and analgesia in laboratory animals* (2nd ed., pp. 385–411). London, UK: Academic Press.

36. Robertson, J. T., & Scicluna, C. (2009). Preoperative evaluation: General consideration. In W. W. Muir & J. A. E. Hubbell (Eds.), *Equine anesthesia* (p. 123). Elsevier.

37. Nann, L. E. (2010). Equine anesthesia. In S. Bryant (Ed.), *Anesthesia for Veterinary Technicians* (p. 357). Danvers, MA: Blackwell Publishing.

38. Hubbell, J. A. E. (2007). Horses. In W. J. Tranquilli, J. C. Thurmon, & K. A. Grimm (Eds.), *Lumb and Jones' veterinary anesthesia and analgesia* (p. 719). Wiley.

39. Doherty, T., & Valverde, A. (2006). Management of sedation and anesthesia. In *Manual of equine anesthesia and analgesia.* (p. 206). Oxford: Blackwell Publishing.

40. Robertson, J. T., & Muir, W. W. (2009). Physical restraint. In W. W. Muir & J. A. E. Hubbell (Eds.), *Equine anesthesia* (p. 110). Elsevier.

41. Bednarski, R. M. (2009). Tracheal and nasal intubation. In W. W. Muir & J. A. E. Hubbell (Eds.), *Equine anesthesia* (p. 282). Elsevier.

42. McNally, E. M., & Pablo, L. S. (2012). Equine Anesthesia. In D. Reeder, S. Miller, D. Wilfong, M. Leitch, & D. Zimmel (Eds.), *AAEVT's equine manual for veterinary technicians* (p. 228). John Wiley & Sons.

43. Yamashita, K., & Muir, W. W. (2009). Intravenous anesthetic and analgesic adjuncts to inhalation anesthesia. In W. W. Muir & J. A. E. Hubbell (Eds.), *Equine anesthesia* (p. 267). Elsevier.

44. Wilson, D. V. (2006). Recognition of pain. In T. Doherty & A. Valverde (Eds.), *Manual of equine anesthesia and analgesia* (pp. 300–301). Blackwell.

45. Lerche, P., & Muir, W. W. (2009). Pain management in horses and cattle. In J. S. Gaynor & W. W. Muir (Eds.), *Handbook of veterinary pain management* (pp. 441–450). Mosby.

46. Dalla Costa, E., Minero, M., Lebelt, D., Stucke, D., Canali, E., & Leach, M. C. (2014). Development of the Horse Grimace Scale (HGS) as a pain assessment tool in horses undergoing routine castration. *PLoS ONE 9*(3), e92281. https://doi.org/10.1371/journal.pone.0092281

47. Taylor, P. M., Hoare, H. R., de Vries, A., Love, E. J., Coumbe, K. M., White, K. L., & Murrell, J. C. (2015). A multicentre, prospective, randomised, blinded clinical trial to compare some perioperative effects of buprenorphine or butorphanol premedication before equine elective general anaesthesia and surgery. *Equine Vet J.*. doi: 10.1111/evj.12442

48. Partial intravenous anesthesia. (2006). In T. Doherty & A. Valverde (Eds.), *Manual of equine anesthesia and analgesia* (p. 217). Blackwell.

49. Ketamine. (2006). In T. Doherty & A. Valverde (Eds.), *Manual of equine anesthesia and analgesia* (p. 145). Blackwell.

50. Valverde, A., Gunkel, C., Doherty, T. J., Giguere, S., & Pollak, A. S. (2005). Effect of a constant rate infusion of lidocaine on the quality of recovery from sevoflurane or isoflurane general anesthesia in horses. *Equine Veterinary Journal, 37*(6), 559–564.

51. Lamont, L. (2006). Intravenous lidocaine. In T. Doherty & A. Valverde (Eds.), *Manual of equine anesthesia and analgesia* (pp. 163–165). Blackwell.

52. Hubbell, J. A. E. (2007). Horses. In W. J. Tranquilli, J. C. Thurmon, & K. A. Grimm (Eds.), *Lumb and Jones' veterinary anesthesia and analgesia* (pp. 722–723). Wiley.

53. Muir, W. W., Hubbell, J. A., Bednarski, R. M., & Lerche, P. (2013). Anesthetic procedures and techniques in pigs. In *Handbook of veterinary anesthesia* (5th ed.). St. Louis, MO: Elsevier/Mosby.

54. Hodgkinson, O. (2007). Practical sedation and anaesthesia in pigs. *Practice, 29,* 34–39.

55. Bollen, P. J. A., Hansen, A. K., & Rasmussen, H. J. (2000). *The laboratory swine: A volume in the Laboratory Animal Pocket Reference Series.* M. A. Suckow (Ed.), pp. 80–81. Boca Raton, FL: CRC Press.

56. Carr, J., & Wilbers, A. (2008). Pet pig medicine: 2. The sick pig. *Practice, 30,* 214–221.

57. Swindle, M. M. (2000). Anesthesia, analgesia, and perioperative care. In *Swine in the laboratory surgery, anesthesia, imaging, and experimental techniques* (2nd ed., pp. 49, 55, 65–66). Boca Raton, FL: CRC Press.

58. Viscardi, A. V., Hunniford, M., Lawlis, P., Leach, M., & Turner, P. V. (2017). Development of a piglet grimace scale to evaluate piglet pain using facial expressions following castration and tail docking: A pilot study. *Front Vet Sci., 4,* 51. doi:10.3389/fvets.2017.00051

59. Di Giminiani, P., Brierley, V. L., Scollo, A., Gottardo, F., Malcolm, E. M., Edwards, S. A., & Leach, M. C. (2016). The assessment of facial expressions in piglets undergoing tail docking and castration: Toward the development of the Piglet Grimace Scale. *Frontiers in Veterinary Science, 3*, 100.

60. Viscardi, A. V., & Turner, P. V. (2018). Use of meloxicam or ketoprofen for piglet pain control following surgical castration. *Frontiers in Veterinary Science, 5*, 299. doi:10.3389/fvets.2018.00299

61. Thurmon, J. C., & Smith, G. W. (2007). Swine. In *Lumb and Jones veterinary anesthesia and analgesia* (4th ed., pp. 747–763). Elsevier.

62. Smith, A. C., & Swindle, M. (2008). Anesthesia and analgesia in swine. In R. Fish, P. J. Danneman, M. Brown, et al. (Eds.), *Anesthesia and analgesia in laboratory animals* (2nd ed., pp. 413–440). London: Academic Press.

63. Mozzachio, K., & Tynes V. V. (2014). Recognition and treatment of pain in pet pigs. In C. Egger, L. Love, & T. Doherty (Eds.), Pain management in veterinary practice (pp. 383–389). Ames, IA: Wiley/Blackwell.

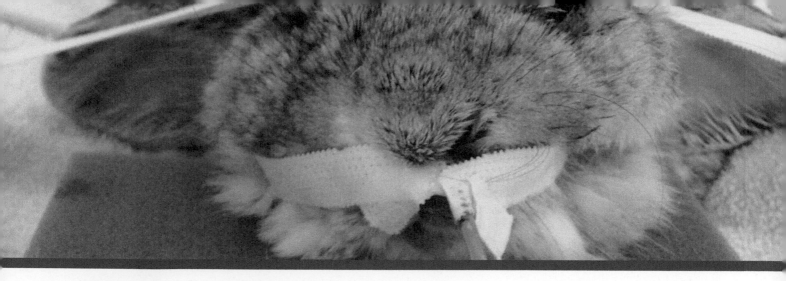

CHAPTER 20

Anesthesia and Analgesia of Pet Birds and Exotic Small Mammals

Lorelei D'Avolio, *LVT, VTS Clinical Practice (Exotics), CVPM*

LEARNING OBJECTIVES

Upon completion of this chapter, it is expected that the reader should be able to:

20.1 Explain anatomical and physiological variances between some of the most common exotic pets seen in clinical practice

20.2 Describe the appropriate equipment and supplies needed when working with exotic species

20.3 Compare and contrast exotic pet anesthesia/analgesia with small animal anesthesia/analgesia

20.4 Identify preanesthetic preparations for pet birds

20.5 Explain premedication strategies for pet birds

20.6 Explain strategies of anesthesia induction for pet birds

20.7 Explain strategies of anesthesia maintenance for pet birds

20.8 Describe how to safely monitor pet birds when using local, regional, and general anesthesia

20.9 Describe how to respond to and/or prevent anesthetic complications specific to pet birds

20.10 Explain strategies of anesthesia recovery for pet birds

20.11 Identify methods of assessing pain in pet birds

20.12 Identify available options for pain management in pet birds

20.13 Identify preanesthetic preparations for exotic small mammals (rabbits, pet rodents, and ferrets)

20.14 Explain premedication strategies for exotic small mammals (rabbits, pet rodents, and ferrets)

20.15 Explain strategies of anesthesia induction for exotic small mammals (rabbits, pet rodents, and ferrets)

20.16 Explain strategies of anesthesia maintenance for exotic small mammals (rabbits, pet rodents, and ferrets)

20.17 Describe how to safely monitor exotic small mammals (rabbits, pet rodents, and ferrets) when using local, regional, and general anesthesia

20.18 Describe how to respond to and/or prevent anesthetic complications specific to exotic small mammals (rabbits, pet rodents, and ferrets)

20.19 Explain strategies of anesthesia recovery for exotic small mammals (rabbits, pet rodents, and ferrets)

20.20 Identify methods of assessing pain in exotic small mammals (rabbits, pet rodents, and ferrets)

20.21 Identify available options for pain management in exotic small mammals (rabbits, pet rodents, and ferrets)

20.22 Explain the basic principles of performing technical procedures such as venous and intraosseous catheterization, air sac cannulation, and other unique techniques used in exotic species

INTRODUCTION

Most veterinary technicians have heard frightening stories of exotic animals being notoriously difficult to anesthetize and even "sponta-neously" dying. These tales have created a sense of fear and anxiety for many veterinary professionals when faced with the task of work-ing with exotic pets. In fact, an Association of Veterinary Anesthetists (AVA) survey found that there is a significantly higher mortality rate in exotic animals undergoing anesthesia versus dogs and cats.[1] These higher mortality rates are believed to stem from four main factors: unfamiliarity with a given species, misinterpretation of the severity of the patient's condition, lack of appropriate equipment, and inexperi-ence in providing difficult critical nursing and supportive care.[1] These factors are mostly preventable and manageable if veterinary techni-cians are taught to recognize and manage them properly. Knowing how to amend these factors will increase the veterinary technician's ability to provide a safer anesthetic event for exotic species. The aim of this chapter is to provide a foundation for the basic skills and knowl-edge needed to provide safe anesthesia and appropriate analgesia to some of the most common exotic species seen in practice. Many of the medications, techniques, and monitoring devices discussed in previous chapters will be described here, but modified to fit the special needs of these unique patients.

PREPARING FOR ANESTHESIA OF COMMONLY SEEN EXOTIC PETS

Providing anesthesia and perioperative care to pet birds and small exotic mammals have similarities to traditional small animal cases; therefore, many of the same questions posed for exotic animal cases mirror those used for small animal cases. The reader should review the small animal questions and consider their application to the exotic animal setting. Questions unique for exotic animals include the following:

- *For the safety of the anesthetist, supporting personnel, and patient, how will the patient be handled?* Exotic species may be small and cute, but they all have defensive and offensive capabilities such as biting and clawing that can cause serious harm. For this reason, proper precaution to prevent injury to both patient and personnel when interacting with or handling them is crucial.
  - Some species of small mammals, such as gerbils and chinchillas, have tails with delicate skin that can tear and slough easily if handled roughly.
  - Hamsters and other small rodents can inflict an unexpected painful bite causing the victim to jerk away, thus flinging the tiny animal and causing potentially fatal trauma to the rodent.
  - Parrots and ferrets have far more powerful jaws than most small animals, and can inflict a serious, often unexpected wound. Many will bite and lock on, creating a situation where the handler may accidentally injure the animal in an effort to release themselves from the bite.
  - Birds have unique anatomical differences that make restraint difficult: their necks can turn almost 180 degrees, they can fly, they have powerful and sharp beaks, and they lack a diaphragm. Improper restraint is not only dangerous for the handler, but also to the bird.
  - Rabbits have incredibly powerful hind legs and are likely to kick wildly when afraid, causing serious scratches to the handler and increasing the likelihood that the rabbit could be dropped or kick off of a table, injuring itself.
- *What physiological considerations must be addressed for exotic animal species?* Exotic animals have a broad range of different physiological traits. Rabbits are very sensitive to heat and are prone to heat stroke, small rodents can become hypothermic extremely fast, and birds have downy insulating feathers that create heating and cooling challenges. The gastrointestinal (GI) transit times of these species differ, each has unique integument, and even the opioid receptors of birds are different than those of other animal species. The key is researching and understanding the patient's biological needs prior to any anesthetic event and making arrangements for keeping them safe in advance.

- *How will stress be reduced when anesthetizing exotic animal species?* Stressed exotic animals release significant amounts of catecholamines such as norepinephrine, which can lead to complications such as arrhythmias, hypertension, hyperthermia, and potentially death. Because most exotic species described here are prey species, stress can be induced simply by a human looking at the animal if they do not have a place to hide. Setting up environments to give patients privacy, comfort, and familiarity to home are crucial. Having a quiet, dimly lit, calm environment will also help these sensitive animals. And proper handling that is practiced and pinpointed to be effective rather than aggressive is paramount. Premedicating with analgesics and sedatives is also important for many exotic species to reduce stress. Premedication typically reduces inhalant anesthetic agent requirements, may provide preemptive analgesia, and generally helps to reduce stress in exotic patients.[2]

- *What is the patient's physical status? Is assigning physical status the same as for assessing small animals? What co-morbidities are present? What medications is the patient on? What have past anesthesia experiences been?* Many veterinary technicians may not think to extrapolate the American Society of Anesthesiologists (ASA) Physical Status Scale for exotic pets. But this important scale is the first step in considering how to create a proper anesthesia and analgesia plan for all animals. Veterinary technicians should be comfortable assessing the ASA physical status of any patient before considering how to proceed with anesthesia. A quick review of ASA physical status is as follows:
  - ASA 1: A normal, healthy patient
  - ASA 2: A patient with mild systemic disease, for example, a bird with a skeletal fracture, obesity of any species, or vitamin A deficiency of birds
  - ASA 3: A patient with severe systemic disease, for example, a ferret with regulated insulinoma or a rodent with chronic *Mycoplasma* infection
  - ASA 4: A patient with severe systemic disease that is a constant threat to life, for example, an egg bound bird, ferret with a foreign body, or rabbit with gastric trichobezoar
  - ASA 5: A moribund patient who is not expected to survive with or without care, for example, patients from category 4 that are in terminal stage of disease or with complicating factors (egg bound bird with oviductal prolapse and anemia, a ferret with a foreign body and severe azotemia and hypoglycemia, or a rabbit with gastric trichobezoar and is bloated and in shock)

- *Is any special species-specific equipment needed?* Being prepared for an exotic animal anesthetic event involves organization and gathering all potentially needed supplies for a successful procedure. Special supplies needed for successful exotic animal anesthesia are described in Table 20-1. Vital signs for pet birds and select small mammals are found in Tables 20-2 and 20-3.

AVIAN ANESTHESIA

The avian patient is by far one of the most interesting species to anesthetize. There are close to 10,000 different species of birds, and their anatomy and size varies greatly. In general practice, the most commonly seen birds are psittacines (parrots) such as budgies, cockatiels, conures, amazons, cockatoos, and macaws. Anesthesia and analgesia of these birds will be the focus of this section.

Birds have a unique respiratory system that helps them meet their oxygen needs while remaining light enough to fly. Starting with the upper respiratory tract, birds have a unique structure called the **choana**. This is an opening located on the dorsal palate that connects to the nares (nostrils) in birds. Additionally, birds do not have an epiglottis but instead have a choanal slit that rests directly onto the glottis creating a direct breathing route through the nares into the trachea (Figure 20-1). Differences in avian respiratory anatomy continue beyond the upper respiratory tract. For example, birds do not have a diaphragm. Instead, birds use their great pectoral and other skeletal muscles to bring air into their bodies. This is important for veterinary technicians to be aware of during restraint. A bird should never be grasped tightly around the body as this inhibits the movement of the keel (sternum) and rib cage which can decrease inspiratory tidal volumes. Placing a bird in ventral recumbency can also inhibit normal tidal volumes by creating pressure on the sternum.

Tech Tip 🐾

If a procedure necessitates that a bird be placed in ventral recumbency, use padding to create a donut shape to place the bird's breast in to alleviate pressure and allow better expansion and proper ventilation or just mechanically ventilate the patient.

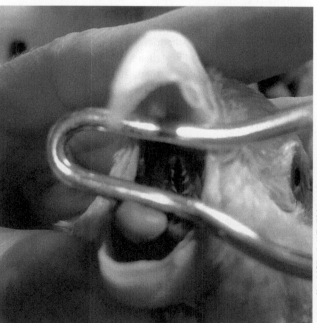

Courtesy of Lorelei D'Avolio, LVT, VTS Clinical Practice (Exotics)

FIGURE 20-1 The choanal slit in the oral cavity of birds. The bird's beak is held open by an oral speculum that provides a safe and effective way to examine the bird's oral cavity.

TABLE 20-1 Special Supplies for Anesthesia of Pet Birds and Small Exotic Mammals

Equipment	Comment
Endotracheal tubes (ETT)	• Stocking cuffed and uncuffed tubes in sizes 1–4.0 mm ensures an appropriate size ETT is available. Birds have complete tracheal rings; therefore, intubating with a cuffed ETT may cause serious damage. For this reason, uncufffed tubes should always be used in birds. Tiny ETT can be made by modifying 16, 18, and 22 gauge IV catheters in patients so small that even a size 1 mm ETT is too large or not available.

Warming equipment	• Exotic animal emergency or surgical patients are frequently hypothermic, and maintaining normal body temperatures can be challenging. Warming units with stagnant heat, such as heating pads, carry a risk of thermal burns. Suggested warming techniques/devices for exotic pets include the following:
	  • Forced air blankets (Bair® Hugger)
	  • Circulating warm water blankets
	  • Radiant heat: heat lamps, infrared bulbs, or ceramic non-light producing bulbs
	  • Electrical resistance with conductive fabric warming blankets (HotDog™ Patient Warming)
	  • Bubble wrap: Bubble wrap acts as a great insulator and is lightweight.
	  • Warmed chlorhexidine scrub and saline for aseptic preparation
	  • Ambient heating: Whether in an incubator or a surgical suite, increasing the heat in the room/enclosure

Masks and chambers for O_2 delivery or anesthesia induction or maintenance	• There is a variety of appropriately sized oxygen masks available for exotic animals. Commercially available small feline masks and rodent specific masks work well for a variety of small mammals, reptiles, and birds. Exotic animal oxygen masks can also be made from water bottles, hard plastic syringe cases, or the actual syringe minus the plunger and needle to suit a variety of sized exotics. • Using large canine oxygen masks can work well as an induction chamber for small rodents. There are also a few options for small induction chambers suitable for exotics from small guinea pigs to large rabbits.
Small laryngoscope blades	• Size #0 to #00 neonate laryngoscope blades are the only sizes that may be helpful to intubate exotic mammals. They can be difficult to find, and even if they are available, many of these patients are still difficult to intubate.
26–30 gauge IV catheters and small hypodermic needles	• Obtaining IV access on small exotics is a challenge. Having 26–30 gauge IV catheters will make catheterization of small birds, chinchillas, and other small rodents more realistic. • Insulin syringes are ideal for repeated IM/IV dosing and phlebotomy in exotic animals under 100 grams.
Clinical textbook for exotic species	• There are many good books on the market that are written to assist less experienced veterinary professionals with treating and managing exotic animals in practice. Select one that teaches species-specific husbandry. • An exotics-specific drug formulary is mandatory for anyone performing dose calculations for exotic pets.
Thermometers	• Ambient, esophageal, and small sized rectal thermometers are necessary to monitor exotic body temperatures depending on the species.
Gram scale	• The ability to accurately weigh patients is critical for medication/fluid dosing
Doppler ultrasound monitor	• Dopplers play a key role in exotic animal anesthesia monitoring both to measure blood pressure and heart rate.
Syringe pump/driver	• These pumps are useful for smaller volumes and lower rate CRIs and fluid therapy.
Non-rebreathing anesthesia circuits	• The force generated by the respiratory muscles during breathing in very small patients is typically not adequate to move gas through areas of high resistance; therefore, non-rebreathing systems are ideal for small exotics because they offer little resistance to air movement.
Safe and escape proof housing for the recovering patient	• Choosing recovery cages that have bars or solid glass/plastic walls to prevent small rodents or other creatures from escaping is necessary. • For climbing species, having a recovery cage that is not tall and does not have bars/perches to climb can prevent sedated or recovering patients from falling and injuring themselves.
Foam wedges	• Small angled wedges should be used to prop up the head and chest to prevent regurgitation in birds and to decrease intrathoracic pressure in rabbits and rodents. Additional wedges can be used to help keep the necks of these patients long and extended.

TABLE 20-2 Biologic Values*

Parameter	Birds	Ferrets	Rabbits	Guinea Pigs	Chinchillas	Rodents
Temperature	102.2–109.4°F (39–43°C)	100–104°F (37.8–40°C)	101–103°F (37.8–39.5°C)	99–103°F (37.2–39.5°C)	98.6–100.4°F (36.9–38°C)	98.8–100.7°F (37.1–38.2°C)
Respiration Rate (breaths per minute)	20–50 breaths/min (see Table 20-3)	33–36	30–60	42–105	45–80	40–200
Heart Rate (beats per minute)	200–600	200–400	120–300	240–310	230–380	260–670
Systolic Blood Pressure (mm/hg)	110–200	90–120	90–120	90–120	90–120	90–120
Tidal Volume (mL/kg)	15–30 (tidal volume is 1.5–2 times higher in birds than mammals based on weight)	10–15	10–15	10–15	10–15	10–15
Range of tidal volumes based on average weights for the species (mL)	0.3–30 (pet birds)	10–11	4–6	2.3–5.3	2.3–5.3	0.8–1.2
Lifespan (years)	7–10 (budgies) 15–20 (conures, cockatiels) 40–70 (cockatoos, amazons, macaws)	7–9	8–10	7–9	15–20	2–3 (rats) 1.5–2.5 (mice and hamsters) 3–5 (gerbils)

* Please note that many of these values will vary based on species, weight, and health parameters. Additionally, many of these numbers, such as blood pressure and tidal volume, have not been definitively researched for each species These values are based on available data and anecdotal resources.

TABLE 20-3 Avian Weight-Based Respiratory Rate Chart

Weight in Grams	Respiratory Rate (Breaths per Minute)
100	40–52
200	35–50
300	30–45
400	25–30
500	20–30
1000	15–20

The complexity of the avian respiratory tract continues with the lungs. They are relatively small in comparison to mammals and there are no lobes or alveoli. Not having a diaphragm, birds have one body cavity called the **coelom** where all the visceral organs lie. Additionally, the respiratory system contains nine air sacs. These air sacs are hollow spaces with thin walls and are connected to the lungs by groups of bronchi and parabronchi. Without alveoli, gas exchange occurs between air capillaries and blood capillaries that lie within these parabronchi. What makes this relevant to the anesthetist is how efficient birds are at gas exchange. Air takes

two complete breathing cycles to pass through the entire respiratory system (Figure 20-2). First inspiration pulls air into the caudal air sacs. Then the first expiration moves that air to the lungs. On second inspiration air moves from the lungs to the cranial air sacs. Lastly, the second expiration expels the air. As the bird breathes, the second inspiration of this cycle is also the first inspiration of another cycle, which creates a unidirectional flow (Figure 20-3). Because the parabronchi are present in each of these connections to and from the lungs, gas exchange is constant. It is estimated that this process allows for 10 times greater oxygen absorption in birds than in mammals.[3] This unique anatomy creates a more efficient respiratory system; however, because of it, birds have less volume of air present in the lungs at the end of expiration, which makes them prone to rapid decreases in oxygenation. Therefore, oxygen should be readily available and provided to them whenever sedatives, tranquilizers, or anesthetics are given.

Tech Tip

Inhalant anesthetic agents such as sevoflurane and isoflurane have a much faster onset of action and recovery with birds than mammals, often inducing anesthesia within seconds!

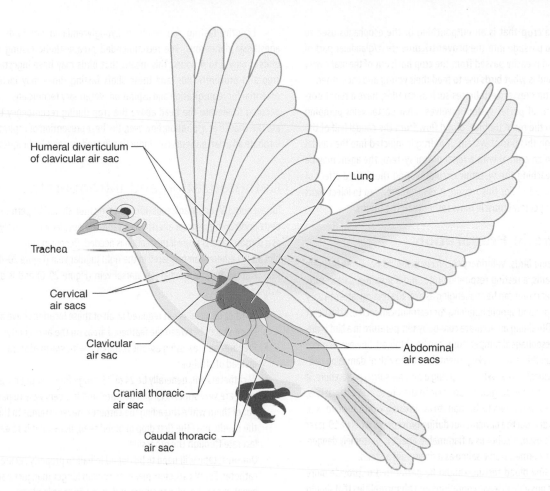

FIGURE 20-2 The respiratory system in birds. Most birds have 9 air sacs (2 cervical, 1 clavicular, 2 cranial thoracic, 2 caudal thoracic, and 2 abdominal).

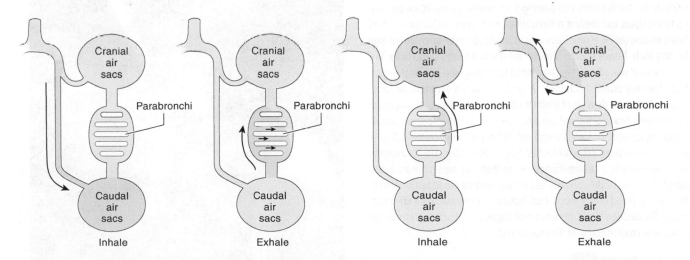

FIGURE 20-3 The movement of air in the avian respiratory system. Inspiration 1: Air enters the trachea and moves into the primary bronchus and neopulmonic (branch work of parabronchial lung tissue that is species variable) region. The air then moves to the caudal air sacs. There may be some movement of air into the paleopulmonic (anatomically paralleled lung tissue found in all bird species that has little anastomosing parabronchial tissue) regions where gas exchange can occur. Both the neopulmonic and paleopulmonic regions are part of the tertiary bronchi (parabronchi) and para-peribronchial gas exchange components of the avian lung. Expiration 1: Air in the caudal air sacs largely moves to the paleopulmonic region with about 12% loss escaping out the trachea through bidirectional neopulmonic regions. Inspiration 2: Remaining air from the paleopulmonic regions moves into the cranial air sacs. Expiration 2: Air moves from the cranial air sacs out through the bronchi and lastly through the trachea. Again, about another 12% of air escapes in this cycle through the neopulmonic regions out the trachea without full gas exchange.[6]

Birds have a **crop** that is an outpouching of the esophagus used to store food prior to passage into the **proventriculus** (first/glandular part of the stomach). Food is easily passed from the crop back out of the oral cavity as regurgitation and is what birds use to feed their young and each other.

Birds, like other lesser vertebrates such as reptiles, have a **renal portal system**. The renal portal system involves renal portal veins pumping blood directly into the renal tubules. Blood flow from the caudal half of the bird passes through this portal system. If a drug is injected into the caudal half of the body in an animal with a renal portal system, the agent may be excreted before reaching the systemic circulation and the desired effect of the drug not achieved.[4,5] For this reason, it is recommended to inject birds parenterally in the pectoral muscle mass rather than the thighs.

Preanesthetic Preparation

Prior to anesthetizing birds, veterinary technicians should obtain an accurate body weight in grams, a resting respiration rate, and if possible, a heart rate. Auscultating the heart rate can be challenging as average resting rates can vary from 200 to 400 bpm and simple handling for restraint can increase the heart rate significantly. Obtaining an accurate core body temperature in a bird is difficult. Cloacal and esophageal temperatures can be obtained; however, it is not recommended to do this in an awake patient due to the risk of damaging delicate cloacal/esophageal tissue. When prepping a bird for surgery/procedure, it is important to use warmed surgical scrub to clean the site to prevent heat loss, and it is important for veterinary technicians to avoid using alcohol on bird skin. Not only will it rapidly cool the patient, but during procedures where a CO_2 laser or electrocautery is used, alcohol as a flammable agent is extremely dangerous. Consider using warmed sterile saline as a rinsing aid instead.

Ideally, baseline blood testing should be performed to provide more information about possible disease conditions and abnormalities that should be addressed before or during anesthesia. Having baseline uric acid levels to evaluate kidney function, packed cell volume (PCV) and total protein (TP) to evaluate hydration status, and blood glucose levels as a minimum are helpful to anticipate and prevent complications. Keep in mind that up to 10% of blood volume (or 1% of total body weight in grams) can be safely drawn from healthy birds. However, depending on the size and health status of the bird and intended procedure, collecting blood for testing and potential surgical blood loss may exceed this amount. As birds will rapidly decompensate after blood loss of more than 1% of their total body weight grams, it is important to determine how much blood can safely be collected prior to venipuncture. For example, most cockatiels weigh only 90 grams, which means 0.9 mL of blood can be collected if they are healthy (90 g × 1mL/100 g). During a surgical procedure such as a salpingohysterectomy there can be significant anticipated blood loss. If the cockatiel had 0.5 mL used for blood testing prior to the surgery, blood loss of over 0.4 mL intraoperatively would be life threatening. The decision about importance of diagnostic laboratory tests should be made in conjunction with the veterinarian.

Tech Tip 🐾

The primary nitrogenous waste product in the avian kidney is uric acid, rather than urea nitrogen and creatinine as in mammals. When estimating renal safety of anesthesia and analgesic drugs in birds, testing uric acid levels and a PCV are the tests of choice.

Lengthy fasting can result in hypoglycemia in pet birds, making anesthesia dangerous. The recommended preanesthetic fasting time for birds is only 2 to 4 hours. This means that birds may have ingesta in their crop and proventriculus, and these short fasting times may increase the potential for regurgitation and aspiration. Veterinary technicians should use padding to elevate the head above the crop during recumbency to alleviate the chance of aspiration; however, the best prevention of aspiration is to intubate all avian patients (see Induction Procedures and Techniques below).

Catheterization and Intraoperative Fluids

Intravenous (IV) or intraosseous (IO) catheterization should be performed in all birds undergoing general anesthesia to provide euvolemia, electrolytes, and as a constant access route if medication is needed. Depending on the size of the bird an IV catheter can be placed in the right jugular vein (Figure 20-4), basilic vein (Figure 20-5), or medial metatarsal vein (Figure 20-6) and is described as follows:

- Feather removal is not required as all of these locations have **apteria** (bare spaces between the feathered areas on the bird's body).
- Prepare the area using aseptic technique with warmed sterile saline instead of alcohol.
- IV catheters will generally be 24 or 26 gauge for most pet birds. These veins are very superficial and delicate, and it is very easy to pass out of them while threading the catheter; never attempt to thread the needle past the first drop of blood seen, because it is so easy to lacerate through the other side.
- One-inch tape will need to be ripped in half to properly secure the catheter. For IV catheters meant to remain longer than just a surgical event, it may be wise to suture and then bandage in place.

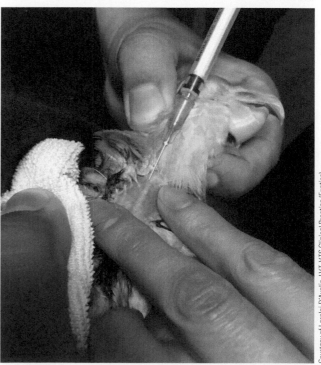

FIGURE 20-4 The right jugular vein; the most easily accessible site for IV catheterization and phlebotomy in psittacines.

Courtesy of Lorelei D'Avolio, LVT, VTS Clinical Practice (Exotics)

An alternative to IV catheterization is the IO route, which involves placement of a needle directly into the center of bone. When correctly seated into the bone marrow, fluids or drugs are rapidly taken up by the rich blood supply and reach systemic circulation almost immediately. Generally, all fluid types and drugs that can be administered IV can be given IO. Proper IO placement locations in birds are the proximal tibiotarsus (Figure 20-7) or the distal ulna (Figure 20-8). IO catheterization can be done using a spinal needle, IV catheter stylet, or regular hypodermic needle, but if using a needle, a sterile piece of cerclage wire may be needed to clear out boney material. Aseptic technique and appropriate local anesthetics are mandatory when placing IO catheters.

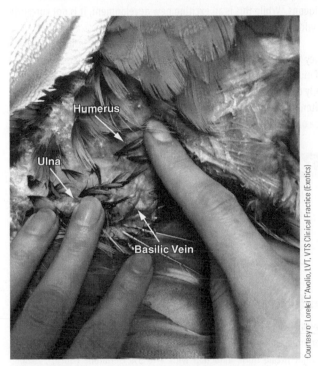

Courtesy of Lorelei D'Avolio, LVT, VTS Clinical Practice (Exotics)

FIGURE 20-5 The basilic vein; an accessible site for IV catheterization and phlebotomy, particularly in larger parrots.

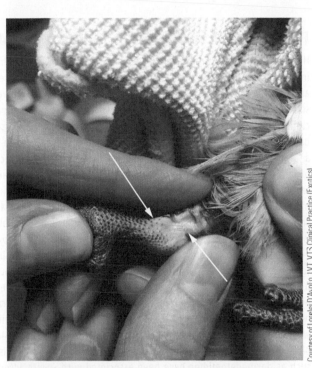

Courtesy of Lorelei D'Avolio, LVT, VTS Clinical Practice (Exotics)

FIGURE 20-6 The medial metatarsal vein; an accessible site for IV catheterization and phlebotomy, particularly in larger parrots.

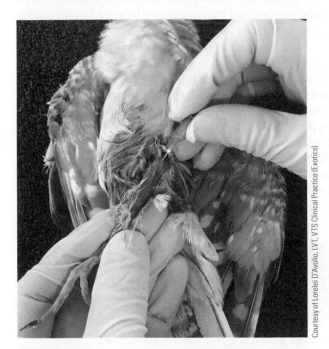

Courtesy of Lorelei D'Avolio, LVT, VTS Clinical Practice (Exotics)

FIGURE 20-7 IO catheter in the proximal tibiotarsus.

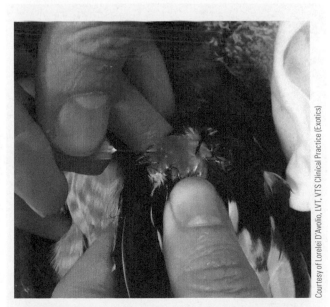

Courtesy of Lorelei D'Avolio, LVT, VTS Clinical Practice (Exotics)

FIGURE 20-8 IO catheterization using a hypodermic needle in the distal ulna. The thumb is placed in the center of the ulna as a guide. The needle is inserted slightly ventral to the dorsal condyle of the distal ulna. The radius and radial carpal bone can be used for orientation.

Intraoperative fluid rates for crystalloids range between 5 and 20 mL/kg/hr depending on the hydration status of the bird, expected and ongoing blood loss, and the length of the anesthetic procedure. For example, starting and continuing a patient at 20 mL/kg/hr may be excessive if the procedure were to last 4 hours, as this could cause fluid overload in the patient. Use of a syringe pump/driver is very helpful in order to provide accurate fluid rates in small birds (Figure 20-9).

Premedication, Sedation, and Restraint

Historically, premedicants have been used less frequently in birds than in other species because there is an increased fear of delayed recovery and respiratory depression. Whether these fears are valid is controversial and unfortunately limited data exist to clarify the debate. However, clinical use of limited premedications, particularly in highly stressed birds, is becoming more standard. Catecholamine release alone can be a fatal event in a highly anxious bird, and it is vital that veterinary technicians learn and practice gentle and calm avian capture and restraint techniques. Working in a darkened room without noise can be calming to birds. Speaking quietly may soothe them. Veterinary technicians should recognize that aggressively throwing a towel over a scared bird is very traumatic. Every attempt should be made to gently drape towels over birds and reward them with gentle sounds or **allopreening** (grooming/rubbing feathers of a bird) while toweled. The goal should be to make this process as enjoyable as possible for the bird. Once properly restrained, benzodiazepines such as midazolam (Versed®) can be given to reduce anxiety in the patient. When used in conjunction with opioids such as butorphanol (Torbugesic®, Torbutrol®), many birds become quite relaxed and sedate. Alpha-2 adrenergic agonists such as dexmedetomidine have been associated with severe adverse effects such as bradycardia, bradyarrhythmias, respiratory depression, hypoxemia, hypercapnia, and death when used alone. While these drugs are widely used in other species and are reversible, the risks in birds do not outweigh the benefits. A preliminary trial was conducted of sedation induced by intranasal administration of midazolam alone or in combination with dexmedetomidine (Dexdomitor®) and reversal by atipamezole (Antisedan®) for a short-term immobilization in pigeons.[7] While this is an exciting new development, it is not currently a widespread practice in pet medicine.

Anticholinergic agents reduce salivary excretions in birds; however, they have a tendency to thicken the excretions, which increases the chance of mucous plug formation and occlusion of the airway. This can be particularly dangerous in birds where tiny diameter endotracheal tubes (ETTs) are often utilized. Anticholinergic drugs (e.g., atropine) should be reserved for use only in bradycardic patients as needed and rarely as a premedicant.

Tech Tip

While midazolam is traditionally given IM, in birds use of intra-nasal midazolam has shown to be equally effective and often requires less restraint, so is therefore less stressful.

Induction Procedures and Techniques

Despite the exposure of the staff to inhalant anesthetic agents, the most common and safest way to induce general anesthesia in pet birds is via a tight fitting mask induction with 4 to 5% isoflurane (IsoFlo®, Isosol®, Aerrane®) or 5 to 8% sevoflurane (Ultane®, SevoFlo®, Petrem®, Flurovess™) in 100% oxygen. Due to the bird's more efficient respiratory system, this process is relatively rapid taking anywhere from 20 to 120 seconds. This fast and simple induction is generally preferred to IV, IO, or IM inductions which require longer and more expert restraint and technique, have longer induction and recovery times, and can be painful.

For longer surgeries or procedures, birds should be intubated to prevent aspiration. Prior to induction, an appropriately sized ETT should be selected. The avian trachea has complete tracheal rings so are not expandable like in dogs and cats. Because of this, they are highly susceptible to pressure necrosis if cuffed ETTs are inflated or if the bird's neck is positioned in ventroflexion, causing the tip of the tube to rub against the tracheal lining. These pressure lesions can cause severe strictures postoperatively that will impede air flow. Cole ETTs (Figure 20-10) are graduated to provide a secure airway at the opening of the glottis without the risks of an inflatable cuff and are highly recommended. Tiny 1.5 and 1 mm ETTs are also available (JorVet Jorgensen Labs) with a stylet to assist in placement, but using various sized IV catheters are viable options

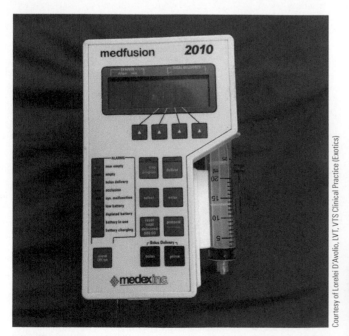

FIGURE 20-9 A syringe pump/driver facilitates very small volumes and low rates of CRI infusions.

Courtesy of Lorelei D'Avolio, LVT, VTS Clinical Practice (Exotics)

FIGURE 20-10 Uncuffed Cole endotracheal tubes.

Courtesy of Lorelei D'Avolio, LVT, VTS Clinical Practice (Exotics)

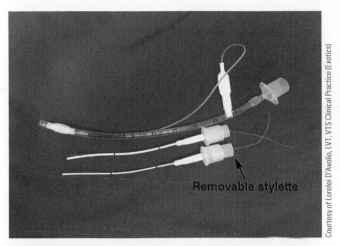

Courtesy of Lorelei D'Avolio, LVT, VTS Clinical Practice (Exotics)

Removable stylette

FIGURE 20–11 Examples of 1.0 and 1.5 mm endotracheal tubes with placement stylets to assist with rigidity in these highly flexible tubes. The 4.0 mm cuffed endotracheal tube in the image is for a size reference.

for small birds such as budgies (Figure 20-11). Intubation will allow veterinary technicians the option to ventilate birds in the event they become bradypneic or apneic, which is a common occurrence especially during prolonged procedures or open coelomic procedures.

Intubation of Parrots

Endotracheal intubation with an uncuffed ETT provides airway access for oxygen and anesthetic delivery, allows for assisted ventilation, and prevents aspiration of saliva and crop contents.

- Have necessary tools ready: appropriate sized uncuffed ETT, $\frac{1}{2}$ inch tape, oral speculum (see Figure 20-1), cotton swabs, mouth gag (can use a piece of tongue depressor wrapped in vet wrap).
- Use a small amount of sterile water-based lubricant to lightly coat the ETT.
- While someone is extending the neck, use the oral speculum to prop open the beak.
- The glottis is easy to see toward the rear of the tongue; however, some species such as budgies and cockatiels have tiny oral cavities, and using a swab to push the tongue down may facilitate visualization.
- Gently insert ETT into trachea and place mouth gag between upper and lower beak to prevent chewing of the tube during recovery. The oral speculum should be removed to prevent the bird from damaging their beak if they were to bite down on it during recovery. The mouth gag is soft and if the bird bit down on it, they would not damage their beak.
- Tape ETT to the beak and/or mouth gag.

Maintenance of Anesthesia

Inhalant anesthesia is the preferred anesthetic maintenance in birds. Preoxygenation for 1 to 2 minutes is beneficial to mitigate the risk of hypoxemia. Oxygen flow rates for anesthetized birds vary; the general rate of 100 to 200 mL/kg/min is acceptable for non-rebreathing circuits. Regardless of the fact that birds do not have an alveolar lung, the MAC of inhalant anesthetic agent is still the accepted way to determine how much gas to administer. Birds have an isoflurane MAC value of approximately 1.3 to 1.5% and

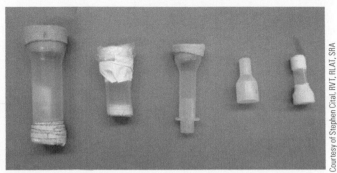

Courtesy of Stephen Cital, RVT, RLAT, SRA

FIGURE 20–12 Various masks made of syringe cases, mole skin tape, bubble wrap, elastic wrap, and other supplies found in the veterinary hospital.

Courtesy of Lorelei D'Avolio, LVT, VTS Clinical Practice (Exotics)

FIGURE 20–13 Air sac tube placement in a pet European Starling. The tube has been placed in the abdominal air sac and will be cut just below the Chinese-finger-trap suture pattern that holds the tube in place.

sevoflurane of approximately 2.5%. These levels should be determined based on vigilant monitoring of the patient while anesthetized. For brief procedures such as taking radiographs, mask anesthesia may be all that is required. If a tight fitting rubber mask is not available, creating a suitable anesthesia mask can be done. For example, a plastic syringe case with a rubber glove on the end can be designed for individual birds (Figure 20-12). For any procedures that are longer than just a few minutes, patients should be intubated and ventilated. Because of their complicated respiratory system, even when birds are still having spontaneous respirations, veterinary technicians should be supplementing them with mechanical ventilation to ensure proper oxygenation. It is very common for birds to become apneic during anesthesia, particularly during coelomic surgery, and mechanical ventilation is vital for their survival.

Assessing anesthetic depth can be difficult, but mimics the same methods used in small animals with the addition of pulling a few down feathers to see if the patient responds to the stimulus. Slight corneal reflexes should be maintained at a surgical plane of anesthesia. In incidences of tracheal obstruction or if head, neck, or mouth surgery needs to be performed, a catheter or cannula can be placed into the caudal thoracic air sac. This unique anatomical feature allows for inhalant anesthetic agents and oxygen to be provided to the avian patient despite the trachea being blocked or unusable. This technique requires aseptic surgical placement of an uncuffed ETT or red ETT rubber catheter (Figure 20-13).

Tech Tip

If air sacs are opened during surgery, such as in a coeliotomy, anesthetic gas and oxygen will escape into the room, limiting patient absorption. It is important to increase the oxygen flow rate and percent of inhalant anesthetic agent in these instances.

Monitoring

Birds can quickly decompensate under anesthesia and should be closely monitored using a variety of techniques. Profound hypothermia can happen relatively quickly. The average body temperature for a bird is 99 to 107° F, and rapid drops in body temperature from the use of anesthetic agents can be associated with decreased heart rate and cardiac instability. Ideal ways to prevent heat loss in birds include radiant heat from above or the use of forced air blankets. Circulating warm water blankets and heated pads can also be used with caution because they can develop hot spots due to uneven heating of the device. Heat loss will be compounded by an open coelomic cavity and use of cold aseptic cleaning fluids. Using a thermometer probe placed in the cloaca, esophagus, or close to the skin in a tucked wing or leg will help gauge trends in the patient's approximate body temperature and allow for earlier responses to fluctuations.

In birds, monitoring the respiration rate may not be possible by watching the reservoir bag, as tidal volume may be so low as to not noticeably move it. But careful assessment of the rising and falling of the chest should be visible. Utilizing a transparent surgical drape is very helpful in monitoring birds for this reason (Veterinary Specialty Products, Overland Park, KS). A clicking or gurgling sound heard with respirations may indicate that the ETT should be checked, suctioned, and potentially replaced. Positive pressure ventilation (PPV) is very useful in birds due to poor ventilation and oxygenation that can rapidly occur if the bird is positioned on its sternum or if the coelomic cavity and associated air sacs are open. Periods of apnea can quickly cause adverse acid-base disorders in birds not given PPV.[6] Whenever available, mechanical ventilation should be used with settings varying from 1 mL up to 1 L based on size of pet bird. Mechanical pressure settings should be set from 4 to 10 cm H_2O in most pet birds, and respiration rates of 6 to 20 breaths per minute. The respiratory rate will increase the smaller the bird species gets.

Tech Tip

Birds have significantly larger tidal volumes for their mass compared to dogs. If a hummingbird could be made to be the same size as a Labrador Retriever dog, the hummingbird would require two to four times larger tidal volume compared to the Labrador Retriever.

Capnography can be a very useful monitoring device for birds. Mainstream capnography such as the EMMA® (Masimo) is preferred over sidestream machines for birds and most exotic pets because of the increased

dead space of the tubing making readings less reliable in the sidestream units. The small monitor of the EMMA® only has 1 mL of dead space when using the pediatric sensor (Figure 20-14). If only sidestream capnography is available, dead space can be reduced by putting an 18 gauge needle directly into the hub of the endotracheal tube connector and then attaching the sidestream line or commercial endotracheal tube adapter (Figures 20-15 and 20-16). A downside to mainstream capnography in such small patients is it may spread disease from one patient to another more easily, can be easily broken, and may put additional weight on the ETT, which could result in it becoming kinked or accidently removed. The end-tidal carbon dioxide ($EtCO_2$) range for most birds is generally 30 to 45 mm Hg; however, readings < 30 mm Hg are not uncommon in small birds. Expired CO_2 levels in birds are not always equal to arterial CO_2 partial pressures because airflow through the avian parabronchi is unidirectional. This one directional flow of air causes all

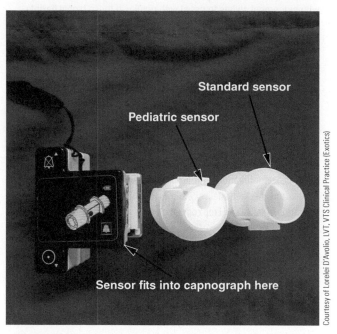

Courtesy of Lorelei D'Avolio, LVT, VTS Clinical Practice (Exotics)

FIGURE 20-14 Capnography used in avian anesthesia. EMMA® by Masimo is an example of mainstream capnography.

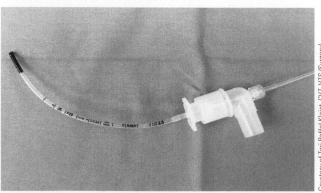

Courtesy of Teri Raffel Kleist, CVT, VTS (Surgery)

FIGURE 20-15 Capnography using traditional elbow sidestream adapter.

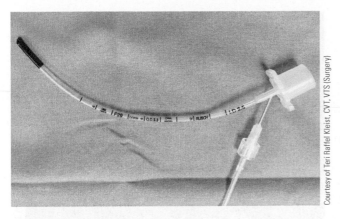

Courtesy of Teri Raffel Kleist, CVT, VTS (Surgery)

FIGURE 20-16 Capnography using 18 gauge needle for a sidestream adapter.

Courtesy of Lorelei D'Avolio, LVT, VTS Clinical Practice (Exotics)

FIGURE 20-17 Doppler probe on the radial artery of a bird used to monitor the heart rate and rhythm.

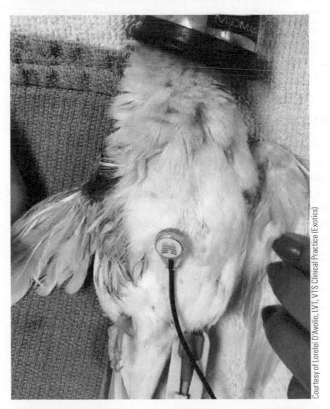

Courtesy of Lorelei D'Avolio, LVT, VTS Clinical Practice (Exotics)

FIGURE 20-18 Doppler probe placement is commonly used in small birds during anesthesia to monitor the heart rate and rhythm directly over the heart.

exhaled gases to pass through the cranial air sacs and trachea without being diluted; therefore, the actual $EtCO_2$ level may be underestimated.

Pulse oximetry (SpO_2) readings are considered unreliable for birds due to the lack of specific calibration for avian red blood cells, which are nucleated. The monitor can, however, provide general information about the patient's oxygenation trends and an audible clue of a patient's pulse rate and quality. Most SpO_2 clamps detect readings when used on the legs or feet of birds and should be used as an accessory monitoring tool.

A monitoring staple in avian anesthesia is the ultrasonic Doppler. The crystal can be placed on many different sites to provide an audible indicator of pulse rate and quality. The crystal is most commonly placed on the radial artery (Figure 20-17) or tibiotarsus artery because the blood pressure cuff can then be placed proximally to the crystal for non-invasive blood pressure monitoring. Other sites include the carotid artery, the ventral keel bone, and directly over the heart (Figure 20-18). Direct arterial blood pressure monitoring is not routinely performed in most pet birds due to the requirements of specific technical skills, cost of required equipment, and risks of the invasive nature of the procedure in birds. If required for a specific procedure, it can be achieved in large birds with placement of the arterial catheter in the brachial or carotid arteries. Indirect monitoring, however, can be safely achieved in most sizes of pet birds. Infant and penile blood pressure cuffs can be placed as described previously, proximal to the Doppler crystal on the radial or tibiotarsus vessels. As with mammals, these systolic readings are not considered 100% accurate and the purpose of monitoring it is to follow trends for each individual patient throughout a procedure. Birds tend to have higher blood pressures than mammals and in general, a systolic blood pressure below 90 mm Hg is considered hypotensive while over 200 mm Hg may be considered hypertensive.[8]

ECG monitoring is another helpful monitoring tool for birds. Using a three lead set with electrode placement in a similar fashion to mammals will demonstrate comparable QRS complexes. Veterinary technicians should use caution and instead of using alligator clips or sticky pads that can injure feathered patients, a stainless steel suture through the skin or needle attached to the ECG leads can be more effective and less traumatic. Esophageal ECG probes are a great atraumatic option and can alleviate multiple monitoring wires and sensors from becoming a nuisance during surgery.

Recovery

Receiving inhalant anesthetic agents for long periods may depress a bird's normal physiologic functions, which slows metabolism of drugs. Birds also have an increased risk of becoming hypoglycemic and hypothermic under anesthesia which may delay recovery. The avian patient should be placed in a lateral or ventral position as soon as possible to facilitate easier respiration. Birds are extubated when they start to show signs of righting themselves and their jaw tone and palpebral reflex returns. During anesthesia birds tend to accumulate a large amount of mucus in their trachea and the area surrounding the ETT. When extubating, it is important to immediately swab around the glottis to prevent aspiration of mucus that may have built up during anesthesia. It is also common for birds to recover in a panicked state where they awaken abruptly, thrashing around. Veterinary technicians should be prepared for these situations by having ETTs loosened, ECG leads removed, and birds loosely wrapped in a warm towel to prevent self-injury. Due to the higher metabolic demands of oxygen in birds, postoperative oxygen therapy is ideal via mask or **flow-by oxygen** (placing the end of the oxygen hose from the anesthesia machine by the patient's nares/mouth using high flow rates) once extubated. However, veterinary technicians should be cognizant of the risks of oxygen toxicity in birds because cellular changes have been seen experimentally in those left in 100% oxygen environments for as little as over 3 hours.[8]

If the patient has external sutures, it is wise to place a modified Elizabethan collar to prevent chewing (Figure 20-19). Nonsteroidal anti-inflammatory drugs (NSAIDs) other than aspirin can be given after anesthesia (carprofen [Rimadyl®] and meloxicam [Metacam®] are the most studied and commonly used NSAIDs in birds), and if either butorphanol or buprenorphine (Buprenex®) were administered as premedicants, they can potentially be redosed during the recovery period depending on when administered as a premedicant. In this delicate recovery time, veterinary technicians should constantly assess birds for pain, discomfort, and stress. A heated recovery cage should be prepared in a quiet and dimly lit room to minimize stress to the bird (Figure 20-20).

AVIAN ANALGESIA

Historically, the results of research about avian analgesia has been confusing, and current therapy relies more on clinical application than proven evidence-based medicine. With increased clinical use of analgesic drugs, many practitioners are reporting improved pain management in their avian patients. When used at current published dosages, most of the avian analgesics show little to no signs of toxicity or undesirable adverse effects. Because of these factors, there is little argument for not implementing a comprehensive pain management protocol for birds, regardless of the lack of proven research.

Noxious stimuli cause pain in birds similar to that in mammals; however, how they display their pain can make it challenging for those unfamiliar with avian behavior to interpret. Signs of pain in birds include:

- Change in normal behaviors such as grooming (could be increased or decreased)
- Change in activity level (could be hyperactive or lethargic)
- Change in vocalization (could be excessive or diminished)
- Favoring one leg or wing, or squinting one eye
- Anorexia
- Excessive fluffing of the feathers (this conserves heat)
- Irritability (assuming the bird is normally docile)

FIGURE 20–19 Parrot with an example of an Elizabethan collar used to prevent chewing on wounds, incisions, or bandages. This collar uses soft material that will both prevent the bird from having access to its body while giving it exciting colors and textures to see and touch.

Courtesy of Lorelei D'Avolio, LVT, VTS Clinical Practice (Exotics)

FIGURE 20–20 A bird recovering from anesthesia. Placing the bird into a small incubator or enclosure with low perches will prevent the bird from injuring itself. The cage also has a plexiglass front to offer a quieter environment as well as keep heat and oxygen inside if they are required.

Courtesy of Lorelei D'Avolio, LVT, VTS Clinical Practice (Exotics)

Past studies show variable to low analgesic effects in birds when given full mu agonist opioids such as fentanyl (Sublimaze®) and morphine (Duramorph®, Astramorph®). Pharmacodynamics studies in pigeons have shown that they have more kappa opioid receptors than mu.[9] Extrapolating that other species of birds share this trait, current theory is that because butorphanol has kappa agonist activity in addition to mu antagonist activity, it is the best option for moderate to severe pain relief in birds. Unfortunately, the duration of effect has shown to be very short and must be administered IM in dosages of 1 to 2 mg/kg every 1 to 4 hours for sustained analgesia.

It is assumed that pharmacological activity of NSAIDs have a similar action in birds as in other animals. Most commonly used NSAIDs, such as meloxicam, are cyclooxygenase (COX) 1 sparing and COX 2 inhibiting with the goal of suppressing inflammation without inhibiting physiologically important prostaglandins. As in other species, gastrointestinal, renal, and coagulation adverse effects are of concern in birds. Meloxicam, however, is considered a safe form of analgesia and is commonly used for chronic and acute pain in birds. Effective dosages for birds have been shown to be 1 to 2 mg/kg IM/PO Q12–24 hours, which is significantly higher than in other animals.[13]

Tech Tip

Birds metabolize meloxicam differently from mammals and require a much higher dosage for effective pain control.[9]

Local anesthetics block sodium channels in the nerve axon, interfering with the conduction of action potential and inhibiting the transmission of pain sensation. Unlike other vertebrates, there is limited research and knowledge about the efficacy and effect of local anesthetics in birds; however, clinical use is common and dosages are extrapolated from published mammalian dosages. Lidocaine (Xylocaine®) and bupivacaine (Marcaine®) are the most commonly used local anesthetics and are MAC sparing. When used synergistically with other analgesics, they may even remove the need for general anesthesia altogether. Because birds may be very small, it is easy to grossly overdose them with local anesthetics, which can result in adverse signs such as excitation, convulsions, cardiopulmonary depression, or even death. Careful dilution with sterile water may be needed to allow for adequate administration volume. For example, if a bird weighing 110 g needed a 2 mg/kg lidocaine infiltrative block of a wound on the thigh with a 2% lidocaine solution, the required dose would be 0.011 mL (110 g × 1 kg/1000 g × 2 mg/kg × 1 mL/20 mg = 0.011 mL). This would be too small of a quantity to adequately infiltrate the area; however, diluting the 0.011 mL lidocaine with 0.2 mL sterile water will provide a good carrying agent to disperse the proper dose to the damaged tissue.

Gabapentin (Neurontin®), an anticonvulsant drug that has been used for treating chronic pain and nerve-pain illnesses such as herpes zoster in humans (see Chapter 5 [Anesthetic and Analgesic Pharmacology] and Chapter 16 [Chronic Pain Management]), appears to be involved with the interaction of calcium channels in the central nervous system; however, its mechanism of action is not fully understood. It has become widely used in veterinary medicine to treat not only seizure disorders, but also chronic pain. In birds, practitioners began using it for potential chronic pain such as in birds with feather destructive behavior or self-mutilation disorders. Currently, avian veterinarians are also using it in similar fashion to mammalian pets to treat chronic pain or postsurgical discomfort. It must be compounded into a liquid, and dosages from 10 to 80 mg/kg given every 6 to 12 hours have been used.

Current multimodal analgesia for birds could include the use of photobiomodulation (Figure 20-21), which uses light to penetrate deep tissues causing the cells to produce more ATP and a repairing effect on DNA as well as increasing cell membrane permeability. This leads to normalization of cell function, pain relief, and healing. Like many other aspects of avian analgesia, there is still a need for more evidence-based research in the use of lasers in birds. However, there does not seem to be any adverse effects to their use, and many clinicians and clients are finding lasers a good addition to the pain management plan.

RABBIT ANESTHESIA

These herbivorous pets are rapidly growing as one of the most common house pets in the United States. There are a huge variety of breeds, from tiny dwarfs weighing only a pound up to giant breeds weighing almost 20 pounds. The anatomy, metabolic rates, and physical characteristics of rabbits are very different from other small mammals, which creates unique challenges for anesthetists.

Preanesthetic Preparation

The chest cavity of rabbits is small with decreased lung capacity. Any rabbits with pre-existing respiratory disease should be thoroughly evaluated prior to anesthesia. This should include radiography of the chest and presurgical

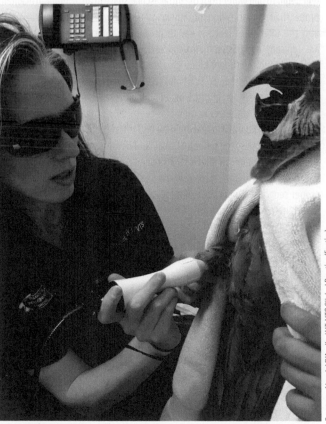

FIGURE 20-21 Use of a Class 4 laser on a bird as part of a multimodal analgesic plan.

bloodwork to determine the physiological status of the rabbit. Veterinary technicians should check vital signs before any medications are given. Temperature, respiration rate, and heart rate should all be monitored and within normal ranges: 101 to 103°F, 120 to 300 bpm, and 30 to 60 breaths per minute respectively.

> **Tech Tip** 🐾
>
> Counting a rabbit's respiratory rate in the cage prior to handling may prove more accurate than after the stress of restraint.

Rabbits do not vomit due to an anatomical valve restricting food from passing from the stomach into the esophagus. As herbivores, it is very important for them to always have food in their stomach, cecum, and intestines to prevent hypoglycemia, dehydration, and ileus. For these reasons, fasting rabbits is strongly discouraged. Removing food an hour or two prior to anesthesia may help keep the oral cavity clean; however, longer fasting is considered deleterious to gastrointestinal motility.

Catheterization and Intravenous Fluids

Intravenous access is critical for rabbits both for the ability to deliver emergency medications and to provide homeostasis, as hypovolemia is common in anesthetized small mammals. Rabbits that are greater than 5% dehydrated will begin to pull water from the intravascular space in an attempt to maintain proper fluid volume. Filling the intravascular space with fluid that can be pulled into the interstitial space is key to preventing poor perfusion. Crystalloids are typically administered to dehydrated patients; however, in a properly hydrated rabbit with low blood pressure, colloids can significantly raise the blood pressure when given as a bolus (2.5–5mL/kg) or as a CRI to increase the intravascular space and improve blood pressure when clinically indicated.

Placing IV catheters in rabbits is technically similar to placing them in cats (refer to Chapter 4 [Fluid Therapy and Intravenous Catheterization] and Appendix E). Unless the patient is weak and decompensated, catheterization should be done after premedications have taken effect. This will minimize a stress response, and the patient is less likely to buck or twitch once receiving proper sedatives. Application of a topical numbing cream such as lidocaine/prilocaine (EMLA®) cream (see Chapter 17 [Analgesic Techniques]) may also increase patient compliance. The easiest locations for IV catheter placement in rabbits are the cephalic veins. The anatomy of the front legs in rabbits is similar to cats, and the vein courses proximally up the center of the anterior surface (Figure 20-22). The restrainer does not need to twist the fascia and skin laterally as much as in cats. These veins also tend to be small, superficial, and delicate. Treating them gently will avoid accidental failure to pass the catheter. Once placed, the catheter can be secured in a similar fashion as other animals.

> **Tech Tip** 🐾
>
> It would be wise to consider using self-adherent bandaging tape (such as VetRap®) or tape to cover the IV lines leading to catheters in rabbits. They love to chew them out once they are feeling better!

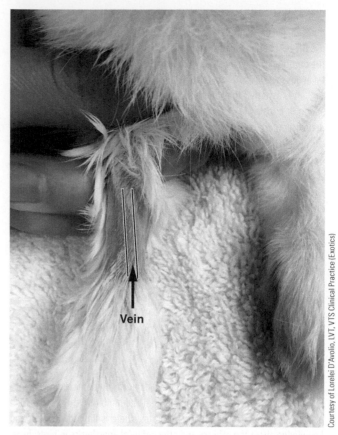

Courtesy of Lorelei D'Avolio, LVT, VTS Clinical Practice (Exotics)

FIGURE 20–22 The cephalic vein of a rabbit is a common site for IV catheter placement.

It may be tempting to place catheters in the large marginal ear veins in rabbits. This technique has been used traditionally in laboratory settings and is sometimes used in clinical practice. Possible problems are the potential for sloughing of the ear if certain medications should leak perivascularly and the rabbit's discomfort from having something unnaturally heavy attached to their ear. Because of these risks and the ease of cephalic catheter placement, most practitioners seeing pet rabbits do not use the ears.

While the lateral saphenous vein is ideal for phlebotomy, it is not ideal for catheter placement. The vein has many angled bends making it difficult to thread a catheter (Figure 20-23). Also, the hind legs are extremely powerful, making proper bandaging and maintenance of the catheter difficult. However, it is possible to place and secure for brief procedures.

> **Tech Tip** 🐾
>
> Rabbit fur is generally much thicker than that of dogs and cats, so it is tempting to aggressively shave. Veterinary technicians should use caution when clipping rabbits—it is very easy to accidentally clip off a nipple or tear the skin.

Premedication, Sedation, and Restraint

One of the key elements of successful rabbit anesthesia is to provide a calm and quiet environment for the patient. Stress in rabbits can complicate routine procedures, causing accidents such as the rabbit bolting/kicking

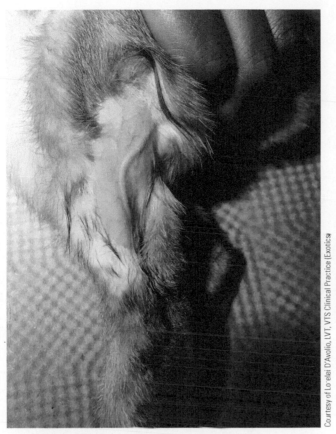

FIGURE 20-23 The lateral saphenous vein of a rabbit is a common phlebotomy site; however, due to the powerful kicking back legs and the curvatures of the vein, it is not a common IV catheter site.

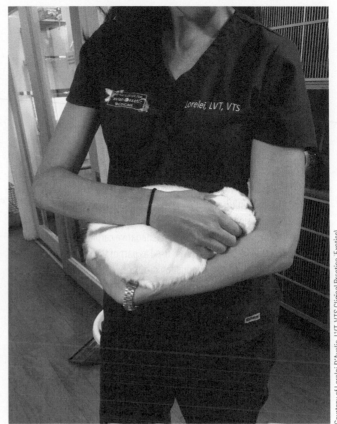

FIGURE 20-24 Appropriate rabbit restraint.

and injuring itself, or causing the rabbit not to eat with subsequent gastrointestinal stasis. Rabbits should be housed in a kennel with something soft and non-slip for them to ambulate normally, away from barking dogs or prowling predatory species, and preferably out of direct bright light. Handlers should be trained to restrain rabbits tightly against their body while supporting the hind end when carrying them (Figure 20-24), and utilizing towels as visual barriers. Sometimes gently covering their eyes can calm them. Keeping this sort of peaceful environment will optimize successful patient anesthesia.

Premedication is recommended for rabbits as they release systemic catecholamines and cortisol under stress, which can result in anesthetic complications. Premedication in rabbits and other small mammal species often means heavy sedation or tranquilization to allow IV catheterization and endotracheal intubation. One unique attribute in rabbits is that many of them produce atropinase, an enzyme that renders atropine useless, and other anticholinergic drugs less effective. If a rabbit requires an anticholinergic drug, it is recommended to try glycopyrrolate (Robinul®). Each patient should be evaluated to determine what the appropriate premedication may include rather than a standard "cocktail" for all rabbits. The drugs mentioned below can be breed dependent, making continual research into particular medication protocols a must.

- Teaming a full mu opioid agonist such as hydromorphone (Dilaudid®) or fentanyl with a benzodiazepine such a midazolam can provide effective and safe analgesia and sedation in even the sickest patient.

These combinations can be given IM or SQ; however, diazepam can have erratic absorption and is painful when given via the SQ or IM route. Therefore, choosing a more water-soluble benzodiazepine such as midazolam is preferred.

- Alpha-2 adrenergic agonists such as xylazine (Rompum®, AnaSed®) or the newer drug dexmedetomidine are used in exotic animal species; however, just as in small animals, they have profound cardiovascular adverse effects and are not appropriate for all patients. They are best used in young and/or healthy rabbits.
- Mixtures such as dexmedetomidine and ketamine (Ketaset®, VetaKet®, Ketalar®, Vetalar®) are commonly used in combination with midazolam to premedicate rabbits. Ketamine should be used with caution in animals with cardiac or renal issues. Ketamine should not be used alone as convulsions can occur.[11]
- Acepromazine (Promace®, generic) can be combined with an opioid to premedicate rabbits; however, it is used less frequently than in dogs and cats due to concern over its side effects and reliability.

Induction Procedures and Techniques

Induction options for rabbits are based on whether a catheter is placed prior to induction and whether the rabbit will be intubated. If a rabbit has an IV catheter and is going to be intubated, propofol (Rapinovet®, PropoFlo®, Propovan®) is a good option for induction. Etomidate (Amidate®) may also be a good option if a catheter is patent, particularly for rabbits suffering from cardiac disease. The neuroactive steroid alfaxalone (Alfaxan®, Alfaxan®Multidose, Alfaxan®Multidose IDX) has a growing body of literature on its use in small mammals, and dosing for rabbits is currently being extrapolated from cats and dogs. Currently, Alfaxan® and Alfaxan®Multidose are only labeled for use in the United States for dogs and cats, but extra label use in rabbits and other exotic species is becoming more common. Alfaxan®Multidose IDX is labeled as an injectable sedative and anesthetic for non-food producing minor species, which does not include rabbits because they are considered food-producing animals in the United States. The requirements for indexing of food animal drug labels are similar to those of the standard Food and Drug Administration (FDA) requirements, which include extensive pharmacokinetic studies and determination of withdrawal times. The use of regular, multidose, or the IDX labeled alfaxalone is considered extra label in rabbits. Reports on alfaxalone use in rabbits indicate minimal negative cardiovascular effects and rapid recoveries.[12] If ketamine was not used as part of the premedication, it can be given IV or IM as part of the induction protocol. Many rabbits are extremely nervous and jumpy, and the stress of restraining them to place an IV catheter may be dangerous or not possible. Sometimes even premedicating them will not achieve a sedate enough rabbit to place a catheter. Including ketamine or alfaxalone to the "premedication cocktail" may not only sedate the patient, but also induce anesthesia, allowing for IV catheter placement. For these patients, while propofol and etomidate may be safe induction agents, they may not be necessary. If the rabbit is not going to be intubated, use of propofol would be contraindicated because it could cause apnea and there would be no way to ventilate the patient. If the patient is not completely anesthetized by the premedication, and IV induction is not being used, mask induction is a common technique. Despite the potential stress involved, it is considered very safe in premedicated rabbits. Veterinary technicians should gently wrap a rabbit in a towel while administering inhalant anesthetic agents to keep the rabbit calm and restrained (Figure 20-25). Sevoflurane is the recommended inhalant anesthetic induction agent due to its rapid onset of action and odor, which is less noxious than isoflurane. Every effort should be made to keep rabbits calm during these inductions.

The oral cavity of rabbits is very long and narrow with a large fleshy tongue and sharp angular teeth. These features make it almost impossible to visualize the glottis, making endotracheal intubation technically difficult in rabbits. Because they primarily breathe through their nose, many practitioners do not routinely intubate rabbits and instead use a mask covering the nose and mouth, or just the nose to deliver inhalant anesthetic agent (Figure 20-26). If there are no anesthetic complications, this technique will suffice. However, in the event of respiratory arrest, there is very little one can do to ventilate the patient. It is possible that by pushing oxygen through a tight-fitting mask, the lungs may become inflated. But more likely, the stomach will become dilated with oxygen rather than the lungs. This "forced mask" ventilation technique is described below:

- Choose a mask that covers the nose and mouth with as few leaks as possible. There are various premanufactured masks, but it may be necessary to create a homemade mask out of syringe cases, small bottles, or tubing.
- Position the head and neck so that the trachea is fully extended and as straight as possible to allow easier air movement during ventilation.
- Stomach gas may need to be released postoperatively by passing a gastric tube or carefully expressing the air out of the stomach postsurgically.

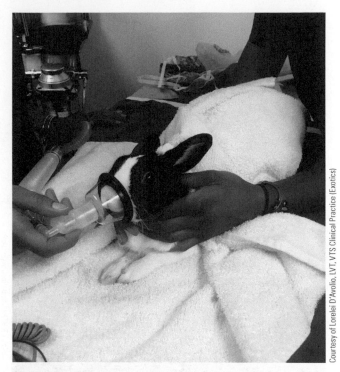

FIGURE 20-25 A rabbit wrapped in a towel while receiving induction via a face mask.

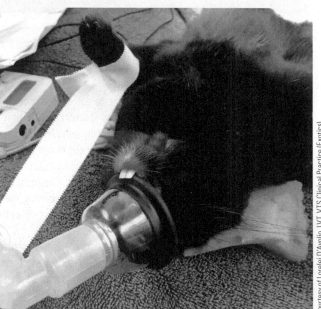

FIGURE 20-26 A rabbit under general anesthesia with a mask covering the nose. As obligate nasal breathers, rabbits can be anesthetized by using a small mask in this fashion.

The standard for rabbit anesthesia is moving toward endotracheal intubation for all anesthesia events. Having a patent airway provides the only guaranteed way veterinary technicians can perform much needed emergency positive pressure ventilations that may both prevent and resolve an apneic event.

Rabbits should be preoxygenated if they are suffering any respiratory compromise, such as an upper respiratory infection or a space occupying mass. Remember that they have a large abdominal cavity with a full stomach, cecum, and intestines. This decreases tidal volume and can compromise respiratory function, making preoxygenation even more valuable.

Endotracheal Intubation of Rabbits

Rabbits are obligate nasal breathers, which means that the normal anatomic position of the epiglottis causes it to be engaged over the caudal rim of the soft palate, sealing the oral pharynx from the lower airways. They also have long, narrow oral cavities with thick tongues, making visualization of the larynx almost impossible.

Since endotracheal intubation in rabbits is commonly done blindly, this skill requires practice. Endotracheal intubation performed blindly in rabbits requires the animal to be very sedate or anesthetized. This procedure generally requires at least two people.

- While administering passive supplemental oxygen, hold the head up to straighten the trachea as much as possible.
- Apply dilute local anesthesia to the supraglottic region. It is recommended to use low dosages of lidocaine and avoid using benzocaine due to its methemoglobinemia-producing effect. Allow the local anesthetic to take effect to reduce laryngeal spasm.
- Insert an appropriate-sized ETT (Figure 20-27).
- Listen for the loudest and clearest breath sounds through the ETT. While listening, attempt to feed the tube into the presumed trachea (Figure 20-28).
  - Some anesthetists prefer to use a modified esophageal stethoscope connected to the end of the ETT to listen for breath sounds, rather than using the ear to ETT method.
- Attaching a capnograph to the ETT to detect expelled CO_2 will confirm proper placement. If this device is not available, hair plucked from the patient can be used to detect whether or not the ETT is in the trachea (Figure 20-29). Visualization of condensation created by a breath on the walls of a clear ETT is also confirmation of proper placement.

Nasal intubation can also be done by inserting a slightly smaller, lubricated ETT into the nasal passages (Figure 20-30).

Using rigid or flexible endoscopes is also an option for practices with this equipment. These techniques often require a third person to drive the camera and can take longer than an experienced technician with excellent blind intubation skills (Figure 20-31).

A commercially available modified laryngeal mask that covers the supraglottic region (V-gel® [Docsinnovent, Inc.]) has been developed for rabbits and felines. Although endotracheal intubation is preferred, V-gel® offers an easier approach to the airway when endotracheal intubation

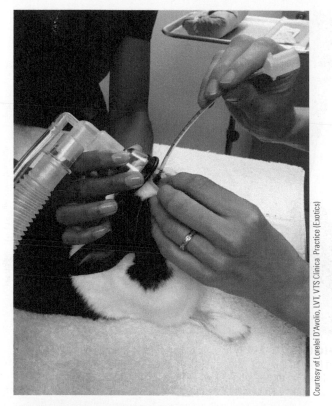

FIGURE 20–27 Positioning a rabbit for endotracheal intubation.

Courtesy of Lorelei D'Avolio, LVT, VTS Clinical Practice (Exotics)

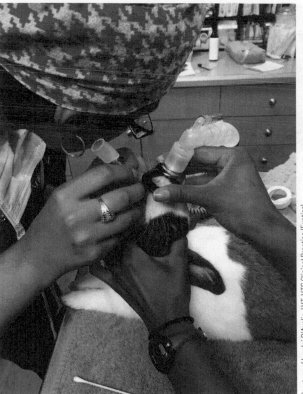

FIGURE 20–28 Veterinary technician is listening through the ETT for loud breath sounds. Once she is able to hear that the tube is placed directly over the glottis, she can advance it into the trachea upon inhalation.

Courtesy of Lorelei D'Avolio, LVT, VTS Clinical Practice (Exotics)

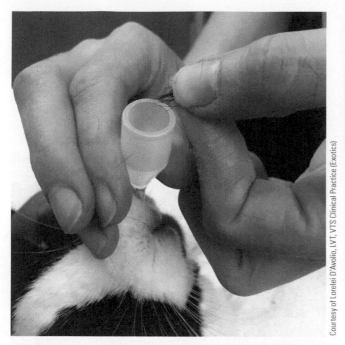

FIGURE 20-29 By holding a few delicate strands of rabbit hair in front of the tube, movement will indicate proper endotracheal intubation as the rabbit breathes through the tube. This technique also works well in birds with downy feathers.

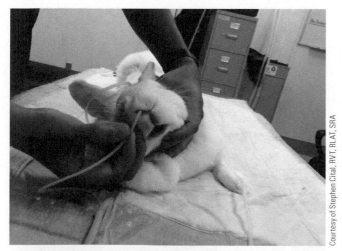

FIGURE 20-30 Nasal intubation in a rabbit.

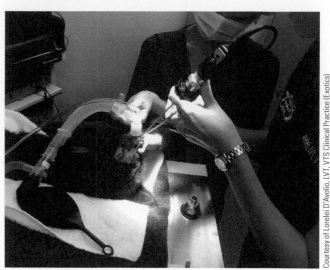

FIGURE 20-31 Endoscopic intubation of a rabbit.

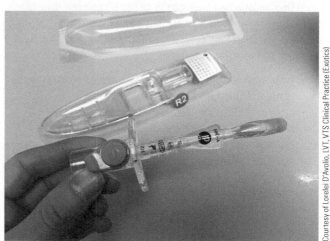

FIGURE 20-32 Example of V-Gel®.

Tech Tip

Avoid using excessive force and trying multiple attempts when intubating rabbits, as soft tissue damage can lead to inflammation, ulceration, and abrasions as well as laryngospasm.

Maintenance

For shorter procedures, injectable anesthetic premedication and induction agents may produce anesthesia of sufficient duration, especially when utilizing proper multimodal analgesia. For longer procedures, delivery of inhalant anesthetic agents such as isoflurane and sevoflurane via a non-rebreathing system is best. Oxygen flow rates for non-rebreathing systems should be between 150 to 300 mL/kg/min to avoid rebreathing

proves too difficult or time consuming. V-gel® is not ideal for animals requiring oral surgery as it takes up a significant amount of space in the mouth and there is potential for fluid leakage around the ETT if the patient is undergoing a dental procedure (Figures 20-32 and 20-33). Providing positive pressure ventilation is also not considered a guarantee with the V-gel®, as there is potential for forced air slipping around the seal. Other endotracheal intubation options are to pass a guide tube into the trachea and then slide an ETT over it or place a V-gel® and then run a guide tube through the V-gel® into the trachea, remove the V-gel®, and then slide an ETT over the guide tube.

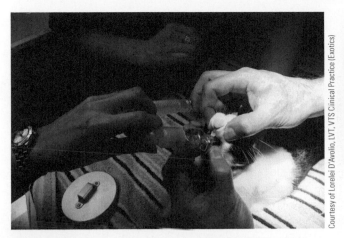

FIGURE 20-33 Placement of the supraglottal airway device V-Gel® in a rabbit.

Courtesy of Lorelei D'Avolio, LVT, VTS Clinical Practice (Exotics)

of carbon dioxide. The vaporizer settings will vary based on premedications and the procedure being performed. Rabbits have an isoflurane MAC value of 2.05 +/− 0.18% and a sevoflurane MAC value between 3.7 and 4.1%, generally requiring higher levels of inhalant anesthetic agent than dogs and cats. Premedications such as ketamine and dexmedetomidine will decrease inhalant anesthetic agent requirements, while more painful procedures may increase the need for higher dosages. The key is for the veterinary technician to constantly be monitoring the patient and adjusting as needed.

Monitoring

A good veterinary technician should be comfortable assessing anesthetic depth by observing and recording reflexes. In rabbits, loss of pedal (digital withdrawal), auricular (ear pinch), and palpebral reflexes indicate appropriate depth, while loss of corneal reflex and dilated pupils can be a sign of deep anesthesia and cerebral hypoxia. Mucous membrane color, capillary refill time, and temperature of extremities can be used to gauge cardiovascular function and tissue perfusion. Another valuable and easy monitoring technique is to use an ultrasonic Doppler to monitor the heart rate and rhythm. Placing the probe directly over the heart and securing it with a piece of tape works well (Figure 20-34).

Tech Tip 🐾

Spending time setting up monitoring equipment with every bell and whistle may prolong anesthetic time. Basic equipment such as a stethoscope and the anesthetist's eyes, ears, touch, and intuition are just as vital as any expensive monitoring equipment.

Conventional monitoring equipment can be used for rabbits; however, pediatric settings that can read a heart rate up to at least 350 bpm should be selected. A SpO_2 monitor that reads higher heart rates will work well on rabbit ears or thin extremities. They do not work well on feet, as rabbits have very thick fur and no hairless pads. As with birds, alligator clamps on ECG leads can cause pressure necrosis or tear thin skin. Using needles, specially manufactured flattened clips, or modified alligator clips is important. Mainstream capnography

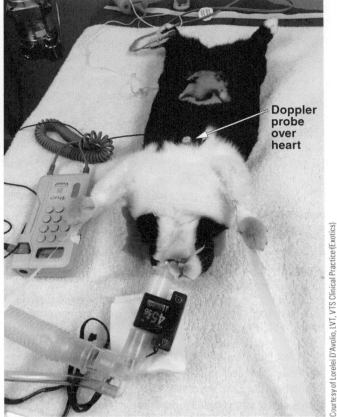

Doppler probe over heart

Courtesy of Lorelei D'Avolio, LVT, VTS Clinical Practice (Exotics)

FIGURE 20-34 Placement of a Doppler probe over the heart of a rabbit prepped for an ovariohysterectomy (OHE).

can be useful in monitoring trends in intubated rabbits. The normal range for end tidal carbon dioxide ($EtCO_2$) is between 35 and 45 mm Hg. Obtaining direct arterial blood is the preferred method for monitoring pH, oxygen, and carbon dioxide levels and it only requires small sample volumes; however, difficulty in obtaining arterial samples may deter frequent monitoring.

Maintaining normothermia is also important, and traditional warming devices such as forced air blankets, water circulating pads, or others that provide heat while preventing burns are ideal. Some other techniques to keep patients warm include using aluminum foil or bubble wrap around the extremities, delivering heated fluids, and radiant heat sources. Pediatric rectal thermometers work well to monitor body temperature.

Tech Tip 🐾

Humid-Vent® is a small heat and moisture exchange device that fits between the ETT and the breathing hose. It has thin filtration paper that keeps inspired gases warm and moist while protecting the anesthesia machine from aerosolized bacterial and viral particles the patient may spread during expiration. This device, while adding a small amount of dead space, may work well during longer procedures to keep the lungs warm and moist (Figure 20-35).

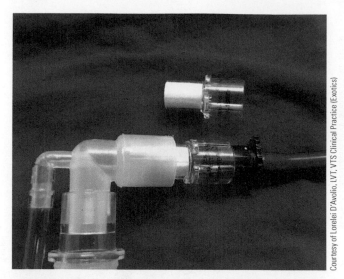

Courtesy of Lorelei D'Avolio, LVT, VTS Clinical Practice (Exotics)

FIGURE 20-35 Humid-Vent is a device that humidifies oxygen and inhalant anesthetic agent prior to passing through an ETT into the patient. The paper filter keeps warm moist air in the chest cavity and protects the anesthesia machine from aerosolized bacteria and viruses when the patient expires.

Blood pressure monitoring is a great tool to assess cardiac function under anesthesia. Due to their small size, placing direct arterial lines to invasively monitor blood pressure is very challenging in rabbits. However, using an appropriately sized blood pressure cuff, sphygmomanometer, and a Doppler to noninvasively monitor peripheral blood pressure is easy and should be part of any long anesthetic procedure. An appropriately sized cuff placed on an appendage or tail works well, as long as the peripheral pulse is strong enough for the equipment to obtain a reading. Shaving small patches of hair will help obtain better results; however, accurate readings are often limited. As with many other devices, veterinary technicians should pay attention to trends rather than exact numbers, recognizing that mean arterial pressure (MAP) should remain over 60 mm Hg and systolic blood pressure (SAP) over 90 mm Hg.

Recovery

Postoperative oxygen therapy is recommended if tolerated by the recovering patient. Once jaw tone returns, the patient can be extubated after examination of the glottis to make sure it is clear of debris or secretions that could be aspirated or obstruct the airway. Rabbits should be placed in sternal recumbency to help aid ventilation. A quiet and dimly lit room away from predators will aid in minimizing stress in pet rabbits.

RABBIT ANALGESIA

Like most exotic pets, analgesia has been underused in pet rabbits for various reasons including veterinary professionals being less familiar with the species, difficulty in assessing pain, and lack of knowledge regarding drug dosage, efficacy, safety, and concerns of adverse effects. But current advances have provided rebuttals to these concerns. It is now commonly accepted that rabbits feel pain like other, more familiar mammals. In fact, it is theorized that as a prey species, rabbits suffer a higher morbidity and mortality in comparison to other mammals when pain is not treated properly. Untreated pain activates the sympathetic nervous system and can cause many dangerous conditions in rabbits such as tachycardia, vasoconstriction,

altered respiratory rate, fluid and electrolyte acid-base imbalance, and cardiac arrhythmias. When combining these possibilities with a reduction in gastrointestinal motility, decreased appetite, dehydration, and a general decreased immune response, rabbits are at significant risk for developing life threatening ileus, commonly referred to as gastrointestinal stasis.

Tech Tip 🐾

Some signs of pain in rabbits may be an audible grinding of the teeth, reduced activity, abnormal posturing, heightened sensitivity to touch, and wincing. But remember, rabbits almost NEVER vocalize, so do not expect howls or cries from rabbits.

How should a veterinary technician develop a proper "pain score" for rabbits? Traditional methods for assessing pain for dogs and cats unfortunately do not correlate well with rabbits. It is very difficult to assess normal behavior in a rabbit once they are taken out of their home environment. Like other prey species, rabbits will hide any signs of pain or injury in order to avoid a perceived predator. While a cat may sit in a corner of a cage vocalizing, squinting its eyes, and holding up a paw to indicate an injury, a rabbit with the same injury may sit silently in a stoic position with wide eyes and rapid breathing, which is the exact appearance as a perfectly normal, healthy rabbit due to normal stress of being in a hospital cage. Alterations in heart rate, respiration rate, and blood pressure may increase with pain, but they may also increase with basic handling and restraint. While a decreased appetite may be apparent in a rabbit experiencing pain, other non-painful factors may cause lack of appetite, such as environmental stress, time of day (rabbits are crepuscular species and tend to eat most at dusk and dawn), or dental malocclusion. Because of these difficulties, it is important that veterinary technicians become familiar with rabbit behavior and body language. It is reasonable to assume procedures that are known to be painful to other mammals are equally painful to rabbits. Problems unique to rabbits that are painful also require appropriate analgesia, such as:

- Gastrointestinal stasis
- Vestibulitis (Figure 20-36)
- Oral ulcerations caused by points on the cheek teeth (Figures 20-37 and 20-38)
- Chronic issues such as spinal spondylosis or degenerative joint disease, which is very prevalent in the backs, stifles, and hips of rabbits

Tech Tip 🐾

The Rabbit Grimace Scale (RbtGS) is a validated scoring system to evaluate acute pain in laboratory rabbits and may serve as a resource for clinical practice and owners. It uses five facial action units (AU) (orbital tightening, cheek flattening, nose shape, whisker position, and ear position) that are scored according to whether it is not present (0), moderately present (1), or obliviously present (2). The RbtGS can be found on The National Centre for the Replacement, Refinement, and Reduction of Animals in Research website at https://www.nc3rs.org.uk/grimacescales.

Courtesy of Lorelei D'Avolio, LVT, VTS Clinical Practice (Exotics)

FIGURE 20-36 Vestibulitis in a rabbit. Persistent head tilt and loss of balance are the classic signs of **vestibulitis** (also known as torticollis or wry neck). Rabbits with vestibular disease can have a head position that ranges from a few degrees to 180 degrees from its normal position. They can circle, fall over, have difficulties standing, and develop eye injuries because a rabbit's prominent eye globe (especially of the eye facing the ground when the head is tilted) is prone to trauma. Vestibulitis can be caused by conditions such as an ear infection, brain disease, or trauma.

Courtesy of Lorelei D'Avolio, LVT, VTS Clinical Practice (Exotics)

FIGURE 20-38 Severely overgrown incisors in a rabbit.

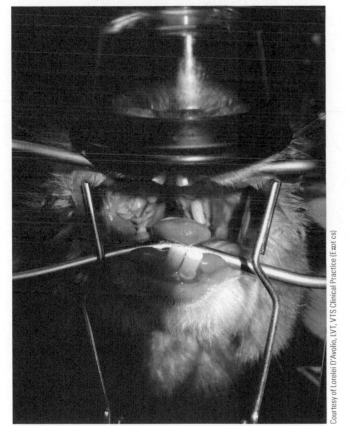

Courtesy of Lorelei D'Avolio, LVT, VTS Clinical Practice (Exotics)

FIGURE 20-37 Severely overgrown cheek teeth (molars) in a rabbit.

As in other mammals, opioids are considered the most effective treatment for moderate to severe pain in rabbits. Because of the dangers associated with gastrointestinal ileus, practitioners have historically been afraid to utilize full mu opioid agonists in rabbits. Both mu and kappa opioid receptor agonists have been shown to inhibit motility in the intestinal tract of some rabbits. However, clinically relevant evidence shows that the effects of untreated pain can severely inhibit motility, in addition to the previously mentioned problems that pain can cause. It is currently considered more important to properly manage pain than to worry about the possibility of an opioid induced slowing of the intestines. There are drugs commonly and safely used to combat slowing of the gastrointestinal tract such as cisapride (Proplusid®) or metoclopramide (Reglan®) that can be compounded into a palatable liquid or given parenterally. Rather than allow rabbits to feel pain, it is considered better practice to treat with opioids as needed and address any possible motility issues as they arise. Drugs such as morphine, hydromorphone, and methadone (Dolophine®) should be considered for any painful procedure. They can be administered IM, IV, SQ, epidurally, or as a CRI. Buprenorphine is also widely used due to its longer duration of effect, and should be used for cases of mild to moderate pain.

Tech Tip

No need to worry that using full mu opioid agonists will make your rabbit patient vomit—rabbits are not able to! They have a very powerful sphincter blocking regurgitation from the stomach to the esophagus.

NSAIDs such as meloxicam work synergistically with opioids and are useful as part of the multimodal analgesic plan for rabbits. Adverse effects are the same as with other species, and include the possibility that by blocking COX enzymes that cause inflammation, they can also block important gastric protective enzymes and therefore cause gastric bleeding or upset. There are also concerns that NSAIDs may inhibit platelet aggregation and exacerbate pre-existing renal insufficiency. Because of this, preoperative use remains controversial. However, NSAID use postoperatively is common. With monitoring of serum biochemical profiles, long term use is considered safe for chronic pain.

Local anesthetics should also be used as part of any good perioperative analgesic plan. Lidocaine and bupivacaine can be administered topically, via infiltration, intra-articular, as a regional nerve block, or epidurally. Simple incisional line blocks and wound infiltration can be used to completely desensitize an area as well as reduce the amount of required anesthetic drugs. Providing maxillary, mandibular, and infraorbital nerve blocks will decrease the amount of post-procedural analgesia required for dental procedures. Epidural administration of lidocaine results in a rapid onset of sensory and motor block to the hind quarters and should be considered for orthopedic or other painful procedures of the hind end.

Alpha-2 adrenergic agonists such as dexmedetomidine are being utilized much more commonly in rabbits. As with other mammals, alpha-2 adrenergic agonists provide not only analgesia but also sedation and muscle relaxation, and are reversible. Their use is recommended for clinically healthy patients who can tolerate the cardiovascular impact they tend to have by causing bradycardia and decreasing cardiac output. When used in proper conditions and with proper monitoring, they are excellent components in multimodal analgesia.

Ketamine is one of the most commonly used analgesics in rabbits due to its wide safety and dissociative properties. As an N-methyl-D-aspartate (NMDA) receptor antagonist, its role is to reduce the development of central sensitization, causing a dissociative sensation in the patient. Ketamine can be given to rabbits IM, SQ, IV, or via CRI.

Thanks to the advent of flavored compounding, tramadol (Ultram®) can be used to treat postoperative and chronic pain in rabbits. While it is a mu opioid receptor agonist, it also inhibits the reuptake of serotonin and norepinephrine. Studies have not provided concrete evidence as to exactly how well this analgesic works in rabbits; however, extrapolating dosages and effects from other mammals has yielded positive clinical results.

As previously described in birds, use of photobiomodulation with class 4 laser therapy is another adjunct to pain management in rabbits. Many rabbits are living longer and experiencing more age-related chronic pain. Photobiomodulation treatment to affected joints is being used more frequently to provide clients with another tool in managing the comfort of their patients. Clinicians are also using photobiomodulation to work on the nerves along the spine to provide analgesia to rabbits with motility issues who are suffering from painful gas and bloat. Use of photobiomodulation for wound management and incisions post-surgery are also becoming more routine.

Tech Tip

Do not be alarmed—drug dosages for rabbits tend to be significantly higher than those used for dogs and cats. Be sure to consult a current exotics formulary, such as *Exotic Animal Formulary* by James Carpenter, for accurate dosing.

PET RODENT ANESTHESIA

Rodents are challenging to anesthetize primarily because of their small sizes. Difficulties include obtaining airways, IV access, maintaining body temperature, and providing proper doses of drugs. With clients expecting higher levels of care now more than ever, veterinary teams must be able to offer surgical and anesthetic services with the highest standards. With the proper equipment, patience, and lots of practice, rodent anesthesia can be safe, interesting, and even fun.

Preanesthetic Preparation

Subclinical respiratory disease is common in rodents. Rats commonly carry *Mycoplasma pulmonis* while guinea pigs carry *Bordetella bronchiseptica*, *Streptococcus pneumoniae,* and *Streptococcus equi* subspecies *zooepidemicus*, which are bacterial diseases often present without obvious clinical signs. Guinea pigs can also carry subclinical adenovirus in their respiratory tract. If clients permit, obtaining radiographs and serology testing prior to an extended anesthetic event is ideal. In older patients, obtaining a complete blood count (CBC) and biochemical profile to evaluate glucose, kidney and liver enzymes, and electrolytes can help avoid pitfalls during anesthesia. Unfortunately, many clients are not willing to permit these potentially costly tests. In instances where preanesthetic testing is not possible, veterinary technicians must be prepared for possible complications and be ready to prevent and manage them if they arise. Keeping patients in a quiet, warm, dark enclosure prior to surgery is important. Allowing cagemates to be present prior to and after a procedure is also helpful for reducing stress. Having a high quality scale that weighs to the tenth of a gram is important to obtain an exact weight and for calculating accurate drug doses. An infant stethoscope will help auscultate rapid heart rates, which can be up to 600 bpm in a mouse. An infant rectal thermometer should be used to obtain proper body temperatures, which are normally around 98°F, except for the guinea pig which can be up to 103°F (see Table 20-2 for biologic values and life expectancies).

Due to their high metabolic rate, rodents should not be fasted prior to surgery. Dehydration and hypoglycemia are a serious concern for these tiny patients. Similar to rabbits, they also lack a vomit reflex and rarely regurgitate. The **hystricomorphs** (guinea pigs and chinchillas) are herbivores, and they require food in their GI tracts at all times. While GI stasis is not a significant concern for most rodents, the hystricomorphs are at serious and grave risk if they do not eat for an extended period of time.

Catheters and Intraoperative Fluids

Fluid stabilization is vital for all patients under anesthesia, especially if they are undergoing a surgical procedure. They should ideally be given preoperatively to optimize hydration, fluid balance, and blood volume; however, administration routes may vary based on the species and size. Placing IV catheters in rodents is achieved with the same basic techniques as with a rabbit using the cephalic vein. But due to patient size, IV catheterization can be difficult to place based on veterinary technician experience, equipment available, and patient status. Guinea pigs have fairly prominent cephalic veins; however, the veins are not straight and their skin tends to be very tough (Figures 20-39 and 20-40). Chinchillas have very tiny cephalic veins and may require catheters sized 26 gauge or narrower. Unlike guinea pigs, their skin is extremely thin and delicate and can be easily torn by clipper blades during routine shaving. Rats and smaller rodents will require tiny laboratory-sized catheters and require deep sedation/anesthesia prior to catheterization. Veterinary technicians should use caution when taping rodent catheters in place as porous tape applied directly on the skin can damage it during removal, and if the skin is wet, these patients can easily slip out of them.

Tech Tip

Gently warming veins with a warm water compress can help dilate tiny veins and facilitate IV access. Using a topical analgesic such as lidocaine/prilocaine cream may help with local discomfort and decrease chances of the patient jerking or moving during catheter placement.

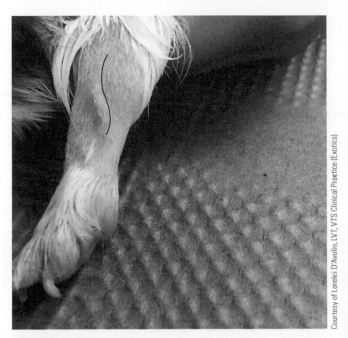

FIGURE 20-39 The cephalic vein of a guinea pig.

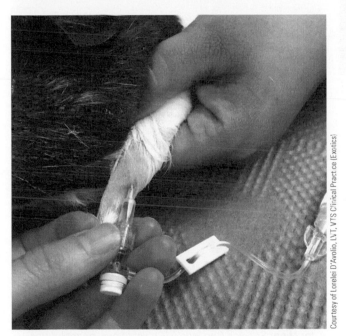

FIGURE 20-40 Placement of an IV catheter into the cephalic vein of a guinea pig.

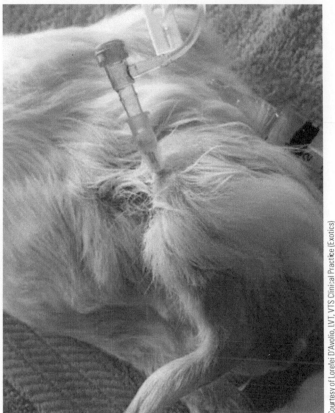

FIGURE 20-41 Guinea pig with an IO catheter of the proximal tibia.

Placement also requires proper aseptic technique, and IO catheters should be sutured in place for stability. When placed properly, risks of complications are rare, but could include infection or emboli of fat or bone marrow. The sites for placement in rodents include the proximal humerus, proximal tibia, and proximal femur. IO catheters should not remain in place for more than 72 hours.

If IV or IO fluids are not possible, SQ fluids should be administered before any surgical procedure prior to or immediately after induction. Fluids should be warmed to prevent hypothermia and can easily be given in the subscapular area. Veterinary technicians will need to draw predetermined doses into syringes rather than using a hanging bag in order to facilitate doses that are sometimes only a few milliliters.

Premedication, Sedation, and Restraint

There are some controversial opinions about whether or not to use multimodal anesthesia with rodents because of their small size. In the event one of the potential adverse effects of a premedication occurs, one must be able to detect and manage the problem. For example, 20 to 40% of Mongolian gerbils have congenital epilepsy and are at a high risk of seizuring if given drugs that reduce the seizure threshold. If a gerbil is given acepromazine and ketamine and begins to seizure, there is little a veterinary technician can do without having IV/IO access. Or if methadone (Dolophine®), midazolam, and ketamine are used and the patient becomes apneic, there is little to do if the patient is not intubated. Conversely, there are profound benefits of

Another fluid administration option for rodents is intraosseous catheterization. IO catheters are fairly easy to place using a spinal needle, IV catheter stylet, or regular hypodermic needle (Figures 20-41 and 20-42). Most guinea pigs or chinchillas require a 20 to 22 gauge needle while a gerbil or hamster may require 25 to 27 gauge. The length of the needle should be long enough to extend $\frac{1}{3}$ to $\frac{1}{2}$ the length of the bone. A sterile piece of cerclage wire or stainless steel suture may be needed to clear out boney material if a stylet is not used. IO catheter placement is considered painful, and appropriate local analgesia or anesthesia/sedation are mandatory.

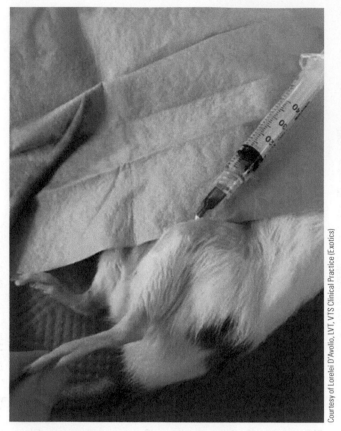

FIGURE 20-42 Rat with an IO catheter of the proximal tibia.

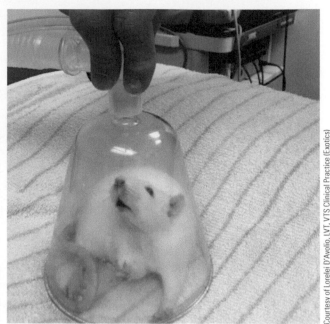

FIGURE 20-43 Rodent in an induction chamber (canine mask).

premedicating animals that are so prone to stress-related adverse effects. Decisions should be made based on the following parameters:

- *Does the facility have proper tools?* For example, 26 gauge or narrower catheters, infusion pumps, or endoscopes.
- *Are veterinary technicians trained/experienced to perform techniques?* For example, blind intubation, placing IO catheters, and titrating dose calculations.
- *Does the facility have the proper reversal agents?* For example, atipamezole (Antisedan®), flumazenil (Romazicon®), and naloxone (Narcan®).
- *Does the facility have the proper monitoring devices (see later in this section) and do the veterinary technicians know how to interpret and place them?*

If the answers to these questions are no, it may be safer to premedicate with analgesics and benzodiazepines alone. Drugs such as midazolam tend to be very safe when given IM in rodents and the anti-anxiety/sedative properties are helpful in these highly stressed patients. If the procedure is not expected to be very painful, butorphanol also has sedating properties and works well to calm these animals prior to mask induction without causing severe adverse effects. After allowing these analgesics/anxiolytics to have effect, patients can either be placed in a small induction chamber (Figure 20-43) or a small mask can be used.

If the answers to these questions are all yes, premedicants can be similar to other mammals. Combinations of analgesics with benzodiazepines, acepromazine, ketamine, dexmedetomidine, and alfaxalone can all be used for premedication/induction based on the procedure and preference of the veterinarians. For the smaller rodents such as mice and hamsters, these doses will often be very small and may need to be diluted to achieve proper dosing.

Patients should be preoxygenated for a few minutes prior to endotracheal intubation or mask induced anesthesia due to their predilection for respiratory diseases and small tidal volumes. If injections of sedatives have been given, veterinary technicians should be prepared that the level of inhalant anesthetic agent may not need to be very high, or may not be needed at all. Diligent monitoring is important, and oxygen should be administered throughout the procedure regardless of whether or not inhalant anesthetic agents are given. Anticholinergic drugs such as atropine or glycopyrrolate are not routinely used as part of a premedication protocol. These parasympatholytic drugs should only be given during anesthesia to treat bradycardia that is causing low cardiac output and low blood pressure and in cases of cardiac arrest.

Tech Tip

For minor species, the standard drug approval process is too expensive and many species are too rare or varied for use in traditional safety and effectiveness studies. The Index of Legally Marketed Unapproved New Animal Drugs for Minor Species (the Index) is a list of new animal drugs intended for use in minor species that have had their safety and effectiveness affirmed through an alternative FDA process. The Index helps veterinarians prescribing drugs to animals or classes of animals representing markets too small to justify the costs of the drug approval process, despite incentives of the Minor Use and Minor Species (MUMS) Animal Health Act of 2004. Indexing provides a faster and less expensive way to obtain legal marketing status for eligible products for minor species. The indexed drug is chemically identical to the non-indexed drug; it is only the label that is different.

Induction Procedures and Techniques

As described in the previous section, parenteral injections and induction can carry significant risks for these small animals. Induction for rodents is usually safest to achieve by mask or chamber induction. Sevoflurane has a faster rate of onset and is less noxious; however, isoflurane is also commonly and safely used.

Despite their inability to vomit, endotracheal intubation is preferred whenever possible. There are several factors that contribute to possible reasons rodents could succumb to respiratory arrest if not intubated including:

- Rodents have very limited chest capacity in comparison to their abdominal cavity.
- Rodents are at risk of harboring undiagnosed respiratory disease as mentioned earlier in this chapter.
- Many analgesics and anesthetic drugs can further compromise the respiratory system.
- Laying in dorsal recumbency puts added pressure on the thoracic cavity from the large abdominal viscera.

Unfortunately, most small rodents are extremely difficult to intubate.

While guinea pigs are the largest of the common pet rodents seen, they are one of the most difficult to intubate. Their anatomy is unique with the presence of a **palatal ostium** (Figure 20-44). This is a tiny opening in the caudal portion of the oral pharynx which is made of tissue that is easily damaged by prodding of an ETT and can bleed profusely, even causing asphyxiation. Additionally, the oral cavity is long and narrow with prominent check teeth. One way to potentially intubate a guinea pig is by using a tiny laryngoscope blade such as a size 0 Miller to provide visualization of the palatal ostium. If an ETT is passed through the opening, blind intubation of the trachea may be possible. Alternatively, an appropriately sized ear cone can be used in a similar fashion (Figure 20-45). A polypropylene catheter can then be directed through the palatal ostium and glottis into the airway, and then an ETT can be fed over the catheter. In hospitals where small sized endoscopes are utilized, they can also be used to assist in visualization.

The smallest patients most veterinary hospitals have the ability and equipment to intubate are chinchillas and rats. These rodents can be intubated in a similar fashion to the rabbit, in that the technique is mostly blind. Proper positioning with the neck dorsally flexed is essential. One technique that may help visualize the glottis is use of a syringe tube with the plunger taken out and cut at a 25 degree angle (Figures 20-46a and 20-46b). Once visualization is made, a tiny tube or even an IV catheter can be used to establish a patent airway (Figure 20-47).

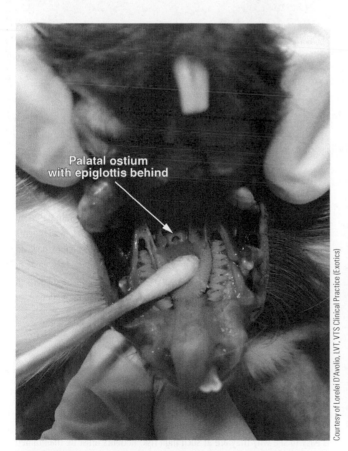

FIGURE 20-44 The palatal ostium in a guinea pig. The palatal ostium is a tiny aperture in the soft palate that separates the oropharynx from the rest of the pharynx, making intubation very difficult.

Courtesy of Lorelei D'Avolio, LVT, VTS Clinical Practice (Exotics)

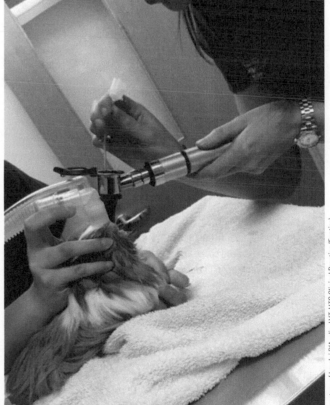

FIGURE 20-45 Using an otoscope cone to assist with intubating a guinea pig. In guinea pigs, special tools may be needed to move the proximal tongue and other soft tissue out of the way to allow visualization of the palatal ostium and perhaps the tracheal opening.

Courtesy of Lorelei D'Avolio, LVT, VTS Clinical Practice (Exotics)

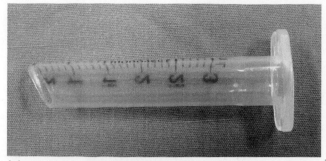

(a)

(b)

Courtesy of Teri Raffel Kleist, CVT, VTS (Surgery)

FIGURE 20-46 Tools to help with visualization of the glottis during endotracheal intubation of rodents. (a) One example is to use a syringe tube with the plunger taken out and cut at a 25 degree angle. (b) The ETT can then be passed through the syringe tube.

In the likely event that endotracheal intubation is not possible, veterinary technicians should do the best they can to provide respiratory assistance using these tips:

- Use an analgesic/anesthetic drug combination that causes the least amount of respiratory depression.
- Preoxygenate rodents with 100% oxygen for 3 to 5 minutes prior to induction.
- Position patients in dorsal recumbency with their neck extended and chest propped up higher than their abdominal cavity to decrease pressure on the thorax (Figure 20-48).
- Monitor respirations for any sudden changes and adjust vaporizer settings accordingly.
- Use a small, tight fitting mask covering the nose and mouth. This will decrease dead space and may be helpful in providing positive pressure ventilation in the event of respiratory collapse. These can be made with rubber gloves covering a syringe cap if commercial rodent masks are not available.

Tech Tip

In guinea pigs, veterinary technicians should swab the oral cavity before, during, and after surgery to ensure it has not filled with saliva or regurgitated materials. Aspiration can happen very quickly in anesthetized guinea pigs; therefore, the use of anticholinergic drugs may be particularly beneficial in this species.

Courtesy of Lorelei D'Avolio, LVT, VTS Clinical Practice (Exotics)

FIGURE 20-47 Method of endotracheal intubation in a rat. Proper positioning with the neck dorsally flexed is essential to passing the ETT through the syringe tube.

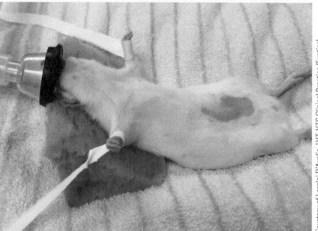

Courtesy of Lorelei D'Avolio, LVT, VTS Clinical Practice (Exotics)

FIGURE 20-48 A pet rat in proper anesthetic position for an abdominal surgery with the thorax elevated on a foam pad.

Maintenance

Anesthesia should be maintained by adjusting the anesthetic rate based on patient monitoring of anesthetic depth and physical parameters. Most rodents with, at minimum, a premedication of midazolam and butorphanol, have an isoflurane MAC value of approximately 1.3% and a sevoflurane MAC value of approximately 2.3%. However, veterinary technicians should be diligent in monitoring and observing the patient in order to adjust levels as needed.

Monitoring

Hypothermia is a dangerous and preventable complication for rodents under anesthesia. Anesthesia can cause hypothermia by vasodilation, redistributing core heat to peripheral limbs, and reducing the ability to shiver/have normal thermogenesis. Combine this with inhalation of cold gases and an open incision site, plus the high surface area to body mass ratio of rodents, and it is easy to see why they are at such high risk. Rodents must be maintained with a variety of heating elements: radiant, environmental (the temperature of the room), forced air blankets, circulating water pads, or heated gel mats. A well lubricated pediatric thermometer is safe to use in almost any rodent.

Additionally, surgical scrub should be warmed prior to use, and using warmed saline instead of alcohol to rinse or lavage is strongly recommended.

Monitoring equipment for rodents is often minimal due to small patient size. Any sized rodent will benefit from the use of Doppler ultrasound probe to monitor heart rate and rhythm. The crystal works best when placed directly over the heart, but in larger rodents may work on the tail base, tarsal, or pedal artery. This will provide, at minimum, a good sense of pain response or light anesthetic plane (increase in heart rate) or if a patient is becoming too deeply anesthetized (decrease in heart rate). While many newer SpO$_2$ probes may work on the extremities of rodents, most equipment cannot read a heart rate over 400 bpm. While this may limit its use in hamsters, mice, or gerbils, it can give accurate levels of blood oxygen saturation and should be included in the monitoring plan. Capnography could be useful in the larger rodents if endotracheal intubation is obtained. Most likely, a mainstream capnography with an exotics converter (EMMA®, Massimo) unit will be required to reduce dead space. An ECG would also be beneficial; however, adjustments to keep the skin safe from trauma as described previously in this chapter are needed.

Beyond these instruments, the most important monitoring for rodents is the veterinary technician's observational skills. Use of a clear surgical drape (Veterinary Specialty Products) is very helpful in keeping the patient visible. Respirations should be regular and consistent. Mucous membranes should remain pink. Reflex responses such as tail pinch, pedal reflex, and ear pinch should be abolished when under an appropriate surgical plane. The patient's position should be monitored, with the head and chest slightly elevated above the abdomen to prevent pressure on the chest, with the face mask or cone fit snugly around the mouth and nose. It is easy for masks to slip and move during manipulation of the rodent during a surgical procedure, and veterinary technicians should be monitoring the patient at all times.

Tech Tip

Many rodents, such as hamsters, have an increased likelihood of corneal ulceration and proptosis of the eyes while under anesthesia. Veterinary technicians should pay attention to this and use lubrication and gentle pressure to keep eyes lubricated to prevent corneal ulcers and closed to prevent proptosis of the eyes.

Recovery

Depending on what was used for premedication, recovery may be very rapid for rodents. A heated, dimly lit, oxygen rich, quiet enclosure should be prepared and ready for recovery. This will prevent hypoxemia, hypothermia, and stress. Any IO or IV catheters should be removed prior to complete recovery, unless they have been properly sutured and secured for long-term use. Rodents will begin to chew at anything that is irritating to them, and if IV/IO lines are uncomfortable and not padded, they will be chewed. Rodents do not tolerate Elizabethan collars well and tend to become highly stressed or wriggle out of them when used, and hystricomorphs may not eat if forced to wear Elizabethan collars, which can lead to dangerous GI stasis. Careful monitoring of incision sites is crucial as rodents may chew on exposed suture, staples, or painful tissue. Additional analgesics should be administered or added, such as NSAIDs or repeated doses of short acting opioids that may have been eliminated from the body. Rodents can and should be offered food immediately after awakening to prevent/correct potential hypoglycemia as well as distract the animal from chewing or barbering its incision.

Tech Tip

Rodents and rabbits regularly suffer from ileus and GI stasis. Postoperative ileus is a major concern in hystricomorphs (guinea pigs and chinchillas) and rabbits. Offering them grass hay, high fiber pellets, and water immediately after recovery is important to prevent ileus.

PET RODENT ANALGESIA

Pain recognition in rodents can be difficult for new veterinary technicians learning about rodent body language. There are the typical hallmarks of pain that all mammals display, such as a decreased appetite and weight loss when experiencing chronic pain. Heart rate and respiratory rate will also elevate during pain; however, once rodents are handled it is impossible to know if the heart rate is elevated from pain or from the stress of being restrained. Rodents may become hypersensitive to even the slightest stimulation, such as a human looking at them in a cage, which can alter respiration and heart rates. Like most animals with an injury, rodents will protect the painful body part (e.g., some rodents with an injured limb will hold it up while others may hold and guard it).

There are other important indicators of pain in some rodents that veterinary technicians should know, which starts by understanding normal behavior for the patient. Getting a detailed history from clients, or even using a web cam to watch patients without causing them stress, is a great way to know what they are doing naturally. Any change from what is natural could be considered a response to pain. Are they pressing or rolling their abdomen? Are they separate from the group they are normally housed with? Are they hiding when normally they would be playing?

Laboratory studies have shown that when rats feel abdominal pain, they walk with an arched back. They, as well as mice, will also have porphyrin staining around the eyes when they are experiencing severe stress and/or pain (see Chapter 21 [Pain Management in Laboratory Mice and Rats]). Rodents that are hunched over, have piloerection (hair standing up), or an unkempt coat may be feeling pain. Some rodents may become very aggressive to humans or cage mates while others may cower in a corner. Most guinea pigs will normally "wheek" loudly when picked up or handled, and many will be silent when they are feeling pain. Mice will often over-groom a painful part of their body, even to the point of mutilation. Laboratory studies have also shown that mice will walk with their tails raised after a painful abdominal procedure.

Tech Tip

Porphyrin secretions from the Harderian glands in many rodents can mimic blood, and many clients describe concern about "bloody tears" or bloody nasal discharge. While excessive porphyrin secretions are a sign of stress, poor diet, pain, or illness, veterinary technicians should assure their clients that their pets are not crying bloody tears.

Interestingly, some rodents are herbivorous, while others are omnivorous. Pets such as guinea pigs and chinchillas are strict herbivores. For them, uncontrolled pain resulting in anorexia will cause similar complications to rabbits, with gastrointestinal stasis being of particular concern. Pets such as rats, hamsters, gerbils, and mice are omnivores. While not eating due to pain is dangerous because it can cause severe hypoglycemia, these rodents do not tend to have the GI complications like herbivores. In addition to proper analgesia, encouraging all rodents to eat after a painful procedure is important. Offering soft/soaked pellets and favorite treat foods should be a mandatory part of recovery.

As with other mammalian species, opioids are the standard for moderate to severe pain management. Full mu opioid agonists such as morphine and hydromorphone, as well as partial mu opioid agonists such as buprenorphine can and should be used to control pain. Due to their small body size and fast metabolic rate, dosages will be significantly higher than in other mammals. Potential adverse effects such as respiratory depression, gastrointestinal stasis, and profound sedation are similar in rodents as other mammals, and patients should be monitored for these adverse effects. Reversal agents work the same in these species and can be used in a similar fashion as dogs and cats based on the unique dosages for rodents listed in current exotics animal formularies. IM injections are easy to administer in the epaxial or gluteal muscles, and administration via SQ or IP (intraperitoneal) route is also viable.

Tech Tip

Some opioids such as buprenorphine have been known to cause pica in rodents, particularly rats. It may be prudent to recover these patients in an enclosure without bedding they are capable of ingesting to prevent potential gastric obstruction.

NSAIDs are used commonly in rodents. Like other mammals, NSAIDs inhibit COX enzymes that cause inflammation and pain. With the newer NSAIDs having improved potency, COX selectivity, and wide safety margin, drugs such as meloxicam are used quite frequently and appear to be effective for more painful conditions than older NSAIDs. They also provide longer duration of action than opioids. The risks of GI and renal damage of NSAIDs exist in rodents similar to that in other species; therefore, these animals should be carefully monitored.

Other safe and easy to administer analgesics in rodents are local anesthetics. While small sizes of patients may create dosing difficulties, use of topical, local infiltration, peripheral nerve blocks, and even epidural injections should all be considered (Figure 20-49). Veterinary technicians should be careful with dosing these tiny animals, as overdosing even with local blocks can be lethal. When quantities are tiny, diluting with saline is recommended in order to achieve a quantity that will sufficiently affect the desired area.

Ketamine and dexmedetomidine are also good options for analgesia in rodents, and are described in the preceding anesthesia section. These drugs function similarly as in other mammals and can be an important part of multimodal analgesia.

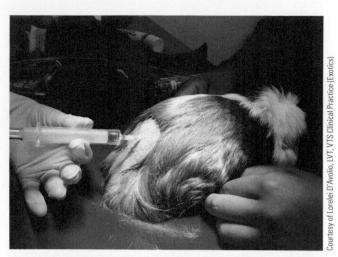

FIGURE 20-49 Epidural needle placement in a guinea pig.

Tech Tip

Multimodal analgesia, administering different analgesics that act in different ways to enhance pain relief, can result in overall reduction of drug dosages. An example of multimodal analgesia for a routine neuter would be to premedicate a pet rat with buprenorphine and midazolam, administer an incisional line block to the surgical site and a testicular block, recover the patient with an injection of meloxicam, and continue postoperative oral tramadol compounded as a suspension and meloxicam for 5 to 7 days.

FERRET ANESTHESIA

For veterinary technicians adept at feline anesthesia, ferret anesthesia may be a more familiar experience than with other exotic species. Their small size presents some additional challenges, and being creative to modify equipment to accommodate them is important. Some key differences between ferrets and cats include the ferret's very long, tubular body with a very flexible spine. They tend to be masters of escaping traditional restraint, capable of wriggling out of almost any towel or grip. While their lanky anatomy can make restraint difficult, most ferrets can be subdued while simply being hung by the scruff of the neck (Figure 20-50). Administering injections, SQ fluids, or even blood draws can be done while in this restraint. Ferrets can deliver a powerful bite if they have not been well socialized; however, most pet ferrets are friendly and docile. While they do have long toenails, they are not retractable and ferrets do not generally use them as weapons like cats do.

Preanesthetic Preparation

It is common for ferrets to have one to several concurrent subclinical diseases that can manifest in ferrets of all ages. At 4 years of age ferrets are considered middle aged and can be expected to have some underlying illnesses. Many ferret owners and inexperienced ferret veterinarians

FIGURE 20–50 A proper "scruff" restraint of a ferret that is often used to facilitate giving injections, performing a physical exam, or other procedures. This restraint works well to temporarily immobilize the ferret.

will miss some key clinical signs of illness, and if these conditions are not diagnosed prior to anesthesia, the results could be fatal. Baseline bloodwork prior to anesthesia is crucial, including those performed on the day of the anesthetic event. At minimum, it should include blood glucose (BG) levels, blood urea nitrogen (BUN), creatinine, alanine aminotransferase (ALT), and a packed cell volume (PCV). Ideally, ferrets should be fasted for 2 hours prior to phlebotomy in order to obtain an accurate glucose level. Chronic disease and gastric ulcerations can cause anemia in a ferret that may appear clinically normal. Cardiac disease is often underdiagnosed in ferrets, as signs may be subtle. Owners may not recognize what coughing sounds or looks like in a ferret. Additionally, seasonal weight gain and loss is very common in ferrets, so clinical signs such as ascites may be overlooked by an owner who thinks it is normal cold weather weight gain. Radiographs and ultrasound of both abdominal and thoracic cavities should be considered as part of a comprehensive preanesthetic work up.

Hypoglycemia is a significant concern with ferrets for two main reasons. First, an insulinoma is a common finding in ferrets of all ages. This cancer of the beta cells in the pancreas causes low blood glucose levels and can be underdiagnosed in ferrets who typically sleep most of the day, so owners may not notice "lethargy." Many inexperienced veterinarians may not know that hind limb weakness, while also a sign of other problems, can be a clinical sign of insulinoma. Second, ferrets have a high metabolic rate compared to cats with a gastrointestinal transit time of only 3 to 4 hours. Safe fasting times for

ferrets should be no longer than that to avoid serious hypoglycemic episodes. This can seem conflicting because unlike rabbits and rodents, ferrets can vomit and should have a brief fast to prevent aspiration. Careful monitoring is important if given opioids or other agents causing nausea.

Like all other exotic species with a rapid metabolism, ferrets are at high risk of becoming hypothermic. Their long, thin bodies denote a high surface area to body mass ratio and clipping fur for surgical procedures, opening the body cavity, and delivering cold inhalant anesthetic agents all contribute to rapid cooling. Veterinary technicians should be prepared, as with other exotic animals, to use warmed surgical scrub and saline as well as provide ambient, radiant, and forced air heat support. For longer procedures, use of a Humid-Vent®, plastic surgical drapes, and heated fluid lines will help prevent hypothermia. The average body temperature for ferrets is 100 to 104°F.

IV Catheterization and Intraoperative Fluids

In order to maintain homeostasis and treat emergency situations that could occur, IV access is critical. Ferret IV catheter placement has some unique challenges for veterinary technicians including being difficult to restrain due to their long, malleable body. They are fast and strong, even when not feeling well, and some form of sedation may be required prior to catheterization. Low dosages of midazolam and butorphanol may suffice; however, veterinary technicians should always be prepared in the event a ferret may buck, nip, or jump. Cephalic veins work well with 22 to 24 gauge catheters; however, due to the thick skin of ferrets, it is recommended that veterinary technicians make a nick in the skin prior to passing the catheter. Without making an entry hole, catheters frequently become burred and may not pass into the vein or could cause hematoma formation. Once this happens, the vein is usually not viable for further attempts. Another drawback in using the front legs is that ferrets tend to dig and burrow. IV lines that are not secured/bandaged well can easily become kinked or pulled out unless the veterinary technician uses copious amounts of tape and bandage material, which often results in the ferret becoming agitated at the bulk of the catheter. They may obsessively gnaw, kick, and writhe until the catheter is destroyed or the line broken. This delicate balance of securing these catheters comes with experience and practice. IV catheters can also be placed in the lateral saphenous or jugular veins; however, these sites tend to be extremely irritating to ferrets. Jugular veins in particular are generally very poorly tolerated by ferrets. It is important for veterinary technicians to remember the role stress plays in their patients when deciding where to place indwelling catheters.

Intraoperative IV fluid maintenance rates for healthy ferrets should be 5 mL/kg/hr, and most crystalloids are appropriate for routine use. Based on blood glucose measurements, veterinary technicians should be prepared to deliver CRIs of glucose to ferrets or "spike" the IV fluid bag with dextrose, or at least have glucose available in the event of an emergency.

Premedication, Sedation, and Restraint

Drugs and multimodal drug combinations used in cats are similarly used in ferrets, at appropriate dosages based on a current exotic animal formulary. Muscle relaxants such as benzodiazepines work well for light tranquilization with minimal effect on cardiopulmonary function. These are often combined with dissociative agents such as ketamine and opioids such as buprenorphine, methadone, hydromorphone, or morphine, which potentiate sedation as well. Acepromazine is also commonly used in ferrets for light sedation, however, should be avoided in sick or hypovolemic patients. Similarly, alpha-2 adrenergic agonists are great for sedation and analgesia, but they can cause bradycardia, arrhythmias, and hypotension. Most formularies offer several options that are based on the patient's health and the practitioners experience; therefore, many combinations can be used. As with other species, anticholinergic drugs should only be given as part of the premedication if indicated by individual patient needs.

Tech Tip

Ketamine is a safe and widely used anesthetic and analgesic in ferrets. It can, however, cause muscle rigidity and salivation. Veterinary technicians should be prepared to use both muscle relaxants and anticholinergic drugs if ketamine is incorporated into the premedication cocktail.

Induction Procedures and Techniques

Similar to rabbits, it is very challenging to place IV catheters in strong, alert ferrets unless they are deeply sedated. Therefore, using IV induction agents such as propofol or etomidate are often not utilized; however, if an IV catheter is achieved prior to induction, these are safe options. An appropriate premedication "cocktail" including ketamine is often enough to induce anesthesia and facilitate IV access and intubation. As with other exotics, using a sevoflurane or isoflurane mask for induction is often utilized if IV induction agents cannot be given. Alfaxalone is a great option for IM induction, although its use is more recent and because of that dosing is based on anecdotal clinical use.

It is recommended to intubate ferrets for all anesthetic procedures to prevent aspiration as well as provide appropriate ventilation. Thankfully, ferrets are relatively easy to intubate in comparison to rabbits and rodents.

- Oral anatomy is similar to that of a cat, and the larynx can be visualized by holding the mouth open with gauze (Figure 20-51) and using a laryngoscope with a pediatric Miller blade.
- Veterinary technicians will need to use 2.5 to 4.0 mm cuffed ETTs for most ferrets.
- Like cats, ferrets are prone to laryngospasms. A drop or spray of 2% lidocaine to the arytenoids will be effective at controlling these spasms.

Tech Tip

Even after induction, ferrets tend to maintain significant jaw tone. Veterinary technicians should never put their fingers in the oral cavity when attempting to intubate, as this is a common time for an inadvertent, yet severe, bite to occur.

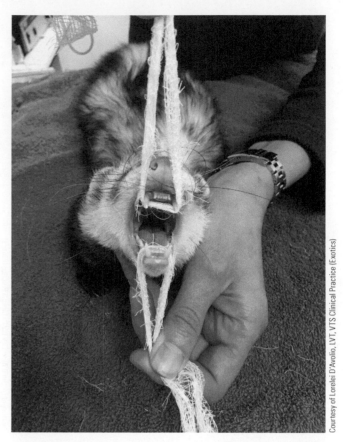

Courtesy of Lorelei D'Avolio, LVT, VTS Clinical Practice (Exotics)

FIGURE 20–51 Using gauze to hold open a ferret's mouth during intubation.

Maintenance

Ferrets should be maintained on inhalant anesthetic agents such as isoflurane or sevoflurane. MAC values for isoflurane are approximately 1.9 to 2.2% and 2.5 to 4.2% for sevoflurane. Administration rates will vary based on premedication and use of analgesics during procedures. If use of an inhalant anesthetic agent is not required, it is still recommended to keep patients on oxygen until recovered.

Monitoring

Monitoring anesthesia for ferrets will vary based on the equipment of the facility. Veterinary technicians should be cognizant at all times of the patient's appropriate reflexes. At a surgical plane, the palpebral reflex should be abolished, and there should also be a loss of pedal (digital withdrawal) reflex and generalized decreased muscle tone. Respiratory rates should be easy to monitor by watching the movement of the reservoir bag, and heart rates can be counted with a simple stethoscope. Normal respiration for a ferret is 33 to 36 breaths per minute, and average heart rates range from 200 to 400 bpm. Use of a Doppler ultrasound probe can be secured to the foot or leg, or placed directly over the heart for constant monitoring of the heart rate and rhythm. If using the foot or distal leg, veterinary technicians should apply a blood pressure cuff and obtain peripheral blood pressure readings. Obtaining arterial blood pressures will greatly depend on staff expertise and available equipment. For most general practices, diligent monitoring using a Doppler

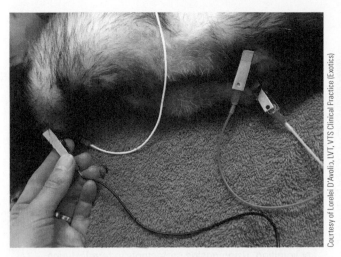

Courtesy of Lorelei D'Avolio, LVT, VTS Clinical Practice (Exotics)

FIGURE 20-52 Electrocardiograph application in a ferret. Using needles instead of the clips when attaching an ECG can prevent tissue trauma.

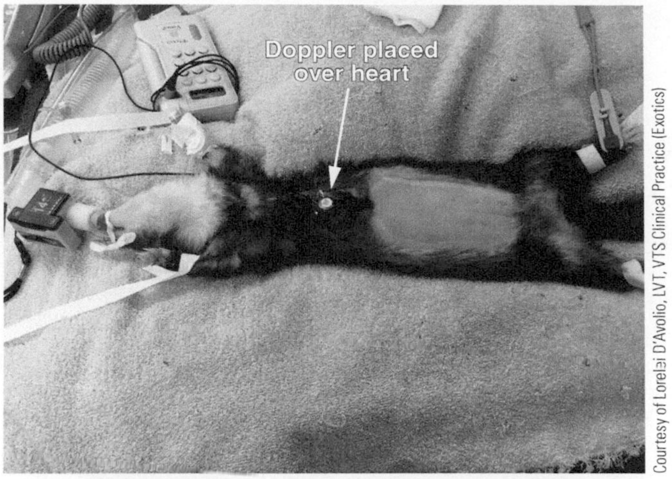

FIGURE 20-53 A ferret anesthetized with a Doppler crystal placed over the heart to monitor heart rate and rhythm.

Courtesy of Lorelei D'Avolio, LVT, VTS Clinical Practice (Exotics)

Doppler placed over heart

blood flow or oscillometric monitor (see Chapter 7 [Anesthesia Monitoring]) for peripheral blood pressuring monitoring shows adequate and important trends/measurements without the potential risks of placing an intra-arterial catheter in a ferret. Use of a mainstream capnograph is preferable to sidestream due to the decrease in dead space and therefore provides more accurate results. Understanding $EtCO_2$ will enable veterinary technicians to know if/when the patient requires ventilatory support, as well as manage plugs or problems with the ETT. Use of ECG is invaluable for veterinary technicians to monitor for signs of arrhythmia or other cardiac abnormalities. While ferret skin is significantly tougher than other exotic animals, their small size may require use of modified clips or needles placed with clips attached to the non-hub side (Figure 20-52). SpO_2 monitors work well with clips on the feet, tongue, or in the rectum to provide information about oxygen saturation of hemoglobin and pulse rate. Constant temperature monitoring is crucial for ferrets. Ideally, an esophageal probe can be set to monitor throughout the procedure; however, regular checks with a rectal thermometer will be sufficient.

Tech Tip

While rabbits have a very small thoracic cavity and the heart that is more cranial, ferrets are the opposite. They have a very large thoracic cavity and proper cardiac auscultation or placement of a Doppler probe intended to be above the heart is much further caudal in the body (Figure 20-53).

Recovery

Successful anesthetic recovery for a ferret will be achieved by veterinary technicians paying close attention to detail. While hypothermia is of great concern during anesthesia, hyperthermia can occur rapidly, especially if they are left to recover on heat. Ferrets cannot sweat and do not pant like dogs, so while a warm recovery cage is ideal, it is important to continue monitoring

them. Veterinary technicians should pay attention to potential signs of hypoglycemia such as extended recovery time, weakness, tremors, and non-responsiveness and offer food soon after the ferret is awake (depending on the surgical procedure and instructions of the veterinarian). Nutritional support is important, often necessitating assistance or force feeding if the ferret is resistant to eating on its own. Continuing IV fluids and analgesia will also help ensure a smooth and successful recovery. Ferrets will chew and lick painful incisions and do not tolerate Elizabethan collars well; therefore, local incisional line blocks are important as well as other analgesics to keep patients comfortable and able to rest. They are also much more likely to eat if they are not in pain.

FERRET ANALGESIA

Ferrets differ from the other species of animals discussed in this chapter, primarily because they are strict carnivores; the top of the food chain, rather than the prey species at the bottom. It is important to recognize what are normal behaviors in ferrets and how they differ from other exotic animals in order to recognize their signs of pain. While prey species spend tremendous effort to hide when they are ill in order to avoid predator attentions, ferrets do not share this instinct. Much like dogs and cats, ferrets will demonstrate signals such as lack of exploring behavior in a new environment (like a veterinary exam room), altered gate/limping, hiding in the back of an enclosure facing away, lack of grooming behaviors/unkempt appearance, decreased food/water intake, and adverse response to palpations when they are experiencing pain. Unique to ferrets is that while most pets will walk with a hunched-back posture when painful, ferrets do the opposite, and their normal hunched posture will be absent when painful. It is also important to recognize that ferrets normally spend up to 70% of the day sleeping; so "lethargy" or excessive sleeping may not be a sign of discomfort for them. Pain can cause a ferret's tail to bristle (Figure 20-54), pawing at the mouth (particularly if having dental pain or nausea), and an alteration in personality from aggressive to gentle or vice versa. They may also appear obtunded (depressed), have their eyes half closed, and grind their teeth. There are many subtle signs of pain; however, that will only be perceptible to someone who is closely monitoring the animals, and who has a good grasp of normal

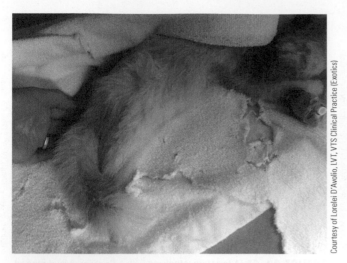

Courtesy of Lorelei D'Avolio, LVT, VTS Clinical Practice (Exotics)

FIGURE 20-54 A ferret tail that is bristled and wiry due to pain. This patient needs proper analgesia quickly.

ferret behavior. This could easily be the role of a veterinary technician, who could also then be capable of assessing the efficacy of chosen analgesics.

Tech Tip

The Ferret Grimace Scale (FGS) is a pain assessment tool that uses five facial AU (orbital tightening, nose bulging, cheek bulging, ear changes, and whisker retraction) that are scored from 0–2 (0 = not present, 1 = moderately present, 2 = obviously present). It can be accessed at https://www.researchgate.net/figure /The-ferret-grimace-scale-Photographs-visualizing-the -normal-appearance-and-changes-0_fig2_321057507.

One important potential complication of untreated pain in ferrets is severe hypoglycemia. Ferrets have a very high a metabolic rate and a very short GI transit time. They also are prone to having clinical or subclinical insulinoma, which can cause hypoglycemia. Combining these three factors along with the likelihood that acute or chronic pain can result in a catabolic state and anorexia creates a strong possibility that ferrets may develop a life threatening hypoglycemia with inadequate pain management, especially during a peri-anesthetic event.

Similar to dogs and cats, a broad range of analgesic drugs can be used to achieve multimodal analgesia in ferrets. Unfortunately, as with other exotics, pharmacokinetics, efficacy, and safety studies are limited, and most dosages are based on the anecdotal experience made by practitioners working regularly with ferrets and not on evidence-based medicine. Because of their anatomy and physiology, many dosages and potential adverse effects are presumed to be similar to those in cats; however, it is still recommended to use an exotic animal formulary when choosing dosages for ferrets. Use of opioids such as butorphanol, buprenorphine, hydromorphone, methadone, morphine, and fentanyl is common and some can cause vomiting. Combining these drugs with alpha-2 adrenergic agonists, NMDA agonists, local analgesics, and NSAIDs should all be part of a comprehensive analgesic plan.

Many clinicians are wary of NSAID use in ferrets due to the role COX 1 inhibitors play in inhibiting important protective mucosal secretions in the stomach. Ferrets are uniquely prone to *Helicobacter* gastritis, and it is thought that concurrent stressors may lead to development of the disease, which NSAIDs may exacerbate. However, there are no references to support this theory, and due to its overwhelming successful clinical use, it is considered a valuable addition to multimodal analgesia in ferrets. Choosing NSAIDs, which are more effective on the COX 2 enzymes, such as meloxicam, is recommended.

Tech Tip

Be conscious of the ferret's sensitive stomach and monitor for signs of gastric ulcers, such as anemia, melena, or vomiting. Using gastric protectants in ferrets during times of stress, pain, or during treatment, may be a good preventative tactic.

Ferrets tend to be very resistant to taking oral medications; however, compounded oral tramadol and butorphanol are effective analgesics to send home with ferrets. With the advent of more specialty compounding pharmacies, many traditionally non-palatable formulations can be made into flavors ferrets may more readily accept. Combining these opioids with gabapentin and use of photobiomodulation is becoming standard in cases of chronic pain or instances of delayed healing such as burns, large wounds, or trauma.

SUMMARY

Anesthetizing and providing pain relief for exotic animals is based on the same principles used in other companion animals; however, the uniqueness of each species makes it necessary to adapt equipment, modify anesthetic protocols, and adjust handling methods to accommodate their differences. Veterinary medicine is constantly evolving and so should methodologies and techniques. Although exotic animal anesthesia may have less evidence-based research to support protocols at this time, and be initially more challenging, knowing the fundamentals of exotic animal anesthesia/analgesia will help ease the fear of working with unfamiliar species and make for smooth and safe experiences. Veterinary technicians that are stressed about exotic animal anesthesia will not benefit the patient, so remember to breathe deeply and apply anesthesia fundamentals when faced with the challenge of anesthetizing unfamiliar patients.

CRITICAL THINKING POINTS

- When working with exotic animal species, collecting as much information as possible about the patient's health status will improve your anesthetic plan and help team members be prepared.

- Special species require special equipment. Supplies and tools specifically designed to perform some of the most basic anesthetic tasks are indicated in anatomically challenging or tiny animals. Equipment should always be made available before the start of an anesthesia case.

- It is important to know the animal you are working with. For example, mice and guinea pigs are both rodents, but their physiology is much different. Their metabolism and reactions to medications will vary. Species physiologic and pharmacologic differences must be recognized so that each species receives a specialized anesthetic plan.

- Hypothermia is one of the biggest threats to anesthetized exotic small animals. Always consider ways to keep patients in an optimal temperature range.

- Although we hope anesthesia will run smoothly, knowing what can happen and how to react in an emergency situation is vital to the patient's safety.

REVIEW QUESTIONS

Multiple Choice

1. What is the best type of endotracheal tube for intubating birds?
   a. A cuffed endotracheal tube that Is fully inflated
   b. A cuffed endotracheal tube that is not inflated
   c. A non-cuffed endotracheal tube
   d. No endotracheal tube is best; use a mask instead

2. A bird is having a coelomotomy. What anesthetic adjustments need to be made in this patient?
   a. Increasing the oxygen flow rate and vaporizer setting due to gas loss through the open air sacs
   b. Decreasing the oxygen flow rate and vaporizer setting due to gas loss through the air sacs
   c. Adding an air sac tube for anesthetic maintenance
   d. No special considerations are needed

3. What is the most effective analgesic opioid in birds, and why?
   a. Tramadol because it can be given orally
   b. Buprenorphine because it has the longest duration
   c. Fentanyl because it is a full mu agonist opioid
   d. Butorphanol because of the high distribution of kappa receptors in birds

4. Which of these exotic species should be fasted 24 hours prior to anesthesia?
   a. Ferrets and rodents
   b. Birds and rabbits
   c. Rabbits and rodents
   d. None of these exotic animals should be fasted 24 hours prior to anesthesia

5. Which animal cannot vomit?
   a. Ferrets
   b. Rabbits
   c. Parrots
   d. Guinea pigs

6. What does providing analgesia to a rabbit prevent?
   a. Over-grooming/barbering
   b. Dangerously rapid recovery
   c. Severe hypoglycemia
   d. Anorexia and subsequent ileus

7. Which is the greatest risk to small exotic patients during anesthesia?
   a. Hypothermia
   b. Hypercapnia
   c. Hypocalcemia
   d. Hypertension

8. Which veins are possible IV catheter sites in parrots?
   a. Right jugular, basilic, medial metatarsal
   b. Cephalic, right jugular, saphenous
   c. Basilic, metacarpal, lateral metatarsal
   d. Femoral, medial metacarpal, right jugular

9. A guinea pig comes in on emergency for being attacked by a dog. The patient is extremely painful and requires surgery as soon as possible to close some lacerations. What would be one of the first things the patient should receive prior to anesthesia induction?
   a. Glucocorticoids to counteract the effects of catecholamines
   b. Dextrose solution to prevent hypoglycemia
   c. Analgesics to relieve some discomfort and ease anesthesia Induction
   d. Acepromazine to relax the patient and cause vasodilation

10. What is the most common induction method in birds?
    a. IV propofol
    b. Mask or chamber induction
    c. IM dexmedetomidine and valium
    d. IM ketamine and midazolam

11. Which monitoring device is considered a staple for exotic pets?
    a. SpO_2
    b. Doppler ultrasound
    c. ECG
    d. Capnography

12. Why is atropine not indicated for use in rabbits?
    a. Atropine is toxic to rabbits.
    b. Some rabbits produce atropinase, which inactivates atropine.
    c. causes severe tachycardia in rabbits.
    d. Atropine burns on IM injection.

13. Which of these species has a palatal ostium, making intubation almost impossible?
    a. Ferret
    b. Parrot
    c. Guinea pig
    d. Rat

14. What may cause severe hypoglycemia in a ferret post-surgery?

    a. Fasting for longer than 4 hours, underlying diabetes mellitus, and high metabolism and long GI transit time

    b. Fasting for longer than 2 hours, underlying pituitary disease, and their very long, tubular gastrointestinal tract that allows food to be absorbed more slowly

    c. Fasting for longer than 2 hours, underlying thyroid disease, and high metabolism and long GI transit time

    d. Fasting for longer than 4 hours, underlying insulinoma, and high metabolism and short GI transit time

15. Which analgesic has been known to cause pica in rats?

    a. Ketamine

    b. Meloxicam

    c. Buprenorphine

    d. Hydromorphone

16. Which of the following statements is true for exotic pets?

    a. Rabbits have the highest morbidity and mortality when experiencing pain because as a prey species, the stress response can contribute to GI stasis.

    b. Ferrets have the highest morbidity and mortality when experiencing pain because they hide their signs of pain making it difficult to assess.

    c. Parrots have the highest morbidity and mortality when experiencing pain because analgesics do not work in non-mammalian species.

    d. Mice, gerbils, and rats have the highest morbidity and mortality when experiencing pain because they are so small that it is impossible to dose them properly with effective analgesia.

17. Which vital sign is typically not feasible to monitor in small rodents?

    a. ECG

    b. Temperature probe

    c. Pulse oximeter

    d. Non-invasive blood pressure

18. Which heat retaining and warming implements are *not* recommended in small exotic species?

    a. Bubble wrap

    b. Forced air blankets

    c. Heavy quilts

    d. Heated tables

19. What precautions should be taken when clipping rabbits and rodents for surgical preparation?

    a. They should not be clipped due to risk of hypothermia.

    b. Taking care not to lacerate their thin, delicate skin.

    c. Straight edge razors should be used instead of electric to prevent stress due to noise.

    d. Hair should be lubricated well prior to clipping.

20. Which species tends to maintain jaw tone even after given induction drugs?

    a. Mice

    b. Rabbits

    c. Birds

    d. Ferrets

Case Studies

Case Study 1: Brigette is a 25-year-old intact female blue and gold macaw. She presents with acute trauma from an attack by a dog. She sustained significant beak trauma, a suspected broken wing, and a laceration to the pectoral region, and is given an ASA level of 3. She is tachypneic and, due to the trauma and suspected blood loss, the DVM has decided not to obtain blood for testing. The plan is to take full body radiographs to assess skeletal and coelomic damage, close the pectoral injury, and assess/repair the beak damage. Brigette is normally a nervous bird and wing flaps/struggles heavily when restrained. The key issues to providing safe analgesia and anesthesia for her include premedicating with pain relief and allowing her to calm down prior to handling and anesthesia.

1. What drug(s) do you suggest as premedication of Brigette?

2. What specific concerns might you have with anesthetizing her?

3. What alternatives could be used to maintain an airway with this patient while working on the beak?

Case Study 2: Sophie is a 5 year-old F/S ferret with insulinoma, confirmed by ultrasound. She has been very lethargic for the last few weeks and lost weight due to eating poorly. She is very weak, almost obtunded, at presentation to the hospital. Despite a guarded prognosis, the owners would like to attempt surgical removal of the pancreatic tumor/s, and she is given an ASA level of 4. Preanesthetic bloodwork shows a slight azotemia with a PCV of 32% and a BG of 43 mg/dL. The key issue in keeping Sophie safe during anesthesia is maintaining her blood glucose levels and maintaining homeostasis.

1. Describe how and why an IV catheter should be placed prior to surgery.

2. Which parameter(s) should be carefully monitored during Sophie's anesthetic procedure and why?

3. What are some ways to correct a hypoglycemic episode during recovery?

Case Study 3: Darlene, a 5-year-old F/S house rabbit, presented to the hospital with a mandibular jaw abscess and malocclusion/overgrown crowns of several molars. She weighed 3.1 kg and her preanesthetic lab work was WNL. She was given an ASA level of 2. The plan was to anesthetize Darlene, obtain skull and dental radiographs, surgically lance and drain the abscess, and correct the dental issues. The key issues for Darlene's procedures were to provide adequate analgesia for what could be a painful procedure, and allow for work to be done in the oral cavity while maintaining a patent airway. She was premedicated with IM buprenorphine, dexmedetomidine, and ketamine. An IV catheter was placed in her right cephalic vein. Darlene was already at an appropriate anesthetic plane, so further induction was not needed. She was started on a CRI of Plasmalyte® at 10 mL/kg/hr. Blind intubation was attempted, but not successful. The hospital did not have an endoscope, and due to concerns over anesthetic duration, the veterinarian decided to continue the procedure with a small mask placed over the nose while working in the oral cavity. Darlene was maintained on oxygen and isoflurane 1%. She was monitored throughout the procedure with an SpO$_2$ monitor clipped to her ear pinna, which showed her oxygen saturation levels to remain between 97 and 100% throughout the procedure. She also had a Doppler ultrasound probe secured to her chest, which showed a consistent heart rate between 180 and 220 bpm, and her ECG readings were WNL. The veterinarian determined that the abscess was related to an infected tooth root and in addition to cleaning the abscess site and removing the capsule, her first premolar was extracted. There was minimal bleeding, and the other teeth were filed using a high-speed drill. It was decided not to administer the reversal atipamezole, and meloxicam was given SQ during recovery. Darlene recovered slowly, but without complications.

1. What other drug(s) could have been used to premedicate Darlene?
2. Why wasn't a supraglottic airway device used when intubation was unsuccessful?
3. What reasons and under what circumstances would you choose to reverse or not to reverse dexmedetomidine?

Case Study 4: Cheese is a 16-month-old intact male pet hamster with overgrown incisor teeth. The owner admits that several weeks prior, her young son dropped the hamster and the top incisors had broken. Now the teeth are growing in crooked, and the hamster may not be able to eat. The owner had been offering soft foods, but she is not sure if the hamster had eaten anything for the last few days. On physical exam, the hamster is aggressive and trying to bite, making a complete oral exam impossible. He is slightly thin with a BCS (body condition score) of 2/5. It is decided that he needs to be anesthetized in order to evaluate and fix the maloccluded incisors. The key issues for anesthetizing this hamster are the limitations of working with such a small patient: there is no lab work to diagnose any potential underlying illness, no way to intubate, and no IV catheter. Due to his potential history of long-term anorexia/hyporexia, he is given an ASA level of 3.

1. Without knowing if the hamster had been eating, what are the potential complications of anesthesia and how could you prevent them?
2. What kind of premedication is recommended for this brief procedure?
3. What are the options to induce anesthesia for this hamster?

Case Study 5: Ditto is a 19-year-old female African Gray Parrot. She is egg-bound, a condition that occurs in birds when they are unable to pass an egg through the oviduct and may be caused by malnutrition, hypocalcemia, oviductal cancer, oviductal torsion, or genetic anomaly. She has a history of egg-binding, and this is the third time she has presented this way. It is decided that due to the previous damage to the oviduct and likelihood that this will be a chronic problem, she should have a salpingohysterectomy, or removal of the oviduct and uterus. Ditto is having significant respiratory distress due to the egg, which is occupying a large amount of space in her coelomic cavity, and she is given an ASA level of 4. The key issues in her anesthetic plan are to provide her proper analgesia, homeostasis, proper monitoring of her vital signs, and being prepared in the event there is an emergency during or after the procedure. Her presurgical bloodwork is WNL, other than an elevated calcium and creatine phosphokinase (CPK), which is expected in an egg-laying bird. It is decided to premedicate Ditto with only butorphanol. She is induced with sevoflurane and an IO catheter is placed in the distal ulna of her right wing. She is started on warm lactated ringer's solution at 10 mL/kg/hr. Emergency drugs are drawn up ahead of time in case they are needed, including a Hetastarch® bolus and the post-surgical meloxicam. Monitoring devices are secured, and the procedure begins.

1. Which monitoring devices should be used to ensure Ditto is ventilating well and maintaining euvolemia?
2. What would influence the decision to administer a bolus of Hetastarch®?
3. Why wasn't Ditto premedicated with dexmedetomidine, ketamine, or other commonly used premedications?

Critical Thinking Questions

1. Leslie is new to exotics only practice. She has had very little experience with avian anesthesia and notices she is scheduled to help Dr. Summers with an avian anesthetic procedure (X-ray and beak trim). Dr. Summers asked Leslie to setup for these brief procedures. Leslie knows the patient is a large, animated, and sometimes aggressive Amazon parrot.
   a. Which special supplies will Leslie need to successfully anesthetize this bird?
   b. What safety precautions should Leslie take working with a difficult patient that can fly?
   c. Would you recommend a preanesthetic drug for this bird that is undergoing just a series of X-rays and a beak trim?

2. Mateo has a guinea pig scheduled for a mass removal. He has never worked with rodents and is excited to help with the procedure. The guinea pig has not been eating or drinking well at home. Mateo understands IV fluids will be crucial for this patient.
   a. Where might Mateo place an IV catheter in this guinea pig?
   b. What lab values would Mateo be interested in, since this guinea pig has recently not been eating or drinking well?
   c. What special precautions will Mateo need to take when monitoring this small patient?

ENDNOTES

1. Lee L. (2006). *Anesthesia for exotic species* (p. 1). Oklahoma: Center for Veterinary Health Sciences.

2. Gunkel, C., & Lafortune, M. (2005). Current techniques in avian anesthesia. *Seminars in Avian and Exotic Pet Medicine, 14*(4), 263–276.

3. O'Malley, B. (2005). *Clinical anatomy and physiology of exotic species: Structure and function of mammals, birds, reptiles, and amphibians* (pp. 3, 5–10, 18, 19, 21, 26, 118, 166). Edinburgh: Elsevier Saunders.

4. Ludders, J., & Matthews, N. (2007). Birds. In W. Tranquilli, J. Thurmon, & K. Grimm (Eds.), Lumb & Jones' veterinary anesthesia and analgesia (4th ed., pp. 841–862). Ames, IA: Wiley-Blackwell.

5. Lumeij, J. T. (1994). Avian medicine: Principles and applications (p. 539). Lake Worth, FL: Wingers Publishing.

6. Ludders, J., & Matthews, N. (2007). Birds. In W. Tranquilli, J. Thurmon, & K. Grimm (Eds.), Lumb & Jones' veterinary anesthesia and analgesia (4th ed., pp. 841–862). Ames, IA: Wiley-Blackwell.

7. Hornak, S., Liptak, T., Ledecky, V., Hromada, R., Bilek, J., Mazensky, D., & Petrovic, V. (2015). A preliminary trial of the sedation induced by intranasal administration of midazolam alone or in combination with dexmedetomidine and reversal by atipamezole for a short-term immobilization in pigeons. *Veterinary Anaesthesia and Analgesia, 42*(2), 192–196. doi: 10.1111/vaa.12187

8. Mitchel, M., & Tully, T. J., (2016). *Current therapy in exotic pet practice* (p. 126). Elsevier Health Sciences.

9. Mansour, A., Khachaturian, H., Lewis, M., et al. (1988). Anatomy of CNS opioid receptors. *Trends in Neurosciences*, 11, 308–314.

10. Sinclair, K. M., Church, M. E., Farver, T. B., Lowenstine, L. J., Owens, S. D., & Paul-Murphy, J. (2012). Effects of meloxicam on hematologic and plasma biochemical analysis variables and results of histologic examination of tissue specimens of Japanese quail (*Coturnix japonica*). *American Journal of Veterinary Research, 73*(11), 1720–1727.

11. Lichtenberger, M. (2008). Proceedings for North American Veterinary Conference (p. 1832).

12. Gil, G., Silcan, G., Villa, A., & Illera, J. (2012), Heart and respiratory rates and adrenal response to propofol of alfaxalone in rabbits. *Veterinary Record, 170*, 444.

13. Desmarchelier, M., Troncy, E., Fitzgerald, G., Lair, S. (2012). Analgesic effects of meloxicam administration on postoperative orthopedic pain in domestic pigeons (*Columba livia*). *American Journal of Veterinary Research, 73*(3), 361–367.

CHAPTER 21

Pain Management in Laboratory Mice and Rats

Mary Ellen Goldberg, *BS, CVT, SRA, CCRA, CCRVN, CVPP, VTS (Lab Animal Medicine, Research/Anesthesia, Physical Rehabilitation)*

LEARNING OBJECTIVES

Upon completion of this chapter, it is expected that the reader should be able to:

21.1 Describe the importance of the Institutional Animal Care and Use Committee (IACUC) in animal research

21.2 Describe how to assess pain in laboratory mice and rats

21.3 Describe pain scoring scales used in laboratory mice and rats

21.4 Identify different analgesics prescribed for laboratory mice and rats

21.5 List behaviors and illnesses associated with pain in mice and rats

INTRODUCTION

Animal research benefits the medical and scientific communities to improve quality of life for wide populations of humans and animals alike. For both animal welfare and humane reasons, laboratory animals must be regularly assessed for the presence of pain. Additionally, pain itself can bring undesirable physiological changes in experimental animals that can ultimately affect and bias research outcomes thereby decreasing the potential benefit and validity of the research. Alleviation of pain, therefore, should be viewed in laboratory animals as a way to reduce the potential for development of additional cofactors, variables, and biases in research projects that could invalidate an entire research project.

Although many types of animals such as dogs, cats, horses, ruminants, and other species can be used within the laboratory research setting, these species have already been described at length in other chapters. The focus of this chapter, therefore, is to help the reader identify and manage pain in laboratory mice and rats. Mice and rat anesthesia is described in Chapter 20 (Anesthesia and Analgesia of Pet Birds and Exotic Small Mammals).

THE ROLE OF THE INSTITUTIONAL ANIMAL CARE AND USE COMMITTEE IN ANIMAL RESEARCH

Before animal research can start, a Principle Investigator (PI) must submit a research proposal including potential pain that will be experienced by laboratory animals during the course of the study to an **Institutional Animal Care and Use Committee** (IACUC). IACUCs are present at institutions such as universities and colleges but can also include private nonacademic research facilities. IACUCs are typically made up of experts such as researchers, professors, veterinarians, and other animal care professionals experienced with animal research as well as a public member not affiliated with the institution to represent general community interests in the proper care and use of animals. The IACUC then reviews the methodologies described in the protocol to ensure that they are ethical and humane for the animals involved in the research. The IACUC may request revisions to the original research proposal and only when IACUC approval is met can research officially begin. IACUCs then serve to monitor research as it progresses and may review final research publication to ensure that all expectations to animal welfare have been met.

Within IACUC proposals, most test subjects (animals used in research projects are referred to as test subjects instead of "patients") are vertebrate animals that must be assigned a USDA pain or distress category. These specific categories help define the degree of pain or distress a research animal may experience (Table 21-1). Interestingly, the most commonly used laboratory species are mice and rats, and these species are not required to be assigned USDA pain or distress categories on IACUC submissions; however, many institutions still require researchers to individually assign them a category and within an annual report must document how many animal subjects are used and to what degree pain or distress was experienced in their mice and rats.

Despite the goal of treating laboratory animals humanely, analgesics sometimes may not be administered in a research setting if analgesic use is perceived to specifically adversely affect experimental results. In these specific cases, justification for not administering analgesics must be made to the IACUC and approved by the committee. Nonanimal alternatives are also required to be considered within IACUC proposals.

IDENTIFICATION OF PAIN IN LABORATORY ANIMALS

Before one can effectively provide analgesia to laboratory rats and mice, it is important to understand how to identify their pain, when pain is likely to occur, how long pain may last, and how well the pain will respond to appropriate therapy. As with other species, measuring laboratory rat and mice responses to painful stimuli fall into three general categories: behavioral, biochemical, and physiological.

- *Behavioral changes* are often the earliest signs of pain and are most frequently observed by animal care staff. Behavioral patterns are therefore effective tools for detecting and grading pain. An advantage to recognizing behavioral changes in laboratory animals is that it does not produce further pain or distress that could occur when collecting biochemical or even physiological data. For example, if an animal care staff member is concerned that a laboratory animal is showing behavioral signs of pain, treatment could be started without having to directly handle the subject. The disadvantage of observing behavioral changes is that they can be subjective assessments and may lead to imprecision and variability between observers. For example, animal care giver A observes a subject and believes it is exhibiting behavior changes; however, animal care giver B disagrees. Because animals cannot verbalize pain or distress, animal care personnel are often forced to make assessments of pain and discomfort based solely on behavioral observations; however, these judgments tend to become more repeatable and reliable with appropriate experience and training in the use of pain scoring scales (described in Pain Scales for Pain Assessment section). Table 21-2 summarizes behavioral signs of pain in mice and rats.
- *Biochemical parameters* are not commonly assessed in clinical practice to determine if an animal is in pain, but they may be performed more commonly in a laboratory setting where samples are already being collected. Determining levels of cortisol, catecholamines, and various hormones such as insulin are frequently employed to assess pain or distress in laboratory animals.
- *Physiological factors* such as increases in heart rate, respiratory rate, and temperature and decreased locomotion have been used as an indirect measure of pain.[1] Unfortunately these physiological factors are not sensitive or specific for pain as stress, excitement, and even handing the animal may influence these parameters. It is recommended, therefore, that these parameters not be trusted as reliable indictors of pain and only used with educated caution in conjunction with other parameters.

Safety Alert

Care should be taken when working with laboratory animals that may be in pain because their behavior may be altered. To avoid injury, be aware that the animal may be more aggressive than normal.

TABLE 21-1 Definition and Examples of USDA Pain and Distress Categories in Laboratory Animals Used in Research

Category B	Category C	Category D	Category E
Animals being bred, acclimatized, or held for use in teaching, testing, experiments, research, or surgery *but not yet used* for such purposes. Noninvasive observation only of animals in the wild.	Animals subjected to procedures that cause no pain or distress, or only momentary or slight pain or distress, and do not require the use of pain-relieving drugs.	Animals subjected to potentially painful or stressful procedures for which they receive appropriate anesthetics, analgesics, and/or tranquilizers.	Animals subjected to potentially painful or stressful procedures that are *not* relieved with anesthetics, analgesics, and/or tranquilizers. Withholding anesthesia/analgesia must be scientifically justified in writing and approved by the IACUC.
Examples	**Examples**	**Examples**	**Examples**
1. Animals being bred or housed, without any research manipulation, prior to euthanasia or transfer to another protocol 2. Observation of animal behavior in the wild without manipulating the animal or its environment	1. Holding or weighing animals in teaching, outreach, or research activities 2. Observation of animal behavior in the lab 3. Ear punching of rodents 4. Tail snips in mice ≤ 21 days old 5. Peripheral injections, blood collection, or catheter implantation 6. Feed studies that do not result in clinical health problems 7. Routine agricultural husbandry procedures approved by the IACUC in a protocol or SOP 8. Live trapping 9. Positive reward training or research 10. Chemical restraint 11. Research procedures that involve no potential increase in pain or distress on client owned animals that are undergoing clinical procedures (e.g., drawing extra blood, choice of antibiotics) 12. Exposure to alterations in environmental conditions (not extreme) with appropriate conditioning and microenvironment 13. Food restriction that reduces the animal's weight by no more than 20% of normal age matched controls 14. AVMA-approved euthanasia procedures 15. Euthanasia of breeding animals or unused offspring 16. Exsanguination with anesthesia 17. Perfusion with anesthesia 18. Unknown genetically engineered phenotype	1. Survival surgery 2. Nonsurvival surgical procedures 3. Laparoscopy or needle biopsies 4. Retro-orbital blood collection 5. Exposure of blood vessels for catheter implantation 6. Induced infections or antibody production 7. Tattooing 8. Exposure of skin to UV light to induce sunburn 9. Tail snips in mice > 21 days old 10. Research procedures that could potentially increase pain or distress (e.g., anesthesia/analgesia studies) on client owned animals that are undergoing clinical procedures 11. Genetically engineered phenotype that causes pain or distress that will be alleviated	1. Toxicological or microbiological testing, cancer research, or infectious disease research that requires continuation after clinical symptoms are evident without medical relief or require death as an endpoint 2. Ocular or skin irritancy testing 3. Food or water deprivation beyond that necessary for ordinary presurgical preparation 4. Application of noxious stimuli such as electrical shock that the animal cannot avoid/escape 5. Any procedures for which needed analgesics, tranquilizers, sedatives, or anesthetics must be withheld for justifiable study purposes 6. Exposure to extreme environmental conditions 7. Euthanasia by procedures not approved by the AVMA 8. Paralysis or immobilization of a conscious animal 9. Genetically engineered phenotype that causes pain or distress that will not be alleviated

TABLE 21-2 Behavioral Signs of Pain in Mice and Rats*

Vocalizing	Posture	Locomotion	Temperament
• Squeaks • Squeals	• Dormouse posture (hunched posture) • Rounded back • Head tilted • Back rigid	• Ataxia • Circling	• Docile or aggressive depending on severity of pain • Ingestion of neonates

This chart is meant to display some of the different signs species may exhibit if in pain. Individuals may not show any of these signs or show signs not listed. This is meant as a general guide.
*Karas, A. Z., Danneman, P. J., & Cadillac, J. M. (2008). Strategies for assessing and minimizing pain. In *Anesthesia and analgesia in laboratory animals* (2nd ed, pp. 205–211). London: Elsevier Inc.

PAIN, STRESS, AND DISTRESS

Recall from previous chapters that acute pain normally serves as a protective function to warn of impending danger and is therefore considered an adaptive response. **Stress** is a broad term used to describe the biological response of one's ability to cope with a threat to physical homeostasis or emotional well-being. Stress occurs when an animal perceives a threat, either consciously or unconsciously, including threats to its well-being such hunger, thirst, growth of a tumor, anemia, poor housing, and countless other potential factors. An animal under stress uses a variety of behavioral or physical mechanisms to counteract **stressors** (the events that lead to stress) in an attempt to return to normal. For example, a mouse with a tumor on its face may avoid eating or have decreased food intake (Figure 21-1). In most cases these adaptations do little to resolve the initial stressor and when an animal cannot make adaptations to resolve the initial pain or stress, it experiences distress. **Distress** is an aversive state in which an animal is unable to adapt completely to stressors and the resulting stress is displayed in maladaptive behaviors.[2] The duration and intensity of the stress and the capability of the individual animal to respond to its stressors affects the likelihood of an animal displaying behavioral or physical signs of distress.[3] For example, if the facial tumor on the mouse continues to grow and the mouse does not eat for an extended period of time it will lose weight, become weaker, and its overall welfare is threatened. Although some stress in life is normal, it is considered distress if an animal reacts to it in such a way that its health is compromised (e.g., anorexia in response to a tumor or self-aggression in response to

psychological stress). Unless scientifically proven otherwise, it is assumed in vertebrates that any procedure that causes more than slight or momentary pain or distress in a human will cause similar pain or distress in an animal.

Research procedures that are assumed to be painful or distressful include:[4]

• Surgery
• Prolonged physical restraint
• Malignant neoplasms (Figure 21-2a)
• Prolonged food or water deprivation
• Adverse stimuli from which the animal cannot escape
• Paralysis or immobility in a conscious animal
• Inflammatory disease
• Organ failure resulting in clinical signs
• Nonhealing skin lesions (Figure 21-2b)
• Whole body irradiation at high dosages
• Withdrawal of more than 10% of an animal's blood volume
• Either repeated use of, large volumes of, or intradermal injections of **Freund's complete adjuvant** (a mineral oil used to boost the immune system)
• Intraperitoneal implantation of cells for monoclonal antibody production
• Experiments that require the animal to reach a moribund state or allow the animal to die spontaneously.

Establishing **endpoint criteria** (one or more predetermined physiological or behavioral signs that identify the point at which an experimental animal's pain and/or distress is terminated, minimized, or reduced) before the start of an experiment allows PIs to prevent unnecessary animal pain and/or distress. Examples of endpoint criteria include euthanizing the animal humanely, terminating a painful procedure, or giving treatment to relieve pain and/or distress. The earliest endpoint criteria should be used to prevent pain or distress and experiments ended as soon as animals reach endpoint criteria.

PAIN SCALES FOR PAIN ASSESSMENT

Pain scales are helpful tools for assessing pain in veterinary patients and should be used regularly in laboratory settings. Pain scoring in laboratory animals has been used since the first behavioral paper was published by Morton and Griffiths in 1985.[5] The Morton and Griffiths scoring scheme is a structured method for assessing pain in animals and has been used in laboratory settings to develop endpoint criteria for experimental studies. There have been modifications to the original scoring scheme; however, the main theory of identifying specific pain behaviors and rating them repeatedly and consistently using the same scale has helped address pain concerns in laboratory species.

Courtesy of Janet Amundson Romich, DVM, MS

FIGURE 21-1 In an attempt to return to normal, an animal under stress uses a variety of behavioral or physical mechanisms to counteract stressors. This mouse has a stressor (facial tumor) and may avoid eating or have decreased food intake to counteract this stressor.

(a)

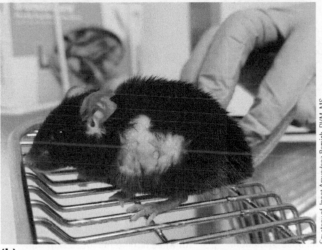

Courtesy of Janet Amundson Romich, DVM, MS

(b)

FIGURE 21-2 Painful conditions in laboratory animals. (a) Tumors are a source of pain in animals. Laboratory animals are frequently assessed for tumor type, tumor size, cumulative tumor burden, and body system most affected by the tumor when determining endpoint criteria of an experiment. (b) Ulcerative dermatitis in mice is a painful condition that commonly appears on the dorsal scapulae, torso, shoulder, and face. Ulcerative lesions may be single or multifocal in distribution and are caused by severe pruritus-induced self-trauma, progressing from superficial wounds to deep ulcerations. Simply trimming the nails of mice can help reduce the incidence of ulcerative dermatitis due to trauma.

Based on their use in assessing pain in laboratory animals, pain scales should include at least three requirements:[1]

- Minimize observer bias and variability of scores among different observers
- Distinguish varying levels of pain intensity based on a particular species and situation
- Detect the degree of "relevance" of pain to the subject

Pain scales can be visual analog scales (VAS), numerical rating scales (NRS), and simple descriptive scales (SDS).[6] A variety of pain scoring systems have been described in Chapter 15 (Acute Pain Management) and Chapter 16 (Chronic Pain Management).

Tech Tip

Since assessing pain is subjective, using the same scale over time provides useful information such as vital sign trends and changes in the patient's response to medications or other treatments.

ANALGESIC CONSIDERATION

In mice and rats, concerns such as lack of evidence-based data on the effects of analgesics, limited number of FDA-approved analgesic drugs, difficulty in extrapolating available dosages from one species to another, and the wide range of species-specific signs of pain makes providing adequate pain management for laboratory animals a challenging task that has hindered the use of analgesics in these species. Despite these concerns, it is important to remember that almost every available analgesic agent has undergone extensive testing in animals and often at one time or another in rodents. There is also advanced veterinary training in laboratory animal medicine that has helped in developing dosage ranges of drugs for mice and rats that are available through a variety of sources. With increased use of analgesic drugs in laboratory settings, many researchers and their staff are reporting improved pain management in their animals and are sharing this information with other researchers. It is the ethical responsibility of researchers to provide reasonable, scientifically justified analgesic drugs to animals in their experiments.

Choosing the proper analgesic for laboratory species is similar to other animal species, and the reader is referred to Chapter 17 (Analgesic Techniques) for questions to answer prior to choosing an analgesic and ways to administer them. Additional questions when considering analgesics for laboratory animals are based on several factors including:[7]

- *Will the analgesic drug interfere with the research project if administered to animals that are part of an experiment?* It is important for the researcher to consider the potential interactions between analgesic therapy and its effects on the research study. For example, opioids may cause respiratory depression and their use may negatively affect a study being conducted on rats involved in a respiratory drug trial or where respiratory parameters are being measured. When reviewers later evaluate research for potential publication, drugs that have been used throughout the research project are considered to determine their potential impact on research outcomes.

- *What facilities and level of animal care are available in the laboratory setting?* Many research facilities have separate holding units for laboratory species that are managed by animal care staff who have daily access to the animals and can provide treatments as needed. For example, ulcerative dermatitis lesions on mice are typically first identified by animal care staff when they are cleaning cages, and these caretakers can also provide treatment as directed by the facility's veterinarian. When designing experiments and submitting potential research for IACUC approval, a researcher must consider who will be responsible for animal care and what said caregiver's training is. Additionally, the researcher must identify the type of postoperative or postprocedural analgesia that is best suited for the test subject. For example, if a procedure is done in a designated laboratory, researchers may need to provide more aggressive analgesia to their animals in the laboratory before returning the animals to the holding unit.

- *What types of analgesics have been used successfully and are available in the facility?* Researchers and veterinarians working in a laboratory facility typically share information about successes and failures of providing analgesia to their animals. Attending conferences, networking with other researchers, and reading drug formularies from other institutions are other ways researchers can learn about providing analgesia to their subjects. Adequate levels of analgesics must be maintained in a laboratory facility

and said drugs must be stored properly (e.g., controlled substances are properly locked, drugs are stored in a cool, dry place).

- *Is any specially adapted equipment needed to administer the analgesic agent?* Most analgesia drugs are administered to laboratory species orally or via injection; however, specialized techniques may be used in a research setting (Figures 21-3a through 21-3c; Figures 21-4a and 21-4b; Figures 21-5a and 21-5b; Figures 21-6a and 21-6b).

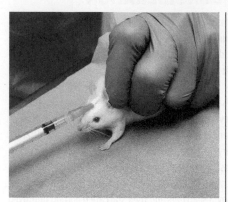

(a)

▲ In mice, a 23 to 27 gauge needle and 1 mL syringe are used for SQ injections. Scruff and restrain the mouse with the nondominant hand allowing access to an area with loose skin

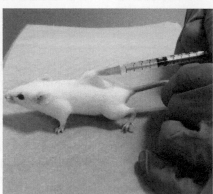

(b)

▲ Or alternatively, restrain by gently holding the tail. Clean the injection site with alcohol. Holding the syringe in the dominant hand, and with the bevel of the needle facing upward and away from your fingers, insert the needle through the skin at the base of the tent or if restraining by the tail inject into an area of loose skin. Pull back on the syringe plunger to aspirate; if a vacuum is not created, the needle may not be below the skin. If blood is seen in the needle hub, it indicates improper needle placement, and the needle must be repositioned. Administer the substance in a steady, fluid motion. Maximum volumes for the mouse = 40 mL/kg total, not > 1.0 mL per site.

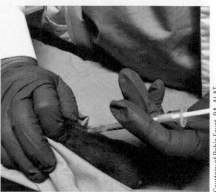

Courtesy of Robin Faust, BA, ALAT

(c)

▲ In rats, a 20 gauge or smaller diameter needle and 1 to 3 mL syringe are used for SQ injections. Scruff and restrain the rat with the nondominant hand allowing access to an area with loose skin. Use a small towel to restrain the rat while tenting the loose skin with the nondominant hand. Clean the injection site with alcohol. Holding the syringe in the dominant hand, and with the bevel of the needle facing upward and away from your fingers, insert the needle at the base of the tented folds of the skin. Pull back on the syringe plunger to aspirate; if a vacuum is not created, the needle may not be below the skin. If blood is seen in the needle hub, it indicates improper needle placement, and the needle must be repositioned. Administer the substance in a steady, fluid motion. Maximum volumes for the rat = 25 mL/kg total, not > 5.0 mL per site.

FIGURE 21-3 SQ injections are used to administer larger volumes of fluid and are administered in the space between the skin and the underlying muscles. Nonirritating substances can be administered subcutaneously in almost any area of the body where the skin overlying the site is loose enough to allow for volume expansion (e.g., flank and dorsal shoulder regions). When administering a solution SQ, the viscosity, concentration, tonicity, and pH of the solution need to be taken into account.

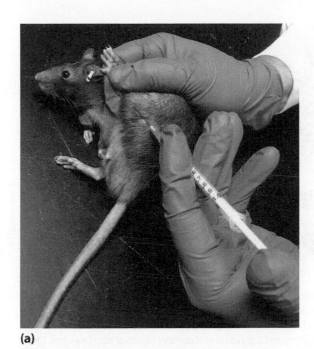

(a)

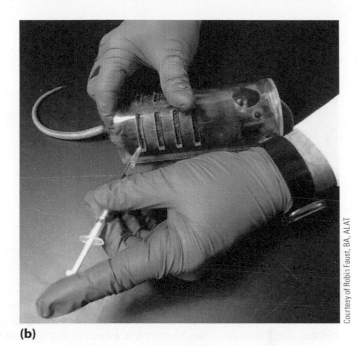

(b)

Courtesy of Robin Faust, BA, ALAT

FIGURE 21-4 IM injections are administered in the thigh muscles of the hind limb. Caution must be used to avoid injury to the sciatic nerve, which is located in close proximity to the femur. When administering a solution IM, the viscosity, concentration, tonicity, and pH of the solution need to be taken into account. IM injections in mice are usually avoided due to their small muscle mass. (a) In rats, a 21-gauge or smaller diameter needle and 1 mL syringe are used for IM injections. Restrain the rat by grasping the skin along its back with the nondominant hand. Clean the injection site with alcohol. With the dominant hand, insert the needle into thigh muscles, and directed away from the femur avoiding the sciatic nerve. Pull back the syringe plunger to aspirate; if blood is seen in the needle hub, it indicates improper needle placement, and the needle must be repositioned. Administer the substance in a steady, fluid motion taking care not to administer fluid too rapidly. (b) IM injections in rats can also be done in a tube restraint. Maximum volumes for the rat = 0.2 mL/site; maximum of 2 to 4 sites.

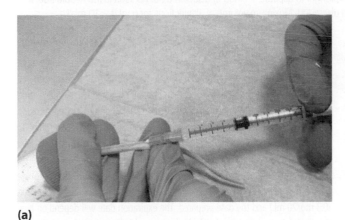

(a)

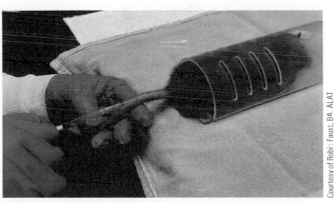

(b)

Courtesy of Robin Faust, BA, ALAT

FIGURE 21-5 IV injections are most commonly administered via the tail vein in rodents. Tail veins are located superficially on the lateral surfaces of the tail at the 10 and 2 o'clock positions. Using a heat source (e.g., gel pack, heated water blanket) can help dilate the blood vessel making it easier to find the vein. (a) For mice, a 27-gauge needle and 1 mL syringe are used for IV injections. Restrain the animal using a mechanical restrainer (e.g., centrifuge tube with a 3 mm hole cut in the end or a mouse specific restraint). Insert the mouse head first into the restraint tube, using the middle finger to keep the mouse from backing out. Clean the injection site with alcohol. The index and middle fingers of the nondominant hand are placed around the tail cranial to where the needle will be inserted (digital pressure will act as a tourniquet). The distal end of the tail is held between the thumb and ring finger below the injection site. With tail under slight tension, insert the needle into lumen of the vein and look for blood to flash back into the needle hub. Do not aspirate the syringe because this will collapse the vessel. Begin injection; if the blood proximal to the needle leaves the vein continue to inject. There will be virtually no resistance when correctly injecting into the vein. If a bubble of solution appears under the skin, stop injecting, remove the needle, and choose a site closer to the base of the tail. Administer the substance in a slow, fluid motion to avoid rupture of the vessel. Remove needle, and apply pressure to stop bleeding before returning animal to cage. Maximum volume of injection for mouse via tail vein = 0.2 mL. (b) For rats, a 23-gauge or smaller diameter needle and a 1 to 3 mL syringe are used for IV injections. Restrain the animal using a mechanical restrainer, clean the injection site with alcohol, and hold the tail with the nondominant hand. With the beveled edge of the needle facing up, align the needle with the vein. Administer the substance in a slow, fluid motion to avoid rupture of the vessel. Remove needle, and apply pressure to stop bleeding before returning animal to cage. Maximum volume of injection (bolus) for a rat via tail vein = 5 mL/kg with bolus IV injection and 20 mL/kg with slow IV injection.

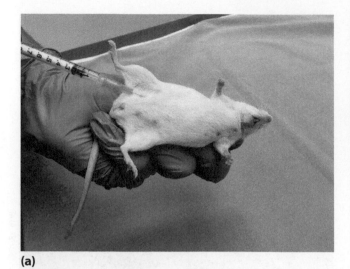

(a)

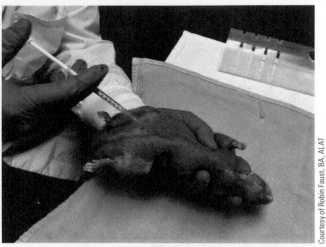

(b)

Courtesy of Robin Faust, BA, ALAT

FIGURE 21-6 IP injections are used in small species for which IV access is challenging and for administration of large fluid volumes. When administering a solution IP, the viscosity, concentration, tonicity, and pH of the solution need to be taken into account. (a) For mice, a 25 to 27 gauge needle and 1 mL syringe are used for IP injections. Restrain the mouse by tilting its head slightly toward the ground so that its head is lower than its hind quarters. Using this position allows the abdominal viscera to shift cranially and minimize accidental puncture of abdominal organs at site of injection. Clean the injection site with alcohol. Insert needle, beveled edge up at a 45 to 90° angle, between ventral midline and top of right thigh to avoid cecum on left side. Keep in mind that the needle must pass through skin and muscle layers to reach the peritoneal cavity. Pull back the syringe plunger to aspirate; if blood or discolored fluid is seen in the needle hub, it indicates improper needle placement, and the needle must be withdrawn and discarded. Maximum volume of IP injections for a mouse is < 10 mL/kg. (b) For a rat, a 23 to 25 gage needle and 1 to 3 mL syringe are used for IP injections. Restrain the rat by either wrapping fingers over the shoulders causing the legs to cross over the chest or wrapping them in a towel. Gently rotate the rat so it is in dorsal recumbency. Place the rat's head/body along the arm and in the crook of the elbow of the non-dominant arm to gently restrain the rat against the body. Restrain the feet/tail with the nondominant hand. Typically, IP injections are given in the rat's caudal right quadrant of the abdomen to avoid damage to the urinary bladder, cecum, and other abdominal organs. Clean the injection site with alcohol. Insert needle, beveled edge up at a 45° angle, between ventral midline and top of right thigh to avoid cecum on left side. Pull back the syringe plunger to aspirate; if blood or discolored fluid is seen in the needle hub, it indicates improper needle placement, and the needle must be withdrawn and discarded. Maximum volume of IP injections for a rat is < 10 mL/kg.

Recall from Chapter 20 (Anesthesia and Analgesia of Pet Birds and Exotic Small Mammals) that specially adapted equipment may be needed to anesthetize rodents and rabbits and the same may be true for providing analgesia to these species. For example, if a constant rate infusion (CRI) is given to a laboratory animal it may require smaller tubing and syringes to ensure the proper dose is given (see Table 20-1).

Tech Tip 🐾

Keep in mind that controlling pain not only involves drug administration but also should include physical, environmental, and behavior management. Acclimation of animals prior to surgery (at least 7 days) allows time for metabolic and hormonal changes caused by the stress of transportation to return to normal and animals to be monitored for signs of illness. A skilled surgeon reduces unintentional surgical trauma. Proper care and non-pharmacologic methods of analgesia include support or bandaging of the traumatized area, appropriate environmental modification (such as nesting material or housing huts for rodents), group housing for socially compatible animals following recovery, easy access to food and water, and provision of appropriate bedding, husbandry, nursing care, and a dry, warm, quiet, stress-free environment.

Drug Dosing in Laboratory Animals

Many laboratory animals are tiny; therefore, accurately obtaining patient weight and correct dosing is crucial to prevent overdose. Since a drug may need to be diluted in order to give as accurate a dose as possible, proper calculation for diluting a drug plus proper storage of the prepared drug is necessary to ensure its concentration and asepsis. For example, in order to give an accurate dose of buprenorphine to a mouse, it needs to be diluted. To prepare a dilution for smaller rodents, one 1 mL vial of buprenorphine (0.3 mg/mL) is drawn into a sterile syringe and added to 19 mL of sterile 0.9% NaCl and dispensed into a sterile vial. The vial of buprenorphine needs to be labeled to include its diluted concentration 0.015 mg/mL solution (0.3 mg/20 mL) and the date prepared. The expiration date will depend on how the diluted drug was prepared and whether or not the diluent contains preservative (drugs in single dose vials, such as the buprenorphine in this example, typically do not contain preservative). For example, diluted buprenorphine made with bacteriostatic 0.9% sodium chloride and prepared under sterile conditions is sterile for 21 days if stored in a sterile glass vial.[8] The solution must also be protected from light and stored in the controlled drug locked box.

Tech Tip 🐾

Drugs intended for parenteral administration purchased in a single dose vial or ampule are labeled for use in a single patient for a single case and typically do not contain preservatives. Using drugs from a single dose vial or ampule for dilution will alter their expiration date.

PAIN MANAGEMENT IN MICE AND RATS

Rodents are a diverse group of mammals characterized by a single pair of growing incisors, robust bodies, short limbs, and long tails. The most common rodents seen in laboratory settings are mice and rats. Pain may be very difficult to detect in rodents because they are prey species and instinctively hide behaviors that may be detrimental to their overall survival if noted by potential predators. Table 21-3 summarizes behavior indicators of pain in rodents as well as specific signs of pain in mice and rats.

Rodent species have the physiological capability to experience pain.[9] The undesirable effects of pain extend beyond local discomfort for the patient and include adverse systemic changes such as inflammation, sympathetic nervous system activation, negative effects on the cardiovascular system, and suppression of the immune system. These systemic effects may slow the animal's recovery and adversely affect prognosis.

TABLE 21-3 Behaviors and Illnesses Associated with Pain in Mice and Rats

Painful Rodent Behaviors	Signs of Pain in Mice	Signs of Pain in Rats
• Decreased food and water[+] intake • An unkempt look (Figure 21-7a) • Lack of grooming • Weight loss • Piloerection • Listlessness • Salivation • **Chromodacryorrhea** (red tears or porphyrin staining) (Figures 21-7b and 21-7c) • Chewing • Licking • Teeth grinding • Hunched posture • Failure to explore cage when disturbed • Vocalization • Self-mutilation • Aggression	• Anorexia • Partial/complete closing of the eyelids; sunken eyes • Changes in respiration, which may include increased or decreased, shallow or labored respiratory patterns • Rough hair coat from lack of grooming; incontinence with soiled hair coat • Increased or decreased vibrissal (whisker) movements • Severe pain or distress indicated by decreased responsiveness to handling, or withdrawal from other mice in the group (Figure 21-7d) • Writhing, scratching, biting, or self-mutilation • Hunched posture • Sudden, sharp movement, such as running • Vocalization when being handled or palpated • Dehydration or weight loss, with wasting of the muscles on the back and a sunken or distended abdomen • Ataxia or circling • Hypothermia	• Anorexia • Markedly reduced breathing frequency • Shallow respiration with visible effort • Expiratory grunts • Exaggerated guarding of potentially painful area • Vocalizations (unsolicited, increased frequency) • Exaggerated lameness • Exaggerated twitching, tremors, or writhing • Unresponsive to external stimuli • Overreacts to external stimuli • Lack of grooming • Mild porphyrin staining • Minor depression • Roughened haircoat, **piloerection** (involuntary erection or bristling of hairs due to a sympathetic reflex or otherwise known as "goose bumps") • Porphyrin staining of nose and eyes • Decreased food consumption • Weight loss • Hunched posture, sitting • Less mobile and alert, responsive only after moderate stimulation • Eyes closed or squinted • Hypothermia, body temperature reduced 1–2°C (1.8–3.6°F) • Tachypnea, increased 30% over baseline • Shallow respirations, abdominal component • Guarding potentially painful site • Licking or scratching of potentially painful site • Bruxism, teeth chattering • **Ptyalism** (excessive flow of saliva) • Self-mutilation

[+]Decreased water intake is especially significant because fluid loss often accompanies anesthesia and surgery. Postoperative patients with pain may require supplemental hydration.

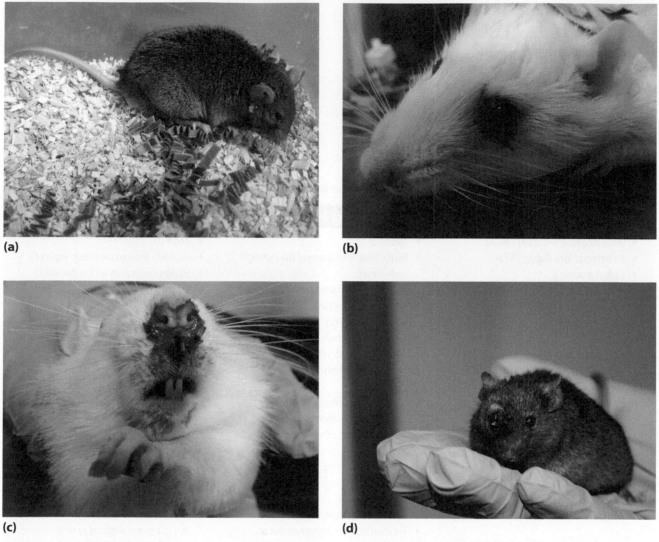

(a)

(b)

(c)

(d)

Courtesy of Janet Amundson Romich, DVM, MS

FIGURE 21–7 Behavioral indicators and signs of pain in rodents. (a) This mouse has an unkept look due to his ruffled coat and hunched posture. (b) Chromodacryorrhea (red tears) and (c) red nasal secretions may indicate pain in rats. This rat has the typical red staining around the nostrils due to release of porphyrins from the Harderian lacrimal glands that travel through the nasolacrimal duct to the nostrils. "Red tears" frequently indicates that the animal has been subjected to some form of stress (e.g., chronic respiratory disease, changes in environment). These porphyrins will fluoresce pink under ultraviolet light. (d) This mouse has an ocular tumor that causes decreased responsiveness to handling.

Tech Tip

Pain assessment is important in a laboratory setting and is sometimes delegated to the veterinary staff in addition to principle investigators and research staff. Observing procedural and surgical methods and the animals themselves help improve post-approval monitoring of compliance. Occasionally, members of the IACUC will also observe study methods and the animals to offer suggestions on ways to monitor and manage pain and ensure protocol compliance.

Pain Scoring Systems in Laboratory Mice and Rats

Grimace scales have been developed for mice and rats as well as other laboratory species such as rabbits and consist of a series of images that show the species in various stages of pain. In the Mouse Grimace Scale (MGS), intensity of each feature is coded on a three-point scale (Figure 21-8). For each of the five features, images of mice exhibiting behavior corresponding to the three values are shown.[10] The Rat Grimace Scale (RGS) also has the intensity of each feature coded on a three-point scale (Figure 21-9).[11] For each of four features, images of rats exhibiting behavior that corresponds to the three values are shown.

	Not present 0	Moderate 1	Severe 2

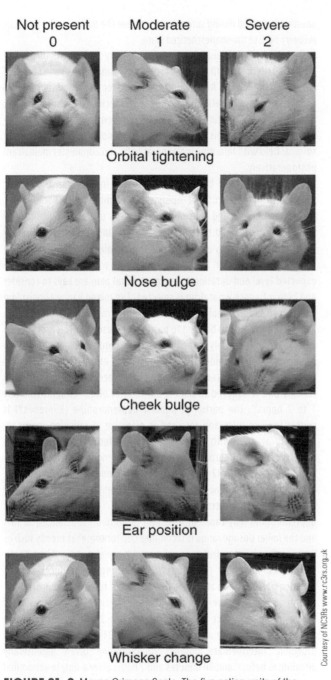

Orbital tightening

Nose bulge

Cheek bulge

Ear position

Whisker change

Courtesy of NC3Rs www.rc3rs.org.uk

FIGURE 21-8 Mouse Grimace Scale. The five action units of the Mouse Grimace Scale include orbital tightening, nose bulge, cheek bulge, ear position, and whisker change. An increase in score indicates an increase response to pain.

Another example of a pain scale that could be used in rodents is described in Table 21-4.[12] Keep in mind that pain can affect a rodent's overall health; therefore, body condition scoring scales such as the Ulman-Cullere and Foltz (mouse) and Hickman (rat) scoring scales can be used to assess the animal's overall health (Figures 21-10a and 21-10b).[13,14] Body condition scoring (BCS) is particularly helpful in cases where pregnancy, organ enlargement, or tumor growth (particularly intra-abdominal growth) may interfere with body weight assessment.

	Not present "0"	Moderate "1"	Obvious "2"

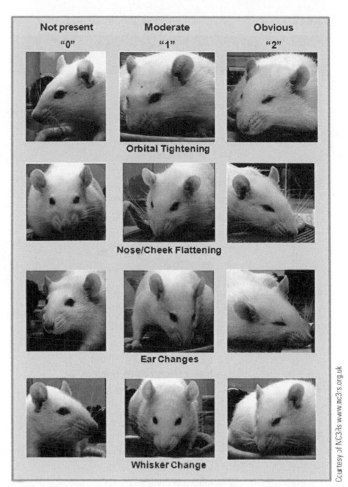

Orbital Tightening

Nose/Cheek Flattening

Ear Changes

Whisker Change

Courtesy of NC3Rs www.nc3rs.org.uk

FIGURE 21-9 Rat Grimace Scale. The four action units of the Rat Grimace Scale include orbital tightening, nose/cheek flattening, ear change, and whisker change. An increase in score indicates an increase response to pain.

Safety Alert ⚠

To avoid being bitten when handling rodents, carry a clean "chuck" towel or gloves to wrap the rodent in for injections.

Treating Rodent Pain

Despite shortcomings of assessment and scoring of pain in rodents, analgesia should be administered if a painful condition is suspected or a painful procedure has been performed. In a research setting, researchers together with laboratory animal veterinarians evaluate individual animals to determine which agent and dosage is appropriate for a particular species and protocol. Resources such as the *Guide for the Care and Use of Laboratory Animals*, 8th ed. (National Research Council of the National Academies) and *Laboratory Animal Anaesthesia*, 4th ed. (Paul Flecknell, Academic Press) as

TABLE 21-4 Pain Scale for Rodents in an Arthritic Study*

Parameter	Scoring Guide	Patient Score
Body Weight (based on pre-study levels)	0 = ≤ 5% decrease 1 = 6–10% decrease 2 = 11–20% decrease 3 = 21–25% decrease 4 > 25% decrease	
Lameness	0 = none 1 = mild, single limb lameness 2 = moderate, multiple limb lameness 3 = severe, non-weight bearing on any limb	
Appearance	0 = normal 1 = huddled, mild piloerection, moves when stimulated 2 = huddled, moderate piloerection, reluctant to move 3 = huddled, ungroomed, severe piloerection, no movement or moribund	
Arthritis Score	0 = normal 1 = mild erythema, no swelling or limb deformity 2 = moderate erythema, mild swelling, no limb deformity 3 = moderate erythema, moderate swelling, mild limb deformity 4 = severe erythema, severe swelling, moderate to severe limb deformity	
		Total:

The pain scale has a corresponding action plan based on the numeric total as follows:

- 0–3 total score or < 1 score in a category: No intervention
- 4–9 total score or > 1 score in a category: Administer analgesic and reevaluate pain score every 8–12 hours.
- 10 – 11: Administer analgesic and reevaluate pain score in 1 hour. If pain still not controlled, then consider additional analgesics or switching analgesic. Reevaluate pain at each administration.

*French, E., Vande Woude, S., Granowski, J., & Maul, D. (2000). Assessment of pain in laboratory animals. *Contemporary Topics, 39*, 85. Presented at 2000 National AALAS meeting, San Diego CA.

well as many institutions' IACUC documents provide dosages for a variety of anesthetic and analgesic agents. Analgesic dosage rates based on body weight in rodents tend to be relatively high compared with other mammals, largely because of their small body size and fast metabolic rate.[15] When given a drug dosage range, smaller patients typically require dosages on the upper end of the scale and larger patients require drug dosages on the lower end (refer to description of allometric scaling in Chapter 8 [Premedication]). Preprocedural administration of analgesics may reduce the amount of anesthesia needed during surgery and improve the quality of the subject's recovery back to the unanesthetized state.

Opioids, nonsteroidal anti-inflammatory drugs (NSAIDs), and local anesthetics are likely the three most commonly used classes of analgesics in rodents. As in other species, preemptive and multimodal analgesia can be beneficial and reduce the incidence of chronic pain. An example of a multimodal analgesic regimen for rodents undergoing surgery is to administer buprenorphine preoperatively, infiltrate the surgical field with lidocaine intraoperatively, and administer meloxicam postoperatively.

All opioids have been used and tested in rodents. As in other species pre-emptive use of opioids is recommended. Additionally, the same concepts and considerations for the use of opioids in other species can be applied to opioid use in mice and rats as well. Opioids vary greatly in potency and efficacy but generally produce few adverse effects in animals or adverse interactions with other drugs. The expected level and duration of postprocedural pain are keys to consider and help influence what drug is selected. The mixed kappa agonist/mu antagonist butorphanol (Torbugesic®, Torbutrol®) can provide weak analgesia and can also be used to partially reverse the effects of full (μ) opioids without entirely reversing kappa receptor-mediated analgesia. Butorphanol has minimal cardiovascular effects and produces mild respiratory depression. Butorphanol decreases food intake in rodents and is believed to have a duration of effect of approximately 1 to 2 hours[16]. The partial μ agonist buprenorphine (Buprenex®) is widely used in mice and rats and is effective in relieving moderate postsurgical pain with minimal adverse effects. Buprenorphine is more potent than morphine and also has a longer duration of action typically lasting 8 to 12 hours.[17] In rats, buprenorphine produces a low level of analgesia at low dosages and much greater analgesia at higher dosages, with little increase in adverse effects.[17] There are reports that buprenorphine may cause pica (eating of nonfood substances) in rats and the lower dosage range is recommended for some rat breeds such as Sprague-Dawley rats. Full μ agonist opioids are also utilized in rodents. In rodents, morphine (Duramorph®, Astramorph®) produces marked sedation in addition to analgesia. Regardless of route of administration, morphine relieves severe pain and has a relatively short duration of action in rats and mice of approximately 2 to 3 hours,[18] thus limiting its use in a laboratory animal setting unless 24-hour care is routinely provided. Intermediate dosages of morphine given parenterally 30 minutes before, and 30 minutes and 2 hours after a simple abdominal surgery provide adequate pain relief in rodents over a 2-day period.[17,19] Fentanyl's duration of action is believed to be very short in rodents, lasting only 30 to 60 minutes. When using fentanyl (Sublimaze®) at higher dosages, particularly if combined with sedatives, be prepared to tracheally intubate and provide positive pressure ventilation to rodents.[20]

NSAIDs are useful alone for mild to moderate inflammatory pain and can be used as part of a multimodal analgesic protocol for more severe pain.[20] Commonly used NSAIDs in rodents include meloxicam (Metacam®), ketoprofen (Ketofen®), and carprofen (Rimadyl®). Reported adverse effects of NSAID use in rodents are similar to that of other species and include gastric irritation or ulceration and potential renal injury.[20]

BC 1

Mouse is emaciated.
- Skeletal structure extremely prominent; little or no flesh cover.
- Vertebrae distinctly segmented.

BC 2

Mouse is underconditioned.
- Segmentation of vertebral column evident.
- Dorsal pelvic bones are readily palpable.

BC 3

Mouse is well-conditioned.
- Vertebrae and dorsal pelvis not prominent; palpable with slight pressure.

BC 4

Mouse is overconditioned.
- Spine is a continuous column.
- Vertebrae palpable only with firm pressure.

BC 5

Mouse is obese.
- Mouse is smooth and bulky.
- Bone structure disappears under flesh and subcutaneous fat.

A "+" or a "-" can be added to the body condition score if additional increments are necessary (i.e., ...2+, 2, 2-...)

(a)

BC 1

Rat is emaciated
- Segmentation of vertebral column prominent if not visible.
- Little or no flesh cover over dorsal pelvis. Pins of pelvis prominent if not visible.
- Segmentation of caudal vertebrae prominent.

BC 2

Rat is underconditioned
- Segmentation of vertebral column prominent.
- Thin flesh cover over dorsal pelvis, little subcutaneous fat. Pins of pelvis easily palpable.
- Thin flesh cover over caudal vertebrae, segmentation palpable with slight pressure.

BC 3

Rat is well-conditioned
- Segmentation of vertebral column easily palpable.
- Moderate subcutaneous fat store over pelvis. Pins of pelvis easily palpable with slight pressure.
- Moderate fat store around tail base, caudal vertebrae may be palpable but not segmented.

BC 4

Rat is overconditioned
- Segmentation of vertebral column palpable with slight pressure.
- Thick subcutaneous fat store over dorsal pelvis. Pins of pelvis palpable with firm pressure.
- Thick fat store over tail base, caudal vertebrae not palpable.

BC 5

Rat is obese
- Segmentation of vertebral column palpable with firm pressure; may be a continuous column
- Thick subcutaneous fat store over dorsal pelvis. Pins of pelvis not palpable with firm pressure.
- Thick fat store over tail base, caudal vertebrae not palpable.

(b)

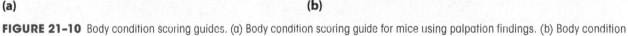

FIGURE 21–10 Body condition scoring guides. (a) Body condition scoring guide for mice using palpation findings. (b) Body condition scoring guide for rats. The chart developed for rat body condition scoring is based on the mouse guide; however, fat deposition in rats was more reliably assessed by palpation of fat overlying the dorsal pelvic protuberances instead of that overlying the vertebral column, as is recommended for mice.

Alpha-2 adrenergic agonists, such as xylazine (Rompum®, AnaSed®, Gemini®), medetomidine (Domitor®), or dexmedetomidine (Dexdomitor®) have analgesic effects in rodents, but they should not be used as the sole analgesic agent because of the level of sedation they produce. These agents combined with an opioid or NSAID result in a synergistic effect, extending both the duration and potency of the total analgesic effect.[17] Ketamine (Ketaset®, VetaKet®, Ketaject®, Vetalar®)) has also been used in conjunction with other agents, especially opioids, to provide analgesia in rodents. CRIs have been used in the research environment for rodents to avoid "peaks and valleys" in drug concentration.[21]

Local anesthetic agents may be administered topically (e.g., topical lidocaine and prilocaine [EMLA®] cream to facilitate venipuncture), locally by infiltration of the surgical site, in peripheral nerve blocks, and in epidural or spinal anesthesia.[22] Lidocaine (Xylocaine®) and bupivacaine (Marcaine®) are the most frequently used local anesthetics, with bupivacaine having a longer duration of action but greater risk for toxicity. The small size of many rodent species makes inadvertent overdose a much more significant hazard. To minimize this risk when using local anesthetics, it is advisable to draw up the maximum safe dose before administration.[23] This volume can, if needed, be diluted 1:2 or 1:4 with sterile 0.9% saline prior to administration to increase the volume.[20]

Gabapentin (Neurontin®) has been evaluated in rats and has been shown to have anti-hyperalgesic and anti-allodynic effects in chronic pain models with repeated dosing.[24]

Dosages of select analgesic drugs used in mice and rats are described in Table 21-5.

Tech Tip

Administration of analgesics to laboratory animals must be properly documented. Without documentation, federal regulators and inspectors consider that analgesics were not provided and would classify that event as noncompliance.

SUMMARY

Every person in the laboratory setting has a vital role to play in the pain management of laboratory animals. It is the ethical duty of veterinarians, veterinary and laboratory animal technicians, animal care staff, researchers, and laboratory personnel to ensure that each research animal receives adequate veterinary care and pain management in accordance with recommendations of the National Research Council's *Guide for the Care and Use of Laboratory Animals*, 8th ed. and the institution's IACUC. Understanding and implementing quality pain management in laboratory species ensures their well-being and the validity of experimental data collected from these animals; therefore, it is difficult to justify withholding analgesic agents in laboratory species because it will alter the experimental results. The research community has a vested interest in assessing pain and providing analgesia to the animals in their experiments in order to provide scientifically sound conclusions from their studies.

CRITICAL THINKING POINTS

- Stress and distress are important concepts in laboratory species which need to be controlled. Both the duration and intensity of the animal's state are important considerations of assessing distress and stress. For example, giving an injection that requires brief restraint may produce acute stress for a few seconds while housing a social species individually in a metabolic cage may lead to chronic distress.

- Closely monitoring laboratory animals will help minimize distress and allow intervention to reduce stressors and provide enrichment for these animals.

- Pain scales should be used regularly in laboratory species to assess pain accurately.

- Preemptive and multimodal analgesia should be used as often as possible in laboratory species just as in other species.

- How a laboratory species responds to analgesic medication should be recorded in a laboratory data book in order to monitor its success or failure.

- Even species without an extensive literature database of information on pain control need to be assessed and given relief for pain by extrapolating data from other species.

- If the laboratory animal has the ability to have nociception, it can feel pain and that pain needs to be managed.

REVIEW QUESTIONS

Multiple Choice

1. Who wrote the first behavioral paper about pain scales for laboratory animals in 1985?
   a. Watson and Crick
   b. Hubel and Wiesel
   c. Morton and Griffiths
   d. Laurel and Hardy

2. What are the commonly used types of pain scales?
   a. VAS, NRS, SDS
   b. NRS, behavioral, SDS
   c. Visual only
   d. Descriptive only

3. What causes stress in laboratory mice and rats?
   a. When an animal anticipates a change in a research protocol such as changing a drug's dosage or administration frequency
   b. When an animal experiences periods of noise reduction in a facility, which may signal its transportation to another facility
   c. When an animal notices increases in food availability as it may be perceived that food reduction will soon follow
   d. When an animal consciously or unconsciously perceives a threat including threats to its well-being

4. Which laboratory species are not currently covered under USDA pain or distress categories?
   a. Rabbits
   b. Livestock
   c. Rats and mice
   d. All species are covered.

TABLE 21–5 Select Analgesic Drug Dosages for Rats and Mice*

Drug Category	Drug	Dosages	Comments
Opioids	butorphanol (Torbugesic®, Torbutrol®)	Rats: 0.2–2 mg/kg given SQ, IM, or IP every 2–4 hrs Mice: 1–5 mg/kg given SQ every 4 hours	Provides mild analgesia Duration of action approximately 1–2 hours
	buprenorphine (Simbadol®, Buprenex®)	Rats: 0.01 – 0.05 mg/kg given SQ or IP every 8–12 hours Mice: 0.05 – 0.1 mg/kg given SQ or IP every 8–12 hours	More potent than morphine Duration of action typically lasts 8–12 hours
	morphine (Duramorph®, Astramorph®)	Rats: Mice: 2–5 mg/kg given SQ or IM every 4 hours Mice: 2–5 mg/kg given SQ or IM every 2–4 hours	Marked sedation in addition to analgesia Duration of action approximately 2–4 hours
NSAIDs	meloxicam (Metacam®)	Rats: 1 mg/kg once daily up to 3 days Mice: 5 mg/kg once daily up to 3 days	Currently difficult to source Recommended to give concurrent SQ fluids
	ketoprofen (Ketofen®)	Rats: 5 mg/kg given SQ, IM, or PO every 24 hours Mice: 2–5 mg/kg every 24 hours	Recommended to give concurrent SQ fluids
	carprofen (Rimadyl®)	Rats: 5 mg/kg IP or SQ every 6–8 hours Mice: 5 mg/kg IP or SQ every 6–8 hours	Drug is viscous and needs to be diluted in sterile water 1:10 or more
Alpha-2 adrenergic agonists mixtures	xylazine (Rompum®, AnaSed®, Gemini®) + Ketamine (Ketaset®, Ketalar®, Vetalar®)	Rats: xylazine: 5–10 mg/kg + ketamine: 75–90 mg/kg IP or SQ (in same syringe) Mice: xylazine: 16–20 mg/kg + ketamine: 75–150 mg/kg + IP or SQ (in same syringe)	May not produce surgical-plane anesthesia for major procedures Xylazine may be reversed with atipamezole (Antisedan®) (1–2.5 mg/kg SC, IP, or IV for rats and mice) or yohimbine (Yobine®, Antagonil®) (1.0–2.0 mg/kg SC, IP or IV) (remember analgesic properties will be reversed)
	dexmedetomidine (Dexdomitor®)	Rats: dexmedetomidine: 0.5–0.75 mg/kg + ketamine: 75–90 mg/kg IP or SQ (in same syringe) Mice: dexmedetomidine: 0.5–1 mg/kg + ketamine: 75–150 mg/kg IP or SQ (in same syringe)	May not produce surgical-plane anesthesia for major procedures Dexmedetomidine may be reversed with atipamezole 1–2.5 mg/kg SC, IP, or IV for rats and mice (remember analgesic properties will be reversed)
Local anesthetics	lidocaine (Xylocaine®)	Rats and mice: Dilute 1:2 or 1:4; do not exceed 7 mg/kg total dose, given SQ or intra-incisional	Use locally before making surgical incision Faster onset than bupivacaine but short (< 1 hour) duration of action
	bupivacaine (Marcaine®)	Rats and mice: Dilute 1:2 or 1:4, do not exceed 8 mg/kg total dose, given SC or intra-incisional	Use locally before making surgical incision Slower onset than lidocaine but longer (4–8 hour) duration of action

*Adapted from UW-Madison Research Animal Resources and Compliance Drug Formulary and University of British Columbia Drug Formulary

5. Which institutional group monitors pain in laboratory species?

   a. ASPCA
   b. FDA
   c. AVMA
   d. IACUC

6. What is porphyrin staining?

   a. A stain that is used in microbiology
   b. A stain that occurs in rodents that are stressed or ill
   c. A stain used to identify corneal ulcers in small mammals
   d. A stain that occurs in urine droppings of birds

7. What is the purpose of the Mouse and Rat Grimace Scales?

   a. To develop an appreciation of rodent emotions
   b. To educate veterinary professionals that rodents have feelings
   c. To show various facial expressions during pain
   d. To show how mice and rates are different from each other

8. Why do smaller animals such as rodents require higher dosages (on a mg/kg basis) than larger animals?

   a. Because of their higher metabolic rate
   b. Because it is better to give more drug than less drug
   c. Because rodents have more plasma proteins, which bind up active drug
   d. Because rodents are not carnivores and therefore they lack enzymes to metabolize drugs

9. How can the accuracy of injectable drug administration to small laboratory animals such as mice be improved?

   a. Use a lower numeric gauge needle so that the drug is delivered to the animal through a wider diameter lumen
   b. Use heparinized syringes to help the drug not adhere to the wall of the syringe during administration
   c. Dilute the drug so that a more precise amount of drug can be measured and administered
   d. Store all drugs in containers that protect them from light exposure

10. What are indirect measures that can indicate pain in rodents?

    a. Muscle atrophy and paralysis
    b. Increased heart rate and respiratory rate
    c. Whisker twitching and excessive grooming
    d. Increased food intake and defecation

11. What are the common adverse effects of NSAID use in rodents?

    a. Gastric irritation and potential renal injury
    b. Renal necrosis and skin rashes
    c. Cardiac arrhythmia and tachypnea
    d. Vomiting and ataxia

12. What is the name for one sign of pain in rodents that causes involuntary bristling of hairs?

    a. Piloerection
    b. Porphyrin staining
    c. Chromodacryorrhea
    d. Vibrissial movement

13. Which opioid provides moderate postsurgical pain relief while producing the least amount of adverse effects in rodents?

    a. Buprenorphine
    b. Morphine
    c. Fentanyl
    d. Butorphanol

14. Which of the following can be seen with morphine administration in rodents?

    a. Polydipsia and polyuria
    b. Tachypnea and tachycardia
    c. Nausea and vomiting
    d. Marked sedation

15. Which of the following can be applied topically to desensitize the skin before placing IV catheters?

    a. Ketamine
    b. Cream with lidocaine and prilocaine
    c. Dexmedetomidine
    d. Tramadol ointment

16. Which opioid may cause pica in some breeds of rats?

    a. Buprenorphine
    b. Morphine
    c. Fentanyl
    d. Butorphanol

17. Which type of pain assessment is particularly helpful in rodents that are pregnant or have organ enlargement or tumor growth that may interfere with body weight assessment?

    a. Grimace scales
    b. Body condition scoring scales
    c. Acute pain scales
    d. Chronic pain scales

18. What are often the earliest signs of pain observed in laboratory mice and rats?

    a. Physiological factor alterations
    b. Biochemical parameter variability
    c. Drug dosing modifications
    d. Behavioral changes

19. Within IACUC proposals, what are most research animals assigned to help define the degree of pain or distress they may experience?

    a. FDA modification score
    b. USDA pain or distress category
    c. MUMS distress identification score
    d. GPS score

20. When may analgesics *not* be administered to mice and rats in a research setting?

    a. If the analgesic is perceived to adversely affect experimental results
    b. If there is no dosage of analgesic available for laboratory species
    c. If a PI believes the cost of the analgesic drug is a financial burden on the study
    d. If the institution is a pain-free research facility

Case Studies

Case Study 1: A 2 ½ year old, M Sprague Dawley rat (life span of a rat is 2–3 years) is part of a research project investigating the effectiveness of a new diabetes drug. This rat weighed 200 g when he first arrived at the facility, and his current weight is 185 g. The rat usually eats standard Adult Rat Food, but he has not eaten for 2 days although he has been drinking water. He sits with a hunched position and is reluctant to move due to a mild limb deformity. He has increased respirations, piloerection, and porphyrin staining. The veterinarian asks you to fill out the Pain Scale for Rodents (Table 21-4) to help determine level of pain.

1. What score do you give this rat?
2. Is intervention needed for this rat?

Case Study 2: As an IACUC member at a large research facility, you are responsible for reading protocols and determining whether or not the laboratory animals will be experiencing pain during research projects. While preparing for the IACUC meeting, you read through a protocol in which the researcher wants to perform tail snips (removal of soft tissue from the tip of the tail using a scalpel) on mice to collect blood.

1. What information about the mice would you like to know prior to determining if the tail snip will be painful?
2. If you determine the tail snips to be painful, what should you do?

Case Study 3: You are working in a laboratory facility and you need to administer a dose of butorphanol to a mouse. In order to give an accurate dose, you need to dilute the butorphanol, which comes in a 10 mL multidose vial and has a concentration of 0.5 mg/mL. The veterinarian asks you to take 1 mL of butorphanol from the vial and add it to 19 mL of sterile saline (0.9% NaCl).

1. What is the final concentration of butorphanol when diluted as described above?
2. What else do you need to do to the vial in which you put butorphanol and sterile saline?
3. Per the research protocol, a mouse is to receive 0.05 mg/kg buprenorphine SQ. Calculate a dose for this mouse using the diluted drug you prepared in question 1.

Case Study 4: You are applying for a job at a research laboratory and are preparing for an interview. You suspect you will be asked how you as a laboratory animal technician can assess whether a laboratory animal is in pain. Describe how you would answer this interview question.

Critical Thinking Questions

1. You have taken a job in a laboratory animal research facility in the Veterinary Department. One of your primary duties is to observe animals in their housing units (cages) to see if they are behaving normally or if there is a problem occurring from their use in the PI's protocol. You come upon this cage in a rat room (Figure 21-11).

FIGURE 21–11 Rat and pups in cage.

Answer the following questions regarding what you see in this image.
   a. Does the mother rat appear normal?
   b. Is she feeding her pups?
   c. Does anything look amiss in the cage?
   d. Are the babies eating?
   e. Would you report this as a normal cage?

2. Many researchers do not want to add any uncontrolled variables to their research projects because they believe this could affect the repeatability and reliability of their data. In other words, if some animals were given analgesic agents and others were not, it would be difficult to compare the data between these groups of animals. Some researchers argue that pain itself could add uncontrolled variables. Describe how you would defend the position that not controlling pain in laboratory species could add uncontrolled variables to a study.

ENDNOTES

1. Karas, A. Z., Danneman, P. J., & Cadillac, J. M. (2008). Strategies for assessing and minimizing pain. In *Anesthesia and analgesia in laboratory animals* (2nd ed, pp. 205–211). London: Elsevier Inc.
2. Aronson, A. L. et al. (1992). *Recognition and alleviation of pain and distress in laboratory animals* (p. 4). Washington, D.C.: National Academy Press.
3. Ward, P. A. (2008). *Recognition and alleviation of pain and distress in laboratory animals* (p. 3). Washington, D.C.: National Academies Press.
4. http://www.ahc.umn.edu/rar/pain&distress.html, accessed May 1, 2020.
5. Morton, D. B., & Griffiths, P. H. M. (1985). Guidelines on the recognition of pain, distress and discomfort in experimental animals and an hypothesis for assessment. *Veterinary Record,* 116, 431–436.
6. Goldberg, M. E. (2010). The fourth vital sign in all creatures great and small. *The NAVTA Journal,* 31–54.
7. Flecknell, P. (1999). Pain-assessment, alleviation and avoidance in laboratory animals. *ANZCCART News, 12*(4), 1–10, 4.
8. DenHerder, J. M., et al. (2017). Effects of time and storage conditions on the chemical and microbiologic stability of diluted buprenorphine for injection. *Journal of the American Association for Laboratory Animal Science, 56*(4), 457–461.
9. Thompson, L. (2014). Recognition and assessment of pain in small exotic mammals. In C. Egger, L. Love, & T. Doherty (Eds.), *Pain management in veterinary practice* (pp. 391–397). Ames, IA: Wiley/Blackwell.
10. Langford, D. J., et al. (2010). Coding facial expressions of pain in the laboratory mouse. *Nature Methods, 7*(6), 448.
11. Sotocinal, S. G. et. al. (2011). The Rat Grimace Scale: A partially automated method for quantifying pain in the laboratory rat via facial expressions. *Molecular Pain, 7*(55), p. 5.
12. French, E., Vande Woude, S., Granowski, J., & Maul, D. (2000). Assessment of pain in laboratory animals. *Contemporary Topics, 39*, 85. Presented at 2000 National AALAS meeting, San Diego CA.
13. Ullman-Culleré, M. H., & Foltz, C. J. (1999). Body condition scoring: A rapid and accurate method for assessing health status in mice. *Laboratory Animal Science, 49*(3).
14. Hickman, D., & Swan, M. (2010). Use of a body condition score technique to assess health status in a rat model of polycystic kidney disease, *Journal of the American Association for Laboratory Animal Science, 49*(2), 155–159.
15. Miller, A. L., & Richardson, C. A. (2011). Rodent analgesia in veterinary clinics of North America. *Exotic Animal Practice, 14*(1), 84
16. Harkness, J. E., Turner, P. V., VandeWoude, S., & Wheler, C. L. (2010). Clinical procedures. In *Harkness and Wagner's biology and medicine of rabbits and rodents* (5th ed., pp. 167–170). Ames, IA: Wiley/Blackwell.
17. Gaertner, D. J., et al. (2008). Anesthesia and analgesia for laboratory rodents. In R. E. Fish, M. J. Brown, P. J. Danneman, & A. Z. Karas (Eds.), Anesthesia and analgesia in laboratory animals (pp. 239–297). American College of Laboratory Animal Medicine Series. San Diego, CA: Academic Press.
18. Gades, N. M., Danneman, P. J., Wixson, S. K., & Tolley, E.A. (2000). The magnitude and duration of the analgesic effect of morphine, butorphanol, and buprenorphine in rats and mice. *Contemporary Topics in Laboratory Animal Science, 39*(2), 8–13.
19. Gonzalez, M. I., Field, M. J., Bramwell, S., McCleary, S., & Singh, L. (2000). Ovariohysterectomy in the rat: A model of surgical pain for evaluation of pre-emptive analgesia? *Pain, 88*(1), 79–88.
20. Greenacre, C. B. (2014). Treatment of pain in small exotic mammals. In C. Egger, L. Love, & T. Doherty (Eds.), *Pain management in veterinary practice* (pp. 399–406). Ames, IA: Wiley/Blackwell
21. Franken, N. D., van Oostrom, H., Stienen, P. J., et al. (2008). Evaluation of analgesic and sedative effects of continuous infusion of dexemetomidine by measuring somatosensory and auditory-evoked potentials in the rat. *Veterinary Anaesthesia and Analgesia, 35*, 424–431.
22. Miller, A. L., & Richardson, C. A. (2011). Rodent analgesia in veterinary clinics of North America. *Exotic Animal Practice 14*(1), 88.
23. Flecknell, P. (2009). Analgesia and post-operative care. In *Laboratory animal anaesthsia* (3rd ed., pp. 139–179). London, UK: Academic Press.
24. Fox, A., Gentry, C., Patel, S., Kesingland, A., & Bevan, S. (2003). Comparative activity of the anticonvulsants oxycarbazepine, carbamazepine, lamotrigine and gabapentin in a model of neuropathic pain in the rat and guinea pig. *Pain, 105*(1–2), 355–362.

Appendix A

OCCUPATIONAL SAFETY

Many aspects of the work performed by veterinary technicians are dangerous. Not all of the dangers are readily apparent at the time of the exposure. Developing and following safety protocols to prevent injuries in the clinic is the job of both employers and employees. Occupational Safety and Health Administration (OSHA) and regulatory agencies in other countries set standards to provide a safe working environment to reduce the amount of injuries occurring within the practice and thereby reducing staff absenteeism due to injuries. It is the employee's responsibility to abide by the safe working policies of the practice. If an injury occurs, an Incident Report should be filled out detailing the injury. In most workplaces, the Incident Report has to be done on the day the incident occurs (if possible). Incident Reports are used to assess risk and determine if specific types of injury can be avoided in the future.

Before, during, and after an anesthetic event there are several items that staff members should be aware of in regard to OSHA compliance including (but not limited to):

- Anesthetic monitoring and scavenging of inhalant anesthetic agents
- Clipping and preparing of the surgical site
- Dispensing drugs
- Electrical equipment use (cords, etc.)
- Fire precautions
- Induction of anesthesia
- Maintaining a clean surgical environment through disinfection of the environment and anesthetic equipment
- Manual handling and transporting of anesthetised patients through proper patient restraint and lifting/moving of patients

It is important to keep safety in mind while performing each task and procedure. As time passes, complacency becomes the norm. Procedures or steps of a procedure that were once executed in a safe manner are either forgotten or modified to a point that personnel are at increased risk of injury or harm.

Some of the safety concerns that are present before, during, and after anesthetic procedures include:

- *Restraining patients*. Working with animals, regardless of size, requires that safety be maintained not only for the patient but for personnel as well. Each species has its own list of restraint concerns. Leashes, halters, ropes, fences, and stanchions must be sturdy and in proper working order. Prep or surgery tables must be available and at a height that is appropriate for personnel to work without excessive bending and stooping. Tables should be movable to allow for positioning for various surgical procedures.

- *Lifting or moving patients*. Take care moving small animals from cage to table ensuring that the patient is properly restrained. When moving large animals from stall to induction, only trained personnel must use proper, fully functional restraint devices. During the induction period, animals are experiencing a new situation and are at heightened stress levels. Expect that they will not act as they would under normal circumstances. Trained personnel, not owners or agents, must handle animals for these procedures.
  - When lifting, pulling, or pushing animals, proper technique can help reduce the incidence of injury. Keeping the back straight when lifting can decrease back injuries. Pulling or pushing large animals during induction or recovery should be done with assistance, either with more personnel or with chain or electric hoists or other assistive devices.

- *Using or storing compressed gas cylinders*. Any size compressed gas cylinder can be dangerous if it is not properly moved or secured. Care must be taken whether the cylinder is full or empty. Injury can occur if the cylinder falls. If a compressed gas cylinder has neck damage, it can become a projectile and cause serious injury to personnel and damage to buildings. If transporting a large compressed gas cylinder, use a hand truck or cart, utilizing the attached strap or chain to keep it in place while it is moving.

- *Using inhalant anesthetic agents*. Veterinary workers involved in surgery and anesthesia are exposed to waste anesthetic gases (WAGs). The amount of WAG exposure is dependent on the job performed and how WAGs are removed from the working environment via a functional scavenging system. OSHA guidelines requires that exposure limit for all halogenated agents is not more than two parts per million.
  - Several scavenging systems are described in Chapter 2 (The Anesthesia Machine). Whether the system is active or passive, the most important aspect of these systems is that they are professionally installed, fully functional, and tested regularly.
  - Another aspect of inhalant anesthetic agent exposure is faulty equipment. Leak testing the anesthetic machine is one of the most influential ways that WAGs can be controlled. Leak tests should be performed before every anesthetic event including checking for faulty reservoir bags and Y-pieces. Clean reservoir bags and Y-pieces should be used for every new patient.
  - During the recovery period, patients can expire a level of inhalant anesthetic agent that is above OSHA acceptable levels. Immediately after a patient is removed from the machine, they are

expiring the highest levels of inhalant anesthetic agent, and personnel are in close proximity to the patient, making their exposure to WAGs greater. It is extremely important that personnel keep themselves away from the end of the endotracheal tube or if the patient is in a cage, that personnel not put their heads into the cage with the patient. The veterinary practice must have proper ventilation in the building so that room air exchanges provide enough fresh air to keep WAG levels as low as possible.

○ Some chemicals, such as inhalant anesthetic agents, are supplied in small glass bottles. Not only is there a splash concern but since they are volatile agents, they also can be dangerous to the mucous membranes and respiratory tract. Having supplies ready if needed to safely clean up a spill and having sufficient ventilation in the area is important when working with inhalant anesthetic agents. Proper ventilation and personal protective equipment (PPE) are important when refilling vaporizers. It is good practice to "announce" to the room that someone is filling a vaporizer in case anyone who is pregnant is in the vicinity and would like to step out of the room.

- *Using chemicals*. Sterilizing and disinfecting agents are used to ensure patient equipment is free from or has significantly reduced numbers of infectious microorganisms. It is imperative that staff handle these agents in a safe manner every time they are used.

○ Workers in the United States have a "right to know." The federal government, through the OSHA, provides this right. Workers must be informed of chemical exposures they face in the workplace. Proper safety equipment must be provided for personnel to work with any substance that is or contains a hazardous chemical. For this system to work, it is important that personnel ask for and use the safety equipment provided by their employer.

○ Chemicals shipped to veterinary practices must have documentation on proper storage and use, and exposure and disposal guidelines. Safety Data Sheets (SDS) come with purchased reagents. If no SDS form is provided, they are accessible online at the company website. These documents must be kept in a way that personnel can access the documents in an emergency situation. Binders with printed SDS are typically kept in veterinary practices in areas where chemicals are being used, or nearby. SDS can also be stored digitally on an accessible space on a computer or shared network. Learn how to read these documents so that you can quickly access life-saving information in emergency situations.

○ Chemicals may be purchased in large containers (the original container) and transferred to smaller containers (secondary or workplace containers) that are used at point-of-use areas. Every time a chemical, such as alcohol or peroxide, is poured from a large bottle into a spray bottle, a secondary container is created. Any chemical that is reconstituted or diluted, such as chlorhexidine, is also deemed to be in a secondary container. These secondary containers need to be labeled to show the product name, the hazardous chemicals it contains, and words or pictures that show the key hazards (e.g., inhalation hazard, ingestion hazard, skin absorption hazard, skin irritant, eye corrosion hazard, etc). There should also be warnings for flammability, as well as exposure concerns for eyes, skin, and lungs.

○ Chemicals must be used or diluted properly to be effective. Checking the instructions frequently helps provide a consistent outcome during disinfection and sterilization procedures. This consistency assures that anesthetic events and patient care are at the highest level of quality as well as increasing safety. When using chemicals, remember to wear proper PPE. Liquids can suddenly splash into the eyes, onto skin and clothing, and could cause chemical burns. Wearing proper protective gear can reduce or eliminate permanent damage. Make sure that bottles have proper fitting caps and that they are placed on the bottle correctly and securely.

Appendix B

ANESTHETIC EQUIPMENT MAINTENANCE AND CLEANING*

Equipment	Suggested Maintenance/ Cleaning
Anesthetic machine	Have professionally serviced 1/yr. This should include a vaporizer calibration and a check of all tubing; check and replace all parts as needed. If exterior of the machine gets soiled, it should be wiped down with a mild soap and water.
Breathing circuits	Clean after every use with 0.2% chlorhexidine solution.
Reservoir (rebreathing) bags	Clean after every use with 0.2% chlorhexidine solution.
CO_2 absorbent	CO_2 granules can be changed on any one of the following schedules: • every two weeks (or some other predetermined length of time), regardless of use • based on a log of machine use and clean when granules start to change color+ (the log will provide historical data to determine a more accurate schedule) • when 2/3 of canister has changed color • based on a log of machine use and choosing an arbitrary length of time (such as 8 hours, 10 hours, 20 hours, etc.) to change the absorber granules Regardless of the method used, it is important to pay attention to the capnograph, the hardening of the granules as they exhaust, and changes in the patient such as increased respiratory rate or red mucous membranes to determine if CO_2 granules are still effective.
Endotracheal tubes	Inflate cuff prior to cleaning. Clean after every use with gentle scrub brush. Clean with 0.2% chlorhexidine solution. ETTs that have a "mucus plug," other fluid remnants, or infectious agent present upon extubation should be discarded.
Machine parts	Careful removal of inhalation/exhalation domes, flutter valves, CO_2 absorber. Clean with warm soapy water made with mild detergent. Chlorhexidine scrub can also be used. If parts of the machine have become grossly contaminated, they should be sterilized (e.g., breathing tubes used for a dog with parvovirus).
ECG leads/machine	Wipe off excess gel with a dry gauze after use. Store attached to a cotton-tipped applicator stick or tongue depressor to avoid tangling of the leads.
Capnograph	Clean unit and sensors prior to each use by wiping with a mild soap solution (e.g., dilute chlorhexidine 1oz/1 gal water) and soft cloth. Disinfect surfaces (but not the screen) prior to use by wiping with isopropyl alcohol. The sample line is intended to be used a single time and then disposed. Adapters can be disinfected so do not need to be treated as a disposable item.[1]
Pulse oximeter – machine	Wipe off outside of machine with a soft towel moistened with disinfectant (e.g., chlorhexidine). NEVER spray disinfectant onto machine.[2]
Doppler	To clean the main unit, use a little water and wipe with a soft dry cloth. Remove the Doppler gel from the probe head after use. It is not recommended to clean the crystal probe of the Doppler with anything other than a gauze pad moistened with water.

Equipment	Suggested Maintenance/ Cleaning
Probes	After each use, wipe off with a gauze sponge moistened with 70% isopropyl alcohol. Do NOT autoclave, ethylene oxide sterilize, or immerse probes.
Esophageal stethoscope	Cap end that attaches to stethoscope with syringe cap (prevents water from entering lumen of the device). Disinfect as is done for endotracheal tubes.
Blood pressure cuffs	Wipe off with towel/gauze moistened with disinfectant. Do NOT spray cuffs. Do NOT allow water to get into the tubing of the cuff.
Arterial catheter – transducer	Calibrate after each use if not a disposable unit. Refer to manufacturer's recommendations for calibration procedure.
Rectal/esophageal thermometer	If probe cover was used, little maintenance is needed. Otherwise wipe off outside of thermometer with 70% isopropyl alcohol and a gauze sponge.
Ventilator	Supply hose, bellows housing, and bellows should be cleaned with a mild soapy solution, rinsed well, and dried thoroughly before reassembly. Frequency should be regularly based on frequency of use. Bellows diaphragm should be inspected for tears or cracks and replaced as needed; bellows flutter valve and ring seals should also be inspected for cracks, tears, and proper fit/seal. NO part should be cleaned with alcohol or steam or ethylene oxide sterilized.[3]
Laryngoscope blade and handle	Blade and handle should undergo low level disinfection with alcohol. Wiping off the blade and outside of the handle with 70% isopropyl alcohol is sufficient. In the event of an exposure to an infectious organism (calici virus, *Chlamydia* or *Bordetella* bacteria, or any other respiratory infectious agent), the laryngoscope should be sterilized. Removal of the batteries from the handle and the lamp from the blade will permit the blade and handle to then be steam autoclaved.

*Courtesy of Teri Raffel Kleist, CVT, VTS (Surgery)
+The color can change back to white over time, so color change becomes less reliable if the granules set for a prolonged time or multiple people are checking them.

ENDNOTES

1. Personal communication, Mike Albano, Service Tech with BioMedic Inc., November 2014. petMAP user's manual, September 2009, p. 22.
2. SurgiVet V3402 handheld digital pulse oximeter operation manual, version 2, January 2000, pp. 4–6.
3. SurgiVet SAV2500 ventilator V725000 operation manual, version 2, June 2007, pp. 5–10.

Appendix C

PATIENT EVALUATION

PATIENT HISTORY

Gathering historical information about a patient's health helps the anesthesia team anticipate complications that may arise during the anesthesia period. For example, vaccine and drug reaction history is important to document in the patient's record and relay to the anesthesia team. Confirming the type of procedure being performed, compliance with fasting recommendations, and the patient's condition the day of the procedure ensure that the patient is prepared for anesthesia.

The client should be informed about the type of procedure that will be performed, the expected outcome and potential complications, and estimated time of the procedure. Consent forms and fee estimates need to be approved/signed by the client. Obtaining emergency contact information and answering client questions is also important.

PHYSICAL EXAMINATION

A thorough physical examination (PE) must be performed on each patient to determine the best possible anesthesia protocol, with each body system methodically examined and further diagnostics performed if indicated by anomalies detected. If abnormalities are found on PE, the veterinarian should be informed and further discussion is warranted on how to best proceed. Many times, further testing will be necessary, whether it be performing hematological analyses, diagnostic imaging, or further medical testing. The goal is to have as much information as possible on the animal's physiological status, and to have them in the best physical condition possible prior to administration of any anesthesia drugs.

The PE provides a "snapshot in time" of a patient's status.[1] Thorough and consistent execution of the PE is necessary for the veterinary team to determine which procedures will be performed on a patient. For example, if coagulation tests reveal some type of clotting abnormality, a patient may not undergo a liver biopsy until the condition is further investigated. The veterinarian will use the information gained from the PE to determine if the concerns are either too numerous or any single concern is too critical to anesthetize the patient on that day. Many states require that the veterinarian perform a PE before surgery is performed. Regardless of who performs the PE, the method of acquiring objective information should be consistent and in line with how the veterinarian wants it collected.

It is important to be sure that the animal being seen is the actual patient to be anesthetized. Horses and especially cattle in herds may not be individually tagged or the permanent identification may be hard to find or read. Writing the signalment (species, breed, age and date of birth, sex and reproductive status) in a medical record gives very specific identification information that provides legal documentation in case of litigation, disease outbreak, and negligent death or neglect cases.[2] Weighing the animal is extremely important along with a Body Condition Score (BCS) assessment as this information can affect the anesthetic protocol decision based on fat distribution or determining the type of anesthetic circuit system that will be used.[3] Pounds should be converted to kilograms since anesthetic drugs are calculated in this manner. Both animal weight and BCS should be written in the medical record.

After signalment information is gathered, the animal should be viewed as a "whole." If possible, look at the animal from a distance. Notice any type of symmetry concerns, obvious skin lesions or physical deformities, and respiratory effort. This is a good time to determine the level of pain and stress currently present. Temperament of the patient plays a significant role in selecting a proper sedation or premedication protocol.

One method of performing a PE is a cranial/caudal approach. Some people prefer this method due to the fact that some species are not as fastidious as others in elimination of bodily wastes.

The PE using a cranial/caudal approach starts with examination of the head. Cranial nerves should be assessed first. Further evaluation must include symmetry of the eyes, ears, and nares, as any abnormalities may indicate a neurologic condition. There should be no head tilt. In addition, eyes should not have any type of exudate and should symmetrically react to a light source and be clear and in a normal position within the orbit. Ears should be clean and dry inside the canal, pinnae should function normally as they react to sound, and there should be no exudate or strong smell. Dependent on species, the nares may be dry or moist, but none should have discharge and the surface should be clean. Parotid, submandibular, and mandibular lymph nodes should be palpated and evaluated.

Mucous membranes, whether evaluated at the ocular conjunctiva or the mouth, should be pink and moist.[4] Individual animals may have various shades of pink that are considered normal. Examples of abnormal mucous membrane colors include:

- Pale mucous membranes, which may indicate anemia, bleeding, shock due to hypovolemia, or pain. In cases of hypovolemia (shock or poor circulation from pre-existing disease processes such as diabetes mellitus or heart disease), the extremities might also be cold.
- Brick red mucous membranes, which may indicate sepsis or endotoxic shock.
- Yellow (icteric) mucous membranes, which may indicate bilirubin accumulation and possible hepatic disease/dysfunction.
- Chocolate brown mucous membranes, which may be due to methemoglobinemia and could indicate acetaminophen toxicity.
- Petechiae (red dots or splotching) on mucous membranes, which may indicate a coagulation disorder.

- Cyanotic mucous membranes, which may be the result of severe or prolonged shock. This is a very poor prognostic indicator and the animal must receive immediate attention and the veterinarian notified.

A generally accepted location to evaluate capillary refill time (CRT) is the gingiva, regardless of species. After blanching the mucous membranes against the gingiva, the tissue should refill in less than 2 seconds. Increased CRT may indicate perfusion problems and is a significant concern for patients undergoing anesthesia.

The patient's mouth should be carefully examined. Any loose or missing teeth or pathology should be reported on the form. Any concerns in this area can have a significant influence on anesthetic induction. Endotracheal intubation may be complicated by variations in anatomy and pathologic changes in the oral cavity, as passage of the endotracheal tube needs to be performed quickly during the induction period.

When following the cranial to caudal approach, the next area to examine is the neck. Any observed abnormalities, such as a jugular pulse or a jugular vein that is abnormal in appearance, could indicate cardiac problems or concerns on venous access. In small animals, jugular access may not be a major concern, but for large animal species this is the preferred vein for administering induction drugs. Check hydration status at the neck or shoulder of the patient. Tenting the skin and its response to releasing the tent can give a general indication on how well the patient is hydrated. Blood testing, with a packed cell volume (PCV)/total protein (TP), is more definitive. Prescapular lymph nodes should be palpated and evaluated.

Auscultation of the thorax helps evaluate the respiratory and cardiac systems. Heart rate and rhythm can be assessed on the left wall of the thorax at the point of the elbow, usually with the stethoscope bell behind the elbow. Normal heart sounds should be a regular rhythm of "lub-dub." Any murmurs, muffled, or dampened sounds can be indications of cardiac or respiratory disease and may affect any anesthetic event.

While auscultating the heart, the respiratory rate can be assessed. During this time observe for any edema in the ventral neck and thorax region. Move to the thoracic regions on both sides of the animal and assess the sounds. Normal lung sounds are quiet, though inspiration is louder than expiration and ventral lung areas are louder than dorsal. Listen for crackles, wheezes, and rales in each lung field. It may be necessary to listen to tracheal sounds if there was any observation of nasal discharge, cough, or history of respiratory disease. Anesthesia puts significant stress on both the respiratory and cardiac body systems; therefore, performing a thorough examination of these two systems helps assess patient anesthetic risk.

Abdominal auscultation can provide some clues of anesthetic risk. A normally functioning gastrointestinal tract means that the patient should be well hydrated and in a catabolic nutritional state. Any abnormalities could mean perfusion problems or a lengthy surgical recovery period. Listen for up to one minute in each of the four abdominal quadrants in large animal species for borborygmi. There should be at least one and not more than four instances of borborygmi sounds. More than four sounds per minute could indicate current or impending increased gastrointestinal (GI) motility possibly leading to diarrhea. Dependent on species, the prefemoral and popliteal lymph nodes should be palpated and evaluated.

The reproductive organs should be assessed in female, male, spayed, and castrated animals. Any abnormalities could indicate an infectious disease or chronic disease condition. While assessing these regions, remember to observe any abnormalities of the urogenital tracts. Urine staining could indicate an ambulation concern or increased recumbency.

Examination of the legs and feet should have started when first beginning to work with the patient. The animal can be observed when moving to the treatment area or from stall to stanchion. Ambulation problems can lengthen the recovery period, especially when moving large animal species from lateral to sternal and standing during the recovery period. Any observed swelling or stiffness could interfere with proper positioning during anesthesia and recovery. All abnormalities must be recorded accurately on the PE form.

A final step for a PE is body temperature, and when to take a patient's temperature depends on the patient. Some individual animals become agitated when a rectal temperature is taken; therefore, leaving this task for last keeps the patient as quiet as possible and decreases the chance that agitation will change any observations or auscultation findings. If the patient is agitated during the PE, it is common to see a spike in body temperature. When there is an elevation in an animal's body temperature approximately one degree above the high end of normal, it may be prudent to allow the patient scheduled for anesthesia to relax for approximately 30 minutes to an hour and then recheck it prior to determining American Society of Anesthesiologists (ASA) physical status (i.e., determining if the patient has a true fever). It is important to get the temperature prior to collecting diagnostic samples when possible.

LABORATORY TESTS

The minimum laboratory tests that should be performed varies with age and health status of the patient. Most young and otherwise apparently healthy patients can have the "Big 4" or "QATs" (quick assessment tests) performed prior to anesthesia. The QATs include PCV/TP/Blood Glucose/BUN (from Azostick®). If an invasive surgery is planned, a complete blood count (CBC) is warranted to determine if adequate number of platelets is present. Additional tests may be warranted based on patient status, which could include serum chemistry panel, electrolyte tests, urinalysis, coagulation panel, blood gas analysis, thoracic radiographs, or electrocardiogram. Many hospitals have specific blood work panels for anesthetic patients based on their physical status or age. Grouping of these panels may follow the manufacturer's blood analyzer's panels that are grouped to keep costs low. The ASA physical status classifications can provide guidance with categorizing which status of patients need specific tests performed.

Hydration, anemia, systemic infection, electrolyte balances, and respiratory, cardiac, kidney, and liver status are all concerns that are addressed with preanesthetic laboratory testing.

Packed Cell Volume/Total Protein

The PCV can be run on an automated machine as part of the CBC or is run from a single heparinized microhematocrit tube of whole blood spun in a centrifuge. This test reveals the ratio of red blood cells to total amount of blood. The level of red cells gives an indication as to the oxygen carrying capacity of the blood. If the PCV is elevated outside of normal parameters, this may indicate dehydration, but further testing is warranted. If the PCV is decreased below normal limits, this may indicate anemia. Again, further testing is warranted.

A subjective evaluation can be made from observing the buffy coat of white blood cells that lies between the plasma and red blood cells. With practice, the anesthetist can learn to evaluate an increased amount of white blood cells. This finding may warrant further testing, such as a CBC. Lastly, the TP is determined using the same microhematocrit blood tube. After breaking the tube above the white blood cell layer, the plasma can be placed on a refractometer and a reading in g/dL can be observed. When this level is within normal limits, the blood has sufficient levels of protein to carry anesthetic drugs in the bloodstream to and from the cells. If the value is below normal limits, this, along with a decreased PCV, may indicate whole blood loss. If the value is above normal limits, this may indicate dehydration and can adversely affect blood pressure, drug elimination, renal and cardiac output, and tissue perfusion.

Complete Blood Cells

A CBC provides information on the amount of red and white blood cells and platelets per μL and also the ratio of white blood cells in the differential. In addition, the size and oxygen carrying ability of the red blood cells is provided. While a machine-generated CBC is helpful, the counting method of the machine can determine the quality of the results. Keeping this in mind, it is helpful to make and stain a blood smear then manually evaluate the morphology of the red and white blood cells and platelets. With practice this can be performed quickly, and the information provided by this task can be invaluable to patient outcomes.

Blood Chemistry and Electrolytes

Blood chemistry and electrolyte tests can provide a good baseline for basic body functions. Most often a chemistry panel determines enzyme levels from the liver and kidneys, as well as pancreatic function and other metabolites. Electrolyte tests provide insight on the levels of substances that are important in maintaining fluid balance, regulating body functions, and conducting electrical impulses in the body. Potassium facilitates muscle contraction and nerve electroconductivity. Calcium also helps with muscle contraction and nerve function as well as blood clotting. Sodium plays a role in nerve function but is important in balancing extracellular fluid. Some analyzers have preloaded panels so the test choices may be limited, whereas other machines allow a wider variety of testing options. Parameters are typically set on the machine and may or may not be species-specific beyond the basics of canine, feline, and equine. Any patient values outside of normal parameters need to be evaluated by the veterinarian before an anesthetic protocol is initiated.

Coagulation Tests

Blood tests for an anesthetic patient may include coagulation tests. These tests can include prothrombin time (PT), activated partial thromboplastin time (aPTT), d-dimers, fibrin degradation products (FDP), and activated clotting time (ACT). Choosing which coagulation tests to perform is determined by whether the concern is about acquired or inherited coagulation problems, if information is needed about the intrinsic or extrinsic pathways, the test's cost, and availability of specific tests. The veterinarian must review any abnormal results.

Urinalysis

Urinalysis (UA) can provide a wealth of information on body function. The three parts of the UA are the macroscopic (gross) examination, biochemical analysis, and microscopic (sediment) evaluation. Each of these parts can not only provide information on the function of the renal system but also other body systems such as the gastrointestinal tract, liver, and pancreas as well as water balance and electrolytes. For example, glucose found on the biochemical analysis may indicate diabetes mellitus in a patient while high levels of protein may indicate kidney disease. A properly functioning urinary system helps the patient eliminate anesthetic agents, maintain blood pressure, and regulate electrolyte and water balance. Any abnormal results should be reported to the veterinarian.

Radiography

Thoracic radiographs may be necessary in anesthetic patients that have any type of abnormal respiratory or cardiac auscultation, history of respiratory disease, trauma, or any other disease process that affect the pulmonary or cardiovascular systems. Radiography can provide information on these systems so that preventive measures can be taken before initiating anesthesia, whether that may be using a ventilator, changing surgical approach, or delaying a procedure until the patient is stable.

Electrocardiogram

An electrocardiogram (ECG) may be indicated for patients that have an abnormal heart rhythm, murmurs or other heart sounds on PE, geriatric patients, and in patients with a known or suspected heart condition, blood electrolyte abnormalities, or other life threatening trauma or condition. Evaluating the electroconductivity of the heart can give life-saving information to the anesthetist so that adjunctive medications, supplemental oxygen, or IV fluids can be administered before, during, or after surgery (see Chapter 7 [Anesthesia Monitoring] and Chapter 12 [Anesthetic Complications]).

ENDNOTES

1. Lien, L., Loly, S., & Ferguson, S. (2014). *Large animal medicine for veterinary technicians*. Ames, IA: Wiley, pp 78–79.
2. Bassert, J., & McCurnin, D. (2017). *McCurnin's clinical textbook for veterinary technicians*, 9th ed. St. Louis, MO: Elsevier, pp. 83–94.
3. Thomas, J., & Lerche, P. (2016). *Anesthesia and analgesia for veterinary technicians*, 5th ed. St. Louis, MO: Mosby Elsevier, pp. 15–16.
4. Jack, C., Watson, P., & Donovan, M. (2014). *Veterinary technician's daily reference guide, canine and feline*, 3rd ed. Ames, IA: Blackwell Publishing, pp. 65–69.

Appendix D

ASA PHYSICAL STATUS CLASSIFICATION SYSTEM

ASA Physical Status Classification System	Definition	Risk Level and Description
ASA I	A normal healthy patient	• Minimal risk • Normal healthy animal, no underlying disease. For example, a healthy patient undergoing an ovariohysterectomy or neuter.
ASA II	A patient with mild systemic disease	• Slight risk, mild disease present • Animal with slight to mild single systemic disturbance for which the animal is able to compensate. For example, obesity, dental disease, and cranial cruciate ligament rupture.
ASA III	A patient with severe systemic disease	• Moderate risk, obvious disease present affecting one or more body systems • Systemic disease or disturbances that are compensated and not causing severe functional limitations. For example, chronic kidney disease, controlled diabetes mellitus, compensated cardiac disease.
ASA IV	A patient with severe systemic disease that is a constant threat to life	• High risk, significantly compromised by disease and functional limitations • Current condition is a threat to life. For example, a horse with colic, a dog with gastric dilatation volvulus (GDV), a dog with congestive heart failure.
ASA V	A moribund patient who is not expected to survive without the operation	• Extreme risk • Patient not expected to survive without the operation. For example, patients from category IV who are in terminal stage of disease (a horse with colic in profound shock, an unresponsive GDV patient, a dog with end-stage heart failure).

"E" denotes emergency and is assigned to any of the I–V categories when the anesthesia is an emergency (for example, IIIE).

Appendix E

IV CATHETER TECHNIQUES

IV CATHETER TECHNIQUE IN DOGS AND CATS

- Palpate area to locate vessel (the cephalic vein in this example); generously shave the area (see Figure 4-10). Be sure to keep the clipper blade flat to the skin to avoid clipper burn and shave against the hair.
- Clean area of loose fur.
- Wash hands, and prep the site by cleaning the skin with an antiseptic scrub using a target scrubbing pattern (see Figure 4-17).
- Use proper aseptic technique when placing an IV catheter. Consider wearing sterile gloves when placing IV catheters (photos show placing a catheter with and without gloves for reference). Never touch the IV catheter lumen with bare hands as this portion is inserted into the vein.
- Unwrap the IV catheter and gently move it off the stylet to ensure the catheter will slide off the stylet without resistance. Pull the IV catheter back onto stylet.
- Open the injection cap and place on a sterile surface (alternately open the T-port and flush with saline; then clamp it off to keep saline in the IV line to prevent injecting air into the vessels once the T-port is connected and IV fluids are started).

- Hold paw/limb with nondominant hand placing your thumb alongside the vessel (if possible) to stabilize it.
- Remove the cap on the end of the catheter and hold it with the bevel up so it is comfortable in your hand.
- Approach the top of the vessel at a 30 degree angle, then insert catheter tip through the skin and into the vessel in one smooth motion (Figure a).
- Once a "flash back" of blood is seen in the catheter hub, reduce the angle of the catheter and advance it approximately 1 mm to make sure the entire bevel of the catheter is within the lumen of the vein (Figure b).
- If blood is still flowing back into the catheter hub, reduce the catheter angle so it is parallel to the limb, and advance it into the vein while keeping the stylet stationary (Figure c). When moving the catheter always keep the catheter and stylet together and check to make sure blood is still flowing back into the hub of the stylet indicating it is still within the lumen of the vein.
- The catheter is then deployed off the stylet (making sure not to pull back on the stylet) and into the vessel (Figure d). When the catheter is moved off the stylet, a column of blood can be seen in the catheter.

(a)

Courtesy of Teri Raffel Kleist, CVT, VTS (Surgery)

(b)

Courtesy of Teri Raffel Kleist, CVT, VTS (Surgery)

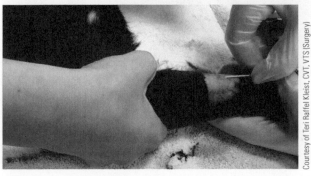

(c)

Courtesy of Teri Raffel Kleist, CVT, VTS (Surgery)

(d)

Courtesy of Teri Raffel Kleist, CVT, VTS (Surgery)

- Once the catheter is fully threaded into the vessel, remove the stylet (Figure e).
- Ask the restrainer to stop holding off the vein and place a finger or thumb over the area of the limb where the end of the catheter enters the skin to stop blood from flowing out of the catheter.
- Place either an injection cap/prn (Figure f), T-port, or administration set on the hub of the catheter (Figure g).
- Place thumb over the injection cap and gently push it toward the catheter while carefully cleaning blood from the area.
- The thumb of the nondominant hand is used to control the catheter during taping. The first piece of tape is thin (tear a 1 inch piece of tape length-wise in half) and placed "sticky-side-up" underneath the hub

of the catheter leaving only ½ to 1 inch on one side and the remainder on the other (Figure h).
- Pull the long portion of the tape over the hub of the catheter to meet the shorter piece and continue to gently wrap it around the limb (Figure i) so that the tape is well adhered to the catheter hub. Pulling the "long over short" avoids a fold in the tape that could compromise the stability of the catheter.
- Gently pinch the two pieces together making the tape snug around the hub. Wind the remaining length of tape around the limb to meet the tape on the catheter hub, but not too tightly to prevent swelling of the paw (Figure j). Swelling is due to a decrease in venous return from the distal limb.

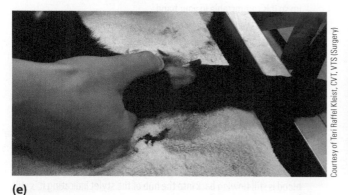

(e)

Courtesy of Teri Raffel Kleist, CVT, VTS (Surgery)

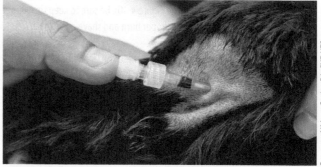

(f)

Courtesy of Kristen Cooley, BA, CVT, VTS (Anesthesia/Analgesia)

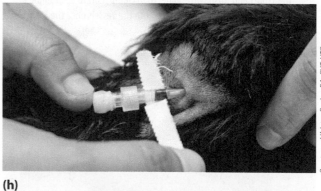

(g)

Courtesy of Teri Raffel Kleist, CVT, VTS (Surgery)

(h)

Courtesy of Kristen Cooley, BA, CVT, VTS (Anesthesia/Analgesia)

(i)

Courtesy of Teri Raffel Kleist, CVT, VTS (Surgery)

(j)

Courtesy of Teri Raffel Kleist, CVT, VTS (Surgery)

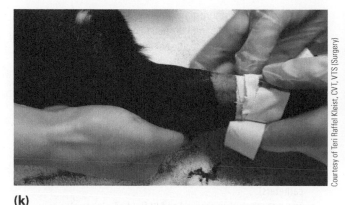

Courtesy of Teri Raffel Kleist, CVT, VTS (Surgery)

(k)

Courtesy of Teri Raffel Kleist, CVT, VTS (Surgery)

(l)

For a short-term anesthesia catheter, the thick piece of tape can now be placed sticky-side-down under the hub of the catheter and snug against the first piece of tape with about ½ to 1 inch on one side of the catheter and the remainder on the other side. Wrap this piece around with gentle pressure that will push the catheter into the vessel (Figure k). When the long piece meets the shorter piece, bring it up to cover the insertion site. Flush the IV catheter with saline. If administering IV fluids, a small loop should be made with the IV tubing (making sure it is not twisted or kinked). While holding the IV tubing in place, use tape to secure it to the patient's limb (Figure l).

For a long-term peripheral catheter, place a T-port onto the catheter hub and take the second piece of narrow tape and begin wrapping it around the T-port distal to the tubing. Place a Telfa® pad or nonstick dressing over the insertion site and gently wrap Kling® gauze around the entire catheter starting under the T-port and wrapping it proximally. Then take the elastic cloth tape (Elasticon®) and wrap it in the same manner making sure it is adhered to the hair below and above the Kling® gauze to keep it from bunching. Take the T-port and secure it to the elastic cloth tape without occluding the tubing. Place a piece of tape with your initials and the date and time the catheter was placed on the elastic cloth tape.

OVER-THE-NEEDLE JUGULAR CATHETER TECHNIQUE IN LARGE ANIMALS

- Shave the area over the jugular groove with a #40 clipper blade (choose an area in the middle ⅓ of the neck whenever possible).
- Brush away hair or wool.
- Perform a surgical scrub of this area, using a target scrubbing technique. Scrubbing time should take at least 5 minutes and should be performed wearing clean exam gloves. Just before the final scrub, a 1 to 3 mL bleb of 2% lidocaine or other local anesthetic can be instilled intradermally over the area of the jugular vein where the catheter will be placed. Use a needle appropriate to the species and age of the patient. Finish the surgical scrub using chlorhexidine or betadine scrub followed by sterile saline. Do not touch this area without wearing sterile gloves.
- Open the sterile glove pack, folding back the edges so that the paper creates a sterile field. Open the IV catheter so that the catheter can be

dropped aseptically on the sterile field. Do likewise for any adaptics (T-port, infusion cap, suture, etc.).
- Don sterile gloves using the open gloving technique.
- With the stylet in place and bevel up, insert the IV catheter into the vein until a "flash back" of blood is seen at the top of the hub. Occlusion of the vessel should either be done by an assistant or with the back of the nondominant clenched hand.
- Thread the IV catheter gently into the jugular vein until the hub reaches the level of the skin; do not advance the stylet (Figure m). Excessive drag can occur in animals with thick skin (such as bulls, camelids, and severely dehydrated patients). In these cases the stylet/catheter must be advanced into the vein further than the "flash back" position before threading the catheter into the vein.

(m)

- Remove the stylet.
- Place any adaptics on the end of the catheter. These adaptics, such as T-ports, must be of sufficient inside diameter. If they are not, they may restrict flow of fluids and should not be used. These additional items should be also be sutured in place (Figure n; see description below).
- Flush the catheter with several mL of normal saline.
- Suture the catheter as described:
  - Pinch a small amount of skin vertically and directly under the catheter hub; lift the skin.
    - If using a swaged suture, pierce a length of swaged 00 suture on a straight needle through the pinched skin.
    - If using a non-swaged suture, pierce a 12 gauge 1 ½" needle through the pinched skin. Release the skin, thread a length of large sized suture material (such as 00) through the 12 gauge needle starting from the hub toward the needle tip, and carefully pull the needle out of the skin and off the suture. Safely discard the needle.
  - Release the skin, tie a surgeon's knot to secure the catheter. Do not overly pinch the skin and cause patient discomfort. Trim excess material.
- Super glue may also be used to further secure the catheter but on the catheter hub ONLY (Figure o). Glue may disintegrate the catheter material producing a hole or detaching the hub from the catheter (polypropylene or polyurethane type catheters).
- The catheter is now secured and IV fluid administration may begin.

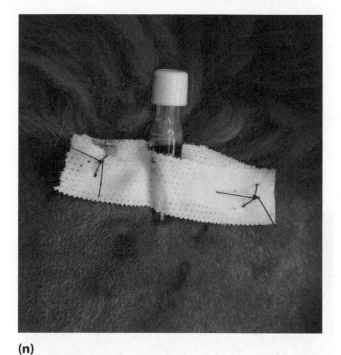

(n)

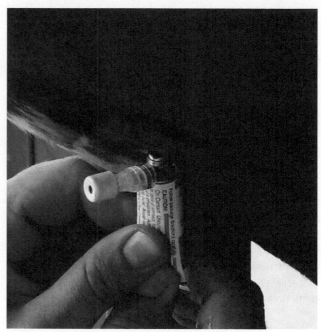

(o)

Appendix F

MATH REVIEW

MEASUREMENT SYSTEMS

Measurement is the use of standard units to determine the weight, length, or volume of a substance. The units of measurements used to calculate drug doses are the household, apothecary (not described), and metric systems. The metric system consists of units of measure that differ from each other by powers of ten. The base units of the metric system are the meter (length), liter (volume), and gram (weight). Prefixes are added to the base unit to denote the size of the metric unit (Table F-1).

Metric Conversions Guide

To convert (change) from one unit to another unit within the same measurement system, equivalents must be memorized and then multiplied or divided based on the desired unit of measure. Conversion between one unit and another involves the use of a conversion factor (number used to either multiple or divide to change the measurement from one unit of measure to another). A conversion factor always has a value of 1. To convert from a larger to a smaller unit of measure, multiply by the conversion factor using dimensional analysis (commonly called unit cancellation). To convert from a smaller to a larger unit of measure, divide by the conversion factor. For example, to convert 1.5 grams to milligrams involves multiplication:

$$(1.5 \text{ g})(1,000 \text{ mg/g}) = 1,500 \text{ mg}$$

To convert 4500 milligrams to grams involves division:

$$(4,500 \text{ mg})(1 \text{ g}/1,000 \text{ mg}) = 4.5 \text{ g}$$

TABLE F-1 Metric Prefixes and Values

Prefix	Value	Decimal Representation	Fractional Representation
micro- (mc or μ)	One millionth of a unit	0.000001 base unit	1/1,000,000 base unit
milli- (m)	One thousandth of a unit	0.001 base unit	1/1,000 base unit
centi- (c)	One hundredth of a unit	0.01 base unit	1/100 base unit
kilo- (k)	One thousand units	1,000 base unit	

Sometimes calculating drug doses or preparing solutions calls for conversion between measurement systems (Table F-2 and Table F-3). The differences between systems can be bridged with conversion to a single system. Dimensional analysis is used to convert between systems using conversion factors. Conversions factor for different measurement systems are listed in the tables below:

For example, the weight of an animal needs to be converted from pounds to kilograms in order to calculate many drug doses. To convert between pounds and kilograms for a 66 pound dog, the following conversion is made:

$$\left(\frac{66 \text{ lb}}{1}\right)\left(\frac{1 \text{ kg}}{2.2 \text{ lb}}\right) = 30 \text{ kg}$$

TABLE F-2 Conversions within the Household System

Unit	Abbreviation	Equivalents
Cup	C	1 C = 8 fl oz
Drop	gt (plural gtt)	60 gtt = 1 t; 180 gtt = 1 T
Ounce (fluid)	oz	2 T = 1 fl oz
Ounce (weight)	oz	16 oz = 1 lb
Pint	pt	1 pt = 2 C
Pound	lb	1 lb = 16 oz
Quart	qt	1 qt = 32 oz
Tablespoon	T (tbs)	1 T = 3 t
Teaspoon	t (tsp)	3 t = 1 T

TABLE F-3 Conversions between the Household and Metric Systems

Household Unit		Metric Unit
1 inch	=	2.54 centimeters
1 ounce	=	30 milliliters
2.2 pounds	=	1 kilogram
1 pint	=	473 milliliters
1 gallon	=	128 fluid ounces
1 teaspoon	=	5 milliliters

DOSE CALCULATIONS

To calculate the total dose of a drug, you need the animal's weight and the amount of a drug needed per unit of the animal's weight (the drug dosage). The dose is the amount of drug given to the animal in one administration.

Dose in mg

An example of calculating a dose in mg is as follows: The induction dosage of Telazol® in dogs and cats is 2 to 5 mg/kg IV and in horses is 1 to 3 mg/kg IV. Using a dosage of 2 mg/kg, how many mg should be given to a female English Cocker Spaniel dog weighing 22 lb?

Step 1: If the animal's weight and the medication dosage are not in the same system of measure, convert the weight so that both measures are expressed in the same unit. In this example, pounds must be converted to kg. From Table F-3, choose the appropriate conversion factor of 1 kg = 2.2 lb. The animal weighs 22 lb; therefore, use the conversion factor 2.2 lb = 1 kg to convert the weight into kg units.

Step 2: Knowing that 2.2 lb = 1 kg, create a conversion factor. *Conversion factors are used to move between units and always have a value of one.* Because they have a value of one, conversion factors do not change the value of the end product. Write this conversion factor as 2.2 lb/1 kg.

Step 3: The next step is to determine in which format to write the conversion factor. There are two choices: 2.2 lb/1 kg or 1 kg/2.2 lb. In this case, cancel out lb to end up with kg. Therefore, lb is on the bottom, and kg is on the top of the conversion factor.

Step 4: Set up the conversions in an equation. Start with what is known (22 lb), and use the conversion factor to set up the equation.

$$\left(\frac{22\text{ lb}}{}\right)\left(\frac{1\text{ kg}}{2.2\text{ lb}}\right) = 10\text{ kg}$$

Step 5: Now perform the calculation by taking the weight of the animal in kg and multiplying it by the dosage. Remember, values on the bottom of the conversion factors are divided into the numbers on the top. Values on the top of the conversion factors are multiplied by the numbers on the top. In this case, take 10 kg, multiply it by 2 mg, and then divide by 1 kg. Keep in mind that units should cancel as well to give the desired unit at the end.

$$\left(\frac{10\text{ kg}}{}\right)\left(\frac{2\text{ mg}}{1\text{ kg}}\right) = 20\text{ mg dose}$$

Step 6: To make sure the answer is correct, prove the work by looking at the units involved. If the animal needs 2 mg of drug for every 1 kg of body weight, and the animal weighs 10 kg, it makes sense that this patient should receive a 20 mg dose of drug. Always double check the answer to avoid errors in calculations.

The entire formula used when an animal's weight is in pounds and the dosage is based in kilograms is to take the animal's weight (in lb) divide by 2.2 (conversion factor to convert to lb to kg), then multiply by amount of drug (in this case, in mg) per kilogram.

$$\frac{Wt\,(\text{lb})}{2.2} \times \text{dose}\,(\text{units/kg}) = \text{dose to animal}$$

This calculation determined the strength of each dose and the total amount of medicine to administer/dispense. To measure the dose for the patient, it needs to be in a measurable unit. For example, milliliters can be dispensed as a unit.

Using a dosage of 2 mg/kg, how many mg should be given to a F/S DSH cat weighing 8 lb?

Step 1: If the animal's weight and the medication dosage are not in the same system of measure, convert the weight so that both measures are expressed in the same unit. In this example, pounds must be converted to kg. From Table F-3, choose the appropriate conversion factor of 1 kg = 2.2 lb. The animal weighs 8 lb; therefore, use the conversion factor 2.2 lb = 1 kg to convert the weight into kg units.

Step 2: Knowing that 2.2 lb = 1 kg, create a conversion factor. *Conversion factors are used to move between units and always have a value of one.* Because they have a value of one, conversion factors do not change the value of the end product. Write this conversion factor as 2.2 lb/1 kg.

Step 3: The next step is to determine in which format to write the conversion factor. There are two choices: 2.2 lb/1 kg or 1 kg/2.2 lb. In this case, cancel out lb to end up with kg. Therefore, lb is on the bottom, and kg is on the top of the conversion factor.

Step 4: Set up the conversions in an equation. Start with what is known (8 lb), and use the conversion factor to set up the equation.

$$\left(\frac{8\text{ lb}}{}\right)\left(\frac{1\text{ kg}}{2.2\text{ lb}}\right) = 3.6\text{ kg}$$

Step 5: Now perform the calculation by taking the weight of the animal in kg and multiplying it by the dosage. Remember, values on the bottom of the conversion factors are divided into the numbers on the top. Values on the top of the conversion factors are multiplied by the numbers on the top. In this case, take 3.6 kg, multiply it by 2 mg, and then divide by 1 kg. Keep in mind that units should cancel as well to give the desired unit at the end.

$$\left(\frac{3.6\text{ kg}}{}\right)\left(\frac{2\text{ mg}}{1\text{ kg}}\right) = 7.2\text{ mg}$$

Step 6: To make sure the answer is correct, prove the work by looking at the units involved. If the animal needs 2 mg of drug for every 1 kg of body weight, and the animal weighs 3.6 kg, it makes sense that this patient should receive a 7.2 mg dose of drug. Always double check the answer to avoid errors in calculations.

Tech Tip

Veterinary professionals should always question dose calculations that seem unreasonably large or small for the size animal for which the drug is prescribed.

Using a dosage of 2 mg/kg, how many mg should be given to an Arabian gelding weighing 1,000 lb?

Step 1: If the animal's weight and the medication dosage are not in the same system of measure, convert the weight so that both measures are expressed in the same unit. In this example, pounds must be converted to kg. From Table F-3, choose the appropriate conversion factor of 1 kg = 2.2 lb. The animal weighs 1,000 lb; therefore, use the conversion factor 2.2 lb = 1 kg to convert the weight into kg units.

Step 2: Knowing that 2.2 lb = 1 kg, create a conversion factor. *Conversion factors are used to move between units and always have a value of one.* Because they have a value of one, conversion factors do not change the value of the end product. Write this conversion factor as 2.2 lb/1 kg.

Step 3: The next step is to determine in which format to write the conversion factor. There are two choices: 2.2 lb/1 kg or 1 kg/2.2 lb. In this case, cancel out lb to end up with kg. Therefore, lb is on the bottom, and kg is on the top of the conversion factor.

Step 4: Set up the conversions in an equation. Start with what is known (1,000 lb), and use the conversion factor to set up the equation.

$$\left(\frac{1,000 \text{ lb}}{}\right)\left(\frac{1 \text{ kg}}{2.2 \text{ lb}}\right) = 454.5 \text{ kg}$$

Step 5: Now perform the calculation by taking the weight of the animal in kg and multiplying it by the dosage. Remember, values on the bottom of the conversion factors are divided into the numbers on the top. Values on the top of the conversion factors are multiplied by the numbers on the top. In this case, take 454.5 kg, multiply it by 2 mg, and then divide by 1 kg. Keep in mind that units should cancel as well to give the desired unit at the end.

$$\left(\frac{454.5 \text{ kg}}{}\right)\left(\frac{2 \text{ mg}}{1 \text{ kg}}\right) = 909 \text{ mg}$$

Step 6: To make sure the answer is correct, prove the work by looking at the units involved. If the animal needs 2 mg of drug for every 1 kg of body weight, and the animal weighs 454.5 kg, it makes sense that this patient should receive a 909 mg dose of drug. Always double check the answer to avoid errors in calculations.

Dose in mL

Using the example of the 22-lb female English Cocker Spaniel dog needing 20 mg of Telazol® again, how many mL of drug will the anesthetist administer?

Step 1: To determine the amount of each dose in mL, it is necessary to know the concentration of drug available. The clinic inventory contains a vial of drug that states the concentration of drug is 100 mg/mL.

Step 2: Determine the amount of mL per dose by dividing the 20 mg dose by 100 mg/mL (the concentration of drug).

$$\left(\frac{20 \text{ mg}}{}\right)\left(\frac{1 \text{ mL}}{100 \text{ mg}}\right) = 0.2 \text{ mL}$$

A 1-mL syringe is used to measure and administer 0.2 mL of drug.

Using the example of the 8 lb DSH cat needing 7.2 mg of Telazol® again, how many mL of drug will the anesthetist administer?

Step 1: To determine the amount of each dose in mL, it is necessary to know the concentration of drug available. The clinic inventory contains a vial of drug that states the concentration of drug is 100 mg/mL.

Step 2: Determine the amount of mL per dose by dividing the 7.2 mg dose by 100 mg/mL (the concentration of drug).

$$\left(\frac{7.2 \text{ mg}}{}\right)\left(\frac{1 \text{ mL}}{100 \text{ mg}}\right) = 0.07 \text{ mL}$$

A 1-mL syringe is used to measure and administer 0.07 mL of drug.

Using the example of the 1,000 lb Arabian horse needing 909 mg of Telazol® again, how many mL of drug will the anesthetist administer?

Step 1: To determine the amount of each dose in mL, it is necessary to know the concentration of drug available. The clinic inventory contains a vial of drug that states the concentration of drug is 100 mg/mL.

Step 2: Determine the amount of mL per dose by dividing the 909 mg dose by 100 mg/mL (the concentration of drug).

$$\left(\frac{909 \text{ mg}}{}\right)\left(\frac{1 \text{ mL}}{100 \text{ mg}}\right) = 9.0 \text{ mL}$$

A 10-mL syringe is used to measure and administer 9.0 mL of drug.

Practice calculating doses in mL for the following patients:
The dosage for dexmedetomidine for dogs, cats, and horses is 0.001 mg/kg IV. Two concentrations of dexmedetomidine are currently available (0.5 mg/mL and 0.1 mg/mL); your clinic stocks the 0.5 mg/mL concentration. Calculate the dose of dexmedetomidine for a 50 pound dog, 10 pound cat, and 1,200 pound horse.

For the 50 lb dog:

$$\left(\frac{50 \text{ lb}}{}\right)\left(\frac{1 \text{ kg}}{2.2 \text{ lb}}\right) = 22.73 \text{ kg}$$

$$\left(\frac{22.73 \text{ kg}}{}\right)\left(\frac{0.001 \text{ mg}}{1 \text{ kg}}\right) = 0.023 \text{ mg}$$

$$\left(\frac{0.023 \text{ mg}}{}\right)\left(\frac{1 \text{ mL}}{0.5 \text{ mg}}\right) = 0.046 \text{ mL}$$

(round up to 0.05 mL)

For the 10 lb cat:

$$\left(\frac{10 \text{ lb}}{}\right)\left(\frac{1 \text{ kg}}{2.2 \text{ lb}}\right) = 4.55 \text{ kg}$$

$$\left(\frac{4.55 \text{ kg}}{}\right)\left(\frac{0.001 \text{ mg}}{1 \text{ kg}}\right) = 0.0046 \text{ mg}$$

$$\left(\frac{0.0046 \text{ mg}}{}\right)\left(\frac{1 \text{ mL}}{0.5 \text{ mg}}\right) = 0.009 \text{ mL}$$

(round up to 0.01 mL)

For the 1,200 lb horse:

$$\left(\frac{1,200 \text{ lb}}{}\right)\left(\frac{1 \text{ kg}}{2.2 \text{ lb}}\right) = 545.45 \text{ kg}$$

$$\left(\frac{545.45 \text{ kg}}{}\right)\left(\frac{0.001 \text{ mg}}{\text{kg}}\right) = 0.545 \text{ mg}$$

$$\left(\frac{0.545 \text{ mg}}{}\right)\left(\frac{1 \text{ mL}}{0.5 \text{ mg}}\right) = 1.09 \text{ mL}$$

(round up to 1.1 mL)

The dosage for buprenorphine for dogs, cats, mice, and rats is 0.03 mg/kg IV. The concentration of buprenorphine is 0.3 mg/mL. Calculate the dose of buprenorphine for a 15 pound dog, 12 pound cat, 25 gram mouse, and 200 gram rat.

For the 15 lb dog:

$$\left(\frac{15 \text{ lb}}{}\right)\left(\frac{1 \text{ kg}}{2.2 \text{ lb}}\right) = 6.82 \text{ kg}$$

$$\left(\frac{6.82 \text{ kg}}{}\right)\left(\frac{0.03 \text{ mg}}{\text{kg}}\right) = 0.2046 \text{ mg}$$

$$\left(\frac{0.2046 \text{ mg}}{}\right)\left(\frac{1 \text{ mL}}{0.3 \text{ mg}}\right) = 0.682 \text{ mL}$$

(round up to 0.7 mL)

For the 12 lb cat:

$$\left(\frac{12 \text{ lb}}{}\right)\left(\frac{1 \text{ kg}}{2.2 \text{ lb}}\right) = 5.45 \text{ kg}$$

$$\left(\frac{5.45 \text{ kg}}{}\right)\left(\frac{0.03 \text{ mg}}{\text{kg}}\right) = 0.1635 \text{ mg}$$

$$\left(\frac{0.1635 \text{ mg}}{}\right)\left(\frac{1 \text{ mL}}{0.3 \text{ mg}}\right) = 0.545 \text{ mL}$$

(round up to 0.5 mL)

For the 25 g mouse (notice a different conversion factor is needed because the weight of the mouse is already in the metric system [see Table F-1]):

$$\left(\frac{25 \text{ g}}{}\right)\left(\frac{1 \text{ kg}}{1,000 \text{ g}}\right) = 0.025 \text{ kg}$$

$$\left(\frac{0.025 \text{ kg}}{}\right)\left(\frac{0.03 \text{ mg}}{\text{kg}}\right) = 0.00075 \text{ mg}$$

$$\left(\frac{0.00075 \text{ mg}}{}\right)\left(\frac{1 \text{ mL}}{0.3 \text{ mg}}\right) = 0.0025 \text{ mL}$$

(round up to 0.003 mL)

For the 200 g rat (notice a different conversion factor is needed because the weight of the mouse is already in the metric system):

$$\left(\frac{200 \text{ g}}{}\right)\left(\frac{1 \text{ kg}}{1,000 \text{ g}}\right) = 0.20 \text{ kg}$$

$$\left(\frac{0.20 \text{ kg}}{}\right)\left(\frac{0.03 \text{ mg}}{\text{kg}}\right) = 0.006 \text{ mg}$$

$$\left(\frac{0.006 \text{ mg}}{}\right)\left(\frac{1 \text{ mL}}{0.3 \text{ mg}}\right) = 0.02 \text{ mL}$$

One analgesic FDA-approved for use in cattle is flunixin meglumine. The dosage for cattle is 2 mg/kg IV (slowly). The concentration of flunixin meglumine is 50 mg/mL. Calculate the dose of flunixin meglumine for a 1,500 pound Holstein cow.

$$\left(\frac{1,500 \text{ lb}}{}\right)\left(\frac{1 \text{ kg}}{2.2 \text{ lb}}\right) = 681.8 \text{ kg}$$

$$\left(\frac{681.8 \text{ kg}}{}\right)\left(\frac{2 \text{ mg}}{\text{kg}}\right) = 1,363.6 \text{ mg}$$

$$\left(\frac{1,363.6 \text{ mg}}{}\right)\left(\frac{1 \text{ mL}}{50 \text{ mg}}\right) = 27.27 \text{ mL}$$

(round up to 27.3 mL)

Dose in Tablets

An example of calculating a dose in tablets is as follows: The dosage of the analgesic robenacoxib for postoperative pain in dogs and cats is 1 mg/kg PO q24h. Using a dosage of 1 mg/kg, how many mg should be given to an English Cocker Spaniel dog weighing 22 lb?

Step 1: If the animal's weight and the medication dosage are not in the same system of measure, convert the weight so that both measures are expressed in the same unit. In this example, pounds must be converted to kg. From Table F-3, choose the appropriate conversion factor of 1 kg = 2.2 lb. The animal weighs 22 lb; therefore, use the conversion factor 2.2 lb = 1 kg to convert the weight into kg units.

Step 2: Knowing that 2.2 lb = 1 kg, create a conversion factor. *Conversion factors are used to move between units and always have a value of one.* Because they have a value of one, conversion factors do not change the value of the end product. Write this conversion factor as 2.2 lb/1 kg.

Step 3: The next step is to determine in which format to write the conversion factor. There are two choices: 2.2 lb/1 kg or 1 kg/2.2 lb. In this case, cancel out lb to end up with kg. Therefore, lb is on the bottom, and kg is on the top of the conversion factor.

Step 4: Set up the conversions in an equation. Start with what is known (22 lb), and use the conversion factor to set up the equation.

$$\left(\frac{22 \text{ lb}}{}\right)\left(\frac{1 \text{ kg}}{2.2 \text{ lb}}\right) = 10 \text{ kg}$$

Step 5: Now perform the calculation by taking the weight of the animal in kg and multiplying it by the dosage. Remember, values on the bottom of the conversion factors are divided into the numbers on the top. Values on the top of the conversion factors are multiplied by the numbers on the top. In this case, take 10 kg, multiply it by 2 mg, and then divide by 1 kg. Keep in mind that units should cancel as well to give the desired unit at the end.

$$\left(\frac{10 \text{ kg}}{}\right)\left(\frac{1 \text{ mg}}{1 \text{ kg}}\right) = 10 \text{ mg dose}$$

Step 6: To make sure the answer is correct, prove the work by looking at the units involved. If the animal needs 1 mg of drug for every 1 kg of body weight, and the animal weighs 10 kg, it makes sense that this patient should receive a 10 mg dose of drug. Always double check the answer to avoid errors in calculations.

The entire formula used when an animal's weight is in pounds and the dosage is based in kilograms is to take the animal's weight (in lb) divide by 2.2 (conversion factor to convert to lb to kg), then multiply by amount of drug (in this case, in mg) per kilogram.

How many tablets will this patient need per dose?

Step 1: To determine the amount of each dose in tablets, it is necessary to know the strength of tablets available. If the clinic inventory contains 10 mg, 20 mg, and 40 mg tablets of this drug, the tablet chosen for the patient is typically the tablet strength closest to the dose or the one more easily divided in half to achieve the dose. Since it was already determined that the animal needs 10 mg per dose, the veterinary technician would choose the 10 mg tablet to perform the calculation.

Step 2: Determine the amount of tablets per dose by dividing the 10 mg dose by 10 mg (tablet strength).

$$\left(\frac{10\ mg}{}\right)\left(\frac{1\ tablet}{10\ mg}\right) = 1\ tablet\ per\ dose$$

Practice calculating doses in tablets for an 11 pound cat:

For the 11 lb cat:

$$\left(\frac{11\ lb}{}\right)\left(\frac{1\ kg}{2.2\ lb}\right) = 5.0\ kg$$

$$\left(\frac{5.0\ kg}{}\right)\left(\frac{1\ mg}{kg}\right) = 15.0\ mg$$

$$\left(\frac{5.0\ mg}{}\right)\left(\frac{1\ tablet}{10\ mg}\right) = 0.5\ tablet\ per\ dose$$

Calculating Total Dose

Knowing how to determine the total dose to dispense to the client for the patient's entire treatment regimen is important when sending a patient home with analgesics or other medications (e.g. antibiotics, antiemetic drugs) following anesthesia and surgery/procedures. Using the example of the English Cocker Spaniel dog that needs 10 mg per dose, how much drug would need to be dispensed if this animal needed the drug q24h for seven days? How many tablets would be dispensed?

Step 1: It was previously determined that each dose in tablet form was 1 tablet of the 10 mg tablet. Because the animal needs the drug q24h (once daily), the animal will get 1 tablet per day (1 tablet × 1 dose per day = 1 tablet per day).

Step 2: The animal needs to take the medication for one week. Multiply 1 tablet per day by 7 days: 1 tablet × 7 days = 7 tablets total to be dispensed.

Step 3: Sometimes the same dose of a drug may be cheaper when a scored tablet of a higher dose is divided in half rather than using a full tablet of the lower dose. For example, splitting a 20 mg tablet in half to get a 10 mg dose may be cheaper than using one 10 mg tablet. Enteric-coated or SR (extended-release) tablets should not be broken as this would expose the drug to the acid environment of the stomach, which can alter their effectiveness or remove the coating that extends the release of the drug. Other drugs that should not be split include those with precise dosing requirements, very small tablets, asymmetrical tablets, capsules, hormones, and medications that are carcinogenic or teratogenic.

SOLUTIONS

Solutions are mixtures of substances not chemically combined with each other. The dissolving substance of a solution is referred to as the solvent and is usually a liquid. The dissolved substance of a solution is referred to as the solute and is usually the solid or particulate part of the mixture. Substances that form solutions are called miscible and those that do not are called immiscible.

Tech Tip

Ratio strengths (parts per some amount) are used primarily in solutions. They represent parts of drug per parts of solution.

When working with solutions, the amount of solute (particles) in the solvent provides information about the relative "strength" of the solution. The amount of solute dissolved in solvent is known as the *concentration*. Concentrations may be expressed as parts (per some amount), weight per volume, volume per volume, and weight per weight. They are usually reported as a percent or percent solution. Remember that a percent is parts per the total times 100. Here are some rules of thumb for working with concentrations.

- *Parts:* Parts per million means 1 mg of solute in a kg (or L) of solvent. Ratios or fractions must be translated into percents of solution to perform many of the necessary calculations. For example, to determine the percent concentration of 1:1,000 epinephrine (see Figure 13-3), divide the parts (1) by the total parts (1,000) and multiply by 100 to determine the percent.

$$\left(\frac{1}{1000}\right)\left(\frac{100}{}\right) = 0.1\%$$

- *Liquid in liquid:* The percent concentration is the volume per 100 volumes of the total mixture. The two volumes may be expressed in any unit as long as they are the same unit within the percent. For example, 1 mL/100 mL, 5 oz/100 oz, and 15 L/100 L.
- *Solids in solids:* The percent concentration is the weight per 100 weights of total mixture. The two weights may be expressed in any units as long as they are the same unit within the percent. For example, 60 mg/100 mg, 55 oz/100 oz, and 4.5 g/100 g.
- *Solids in liquid:* The percent concentration is the weight in grams per 100 volume parts in milliliters. The weight must be in grams and the volume must be in milliliters. For example, dextrose 5% = 5 g/100 mL. Dextrose 5% can also be expressed as 5,000 mg/100 mL or 50 mg/mL, as long as the number is based originally on grams.

Tech Tip

When using ratios, it is important to understand what the notation represents. A 1:2 ratio signifies of the total amount (2 parts), 1 part is the medication. A 1:1,000 ratio signifies of the total amount (1,000 parts), 1 part is medication. Therefore, a 1:1 ratio signifies of the total amount (1 part), the entire 1 part is medication. In other words, a 1:1 ratio contains all medication.

Percent Concentration Calculations

Occasionally the veterinary technician may have to prepare a drug solution from a pure drug or stock solution. *Pure drugs* are substances, in solid or liquid form, that are 100% pure. A *stock solution* is a relatively concentrated solution from which more dilute (weaker) solutions are made.

One method of determining the amount of pure drug needed to make a solution is the ratio–proportion method. The formula used to determine the amount of pure drug needed to make the solution is

$$\frac{\text{Amount of drug}}{\text{Amount of finished solution}} = \frac{\text{Percentage of finished solution}}{100\% \left(\text{based on a pure drug}\right)}$$

For example, the amount of sodium chloride needed to make 500 mL of a 0.9% solution can be determined by performing the following calculation:

Step 1: Set up the ratio of each solution based on the information provided. Recall that a 0.9% solution has the weight in grams per 100 volume parts in milliliters. For this example, 0.9 g NaCl is in 100 mL of solution. The new solution has an unknown amount of NaCl in 500 mL of solution.

$$\frac{X\,g}{500\;mL} = \frac{0.9}{100\;mL}$$

Step 2: Cross multiply to get X on one side of the equation. In this example, 500 mL multiplied by 0.9 g equals 450 (mL)(g). 100 mL multiplied by X g is 100 (mL)(g).

$$100\;(mL) \times X\;(g) = 450\;(mL)(g)$$

Step 3: Isolate X by cross multiplying. In this example, 450 (mL)(g) is divided by 100 mL. The other side of the equation is also divided by 100 mL so that X g is isolated on one side of the equation.

$$\frac{100\;(mL) \times X\;(g)}{100\;mL} = \frac{450\;(mL)(g)}{100\;mL}$$

$$X = 4.5\;g\text{ of sodium chloride}$$

(The answer technically is 4.5 mL; however, 4.5 mL of sodium chloride can be multiplied by the density of solution, which is 1 g/mL. This gives a final answer of 4.5 g.)

The amount of drug used to prepare a solution adds to the total volume of the solvent. The amount of volume contributed by a dry drug varies by its individual structure and is usually not accounted for when preparing a solution. However, it is fairly easy to determine the volume a liquid drug will add to the solvent volume. To determine the amount of solvent needed to make the finished solution, subtract the volume of liquid drug from the amount of finished solution.

For example, to prepare a liter of 4% formaldehyde fixative solution from a 37% stock solution, use the ratio–proportion method as follows:

Step 1: Set up the ratio of each solution based on the information provided. For this example, to determine the volume of a 37% solution needed to make a 4% solution, set up a ratio of the percent solution needed (4%) over the more concentrated solution (37%). Then set up a ratio of the amount of liquid needed (X mL) of the total solution (1,000 mL). Make sure that the concentrated solution (37%) is across from the total volume (1,000 mL) and the new concentration (4%) is across from the new volume (X mL).

$$\frac{X\;mL}{1,000\;mL} = \frac{4\%}{37\%}$$

Step 2: Solve for X by cross multiplying.

$$37\;(\%) \times X\;(mL) = 4,000\;(\%)\;(mL)$$

Step 3: Solve for X by dividing each side by 37% to isolate X.

$$\frac{37\;(\%) \times X\;(mL)}{37\%} = \frac{4,000\;(\%)(mL)}{37\%}$$

$$X = 108\;mL\text{ of stock solution}$$

Step 4: To determine the amount of solvent to add to the stock solution, take the total desired amount (in this case, 1 L), convert 1 L to 1,000 mL (since the volume of stock solution is in mL, this volume of stock solution also needs to be in mL), and subtract the amount of stock solution in mL

$$1,000\;mL - 108\;mL = 892\;mL\text{ of solvent should be added}$$

Another way to make the final solution is add diluent (solvent) quantity sufficient (qs) to 1,000 mL. *Quantity sufficient* is the amount of diluent needed to yield the final volume. In this case, the 108 mL of stock solution would be placed in a container that held at least 1,000 mL (1 L) and solvent would be added to reach 1,000 mL, thereby yielding the desired quantity and concentration.

Another method of determining volume for a desired final volume is via the volume concentration method. The equation for this is as follows:

$$V_s \times C_s = V_d \times C_d$$

V_s = volume of the beginning or stock solution
C_s = concentration of the beginning or stock solution
V_d = volume of the final solution
C_d = concentration of the final solution

An example of using this equation is as follows: How much water must be added to a liter of 90% alcohol to change it to a 40% solution?

Step 1: Consider what is known. There is 1 L of 90% alcohol. Now consider what is wanted. A 40% solution is desired, and whatever amount of water needed to make this solution can be used. Set up an equation based on this information in which V_s is 1,000 mL, C_s is 90%, C_d is 40%, and V_d is unknown.

$$1,000\;mL \times 90\% = V \times 40\%$$

Step 2: Solve for V_d by dividing each side by 40%.

$$\frac{90,000\;mL\;(\%)}{40\%} = \frac{V_d\;(40\%)}{40\%}$$

$$2,250\;mL = V_d$$

It is also possible to convert the percent of a solution to a decimal before solving the preceding equation. In decimal form, 90% equals 0.90 and 40% equals 0.40.

$$1,000\;mL \times 0.90 = V_d \times 0.40$$

$$900\;mL = 0.40\;V_d$$

$$2,250\;mL = V_d$$

Step 3: Remember that 2,250 mL is the final volume. Therefore, it is necessary to subtract the original amount of 90% alcohol from the final volume to know how much water needs to be added to the 1,000 mL of 90% alcohol to make a 40% solution.

$$2,250 \text{ mL} - 1,000 \text{ mL} = 1,250 \text{ mL}$$

Sometimes drug concentrations are listed in percents. The percent concentrations are usually found on the front of the drug vial or bottle. These containers should also have the concentration listed in mg/mL.

An example is as follows: Lidocaine is a drug used as a topical anesthetic and as an antiarrhythmic. The dosage for a dog is 3 mg/kg. Calculate the dose of lidocaine in mL for a 15 lb Poodle dog.

Step 1: The animal's weight and the medication dosage are not in the same system of measure; therefore, the weight must be converted so that both measures are expressed in the same unit. In this example, pounds must be converted to kg. From Table F-3, choose the appropriate conversion factor of 1 kg = 2.2 lb. The animal weighs 15 lb; therefore, use the conversion factor 2.2 lb = 1 kg to convert the weight into kg units.

Step 2: Knowing that 2.2 lb = 1 kg, create a conversion factor. *Conversion factors are used to move between units and always have a value of one.* Because they have a value of one, conversion factors do not change the value of the end product. Write this conversion factor as 2.2 lb/1 kg.

Step 3: The next step is to determine in which format to write the conversion factor. There are two choices: 2.2 lb/1 kg or 1 kg/2.2 lb. In this case, cancel out lb to end up with kg. Therefore, lb is on the bottom, and kg is on the top of the conversion factor.

Step 4: Set up the conversions in an equation. Start with what is known (15 lb), and use the conversion factor to set up the equation.

$$\left(\frac{15 \text{ lb}}{1}\right)\left(\frac{1 \text{ kg}}{2.2 \text{ lb}}\right) = 6.81 \text{ kg}$$

Step 5: Calculate the dose needed in mg by taking the weight of the animal in kg and multiplying it by the dosage. Remember, values on the bottom of the conversion factors are divided into the numbers on the top. Values on the top of the conversion factors are multiplied by the numbers on the top. In this case, take 6.81 kg, multiply it by 3 mg, and then divide by 1 kg. Keep in mind that units should cancel as well to give the desired unit at the end.

$$\left(\frac{6.81 \text{ kg}}{1}\right)\left(\frac{3 \text{ mg}}{\text{kg}}\right) = 20.45 \text{ mg}$$

Step 6: Determine the concentration of the drug. The front of the vial lists the concentration of lidocaine as 2% (see Figure 5-13a). Because percents are parts per the total, 2% is 2 parts per 100. In solutions, percent represents the number of grams of drug per 100 mL of solution. Remember that percents are g/100 mL; therefore, lidocaine 2% equals 2 g/100 mL.

Step 7: In this example, the units of weight (mg and g) need to be the same to perform the calculation. The dose of 20.45 mg needs to be converted to g. Knowing that 1,000 mg = 1 g, create a conversion factor. Write this conversion factor as 1,000 g/1 kg.

Step 8: The next step is to determine in which format to write the conversion factor. There are two choices: 1,000 g/1 kg or 1 kg/1,000 g.

The choice is made based on which unit of measure to cancel out and which unit of measure to end up with. In this case, cancel out mg to end up with g. Therefore, mg is on the bottom and g is on the top of the conversion factor.

$$\left(\frac{20.45 \text{ mg}}{1}\right)\left(\frac{1 \text{ g}}{1,000 \text{ mg}}\right) = 0.02045 \text{ g}$$

Step 9: The last step is to calculate the dose in mL using the concentration of lidocaine (2 g/100 mL). The concentration should be written as 100 mL/2 g so that the grams cancel out and the milliliters are on the top to provide a final answer in mL.

$$\left(\frac{0.02045 \text{ g}}{1}\right)\left(\frac{100 \text{ mL}}{2 \text{ g}}\right) = 1.02 \text{ mL}$$

This dog would get 1.02 mL per dose (which would be rounded to 1.0 mL). This calculation demonstrates the rationale for the Food and Drug Administration (FDA) mandating that drug labels must not only have percents (2%) or parts per the whole (1:1,000) listed on them but must also have the concentration listed. There is an increased risk of mathematical errors if the drug strength is not listed in metric concentrations such as mg/mL (as demonstrated in Figure 5-13a). Percents or parts per the whole require extra mathematical steps (steps 6–8 above) that are not needed if the metric concentration such as mg/mL is provided on the label.

Percentages

Percentages are used frequently in delivering medications to patients. For example, 0.9% saline is used for IV administration, 2% lidocaine is used for regional anesthesia, and 5% dextrose for IV supplementation. The one clue that is central to understanding and remembering how to deal with percentage problems is that percentage is always based on 100. The value is parts of one hundred. When thinking about 5% dextrose, the dextrose within the fluid is 5 grams per 100 mL of fluid. If we use this knowledge, then 0.9% saline contains 0.9 grams of NaCl per 100 mL of fluid.

In the case of 2% lidocaine, we now know that it is 2 grams of lidocaine per 100 mL of solution. Clinically, a vial of 2% solution of lidocaine is labeled as 20 mg/mL ([2 g/100 mL] × [1,000 mg/1 g] = 2,000 mg/100 mL = 20 mg/mL). Converting to a decimal it looks like this:

$$\frac{2}{100}$$

To convert the fraction to a decimal, move the decimal point two places to the left of "2." So, 2% becomes 0.02.

Percentage values can be added together, for example,
20% + 35% + 2% = 57%.

Percent Concentration Calculations

To prepare a drug solution from a pure drug or stock solution, two mathematical methods can be used: the ratio-proportion method or the percent solution method. To properly dilute solutions the ratio-proportion method uses the following equation:

$$\frac{\text{Amount of drug}}{\text{Amount of finished solution}} = \frac{\text{Percentage of finished solution}}{100\% \left(\text{based on a pure drug}\right)}$$

For example, to determine how much sodium chloride is needed to make 500 mL of a 0.9% solution (0.9% NaCl = 0.9 g NaCl per 100 mL solution), the following calculation would be performed:

$$\frac{X\,g}{500\,mL} = \frac{0.9}{100\,mL}$$

To determine X, cross multiply to get

$$100\,(mL)\,X(g) = 450\,(mL)(g)$$

Then divide each side by 100% to isolate X

$$X = 4.5\,g$$

(The answer is technically 4.5 mL; however, 4.5 mL sodium chloride can be multiplied by its density (1 g/mL) which gives a final answer of 4.5 g.)

Another way to determine volume of a desired final volume is with the percent solution method. The equation for this is:

$$V_1 \times C_1 = V_2 \times C_2$$

The volume is shown with a V and concentration is shown with a C.

V_1 = volume of the first solution
C_1 = concentration of the first solution
V_2 = volume of the second solution
C_2 = concentration of the second solution

Using this formula we need to have three values and must solve for only one. Consider the following example:

The clinic purchases a stock solution of 70% alcohol. We would like 100 mL of 5% alcohol. For this problem 70% is C_1; 5% is C_2; 100 mL is V_2. V_1 is unknown.

$$V_1 \times 70\% = 5\% \times 100\,mL$$

$$V_1 = \frac{5\% \times 100\,mL}{70\%}$$

We can cancel out the percentage symbol, which then leaves mL as the value for our final product.

$$V_1 = \frac{5\% \times 100\,mL}{70\%}$$

$$V_1 = \frac{500\,mL}{70}$$

$$V_1 = 7.14\,mL\ (\text{round down to 7.1 mL})$$

7.1 mL is the amount of the original 70% solution that must be used to make the diluted solution.

The total amount of the diluted solution is 100 mL. The amount of the stock solution is subtracted from the total diluted solution so that the percentage solution is the correct final amount.

$$\begin{array}{r} 100\,mL \\ -\ 7.1\,mL \\ \hline 92.9\,mL \end{array}$$

92.9 mL is the amount of water that is added to the 7.1 mL of stock alcohol to make a 5% solution.

Consider the following example:

Acepromazine is typically diluted to 2 mg/mL for use in small animals from the 10 mg/mL concentration used for large animals. How much 10 mg/mL acepromazine and how much sterile water would you need to make 20 mL of the 2 mg/mL concentration (enough to fit into a 20 mL sterile vial)?

$$V_1 \times C_1 = V_2 \times C_2$$

$$V_1 \times 10\,mg/mL = 20\,mL \times 2\,mg/mL$$

$$V_1 \times 10\,mg/mL = 40\,mg$$

$$V_1 = \frac{40\,mg}{10\,mg/mL}$$

$$V_1 = 4\,mL\ \text{of the 10 mg/mL acepromazine}$$

The total amount desired of the diluted acepromazine is 20 mL. The amount of the 10 mg/mL acepromazine is subtracted from the total amount desired of the diluted solution so that the 2 mg/mL acepromazine solution is the correct final concentration.

20 mL total − 4 mL (10mg/mL) acepromazine = 16 mL sterile water

16 mL is the amount of diluent that is added to 4 mL of 10 mg/mL acepromazine to make 20 mL of 2 mg/mL acepromazine.

Calculating Constant Rate Infusion (CRI)

When administering drugs via CRI, the constant dose delivered over a specified period of time is calculated. For example, the drug dosage for CRI is often expressed in micrograms (mcg or µg) per kilogram per minute (*mcg/kg/min*), micrograms per kilogram per hour (*mcg/kg/min*), milligrams per kilogram per minute (*mcg/kg/min*), or milligrams per kilogram per hour (*mcg/kg/min*). The drug concentration is usually expressed as milligrams per milliliter (*mg/mL*) or micrograms per milliliter (*mcg/mL*) and then ultimately the drug is delivered at a rate expressed in volume (usually milliliters per hour [*mL/hr*]). All of the different units need to be the same so that the desired unit is left at the end of the calculation.

To perform CRI calculations, the drug dosage (e.g., 0.05 mg/kg/hr), drug concentration (e.g., hydromorphone is 2 mg/mL), patient's weight in kg (e.g., the dog weighs 30 lb), and fluid rate in mL per hour (e.g., 25 mL/hr) are needed. Remember that all of the above information is not always provided; sometimes you will need to look it up in a drug formulary or solve for an answer.

Consider the following CRI calculation: A female spayed Beagle dog weighing 30 pounds is prescribed a hydromorphone CRI at a dosage of 0.05 mg/kg/hr following surgical repair of a femoral fracture. The concentration of hydromorphone is 2 mg/mL. What is the drug concentration (mL/hr) at which the syringe pump should be set?

- First make sure the information is in the same units. In this example, the dosage in mg/kg/hr, the concentration of hydromorphone in mg/mL, and the desired rate in mL/hr are all in common units (see Table F-3).
  - The dog's weight is in pounds and needs to be converted to kg.

$$\left(\frac{30\,lb}{1}\right)\left(\frac{1\,kg}{2.2\,lb}\right) = 13.6\,kg$$

  - Determine the amount of drug to administer based on the patient's weight (0.05 mg/kg/hr) (13.6 kg) = 0.68 mg/hr
  - Use the drug concentration to determine the administration rate in mL/hr

$$\left(\frac{0.68\,mg}{hr}\right)\left(\frac{1\,mL}{2\,mg}\right) = 0.34\,mL/hr$$

If using a syringe pump/driver, it should be set to administer hydromorphone at 0.34 mL/hr.

Consider a different way of administering the hydromorphone. The veterinarian would like to administer hydromorphone in a 250 mL bag of fluids (0.9% NaCl) at a rate of 25 mL/hr. How much 2mg/mL hydromorphone will you add to the 250 mL bag of NaCl?

- Determine how long the bag of fluids will last.

$$(250 \text{ mL}) \left(\frac{1 \text{ hr}}{25 \text{ mL}} \right) = 10 \text{ hr}$$

- We know the administration rate is 0.34 mL/hr.

$$(10 \text{ hr}) \left(\frac{0.34 \text{ mL}}{1 \text{ hr}} \right) = 3.4 \text{ mL}$$

- To ensure we do not have a volume greater than 250 mL in the 250 mL bag, we must first remove 3.4 mL of 0.9% NaCl from the bag and then add 3.4 mL hydromorphone.
- If we administer this bag of 0.9% NaCl at 25 mL/hr, the hydromorphone will run at 0.34 mL/hr for 10 hr.

Another equation to use for this problem is as follows:

○ To determine how much hydromorphone to add to the 250 mL bag, use the following equation:

$$M = \frac{D \times W \times V}{R}$$

M = amount of drug (in units used to solve the equation so everything cancels out) to be added to the solution = unknown (?)
D = dosage = 0.05 mg/kg/hr
W = weight of the patient in kg − 13.6 kg
V = volume of the solution = 250 mL
R = infusion rate = 25 mL/hr

- In this example, the equation now looks like this:

$$M = \frac{(0.05 \text{ mg/kg/hr}) (13.6 \text{ kg}) (250 \text{ mL})}{(25 \text{ mL/hr})}$$

- Perform the calculation and unit calculation:

$$M = \frac{170 \text{ mg}}{25}$$

$$M = 6.8 \text{ mg}$$

- Use the concentration of the drug (hydromorphone = 2 mg/mL) to determine the amount of mL to add to the bag

$$6.8 \text{ mg} \times 1 \text{ mL}/2 \text{ mg} = 3.4 \text{ mL}$$

- In this example, 3.4 mL of hydromorphone (2 mg/mL) needs to be added to the 250 mL bag. Again remember to remove 3.4 mL of 0.9% NaCl from the bag before adding the hydromorphone to not exceed 250 mL of total volume, which will alter the concentration.

Veterinarians sometimes prescribe several analgesic drugs for patients undergoing more painful procedures. Each drug is calculated separately and is based on the dosages provided by the veterinarian. For example, a veterinarian requests a CRI of morphine, lidocaine, and ketamine (MLK). The dosages are 200 mcg/kg/hr for morphine; 30 mcg/kg/min for lidocaine, and 600 mcg/kg/hr for ketamine. The concentration of morphine is 15 mg/mL, lidocaine is 20 mg/mL, and ketamine is 100 mg/mL. The veterinarian determines that the administration rate is 5 mL/kg/hr. Based on the administration rate

of 5 mL/kg/hr in a 60 lb (60 lb × 1 kg/2.2 lb = 27.3 kg) female spayed Labrador Retriever dog, determine the amount of each drug needed for a 500 mL bag of fluids.

Fluid rate:

- 5 mL/kg/hr × 27.3 kg = 136.5 mL/hr

For morphine:

- First make sure the information is in the same units. In this example the dosage is in mcg/kg/min while the concentration of morphine is in mg/mL and the desired rate is in mL/hr.
  ○ 200 mcg/kg/hr × 27.3 kg = 5,460 mcg/hr
  ○ 5,460 mcg/hr × 1 mg/1,000 mcg = 5.46 mg/hr
  ○ 5.46 mg/hr ÷ 136.5 mL/hr = 0.04 mg/mL
  ○ 0.04 mg/mL × 500 mL bag = 20 mg

$$\left(20 \text{ mg} = \frac{D \times W \times V}{R} \right)$$

  ○ 20 mg ÷ 15 mg/mL (concentration of morphine) = 1.3 mL

- In other words,

$$M = \frac{D \times W \times V}{R} = 20 \text{ mg}$$

M ÷ concentration of morphine = volume of morphine
20 mg ÷ 15 mg/mL = 1.3 mL

For lidocaine:

- First make sure the information is in the same units. In this example the dosage is in mcg/kg/min while the concentration of lidocaine is in mg/mL and the desired rate is in mL/hr.
  ○ 30 mcg/kg/min × 27.3 kg = 819 mcg/min
  ○ 819 mcg/min × 1 mg/1,000 mcg = 0.819 mg/min
  ○ 0.819 mg/min × 60 min/hr = 49.14 mg/hr
  ○ 49.14 mg/hr ÷ 136.5 mL/hr = 0.36 mg/mL
  ○ 0.36 mg/mL × 500 mL bag = 180 mg

$$\left(180 \text{ mg} = \frac{D \times W \times V}{R} \right)$$

  ○ 180 mg ÷ 20 mg/mL (concentration of lidocaine) = 9 mL

- In other words,
  ○ $M = \frac{D \times W \times V}{R} = 180 \text{ mg}$
  ○ M ÷ concentration of lidocaine = volume of lidocaine
  ○ 180 mg ÷ 20 mg/mL = 9 mL

For ketamine:

- First make sure the information is in the same units. In this example the dosage is in mcg/kg/hr while the concentration of ketamine is in mg/mL and the desired rate is in mL/hr.
  ○ 600 mcg/kg/hr × 27.3 kg = 16,380 mcg/hr
  ○ 16,380 mcg/hr × 1 mg/1,000 mcg = 16.38 mg/hr
  ○ 16.38 mg/hr ÷ 136.5 mL/hr = 0.12 mg/mL
  ○ 0.12 mg/mL × 500 mL bag = 60 mg

$$\left(60 \text{ mg} = \frac{D \times W \times V}{R} \right)$$

- ○ 60 mg ÷ 100 mg/mL (concentration of ketamine) = 0.6 mL
- ○ In other words,
- ○ $M = \dfrac{D \times W \times V}{R} = 60$ mg
- ○ M ÷ concentration of ketamine = volume of ketamine
- ○ 60 mg ÷ 100 mg/mL = 0.6 mL

In this example, 1.3 mL morphine (15 mg/mL), 9 mL lidocaine (20 mg/mL), and 0.6 mL ketamine (100 mg/mL) needs to be added to the 500 mL bag. Remember to remove 10.9 mL of fluid (1.3 mL + 9 mL + 0.6 mL) before adding the drugs to not exceed a total volume of 500 mL, which will alter the concentration.

Based on the administration rate of 5 mL/kg/hr in the 27.3 kg patient, determine how long a 500 mL fluid bag will last.

- Calculate the rate of fluids needed per hour for the weight of the patient.

$$5 \text{ mL/kg/hr} \times 27.3 \text{ kg} = 136.5 \text{ mL/hr}$$

(This is the delivery rate set on the infusion pump or syringe pump.)

- To determine how long the bag will last, take the fluid bag size and divide it by the hourly rate.
  - ○ $500 \text{ mL} \div 136.5 \text{ mL/hr} = 500 \text{ mL} \times \dfrac{1 \text{ hr}}{136.5 \text{ mL}} = 3.66 \text{ hr}$

Appendix G

FLUID DRIP RATES

Patient Weight (kg)	Fluid rate mL/kg/hr	mL/hr	10 gtt/mL (... drops per second)	15 gtt/mL (... drops per second)	20 gtt/mL (... drops per second)	60 gtt/mL (1 drop/... seconds)
1	2.5	2.5				24.0
2	2.5	5				12.0
3	2.5	7.5				8.0
4	2.5	10				6.0
5	2.5	12.5				4.8
6	2.5	15				4.0
7	2.5	17.5				3.4
8	2.5	20				3.0
9	2.5	22.5				2.7
10	3	30	0.08	0.13	0.17	2.0
12	3	36	0.10	0.15	0.20	
14	3	42	0.12	0.18	0.23	
16	3	48	0.13	0.20	0.27	
18	3	54	0.15	0.23	0.30	
20	3	60	0.17	0.25	0.33	
22	3	66	0.18	0.28	0.37	
24	3	72	0.20	0.30	0.40	
26	3	78	0.22	0.33	0.43	
28	3	84	0.23	0.35	0.47	
30	3	90	0.25	0.38	0.50	
35	3	105	0.29	0.44	0.58	
40	3	120	0.33	0.50	0.67	
45	3	135	0.38	0.56	0.75	
50	3	150	0.42	0.63	0.83	
55	3	165	0.46	0.69	0.92	
60	3	180	0.50	0.75	1.00	
65	3	195	0.54	0.81	1.08	
70	3	210	0.58	0.88	1.17	

Gravity flow drip rates using the 10, 15, and 20 drops per ml (... drops per second) and the 60 drops per ml (1 drop per ... seconds) administration sets.

Appendix H

EXAMPLE OF AN ANESTHESIA RECORD

Patient Name	(First & Last)			Client #		Procedure(s)					
		Age	Wt/Kg	Temp	Pulse	Resp	PCV	TP	BUN		
Species/Breed											

Date	Case Clinician		ASA Status		ER	System Type:	Vent □	ETTSIZE:	
M T W R F S SN			I II III IV V		Y N	Circle □ Non-Reb □			

PREMEDS	DOSE	ROUTE	TIME	MONITORS:	FLUID PUMP □		
	mg			PULSE OK □	VENOUS CATHETER(S):	□ STERILE □ NON-STERILE	
	mg			ECG □	□ PLACED PRIOR TO ANESTHESIA		
	mg			ETCO2 □	LOCATION:	GA.	

PREMED RESPONSE:

HISTORY:

IBP □ LOCATION_____

NIBP □ DOPPLER □ OSCILLOMETRIC Cuff Size_____ Location_____

CPR/DNR

TIME	00	15	30	45	00	15	30	45	00	15	30	45	00	15	30	45	TOTALS
FLUIDS																	mLs
																	mLs
																	mLs
																	mLs
INDUCTION																	
MEDS																	

TEMPERATURE

COMMENTS

ISOFLURANE □ 6 5 4 3 2 1
SEVOFLURANE □
O₂ L/Min

SYMBOLS
V Systolic
+ Mean
^ Diastolic
• Pulse
× CO₂
O Respirations
A Assisted
C Controlled
★ SpO₂

CODES
S Surgery
S End Surgery
T Tourniquet
T Tourniquet Off

SPECIAL PROCEDURES
Epidural □
Nerve Block □
Site_____

time tgh=

post-op pain score

/10

200 180 160 140 120 100 90 80 70 60 50 40 30 20 10

200 180 160 140 120 100 90 80 70 60 50 40 30 20 10

ANESTHETIST(S)

Codes and Remark Numbers			RECOVERY NOTES	
LA Assisted □	Extubation Time	Post op Temp.	Recovery Notes & Analgesic Recommendations	
Sternal _____				
Standing _____				

VMTH-21	ANESTHESIA RECORD	Revised 05/2021

Appendix I

NORMAL VITAL SIGNS*

NORMAL BODY TEMPERATURES (RECTAL)

Species	°C	°F
Cats	37.8–39.2	100–102.5
Dogs	37.5–39.2	99.5–102.5
Horses	37.2–38.6 (adults)	99–101.5
	37.5–39.2 (foals)	99.5–102.7
Cattle	37.8–39.2 (> 1 year of age)	100–102.5
	38.6–39.4 (< 1 year of age)	101.5–103.5
Goats	38.5–40.2	101.3–104.5
Sheep	38.5–40	101.3–104
Swine	37.8–38.9 (adult)	100–102
	38.9–40 (piglet)	102–104

Temperatures may fluctuate over the course of the day. Normal increases in temperature may occur in the evening, with food intake, muscular activity, approaching estrus, and gestation; normal decreases in temperature may occur in the morning, following intake of large volumes of cool fluids, and low ambient temperature.

NORMAL HEART RATES IN ANESTHETIZED ANIMALS

While auscultating the heart, assess and record *rate* (record as beats per minute to determine if the patient's heart rate is bradycardic, tachycardic, or normal), *rhythm* (record whether or not the heart is beating at regular intervals), and *strength and quality* (record strength as strong, bounding, weak, or thready/wiry). Pulse is a subjective assessment of cardiac output and tissue perfusion. A good quality pulse will be regular, easily located, and strong. Conversely, a poor quality pulse is irregular, very weak, or too rapid.

Species	Heart Rate in Beats per Minute
Cats	120–200
Dogs	60–140
Horses	28–40
Cattle	60–100
Goats	80–150
Sheep	80–150
Swine	80–130

Checking mucous membrane (MM) color provides the anesthetist with a subjective assessment of peripheral blood flow. The color should be pink with a capillary refill time (CRT) of less than 2 seconds.

NORMAL RESTING RESPIRATORY RATES IN ANESTHETIZED ANIMALS**

Respiratory rate, rhythm, and quality can also be assessed visually and via auscultation. It is also important to determine if there is a regular pattern to the respirations, what is the amount of effort being used to breathe, and are the breaths labored, normal, or follow an abnormal pattern.

Species	Respiratory Rate in Breaths per Minute
Cats	8–15
Dogs	8–15
Horses	10–30
Cattle	20–40
Goats	20–40
Sheep	20–40
Swine	10–25

**It is difficult to put a set range of normal on respiratory rates because it can really vary between patients. Ideally, the EtCO$_2$ should be used in conjunction with monitoring the respiratory rate to ensure carbon dioxide levels are within normal range.

MONITORING PARAMETER VALUES IN ANESTHETIZED PATIENTS

Monitoring Parameter	Normal Range
Blood pressure	Systolic: > 100 mm Hg and < 160 mm Hg
Mean arterial pressure	> 60 mm Hg
	> 70 mm Hg in horses
SpO$_2$	95–100%
EtCO$_2$	35–45 mm Hg
PaCO$_2$	35–45 mm Hg[+]
PaO$_2$	> 80 mm Hg
Arterial pH	7.35–7.45
Capillary refill time	< 2 sec

[+]Some anesthesiologists allow a certain degree of "permissive hypercapnia" under anesthesia, particularly in species such as equine. There is usually a 2–5 mm Hg difference between PaCO$_2$ and EtCO$_2$.

*Adapted from *Plumb's Veterinary Drug Handbook*, 9th ed.; values for small mammals and birds can be found in Chapter 20 (Anesthesia and Analgesia of Pet Birds and Exotic Small Mammals).

Appendix J

PHYSICAL REHABILITATION HISTORY, REGULATION, AND CREDENTIALING

HISTORY OF PHYSICAL THERAPY AND ITS PRINCIPLES

Physical therapy (PT) is an established profession in humans and companion animals that primarily focuses on restoring, maintaining, and promoting optimal function, fitness, wellness, and quality of life to patients.[1] Physical therapists achieve these goals by identifying risk factors in order to prevent or slow the progression of functional decline and disability and to enhance participation in the daily activities of their patients. Despite physical therapists using therapeutic interventions dating back to Hippocrates, physical therapy did not emerge as a profession in Europe until the 1500s when exercise and muscle reeducation programs were used to treat a variety of orthopedic diseases and injuries.[2] Physical therapy programs in the United States evolved in 1916 during the polio epidemic and were officially acknowledged in 1917 when physical therapists were needed for the rehabilitation of injured soldiers during World War I.[3]

HISTORY OF VETERINARY PHYSICAL REHABILITATION AND ITS PRINCIPLES

The use of physical therapy techniques on animals has benefited from large amounts of evidence-based medicine from the human PT profession, and many human therapies can be easily applied to veterinary medicine.[4] Physical modalities are treatments by physical means, which may be in the form of a mechanical or thermal device or the prescription of a specific therapeutic exercise. In humans, physical modalities have proven effective in managing chronic pain conditions, and therapeutic exercise has helped patients improve functional capabilities. Veterinary physical rehabilitative medicine began in the 1960s with the boom of equine sporting events and the subsequent increase in sporting related injuries. Professional interest grew in the 1990s when the American Veterinary Medical Association (AVMA) added "veterinary physical therapy" to its guidelines in 1996, and within two years, the first canine physical rehabilitation certification program was launched in the United States. The American Association of Rehabilitation Veterinarians (AARV) defined veterinary rehabilitation and physical medicine in 2007 and set standards

for treatment of physical injury or illness in an animal to decrease pain and restore function.[5] The AVMA guidelines for Complementary and Alternative Veterinary Medicine (CAVM) were initially approved in 2001 and revised in 2007 and 2019. CAVM guidelines help veterinarians make informed decisions about medical approaches known as "complementary," "alternative," and "integrative." CAVM is held to the same standards as all veterinary medicine and examples include, but are not limited to, veterinary acupuncture and veterinary manual or manipulative therapy (similar to osteopathy, chiropractic, or physical medicine and therapy).

Veterinarians wishing to provide the CAVM therapies on animals should have the requisite skills and knowledge for any treatment they perform and are encouraged, but are not required, to obtain formal training. The AVMA has not evaluated the training or curricula of veterinary physical rehabilitation certification programs, but has recognized the American College of Veterinary Sports Medicine and Rehabilitation (ACVSMR) since 2010.[6] The (AVMA) officially accepted the American College of Veterinary Sports Medicine and Rehabilitation (ACVSMR) as a board certification specialty in 2010. The following year, the AVMA and the American Animal Hospital Association (AAHA) set new standards for annual wellness exams that included pain assessment and osteoarthritis management, which shifted from a solely pharmacological approach to one combined with physical and other ancillary methods such as the use of cold, heat, light, water, transcutaneous electrical nerve stimulation (TENS), massage, and exercise to ease an animal's pain.

REGULATORY CONSIDERATIONS FOR VETERINARY PHYSICAL REHABILITATION PROVIDERS

Determining exactly who is legally qualified to practice veterinary physical rehabilitation is complicated and controversial but is ultimately determined by individual state Physical Therapy and Veterinary Practice Acts. Since 1993, professional organizations such as the AVMA and the American Physical Therapy Association (APTA) have been trying to determine who is qualified to provide veterinary rehabilitation services. The APTA believes physical therapists are the providers of choice regardless of whether the client is human or animal, while the AVMA suggests physical therapy services should be performed by a licensed veterinarian or physical therapist who is educated in nonhuman animal anatomy and physiology. To avoid potential conflict, it is best to use the term *animal physical rehabilitation* or *veterinary physical rehabilitation* when describing or promoting physical rehabilitation services. The terms *physical therapy* and *physical therapist* are protected in many states for use on *humans, people, or persons* by a credentialed physical therapist. Physical therapy practice acts with the written terms *human* or *person* restrict credentialed physical therapists from using their professional title when working with animals, requiring them to act as a skilled veterinary assistant.[7] Ideally, a veterinarian credentialed in physical rehabilitation should collaborate with physical therapists to develop therapeutic plans that allow the credentialed veterinary technician to administer treatments as directed.[8]

In most clinical settings, the credentialed veterinary technician carries out the physical rehabilitation treatment plan developed by a veterinarian or credentialed physical rehabilitation professional. Individual state veterinary practice acts should be reviewed for specific limitations, if any, regarding veterinary physical rehabilitation. Some veterinary practice acts limit application of veterinary physical rehabilitation to *direct* or *indirect* supervision by a state licensed veterinarian for both the physical therapist (acting as a skilled veterinary assistant) and credentialed veterinary technician. The veterinary practice act may further limit all services and therapies to a state licensed veterinarian. Those states permitting credentialed veterinary technicians to work through indirect veterinary supervision are subject to strict guidelines set forth by the state's veterinary medical board. These guidelines often include standard medical record keeping requirements and maintaining a veterinary referral database, progress note protocol, and emergency plan. In most cases the credentialed veterinary technician is best suited to provide physical rehabilitation services in the presence of a licensed veterinarian for patient and client safety.

Veterinarians are not subject to such restrictions and may provide physical rehabilitation by omitting the term "physical therapy" from the list of services provided without violating individual state practice acts. In most states' veterinary practice acts there is no mention of physical therapy or animal physical rehabilitation in regard to what constitutes the practice of veterinary medicine, leaving veterinary professionals with a rather vague understanding of the law or leaving it open to individual interpretation. Regardless of legislative wording, the AVMA states that veterinarians should have adequate skills and knowledge for any treatment modality they are providing.

VETERINARY PHYSICAL REHABILITATION CREDENTIALING

Since 1999 hundreds of veterinarians, credentialed veterinary technicians, physical therapists, and physical therapy assistants have become credentialed in canine or equine physical rehabilitation through one of the two programs in the United States (the first program established in 1999 is sponsored through the University of Tennessee, Knoxville, and offers certification in either canine or equine physical rehabilitation; the second program established in 2002 is the Canine Rehabilitation Institute [CRI] based in Wellington, Florida, and works jointly with Colorado State University's

College of Veterinary Medicine and Biomedical Sciences to offer a Certified Canine Rehabilitation Assistant [CCRA] certificate). Each credentialing program is recognized by American Association of Veterinary State Boards (AAVSB) and the Registry of Approved Continuing Education (RACE) and limits program attendance to licensed veterinarians, credentialed veterinary technicians, physical therapists, physical therapy assistants, or students of the aforementioned programs. Credentialing awards differ between the two schools with graduates from the Tennessee program earning a Certified Canine Rehabilitation Practitioner (CCRP) or Certified Equine Rehabilitation Practitioner (CERP) title while graduates from the CRI program earn a Certified Canine Rehabilitation Therapist (CCRT) if they are a veterinarian or physical therapist, a Certified Canine Rehabilitation Veterinary Nurse (CCRVN) if they are a Certified, Licensed, or Registered Veterinary Technician (CVT, LVT, RVT), Animal Health Technician (AHT) for those who have completed a 2-year veterinary technician program, or Certified Canine Rehabilitation Assistant (CCRA) if they are physical therapy assistants or veterinary staff without formal credentials. For veterinary technicians residing in Canada, the Northern College of Applied Arts and Technology offers a 1-year certification program to graduates of a veterinary technology.

Curricula for programs in the United States focus on anatomy and pathophysiology for common conditions benefiting from physical rehabilitation, assessment techniques, manual therapy techniques, physical modalities, therapeutic exercise, pain management, the business of canine rehabilitation, supervised externships, case reports, and final examinations (including written and lab practical). Additional coursework is available for students wishing to expand their studies in areas of orthotics and prosthetics, canine sports medicine, neurological rehabilitation, orthopedics, low level laser techniques, geriatrics, nutrition, and joint mobilization techniques.[9] Most students enrolled in physical rehabilitation programs are able to complete coursework within the first year of enrollment and complete the certification program within the second year.

Additional specialization after rehabilitation credentialing is available to veterinarians, who may become a diplomat of the American College of Sports Medicine and Rehabilitation once they have met credentialing criteria. Board certification was established and recognized by the AVMA in 2010 after the development of the AARV in 2007. The AARV sets guiding principles for the ideal practice of veterinary physical rehabilitative medicine.[10]

Credentialed veterinary technicians in veterinary physical rehabilitation may become associate members of the AARV whose mission encourages professional development in the field of veterinary physical rehabilitation and provides assistance to veterinarians specializing in physical rehabilitation to improve the quality of animal's lives.[11] The AARV serves as a basis for newly credentialed rehabilitation veterinary technicians wishing to learn more about the specialty field. Credentialed veterinary technicians wishing to enroll in a veterinary physical rehabilitation program should check the requirements section for eligibility. Depending on the program, enrollment may require concurrent participation with a licensed veterinarian and proof of graduation in a veterinary technology program or state licensure.

In 2012, a group of Credentialed Veterinary Technicians certified in veterinary physical rehabilitation formed a Board of Directors committee, the American Association of Rehabilitation Veterinary Technicians (AARVT), which was endorsed and supervised by the AARV. In late 2012, it formed a separate organization, the Academy of Physical Rehabilitation Veterinary Technicians (APRVT), to encourage specialization of veterinary technicians in the field and follow existing NAVTA guidelines. This change was finalized in 2014 by the state of Missouri and the AARVT Board of Directors became the APRVT. The Charter Members of APRVT worked on establishing an Academy following the NAVTA Committee on Veterinary Technicians Specialties (CVTS) guidelines. The APRVT was granted provisional accreditation by the CVTS in February 2017 and as a recognized VTS Academy, the APRVT focuses on providing credentialed veterinary technicians expert level skills, knowledge and assistance for the AARV veterinary members and diplomates of the ACVSMR who encourage and promote research within the profession to validate the science of veterinary sports medicine and rehabilitation. The first certifying examination was held August 9, 2018. At the publication date of this text, the APRVT has met the NAVTA CVTS requirements for provisional recognition.[12] Specific requirements for becoming a VTS (Physical Rehabilitation) can be found on APRVT website at https://www.aprvt.com

ENDNOTES

1 Levine, D., Millis, D. L., & Marcellin-Little, D. J. (2015). Introduction to veterinary physical rehabilitation. *Vet Clinics of North America: Small Animal Practice, 1247*.

2 Murphy, W. (1995). Through a glass darkly. In: *Healing the generations: A history of physical therapy and the American Physical Therapy Association* (pp. 7–11). Lyme, CT: Greenwich Publishing Group Inc.

3 APTA (2011). Today's physical therapist: A comprehensive review of 21st-century health care profession. *Today's Physical Therapist, 1*, 6–9

4 Millis, D. L. (2009). Physical therapy and rehabilitation in dogs. In J. S. Gaynor & W. W. Muir (Eds.), *Handbook of veterinary pain management* (2nd ed., p. 507). St. Louis, MO: Mosby.

5 American Association of Rehabilitation Veterinarians. (2011). *Model standards for veterinary rehabilitation practice.* http://www.rehabvets.org/_docs/AARV_Model_Guidelines_2011-02-07.pdf

6 Journal of the American Veterinary Medical Association News. (2010, June 1). https://www.avma.org/News/JAVMANews/Pages/100601h.aspx

7 Levine, D., & Millis D. (2014). Regulatory and practice issues for veterinary and physical therapy professions. In D. Millis & D. Levine (Eds.), *Canine rehabilitation and physical therapy* (pp. 11–12). Elseveir-Saunders.

8 Van Dyke, J. (2009). Canine rehabilitation: An inside look at a fast-growing market segment. *DVM News Magazine*.

9 Animal Physical Rehabilitation Certification Program Resources. http://www.canineequinerehab.com and http://www.caninerehabilitation.com

10 American Association of Rehabilitation Veterinarians. (2011, February 7). *Model standards for veterinary rehabilitation practice.* http://www.rehabvets.org/_docs/AARV_Model_Guidelines_2011-02-07.pdf

11 American Association of Rehabilitation Veterinarians. Mission statement. http://rehabvets.org/AARV-Membership.lasso

12 Academy of Physical Rehabilitation Veterinary Technicians. Our History. https://www.aprvt.com/history.html

ABBREVIATIONS

α: Alpha.

$α_1$: Alpha-1 (in reference to adrenergic receptors).

$α_2$: Alpha-2 (in reference to adrenergic receptors).

β: Beta.

$β_1$: Beta-1 (in reference to adrenergic receptors).

$β_2$: Beta-2 (in reference to adrenergic receptors).

f: Respiratory rate.

k: Kappa.

μ: Mu.

°: Degree.

%: Percent.

A

A-a: Alveolar-arterial.

AAFP: American Association of Equine Practitioners.

AAFP: American Association of Feline Practitioners.

AAHA: American Animal Hospital Association.

AARV: American Association of Rehabilitation Veterinarians.

AAVSB: American Association of Veterinary State Boards.

AC: Anterior cingulated cortex.

ACD: Acid citrate dextrose.

ACEi: Angiotensin converting enzyme inhibitor.

ACh: Acetylcholine.

ACT: Activated clotting time.

ACVA: American College of Veterinary Anesthesia (not currently used).

ACVAA: American College of Veterinary Anesthesia and Analgesia.

ACVSMR: American College of Veterinary Sports Medicine and Rehabilitation.

AD: Right ear.

ADEs: Adverse drug effects.

ADME: Absorption, distribution, metabolism, and elimination.

A-fib: Atrial fibrillation.

AHT: Animal Health Technologist.

AKI: Acute kidney injury.

ALKP: Alkaline phosphatase; sometimes abbreviated Alk phos.

ALS: Advanced life support.

ALT: Alanine aminotransferase.

AMDUCA: Animal Medicinal Drug Use Clarification Act.

AMPA: α-Amino-3-hydroxy-5-methyl-4-isoxazolepropionic acid.

ANS: Autonomic nervous system.

APC: Atrial premature contractions.

APL: Adjustable pressure limiting (valve).

APTA: American Physical Therapy Association.

aPTT: Activated partial thromboplastin time.

ARDS: Acute respiratory distress syndrome.

AROM: Active range of motion.

ASA: American Society of Anesthesiologists.

AST: aspartate aminotransferase.

ASTM: American Society for Testing and Materials.

ATP: Adenosine triphosphate.

AU: Action unit.

AV: atrioventricular.

AVMA: American Veterinary Medical Association.

AVTA: Academy of Veterinary Technicians in Anesthesia (not currently used).

AVTAA: Academy of Veterinary Technicians in Anesthesia and Analgesia.

B

BBB: Blood-brain barrier.

BCS: Body condition scoring.

BG: Blood glucose.

BGPC: Blood-gas partition coefficient.

BLS: Basic life support.

BP: Blood pressure.

bpm: Beats per minute.

BUN: Blood urea nitrogen.

C

C: Celsius.

C-I: Schedule 1 controlled substance.

C-II: Schedule 2 controlled substance.

C-III: Schedule 3 controlled substance.

C-IV: Schedule 4 controlled substance.

C-V: Schedule 5 controlled substance.

CaO_2: oxygen content in arterial blood.

CAVM: Complementary and Alternative Veterinary Medicine

CBC: Complete blood count.

CBD: Cannabidiol.

CBF: Cerebral blood flow.

CBPI: Canine Brief Pain Inventory.

CBW: Canine Body Worker.

CCRA: Certified Canine Rehabilitation Assistant.

CCRP: Certified Canine Rehabilitation Practitioner.

CCRT: Certified Canine Rehabilitation Therapist.

CCRVN: Certified Canine Rehabilitation Veterinary Nurse.

CE: Continuing education.

CERP: Certified Equine Rehabilitation Practitioner.

Cl^-: Chloride.

$cm\ H_2O$: Centimeters of water.

CMPS: Composite measure pain scale.

CNS: Central nervous system.

CO: Cardiac output.

CO_2: Carbon dioxide.

CODI: Cincinnati Orthopedic Disability Index.

COP: Colloidal oncotic pressure.

COX: Cyclooxygenase.

CPA: Cardiopulmonary arrest.

CPCR: Cardiopulmonary cerebral resuscitation.

CPDA: citrate phosphate dextrose adenine.

CPK: creatine phosphokinase.

CPROM: Continuous passive range of motion.

CRI: Canine Rehabilitation Institute.

CRI: Constant rate infusion.

CRT: Capillary refill time.

CRTZ: Chemoreceptor trigger zone.

CSA: Canadian Standards Association.

C-section: Cesarean section.

CSU: Colorado State University.

CT: Computed tomography.

CVM: Center for Veterinary Medicine.

CVP: Central venous pressure.

CVPP: Certified Veterinary Pain Practitioner.

CVT: Certified Veterinary Technician.

D

D_5W: 5% dextrose in water.

DA: Dopamine.

DAP: Diastolic arterial pressure.

DEA: Drug Enforcement Administration.

DHA: Docosahexaenoic acid.

DIVAS: Dynamic interactive visual analogue scales.

DLH: Domestic longhair (cat).

DM: Diabetes mellitus.

DNR: do not resuscitate.

DO_2:   Oxygen delivery. The rate of oxygen transported from the lungs to the peripheral tissues.

DSH: Domestic shorthair (cat).

E

ECF: Extracellular fluid.

ECG: Electrocardiogram.

ECVAA: European College of Veterinary Anaesthesia and Analgesia.

EMD: Electromechanical dissociation.

EPA: Eicosapentaenoic acid.

ERV: expiratory reserve volume.

ESWT: Extracorporeal shockwave therapy.

$EtCO_2$:   End tidal carbon dioxide.

ETT: Endotracheal tube; also abbreviated ET tube.

ET tube: Endotracheal tube; also abbreviated ETT.

F

F: Bioavailability.

F: Fahrenheit.

FARAD: Food Animal Residue Avoidance Database.

FDA: Food and Drug Administration.

FDA-CVM: Food and Drug Administration Center for Veterinary Medicine.

FDP: Fibrin degradation products.

FEI: Federation Equestrian International.

FGS: Ferret Grimace Scale.

FIC: Feline interstitial cystitis.

FIM: Functional Independence Measure.

FiO_2:   Fraction of inspired oxygen.

FK: fentanyl and ketamine.

FL: Fentanyl and lidocaine.

FLK: Fentanyl, lidocaine, and ketamine.

FOPS: Feline orofacial pain syndrome.

Fr: French.

FRC: Functional residual capacity.

G

g: Gram.

G: Gauge; also abbreviated ga.

ga: Gauge; also abbreviated G.

GaAa: Gallium arsenate.

GaAlAs: Gallium aluminum arsenate.

GABA: Gamma-aminobutyric acid.

GCMS: Glasgow Composite Measure Pain Score.

GCPS: Glasgow Composite Pain Scale.

GDV: Gastric dilatation volvulus.

GG: Glyceryl guaiacolate.

GI: Gastrointestinal.

GKD: Guaifenesin-ketamine-detomidine.

GKR: Guaifenesin-ketamine-romifidine.

GKX: Guaifenesin-ketamine-xylazine.

gt: Drop.

gtt: Drops.

H

h: Hour; also abbreviated hr.

H_2O: Water.

Hb: Hemoglobin.

HCO_3^-:   Bicarbonate.

HCPI: Helsinki Chronic Pain Index.

He-Ne: helium-neon.

HES: Hydroxyethyl.

HGS: Horse Grimace Scale.

HPA: Hypothalamic-pituitary-adrenal.

HQHV: High quality high volume.

HQHV: High quality/high volume.

HR: Heart rate.

hr: Hour; also abbreviated h.

HYPP: Hyperkalemic periodic paralysis.

I

IA: Intra-articular.

IACUC: Institutional Animal Care and Use Committee.

IASP: International Association for the Study of Pain.

IC: Insular cortex.

ICF: Intracellular fluid.

iCO_2:   Inspired carbon dioxide.

ICP: Intracranial pressure.

ICU: Intensive care unit.

IM: Intramuscular.

IO: Intraosseous.

IOP: Intraocular pressure.

IPPV: Intermittent positive pressure ventilation.

Ipsilateral: Same side.

IRV: Inspiratory reserve volume.

ITD: Impedance threshold device.

IV: Intravenous.

IVAA: Intravenous analgesic adjunct.

IVAPM: International Veterinary Academy of Pain Management.
IVDD: Intervertebral disc disease.
IVECCS: International Veterinary Emergency & Critical Care Symposium.
IVRA: Intravenous regional anesthesia.

J

J: Joule.
JAVMA: Journal of the American Veterinary Medical Association.

K

K^+: Potassium.
KXM: Ketamine-xylazine-midazolam.
kg: Kilograms.

L

L/min: Liters per minute.
LASER: Light amplification by stimulated emission of radiation.
LLLT: Low level laser therapy.
LOAD: Liverpool Osteoarthritis in Dog.
LOX: Lipoxygenase.
LRS: Lactated Ringer's solution.
LSD: Lysergic acid diethylamide; an hallucinogenic drug.
LVT: Licensed Veterinary Technician.

M

M1: O-desmethyltramadol.
mA: milliampere.
MAC: Minimum alveolar concentration.
MAP: Mean arterial pressure.
mcg: Microgram.
mcg/kg/min: Micrograms per kilogram per minute.
MCPS: Multidimensional composite pain scale.
MDR-1: Multidrug resistance mutation 1.
M_E: Minute volume.
mg: Milligram.
Mg^{+2}: Magnesium.
MGS: Mouse Grimace Scale.
Miosis: Constriction of the pupil.
MK: Morphine and ketamine.
mL: Milliliter.
mL/kg: Milliliter per kilogram.
mL/min: Milliliter per minute.
ML: Morphine and lidocaine.
MLK: Morphine, lidocaine, and ketamine.
mm: Millimeter.
mm Hg: Millimeters of mercury.
MM: Mucous membrane.
MRI: Magnetic resonance imaging.
MSM: Methylsulfonylmethane.
MUMS: Minor use and minor species.
MV: Minute volume; also abbreviated VE.

N

N_2O: Nitrous oxide.
Na^+: Sodium.

NaCl: Sodium chloride.
NAVAS: North American Veterinary Anesthesia Society.
NAVEL: Naloxone, atropine, vasopressin, epinephrine, and lidocaine.
NAVTA: National Association of Veterinary Technicians in America.
NAVTA-CVTS: National Association of Veterinary Technicians in America— Committee on Veterinary Technician Specialties.
NGF: Nerve growth factor.
NIBP: Non-invasive blood pressure.
NIOSH: National Institute for Occupational Safety and Health.
NK: Neurokinin.
NMDA: N-methyl-D-aspartate.
NMES: Neuromuscular electrical stimulation.
NRS: Numerical rating scale.
NSAID: Non-steroidal anti-inflammatory drug.
NT: Nasotracheal.

O

O_2: Oxygen.
OA: Osteoarthritis.
OACM: Osteoarthritis case manager.
OHE: Ovariohysterectomy.
OHSA: Occupational Health and Safety Administration.
OTM: Oral transmucosal.

P

$PaCO_2$: Partial pressure of carbon dioxide molecules dissolved in the plasma phase of an arterial sample (i.e. not bound to Hb).
PaO_2: Partial pressure of oxygen molecules dissolved in the plasma phase of an arterial sample (i.e. not bound to Hb).
PaO_2/FiO_2 (P/F) ratio: Partial pressure of oxygen/ Fraction of inspired oxygen ratio.
PAO_2: Partial pressure of oxygen in the alveoli.
P_{bar}: The barometric pressure (at sea level, 760 mm Hg).
PCV: Packed cell volume.
PDA: Patient ductus arteriosis.
PF: Physical examination.
PEA: Pulseless electrical activity.
PEMF: Pulsed electromagnetic field.
P/F ratio: Partial pressure of oxygen/ Fraction of inspired oxygen ratio; also abbreviated PaO_2/FiO_2.
PFC: Prefrontal cortex.
PGE_2: Prostaglandin E_2.
PGS: Piglet Grimace Scale.
pH: Scale of acidity from 0 to 14 with lower numbers being more acidic and higher numbers being more basic.
P_{H2O}: The water vapor pressure in the atmosphere (47 mm Hg).
PI: Principle Investigator.
PISS: Pin index safety system.
PIVA: Partial intravenous anesthesia.
PLR: Pupillary light reflex.
PNS: Peripheral nervous system.
PO: Orally.
pO_2: Partial pressure of oxygen molecules dissolved in the plasma phase.
PO_4^{-3}: Phosphate.
PONV: Post-operative nausea and vomiting.
PP: Pulse pressure.

ppm: Parts per million.

PPV: Positive pressure ventilation.

pRBC: Packed red blood cells.

PROM: Passive range of motion.

PRT: Palliative radiation therapy.

PSGAG: Polysulfated glycosaminoglycan.

psi: Pounds per square inch.

PT: Physical therapy; also prothrombin time.

PVC: Polyvinylchloride; also premature ventricular contraction (also abbreviated VPC).

PvCO$_2$: carbon dioxide dissolved in venous blood plasma.

PvO$_2$: oxygen dissolved in venous blood plasma.

Q

q: Every.

QATs: Quick assessment tests.

R

RAC: Reticular activation center.

RACE: Registry of Approved Continuing Education.

RBC: Red blood cell.

RbtGS: Rabbit Grimace Scale.

RECOVER: Reassessment Campaign on Veterinary Resuscitation.

RGS: Rat Grimace Scale.

RICE: Rest, Ice, Compression, and Elevation.

ROM: Range of motion.

ROSC: Return of spontaneous circulation.

RR: Respiratory rate.

RUM: Radial, ulnar, and median (nerves).

RUMM: Radial, ulnar, median, and musculocutaneous (nerves).

RV: Residual volume.

RVT: Registered Veterinary Technician.

S

SA: Sinoatrial.

SAGM: saline adenine glucose mannitol.

SAMe: S-adenosylmethionine.

SaO$_2$: Arterial oxygen saturation.

SAP: Systolic arterial pressure.

SC: Subcutaneous; also abbreviated SQ and SubQ.

SDMA: Symmetric dimethylarginine.

SDS: Simple descriptive scale; also Safety Data Sheets.

Sec: Second.

SF: Short form.

SGS: Sheep Grimace Scale.

SNRI: Serotonin-norepinephrine reuptake inhibitor.

SOAP: Subjective, objective, assessment, and plan (method of record keeping).

SPFES: Sheep Pain Facial Expression Scale.

SpO$_2$: Peripheral oxygen saturation as determined by the pulse-oximeter, reported as a percentage.

SQ: Subcutaneous; also abbreviated SubQ and SC.

SSRI: Selective serotonin reuptake inhibitor.

SSS: Sick sinus syndrome.

SubQ: Subcutaneous; also abbreviated SQ and SC.

SV: Stroke volume.

SVR: Systemic vascular resistance.

SVT: Supraventricular tachycardia.

T

TACO: Transfusion-associated circulatory overload.

TCA: Tricyclic antidepressant.

TCM: Traditional Chinese Medicine.

TENS: Transcutaneous electrical nerve stimulator.

TI: Therapeutic index.

TIVA: Total intravenous anesthesia.

TLC: Total lung capacity; also tender loving care.

TP: Total protein.

TPN: Total parenteral nutrition.

TPR: Temperature, pulse, and respiration.

TRALI: Transfusion-related acute lung injury.

trkA: Tropomyosin receptor kinase A.

TrpV1: Transient receptor potential cation channel subfamily V member 1.

TV: Tidal volume; also abbreviated V$_T$.

U

UNESP: Universidade Estadual Paulista (São Paulo State University).

UA: Urinalysis.

US: Ultrasound.

USDA: United States Department of Agriculture.

V

V/Q mismatch: Ventilation-perfusion mismatch.

VAS: Visual analog scale.

V$_C$: Vital capacity.

V$_E$: Minute volume; also abbreviated MV.

VF: Ventricular fibrillation; also abbreviated V-fib.

V-fib: Ventricular fibrillation; also abbreviated VF.

VPC: Ventricular premature contraction; also abbreviated PVC.

V$_T$: Tidal volume; also abbreviated TV.

V-tach: Ventricular tachycardia.

VTAS: Veterinary Technician Anesthetist Society.

VTBI: Volume to be infused.

VTS: Veterinary Technician Specialist.

VTS (Anesthesia/Analgesia): Veterinary Technician Specialist (VTS) in anesthesia/analgesia.

W

W: Watt.

WAG: Waste anesthetic gas.

WBC: White blood cell.

WHO: World Health Organization.

WNL: Within normal limits.

GLOSSARY

1st-degree AV block Type of AV block that results in a prolonged but consistent P-R interval on the ECG, but all impulses are conducted through the AV node to the ventricles and QRS complexes appear consistently on the ECG.

2nd-degree AV block Type of AV block that is characterized by a delay and intermittent lack of transmission of impulse through the AV node, resulting in intermittently absent QRS complexes and "dropped beats." There are two subtypes of 2nd-degree AV block: Mobitz Type 1 and Mobitz Type 2.

3rd degree AV block Type of AV block that is due to a defect in the cardiac conduction system in which there is no conduction through the atrioventricular node, therefore causing a complete dissociation between the sinoatrial (SA) and AV nodes.

A

Active assisted range of motion Movement where external forces are used to guide a joint through motion *voluntarily generated* by the patient.

Active range of motion (AROM) Movement around a joint *without external intervention*; therefore, the patient causes the movement.

Active restricted range of motion Movement where a device or equipment such as a weighted vest is used to increase voluntary effort by the patient.

Active scavenging systems Hoses that use a fan or vacuum to remove waste gas after it has passively moved from the pop-off valve to the scavenging duct system.

Acupuncture One of the five components of Traditional Chinese Medicine that involves inserting fine needles through the skin at specific points of the body (acupoints), often corresponding with motor points in the muscle to affect bodily systems and healing processes.

Acute pain Pain that arises suddenly in response to a specific injury and is usually treatable; is adaptive (physiologic) pain that serves as a warning signal to notify the body of injury or tissue damage.

Adaptive pain Acute pain that acts as a warning signal to the body to notify it of injury and/or tissue damage; also known as physiologic pain.

Adjustable pressure limiting (APL) valve Safety valve that allows excess gases to vent from the breathing circuit to avoid a build-up of pressure within the system; also called a pop-off valve.

Afferent Conducted inward or toward something; e.g., afferent neurons carry nerve impulses from receptors or sense organs toward the central nervous system.

Affinity The measure of a drug's strength with which it binds to its receptor.

Afterload The tension in the left ventricle immediately before the aortic valve opens.

Agent gas analysis Instrument that determines the amount of inhalant anesthetic agent present at inspiration (coming from machine) and expiration (having just left the alveoli).

Agonal breathing An abnormal pre-death breathing pattern characterized by short, gasping breaths.

Agonist Drug that binds to a cell receptor and stimulates a response characteristic of that receptor.

Allodynia Exaggerated and prolonged responsiveness of neurons to even normal input after tissue damage; in other words pain due to a stimulus that does not normally provoke pain.

Allometric scaling Approach to choosing drug dosages which considers the influence of patient size in dosage selection; choosing drug dosages based on the size of the patient so that smaller patients receive doses calculated from the higher end of the drug dosage range and larger patients receive doses calculated from the lower end of the drug dosage range; also known as metabolic scaling.

Allopreening Grooming/rubbing feathers of a bird.

Alternative medicine Many types of health care practices not typically taught in western medicine and may be combined with conventional western medical treatments; when used together it is known as complementary or integrative medicine.

Alveolar dead space The volume of gas in alveoli that is ventilated but not perfused.

Alveolar oxygen equation Mathematical determination of the partial pressure of oxygen in the alveoli; $PAO_2 = ((P_{bar} - P_{H_2O}) \times FiO_2) - (PaCO_2/0.8)$.

Analgesia Relief of pain without loss of consciousness; absence of pain or noxious stimulation.

Analgesic An agent that has pain relieving properties.

Analgesic adjunctives Drugs that on their own have minimal pain relieving effects, but are included to enhance the effect of another analgesic agent.

Anatomical dead space At inspiration the volume of air occupying the space in the nose, pharynx, larynx, trachea, bronchi, and bronchioles.

Anesthesia induction Careful administration of appropriate agents to induce unconsciousness and "bring about" anesthesia.

Anesthesia Without pain; part or entire loss of sensation to the body.

Anesthetic depth The degree to which the central nervous system (CNS) is depressed by a general anesthetic agent.

Anion Negatively charged ion.

Antagonist Drug that binds to a cell receptor and prevents the cell from performing some function.

Antiemetic An agent that reduces or prevents nausea and vomiting.

Anxiolysis Having anti-anxiety ability or relieves anxiety.

Anxiolytic Anti-anxiety; term is used to describe anti-anxiety drugs.

Apnea Cessation of breathing.

Apneustic breathing Breathing pattern in which a patient takes a prolonged inspiratory breath, then holds the breath, and then quickly exhales. Seen most often with ketamine administration.

Apneustic ventilation Prolonged inspirations with subsequent short exhalations; patient appears to be holding its breath.

Apparatus dead space The area where bidirectional flow takes place within a breathing circuit or endotracheal tube but no gas exchange occurs; also called mechanical dead space.

Arterial oxygen saturation (SaO$_2$) The amount of oxygen that binds to hemoglobin in the red blood cells; it is determined invasively via an arterial blood gas analysis.

Arterial oxygen tension (PaO$_2$) The amount of oxygen molecules dissolved in the plasma phase of an arterial sample (i.e., not bound to Hb); it is determined invasively via an arterial blood gas analysis.

American Society of Anesthesiologists (ASA) physical status The American Society of Anesthesiologists classification system rating patient risk during anesthesia based on the patient's health.

Apteria Bare spaces between the feathered areas on the bird's body

Atelectasis Alveolar collapse

Atrial fibrillation (A-fib) Dysrhythmia in which atrial depolarization and repolarization occur, but there is no coordinated atrial contraction; on ECG complexes are characterized by a fast, "irregularly irregular" pattern of normal QRS complexes without discernible P waves.

Atrial premature contractions Extra heart beat arising from a focus of electrical activity in the atria other than the sinoatrial node; on ECG they appear as normal-looking complexes that appear early in the normal cadence of the underlying rhythm.

Atrioventricular (AV) block Cardiac dysrhythmia that results from a delay in conduction between the SA node and the AV node.

Atrioventricular (AV) node A group of specialized cardiac muscle cells at the junction between the atrium and ventricle.

Autoregulation of blood flow Protective mechanism that maintains constant blood flow and perfusion pressure to the kidneys and brain despite changes in systemic blood pressure.

Autoregulation of cerebral blood flow Protective mechanism that maintains constant cerebral blood flow via contraction and relaxation of cerebral arterioles in response to changes in cerebral perfusion pressures.

B

Bagging Slang term for providing mechanical or manual ventilation to a patient.

Balanced analgesia Concept of using a combination of different analgesic drug classes to target different sites along the nociceptive pain pathway in an effort to maximize patient comfort and recovery; often allows for the reduction in dosage and frequency of individual drugs, thereby reducing adverse effects; also called multimodal analgesia.

Balanced anesthesia Theory in which general anesthesia is produced by administering several drugs with the goal of exploiting each drug's positive actions while avoiding potential adverse effects associated with large doses of a single drug. The philosophy encourages the use of several agents, each designed to affect a different function (e.g., providing muscle relaxation and amnesia or blocking pain and motor function); also known as multimodal anesthesia.

Balanced solution Electrolyte solution that is buffered with precursors of bicarbonate and more closely mimic plasma electrolyte composition, particularly in regards to chloride content.

Barotrauma Damage to lung tissue from overpressurization and overstretching of the delicate lung tissue.

Bioavailability (F) The degree to which a drug is absorbed and reaches the circulation; drugs that are 100% available have a bioavailability of 1.

Blood products Types of colloid that contain red blood cells, plasma proteins, or both; sometimes referred to as natural colloids; examples include whole blood, packed red blood cells, plasma, and albumin.

Blood-brain barrier (BBB) Partition formed by tight junctions between the capillary endothelial cells and also by the glial cells that surround the capillaries, which restricts the penetration of polar and ionized molecules into brain neurons.

Blood-gas partition coefficient (BGPC) Ratio of gas concentration in blood compared to the gas concentration in contact with blood when partial pressures in both compartments are equal; it is a measure of the tendency of an inhalation anesthetic agent to exist as a gas or to dissolve in blood.

Bolus Volume of fluid given rapidly, typically by the IV route.

Bradypnea Slow, regular breathing.

Bronchial intubation Inadvertent placement of the endotracheal tube into one of the mainstem bronchi; usually caused by an excessively long ETT. This only allows for one lung ventilation and may lead to hypoxemia.

Buffer Additives that help fluid resist changes in pH; fluids given to animals may be modified with bicarbonate buffers, such as gluconate, lactate or acetate.

Bundle of His Collection of heart muscle cells that transmit electrical impulses from the AV node to the apex of the heart.

C

Capnogram A record of a graph that has time on the x-axis and CO_2 levels on the y-axis.

Capnography A graphical record of expired CO_2 concentrations during a respiratory cycle.

Carbon dioxide (CO_2) absorber device containing absorbent chemicals that removes CO_2 from the exhaled gases before they are reintroduced to the patient.

Cardiac contractility The intrinsic strength of the heart's contraction; also called inotropy.

Cardiac contractility The intrinsic strength of the heart's contraction; also called inotropy.

Cardiac output (CO) Volume of blood ejected by each ventricle per minute; it is the product of stroke volume (SV) and heart rate (HR); abbreviated CO.

Cardiac pump theory Concept that external chest compressions lead to blood flow during cardiopulmonary cerebral resuscitation (CPCR) because the ventricles of the heart are directly compressed between the ribs.

Cardiopulmonary arrest (CPA) A sudden total failure of both the circulatory and respiratory systems.

Catalepsy Condition characterized by a loss of sensation and consciousness accompanied by rigidity of the body.

Cataleptic "Dissociative-like" state where patients become detached from and not responsive to their surroundings.

Cation Positively charged ion.

Ceiling effect Property of a drug in which increasing dosages will not produce a greater effect.

Central sensitization Increased responsiveness of nociceptive neurons and changes to the central nervous system that alters the body's normal response to pain in other words, persistent post-injury changes in the central nervous system that result in pain hypersensitivity.

Central venous catheter (central line) An indwelling catheter placed into a central vein, typically the jugular or caudal vena cava, for long term fluid administration. These catheters allow for administration of IV fluids and medication, sampling of blood, use of hyperosmolar solution (> 600 mOsm/L), parenteral nutrition, potentially irritating drugs known to cause phlebitis and tissue sloughing, and central venous pressure measurement.

Central venous oxygen saturation The percentage of hemoglobin bound to oxygen in a central vein (e.g., cranial vena cava). This percentage is used clinically as a marker of inadequate oxygen delivery.

Central venous pressure (CVP) The pressure within a central vein (e.g., cranial vena cava).

Chemoreceptor trigger zone (CRTZ) Area of the brain that lies outside of the blood-brain barrier near the fourth ventricle of the brain; this area sends information to the vomiting/emetic center located in the medulla, which integrates many inputs that result in vomiting.

Choana An opening located on the dorsal palate that connects to the nares (nostrils) in birds.

Chromodachryorrhea Red tears or porphyrin staining; sign of pain in rats.

Chronic pain Pathologic (maladaptive) pain that serves no useful biological purpose that persists for longer than the expected time frame for normal healing.

Closed circle system Rebreathing system in which the oxygen flow rate approximates the patient's metabolic oxygen consumption.

Cluster care Coordinating treatment times with other procedures to minimize patient stress and allow rest in between necessary medical evaluations.

Coaxial A smaller tube located within a larger tube.

Coelom One body cavity in birds where all the visceral organs lie.

Colloid Solutions that contain higher molecular weight molecules that contribute to oncotic pressure.

Common gas outlet Point at which oxygen and anesthetic gas enter the breathing circuit; also called the fresh gas inlet.

Compensation An abnormal movement or posture from abnormal skeletal alignment or associated movement.

Competitive antagonist Drug that competes with the agonist for the same receptors and is reversible by giving high dosages of agonist (reversible or surmountable antagonism).

Conduction Transfer of heat by the direct interaction of the molecules in one area with those in another area.

Conductive heat loss Heat loss through transfer of heat from the warmer patient body to cooler objects/surfaces in contact with the animal.

Contralateral Opposite side.

Controlled substance Drug or chemical whose manufacturing, sale, possession, and use is strictly regulated by the federal government due to its risk of human abuse and addiction.

Convection Heat transfer by movement of air or fluid from a warm area to a cooler area.

Convective heat loss Heat loss to cooler objects not in direct contact with the patient's body (such as air current or ambient temperature).

Crop In birds the outpouching of the esophagus used to store food prior to passage into the proventriculus.

Cross-matching Test performed prior to a blood transfusion in order to determine if the donor's blood is compatible with the blood of an intended recipient.

Cryotherapy Application of cold that is used during the acute inflammatory phase of tissue healing and after exercise to minimize the metabolic rate of reactions involved in tissue injury and healing; also known as cold therapy.

Crystalloid Solutions that consist of a sodium or dextrose base dissolved in water. Some crystalloids have other electrolytes and/or buffers added to them.

Cumulative Property in which administration of the drug may produce effects that are more pronounced than those produced by the first dose. In other words, an animal is unable to metabolize and excrete a normal drug dose before the next dose is given. When the second dose of the drug is given, some drug from the first dose remains in the body, which results in too much drug in the animal, potentially leading to toxicity.

Cushing reflex Physiological nervous system response to increased intracranial pressure that results from increased sympathetic discharge to the peripheral vasculature (sometimes resulting in hypertension) with a reflex increase in parasympathetic discharge to the heart (resulting in bradycardia) in an attempt to maintain cerebral perfusion; also known as Cushing response or Cushing reflex vasopressor response.

Cyclooxygenase Enzymes responsible for formation of prostaglandins, prostacyclin, and thromboxane; abbreviated COX. Inhibition of COX can provide relief from inflammation and pain.

D

Dead space The volume of inhaled air that does not take part in gas exchange.

Dead space ventilation Areas where there is ventilated air but no gas exchange takes place.

Defibrillation The stopping of fibrillation of the heart by administering a controlled electric shock in order to allow restoration of the normal rhythm.

Dehydration Loss of body water; refers to lack of fluid in the interstitial compartment.

Diabetes mellitus (DM) Endocrinopathy that results from deficient serum insulin levels (hypoinsulinemia; Type I) or decreased insulin secretion and sensitivity of peripheral tissues to its effects (Type II). Dogs and cats with DM present with polyuria, polydipsia, polyphagia, and weight loss.

Diastolic pressure Lowest pressure during relaxation of the heart, it is comprised of heart rate and systemic vascular resistance (vessel size) and volume.

Diffusion impairment Impairment of gas exchange that can cause hypoxemia; may be caused by conditions such as chronic fibrosis of the lungs.

Diffusion Particles move from an area of high concentration to an area where particles are in low concentration.

Dissociative anesthesia Anesthesia that produces a catalepsy-like state, in which the patient feels dissociated from its environment, and good somatic analgesia but poor visceral analgesia.

Distress An aversive state in which an animal is unable to adapt completely to stressors and the resulting stress is displayed in maladaptive behaviors.

Dosage Amount of drug per animal species' body weight or measure; e.g., mg/kg or g/lb.

Dose Amount of drug administered at one time to achieve the desired effect; e.g., mL, mg, or tablets.

Dysphoria The state of feeling unwell, unhappy, or emotional or mental discomfort. In pets, can include restlessness, anxiety, disorientation, vocalization, and even aggression or attempts to escape.

Drug withdrawal time An estimate of the amount of time it takes 99.9% of the drug to be eliminated from the plasma.

Dyspnea Difficult or labored breathing.

E

Efficacy The effectiveness of the drug to work once it is in the patient.

Electrolyte Particles or ions that can conduct an electrical charge; in body systems they are defined as charged particles such as Na^+ or Cl^-.

Emergence delirium State in which the patient is agitated and inconsolable, uncooperative and typically thrashing, vocalizing and defecating. This condition can be seen during anesthesia recovery and is self-limiting.

End-tidal carbon dioxide (EtCO$_2$) Value used to non-invasively monitor the adequacy of ventilation and subsequent circulation on a continual basis. $EtCO_2$ reflects systemic arterial CO_2 and is obtained via capnography that correlates well with the CO_2 in blood; reported in mm Hg.

Endorphins Any of a group of hormones secreted within the brain and nervous system that activate the body's opiate receptors causing an analgesic effect.

Endpoint criteria One or more predetermined physiological or behavioral signs that identify the point at which an experimental animal's pain and/or distress is terminated, minimized, or reduced. Examples of endpoint criteria include euthanizing the animal humanely, terminating a painful procedure, or giving treatment to relieve pain and/or distress.

Epidural analgesia Segmental pain relief by providing a drug at its target (e.g., a drug epidurally can migrate to another location).

Epidural anesthesia Complete sensory, motor, and possible autonomic blockade produced by local anesthetic agents injected epidurally.

Eucapnia   Normal CO_2 tension in blood.

Eupnea   Ordinary and quiet breathing.

Euvolemia   Adequate circulating plasma volume.

Evaporation   Act of conversion from liquid to gas, providing cooling of the surface.

Evaporative heat loss   Heat loss due to evaporation of liquids from the surface of the body or within an open body cavity.

Expiratory pause   Break in breathing following expiration.

Expiratory reserve volume (ERV)   Amount of air expired over the tidal volume (extra amount that could be exhaled after normal expiration).

Extracorporeal shockwave therapy (ESWT)   Application of high-energy, high-amplitude pressure waves to tissues to facilitate tissue healing.

Extubation   Process of removing the endotracheal tube.

F

First-order neurons   Primary afferent (receiving) fibers that originate in the periphery and projects to the spinal cord.

First-pass metabolism   The process by which an orally administered drug passes to the liver first, which reduces the amount of active drug in systemic circulation.

Flow-by oxygen   Method of delivering oxygen by placing the end of the oxygen hose from the anesthesia machine by the patient's nares/mouth using high flow rates.

Flowmeter   A glass cylinder that controls the flow of oxygen through the vaporizer to the breathing system at a specific flow rate shown in liters per minute (L/min) or milliliters per minute (mL/min); it is located downstream from the pressure regulator.

Fluid overload   Condition in which the body's fluid requirements are met and the administration of fluid occurs at a rate that is greater than the rate at which the body can use or eliminate the fluid; in other words, administration of more fluid than the body can handle; also known as circulatory overload.

Focal erosive lesions   Local destruction that occurs at the joint margins and in subchondral bone of patients with arthritis. These erosions progress throughout the course of disease and generally correlate with disease severity.

French (Fr)   Unit of measure that describes a catheter's circumference and is a series of whole numbers that represents three times the catheter's outer diameter in millimeters.

Fresh gas inlet   Point at which oxygen and anesthetic gas enter the breathing circuit; also called the common gas outlet.

Freund's complete adjuvant   A mineral oil used to boost the immune system.

Full agonist   Drug that binds to a cell receptor and stimulates a maximal response characteristic of that receptor.

Functional Independence Measure (FIM)   Tool used to assess a patient's ability to carry out activities of daily living safely and autonomously.

Functional residual capacity (FRC)   Amount of air remaining in the lungs after a normal; comprised of expiratory reserve volume + residual volume.

G

Gauge   Unit of measure that describe a catheter's diameter in which larger gauge numbers have narrower diameters than those with smaller gauge numbers.

General anesthesia   A reversible state of unconsciousness produced by anesthetic agents, with absence of pain perception over the entire body and a greater or lesser degree of muscular relaxation; the drugs producing this state are most commonly administered by inhalation, intravenously, or intramuscularly.

Goniometer   An instrument that objectively measures the range of motion at a joint.

Goniometry   Measurement of a joint's range of motion.

H

Hanger yokes   The part of the anesthesia machine that allows the portable compressed gas cylinder to be connected with the anesthesia machine.

High-pressure system   Area of the anesthetic machine responsible for taking gases at cylinder pressure and reducing and regulating them to be used safely within the machine; includes the compressed gas cylinder or piped oxygen, hanger yokes, high-pressure hoses, pressure gauges, and pressure regulators. Pressures in the high-pressure system may be as high as 2200 psi.

Hudson demand valve   A special valve that when attached to an oxygen source can provide a large volume of oxygen at the push of a button (75 L/min). This device can be used to briefly ventilate a large animal or provide oxygen.

Hydrophilic

Hyperadrenocorticism water soluble (Cushing's disease)   Endocrinopathy that results from either over secretion of cortisol from the adrenal gland due to a pituitary gland tumor (most common) or adrenal gland tumor or excessive cortisol from overuse of corticosteroids. This condition is seen more commonly in dogs who often present with a pendulous abdomen, alopecia, and muscle weakness.

Hyperalgesia   Increased pain from a stimulus that should normally provoke pain.

Hypercapnia   Elevated CO_2 tension in blood.

Hyperesthesia   An abnormal increase in sensitivity to sensory stimuli.

Hyperosmolar   Term used most commonly to describe a solution that has a higher dissolved particle composition than blood.

Hyperpnea   Rapid breathing that may be increased in depth; overrespiration.

Hyperthermia   A core body temperature elevation that is above the accepted normal range for each species.

Hyperthyroidism   Endocrinopathy caused by overproduction of thyroid hormone by the thyroid glands and is seen more commonly in cats; excessive thyroid hormones result in increases in metabolic rate and oxygen consumption, weight loss, hyperactivity, tachycardia, and tachypnea.

Hypertonic   Containing more dissolved particles (such as salt and other electrolytes) than is found in normal cells and plasma.

Hyperventilation   Increase in the ventilation rate and/or tidal volume that leads to hypocapnia.

Hypoadrenocorticism (Addison's disease)   Endocrinopathy resulting from decreased production of glucocorticoids and mineralocorticoids from the adrenal gland. This condition is seen more commonly in dogs who present with weakness, dehydration, hypotension, vomiting, and weight loss associated with electrolyte abnormalities such as hyperkalemia, hyponatremia, and hypochloremia.

Hypocapnia   Lowered CO_2 tension in blood.

Hypopnea   Slow breathing that may be shallow in depth; underrespiration.

Hypotension   A decrease in blood pressure outside of the normal range; mean arterial blood pressure less than 60 mm Hg or systolic arterial blood pressure less than 90 mm Hg if taken indirectly (less than 80 mm Hg if taken directly).

Hypothermia   A core body temperature reduction that is below the accepted normal range for each species.

Hypothyroidism   Endocrinopathy caused by less than normal levels of thyroid hormone due to underproduction of thyroid hormone by the thyroid gland and is seen more commonly in canines; untreated patients may have reduced cardiac output, decreased circulating blood volume, and the potential for decreased hepatic metabolism and renal excretion of drugs.

Hypotonic   Containing less dissolved particles (such as salt and other electrolytes) than is found in normal cells and plasma.

Hypoventilation Reduction in the ventilation rate and/or tidal volume that leads to hypercapnia.

Hypovolemia Reduction of extracellular fluid volume resulting in decreased tissue perfusion; describes lack of fluid in the intravascular compartment.

Hypoxemia Low partial pressure of dissolved oxygen in arterial blood; a condition in which the blood is not sufficiently oxygenated to meet metabolic requirements of the body or tissues.

Hypoxia Reduced level of tissue oxygenation; a condition in which the body as a whole or a region of the body is deprived of adequate oxygen supply.

Hypoxic mixture A mixture of gases that does not provide enough oxygen for an individual patient.

Hystricomorph Any rodent of the suborder *Hystricomorpha*, such as guinea pigs and chinchillas.

I

Idioventricular rhythm Slow regular ventricular rhythm that have wide QRS complexes that can be mistaken for VPCs.

Impedance threshold device A valve preset to open at a specific pressure (cracking pressure) that limits the entry of air into the lungs during chest recoil between chest compressions (reduces intrathoracic pressure) to improve venous return to the heart.

Inotropy Cardiac contraction force

Insensible body water losses Body water lost as a result of normal metabolic processes but not easily measured; such losses occur through sweating, breathing, and mucous membrane evaporation.

Inspiratory reserve volume (IRV) Amount of air inspired over the tidal volume (extra amount that could be inhaled after normal inspiration).

Insufflation Inflating the stomach with air for internal visualization using an endoscope.

Institutional Animal Care and Use Committee (IACUC) Committee present at institutions that perform animal research whose function is to review and monitor the methodologies described in the research protocol to ensure that they provide ethical and humane for the animals involved in the research; they are typically made up of experts such as researchers, professors, veterinarians, and other animal care professionals experienced with the use of animal research as well as a public member not affiliated with the institution to represent general community interests in the proper care and use of animals.

Intermediate-pressure system Part of the anesthetic machine that receives gas from the pressure regulator and moves it into the flush valve and flowmeter; includes the pipeline inlet connections, pipeline pressure gauge (may not be present on portable anesthetic machines), conduits from pipeline to flowmeter, conduits from regulator to flowmeter, flowmeter assembly, oxygen fail-safe valve, oxygen supply failure valve, and oxygen flush valve; pressures in the intermediate system range from 35 to 75 psi but is typically set between 50-55 psi in most anesthesia machines with one regulator.

Intermittent positive pressure ventilation (IPPV) Process when a dedicated person administers breaths or mechanically (via a ventilator) ventilates a patient with the use of an endotracheal or tracheostomy tube, and an anesthetic breathing system.

Interposed abdominal compressions Abdominal compressions over the cranial abdomen used during CPCR to facilitate venous return from the abdomen.

Interventricular septum Wall separating the left and right ventricles.

Ion Charged particle.

Ionization Process by which an atom or molecule becomes charged.

Ipsilateral Same side.

Isotonic Term used to describe crystalloids that have an osmolality and sodium concentration similar to plasma.

IV bolus Rapid administration of a drug via the intravenous route.

L

Lancinating Stabbing.

Laryngospasm Reflexive closure of the laryngeal cartilage usually caused by stimulation of the larynx and an inadequate anesthetic depth.

Linear tracking Maintaining a plane of motion consistent with the sagittal plane. During ambulation, linear tracking of limbs should be preserved when providing neuromuscular reeducation to the joints.

Lipophilic Lipid soluble.

Local anesthesia Anesthesia produced in a limited area, as by injection of a local anesthetic, topical application, or freezing. Examples include infiltration, topical, regional block, spinal, epidural techniques.

Locoregional anesthesia Local and regional anesthesia techniques that desensitize a localized area of the body so that surgical procedures can be performed in conscious animals or to decrease the level of general anesthetic needed in anesthetized animals.

Low level laser therapy Form of phototherapy that involves the application of low light intensities to injuries and lesions to stimulate healing and induce minimal temperature elevation; commonly known as LLLT, cold laser therapy, or phototherapy.

Low-flow circle system Rebreathing system in which oxygen flow rates are greater than the patient's metabolic oxygen needs.

Low-pressure system Area of the anesthetic machine comprised of the components downstream of the flow control valve (flowmeter, piping from the flowmeter to the vaporizer, the vaporizer, the channel from the vaporizer to the common gas outlet, and the channel from the common gas outlet to the breathing system). Pressures within this system are near ambient pressure and will vary once they reach the breathing system; between 0 and 30 cm H_2O.

M

Maintenance fluid Type of fluid therapy that meets the daily water and electrolyte requirements of the patient.

Maladaptive pain Chronic pain that serves no useful biologic purpose; also called pathologic pain.

Malignant hyperthermia Rare genetic condition in which uncontrolled skeletal muscle activity and increased metabolism lead to rapid increases in body temperature during general anesthesia; more commonly occurs in pigs but can also rarely occur in humans, dogs, cats, and horses. The condition results in dramatic increased carbon dioxide production and heat that result in hypercapnia, hyperthermia, and death.

Mean arterial pressure (MAP) The average pressure in the arteries during one cardiac cycle; it is the product of systemic vascular resistance (SVR) and cardiac output (CO); it is the driving pressure for organ perfusion and can be calculated using this equation: (systolic − diastolic)/3 + diastolic.

Mechanical dead space The area where bidirectional flow takes place within a breathing circuit or endotracheal tube but no gas exchange occurs; also called apparatus dead space.

Mentation Level of consciousness and the patient's reaction to its environment.

Metabolic oxygen consumption The amount of oxygen consumed by an individual based on that individual's metabolism, disease state, and drug history.

Minimum alveolar concentration (MAC) Lowest alveolar concentration of inhalant anesthetic agent required to prevent movement in response to noxious stimuli in 50% of patients; used to compare the relative "potency" of inhalant anesthetic agents.

Minute volume (V_E) The volume of air breathed over one minute (200 mL/kg/min).

Mixed agonist/antagonist Drug that binds to more than one type of receptor and simultaneously stimulates one and blocks another.

Mobitz Type 1 (Wenckebach) Type of 2nd degree AV block that is characterized by a gradually increasing P-R interval with an eventual complete block of the signal through the AV node, dropping the QRS completely.

Mobitz Type 2 Type of 2nd degree AV block that shows a consistent and normal PR interval with occasional complete block of the signal through the AV node, again completely dropping QRS complexes. This arrhythmia has the potential for deteriorating into 3rd-degree AV block and should always be treated.

Modulation Process that occurs when first-order neurons synapse with second-order neurons in the dorsal horn cells of the spinal cord; modulation alters the traveling pain by a complex balance of stimulatory and inhibitory substances so that the pain signal is either enhanced or decreased.

Morbidity A diseased state, disability, or poor health; ratio of diseased animals to well animals in a population.

Mortality The state of death; ratio of diseased animals that die to diseased animals.

Multimodal analgesia Concept of using a combination of different analgesic drug classes to target different sites along the nociceptive pain pathway in an effort to maximize patient comfort and recovery; often allows for the reduction in dosage and frequency of individual drugs, thereby reducing adverse effects; also called balanced analgesia.

Multimodal anesthesia Theory in which general anesthesia is produced by administering several drugs with the goal of exploiting each drug's positive actions while avoiding potential adverse effects associated with large doses of a single drug. The philosophy encourages the use of several agents, each designed to affect a different function (e.g., providing muscle relaxation and amnesia or blocking pain and motor function); also known as balanced anesthesia

Myoclonus Involuntary jerking of a muscle or group of muscles.

N

Neuroleptanalgesia Mixing a sedative or tranquilizer with an analgesic drug (typically an opioid) to provide more profound sedation and potentially analgesia then if either drug was used alone.

Neuromuscular electrical stimulation (NMES) Administration of an electrical current generated by a stimulator that travels through leads to electrodes placed over the affected area via contact gel.

Neuropathic pain Pain caused by direct damage to or indirect alteration of the nervous system.

Neuroplasticity The central nervous system's ability to reorganize itself by forming new neural connections.

Nociception The neural process of encoding noxious stimuli; the processing of a noxious (tissue-damaging) stimulus resulting in the perception of pain by the brain.

Nociceptive pain Pain caused by direct activation of special sensory pain neurons

Nociceptor A high threshold sensory receptor of the peripheral nervous system that is capable of transducing and encoding noxious stimuli; a receptor of a sensory neuron (nerve cell) that responds to potentially damaging stimuli by sending signals to the spinal cord and brain.

Non-cumulative Property of a drug that does not produce effects that are more pronounced in serial doses than those produced by the first dose.

Non-rebreathing systems Type of anesthetic system that utilizes high fresh gas flow rates to avoid rebreathing of expired gases. This system bypasses the unidirectional valves and CO_2 absorber to reduce the resistance to breathing for small (< 7kg), spontaneously breathing patients.

Noncompetitive antagonist Drug that binds to a different site than the agonist's binding site, which changes the agonist's receptor and is irreversible (irreversible or insurmountable antagonism).

Noxious Harmful or destructive.

Nutraceuticals Foods or food products that theoretically provide health and medical benefits, including the prevention and treatment of disease; the term nutraceutical is made by combining the words "nutrition" and "pharmaceutical."

O

Objective Based on actual data and not influenced by emotions or personal prejudices; objective information in a SOAP includes body weight, vital parameters, physical exam findings, and results from diagnostic testing.

Oncotic pressure Form of osmotic pressure exerted by proteins, mainly albumin, in plasma that maintains volume within the vascular space by preventing fluid from leaving it.

One-way valves Valves that directs the flow of gases in a circle to prevent the direct return of carbon dioxide to the patient; also called unidirectional valves.

Ongoing fluid loss phase Phase of fluid therapy that replaces body water lost through things like continued vomiting, diarrhea, oozing skin, and excessive drooling.

Ongoing maintenance phase Phase of fluid therapy used to replace body water lost continuously during daily normal body functions.

Oscillometric Electronic blood pressure sensor with a numerical readout.

Osmolality Osmotic pressure of a solution based on the number of particles per kilogram of solution; is independent of the size of the particles.

Osmole Molecular weight of a solute, in grams, divided by the number of ions or particles into which it dissociates in solution.

Osmotic pressure Pressure or force that develops when two solutions of different concentrations are separated by a selectively permeable membrane.

Over-the-needle catheter Catheter with a stylet with a sharp needle tip, which is used to penetrate the skin and introduce the catheter into the blood vessel.

Over-the-wire catheter Catheter with a lumen so that a guide wire can be used to establish the path for the catheter.

Overpressurizing the vaporizer Technique that is done by setting the vaporizer at a higher than the desired concentration so the target concentration is met in a shorter period of time; it is done to increase the anesthetic concentration in the circuit.

Oxygen concentrator Cost-effective method to supply oxygen to an anesthesia machine; this method collects room air and separates the oxygen from the other gases; namely, nitrogen, argon, and carbon dioxide.

Oxygen consumption The rate at which oxygen is removed from the blood for use by the tissues.

Oxygen delivery (DO_2) The rate of oxygen transport from the lungs to the peripheral tissues; it is a function of both cardiac output and arterial oxygen content (CaO_2), and it cannot be directly measured.

Oxygen flush valve (oxygen fast flush valve) A button that can be utilized to supply large volumes of oxygen quickly to the patient; when depressed, this valve rapidly delivers large volumes of 100% oxygen to the breathing circuit and bypasses the flowmeter and the vaporizer.

Oxygenation The process of taking oxygen from inspired air and delivering it to the tissues.

Oxyhemoglobin dissociation curve A graph displaying the nonlinear tendency for oxygen to bind to hemoglobin. It describes the relation between the partial pressure of oxygen (x axis) and the percent oxygen saturation (y axis).

P

PaCO$_2$ Partial pressure of carbon dioxide molecules dissolved in the plasma phase of an arterial sample (i.e., not bound to Hb).

Palatal ostium Tiny opening in the soft palate of guinea pigs that separates the oral pharynx from the rest of the pharynx, making intubation very difficult.

Palliative radiation therapy (PRT) A type of radiation therapy administered to cancer patients with the goal to alleviate pain associated with the condition.

P$_a$O$_2$ Partial pressure of oxygen molecules dissolved in the plasma phase of an arterial sample (i.e. not bound to Hb).

P$_A$O$_2$ Partial pressure of oxygen in the alveoli.

P/F ratio P$_a$O$_2$/FiO$_2$ Objective number that can be utilized to assess the adequacy of a patient's response to oxygen therapy; this number can help predict the likelihood of a patient developing hypoxemia once taken off 100% oxygen and moved to recovery.

Parasympatholytic Drug that blocks the effects of the parasympathetic nervous system by competitively blocking the binding of acetylcholine at muscarinic receptors; effects include increased heart rate, bronchodilation, decreased gastric motility and secretions, and mydriasis (pupillary dilation).

Parenteral Route other than the gastrointestinal tract; e.g., IV, IM, and SQ.

Paroxysmal A sudden attack or increase in clinical signs.

Partial agonist Drug that binds to the receptor to produce a submaximal response relative to the full agonist.

Partial pressure The pressure that each gas in a mixture will exert on a column of mercury; expressed in mm Hg.

Partial pressure difference A measure of the difference in partial pressures between anesthetic gas in the alveoli versus that in the blood affects the rate of uptake (absorption) of the inhaled anesthetic agent.

Partial pressure of oxygen in the alveoli (P$_A$O$_2$) Partial pressure of inspired oxygen minus consumed oxygen; value determined by the alveolar oxygen equation because it is not possible to collect gas from the alveoli.

Partial pressure of oxygen in the arterial blood (P$_a$O$_2$) A measurement of oxygen molecules dissolved in the plasma phase of an arterial sample (i.e., not bound to hemoglobin); reflects how well oxygen is able to move from the lungs to the blood; it is often altered by severe illnesses.

Passive range of motion (PROM) Movement around a joint where *no voluntary movement* is required by the patient because the motion is performed by an external force such as the therapist's hands.

Passive scavenging systems Hoses that rely on positive pressure generated during patient exhalation and gravity to move gases into a disposal assembly; an example is an activated charcoal canister that absorbs anesthetic gases.

Patent airway Open passageway into and out of the lungs.

Pathologic pain Chronic pain that serves no useful biologic purpose; also called maladaptive pain.

Peel-away catheter Type of through-the-needle catheter that uses a large gauge needle and a peel-away sheath to facilitate catheter placement.

Perception The cerebral cortical response to nociceptive signals that are projected by third-order neurons to the brain; clinically, pain realization in the patient.

Perfusion The flow of blood through a specific organ or tissue.

Perioperative period The time period describing the entire duration of a patient's surgical/anesthetic experience including preoperative, intraoperative, and postoperative time periods.

Peripheral capillary oxygen saturation (SpO$_2$) An estimate of the oxygen saturation level; it is determined non-invasively by a pulse oximeter.

Peripheral sensitization An increased sensitivity of peripheral nociceptors at or around the point of the original noxious stimulation; results in alteration of the body's normal response to pain.

Pharmacodynamics Study of the drug's mechanism of action and its effect on the animal. In other words, what the drug does to the body.

Pharmacokinetics The study of physiological movement of a drug within the body after it is administered and includes its absorption, distribution, metabolism, and elimination of drugs. In other words, what the body does to the drug.

Photobiostimulation Light energy modulating or stimulating cellular function.

Photoplethysmography A technique used to detect blood volume changes in tissue beds; commonly called pleth.

Physical rehabilitation The use of noninvasive techniques, excluding veterinary chiropractic, for the rehabilitation of injuries, disease processes, and congenital deformities in nonhuman animals.

Physiological dead space The anatomical dead space and the alveolar dead space.

Physiologic pain Acute pain that acts as a warning signal to the body to notify it of injury and/or tissue damage; also known as adaptive pain.

Piloerection Involuntary erection or bristling of hairs due to a sympathetic reflex; otherwise known as "goose bumps."

Pin Index Safety System (PISS) precise configuration of a series of holes located just beneath the outlet port on the neck of the compressed gas cylinder and its corresponding pins located on the hanger yoke or pressure regulator; this system prevents the unintentional placement of a cylinder of the wrong medical gas onto the hanger yoke or pressure regulator of the anesthesia machine that was designed for another gas.

Plasma esterase Enzyme found in the plasma that cleave apart compounds and make them inactive.

Point-of-care instrumentation Medical diagnostic instruments that can be used at the time and place of patient care.

Polypnea Rapid, shallow panting.

Pop-off valve Safety valve that allows excess gases to vent from the breathing circuit to avoid a build-up of pressure within the system; also called an adjustable pressure limiting (APL) valve.

Potency The measure of drug activity in terms of the amount of the drug required to produce an effect; refers to the dosage required to produce maximum effect.

Pre-emptive analgesia Providing pain control before a painful stimulus occurs; the goal is to decrease pain in the immediate postoperative period and reduce the incidence of chronic pain by preventing the establishment of central sensitization and hyperexcitability; also known as preventative analgesia.

Precordial thump A technique where one delivers a strong blow using the heel of the hand directly over the heart in an effort to convert ventricular fibrillation to a sinus rhythm in the absence of a defibrillator.

Preload The volume of blood entering the right side of the heart.

Premedication Any agent that is given before anesthesia begins.

Pressure manometer Pressure gauge located after the expiratory unidirectional valve that reflects the pressure in the breathing system and the patient's lungs; it is calibrated in cm H$_2$O (centimeters of water) or mm Hg (millimeters of mercury).

Pressure regulators (pressure reducing valves) Units that have two gauges (one that reads the tank pressure and one that reads the line pressure) to control the pressure at a steady level and reduce the compressed gas cylinder's pressure to a level that is appropriate and safe for use inside the anesthesia machine.

Protein binding Binding of drugs to proteins in blood plasma, which determines how effective the drug is in the body. The bound drug is kept in the blood stream while the unbound component may be metabolized or excreted; therefore, if a drug is 95% bound to a protein and 5% is free, that means 5% is active in the system and causing pharmacological effects.

Proventriculus First/glandular part of the stomach.

Ptyalism Excessive flow of saliva.

Pulse oximeter A non-invasive monitor that measures oxygenation of blood hemoglobin by providing a continuous, automatic, and audible test of cardiopulmonary function; also called pulse ox.

Pulse therapy Method of drug administration in which the patient is given the drug for several days then off of the drug for several days in order to limit potential adverse effects.

Pulsed electromagnetic field therapy (PEMF) Type of electromagnetic therapy in which small electrical currents are intermittently applied to the body.

Purkinje fibers Specialized cardiac muscle cells that conduct impulses to facilitate synchronized contractions of the ventricles.

Q

Quick assessment tests (QATs) Minimum laboratory tests (PCV/TP/Glu/BUN [from an Azostick®]) recommended prior to anesthesia.

R

Radiant heat loss Heat loss when heat generated within the patient's body is given off to the atmosphere.

Range of motion (ROM) Measurement of movement around a joint.

Reassessment Campaign On Veterinary Resuscitation (RECOVER) Cardiopulmonary Cerebral Resuscitation guidelines that are the standard of care for providing veterinary CPCR.

RECOVER Reassessment Campaign on Veterinary Resuscitation.

Referred pain Pain that is perceived to originate from a location other than its true source.

Regional anesthesia Anesthesia caused by interrupting the sensory nerve conductivity of any region of the body. Regional anesthesia may be produced by a field block (encircling the operative field by means of injections of a local anesthetic) or by a nerve block (making multiple injections in close proximity to the nerves supplying the area).

Renal portal system Portal venous system in which the renal portal veins pump blood directly into the renal tubules. In animals with a renal portal system such as birds, blood flow from their caudal half enters the renal portal system resulting in a drug being excreted before reaching the systemic circulation; therefore, the desired effect of the drug is not achieved.

Replacement fluid Type of fluid therapy that replaces water and electrolyte deficits caused by dehydration.

Replacement phase Phase of fluid therapy which serves to correct dehydration and replace water previously lost.

Reservoir bag Bag that attaches to a port located on the CO_2 absorber side of the circle near the expiratory unidirectional valve to provide extra gas holding capacity within the system during exhalation; sometimes referred to as a rebreathing bag for rebreathing systems.

Residual volume (RV) Air remaining in the lungs after a forced expiration (amount of air trapped in alveoli).

Respiratory sinus arrhythmia Heart rhythm in which heart rate increases slightly during inspiration and decreases during exhalation; caused by fluctuations in vagal tone associated with respiration.

Resuscitation phase Phase of fluid therapy which serves to expand intravascular volume.

Reticular activation center (RAC) Area of the brain responsible for maintaining consciousness in animals.

Right-to-left shunt Physiologic shunt due to unusual pulmonary blood flow that can cause deoxygenated blood to mix with oxygenated blood going back into systemic circulation through the left ventricle to cause hypoxemia.

ROM Passive range of motion.

S

Second order neurons Afferent fibers that originate in the the dorsal horn cells of the spinal cord and carry signals from the spinal cord to the brain.

Sedation The state characterized by CNS depression accompanied by sleepiness/drowsiness and some degree of relaxation.

Sedative Any drug which results in central nervous system depression and sleepiness.

Semiclosed circle system Rebreathing system that utilizes fresh gas flow rates that exceed the metabolic oxygen consumption rate of the patient.

Sensible body water losses Body water lost in urine and feces and can actually be measured.

Shock Any state where the body's consumption of oxygen is not met by the body's delivery of oxygen.

Sick sinus syndrome (SSS) Cardiac condition characterized by bradycardia and a history of intermittent weakness and collapse due to a heart rhythm in which the SA node does not discharge an impulse to trigger the heart to contract.

Signalment Description of the animal containing information such as species, breed, sex, age, and sexual status (intact or neutered).

Sinoatrial (SA) node A bundle of specialized cardiac muscle cells that lie in the cranial portion of the right atrium. The SA node is called the pacemaker of the heart because it keeps the pace of the heart rate by automatically generating the electrical impulses that trigger the heartbeat.

Sinus bradycardia A decrease in heart rate outside of the normal range characterized by slow but normal complexes on the ECG.

Sinus tachycardia An increase in heart rate outside the normal range characterized by fast but normal complexes on the ECG; usually defined as originating somewhere other than the SA node but above the AV junction.

SOAP Subjective, objective, assessment, and plan (method of record keeping).

Solute Dissolved substance of a solution (e.g., sodium).

Solvent The dissolving substance of a solution (water is the biological solvent).

Somatic pain Pain originating from ligaments, tendons, bones, blood vessels, and muscles and in people is described as cramping, aching, throbbing, and sharp.

Splitting ratio The ratio of the gas that picks up inhalant anesthetic agent to the gas that does not pick up inhalant anesthetic agent determines the concentration of gas leaving the vaporizer.

Stage I of anesthesia Stage of anesthesia in which the patient is gradually losing consciousness and may exhibit signs of fear or excitement; also called the induction stage.

Stage II of anesthesia Stage of anesthesia in which the patient may exhibit involuntary reactions like vocalizing and paddling; also known as the excitement stage.

Stage III of anesthesia Stage of anesthesia in which the patient is unconscious and has progressive muscle relaxation and loss of reflexes as the animal transitions from a light to a deep plane of anesthesia; also known as the surgical anesthesia (operative) stage. Stage III is divided into three planes; light or plane 1, medium or plane 2, and deep or plane 3. (It should be noted that some sources describe four planes in Stage III.)

Stage IV of anesthesia   Stage of anesthesia in which the patient's body is extremely depressed, reflexes are absent, muscle tone is flaccid, pupils are dilated, and the patient experiences marked cardiopulmonary depression; also known as the anesthetic overdose (danger) stage.

Stress   Broad term used to describe the biological response of an animal to cope with a threat to its physical homeostasis or emotional well-being.

Stressor   Event that leads to stress.

Stroke volume (SV)   Amount of blood ejected in one cardiac cycle.

Subanesthetic dosage   A lower drug dosage than would cause anesthesia.

Subchondral sclerosis   Increased bone density or thickening in the subchondral layer.

Subjective   Based on a given person's experience, understanding, and feelings; subjective information in a SOAP include patient history, clinical signs, and behavioral/temperament assessment.

Supraventricular tachycardia   An increase in heart rate outside the normal range characterized by a fast rhythm with an origin somewhere above the ventricles; fast complexes usually without P waves on the ECG.

Surgical anesthesia   The degree of anesthesia at which a response to surgical stimulus does not occur.

Surgical MAC   Concentration of drug necessary to keep 95% of patients immobile during surgical stimulation; calculated by multiplying 1.5 X the MAC value.

Systemic vascular resistance (SVR)   Resistance to blood flow by the peripheral circulation.

Systolic blood pressure   Peak pressure during contraction of the heart, it is comprised of stroke volume and arterial compliance.

T

T-piece   Fresh gas port that sits at an angle between 45 and 90° to the patient end of the circuit for a non-rebreathing system.

Tachypnea   Rapid breathing that is not labored.

Therapeutic index (TI)   Comparison of the amount of a therapeutic agent that causes the therapeutic effect to the amount that causes toxicity; also called margin of safety.

Therapeutic ultrasound   Treatment modality that delivers energy in the form of acoustic vibrations to produce thermal and non-thermal effects.

Third-order neurons   Afferent fibers that carry signals to the cerebral cortex where perception occurs.

Thoracic pump theory   Concept that external chest compressions lead to blood flow during cardiopulmonary cerebral resuscitation (CPCR) because chest compressions increase overall intrathoracic pressure, thereby compressing the aorta and collapsing the vena cava leading to blood flow out of the thorax.

Threshold   The minimum intensity of a stimulus; e. g., pain threshold is the minimum intensity of a stimulus that is perceived as painful (i.e., you need to have a certain amount of stimulation for it to be perceived).

Thrombogenic   The tendency to initiate blood clot formation.

Through-the-needle catheter   Catheter with a larger gauge needle through which a smaller gauge catheter is fed and is used for placement of central lines.

Tidal volume (V_T)   Amount of air exchanged during normal respiration (air inhaled and exhaled in one breath); calculated in mL/kg.

Titrated   Drug is given in a series of IV bolus injections.

To effect   Something is given in order to obtain a desired result; the desired result of anesthesia induction is unconsciousness.

Tonicity   the concentration of a solution as compared to another solution; is based on osmolality, which is the osmotic pressure of a solution based on the number of particles per kilogram of solution.

Total lung capacity (TLC)   Maximum volume to which the lungs can be expanded tidal volume + expiratory reserve volume + residual volume.

Toxic dosage   Drug dosage at which a substance can damage an organism.

Tranquilization   Administration of any of a group of compounds that calm and relax an anxious patient but do not induce sleep.

Tranquilizer   Any drug that calms an animal; is used to reduce anxiety and aggression in animals.

Transducer   A device that converts one form of energy to another.

Transduction   The initial reaction to and conversion of a noxious stimulus (mechanical, chemical, or thermal) into electrical energy by a peripheral nociceptor (free afferent nerve ending).

Transient receptor potential cation channel subfamily V member 1 (TrpV1) receptor   Receptor mainly found in nociceptive neurons of the peripheral nervous system and in lesser amounts in other tissues such as the central nervous system that detects and regulates body temperature and provides pain and scalding sensation; previously named the capsaicin receptor and the vanilloid receptor 1.

Transmission   The propagation through the peripheral nervous system via first-order neurons; second step in the nociceptive pathway where the electrochemical signal travels through the peripheral nervous system via first order neurons to the spinal cord.

Transmucosal   A route of administration that is "across the mucosa." Transmucosal drug administration involves a drug that is squirted or applied to an oral mucous membrane where it is absorbed rather than swallowed.

U

Unidirectional valves   Valve that directs the flow of gases in a circle to prevent the direct return of carbon dioxide to the patient; also called one-way valves.

V

Vapor pressure   Measure of the tendency for the molecules in the liquid state to enter the gaseous state.

Vaporizers   Units that convert anesthetic agents from a volatile liquid into a vapor that is delivered in the carrier gas.

Vasomotor tone   The degree of vasodilation or vasoconstriction.

Ventilation-perfusion (V/Q) mismatch   Unequal perfusion of the alveoli due to gravity.

Ventilation-perfusion mismatch (V/Q mismatch)   Unequal perfusion of the alveoli due to gravity.

Ventricular fibrillation (V-fib or VF)   Chaotic electrical activity of the heart with no mechanical activity.

Ventricular premature contractions (VPC or PVC)   Abnormal impulse that originates within the ventricle causing the ventricles to contract prematurely, before the chamber is full of blood, and can lead to a decrease of blood flow to the body.

Ventricular tachycardia (V-tach)   Dysrhythmia classified as having a run of greater than three ventricular premature contractions in a row or a sustained elevated ventricular rate.

Vestibulitis   Inflammation of the vestibular nerve that causes persistent head tilt and loss of balance in rabbits; can be caused by conditions such as an ear infection, brain disease, or trauma; also known as torticollis or wry neck.

Visceral pain   Pain that originates from internal organs and in people is often described as dull, aching, and can be difficult to specifically localize.

Vital capacity (V_c)   Largest amount of air that can be moved in the lungs (tidal volume + inspiratory and expiratory reserve volumes).

Volatile   Easily evaporated at room temperature.

Volatile anesthetics   Anesthetic agents that are typically liquids at room temperature and are vaporized in the presence of oxygen or a carrier gas and is delivered to a patient via the respiratory tract.

W

Waiver   A legal form that someone fills out with the intent of releasing someone else from liability.

Waste anesthetic gas (WAG)   Waste anesthetic gas.

Weaning from a mechanical ventilator   The process of getting the patient to breathe spontaneously so it no longer needs mechanical ventilation, and ensures that the patient is clinically stable and able to maintain its expired CO_2 levels within a stable range.

Wind up   Phenomenon that occurs in response to a barrage of afferent nociceptive impulses, causing *peripheral* nociceptor sensitivity and widening or expansion of the peripheral receptive fields; in other words, a heightened sensitivity that results in altered pain thresholds, both peripherally and centrally, so that pain is experienced in places unrelated to the original source.

Wye-piece   Joining of the inspiratory limb with the expiratory limb on a circle breathing system; the wye-piece also connects to the endotracheal tube or facemask; also called the Y-piece.

Y

Y-piece   Joining of the inspiratory limb with the expiratory limb on a circle breathing system; the Y-piece also connects to the endotracheal tube or facemask; also called the wye-piece.

Yoke block   A solid cylindrical piece of material that fits between the retaining screw and nipple of the yoke; prevents dust and dirt from entering the yoke area and are typically chained to the anesthesia machine.

INDEX

Note: Page numbers followed by "*f*" and "*t*" refer to figures and tables, respectively.